Procedure Index

Following is an index that provides page-number references for the various statistical procedures discussed in the book. *Note:* This index includes only numbered procedures (i.e., Procedure x.x), not all procedures.

Introductory
STATISTICS
10TH EDITION

Introductory
STATISTICS
10TH EDITION

Neil A. Weiss, Ph.D.

School of Mathematical and Statistical Sciences
Arizona State University

Biographies by **Carol A. Weiss**

PEARSON

Boston Columbus Indianapolis New York San Francisco
Amsterdam Cape Town Dubai London Madrid Milan Munich Paris Montreal Toronto
Delhi Mexico City Sao Paulo Sydney Hong Kong Seoul Singapore Taipei Tokyo

On the cover: Pelicans are large, social water birds found primarily along coastlines. They are best known for their aerial dives and elastic throat pouch with which they catch fish. According to the National Geographic Society, pelicans live for about 10 to 25 years in the wild. On average, they are 5.8 ft long, have a wingspan of 10 ft, and weigh 30 lb. Although, in North America, the brown pelican is endangered, populations have somewhat recovered after decades of population decline from the deleterious effects of pesticides, such as DDT, on egg survival.

Cover photograph: Tirc83/Getty Images

Editor in Chief: Deirdre Lynch
Senior Acquisitions Editor: Suzanna Bainbridge
Editorial Assistants: Justin Billing and Salena Casha
Program Team Lead: Marianne Stepanian
Program Manager: Chere Bemelmans
Project Team Lead: Christina Lepre
Project Manager: Shannon Steed
Senior Designer: Barbara T. Atkinson
Manager, Multimedia Production: Christine Stavrou
Multimedia Producer: Stephanie Green

Software Development: Bob Carroll, Marty Wright
Senior Marketing Manager: Erin Kelly
Marketing Coordinator: Kathleen DeChavez
Senior Author Support/Technology Specialist: Joe Vetere
Rights and Permissions Advisor: Diahanne Lucas
Senior Procurement Specialist: Carol Melville
Cover Designer: Jenny Willingham
Text Design: Rokusek Design, Inc.
Production Coordination, Composition, and
    Illustrations: Aptara Corporation

Acknowledgements of third party content appear on page C-1, which constitutes an extension of this copyright page.

PEARSON, ALWAYS LEARNING, MyStatLab and MathXL are exclusive trademarks in the U.S. and/or other countries owned by Pearson Education, Inc. or its affiliates.

Unless otherwise indicated herein, any third-party trademarks that may appear in this work are the property of their respective owners and any references to third-party trademarks, logos or other trade dress are for demonstrative or descriptive purposes only. Such references are not intended to imply any sponsorship, endorsement, authorization, or promotion of Pearson's products by the owners of such marks, or any relationship between the owner and Pearson Education, Inc. or its affiliates, authors, licensees or distributors.

MICROSOFT AND/OR ITS RESPECTIVE SUPPLIERS MAKE NO REPRESENTATIONS ABOUT THE SUITABILITY OF THE INFORMATION CONTAINED IN THE DOCUMENTS AND RELATED GRAPHICS PUBLISHED AS PART OF THE SERVICES FOR ANY PURPOSE. ALL SUCH DOCUMENTS AND RELATED GRAPHICS ARE PROVIDED "AS IS" WITHOUT WARRANTY OF ANY KIND. MICROSOFT AND/OR ITS RESPECTIVE SUPPLIERS HEREBY DISCLAIM ALL WARRANTIES AND CONDITIONS WITH REGARD TO THIS INFORMATION, INCLUDING ALL WARRANTIES AND CONDITIONS OF MERCHANTABILITY, WHETHER EXPRESS, IMPLIED OR STATUTORY, FITNESS FOR A PARTICULAR PURPOSE, TITLE AND NON-INFRINGEMENT. IN NO EVENT SHALL MICROSOFT AND/OR ITS RESPECTIVE SUPPLIERS BE LIABLE FOR ANY SPECIAL, INDIRECT OR CONSEQUENTIAL DAMAGES OR ANY DAMAGES WHATSOEVER RESULTING FROM LOSS OF USE, DATA OR PROFITS, WHETHER IN AN ACTION OF CONTRACT, NEGLIGENCE OR OTHER TORTIOUS ACTION, ARISING OUT OF OR IN CONNECTION WITH THE USE OR PERFORMANCE OF INFORMATION AVAILABLE FROM THE SERVICES. THE DOCUMENTS AND RELATED GRAPHICS CONTAINED HEREIN COULD INCLUDE TECHNICAL INACCURACIES OR TYPOGRAPHICAL ERRORS. CHANGES ARE PERIODICALLY ADDED TO THE INFORMATION HEREIN. MICROSOFT AND/OR ITS RESPECTIVE SUPPLIERS MAY MAKE IMPROVEMENTS AND/OR CHANGES IN THE PRODUCT(S) AND/OR THE PROGRAM(S) DESCRIBED HEREIN AT ANY TIME. PARTIAL SCREEN SHOTS MAY BE VIEWED IN FULL WITHIN THE SOFTWARE VERSION SPECIFIED.

MICROSOFT® WINDOWS®, and MICROSOFT OFFICE® ARE REGISTERED TRADEMARKS OF THE MICROSOFT CORPORATION IN THE U.S.A. AND OTHER COUNTRIES. THIS BOOK IS NOT SPONSORED OR ENDORSED BY OR AFFILIATED WITH THE MICROSOFT CORPORATION.

Library of Congress Cataloging-in-Publication Data

Weiss, N. A. (Neil A.)
    Introductory statistics / Neil A. Weiss, Arizona State University; biographies by Carol A. Weiss. – 10th edition.
        pages cm
    Includes index.
        ISBN 978-0-321-98917-8
    1. Statistics–Textbooks.   I. Weiss, Carol A.   II. Title.
    QA276.12.W45 2016
    519.5–dc23

                                                                    2014019105

4 16

www.pearsonhighered.com

ISBN 13: 978-0-321-98917-8
ISBN 10: 0-321-98917-1

About the Author

Neil A. Weiss received his Ph.D. from UCLA and subsequently accepted an assistant professor position at Arizona State University (ASU), where he was ultimately promoted to the rank of full professor. Dr. Weiss has taught statistics, probability, and mathematics—from the freshman level to the advanced graduate level—for more than 30 years.

In recognition of his excellence in teaching, Dr. Weiss received the *Dean's Quality Teaching Award* from the ASU College of Liberal Arts and Sciences. He has also been runner-up twice for the *Charles Wexler Teaching Award* in the ASU School of Mathematical and Statistical Sciences. Dr. Weiss's comprehensive knowledge and experience ensures that his texts are mathematically and statistically accurate, as well as pedagogically sound.

In addition to his numerous research publications, Dr. Weiss is the author of *A Course in Probability* (Addison-Wesley, 2006). He has also authored or coauthored books in finite mathematics, statistics, and real analysis, and is currently working on a new book on applied regression analysis and the analysis of variance. His texts—well known for their precision, readability, and pedagogical excellence—are used worldwide.

Dr. Weiss is a pioneer of the integration of statistical software into textbooks and the classroom, first providing such integration in the book *Introductory Statistics* (Addison-Wesley, 1982). He and Pearson Education continue that spirit to this day.

In his spare time, Dr. Weiss enjoys walking, studying and practicing meditation, and playing hold'em poker. He is married and has two sons.

Dedicated to Aaron and Greg

Contents

PART I   Introduction

CHAPTER 1   The Nature of Statistics   1

PART II   Descriptive Statistics

CHAPTER 2   Organizing Data   36

CHAPTER 3   Descriptive Measures   93

*Indicates optional material.

PART III   Probability, Random Variables, and Sampling Distributions

PART IV   Inferential Statistics

*Indicates optional material.

*Indicates optional material.

**Indicates optional material on the WeissStats site.

*Indicates optional material.

**Indicates optional material on the WeissStats site.

MODULE C Design of Experiments and Analysis of Variance   C-0

Appendixes

WeissStats Resource Site (brief contents)

Note: Visit the WeissStats Resource Site at **www.pearsonhighered.com/weiss-stats** for detailed contents.

Additional Sections

Additional Statistical Tables

Applets

Data Sets

Data Sources

Focus Database

Formulas

JMP Concept Discovery Modules

Minitab Macros

Procedures Booklet

Regression-ANOVA Modules

StatCrunch Reports

Technology Basics

TI Programs

*indicates optional material

Preface

Using and understanding statistics and statistical procedures have become required skills in virtually every profession and academic discipline. The purpose of this book is to help students master basic statistical concepts and techniques and to provide real-life opportunities for applying them.

Audience and Approach

Introductory Statistics is intended for one- or two-semester courses or for quarter-system courses. Instructors can easily fit the text to the pace and depth they prefer. Introductory high school algebra is a sufficient prerequisite.

Although mathematically and statistically sound (the author has also written books at the senior and graduate levels), the approach does not require students to examine complex concepts. Rather, the material is presented in a natural and intuitive way. Simply stated, students will find this book's presentation of introductory statistics easy to understand.

About This Book

Introductory Statistics presents the fundamentals of statistics, featuring data production and data analysis. Data exploration is emphasized as an integral prelude to statistical inference.

This edition of *Introductory Statistics* continues the book's tradition of being on the cutting edge of statistical pedagogy, technology, and data analysis. It includes hundreds of new and updated exercises with real data from journals, magazines, newspapers, and websites.

The following Guidelines for Assessment and Instruction in Statistics Education (GAISE), funded and endorsed by the American Statistical Association, are supported and adhered to in *Introductory Statistics*:

- Emphasize statistical literacy and develop statistical thinking.
- Use real data.
- Stress conceptual understanding rather than mere knowledge of procedures.
- Foster active learning in the classroom.
- Use technology for developing conceptual understanding and analyzing data.
- Use assessments to improve and evaluate student learning.

Changes in the Tenth Edition

The goal for this edition was to create an even more flexible and user-friendly book, to provide several new step-by-step procedures for making statistical analyses easier to apply, to add a fourth category of exercises, to expand the use of technology for developing understanding and analyzing data, and to refurbish the exercises. Several important revisions are presented as follows.

New! **New Case Studies.** Fifty percent of the chapter-opening case studies have been replaced.

New! **New and Revised Exercises.** This edition contains more than **3000** high-quality exercises, which far exceeds what is found in typical introductory statistics books. Over 35% of the exercises are new, updated, or modified.

New! **WeissStats Resource Site.** The WeissStats Resource Site (aka WeissStats site) provides an extensive array of resources for both instructors and students, including additional topics, applets, all data sets from the book in multiple formats, a procedures booklet, and technology appendixes. In addition to several new items, the site offers universal access to those items formerly included on the WeissStats CD. Refer to the table of contents for a brief list of the contents of the WeissStats site or visit the site at www.pearsonhighered.com/weiss-stats. *Note:* Resources for instructors only are available on the Instructor Resource Center at www.pearsonhighered.com/irc.

New! **Chebyshev's Rule and the Empirical Rule.** A new (optional) section of Chapter 3 has been added that is dedicated to an examination of Chebyshev's rule and the empirical rule. The empirical rule is further examined in Chapter 6 when the normal distribution is discussed.

New! **Quartiles.** The method for calculating quartiles has been modified to make it more easily accessible to students. Furthermore, a dedicated procedure that provides a step-by-step method for finding the quartiles of a data set has been included.

Revised! **Distribution Shapes.** The material on distribution shapes in Section 2.4 has been significantly modified and clarified. Students will find this revised approach easier to understand and apply.

Revised!   **Regression Analysis.**   Major improvements have been made to the chapter on Descriptive Methods in Regression and Correlation. These improvements include a comprehensive discussion of scatterplots, a simpler introduction to the least-squares criterion, and easier introductory examples for the regression equation, the sums of squares and coefficient of determination, and the linear correlation coefficient.

Expanded!   **Warm-up Exercises.**   In this edition, hundreds of "warm-up" exercises have been added. These exercises provide context-free problems that allow students to concentrate solely on the relevant concepts before moving on to applied exercises.

Expanded!   **Density Curves.**   The discussion of density curves has been significantly expanded and now includes several examples and many more exercises.

Expanded!   **Type II Error Probabilities and Power.** Section 9.7, which covers Type II error probabilities and power, has undergone major revision, including increased visuals and the addition of procedures for calculating Type II error probabilities and for constructing power curves.

Note: See the Technology section of this preface for a discussion of technology additions, revisions, and improvements.

Hallmark Features and Approach

Chapter-Opening Features.   Each chapter begins with a general description of the chapter, an explanation of how the chapter relates to the text as a whole, and a chapter outline. A classic or contemporary case study highlights the real-world relevance of the material.

End-of-Chapter Features. Each chapter ends with features that are useful for review, summary, and further practice.

- *Chapter Reviews.* Each chapter review includes *chapter objectives*, a list of *key terms* with page references, and *review problems* to help students review and study the chapter. Items related to optional materials are marked with asterisks, unless the entire chapter is optional.
- *Focusing on Data Analysis.* This feature lets students work with large data sets, practice technology use, and discover the many methods of exploring and analyzing data. For details, see the introductory Focusing on Data Analysis section on page 34 of Chapter 1.
- *Case Study Discussion.* At the end of each chapter, the chapter-opening case study is reviewed and discussed in light of the chapter's major points, and then problems are presented for students to solve.
- *Biographical Sketches.* Each chapter ends with a brief biography of a famous statistician. Besides being of general interest, these biographies teach students about the development of the science of statistics.

Formula/Table Card. The book's detachable formula/table card (FTC) contains all the formulas and many of the tables that appear in the text. The FTC is helpful for quick-reference purposes; many instructors also find it convenient for use with examinations.

Procedure Boxes, Index, and Booklet. To help students learn how to perform statistical analyses, easy-to-follow, step-by-step procedures have been provided. Each step is highlighted and presented again within the illustrating example. This approach shows how the procedure is applied and helps students master its steps. Additionally:

- A *Procedure Index* (located near the front of the book) provides a quick and easy way to find the right procedure for performing any statistical analysis.
- A *Procedures Booklet* (available in the Procedures Booklet section of the WeissStats Resource Site) provides a convenient way to access any required procedure.

ASA/MAA–Guidelines Compliant. *Introductory Statistics* follows American Statistical Association (ASA) and Mathematical Association of America (MAA) guidelines, which stress the interpretation of statistical results, the contemporary applications of statistics, and the importance of critical thinking.

Populations, Variables, and Data. Through the book's consistent and proper use of the terms *population, variable*, and *data*, statistical concepts are made clearer and more unified. This strategy is essential for the proper understanding of statistics.

Data Analysis and Exploration. Data analysis is emphasized, both for exploratory purposes and to check assumptions required for inference. Recognizing that not all readers have access to technology, the book provides ample opportunity to analyze and explore data without the use of a computer or statistical calculator.

Parallel Critical-Value/*P*-Value Approaches. Through a parallel presentation, the book offers complete flexibility in the coverage of the critical-value and *P*-value approaches to hypothesis testing. Instructors can concentrate on either approach, or they can cover and compare both approaches. The dual procedures, which provide both the critical-value and *P*-value approaches to a hypothesis-testing method, are combined in a side-by-side, easy-to-use format.

Interpretations. This feature presents the meaning and significance of statistical results in everyday language and highlights the importance of interpreting answers and results.

You Try It! This feature, which follows most examples, allows students to immediately check their understanding by working a similar exercise.

What Does It Mean? This margin feature states in "plain English" the meanings of definitions, formulas, key facts, and some discussions—thus facilitating students' understanding of the formal language of statistics.

Examples and Exercises

Real-World Examples. Every concept discussed in the text is illustrated by at least one detailed example. Based on real-life situations, these examples are interesting as well as illustrative.

Real-World Exercises. Constructed from an extensive variety of articles in newspapers, magazines, statistical abstracts, journals, and websites, the exercises provide current, real-world applications whose sources are explicitly cited.

New to this edition, a fourth category of exercises has been added, namely, Applying the Concepts and Skills. As a consequence, the exercise sets are now divided into the following four categories:

- *Understanding the Concepts and Skills* exercises help students master the basic concepts and skills explicitly discussed in the section. These exercises consist of two types: (1) Non-computational problems that test student understanding of definitions, formulas, and key facts; (2) "warm-up" exercises, which require only simple computations and provide context-free problems that allow students to concentrate solely on the relevant concepts before moving on to applied exercises. For pedagogical reasons, it is recommended that warm-up exercises be done without the use of a statistical technology.
- *Applying the Concepts and Skills* exercises provide students with an extensive variety of applied problems that hone student skills with real-life data. These exercises can be done with or without the use of a statistical technology, at the instructor's discretion.
- *Working with Large Data Sets* exercises are intended to be done with a statistical technology and let students apply and interpret the computing and statistical capabilities of Minitab®, Excel®, the TI-83/84 Plus®, or any other statistical technology.
- *Extending the Concepts and Skills* exercises invite students to extend their skills by examining material not necessarily covered in the text. These exercises include many critical-thinking problems.

Notes: An exercise number set in cyan indicates that the exercise belongs to a group of exercises with common instructions. Also, exercises related to optional materials are marked with asterisks, unless the entire section is optional.

Data Sets. In most examples and exercises, both raw data and summary statistics are presented. This practice gives a more realistic view of statistics and lets students solve problems by computer or statistical calculator. More than **1000** data sets are included, many of which are new or updated. All data sets are available in multiple formats in the Data Sets section of the WeissStats Resource Site, www.pearsonhighered.com/weiss-stats.

Technology

Parallel Presentation. The book's technology coverage is completely flexible and includes options for use of Minitab, Excel, and the TI-83/84 Plus. Instructors can concentrate on one technology or cover and compare two or more technologies.

Updated! **The Technology Center.** This in-text, statistical-technology presentation discusses three of the most popular applications—Minitab, Excel, and the TI-83/84 Plus graphing calculators—and includes step-by-step instructions for the implementation of each of these applications. The Technology Centers are integrated as optional material and reflect the latest software releases.

Updated! **Technology Appendixes.** The appendixes for Excel, Minitab, and the TI-83/84 Plus have been updated to correspond to the latest versions of these three statistical technologies. These appendixes introduce the three statistical technologies, explain how to input data, and discuss how to perform other basic tasks. They are entitled *Getting Started with …* and are located in the Technology Basics section of the WeissStats Resource Site, www.pearsonhighered.com/weiss-stats.

Expanded! **Built-in Technology Manuals.** The Technology Center features (in the book) and the technology appendixes (on the WeissStats site) make it unnecessary for students to purchase technology manuals. Students who will be using Minitab, Excel, or the TI-83/84 Plus to solve exercises should study the appropriate technology appendix(es) before commencing with The Technology Center sections.

Expanded! **TI Programs.** The TI-83/84 Plus does not have built-in applications for a number of the statistical analyses discussed in the book. So that users of the TI-83/84 Plus can do such analyses with their calculators, the author has made available TI programs. Those programs are obtainable from the TI Programs section of the WeissStats Resource Site.

Computer Simulations. Computer simulations, appearing in both the text and the exercises, serve as pedagogical aids for understanding complex concepts such as sampling distributions.

StatCrunch **Interactive StatCrunch Reports.** Sixty-four StatCrunch reports have been written specifically for *Introductory Statistics*. Each report corresponds to a statistical analysis covered in the book. These interactive reports, keyed to the book with a StatCrunch icon, explain how to use StatCrunch online statistical software to solve problems previously solved by hand in the book. Go to www.statcrunch.com, choose **Explore ▼ Groups**, and search "Weiss Introductory Statistics 10/e" to access the

StatCrunch Reports. Alternatively, you can access these reports from the document Access to StatCrunch Reports.pdf, which is in the StatCrunch section of the WeissStats Resource Site. *Note:* Analyzing data in StatCrunch requires a MyStatLab or StatCrunch account.

APPLET **Java Applets.** Twenty-one Java applets have been custom written for *Introductory Statistics*. These applets, keyed to the book with an applet icon, give students additional interactive activities for the purpose of clarifying statistical concepts in an interesting and fun way. The applets are available from the Applets section of the WeissStats Resource Site.

Organization

Introductory Statistics offers considerable flexibility in choosing material to cover. The following flowchart indicates different options by showing the interdependence among chapters; the prerequisites for a given chapter consist of all chapters that have a path that leads to that chapter.

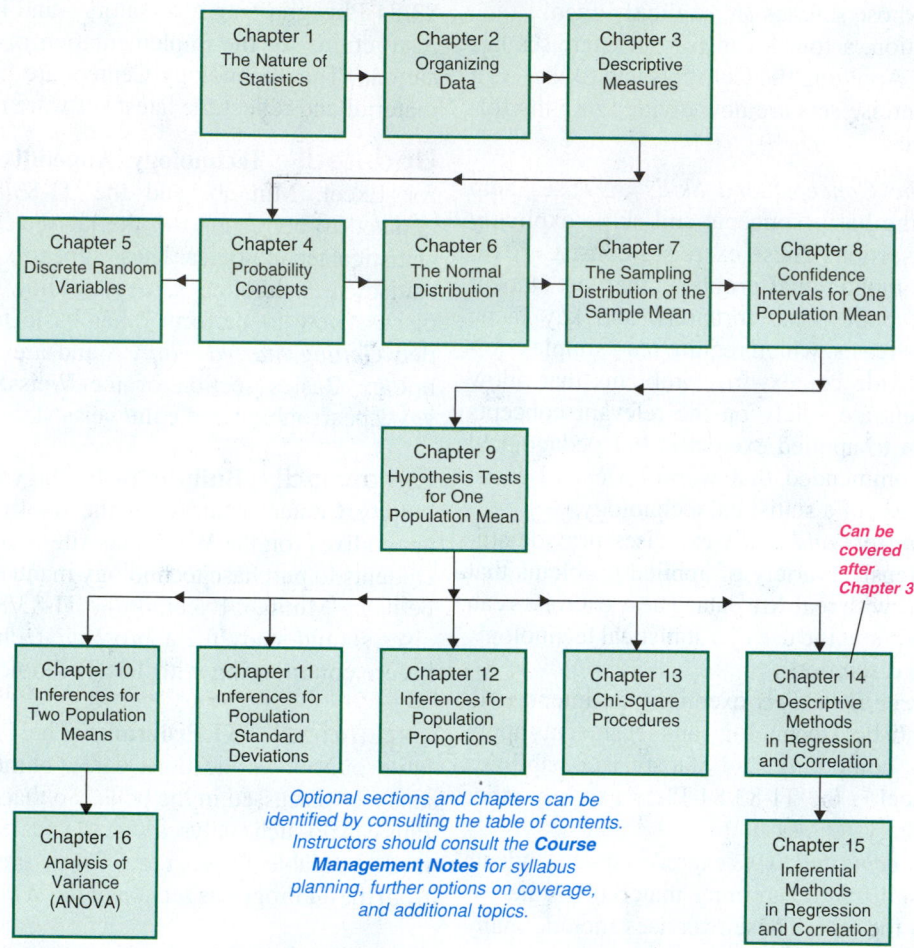

Acknowledgments

For this and the previous few editions of the book, it is our pleasure to thank the following reviewers, whose comments and suggestions resulted in significant improvements:

Olcay Akman, *Illinois State University*
James Albert, *Bowling Green State University*
John F. Beyers, II, *University of Maryland, University College*
David K. Britz, *Raritan Valley Community College*
Josef Brown, *New Mexico Tech*
Yvonne Brown, *Pima Community College*
Beth Chance, *California Polytechnic State University*
Brant Deppa, *Winona State University*
Carol DeVille, *Louisiana Tech University*

Jacqueline Fesq, *Raritan Valley Community College*
Robert Forsythe, *Frostburgh State University*
Richard Gilman, *Holy Cross College*
Donna Gorton, *Butler Community College*
David Groggel, *Miami University*
Joel Haack, *University of Northern Iowa*
Bernard Hall, *Newbury College*
Jessica Hartnett, *Gannon College*
Jane Harvill, *Baylor University*
Lance Hemlow, *Raritan Valley Community College*

Susan Herring, *Sonoma State University*
David Holmes, *The College of New Jersey*
Lorraine Hughes, *Mississippi State University*
Michael Hughes, *Miami University*
Satish Iyengar, *University of Pittsburgh*
Yvette Janecek, *Blinn College*
Jann-Huei Jinn, *Grand Valley State University*
Jeffrey Jones, *County College of Morris*
Thomas Kline, *University of Northern Iowa*
Lynn Kowski, *Raritan Valley Community College*
Christopher Lacke, *Rowan University*
Sheila Lawrence, *Rutgers University*
Tze-San Lee, *Western Illinois University*
Ennis Donice McCune, *Stephen F. Austin State University*
Jackie Miller, *The Ohio State University*
Luis F. Moreno, *Broome Community College*
Bernard J. Morzuch, *University of Massachusetts, Amherst*
Dennis M. O'Brien, *University of Wisconsin, La Crosse*
Dwight M. Olson, *John Carroll University*
Bonnie Oppenheimer, *Mississippi University for Women*
JoAnn Paderi, *Lourdes College*
Melissa Pedone, *Valencia Community College*
Alan Polansky, *Northern Illinois University*
Cathy D. Poliak, *Northern Illinois University*
Kimberley A. Polly, *Indiana University*

Geetha Ramachandran, *California State University*
B. Madhu Rao, *Bowling Green State University*
Gina F. Reed, *Gainesville College*
Steven E. Rigdon, *Southern Illinois University, Edwardsville*
Kevin M. Riordan, *South Suburban College*
Sharon Ross, *Georgia Perimeter College*
Edward Rothman, *University of Michigan*
Rina Santos, *College of Alameda*
George W. Schultz, *St. Petersburg College*
Arvind Shah, *University of South Alabama*
Sean Simpson, *Westchester Community College, SUNY*
Cid Srinivasan, *University of Kentucky, Lexington*
W. Ed Stephens, *McNeese State University*
Kathy Taylor, *Clackamas Community College*
Alane Tentoni, *Northwest Mississippi Community College*
Bill Vaughters, *Valencia Community College*
Roumen Vesselinov, *University of South Carolina*
Brani Vidakovic, *Georgia Institute of Technology*
Jackie Vogel, *Austin Peay State University*
Donald Waldman, *University of Colorado, Boulder*
Daniel Weiner, *Boston University*
Dawn White, *California State University, Bakersfield*
Marlene Will, *Spalding University*
Latrica Williams, *St. Petersburg College*
Matthew Wood, *University of Missouri, Columbia*
Nicholas A. Zaino Jr., *University of Rochester*

Our thanks are also extended to Joe Fred Gonzalez, Jr., for his many suggestions over the years for improving the book; and to Daniel Collins, Fuchun Huang, Charles Kaufman, Sharon Lohr, Richard Marchand, Shahrokh Parvini, Kathy Prewitt, Walter Reid, and Bill Steed, with whom we have had several illuminating consultations. Thanks also go to Matthew Hassett and Ronald Jacobowitz for their many helpful comments and suggestions.

Several other people provided useful input and resources. They include Thomas A. Ryan, Jr., Webster West, William Feldman, Frank Crosswhite, Lawrence W. Harding, Jr., George McManus, Greg Weiss, Jeanne Sholl, R. B. Campbell, Linda Holderman, Mia Stephens, Howard Blaut, Rick Hanna, Alison Stern-Dunyak, Dale Phibrick, Christine Sarris, and Maureen Quinn. Our sincere thanks go to all of them for their help in making this a better book.

Thanks to Larry Griffey for his formula/table card. Our gratitude also goes to Toni Garcia for writing the *Instructor's Solutions Manual* and the *Student's Solutions Manual*.

We express our appreciation to Dennis Young for his linear models modules and for his collaboration on numerous statistical and pedagogical issues. For checking the accuracy of the entire text and answers to the exercises, we extend our gratitude to Todd Hendricks and Susan Herring.

We are also grateful to David Lund and Patricia Lee for obtaining the database for the Focusing on Data Analysis sections. Our thanks are extended to the following people for their research in finding myriad interesting statistical studies and data for the examples, exercises, and case studies: Toni Garcia, Traci Gust, David Lund, Jelena Milovanovic, and Greg Weiss.

Many thanks go to Christine Stavrou and Stephanie Green for directing the development of the WeissStats Resource Site and to Cindy Scott, Carol Weiss, and Dennis Young for constructing the data files. Our appreciation also goes to our software editors, Bob Carroll and Marty Wright.

We are grateful to Kelly Ricci of Aptara Corporation, who, along with Marianne Stepanian, Shannon Steed, Chere Bemelmans, Christina Lepre, Joe Vetere, and Sonia Ashraf of Pearson Education, coordinated the development and production of the book. We also thank our copyeditor, Bret Workman, and our proofreaders, Carol Weiss, Greg Weiss, Danielle Kortan, and Cindy Scott.

To Barbara Atkinson (Pearson Education) and Rokusek Design, Inc., we express our thanks for awesome interior and cover designs. Our sincere thanks also go to all the people at Aptara for a terrific job of composition and illustration. We thank Aptara Corporation for photo research.

Without the help of many people at Pearson Education, this book and its numerous ancillaries would not have been possible; to all of them go our heartfelt thanks. In addition to the Pearson Education people mentioned above, we give special thanks to Greg Tobin and Deirdre Lynch, and to the following other people at Pearson Education: Suzanna Bainbridge, Ruth Berry, Justin Billing, Salena Casha, Erin Kelly, Kathleen DeChavez, Diahanne Lucas, Caroline Fell, and Carol Melville.

Finally, we convey our appreciation to Carol A. Weiss. Apart from writing the text, she was involved in every aspect of development and production. Moreover, Carol did a superb job of researching and writing the biographies.

N.A.W.

Supplements

Student Supplements

Student's Edition

- This version of the text includes the answers to the odd-numbered Understanding the Concepts and Skills exercises, the odd-numbered Applying the Concepts and Skills exercises, and all Review Problems of those two exercise categories. (The Instructor's Edition contains the answers to all of those exercises.)
- ISBN: 0-321-98917-1 / 978-0-321-98917-8

Student's Solutions Manual

- Written by Toni Garcia, this supplement contains detailed, worked-out solutions to the odd-numbered section exercises (Understanding the Concepts and Skills, Applying the Concepts and Skills, Working with Large Data Sets, and Extending the Concepts and Skills) and all Review Problems.
- ISBN: 0-321-98928-7 / 978-0-321-98928-4

WeissStats Resource Site (aka WeissStats site)

- This website offers universal access to an extensive array of resources: additional topics, applets, all data sets from the book in multiple formats, a procedures booklet, technology appendixes, and much more.
- URL: www.pearsonhighered.com/weiss-stats.

Instructor Supplements

Instructor's Edition

- This version of the text includes the answers to all of the Understanding the Concepts and Skills exercises and Applying the Concepts and Skills exercises. (The Student's Edition contains the answers to only the odd-numbered ones in the sections.)
- ISBN: 0-321-98982-1 / 978-0-321-98982-6

Instructor's Solutions Manual (download only)

- Written by Toni Garcia, this supplement contains detailed, worked-out solutions to all of the section exercises (Understanding the Concepts and Skills, Applying the Concepts and Skills, Working with Large Data Sets, and Extending the Concepts and Skills), the Review Problems, the Focusing on Data Analysis exercises, and the Case Study Discussion exercises.
- Available for download within MyStatLab or at www.pearsonhighered.com/irc.

Online Test Bank

- Written by Michael Butros, this supplement provides three examinations for each chapter of the text.
- Answer keys are included.
- Available for download within MyStatLab or at www.pearsonhighered.com/irc.

TestGen®

TestGen® (www.pearsoned.com/testgen) enables instructors to build, edit, print, and administer tests using a computerized bank of questions developed to cover all the objectives of the text. TestGen is algorithmically based, allowing instructors to create multiple but equivalent versions of the same question or test with the click of a button. Instructors can also modify test bank questions or add new questions. The software and testbank are available for download from Pearson Education's online catalog.

PowerPoint Lecture Presentation

- Classroom presentation slides are geared specifically to the sequence of this textbook.
- These PowerPoint slides are available within MyStatLab or at www.pearsonhighered.com/irc.

Technology Resources

Minitab®

This software is a condensed version of the Professional release of MINITAB statistical software. It offers the full range of statistical methods and graphical capabilities, along with worksheets that can include up to 10,000 data points. Individual copies of the software can be bundled with the text (CD only) ISBN: 0-13-143661-9 / 978-0-13-143661-9.

XLSTAT™ (access code required)

The XLSTAT statistical analysis add-in offers a wide variety of functions to enhance analytical capabilities of Microsoft Excel®, making it an ideal tool for your everyday data analysis and statistics requirements. This version has been specifically built for your course. XLSTAT is compatible with all Excel versions (except 2008 for Mac). To register, visit www.pearsonhighered.com/xlstat.

JMP® Student Edition

JMP Student Edition is an easy-to-use, streamlined version of JMP desktop statistical discovery software from SAS Institute Inc. and is available for bundling with the text (ISBN: 0-321-89164-3 / 978-0-321-89164-8).

IBM® SPSS® Statistics Student Version

SPSS, a statistical and data management software package, is also available for bundling with the text (ISBN: 0-321-97825-0 / 978-0-321-97825-7).

MathXL® for Statistics Online Course (access code required)

MathXL® is the homework and assessment engine that runs MyStatLab. (MyStatLab is MathXL plus a learning management system.)

With MathXL for Statistics, instructors can:

- Create, edit, and assign online homework and tests using algorithmically generated exercises correlated at the objective level to the textbook.
- Create and assign their own online exercises and import TestGen tests for added flexibility.
- Maintain records of all student work, tracked in MathXL's online gradebook.

With MathXL for Statistics, students can:

- Take chapter tests in MathXL and receive personalized study plans and/or personalized homework assignments based on their test results.
- Use the study plan and/or the homework to link directly to tutorial exercises for the objectives they need to study.
- Students can also access supplemental animations and video clips directly from selected exercises.

- Knowing that students often use external statistical software, we make it easy to copy our data sets, both from the eText and the MyStatLab questions, into software like StatCrunch™, Minitab, Excel, and more.

MathXL for Statistics is available to qualified adopters. For more information, visit our website at www.mathxl.com, or contact your Pearson representative.

MyStatLab™ Online Course (access code required)

MyStatLab from Pearson is the world's leading online resource in statistics, integrating interactive homework, assessment, and media in a flexible, easy-to-use format. MyStatLab is a course management system that delivers proven results in helping individual students succeed.

- MyStatLab can be implemented successfully in any environment—lab-based, hybrid, fully online, traditional—and demonstrates the quantifiable difference that integrated usage has on student retention, subsequent success, and overall achievement.
- MyStatLab's comprehensive online gradebook automatically tracks students' results on tests, quizzes, homework, and in the study plan. Instructors can use the gradebook to provide positive feedback or intervene if students have trouble. Gradebook data can be easily exported to a variety of spreadsheet programs, such as Microsoft Excel. Instructors can determine which points of data to export, and then analyze the results to determine success.

MyStatLab provides engaging experiences that personalize, stimulate, and measure learning for each student. In addition to the resources below, each course includes a full interactive online version of the accompanying textbook.

- *Tutorial Exercises with Multimedia Learning Aids*: The homework and practice exercises in MyStatLab align with the exercises in the textbook, and most regenerate algorithmically to give students unlimited opportunity for practice and mastery. Exercises offer immediate helpful feedback, guided solutions, sample problems, animations, videos, and eText clips for extra help at point-of-use.
- *MyStatLab Accessibility*: MyStatLab is compatible with the JAWS 12/13 screen reader, and enables multiple-choice and free-response problem-types to be read and interacted with via keyboard controls and math notation input.
- *StatTalk Videos*: Fun-loving statistician Andrew Vickers takes to the streets of Brooklyn, NY, to demonstrate important statistical concepts through interesting stories and real-life events. This series of 24 fun and engaging videos will help students actually understand statistical concepts. Available with an instructor's user guide and assessment questions.

- *Additional Question Libraries*: In addition to algorithmically regenerated questions that are aligned with your textbook, MyStatLab courses come with two additional question libraries:

  - 450 exercises in Getting Ready for Statistics cover the developmental math topics students need for the course. These can be assigned as a prerequisite to other assignments, if desired.
  - 1000 exercises in the Conceptual Question Library require students to apply their statistical understanding.

- *StatCrunch*™: MyStatLab integrates the web-based statistical software, StatCrunch, within the online assessment platform so that students can easily analyze data sets from exercises and the text. In addition, MyStatLab includes access to www.statcrunch.com, a website where users can access tens of thousands of shared data sets, create and conduct online surveys, perform complex analyses using the powerful statistical software, and generate compelling reports.

- *Statistical Software, Support and Integration*: We make it easy to copy our data sets, both from the eText and the MyStatLab questions, into software such as StatCrunch, Minitab, Excel, and more. Students have access to a variety of support tools—Technology Tutorial Videos, Technology Study Cards, and Technology Manuals for select titles—to learn how to effectively use statistical software. And, MyStatLab comes from an experienced partner with educational expertise and an eye on the future.

- Knowing that you are using a Pearson product means knowing that you are using quality content. That means that our eTexts are accurate and our assessment tools work. It means we are committed to making MyStatLab as accessible as possible.

- Whether you are just getting started with MyStatLab, or have a question along the way, we're here to help you learn about our technologies and how to incorporate them into your course.

To learn more about how MyStatLab combines proven learning applications with powerful assessment, visit www.mystatlab.com or contact your Pearson representative.

StatCrunch® StatCrunch

StatCrunch is powerful web-based statistical software that allows users to perform complex analyses, share data sets, and generate compelling reports of their data. The vibrant online community offers tens of thousands of shared data sets for students to analyze.

- *Collect.* Users can upload their own data to StatCrunch or search a large library of publicly shared data sets, spanning almost any topic of interest. Also, an online survey tool allows users to quickly collect data via web-based surveys.
- *Crunch.* A full range of numerical and graphical methods allow users to analyze and gain insights from any data set. Interactive graphics help users understand statistical concepts, and are available for export to enrich reports with visual representations of data.
- *Communicate.* Reporting options help users create a wide variety of visually-appealing representations of their data.

Full access to StatCrunch is available with a MyStatLab kit, and StatCrunch is available by itself to qualified adopters. StatCrunch Mobile is now available, just visit www.statcrunch.com from the browser on your smartphone or tablet. For more information, visit our website at www.statcrunch.com, or contact your Pearson representative.

Data Sources

1stock1
A Handbook of Small Data Sets
A. C. Nielsen Company
AAA Foundation for Traffic Safety
AAMC Faculty Roster
AAUP Annual Report on the Economic Status of the Profession
ABC Global Kids Study
ABCNEWS Poll
ABCNews.com
About.com Pediatrics
Accident Facts
ACT High School Profile Report
ACT, Inc.
Acta Opthalmologica
AFI's 100 Years… 100 Movies — 10th Anniversary Edition
Agricultural Marketing Service
Agricultural Research Service
AHA Hospital Statistics
Air Travel Consumer Report
Alcohol Consumption and Related Problems: Alcohol and Health Monograph 1
All About Diabetes
Alliance for Cervical Cancer Protection
Alzheimer's Care Quarterly
American Association of University Professors
American Community Survey
American Council of Life Insurers
American Demographics
American Diabetes Association
American Express Retail Index
American Film Institute
American Hospital Association
American Hospital Association Annual Survey
American Housing Survey for the United States
American Industrial Hygiene Association Journal
American Journal of Applied Sciences
American Journal of Clinical Nutrition
American Journal of Obstetrics and Gynecology
American Journal of Physical Anthropology
American Journal of Political Science
American Laboratory
American Scientist
American Statistical Association

American Statistician
American Veterinary Medical Association
American Wedding Study
America's Families and Living Arrangements
America's Network Telecom Investor Supplement
Amstat News
Amusement Business
Analytical Chemistry
Analytical Services Division Transport Statistics
Animal Action Report
Animal Behaviour
Annals of Epidemiology
Anthropometric Reference Data for Children and Adults
Appetite
Applied Psychology in Criminal Justice
Aquaculture
Aquatic Biology
Arbitron
Archives of Physical Medicine and Rehabilitation
Arizona Chapter of the American Lung Association
Arizona Department of Revenue
Arizona Republic
Arizona Residential Property Valuation System
Arizona State University
Arizona State University Enrollment Summary
Arthritis Today
Asian Import
Associated Newspapers Ltd
Associated Press
Association of American Medical Colleges
Association of American Universities
Atlantic Oceanographic & Meteorological Laboratory
Atlantic Hurricane Database
Auckland University of Technology
Augusta National Golf Club
Australian Journal of Rural Health
Australian Journal of Zoology
Auto Trader
Avis Rent-A-Car
Baltimore Ravens
BARRON'S
Baseball Almanac

BBC News Magazine
Beachbody, LLC
Beer Institute Annual Report
Behavior Research Center
Behavioral Ecology and Sociobiology
Behavioral Risk Factor Surveillance System Summary Prevalence Report
Behavioural Pharmacology
Bell Systems Technical Journal
Biofuel Transportation Database
Biological Conservation
Biology of Sex Differences
Biometrics
Biometrika
BioScience
Board of Governors of the Federal Reserve System
Boston Athletic Association
Boston Globe
Box Office Mojo
Boyce Thompson Southwestern Arboretum
Brewer's Almanac
Bride's Magazine
British Bankers' Association
British Journal of Educational Psychology
British Journal of Haematology
British Journal of Visual Impairment
British Medical Journal
Brokerage Report
Bureau of Crime Statistics and Research of Australia
Bureau of Economic Analysis
Bureau of Educational and Cultural Affairs
Bureau of Justice Statistics
Bureau of Justice Statistics Special Report
Bureau of Labor Statistics
Business Times
Buyers of New Cars
California Agriculture
California Wild: Natural Sciences for Thinking Animals
Car Shopping Trends Report
CareerBuilder
CBS News
Celebrity Net Worth
Cellular Telecommunications & Internet Association
Census of Agriculture
Centers for Disease Control and Prevention
Central Election Commission of the Russian Federation

Central Intelligence Agency
Chance
Characteristics of New Housing
Chemical & Pharmaceutical Bulletin
Chesapeake Biological Laboratory
Climates of the World
Climatography of the United States
Clinical Journal of Sports Medicine
Clinical Linguistics and Phonetics
CNBC
CNN/USA TODAY
Coleman & Associates, Inc.
College Board
College Entrance Examination Board
College-Bound Seniors Total Group Profile Report
Communications Industry Forecast & Report
Compendium of Federal Justice Statistics
Conde Nast Bridal Group
Congressional Directory
Consumer Expenditure Survey
Consumer Reports
Contributions to Boyce Thompson Institute
Controlling Road Rage: A Literature Review and Pilot Study
Crime in the United States
Criminal Justice and Behavior
CTIA–The Wireless Association
Current Housing Reports
Current Population Reports
Current Population Survey
Daily Mail
Daily Racing Form
Dallas Mavericks Roster
Data from the National Health Interview Survey
DataGenetics
Dave Leip's Atlas of U.S. Presidential Elections
Deep Sea Research Part I: Oceanographic Research Papers
Demographic Profiles
Demography
Department of Information Resources and Communications
Desert Samaritan Hospital
Dietary Guidelines for Americans
Dietary Reference Intakes
Digest of Education Statistics
Discover
Early Medieval Europe
Eastern Mediterranean Health Journal
Ecology
Economic Development Corporation Report
Edinburgh Medical and Surgical Journal
Edison Research
Edmunds.com
Educational Attainment in the United States
Educational Research
Educational Testing Service
eMarketer
Employment and Earnings
Energy Information Administration

Environmental Geology
ESPN
ESPN MLB Scoreboard
Estimates of School Statistics Database
Everyday Health Network
Experimental Agriculture
Experimental Brain Research
Family & Intimate Partner Violence Quarterly
Family Planning Perspectives
Federal Bureau of Investigation
Federal Highway Administration
Federal Highway Administration Annual Highway Statistics
Federal Reserve System
Federal Bureau of Prisons
Financial Planning
Fixed-Site Amusement Ride Injury Survey
FlightStats On-time Performance Report Summary
Florida Department of Environmental Protection
Florida State Center for Health Statistics
Food Consumption, Prices, and Expenditures
Footwear News
Forbes
Forest Mensuration
Fortune Magazine
Friends of the Earth
Fuel Economy Guide
Gallup
Gallup Poll
Gallup, Inc.
Geography
Georgia State University
Global Attractions Attendance Report
Global Index of Religiosity and Atheism
Golf Laboratories, Inc.
Golf.com
Governors' Political Affiliations & Terms of Office
Graduating Student and Alumni Survey
GRE Guide to the Use of Scores
Hanna Properties
Harris Interactive
Harris Poll
Harvard University
Heredity
Higher Education Research Institute
Highway Construction Safety and the Aging Driver
Highway Statistics
Hilton Hotels Corporation
Hirslanden Clinic
Historical Income Tables
HIV Surveillance Report
Hollywood Demographics
Homestyle Pizza
Hospital Statistics
HuffPost
HuffPost Style
Human Biology

Hydrobiologia
Income, Individual Income Tax Returns
Income, Poverty and Health Insurance Coverage in the United States
Indiewire
Industry Research
Infoplease
Information Please Almanac
Injury Facts
Inside MS
Institute of Medicine of the National Academy of Sciences
Insurance Institute for Highway Safety
Internal Revenue Service
International Association of Amusement Parks and Attractions.
International Classification of Diseases
International Communications Research
International Data Base
International Journal of Obesity
International Journal of Public Health
Iowa Agriculture Experiment Station
Iowa State University
Japan Automobile Manufacturer's Association
Japan Statistics Bureau
JAVMA News
Joint Committee on Printing
Journal of Abnormal Psychology
Journal of Advertising Research
Journal of American College Health
Journal of Anaesthesiology Clinical Pharmacology
Journal of Anatomy
Journal of Applied Behavioral Analysis
Journal of Applied Ecology
Journal of Applied Ichthyology
Journal of Applied Psychology
Journal of Applied Research in Higher Education
Journal of Applied Social Psychology
Journal of Applied Sport Psychology
Journal of Bone and Joint Surgery
Journal of Chemical Ecology
Journal of Child Nutrition and Management
Journal of Chronic Diseases
Journal of Clinical Endocrinology & Metabolism
Journal of Clinical Oncology
Journal of Early Adolescence
Journal of Environmental Psychology
Journal of Environmental Science and Health
Journal of Experimental Biology
Journal of Experimental Social Psychology
Journal of Family Violence
Journal of Forensic Identification
Journal of Gerontology Series A: Biological Sciences and Medical Sciences
Journal of Health, Population and Nutrition
Journal of Herpetology
Journal of Human Evolution
Journal of Mammalogy

*Journal of Nursing and Healthcare of
    Chronic Illness*
Journal of Nutrition
Journal of Organizational Behavior
Journal of Paleontology
Journal of Pediatrics
Journal of Pharmaceutical Sciences
Journal of Poverty & Social Justice
Journal of Pregnancy
Journal of Prosthetic Dentistry
Journal of Real Estate and Economics
Journal of Statistics Education
Journal of Sustainable Tourism
*Journal of the American Academy of Child
    and Adolescent Psychiatry*
Journal of the American Geriatrics Society
*Journal of the American Medical
    Association*
*Journal of the American Public Health
    Association*
Journal of the Royal Statistical Society
Journal of Tropical Ecology
*Journal of Water Resources Planning and
    Management*
Journal of Wildlife Management
Journal of Zoology, London
*Journalism & Mass Communication
    Quarterly*
Kelley Blue Book
Kelley Blue Book Company
Kennedy: The Classical Biography
Labor Force Statistics
Land Economics
Lawlink
Leonard Maltin Movie Guide
Life Expectancy at Birth
Life Insurers Fact Book
Limnology and Oceanography
Literary Digest
Los Angeles Dodgers
Los Angeles Times
Main Economic Indicators
Mammalia
Management
Manufactured Housing Statistics
Marine Ecology Progress Series
Marine Mammal Science
Market Survey of Long-Term Care Costs
Mayo Clinical Proceedings
*Median Sales Price of Existing
    Single-Family Homes for Metropolitan
    Areas*
Medical Biology and Etruscan Origins
Medical College of Wisconsin Eye Institute
Medical Principles and Practice
Medicine and Science in Sports & Exercise
Mega Millions
Mellman Group
Merck Manual
MetLife Mature Market Institute
Minitab Inc.
MLB.com
Money Stock Measures

Monitoring the Future
Monthly Labor Review
Monthly Tornado Statistics
Morningstar
Morrison Planetarium
Motor Vehicle Statistics of Japan
Motorcycle USA
National Aeronautics and Space
    Administration
National Agricultural Statistics Service
National Anti-Vivisection Society
National Association of Colleges and
    Employers
National Association of Realtors
*National Association of State Racing
    Commissioners*
National Basketball Association
National Cancer Institute
National Center for Education Statistics
National Center for Health Statistics
National Center on Addiction and Substance
    Abuse at Columbia University
National Collegiate Athletic Association
National Constitution Center
National Corrections Reporting Program
National Education Association
National Football League
National Geographic
National Geographic Society
National Golf Foundation
National Governors Association
*National Health and Nutrition Examination
    Survey*
National Health Interview Study
National Health Interview Survey
National Highway Traffic Safety
    Administration
*National Household Travel Survey, Summary
    of Travel Trends*
National Hurricane Center
National Institute of Aging
National Institute of Child Health and
    Human Development Neonatal Research
    Network
National Institute of Hygiene
National Institute of Mental Health
National Institute on Drug Abuse
National Interagency Fire Center
National Longitudinal Survey of Youth
National Low Income Housing Coalition
National Mortgage News
National Oceanic & Atmospheric
    Administration
National Safety Council
National Science Foundation
National Survey on Drug Use and Health
National Vital Statistics Reports
Nature
NCAA
NCAA.com
New England Journal of Medicine
New England Patriots Roster
New Scientist

New York Times
Newsweek
Newsweek, Inc
NewYork Times
Nielsen Media Research
Nielsen Report on Television
Nigerian Medical Journal
Nutrition
Obstetrics & Gynecology
Occupant Restraint Use
OECD in Figures
Office of Aviation Enforcement and
    Proceedings
Office of Immigration Statistics
Office of Justice Programs
Opinion Dynamics Poll
Opinion Research Corporation
Organisation for Economic Co-operation
    and Development
Origin of Species
Osteoporosis International
Out of Reach
Parade Magazine
Payless ShoeSource
Peacecorps.org
Pediatric Research
Pediatrics
Penn Schoen Berland
Pew Forum on Religion and Public Life
Pew Internet & American Life Project
Pew Research Center
PGA TOUR
Philosophical Magazine
Phoenix Gazette
Physician Specialty Data Book
PIN analysis
Player Roster
PLOS Biology
PLOS ONE
Pollstar
Popular Mechanics
*Population-at-Risk Rates and Selected
    Crime Indicators*
Preventative Medicine
pricewatch.com
Primetime Broadcast Programs
Prison Statistics
*Proceedings of the 6th Berkeley Symposium
    on Mathematics and Statistics, VI*
*Proceedings of the American Zoo and
    Aquarium Association Nutrition Advisory
    Group*
*Proceedings of the National Academy of
    Science USA*
Proceedings of the Royal Society of London
Professional Golfers' Association of
    America
Profile of Jail Inmates
Psychology of Addictive Behaviors
Pulse Opinion Research, LLC
Quality Engineering
Quinnipiac University
R. L. Polk & Co.

R. R. Bowker Company
Ranking of the States and Estimates of School Statistics
Rasmussen Reports
Recording Industry Association of America
Religious Landscape Survey
Reports
Research Quarterly for Exercise and Sport
Research Resources, Inc.
Residential Energy Consumption Survey: Consumption and Expenditures
Richard's Heating and Cooling
Robson Communities, Inc.
Roche
Roper Starch Worldwide, Inc.
Rubber Age
Runner's World
Salary Survey
Scarborough Research
Science
Science and Engineering Degrees
Science and Engineering Doctorate Awards
Science and Engineering Indicators
Science News
Scientific American
Scottish Executive
Selected Manpower Statistics
Semi-annual Wireless Survey
Sexually Transmitted Disease Surveillance
Significance Magazine
Smartphone Ownership
Sneak Previews
Snell, Perry and Associates
Soccer & Society
Social Forces
Social Indicators Research
South Carolina Budget and Control Board
South Carolina Statistical Abstract
Southwest Airlines
Sports Illustrated
Sports Illustrated Sites
SportsCenturyRetrospective
Stanford Revision of the Binet–Simon Intelligence Scale
Statistical Abstract of the United States
Statistical Report
Statistical Summary of Students and Staff
Statistical Yearbook
Statistics Norway
Statistics of Income, Individual Income Tax Returns
STATS
Status of the Profession
Stock Performance Guide
Stockholm Transit District
Storm Prediction Center
Summary of Travel Trends
Surveillance Epidemiology and End Results Fact Sheet
Survey of Consumer Finances
Survey of Current Business
Survey of Graduate Science Engineering Students and Postdoctorates
TalkBack Live

Teaching Issues and Experiments in Ecology
Technometrics
Television Bureau of Advertising
Tempe Daily News
Tesla Motors
Texas Comptroller of Public Accounts
The AMATYC Review
The American Freshman
The American Statistician
The Bowker Annual Library and Book Trade Almanac
The Business Journal
The Cross-Platform Report
The Design and Analysis of Factorial Experiments
The Earth: Structure, Composition and Evolution
The Geyser Observation and Study Association
The History of Statistics
The Infinite Dial
The Journal of Arachnology
The Lancet
The Lobster Almanac
The Marathon: Physiological, Medical, Epidemiological, and Psychological Studies
The Methods of Statistics
The Nielsen Company
The Open University
The Plant Cell
The Street
The Washington Post
The World Bank
Themed Entertainment Association
Thoroughbred Times
TIME
Time Spent Viewing
Times Higher Education
TIMS
TNS Intersearch
Trade & Environment Database (TED) Case Studies
Trademark Reporter
Travel + Leisure Golf
Trends in Television
Tropical Biodiversity
Tropical Cyclone Report
TV Basics
TVbytheNumbers
U.S. Agency for International Development
U.S. Agricultural Trade Update
U.S. Bureau of Citizenship and Immigration Services
U.S. Bureau of Economic Analysis
U.S. Census Bureau
U.S. Coast Guard
U.S. Congress, Joint Committee on Printing
U.S. Department of Agriculture
U.S. Department of Commerce
U.S. Department of Defense
U.S. Department of Education
U.S. Department of Energy

U.S. Department of Health and Human Services
U.S. Department of Housing and Urban Development
U.S. Department of Justice
U.S. Energy Information Administration
U.S. Environmental Protection Agency
U.S. Federal Highway Administration
U.S. Geological Survey
U.S. National Center for Health Statistics
U.S. National Oceanic and Atmospheric Administration
U.S. Naturalizations
U.S. News and World Report
U.S. Overseas Loans and Grants
U.S. Postal Service
U.S. Substance Abuse and Mental Health Services Administration
U.S. Women's Open
Ultrasound in Medicine and Biology
Uniform Crime Reports
United States Pharmacopeia
University of Delaware
University of Helsinki
University of Malaysia
University of Maryland
University of Nevada, Las Vegas
Urban Studies
USA TODAY
Usability News
Utah Behavioral Risk Factor Surveillance System (BRFSS) Local Health District Report
Vegetarian Journal
Vegetarian Resource Group
VentureOne Corporation
Veronis Suhler Stevenson
Vital and Health Statistics
Vital Statistics of the United States
Wall Street Journal
Washington Post
Weatherwise
Weekly Retail Gasoline and Diesel Prices
Wichita Eagle
Wikipedia
WIN-Gallup International
Women and Cardiovascular Disease Hospitalizations
Women's Health Initiative
WONDER database
World Almanac
World FactBook
World Meteorological Association
World Radiation Center
World Series History
www.house.gov
YAHOO News
Yahoo! Contributor Network
Year-End Industry Shipment and Revenue Statistics
YouGov
Zillow.com
Zogby International

The Nature of Statistics

CHAPTER OBJECTIVES

What does the word *statistics* bring to mind? To most people, it suggests numerical facts or data, such as unemployment figures, farm prices, or the number of marriages and divorces. Two common definitions of the word *statistics* are as follows:

1. [used with a plural verb] facts or data, either numerical or nonnumerical, organized and summarized so as to provide useful and accessible information about a particular subject.
2. [used with a singular verb] the science of organizing and summarizing numerical or nonnumerical information.

Statisticians also analyze data for the purpose of making generalizations and decisions. For example, a political analyst can use data from a portion of the voting population to predict the political preferences of the entire voting population, or a city council can decide where to build a new airport runway based on environmental impact statements and demographic reports that include a variety of statistical data.

In this chapter, we introduce some basic terminology so that the various meanings of the word *statistics* will become clear to you. We also examine two primary ways of producing data, namely, through sampling and experimentation. We discuss sampling designs in Sections 1.2 and 1.3 and experimental designs in Section 1.4.

CASE STUDY

Top Films of All Time

Honoring the 10th anniversary of its award-winning series, the American Film Institute (AFI) again conducted a poll of 1500 film artists, critics, and historians, asking them to pick their 100 favorite films from a list of 400. The films on the list were made between 1915 and 2005.

After tallying the responses, AFI compiled a list representing the top 100 films. *Citizen Kane,* made in 1941, again finished in first place, followed by *The Godfather,* which was made in 1972. The following table shows the top 40 finishers in the poll. [SOURCE: Data from *AFI's 100 Years...100 Movies — 10th Anniversary Edition*. Published by the American Film Institute.]

Rank	Film	Year	Rank	Film	Year
1	Citizen Kane	1941	21	Chinatown	1974
2	The Godfather	1972	22	Some Like It Hot	1959
3	Casablanca	1942	23	The Grapes of Wrath	1940
4	Raging Bull	1980	24	E.T. The Extra-Terrestrial	1982
5	Singin' in the Rain	1952	25	To Kill a Mockingbird	1962
6	Gone with the Wind	1939	26	Mr. Smith Goes to Washington	1939
7	Lawrence of Arabia	1962	27	High Noon	1952
8	Schindler's List	1993	28	All About Eve	1950
9	Vertigo	1958	29	Double Indemnity	1944
10	The Wizard of Oz	1939	30	Apocalypse Now	1979
11	City Lights	1931	31	The Maltese Falcon	1941
12	The Searchers	1956	32	The Godfather Part II	1974
13	Star Wars	1977	33	One Flew Over the Cuckoo's Nest	1975
14	Psycho	1960	34	Snow White and the Seven Dwarfs	1937
15	2001: A Space Odyssey	1968	35	Annie Hall	1977
16	Sunset Blvd.	1950	36	The Bridge on the River Kwai	1957
17	The Graduate	1967	37	The Best Years of Our Lives	1946
18	The General	1927	38	The Treasure of the Sierra Madre	1948
19	On the Waterfront	1954	39	Dr. Strangelove	1964
20	It's a Wonderful Life	1946	40	The Sound of Music	1965

Armed with the knowledge that you gain in this chapter, you will be asked to further analyze this AFI poll at the end of the chapter.

1.1   Statistics Basics

You probably already know something about statistics. If you read newspapers, surf the Web, watch the news on television, or follow sports, you see and hear the word *statistics* frequently. In this section, we use familiar examples such as baseball statistics and voter polls to introduce the two major types of statistics: **descriptive statistics** and **inferential statistics.** We also introduce terminology that helps differentiate among various types of statistical studies.

Descriptive Statistics

Each spring in the late 1940s, President Harry Truman officially opened the major league baseball season by throwing out the "first ball" at the opening game of the Washington Senators. We use the 1948 baseball season to illustrate the first major type of statistics, descriptive statistics.

EXAMPLE 1.1   Descriptive Statistics

The 1948 Baseball Season In 1948, the Washington Senators (Nationals) played 153 games, winning 56 and losing 97. They finished seventh in the American League and were led in hitting by Bud Stewart, whose batting average was .279. Baseball statisticians compiled these and many other statistics by organizing the complete records for each game of the season.

Although fans take baseball statistics for granted, much time and effort is required to gather and organize them. Moreover, without such statistics, baseball would be much harder to follow. For instance, imagine trying to select the best hitter in the American League given only the official score sheets for each game. (More than 600 games were played in 1948; the best hitter was Ted Williams, who led the league with a batting average of .369.)

The work of baseball statisticians is an illustration of *descriptive statistics*.

DEFINITION 1.1

Descriptive Statistics

Descriptive statistics consists of methods for organizing and summarizing information.

Descriptive statistics includes the construction of graphs, charts, and tables and the calculation of various descriptive measures such as averages, measures of variation, and percentiles. We discuss descriptive statistics in detail in Chapters 2 and 3.

Inferential Statistics

We use the 1948 presidential election to introduce the other major type of statistics, inferential statistics.

EXAMPLE 1.2 **Inferential Statistics**

The 1948 Presidential Election In the fall of 1948, President Truman was concerned about statistics. The Gallup Poll taken just prior to the election predicted that he would win only 44.5% of the vote and be defeated by the Republican nominee, Thomas E. Dewey. But the statisticians had predicted incorrectly. Truman won more than 49% of the vote and, with it, the presidency. The Gallup Organization modified some of its procedures and has correctly predicted the winner ever since.

Political polling provides an example of inferential statistics. Interviewing everyone of voting age in the United States on their voting preferences would be expensive and unrealistic. Statisticians who want to gauge the sentiment of the entire **population** of U.S. voters can afford to interview only a carefully chosen group of a few thousand voters. This group is called a **sample** of the population. Statisticians analyze the information obtained from a sample of the voting population to make inferences (draw conclusions) about the preferences of the entire voting population. Inferential statistics provides methods for drawing such conclusions.

The terminology just introduced in the context of political polling is used in general in statistics.

DEFINITION 1.2

Population and Sample

Population: The collection of all individuals or items under consideration in a statistical study.

Sample: That part of the population from which information is obtained.

Figure 1.1 on the following page depicts the relationship between a population and a sample from the population.

Now that we have discussed the terms *population* and *sample,* we can define *inferential statistics.*

DEFINITION 1.3

Inferential Statistics

Inferential statistics consists of methods for drawing and measuring the reliability of conclusions about a population based on information obtained from a sample of the population.

FIGURE 1.1
Relationship between population and sample

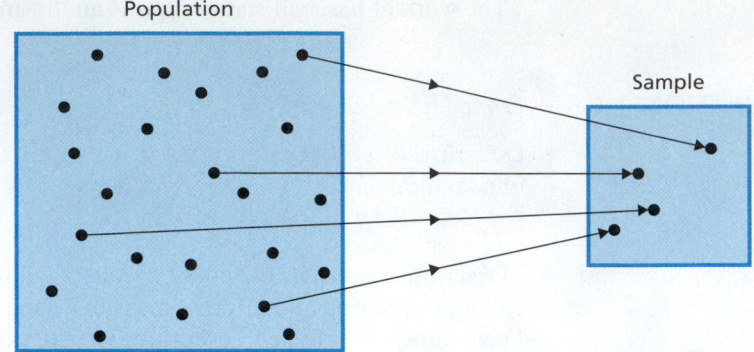

Descriptive statistics and inferential statistics are interrelated. You must almost always use techniques of descriptive statistics to organize and summarize the information obtained from a sample before carrying out an inferential analysis. Furthermore, as you will see, the preliminary descriptive analysis of a sample often reveals features that lead you to the choice of (or to a reconsideration of the choice of) the appropriate inferential method.

Classifying Statistical Studies

As you proceed through this book, you will obtain a thorough understanding of the principles of descriptive and inferential statistics. In this section, you will classify statistical studies as either descriptive or inferential. In doing so, you should consider the purpose of the statistical study.

If the purpose of the study is to examine and explore information for its own intrinsic interest only, the study is descriptive. However, if the information is obtained from a sample of a population and the purpose of the study is to use that information to draw conclusions about the population, the study is inferential.

Thus, a descriptive study may be performed either on a sample or on a population. Only when an inference is made about the population, based on information obtained from the sample, does the study become inferential.

Examples 1.3 and 1.4 further illustrate the distinction between descriptive and inferential studies. In each example, we present the result of a statistical study and classify the study as either descriptive or inferential. Classify each study yourself before reading our explanation.

■■■   **EXAMPLE 1.3**   **Classifying Statistical Studies**

The 1948 Presidential Election Table 1.1 displays the voting results for the 1948 presidential election.

TABLE 1.1
Final results of the 1948 presidential election

You try it!

Ticket	Votes	Percentage
Truman–Barkley (Democratic)	24,179,345	49.7
Dewey–Warren (Republican)	21,991,291	45.2
Thurmond–Wright (States Rights)	1,176,125	2.4
Wallace–Taylor (Progressive)	1,157,326	2.4
Thomas–Smith (Socialist)	139,572	0.3

Exercise 1.7 on page 7

Classification This study is descriptive. It is a summary of the votes cast by U.S. voters in the 1948 presidential election. No inferences are made.   ■

■■■   EXAMPLE 1.4   Classifying Statistical Studies

Testing Baseballs   For the 101 years preceding 1977, the major leagues purchased baseballs from the Spalding Company. In 1977, that company stopped manufacturing major league baseballs, and the major leagues then bought their baseballs from the Rawlings Company.

Early in the 1977 season, pitchers began to complain that the Rawlings ball was "livelier" than the Spalding ball. They claimed it was harder, bounced farther and faster, and gave hitters an unfair advantage. Indeed, in the first 616 games of 1977, 1033 home runs were hit, compared to only 762 home runs hit in the first 616 games of 1976.

Sports Illustrated magazine sponsored a study of the liveliness question and published the results in the article "They're Knocking the Stuffing Out of It" (*Sports Illustrated*, June 13, 1977, pp. 23–27) by L. Keith. In this study, an independent testing company randomly selected 85 baseballs from the current (1977) supplies of various major league teams. It measured the bounce, weight, and hardness of the chosen baseballs and compared these measurements with measurements obtained from similar tests on baseballs used in 1952, 1953, 1961, 1963, 1970, and 1973.

The conclusion was that "...the 1977 Rawlings ball is livelier than the 1976 Spalding, but not as lively as it could be under big league rules, or as the ball has been in the past."

Classification   This study is inferential. The independent testing company used a sample of 85 baseballs from the 1977 supplies of major league teams to make an inference about the population of all such baseballs. (An estimated 360,000 baseballs were used by the major leagues in 1977.)   ■

You try it!

Exercise 1.9 on page 7

The *Sports Illustrated* study also shows that it is often not feasible to obtain information for the entire population. Indeed, after the bounce and hardness tests, all of the baseballs sampled were taken to a butcher in Plainfield, New Jersey, to be sliced in half so that researchers could look inside them. Clearly, testing every baseball in this way would not have been practical.

The Development of Statistics

Historically, descriptive statistics appeared before inferential statistics. Censuses were taken as long ago as Roman times. Over the centuries, records of such things as births, deaths, marriages, and taxes led naturally to the development of descriptive statistics.

Inferential statistics is a newer arrival. Major developments began to occur with the research of Karl Pearson (1857–1936) and Ronald Fisher (1890–1962), who published their findings in the early years of the twentieth century. Since the work of Pearson and Fisher, inferential statistics has evolved rapidly and is now applied in a myriad of fields.

Familiarity with statistics will help you make sense of many things you read in newspapers and magazines and on the Internet. For instance, could the *Sports Illustrated* baseball test (Example 1.4), which used a sample of only 85 baseballs, legitimately draw a conclusion about 360,000 baseballs? After working through Chapter 9, you will understand why such inferences are reasonable.

? What Does It Mean?

An understanding of statistical reasoning and of the basic concepts of descriptive and inferential statistics has become mandatory for virtually everyone, in both their private and professional lives.

Observational Studies and Designed Experiments

Besides classifying statistical studies as either descriptive or inferential, we often need to classify them as either *observational studies* or *designed experiments*. In an **observational study,** researchers simply observe characteristics and take measurements, as in a sample survey. In a **designed experiment,** researchers impose treatments and controls (discussed in Section 1.4) and then observe characteristics and take measurements. Observational studies can reveal only *association,* whereas designed experiments can help establish *causation.*

Note that, in an observational study, someone is observing data that already exist (i.e., the data were there and would be there whether someone was interested in them or not). In a designed experiment, however, the data do not exist until someone does something (the experiment) that produces the data. Examples 1.5 and 1.6 illustrate some major differences between observational studies and designed experiments.

EXAMPLE 1.5   An Observational Study

Vasectomies and Prostate Cancer   Approximately 450,000 vasectomies are performed each year in the United States. In this surgical procedure for contraception, the tube carrying sperm from the testicles is cut and tied.

Several studies have been conducted to analyze the relationship between vasectomies and prostate cancer. The results of one such study by E. Giovannucci et al. appeared in the paper "A Retrospective Cohort Study of Vasectomy and Prostate Cancer in U.S. Men" (*Journal of the American Medical Association*, Vol. 269(7), pp. 878–882).

Dr. Giovannucci, study leader and epidemiologist at Harvard-affiliated Brigham and Women's Hospital, said that "...we found 113 cases of prostate cancer among 22,000 men who had a vasectomy. This compares to a rate of 70 cases per 22,000 among men who didn't have a vasectomy."

The study shows about a 60% elevated risk of prostate cancer for men who have had a vasectomy, thereby revealing an association between vasectomy and prostate cancer. But does it establish causation: that having a vasectomy causes an increased risk of prostate cancer?

The answer is no, because the study was observational. The researchers simply observed two groups of men, one with vasectomies and the other without. Thus, although an association was established between vasectomy and prostate cancer, the association might be due to other factors (e.g., temperament) that make some men more likely to have vasectomies and also put them at greater risk of prostate cancer.

You try it!

Exercise 1.19 on page 8

EXAMPLE 1.6   A Designed Experiment

Folic Acid and Birth Defects   For several years, evidence had been mounting that folic acid reduces major birth defects. Drs. A. E. Czeizel and I. Dudas of the National Institute of Hygiene in Budapest directed a study that provided the strongest evidence to date. Their results were published in the paper "Prevention of the First Occurrence of Neural-Tube Defects by Periconceptional Vitamin Supplementation" (*New England Journal of Medicine*, Vol. 327(26), p. 1832).

For the study, the doctors enrolled 4753 women prior to conception and divided them randomly into two groups. One group took daily multivitamins containing 0.8 mg of folic acid, whereas the other group received only trace elements (minute amounts of copper, manganese, zinc, and vitamin C). A drastic reduction in the rate of major birth defects occurred among the women who took folic acid: 13 per 1000, as compared to 23 per 1000 for those women who did not take folic acid.

In contrast to the observational study considered in Example 1.5, this is a designed experiment and does help establish causation. The researchers did not simply observe two groups of women but, instead, randomly assigned one group to take daily doses of folic acid and the other group to take only trace elements.

You try it!

Exercise 1.21 on page 8

Exercises 1.1

Understanding the Concepts and Skills

1.1 Define the following terms:
a. Population
b. Sample

1.2 What are the two major types of statistics? Describe them in detail.

1.3 Identify some methods used in descriptive statistics.

1.4 Explain two ways in which descriptive statistics and inferential statistics are interrelated.

1.5 Define the following terms:
a. Observational study
b. Designed experiment

1.6 Fill in the following blank: Observational studies can reveal only association, whereas designed experiments can help establish _____.

Applying the Concepts and Skills

In Exercises 1.7–1.12, classify each of the studies as either descriptive or inferential. Explain your answers.

1.7 **TV Viewing Times.** Data from a sample of Americans yielded the following estimates of average TV viewing time per month for all Americans 2 years old and older. The times are in hours and minutes; Q1 stands for first quarter. [SOURCE: *The Cross-Platform Report*, Quarter 1, 2011. Published by The Nielsen Company, © 2011.]

Viewing method	Q1 2011	Q1 2010	Change (%)
Watching TV in the home	158:47	158:25	0.2
Watching timeshifted TV	10:46	9:36	12.2
DVR playback	26:14	25:48	1.7
Using the Internet on a computer	25:33	25:54	−1.4
Watching video on the Internet	4:33	3:23	34.5
Mobile subscribers watching video on a mobile phone	4:20	3:37	20.0

1.8 **Professional Athlete Salaries.** From the *Statistical Abstract of the United States* and the article "Average Salaries in the NBA, NFL, MLB and NHL" by J. Dorish, published on the Yahoo! Contributor Network, we obtained the following data on average professional athletes' salaries for the years 2005 and 2011.

Sport	Average salary ($millions)	
	2005	2011
Baseball (MLB)	2.48	3.31
Basketball (NBA)	4.04	5.15
Football (NFL)	1.40	1.90

1.9 **Home Sales.** Zillow.com is an online database that provides real estate information for U.S. homes that are for rent or sale. It also presents statistics on recently sold homes. The following table gives various information on all homes sold in several different cities across the United States for the month of September 2012.

City	Price per square foot	Sale to list price ratio	% foreclosure re-sales
Scottsdale, AZ	$167	0.973	12.43%
Washington, DC	$436	0.990	2.88%
San Francisco, CA	$636	1.026	6.55%
Las Vegas, NV	$ 74	1.000	19.45%
Nashville, TN	$106	0.973	18.09%

1.10 **Drug Use.** The U.S. Substance Abuse and Mental Health Services Administration collects and publishes data on nonmedical drug use, by type of drug and age group, in *National Survey on Drug Use and Health*. The following table provides data for the years 2003 and 2008. The percentages shown are estimates for the entire nation based on information obtained from a sample (NA, not available).

Type of drug	Percentage, 18–25 years old			
	Ever used		Current user	
	2003	2008	2003	2008
Any illicit drug	60.5	56.6	20.3	19.6
Marijuana and hashish	53.9	50.4	17.0	16.5
Cocaine	15.0	14.4	2.2	1.5
Hallucinogens	23.3	17.7	1.7	1.7
Inhalants	14.9	10.4	0.4	0.3
Any psychotherapeutic	29.0	29.2	6.0	5.9
Alcohol	87.1	85.6	61.4	61.2
"Binge" alcohol use	NA	NA	41.6	41.8
Cigarettes	70.2	64.2	40.2	35.7
Smokeless tobacco	22.0	20.3	4.7	5.4
Cigars	45.2	41.4	11.4	11.3

1.11 **Dow Jones Industrial Averages.** From the *Stock Performance Guide*, published online by 1stock1 on the website 1Stock1.com, we found the closing values of the Dow Jones Industrial Averages as of the end of December for the years 2004 through 2013.

Year	Closing
2004	10,783.01
2005	10,717.50
2006	12,463.15
2007	13,264.82
2008	8,776.39
2009	10,428.05
2010	11,577.51
2011	12,217.56
2012	13,104.14
2013	16,576.66

1.12 In-Demand College Majors. In a June 2013 article, published online by The Street, B. O'Connell discussed the results of a survey on opportunities for graduating college students. In one aspect of the survey, the following percentage estimates were reported on which college majors were in demand among U.S. firms. [SOURCE: "The Most In-Demand College Majors This Year." Published by Career-Builder, LLC, © 2013.]

Major	Percentage of U.S. firms
Business studies	31%
Computer sciences	24%
Engineering	17%
Health care sciences	10%
Engineering technologies	9%
Math and statistics	9%
Communications	7%
Education	7%
Science technology	6%
Liberal arts	6%

1.13 Thoughts on Evolution. In an article titled "Who has designs on your student's minds?" (*Nature*, Vol. 434, pp. 1062–1065), author G. Brumfiel postulated that support for Darwinism increases with level of education. The following table provides percentages of U.S. adults, by educational level, who believe that evolution is a scientific theory well supported by evidence.

Education	Percentage
Postgraduate education	65%
College graduate	52%
Some college education	32%
High school or less	20%

a. Do you think that this study is descriptive or inferential? Explain your answer.
b. If, in fact, the study is inferential, identify the sample and population.

1.14 Big-Banks Break-up. A nationwide survey of 1000 U.S. adults, conducted in March 2013 by Rasmussen Reports (field work by Pulse Opinion Research, LLC), found that 50% of respondents favored a plan to break up the 12 megabanks, which then controlled about 69% of the banking industry.
a. Identify the population and sample for this study.
b. Is the percentage provided a descriptive statistic or an inferential statistic? Explain your answer.

1.15 Genocide. The document "American Attitudes about Genocide" provided highlights of a nationwide poll with 1000 participants. The survey, conducted by Penn Schoen Berland between June 30 and July 10, 2012, revealed that "66% of respondents believe that genocide is preventable."
a. Is the statement in quotes an inferential or a descriptive statement? Explain your answer.
b. Based on the same information, what if the statement had been "66% of Americans believe that genocide is preventable"?

1.16 Vasectomies and Prostate Cancer. Refer to the vasectomy/prostate cancer study discussed in Example 1.5 on page 6.
a. How could the study be modified to make it a designed experiment?
b. Comment on the feasibility of the designed experiment that you described in part (a).

In Exercises 1.17–1.22, state whether the investigation in question is an observational study or a designed experiment. Justify your answer in each case.

1.17 The Salk Vaccine. In the 1940s and early 1950s, the public was greatly concerned about polio. In an attempt to prevent this disease, Jonas Salk of the University of Pittsburgh developed a polio vaccine. In a test of the vaccine's efficacy, involving nearly 2 million grade-school children, half of the children received the Salk vaccine; the other half received a placebo, in this case an injection of salt dissolved in water. Neither the children nor the doctors performing the diagnoses knew which children belonged to which group, but an evaluation center did. The center found that the incidence of polio was far less among the children inoculated with the Salk vaccine. From that information, the researchers concluded that the vaccine would be effective in preventing polio for all U.S. school children; consequently, it was made available for general use.

1.18 Do Left-Handers Die Earlier? According to a study published in the *Journal of the American Public Health Association*, left-handed people do not die at an earlier age than right-handed people, contrary to the conclusion of a highly publicized report done 2 years earlier. The investigation involved a 6-year study of 3800 people in East Boston older than age 65. Researchers at Harvard University and the National Institute of Aging found that the "lefties" and "righties" died at exactly the same rate. "There was no difference, period," said Dr. J. Guralnik, an epidemiologist at the institute and one of the coauthors of the report.

1.19 Sex, Sleep, and PTSD. In the article, "One's Sex, Sleep, and Posttraumatic Stress Disorder" (*Biology of Sex Differences*, Vol. 3, No. 29, pp. 1–7), I. Kobayashi et al. study the relationship between one's sex, sleep patterns, and posttraumatic stress disorder (PTSD) after trauma exposure. The authors report that women have a higher lifetime prevalence of PTSD as well as a greater risk of developing PTSD following trauma exposure. Relationships between sleep and physical health have been documented in a number of studies, and the authors explore the possibility that disruptive sleep habits are common among people with PTSD and also a possible risk factor for the development of PTSD. A questionnaire of men and women with and without PTSD produced data on their sleep habits.

1.20 Aspirin and Cardiovascular Disease. In the article by P. Ridker et al. titled "A Randomized Trial of Low-dose Aspirin in the Primary Prevention of Cardiovascular Disease in Women" (*New England Journal of Medicine*, Vol. 352, pp. 1293–1304), the researchers noted that "We randomly assigned 39,876 initially healthy women 45 years of age or older to receive 100 mg of aspirin or placebo on alternate days and then monitored them for 10 years for a first major cardiovascular event (i.e., nonfatal myocardial infarction, nonfatal stroke, or death from cardiovascular causes)."

1.21 Heart Failure. In the paper "Cardiac-Resynchronization Therapy with or without an Implantable Defibrillator in Advanced Chronic Heart Failure" (*New England Journal of Medicine*, Vol. 350, pp. 2140–2150), M. Bristow et al. reported the results of a study of methods for treating patients who had advanced heart failure due to ischemic or nonischemic cardiomyopathies. A total of 1520 patients were randomly assigned in a 1:2:2 ratio to receive optimal pharmacologic therapy alone or in combination with either a pacemaker or a pacemaker–defibrillator combination. The patients were then observed until they died or were hospitalized for any cause.

1.22 Starting Salaries. The National Association of Colleges and Employers (NACE) compiles information on salary offers to new college graduates and publishes the results in *Salary Survey*.

Extending the Concepts and Skills

1.23 Ballistic Fingerprinting. In an on-line press release, ABCNews.com reported that "...73 percent of Americans...favor a law that would require every gun sold in the United States to be test-fired first, so law enforcement would have its fingerprint in case it were ever used in a crime."

a. Do you think that the statement in the press release is inferential or descriptive? Can you be sure?

b. Actually, ABCNews.com conducted a telephone survey of a random national sample of 1032 adults and determined that 73% of them favored a law that would require every gun sold in the United States to be test-fired first, so law enforcement would have its fingerprint in case it were ever used in a crime. How would you rephrase the statement in the press release to make clear that it is a descriptive statement? an inferential statement?

1.24 Causes of Death. The National Center for Health Statistics published the following data on the leading causes of death in 2010 in *National Vital Statistics Reports*. Deaths are classified according to the tenth revision of the *International Classification of Diseases*. Rates are per 100,000 population.

Rank	Cause of death	Rate
1	Diseases of the heart	193.6
2	Malignant neoplasms	186.2
3	Chronic lower respiratory diseases	44.7
4	Cerebrovascular diseases	41.9
5	Accidents (unintentional injuries)	39.1
6	Alzheimer's disease	27.0
7	Diabetes mellitus	22.4

Do you think that these rates are descriptive statistics or inferential statistics? Explain your answer.

1.25 Medical Testing on Animals. In its Summer 2013 *Animal Action Report*, the National Anti-Vivisection Society stated that "59% of Americans between the ages of 18 and 29 oppose medical testing on animals." The percentage of 59% was computed from sample data.

a. Identify the population under consideration.

b. Identify the sample under consideration.

c. Is the statement in quotes descriptive or inferential?

d. If you wanted to make it clear that the the percentage of 59% was computed from sample data, how would you rephrase the statement in quotes?

1.26 Lobbying Congress. In the special report, "Bitter Pill: Why Medical Bills Are Killing Us" (*TIME*, Vol. 181, No. 8, 2013), S. Brill presented an in-depth investigation of hospital billing practices that reveals why U.S. health care spending is out of control. One of the many statistics provided in the report is that, during the period from 1998 through 2012, the pharmaceutical and health-care-products industries and organizations representing doctors, hospitals, nursing homes, health services, and HMOs spent $5.36 billion lobbying Congress.

a. Under what conditions would the $5.36 billion lobbying-expenditure figure be a descriptive statistic? Explain your answer.

b. Under what conditions would the $5.36 billion lobbying-expenditure figure be an inferential statistic? Explain your answer.

1.2   Simple Random Sampling

Throughout this book, we present examples of organizations or people conducting studies: A consumer group wants information about the gas mileage of a particular make of car, so it performs mileage tests on a sample of such cars; a teacher wants to know about the comparative merits of two teaching methods, so she tests those methods on two groups of students. This approach reflects a healthy attitude: To obtain information about a subject of interest, plan and conduct a study.

Suppose, however, that a study you are considering has already been done. Repeating it would be a waste of time, energy, and money. Therefore, before planning and conducting a study, do a literature search. You do not necessarily need to go through the entire library or make an extensive Internet search. Instead, you might use an information collection agency that specializes in finding studies on specific topics.

? What Does It Mean?

You can often avoid the effort and expense of a study if someone else has already done that study and published the results.

Census, Sampling, and Experimentation

If the information you need is not already available from a previous study, you might acquire it by conducting a **census**—that is, by obtaining information for the entire population of interest. However, conducting a census may be time consuming, costly, impractical, or even impossible.

Two methods other than a census for obtaining information are **sampling** and **experimentation.** In much of this book, we concentrate on sampling. However, we introduce experimentation in Section 1.4, discuss it sporadically throughout the text, and examine it in detail in the chapter *Design of Experiments and Analysis of Variance* (Module C) in the Regression-ANOVA Modules section on the WeissStats site.

If sampling is appropriate, you must decide how to select the sample; that is, you must choose the method for obtaining a sample from the population. Because the sample will be used to draw conclusions about the entire population, it should be a **representative sample**—that is, it should reflect as closely as possible the relevant characteristics of the population under consideration.

For instance, using the average weight of a sample of professional football players to make an inference about the average weight of all adult males would be unreasonable. Nor would it be reasonable to estimate the median income of California residents by sampling the incomes of Beverly Hills residents.

To see what can happen when a sample is not representative, consider the presidential election of 1936. Before the election, the *Literary Digest* magazine conducted an opinion poll of the voting population. Its survey team asked a sample of the voting population whether they would vote for Franklin D. Roosevelt, the Democratic candidate, or for Alfred Landon, the Republican candidate.

Based on the results of the survey, the magazine predicted an easy win for Landon. But when the actual election results were in, Roosevelt won by the greatest landslide in the history of presidential elections! What happened?

- The sample was obtained from among people who owned a car or had a telephone. In 1936, that group included only the more well-to-do people, and historically such people tend to vote Republican.
- The response rate was low (less than 25% of those polled responded), and there was a nonresponse bias (a disproportionate number of those who responded to the poll were Landon supporters).

The sample obtained by the *Literary Digest* was not representative.

Most modern sampling procedures involve the use of **probability sampling.** In probability sampling, a random device—such as tossing a coin, consulting a table of random numbers, or employing a random-number generator—is used to decide which members of the population will constitute the sample instead of leaving such decisions to human judgment.

The use of probability sampling may still yield a nonrepresentative sample. However, probability sampling helps eliminate unintentional selection bias and permits the researcher to control the chance of obtaining a nonrepresentative sample. Furthermore, the use of probability sampling guarantees that the techniques of inferential statistics can be applied. In this section and the next, we examine the most important probability-sampling methods.

Simple Random Sampling

The inferential techniques considered in this book are intended for use with only one particular sampling procedure: **simple random sampling.**

DEFINITION 1.4

? **What Does It Mean?**

Simple random sampling corresponds to our intuitive notion of random selection by lot.

Simple Random Sampling; Simple Random Sample

Simple random sampling: A sampling procedure for which each possible sample of a given size is equally likely to be the one obtained.

Simple random sample: A sample obtained by simple random sampling.

There are two types of simple random sampling. One is **simple random sampling with replacement (SRSWR),** whereby a member of the population can be selected more than once; the other is **simple random sampling without replacement (SRS),** whereby a member of the population can be selected at most once. *Unless we specify otherwise, assume that simple random sampling is done without replacement.*

In Example 1.7, we chose a very small population—the five top Oklahoma state officials—to illustrate simple random sampling. In practice, we would not sample from such a small population but would instead take a census. Using a small population here makes understanding the concept of simple random sampling easier.

Notation

n = sample size

$\bar{x}$ = sample mean

s = sample stdev

Q_j = jth quartile

N = population size

μ = population mean

σ = population stdev

d = paired difference

$\hat{p}$ = sample proportion

p = population proportion

O = observed frequency

E = expected frequency

Chapter 3   Descriptive Measures

- Sample mean: $\bar{x} = \dfrac{\Sigma x_i}{n}$

- Range: Range = Max − Min

- Sample standard deviation:

$$s = \sqrt{\frac{\Sigma(x_i - \bar{x})^2}{n-1}} \quad \text{or} \quad s = \sqrt{\frac{\Sigma x_i^2 - (\Sigma x_i)^2/n}{n-1}}$$

- Interquartile range: IQR = $Q_3 - Q_1$

- Lower limit = $Q_1 - 1.5 \cdot$ IQR,   Upper limit = $Q_3 + 1.5 \cdot$ IQR

- Population mean (mean of a variable): $\mu = \dfrac{\Sigma x_i}{N}$

- Population standard deviation (standard deviation of a variable):

$$\sigma = \sqrt{\frac{\Sigma(x_i - \mu)^2}{N}} \quad \text{or} \quad \sigma = \sqrt{\frac{\Sigma x_i^2}{N} - \mu^2}$$

- Standardized variable: $z = \dfrac{x - \mu}{\sigma}$

Chapter 4   Probability Concepts

- Probability for equally likely outcomes:

$$P(E) = \frac{f}{N}$$

where f denotes the number of ways event E can occur and N denotes the total number of outcomes possible.

- Special addition rule:

$$P(A \text{ or } B \text{ or } C \text{ or } \cdots) = P(A) + P(B) + P(C) + \cdots$$

($A, B, C, \ldots$ mutually exclusive)

- Complementation rule: $P(E) = 1 - P(\text{not } E)$

- General addition rule: $P(A \text{ or } B) = P(A) + P(B) - P(A \text{ \& } B)$

- Conditional probability rule: $P(B \mid A) = \dfrac{P(A \text{ \& } B)}{P(A)}$

- General multiplication rule: $P(A \text{ \& } B) = P(A) \cdot P(B \mid A)$

- Special multiplication rule:

$$P(A \text{ \& } B \text{ \& } C \text{ \& } \cdots) = P(A) \cdot P(B) \cdot P(C) \cdots$$

($A, B, C, \ldots$ independent)

- Rule of total probability:

$$P(B) = \sum_{j=1}^{k} P(A_j) \cdot P(B \mid A_j)$$

($A_1, A_2, \ldots, A_k$ mutually exclusive and exhaustive)

- Bayes's rule:

$$P(A_i \mid B) = \frac{P(A_i) \cdot P(B \mid A_i)}{\sum_{j=1}^{k} P(A_j) \cdot P(B \mid A_j)}$$

($A_1, A_2, \ldots, A_k$ mutually exclusive and exhaustive)

- Factorial: $k! = k(k - 1) \cdots 2 \cdot 1$

- Permutations rule: $_mP_r = \dfrac{m!}{(m-r)!}$

- Special permutations rule: $_mP_m = m!$

- Combinations rule: $_mC_r = \dfrac{m!}{r!(m-r)!}$

- Number of possible samples: $_NC_n = \dfrac{N!}{n!(N-n)!}$

Chapter 5   Discrete Random Variables

- Mean of a discrete random variable X: $\mu = \Sigma x P(X = x)$

- Standard deviation of a discrete random variable X:

$$\sigma = \sqrt{\Sigma(x - \mu)^2 P(X = x)} \quad \text{or} \quad \sigma = \sqrt{\Sigma x^2 P(X = x) - \mu^2}$$

- Factorial: $k! = k(k - 1) \cdots 2 \cdot 1$

- Binomial coefficient: $\dbinom{n}{x} = \dfrac{n!}{x!(n-x)!}$

- Binomial probability formula:

$$P(X = x) = \binom{n}{x} p^x (1 - p)^{n-x}$$

where n denotes the number of trials and p denotes the success probability.

- Mean of a binomial random variable: $\mu = np$

- Standard deviation of a binomial random variable:

$$\sigma = \sqrt{np(1 - p)}$$

- Poisson probability formula: $P(X = x) = e^{-\lambda} \dfrac{\lambda^x}{x!}$

- Mean of a Poisson random variable: $\mu = \lambda$

- Standard deviation of a Poisson random variable: $\sigma = \sqrt{\lambda}$

Chapter 6   The Normal Distribution

- z-score for an x-value: $z = \dfrac{x - \mu}{\sigma}$

- x-value for a z-score: $x = \mu + z \cdot \sigma$

Formula/Table Card for Weiss's *Introductory Statistics*, 10/e
Larry R. Griffey

Table VIII — Values of F_α

dfd	α	1	2	3	4	5	6	7	8	9
1	0.10	39.86	49.50	53.59	55.83	57.24	58.20	58.91	59.44	59.86
	0.05	161.45	199.50	215.71	224.58	230.16	233.99	236.77	238.88	240.54
	0.025	647.79	799.50	864.16	899.58	921.85	937.11	948.22	956.66	963.28
	0.01	4052.2	4999.5	5403.4	5624.6	5763.6	5859.0	5928.4	5981.1	6022.5
	0.005	16211	20000	21615	22500	23056	23437	23715	23925	24091
2	0.10	8.53	9.00	9.16	9.24	9.29	9.33	9.35	9.37	9.38
	0.05	18.51	19.00	19.16	19.25	19.30	19.33	19.35	19.37	19.38
	0.025	38.51	39.00	39.17	39.25	39.30	39.33	39.36	39.37	39.39
	0.01	98.50	99.00	99.17	99.25	99.30	99.33	99.36	99.37	99.39
	0.005	198.50	199.00	199.17	199.25	199.30	199.33	199.36	199.37	199.39
3	0.10	5.54	5.46	5.39	5.34	5.31	5.28	5.27	5.25	5.24
	0.05	10.13	9.55	9.28	9.12	9.01	8.94	8.89	8.85	8.81
	0.025	17.44	16.04	15.44	15.10	14.88	14.73	14.62	14.54	14.47
	0.01	34.12	30.82	29.46	28.71	28.24	27.91	27.67	27.49	27.35
	0.005	55.55	49.80	47.47	46.19	45.39	44.84	44.43	44.13	43.88
4	0.10	4.54	4.32	4.19	4.11	4.05	4.01	3.98	3.95	3.94
	0.05	7.71	6.94	6.59	6.39	6.26	6.16	6.09	6.04	6.00
	0.025	12.22	10.65	9.98	9.60	9.36	9.20	9.07	8.98	8.90
	0.01	21.20	18.00	16.69	15.98	15.52	15.21	14.98	14.80	14.66
	0.005	31.33	26.28	24.26	23.15	22.46	21.97	21.62	21.35	21.14
5	0.10	4.06	3.78	3.62	3.52	3.45	3.40	3.37	3.34	3.32
	0.05	6.61	5.79	5.41	5.19	5.05	4.95	4.88	4.82	4.77
	0.025	10.01	8.43	7.76	7.39	7.15	6.98	6.85	6.76	6.68
	0.01	16.26	13.27	12.06	11.39	10.97	10.67	10.46	10.29	10.16
	0.005	22.78	18.31	16.53	15.56	14.94	14.51	14.20	13.96	13.77
6	0.10	3.78	3.46	3.29	3.18	3.11	3.05	3.01	2.98	2.96
	0.05	5.99	5.14	4.76	4.53	4.39	4.28	4.21	4.15	4.10
	0.025	8.81	7.26	6.60	6.23	5.99	5.82	5.70	5.60	5.52
	0.01	13.75	10.92	9.78	9.15	8.75	8.47	8.26	8.10	7.98
	0.005	18.63	14.54	12.92	12.03	11.46	11.07	10.79	10.57	10.39
7	0.10	3.59	3.26	3.07	2.96	2.88	2.83	2.78	2.75	2.72
	0.05	5.59	4.74	4.35	4.12	3.97	3.87	3.79	3.73	3.68
	0.025	8.07	6.54	5.89	5.52	5.29	5.12	4.99	4.90	4.82
	0.01	12.25	9.55	8.45	7.85	7.46	7.19	6.99	6.84	6.72
	0.005	16.24	12.40	10.88	10.05	9.52	9.16	8.89	8.68	8.51
8	0.10	3.46	3.11	2.92	2.81	2.73	2.67	2.62	2.59	2.56
	0.05	5.32	4.46	4.07	3.84	3.69	3.58	3.50	3.44	3.39
	0.025	7.57	6.06	5.42	5.05	4.82	4.65	4.53	4.43	4.36
	0.01	11.26	8.65	7.59	7.01	6.63	6.37	6.18	6.03	5.91
	0.005	14.69	11.04	9.60	8.81	8.30	7.95	7.69	7.50	7.34

Table VIII (cont.) — Values of F_α

dfd	α	1	2	3	4	5	6	7	8	9
9	0.10	3.36	3.01	2.81	2.69	2.61	2.55	2.51	2.47	2.44
	0.05	5.12	4.26	3.86	3.63	3.48	3.37	3.29	3.23	3.18
	0.025	7.21	5.71	5.08	4.72	4.48	4.32	4.20	4.10	4.03
	0.01	10.56	8.02	6.99	6.42	6.06	5.80	5.61	5.47	5.35
	0.005	13.61	10.11	8.72	7.96	7.47	7.13	6.88	6.69	6.54
10	0.10	3.29	2.92	2.73	2.61	2.52	2.46	2.41	2.38	2.35
	0.05	4.96	4.10	3.71	3.48	3.33	3.22	3.14	3.07	3.02
	0.025	6.94	5.46	4.83	4.47	4.24	4.07	3.95	3.85	3.78
	0.01	10.04	7.56	6.55	5.99	5.64	5.39	5.20	5.06	4.94
	0.005	12.83	9.43	8.08	7.34	6.87	6.54	6.30	6.12	5.97
11	0.10	3.23	2.86	2.66	2.54	2.45	2.39	2.34	2.30	2.27
	0.05	4.84	3.98	3.59	3.36	3.20	3.09	3.01	2.95	2.90
	0.025	6.72	5.26	4.63	4.28	4.04	3.88	3.76	3.66	3.59
	0.01	9.65	7.21	6.22	5.67	5.32	5.07	4.89	4.74	4.63
	0.005	12.23	8.91	7.60	6.88	6.42	6.10	5.86	5.68	5.54
12	0.10	3.18	2.81	2.61	2.48	2.39	2.33	2.28	2.24	2.21
	0.05	4.75	3.89	3.49	3.26	3.11	3.00	2.91	2.85	2.80
	0.025	6.55	5.10	4.47	4.12	3.89	3.73	3.61	3.51	3.44
	0.01	9.33	6.93	5.95	5.41	5.06	4.82	4.64	4.50	4.39
	0.005	11.75	8.51	7.23	6.52	6.07	5.76	5.52	5.35	5.20
13	0.10	3.14	2.76	2.56	2.43	2.35	2.28	2.23	2.20	2.16
	0.05	4.67	3.81	3.41	3.18	3.03	2.92	2.83	2.77	2.71
	0.025	6.41	4.97	4.35	4.00	3.77	3.60	3.48	3.39	3.31
	0.01	9.07	6.70	5.74	5.21	4.86	4.62	4.44	4.30	4.19
	0.005	11.37	8.19	6.93	6.23	5.79	5.48	5.25	5.08	4.94
14	0.10	3.10	2.73	2.52	2.39	2.31	2.24	2.19	2.15	2.12
	0.05	4.60	3.74	3.34	3.11	2.96	2.85	2.76	2.70	2.65
	0.025	6.30	4.86	4.24	3.89	3.66	3.50	3.38	3.29	3.21
	0.01	8.86	6.51	5.56	5.04	4.69	4.46	4.28	4.14	4.03
	0.005	11.06	7.92	6.68	6.00	5.56	5.26	5.03	4.86	4.72
15	0.10	3.07	2.70	2.49	2.36	2.27	2.21	2.16	2.12	2.09
	0.05	4.54	3.68	3.29	3.06	2.90	2.79	2.71	2.64	2.59
	0.025	6.20	4.77	4.15	3.80	3.58	3.41	3.29	3.20	3.12
	0.01	8.68	6.36	5.42	4.89	4.56	4.32	4.14	4.00	3.89
	0.005	10.80	7.70	6.48	5.80	5.37	5.07	4.85	4.67	4.54
16	0.10	3.05	2.67	2.46	2.33	2.24	2.18	2.13	2.09	2.06
	0.05	4.49	3.63	3.24	3.01	2.85	2.74	2.66	2.59	2.54
	0.025	6.12	4.69	4.08	3.73	3.50	3.34	3.22	3.12	3.05
	0.01	8.53	6.23	5.29	4.77	4.44	4.20	4.03	3.89	3.78
	0.005	10.58	7.51	6.30	5.64	5.21	4.91	4.69	4.52	4.38

Note: dfn = degrees of freedom, numerator; dfd = degrees of freedom, denominator. The figure above shows the F-distribution curve with shaded area α to the right of F_α.

Chapter 7   The Sampling Distribution of the Sample Mean

- Mean of the variable $\bar{x}$: $\mu_{\bar{x}} = \mu$

- Standard deviation of the variable $\bar{x}$: $\sigma_{\bar{x}} = \sigma/\sqrt{n}$

Chapter 8   Confidence Intervals for One Population Mean

- Standardized version of the variable $\bar{x}$:

$$z = \frac{\bar{x} - \mu}{\sigma/\sqrt{n}}$$

- z-interval for μ (σ known, normal population or large sample):

$$\bar{x} \pm z_{\alpha/2} \cdot \frac{\sigma}{\sqrt{n}}$$

- Margin of error for the estimate of μ: $E = z_{\alpha/2} \cdot \dfrac{\sigma}{\sqrt{n}}$

- Sample size for estimating μ:

$$n = \left(\frac{z_{\alpha/2} \cdot \sigma}{E}\right)^2$$

rounded up to the nearest whole number.

- Studentized version of the variable $\bar{x}$:

$$t = \frac{\bar{x} - \mu}{s/\sqrt{n}}$$

- t-interval for μ (σ unknown, normal population or large sample):

$$\bar{x} \pm t_{\alpha/2} \cdot \frac{s}{\sqrt{n}}$$

with df $= n - 1$.

Chapter 9   Hypothesis Tests for One Population Mean

- z-test statistic for H_0: $\mu = \mu_0$ (σ known, normal population or large sample):

$$z = \frac{\bar{x} - \mu_0}{\sigma/\sqrt{n}}$$

- t-test statistic for H_0: $\mu = \mu_0$ (σ unknown, normal population or large sample):

$$t = \frac{\bar{x} - \mu_0}{s/\sqrt{n}}$$

with df $= n - 1$.

- Symmetry property of a Wilcoxon signed-rank distribution:

$$W_{1-A} = n(n + 1)/2 - W_A$$

- Wilcoxon signed-rank test statistic for H_0: $\mu = \mu_0$ (symmetric population):

$$W = \text{sum of the positive ranks}$$

Chapter 10   Inferences for Two Population Means

- Pooled sample standard deviation:

$$s_p = \sqrt{\frac{(n_1 - 1)s_1^2 + (n_2 - 1)s_2^2}{n_1 + n_2 - 2}}$$

- Pooled t-test statistic for H_0: $\mu_1 = \mu_2$ (independent samples, normal populations or large samples, and equal population standard deviations):

$$t = \frac{\bar{x}_1 - \bar{x}_2}{s_p\sqrt{(1/n_1) + (1/n_2)}}$$

with df $= n_1 + n_2 - 2$.

- Pooled t-interval for $\mu_1 - \mu_2$ (independent samples, normal populations or large samples, and equal population standard deviations):

$$(\bar{x}_1 - \bar{x}_2) \pm t_{\alpha/2} \cdot s_p\sqrt{(1/n_1) + (1/n_2)}$$

with df $= n_1 + n_2 - 2$.

- Degrees of freedom for nonpooled t-procedures:

$$\Delta = \frac{[(s_1^2/n_1) + (s_2^2/n_2)]^2}{\dfrac{(s_1^2/n_1)^2}{n_1 - 1} + \dfrac{(s_2^2/n_2)^2}{n_2 - 1}}$$

rounded down to the nearest integer.

- Nonpooled t-test statistic for H_0: $\mu_1 = \mu_2$ (independent samples, and normal populations or large samples):

$$t = \frac{\bar{x}_1 - \bar{x}_2}{\sqrt{(s_1^2/n_1) + (s_2^2/n_2)}}$$

with df $= \Delta$.

- Nonpooled t-interval for $\mu_1 - \mu_2$ (independent samples, and normal populations or large samples):

$$(\bar{x}_1 - \bar{x}_2) \pm t_{\alpha/2} \cdot \sqrt{(s_1^2/n_1) + (s_2^2/n_2)}$$

with df $= \Delta$.

- Symmetry property of a Mann–Whitney distribution:

$$M_{1-A} = n_1(n_1 + n_2 + 1) - M_A$$

- Mann–Whitney test statistic for H_0: $\mu_1 = \mu_2$ (independent samples and same-shape populations):

$$M = \text{sum of the ranks for sample data from Population 1}$$

- Paired t-test statistic for H_0: $\mu_1 = \mu_2$ (paired sample, and normal differences or large sample):

$$t = \frac{\bar{d}}{s_d/\sqrt{n}}$$

with df $= n - 1$.

Formula/Table Card for Weiss's *Introductory Statistics, 10/e*
Larry R. Griffey

Table I Random numbers

Line number	00–09		10–19		20–29		30–39		40–49	
00	15544	80712	97742	21500	97081	42451	50623	56071	28882	28739
01	01011	21285	04729	39986	73150	31548	30168	76189	56996	19210
02	47435	53308	40718	29050	74858	64517	93573	51058	68501	42723
03	91312	75137	86274	59834	69844	19853	06917	17413	44474	86530
04	12775	08768	80791	16298	22934	09630	98862	39746	64623	32768
05	31466	43761	94872	92230	52367	13205	38634	55882	77518	36252
06	09300	43847	40881	51243	97810	18903	53914	31688	06220	40422
07	73582	13810	57784	72454	68997	72229	30340	08844	53924	89630
08	11092	81392	58189	22697	41063	09451	09789	00637	06450	85990
09	93322	98567	00116	35605	66790	52965	62877	21740	56476	49296
10	80134	12484	67089	08674	70753	90959	45842	59844	45214	36505
11	97888	31797	95037	84400	76041	96668	75920	68482	56855	97417
12	92612	27082	59459	69380	98654	20407	88151	56263	27126	63797
13	72744	45586	43279	44218	83638	05422	00995	70217	78925	39097
14	96256	70653	45285	26293	78305	80252	03625	40159	68760	84716
15	07851	47452	66742	83331	54701	06573	98169	37499	67756	68301
16	25594	41552	96475	56151	02089	33748	65289	89956	89559	33687
17	65358	15155	59374	80940	03411	94656	69440	47156	77115	99463
18	09402	31008	53424	21928	02198	61201	02457	87214	59750	51330
19	97424	90765	01634	37328	41243	33564	17884	94747	93650	77668

Table III Normal scores

Ordered position	n = 5	6	7	8	9	10	11	12	13
1	−1.18	−1.28	−1.36	−1.43	−1.50	−1.55	−1.59	−1.64	−1.68
2	−0.50	−0.64	−0.76	−0.85	−0.93	−1.00	−1.06	−1.11	−1.16
3	0.00	−0.20	−0.35	−0.47	−0.57	−0.65	−0.73	−0.79	−0.85
4	0.50	0.20	0.00	−0.15	−0.27	−0.37	−0.46	−0.53	−0.60
5	1.18	0.64	0.35	0.15	0.00	−0.12	−0.22	−0.31	−0.39
6		1.28	0.76	0.47	0.27	0.12	0.00	−0.10	−0.19
7			1.36	0.85	0.57	0.37	0.22	0.10	0.00
8				1.43	0.93	0.65	0.46	0.31	0.19
9					1.50	1.00	0.73	0.53	0.39
10						1.55	1.06	0.79	0.60
11							1.59	1.11	0.85
12								1.64	1.16
13									1.68

Table VI Values of M_α

n_2	α	n_1 = 3	4	5	6	7	8	9	10
3	0.10	14	20	27	36	45	55	66	78
	0.05	15	21	29	37	46	57	68	80
	0.025	—	22	30	38	48	58	70	82
	0.01	—	—	—	39	49	59	71	83
	0.005	—	—	—	—	—	60	72	85
4	0.10	16	23	31	40	49	60	72	85
	0.05	17	24	32	41	51	62	74	87
	0.025	18	25	33	43	53	64	76	89
	0.01	—	26	35	44	54	65	78	91
	0.005	—	—	—	45	55	66	79	93
5	0.10	18	26	34	44	54	65	78	91
	0.05	20	27	36	46	56	68	80	94
	0.025	21	28	37	47	58	70	83	96
	0.01	—	30	39	49	60	72	85	99
	0.005	—	—	40	50	61	73	86	101
6	0.10	21	29	38	48	59	71	84	98
	0.05	22	30	40	50	61	73	87	101
	0.025	23	32	41	52	63	76	89	103
	0.01	24	33	43	54	65	78	92	106
	0.005	—	34	44	55	67	80	94	108
7	0.10	23	31	41	52	63	76	89	104
	0.05	24	33	43	54	66	79	93	107
	0.025	26	35	45	56	68	81	95	110
	0.01	27	36	47	58	71	84	98	114
	0.005	—	37	48	60	72	86	101	116
8	0.10	25	34	44	56	68	81	95	110
	0.05	27	36	47	58	71	84	99	114
	0.025	28	38	49	61	73	87	102	117
	0.01	29	39	51	63	76	90	105	121
	0.005	30	40	52	65	78	92	108	124
9	0.10	27	37	48	60	72	86	101	116
	0.05	29	39	50	63	76	90	105	121
	0.025	31	41	53	65	78	93	108	124
	0.01	32	43	55	68	81	96	112	129
	0.005	33	44	56	70	84	99	114	131
10	0.10	29	40	51	64	77	91	106	123
	0.05	31	42	54	67	80	95	111	127
	0.025	33	44	56	69	83	98	114	131
	0.01	34	46	59	72	87	102	119	136
	0.005	36	48	61	74	89	105	121	139

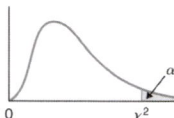

Table VII Values of χ_α^2

$\chi_{0.10}^2$	$\chi_{0.05}^2$	$\chi_{0.025}^2$	$\chi_{0.01}^2$	$\chi_{0.005}^2$	df
2.706	3.841	5.024	6.635	7.879	1
4.605	5.991	7.378	9.210	10.597	2
6.251	7.815	9.348	11.345	12.838	3
7.779	9.488	11.143	13.277	14.860	4
9.236	11.070	12.833	15.086	16.750	5
10.645	12.592	14.449	16.812	18.548	6
12.017	14.067	16.013	18.475	20.278	7
13.362	15.507	17.535	20.090	21.955	8
14.684	16.919	19.023	21.666	23.589	9
15.987	18.307	20.483	23.209	25.188	10
17.275	19.675	21.920	24.725	26.757	11
18.549	21.026	23.337	26.217	28.300	12
19.812	22.362	24.736	27.688	29.819	13
21.064	23.685	26.119	29.141	31.319	14
22.307	24.996	27.488	30.578	32.801	15
23.542	26.296	28.845	32.000	34.267	16
24.769	27.587	30.191	33.409	35.718	17
25.989	28.869	31.526	34.805	37.156	18
27.204	30.143	32.852	36.191	38.582	19
28.412	31.410	34.170	37.566	39.997	20
29.615	32.671	35.479	38.932	41.401	21
30.813	33.924	36.781	40.290	42.796	22
32.007	35.172	38.076	41.638	44.181	23
33.196	36.415	39.364	42.980	45.559	24
34.382	37.653	40.647	44.314	46.928	25
35.563	38.885	41.923	45.642	48.290	26
36.741	40.113	43.195	46.963	49.645	27
37.916	41.337	44.461	48.278	50.994	28
39.087	42.557	45.722	49.588	52.336	29
40.256	43.773	46.979	50.892	53.672	30
51.805	55.759	59.342	63.691	66.767	40
63.167	67.505	71.420	76.154	79.490	50
74.397	79.082	83.298	88.381	91.955	60
85.527	90.531	95.023	100.424	104.213	70
96.578	101.879	106.628	112.328	116.320	80
107.565	113.145	118.135	124.115	128.296	90
118.499	124.343	129.563	135.811	140.177	100

- Paired t-interval for $\mu_1 - \mu_2$ (paired sample, and normal differences or large sample):

$$\bar{d} \pm t_{\alpha/2} \cdot \frac{s_d}{\sqrt{n}}$$

with df $= n - 1$.

- Paired Wilcoxon signed-rank test statistic for H_0: $\mu_1 = \mu_2$ (paired sample and symmetric differences):

$$W = \text{sum of the positive ranks}$$

Chapter 11   Inferences for Population Standard Deviations

- χ^2-test statistic for H_0: $\sigma = \sigma_0$ (normal population):

$$\chi^2 = \frac{n-1}{\sigma_0^2} s^2$$

with df $= n - 1$.

- χ^2-interval for σ (normal population):

$$\sqrt{\frac{n-1}{\chi_{\alpha/2}^2}} \cdot s \quad \text{to} \quad \sqrt{\frac{n-1}{\chi_{1-\alpha/2}^2}} \cdot s$$

with df $= n - 1$.

- F-test statistic for H_0: $\sigma_1 = \sigma_2$ (independent samples and normal populations):

$$F = s_1^2/s_2^2$$

with df $= (n_1 - 1, n_2 - 1)$.

- F-interval for σ_1/σ_2 (independent samples and normal populations):

$$\frac{1}{\sqrt{F_{\alpha/2}}} \cdot \frac{s_1}{s_2} \quad \text{to} \quad \frac{1}{\sqrt{F_{1-\alpha/2}}} \cdot \frac{s_1}{s_2}$$

with df $= (n_1 - 1, n_2 - 1)$.

Chapter 12   Inferences for Population Proportions

- Sample proportion:

$$\hat{p} = \frac{x}{n}$$

where x denotes the number of members in the sample that have the specified attribute.

- z-interval for p:

$$\hat{p} \pm z_{\alpha/2} \cdot \sqrt{\hat{p}(1 - \hat{p})/n}$$

(*Assumption:* both x and $n - x$ are 5 or greater)

- Margin of error for the estimate of p:

$$E = z_{\alpha/2} \cdot \sqrt{\hat{p}(1 - \hat{p})/n}$$

- Sample size for estimating p:

$$n = 0.25 \left(\frac{z_{\alpha/2}}{E}\right)^2 \quad \text{or} \quad n = \hat{p}_g(1 - \hat{p}_g)\left(\frac{z_{\alpha/2}}{E}\right)^2$$

rounded up to the nearest whole number (g = "educated guess")

- z-test statistic for H_0: $p = p_0$:

$$z = \frac{\hat{p} - p_0}{\sqrt{p_0(1 - p_0)/n}}$$

(*Assumption:* both np_0 and $n(1 - p_0)$ are 5 or greater)

- Pooled sample proportion: $\hat{p}_p = \dfrac{x_1 + x_2}{n_1 + n_2}$

- z-test statistic for H_0: $p_1 = p_2$:

$$z = \frac{\hat{p}_1 - \hat{p}_2}{\sqrt{\hat{p}_p(1 - \hat{p}_p)}\sqrt{(1/n_1) + (1/n_2)}}$$

(*Assumptions:* independent samples; x_1, $n_1 - x_1$, x_2, $n_2 - x_2$ are all 5 or greater)

- z-interval for $p_1 - p_2$:

$$(\hat{p}_1 - \hat{p}_2) \pm z_{\alpha/2} \cdot \sqrt{\hat{p}_1(1 - \hat{p}_1)/n_1 + \hat{p}_2(1 - \hat{p}_2)/n_2}$$

(*Assumptions:* independent samples; x_1, $n_1 - x_1$, x_2, $n_2 - x_2$ are all 5 or greater)

- Margin of error for the estimate of $p_1 - p_2$:

$$E = z_{\alpha/2} \cdot \sqrt{\hat{p}_1(1 - \hat{p}_1)/n_1 + \hat{p}_2(1 - \hat{p}_2)/n_2}$$

- Sample size for estimating $p_1 - p_2$:

$$n_1 = n_2 = 0.5 \left(\frac{z_{\alpha/2}}{E}\right)^2$$

or

$$n_1 = n_2 = (\hat{p}_{1g}(1 - \hat{p}_{1g}) + \hat{p}_{2g}(1 - \hat{p}_{2g}))\left(\frac{z_{\alpha/2}}{E}\right)^2$$

rounded up to the nearest whole number (g = "educated guess")

Chapter 13   Chi-Square Procedures

- Expected frequencies for a chi-square goodness-of-fit test:

$$E = np$$

- Test statistic for a chi-square goodness-of-fit test:

$$\chi^2 = \Sigma(O - E)^2/E$$

with df $= c - 1$, where c is the number of possible values for the variable under consideration.

- Expected frequencies for a chi-square independence test or a chi-square homogeneity test:

$$E = \frac{R \cdot C}{n}$$

where R = row total and C = column total.

- Test statistic for a chi-square independence test:

$$\chi^2 = \Sigma(O - E)^2/E$$

with df $= (r - 1)(c - 1)$, where r and c are the number of possible values for the two variables under consideration.

- Test-statistic for a chi-square homogeneity test:

$$\chi^2 = \Sigma(O - E)^2/E$$

with df $= (r - 1)(c - 1)$, where r is the number of populations and c is the number of possible values for the variable under consideration.

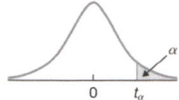

Table IV   Values of t_α

df	$t_{0.10}$	$t_{0.05}$	$t_{0.025}$	$t_{0.01}$	$t_{0.005}$	df
1	3.078	6.314	12.706	31.821	63.657	1
2	1.886	2.920	4.303	6.965	9.925	2
3	1.638	2.353	3.182	4.541	5.841	3
4	1.533	2.132	2.776	3.747	4.604	4
5	1.476	2.015	2.571	3.365	4.032	5
6	1.440	1.943	2.447	3.143	3.707	6
7	1.415	1.895	2.365	2.998	3.499	7
8	1.397	1.860	2.306	2.896	3.355	8
9	1.383	1.833	2.262	2.821	3.250	9
10	1.372	1.812	2.228	2.764	3.169	10
11	1.363	1.796	2.201	2.718	3.106	11
12	1.356	1.782	2.179	2.681	3.055	12
13	1.350	1.771	2.160	2.650	3.012	13
14	1.345	1.761	2.145	2.624	2.977	14
15	1.341	1.753	2.131	2.602	2.947	15
16	1.337	1.746	2.120	2.583	2.921	16
17	1.333	1.740	2.110	2.567	2.898	17
18	1.330	1.734	2.101	2.552	2.878	18
19	1.328	1.729	2.093	2.539	2.861	19
20	1.325	1.725	2.086	2.528	2.845	20
21	1.323	1.721	2.080	2.518	2.831	21
22	1.321	1.717	2.074	2.508	2.819	22
23	1.319	1.714	2.069	2.500	2.807	23
24	1.318	1.711	2.064	2.492	2.797	24
25	1.316	1.708	2.060	2.485	2.787	25
26	1.315	1.706	2.056	2.479	2.779	26
27	1.314	1.703	2.052	2.473	2.771	27
28	1.313	1.701	2.048	2.467	2.763	28
29	1.311	1.699	2.045	2.462	2.756	29
30	1.310	1.697	2.042	2.457	2.750	30
31	1.309	1.696	2.040	2.453	2.744	31
32	1.309	1.694	2.037	2.449	2.738	32
33	1.308	1.692	2.035	2.445	2.733	33
34	1.307	1.691	2.032	2.441	2.728	34
35	1.306	1.690	2.030	2.438	2.724	35
36	1.306	1.688	2.028	2.434	2.719	36
37	1.305	1.687	2.026	2.431	2.715	37
38	1.304	1.686	2.024	2.429	2.712	38
39	1.304	1.685	2.023	2.426	2.708	39
40	1.303	1.684	2.021	2.423	2.704	40
41	1.303	1.683	2.020	2.421	2.701	41
42	1.302	1.682	2.018	2.418	2.698	42
43	1.302	1.681	2.017	2.416	2.695	43
44	1.301	1.680	2.015	2.414	2.692	44
45	1.301	1.679	2.014	2.412	2.690	45
46	1.300	1.679	2.013	2.410	2.687	46
47	1.300	1.678	2.012	2.408	2.685	47
48	1.299	1.677	2.011	2.407	2.682	48
49	1.299	1.677	2.010	2.405	2.680	49

Table IV (cont.)   Values of t_α

df	$t_{0.10}$	$t_{0.05}$	$t_{0.025}$	$t_{0.01}$	$t_{0.005}$	df
50	1.299	1.676	2.009	2.403	2.678	50
51	1.298	1.675	2.008	2.402	2.676	51
52	1.298	1.675	2.007	2.400	2.674	52
53	1.298	1.674	2.006	2.399	2.672	53
54	1.297	1.674	2.005	2.397	2.670	54
55	1.297	1.673	2.004	2.396	2.668	55
56	1.297	1.673	2.003	2.395	2.667	56
57	1.297	1.672	2.002	2.394	2.665	57
58	1.296	1.672	2.002	2.392	2.663	58
59	1.296	1.671	2.001	2.391	2.662	59
60	1.296	1.671	2.000	2.390	2.660	60
61	1.296	1.670	2.000	2.389	2.659	61
62	1.295	1.670	1.999	2.388	2.657	62
63	1.295	1.669	1.998	2.387	2.656	63
64	1.295	1.669	1.998	2.386	2.655	64
65	1.295	1.669	1.997	2.385	2.654	65
66	1.295	1.668	1.997	2.384	2.652	66
67	1.294	1.668	1.996	2.383	2.651	67
68	1.294	1.668	1.995	2.382	2.650	68
69	1.294	1.667	1.995	2.382	2.649	69
70	1.294	1.667	1.994	2.381	2.648	70
71	1.294	1.667	1.994	2.380	2.647	71
72	1.293	1.666	1.993	2.379	2.646	72
73	1.293	1.666	1.993	2.379	2.645	73
74	1.293	1.666	1.993	2.378	2.644	74
75	1.293	1.665	1.992	2.377	2.643	75
80	1.292	1.664	1.990	2.374	2.639	80
85	1.292	1.663	1.988	2.371	2.635	85
90	1.291	1.662	1.987	2.368	2.632	90
95	1.291	1.661	1.985	2.366	2.629	95
100	1.290	1.660	1.984	2.364	2.626	100
200	1.286	1.653	1.972	2.345	2.601	200
300	1.284	1.650	1.968	2.339	2.592	300
400	1.284	1.649	1.966	2.336	2.588	400
500	1.283	1.648	1.965	2.334	2.586	500
600	1.283	1.647	1.964	2.333	2.584	600
700	1.283	1.647	1.963	2.332	2.583	700
800	1.283	1.647	1.963	2.331	2.582	800
900	1.282	1.647	1.963	2.330	2.581	900
1000	1.282	1.646	1.962	2.330	2.581	1000
2000	1.282	1.646	1.961	2.328	2.578	2000

1.282	1.645	1.960	2.326	2.576
$z_{0.10}$	$z_{0.05}$	$z_{0.025}$	$z_{0.01}$	$z_{0.005}$

Table V   Values of W_α

n	$W_{0.10}$	$W_{0.05}$	$W_{0.025}$	$W_{0.01}$	$W_{0.005}$	n
7	22	24	26	28	—	7
8	28	30	32	34	36	8
9	34	37	39	42	43	9
10	41	44	47	50	52	10
11	48	52	55	59	61	11
12	56	64	64	68	71	12
13	65	70	74	78	81	13
14	74	79	84	89	92	14
15	83	90	95	100	104	15
16	94	100	106	112	117	16
17	104	112	118	125	130	17
18	116	124	131	138	143	18
19	128	136	144	152	158	19
20	140	150	158	167	173	20

Chapter 14   Descriptive Methods in Regression and Correlation

- S_{xx}, S_{xy}, and S_{yy}:

$$S_{xx} = \Sigma(x_i - \bar{x})^2 = \Sigma x_i^2 - (\Sigma x_i)^2/n$$

$$S_{xy} = \Sigma(x_i - \bar{x})(y_i - \bar{y}) = \Sigma x_i y_i - (\Sigma x_i)(\Sigma y_i)/n$$

$$S_{yy} = \Sigma(y_i - \bar{y})^2 = \Sigma y_i^2 - (\Sigma y_i)^2/n$$

- Regression equation: $\hat{y} = b_0 + b_1 x$, where

$$b_1 = \frac{S_{xy}}{S_{xx}} \quad \text{and} \quad b_0 = \frac{1}{n}(\Sigma y_i - b_1 \Sigma x_i) = \bar{y} - b_1 \bar{x}$$

- Total sum of squares: $SST = \Sigma(y_i - \bar{y})^2 = S_{yy}$

- Regression sum of squares: $SSR = \Sigma(\hat{y}_i - \bar{y})^2 = S_{xy}^2/S_{xx}$
- Error sum of squares: $SSE = \Sigma(y_i - \hat{y}_i)^2 = S_{yy} - S_{xy}^2/S_{xx}$
- Regression identity: $SST = SSR + SSE$
- Coefficient of determination: $r^2 = \dfrac{SSR}{SST}$

- Linear correlation coefficient:

$$r = \frac{\frac{1}{n-1}\Sigma(x_i - \bar{x})(y_i - \bar{y})}{s_x s_y} \quad \text{or} \quad r = \frac{S_{xy}}{\sqrt{S_{xx}S_{yy}}}$$

Chapter 15   Inferential Methods in Regression and Correlation

- Population regression equation: $y = \beta_0 + \beta_1 x$

- Standard error of the estimate: $s_e = \sqrt{\dfrac{SSE}{n-2}}$

- Test statistic for H_0: $\beta_1 = 0$:

$$t = \frac{b_1}{s_e/\sqrt{S_{xx}}}$$

with df $= n - 2$.

- Confidence interval for β_1:

$$b_1 \pm t_{\alpha/2} \cdot \frac{s_e}{\sqrt{S_{xx}}}$$

with df $= n - 2$.

- Confidence interval for the conditional mean of the response variable corresponding to x_p:

$$\hat{y}_p \pm t_{\alpha/2} \cdot s_e \sqrt{\frac{1}{n} + \frac{(x_p - \bar{x})^2}{S_{xx}}}$$

with df $= n - 2$.

- Prediction interval for an observed value of the response variable corresponding to x_p:

$$\hat{y}_p \pm t_{\alpha/2} \cdot s_e \sqrt{1 + \frac{1}{n} + \frac{(x_p - \bar{x})^2}{S_{xx}}}$$

with df $= n - 2$.

- Test statistic for H_0: $\rho = 0$:

$$t = \frac{r}{\sqrt{\dfrac{1 - r^2}{n - 2}}}$$

with df $= n - 2$.

- Test statistic for a correlation test for normality:

$$R_p = \frac{\Sigma x_i w_i}{\sqrt{S_{xx}\Sigma w_i^2}}$$

where x and w denote observations of the variable and the corresponding normal scores, respectively.

Chapter 16   Analysis of Variance (ANOVA)

- Notation in one-way ANOVA:

  k = number of populations
  n = total number of observations
  $\bar{x}$ = mean of all n observations
  n_j = size of sample from Population j
  $\bar{x}_j$ = mean of sample from Population j
  s_j^2 = variance of sample from Population j
  T_j = sum of sample data from Population j

- Defining formulas for sums of squares in one-way ANOVA:

$$SST = \Sigma(x_i - \bar{x})^2$$
$$SSTR = \Sigma n_j(\bar{x}_j - \bar{x})^2$$
$$SSE = \Sigma(n_j - 1)s_j^2$$

- One-way ANOVA identity: $SST = SSTR + SSE$
- Computing formulas for sums of squares in one-way ANOVA:

$$SST = \Sigma x_i^2 - (\Sigma x_i)^2/n$$
$$SSTR = \Sigma(T_j^2/n_j) - (\Sigma x_i)^2/n$$
$$SSE = SST - SSTR$$

- Mean squares in one-way ANOVA:

$$MSTR = \frac{SSTR}{k - 1} \qquad MSE = \frac{SSE}{n - k}$$

- Test statistic for one-way ANOVA (independent samples, normal populations, and equal population standard deviations):

$$F = \frac{MSTR}{MSE}$$

with df $= (k - 1, n - k)$.

- Confidence interval for $\mu_i - \mu_j$ in the Tukey multiple-comparison method (independent samples, normal populations, and equal population standard deviations):

$$(\bar{x}_i - \bar{x}_j) \pm \frac{q_\alpha}{\sqrt{2}} \cdot s \sqrt{(1/n_i) + (1/n_j)}$$

where $s = \sqrt{MSE}$ and q_α is obtained for a q-curve with parameters k and $n - k$.

- Test statistic for a Kruskal–Wallis test (independent samples, same-shape populations, all sample sizes 5 or greater):

$$K = \frac{SSTR}{SST/(n - 1)} \quad \text{or} \quad K = \frac{12}{n(n + 1)}\sum_{j=1}^{k} \frac{R_j^2}{n_j} - 3(n + 1)$$

where $SSTR$ and SST are computed for the ranks of the data, and R_j denotes the sum of the ranks for the sample data from Population j. K has approximately a chi-square distribution with df $= k - 1$.

Formula/Table Card for Weiss's *Introductory Statistics*, 10/e
Larry R. Griffey

Table II (cont.) Areas under the standard normal curve

z	0.00	0.01	0.02	0.03	0.04	0.05	0.06	0.07	0.08	0.09
0.0	0.5000	0.5040	0.5080	0.5120	0.5160	0.5199	0.5239	0.5279	0.5319	0.5359
0.1	0.5398	0.5438	0.5478	0.5517	0.5557	0.5596	0.5636	0.5675	0.5714	0.5753
0.2	0.5793	0.5832	0.5871	0.5910	0.5948	0.5987	0.6026	0.6064	0.6103	0.6141
0.3	0.6179	0.6217	0.6255	0.6293	0.6331	0.6368	0.6406	0.6443	0.6480	0.6517
0.4	0.6554	0.6591	0.6628	0.6664	0.6700	0.6736	0.6772	0.6808	0.6844	0.6879
0.5	0.6915	0.6950	0.6985	0.7019	0.7054	0.7088	0.7123	0.7157	0.7190	0.7224
0.6	0.7257	0.7291	0.7324	0.7357	0.7389	0.7422	0.7454	0.7486	0.7517	0.7549
0.7	0.7580	0.7611	0.7642	0.7673	0.7704	0.7734	0.7764	0.7794	0.7823	0.7852
0.8	0.7881	0.7910	0.7939	0.7967	0.7995	0.8023	0.8051	0.8078	0.8106	0.8133
0.9	0.8159	0.8186	0.8212	0.8238	0.8264	0.8289	0.8315	0.8340	0.8365	0.8389
1.0	0.8413	0.8438	0.8461	0.8485	0.8508	0.8531	0.8554	0.8577	0.8599	0.8621
1.1	0.8643	0.8665	0.8686	0.8708	0.8729	0.8749	0.8770	0.8790	0.8810	0.8830
1.2	0.8849	0.8869	0.8888	0.8907	0.8925	0.8944	0.8962	0.8980	0.8997	0.9015
1.3	0.9032	0.9049	0.9066	0.9082	0.9099	0.9115	0.9131	0.9147	0.9162	0.9177
1.4	0.9192	0.9207	0.9222	0.9236	0.9251	0.9265	0.9279	0.9292	0.9306	0.9319
1.5	0.9332	0.9345	0.9357	0.9370	0.9382	0.9394	0.9406	0.9418	0.9429	0.9441
1.6	0.9452	0.9463	0.9474	0.9484	0.9495	0.9505	0.9515	0.9525	0.9535	0.9545
1.7	0.9554	0.9564	0.9573	0.9582	0.9591	0.9599	0.9608	0.9616	0.9625	0.9633
1.8	0.9641	0.9649	0.9656	0.9664	0.9671	0.9678	0.9686	0.9693	0.9699	0.9706
1.9	0.9713	0.9719	0.9726	0.9732	0.9738	0.9744	0.9750	0.9756	0.9761	0.9767
2.0	0.9772	0.9778	0.9783	0.9788	0.9793	0.9798	0.9803	0.9808	0.9812	0.9817
2.1	0.9821	0.9826	0.9830	0.9834	0.9838	0.9842	0.9846	0.9850	0.9854	0.9857
2.2	0.9861	0.9864	0.9868	0.9871	0.9875	0.9878	0.9881	0.9884	0.9887	0.9890
2.3	0.9893	0.9896	0.9898	0.9901	0.9904	0.9906	0.9909	0.9911	0.9913	0.9916
2.4	0.9918	0.9920	0.9922	0.9925	0.9927	0.9929	0.9931	0.9932	0.9934	0.9936
2.5	0.9938	0.9940	0.9941	0.9943	0.9945	0.9946	0.9948	0.9949	0.9951	0.9952
2.6	0.9953	0.9955	0.9956	0.9957	0.9959	0.9960	0.9961	0.9962	0.9963	0.9964
2.7	0.9965	0.9966	0.9967	0.9968	0.9969	0.9970	0.9971	0.9972	0.9973	0.9974
2.8	0.9974	0.9975	0.9976	0.9977	0.9977	0.9978	0.9979	0.9979	0.9980	0.9981
2.9	0.9981	0.9982	0.9982	0.9983	0.9984	0.9984	0.9985	0.9985	0.9986	0.9986
3.0	0.9987	0.9987	0.9987	0.9988	0.9988	0.9989	0.9989	0.9989	0.9990	0.9990
3.1	0.9990	0.9991	0.9991	0.9991	0.9992	0.9992	0.9992	0.9992	0.9993	0.9993
3.2	0.9993	0.9993	0.9994	0.9994	0.9994	0.9994	0.9994	0.9995	0.9995	0.9995
3.3	0.9995	0.9995	0.9996	0.9996	0.9996	0.9996	0.9996	0.9996	0.9996	0.9997
3.4	0.9997	0.9997	0.9997	0.9997	0.9997	0.9997	0.9997	0.9997	0.9997	0.9998
3.5	0.9998	0.9998	0.9998	0.9998	0.9998	0.9998	0.9998	0.9998	0.9998	0.9998
3.6	0.9998	0.9998	0.9999	0.9999	0.9999	0.9999	0.9999	0.9999	0.9999	0.9999
3.7	0.9999	0.9999	0.9999	0.9999	0.9999	0.9999	0.9999	0.9999	0.9999	0.9999
3.8	0.9999	0.9999	0.9999	0.9999	0.9999	0.9999	0.9999	0.9999	0.9999	0.9999
3.9	1.0000†									

† For $z \geq 3.90$, the areas are 1.0000 to four decimal places.

Table II Areas under the standard normal curve

z	0.09	0.08	0.07	0.06	0.05	0.04	0.03	0.02	0.01	0.00
−3.9	0.0001	0.0001	0.0001	0.0001	0.0001	0.0001	0.0001	0.0001	0.0001	0.0000†
−3.8	0.0001	0.0001	0.0001	0.0001	0.0001	0.0001	0.0001	0.0001	0.0001	0.0001
−3.7	0.0001	0.0001	0.0001	0.0001	0.0001	0.0001	0.0001	0.0001	0.0001	0.0001
−3.6	0.0001	0.0001	0.0001	0.0001	0.0002	0.0002	0.0002	0.0002	0.0002	0.0002
−3.5	0.0002	0.0002	0.0002	0.0002	0.0002	0.0002	0.0002	0.0002	0.0002	0.0002
−3.4	0.0002	0.0003	0.0003	0.0003	0.0003	0.0003	0.0003	0.0003	0.0003	0.0003
−3.3	0.0003	0.0004	0.0004	0.0004	0.0004	0.0004	0.0004	0.0005	0.0005	0.0005
−3.2	0.0005	0.0005	0.0005	0.0006	0.0006	0.0006	0.0006	0.0006	0.0007	0.0007
−3.1	0.0007	0.0007	0.0008	0.0008	0.0008	0.0008	0.0009	0.0009	0.0009	0.0010
−3.0	0.0010	0.0010	0.0011	0.0011	0.0011	0.0012	0.0012	0.0013	0.0013	0.0013
−2.9	0.0014	0.0014	0.0015	0.0015	0.0016	0.0016	0.0017	0.0018	0.0018	0.0019
−2.8	0.0019	0.0020	0.0021	0.0021	0.0022	0.0023	0.0023	0.0024	0.0025	0.0026
−2.7	0.0026	0.0027	0.0028	0.0029	0.0030	0.0031	0.0032	0.0033	0.0034	0.0035
−2.6	0.0036	0.0037	0.0038	0.0039	0.0040	0.0041	0.0043	0.0044	0.0045	0.0047
−2.5	0.0048	0.0049	0.0051	0.0052	0.0054	0.0055	0.0057	0.0059	0.0060	0.0062
−2.4	0.0064	0.0066	0.0068	0.0069	0.0071	0.0073	0.0075	0.0078	0.0080	0.0082
−2.3	0.0084	0.0087	0.0089	0.0091	0.0094	0.0096	0.0099	0.0102	0.0104	0.0107
−2.2	0.0110	0.0113	0.0116	0.0119	0.0122	0.0125	0.0129	0.0132	0.0136	0.0139
−2.1	0.0143	0.0146	0.0150	0.0154	0.0158	0.0162	0.0166	0.0170	0.0174	0.0179
−2.0	0.0183	0.0188	0.0192	0.0197	0.0202	0.0207	0.0212	0.0217	0.0222	0.0228
−1.9	0.0233	0.0239	0.0244	0.0250	0.0256	0.0262	0.0268	0.0274	0.0281	0.0287
−1.8	0.0294	0.0301	0.0307	0.0314	0.0322	0.0329	0.0336	0.0344	0.0351	0.0359
−1.7	0.0367	0.0375	0.0384	0.0392	0.0401	0.0409	0.0418	0.0427	0.0436	0.0446
−1.6	0.0455	0.0465	0.0475	0.0485	0.0495	0.0505	0.0516	0.0526	0.0537	0.0548
−1.5	0.0559	0.0571	0.0582	0.0594	0.0606	0.0618	0.0630	0.0643	0.0655	0.0668
−1.4	0.0681	0.0694	0.0708	0.0721	0.0735	0.0749	0.0764	0.0778	0.0793	0.0808
−1.3	0.0823	0.0838	0.0853	0.0869	0.0885	0.0901	0.0918	0.0934	0.0951	0.0968
−1.2	0.0985	0.1003	0.1020	0.1038	0.1056	0.1075	0.1093	0.1112	0.1131	0.1151
−1.1	0.1170	0.1190	0.1210	0.1230	0.1251	0.1271	0.1292	0.1314	0.1335	0.1357
−1.0	0.1379	0.1401	0.1423	0.1446	0.1469	0.1492	0.1515	0.1539	0.1562	0.1587
−0.9	0.1611	0.1635	0.1660	0.1685	0.1711	0.1736	0.1762	0.1788	0.1814	0.1841
−0.8	0.1867	0.1894	0.1922	0.1949	0.1977	0.2005	0.2033	0.2061	0.2090	0.2119
−0.7	0.2148	0.2177	0.2206	0.2236	0.2266	0.2296	0.2327	0.2358	0.2389	0.2420
−0.6	0.2451	0.2483	0.2514	0.2546	0.2578	0.2611	0.2643	0.2676	0.2709	0.2743
−0.5	0.2776	0.2810	0.2843	0.2877	0.2912	0.2946	0.2981	0.3015	0.3050	0.3085
−0.4	0.3121	0.3156	0.3192	0.3228	0.3264	0.3300	0.3336	0.3372	0.3409	0.3446
−0.3	0.3483	0.3520	0.3557	0.3594	0.3632	0.3669	0.3707	0.3745	0.3783	0.3821
−0.2	0.3859	0.3897	0.3936	0.3974	0.4013	0.4052	0.4090	0.4129	0.4168	0.4207
−0.1	0.4247	0.4286	0.4325	0.4364	0.4404	0.4443	0.4483	0.4522	0.4562	0.4602
−0.0	0.4641	0.4681	0.4721	0.4761	0.4801	0.4840	0.4880	0.4920	0.4960	0.5000

† For $z \leq -3.90$, the areas are 0.0000 to four decimal places.

EXAMPLE 1.7   Simple Random Samples

TABLE 1.2
Five top Oklahoma state officials

Governor (G)
Lieutenant Governor (L)
Secretary of State (S)
Attorney General (A)
Treasurer (T)

TABLE 1.3
The 10 possible samples of two officials

G, L	G, S	G, A	G, T	L, S
L, A	L, T	S, A	S, T	A, T

TABLE 1.4
The five possible samples of four officials

G, L, S, A	G, L, S, T
G, L, A, T	G, S, A, T
L, S, A, T	

You
try it!

Exercise 1.43
on page 16

Sampling Oklahoma State Officials As reported by the *World Almanac*, the top five state officials of Oklahoma are as shown in Table 1.2. Consider these five officials a population of interest.

a. List the possible samples (without replacement) of two officials from this population of five officials.
b. Describe a method for obtaining a simple random sample of two officials from this population of five officials.
c. For the sampling method described in part (b), what are the chances that any particular sample of two officials will be the one selected?
d. Repeat parts (a)–(c) for samples of size 4.

Solution For convenience, we represent the officials in Table 1.2 by using the letters in parentheses.

a. Table 1.3 lists the 10 possible samples of two officials from this population of five officials.
b. To obtain a simple random sample of size 2, we could write the letters that correspond to the five officials (G, L, S, A, and T) on separate pieces of paper. After placing these five slips of paper in a box and shaking it, we could, while blindfolded, pick two slips of paper.
c. The procedure described in part (b) will provide a simple random sample. Consequently, each of the possible samples of two officials is equally likely to be the one selected. There are 10 possible samples, so the chances are $\frac{1}{10}$ (1 in 10) that any particular sample of two officials will be the one selected.
d. Table 1.4 lists the five possible samples of four officials from this population of five officials. A simple random sampling procedure, such as picking four slips of paper out of a box, gives each of these samples a 1 in 5 chance of being the one selected.

Random-Number Tables

Obtaining a simple random sample by picking slips of paper out of a box is usually impractical, especially when the population is large. Fortunately, we can use several practical procedures to get simple random samples. One common method involves a **table of random numbers**—a table of randomly chosen digits, as illustrated in Example 1.8.

EXAMPLE 1.8   Random-Number Tables

Sampling Student Opinions Student questionnaires, known as "teacher evaluations," gained widespread use in the late 1960s and early 1970s. Generally, professors hand out evaluation forms a week or so before the final.

That practice, however, poses several problems. On some days, less than 60% of students registered for a class may attend. Moreover, many of those who are present complete their evaluation forms in a hurry in order to prepare for other classes. A better method, therefore, might be to select a simple random sample of students from the class and interview them individually.

During one semester, Professor Hassett wanted to sample the attitudes of the students taking college algebra at his school. He decided to interview 15 of the 728 students enrolled in the course. Using a registration list on which the 728 students were numbered 1–728, he obtained a simple random sample of 15 students by randomly selecting 15 numbers between 1 and 728. To do so, he used the random-number table that appears in Appendix A as Table I and here as Table 1.5.

TABLE 1.5
Random numbers

Line number	Column number									
	00–09		10–19		20–29		30–39		40–49	
00	15544	80712	97742	21500	97081	42451	50623	56071	28882	28739
01	01011	21285	04729	39986	73150	31548	30168	76189	56996	19210
02	47435	53308	40718	29050	74858	64517	93573	51058	68501	42723
03	91312	75137	86274	59834	69844	19853	06917	17413	44474	86530
04	12775	08768	80791	16298	22934	09630	98862	39746	64623	32768
05	31466	43761	94872	92230	52367	13205	38634	55882	77518	36252
06	09300	43847	40881	51243	97810	18903	53914	31688	06220	40422
07	73582	13810	57784	72454	68997	72229	30340	08844	53924	89630
08	11092	81392	58189	22697	41063	09451	09789	00637	06450	85990
09	93322	98567	00116	35605	66790	52965	62877	21740	56476	49296
10	80134	12484	67089	08674	70753	90959	45842	59844	45214	36505
11	97888	31797	95037	84400	76041	96668	75920	68482	56855	97417
12	92612	27082	59459	69380	98654	20407	88151	56263	27126	63797
13	72744	45586	43279	44218	83638	05422	00995	70217	78925	39097
14	96256	70653	45285	26293	78305	80252	03625	40159	68760	84716
15	07851	47452	66742	83331	54701	06573	98169	37499	67756	68301
16	25594	41552	96475	56151	02089	33748	65289	89956	89559	33687
17	65358	15155	59374	80940	03411	94656	69440	47156	77115	99463
18	09402	31008	53424	21928	02198	61201	02457	87214	59750	51330
19	97424	90765	01634	37328	41243	33564	17884	94747	93650	77668

TABLE 1.6
Registration numbers of students interviewed

69	303	458	652	178
386	97	9	694	578
539	628	36	24	404

StatCrunch

Report 1.1

You try it!

Exercise 1.49(a) on page 16

To select 15 random numbers between 1 and 728, we first pick a random starting point, say, by closing our eyes and placing a finger on Table 1.5. Then, beginning with the three digits under the finger, we go down the table and record the numbers as we go. Because we want numbers between 1 and 728 only, we discard the number 000 and numbers between 729 and 999. To avoid repetition, we also eliminate duplicate numbers. If we have not found enough numbers by the time we reach the bottom of the table, we move over to the next column of three-digit numbers and go up.

Using this procedure, Professor Hassett began with 069, circled in Table 1.5. Reading down from 069 to the bottom of Table 1.5 and then up the next column of three-digit numbers, he found the 15 random numbers displayed in Fig. 1.2 and in Table 1.6. Professor Hassett then interviewed the 15 students whose registration numbers are shown in Table 1.6.

FIGURE 1.2
Procedure used by Professor Hassett to obtain 15 random numbers between 1 and 728 from Table 1.5

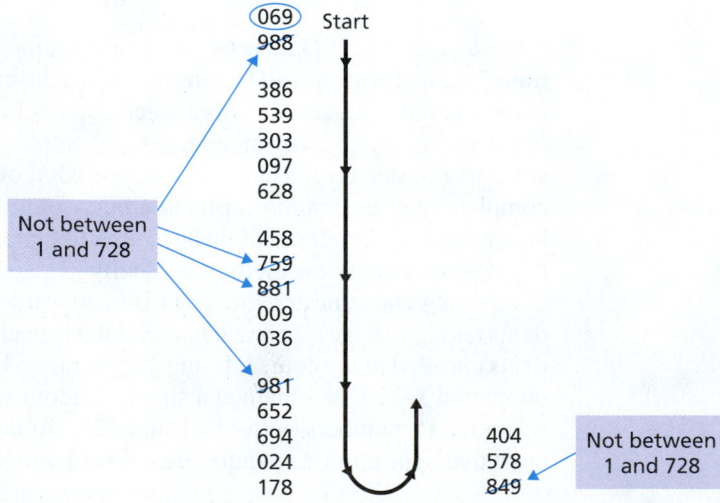

Simple random sampling, the basic type of probability sampling, is also the foundation for the more complex types of probability sampling, which we explore in Section 1.3.

Random-Number Generators

Nowadays, statisticians prefer statistical software packages or graphing calculators, rather than random-number tables, to obtain simple random samples. The built-in programs for doing so are called **random-number generators.** When using random-number generators, be aware of whether they provide samples with replacement or samples without replacement. We discuss the use of random-number generators for obtaining simple random samples in the Technology Center at the end of this section.

Sources of Bias in Sample Surveys

Probability sampling eliminates bias in the selection of a sample from a complete and accurate list of the population. Nonetheless, many sources of bias often work their way into sample surveys, especially those involving large, human populations. We present a few of the many such sources in the exercises. For more details about this important topic, we recommend the book *Sampling: Design and Analysis, 2/e,* by Sharon L. Lohr (Boston: Brooks/Cole, 2010).

THE TECHNOLOGY CENTER

Today, programs for conducting statistical and data analyses are available in dedicated statistical software packages, general-use spreadsheet software, and graphing calculators. In this book, we discuss three of the most popular statistical technologies: Minitab, Excel, and the TI-83/84 Plus.

For Excel, we mostly use XLSTAT from Addinsoft, a statistics add-in that complements Excel's standard statistics capabilities. The version of XLSTAT that we use in this book is XLSTAT for Pearson, a special edition of XLSTAT that has been specifically built for Pearson Education.

"TI-83/84 Plus" is a shorthand for "TI-83 Plus and/or TI-84 Plus." The TI-calculator output that we show in this book is actually from the TI-84 Plus *C* Silver Edition. The style of the output from this calculator differs somewhat from that of the TI-83 Plus and earlier versions of the TI-84 Plus. Nonetheless, in most cases, keystrokes and output-content remain essentially the same for all three calculators.

At the end of most sections of this book, in subsections titled "The Technology Center," we present and interpret output from the three technologies that provides technology solutions to problems solved by hand earlier in the section. For this aspect of The Technology Center, you need neither a computer nor a graphing calculator, nor do you need working knowledge of any of the technologies.

Another aspect of The Technology Center provides step-by-step instructions for using the three technologies to obtain the output presented. When studying this material, you will get the best results by performing the steps described. *Note that, depending on the version of a statistical technology that you use, the appropriate instructions and the output you obtain may differ somewhat from that shown in the book.*

Successful use of technology requires knowing how to input data. We discuss that and other basic tasks for Minitab, Excel, and the TI-83/84 Plus in documents contained in the Technology Basics section of the WeissStats Resource Site (or, briefly, WeissStats site). The URL for the WeissStats site is www.pearsonhighered.com/weiss-stats. Note also that files for all appropriate data sets in the book can be found in multiple formats in the Data Sets section of the WeissStats site.

Using Technology to Obtain an SRS

In this Technology Center, we present output and step-by-step instructions for using technology to obtain a simple random sample without replacement.

Note to TI-83/84 Plus users:

- At the time of this writing, the TI-83/84 Plus does not have a built-in program for simple random sampling without replacement. However, a TI program called SRS, supplied in the TI Programs section of the WeissStats site, allows you to perform this procedure. Your instructor can show you how to download the program to your calculator.
- We recommend that you first *seed* the TI random-number generator, a task that only needs to be done once. Your instructor can also show you how to accomplish that task.
- After you apply the SRS program, the required SRS will be in List 1 (L1). *Warning:* Any data that you may have previously stored in List 1 will be erased during program execution, so copy those data to another list prior to program execution if you want to retain them.

EXAMPLE 1.9   Using Technology to Obtain an SRS

Sampling Student Opinions Recall that, during one semester, Professor Hassett wanted to sample the attitudes of the students taking college algebra at his school. He decided to interview 15 of the 728 students enrolled in the course. Use Minitab, Excel, or the TI-83/84 Plus to obtain a simple random sample without replacement (SRS) of 15 of the 728 students.

Solution Recall that Professor Hassett had a registration list on which the 728 students enrolled in the course were numbered 1–728. Thus, to obtain a simple random sample without replacement of 15 of the 728 students, we need only get an SRS of 15 numbers from the numbers 1–728.

To accomplish that, we applied the appropriate random-number programs. The results are shown in Output 1.1. Steps for generating that output are presented in Instructions 1.1.

OUTPUT 1.1   SRS of 15 numbers from 1–728

MINITAB

	REGNO	SAMPLE
1	1	321
2	2	430
3	3	241
4	4	11
5	5	545
6	6	185
7	7	16
8	8	534
9	9	428
10	10	596
11	11	460
12	12	641
13	13	331
14	14	46
15	15	203
16	16	
17	17	

EXCEL

Sampled data:

REGNO
541
325
278
387
596
109
563
69
300
142
163
399
416
370
157

TI-83/84 PLUS

```
NORMAL FLOAT AUTO REAL RADIAN MP

L1      L2     L3     L4     L5     1
86      ------ ------ ------ ------
176
622
272
435
442
57
13
10
507
511

L1(1)=86
```

Note: Only the first 11 numbers are visible in this output.

Of course, the 15 numbers that we obtained differ for the three technologies and will most likely differ from the 15 numbers that you would get by applying any of the programs.

INSTRUCTIONS 1.1   Steps for generating Output 1.1

MINITAB

1. Store the numbers 1–728 in a column named REGNO†
2. Choose **Calc ➤ Random Data ➤ Sample From Columns…**
3. Press the F3 key to reset the dialog box
4. Type 15 in the **Number of rows to sample** text box
5. Specify REGNO in the **From columns** text box
6. Type SAMPLE in the **Store samples in** text box
7. Click **OK**
8. The required sample is in the column named SAMPLE

EXCEL

1. Store the numbers 1–728 in a column named REGNO‡
2. Choose **XLSTAT ➤ Preparing data ➤ Data sampling**
3. Click the reset button (↻) in the lower left corner of the dialog box
4. Click in the **Data** selection box
5. Select the column of the worksheet that contains the REGNO data

6. Click the arrow button at the right of the **Sampling** drop-down list box and select **Random without replacement**
7. Type 15 in the **Sample size** text box
8. Click **OK**
9. Click the **Continue** button in the **XLSTAT – Selections** dialog box

TI-83/84 PLUS

1. Press **PRGM**
2. Arrow down to SRS and press **ENTER** twice
3. Type 1 for **MIN** (the smallest possible value) and press **ENTER**
4. Type 728 for **MAX** (the largest possible value) and press **ENTER**
5. Type 15 for **SAMPLE SIZE** and press **ENTER**
6. After the program completes, press **STAT** and then **ENTER**
7. The required sample is in **L1** (List 1)

† There are several ways to accomplish this task. Consult your instructor or search "Patterned data" in the Index tab of Minitab's Help facility.

‡ There are several ways to accomplish this task. Consult your instructor for his or her preferred method.

Exercises 1.2

Understanding the Concepts and Skills

1.27 Explain why a census is often not the best way to obtain information about a population.

1.28 Identify two statistical methods other than a census for obtaining information.

1.29 In sampling, explain why obtaining a representative sample is important.

1.30 Provide a scenario of your own in which a sample is not representative.

1.31 Regarding probability sampling:
a. What is it?
b. Does probability sampling always yield a representative sample? Explain your answer.
c. Identify some advantages of probability sampling.

1.32 Regarding simple random sampling:
a. What is simple random sampling?
b. What is a simple random sample?
c. Identify two forms of simple random sampling and explain the difference between the two.

1.33 The inferential procedures discussed in this book are intended for use with only one particular sampling procedure. What sampling procedure is that?

1.34 Identify two methods for obtaining a simple random sample.

1.35 What is the acronym used for simple random sampling without replacement?

1.36 The members of a population are numbered 1–5.
a. List the 10 possible samples (without replacement) of size 3 from this population.
b. If an SRS of size 3 is taken from the population, what are the chances of selecting 1, 3, and 5? Explain your answer.
c. Use Table I in Appendix A to obtain an SRS of size 3 from the population. Start at the single-digit number in line number 5 and column number 20, read down the column, up the next, and so on.

1.37 The members of a population are numbered 1–4.
a. List the 6 possible samples (without replacement) of size 2 from this population.
b. If an SRS of size 2 is taken from the population, what are the chances of selecting 2 and 3? Explain your answer.
c. Use Table I in Appendix A to obtain an SRS of size 2 from the population. Start at the single-digit number in line number 17 and column number 7, read down the column, up the next, and so on.

1.38 The members of a population are numbered 1–90.
a. Use Table I in Appendix A to obtain an SRS of size 5 from the population. Start at the two-digit number in line number 15 and column numbers 25–26, read down the column, up the next, and so on.
b. If you have access to a random-number generator, use it to solve part (a).

1.39 The members of a population are numbered 1–50.

a. Use Table I in Appendix A to obtain an SRS of size 6 from the population. Start at the two-digit number in line number 10 and column numbers 10–11, read down the column, up the next, and so on.

b. If you have access to a random-number generator, use it to solve part (a).

Applying the Concepts and Skills

1.40 Memorial Day Poll. In the year 2000, an on-line poll was conducted over Memorial Day weekend that asked people what they were doing to observe the holiday. The choices were: (1) stay home and relax, (2) vacation outdoors over the weekend, or (3) visit a military cemetery. More than 22,000 people participated in the poll, with 86% selecting option 1. Discuss this poll with regard to its suitability.

1.41 Estimating Median Income. Explain why a sample of 30 dentists from Seattle taken to estimate the median income of all Seattle residents is not representative.

1.42 Oklahoma State Officials. The five top Oklahoma state officials are displayed in Table 1.2 on page 11. Use that table to solve the following problems.

a. List the possible samples of size 1 that can be obtained from the population of five officials.

b. What is the difference between obtaining a simple random sample of size 1 and selecting one official at random?

c. List the possible samples (without replacement) of size 5 that can be obtained from the population of five officials.

d. What is the difference between obtaining a simple random sample of size 5 and taking a census of the five officials?

1.43 Oklahoma State Officials. The five top Oklahoma state officials are displayed in Table 1.2 on page 11. Use that table to solve the following problems.

a. List the 10 possible samples (without replacement) of size 3 that can be obtained from the population of five officials.

b. If a simple random sampling procedure is used to obtain a sample of three officials, what are the chances that it is the first sample on your list in part (a)? the second sample? the tenth sample?

1.44 Best-Selling Albums. The Recording Industry Association of America provides data on the best-selling albums of all time. As of May 28, 2013, the top six best-selling albums of all time (U.S. sales only), are by the artists the Eagles (E), Michael Jackson (M), Pink Floyd (P), Led Zeppelin (L), AC/DC (A), and Billy Joel (B).

a. List the 15 possible samples (without replacement) of two artists that can be selected from the six. For brevity, use the initial provided.

b. Describe a procedure for taking a simple random sample of two artists from the six.

c. If a simple random sampling procedure is used to obtain two artists, what are the chances of selecting P and A? M and E?

1.45 Best-Selling Albums. Refer to Exercise 1.44.

a. List the 15 possible samples (without replacement) of four artists that can be selected from the six.

b. Describe a procedure for taking a simple random sample of four artists from the six.

c. If a simple random sampling procedure is used to obtain four artists, what are the chances of selecting E, A, L, and B? P, B, M, and A?

1.46 Best-Selling Albums. Refer to Exercise 1.44.

a. List the 20 possible samples (without replacement) of three artists that can be selected from the six.

b. Describe a procedure for taking a simple random sample of three artists from the six.

c. If a simple random sampling procedure is used to obtain three artists, what are the chances of selecting M, A, and L? P, L, and E?

1.47 Social Networking Websites. From Wikipedia.com, we obtained the top seven major active social networking websites in the United States, excluding dating websites. Ranked according to registered users, as of April 2013, from most popular to least popular, they are Facebook (F), Twitter (T), Goggle+ (G), Habbo (H), LinkedIn (L), Bebo (B), and Tagged (A).

a. List the 21 possible samples (without replacement) of two social media websites that can be selected from the seven. For brevity, use the initial provided.

b. If a simple random sampling procedure is used to obtain two of these social media websites, what are the chances of selecting B and A? T and G?

1.48 Keno. In the game of keno, 20 balls are selected at random from 80 balls numbered 1–80.

a. Use Table I in Appendix A to simulate one game of keno by obtaining 20 random numbers between 1 and 80. Start at the two-digit number in line number 5 and column numbers 31–32, read down the column, up the next, and so on.

b. If you have access to a random-number generator, use it to solve part (a).

1.49 The International 500. Each year, *Fortune Magazine* publishes an article titled "The International 500" that provides a ranking by sales of the top 500 firms outside the United States. Suppose that you want to examine various characteristics of successful firms. Further suppose that, for your study, you decide to take a simple random sample of 10 firms from *Fortune Magazine*'s list of "The International 500."

a. Use Table I in Appendix A to obtain 10 random numbers that you can use to specify your sample. Start at the three-digit number in line number 14 and column numbers 10–12, read down the column, up the next, and so on.

b. If you have access to a random-number generator, use it to solve part (a).

1.50 Megacities Risk. In an issue of *Discover* (Vol. 26, No. 5, p. 14), A. Casselman looked at the natural-hazards risk index of megacities to evaluate potential loss from catastrophes such as earthquakes, storms, and volcanic eruptions. Urban areas have more to lose from natural perils, technological risks, and environmental hazards than rural areas. The top 10 megacities in the world are Tokyo, San Francisco, Los Angeles, Osaka, Miami, New York, Hong Kong, Manila, London, and Paris.

a. Suppose that you decide to take a simple random sample of five of these 10 megacities. Use Table I in Appendix A to obtain five random numbers that you can use to specify your sample.

b. If you have access to a random-number generator, use it to solve part (a).

1.51 Element Hunters. In the article "Element Hunters" (*National Geographic*, Vol. 233, No. 5, pp. 112–120), R. Dunn reports about the search for new undiscovered elements. Since 1940, scientists have been synthesizing elements one by one. The first was neptunium (Np), element number 93. There are, as of this writing, a total of 26 new synthetic elements. The following table provides their element numbers and symbols.

Number	Symbol	Number	Symbol
93	Np	106	Sg
94	Pu	107	Bh
95	Am	108	Hs
96	Cm	109	Mt
97	Bk	110	Ds
98	Cf	111	Rg
99	Es	112	Cn
100	Fm	113	Uut
101	Md	114	Fl
102	No	115	Uup
103	Lr	116	Lv
104	Rf	117	Uus
105	Db	118	Uuo

a. Suppose that you decide to take a simple random sample of eight of these new elements. Use Table I in Appendix A to obtain eight random numbers that you can use to specify your sample.

b. If you have access to a random-number generator, use it to solve part (a).

Extending the Concepts and Skills

Sources of Bias in Sample Surveys. As you know, probability sampling eliminates bias in choosing a sample from a (list of the entire) population. Nonetheless, many sources of bias often work their way into sample surveys of large human populations, such as those done by government agencies and in opinion polls. Exercises 1.52–1.54 present some aspects of sources of bias in sample surveys.

1.52 Undercoverage. Oftentimes, an accurate and complete list of the population is unavailable. In such cases, one or more groups will be omitted from the sampling process because they are not listed as part of the population. This type of bias is called *undercoverage.*
a. Explain why a sample survey of households will generally suffer from undercoverage.
b. Provide another example where bias due to undercoverage is likely to occur.

1.53 Nonresponse. When responses are not obtained from some of the individuals in the sample because either those individuals cannot be reached or refuse to participate, we have *nonresponse bias.*
a. Discuss some of the dangers of nonresponse.
b. Many sample surveys that are reported in the media have response rates as low as 10%. Explain the consequences of such low response rates in trying to generalize the results to the entire population.

1.54 Response bias. When the behavior of the interviewer or respondent results in inaccurate responses, we have *response bias.*
a. Explain why a survey question "Do you smoke marijuana" might result in response bias?
b. Provide some additional survey situations that might be conducive to response bias.
c. Provide some additional factors that might lead to response bias.

1.3 Other Sampling Designs*

Simple random sampling is the most natural and easily understood method of probability sampling—it corresponds to our intuitive notion of random selection by lot. However, simple random sampling does have drawbacks. For instance, it may fail to provide sufficient coverage when information about subpopulations is required and may be impractical when the members of the population are widely scattered geographically.

In this section, we examine some commonly used sampling procedures that are often more appropriate than simple random sampling. Remember, however, that the inferential procedures discussed in this book must be modified before they can be applied to data that are obtained by sampling procedures other than simple random sampling.

Systematic Random Sampling

One method that takes less effort to implement than simple random sampling is **systematic random sampling**. Procedure 1.1 presents a step-by-step method for implementing systematic random sampling.

■■■ PROCEDURE 1.1   Systematic Random Sampling

Step 1   Divide the population size by the sample size and round the result down to the nearest whole number, *m.*

Step 2   Use a random-number table or a similar device to obtain a number, *k,* between 1 and *m.*

Step 3   Select for the sample those members of the population that are numbered $k, k + m, k + 2m, \ldots$.

EXAMPLE 1.10   Systematic Random Sampling

Sampling Student Opinions Recall Example 1.8, in which Professor Hassett wanted a sample of 15 of the 728 students enrolled in college algebra at his school. Use systematic random sampling to obtain the sample.

Solution   We apply Procedure 1.1.

Step 1   Divide the population size by the sample size and round the result down to the nearest whole number, m.

The population size is the number of students in the class, which is 728, and the sample size is 15. Dividing the population size by the sample size and rounding down to the nearest whole number, we get 728/15 = 48 (rounded down). Thus, $m = 48$.

Step 2   Use a random-number table or a similar device to obtain a number, k, between 1 and m.

Referring to Step 1, we see that we need to randomly select a number between 1 and 48. Using a random-number table, we obtained the number 22 (but we could have conceivably gotten any number between 1 and 48, inclusive). Thus, $k = 22$.

Step 3   Select for the sample those members of the population that are numbered $k, k + m, k + 2m, \ldots$.

From Steps 1 and 2, we see that $k = 22$ and $m = 48$. Hence, we need to list every 48th number, starting at 22, until we have 15 numbers. Doing so, we get the 15 numbers displayed in Table 1.7.

Interpretation   If Professor Hassatt had used systematic random sampling and had begun with the number 22, he would have interviewed the 15 students whose registration numbers are shown in Table 1.7.

TABLE 1.7

Numbers obtained by systematic random sampling

22	166	310	454	598
70	214	358	502	646
118	262	406	550	694

You try it!

Exercise 1.71 on page 24

Systematic random sampling is easier to execute than simple random sampling and usually provides comparable results. The exception is the presence of some kind of cyclical pattern in the listing of the members of the population (e.g., male, female, male, female, ...), a phenomenon that is relatively rare.

Cluster Sampling

Another sampling method is **cluster sampling,** which is particularly useful when the members of the population are widely scattered geographically. Procedure 1.2 provides a step-by-step method for implementing cluster sampling.

PROCEDURE 1.2   Cluster Sampling

Step 1   Divide the population into groups (clusters).

Step 2   Obtain a simple random sample of the clusters.

Step 3   Use all the members of the clusters obtained in Step 2 as the sample.

EXAMPLE 1.11   Cluster Sampling

The 300 members of a population have been divided into clusters of equal size 20. Use cluster sampling to obtain a sample of size 60 from the population.

Solution   We apply Procedure 1.2.

Step 1   Divide the population into groups (clusters).

We know that the population size is 300 and that the population has been divided into clusters of equal size 20. Therefore, there are 15 (300/20) clusters. We number the 20 members of the population in cluster #1 from 1–20, those in cluster #2 from 21–40, and so forth.

Step 2   Obtain a simple random sample of the clusters.

Because each cluster contains 20 members of the population and we want a sample of size 60, we need a simple random sample of 3 (60/20) of the 15 clusters. Using a random-number generator, we obtained clusters #3, #4, and #10.

Step 3   Use all the members of the clusters obtained in Step 2 as the sample.

From Step 2, we see that the 60 members of the population comprising the sample are those whose numbers are 41–60, 61–80, and 181–200. *Note:* Of course, the 60 members sampled would be different if one or more of the three clusters that were randomly selected in Step 2 were different.

You try it!

Exercise 1.61
on page 23

Many years ago, citizens' groups pressured the city council of Tempe, Arizona, to install bike paths in the city. The council members wanted to be sure that they were supported by a majority of the taxpayers, so they decided to poll the city's homeowners.

Their first survey of public opinion was a questionnaire mailed out with the city's 18,000 homeowner water bills. Unfortunately, this method did not work very well. Only 19.4% of the questionnaires were returned, and a large number of those had written comments that indicated they came from avid bicyclists or from people who strongly resented bicyclists. The city council realized that the questionnaire generally had not been returned by the average homeowner.

An employee in the city's planning department had sample survey experience, so the council asked her to do a survey. She was given two assistants to help her interview 300 homeowners and 10 days to complete the project.

The planner first considered taking a simple random sample of 300 homes: 100 interviews for herself and for each of her two assistants. However, the city was so spread out that an interviewer of 100 randomly scattered homeowners would have to drive an average of 18 minutes from one interview to the next. Doing so would require approximately 30 hours of driving time for each interviewer and could delay completion of the report. The planner needed a different sampling design.

■■ ■   EXAMPLE 1.12   Cluster Sampling

Bike Paths Survey   To save time, the planner decided to use cluster sampling. The residential portion of the city was divided into 947 blocks, each containing 20 homes, as shown in Fig. 1.3. Explain how the planner used cluster sampling to obtain a sample of 300 homes.

FIGURE 1.3
A typical block of homes

Solution   We apply Procedure 1.2.

Step 1   Divide the population into groups (clusters).

The planner used the 947 blocks as the clusters, thus dividing the population (residential portion of the city) into 947 groups.

Step 2 **Obtain a simple random sample of the clusters.**

Because each block (cluster) contained 20 homes and the planner needed a sample of 300 homes, she required 15 (300/20) blocks. The planner numbered the blocks (clusters) from 1 to 947 and then used a table of random numbers to obtain a simple random sample of 15 of the 947 blocks.

Step 3 **Use all the members of the clusters obtained in Step 2 as the sample.**

The sample consisted of the 300 homes comprising the 15 sampled blocks:

$$15 \text{ blocks} \times 20 \text{ homes per block} = 300 \text{ homes.}$$

Interpretation The planner used cluster sampling to obtain a sample of 300 homes: 15 blocks of 20 homes per block. Each of the three interviewers was then assigned 5 of these 15 blocks. This method gave each interviewer 100 homes to visit (5 blocks of 20 homes per block) but saved much travel time because an interviewer could complete the interviews on an entire block before driving to another neighborhood. The report was finished on time.

You try it!

Exercise 1.73(a) on page 24

Although cluster sampling can save time and money, it does have disadvantages. Ideally, each cluster should mirror the entire population. In practice, however, members of a cluster may be more homogeneous than the members of the entire population, which can cause problems.

For instance, consider a simplified small town, as depicted in Fig. 1.4. The town council wants to build a town swimming pool. A town planner needs to sample homeowner opinion about using public funds to build the pool. Many upper-income and middle-income homeowners may say "No" if they own or can access pools. Many low-income homeowners may say "Yes" if they do not have access to pools.

FIGURE 1.4

Clusters for a small town

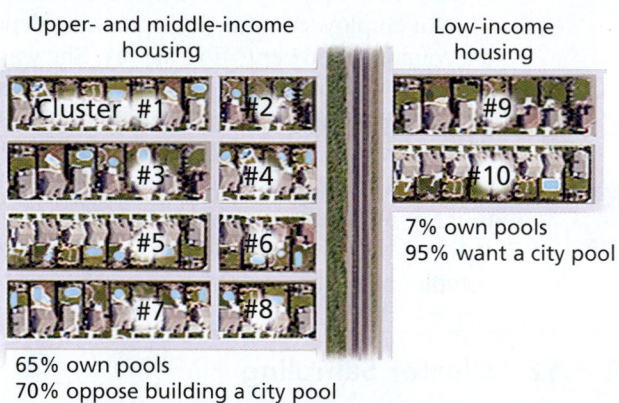

If the planner uses cluster sampling and interviews the homeowners of, say, three randomly selected clusters, there is a good chance that no low-income homeowners will be interviewed.[†] And if no low-income homeowners are interviewed, the results of the survey will be misleading. If, for instance, the planner surveyed clusters #3, #5, and #8, then his survey would show that only about 30% of the homeowners want a pool. However, that is not true, because more than 40% of the homeowners actually want a pool. The clusters most strongly in favor of the pool would not have been included in the survey.

In this hypothetical example, the town is so small that common sense indicates that a cluster sample may not be representative. However, in situations with hundreds of clusters, such problems may be difficult to detect.

[†]There are 120 possible three-cluster samples, and 56 of those contain neither of the low-income clusters, #9 and #10. In other words, 46.7% of the possible three-cluster samples contain neither of the low-income clusters.

Stratified Sampling

Another sampling method, known as **stratified sampling,** is often more reliable than cluster sampling. In stratified sampling, the population is first divided into subpopulations, called **strata,** and then sampling is done from each stratum. Ideally, the members of each stratum should be homogeneous relative to the characteristic under consideration.

In stratified sampling, the strata are often sampled in proportion to their size, which is called **proportional allocation.** Procedure 1.3 presents a step-by-step method for implementing stratified (random) sampling with proportional allocation.

■■■ PROCEDURE 1.3   Stratified Random Sampling with Proportional Allocation

Step 1   Divide the population into subpopulations (strata).

Step 2   From each stratum, obtain a simple random sample of size proportional to the size of the stratum; that is, the sample size for a stratum equals the total sample size times the stratum size divided by the population size.

Step 3   Use all the members obtained in Step 2 as the sample.

■■■ EXAMPLE 1.13   Stratified Sampling with Proportional Allocation

The 2000 members of a population have been divided into four strata of sizes 400, 600, 800, and 200. Use stratified sampling with proportional allocation to obtain a sample of size 10 from the population.

Solution   We apply Procedure 1.3.

Step 1   Divide the population into subpopulations (strata).

We know that the population size is 2000 and that the population has been divided into four strata of sizes 400, 600, 800, and 200. We number the 400 members of the population in strata #1 from 1–400, those in strata #2 from 401−1000, and so forth.

Step 2   From each stratum, obtain a simple random sample of size proportional to the size of the stratum; that is, the sample size for a stratum equals the total sample size times the stratum size divided by the population size.

The sample size for stratum #1 is

$$\text{Total sample size} \times \frac{\text{Size of stratum \#1}}{\text{Population size}} = 10 \times \frac{400}{2000} = 2.$$

Similarly, we find that the sample sizes for strata #2, #3, and #4 are 3, 4, and 1, respectively.

We used a random-number generator to get a simple random sample of size 2 from the 400 members of stratum #1 (numbered 1−400) and obtained 166 and 264. Proceeding similarly, we obtained the required simple random samples shown in the final column of the following table.

Stratum	Size	Numbered	Sample size	Sample
#1	400	1–400	2	166, 264
#2	600	401–1000	3	454, 511, 620
#3	800	1001–1800	4	1246, 1420, 1759, 1793
#4	200	1801–2000	1	1938

Exercise 1.63
on page 23

Step 3 Use all the members obtained in Step 2 as the sample.

From Step 2, we see that the 10 members of the population comprising the sample are those whose numbers are 166, 264, 454, 511, 620, 1246, 1420, 1759, 1793, and 1938. *Note:* Of course, the 10 members sampled would be different if one or more of the four samples in the last column of the preceding table were different.

EXAMPLE 1.14   Stratified Sampling with Proportional Allocation

Town Swimming Pool Consider again the town swimming pool situation discussed on page 20. The town has 250 homeowners of which 25, 175, and 50 are upper income, middle income, and low income, respectively. Explain how we can obtain a sample of 20 homeowners, using stratified sampling with proportional allocation, stratifying by income group.

Solution We apply Procedure 1.3.

Step 1 Divide the population into subpopulations (strata).

We divide the homeowners in the town into three strata according to income group: upper income, middle income, and low income.

Step 2 From each stratum, obtain a simple random sample of size proportional to the size of the stratum; that is, the sample size for a stratum equals the total sample size times the stratum size divided by the population size.

Of the 250 homeowners, 25 are upper income, 175 are middle income, and 50 are lower income. The sample size for the upper-income homeowners is, therefore,

$$\text{Total sample size} \times \frac{\text{Number of upper-income homeowners}}{\text{Total number of homeowners}} = 20 \cdot \frac{25}{250} = 2.$$

Similarly, we find that the sample sizes for the middle-income and lower-income homeowners are 14 and 4, respectively. Thus, we take a simple random sample of size 2 from the 25 upper-income homeowners, of size 14 from the 175 middle-income homeowners, and of size 4 from the 50 lower-income homeowners.

Step 3 Use all the members obtained in Step 2 as the sample.

The sample consists of the 20 homeowners selected in Step 2, namely, the 2 upper-income, 14 middle-income, and 4 lower-income homeowners.

Exercise 1.73(c)
on page 24

Interpretation This stratified sampling procedure ensures that no income group is missed. It also improves the precision of the statistical estimates (because the homeowners within each income group tend to be homogeneous) and makes it possible to estimate the separate opinions of each of the three strata (income groups).

In both Examples 1.13 and 1.14, the calculated sample sizes for all the strata turned out to be whole numbers. Sometimes, however, things don't work out that easily.

For instance, suppose that the population size is 100, the total sample size is 10, and the strata sizes are 29, 41, and 30. Applying the formula

$$\text{Stratum sample size} = \text{Total sample size} \times \frac{\text{Size of stratum}}{\text{Population size}},$$

we find the strata "sample sizes" to be 2.9, 4.1, and 3. Of course, sample sizes must be whole numbers. To remedy the situation here, we can simply round to get the strata sample sizes 3, 4, and 3. Note that, in this case, rounding preserves the condition that the total sample size (10) equals the sum of the strata sample sizes (3 + 4 + 3).

Simple rounding, however, doesn't always remedy the situation. For instance, suppose that the population size is 100, the total sample size is 10, and the strata sizes are 26, 47, and 27. Applying the preceding formula, we find the strata "sample sizes" to be 2.6, 4.7, and 2.7. Rounding these "sample sizes" yields the whole numbers 3, 5, and 3, which do not sum to the total sample size of 10. One way to remedy the situation here would be to round up the two strata "sample sizes" that are closest to their rounded-up values (i.e., round 4.7 and 2.7 up to 5 and 3, respectively) and round down the other strata "sample size" (i.e., round 2.6 down to 2).

Generally, we often need to get creative in deciding how to deal with fractional strata "sample sizes." You will be asked to do so in some of the exercises at the end of this section.

Multistage Sampling

Most large-scale surveys combine one or more of simple random sampling, systematic random sampling, cluster sampling, and stratified sampling. Such **multistage sampling** is used frequently by pollsters and government agencies.

For instance, the U.S. National Center for Health Statistics conducts surveys of the civilian noninstitutional U.S. population to obtain information on illnesses, injuries, and other health issues. Data collection is by a multistage probability sample of approximately 42,000 households. Information obtained from the surveys is published in the *National Health Interview Survey*.

Exercises 1.3

Understanding the Concepts and Skills

In each of Exercises 1.55–1.58, fill in the blank(s).

1.55 Systematic random sampling is easier to execute than simple random sampling and usually provides comparable results. The exception is the presence of some kind of _____ _____ in the listing of the members of the population.

1.56 Ideally, in cluster sampling, each cluster should _____ the entire population.

1.57 Ideally, in stratified sampling, the members of each stratum should be _____ relative to the characteristic under consideration.

1.58 Surveys that combine one or more of simple random sampling, systematic random sampling, cluster sampling, and stratified sampling employ what is called _____ sampling.

1.59 The members of a population have been numbered 1–372. A sample of size 5 is to be taken from the population, using systematic random sampling.
a. Apply Procedure 1.1 on page 17 to determine the sample (i.e., the numbers corresponding to the members of the population that are included in the sample).
b. Suppose that, in Step 2 of Procedure 1.1, the random number chosen is 10 (i.e., $k = 10$). Determine the sample.

1.60 The members of a population have been numbered 1–500. A sample of size 9 is to be taken from the population, using systematic random sampling.
a. Apply Procedure 1.1 on page 17 to determine the sample (i.e., the numbers corresponding to the members of the population that are included in the sample).
b. Suppose that, in Step 2 of Procedure 1.1, the random number chosen is 48 (i.e., $k = 48$). Determine the sample.

1.61 The members of a population have been numbered 1–50. A sample of size 20 is to be taken from the population, using cluster sampling. The clusters are of equal size 10, where cluster #1 consists of the members of the population numbered 1–10, cluster #2 consists of the members of the population numbered 11–20, and so forth.
a. Apply Procedure 1.2 on page 18 to determine the sample (i.e., the numbers corresponding to the members of the population that are included in the sample).
b. Suppose that, in Step 2 of Procedure 1.2, clusters #1 and #3 are selected. Determine the sample.

1.62 The members of a population have been numbered 1–100. A sample of size 30 is to be taken from the population, using cluster sampling. The clusters are of equal size 10, where cluster #1 consists of the members of the population numbered 1–10, cluster #2 consists of the members of the population numbered 11–20, and so forth.
a. Apply Procedure 1.2 on page 18 to determine the sample (i.e., the numbers corresponding to the members of the population that are included in the sample).
b. Suppose that, in Step 2 of Procedure 1.2, clusters #2, #6, and #9 are selected. Determine the sample.

1.63 The members of a population have been numbered 1–1000. A sample of size 20 is to be taken from the population, using stratified random sampling with proportional allocation. The strata are of sizes 300, 200, 400, and 100, where stratum #1 consists of the members of the population numbered 1–300, stratum #2 consists of the members of the population numbered 301–500, and so forth.
a. Determine the sample sizes that will be taken from the strata.
b. Apply Procedure 1.3 on page 21 to determine the sample (i.e., the numbers corresponding to the members of the population that are included in the sample).

1.64 The members of a population have been numbered 1–500. A sample of size 10 is to be taken from the population, using stratified random sampling with proportional allocation. The strata are of sizes 200, 150, and 150, where stratum #1 consists of the members of the population numbered 1–200, stratum #2 consists of the members of the population numbered 201–350, and so forth.

a. Determine the sample sizes that will be taken from the strata.

b. Apply Procedure 1.3 on page 21 to determine the sample (i.e., the numbers corresponding to the members of the population that are included in the sample).

Applying the Concepts and Skills

1.65 Ghost of Speciation Past. In the article, "Ghost of Speciation Past" (*Nature*, Vol. 435, pp. 29–31), T. Kocher looked at the origins of a diverse flock of cichlid fishes in the lakes of southeast Africa. Suppose that you wanted to select a sample from the hundreds of species of cichlid fishes that live in the lakes of southeast Africa. If you took a simple random sample from the species of each lake and combined all the simple random samples into one sample, which type of sampling design would you have used? Explain your answer.

1.66 Number of Farms. The National Agricultural Statistics Service (NASS) conducts studies of the number of acres devoted to farms in each county of the United States. Suppose that we divide the United States into the four census regions (Northeast, North Central, South, and West), take a simple random sample of counties from each of the four regions, and combine all four simple random samples into one sample. What type of sampling design have we used? Explain your answer.

1.67 John F. Kennedy. In one of his books, Ted Sorenson, Special Counsel to President John F. Kennedy, presents an intimate biography of the extraordinary man. According to Sorenson, Kennedy "read every fiftieth letter of the thirty thousand coming weekly to the White House." What type of sampling design was Kennedy using in this case? Explain your answer. [SOURCE: From *Kennedy: The Classical Biography* by Ted Sorenson. Published by Harper Perennial, © 2009.]

1.68 Litigation Surveys. In the article, "Non-probability Sampling Designs for Litigation Surveys (*Trademark Reporter*, Vol. 81, pp. 169–179), J. Jacoby and H. Handlin discussed the controversy about whether nonprobability samples are acceptable as evidence in litigation. The authors randomly selected 26 journals from a list of 1285 scholarly journals in the social and behavioral sciences. They examined all articles published during one year in each of the 26 journals selected with regard to sampling methods. What type of sampling design was used by these two authors in their investigation? Explain your answer.

1.69 Immunization of Schoolchildren. In the article, "Reasons for Non-uptake of Measles, Mumps, and Rubella Catch Up Immunisation in a Measles Epidemic and Side Effects of the Vaccine" (*British Medical Journal*, Vol. 310, pp. 1629–1632), R. Roberts et al. discussed a follow-up survey to examine why almost 10,000 children, ages 11–15, whose records showed no previous immunization were not immunized. In the survey, 10 of the 46 schools participating in the immunization campaign were randomly chosen and then the parents of all the nonimmunized children at the 10 selected schools were sent a questionnaire. What type of sampling design was used by these authors in their survey? Explain your answer.

1.70 University Parking Facilities. During one year, a university wanted to gauge the sentiment of the people using the university's parking facilities. Each of the 8493 people that used the parking facilities had a sticker with a unique number between 1 and 8493. The university committee on parking decided to sample 30 users of the parking facilities and obtain their views on those facilities. The committee selected a number at random between 1 and 283 and got the number 10. The people interviewed were the ones whose stickers had numbers 10, 293, 576, ..., 8217. What type of sampling design was used by the university committee on parking? Explain your answer.

1.71 The International 500. In Exercise 1.49 on page 16, you used simple random sampling to obtain a sample of 10 firms from *Fortune Magazine*'s list of "The International 500."

a. Use systematic random sampling to accomplish that same task.

b. Which method is easier: simple random sampling or systematic random sampling?

c. Does it seem reasonable to use systematic random sampling to obtain a representative sample? Explain your answer.

1.72 Keno. In the game of keno, 20 balls are selected at random from 80 balls numbered 1–80. In Exercise 1.48 on page 16, you used simple random sampling to simulate one game of keno.

a. Use systematic random sampling to obtain a sample of 20 of the 80 balls.

b. Which method is easier: simple random sampling or systematic random sampling?

c. Does it seem reasonable to use systematic random sampling to simulate one game of keno? Explain your answer.

1.73 Sampling Dorm Residents. Students in the dormitories of a university in the state of New York live in clusters of four double rooms, called *suites*. There are 48 suites, with eight students per suite.

a. Describe a cluster sampling procedure for obtaining a sample of 24 dormitory residents.

b. Students typically choose friends from their classes as suitemates. With that in mind, do you think cluster sampling is a good procedure for obtaining a representative sample of dormitory residents? Explain your answer.

c. The university housing office has separate lists of dormitory residents by class level. The number of dormitory residents in each class level is as follows.

Class level	Number of dorm residents
Freshman	128
Sophomore	112
Junior	96
Senior	48

Use the table to design a procedure for obtaining a stratified sample (with proportional allocation) of 24 dormitory residents.

1.74 Best High Schools. In an issue of *Newsweek* (Vol. CXLV, No. 20, pp. 48–57), B. Kantrowitz listed "The 100 best high schools in America" according to a ranking devised by J. Mathews. Another characteristic measured from the high school is the percent free lunch, which is the percentage of student body that is eligible for free and reduced-price lunches, an indicator of socioeconomic status. A percentage of 40% or more generally signifies a high concentration of children in poverty. The top 100 schools, grouped according to their percent free lunch, is as follows.

Percent free lunch	Number of top 100 ranked high schools
0–under 10	50
10–under 20	18
20–under 30	11
30–under 40	8
40 or over	13

a. Use the table to design a procedure for obtaining a stratified sample (with proportional allocation) of 25 high schools from the list of the top 100 ranked high schools.

b. If stratified random sampling with proportional allocation is used to select the sample of 25 high schools, how many would be selected from the stratum with a percent-free-lunch value of 30–under 40?

1.75 U.S. House of Representatives. There are 435 representatives in the 113th session of the U.S. House of Representatives. On the website www.house.gov, you can find an alphabetized list of the 435 congresspersons. In 2013, the first representative listed is Robert Aderholt, a Republican from Alabama, and the last representative listed is Todd Young, a Republican from Indiana. Suppose that the alphabetized list is indexed 1 through 435.

a. Use systematic random sampling to obtain a sample of 15 of the 435 representatives.

b. Suppose that, in Step 2 of Procedure 1.1, the random number chosen is 12 (i.e., $k = 12$). Determine the sample.

1.76 Peace Corps Volunteers. The Peace Corps is an independent U.S. government agency that provides trained volunteers for countries requesting assistance. According to Peacecorps.org, as of September 2012, volunteers currently serve in about 76 different host countries. The average age of a volunteer is about 28 years old. The following table reports the percentage of total volunteers serving by geographic region.

Region	Percent of volunteers
Africa	43%
Latin America	21%
Eastern Europe/Central Asia	15%
Asia	10%
Caribbean	4%
North Africa/Middle East	4%
Pacific Islands	3%

a. Use the table to design a procedure for obtaining a stratified sample (with proportional allocation) of 50 Peace Corps volunteers.

b. If stratified random sampling with proportional allocation is used to select the sample of 50 Peace Corps volunteers, how many would be selected from the Caribbean?

Extending the Concepts and Skills

1.77 The Terri Schiavo Case. In the early part of 2005, the Terri Schiavo case received national attention as her husband sought to have life support removed, and her parents sought to maintain that life support. The courts allowed the life support to be removed, and her death ensued. A Harris Poll of 1010 U.S. adults was taken by telephone on April 21, 2005, to determine how common it is for life support systems to be removed. Those questioned in the sample were asked: (1) Has one of your parents, a close friend, or a family member died in the last 10 years? (2) Before (this death/these deaths) happened, was this person/were any of these people, kept alive by any support system? (3) Did this person die while on a life support system, or had it been withdrawn? Respondents were also asked questions about age, sex, race, education, region, and household income to ensure that results represented a cross section of U.S. adults.

a. What kind of sampling design was used in this survey? Explain your answer.

b. If 78% of the respondents answered the first question in the affirmative, what was the approximate sample size for the second question?

c. If 28% of those responding to the second question answered "yes," what was the approximate sample size for the third question?

1.78 In simple random sampling, all samples of a given size are equally likely. Is that true in systematic random sampling? Explain your answer.

1.79 In simple random sampling, it is also true that each member of the population is equally likely to be selected, the chance for each member being equal to the sample size divided by the population size.

a. Under what circumstances is that fact also true for systematic random sampling? Explain your answer.

b. Provide an example in which that fact is not true for systematic random sampling.

1.80 In simple random sampling, it is also true that each member of the population is equally likely to be selected, the chance for each member being equal to the sample size divided by the population size. Show that this fact is also true for stratified random sampling with proportional allocation.

1.4   Experimental Designs*

As we mentioned earlier, two methods for obtaining information, other than a census, are sampling and experimentation. In Sections 1.2 and 1.3, we discussed some of the basic principles and techniques of sampling. Now, we do the same for experimentation.

Principles of Experimental Design

The study presented in Example 1.6 on page 6 illustrates three basic principles of experimental design: **control, randomization,** and **replication.**

- *Control:* The doctors compared the rate of major birth defects for the women who took folic acid to that for the women who took only trace elements.
- *Randomization:* The women were divided randomly into two groups to avoid unintentional selection bias.
- *Replication:* A large number of women were recruited for the study to make it likely that the two groups created by randomization would be similar and also to increase the chances of detecting any effect due to the folic acid.

In the language of experimental design, each woman in the folic acid study is an **experimental unit,** or a **subject.** More generally, we have the following definition.

DEFINITION 1.5

Experimental Units; Subjects

In a designed experiment, the individuals or items on which the experiment is performed are called **experimental units.** When the experimental units are humans, the term **subject** is often used in place of experimental unit.

In the folic acid study, both doses of folic acid (0.8 mg and none) are called *treatments* in the context of experimental design. Generally, each experimental condition is called a **treatment,** of which there may be several.

Now that we have introduced the terms *experimental unit* and *treatment*, we can present the three basic principles of experimental design in a general setting.

KEY FACT 1.1

Principles of Experimental Design

The following principles of experimental design enable a researcher to conclude that differences in the results of an experiment not reasonably attributable to chance are likely caused by the treatments.

- **Control:** Two or more treatments should be compared.
- **Randomization:** The experimental units should be randomly divided into groups to avoid unintentional selection bias in constituting the groups.
- **Replication:** A sufficient number of experimental units should be used to ensure that randomization creates groups that resemble each other closely and to increase the chances of detecting any differences among the treatments.

One of the most common experimental situations involves a specified treatment and *placebo,* an inert or innocuous medical substance. Technically, both the specified treatment and placebo are treatments. The group receiving the specified treatment is called the **treatment group,** and the group receiving placebo is called the **control group.** In the folic acid study, the women who took folic acid constituted the treatment group, and those women who took only trace elements constituted the control group.

Terminology of Experimental Design

In the folic acid study, the researchers were interested in the effect of folic acid on major birth defects. Birth-defect classification (whether major or not) is the **response variable** for this study. The daily dose of folic acid is called the **factor.** In this case, the factor has two **levels,** namely, 0.8 mg and none.

When there is only one factor, as in the folic acid study, the treatments are the same as the levels of the factor. If a study has more than one factor, however, each treatment is a combination of levels of the various factors.

DEFINITION 1.6

Response Variable, Factors, Levels, and Treatments

Response variable: The characteristic of the experimental outcome that is to be measured or observed.

Factor: A variable whose effect on the response variable is of interest in the experiment.

Levels: The possible values of a factor.

Treatment: Each experimental condition. For one-factor experiments, the treatments are the levels of the single factor. For multifactor experiments, each treatment is a combination of levels of the factors.

■■■ **EXAMPLE 1.15**   Experimental Design

Weight Gain of Golden Torch Cacti The golden torch cactus (*Trichocereus spachianus*), a cactus native to Argentina, has excellent landscape potential. W. Feldman and F. Crosswhite, two researchers at the Boyce Thompson Southwestern Arboretum, investigated the optimal method for producing these cacti.

The researchers examined, among other things, the effects of a hydrophilic polymer and irrigation regime on weight gain. Hydrophilic polymers are used as soil additives to keep moisture in the root zone. For this study, the researchers chose Broadleaf P-4 polyacrylamide, abbreviated P4. The hydrophilic polymer was either used or not used, and five irrigation regimes were employed: none, light, medium, heavy, and very heavy. Identify the

a. experimental units.   **b.** response variable.   **c.** factors.
d. levels of each factor.   **e.** treatments.

Solution

a. The experimental units are the cacti used in the study.
b. The response variable is weight gain.
c. The factors are hydrophilic polymer and irrigation regime.
d. Hydrophilic polymer has two levels: with and without. Irrigation regime has five levels: none, light, medium, heavy, and very heavy.
e. Each treatment is a combination of a level of hydrophilic polymer and a level of irrigation regime. Table 1.8 depicts the 10 treatments for this experiment. In the table, we abbreviated "very heavy" as "Xheavy."

You try it!

Exercise 1.93
on page 30

TABLE 1.8

Schematic for the 10 treatments in the cactus study

		Irrigation regime				
		None	Light	Medium	Heavy	Xheavy
Polymer	No P4	No water No P4 (Treatment 1)	Light water No P4 (Treatment 2)	Medium water No P4 (Treatment 3)	Heavy water No P4 (Treatment 4)	Xheavy water No P4 (Treatment 5)
	With P4	No water With P4 (Treatment 6)	Light water With P4 (Treatment 7)	Medium water With P4 (Treatment 8)	Heavy water With P4 (Treatment 9)	Xheavy water With P4 (Treatment 10)

Statistical Designs

Once we have chosen the treatments, we must decide how the experimental units are to be assigned to the treatments (or vice versa). The women in the folic acid study were randomly divided into two groups; one group received folic acid and the other only trace elements. In the cactus study, 40 cacti were divided randomly into 10 groups of 4 cacti each, and then each group was assigned a different treatment from among the 10 depicted in Table 1.8. Both of these experiments used a **completely randomized design.**

DEFINITION 1.7   **Completely Randomized Design**

In a **completely randomized design,** all the experimental units are assigned randomly among all the treatments.

Although the completely randomized design is commonly used and simple, it is not always the best design. Several alternatives to that design exist.

For instance, in a **randomized block design,** experimental units that are similar in ways that are expected to affect the response variable are grouped in **blocks.** Then the random assignment of experimental units to the treatments is made block by block.

DEFINITION 1.8

> ### Randomized Block Design
>
> In a **randomized block design,** the experimental units are assigned randomly among all the treatments separately within each block.

Example 1.16 contrasts completely randomized designs and randomized block designs.

EXAMPLE 1.16   Statistical Designs

Golf Ball Driving Distances   Suppose we want to compare the driving distances for five different brands of golf ball. For 40 golfers, discuss a method of comparison based on

a. a completely randomized design.
b. a randomized block design.

Solution   Here the experimental units are the golfers, the response variable is driving distance, the factor is brand of golf ball, and the levels (and treatments) are the five brands.

a. For a completely randomized design, we would randomly divide the 40 golfers into five groups of 8 golfers each and then randomly assign each group to drive a different brand of ball, as illustrated in Fig. 1.5.

FIGURE 1.5   Completely randomized design for golf ball experiment

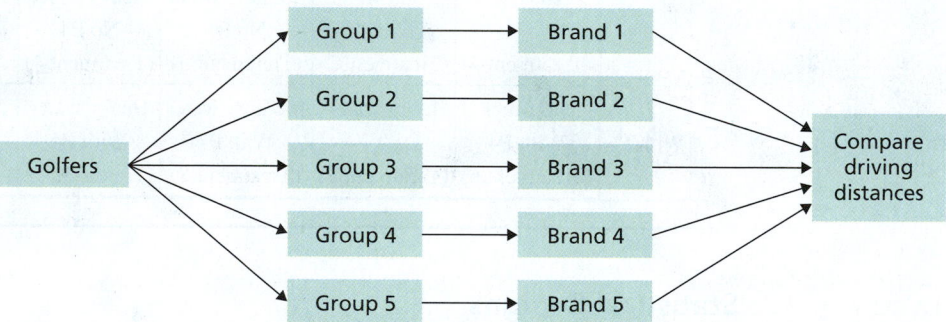

b. Because driving distance is affected by gender, using a randomized block design that blocks by gender is probably a better approach. We could do so by using 20 men golfers and 20 women golfers. We would randomly divide the 20 men into five groups of 4 men each and then randomly assign each group to drive a different brand of ball, as shown in Fig. 1.6. Likewise, we would randomly divide the 20 women into five groups of 4 women each and then randomly assign each group to drive a different brand of ball, as also shown in Fig. 1.6.

By blocking, we can isolate and remove the variation in driving distances between men and women and thereby make it easier to detect any differences in driving distances among the five brands of golf ball. Additionally, blocking permits us to analyze separately the differences in driving distances among the five brands for men and women.

You try it!

Exercise 1.99 on page 30

As illustrated in Example 1.16, blocking can isolate and remove systematic differences among blocks, thereby making any differences among treatments easier to detect. Blocking also makes possible the separate analysis of treatment effects on each block.

FIGURE 1.6 Randomized block design for golf ball experiment

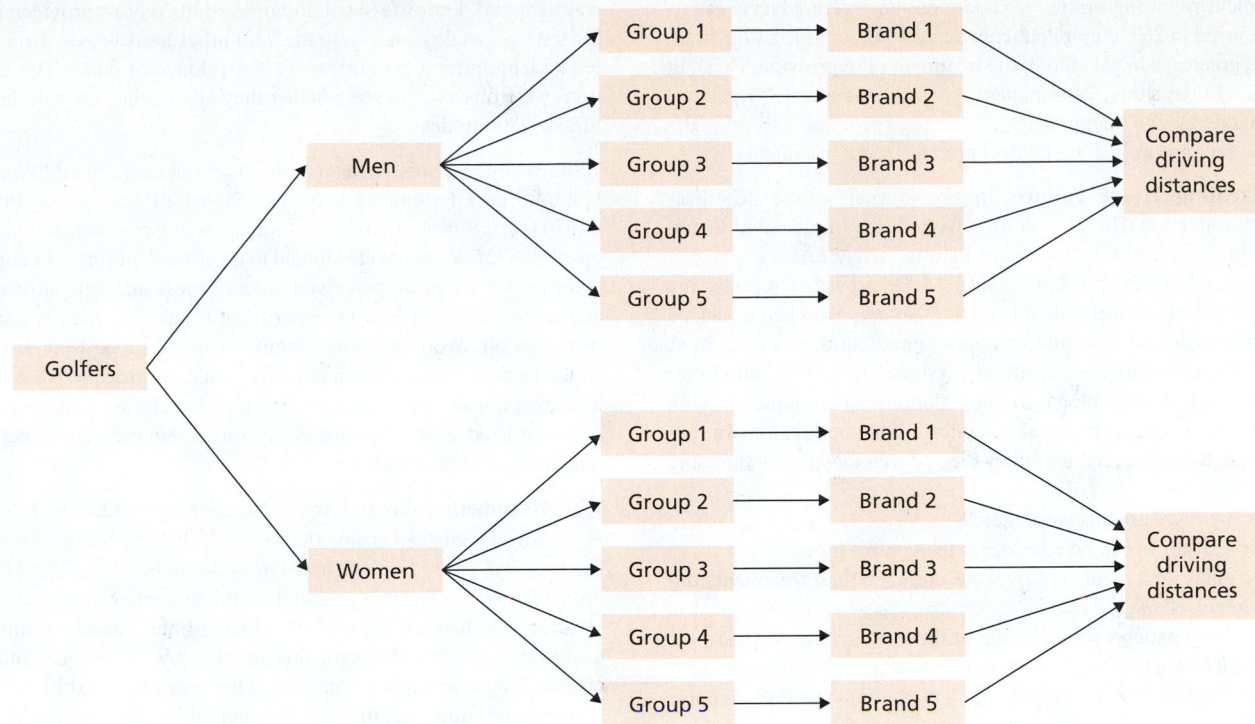

In this section, we introduced some of the basic terminology and principles of experimental design. However, we have just scratched the surface of this vast and important topic to which entire courses and books are devoted. Further discussion of experimental design is provided in the chapter *Design of Experiments and Analysis of Variance* (Module C) in the Regression-ANOVA Modules section on the WeissStats site.

Exercises 1.4

Understanding the Concepts and Skills

1.81 In a designed experiment,
a. what are the experimental units?
b. if the experimental units are humans, what term is often used in place of experimental unit?

1.82 State and explain the significance of the three basic principles of experimental design.

1.83 Define each of the following terms in the context of experimental design.
a. Response variable **b.** Factor
c. Levels **d.** Treatments

1.84 In this section, we discussed two types of statistical designs. Identify and explain the meaning of each one.

1.85 In a designed experiment, there is one factor with four levels. How many treatments are there?

1.86 In a designed experiment, there is one factor with five levels. How many treatments are there?

1.87 In a designed experiment, there are two factors, say, Factor A and Factor B. Factor A has three levels, say, a_1, a_2, and a_3; Factor B has four levels, say, b_1, b_2, b_3, and b_4.

a. Construct a schematic for the treatments similar to Table 1.8 on page 27.
b. Use part (a) to determine the number of treatments.
c. Could you have determined the answer for part (b) without constructing the schematic in part (a)? Explain your answer.

1.88 In a designed experiment, there are two factors, say, Factor A and Factor B. Factor A has four levels, say, a_1, a_2, a_3, and a_4; Factor B has two levels, say, b_1 and b_2.
a. Construct a schematic for the treatments similar to Table 1.8 on page 27.
b. Use part (a) to determine the number of treatments.
c. Could you have determined the answer for part (b) without constructing the schematic in part (a)? Explain your answer.

1.89 In a designed experiment, there are two factors. One factor has m levels and the other factor has n levels. Determine the number of treatments. *Hint:* Refer to Exercise 1.87 or Exercise 1.88.

Applying the Concepts and Skills

1.90 Adverse Effects of Prozac. Prozac (fluoxetine hydrochloride), a product of Eli Lilly and Company, is used for the treatment of depression, obsessive–compulsive disorder (OCD), and bulimia nervosa. An issue of the magazine *Arthritis Today* contained an advertisement reporting on the "...treatment-emergent adverse events that occurred in 2% or more patients treated with Prozac and with incidence greater than placebo in the treatment of depression, OCD, or bulimia." In the study, 2444 patients took Prozac and 1331 patients were given placebo. Identify the

a. treatment group.   b. control group.   c. treatments.

1.91 Treating Heart Failure. In the journal article "Cardiac-Resynchronization Therapy with or without an Implantable Defibrillator in Advanced Chronic Heart Failure" (*New England Journal of Medicine*, Vol. 350, pp. 2140–2150), M. Bristow et al. reported the results of a study of methods for treating patients who had advanced heart failure due to ischemic or nonischemic cardiomyopathies. A total of 1520 patients were randomly assigned in a 1:2:2 ratio to receive optimal pharmacologic therapy alone or in combination with either a pacemaker or a pacemaker–defibrillator combination. The patients were then observed until they died or were hospitalized for any cause.

a. How many treatments were there?
b. Which group would be considered the control group?
c. How many treatment groups were there? Which treatments did they receive?
d. How many patients were in each of the three groups studied?
e. Explain how a table of random numbers or a random-number generator could be used to divide the patients into the three groups.

In Exercises 1.92–1.97, we present descriptions of designed experiments. In each case, identify the
a. experimental units.
b. response variable.
c. factor(s).
d. levels of each factor.
e. treatments.

1.92 Increasing Unit Sales. Supermarkets are interested in strategies to increase temporarily the unit sales of a product. In one study, researchers compared the effect of display type and price on unit sales for a particular product. The following display types and pricing schemes were employed.

- Display types: normal display space interior to an aisle, normal display space at the end of an aisle, and enlarged display space.
- Pricing schemes: regular price, reduced price, and cost.

1.93 Highway Signs. A driver's ability to detect highway signs is an important consideration in highway safety. In his dissertation, *Highway Construction Safety and the Aging Driver*, S. Younes investigated the distance at which drivers can first detect highway caution signs. This distance is called the *detection distance*. Younes analyzed the effect that sign size and sign material have on detection distance. Drivers were randomly assigned to one combination of sign size (small, medium, and large), and sign material (1, 2, and 3). Each driver covered the same stretch of highway at a constant speed during the same time of day, and the detection distance, in feet, was determined for the driver's assigned caution sign.

1.94 Oat Yield and Manure. In a classic study, described by F. Yates in *The Design and Analysis of Factorial Experiments* (Commonwealth Bureau of Soils, Technical Communication No. 35), the effect on oat yield was compared for three different varieties of oats and four different concentrations of manure (0, 0.2, 0.4, and 0.6 cwt per acre).

1.95 The Lion's Mane. In a study by P. M. West titled "The Lion's Mane" (*American Scientist*, Vol. 93, No. 3, pp. 226–236), the effects

of the mane of a male lion as a signal of quality to mates and rivals was explored. Four life-sized dummies of male lions provided a tool for testing female response to the unfamiliar lions whose manes varied by length (long or short) and color (blonde or dark). The female lions were observed to see whether they approached each of the four life-sized dummies.

1.96 Sexual Signals. In a study by A. Elliot et al., titled "Women's Use of Red Clothing as a Sexual Signal in Intersexual Interaction" (*Journal of Experimental Social Psychology*, Vol. 49, Issue 3, pp. 599–602), women were studied to determine the effect of apparel-color choice based on perceived attractiveness and gender of a new acquaintance. The following experiment is based on the researchers' investigation. Women are randomly assigned to be told that they would be conversing with an attractive male, an unattractive male, an attractive female, or an unattractive female. The women must wear a standardized shirt before the conversation, but can choose between red, green, and blue.

1.97 Dexamethasone and IQ. In the paper "Outcomes at School Age After Postnatal Dexamethasone Therapy for Lung Disease of Prematurity" (*New England Journal of Medicine*, Vol. 350, No. 13, pp. 1304–1313), T. Yeh et al. studied the outcomes at school age in children who had participated in a double-blind, placebo-controlled trial of early postnatal dexamethasone therapy for the prevention of chronic lung disease of prematurity. One result reported in the study was that, at school age, the control group of 74 children had an average IQ score of 84.4, whereas the dexamethasone group of 72 children had an average IQ score of 78.

1.98 Lifetimes of Flashlight Batteries. Two different options are under consideration for comparing the lifetimes of four brands of flashlight battery, using 20 flashlights.
a. One option is to randomly divide 20 flashlights into four groups of 5 flashlights each and then randomly assign each group to use a different brand of battery. Would this statistical design be a completely randomized design or a randomized block design? Explain your answer.
b. Another option is to use 20 flashlights—five different brands of 4 flashlights each—and randomly assign the 4 flashlights of each brand to use a different brand of battery. Would this statistical design be a completely randomized design or a randomized block design? Explain your answer.

1.99 Dental Hygiene: Which Toothbrush? In an experiment reported by J. Singer and D. Andrade in the article "Regression Models for the Analysis of Pretest/Posttest Data" (*Biometrics*, Vol. 53, pp. 729–735), the effect of using either a conventional or experimental (hugger) toothbrush was investigated. Twelve female and 12 male preschoolers were selected. Within each gender group, six were randomly assigned to the conventional toothbrush and the remaining six to the experimental toothbrush. After each subject brushed with the assigned toothbrush, a dental plaque index was measured. The higher the dental plaque index, the greater was the amount of plaque on an individual's teeth.
a. Is the statistical design described here a completely randomized design or a randomized block design? Explain your answer.
b. If the statistical design is a randomized block design, what are the blocks?

Extending the Concepts and Skills

1.100 The Salk Vaccine. In Exercise 1.17 on page 8, we discussed the Salk vaccine experiment. The experiment utilized a technique called *double-blinding* because neither the children nor the doctors involved knew which children had been given the vaccine and which had been given placebo. Explain the advantages of using double-blinding in the Salk vaccine experiment.

1.101 In sampling from a population, state which type of sampling design corresponds to each of the following experimental designs:
a. Completely randomized design
b. Randomized block design

CHAPTER IN REVIEW

You Should Be Able to

1. classify a statistical study as either descriptive or inferential.

2. identify the population and the sample in an inferential study.

3. explain the difference between an observational study and a designed experiment.

4. classify a statistical study as either an observational study or a designed experiment.

5. explain what is meant by a representative sample.

6. describe simple random sampling.

7. use a table of random numbers to obtain a simple random sample.

*8. describe systematic random sampling, cluster sampling, and stratified sampling.

*9. state the three basic principles of experimental design.

*10. identify the treatment group and control group in a study.

*11. identify the experimental units, response variable, factor(s), levels of each factor, and treatments in a designed experiment.

*12. distinguish between a completely randomized design and a randomized block design.

Key Terms

blocks,* *28*
census, *9*
cluster sampling,* *18*
completely randomized design,* *27*
control,* *26*
control group,* *26*
descriptive statistics, *3*
designed experiment, *5*
experimental unit,* *26*
experimentation, *9*
factor,* *26*
inferential statistics, *3*
levels,* *26*
multistage sampling,* *23*

observational study, *5*
population, *3*
probability sampling, *10*
proportional allocation,* *21*
randomization,* *26*
randomized block design,* *28*
random-number generator, *13*
replication,* *26*
representative sample, *10*
response variable,* *26*
sample, *3*
sampling, *9*
simple random sample, *10*
simple random sampling, *10*

simple random sampling
    with replacement (SRSWR), *10*
simple random sampling
    without replacement (SRS), *10*
strata,* *21*
stratified random sampling with
    proportional allocation,* *21*
stratified sampling,* *21*
subject,* *26*
systematic random sampling,* *17*
table of random numbers, *11*
treatment,* *26*
treatment group,* *26*

REVIEW PROBLEMS

Understanding the Concepts and Skills

1. In a newspaper or magazine, or on the Internet, find an example of
a. a descriptive study.   b. an inferential study.

2. Almost any inferential study involves aspects of descriptive statistics. Explain why.

3. Regarding observational studies and designed experiments:
a. Describe each type of statistical study.
b. With respect to possible conclusions, what important difference exists between these two types of statistical studies?

4. Before planning and conducting a study to obtain information, what should be done?

5. Explain the meaning of
a. a representative sample.
b. probability sampling.
c. simple random sampling.

6. Which of the following sampling procedures involve the use of probability sampling?
a. A college student is hired to interview a sample of voters in her town. She stays on campus and interviews 100 students in the cafeteria.
b. A pollster wants to interview 20 gas station managers in Baltimore. He posts a list of all such managers on his wall, closes his eyes, and tosses a dart at the list 20 times. He interviews the people whose names the dart hits.

*7. Describe each of the following sampling methods and indicate conditions under which each is appropriate.
  a. Systematic random sampling
  b. Cluster sampling
  c. Stratified random sampling with proportional allocation

*8. Identify and explain the significance of the three basic principles of experimental design.

Applying the Concepts and Skills

9. **Baseball Scores.** From ESPN MLB Scoreboard, we obtained the following major league baseball scores for August 14, 2013. Is this study descriptive or inferential? Explain your answer.

Games in AL stadiums	Games in NL stadiums
Indians 9, Twins 8	Reds 5, Cubs 0
Tigers 6, White Sox 4	Padres 2, Rockies 4
Marlins 5, Royals 2	Orioles 4, Diamondbacks 5
Angels 3, Yankees 11	Giants 5, Nationals 6
Red Sox 3, Blue Jays 4	Phillies 3, Braves 6
Mariners 4, Rays 5	Pirates 5, Cardinals 1
Brewers 4, Rangers 5	Mets 4, Dodgers 5
Astros 2, Athletics 1	

10. **Working Lottery Winners.** In a national poll taken on August 7–11, 2013, by Gallup, Inc., 1039 adults who were employed full or part time were asked the following question: "If you won 10 million dollars in the lottery, would you continue to work, or would you stop working?" Sixty-eight percent of the respondents said that they would continue to work.
  a. Is this study descriptive or inferential? Explain your answer.
  b. The title of the article discussing the survey was "In U.S., Most Would Still Work Even if They Won Millions." Is the statement in quotes here descriptive or inferential? Explain your answer.

11. **British Backpacker Tourists.** Research by G. Visser and C. Barker in "A Geography of British Backpacker Tourists in South Africa" (*Geography*, Vol. 89, No. 3, pp. 226–239) reflected on the impact of British backpacker tourists visiting South Africa. A sample of British backpackers was interviewed. The information obtained from the sample was used to construct the following table for the age distribution of all British backpackers. Classify this study as descriptive or inferential, and explain your answer.

Age (yr)	Percentage
Less than 21	9
21–25	46
26–30	27
31–35	10
36–40	4
Over 40	4

12. **Peanut Allergies.** In the article "Food Allergy Advice May Be Peanuts" (*Science News*, Vol. 174, No. 12, pp. 8–9), N. Seppa reports that early exposure to peanuts seems to lessen the risk of nut allergy. Of 4000 Jewish children sampled in Britain, 1.85% had peanut allergies; and of 4600 Jewish children sampled in Israel, where early peanut consumption is more common, 0.17% had peanut allergies. The researcher chose Jewish children in both countries to limit genetic differences between groups.
  a. Is this study descriptive or inferential?
  b. Is this study observational or experimental?

13. **Persistent Poverty and IQ.** An article appearing in an issue of the *Arizona Republic* reported on a study conducted by G. Duncan of the University of Michigan. According to the report, "Persistent poverty during the first 5 years of life leaves children with IQs 9.1 points lower at age 5 than children who suffer no poverty during that period…." Is this statistical study an observational study or is it a designed experiment? Explain your answer.

14. **Wasp Hierarchical Status.** In an issue of *Discover* (Vol. 26, No. 2, pp. 10–11), J. Netting described the research of E. Tibbetts of the University of Arizona in the article, "The Kind of Face Only a Wasp Could Trust." Tibbetts found that wasps signal their strength and status with the number of black splotches on their yellow faces, with more splotches denoting higher status. Tibbetts decided to see if she could cheat the system. She painted some of the insects' faces to make their status appear higher or lower than it really was. She then placed the painted wasps with a group of female wasps to see if painting the faces altered their hierarchical status. Was this investigation an observational study or a designed experiment? Justify your answer.

15. **Incomes of College Students' Parents.** A researcher wants to estimate the average income of parents of college students. To accomplish that, he surveys a sample of 250 students at Yale. Is this a representative sample? Explain your answer.

16. **On-Time Airlines.** From the *FlightStats On-time Performance Report Summary*, we found that, in July 2013, the top five North American airlines in terms of percentage of on-time arrivals were Hawaiian (H), Horizon (Z), Compass (C), Alaska (A), and Jazz (J).
  a. List the 10 possible samples (without replacement) of size 3 that can be obtained from the population of five airlines. Use the parenthetical abbreviations in your list.
  b. If a simple random sampling procedure is used to obtain a sample of three of these five airlines, what are the chances that it is the first sample on your list in part (a)? the second sample? the tenth sample?
  c. Describe three methods for obtaining a simple random sample of three of these five airlines.
  d. Use one of the methods that you described in part (c) to obtain a simple random sample of three of these five airlines.

17. **Top North American Athletes.** As part of ESPN's SportsCenturyRetrospective, a panel chosen by ESPN ranked the top 100 North American athletes of the twentieth century. For a class project, you are to obtain a simple random sample of 15 of these 100 athletes and briefly describe their athletic feats.
  a. Explain how you can use Table I in Appendix A to obtain the simple random sample.
  b. Starting at the three-digit number in line number 10 and column numbers 7–9 of Table I, read down the column, up the next, and so on, to find 15 numbers that you can use to identify the athletes to be considered.
  c. If you have access to a random-number generator, use it to obtain the required simple random sample.

18. **QuickVote.** TalkBack Live, a production of CNN, conducted on-line surveys on various issues. One survey, called a *QuickVote*, of 680 people asked "Would you vote for a third-party candidate?" Of the 680 people surveyed, 608 (89.4%) responded "yes" and 72 (10.6%) responded "no." Beneath the vote tally, the following statement regarding the sampling procedure was found:

> This QuickVote is not scientific and reflects the opinions of only those Internet users who have chosen to

participate. The results cannot be assumed to represent the opinions of Internet users in general, nor the public as a whole. The QuickVote sponsor is not responsible for content, functionality or the opinions expressed therein.

Discuss the preceding statement in light of what you have learned in this chapter.

19. Leisure Activities and Dementia. An article appearing in the *Los Angeles Times* discussed the study "Leisure Activities and the Risk of Dementia in the Elderly" (*New England Journal of Medicine*, Vol. 348, pp. 2508–2516) by J. Verghese et al. The article in the *Times*, titled "Crosswords Reduce Risk of Dementia," contained the following statement: "Elderly people who frequently read, do crossword puzzles, practice a musical instrument or play board games cut their risk of Alzheimer's and other forms of dementia by nearly two-thirds compared with people who seldom do such activities." Comment on the statement in quotes, keeping in mind the type of study for which causation can be reasonably inferred.

20. Hepatitis B and Pancreatic Cancer. The article "Study Links Hepatitis B and Cancer of Pancreas" by D. Grady, appeared in the September 29, 2008 issue of the *New York Times*. It reported that, for the first time, a study showed that people with pancreatic cancer are more likely than those without the disease to have been infected with the hepatitis B virus. The study by M. Hassan et al., titled "Association Between Hepatitis B Virus and Pancreatic Cancer" (*Journal of Clinical Oncology*, Vol. 26, No. 28, pp. 4557–4562) compared 476 people who had pancreatic cancer with 879 healthy control subjects. All were tested to see whether they had ever been infected with the viruses that cause hepatitis B or hepatitis C. The results were that no connection was found to hepatitis C, but the cancer patients were twice as likely as the healthy subjects to have had hepatitis B. The researchers noted, however, that "…while the study showed an association, it did not prove cause and effect. More work is needed to determine whether the virus really can cause pancreatic cancer." Explain the validity of the statement in quotes.

*21. Top North American Athletes. Refer to Problem 17.
a. Use systematic random sampling to obtain a sample of 15 athletes.
b. In this case, is systematic random sampling an appropriate alternative to simple random sampling? Explain your answer.

*22. Water Quality. In the article "Randomized Stratified Sampling Methodology for Water Quality in Distribution Systems" (*Journal of Water Resources Planning and Management*, Vol. 130, Issue 4, pp. 330–338), V. Speight et al. proposed the method of stratified sampling to collect water samples for water-quality testing. The following table separates the Durham, North Carolina water distribution system into strata based on distance from the nearest treatment plant.

Distance from treatment center	Stratum size
Less than 1.5 miles	1310
1.5–less than 3.0 miles	3166
3.0–less than 4.5 miles	2825
4.5–less than 6.0 miles	1593
6.0–less than 7.5 miles	1350
7.5 miles or greater	1463

Use the table to design a procedure for obtaining a stratified sample (with proportional allocation) of 80 water samples from Durham.

Hint: Refer to the remarks about strata sample size on pages 22–23 following Example 1.14.

*23. AVONEX and MS. An issue of *Inside MS* contained an article describing AVONEX (interferon beta-1a), a drug used in the treatment of relapsing forms of multiple sclerosis (MS). Included in the article was a report on "…adverse events and selected laboratory abnormalities that occurred at an incidence of 2% or more among the 158 multiple sclerosis patients treated with 30 mcg of AVONEX once weekly by IM injection." In the study, 158 patients took AVONEX and 143 patients were given placebo.
a. Is this study observational or is it a designed experiment?
b. Identify the treatment group, control group, and treatments.

*24. Plant Density and Tomato Yield. In "Effects of Plant Density on Tomato Yields in Western Nigeria" (*Experimental Agriculture*, Vol. 12(1), pp. 43–47), B. Adelana reported on the effect of tomato variety and planting density on yield. Four tomato varieties (Harvester, Pusa Early Dwarf, Ife No. 1, and Ibadan Local) were grown at four densities (10,000, 20,000, 30,000, and 40,000 plants/ha). Identify the
a. experimental units. **b.** response variable.
c. factor(s). **d.** levels of each factor.
e. treatments.

*25. Child-Proof Bottles. Designing medication packaging that resists opening by children, but yields readily to adults, presents numerous challenges. In the article "Painful Design" (*American Scientist*, Vol. 93, No. 2, pp. 113–118), H. Petroski examined the packaging used for Aleve, a brand of pain reliever. Three new container designs were given to a panel of children aged 42 months to 51 months. For each design, the children were handed the bottle, shown how to open it, and then left alone with it. If more than 20% of the children succeeded in opening the bottle on their own within 10 minutes, even if by using their teeth, the bottle failed to qualify as child resistant. Identify the
a. experimental units. **b.** response variable.
c. factor(s). **d.** levels of each factor.
e. treatments.

*26. Doughnuts and Fat. A classic study, conducted in 1935 by B. Lowe at the Iowa Agriculture Experiment Station, analyzed differences in the amount of fat absorbed by doughnuts in cooking with four different fats. For the experiment, 24 batches of doughnuts were randomly divided into four groups of 6 batches each. The four groups were then randomly assigned to the four fats. What type of statistical design was used for this study? Explain your answer.

*27. Comparing Gas Mileages. An experiment is to be conducted to compare four different brands of gasoline for gas mileage.
a. Suppose that you randomly divide 24 cars into four groups of 6 cars each and then randomly assign the four groups to the four brands of gasoline, one group per brand. Is this experimental design a completely randomized design or a randomized block design? If it is the latter, what are the blocks?
b. Suppose, instead, that you use six different models of cars whose varying characteristics (e.g., weight and horsepower) affect gas mileage. Four cars of each model are randomly assigned to the four different brands of gasoline. Is this experimental design a completely randomized design or a randomized block design? If it is the latter, what are the blocks?
c. Which design is better, the one in part (a) or the one in part (b)? Explain your answer.

FOCUSING ON DATA ANALYSIS

UWEC UNDERGRADUATES

The file named Focus.txt in the Focus Database section of the WeissStats site contains information on the undergraduate students at the University of Wisconsin - Eau Claire (UWEC). Those students constitute the population of interest in the *Focusing on Data Analysis* sections that appear at the end of each chapter of the book.[†]

Thirteen variables are considered. Table 1.9 lists the variables and the names used for those variables in the data files. We call the database of information for those variables the **Focus database.**

Also provided in the Focus Database section is a file called FocusSample.txt that contains data on the same 13 variables for a simple random sample of 200 of the undergraduate students at UWEC. Those 200 students constitute a sample that can be used for making statistical inferences in the *Focusing on Data Analysis* sections. We call this sample data the **Focus sample.**

Large data sets are almost always analyzed by computer, and that is how you should handle both the Focus database and the Focus sample. We have supplied the Focus database and Focus sample in several file formats in the Focus Database section of the WeissStats site.

If you use a statistical software package for which we have not supplied a Focus database file, you should (1) input the file Focus.txt into that software, (2) ensure that the variables are named as indicated in Table 1.9, and (3) save the worksheet to

a file named Focus in the format suitable to your software, that is, with the appropriate file extension. Then, any time that you want to analyze the Focus database, you can simply retrieve your Focus worksheet. These same remarks apply to the Focus sample, as well as to the Focus database.

TABLE 1.9

Variables and variable names for the Focus database

Variable	Variable name
Sex	SEX
High school percentile	HSP
Cumulative GPA	GPA
Age	AGE
Total earned credits	CREDITS
Classification	CLASS
School/college	COLLEGE
Primary major	MAJOR
Residency	RESIDENCY
Admission type	TYPE
ACT English score	ENGLISH
ACT math score	MATH
ACT composite score	COMP

[†]We have restricted attention to those undergraduate students at UWEC with complete records for all the variables under consideration.

CASE STUDY DISCUSSION

TOP FILMS OF ALL TIME

At the beginning of this chapter, we discussed the results of a survey by the American Film Institute (AFI). Now that you have learned some of the basic terminology of statistics, we want you to examine that survey in greater detail.

Answer each of the following questions pertaining to the survey. In doing so, you may want to reread the description of the survey given on pages 1–2.

a. Identify the population.
b. Identify the sample.

c. Is the sample representative of the population of all U.S. moviegoers? Explain your answer.
d. Consider the following statement: "Among the 1500 film artists, critics, and historians polled by AFI, the top-ranking film was *Citizen Kane*." Is this statement descriptive or inferential? Explain your answer.
e. Suppose that the statement in part (d) is changed to the following statement: "Based on the AFI poll, *Citizen Kane* is the top-ranking film among all film artists, critics, and historians." Is this statement descriptive or inferential? Explain your answer.

BIOGRAPHY

FLORENCE NIGHTINGALE: LADY OF THE LAMP

Florence Nightingale (1820–1910), the founder of modern nursing, was born in Florence, Italy, into a wealthy English family. In 1849, over the objections of her parents, she entered the Institution of Protestant Deaconesses at Kaiserswerth, Germany, which "…trained country girls of good character to nurse the sick."

The Crimean War began in March 1854 when England and France declared war on Russia. After serving as superintendent of the Institution for the Care of Sick Gentlewomen in London, Nightingale was appointed by the English Secretary of State at War, Sidney Herbert, to be in charge of 38 nurses who were to be stationed at military hospitals in Turkey.

Nightingale found the conditions in the hospitals appalling—overcrowded, filthy, and without sufficient facilities. In addition to the administrative duties she undertook to alleviate those conditions, she spent many hours tending patients. After 8:00 P.M. she allowed none of her nurses in the wards, but made rounds herself every night, a deed that earned her the epithet Lady of the Lamp.

Nightingale was an ardent believer in the power of statistics and used statistics extensively to gain an understanding of social and health issues. She lobbied to introduce statistics into the curriculum at Oxford and invented the coxcomb chart, a type of pie chart. Nightingale felt that charts and diagrams were a means of making statistical information understandable to people who would otherwise be unwilling to digest the dry numbers.

In May 1857, as a result of Nightingale's interviews with officials ranging from the Secretary of State to Queen Victoria herself, the Royal Commission on the Health of the Army was established. Under the auspices of the commission, the Army Medical School was founded. In 1860, Nightingale used a fund set up by the public to honor her work in the Crimean War to create the Nightingale School for Nurses at St. Thomas's Hospital. During that same year, at the International Statistical Congress in London, she authored one of the three papers discussed in the Sanitary Section and also met Adolphe Quetelet (see Chapter 2 biography), who had greatly influenced her work.

After 1857, Nightingale lived as an invalid, although it has never been determined that she had any specific illness. In fact, many speculated that her invalidism was a stratagem she employed to devote herself to her work.

Nightingale was elected an Honorary Member of the American Statistical Association in 1874. In 1907, she was presented the Order of Merit for meritorious service by King Edward VII; she was the first woman to receive that award.

Florence Nightingale died in 1910. An offer of a national funeral and burial at Westminster Abbey was declined, and, according to her wishes, Nightingale was buried in the family plot in East Mellow, Hampshire, England.

CHAPTER

2

Organizing Data

CHAPTER OBJECTIVES

In Chapter 1, we introduced two major interrelated branches of statistics: descriptive statistics and inferential statistics. In this chapter, you will begin your study of descriptive statistics, which consists of methods for organizing and summarizing information.

In Section 2.1, we show you how to classify data by type. Knowing the data type can help you choose the correct statistical method. In Section 2.2, we explain how to group and graph qualitative data so that they are easier to work with and understand. In Section 2.3, we do likewise for quantitative data. In that section, we also introduce stem-and-leaf diagrams—one of an arsenal of statistical tools known collectively as **exploratory data analysis.**

In Section 2.4, we discuss the identification of the shape of a data set. In Section 2.5, we present tips for avoiding confusion when you read and interpret graphical displays.

CASE STUDY

World's Richest People

Each year, *Forbes* magazine publishes a list of the world's richest people. In 2013, over 50 reporters worked to compile *Forbes*'s

27th anniversary "World's Billionaires" rankings. To estimate net worth, *Forbes* values individual assets and accounts for debt. Assets include stakes in public and private companies, real estate, yachts, art, and cash.

The magazine includes wealth belonging to a person's immediate relatives provided that the wealth can ultimately be traced to one living individual. In those cases "& family" is added to indicate that the net worth given includes money belonging to more than one person.

From the *Forbes* article, we constructed the following table showing the 25 richest people (out of 1426 on the complete list), as of March 2013.

Rank	Name	Age	Citizenship	Wealth ($ billions)
1	Carlos Slim Helu & family	73	Mexico	73
2	Bill Gates	57	United States	67
3	Amancio Ortega	77	Spain	57
4	Warren Buffett	82	United States	53.5
5	Larry Ellison	68	United States	43
6	Charles Koch	77	United States	34
6	David Koch	73	United States	34
8	Li Ka-shing	84	Hong Kong	31
9	Liliane Bettencourt & family	90	France	30
10	Bernard Arnault & family	64	France	29
11	Christy Walton & family	58	United States	28.2
12	Stefan Persson	65	Sweden	28
13	Michael Bloomberg	71	United States	27
14	Jim Walton	65	United States	26.7
15	Sheldon Adelson	79	United States	26.5
16	Alice Walton	63	United States	26.3
17	S. Robson Walton	69	United States	26.1
18	Karl Albrecht	93	Germany	26
19	Jeff Bezos	49	United States	25.2
20	Larry Page	40	United States	23
21	Sergey Brin	39	United States	22.8
22	Mukesh Ambani	56	India	21.5
23	Michele Ferrero & family	88	Italy	20.4
24	Lee Shau Kee	85	Hong Kong	20.3
24	David Thomson & family	55	Canada	20.3

At the end of this chapter, you will apply some of your newly learned statistical skills to analyze these data.

2.1 Variables and Data

A characteristic that varies from one person or thing to another is called a **variable.** Examples of variables for humans are height, weight, number of siblings, sex, marital status, and eye color. The first three of these variables yield numerical information and are examples of **quantitative variables;** the last three yield nonnumerical information and are examples of **qualitative variables,** also called **categorical variables.**[†]

Quantitative variables can be classified as either *discrete* or *continuous*. A **discrete variable** is a variable whose possible values can be listed, even though the list may continue indefinitely. This property holds, for instance, if either the variable has only a finite number of possible values or its possible values are some collection of whole numbers.[‡] A discrete variable usually involves a count of something, such as the number of siblings a person has, the number of cars owned by a family, or the number of students in an introductory statistics class.

A **continuous variable** is a variable whose possible values form some interval of numbers. Typically, a continuous variable involves a measurement of something, such as the height of a person, the weight of a newborn baby, or the length of time a car battery lasts.

[†]Values of a qualitative variable are sometimes coded with numbers—for example, zip codes, which represent geographical locations. We cannot do arithmetic with such numbers, in contrast to those of a quantitative variable.

[‡]Mathematically speaking, a discrete variable is any variable whose possible values form a *countable set*, a set that is either finite or countably infinite.

The preceding discussion is summarized graphically in Fig. 2.1 and verbally in the following definition.

DEFINITION 2.1

Variables

Variable: A characteristic that varies from one person or thing to another.

Qualitative variable: A nonnumerically valued variable.

Quantitative variable: A numerically valued variable.

Discrete variable: A quantitative variable whose possible values can be listed. In particular, a quantitative variable with only a finite number of possible values is a discrete variable.

Continuous variable: A quantitative variable whose possible values form some interval of numbers.

What Does It Mean?

A discrete variable usually involves a count of something, whereas a continuous variable usually involves a measurement of something.

FIGURE 2.1

Types of variables

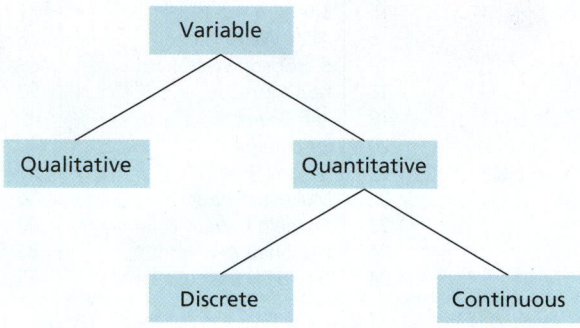

The values of a variable for one or more people or things yield **data.** Thus the information collected, organized, and analyzed by statisticians is data. Data, like variables, can be classified as **qualitative data, quantitative data, discrete data,** and **continuous data.**

DEFINITION 2.2

Data

Data: Values of a variable.

Qualitative data: Values of a qualitative variable.

Quantitative data: Values of a quantitative variable.

Discrete data: Values of a discrete variable.

Continuous data: Values of a continuous variable.

What Does It Mean?

Data are classified according to the type of variable from which they were obtained.

Each individual piece of data is called an **observation,** and the collection of all observations for a particular variable is called a **data set.**[†] We illustrate various types of variables and data in Examples 2.1–2.4.

EXAMPLE 2.1   **Variables and Data**

The 118th Boston Marathon At noon on April 21, 2014, about 35,671 men and women set out to run 26 miles and 385 yards from rural Hopkinton to Boston. Thousands of people lining the streets leading into Boston and millions more on television watched this 118th running of the Boston Marathon.

The Boston Marathon provides examples of different types of variables and data, which are compiled by the Boston Athletic Association and others. The

---

[†]Sometimes *data set* is used to refer to all the data for all the variables under consideration.

classification of each entrant as either male or female illustrates the simplest type of variable. "Gender" is a qualitative variable because its possible values (male or female) are nonnumerical. Thus, for instance, the information that Meb Keflezighi is a male and Rita Jeptoo is a female is qualitative data.

"Place of finish" is a quantitative variable, which is also a discrete variable because it makes sense to talk only about first place, second place, and so on—there are only a finite number of possible finishing places. Thus, the information that, among the women, Rita Jeptoo and Buzunesh Deba finished first and second, respectively, is discrete, quantitative data.

"Finishing time" is a quantitative variable, which is also a continuous variable because the finishing time of a runner can conceptually be any positive number. The information that Meb Keflezighi won the men's competition in 2:08:37 and Rita Jeptoo won the women's competition in 2:18:57 is continuous, quantitative data.

You try it!

Exercise 2.7 on page 40

◼◼◼ **EXAMPLE 2.2   Variables and Data**

Human Blood Types Human beings have one of four blood types: A, B, AB, or O. What kind of data do you receive when you are told your blood type?

Solution Blood type is a qualitative variable because its possible values are nonnumerical. Therefore your blood type is qualitative data.

◼◼◼ **EXAMPLE 2.3   Variables and Data**

Household Size The U.S. Census Bureau collects data on household size and publishes the information in *Current Population Reports*. What kind of data is the number of people in your household?

Solution Household size is a quantitative variable, which is also a discrete variable because its possible values are 1, 2, ... Therefore the number of people in your household is discrete, quantitative data.

◼◼◼ **EXAMPLE 2.4   Variables and Data**

The World's Highest Waterfalls The *Information Please Almanac* lists the world's highest waterfalls. The list shows that Angel Falls in Venezuela is 3281 feet high, or more than twice as high as Ribbon Falls in Yosemite, California, which is 1612 feet high. What kind of data are these heights?

Solution Height is a quantitative variable, which is also a continuous variable because height can conceptually be any positive number. Therefore the waterfall heights are continuous, quantitative data.

Classification and the Choice of a Statistical Method

Some of the statistical procedures that you will study are valid for only certain types of data. This limitation is one reason why you must be able to classify data. The classifications we have discussed are sufficient for most applications, even though statisticians sometimes use additional classifications.

Data classification can be difficult; even statisticians occasionally disagree over data type. For example, some classify amounts of money as discrete data; others say it is continuous data. In most cases, however, data classification is fairly clear and will help you choose the correct statistical method for analyzing the data.

Exercises 2.1

Understanding the Concepts and Skills

2.1 Give an example, other than those presented in this section, of a
a. qualitative variable.
b. discrete, quantitative variable.
c. continuous, quantitative variable.

2.2 Explain the meaning of
a. qualitative variable.
b. discrete, quantitative variable.
c. continuous, quantitative variable.

2.3 Explain the meaning of
a. qualitative data.   **b.** discrete, quantitative data.
c. continuous, quantitative data.

2.4 Provide a reason why the classification of data is important.

2.5 Of the variables you have studied so far, which type yields non-numerical data?

Applying the Concepts and Skills

For each part of Exercises 2.6–2.11, classify the data as either qualitative or quantitative; if quantitative, further classify it as discrete or continuous. Also, identify the variable under consideration in each case.

2.6 World's Highest Temperatures. Information provided by the World Meteorological Association revealed the following data on the highest recorded temperature for each continent.

Rank	Continent	Place	Temp (°F)
1	N. America	Death Valley, CA, USA	134
2	Africa	Kebili, Tunisia	131
3	Asia	Tirat Tsvi, Israel	129
4	Australia	Oodnadatta, S. Australia	123
5	S. America	Rivadavia, Argentina	120
6	Europe	Athens, Greece	118
7	Oceania	Tuguegarao, Philippines	108
8	Antarctica	Vanda Station, Scott Coast	59

a. What type of data is presented in the first column of the table?
b. What type of data is presented in the second column of the table?
c. What type of data is presented in the fourth column of the table?
d. What type of data is provided by the information that Death Valley is in the United States?

2.7 Earthquakes. The U.S. Geological Survey monitors and reports on earthquakes, providing daily real-time, worldwide earthquake lists. Some of the information for four of the 105 earthquakes that occurred on May 10, 2013, is shown in the following table. Magnitude is given on the Richter scale and NST stands for the number of stations that reported the activity on the same earthquake.

Time	Magnitude	Depth (km)	NST	Region
03:46:54	1.6	16.7	25	N. California
15:19:57	1.8	7.7	11	Tennessee
19:57:02	1.5	7.0	15	Nevada
23:12:01	2.4	11.0	6	Puerto Rico

Identify the type of data provided by the information in the
a. first column of the table.
b. second column of the table.
c. third column of the table.
d. fourth column of the table.
e. fifth column of the table.

2.8 Top 10 IPOs. An online article from the *Washington Post*, titled "Facebook Joins Ranks of Largest IPOs in U.S. History," gave the following information on the top 10 IPOs (initial public offerings) in the United States as of May 2012.

Rank	Company	IPO ($billions)
1	Visa	17.86
2	Enel	16.45
3	Facebook	16.00
4	General Motors	15.77
5	Deutsche Telekom	13.00
6	AT&T Wireless	10.62
7	Kraft Foods	8.68
8	France Telecom	7.29
9	Telstra	5.65
10	Swisscom	5.58

a. What type of data is presented in the first column of the table?
b. What type of data is presented in the second column of the table?
c. What type of data is presented in the third column of the table?
d. What type of data is provided by the information that Facebook is a social networking business?

2.9 Earnings from the Crypt. On the Celebrity Net Worth website, we found the article "The 10 Highest Earning Dead Celebrities" by G. Golden. According to the article, the highest earning dead celebrities in 2012 are as shown in the following table.

Rank	Name	Earnings ($millions)
1	Michael Jackson	200
2	Elvis Presley	60
3	Marilyn Monroe	30
4	Charles Schulz	26
5	John Lennon	15
6	Elizabeth Taylor	13
7	Albert Einstein	12
8	Theodor Geisel (Dr. Suess)	11
9	Jimi Hendrix	10
10	Stieg Larsson	9

What type of data is presented in the
a. first column of the table?
b. second column of the table?
c. third column of the table?

2.10 World University Rankings. The *Times Higher Education* "World University Rankings" presents global university performance tables to judge world class universities on the basis of teaching, research, knowledge transfer, and international outlook. The rankings employ 13 carefully calibrated performance indicators to provide comprehensive and balanced comparisons. The top 10 universities for 2012–2013 are as shown in the following table.

Rank	Institution	Overall score
1	California Institute of Technology	95.5
2	Stanford University	93.7
2	University of Oxford	93.7
4	Harvard University	93.6
5	Massachusetts Institute of Technology	93.1
6	Princeton University	92.7
7	University of Cambridge	92.6
8	Imperial College London	90.6
9	University of California, Berkeley	90.5
10	University of Chicago	90.4

What type of data is presented in the
a. first column of the table?   **b.** second column of the table?
c. third column of the table?

2.11 Recording Industry Statistics. The Recording Industry Association of America (RIAA) reports annual statistics on shipments and revenues. For the year 2012, RIAA reported the following data on physical shipments. [SOURCE: *Year-End Industry Shipment and Revenue Statistics*. Published by RIAA, © 2012.]

Product	Units shipped (millions)	Retail dollar value ($millions)
CD	210.9	2532.0
CD single	1.1	3.2
LP/EP	7.1	162.6
Vinyl single	0.4	4.7
Music video	6.2	118.2
DVD audio	0.0	0.2
SA CD	0.1	1.3

Identify the type of data provided by the information in each of the following columns of the table:
a. first   **b.** second   **c.** third

2.12 RBI Kings. As reported on MLB.com, the five players with the highest runs batted in (RBI) during the 2012 Major League Baseball season are listed in the following table. Also included are the teams for which they played, their positions, and their weights. Identify the type of data provided by the information in each column of the table.

Player	Team	Position	RBI	Weight (lb)
M. Cabrera	Detroit	3B	139	240
J. Hamilton	Texas	LF	128	225
C. Headley	San Diego	3B	115	220
R. Braun	Milwaukee	LF	112	205
E. Encarnacion	Toronto	1B	110	230

2.13 Top Broadcast Shows. As reported in *Primetime Broadcast Programs*, a publication of The Nielsen Company, the top three prime broadcast network television shows for the week of August 5, 2013, are as presented in the following table. Ratings are the percentage of TV homes in the United States tuned into television. Identify the type of data provided by the information in each column of the table.

Rank	Program	Network	Rating
1	Under the Dome	CBS	6.4
2	America's Got Talent	NBC	6.2
3	The Bachelorette	ABC	6.0

2.14 The Fulbright Program. The U.S. government's flagship international educational exchange program, the Fulbright program, is sponsored by the Bureau of Educational and Cultural Affairs. Fulbrights are awarded based on "academic and leadership potential, providing the opportunity to exchange ideas and contribute to finding solutions to shared international concerns." For 2012–2013, the top 10 doctoral/research institutions receiving Fulbright awards are as shown in the following table. The success rate gives the number of grants per application, expressed as a percentage. Identify the type of data provided by the information in each column of the table.

Institution	Grants	Applications	Success rate
U. of Michigan, Ann Arbor	40	141	28.4%
Harvard	31	132	23.5%
Brown	29	85	34.1%
U. of Chicago	24	102	23.5%
U. of California, Berkeley	23	97	23.7%
Yale	23	98	23.5%
Arizona State University	23	58	39.7%
Columbia	22	79	27.8%
Northwestern	22	101	21.8%
U. of Texas at Austin	22	77	28.6%

2.15 Top 10 Green Cars. The following table presents information on the Kelley Blue Book's "10 Best Green Cars of 2013." Note that mpg and mpge are abbreviations for miles per gallon and miles per gallon equivalent, respectively. Identify the type of data provided by the information in each column of the table. [SOURCE: "10 Best Green Cars of 2013." Published by the Kelley Blue Book Co., Inc.]

Rank	Make/model	Type	Combined mileage
1	Nissan Leaf	Electric	116 mpge
2	Tesla Model S	Electric	89 mpge
3	Ford Focus	Electric	105 mpge
4	Chevrolet Volt	Hybrid	First 38 miles: 98 mpge; next 344 miles: 37 mpg
5	Toyota Prius Plug-in	Hybrid	First 11 miles: 95 mpge; next 529 miles: 50 mpg
6	Ford C-Max Energi	Hybrid	First 21 miles: 100 mpge; next 599 miles: 43 mpg
7	Volkswagen Jetta	Hybrid	45 mpg
8	Honda Fit	Electric	118 mpge
9	Toyota Avalon	Hybrid	40 mpg
10	Lincoln MKZ	Hybrid	45 mpg

Extending the Concepts and Skills

2.16 Ordinal Data. Another important type of data is *ordinal data*, which are data about order or rank given on a scale such as 1, 2, 3, … or A, B, C, …. Following are several variables. Which, if any, yield ordinal data? Explain your answer.
a. Height   **e.** Number of siblings
b. Weight   **f.** Religion
c. Age   **g.** Place of birth
d. Sex   **h.** High school class rank

2.2 Organizing Qualitative Data

Some situations generate an overwhelming amount of data. We can often make a large or complicated set of data more compact and easier to understand by organizing it in a table, chart, or graph. In this section, we examine some of the most important ways to organize qualitative data. In the next section, we do that for quantitative data.

Frequency Distributions

Recall that qualitative data are values of a qualitative (nonnumerically valued) variable. One way of organizing qualitative data is to construct a table that gives the number of times each distinct value occurs. The number of times a particular distinct value occurs is called its **frequency** (or **count**).

DEFINITION 2.3

What Does It Mean?

A frequency distribution provides a table of the values of the observations and how often they occur.

Frequency Distribution of Qualitative Data

A **frequency distribution** of qualitative data is a listing of the distinct values and their frequencies.

Procedure 2.1 provides a step-by-step method for obtaining a frequency distribution of qualitative data.

PROCEDURE 2.1

To Construct a Frequency Distribution of Qualitative Data

Step 1   List the distinct values of the observations in the data set in the first column of a table.

Step 2   For each observation, place a tally mark in the second column of the table in the row of the appropriate distinct value.

Step 3   Count the tallies for each distinct value and record the totals in the third column of the table.

Note: When applying Step 2 of Procedure 2.1, you may find it useful to cross out each observation after you tally it. This strategy helps ensure that no observation is missed or duplicated.

EXAMPLE 2.5   **Frequency Distribution of Qualitative Data**

Political Party Affiliations   Professor Weiss asked his introductory statistics students to state their political party affiliations as Democratic, Republican, or Other. The responses of the 40 students in the class are given in Table 2.1. Determine the frequency distribution of these data.

TABLE 2.1

Political party affiliations of the students in introductory statistics

Democratic	Other	Democratic	Other	Democratic
Republican	Republican	Other	Other	Republican
Republican	Republican	Republican	Democratic	Republican
Republican	Democratic	Democratic	Other	Republican
Democratic	Democratic	Republican	Democratic	Democratic
Republican	Republican	Other	Other	Democratic
Republican	Democratic	Republican	Other	Other
Republican	Republican	Republican	Democratic	Republican

Solution We apply Procedure 2.1.

Step 1 List the distinct values of the observations in the data set in the first column of a table.

The distinct values of the observations are Democratic, Republican, and Other, which we list in the first column of Table 2.2.

TABLE 2.2

Table for constructing a frequency distribution for the political party affiliation data in Table 2.1

Party	Tally	Frequency
Democratic	⦀⦀ III	13
Republican	⦀⦀⦀ III	18
Other	⦀ IIII	9
		40

Step 2 For each observation, place a tally mark in the second column of the table in the row of the appropriate distinct value.

The first affiliation listed in Table 2.1 is Democratic, calling for a tally mark in the Democratic row of Table 2.2. The complete results of the tallying procedure are shown in the second column of Table 2.2.

Step 3 Count the tallies for each distinct value and record the totals in the third column of the table.

Counting the tallies in the second column of Table 2.2 gives the frequencies in the third column of Table 2.2. The first and third columns of Table 2.2 provide a frequency distribution for the data in Table 2.1.

Interpretation From Table 2.2, we see that, of the 40 students in the class, 13 are Democrats, 18 are Republicans, and 9 are Other.

By simply glancing at Table 2.2, we can easily obtain various pieces of useful information. For instance, we see that more students in the class are Republicans than any other political party affiliation.

StatCrunch

Report 2.1

You try it!

Exercise 2.27(a) on page 50

Relative-Frequency Distributions

In addition to the frequency that a particular distinct value occurs, we are often interested in the **relative frequency,** which is the ratio of the frequency to the total number of observations:

$$\text{Relative frequency} = \frac{\text{Frequency}}{\text{Number of observations}}.$$

For instance, as we see from Table 2.2, the relative frequency of Democrats in Professor Weiss's introductory statistics class is

$$\text{Relative frequency of Democrats} = \frac{\text{Frequency of Democrats}}{\text{Number of observations}} = \frac{13}{40} = 0.325.$$

In terms of percentages, 32.5% of the students in Professor Weiss's introductory statistics class are Democrats. We see that a relative frequency is just a **percentage** expressed as a decimal.

As you might expect, a relative-frequency distribution of qualitative data is similar to a frequency distribution, except that we use relative frequencies instead of frequencies.

? What Does It Mean?

A relative-frequency distribution provides a table of the values of the observations and (relatively) how often they occur.

DEFINITION 2.4

Relative-Frequency Distribution of Qualitative Data

A **relative-frequency distribution** of qualitative data is a listing of the distinct values and their relative frequencies.

To obtain a relative-frequency distribution, we first find a frequency distribution and then divide each frequency by the total number of observations. Thus, we have Procedure 2.2.

■■■ **PROCEDURE 2.2** **To Construct a Relative-Frequency Distribution of Qualitative Data**

Step 1 Obtain a frequency distribution of the data.

Step 2 Divide each frequency by the total number of observations.

■■■ **EXAMPLE 2.6** **Relative-Frequency Distribution of Qualitative Data**

Political Party Affiliations Refer to Example 2.5 on pages 42–43. Construct a relative-frequency distribution of the political party affiliations of the students in Professor Weiss's introductory statistics class presented in Table 2.1 on page 42.

Solution We apply Procedure 2.2.

Step 1 Obtain a frequency distribution of the data.

We obtained a frequency distribution of the data in Example 2.5; specifically, see the first and third columns of Table 2.2 on page 43.

Step 2 Divide each frequency by the total number of observations.

Dividing each entry in the third column of Table 2.2 by the total number of observations, 40, we obtain the relative frequencies displayed in the second column of Table 2.3. The two columns of Table 2.3 provide a relative-frequency distribution for the data in Table 2.1.

TABLE 2.3

Relative-frequency distribution for the political party affiliation data in Table 2.1

Party	Relative frequency	
Democratic	0.325	← *13/40*
Republican	0.450	← *18/40*
Other	0.225	← *9/40*
	1.000	

StatCrunch

Report 2.2

You try it!

Exercise 2.27(b) on page 50

Interpretation From Table 2.3, we see that 32.5% of the students in Professor Weiss's introductory statistics class are Democrats, 45.0% are Republicans, and 22.5% are Other.

Note: Relative-frequency distributions are better than frequency distributions for comparing two data sets. Because relative frequencies always fall between 0 and 1, they provide a standard for comparison.

Pie Charts

Another method for organizing and summarizing data is to draw a picture of some kind. The old saying "a picture is worth a thousand words" has particular relevance in statistics—a graph or chart of a data set often provides the simplest and most efficient display.

Two common methods for graphically displaying qualitative data are *pie charts* and *bar charts*. We begin with pie charts.

DEFINITION 2.5

Pie Chart

A **pie chart** is a disk divided into wedge-shaped pieces proportional to the relative frequencies of the qualitative data.

Procedure 2.3 presents a step-by-step method for constructing a pie chart.

PROCEDURE 2.3

To Construct a Pie Chart

Step 1 Obtain a relative-frequency distribution of the data by applying Procedure 2.2.

Step 2 Divide a disk into wedge-shaped pieces proportional to the relative frequencies.

Step 3 Label the slices with the distinct values and their relative frequencies.

EXAMPLE 2.7

Pie Charts

FIGURE 2.2

Pie chart of the political party affiliation data in Table 2.1

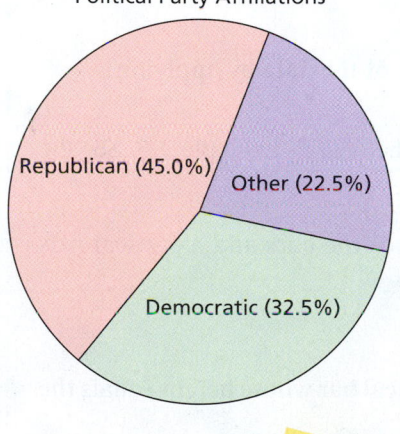
Political Party Affiliations

Republican (45.0%) · Other (22.5%) · Democratic (32.5%)

StatCrunch
Report 2.3

You try it!

Exercise 2.27(c) on page 50

Political Party Affiliations Construct a pie chart of the political party affiliations of the students in Professor Weiss's introductory statistics class presented in Table 2.1 on page 42.

Solution We apply Procedure 2.3.

Step 1 Obtain a relative-frequency distribution of the data by applying Procedure 2.2.

We obtained a relative-frequency distribution of the data in Example 2.6. See the columns of Table 2.3.

Step 2 Divide a disk into wedge-shaped pieces proportional to the relative frequencies.

Referring to the second column of Table 2.3, we see that, in this case, we need to divide a disk into three wedge-shaped pieces that comprise 32.5%, 45.0%, and 22.5% of the disk. We do so by using a protractor and the fact that there are 360° in a circle. Thus, for instance, the first piece of the disk is obtained by marking off 117° (32.5% of 360°). See the three wedges in Fig. 2.2.

Step 3 Label the slices with the distinct values and their relative frequencies.

Referring again to the relative-frequency distribution in Table 2.3, we label the slices as shown in Fig. 2.2. Notice that we expressed the relative frequencies as percentages. Either method (decimal or percentage) is acceptable.

Bar Charts

Another graphical display for qualitative data is the *bar chart*. Frequencies, relative frequencies, or percents can be used to label a bar chart. Although we primarily use relative frequencies, some of our applications employ frequencies or percents.

DEFINITION 2.6

Bar Chart

A **bar chart** displays the distinct values of the qualitative data on a horizontal axis and the relative frequencies (or frequencies or percents) of those values on a vertical axis. The relative frequency of each distinct value is represented by a vertical bar whose height is equal to the relative frequency of that value. The bars should be positioned so that they do not touch each other.

Procedure 2.4 presents a step-by-step method for constructing a bar chart.

PROCEDURE 2.4 **To Construct a Bar Chart**

Step 1 Obtain a relative-frequency distribution of the data by applying Procedure 2.2.

Step 2 Draw a horizontal axis on which to place the bars and a vertical axis on which to display the relative frequencies.

Step 3 For each distinct value, construct a vertical bar whose height equals the relative frequency of that value.

Step 4 Label the bars with the distinct values, the horizontal axis with the name of the variable, and the vertical axis with "Relative frequency."

EXAMPLE 2.8 **Bar Charts**

FIGURE 2.3

Bar chart of the political party affiliation data in Table 2.1

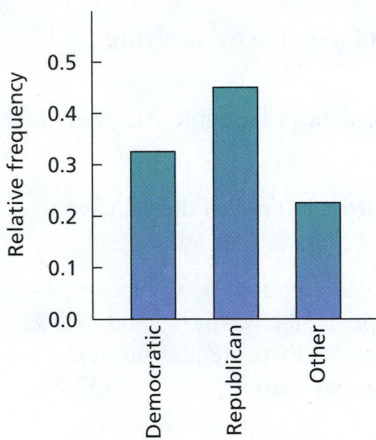

Political Party Affiliations

Political Party Affiliations Construct a bar chart of the political party affiliations of the students in Professor Weiss's introductory statistics class presented in Table 2.1 on page 42.

Solution We apply Procedure 2.4.

Step 1 Obtain a relative-frequency distribution of the data by applying Procedure 2.2.

We obtained a relative-frequency distribution of the data in Example 2.6. See the columns of Table 2.3 on page 44.

Step 2 Draw a horizontal axis on which to place the bars and a vertical axis on which to display the relative frequencies.

See the horizontal and vertical axes in Fig. 2.3.

Step 3 For each distinct value, construct a vertical bar whose height equals the relative frequency of that value.

Referring to the second column of Table 2.3, we see that, in this case, we need three vertical bars of heights 0.325, 0.450, and 0.225, respectively. See the three bars in Fig. 2.3.

Step 4 Label the bars with the distinct values, the horizontal axis with the name of the variable, and the vertical axis with "Relative frequency."

Referring again to the relative-frequency distribution in Table 2.3, we label the bars and axes as shown in Fig. 2.3.

StatCrunch

Report 2.4

You try it!

Exercise 2.27(d) on page 50

THE TECHNOLOGY CENTER

In this Technology Center, we present output and step-by-step instructions for using technology to obtain frequency distributions, relative-frequency distributions, pie charts, and bar charts for qualitative data.

Note to TI-83/84 Plus users: At the time of this writing, the TI-83/84 Plus does not have built-in programs for performing the aforementioned tasks.

EXAMPLE 2.9   **Using Technology to Obtain Frequency and Relative-Frequency Distributions of Qualitative Data**

Political Party Affiliations Use Minitab or Excel to obtain frequency and relative-frequency distributions of the political party affiliation data displayed in Table 2.1 on page 42 (and provided in electronic files in the Data Sets section of the WeissStats site).

Solution We applied the appropriate programs to the data, resulting in Output 2.1. Steps for generating that output are presented in Instructions 2.1.

OUTPUT 2.1   Frequency and relative-frequency distributions of the political party affiliation data

MINITAB

Tally for Discrete Variables: PARTY

PARTY	Count	Percent
Democratic	13	32.50
Other	9	22.50
Republican	18	45.00
N=	40	

EXCEL

Descriptive statistics (Qualitative data):			
Sample	Category	Frequency per category	Rel. frequency per category (%)
PARTY	Democratic	13.0000	32.5000
	Other	9.0000	22.5000
	Republican	18.0000	45.0000

Compare Output 2.1 to Tables 2.2 and 2.3 on pages 43 and 44, respectively. Note that both Minitab and Excel use percents instead of relative frequencies.

INSTRUCTIONS 2.1   Steps for generating Output 2.1

MINITAB

1 Store the data from Table 2.1 in a column named PARTY
2 Choose **Stat ➤ Tables ➤ Tally Individual Variables...**
3 Press the F3 key to reset the dialog box
4 Specify PARTY in the **Variables** text box
5 Check the **Counts** and **Percents** check boxes from the **Display** list
6 Click **OK**

EXCEL

1 Store the data from Table 2.1 in a column named PARTY
2 Choose **XLSTAT ➤ Describing data ➤ Descriptive statistics**
3 Click the reset button in the lower left corner of the dialog box

4 Uncheck the **Quantitative data** check box
5 Click in the **Qualitative data** selection box and then select the column of the worksheet that contains the PARTY data
6 Click the **Options** tab and uncheck the **Charts** check box
7 Click the **Outputs** tab and click the **None** button below the **Qualitative data** check box list
8 Check the **Frequency per category** and **Rel. frequency per category (%)** check boxes in the **Qualitative data** check box list
9 Ensure that the **Display vertically** check box under the **Qualitative data** check box list is unchecked
10 Click **OK**
11 Click the **Continue** button in the **XLSTAT – Selections** dialog box

Note: The steps in Instructions 2.1 are specifically for the data set in Table 2.1 on the political party affiliations of the students in Professor Weiss's introductory statistics class. To apply those steps for a different data set, simply make the necessary changes in the instructions to reflect the different data set—in this case, to steps 1 and 4 in Minitab and to steps 1 and 5 in Excel. Similar comments hold for all technology instructions throughout the book.

EXAMPLE 2.10   **Using Technology to Obtain a Pie Chart**

Political Party Affiliations Use Minitab or Excel to obtain a pie chart of the political party affiliation data in Table 2.1 on page 42.

Solution We applied the pie-chart programs to the data, resulting in Output 2.2. Steps for generating that output are presented in Instructions 2.2 on the next page.

OUTPUT 2.2   Pie charts of the political party affiliation data

MINITAB

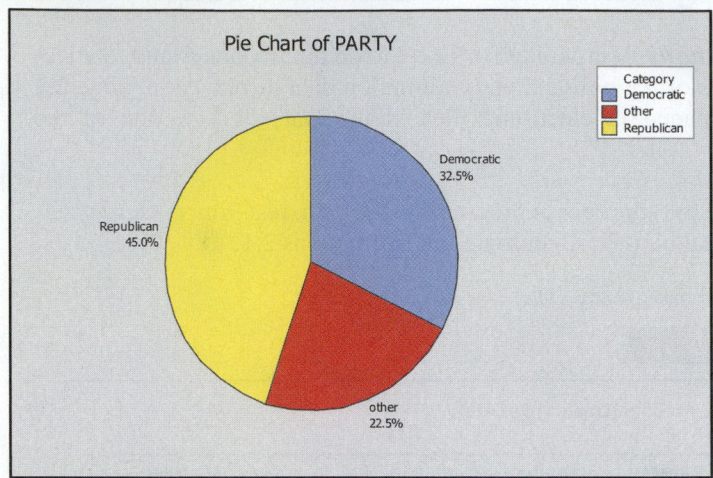

EXCEL

Descriptive statistics (Qualitative data):			
Sample	Category	Frequency per category	Rel. frequency per category (%)
PARTY	Democratic	13.0000	32.5000
	Other	9.0000	22.5000
	Republican	18.0000	45.0000
Pie charts:			

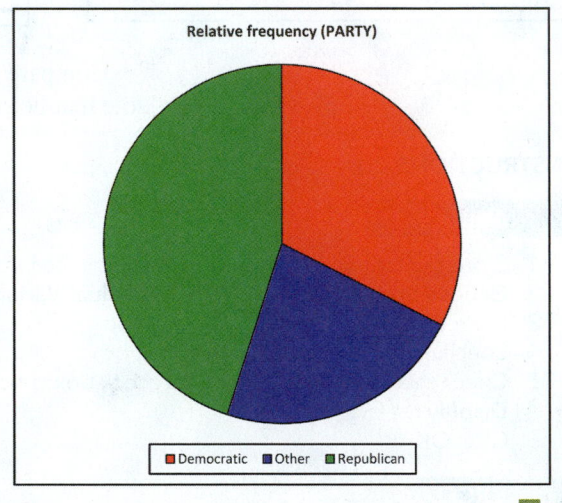

INSTRUCTIONS 2.2   Steps for generating Output 2.2

MINITAB

1 Store the data from Table 2.1 in a column named PARTY
2 Choose **Graph ➤ Pie Chart...**
3 Press the F3 key to reset the dialog box
4 Specify PARTY in the **Categorical variables** text box
5 Click the **Labels...** button
6 Click the **Slice Labels** tab
7 Check the first and third check boxes from the **Label pie slices with** list
8 Click **OK** twice

EXCEL

1 Store the data from Table 2.1 in a column named PARTY
2 Choose **XLSTAT ➤ Visualizing data ➤ Univariate plots**
3 Click the reset button in the lower left corner of the dialog box

4 Uncheck the **Quantitative data** check box
5 Click in the **Qualitative data** selection box and then select the column of the worksheet that contains the PARTY data
6 Click the **Outputs** tab and then click the **None** button below the **Qualitative data** check box list
7 Check the **Frequency per category** and **Rel. frequency per category (%)** check boxes in the **Qualitative data** check box list
8 Ensure that the **Display vertically** check box under the **Qualitative data** check box list is unchecked
9 Click the **Charts (2)** tab
10 In the **Qualitative data** check box list, uncheck the **Bar charts** check box and check the **Pie charts** check box
11 Select the **Relative frequencies** option button in the **Values used** list
12 Click **OK**
13 Click the **Continue** button in the **XLSTAT – Selections** dialog box

EXAMPLE 2.11   Using Technology to Obtain a Bar Chart

Political Party Affiliations Use Minitab or Excel to obtain a bar chart of the political party affiliation data in Table 2.1 on page 42.

Solution We applied the bar-chart programs to the data, resulting in Output 2.3. Steps for generating that output are presented in Instructions 2.3.

OUTPUT 2.3   Bar charts of the political party affiliation data

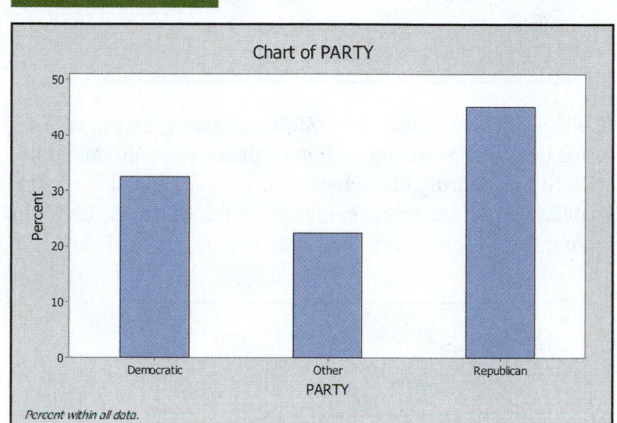

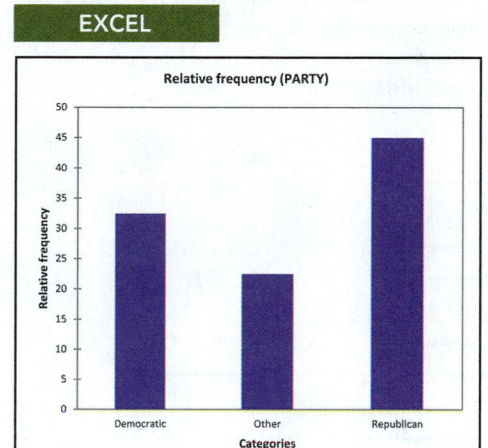

Compare Output 2.3 to the bar chart obtained by hand in Fig. 2.3 on page 46. Notice that, by default, both Minitab and Excel arrange the distinct values of the qualitative data in alphabetical order, in this case, Democratic, Other, and Republican.

INSTRUCTIONS 2.3   Steps for generating Output 2.3

MINITAB

1 Store the data from Table 2.1 in a column named PARTY
2 Choose **Graph ➤ Bar Chart...**
3 Click **OK**
4 Press the F3 key to reset the dialog box
5 Specify PARTY in the **Categorical variables** text box
6 Click the **Chart Options...** button
7 Check the **Show Y as Percent** check box
8 Click **OK** twice

EXCEL

1 Store the data from Table 2.1 in a column named PARTY
2 Choose **XLSTAT ➤ Visualizing data ➤ Univariate plots**
3 Click the reset button in the lower left corner of the dialog box

4 Uncheck the **Quantitative data** check box
5 Click in the **Qualitative data** selection box and then select the column of the worksheet that contains the PARTY data
6 Click the **Outputs** tab and then click the **None** button below the **Qualitative data** check box list
7 Check the **Frequency per category** and **Rel. frequency per category (%)** check boxes in the **Qualitative data** check box list
8 Ensure that the **Display vertically** check box under the **Qualitative data** check box list is unchecked
9 Click the **Charts (2)** tab
10 Select the **Relative frequencies** option button in the **Values used** list
11 Click **OK**
12 Click the **Continue** button in the **XLSTAT – Selections** dialog box

Exercises 2.2

Understanding the Concepts and Skills

2.17 What is a frequency distribution of qualitative data and why is it useful?

2.18 Explain the difference between
a. frequency and relative frequency.
b. percentage and relative frequency.

2.19 Answer true or false to each of the statements in parts (a) and (b), and explain your reasoning.

a. Two data sets that have identical frequency distributions have identical relative-frequency distributions.

b. Two data sets that have identical relative-frequency distributions have identical frequency distributions.

c. Use your answers to parts (a) and (b) to explain why relative-frequency distributions are better than frequency distributions for comparing two data sets.

In Exercises 2.20–2.25, we have presented some simple qualitative data sets for practicing the concepts. For each data set,
a. determine a frequency distribution.
b. obtain a relative-frequency distribution.
c. draw a pie chart.
d. construct a bar chart.

2.20

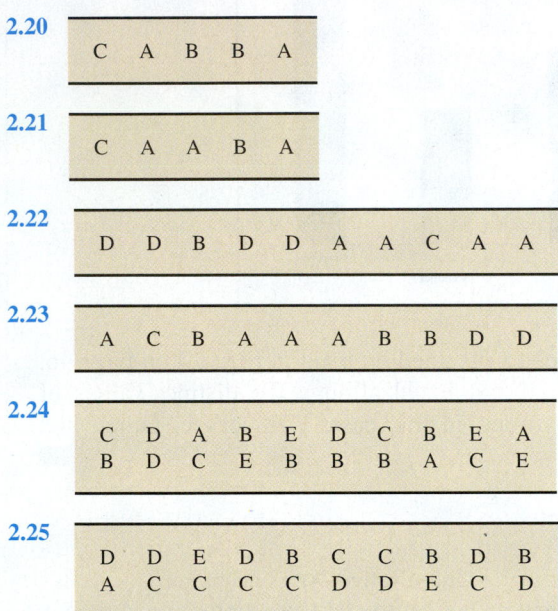

| C | A | B | B | A |

2.21

| C | A | A | B | A |

2.22

| D | D | B | D | D | A | A | C | A | A |

2.23

| A | C | B | A | A | A | B | B | D | D |

2.24

| C | D | A | B | E | D | C | B | E | A |
| B | D | C | E | B | B | B | A | C | E |

2.25

| D | D | E | D | B | C | C | B | D | B |
| A | C | C | C | C | D | D | E | C | D |

Applying the Concepts and Skills

For each data set in Exercises 2.26–2.31,
a. determine a frequency distribution.
b. obtain a relative-frequency distribution.
c. draw a pie chart.
d. construct a bar chart.

2.26 **Primetime Broadcast Shows.** From the TVbytheNumbers website, we obtained the networks for the top 20 primetime broadcast TV shows by total viewership for the week ending August 18, 2013.

CBS	NBC	NBC	ABC	CBS
CBS	CBS	CBS	CBS	FOX
CBS	CBS	CBS	CBS	NBC
NBC	FOX	CBS	NBC	ABC

2.27 **NCAA Wrestling Champs.** From NCAA.com–the official Web site for NCAA sports—we obtained the National Collegiate Athletic Association wrestling champions for the years 1989–2013 in the document "Championship History." They are displayed in the following table.

Year	Champion	Year	Champion
1989	Oklahoma St.	2002	Minnesota
1990	Oklahoma St.	2003	Oklahoma St.
1991	Iowa	2004	Oklahoma St.
1992	Iowa	2005	Oklahoma St.
1993	Iowa	2006	Oklahoma St.
1994	Oklahoma St.	2007	Minnesota
1995	Iowa	2008	Iowa
1996	Iowa	2009	Iowa
1997	Iowa	2010	Iowa
1998	Iowa	2011	Penn State
1999	Iowa	2012	Penn State
2000	Iowa	2013	Penn State
2001	Minnesota		

2.28 **Colleges of Students.** The following table provides data on college for the students in one section of the course Introduction to Computer Science during one semester at Arizona State University. In the table, we use the abbreviations BUS for Business, ENG for Engineering and Applied Sciences, and LIB for Liberal Arts and Sciences.

ENG	ENG	BUS	BUS	ENG
LIB	LIB	ENG	ENG	ENG
BUS	BUS	ENG	BUS	ENG
LIB	BUS	BUS	BUS	ENG
ENG	ENG	LIB	ENG	BUS

2.29 **Class Levels.** Earlier in this section, we considered the political party affiliations of the students in Professor Weiss's introductory statistics course. The class levels of those students are as follows, where Fr, So, Jr, and Sr denote freshman, sophomore, junior, and senior, respectively.

So	So	Jr	Fr	Jr	So	Jr	So
So	So	Sr	So	Jr	Jr	Sr	Fr
Jr	Jr	So	Jr	Fr	Sr	Jr	So
Jr	Fr	Fr	Jr	Sr	So	Sr	Sr
So	Jr	So	Sr	So	So	Fr	So

2.30 **U.S. Regions.** The U.S. Census Bureau divides the states in the United States into four regions: Northeast (NE), Midwest (MW), South (SO), and West (WE). The following table gives the region of each of the 50 states.

SO	WE	WE	MW	NE	WE	WE	SO	MW	SO
WE	NE	WE	SO	MW	MW	NE	WE	SO	WE
WE	SO	MW	SO	MW	WE	SO	NE	SO	SO
SO	SO	MW	NE	SO	NE	MW	NE	WE	MW
WE	SO	MW	SO	MW	NE	MW	SO	NE	WE

2.31 **Road Rage.** The report *Controlling Road Rage: A Literature Review and Pilot Study* was prepared for the AAA Foundation for Traffic Safety by D. Rathbone and J. Huckabee. The authors discuss the results of a literature review and pilot study on how to prevent aggressive driving and road rage. As described in the study, *road rage* is criminal behavior by motorists characterized by uncontrolled anger that results in violence or threatened violence on the road. One of the goals of the study was to determine when road rage occurs most often. The days on which 69 road rage incidents occurred are presented in the following table.

F	F	Tu	Tu	F	Su	F	F	Tu	F
Tu	Sa	Sa	F	Sa	Tu	W	W	Th	Th
Th	Sa	M	Tu	Th	Su	W	Th	W	Tu
Tu	F	Th	Th	F	W	F	Th	F	Sa
F	W	W	F	Tu	W	W	Th	M	M
F	Su	Tu	F	W	Su	W	Th	M	Tu
F	W	Th	M	Su	Sa	Sa	F	F	

In each of Exercises 2.32–2.37, we have presented a frequency distribution of qualitative data. For each exercise,
a. *obtain a relative-frequency distribution.*
b. *draw a pie chart.*
c. *construct a bar chart.*

2.32 Robbery Locations. The Department of Justice and the Federal Bureau of Investigation publish a compilation on crime statistics for the United States in *Crime in the United States*. The following table provides a frequency distribution for robbery type during a one-year period.

Robbery type	Frequency
Street/highway	127,403
Commercial house	37,885
Gas or service station	7,009
Convenience store	14,863
Residence	49,361
Bank	5,777
Miscellaneous	48,878

2.33 M&M Colors. Observing that the proportion of blue M&Ms in his bowl of candy appeared to be less than that of the other colors, R. Fricker, Jr., decided to compare the color distribution in randomly chosen bags of M&Ms to the theoretical distribution reported by M&M/MARS consumer affairs. Fricker published his findings in the article "The Mysterious Case of the Blue M&Ms" (*Chance*, Vol. 9(4), pp. 19–22). For his study, Fricker bought three bags of M&Ms from local stores and counted the number of each color. The average number of each color in the three bags was distributed as shown in the following table.

Color	Frequency
Brown	152
Yellow	114
Red	106
Orange	51
Green	43
Blue	43

2.34 Freshmen Politics. The Higher Education Research Institute of the University of California, Los Angeles, publishes information on characteristics of incoming college freshmen in *The American Freshman*. In 2000, 27.7% of incoming freshmen characterized their political views as liberal, 51.9% as moderate, and 20.4% as conservative. For this year, a random sample of 500 incoming college freshmen yielded the following frequency distribution for political views.

Political view	Frequency
Liberal	147
Moderate	237
Conservative	116

2.35 Medical School Faculty. The Association of American Medical Colleges (AAMC) compiles data on medical school faculty and publishes the results in *AAMC Faculty Roster*. The following table presents a frequency distribution of rank for medical school faculty during one year.

Rank	Frequency
Professor	32,511
Associate professor	28,572
Assistant professor	59,277
Instructor	14,289
Other	3,276

2.36 Drug Types. Drug dealer motivation is traditionally attributed to greed or social pressures. Researchers R. Highland and D. Dabney explore possible personality characteristics as a motivation in the article "Using Adlerian Theory to Shed Light on Drug Dealer Motivations" (*Applied Psychology in Criminal Justice*, Vol. 5, Issue 2, pp. 109–138). The following table lists the types of drug sold from a sample of convicted drug dealers.

Drug sold	Frequency
Marijuana	73
Crack cocaine	62
Powder cocaine	45
Ecstasy	20
Methamphetamine	17
Heroin	5
Other	4

2.37 An Edge in Roulette? An American roulette wheel contains 18 red numbers, 18 black numbers, and 2 green numbers. The following table shows the frequency with which the ball landed on each color in 200 trials.

Number	Red	Black	Green
Frequency	88	102	10

2.38 Health Status. The National Center for Health Statistics collects data on health insurance coverage status and coverage type. The following data from the *National Health Interview Survey* provides the frequency, in millions, of the current health status for persons aged 19–64 and persons aged 65 and over.

Health status	Frequency of persons aged	
	19–64	65 and over
Excellent/very good	119.7	18.1
Good	48.4	14.2
Fair or poor	20.2	9.2

a. Obtain a relative-frequency distribution of health status for persons aged 19–64 and a relative-frequency distribution of health status for persons aged 65 and over.
b. Draw a pie chart for both age groups.
c. Compare your pie charts.

Working with Large Data Sets

In Exercises 2.39–2.41, use the technology of your choice to
a. *determine a frequency distribution.*
b. *obtain a relative-frequency distribution.*
c. *draw a pie chart.*
d. *construct a bar chart.*

If an exercise discusses more than one data set, do parts (a)–(d) for each data set.

2.39 **Japanese Vehicle Exports.** The Japan Automobile Manufacturer's Association provides data on exported vehicles in *Motor Vehicle Statistics of Japan*. In 2010, cars, trucks, and buses constituted 88.3%, 9.3%, and 2.4% of vehicle exports, respectively. A random sample of last year's exports yielded the vehicle-type data on the WeissStats site.

2.40 **Marital Status and Drinking.** Research by W. Clark and L. Midanik (*Alcohol Consumption and Related Problems: Alcohol and Health Monograph 1*. DHHS Pub. No. (ADM) 82–1190) examined, among other issues, alcohol consumption patterns of U.S. adults by marital status. Data for marital status and number of drinks per month, based on the researchers' survey results, are provided on the WeissStats site.

2.41 **Ballot Preferences.** In Issue 338 of the *Amstat News*, then-president of the American Statistical Association, F. Scheuren, reported the results of a survey on how members would prefer to receive ballots in annual elections. On the WeissStats site, you will find data for preference and highest degree obtained for the 566 respondents.

2.3 Organizing Quantitative Data

In the preceding section, we discussed methods for organizing qualitative data. Now we discuss methods for organizing quantitative data.

To organize quantitative data, we first group the observations into **classes** (also known as **categories** or **bins**) and then treat the classes as the distinct values of qualitative data. Consequently, once we group the quantitative data into classes, we can construct frequency and relative-frequency distributions of the data in exactly the same way as we did for qualitative data.

Three commonsense and important guidelines for grouping quantitative data into classes are:

1. The number of classes should be small enough to provide an effective summary but large enough to display the relevant characteristics of the data. A rule of thumb is that the number of classes should be between 5 and 20.
2. Each observation must belong to one, and only one, class. That is, each observation should belong to some class and no observation should belong to more than one class.
3. Whenever feasible, all classes should have the same width. Roughly speaking, this guideline means that, if possible, all classes should cover the same number of possible values. We'll make this guideline more precise later in this section.

> **? What Does It Mean?**
>
> The reason for grouping is to organize the data into a sensible number of classes in order to make the data more accessible and understandable.

The list of guidelines could go on, but for our purposes these three guidelines provide a solid basis for grouping data.

Several methods can be used to group quantitative data into classes. Here we discuss three of the most common methods: *single-value grouping, limit grouping, and cutpoint grouping.*

Single-Value Grouping

In some cases, the most appropriate way to group quantitative data is to use classes in which each class represents a single possible value. Such classes are called **single-value classes,** and this method of grouping quantitative data is called **single-value grouping.**

Thus, in single-value grouping, we use the distinct values of the observations as the classes, a method completely analogous to that used for qualitative data. Single-value grouping is particularly suitable for discrete data in which there are only a small number of distinct values.

 **EXAMPLE 2.12**   Single-Value Grouping

TVs per Household   The Television Bureau of Advertising publishes information on television ownership in *Trends in Television*. Table 2.4 gives the number of TV

TABLE 2.4

Number of TV sets in each of 50 randomly selected households

1	1	1	2	6	3	3	4	2	4
3	2	1	5	2	1	3	6	2	2
3	1	1	4	3	2	2	2	2	3
0	3	1	2	1	2	3	1	1	3
3	2	1	2	1	1	3	1	5	1

sets per household for 50 randomly selected households. Use single-value grouping to organize these data into frequency and relative-frequency distributions.

Solution The (single-value) classes are the distinct values of the data in Table 2.4, which are the numbers 0, 1, 2, 3, 4, 5, and 6. See the first column of Table 2.5.

Tallying the data in Table 2.4, we get the frequencies shown in the second column of Table 2.5. Dividing each such frequency by the total number of observations, 50, we get the relative frequencies in the third column of Table 2.5.

TABLE 2.5

Frequency and relative-frequency distributions, using single-value grouping, for the number-of-TVs data in Table 2.4

Number of TVs	Frequency	Relative frequency
0	1	0.02
1	16	0.32
2	14	0.28
3	12	0.24
4	3	0.06
5	2	0.04
6	2	0.04
	50	1.00

StatCrunch

Report 2.5

You try it!

Exercise 2.81(a)–(b) on page 69

Thus, the first and second columns of Table 2.5 provide a frequency distribution of the data in Table 2.4, and the first and third columns provide a relative-frequency distribution.

Limit Grouping

A second way to group quantitative data is to use **class limits.** With this method, each class consists of a range of values. The smallest value that could go in a class is called the **lower limit** of the class, and the largest value that could go in the class is called the **upper limit** of the class.

This method of grouping quantitative data is called **limit grouping.** It is particularly useful when the data are expressed as whole numbers and there are too many distinct values to employ single-value grouping.

EXAMPLE 2.13   Limit Grouping

TABLE 2.6

Days to maturity for 40 short-term investments

70	64	99	55	64	89	87	65
62	38	67	70	60	69	78	39
75	56	71	51	99	68	95	86
57	53	47	50	55	81	80	98
51	36	63	66	85	79	83	70

Days to Maturity for Short-Term Investments Table 2.6 displays the number of days to maturity for 40 short-term investments. The data are from *BARRON'S* magazine. Use limit grouping, with grouping by 10s, to organize these data into frequency and relative-frequency distributions.

Solution Because we are grouping by 10s and the shortest maturity period is 36 days, our first class is 30–39, that is, for maturity periods from 30 days up to, and including, 39 days. The longest maturity period is 99 days, so grouping by 10s results in the seven classes given in the first column of Table 2.7 on the next page.

Next we tally the data in Table 2.6 into the classes. For instance, the first investment in Table 2.6 has a 70-day maturity period, calling for a tally mark on the line for the class 70–79 in Table 2.7. The results of the tallying procedure are shown in the second column of Table 2.7.

Counting the tallies for each class, we get the frequencies in the third column of Table 2.7. Dividing each such frequency by the total number of observations, 40, we get the relative frequencies in the fourth column of Table 2.7.

Thus, the first and third columns of Table 2.7 provide a frequency distribution of the data in Table 2.6, and the first and fourth columns provide a relative-frequency distribution.

You try it!

Exercise 2.85(a)–(b) on page 70

TABLE 2.7

Frequency and relative-frequency distributions, using limit grouping, for the days-to-maturity data in Table 2.6

Days to maturity	Tally	Frequency	Relative frequency
30–39	III	3	0.075
40–49	I	1	0.025
50–59	LHT III	8	0.200
60–69	LHT LHT	10	0.250
70–79	LHT II	7	0.175
80–89	LHT II	7	0.175
90–99	IIII	4	0.100
		40	1.000

In Definition 2.7, we summarize our discussion of limit grouping and also define two additional terms.

DEFINITION 2.7

Terms Used in Limit Grouping

Lower class limit: The smallest value that could go in a class.

Upper class limit: The largest value that could go in a class.

Class width: The difference between the lower limit of a class and the lower limit of the next-higher class.

Class mark: The average of the two class limits of a class.

For instance, consider the class 50–59 in Example 2.13. The lower limit is 50, the upper limit is 59, the width is $60 - 50 = 10$, and the mark is $(50 + 59)/2 = 54.5$.

Cutpoint Grouping

A third way to group quantitative data is to use **class cutpoints.** As with limit grouping, each class consists of a range of values. The smallest value that could go in a class is called the **lower cutpoint** of the class, and the smallest value that could go in the next-higher class is called the **upper cutpoint** of the class. Note that the lower cutpoint of a class is the same as its lower limit and that the upper cutpoint of a class is the same as the lower limit of the next higher class.

The method of grouping quantitative data by using cutpoints is called **cutpoint grouping**. This method is particularly useful when the data are continuous and are expressed with decimals.

EXAMPLE 2.14   Cutpoint Grouping

TABLE 2.8

Weights, in pounds, of 37 males aged 18–24 years

129.2	185.3	218.1	182.5	142.8
155.2	170.0	151.3	187.5	145.6
167.3	161.0	178.7	165.0	172.5
191.1	150.7	187.0	173.7	178.2
161.7	170.1	165.8	214.6	136.7
278.8	175.6	188.7	132.1	158.5
146.4	209.1	175.4	182.0	173.6
149.9	158.6			

Weights of 18- to 24-Year-Old Males   The U.S. National Center for Health Statistics publishes data on weights and heights by age and sex in the document *Vital and Health Statistics*. The weights shown in Table 2.8, given to the nearest tenth of a pound, were obtained from a sample of 18- to 24-year-old males. Use cutpoint grouping to organize these data into frequency and relative-frequency distributions. Use a class width of 20 and a first cutpoint of 120.

Solution   Because we are to use a first cutpoint of 120 and a class width of 20, our first class is 120–under 140, as shown in the first column of Table 2.9. This class is for weights of 120 lb up to, but not including, weights of 140 lb. The largest weight in Table 2.8 is 278.8 lb, so the last class in Table 2.9 is 260–under 280.

Tallying the data in Table 2.8 gives us the frequencies in the second column of Table 2.9. Dividing each such frequency by the total number of observations, 37, we get the relative frequencies (rounded to three decimal places) in the third column of Table 2.9.

TABLE 2.9

Frequency and relative-frequency distributions, using cutpoint grouping, for the weight data in Table 2.8

Weight (lb)	Frequency	Relative frequency
120–under 140	3	0.081
140–under 160	9	0.243
160–under 180	14	0.378
180–under 200	7	0.189
200–under 220	3	0.081
220–under 240	0	0.000
240–under 260	0	0.000
260–under 280	1	0.027
	37	0.999

You try it!

Exercise 2.89(a)–(b) on page 70

Thus, the first and second columns of Table 2.9 provide a frequency distribution of the data in Table 2.8, and the first and third columns provide a relative-frequency distribution.

Note: Although relative frequencies must always sum to 1, their sum in Table 2.9 is given as 0.999. This discrepancy occurs because each relative frequency is rounded to three decimal places, and, in this case, the resulting sum differs from 1 by a little. Such a discrepancy is called *rounding error* or *roundoff error*.

In Definition 2.8, we summarize our discussion of cutpoint grouping and also define two additional terms. Note that the definition of class width here is consistent with that given in Definition 2.7.

DEFINITION 2.8

Terms Used in Cutpoint Grouping

Lower class cutpoint: The smallest value that could go in a class.

Upper class cutpoint: The smallest value that could go in the next-higher class (equivalent to the lower cutpoint of the next-higher class).

Class width: The difference between the cutpoints of a class.

Class midpoint: The average of the two cutpoints of a class.

For instance, consider the class 160–under 180 in Example 2.14. The lower cutpoint is 160, the upper cutpoint is 180, the width is $180 - 160 = 20$, and the midpoint is $(160 + 180)/2 = 170$.

Choosing the Grouping Method

We have now discussed three methods for grouping quantitative data: single-value grouping, limit grouping, and cutpoint grouping. The following table provides guidelines for deciding which grouping method should be used.

Grouping method	When to use
Single-value grouping	Use with discrete data in which there are only a small number of distinct values.
Limit grouping	Use when the data are expressed as whole numbers and there are too many distinct values to employ single-value grouping.
Cutpoint grouping	Use when the data are continuous and are expressed with decimals.

Choosing the Classes

We have explained how to group quantitative data into specified classes, but we have not discussed how to choose the classes. The reason is that choosing the classes is somewhat subjective, and, moreover, grouping is almost always done with technology.

Hence, understanding the logic of grouping is more important for you than understanding all the details of grouping. For those interested in exploring more details of grouping, we have provided them in the Extending the Concepts and Skills exercises at the end of this section.

Histograms

As we mentioned in Section 2.2, another method for organizing and summarizing data is to draw a picture of some kind. Three common methods for graphically displaying quantitative data are *histograms, dotplots*, and *stem-and-leaf diagrams*. We begin with histograms.

A histogram of quantitative data is the direct analogue of a bar chart of qualitative data, where we use the classes of the quantitative data in place of the distinct values of the qualitative data. However, to help distinguish a histogram from a bar chart, we position the bars in a histogram so that they touch each other. Frequencies, relative frequencies, or percents can be used to label a histogram.

DEFINITION 2.9

? **What Does It Mean?**

A histogram provides a graph of the values of the observations and how often they occur.

Histogram

A **histogram** displays the classes of the quantitative data on a horizontal axis and the frequencies (relative frequencies, percents) of those classes on a vertical axis. The frequency (relative frequency, percent) of each class is represented by a vertical bar whose height is equal to the frequency (relative frequency, percent) of that class. The bars should be positioned so that they touch each other.

- For single-value grouping, we use the distinct values of the observations to label the bars, with each such value centered under its bar.

- For limit grouping or cutpoint grouping, we use the lower class limits (or, equivalently, lower class cutpoints) to label the bars. *Note:* Some statisticians and technologies use class marks or class midpoints centered under the bars.

As expected, a histogram that uses frequencies on the vertical axis is called a **frequency histogram**. Similarly, a histogram that uses relative frequencies or percents on the vertical axis is called a **relative-frequency histogram** or **percent histogram**, respectively.

Procedure 2.5 presents a method for constructing a histogram.

■■■■ **PROCEDURE 2.5** **To Construct a Histogram**

Step 1 Obtain a frequency (relative-frequency, percent) distribution of the data.

Step 2 Draw a horizontal axis on which to place the bars and a vertical axis on which to display the frequencies (relative frequencies, percents).

Step 3 For each class, construct a vertical bar whose height equals the frequency (relative frequency, percent) of that class.

Step 4 Label the bars with the classes, as explained in Definition 2.9, the horizontal axis with the name of the variable, and the vertical axis with "Frequency" ("Relative frequency," "Percent").

■■■■   **EXAMPLE 2.15**   Histograms

TVs, Days to Maturity, and Weights   Construct frequency histograms and relative-frequency histograms for the data on number of televisions per household (Example 2.12), days to maturity for short-term investments (Example 2.13), and weights of 18- to 24-year-old males (Example 2.14).

Solution   We previously grouped the three data sets using single-value grouping, limit grouping, and cutpoint grouping, respectively, as shown in Tables 2.5, 2.7, and 2.9. We repeat those tables here in Table 2.10.

TABLE 2.10   Frequency and relative-frequency distributions for the data on (a) number of televisions per household, (b) days to maturity for short-term investments, and (c) weights of 18- to 24-year-old males

Number of TVs	Frequency	Relative frequency
0	1	0.02
1	16	0.32
2	14	0.28
3	12	0.24
4	3	0.06
5	2	0.04
6	2	0.04

(a) Single-value grouping

Days to maturity	Frequency	Relative frequency
30–39	3	0.075
40–49	1	0.025
50–59	8	0.200
60–69	10	0.250
70–79	7	0.175
80–89	7	0.175
90–99	4	0.100

(b) Limit grouping

Weight (lb)	Frequency	Relative frequency
120–under 140	3	0.081
140–under 160	9	0.243
160–under 180	14	0.378
180–under 200	7	0.189
200–under 220	3	0.081
220–under 240	0	0.000
240–under 260	0	0.000
260–under 280	1	0.027

(c) Cutpoint grouping

We will apply Procedure 2.5 to construct a frequency histogram for the weight data and leave the details of construction of the other five histograms to the reader. All six histograms are shown in Figs. 2.4–2.6 on the next page.

Step 1   Obtain a frequency distribution of the data.

We have already obtained a frequency distribution of the weight data, as shown in the first two columns of Table 2.10(c).

Step 2   Draw a horizontal axis on which to place the bars and a vertical axis on which to display the frequencies.

See the horizontal and vertical axes in Fig. 2.6(a).

Step 3   For each class, construct a vertical bar whose height equals the frequency of that class.

Referring to the second column of Table 2.10(c), we see that, in this case, we need vertical bars of heights 3, 9, 14, 7, 3, 0 , 0, and 1, respectively. See the bars in Fig. 2.6(a).

Step 4   Label the bars with the classes, as explained in Definition 2.9, the horizontal axis with the name of the variable, and the vertical axis with "Frequency."

Because the weight data have been grouped by using cutpoint grouping, we use the lower class cutpoints to label the bars. Referring to the first column of Table 2.10(c), we label the bars as shown on the horizontal axis in Fig. 2.6(a). Because the variable is weight (measured in pounds) and we want a frequency histogram, we label the horizontal axis "Weight (lb)" and the vertical axis "Frequency," as shown in Fig. 2.6(a).

FIGURE 2.4   Single-value grouping. Number of TVs per household: (a) frequency histogram; (b) relative-frequency histogram

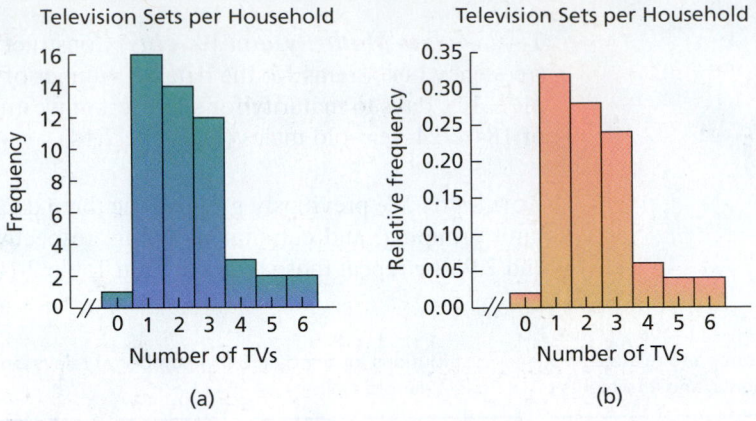

(a)   (b)

FIGURE 2.5   Limit grouping. Days to maturity: (a) frequency histogram; (b) relative-frequency histogram

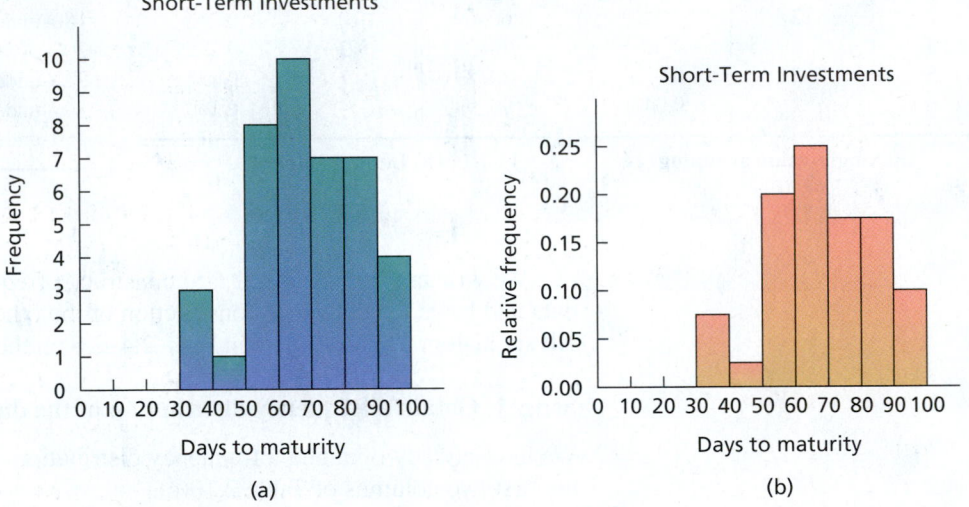

(a)   (b)

FIGURE 2.6   Cutpoint grouping. Weight of 18- to 24-year-old males: (a) frequency histogram; (b) relative-frequency histogram

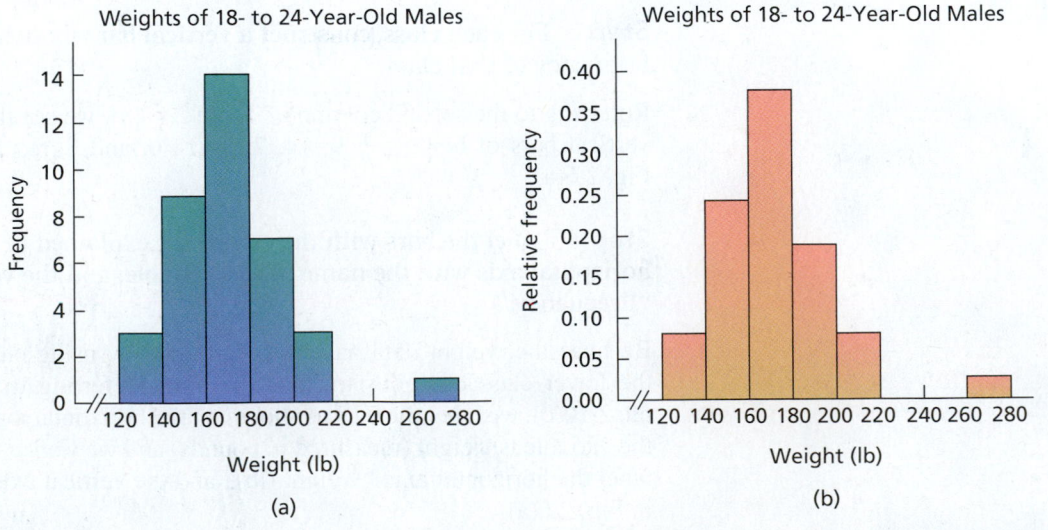

(a)   (b)

Observe the following facts about the histograms in Figs. 2.4, 2.5, and 2.6:

- In each figure, the frequency histogram and relative-frequency histogram have proportionally the same shape, and the same would be true for the percent histogram. This result holds because frequencies, relative-frequencies, and percents are proportional.
- Because the histograms in Fig. 2.4 are based on single-value grouping, the distinct values label the bars, with each such value centered under its bar.
- Because the histograms in Figs. 2.5 and 2.6 are based on limit grouping and cut-point grouping, respectively, the lower class limits (or, equivalently, lower class cutpoints) label the bars.
- We did not show percent histograms in Figs. 2.4, 2.5, and 2.6. However, each percent histogram would look exactly like the corresponding relative-frequency histogram, except that the relative frequencies would be changed to percents (obtained by multiplying each relative frequency by 100) and "Percent," instead of "Relative frequency," would be used to label the vertical axis.
- The symbol // is used on the horizontal axes in Figs. 2.4 and 2.6. This symbol indicates that the zero point on that axis is not in its usual position at the intersection of the horizontal and vertical axes. Whenever any such modification is made, whether on a horizontal axis or a vertical axis, the symbol // or some similar symbol should be used to indicate that fact.

StatCrunch

Report 2.6

You try it!

Exercise 2.85(c)–(d) on page 70

Relative-frequency (or percent) histograms are better than frequency histograms for comparing two data sets. The same vertical scale is used for all relative-frequency histograms—a minimum of 0 and a maximum of 1—making direct comparison easy. In contrast, the vertical scale of a frequency histogram depends on the number of observations, making comparison more difficult.

We note that histograms based on limit or cutpoint grouping can be sensitive to the choice of classes. So, it is sometimes important to experiment with making different choices for the classes in order to see whether and how such choices affect the shape or other aspects of the histogram. You'll want to use technology for such experimentation because doing so by hand will be tedious and time-consuming.

Dotplots

Another type of graphical display for quantitative data is the *dotplot*.

DEFINITION 2.10

Dotplot

A **dotplot** is a graph in which each observation is plotted as a dot at an appropriate place above a horizontal axis. Observations having equal values are stacked vertically.

Dotplots are particularly useful for showing the relative positions of the data in a data set or for comparing two or more data sets. Procedure 2.6 presents a method for constructing a dotplot.

PROCEDURE 2.6   **To Construct a Dotplot**

Step 1   Draw a horizontal axis that displays the possible values of the quantitative data.

Step 2   Record each observation by placing a dot over the appropriate value on the horizontal axis.

Step 3   Label the horizontal axis with the name of the variable.

EXAMPLE 2.16   Dotplots

Prices of DVD Players   One of Professor Weiss's sons wanted to add a new DVD player to his home theater system. He used the Internet to shop and went to price-watch.com. There he found 16 quotes on different brands and styles of DVD players. Table 2.11 lists the prices, in dollars. Construct a dotplot for these data.

TABLE 2.11

Prices, in dollars, of 16 DVD players

210	219	214	197
224	219	199	199
208	209	215	199
212	212	219	210

Solution   We apply Procedure 2.6.

Step 1   Draw a horizontal axis that displays the possible values of the quantitative data.

See the horizontal axis in Fig. 2.7.

FIGURE 2.7

Dotplot for DVD-player prices in Table 2.11

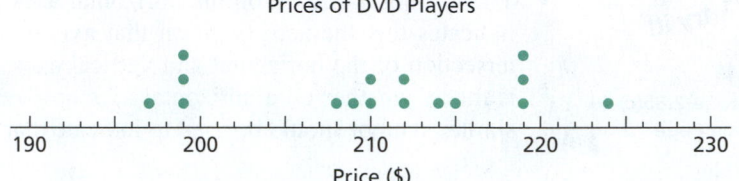

Step 2   Record each observation by placing a dot over the appropriate value on the horizontal axis.

The first price is $210, which calls for a dot over the "210" on the horizontal axis in Fig. 2.7. Continuing in this manner, we get all the dots shown in Fig. 2.7.

StatCrunch

Report 2.7

You try it!

Exercise 2.93
on page 70

Step 3   Label the horizontal axis with the name of the variable.

The variable here is "Price," with which we label the horizontal axis in Fig. 2.7. ■

Dotplots are similar to histograms. In fact, when data are grouped using single-value grouping, a dotplot and a frequency histogram are essentially identical. However, for single-value grouped data that involve decimals, dotplots are generally preferable to histograms because they are easier to construct and use.

Stem-and-Leaf Diagrams

Statisticians continue to invent ways to display data. One method, developed in the 1960s by the late Professor John Tukey of Princeton University, is called a *stem-and-leaf diagram*. This ingenious diagram is often easier to construct than either a frequency distribution or a histogram and generally displays more information.

DEFINITION 2.11

Stem-and-Leaf Diagram

In a **stem-and-leaf diagram** (or **stemplot**), each observation is separated into two parts, namely, a **stem**—consisting of all but the rightmost digit—and a **leaf**, the rightmost digit.

Note that stems may use as many digits as required, but each leaf must contain only one digit. Procedure 2.7 presents a step-by-step method for constructing a stem-and-leaf diagram.

■■■ **PROCEDURE 2.7** **To Construct a Stem-and-Leaf Diagram**

Step 1 Think of each observation as a stem—consisting of all but the rightmost digit—and a leaf, the rightmost digit.

Step 2 Write the stems from smallest to largest in a vertical column to the left of a vertical rule.

Step 3 Write each leaf to the right of the vertical rule in the row that contains the appropriate stem.

Step 4 Arrange the leaves in each row in ascending order.

■■■ **EXAMPLE 2.17** **Stem-and-Leaf Diagrams**

TABLE 2.12

Days to maturity for 40 short-term investments

70	64	99	55	64	89	87	65
62	38	67	70	60	69	78	39
75	56	71	51	99	68	95	86
57	53	47	50	55	81	80	98
51	36	63	66	85	79	83	70

Days to Maturity for Short-Term Investments Table 2.12 repeats the data on the number of days to maturity for 40 short-term investments. Previously, we grouped these data with a frequency distribution (Table 2.7 on page 54) and graphed them with a frequency histogram (Fig. 2.5(a) on page 58). Now let's construct a stem-and-leaf diagram, which simultaneously groups the data and provides a graphical display similar to a histogram.

Solution We apply Procedure 2.7.

Step 1 Think of each observation as a stem—consisting of all but the rightmost digit—and a leaf, the rightmost digit.

Referring to Table 2.12, we note that these observations are two-digit numbers. Thus, in this case, we use the first digit of each observation as the stem and the second digit as the leaf.

Step 2 Write the stems from smallest to largest in a vertical column to the left of a vertical rule.

Referring again to Table 2.12, we see that the stems consist of the numbers 3, 4, . . . , 9. See the numbers to the left of the vertical rule in Fig. 2.8(a).

Step 3 Write each leaf to the right of the vertical rule in the row that contains the appropriate stem.

The first number in Table 2.12 is 70, which calls for a 0 to the right of the stem 7. Reading down the first column of Table 2.12, we find that the second number is 62, which calls for a 2 to the right of the stem 6. We continue in this manner until we account for all of the observations in Table 2.12. The result is the diagram displayed in Fig. 2.8(a).

FIGURE 2.8

Constructing a stem-and-leaf diagram for the days-to-maturity data

Stems   Leaves

(a)	(b)
3 | 8 6 9	3 | 6 8 9
4 | 7	4 | 7
5 | 7 1 6 3 5 1 0 5	5 | 0 1 1 3 5 5 6 7
6 | 2 4 7 3 6 4 0 9 8 5	6 | 0 2 3 4 4 5 6 7 8 9
7 | 0 5 1 0 9 8 0	7 | 0 0 0 1 5 8 9
8 | 5 9 1 7 0 3 6	8 | 0 1 3 5 6 7 9
9 | 9 9 5 8	9 | 5 8 9 9

Step 4 **Arrange the leaves in each row in ascending order.**

The first row of leaves in Fig. 2.8(a) is 8, 6, and 9. Arranging these numbers in ascending order, we get the numbers 6, 8, and 9, which we write in the first row to the right of the vertical rule in Fig. 2.8(b). We continue in this manner until the leaves in each row are in ascending order, as shown in Fig. 2.8(b), which is the stem-and-leaf diagram for the days-to-maturity data.

The stem-and-leaf diagram for the days-to-maturity data is similar to a frequency histogram for those data because the length of the row of leaves for a class equals the frequency of the class. [Turn the stem-and-leaf diagram in Fig. 2.8(b) 90° counterclockwise, and compare it to the frequency histogram shown in Fig. 2.5(a) on page 58.]

StatCrunch
Report 2.8

You try it!

Exercise 2.97
on page 71

In our next example, we describe the use of the stem-and-leaf diagram for three-digit numbers and also introduce the technique of using more than one line per stem.

EXAMPLE 2.18   Stem-and-Leaf Diagrams

TABLE 2.13

Cholesterol levels for 20 high-level patients

210	209	212	208
217	207	210	203
208	210	210	199
215	221	213	218
202	218	200	214

FIGURE 2.9

Stem-and-leaf diagram for cholesterol levels: (a) one line per stem; (b) two lines per stem.

Cholesterol Levels According to the *National Health and Nutrition Examination Survey*, published by the Centers for Disease Control, the average cholesterol level for children between 4 and 19 years of age is 165 mg/dL. A pediatrician tested the cholesterol levels of several young patients and was alarmed to find that many had levels higher than 200 mg/dL. Table 2.13 presents the readings of 20 patients with high levels. Construct a stem-and-leaf diagram for these data by using

a. one line per stem.   **b.** two lines per stem.

Solution Because these observations are three-digit numbers, we use the first two digits of each number as the stem and the third digit as the leaf.

a. Using one line per stem and applying Procedure 2.7, we obtain the stem-and-leaf diagram displayed in Fig. 2.9(a).

```
                                    19 |
                                    19 | 9
                                    20 | 0 2 3
                                    20 | 7 8 8 9
   19 | 9                           21 | 0 0 0 0 2 3 4
   20 | 0 2 3 7 8 8 9               21 | 5 7 8 8
   21 | 0 0 0 0 2 3 4 5 7 8 8       22 | 1
   22 | 1                           22 |
         (a)                               (b)
```

You try it!

Exercise 2.99
on page 71

b. The stem-and-leaf diagram in Fig. 2.9(a) is only moderately helpful because there are so few stems. Figure 2.9(b) is a better stem-and-leaf diagram for these data. It uses two lines for each stem, with the first line for the leaf digits 0–4 and the second line for the leaf digits 5–9.

In Example 2.18, we saw that using two lines per stem provides a more useful stem-and-leaf diagram for the cholesterol data than using one line per stem. When there are only a few stems, we might even want to use five lines per stem, where the first line is for leaf digits 0 and 1, the second line is for leaf digits 2 and 3, . . . , and the fifth line is for leaf digits 8 and 9.

For instance, suppose you have data on the heights, in inches, of the students in your class. Most, if not all, of the observations would be in the 60- to 80-inch range, which would give only a few stems. This is a case where five lines per stem would probably be best.

Although stem-and-leaf diagrams have several advantages over the more classical techniques for grouping and graphing, they do have some drawbacks. For instance, they are generally not useful with large data sets and can be awkward with data containing many digits; histograms are usually preferable to stem-and-leaf diagrams in such cases.

THE TECHNOLOGY CENTER

Grouping or graphing data by hand can be tedious. You can avoid the tedium by using technology. In this Technology Center, we first present output and step-by-step instructions to group quantitative data using single-value grouping.

Note to TI-83/84 Plus users:

- At the time of this writing, the TI-83/84 Plus does not have a built-in program for single-value grouping. However, a TI program called TALLY, supplied in the TI Programs section on the WeissStats site, allows you to perform this procedure. Your instructor can show you how to download the program to your calculator.
- After you apply the TALLY program, the required frequency distribution will be in Lists 1 and 2, and the required relative-frequency distribution will be in Lists 1 and 3. *Warning:* Any data that you may have previously stored in Lists 1–3 will be erased during program execution, so copy those data to other lists prior to program execution if you want to retain them.

EXAMPLE 2.19   **Using Technology to Obtain Frequency and Relative-Frequency Distributions of Quantitative Data Using Single-Value Grouping**

TVs per Household Table 2.4 on page 53 shows data on the number of TV sets per household for 50 randomly selected households. Use Minitab, Excel, or the TI-83/84 Plus to obtain frequency and relative-frequency distributions of these quantitative data using single-value grouping.

Solution We applied the grouping programs to the data, resulting in Output 2.4. Steps for generating that output are presented in Instructions 2.4 on the next page.

OUTPUT 2.4   Frequency and relative-frequency distributions, using single-value grouping, for the number-of-TVs data

MINITAB

Tally for Discrete Variables: TVs

TVs	Count	Percent
0	1	2.00
1	16	32.00
2	14	28.00
3	12	24.00
4	3	6.00
5	2	4.00
6	2	4.00
N=	50	

EXCEL

Descriptive statistics (Qualitative data):			
Sample	Category	Frequency per category	Rel. frequency per category (%)
TVs	0	1.0000	2.0000
	1	16.0000	32.0000
	2	14.0000	28.0000
	3	12.0000	24.0000
	4	3.0000	6.0000
	5	2.0000	4.0000
	6	2.0000	4.0000

TI-83/84 PLUS

NORMAL FLOAT AUTO REAL RADIAN MP					
L1	L2	L3	L4	L5	1
0	1	.02	------	------	
1	16	.32			
2	14	.28			
3	12	.24			
4	3	.06			
5	2	.04			
6	2	.04			
------	------	------			

L1(1)=0

Compare Output 2.4 to Table 2.5 on page 53. Note that both Minitab and Excel use percents instead of relative frequencies.

INSTRUCTIONS 2.4 Steps for generating Output 2.4

MINITAB

1 Store the data from Table 2.4 in a column named TVs
2 Choose **Stat ➤ Tables ➤ Tally Individual Variables...**
3 Press the F3 key to reset the dialog box
4 Specify TVs in the **Variables** text box
5 Check the **Counts** and **Percents** check boxes from the **Display** list
6 Click **OK**

EXCEL

1 Store the data from Table 2.4 in a column named TVs
2 Choose **XLSTAT ➤ Describing data ➤ Descriptive statistics**
3 Click the reset button in the lower left corner of the dialog box
4 Uncheck the **Quantitative data** check box
5 Click in the **Qualitative data** selection box and then select the column of the worksheet that contains the TVs data
6 Click the **Options** tab and uncheck the **Charts** check box

7 Click the **Outputs** tab and click the **None** button below the **Qualitative data** check box list
8 Check the **Frequency per category** and **Rel. frequency per category (%)** check boxes in the **Qualitative data** check box list
9 Ensure that the **Display vertically** check box under the **Qualitative data** check box list is unchecked
10 Click **OK**
11 Click the **Continue** button in the **XLSTAT – Selections** dialog box

TI-83/84 PLUS

1 Store the data from Table 2.4 in a list named TVS
2 Press **PRGM**
3 Arrow down to TALLY and press **ENTER** twice
4 Press **2ND ➤ LIST**
5 Arrow down to TVS and press **ENTER** twice
6 After the program completes, press **STAT** and then **ENTER**
7 The distinct values of the data are in **L1** (List 1), the frequencies are in **L2**, and the relative frequencies are in **L3**

Next, we explain how to use Minitab, Excel, or the TI-83/84 Plus to construct a histogram.

EXAMPLE 2.20 Using Technology to Obtain a Histogram

Days to Maturity for Short-Term Investments Table 2.6 on page 53 gives data on the number of days to maturity for 40 short-term investments. Use Minitab, Excel, or the TI-83/84 Plus to obtain a frequency histogram of those data.

Solution We applied the histogram programs to the data, resulting in Output 2.5. Steps for generating that output are presented in Instructions 2.5.

OUTPUT 2.5 Histograms of the days-to-maturity data

MINITAB

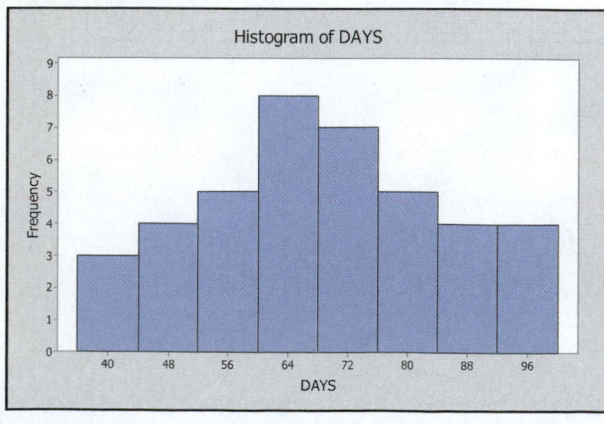

EXCEL

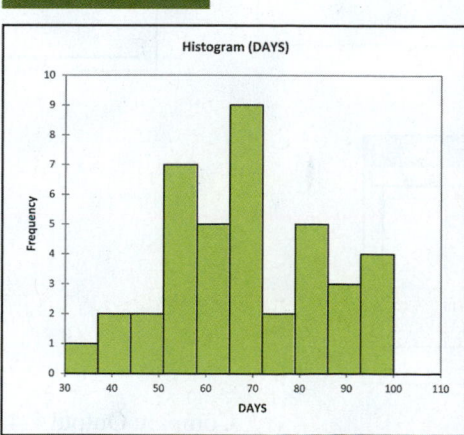

OUTPUT 2.5 (cont.)
Histograms of the days-to-maturity data

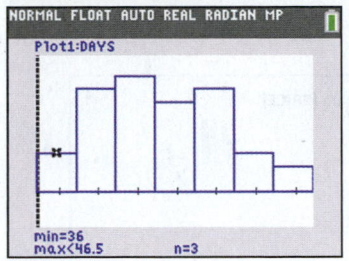

Some technologies require the user to specify a histogram's classes; others automatically choose the classes; others allow the user to specify the classes or to let the program choose them.

We generated all three histograms in Output 2.5 by letting the programs automatically choose the classes, which explains why the three histograms differ from each other and from the histogram we constructed by hand in Fig. 2.5(a) on page 58.

INSTRUCTIONS 2.5   Steps for generating Output 2.5

MINITAB

1 Store the data from Table 2.6 in a column named DAYS
2 Choose **Graph ➤ Histogram...**
3 Click **OK**
4 Press the F3 key to reset the dialog box
5 Specify DAYS in the **Graph variables** text box
6 Click **OK**

EXCEL

1 Store the data from Table 2.6 in a column named DAYS
2 Choose **XLSTAT ➤ Visualizing data ➤ Histograms**
3 Click the reset button in the lower left corner of the dialog box
4 Click in the **Data** selection box and then select the column of the worksheet that contains the DAYS data
5 Click the **Outputs** tab and uncheck the **Descriptive statistics** check box

6 Click the **Charts** tab and select **Frequency** from the **Ordinate of the histograms** drop-down list box
7 Click **OK**
8 Click the **Continue** button in the **XLSTAT – Selections** dialog box

TI-83/84 PLUS

1 Store the data from Table 2.6 in a list named DAYS
2 Ensure that all stat plots and all **Y =** functions are off (see notes below instructions)
3 Press **2ND ➤ STAT PLOT** and then press **ENTER** twice
4 Arrow to the third graph icon and press **ENTER**
5 Press the down-arrow key
6 Press **2ND ➤ LIST**
7 Arrow down to DAYS and press **ENTER** twice
8 Type 1 for **Freq**
9 Press **ZOOM**, then **9**, and then **TRACE**

Note to TI-83/84 Plus users: Step 2 of the TI-83/84 Plus instructions in Instructions 2.5 states that you should ensure that all stat plots and all **Y =** functions are off (deselected).

• To turn off all stat plots: Press **2ND ➤ STAT PLOT**, press **4**, and then press **ENTER**.
• To turn off all **Y =** functions: Press **VARS**, arrow over to **Y-VARS**, press **4**, press **2**, and then press **ENTER**.

In our next example, we show how to use Minitab, Excel, or the TI-83/84 Plus to construct a dotplot.

Note to TI-83/84 Plus users: At the time of this writing, the TI-83/84 Plus does not have a built-in program for a dotplot. However, a TI program called DOTPLOT, supplied in the TI Programs section on the WeissStats site, allows you to perform this procedure. Your instructor can show you how to download the program to your calculator. *Warning:* Any data that you may have previously stored in Lists 1 and 2 will be erased during program execution, so copy those data to other lists prior to program execution if you want to retain them.

EXAMPLE 2.21   Using Technology to Obtain a Dotplot

Prices of DVD Players Table 2.11 on page 60 supplies data on the prices of 16 DVD players. Use Minitab, Excel, or the TI-83/84 Plus to obtain a dotplot of those data.

Solution We applied the dotplot programs to the data, resulting in Output 2.6. Steps for generating that output are presented in Instructions 2.6.

OUTPUT 2.6 Dotplots for the DVD price data

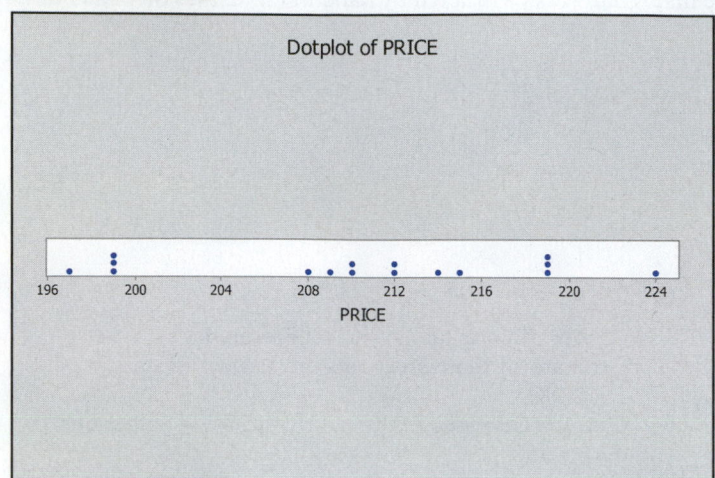

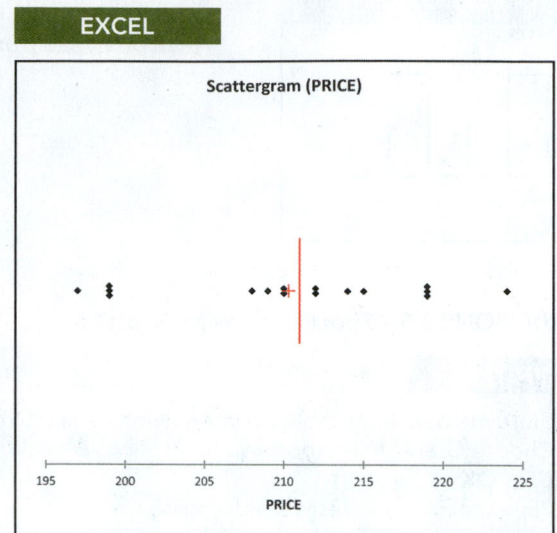

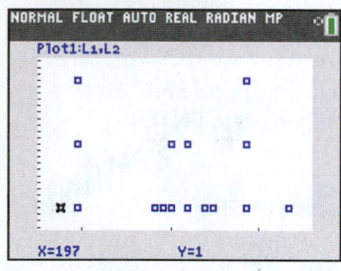

Compare Output 2.6 to the dotplot obtained by hand in Fig. 2.7 on page 60. Note that Excel refers to the diagram as a scattergram rather than as a dotplot.

INSTRUCTIONS 2.6 Steps for generating Output 2.6

MINITAB

1 Store the data from Table 2.11 in a column named PRICE
2 Choose **Graph ➤ Dotplot...**
3 Click **OK**
4 Press the F3 key to reset the dialog box
5 Specify PRICE in the **Graph variables** text box
6 Click **OK**

EXCEL

1 Store the data from Table 2.11 in a column named PRICE
2 Choose **XLSTAT ➤ Visualizing data ➤ Univariate plots**
3 Click the reset button in the lower left corner of the dialog box
4 Click in the **Quantitative data** selection box and then select the column of the worksheet that contains the PRICE data

5 Click the **Options** tab and uncheck the **Descriptive statistics** check box
6 Click the **Charts (1)** tab
7 In the **Quantitative data** check box list, uncheck the **Box plots** check box and check the **Scattergrams** check box
8 Click the **Options** tab next to the **Chart types** tab and then select the **Horizontal** option button
9 Click **OK**
10 Click the **Continue** button in the **XLSTAT – Selections** dialog box

TI-83/84 PLUS

1 Store the data from Table 2.11 in a list named PRICE
2 Press **PRGM**
3 Arrow down to DOTPLOT and press **ENTER** twice
4 Press **2ND ➤ LIST**
5 Arrow down to PRICE and press **ENTER** twice
6 To terminate the program, press **ENTER**

Our final illustration in this Technology Center shows how to use Minitab or Excel to obtain stem-and-leaf diagrams. *Note to TI-83/84 Plus users:* At the time of this writing, the TI-83/84 Plus does not have a built-in program for generating stem-and-leaf diagrams.

EXAMPLE 2.22   Using Technology to Obtain a Stem-and-Leaf Diagram

Cholesterol Levels Table 2.13 on page 62 provides the cholesterol levels of 20 patients with high levels. Apply Minitab or Excel to obtain a stem-and-leaf diagram for those data.

Solution We applied the stem-and-leaf programs to the data, resulting in Output 2.7. Steps for generating that output are presented in Instructions 2.7.

OUTPUT 2.7   Stem-and-leaf diagrams for cholesterol levels

MINITAB

Stem-and-Leaf Display: LEVEL

```
Stem-and-leaf of LEVEL   N  = 20
Leaf Unit = 1.0

   1    19  9
   2    20  0
   4    20  23
   4    20
   5    20  7
   8    20  889
  (4)   21  0000
   8    21  23
   6    21  45
   4    21  7
   3    21  88
   1    22  1
```

EXCEL

Stem-and-leaf plot (LEVEL):

Unit:	10
19	9
20	0237889
21	00002345788
22	1

Observe that Minitab used five lines per stem, whereas Excel used one line per stem. Minitab has an option for specifying the number of lines per stem (through the appropriate use of the **Increment** text box), but Excel does not.

INSTRUCTIONS 2.7   Steps for generating Output 2.7

MINITAB

1 Store the data from Table 2.13 in a column named LEVEL
2 Choose **Graph ➤ Stem-and-Leaf...**
3 Press the F3 key to reset the dialog box
4 Specify LEVEL in the **Graph variables** text box
5 Click **OK**

EXCEL

1 Store the data from Table 2.13 in a column named LEVEL
2 Choose **XLSTAT ➤ Visualizing data ➤ Univariate plots**

3 Click the reset button in the lower left corner of the dialog box
4 Click in the **Quantitative data** selection box and then select the column of the worksheet that contains the LEVEL data
5 Click the **Options** tab and uncheck the **Descriptive statistics** check box
6 Click the **Charts (1)** tab, uncheck the **Box plots** check box, and check the **Stem-and-leaf plots** check box
7 Click **OK**
8 Click the **Continue** button in the **XLSTAT – Selections** dialog box

Exercises 2.3

Understanding the Concepts and Skills

2.42 Identify an important reason for grouping data.

2.43 Do the concepts of class limits, marks, cutpoints, and midpoints make sense for qualitative data? Explain your answer.

2.44 State three of the most important guidelines in choosing the classes for grouping a quantitative data set.

2.45 With regard to grouping quantitative data into classes in which each class represents a range of possible values, we discussed two methods for depicting the classes. Identify the two methods and explain the relative advantages and disadvantages of each method.

2.46 For quantitative data, we examined three types of grouping: single-value grouping, limit grouping, and cutpoint grouping. For

each type of data given, decide which of these three grouping types is usually best. Explain your answers.
a. Continuous data displayed to one or more decimal places
b. Discrete data in which there are relatively few distinct observations

2.47 We used slightly different methods for determining the "middle" of a class with limit grouping and cutpoint grouping. Identify the methods and the corresponding terminologies.

2.48 Explain the difference between a frequency histogram and a relative-frequency histogram.

2.49 Explain the advantages and disadvantages of frequency histograms versus frequency distributions.

2.50 For data that are grouped in classes based on more than a single value, lower class limits (or cutpoints) are used on the horizontal axis of a histogram for depicting the classes. Class marks (or midpoints) can also be used, in which case each bar is centered over the mark (or midpoint) of the class it represents. Explain the advantages and disadvantages of each method.

2.51 Discuss the relative advantages and disadvantages of stem-and-leaf diagrams versus frequency histograms.

2.52 Suppose that you have a data set that contains a large number of observations. Which graphical display is generally preferable: a histogram or a stem-and-leaf diagram? Explain your answer.

2.53 Suppose that you have constructed a stem-and-leaf diagram and discover that it is only moderately useful because there are too few stems. How can you remedy the problem?

In each of Exercises 2.54–2.59, we have presented a "data scenario." In each case, decide which type of grouping (single-value, limit, or cutpoint) is probably the best.

2.54 Number of Bedrooms. The number of bedrooms per single-family dwelling

2.55 Ages of Householders. The ages of householders, given as a whole number

2.56 Sleep Aids. The additional sleep, to the nearest tenth of an hour, obtained by a sample of 100 patients by using a particular brand of sleeping pill

2.57 Number of Cars. The number of automobiles per family

2.58 Gas Mileage. The gas mileages, rounded to the nearest number of miles per gallon, of all new car models

2.59 Giant Tarantulas. The carapace lengths, to the nearest hundredth of a millimeter, of a sample of 50 giant tarantulas

In Exercises 2.60–2.71, we have presented some quantitative data sets and specified a grouping method for practicing the concepts. For each data set,
a. *determine a frequency distribution.*
b. *obtain a relative-frequency distribution.*
c. *construct a frequency histogram based on your result from part (a).*
d. *construct a relative-frequency histogram based on your result from part (b).*

2.60 Use single-value grouping.

3	3	3	4	4	1	3	4	3	1

2.61 Use single-value grouping.

3	3	1	3	2	4	3	1	2	3

2.62 Use single-value grouping.

2	1	1	1	1	1	0	3	2	2
1	1	4	1	2	2	2	3	0	2

2.63 Use single-value grouping.

4	2	4	3	4	3	3	3	4	4
3	0	3	2	2	3	1	4	2	3

2.64 Use limit grouping with a first class of 0–9 and a class width of 10.

9	27	16	24	18
37	20	21	31	31
41	9	32	32	39
4	15	27	37	30

2.65 Use limit grouping with a first class of 0–4 and a class width of 5.

9	8	21	25	16
2	2	8	20	16
12	4	26	25	6
0	17	6	16	2

2.66 Use limit grouping with a first class of 30–36 and a class width of 7.

39	55	31	56	53
74	35	59	37	62
47	33	46	69	47
50	54	31	36	47
65	78	60	36	34

2.67 Use limit grouping with a first class of 50–59 and a class width of 10.

79	61	77	74	54
70	62	83	93	66
70	89	80	98	54
90	83	50	72	82
54	85	51	86	54

2.68 Use cutpoint grouping with a first class of 10–under 15.

12.7	20.9	22.0	10.1	22.0
16.6	26.2	38.8	17.3	19.5
15.8	24.3	13.2	12.9	16.4
24.5	25.2	22.0	20.8	26.9

2.69 Use cutpoint grouping with a first class of 40–under 46.

50.0	53.4	44.5	55.2	51.2
56.8	52.5	51.6	48.6	56.9
54.3	51.0	43.5	57.5	65.4
41.6	57.8	54.2	52.5	50.7

2.70 Use cutpoint grouping with a first class midpoint of 0.5 and a class width of 1.

0.45	1.04	6.80	1.50	1.65
1.31	4.31	4.15	1.57	1.87
2.65	1.39	2.70	0.86	1.41
7.69	2.78	5.32	1.22	1.88
2.00	0.44	6.12	1.89	3.29

2.71 Use cutpoint grouping with a first cutpoint of 25 and a class width of 3.

43.01	40.39	38.55	38.01	42.55
41.21	36.16	35.68	36.98	30.28
30.90	39.06	35.94	25.91	33.57
31.44	42.13	34.26	35.25	33.02
37.26	31.30	39.25	34.93	32.46

In each of Exercises 2.72–2.75, construct a dotplot for the data.

2.72

3	1	3	3	4	1	1	1	2	2

2.73

3	2	1	3	2	3	0	2	1	2

2.74

13	13	10	10	11
10	11	10	15	13
13	12	10	13	10

2.75

32	30	38	33	30
31	36	37	30	34
34	36	32	37	33

In each of Exercises 2.76–2.79, construct a stem-and-leaf diagram for the data, using the specified number of lines per stem.

2.76 Use one line per stem.

112	60	97	99	106
117	108	100	125	118

2.77 Use one line per stem.

55	32	41	20	21
37	38	43	32	62

2.78 Use five lines per stem.

15	11	12	22	17
10	23	16	15	23
10	14	9	16	5
22	14	14	8	20

2.79 Use two lines per stem.

34	46	22	33	25
27	45	25	22	36
28	27	40	32	27
24	37	29	32	31

Applying the Concepts and Skills

For each data set in Exercises 2.80–2.91, use the specified grouping method to
a. determine a frequency distribution.
b. obtain a relative-frequency distribution.
c. construct a frequency histogram based on your result from part (a).
d. construct a relative-frequency histogram based on your result from part (b).

2.80 **Number of Siblings.** Professor Weiss asked his introductory statistics students to state how many siblings they have. The responses are shown in the following table. Use single-value grouping.

1	3	2	1	1	0	1	1
3	0	2	2	1	2	0	2
1	2	2	1	0	1	1	1
1	1	0	2	0	3	4	2
0	2	1	1	2	1	1	0

2.81 **Household Size.** The U.S. Census Bureau conducts nationwide surveys on characteristics of U.S. households and publishes the results in *Current Population Reports*. Following are data on the number of people per household for a sample of 40 households. Use single-value grouping.

2	5	2	1	1	2	3	4
1	4	4	2	1	4	3	3
7	1	2	2	3	4	2	2
6	5	2	5	1	3	2	5
2	1	3	3	2	2	3	3

2.82 **Cottonmouth Litter Size.** In the paper "The Eastern Cottonmouth (*Agkistrodon piscivorus*) at the Northern Edge of Its Range" (*Journal of Herpetology*, Vol. 29, No. 3, pp. 391–398), C. Blem and L. Blem examined the reproductive characteristics of the eastern cottonmouth, a once widely distributed snake whose numbers have decreased recently due to encroachment by humans. A simple random sample of 24 female cottonmouths in Florida yielded the following data on number of young per litter. Use single-value grouping.

8	6	7	7	4	3	1	7
5	6	5	5	6	8	5	5
7	4	6	6	5	5	5	4

2.83 **Household Computers.** Every year since 1998, Arbitron and Edison Research have conducted a nationally representative survey focusing on trends in digital platforms. According to their publication *The Infinite Dial*, in 2002, there were an average of 1.2 working computers in the U.S. home. A random sample of 45 households taken this year yielded the following data on number of working computers. Use single-value grouping.

1	1	1	1	1	2	0	1	1
0	1	4	2	2	3	4	2	5
1	1	1	1	1	2	2	1	0
1	4	4	2	3	0	2	1	0
1	1	2	1	1	1	2	2	1

2.84 **Residential Energy Consumption.** The U.S. Energy Information Administration collects data on residential energy consumption and expenditures. Results are published in the document

Residential Energy Consumption Survey: Consumption and Expenditures. The following table gives one year's energy consumption for a sample of 50 households in the South. Data are in millions of BTUs. Use limit grouping with a first class of 40–49 and a class width of 10.

130	55	45	64	155	66	60	80	102	62
58	101	75	111	151	139	81	55	66	90
97	77	51	67	125	50	136	55	83	91
54	86	100	78	93	113	111	104	96	113
96	87	129	109	69	94	99	97	83	97

2.85 Early-Onset Dementia. Dementia is a person's loss of intellectual and social abilities that is severe enough to interfere with judgment, behavior, and daily functioning. Alzheimer's disease is the most common type of dementia. In the article "Living with Early Onset Dementia: Exploring the Experience and Developing Evidence-Based Guidelines for Practice" (*Alzheimer's Care Quarterly*, Vol. 5, Issue 2, pp. 111–122), P. Harris and J. Keady explored the experience and struggles of people diagnosed with dementia and their families. A simple random sample of 21 people with early-onset dementia gave the following data on age, in years, at diagnosis. Use limit grouping with a first class of 40–44 and a class width of 5.

60	58	52	58	59	58	51
61	54	59	55	53	44	46
47	42	56	57	49	41	43

2.86 Cheese Consumption. The U.S. Department of Agriculture reports in *Food Consumption, Prices, and Expenditures* that the average American consumed about 33.5 lb of cheese in 2011. Cheese consumption has increased steadily since 1960, when the average American ate only 8.3 lb of cheese annually. The following table provides last year's cheese consumption, in pounds, for 35 randomly selected Americans. Use limit grouping with a first class of 20–22 and a class width of 3.

45	28	32	37	41	39	33
32	31	35	27	46	25	41
35	31	44	23	38	27	32
43	32	25	36	26	30	35
36	36	35	21	43	35	28

2.87 Chronic Hemodialysis and Anxiety. Patients who undergo chronic hemodialysis often experience severe anxiety. Videotapes of progressive relaxation exercises were shown to one group of patients and neutral videotapes to another group. Then both groups took the State-Trait Anxiety Inventory, a psychiatric questionnaire used to measure anxiety, on which higher scores correspond to higher anxiety. In the paper "The Effectiveness of Progressive Relaxation in Chronic Hemodialysis Patients" (*Journal of Chronic Diseases*, Vol. 35, No. 10), R. Alarcon et al. presented the results of the study. The following data give score results for the group that viewed relaxation-exercises videotapes. Use limit grouping with a first class of 12–17 and a class width of 6.

30	41	28	14	40	36	38	24
61	36	24	45	38	43	32	28
37	34	20	23	34	47	25	31
39	14	43	40	29	21	40	

2.88 Primetime Broadcast Shows. From the TVbytheNumbers website, we obtained the viewing audiences, in millions, for the top 20 primetime broadcast TV shows for the week ending August 18, 2013. Use cutpoint grouping with a first class of 4–under 5.

10.360	9.508	8.785	8.299	8.089
7.434	6.712	6.572	6.478	6.469
6.331	6.169	6.069	5.877	5.751
5.608	5.493	5.443	5.098	4.865

2.89 Clocking the Cheetah. The cheetah (*Acinonyx jubatus*) is the fastest land mammal and is highly specialized to run down prey. The cheetah often exceeds speeds of 60 mph and, according to the online document "Cheetah Conservation in Southern Africa" (*Trade & Environment Database (TED) Case Studies*, Vol. 8, No. 2) by J. Urbaniak, the cheetah is capable of speeds up to 72 mph. The following table gives the speeds, in miles per hour, over $\frac{1}{4}$ mile for 35 cheetahs. Use cutpoint grouping with 52 as the first cutpoint and classes of equal width 2.

57.3	57.5	59.0	56.5	61.3	57.6	59.2
65.0	60.1	59.7	62.6	52.6	60.7	62.3
65.2	54.8	55.4	55.5	57.8	58.7	57.8
60.9	75.3	60.6	58.1	55.9	61.6	59.6
59.8	63.4	54.7	60.2	52.4	58.3	66.0

2.90 Fuel Tank Capacity. *Consumer Reports* provides information on new automobile models, including price, mileage ratings, engine size, body size, and indicators of features. A simple random sample of 35 new models yielded the following data on fuel tank capacity, in gallons. Use cutpoint grouping with 12 as the first cutpoint and classes of equal width 2.

17.2	23.1	17.5	15.7	19.8	16.9	15.3
18.5	18.5	25.5	18.0	17.5	14.5	20.0
17.0	20.0	24.0	26.0	18.1	21.0	19.3
20.0	20.0	12.5	13.2	15.9	14.5	22.2
21.1	14.4	25.0	26.4	16.9	16.4	23.0

2.91 Oxygen Distribution. In the article "Distribution of Oxygen in Surface Sediments from Central Sagami Bay, Japan: In Situ Measurements by Microelectrodes and Planar Optodes" (*Deep Sea Research Part I: Oceanographic Research Papers*, Vol. 52, Issue 10, pp. 1974–1987), R. Glud et al. explored the distributions of oxygen in surface sediments from central Sagami Bay. The oxygen distribution gives important information on the general biogeochemistry of marine sediments. Measurements were performed at 16 sites. A sample of 22 depths yielded the following data, in millimoles per square meter per day, on diffusive oxygen uptake. Use cutpoint grouping with a first class of 0–under 1.

1.8	2.0	1.8	2.3	3.8	3.4	2.7	1.1
3.3	1.2	3.6	1.9	7.6	2.0	1.5	2.0
1.1	0.7	1.0	1.8	1.8	6.7		

2.92 Exam Scores. Construct a dotplot for the following exam scores of the students in an introductory statistics class.

88	82	89	70	85
63	100	86	67	39
90	96	76	34	81
64	75	84	89	96

2.93 Age of Passenger Cars. According to R. L. Polk & Co., the average age of passenger cars in the United States was 9.1 years in 2000. A sample of 37 passenger cars taken this year provided the ages, in years, displayed in the following table. Construct a dotplot for the ages.

10	13	6	9	12	7	7	10
10	12	8	14	6	15	18	7
12	7	6	24	9	19	6	4
12	11	18	9	9	11	19	6
13	15	2	7	11			

2.94 Stressed-Out Bus Drivers. Frustrated passengers, congested streets, time schedules, and air and noise pollution are just some of the physical and social pressures that lead many urban bus drivers to retire prematurely with disabilities such as coronary heart disease and stomach disorders. An intervention program designed by the Stockholm Transit District was implemented to improve the work conditions of the city's bus drivers. Improvements were evaluated by G. Evans et al., who collected physiological and psychological data for bus drivers who drove on the improved routes (intervention) and for drivers who were assigned the normal routes (control). Their findings were published in the article "Hassles on the Job: A Study of a Job Intervention With Urban Bus Drivers" (*Journal of Organizational Behavior*, Vol. 20, pp. 199–208). Following are data, based on the results of the study, for the heart rates, in beats per minute, of the intervention and control drivers.

Intervention		Control						
68	66	74	52	67	63	77	57	80
74	58	77	53	76	54	73	54	
69	63	60	77	63	60	68	64	
68	73	66	71	66	55	71	84	
64	76	63	73	59	68	64	82	

a. Obtain dotplots for each of the two data sets, using the same scales.
b. Use your result from part (a) to compare the two data sets.

2.95 Acute Postoperative Days. Several neurosurgeons wanted to determine whether a dynamic system (Z-plate) reduced the number of acute postoperative days in the hospital relative to a static system (ALPS plate). R. Jacobowitz, Ph.D., an Arizona State University professor, along with G. Vishteh, M.D., and other neurosurgeons obtained the following data on the number of acute postoperative days in the hospital using the dynamic and static systems.

Dynamic							Static		
7	5	8	8	6	7	7	6	18	9
9	10	7	7	7	7	8	7	14	9

a. Obtain dotplots for each of the two data sets, using the same scales.
b. Use your result from part (a) to compare the two data sets.

2.96 M&Ms. In the article "Sweetening Statistics—What M&M's Can Teach Us" (Minitab Inc., August 2008), M. Paret and E. Martz discussed several statistical analyses that they performed on bags of M&Ms. The authors took a random sample of 30 small bags of peanut M&Ms and obtained the following weights, in grams (g). Construct a stem-and-leaf diagram for these weights.

55.0	50.8	52.1	57.0	52.1	53.5
51.3	51.5	46.4	55.3	45.5	54.1
55.3	50.3	47.2	53.8	50.7	51.5
50.5	51.8	53.6	52.0	51.9	54.3
48.0	53.3	53.5	56.0	49.1	53.9

2.97 Women in the Workforce. In an issue of *Science* (Vol. 308, No. 5721, p. 483), D. Normile reported on a study from the Japan Statistics Bureau of the 30 industrialized countries in the Organization for Economic Co-operation and Development (OECD) titled "Japan Mulls Workforce Goals for Women." Following are the percentages of women in scientific workforces for a sample of 17 countries. Construct a stem-and-leaf diagram for these percentages.

26	12	44	13	40	18	39	21	35
27	34	28	34	28	33	29	29	

2.98 Process Capability. R. Morris and E. Watson studied various aspects of process capability in the paper "Determining Process Capability in a Chemical Batch Process" (*Quality Engineering*, Vol. 10(2), pp. 389–396). In one part of the study, the researchers compared the variability in product of a particular piece of equipment to a known analytic capability to decide whether product consistency could be improved. The following data were obtained for 10 batches of product.

30.1	30.7	30.2	29.3	31.0
29.6	30.4	31.2	28.8	29.8

Construct a stem-and-leaf diagram for these data with
a. one line per stem.   b. two lines per stem.
c. Which stem-and-leaf diagram do you find more useful? Why?

2.99 University Patents. The number of patents a university receives is an indicator of the research level of the university. From a study titled *Science and Engineering Indicators* issued by the National Science Foundation, we found the number of U.S. patents awarded to a sample of 36 private and public universities to be as follows.

93	27	11	30	9	30	35	20	9
35	24	19	14	29	11	2	55	15
35	2	15	4	16	79	16	22	49
3	69	23	18	41	11	7	34	16

Construct a stem-and-leaf diagram for these data with
a. one line per stem.   b. two lines per stem.
c. Which stem-and-leaf diagram do you find more useful? Why?

2.100 San Francisco Giants. From the Baseball Almanac website, we found the heights, in inches, of the players on the 2012 World Series–winning San Francisco Giants baseball team.

76	75	77	75	72	74	73	71	75
71	76	74	72	72	76	75	76	72
70	76	76	74	74	73	71	74	73
77	72	74	73	73	76	76	71	70
71	71	72	73	74	74	71	76	73

a. Construct a stem-and-leaf diagram of these data with five lines per stem.
b. Why is it better to use five lines per stem here instead of one or two lines per stem?

2.101 Detroit Tigers. From the Baseball Almanac website, we found the heights, in inches, of the players on the 2012 World Series runner-up Detroit Tigers baseball team.

72	77	75	75	73	77	72	75
80	71	71	73	77	78	72	75
72	75	77	76	77	72	74	74
71	72	73	72	74	76	77	71
71	71	74	71	73	72	76	72
76	73	76	72	71	75	68	

a. Construct a stem-and-leaf diagram of these data with five lines per stem.

b. Why is it better to use five lines per stem here instead of one or two lines per stem?

2.102 Adjusted Gross Incomes. The *Internal Revenue Service* (IRS) publishes data on adjusted gross incomes in *Statistics of Income, Individual Income Tax Returns*. The following relative-frequency histogram shows one year's individual income tax returns for adjusted gross incomes of less than $50,000.

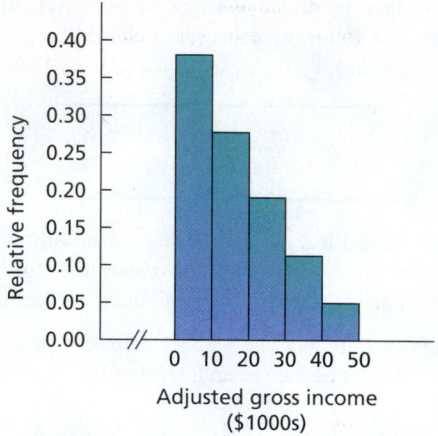

Adjusted gross income
($1000s)

Use the histogram and the fact that adjusted gross incomes are expressed to the nearest whole dollar to answer each of the following questions.

a. Approximately what percentage of the individual income tax returns had an adjusted gross income between $10,000 and $19,999, inclusive?

b. Approximately what percentage had an adjusted gross income of less than $30,000?

c. The IRS reported that 89,928,000 individual income tax returns had an adjusted gross income of less than $50,000. Approximately how many had an adjusted gross income between $30,000 and $49,999, inclusive?

2.103 Cholesterol Levels. According to the *National Health and Nutrition Examination Survey*, published by the *Centers for Disease Control and Prevention*, the average cholesterol level for children between 4 and 19 years of age is 165 mg/dL. A pediatrician who tested the cholesterol levels of several young patients was alarmed to find that many had levels higher than 200 mg/dL. The following relative-frequency histogram shows the readings for some patients who had high cholesterol levels.

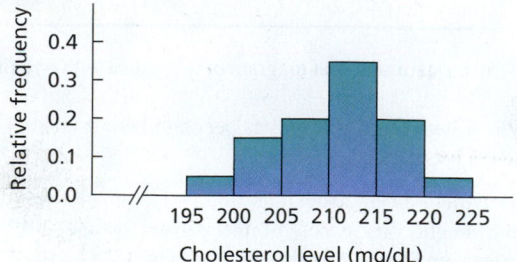

Cholesterol level (mg/dL)

Use the graph to answer the following questions. Note that cholesterol levels are always expressed as whole numbers.

a. What percentage of the patients have cholesterol levels between 205 and 209, inclusive?

b. What percentage of the patients have levels of 215 or higher?

c. If the number of patients is 20, how many have levels between 210 and 214, inclusive?

2.104 Hospital Beds. The number of hospital beds available in community hospitals is decreasing. Through advancement in care and technology, hospitals are getting more efficient. However, the aging and increasingly obese and diabetic population intensifies the need. The *American Hospital Association Annual Survey*, published by the American Hospital Association (AHA), gave the number of hospital beds available per 1000 people for each of the 50 states plus the District of Columbia for 2010. Following is a stem-and-leaf diagram of the data. The stems represent units and the leaves tenths (e.g., the first entry is 1.7).

```
1 | 7 7 8 9 9 9
2 | 0 0 1 1 1 2 2 2 3 3 4 4 4 4 4 4 4
2 | 6 6 6 7 7 7 8 9
3 | 0 0 1 1 1 2 2 3 3 3 4
3 | 5 5 7 9
4 | 0 4
4 |
5 | 0 0
5 | 7
```

How many states (including the District of Columbia) had

a. at least three but less than four hospital beds per 1000 people available?

b. at least four and a half hospital beds per 1000 people available?

2.105 Parkinson's Disease. Parkinson's disease affects internally generated movements such as movements recalled from memory. L-Dopa is a drug that is used in clinical treatment of Parkinson's disease. In the article, "L-Dopa Induces Under-Damped Visually Guided Motor Responses in Parkinson's Disease" (*Experimental Brain Research*, Vol. 202, Issue 3, pp. 553–559), Wing-Lok Au et al. explore the effects of L-Dopa on patients with Parkinson's disease. Following is a stem-and-leaf diagram of the ages, in years, at which a sample of 14 patients experienced the onset of symptoms of Parkinson's disease.

```
3 | 9
4 |
4 | 8 9
5 | 0 2 2 3 3 4 4
5 |
6 | 2
6 | 5
7 | 1 1
```

How many of these 14 patients were

a. under 45 years old at the onset of symptoms?

b. at least 65 years old at the onset of symptoms?

c. between 50 years old and 64 years old, inclusive, at the onset of symptoms?

Working with Large Data Sets

2.106 The Great White Shark. In an article titled "Great White, Deep Trouble" (*National Geographic*, Vol. 197(4), pp. 2–29), Peter Benchley—the author of *JAWS*—discussed various aspects of the Great White Shark (*Carcharodon carcharias*). Data on the number of pups borne in a lifetime by each of 80 Great White Shark females are provided on the WeissStats site. Use the technology of your choice to
a. obtain frequency and relative-frequency distributions, using single-value grouping.
b. construct and interpret either a frequency histogram or a relative-frequency histogram.

2.107 The Beatles. In the article, "Length of The Beatles' Songs" (*Chance*, Vol. 25, No. 1, pp. 30–33), T. Koyama discusses aspects and interpretations of the lengths of songs by The Beatles. Data on the length, in seconds, of 229 Beatles' songs are presented on the WeissStats site. Use the technology of your choice to
a. obtain frequency and relative-frequency distributions.
b. get and interpret a frequency histogram or a relative-frequency histogram.
c. construct a dotplot.
d. Compare your graphs from parts (b) and (c).

2.108 High School Completion. As reported by the U.S. Census Bureau in *Educational Attainment in the United States*, the percentage of adults in each state and the District of Columbia who have completed high school is provided on the WeissStats site. Apply the technology of your choice to construct a stem-and-leaf diagram of the percentages with
a. one line per stem.   **b.** two lines per stem.
c. five lines per stem.
d. Which stem-and-leaf diagram do you consider most useful? Explain your answer.

2.109 Bachelor's Completion. As reported by the U.S. Census Bureau in *Educational Attainment in the United States*, the percentage of adults in each state and the District of Columbia who have completed a bachelor's degree is provided on the WeissStats site. Apply the technology of your choice to construct a stem-and-leaf diagram of the percentages with
a. one line per stem.   **b.** two lines per stem.
c. five lines per stem.
d. Which stem-and-leaf diagram do you consider most useful? Explain your answer.

2.110 Body Temperature. A study by researchers at the University of Maryland addressed the question of whether the mean body temperature of humans is 98.6°F. The results of the study by P. Mackowiak et al. appeared in the article "A Critical Appraisal of 98.6°F, the Upper Limit of the Normal Body Temperature, and Other Legacies of Carl Reinhold August Wunderlich" (*Journal of the American Medical Association*, Vol. 268, pp. 1578–1580). Among other data, the researchers obtained the body temperatures of 93 healthy humans, as provided on the WeissStats site. Use the technology of your choice to obtain and interpret
a. a frequency histogram or a relative-frequency histogram of the temperatures.
b. a dotplot of the temperatures.
c. a stem-and-leaf diagram of the temperatures.
d. Compare your graphs from parts (a)–(c). Which do you find most useful?

Extending the Concepts and Skills

2.111 Exam Scores. The exam scores for the students in an introductory statistics class are as follows.

88	82	89	70	85
63	100	86	67	39
90	96	76	34	81
64	75	84	89	96

a. Group these exam scores, using the classes 30–39, 40–49, 50–59, 60–69, 70–79, 80–89, and 90–100.
b. What are the widths of the classes?
c. If you wanted all the classes to have the same width, what classes would you use?

Choosing the Classes. One way that we can choose the classes to be used for grouping a quantitative data set is to first decide on the (approximate) number of classes. From that decision, we can then determine a class width and, subsequently, the classes themselves. Several methods can be used to decide on the number of classes. One method is to use the following guidelines, based on the number of observations:

Number of observations	Number of classes
25 or fewer	5–6
25–50	7–14
Over 50	15–20

With the preceding guidelines in mind, we can use the following step-by-step procedure for choosing the classes.

Step 1 Decide on the (approximate) number of classes.

Step 2 Calculate an approximate class width as

$$\frac{\text{Maximun observation} - \text{Minimum observation}}{\text{Number of classes}}$$

and use the result to decide on a convenient class width.

Step 3 Choose a number for the lower limit (or cutpoint) of the first class, noting that it must be less than or equal to the minimum observation.

Step 4 Obtain the other lower class limits (or cutpoints) by successively adding the class width chosen in Step 2.

Step 5 Use the results of Step 4 to specify all of the classes.

Exercises 2.112 and 2.113 provide you with some practice in applying the preceding step-by-step procedure.

2.112 Days to Maturity for Short-Term Investments. Refer to the days-to-maturity data in Table 2.6 on page 53. Note that there are 40 observations, the smallest and largest of which are 36 and 99, respectively. Apply the preceding procedure to choose classes for limit grouping. Use approximately seven classes. *Note:* If in Step 2 you decide on 10 for the class width and in Step 3 you choose 30 for the lower limit of the first class, then you will get the same classes as used in Example 2.13; otherwise, you will get different classes (which is fine).

2.113 Weights of 18- to 24-Year-Old Males. Refer to the weight data in Table 2.8 on page 54. Note that there are 37 observations, the smallest and largest of which are 129.2 and 278.8, respectively. Apply the preceding procedure to choose classes for cutpoint grouping. Use approximately eight classes. *Note:* If in Step 2 you decide on 20 for the class width and in Step 3 you choose 120 for the lower cutpoint of the first class, then you will get the same classes as used in Example 2.14; otherwise, you will get different classes (which is fine).

Contingency Tables. The methods presented in this section and the preceding section apply to grouping data obtained from observing values of one variable of a population. Such data are called **univariate data.** For instance, in Example 2.14 on page 54, we examined data obtained from observing values of the variable "weight" for a sample of 18- to 24-year-old males; those data are univariate. We could have considered not only the weights of the males but also their heights. Then, we would have data on two variables, height and weight. Data obtained from observing values of two variables of a population are called **bivariate data.** Tables called **contingency tables** can be used to group bivariate data, as explained in Exercise 2.114.

2.114 Age and Gender. The following bivariate data on age (in years) and gender were obtained from the students in a freshman calculus course. The data show, for example, that the first student on the list is 21 years old and is a male.

Age	Gender	Age	Gender	Age	Gender	Age	Gender	Age	Gender
21	M	29	F	22	M	23	F	21	F
20	M	20	M	23	M	44	M	28	F
42	F	18	F	19	F	19	M	21	F
21	M	21	M	21	M	21	F	21	F
19	F	26	M	21	F	19	M	24	F
21	F	24	F	21	F	25	M	24	F
19	F	19	M	20	F	21	M	24	F
19	M	25	M	20	F	19	M	23	M
23	M	19	F	20	F	18	F	20	F
20	F	23	M	22	F	18	F	19	M

a. Group these data in the following contingency table. For the first student, place a tally mark in the box labeled by the "21–25" column and the "Male" row, as indicated. Tally the data for the other 49 students.

		Age (yr)		
Gender		**Under 21**	**21–25**	**Over 25**	**Total**
	Male		|		
	Female				
	Total				

b. Construct a table like the one in part (a) but with frequencies replacing the tally marks. Add the frequencies in each row and column of your table and record the sums in the proper "Total" boxes.

c. What do the row and column totals in your table in part (b) represent?

d. Add the row totals and add the column totals. Why are those two sums equal, and what does their common value represent?

e. Construct a table that shows the relative frequencies for the data. (*Hint:* Divide each frequency obtained in part (b) by the total of 50 students.)

f. Interpret the entries in your table in part (e) as percentages.

Relative-Frequency Polygons. Another graphical display commonly used is the **relative-frequency polygon.** In a relative-frequency polygon, a point is plotted above each class mark in limit grouping and above each class midpoint in cutpoint grouping at a height equal to the relative frequency of the class. Then the points are connected with lines. For instance, the grouped days-to-maturity data given in Table 2.10(b) on page 57 yields the following relative-frequency polygon.

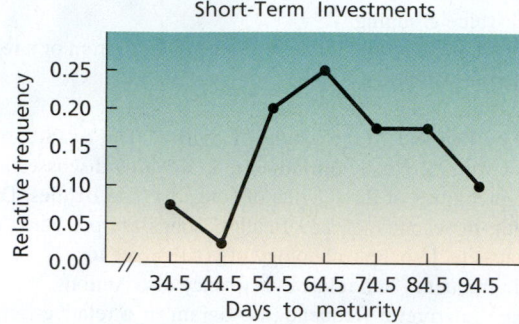

2.115 Residential Energy Consumption. Construct a relative-frequency polygon for the energy-consumption data given in Exercise 2.84. Use the classes specified in that exercise.

2.116 Clocking the Cheetah. Construct a relative-frequency polygon for the speed data given in Exercise 2.89. Use the classes specified in that exercise.

2.117 As mentioned, for relative-frequency polygons, we label the horizontal axis with class marks in limit grouping and class midpoints in cutpoint grouping. How do you think the horizontal axis is labeled in single-value grouping?

Ogives. Cumulative information can be portrayed using a graph called an **ogive** (ō′jīv). To construct an ogive, we first make a table that displays cumulative frequencies and cumulative relative frequencies. A **cumulative frequency** is obtained by summing the frequencies of all classes representing values less than a specified lower class limit (or cutpoint). A **cumulative relative frequency** is found by dividing the corresponding cumulative frequency by the total number of observations.

For instance, consider the grouped days-to-maturity data given in Table 2.10(b) on page 57. From that table, we see that the cumulative frequency of investments with a maturity period of less than 50 days is 4 (3 + 1) and, therefore, the cumulative relative frequency is 0.1 (4/40). Table 2.14 shows all cumulative information for the days-to-maturity data.

TABLE 2.14

Cumulative information for days-to-maturity data

Less than	Cumulative frequency	Cumulative relative frequency
30	0	0.000
40	3	0.075
50	4	0.100
60	12	0.300
70	22	0.550
80	29	0.725
90	36	0.900
100	40	1.000

Using Table 2.14, we can now construct an ogive for the days-to-maturity data. In an ogive, a point is plotted above each lower class limit (or cutpoint) at a height equal to the cumulative relative frequency. Then the points are connected with lines. An ogive for the days-to-maturity data is as follows.

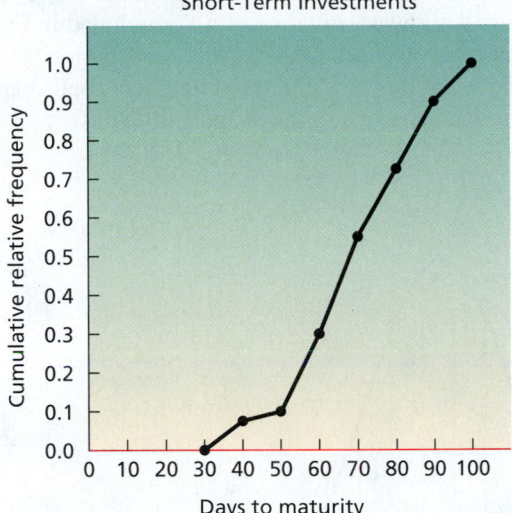

Short-Term Investments

2.118 Residential Energy Consumption. Refer to the energy-consumption data given in Exercise 2.84.
a. Construct a table similar to Table 2.14 for the data, based on the classes specified in Exercise 2.84. Interpret your results.
b. Construct an ogive for the data.

2.119 Clocking the Cheetah. Refer to the speed data given in Exercise 2.89.
a. Construct a table similar to Table 2.14 for the data, based on the classes specified in Exercise 2.89. Interpret your results.
b. Construct an ogive for the data.

Further Stem-and-Leaf Techniques. In constructing a stem-and-leaf diagram, rounding or truncating each observation to a suitable number of digits is often useful. Exercises 2.120 and 2.121 involve rounding and truncating numbers for use in stem-and-leaf diagrams.

2.120 Cardiovascular Hospitalizations. The Florida State Center for Health Statistics reported in *Women and Cardiovascular Disease Hospitalizations* that, for cardiovascular hospitalizations, the mean age of women is 71.9 years. At one hospital, a random sample of 20 female cardiovascular patients had the following ages, in years.

75.9	83.7	87.3	74.5	82.5
78.2	76.1	52.8	56.4	53.8
88.2	78.9	81.7	54.4	52.7
58.9	97.6	65.8	86.4	72.4

a. Round each observation to the nearest year and then construct a stem-and-leaf diagram of the rounded data.
b. Truncate each observation by dropping the decimal part, and then construct a stem-and-leaf diagram of the truncated data.
c. Compare the stem-and-leaf diagrams that you obtained in parts (a) and (b).

2.121 Shoe and Apparel E-Tailers. In the special report "Mousetrap: The Most-Visited Shoe and Apparel E-tailers" (*Footwear News*, Vol. 58, No. 3, p. 18), we found the following data on the average time, in minutes, spent per user per month from January to June of one year for a sample of 15 shoe and apparel retail websites.

13.3	9.0	11.1	9.1	8.4
15.6	8.1	8.3	13.0	17.1
16.3	13.5	8.0	15.1	5.8

The following Minitab output shows a stem-and-leaf diagram for these data. The second column gives the stems, and the third column gives the leaves.

```
Stem-and-Leaf Display: TIME

Stem-and-leaf of TIME   N  = 15
Leaf Unit = 1.0


  1     0   5
  1     0
  7     0   888899
 (1)    1   1
  7     1   333
  4     1   55
  2     1   67
```

Did Minitab use rounding or truncation to obtain this stem-and-leaf diagram? Explain your answer.

2.4 Distribution Shapes

In this section, we discuss *distributions* and their associated properties.

DEFINITION 2.12

Distribution of a Data Set

The **distribution of a data set** is a table, graph, or formula that provides the values of the observations and how often they occur.

Up to now, we have portrayed distributions of data sets by frequency distributions, relative-frequency distributions, frequency histograms, relative-frequency histograms, dotplots, stem-and-leaf diagrams, pie charts, and bar charts.

An important aspect of the distribution of a quantitative data set is its shape. Indeed, as we demonstrate in later chapters, the shape of a distribution frequently plays a role in determining the appropriate method of statistical analysis. To identify the shape of a distribution, the best approach usually is to use a smooth curve that approximates the overall shape.

For instance, Fig. 2.10 displays a relative-frequency histogram for the heights of the 3264 female students who attend a midwestern college. Also included in Fig. 2.10 is a smooth curve that approximates the overall shape of the distribution. Both the histogram and the smooth curve show that this distribution of heights is bell shaped (or mound shaped), but the smooth curve makes seeing the shape a little easier.

FIGURE 2.10

Relative-frequency histogram and approximating smooth curve for the distribution of heights

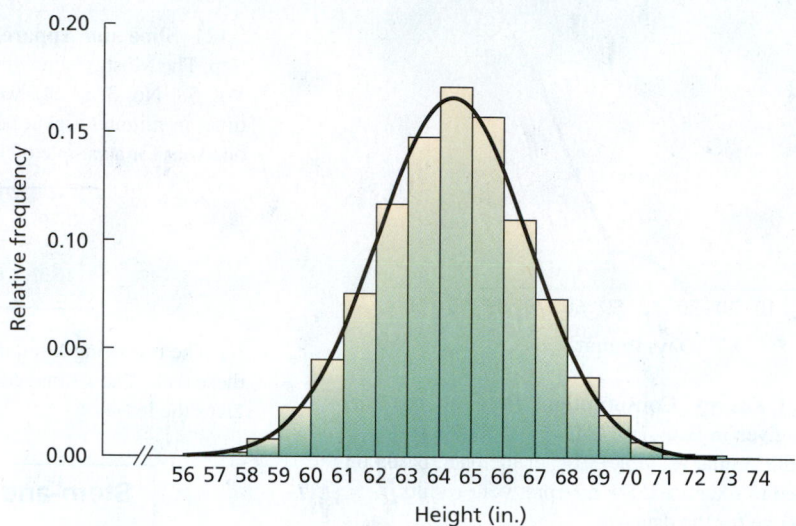

Another advantage of using smooth curves to identify distribution shapes is that we need not worry about minor differences in shape. Instead we can concentrate on overall patterns, which, in turn, allows us to classify most distributions by designating relatively few shapes.

Three important general aspects of the shape of a distribution involve *modality, symmetry,* and *skewness,* which we discuss now.

Modality

When considering the shape of a distribution, you should observe its number of peaks (highest points). A distribution is **unimodal** if it has one peak, **bimodal** if it has two peaks, and **multimodal** if it has three or more peaks. Figure 2.11 shows examples of unimodal, bimodal, and multimodal distributions.

FIGURE 2.11

Examples of (a) unimodal, (b) bimodal, and (c) multimodal distributions

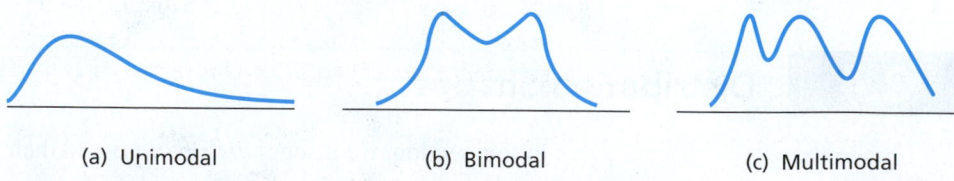

(a) Unimodal   (b) Bimodal   (c) Multimodal

Technically, a distribution is bimodal or multimodal only if the peaks are the same height. However, in practice, distributions with pronounced but not necessarily equal-height peaks are often called bimodal or multimodal.

Symmetry

Another important consideration when examining the shape of a distribution is *symmetry*. A distribution that can be divided into two pieces that are mirror images of one another is called **symmetric.** The distributions in Figs. 2.10 and 2.11(b) are symmetric.

The three distributions in Fig. 2.12, called **bell shaped, triangular,** and **uniform** (or **rectangular**), are specific categories of symmetric distributions. Note that the distribution of heights in Fig. 2.10 is bell shaped.

FIGURE 2.12

Examples of symmetric distributions: (a) bell shaped, (b) triangular, and (c) uniform

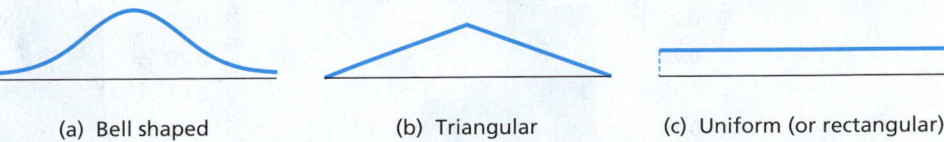

(a) Bell shaped   (b) Triangular   (c) Uniform (or rectangular)

Observe that bell-shaped distributions and triangular distributions are also unimodal. A uniform distribution has either no peaks or infinitely many peaks, depending on how you look at it. In any case, we do not classify a uniform distribution according to modality.

Skewness

A unimodal distribution that is not symmetric is either *right skewed* or *left skewed*. A **right-skewed** distribution rises to its peak rapidly and comes back toward the horizontal axis more slowly—its "right tail" is longer than its "left tail." A **left-skewed** distribution rises to its peak slowly and comes back toward the horizontal axis more rapidly—its "left tail" is longer than its "right tail." Figure 2.13 shows generic right-skewed and left-skewed distributions.

FIGURE 2.13

Generic skewed distributions: (a) right skewed and (b) left skewed

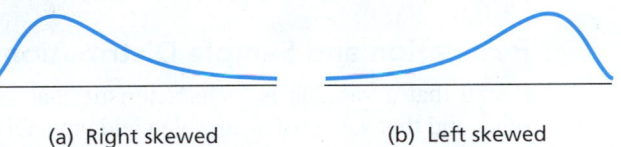

(a) Right skewed   (b) Left skewed

It is important to note the following distinction between general and specific classifications of distribution shape:

- Modality, symmetry, and skewness are *general* classifications of distribution shape.
- Such designations as bell shaped, triangular, and uniform are *specific* classifications of distribution shape.

For instance, the distribution of heights in Fig. 2.10 is classified generally as unimodal and symmetric, and is classified specifically as bell shaped. As another example, consider the distribution shown in Fig. 2.14. Its general classification is unimodal and right skewed, and its specific classification is **reverse J shaped.**

FIGURE 2.14

Reverse-J-shaped distribution

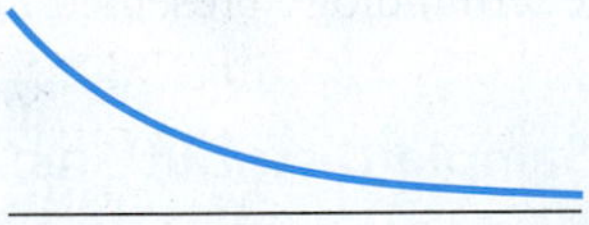

In our study of statistics, the distinction among symmetric, right skewed, and left skewed will play a prominent role, so we often concentrate on that aspect of distribution shape. Our next example provides an illustration.

■■■ **EXAMPLE 2.23**   **Identifying Distribution Shapes**

Household Size A relative-frequency histogram for household size in the United States shown in Fig. 2.15(a) (next page) is based on data contained in *Current Population Reports*, a publication of the U.S. Census Bureau.[†] Classify the distribution of sizes for U.S. households as (roughly) symmetric, right skewed, or left skewed.

[†] Actually, the class 7 portrayed in Fig. 2.15 is for seven or more people. However, we have considered it 7 for illustrative purposes.

FIGURE 2.15   Relative-frequency histogram for household size

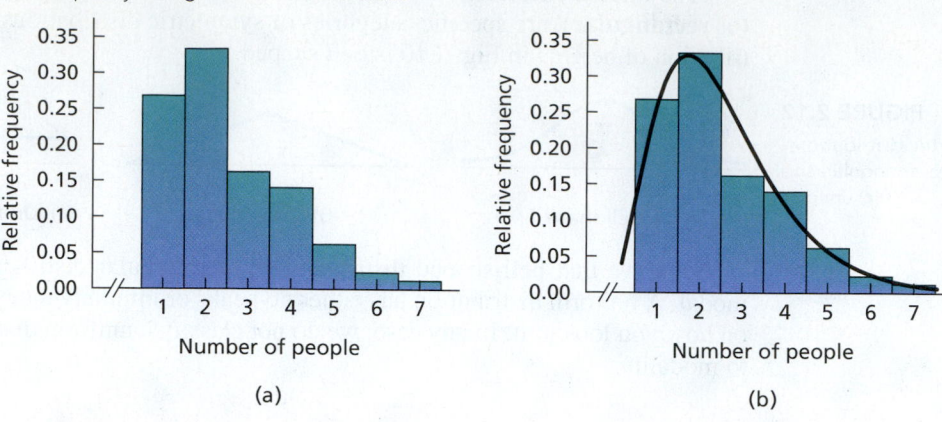

(a)   (b)

Exercise 2.143
on page 81

Solution   First, we draw a smooth curve through the histogram shown in Fig. 2.15(a) to get Fig. 2.15(b). Next, we note that the distribution of household sizes is unimodal and nonsymmetric. Then, by referring to Fig. 2.13 on page 77, we find that the distribution of household sizes is right skewed.   ■

It is important to note that, when classifying distributions, we must be flexible. Thus, for instance, exact symmetry is not required to classify a distribution as symmetric.

Population and Sample Distributions

Recall that a variable is a characteristic that varies from one person or thing to another and that values of a variable yield data. Distinguishing between data for an entire population and data for a sample of a population is an essential aspect of statistics.

DEFINITION 2.13

Population and Sample Data

Population data: The values of a variable for the entire population.
Sample data: The values of a variable for a sample of the population.

Note: Population data are also called **census data.**

To distinguish between the distribution of population data and the distribution of sample data, we use the terminology presented in Definition 2.14.

DEFINITION 2.14

Population and Sample Distributions; Distribution of a Variable

The distribution of population data is called the **population distribution,** or the **distribution of the variable.**

The distribution of sample data is called a **sample distribution.**

For a particular population and variable, sample distributions vary from sample to sample. However, there is only one population distribution, namely, the distribution of the variable under consideration on the population under consideration. The following example illustrates this point and some others as well.

■ ■ ■   EXAMPLE 2.24   Population and Sample Distributions

Household Size   In Example 2.23, we considered the distribution of household size for U.S. households. There the variable is household size, and the population consists of all U.S. households. We repeat the graph for that example in Fig. 2.16(a). This

FIGURE 2.16 Population distribution and six sample distributions for household size

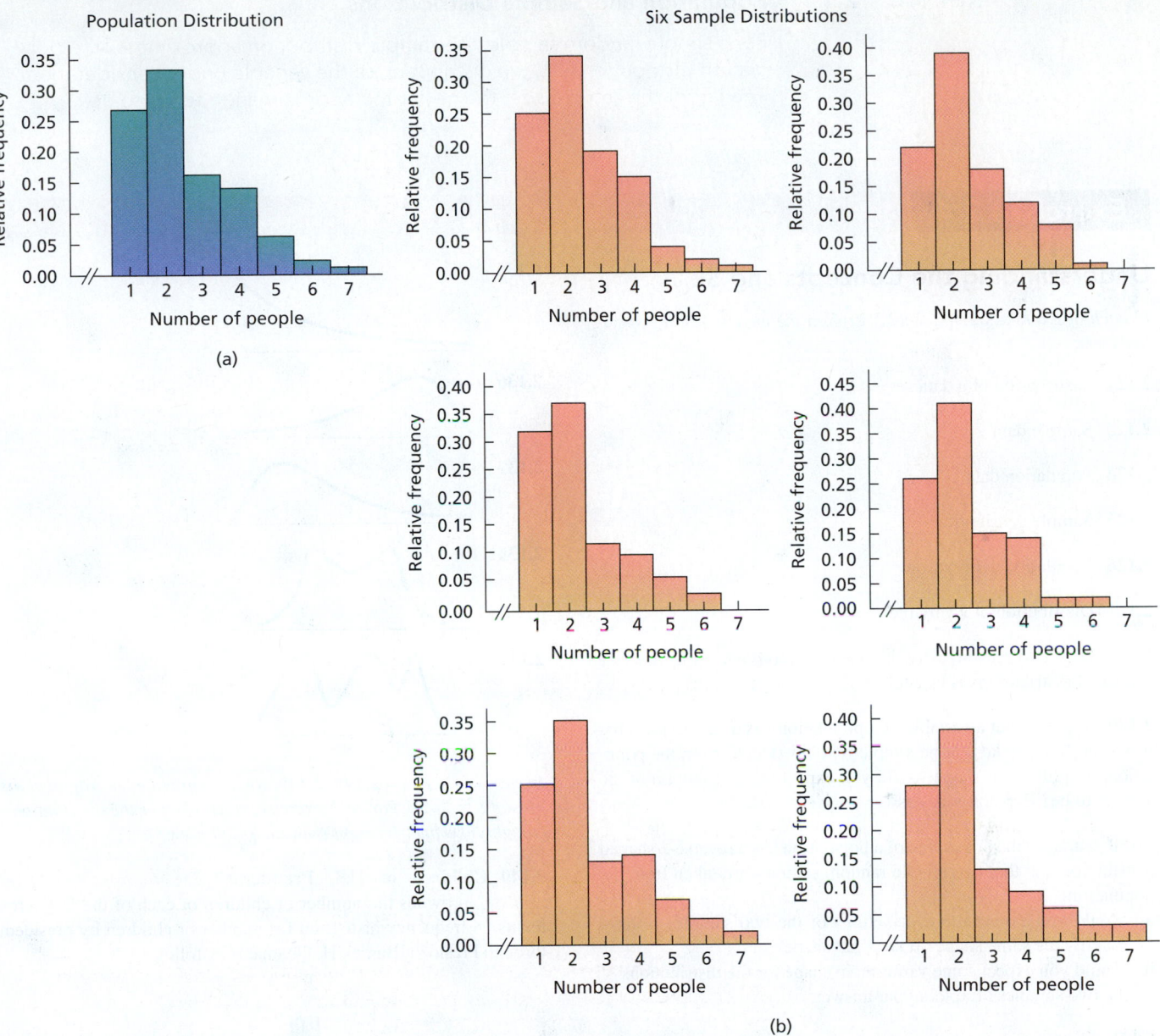

(a)

(b)

graph is a relative-frequency histogram of household size for the population of all U.S. households; it gives the population distribution or, equivalently, the distribution of the variable "household size."

We simulated six simple random samples of 100 households each from the population of all U.S. households. Figure 2.16(b) shows relative-frequency histograms of household size for all six samples. Compare the six sample distributions in Fig. 2.16(b) to each other and to the population distribution in Fig. 2.16(a).

Solution The distributions of the six samples are similar but have definite differences. This result is not surprising because we would expect variation from one sample to another. Nonetheless, the overall shapes of the six sample distributions are roughly the same and also are similar in shape to the population distribution—all of these distributions are right skewed.

In practice, we usually do not know the population distribution. As Example 2.24 suggests, however, we can use the distribution of a simple random sample from the population to get a rough idea of the population distribution.

KEY FACT 2.1 | **Population and Sample Distributions**

For a simple random sample, the sample distribution approximates the population distribution (i.e., the distribution of the variable under consideration). The larger the sample size, the better the approximation tends to be.

Exercises 2.4

Understanding the Concepts and Skills

In each of Exercises 2.122–2.127, explain the meaning of the specified term.

2.122 Distribution of a data set

2.123 Sample data

2.124 Population data

2.125 Sample distribution

2.126 Population distribution

2.127 Distribution of a variable

2.128 Give two reasons why the use of smooth curves to describe shapes of distributions is helpful.

2.129 Suppose that a variable of a population has a bell-shaped distribution. If you take a large simple random sample from the population, roughly what shape would you expect the distribution of the sample to be? Explain your answer.

2.130 Suppose that a variable of a population has a reverse-J-shaped distribution and that two simple random samples are taken from the population.
a. Would you expect the distributions of the two samples to have roughly the same shape? If so, what shape?
b. Would you expect some variation in shape for the distributions of the two samples? Explain your answer.

2.131 Identify and sketch three distribution shapes that are symmetric.

In each of Exercises 2.132–2.139, we have drawn a smooth curve that represents a distribution. In each case, do the following:
a. Identify the shape of the distribution with regard to modality.
b. Identify the shape of the distribution with regard to symmetry (or nonsymmetry).
c. If the distribution is unimodal and nonsymmetric, classify it as either right skewed or left skewed.

2.132

2.133

2.134

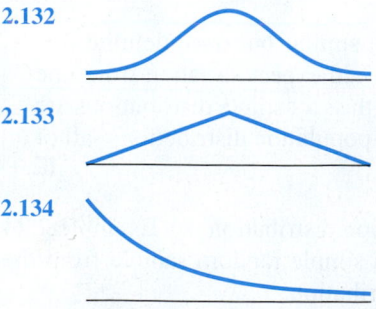

2.135

2.136

2.137

2.138

2.139

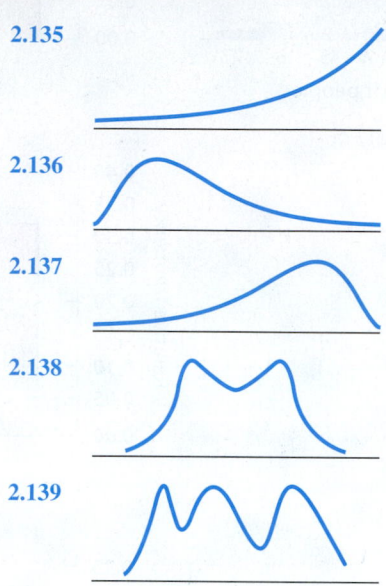

In each of Exercises 2.140–2.149, we have provided a graphical display of a data set. For each exercise, state whether the distribution is (roughly) symmetric, right skewed, or left skewed.

2.140 Children of U.S. Presidents. The *Information Please Almanac* provides the number of children of each of the U.S. presidents. A frequency histogram for number of children by president, through President Barack H. Obama, is as follows.

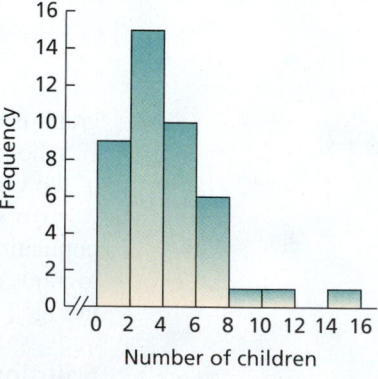

2.141 Clocking the Cheetah. The cheetah (*Acinonyx jubatus*) is the fastest land mammal and is highly specialized to run down prey. The cheetah often exceeds speeds of 60 mph and, according to the on-line document "Cheetah Conservation in Southern Africa" (*Trade & Environment Database (TED) Case Studies*, Vol. 8, No. 2) by J. Urbaniak, the cheetah is capable of speeds up to 72 mph. Following is a frequency histogram for the speeds, in miles per hour, for a sample of 35 cheetahs.

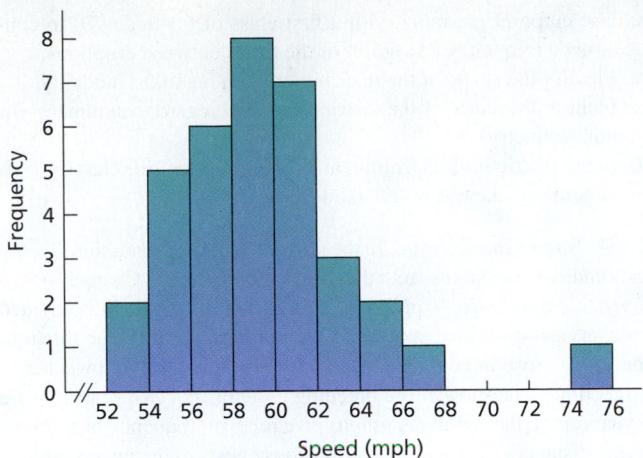

2.144 Baltimore Ravens. From *Player Roster*, the official roster of the 2013 Super Bowl champion Baltimore Ravens, we obtained the heights, in inches, of the players on that team. A dotplot of those heights is as follows.

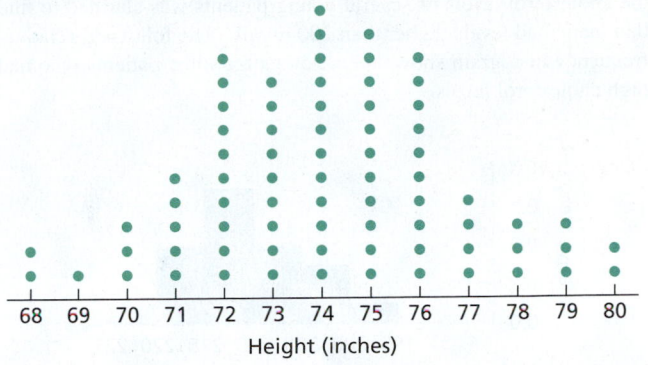

2.142 Malnutrition and Poverty. R. Reifen et al. studied various nutritional measures of Ethiopian school children and published their findings in the paper "Ethiopian-Born and Native Israeli School Children Have Different Growth Patterns" (*Nutrition*, Vol. 19, pp. 427–431). The study, conducted in Azezo, North West Ethiopia, found that malnutrition is prevalent in primary and secondary school children because of economic poverty. A frequency histogram for the weights, in kilograms (kg), of 60 randomly selected male Ethiopian-born school children ages 12–15 years old is as follows.

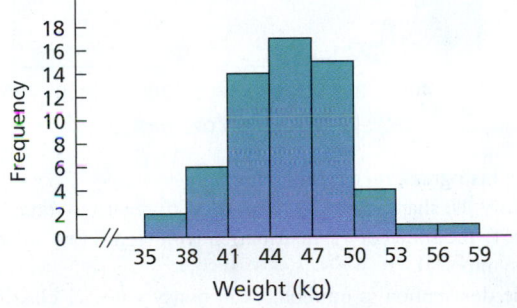

2.145 PCBs and Pelicans. Polychlorinated biphenyls (PCBs), industrial pollutants, are known to be carcinogens and a great danger to natural ecosystems. As a result of several studies, PCB production was banned in the United States in 1979 and by the Stockholm Convention on Persistent Organic Pollutants in 2001. One study, published in 1972 by R. W. Risebrough, is titled "Effects of Environmental Pollutants Upon Animals Other Than Man" (*Proceedings of the 6th Berkeley Symposium on Mathematics and Statistics, VI*, University of California Press, pp. 443–463). In that study, 60 Anacapa pelican eggs were collected and measured for their shell thickness, in millimeters (mm), and concentration of PCBs, in parts per million (ppm). Following is a relative-frequency histogram of the PCB concentration data.

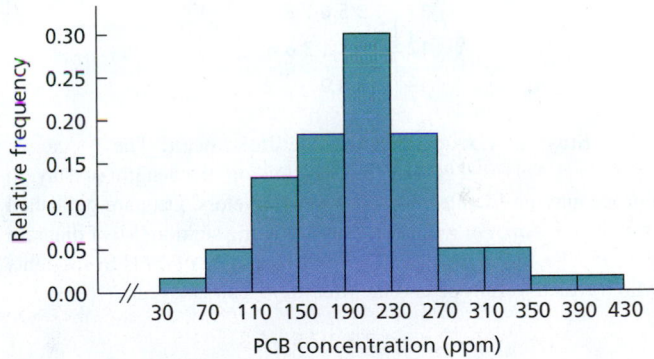

2.143 The Coruro's Burrow. The subterranean coruro (*Spalacopus cyanus*) is a social rodent that lives in large colonies in underground burrows that can reach lengths of up to 600 meters. Zoologists S. Begall and M. Gallardo studied the characteristics of the burrow systems of the subterranean coruro in central Chile and published their findings in the *Journal of Zoology, London* (Vol. 251, pp. 53–60). A sample of 51 burrows, whose depths were measured in centimeters, yielded the following frequency histogram.

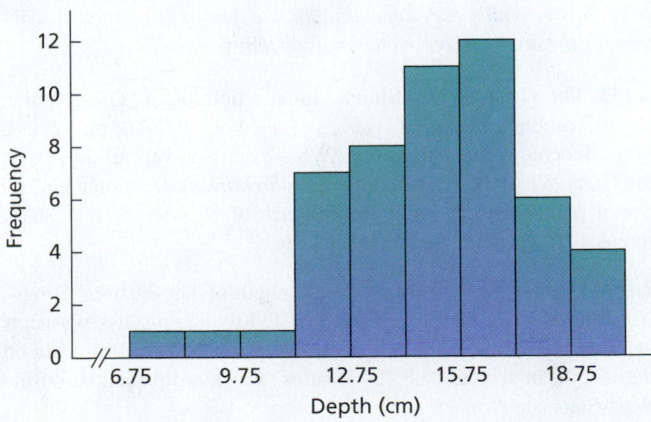

2.146 Adjusted Gross Incomes. The Internal Revenue Service (IRS) publishes data on adjusted gross incomes in the document *Statistics of Income, Individual Income Tax Returns*. The following relative-frequency histogram shows one year's individual income tax returns for adjusted gross incomes of less than $50,000.

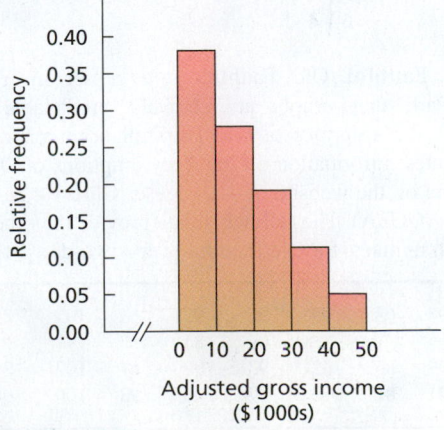

2.147 Cholesterol Levels. According to the *National Health and Nutrition Examination Survey*, published by the Centers for Disease Control and Prevention, the average cholesterol level for children between 4 and 19 years of age is 165 mg/dL. A pediatrician who tested the cholesterol levels of several young patients was alarmed to find that many had levels higher than 200 mg/dL. The following relative-frequency histogram shows the readings for some patients who had high cholesterol levels.

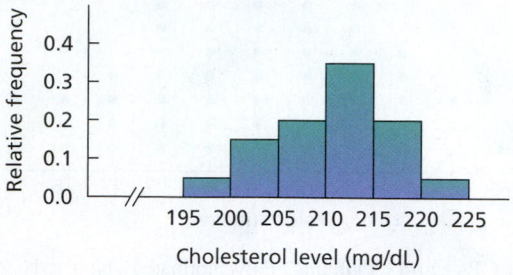

2.148 Sickle Cell Disease. A study published by E. Anionwu et al. in the *British Medical Journal* (Vol. 282, pp. 283–286) measured the steady-state hemoglobin levels of patients with three different types of sickle cell disease. Following is a stem-and-leaf diagram of the data.

```
 7 | 2 7
 8 | 0 1 1 3 4 4 5 6 7
 9 | 1 1 1 2 8
10 | 0 1 3 4 6 7 9
11 | 1 3 5 6 7 8 9
12 | 0 0 1 1 3 6 6
13 | 3 3 8 9
```

2.149 Stays in Europe and the Mediterranean. The Bureau of Economic Analysis gathers information on the length of stay in Europe and the Mediterranean by U.S. travelers. Data are published in *Survey of Current Business*. The following stem-and-leaf diagram portrays the length of stay, in days, of a sample of 36 U.S. residents who traveled to Europe and the Mediterranean last year.

```
0 | 1 1 1 2 3 3 3 5 5 6 8
1 | 0 0 0 1 2 2 2 3 4 5 6 7 8
2 | 0 1 1 1 7
3 | 1 2
4 | 1 4 8
5 | 6
6 | 4
```

2.150 Old Faithful. Old Faithful is a geyser in Yellowstone National Park that erupts at relatively predictable intervals. Since 2000, the eruptions of Old Faithfull occur on average every 90 minutes. Information on the daily eruptions of Old Faithful can be found on the website of The Geyser Observation and Study Association (GOSA). The following data provide the times between eruptions, in minutes, for Old Faithful over a two-day period.

96	68	99	97	92	61	105	89
95	71	93	104	85	107	91	97
99	86	102	93	99	64	102	86
101	81	117	87	106	93	106	106

a. Use cutpoint grouping with a first class of 60–under 70 to construct a frequency histogram of the times between eruptions.
b. Identify the shape of the distribution with regard to modality.
c. Identify the shape of the distribution with regard to symmetry (or nonsymmetry).
d. If the distribution is unimodal and nonsymmetric, classify it as either right skewed or left skewed.

2.151 Snow Goose Nests. In the article "Trophic Interaction Cycles in Tundra Ecosystems and the Impact of Climate Change" (*Bio-Science*, Vol. 55, No. 4, pp. 311–321), R. Ims and E. Fuglei provided an overview of animal species in the northern tundra. One threat to the snow goose in arctic Canada is the lemming. Snowy owls act as protection to the snow goose breeding grounds. For two years that are 3 years apart, the following graphs give relative frequency histograms of the distances, in meters, of snow goose nests to the nearest snowy owl nest.

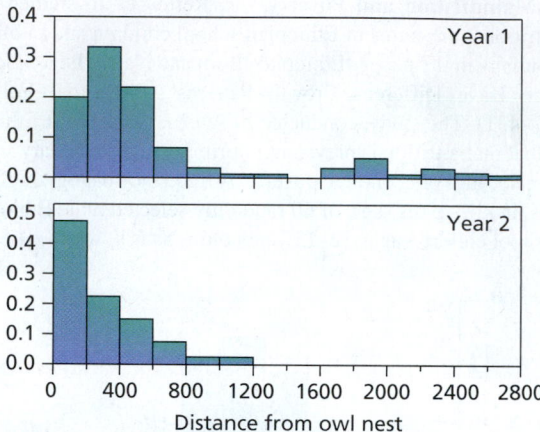

For each histogram, do the following:
a. Identify the shape of the distribution with regard to modality.
b. Identify the shape of the distribution with regard to symmetry (or nonsymmetry).
c. If the distribution is unimodal and nonsymmetric, classify it as either right skewed or left skewed.
d. Compare the two distributions.

Working with Large Data Sets

In each of Exercises 2.152–2.157,
a. use the technology of your choice to identify the modality and symmetry (or nonsymmetry) of the distribution of the data set.
b. if unimodal, classify the distribution as symmetric, right skewed, or left skewed.
Note: Answers may vary depending on the type of graph that you obtain for the data and on the technology that you use.

2.152 The Great White Shark. In an article titled "Great White, Deep Trouble" (*National Geographic*, Vol. 197(4), pp. 2–29), Peter Benchley—the author of JAWS—discussed various aspects of the Great White Shark (*Carcharodon carcharias*). Data on the number of pups borne in a lifetime by each of 80 Great White Shark females are given on the WeissStats site.

2.153 The Beatles. In the article, "Length of The Beatles' Songs" (*Chance*, Vol. 25, No. 1, pp. 30–33), T. Koyama discusses aspects and interpretations of the lengths of songs by The Beatles. Data on the length, in seconds, of 229 Beatles' songs are presented on the WeissStats site.

2.154 High School Completion. As reported by the U.S. Census Bureau in *Educational Attainment in the United States*, the percentage of adults in each state and the District of Columbia who have completed high school is provided on the WeissStats site.

2.155 Bachelor's Completion. As reported by the U.S. Census Bureau in the document *Educational Attainment in the United States*, the percentage of adults in each state and the District of Columbia who have completed a bachelor's degree is provided on the WeissStats site.

2.156 Body Temperature. A study by researchers at the University of Maryland addressed the question of whether the mean body temperature of humans is 98.6°F. The results of the study by P. Mackowiak et al. appeared in the article "A Critical Appraisal of 98.6°F, the Upper Limit of the Normal Body Temperature, and Other Legacies of Carl Reinhold August Wunderlich" (*Journal of the American Medical Association*, Vol. 268, pp. 1578–1580). Among other data, the researchers obtained the body temperatures of 93 healthy humans, as provided on the WeissStats site.

2.157 Forearm Length. In 1903, K. Pearson and A. Lee published the paper "On the Laws of Inheritance in Man. I. Inheritance of Physical Characters" (*Biometrika*, Vol. 2, pp. 357–462). The article examined and presented data on forearm length, in inches, for a sample of 140 men, which we present on the WeissStats site.

Extending the Concepts and Skills

2.158 Class Project: Number of Siblings. This exercise is a class project and works best in relatively large classes.
a. Determine the number of siblings for each student in the class.
b. Obtain a relative-frequency histogram for the number of siblings. Use single-value grouping.
c. Obtain a simple random sample of about one-third of the students in the class.
d. Find the number of siblings for each student in the sample.
e. Obtain a relative-frequency histogram for the number of siblings for the sample. Use single-value grouping.
f. Repeat parts (c)–(e) three more times.
g. Compare the histograms for the samples to each other and to that for the entire population. Relate your observations to Key Fact 2.1.

2.159 Class Project: Random Digits. This exercise can be done individually or, better yet, as a class project.

a. Use a table of random numbers or a random-number generator to obtain 50 random integers between 0 and 9.
b. Without graphing the distribution of the 50 numbers you obtained, guess its specific shape. Explain your reasoning.
c. Construct a relative-frequency histogram based on single-value grouping for the 50 numbers that you obtained in part (a). Is its shape about what you expected?
d. If your answer to part (c) was "no," provide an explanation.
e. What would you do to make getting a "yes" answer to part (c) more plausible?
f. If you are doing this exercise as a class project, repeat parts (a)–(c) for 1000 random integers.

Simulation. For purposes of both understanding and research, simulating variables is often useful. Simulating a variable involves the use of a computer or statistical calculator to generate observations of the variable. In Exercises 2.160 and 2.161, the use of simulation will enhance your understanding of distribution shapes and the relation between population and sample distributions.

2.160 Random Digits. In this exercise, use technology to work Exercise 2.159, as follows:
a. Use the technology of your choice to obtain 50 random integers between 0 and 9.
b. Use the technology of your choice to get a relative-frequency histogram based on single-value grouping for the numbers that you obtained in part (a).
c. Repeat parts (a) and (b) five more times.
d. Are the specific shapes of the distributions that you obtained in parts (a)–(c) about what you expected?
e. Repeat parts (a)–(d), but generate 1000 random integers each time instead of 50.

2.161 Standard Normal Distribution. One of the most important distributions in statistics is the *standard normal distribution*. We discuss this distribution in detail in Chapter 6.
a. Use the technology of your choice to generate a sample of 3000 observations from a variable that has the standard normal distribution.
b. Use the technology of your choice to get a relative-frequency histogram for the 3000 observations that you obtained in part (a).
c. Based on the histogram you obtained in part (b), what specific shape does the standard normal distribution have? Explain your reasoning.

2.5 Misleading Graphs*

Graphs and charts are frequently misleading, sometimes intentionally and sometimes inadvertently. Regardless of intent, we need to read and interpret graphs and charts with a great deal of care. In this section, we examine some misleading graphs and charts.

 **EXAMPLE 2.25** Truncated Graphs

Unemployment Rates The Bureau of Labor Statistics collects data on unemployment rates and publishes its finding in *Labor Force Statistics*. Figure 2.17(a) on the following page shows a bar chart from an article in a major metropolitan newspaper. The graph displays the unemployment rates in the United States from September of one year through March of the next year.

FIGURE 2.17

Unemployment rates: (a) truncated graph; (b) nontruncated graph

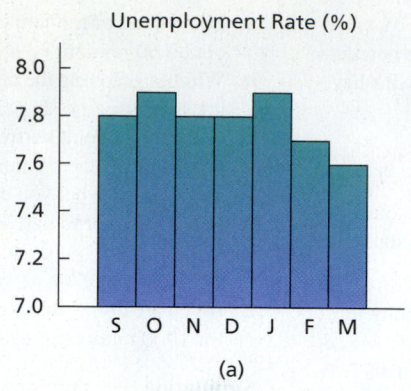
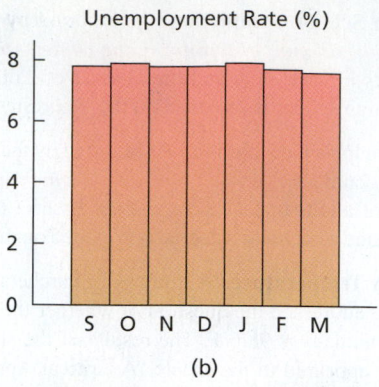

(a)   (b)

Because, in Fig. 2.17(a), the bar for March is about two-thirds as large as the bar for January, a quick look at that figure might lead you to conclude that the unemployment rate dropped by roughly one-third between January and March. In reality, however, the unemployment rate dropped by less than one twenty-fifth, from 7.9% to 7.6%. Let's analyze the graph more carefully to discover what it truly represents.

Figure 2.17(a) is an example of a **truncated graph** because the vertical axis, which should start at 0%, starts at 7% instead. Thus, the part of the graph from 0% to 7% has been cut off, or truncated. This truncation causes the bars to be out of proportion and, hence, creates a misleading impression.

Figure 2.17(b) is a nontruncated version of Fig. 2.17(a). Although the nontruncated version provides a correct graphical display, the "ups" and "downs" in the unemployment rates are not as easy to spot as they are in the truncated graph.

Truncated graphs have long been a target of statisticians, and many statistics books warn against their use. Nonetheless, as illustrated by Example 2.25, truncated graphs are still used today, even in reputable publications.

However, Example 2.25 also suggests that cutting off part of the vertical axis of a graph may allow relevant information to be conveyed more easily. In such cases, though, the illustrator should include a special symbol, such as //, to signify that the vertical axis has been modified.

The two graphs shown in Fig. 2.18 provide an excellent illustration. Both portray the number of new single-family homes sold per month over several months. The graph

FIGURE 2.18

New single-family home sales

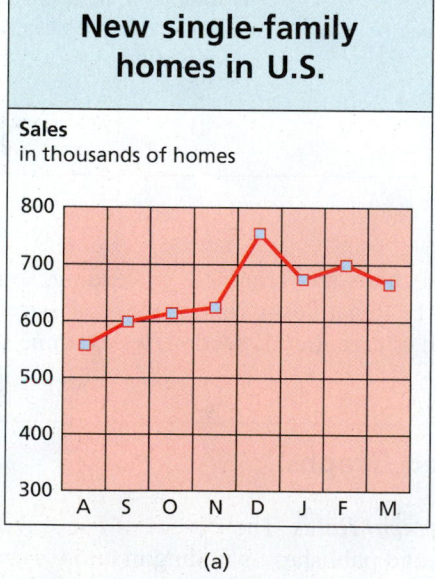
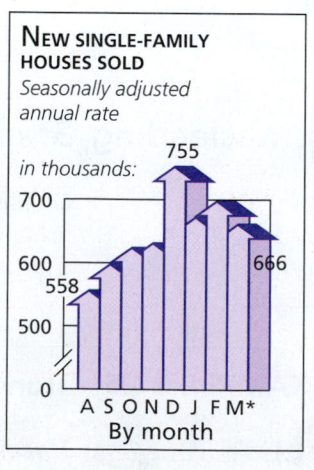

(a)   (b)

SOURCES: Data from U.S. Department of Commerce and U.S. Department of Housing and Urban Development

in Fig. 2.18(a) is truncated—most likely in an attempt to present a clear visual display of the variation in sales. The graph in Fig. 2.18(b) accomplishes the same result but is less subject to misinterpretation; you are aptly warned by the slashes that part of the vertical axis between 0 and 500 has been removed.

Improper Scaling

Misleading graphs and charts can also result from *improper scaling*.

 **EXAMPLE 2.26**   Improper Scaling

FIGURE 2.19
Pictogram for home building

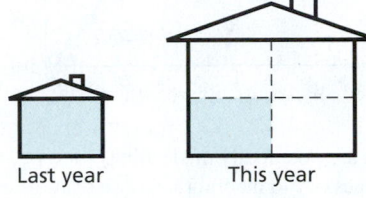

Last year   This year

Home Building A developer is preparing a brochure to attract investors for a new shopping center to be built in an area of Denver, Colorado. The area is growing rapidly; this year twice as many homes will be built there as last year. To illustrate that fact, the developer draws a **pictogram** (a symbol representing an object or concept by illustration), as shown in Fig. 2.19.

The house on the left represents the number of homes built last year. Because the number of homes that will be built this year is double the number built last year, the developer makes the house on the right twice as tall and twice as wide as the house on the left. However, this **improper scaling** gives the visual impression that four times as many homes will be built this year as last. Thus the developer's brochure may mislead the unwary investor. ◼

Graphs and charts can be misleading in countless ways besides the two that we discussed. Many more examples of misleading graphs can be found in the entertaining and classic book *How to Lie with Statistics* by Darrell Huff (New York: Norton, 1993). The main purpose of this section has been to show you to construct and read graphs and charts carefully.

Exercises 2.5

Understanding the Concepts and Skills

2.162 Give one reason why constructing and reading graphs and charts carefully is important.

2.163 This exercise deals with truncated graphs.
a. What is a truncated graph?
b. Give a legitimate motive for truncating the axis of a graph.
c. If you have a legitimate motive for truncating the axis of a graph, how can you correctly obtain that objective without creating the possibility of misinterpretation?

2.164 In a current newspaper or magazine, find two examples of graphs that might be misleading. Explain why you think the graphs are potentially misleading.

Applying the Concepts and Skills

2.165 Reading Skills. Each year the director of the reading program in a school district administers a standard test of reading skills. Then the director compares the average score for his district with the national average. Figure 2.20 was presented to the school board in the year 2013.

FIGURE 2.20
Average reading scores

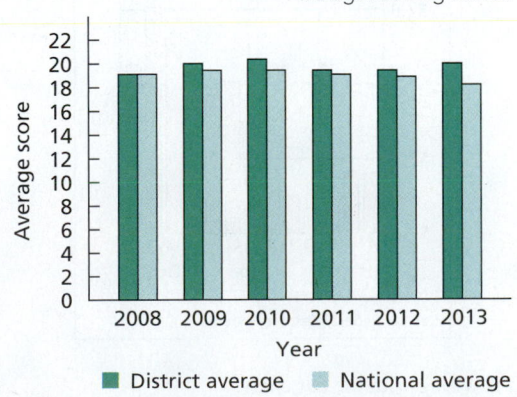

a. Obtain a truncated version of Fig. 2.20 by sliding a piece of paper over the bottom of the graph so that the bars start at 16.
b. Repeat part (a), but have the bars start at 18.
c. What misleading impression about the year 2013 scores is given by the truncated graphs obtained in parts (a) and (b)?

2.166 America's Melting Pot. The U.S. Census Bureau publishes data on the population of the United States by race and Hispanic origin in *American Community Survey*. From that document, we constructed the following bar chart. Note that people who are Hispanic may be of any race, and people in each race group may be either Hispanic or not Hispanic.

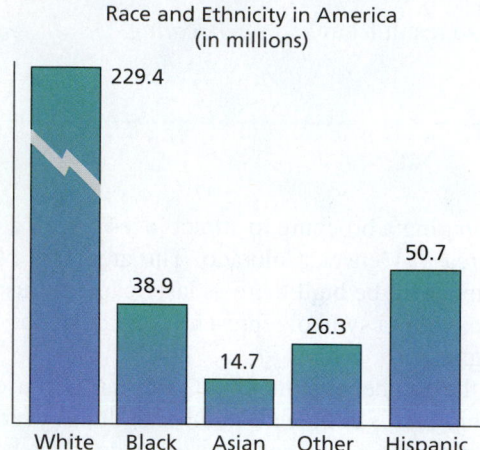

Race and Ethnicity in America
(in millions)

a. Explain why a break is shown in the first bar.
b. Why was the graph constructed with a broken bar?
c. Is this graph potentially misleading? Explain your answer.

2.167 M2 Money Supply. The Federal Reserve System publishes weekly figures of M2 money supply in the document *Money Stock Measures*. M2 includes such things as cash in circulation, deposits in checking accounts, nonbank traveler's checks, accounts such as savings deposits, and money-market mutual funds. The following bar chart provides data on the M2 money supply over 3 months during one year.

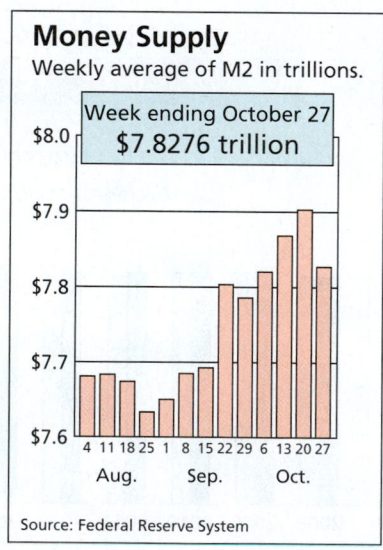

Money Supply
Weekly average of M2 in trillions.

Week ending October 27
$7.8276 trillion

Source: Federal Reserve System

a. What is wrong with the bar chart?
b. Construct a version of the bar chart with a nontruncated and unmodified vertical axis.
c. Construct a version of the bar chart in which the vertical axis is modified in an acceptable manner.

2.168 Drunk-Driving Fatalities. Drunk-driving fatalities represent the total number of people (occupants and non-occupants) killed in motor vehicle traffic crashes in which at least one driver had a blood alcohol content (BAC) of 0.08 or higher. The following graph is based on data published by the National Highway Traffic Safety Administration in *Federal Highway Administration Annual Highway Statistics*.

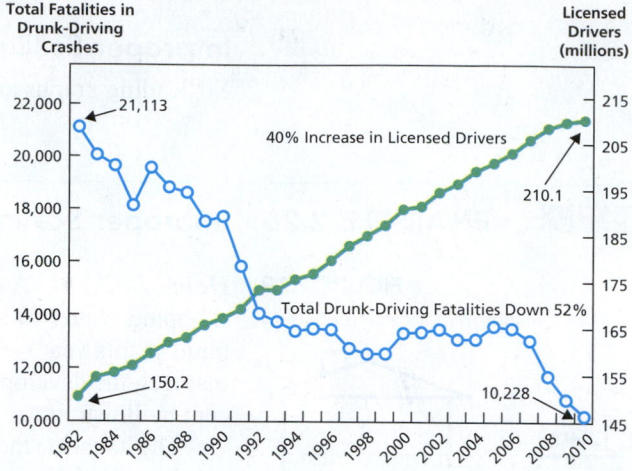

a. What features of the graph are potentially misleading?
b. Do you think that it was necessary to incorporate those features in order to display the data?
c. What could be done to more correctly display the data?

2.169 Happiest Age. An article, titled "We're Happiest at 74; It's All Downhill till 40, Then Life Gets Better, Say Scientists," included a graph, similar to the following one, of a happiness scale versus age. The article argues that happiness decreases until the age of 46 and then increases again, and that people are happiest when they hit 74. Notice the horizontal axis as well as the vertical axis. [SOURCE: Graph from *Daily Mail*. Copyright © 2010 by Associated Newspapers Ltd. Used by permission of Associated Newspapers Ltd.]

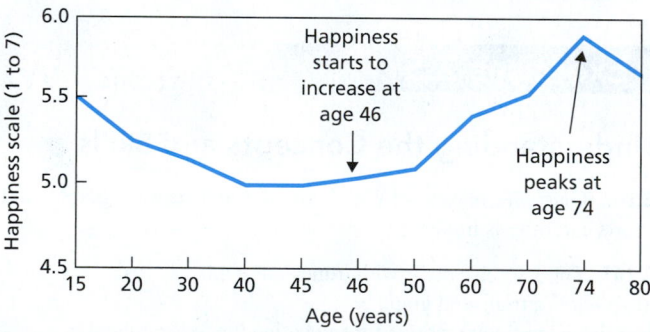

a. Cover the numbers on the vertical axis of the graph with a piece of paper.
b. What impression does the graph convey regarding the percentage drop in happiness between the ages of 15 and 20?
c. Now remove the piece of paper from the graph. Use the vertical scale to find the actual percentage drop in happiness between the ages of 15 and 20.
d. Why is the graph potentially misleading?
e. What can be done to make the graph less potentially misleading?

Extending the Concepts and Skills

2.170 Home Building. Refer to Example 2.26 on page 85. Suggest a way in which the developer can accurately illustrate that twice as many homes will be built in the area this year as last.

2.171 Marketing Golf Balls. A golf ball manufacturer has determined that a newly developed process results in a ball that lasts roughly twice as long as a ball produced by the current process. To illustrate this advance graphically, she designs a brochure showing a "new" ball having twice the radius of the "old" ball.

a. What is wrong with this depiction?

b. How can the manufacturer accurately illustrate the fact that the "new" ball lasts twice as long as the "old" ball?

Old ball   New ball

CHAPTER IN REVIEW

You Should Be Able to

1. classify variables and data as either qualitative or quantitative.

2. distinguish between discrete and continuous variables and between discrete and continuous data.

3. construct a frequency distribution and a relative-frequency distribution for qualitative data.

4. draw a pie chart and a bar chart.

5. group quantitative data into classes using single-value grouping, limit grouping, or cutpoint grouping.

6. identify terms associated with the grouping of quantitative data.

7. construct a frequency distribution and a relative-frequency distribution for quantitative data.

8. construct a frequency histogram and a relative-frequency histogram.

9. construct a dotplot.

10. construct a stem-and-leaf diagram.

11. identify the modality and symmetry (or nonsymmetry) of the distribution of a data set.

12. specify whether a unimodal distribution is symmetric, right skewed, or left skewed.

13. understand the relationship between sample distributions and the population distribution (distribution of the variable under consideration).

14. identify and correct misleading graphs.*

Key Terms

bar chart, *45*
bell shaped, *77*
bimodal, *76*
bins, *52*
categorical variable, *37*
categories, *52*
census data, *78*
class cutpoints, *54*
class limits, *53*
class mark, *54*
class midpoint, *55*
class width, *54, 55*
classes, *52*
continuous data, *38*
continuous variable, *38*
count, *42*
cutpoint grouping, *54*
data, *38*
data set, *38*
discrete data, *38*
discrete variable, *38*
distribution of a data set, *75*
distribution of a variable, *78*
dotplot, *59*

exploratory data analysis, *36*
frequency, *42*
frequency distribution, *42*
frequency histogram, *56*
histogram, *56*
improper scaling,* *85*
leaf, *60*
left skewed, *77*
limit grouping, *53*
lower class cutpoint, *55*
lower class limit, *54*
modality, *76*
multimodal, *76*
observation, *38*
percent histogram, *56*
percentage, *43*
pictogram, *85*
pie chart, *45*
population data, *78*
population distribution, *78*
qualitative data, *38*
qualitative variable, *38*
quantitative data, *38*
quantitative variable, *38*

rectangular, *77*
relative frequency, *43*
relative-frequency distribution, *43*
relative-frequency histogram, *56*
reverse J shaped, *77*
right skewed, *77*
sample data, *78*
sample distribution, *78*
single-value classes, *52*
single-value grouping, *52*
skewness, *77*
stem, *60*
stem-and-leaf diagram, *60*
stemplot, *60*
symmetric, *76*
symmetry, *76*
triangular, *77*
truncated graph,* *84*
uniform, *77*
unimodal, *76*
upper class cutpoint, *55*
upper class limit, *54*
variable, *38*

REVIEW PROBLEMS

Understanding the Concepts and Skills

1. This problem is about variables.
a. What is a variable?
b. Identify two main types of variables.
c. Identify the two types of quantitative variables.

2. This problem is about data.
a. What are data?
b. How is data type determined?

3. For a qualitative data set, what is a
a. frequency distribution?
b. relative-frequency distribution?

4. What is the relationship between a frequency or relative-frequency distribution of a quantitative data set and that of a qualitative data set?

5. Identify two main types of graphical displays that are used for qualitative data.

6. In a bar chart, unlike in a histogram, the bars do not abut. Give a possible reason for that.

7. Some users of statistics prefer pie charts to bar charts because people are accustomed to having the horizontal axis of a graph show order. For example, someone might infer from Fig. 2.3 on page 46 that "Republican" is less than "Other" because "Republican" is shown to the left of "Other" on the horizontal axis. Pie charts do not lead to such inferences. Give other advantages and disadvantages of each method.

8. When is the use of single-value grouping particularly appropriate?

9. A quantitative data set has been grouped by using limit grouping with equal-width classes. The lower and upper limits of the first class are 3 and 8, respectively, and the class width is 6.
a. What is the class mark of the second class?
b. What are the lower and upper limits of the third class?
c. Which class would contain an observation of 23?

10. A quantitative data set has been grouped by using limit grouping with equal-width classes of width 5. The class limits are whole numbers.
a. If the class mark of the first class is 8, what are its lower and upper limits?
b. What is the class mark of the second class?
c. What are the lower and upper limits of the third class?
d. Which class would contain an observation of 28?

11. A quantitative data set has been grouped by using cutpoint grouping with equal-width classes.
a. If the lower and upper cutpoints of the first class are 5 and 15, respectively, what is the common class width?
b. What is the midpoint of the second class?
c. What are the lower and upper cutpoints of the third class?
d. Which class would contain an observation of 32.4?

12. A quantitative data set has been grouped by using cutpoint grouping with equal-width classes of width 8.
a. If the midpoint of the first class is 10, what are its lower and upper cutpoints?

b. What is the class midpoint of the second class?
c. What are the lower and upper cutpoints of the third class?
d. Which class would contain an observation of 22?

13. Explain the relative positioning of the bars in a histogram to the numbers that label the horizontal axis when each of the following quantities is used to label that axis.
a. Lower class limits
b. Lower class cutpoints
c. Class marks
d. Class midpoints

14. Sketch the curve corresponding to each of the following specific distribution shapes.
a. Bell shaped   **b.** Triangular
c. Reverse J shaped   **d.** Uniform

15. Draw a smooth curve that represents a symmetric trimodal (three-peak) distribution.

16. A variable of a population has a left-skewed distribution.
a. If a large simple random sample is taken from the population, roughly what shape will the distribution of the sample have? Explain your answer.
b. If two simple random samples are taken from the population, would you expect the two sample distributions to have identical shapes? Explain your answer.
c. If two simple random samples are taken from the population, would you expect the two sample distributions to have similar shapes? If so, what shape would that be? Explain your answers.

Applying the Concepts and Skills

17. Largest Hydroelectric Plants. According to Wikipedia, the world's five largest hydroelectric plants, based on installed capacity, are as shown in the following table. Capacities are in megawatts.

Rank	Name	Country	Capacity
1	Three Gorges Dam	China	22,500
2	Itaipu Dam	Brazil/Paraguay	14,000
3	Guri	Venezuela	10,200
4	Tucuruí	Brazil	8,370
5	Grand Coulee	United States	6,809

a. What type of data is given in the first column of the table?
b. What type of data is given in the fourth column?
c. What type of data is given in the third column?

18. DVD Players. Refer to Example 2.16 on page 60.
a. Explain why a frequency histogram of the DVD prices with single-value classes would be essentially identical to the dotplot shown in Fig. 2.7.
b. Would the dotplot and a frequency histogram be essentially identical with other than single-value classes? Explain your answer.

19. Inauguration Ages. From the *Information Please Almanac*, we obtained the ages at inauguration for the first 44 presidents of the United States (from George Washington to Barack H. Obama).

President	Age at inaug.	President	Age at inaug.
G. Washington	57	B. Harrison	55
J. Adams	61	G. Cleveland	55
T. Jefferson	57	W. McKinley	54
J. Madison	57	T. Roosevelt	42
J. Monroe	58	W. Taft	51
J. Q. Adams	57	W. Wilson	56
A. Jackson	61	W. Harding	55
M. Van Buren	54	C. Coolidge	51
W. Harrison	68	H. Hoover	54
J. Tyler	51	F. Roosevelt	51
J. Polk	49	H. Truman	60
Z. Taylor	64	D. Eisenhower	62
M. Fillmore	50	J. Kennedy	43
F. Pierce	48	L. Johnson	55
J. Buchanan	65	R. Nixon	56
A. Lincoln	52	G. Ford	61
A. Johnson	56	J. Carter	52
U. Grant	46	R. Reagan	69
R. Hayes	54	G. Bush	64
J. Garfield	49	W. Clinton	46
C. Arthur	50	G. W. Bush	54
G. Cleveland	47	B. Obama	47

a. Identify the classes for grouping these data, using limit grouping with classes of equal width 5 and a first class of 40–44.

b. Identify the class marks of the classes found in part (a).

c. Construct frequency and relative-frequency distributions of the inauguration ages based on your classes obtained in part (a).

d. Draw a frequency histogram for the inauguration ages based on your grouping in part (a).

e. Classify the shape of the distribution of inauguration ages for the first 44 presidents of the United States both generally and specifically.

20. Inauguration Ages. Refer to Problem 19. Construct a dotplot for the ages at inauguration of the first 44 presidents of the United States.

21. Inauguration Ages. Refer to Problem 19. Construct a stem-and-leaf diagram for the inauguration ages of the first 44 presidents of the United States.

a. Use one line per stem.

b. Use two lines per stem.

c. Which of the two stem-and-leaf diagrams that you just constructed corresponds to the frequency distribution of Problem 19(c)?

22. Battery Lifetime. In the article "Comparing the lifetime of two brands of batteries" (*Journal of Statistics Education*, Vol. 21, No. 1, pp. 1–19) by P. Dunn, two brands of AA alkaline batteries were compared. The following table gives the number of pulses required for a sample of 18 Energizer and Ultracell batteries to reach a voltage level of 1.3.

37	57	37	46	56	44
47	54	58	56	38	57
57	34	57	60	54	40

a. Construct a stem-and-leaf diagram for these data with one line per stem.

b. Construct a stem-and-leaf diagram with two lines per stem.

c. Which stem-and-leaf diagram do you find more useful? Why?

23. Busy Bank Tellers. The Prescott National Bank has six tellers available to serve customers. The data in the following table provide the number of busy tellers observed during 25 spot checks.

6	5	4	1	5
6	1	5	5	5
3	5	2	4	3
4	5	0	6	4
3	4	2	3	6

a. Use single-value grouping to organize these data into frequency and relative-frequency distributions.

b. Draw a relative-frequency histogram for the data based on the grouping in part (a).

c. Identify the modality of the distribution of these numbers of busy tellers.

d. State whether the distribution is (roughly) symmetric, right skewed, or left skewed.

e. Construct a dotplot for the data on the number of busy tellers.

f. Compare the dotplot that you obtained in part (e) to the relative-frequency histogram that you drew in part (b).

24. On-Time Arrivals. The *Air Travel Consumer Report* is a monthly product of the Department of Transportation's Office of Aviation Enforcement and Proceedings. The report is designed to assist consumers with information on the quality of services provided by the airlines. Following are the percentages of on-time arrivals for June 2013 by the 16 reporting airlines.

93.1	87.5	77.9	77.3
76.1	73.8	73.0	72.5
71.1	70.6	69.8	69.6
69.0	66.0	65.9	61.8

a. Identify the classes for grouping these data, using cutpoint grouping with classes of equal width 5 and a first lower class cutpoint of 60.

b. Identify the class midpoints of the classes found in part (a).

c. Construct frequency and relative-frequency distributions of the data based on your classes from part (a).

d. Draw a frequency histogram of the data based on your classes from part (a).

e. Round each observation to the nearest whole number, and then construct a stem-and-leaf diagram with two lines per stem.

f. Obtain the greatest integer in each observation, and then construct a stem-and-leaf diagram with two lines per stem.

g. Which of the stem-and-leaf diagrams in parts (e) and (f) corresponds to the frequency histogram in part (d)? Explain why.

25. Old Ballplayers. From the ESPN Web site, we obtained the age of the oldest player on each of the major league baseball teams during one season. Here are the data.

33	37	36	40	36	36
40	36	37	36	40	42
37	42	38	39	35	37
40	44	39	40	46	38
37	40	37	42	41	41

a. Construct a dotplot for these data.

b. Use your dotplot from part (a) to identify the modality and symmetry (or nonsymmetry) of the distribution of these ages.

26. Latent Fingerprints. Firearms, live ammunition, and spent cartridge casings are often submitted to crime laboratories to be processed for latent fingerprints. B. Maldonado explored the chances of successfully recovering fingerprints in the article, "Study on Developing Latent Fingerprints on Firearm Evidence" (*Journal of Forensic Identification*, Vol. 62, Issue 5, pp. 425–429). The following table provides a frequency distribution for the type of evidence submitted to crime laboratories over a 2-year period at the Denver Police Department.

Evidence	Frequency
Firearm	289
Magazine	161
Live Cartridge	2727
Spent Cartridge Casing	259

a. Obtain a relative-frequency distribution.

b. Draw a pie chart.

c. Construct a bar chart.

d. Interpret your results.

27. U.S. Divisions. The U.S. Census Bureau divides the states in the United States into nine divisions: East North Central (ENC), East South Central (ESC), Middle Atlantic (MAC), Mountain (MTN), New England (NED), Pacific (PAC), South Atlantic (SAC), West North Central (WNC), and West South Central (WSC). The following table gives the divisions of each of the 50 states.

ESC	PAC	MTN	WSC	PAC	MTN	NED	SAC	SAC	SAC
PAC	MTN	ENC	ENC	WNC	WNC	ESC	WSC	NED	SAC
NED	ENC	WNC	ESC	WNC	MTN	WNC	MTN	NED	MAC
MTN	MAC	SAC	WNC	ENC	WSC	PAC	MAC	NED	SAC
WNC	ESC	WSC	MTN	NED	SAC	PAC	SAC	ENC	MTN

a. Identify the population and variable under consideration.

b. Obtain both a frequency distribution and a relative-frequency distribution of the divisions.

c. Draw a pie chart of the divisions.

d. Construct a bar chart of the divisions.

e. Interpret your results.

*28. Clean Fossil Fuels. In the article, "Squeaky Clean Fossil Fuels" (*New Scientist*, Vol. 186, No. 2497, p. 26), F. Pearce reported on the benefits of using clean fossil fuels that release no carbon dioxide (CO_2), helping to reduce the threat of global warming. One technique of slowing down global warming caused by CO_2 is to bury the CO_2 underground in old oil or gas wells, coal mines, or porous rocks filled with salt water. Global estimates are that 11,000 billion tonnes of CO_2 could be disposed of underground, several times more than the likely emissions of CO_2 from burning fossil fuels in the coming century. This could give the world extra time to give up its reliance on fossil fuels. The following bar chart shows the distribution of space available to bury CO_2 gas underground.

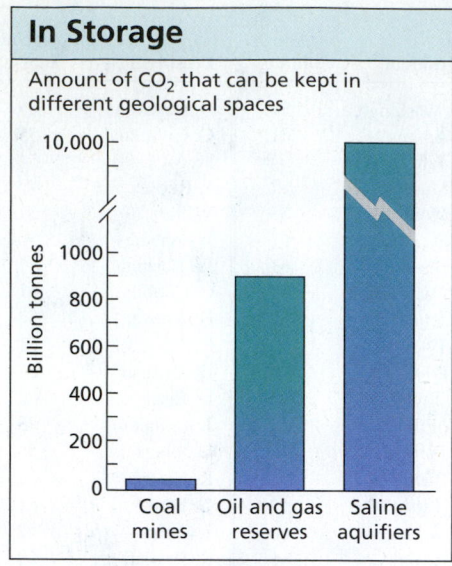

a. Explain why the break is found in the third bar.

b. Why was the graph constructed with a broken bar?

*29. Reshaping the Labor Force. The following graph is based on one that appeared in an *Arizona Republic* newspaper article entitled "Hand That Rocked Cradle Turns to Work as Women Reshape U.S. Labor Force." The graph depicts the labor force participation rates for the years 1960, 1980, and 2000.

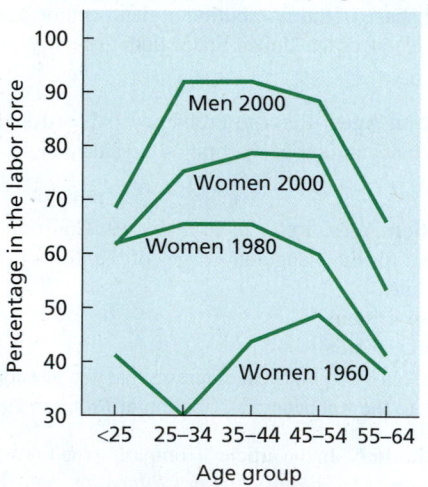

a. Cover the numbers on the vertical axis of the graph with a piece of paper.

b. Look at the 1960 and 2000 graphs for women, focusing on the 35- to 44-year-old age group. What impression does the graph convey regarding the ratio of the percentages of women in the labor force for 1960 and 2000?

c. Now remove the piece of paper from the graph. Use the vertical scale to find the actual ratio of the percentages of 35- to 44-year-old women in the labor force for 1960 and 2000.

d. Why is the graph potentially misleading?

e. What can be done to make the graph less potentially misleading?

Working with Large Data Sets

30. Hair and Eye Color. In the article "Graphical Display of Two-Way Contingency Tables" (*The American Statistician*, Vol. 28, No. 1, pp. 9–12), R. Snee presented data on hair color and eye color among 592 students in an elementary statistics course at the University of Delaware. Raw data for that information are presented on the WeissStats site. Use the technology of your choice to do the following tasks, and interpret your results.
a. Obtain both a frequency distribution and a relative-frequency distribution for the hair-color data.
b. Get a pie chart of the hair-color data.
c. Determine a bar chart of the hair-color data.
d. Repeat parts (a)–(c) for the eye-color data.

In Problems 31–33,
a. *identify the population and variable under consideration.*
b. *use the technology of your choice to obtain and interpret a frequency histogram, a relative-frequency histogram, or a percent histogram of the data.*
c. *use the technology of your choice to obtain a dotplot of the data.*
d. *use the technology of your choice to obtain a stem-and-leaf diagram of the data.*
e. *identify the modality and symmetry (or nonsymmetry) of the distribution.*
f. *if the distribution is unimodal and nonsymmetric, state whether it is right skewed or left skewed.*

31. Agricultural Exports. The U.S. Department of Agriculture collects data pertaining to the value of agricultural exports and publishes its findings in *U.S. Agricultural Trade Update*. For one year, the values of these exports, by state, are provided on the WeissStats site. Data are in millions of dollars.

32. Life Expectancy by Country. Life expectancy is the average number of years to be lived by a group of people born in the same year if mortality at each age remains constant in the future. From the *World FactBook*, published by the Central Intelligence Agency (CIA), we obtained the life expectancy for people in various countries. Those data are presented on the WeissStats site.

33. High and Low Temperatures. The U.S. National Oceanic and Atmospheric Administration publishes temperature data in *Climatography of the United States*. According to that document, the annual average maximum and minimum temperatures for selected cities in the United States are as provided on the WeissStats site. [*Note*: Do parts (a)–(f) for both the maximum and minimum temperatures.]

FOCUSING ON DATA ANALYSIS

UWEC UNDERGRADUATES

Recall from Chapter 1 (see page 34) that the Focus database and Focus sample contain information on the undergraduate students at the University of Wisconsin - Eau Claire (UWEC). Now would be a good time for you to review the discussion about these data sets.

a. For each of the following variables, make an educated guess at its distribution shape in terms of modality, symmetry, and skewness: high school percentile, cumulative GPA, age, ACT English score, ACT math score, and ACT composite score.
b. Open the Focus sample (FocusSample) in the statistical software package of your choice and then obtain and interpret histograms for each of the samples corresponding to the variables in part (a). Compare your results with the educated guesses that you made in part (a).

c. If your statistical software package will accommodate the entire Focus database (Focus), open that worksheet and then obtain and interpret histograms for each of the variables in part (a). Compare your results with the educated guesses that you made in part (a). Also discuss and explain the relationship between the histograms that you obtained in this part and those that you obtained in part (b).
d. Open the Focus sample and then determine and interpret pie charts and bar charts of the samples for the variables sex, classification, residency, and admission type.
e. If your statistical software package will accommodate the entire Focus database, open that worksheet and then obtain and interpret pie charts and bar charts for each of the variables in part (d). Also discuss and explain the relationship between the pie charts and bar charts that you obtained in this part and those that you obtained in part (d).

CASE STUDY DISCUSSION

WORLD'S RICHEST PEOPLE

Recall that, each year, *Forbes* magazine publishes a list of the world's richest people. On page 37, we constructed a table, based on a 2013 *Forbes* article, that shows the 25 richest people in the world, as of March of that year.

a. For each of the five columns of the table, classify the data as either qualitative or quantitative; if quantitative, further classify it as discrete or continuous. Also, identify the variable under consideration in each case.

b. Obtain frequency and relative-frequency distributions of the citizenship data.
c. Construct and interpret a pie chart for the citizenship data.
d. Construct and interpret a bar chart for the citizenship data.
e. Use limit grouping, with grouping by 10s, to organize the age data into frequency and relative-frequency distributions.
f. Construct frequency and relative-frequency histograms of the age data based on your grouping in part (e).

g. Identify and interpret the (rough) shape of your histograms in part (f) in terms of modality, symmetry, and skewness.

h. Obtain stem-and-leaf diagrams of the age data, using both one and two lines per stem. Which of the two stem-and-leaf diagrams corresponds to your frequency histogram in part (f)?

i. Construct a dotplot of the age data.

j. Use cutpoint grouping to organize the wealth data into frequency and relative-frequency distributions. Employ a class width of 5 and a first cutpoint of 20.

k. Construct frequency and relative-frequency histograms of the wealth data based on your grouping in part (j).

l. Identify and interpret the shape of your histograms in part (k) in terms of modality, symmetry, and skewness.

m. Truncate each wealth to a whole number (i.e., find the greatest integer in each wealth), and then obtain a stem-and-leaf diagram of the resulting data, using two lines per stem.

n. Round each wealth to a whole number, and then obtain a dotplot of the resulting data.

BIOGRAPHY

ADOLPHE QUETELET: ON "THE AVERAGE MAN"

Lambert Adolphe Jacques Quetelet was born in Ghent, Belgium, on February 22, 1796. He attended school locally and, in 1819, received the first doctorate of science degree granted at the newly established University of Ghent. In that same year, he obtained a position as a professor of mathematics at the Brussels Athenaeum.

Quetelet was elected to the Belgian Royal Academy in 1820 and served as its secretary from 1834 until his death in 1874. He was founder and director of the Royal Observatory in Brussels, founder and a major contributor to the journal *Correspondance Mathématique et Physique*, and, according to Stephen M. Stigler in *The History of Statistics*, was "...active in the founding of more statistical organizations than any other individual in the nineteenth

century." Among the organizations he established was the International Statistical Congress, initiated in 1853.

In 1835, Quetelet wrote a two-volume set titled *A Treatise on Man and the Development of His Faculties,* the publication in which he introduced his concept of the "average man" and that firmly established his international reputation as a statistician and sociologist. A review in the *Athenaeum* stated, "We consider the appearance of these volumes as forming an epoch in the literary history of civilization."

In 1855, Quetelet suffered a stroke that limited his work but not his popularity. He died on February 17, 1874. His funeral was attended by royalty and famous scientists from around the world. A monument to his memory was erected in Brussels in 1880.

Descriptive Measures

CHAPTER OBJECTIVES

In Chapter 2, you began your study of descriptive statistics. There you learned how to organize data into tables and summarize data with graphs.

Another method of summarizing data is to compute numbers, such as averages and percentiles, that describe the data set. Numbers that are used to describe data sets are called **descriptive measures.** In this chapter, we continue our discussion of descriptive statistics by examining some of the most commonly used descriptive measures.

In Section 3.1, we present *measures of center*—descriptive measures that indicate the center, or most typical value, in a data set. Next, in Sections 3.2 and 3.3, we examine *measures of variation*—descriptive measures that indicate the amount of variation or spread in a data set.

The five-number summary, which we discuss in Section 3.4, includes descriptive measures that can be used to obtain both measures of center and measures of variation. That summary also provides the basis for a widely used graphical display, the boxplot.

In Section 3.5, we examine descriptive measures of populations. We also illustrate how sample data can be used to provide estimates of descriptive measures of populations when census data are unavailable.

CASE STUDY

The Beatles' Song Length

In the article "Length of The Beatles' Songs" (*Chance*, Vol. 25, No. 1, pp. 30–33), T. Koyama discusses aspects and interpretations of the lengths of songs by The Beatles. In this case study, we concentrate on the 175 songs that appear in The Beatles' 12 studio record albums.

The following table provides some basic information about the studio albums. At the end of this chapter, you will find the data on album name and length, in seconds, of each song in an album.

As noted by Professor Koyama, the B-side of Yellow Submarine consists of six instrumental songs composed by George Martin and another song credited to Lennon and McCartney and arranged by George Martin, none of which are considered Beatles' songs.

Album	Release date	Number of songs
Please Please Me	03/22/1963	14
With The Beatles	11/22/1963	14
A Hard Day's Night	06/26/1964	13
Beatles for Sale	12/04/1964	14
Help!	08/06/1965	14
Rubber Soul	12/03/1965	14
Revolver	08/05/1966	14
Sgt. Pepper's Lonely Hearts Club Band	06/01/1967	13
The Beatles (White Album)	11/22/1968	30
Yellow Submarine	01/17/1969	6
Abbey Road	09/26/1969	17
Let It Be	05/08/1970	12

In this chapter, we demonstrate several additional techniques to help you analyze data. At the end of the chapter, you will apply those techniques to analyze the lengths of The Beatles' studio-album songs.

3.1   Measures of Center

Descriptive measures that indicate where the center or most typical value of a data set lies are called **measures of central tendency** or, more simply, **measures of center**. Measures of center are often called *averages*.

In this section, we discuss the three most important measures of center: the *mean, median,* and *mode.* The mean and median apply only to quantitative data, whereas the mode can be used with either quantitative or qualitative (categorical) data.

The Mean

The most commonly used measure of center is the *mean.* When people speak of taking an average, they are most often referring to the mean.

DEFINITION 3.1

What Does It Mean?

The mean of a data set is its arithmetic average.

Mean of a Data Set

The **mean** of a data set is the sum of the observations divided by the number of observations.

EXAMPLE 3.1   The Mean

TABLE 3.1
Data Set I

$300	300	300	940	300
300	400	300	400	
450	800	450	1050	

TABLE 3.2
Data Set II

$300	300	940	450	400
400	300	300	1050	300

Weekly Salaries Professor Hassett spent one summer working for a small mathematical consulting firm. The firm employed a few senior consultants, who made between $800 and $1050 per week; a few junior consultants, who made between $400 and $450 per week; and several clerical workers, who made $300 per week.

The firm required more employees during the first half of the summer than the second half. Tables 3.1 and 3.2 list typical weekly earnings for the two halves of the summer. Find the mean of each of the two data sets.

Solution As we see from Table 3.1, Data Set I has 13 observations. The sum of those observations is $6290, so

$$\text{Mean of Data Set I} = \frac{\$6290}{13} = \$483.85 \text{ (rounded to the nearest cent)}.$$

StatCrunch

Report 3.1

You try it!

Exercise 3.19(a) on page 101

Similarly,

$$\text{Mean of Data Set II} = \frac{\$4740}{10} = \$474.00.$$

Interpretation The employees who worked in the first half of the summer earned more, on average (a mean salary of $483.85), than those who worked in the second half (a mean salary of $474.00).

The Median

Another frequently used measure of center is the median. Essentially, the *median* of a data set is the number that divides the bottom 50% of the data from the top 50%. A more precise definition of the median follows.

DEFINITION 3.2

? What Does It Mean?

The median of a data set is the middle value in its ordered list.

Median of a Data Set

Arrange the data in increasing order.

- If the number of observations is odd, then the **median** is the observation exactly in the middle of the ordered list.
- If the number of observations is even, then the **median** is the mean of the two middle observations in the ordered list.

In both cases, if we let n denote the number of observations, then the median is at position $(n + 1)/2$ in the ordered list.

■■■ **EXAMPLE 3.2** **The Median**

Weekly Salaries Consider again the two sets of salary data shown in Tables 3.1 and 3.2. Determine the median of each of the two data sets.

Solution To find the median of Data Set I, we first arrange the data in increasing order:

300 300 300 300 300 300 **400** 400 450 450 800 940 1050

The number of observations is 13, so $(n + 1)/2 = (13 + 1)/2 = 7$. Consequently, the median is the seventh observation in the ordered list, which is 400 (shown in boldface).

To find the median of Data Set II, we first arrange the data in increasing order:

300 300 300 300 **300** **400** 400 450 940 1050

The number of observations is 10, so $(n + 1)/2 = (10 + 1)/2 = 5.5$. Consequently, the median is halfway between the fifth and sixth observations (shown in boldface) in the ordered list, which is 350.

StatCrunch

Report 3.2

You try it!

Exercise 3.19(b) on page 101

Interpretation Again, the analysis shows that the employees who worked in the first half of the summer tended to earn more (a median salary of $400) than those who worked in the second half (a median salary of $350).

To determine the median of a data set, you must first arrange the data in increasing order. Constructing a stem-and-leaf diagram as a preliminary step to ordering the data is often helpful.

The Mode

The final measure of center that we discuss here is the *mode*.

DEFINITION 3.3

What Does It Mean?

The mode of a data set is its most frequently occurring value.

Mode of a Data Set

Find the frequency of each value in the data set.

- If no value occurs more than once, then the data set has *no mode*.
- Otherwise, any value that occurs with the greatest frequency is a **mode** of the data set.

EXAMPLE 3.3   The Mode

TABLE 3.3

Frequency distribution for Data Set I

Salary	Frequency
300	6
400	2
450	2
800	1
940	1
1050	1

StatCrunch

You try it!

Report 3.3

Exercise 3.19(c)
on page 101

Weekly Salaries   Determine the mode(s) of each of the two sets of salary data given in Tables 3.1 and 3.2 on page 94.

Solution   Referring to Table 3.1, we obtain the frequency of each value in Data Set I, as shown in Table 3.3. From Table 3.3, we see that the greatest frequency is 6, and that 300 is the only value that occurs with that frequency. So the mode is $300.

Proceeding in the same way, we find that, for Data Set II, the greatest frequency is 5 and that 300 is the only value that occurs with that frequency. So the mode is $300.

Interpretation   The most frequent salary was $300 both for the employees who worked in the first half of the summer and those who worked in the second half.   ■

A data set will have more than one mode if more than one of its values occurs with the greatest frequency. For instance, suppose the first two $300-per-week employees who worked in the first half of the summer were promoted to $400-per-week jobs. Then the weekly earnings for the 13 employees would be as follows.

$400	400	300	940	300
300	400	300	400	
450	800	450	1050	

Now, both the value 300 and the value 400 would occur with greatest frequency, 4. This new data set would thus have two modes, 300 and 400.

Comparison of the Mean, Median, and Mode

The mean, median, and mode of a data set are often different. Table 3.4 summarizes the definitions of these three measures of center and gives their values for Data Set I and Data Set II, which we computed in Examples 3.1–3.3.

In both Data Sets I and II, the mean is larger than the median. The reason is that the mean is strongly affected by the few large salaries in each data set. In general, the mean is sensitive to extreme (very large or very small) observations, whereas the median is not. Consequently, when the choice for the measure of center is between the

TABLE 3.4

Means, medians, and modes of salaries in Data Set I and Data Set II

Measure of center	Definition	Data Set I	Data Set II
Mean	$\dfrac{\text{Sum of observations}}{\text{Number of observations}}$	$483.85	$474.00
Median	Middle value in ordered list	$400.00	$350.00
Mode	Most frequent value	$300.00	$300.00

mean and the median, the median is usually preferred for data sets that have extreme observations.

Figure 3.1 shows the relative positions of the mean and median for archetypal right-skewed, symmetric, and left-skewed distributions. Generally, the mean and median are exactly equal for an exactly symmetric distribution and are approximately equal for a roughly symmetric distribution. The mean is *usually* greater than the median for a right-skewed distribution and is *usually* less than the median for a left-skewed distribution; the most common exceptions occur for discrete variables with only a few possible values.

FIGURE 3.1

Relative positions of the mean and median for archetypal (a) right-skewed, (b) symmetric, and (c) left-skewed distributions

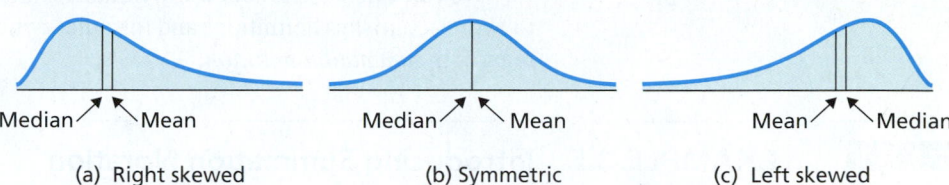

A **resistant measure** is not sensitive to the influence of a few extreme observations. The median is a resistant measure of center, but the mean is not. A *trimmed mean* can improve the resistance of the mean: removing a percentage of the smallest and largest observations before computing the mean gives a **trimmed mean.** In Exercise 3.54, we discuss trimmed means in more detail.

APPLET

Applet 3.1

The mode for each of Data Sets I and II differs from both the mean and the median. Whereas the mean and the median are aimed at finding the center of a data set, the mode is really not—the value that occurs most frequently may not be near the center.

It should now be clear that the mean, median, and mode generally provide different information. There is no simple rule for deciding which measure of center to use in a given situation. Even experts may disagree about the most suitable measure of center for a particular data set.

EXAMPLE 3.4   Selecting an Appropriate Measure of Center

a.   A student takes four exams in a biology class. His grades are 88, 75, 95, and 100. Which measure of center is the student likely to report?

b.   The National Association of REALTORS publishes data on resale prices of U.S. homes. Which measure of center is most appropriate for such resale prices?

c.   The 2014 Boston Marathon had two categories of official finishers: male and female, of which there were 19,579 and 16,092, respectively. Which measure of center should be used here?

Solution

a.   Chances are that the student would report the mean of his scores, which is 89.5. The mean is probably the most suitable measure of center for the student to use because it takes into account the numerical value of each score and therefore indicates his overall performance.

b.   The most appropriate measure of center for resale home prices is the median because it is aimed at finding the center of the data on resale home prices and because it is not strongly affected by the relatively few homes with extremely high resale prices. Thus the median provides a better indication of the "typical" resale price than either the mean or the mode.

You try it!

Exercise 3.27 on page 102

c.   The only suitable measure of center for these data is the mode, which is "male." Each observation in this data set is either "male" or "female." There is no way to compute a mean or median for such data. *Of the mean, median, and mode, the mode is the only measure of center that can be used for qualitative data.*

Many measures of center that appear in newspapers or that are reported by government agencies are medians, as is the case for household income and number of years of school completed. In an attempt to provide a clearer picture, some reports include both the mean and the median. For instance, the National Center for Health Statistics does so for daily intake of nutrients in the publication *Vital and Health Statistics*.

Summation Notation

In statistics, as in algebra, letters such as x, y, and z are used to denote variables. So, for instance, in a study of heights and weights of college students, we might let x denote the variable "height" and y denote the variable "weight."

We can often use notation for variables, along with other mathematical notations, to express statistics definitions and formulas concisely. Of particular importance, in this regard, is *summation notation*.

EXAMPLE 3.5   Introducing Summation Notation

Exam Scores The exam scores for the student in Example 3.4(a) are 88, 75, 95, and 100.

a. Use mathematical notation to represent the individual exam scores.
b. Use summation notation to express the sum of the four exam scores.

Solution Let x denote the variable "exam score."

a. We use the symbol x_i (read as "x sub i") to represent the ith observation of the variable x. Thus, for the exam scores,

$$x_1 = \text{score on Exam 1} = 88;$$
$$x_2 = \text{score on Exam 2} = 75;$$
$$x_3 = \text{score on Exam 3} = 95;$$
$$x_4 = \text{score on Exam 4} = 100.$$

More simply, we can just write $x_1 = 88$, $x_2 = 75$, $x_3 = 95$, and $x_4 = 100$. The numbers 1, 2, 3, and 4 written below the xs are called **subscripts.** Subscripts do not necessarily indicate order but, rather, provide a way of keeping the observations distinct.

b. We can use the notation in part (a) to write the sum of the exam scores as

$$x_1 + x_2 + x_3 + x_4.$$

Summation notation, which uses the uppercase Greek letter Σ (sigma), provides a shorthand description for that sum. The letter Σ corresponds to the uppercase English letter S and is used here as an abbreviation for the phrase "the sum of." So, in place of $x_1 + x_2 + x_3 + x_4$, we can use **summation notation, Σx_i,** read as "summation x sub i" or "the sum of the observations of the variable x." For the exam-score data,

$$\Sigma x_i = x_1 + x_2 + x_3 + x_4 = 88 + 75 + 95 + 100 = 358.$$

Interpretation The sum of the student's four exam scores is 358 points. ∎

Note the following about summation notation:

- When no confusion can arise, we sometimes write Σx_i even more simply as Σx.
- For clarity, we sometimes use **indices** to write Σx_i as $\sum_{i=1}^{n} x_i$, which is read as "summation x sub i from i equals 1 to n," where n stands for the number of observations.

The Sample Mean

In the remainder of this section and in Sections 3.2–3.4, we concentrate on descriptive measures of samples. In Section 3.5, we discuss descriptive measures of populations and their relationship to descriptive measures of samples.

Recall that values of a variable for a sample from a population are called *sample data*. The mean of sample data is called a **sample mean**. The symbol used for a sample mean is a bar over the letter representing the variable. So, for a variable x, we denote a sample mean as $\bar{x}$, read as "x bar." If we also use the letter n to denote the **sample size** or, equivalently, the number of observations, we can express the definition of a sample mean concisely.

DEFINITION 3.4

? What Does It Mean?

A sample mean is the arithmetic average (mean) of sample data.

Sample Mean

For a variable x, the mean of the observations for a sample is called a **sample mean** and is denoted $\bar{x}$. Symbolically,

$$\bar{x} = \frac{\Sigma x_i}{n},$$

where n is the sample size.

■■ ■ ■   **EXAMPLE 3.6**   **The Sample Mean**

TABLE 3.5
Arterial blood pressures of 16 children of diabetic mothers

81.6	84.1	87.6	82.8
82.0	88.9	86.7	96.4
84.6	104.9	90.8	94.0
69.4	78.9	75.2	91.0

You try it!

Exercise 3.31
on page 102

Children of Diabetic Mothers The paper "Correlations Between the Intrauterine Metabolic Environment and Blood Pressure in Adolescent Offspring of Diabetic Mothers" (*Journal of Pediatrics*, Vol. 136, Issue 5, pp. 587–592) by N. Cho et al. presented findings of research on children of diabetic mothers. Past studies showed that maternal diabetes results in obesity, blood pressure, and glucose tolerance complications in the offspring.

Table 3.5 presents the arterial blood pressures, in millimeters of mercury (mm Hg), for a sample of 16 children of diabetic mothers. Determine the sample mean of these arterial blood pressures.

Solution Let x denote the variable "arterial blood pressure." We want to find the mean, $\bar{x}$, of the 16 observations of x shown in Table 3.5. The sum of those observations is $\Sigma x_i = 1378.9$. The sample size (or number of observations) is 16, so $n = 16$. Thus,

$$\bar{x} = \frac{\Sigma x_i}{n} = \frac{1378.9}{16} = 86.18.$$

Interpretation The mean arterial blood pressure of the sample of 16 children of diabetic mothers is 86.18 mm Hg.   ■

THE TECHNOLOGY CENTER

All statistical technologies have programs that automatically compute the mean and median of a data set. In this subsection, we present output and step-by-step instructions for such programs.

EXAMPLE 3.7   **Using Technology to Obtain the Mean and Median**

Weekly Salaries Use Minitab, Excel, or the TI-83/84 Plus to find the mean and median of the salary data for Data Set I, displayed in Table 3.1 on page 94.

Solution We applied the mean and median programs to the data, resulting in Output 3.1. Steps for generating that output are presented in Instructions 3.1.

OUTPUT 3.1 Mean and median for Data Set I

MINITAB

Descriptive Statistics: SALARY

```
Variable    Mean   Median
SALARY     483.8   400.0
```

EXCEL

Descriptive statistics (Quantitative data):

Statistic	SALARY
Median	400.0000
Mean	483.8462

TI-83/84 PLUS

```
NORMAL FLOAT AUTO REAL RADIAN MP
mean(∟SAL)
                        483.8461538
median(∟SAL)
                                 400
```

As shown in Output 3.1, the mean and median of the salary data for Data Set I are 483.8 (to one decimal place) and 400, respectively.

INSTRUCTIONS 3.1 Steps for generating Output 3.1

MINITAB

1 Store the data from Table 3.1 in a column named SALARY
2 Choose **Stat ➤ Basic Statistics ➤ Display Descriptive Statistics...**
3 Press the F3 key to reset the dialog box
4 Specify SALARY in the **Variables** text box
5 Click the **Statistics...** button
6 Select the **None** option button from the **Check statistics** list
7 Check the **Mean** and **Median** check boxes
8 Click **OK** twice

EXCEL

1 Store the data from Table 3.1 in a column named SALARY
2 Choose **XLSTAT ➤ Describing data ➤ Descriptive statistics**
3 Click the reset button in the lower left corner of the dialog box
4 Click in the **Quantitative data** selection box and then select the column of the worksheet that contains the SALARY data
5 Click the **Options** tab and uncheck the **Charts** check box

6 Click the **Outputs** tab and click the **None** button below the **Quantitative data** check box list
7 Check the **Median** and **Mean** check boxes in the **Quantitative data** check box list
8 Ensure that the **Display vertically** check box under the **Quantitative data** check box list is checked
9 Click **OK**
10 Click the **Continue** button in the **XLSTAT – Selections** dialog box

TI-83/84 PLUS

1 Store the data from Table 3.1 in a list named SAL
2 Press **2ND ➤ LIST**
3 Arrow over to **MATH**
4 Press **3**
5 Press **2ND ➤ LIST**
6 Arrow down to SAL and press **ENTER**
7 Press **)** and then **ENTER**
8 Press **2ND ➤ LIST**
9 Arrow over to **MATH**
10 Press **4**
11 Press **2ND ➤ LIST**
12 Arrow down to SAL and press **ENTER**
13 Press **)** and then **ENTER**

Note to Minitab users: Versions 15 and earlier of Minitab do not provide a **Check statistics** list. Users of such versions will need to manually check and/or uncheck check boxes.

Exercises 3.1

Understanding the Concepts and Skills

3.1 Explain in detail the purpose of a measure of center.

3.2 Name and describe the three most important measures of center.

3.3 Of the mean, median, and mode, which is the only one appropriate for use with qualitative data?

3.4 True or false: The mean, median, and mode can all be used with quantitative data. Explain your answer.

3.5 Consider the data set 1, 2, 3, 4, 5, 6, 7, 8, 9.
a. Obtain the mean and median of the data.
b. Replace the 9 in the data set by 99 and again compute the mean and median. Decide which measure of center works better here, and explain your answer.
c. For the data set in part (b), the mean is neither central nor typical for the data. The lack of what property of the mean accounts for this result?

3.6 Complete the following statement: A descriptive measure is *resistant* if ....

3.7 Floor Space. The U.S. Department of Housing and Urban Development compiles information on new, privately owned single-family houses. According to the document *Characteristics of New Housing*, in 2012 the mean floor space of such homes was 2505 sq ft and the median was 2306 sq ft. Which measure of center do you think is more appropriate? Justify your answer.

3.8 Net Worth. The Board of Governors of the Federal Reserve System publishes information on family net worth in the *Survey of Consumer Finances*. In 2010, the mean net worth of families in the United States was $498.8 thousand and the median net worth was $77.3 thousand. Which measure of center do you think is more appropriate? Explain your answer.

In Exercises 3.9–3.14, we have provided simple data sets for you to practice the basics of finding measures of center. For each data set, determine the
a. mean.   *b. median.*   *c. mode(s).*

3.9 4, 0, 5   **3.10** 3, 5, 7

3.11 1, 2, 4, 4   **3.12** 2, 5, 0, −1

3.13 1, 9, 8, 4, 3   **3.14** 4, 2, 0, 2, 2

3.15 Explain what each symbol represents.
a. Σ **b.** n **c.** $\bar{x}$

3.16 For a particular population, is the population mean a variable? What about a sample mean?

3.17 Consider these sample data: $x_1 = 1$, $x_2 = 7$, $x_3 = 4$, $x_4 = 5$, $x_5 = 10$.
a. Find n.   **b.** Compute Σx_i.   **c.** Determine $\bar{x}$

3.18 Consider these sample data: $x_1 = 12$, $x_2 = 8$, $x_3 = 9$, $x_4 = 17$.
a. Find n.   **b.** Compute Σx_i.   **c.** Determine $\bar{x}$.

Applying the Concepts and Skills

In Exercises 3.19–3.26, find the
a. mean.   *b. median.*   *c. mode(s).*
For the mean and the median, round each answer to one more decimal place than that used for the observations.

3.19 Amphibian Embryos. In a study of the effects of radiation on amphibian embryos titled "Shedding Light on Ultraviolet Radiation and Amphibian Embryos" (*BioScience*, Vol. 53, No. 6, pp. 551–561), L. Licht recorded the time it took for a sample of seven different species of frogs' and toads' eggs to hatch. The following table shows the times to hatch, in days.

6	7	11	6	5	5	11

3.20 Hurricanes. An article by D. Schaefer et al. (*Journal of Tropical Ecology*, Vol. 16, pp. 189–207) reported on a long-term study of the effects of hurricanes on tropical streams of the Luquillo Experimental Forest in Puerto Rico. The study shows that Hurricane Hugo had a significant impact on stream water chemistry. The following table shows a sample of 10 ammonia fluxes in the first year after Hugo. Data are in kilograms per hectare per year.

96	66	147	147	175
116	57	154	88	154

3.21 Tornado Touchdowns. Each year, tornadoes that touch down are recorded by the Storm Prediction Center and published in *Monthly Tornado Statistics*. The following table gives the number of tornadoes that touched down in the United States during each month of one year. [SOURCE: National Oceanic and Atmospheric Administration]

3	2	47	118	204	97
68	86	62	57	98	99

3.22 Technical Merit. In one Winter Olympics, Michelle Kwan competed in the Short Program ladies singles event. From nine judges, she received scores ranging from 1 (poor) to 6 (perfect). The following table provides the scores that the judges gave her on technical merit, found in an article by S. Berry (*Chance*, Vol. 15, No. 2, pp. 14–18).

5.8	5.7	5.9	5.7	5.5	5.7	5.7	5.7	5.6

3.23 Billionaires' Club. Each year, *Forbes* magazine compiles a list of the 400 richest Americans. As of September 19, 2012, the top 10 on the list are as shown in the following table.

Person	Wealth ($ billions)
Bill Gates	66.0
Warren Buffett	46.0
Larry Ellison	41.0
Charles Koch	31.0
David Koch	31.0
Christy Walton & family	27.9
Jim Walton	26.8
Alice Walton	26.3
S. Robson Walton	26.1
Michael Bloomberg	25.0

3.24 Tablet Computers. Tablet personal computers such as iPad and Kindle Fire are one-piece portable computers that typically offer a touchscreen. Tablets are available in different sizes but generally offer a screen that is greater than 7 inches diagonally. Consumer Reports reviews tablets of different sizes and Internet capabilities. The battery lives, in hours, for a sample of tablets with screen sizes between 9 and 12 inches are provided in the following table.

8.4	5.6	10.9	10.8
11.5	8.0	8.5	12.9

3.25 The Beatles. The English rock band, The Beatles, was formed in Liverpool in 1960. There are 12 studio albums that are considered part of their core catalogue. In the article, "Length of The Beatles' Songs" (*Chance*, Vol. 25, No. 1, pp. 30–33), T. Koyama lists the album title, date of release, and number of songs on each album. Here are the album names and numbers of songs.

Album	Number of songs
Please Please Me	14
With the Beatles	14
A Hard Day's Night	13
Beatles for Sale	14
Help!	14
Rubber Soul	14
Revolver	14
Sgt. Pepper	13
The Beatles	30
Yellow Submarine	6
Abbey Road	17
Let It Be	12

3.26 Router Horsepower. In the article "Router Roundup" (*Popular Mechanics*, Vol. 180, No. 12, pp. 104–109), T. Klenck reported on tests of seven fixed-base routers for performance, features, and handling. The following table gives the horsepower (hp) for each of the seven routers tested.

1.75	2.25	2.25	2.25	1.75	2.00	1.50

3.27 Medieval Cremation Burials. In the article "Material Culture as Memory: Combs and Cremations in Early Medieval Britain" (*Early Medieval Europe*, Vol. 12, Issue 2, pp. 89–128), H. Williams discussed the frequency of cremation burials found in 17 archaeological sites in eastern England. Here are the data.

83	64	46	48	523	35	34	265	2484
46	385	21	86	429	51	258	119	

a. Obtain the mean, median, and mode of these data.
b. Which measure of center do you think works best here? Explain your answer.

3.28 Monthly Motorcycle Casualties. The *Scottish Executive, Analytical Services Division Transport Statistics*, compiles data on motorcycle casualties. During one year, monthly casualties from motorcycle accidents in Scotland for built-up roads and non–built-up roads were as follows.

Month	Built-up	Non–built-up
January	25	16
February	38	9
March	38	26
April	56	48
May	61	73
June	52	72
July	50	91
August	90	69
September	67	71
October	51	28
November	64	19
December	40	12

a. Find the mean, median, and mode of the number of motorcycle casualties for built-up roads.
b. Find the mean, median, and mode of the number of motorcycle casualties for non–built-up roads.
c. If you had a list of only the month of each casualty, what month would be the modal month for each type of road?

3.29 Daily Motorcycle Accidents. The *Scottish Executive, Analytical Services Division Transport Statistics*, compiles data on motorcycle accidents. During one year, the numbers of motorcycle accidents in Scotland were tabulated by day of the week for built-up roads and non–built-up roads and resulted in the following data.

Day	Built-up	Non–built-up
Monday	88	70
Tuesday	100	58
Wednesday	76	59
Thursday	98	53
Friday	103	56
Saturday	85	94
Sunday	69	102

a. Find the mean and median of the number of accidents for built-up roads.
b. Find the mean and median of the number of accidents for non–built-up roads.
c. If you had a list of only the day of the week for each accident, what day would be the modal day for each type of road?
d. What might explain the difference in the modal days for the two types of roads?

In each of Exercises 3.30–3.33,
a. find n.
b. compute Σx_i.
c. determine the sample mean. Round your answer to one more decimal place than that used for the observations.

3.30 Honeymoons. Popular destinations for the newlyweds of today are the Caribbean and Hawaii. According to a recent *American Wedding Study* by the Conde Nast Bridal Group, a honeymoon, on average, lasts 9.4 days and costs $5111. A sample of 12 newlyweds reported the following lengths of stay of their honeymoons.

5	14	7	10	6	8
12	9	10	9	7	11

3.31 Sleep. In 1908, W. S. Gosset published the article "The Probable Error of a Mean" (*Biometrika*, Vol. 6, pp. 1–25). In this pioneering paper, written under the pseudonym "Student," Gosset introduced what later became known as Student's *t*-distribution, which we discuss in a later chapter. Gosset used the following data set, which shows the additional sleep in hours obtained by a sample of 10 patients given laevohysocyamine hydrobromide.

1.9	0.8	1.1	0.1	−0.1
4.4	5.5	1.6	4.6	3.4

3.32 Pesticides in Pakistan. Pesticides are chemicals often used in agriculture to control pests. In Pakistan, 70% of the population depends on agriculture, and pesticide use there has increased rapidly. In the article, "Monitoring Pesticide Residues in Fresh Fruits Marketed in Peshawar, Pakistan" (*American Laboratory*, Vol. 37, No. 7, pp. 22–24), J. Shah et al. sampled the most commonly used fruit in Pakistan and analyzed the pesticide residues in the fruit. The amounts, in mg/kg, of the pesticide Dichlorovos for a sample of apples, guavas, and mangos were as follows.

0.2	1.6	4.0	5.4	5.7	11.4
0.2	3.4	2.4	6.6	4.2	2.7

3.33 **U.S. Supreme Court Justices.** From *Wikipedia*, we found that the ages of the justices of the U.S. Supreme Court, as of September 9, 2013, are as follows, in years.

58	77	77	65	80
75	63	59	53	

In each of Exercises 3.34–3.41,
a. determine the mode of the data.
b. decide whether it would be appropriate to use either the mean or the median as a measure of center. Explain your answer.

3.34 **Primetime Broadcast Shows.** From the TVbytheNumbers website, we obtained the networks for the top 20 primetime broadcast TV shows by total viewership for the week ending August 18, 2013.

CBS	NBC	NBC	ABC	CBS
CBS	CBS	CBS	CBS	FOX
CBS	CBS	CBS	CBS	NBC
NBC	FOX	CBS	NBC	ABC

3.35 **NCAA Wrestling Champs.** From NCAA.com—the official Web site for NCAA sports—we obtained the National Collegiate Athletic Association wrestling champions for the years 1989–2013 in the document "Championship History". They are displayed in the following table.

Year	Champion	Year	Champion
1989	Oklahoma St.	2002	Minnesota
1990	Oklahoma St.	2003	Oklahoma St.
1991	Iowa	2004	Oklahoma St.
1992	Iowa	2005	Oklahoma St.
1993	Iowa	2006	Oklahoma St.
1994	Oklahoma St.	2007	Minnesota
1995	Iowa	2008	Iowa
1996	Iowa	2009	Iowa
1997	Iowa	2010	Iowa
1998	Iowa	2011	Penn State
1999	Iowa	2012	Penn State
2000	Iowa	2013	Penn State
2001	Minnesota		

3.36 **Road Rage.** The report *Controlling Road Rage: A Literature Review and Pilot Study* was prepared for the AAA Foundation for Traffic Safety by D. Rathbone and J. Huckabee. The authors discussed the results of a literature review and pilot study on how to prevent aggressive driving and road rage. As described in the study, *road rage* is criminal behavior by motorists characterized by uncontrolled anger that results in violence or threatened violence on the road. One of the goals of the study was to determine when road rage occurs most often. The days on which 69 road rage incidents occurred are presented in the following table.

F	F	Tu	Tu	F	Su	F	F	Tu	F
Tu	Sa	Sa	F	Sa	Tu	W	W	Th	Th
Th	Sa	M	Tu	Th	Su	W	Th	W	Tu
Tu	F	Th	Th	F	W	F	Th	F	Sa
F	W	W	F	Tu	W	W	Th	M	M
F	Su	Tu	F	W	Su	W	Th	M	Tu
F	W	Th	M	Su	Sa	Sa	Sa	F	F

3.37 **U.S. Supreme Court Justices.** From *Wikipedia*, we found that the law schools of the justices of the U.S. Supreme Court, as of September 9, 2013, are as follows.

Harvard	Harvard	Harvard
Yale	Columbia	Harvard
Yale	Yale	Harvard

3.38 **Robbery Locations.** The Department of Justice and the Federal Bureau of Investigation publish a compilation on crime statistics for the United States in *Crime in the United States*. The following table provides a frequency distribution for robbery type during a one-year period.

Robbery type	Frequency
Street/highway	127,403
Commercial house	37,885
Gas or service station	7,009
Convenience store	14,863
Residence	49,361
Bank	5,777
Miscellaneous	48,878

3.39 **Freshmen Politics.** The Higher Education Research Institute of the University of California, Los Angeles, publishes information on characteristics of incoming college freshmen in *The American Freshman*. In 2000, 27.7% of incoming freshmen characterized their political views as liberal, 51.9% as moderate, and 20.4% as conservative. For this year, a random sample of 500 incoming college freshmen yielded the following frequency distribution for political views.

Political view	Frequency
Liberal	147
Moderate	237
Conservative	116

3.40 **Medical School Faculty.** The Association of American Medical Colleges (AAMC) compiles data on medical school faculty and publishes the results in *AAMC Faculty Roster*. The following table presents a frequency distribution of rank for medical school faculty during one year.

Rank	Frequency
Professor	32,511
Associate professor	28,572
Assistant professor	59,277
Instructor	14,289
Other	3,276

3.41 **An Edge in Roulette?** An American roulette wheel contains 18 red numbers, 18 black numbers, and 2 green numbers. The following table shows the frequency with which the ball landed on each color in 200 trials.

Number	Red	Black	Green
Frequency	88	102	10

Working with Large Data Sets

In each of Exercises 3.42–3.50, use the technology of your choice to obtain the measures of center that are appropriate from among the mean, median, and mode. Discuss your results and decide which measure of center is most appropriate. Provide a reason for your answer. Note: If an exercise contains more than one data set, perform the aforementioned tasks for each data set.

3.42 Japanese Vehicle Exports. The Japan Automobile Manufacturer's Association provides data on exported vehicles in *Motor Vehicle Statistics of Japan*. In 2010, cars, trucks, and buses constituted 88.3%, 9.3%, and 2.4% of vehicle exports, respectively. A random sample of last year's exports yielded the vehicle-type data on the WeissStats site.

3.43 U.S. Hospitals. The American Hospital Association conducts annual surveys of hospitals in the United States and publishes its findings in *AHA Hospital Statistics*. Data on hospital type for U.S. registered hospitals can be found on the WeissStats site. For convenience, we use the following abbreviations:

- NPC: Nongovernment not-for-profit community hospitals
- IOC: Investor-owned (for-profit) community hospitals
- SLC: State and local government community hospitals
- FGH: Federal government hospitals
- NFP: Nonfederal psychiatric hospitals
- NLT: Nonfederal long-term-care hospitals
- HUI: Hospital units of institutions

3.44 Marital Status and Drinking. Research by W. Clark and L. Midanik (*Alcohol Consumption and Related Problems: Alcohol and Health Monograph 1*. DHHS Pub. No. (ADM) 82–1190) examined, among other issues, alcohol consumption patterns of U.S. adults by marital status. Data for marital status and number of drinks per month, based on the researchers' survey results, are provided on the WeissStats site.

3.45 Ballot Preferences. In Issue 338 of the *Amstat News*, then-president of the American Statistical Association Fritz Scheuren reported the results of a survey on how members would prefer to receive ballots in annual elections. On the WeissStats site, you will find data for preference and highest degree obtained for the 566 respondents.

3.46 The Great White Shark. In an article titled "Great White, Deep Trouble" (*National Geographic*, Vol. 197(4), pp. 2–29), Peter Benchley—the author of *JAWS*—discussed various aspects of the Great White Shark (*Carcharodon carcharias*). Data on the number of pups borne in a lifetime by each of 80 Great White Shark females are provided on the WeissStats site.

3.47 The Beatles. In the article, "Length of The Beatles' Songs" (*Chance*, Vol. 25, No. 1, pp. 30–33), T. Koyama discusses aspects and interpretations of the lengths of songs by The Beatles. Data on the length, in seconds, of 229 Beatles' songs are presented on the WeissStats site.

3.48 High School Completion. As reported by the U.S. Census Bureau in *Educational Attainment in the United States*, the percentage of adults in each state and the District of Columbia who have completed high school is provided on the WeissStats site.

3.49 Bachelor's Completion. As reported by the U.S. Census Bureau in *Educational Attainment in the United States*, the percentage of adults in each state and the District of Columbia who have completed a bachelor's degree is provided on the WeissStats site.

3.50 Body Temperature. A study by researchers at the University of Maryland addressed the question of whether the mean body temperature of humans is 98.6°F. The results of the study by P. Mackowiak et al. appeared in the article "A Critical Appraisal of 98.6°F, the Upper Limit of the Normal Body Temperature, and Other Legacies of Carl Reinhold August Wunderlich" (*Journal of the American Medical Association*, Vol. 268, pp. 1578–1580). Among other data, the researchers obtained the body temperatures of 93 healthy humans, as provided on the WeissStats site.

In each of Exercises 3.51–3.52,
a. *use the technology of your choice to determine the mean and median of each of the two data sets.*
b. *compare the two data sets by using your results from part (a).*

3.51 Treating Psychotic Illness. L. Petersen et al. evaluated the effects of integrated treatment for patients with a first episode of psychotic illness in the paper "A Randomised Multicentre Trial of Integrated Versus Standard Treatment for Patients with a First Episode of Psychotic Illness" (*British Medical Journal*, Vol. 331, (7517):602). Part of the study included a questionnaire that was designed to measure client satisfaction for both the integrated treatment and a standard treatment. The data on the WeissStats site are based on the results of the client questionnaire.

3.52 The Etruscans. Anthropologists are still trying to unravel the mystery of the origins of the Etruscan empire, a highly advanced Italic civilization formed around the eighth century B.C. in central Italy. Were they native to the Italian peninsula or, as many aspects of their civilization suggest, did they migrate from the East by land or sea? The maximum head breadth, in millimeters, of 70 modern Italian male skulls and that of 84 preserved Etruscan male skulls were analyzed to help researchers decide whether the Etruscans were native to Italy. The resulting data can be found on the WeissStats site. [SOURCE: N. Barnicot and D. Brothwell, "The Evaluation of Metrical Data in the Comparison of Ancient and Modern Bones." In *Medical Biology and Etruscan Origins*, G. Wolstenholme and C. O'Connor, eds., Little, Brown & Co., 1959]

Extending the Concepts and Skills

3.53 Food Choice. As you discovered earlier, *ordinal data* are data about order or rank given on a scale such as 1, 2, 3, . . . or A, B, C, ... Most statisticians recommend using the median to indicate the center of an ordinal data set, but some researchers also use the mean. In the paper "Measurement of Ethical Food Choice Motives" (*Appetite*, Vol. 34, pp. 55–59), research psychologists M. Lindeman and M. Väänänen of the University of Helsinki published a study on the factors that most influence people's choice of food. One of the questions asked of the participants was how important, on a scale of 1 to 4 (1 = not at all important, 4 = very important), is ecological welfare in food choice motive, where ecological welfare includes animal welfare and environmental protection. Here are the ratings given by 14 of the participants.

2	4	1	2	4	3	3
2	2	1	2	4	2	3

a. Compute the mean of the data.
b. Compute the median of the data.
c. Decide which of the two measures of center is best.

3.54 Outliers and Trimmed Means. Some data sets contain *outliers*, observations that fall well outside the overall pattern of the data.

(We discuss outliers in more detail in Section 3.4.) Suppose, for instance, that you are interested in the ability of high school algebra students to compute square roots. You decide to give a square-root exam to 10 of these students. Unfortunately, one of the students had a fight with his girlfriend and cannot concentrate—he gets a 0. The 10 scores are displayed in increasing order in the following table. The score of 0 is an outlier.

0	58	61	63	67	69	70	71	78	80

Statisticians have a systematic method for avoiding extreme observations and outliers when they calculate means. They compute *trimmed means,* in which high and low observations are deleted or "trimmed off" before the mean is calculated. For instance, to compute the 10% trimmed mean of the test-score data, we first delete both the bottom 10% and the top 10% of the ordered data, that is, 0 and 80. Then we calculate the mean of the remaining data. Thus the 10% trimmed mean of the test-score data is

$$\frac{58 + 61 + 63 + 67 + 69 + 70 + 71 + 78}{8} = 67.1.$$

The following table displays a set of scores for a 40-question algebra final exam.

2	15	16	16	19	21	21	25	26	27
4	15	16	17	20	21	24	25	27	28

a. Do any of the scores look like outliers?
b. Compute the usual mean of the data.
c. Compute the 5% trimmed mean of the data.
d. Compute the 10% trimmed mean of the data.
e. Compare the means you obtained in parts (b)–(d). Which of the three means provides the best measure of center for the data?

3.55 Explain the difference between the quantities $(\Sigma x_i)^2$ and Σx_i^2. Construct an example to show that, in general, those two quantities are unequal.

3.56 Explain the difference between the quantities $\Sigma x_i y_i$ and $(\Sigma x_i)(\Sigma y_i)$. Provide an example to show that, in general, those two quantities are unequal.

3.2 Measures of Variation

Up to this point, we have discussed only descriptive measures of center, specifically, the mean, median, and mode. However, two data sets can have the same mean, median, or mode and still differ in other respects. For example, consider the heights of the five starting players on each of two men's college basketball teams, as shown in Fig. 3.2.

FIGURE 3.2

Five starting players on two basketball teams

	Team I					Team II				
Feet and inches	6'	6'1"	6'4"	6'4"	6'6"	5'7"	6'	6'4"	6'4"	7'
Inches	72	73	76	76	78	67	72	76	76	84

The two teams have the same mean height, 75 inches (6′ 3″); the same median height, 76 inches (6′ 4″); and the same mode, 76 inches (6′ 4″). Nonetheless, the two data sets clearly differ. In particular, the heights of the players on Team II vary much more than those on Team I. To describe that difference quantitatively, we use a descriptive measure that indicates the amount of variation, or spread, in a data set. Such descriptive measures are referred to as **measures of variation** or **measures of spread.**

Just as there are several different measures of center, there are also several different measures of variation. In this section, we examine two of the most frequently used measures of variation: the *range* and *sample standard deviation.*

The Range

The contrast between the height difference of the two teams is clear if we place the shortest player on each team next to the tallest, as in Fig. 3.3.

FIGURE 3.3
Shortest and tallest starting
players on the teams

| Feet and inches | 6' | 6'6" | | 5'7" | 7' |
| Inches | 72 | 78 | | 67 | 84 |

The **range** of a data set is the difference between the maximum (largest) and minimum (smallest) observations. From Fig. 3.3,

Team I: Range $= 78 - 72 = 6$ inches,

Team II: Range $= 84 - 67 = 17$ inches.

Interpretation The difference between the heights of the tallest and shortest players on Team I is 6 inches, whereas that difference for Team II is 17 inches.

StatCrunch

Report 3.4

Exercise 3.63(a)
on page 113

DEFINITION 3.5

? What Does It Mean?

The range of a data set is the difference between its largest and smallest values.

Range of a Data Set

The **range** of a data set is given by the formula

$$\text{Range} = \text{Max} - \text{Min},$$

where Max and Min denote the maximum and minimum observations, respectively.

The range of a data set is easy to compute, but takes into account only the largest and smallest observations. For that reason, two other measures of variation, the *standard deviation* and the *interquartile range,* are generally favored over the range. We discuss the standard deviation in this section and consider the interquartile range in Section 3.4.

The Sample Standard Deviation

In contrast to the range, the standard deviation takes into account all the observations. It is the preferred measure of variation when the mean is used as the measure of center.

Roughly speaking, the **standard deviation** measures variation by indicating how far, on average, the observations are from the mean. For a data set with a large amount of variation, the observations will, on average, be far from the mean; so the standard deviation will be large. For a data set with a small amount of variation, the observations will, on average, be close to the mean; so the standard deviation will be small.

The formulas for the standard deviations of sample data and population data differ slightly. In this section, we concentrate on the sample standard deviation. We discuss the population standard deviation in Section 3.5.

The first step in computing a sample standard deviation is to find the **deviations from the mean,** that is, how far each observation is from the mean.

■■■ **EXAMPLE 3.8** **The Deviations From the Mean**

Heights of Starting Players The heights, in inches, of the five starting players on Team I are 72, 73, 76, 76, and 78, as we saw in Fig. 3.2 on page 105. Find the deviations from the mean.

Solution The mean height of the starting players on Team I is

$$\bar{x} = \frac{\Sigma x_i}{n} = \frac{72 + 73 + 76 + 76 + 78}{5} = \frac{375}{5} = 75 \text{ inches.}$$

To find the deviation from the mean for an observation x_i, we subtract the mean from it; that is, we compute $x_i - \bar{x}$. For instance, the deviation from the mean for the height of 72 inches is $x_i - \bar{x} = 72 - 75 = -3$. The deviations from the mean for all five observations are given in the second column of Table 3.6 and are represented by arrows in Fig. 3.4.

TABLE 3.6

Deviations from the mean

Height x	Deviation from mean $x - \bar{x}$
72	−3
73	−2
76	1
76	1
78	3

FIGURE 3.4

Observations (shown by dots) and deviations from the mean (shown by arrows)

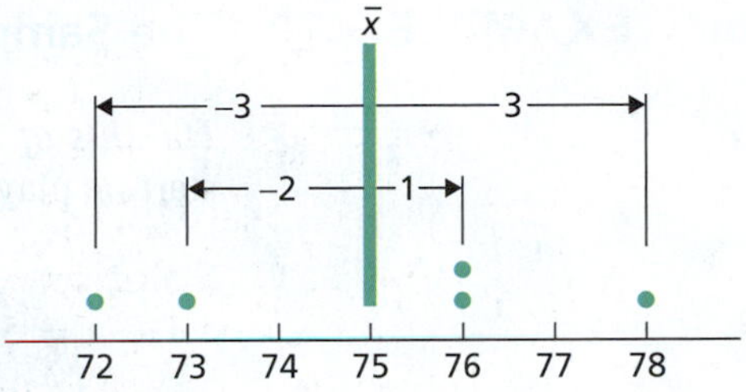

The second step in computing a sample standard deviation is to obtain a measure of the total deviation from the mean for all the observations. Although the quantities $x_i - \bar{x}$ represent deviations from the mean, adding them to get a total deviation from the mean is of no value because their sum, $\Sigma(x_i - \bar{x})$, always equals zero. Summing the second column of Table 3.6 illustrates this fact for the height data of Team I.

To obtain quantities that do not sum to zero, we square the deviations from the mean. The sum of the squared deviations from the mean, $\Sigma(x_i - \bar{x})^2$, is called the **sum of squared deviations** and gives a measure of total deviation from the mean for all the observations. We show how to calculate it next.

EXAMPLE 3.9   **The Sum of Squared Deviations**

Heights of Starting Players Compute the sum of squared deviations for the heights of the starting players on Team I.

Solution To get Table 3.7, we added a column for $(x - \bar{x})^2$ to Table 3.6.

TABLE 3.7

Table for computing the sum of squared deviations for the heights of Team I

Height x	Deviation from mean $x - \bar{x}$	Squared deviation $(x - \bar{x})^2$
72	−3	9
73	−2	4
76	1	1
76	1	1
78	3	9
		24

From the third column of Table 3.7, $\Sigma(x_i - \bar{x})^2 = 24$. The sum of squared deviations is 24 inches^2.

The third step in computing a sample standard deviation is to take an average of the squared deviations. We do so by dividing the sum of squared deviations by $n - 1$, or 1 less than the sample size. The resulting quantity is called a **sample variance** and is denoted s_x^2 or, when no confusion can arise, s^2. In symbols,

$$s^2 = \frac{\Sigma(x_i - \bar{x})^2}{n - 1}.$$

Note: If we divided by n instead of by $n - 1$, the sample variance would be the mean of the squared deviations. Although dividing by n seems more natural, we divide by $n - 1$ for the following reason. One of the main uses of the sample variance is to estimate the population variance (defined in Section 3.5). Division by n tends to underestimate the population variance, whereas division by $n - 1$ gives, on average, the correct value.

EXAMPLE 3.10   **The Sample Variance**

Heights of Starting Players Determine the sample variance of the heights of the starting players on Team I.

Solution From Example 3.9, the sum of squared deviations is 24 inches². Because $n = 5$,

$$s^2 = \frac{\Sigma(x_i - \bar{x})^2}{n - 1} = \frac{24}{5 - 1} = 6.$$

The sample variance is 6 inches².

As we have just seen, a sample variance is in units that are the square of the original units, the result of squaring the deviations from the mean. Because descriptive measures should be expressed in the original units, the final step in computing a sample standard deviation is to take the square root of the sample variance, which gives us the following definition.

DEFINITION 3.6

What Does It Mean?

Roughly speaking, the sample standard deviation indicates how far, on average, the observations in the sample are from the mean of the sample.

Sample Standard Deviation

For a variable x, the standard deviation of the observations for a sample is called a **sample standard deviation.** It is denoted s_x or, when no confusion will arise, simply **s**. We have

$$s = \sqrt{\frac{\Sigma(x_i - \bar{x})^2}{n - 1}},$$

where n is the sample size and $\bar{x}$ is the sample mean.

EXAMPLE 3.11   **The Sample Standard Deviation**

Heights of Starting Players Determine the sample standard deviation of the heights of the starting players on Team I.

Solution From Example 3.10, the sample variance is 6 inches². Thus the sample standard deviation is

$$s = \sqrt{\frac{\Sigma(x_i - \bar{x})^2}{n - 1}} = \sqrt{6} = 2.4 \text{ inches (rounded)}.$$

Interpretation Roughly speaking, on average, the heights of the players on Team I vary from the mean height of 75 inches by about 2.4 inches.

For teaching purposes, we spread our calculations of a sample standard deviation over four separate examples. Now we summarize the procedure with three steps.

Step 1 Calculate the sample mean, $\bar{x}$.

Step 2 Construct a table to obtain the sum of squared deviations, $\Sigma(x_i - \bar{x})^2$.

Step 3 Apply Definition 3.6 to determine the sample standard deviation, s.

■■ ■ ■   **EXAMPLE 3.12**   **The Sample Standard Deviation**

TABLE 3.8

Table for computing the sum of squared deviations for the heights of Team II

x	$x - \bar{x}$	$(x - \bar{x})^2$
67	-8	64
72	-3	9
76	1	1
76	1	1
84	9	81
		156

Exercise 3.63(b)
on page 113

StatCrunch

Report 3.5

Heights of Starting Players The heights, in inches, of the five starting players on Team II are 67, 72, 76, 76, and 84. Determine the sample standard deviation of these heights.

Solution We apply the three-step procedure just described.

Step 1 Calculate the sample mean, $\bar{x}$.

We have

$$\bar{x} = \frac{\Sigma x_i}{n} = \frac{67 + 72 + 76 + 76 + 84}{5} = \frac{375}{5} = 75 \text{ inches.}$$

Step 2 Construct a table to obtain the sum of squared deviations, $\Sigma(x_i - \bar{x})^2$.

Table 3.8 provides columns for x, $x - \bar{x}$, and $(x - \bar{x})^2$. The third column shows that $\Sigma(x_i - \bar{x})^2 = 156$ inches^2.

Step 3 Apply Definition 3.6 to determine the sample standard deviation, s.

Because $n = 5$ and $\Sigma(x_i - \bar{x})^2 = 156$, the sample standard deviation is

$$s = \sqrt{\frac{\Sigma(x_i - \bar{x})^2}{n - 1}} = \sqrt{\frac{156}{5 - 1}} = \sqrt{39} = 6.2 \text{ inches (rounded).}$$

Interpretation Roughly speaking, on average, the heights of the players on Team II vary from the mean height of 75 inches by about 6.2 inches.   ■

In Examples 3.11 and 3.12, we found that the sample standard deviations of the heights of the starting players on Teams I and II are 2.4 inches and 6.2 inches, respectively. Hence Team II, which has more variation in height than Team I, also has a larger standard deviation.

KEY FACT 3.1

Variation and the Standard Deviation

The more variation that there is in a data set, the larger is its standard deviation.

Applet 3.2

Key Fact 3.1 shows that the standard deviation satisfies the basic criterion for a measure of variation; in fact, the standard deviation is the most commonly used measure of variation. However, the standard deviation does have its drawbacks. For instance, it is not resistant: its value can be strongly affected by a few extreme observations.

A Computing Formula for *s*

Next, we present an alternative formula for obtaining a sample standard deviation, which we call the *computing formula* for *s*. We call the original formula given in Definition 3.6 the *defining formula* for *s*.

FORMULA 3.1

Computing Formula for a Sample Standard Deviation

A sample standard deviation can be computed using the formula

$$s = \sqrt{\frac{\Sigma x_i^2 - (\Sigma x_i)^2/n}{n-1}},$$

where n is the sample size.

Note: In the numerator of the computing formula, the division of $(\Sigma x_i)^2$ by n should be performed before the subtraction from Σx_i^2. In other words, first compute $(\Sigma x_i)^2/n$ and then subtract the result from Σx_i^2.

The computing formula for s is equivalent to the defining formula—both formulas give the same answer, although differences owing to roundoff error are possible. However, the computing formula is usually faster and easier for doing calculations by hand and also reduces the chance for roundoff error.

Before illustrating the computing formula for s, let's investigate its expressions, Σx_i^2 and $(\Sigma x_i)^2$. The expression Σx_i^2 represents the sum of the squares of the data; to find it, first square each observation and then sum those squared values. The expression $(\Sigma x_i)^2$ represents the square of the sum of the data; to find it, first sum the observations and then square that sum.

■■■   **EXAMPLE 3.13**   **Computing Formula for a Sample Standard Deviation**

TABLE 3.9

Table for computation of s, using the computing formula

Heights of Starting Players Find the sample standard deviation of the heights for the five starting players on Team II by using the computing formula.

Solution We need the sums Σx_i^2 and $(\Sigma x_i)^2$, which Table 3.9 shows to be 375 and 28,281, respectively. Now applying Formula 3.1, we get

x	x^2
67	4,489
72	5,184
76	5,776
76	5,776
84	7,056
375	28,281

$$s = \sqrt{\frac{\Sigma x_i^2 - (\Sigma x_i)^2/n}{n-1}} = \sqrt{\frac{28,281 - (375)^2/5}{5-1}}$$

$$= \sqrt{\frac{28,281 - 28,125}{4}} = \sqrt{\frac{156}{4}} = \sqrt{39} = 6.2 \text{ inches,}$$

which is the same value that we got by using the defining formula.   **■**

You try it!

Exercise 3.63(c) on page 113

Rounding Basics

Here is an important rule to remember when you use only basic calculator functions to obtain a sample standard deviation or any other descriptive measure.

Rounding Rule: Do not perform any rounding until the computation is complete; otherwise, substantial roundoff error can result.

Another common rounding rule is to round final answers that contain units to one more decimal place than the raw data. Although we usually abide by this convention, occasionally we vary from it for pedagogical reasons. In general, you should stick to this rounding rule as well.

Further Interpretation of the Standard Deviation

Again, the standard deviation is a measure of variation—the more variation there is in a data set, the larger is its standard deviation. Table 3.10 contains two data sets, each with 10 observations. Notice that Data Set II has more variation than Data Set I.

TABLE 3.10

Data sets that have different variation

Data Set I	41	44	45	47	47	48	51	53	58	66
Data Set II	20	37	48	48	49	50	53	61	64	70

TABLE 3.11

Means and standard deviations of the data sets in Table 3.10

Data Set I	Data Set II
$\bar{x} = 50.0$	$\bar{x} = 50.0$
$s = 7.4$	$s = 14.2$

We computed the sample mean and sample standard deviation of each data set and summarized the results in Table 3.11. As expected, the standard deviation of Data Set II is larger than that of Data Set I.

To enable you to compare visually the variations in the two data sets, we produced the graphs shown in Figs. 3.5 and 3.6. On each graph, we marked the observations with dots. In addition, we located the sample mean, $\bar{x} = 50$, and measured intervals equal in length to the standard deviation: 7.4 for Data Set I and 14.2 for Data Set II.

FIGURE 3.5   Data Set I; $\bar{x} = 50$, $s = 7.4$

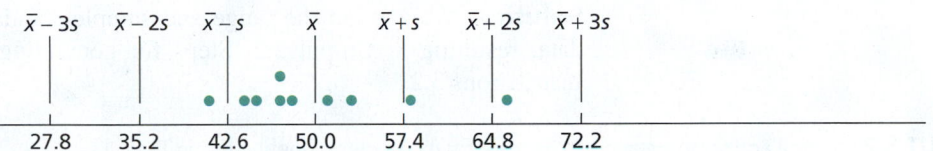

FIGURE 3.6   Data Set II; $\bar{x} = 50$, $s = 14.2$

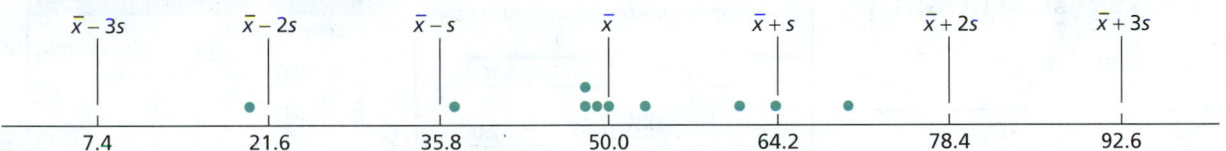

In Fig. 3.5, note that the horizontal position labeled $\bar{x} + 2s$ represents the number that is two standard deviations to the right of the mean, which in this case is

$$\bar{x} + 2s = 50.0 + 2 \cdot 7.4 = 50.0 + 14.8 = 64.8.^{\dagger}$$

Likewise, the horizontal position labeled $\bar{x} - 3s$ represents the number that is three standard deviations to the left of the mean, which in this case is

$$\bar{x} - 3s = 50.0 - 3 \cdot 7.4 = 50.0 - 22.2 = 27.8.$$

Figure 3.6 is interpreted in a similar manner.

The graphs shown in Figs. 3.5 and 3.6 vividly illustrate that Data Set II has more variation than Data Set I. They also show that for each data set, all observations lie within a few standard deviations to either side of the mean. This result is no accident.

KEY FACT 3.2

Three-Standard-Deviations Rule

Almost all the observations in any data set lie within three standard deviations to either side of the mean.

A data set with a great deal of variation has a large standard deviation, so three standard deviations to either side of its mean will be extensive, as shown in Fig. 3.6. A data set with little variation has a small standard deviation, and hence three standard deviations to either side of its mean will be narrow, as shown in Fig. 3.5.

---

†Recall that the rules for the order of arithmetic operations say to multiply and divide before adding and subtracting. So, to evaluate $a + b \cdot c$, find $b \cdot c$ first and then add the result to a. Similarly, to evaluate $a - b \cdot c$, find $b \cdot c$ first and then subtract the result from a.

THE TECHNOLOGY CENTER

Most statistical technologies have programs that automatically compute the range and sample standard deviation of a data set. In this subsection, we present output and step-by-step instructions for such programs.

EXAMPLE 3.14   **Using Technology to Obtain the Range and Sample Standard Deviation**

Heights of Starting Players The first column of Table 3.9 on page 110 gives the heights of the five starting players on Team II. Use Minitab, Excel, or the TI-83/84 Plus to find the range and sample standard deviation of those heights.

Solution We applied the range and sample-standard-deviation programs to the data, resulting in Output 3.2. Steps for generating that output are presented in Instructions 3.2.

OUTPUT 3.2   Range and sample standard deviation for the heights of the players on Team II

MINITAB

Descriptive Statistics: HEIGHT

```
Variable  StDev  Range
HEIGHT    6.24   17.00
```

EXCEL

Descriptive statistics (Quantitative data):

Statistic	HEIGHT
Range	17.0000
Standard deviation (n-1)	6.2450

TI-83/84 PLUS

```
NORMAL FLOAT AUTO REAL RADIAN MP
stdDev(ʟHT)
                    6.244997998
min(ʟHT)
                             67.
max(ʟHT)
                             84.
```

As shown in Output 3.2, the sample standard deviation of the heights for the starting players on Team II is 6.24 inches (to two decimal places). The Minitab and Excel outputs also show that the range of the heights is 17 inches. We can get the range from the TI-83/84 Plus output by subtracting the minimum (min) from the maximum (max): Range $= 84 - 67 = 17$.

INSTRUCTIONS 3.2   Steps for generating Output 3.2

MINITAB

1 Store the data from Table 3.9 in a column named HEIGHT
2 Choose **Stat ➤ Basic Statistics ➤ Display Descriptive Statistics...**
3 Press the F3 key to reset the dialog box
4 Specify HEIGHT in the **Variables** text box
5 Click the **Statistics...** button
6 Select the **None** option button from the **Check statistics** list
7 Check the **Standard deviation** and **Range** check boxes
8 Click **OK** twice

EXCEL

1 Store the data from Table 3.9 in a column named HEIGHT
2 Choose **XLSTAT ➤ Describing data ➤ Descriptive statistics**

3 Click the reset button in the lower left corner of the dialog box
4 Click in the **Quantitative data** selection box and then select the column of the worksheet that contains the HEIGHT data
5 Click the **Options** tab and uncheck the **Charts** check box
6 Click the **Outputs** tab and click the **None** button below the **Quantitative data** check box list
7 Check the **Range** and **Standard deviation (n-1)** check boxes in the **Quantitative data** check box list
8 Ensure that the **Display vertically** check box under the **Quantitative data** check box list is checked
9 Click **OK**
10 Click the **Continue** button in the **XLSTAT – Selections** dialog box

(continued)

Note to Minitab users: Versions 15 and earlier of Minitab do not provide a **Check statistics** list. Users of such versions will need to manually check and/or uncheck check boxes.

Exercises 3.2

Understanding the Concepts and Skills

3.57 Explain the purpose of a measure of variation.

3.58 Why is the standard deviation preferable to the range as a measure of variation?

3.59 When you use the standard deviation as a measure of variation, what is the reference point?

3.60 Darts. The following dartboards represent darts thrown by two players, Tracey and Joan.

Tracey   Joan

For the variable "distance from the center," which player's board represents data with a smaller sample standard deviation? Explain your answer.

3.61 Consider the data set 1, 2, 3, 4, 5, 6, 7, 8, 9.
a. Use the defining formula to obtain the sample standard deviation.
b. Replace the 9 in the data set by 99, and again use the defining formula to compute the sample standard deviation.
c. Compare your answers in parts (a) and (b). The lack of what property of the standard deviation accounts for its extreme sensitivity to the change of 9 to 99?

3.62 Consider the following four data sets.

Data Set I	Data Set II	Data Set III	Data Set IV
1 5	1 9	5 5	2 4
1 8	1 9	5 5	4 4
2 8	1 9	5 5	4 4
2 9	1 9	5 5	4 10
5 9	1 9	5 5	4 10

a. Compute the mean of each data set.
b. Although the four data sets have the same means, in what respect are they quite different?
c. Which data set appears to have the least variation? the greatest variation?
d. Compute the range of each data set.
e. Use the defining formula to compute the sample standard deviation of each data set.
f. From your answers to parts (d) and (e), which measure of variation better distinguishes the spread in the four data sets: the range or the standard deviation? Explain your answer.
g. Are your answers from parts (c) and (e) consistent?

3.63 Age of U.S. Residents. The U.S. Census Bureau publishes information about ages of people in the United States in *Current Population Reports*. A sample of five U.S. residents have the following ages, in years.

21	54	9	45	51

a. Determine the range of these ages.
b. Find the sample standard deviation of these ages by using the defining formula, Definition 3.6 on page 108.
c. Find the sample standard deviation of these ages by using the computing formula, Formula 3.1 on page 110.
d. Compare your work in parts (b) and (c).

3.64 Consider the data set 3, 3, 3, 3, 3, 3.
a. Guess the value of the sample standard deviation without calculating it. Explain your reasoning.
b. Use the defining formula to calculate the sample standard deviation.
c. Complete the following statement and explain your reasoning: If all observations in a data set are equal, the sample standard deviation is _____.
d. Complete the following statement and explain your reasoning: If the sample standard deviation of a data set is 0, then ....

In Exercises 3.65–3.70, we have provided simple data sets for you to practice the basics of finding measures of variation. For each data set, determine the
a. range.   *b. sample standard deviation.*

3.65 4, 0, 5   **3.66** 3, 5, 7

3.67 1, 2, 4, 4   **3.68** 2, 5, 0, −1

3.69 1, 9, 8, 4, 3   **3.70** 4, 2, 0, 2, 2

Applying the Concepts and Skills

In Exercises 3.71–3.78, determine the range and sample standard deviation for each of the data sets. For the sample standard deviation, round each answer to one more decimal place than that used for the observations.

3.71 Amphibian Embryos. In a study of the effects of radiation on amphibian embryos titled "Shedding Light on Ultraviolet Radiation and Amphibian Embryos" (*BioScience*, Vol. 53, No. 6, pp. 551–561), L. Licht recorded the time it took for a sample of seven different species of frogs' and toads' eggs to hatch. The following table shows the times to hatch, in days.

6	7	11	6	5	5	11

3.72 Hurricanes. An article by D. Schaefer et al. (*Journal of Tropical Ecology*, Vol. 16, pp. 189–207) reported on a long-term study of the effects of hurricanes on tropical streams of the Luquillo Experimental Forest in Puerto Rico. The study showed that Hurricane Hugo had a significant impact on stream water chemistry. The following table shows a sample of 10 ammonia fluxes in the first year after Hugo. Data are in kilograms per hectare per year.

96	66	147	147	175
116	57	154	88	154

3.73 Tornado Touchdowns. Each year, tornadoes that touch down are recorded by the Storm Prediction Center and published in *Monthly Tornado Statistics*. The following table gives the number of tornadoes that touched down in the United States during each month of one year. [SOURCE: National Oceanic and Atmospheric Administration.]

3	2	47	118	204	97
68	86	62	57	98	99

3.74 Technical Merit. In one Winter Olympics, Michelle Kwan competed in the Short Program ladies singles event. From nine judges, she received scores ranging from 1 (poor) to 6 (perfect). The following table provides the scores that the judges gave her on technical merit, found in an article by S. Berry (*Chance*, Vol. 15, No. 2, pp. 14–18).

5.8	5.7	5.9	5.7	5.5	5.7	5.7	5.7	5.6

3.75 Billionaires' Club. Each year, *Forbes* magazine compiles a list of the 400 richest Americans. As of September 19, 2012, the top 10 on the list are as shown in the following table.

Person	Wealth ($ billions)
Bill Gates	66.0
Warren Buffett	46.0
Larry Ellison	41.0
Charles Koch	31.0
David Koch	31.0
Christy Walton & family	27.9
Jim Walton	26.8
Alice Walton	26.3
S. Robson Walton	26.1
Michael Bloomberg	25.0

3.76 Tablet Computers. Tablet personal computers such as iPad and Kindle Fire are one-piece portable computers that typically offer a touchscreen. Tablets are available in different sizes but generally offer a screen that is greater than 7 inches diagonally. Consumer Reports reviews tablets of different sizes and Internet capabilities. The battery lives, in hours, for a sample of tablets with screen sizes between 9 and 12 inches are provided in the following table.

8.4	5.6	10.9	10.8
11.5	8.0	8.5	12.9

3.77 The Beatles. The English rock band, The Beatles, was formed in Liverpool in 1960. There are 12 studio albums that are considered part of their core catalogue. In the article, "Length of The Beatles' Songs" (*Chance*, Vol. 25, No. 1, pp. 30–33), T. Koyama lists the album title, date of release, and number of songs on each album. Here are the album names and numbers of songs.

Album	Number of songs
Please Please Me	14
With the Beatles	14
A Hard Day's Night	13
Beatles for Sale	14
Help!	14
Rubber Soul	14
Revolver	14
Sgt. Pepper	13
The Beatles	30
Yellow Submarine	6
Abbey Road	17
Let It Be	12

3.78 Router Horsepower. In the article "Router Roundup" (*Popular Mechanics*, Vol. 180, No. 12, pp. 104–109), T. Klenck reported on tests of seven fixed-base routers for performance, features, and handling. The following table gives the horsepower for each of the seven routers tested.

1.75	2.25	2.25	2.25	1.75	2.00	1.50

3.79 Medieval Cremation Burials. In the article "Material Culture as Memory: Combs and Cremations in Early Medieval Britain" (*Early Medieval Europe*, Vol. 12, Issue 2, pp. 89–128), H. Williams discussed the frequency of cremation burials found in 17 archaeological sites in eastern England. Here are the data.

83	64	46	48	523	35	34	265	2484
46	385	21	86	429	51	258	119	

a. Obtain the sample standard deviation of these data.
b. Do you think that, in this case, the sample standard deviation provides a good measure of variation? Explain your answer.

3.80 Monthly Motorcycle Casualties. The *Scottish Executive, Analytical Services Division Transport Statistics*, compiles data on motorcycle casualties. During one year, monthly casualties resulting from motorcycle accidents in Scotland for built-up roads and non–built-up roads were as follows.

Month	Built-up	Non–built-up
January	25	16
February	38	9
March	38	26
April	56	48
May	61	73
June	52	72
July	50	91
August	90	69
September	67	71
October	51	28
November	64	19
December	40	12

a. Without doing any calculations, make an educated guess at which of the two data sets, built-up or non–built-up, has the greater variation.

b. Find the range and sample standard deviation of each of the two data sets. Compare your results here to the educated guess that you made in part (a).

3.81 Daily Motorcycle Accidents. The *Scottish Executive, Analytical Services Division Transport Statistics*, compiles data on motorcycle accidents. During one year, the numbers of motorcycle accidents in Scotland were tabulated by day of the week for built-up roads and non–built-up roads and resulted in the following data.

Day	Built-up	Non–built-up
Monday	88	70
Tuesday	100	58
Wednesday	76	59
Thursday	98	53
Friday	103	56
Saturday	85	94
Sunday	69	102

a. Without doing any calculations, make an educated guess at which of the two data sets, built-up or non–built-up, has the greater variation.

b. Find the range and sample standard deviation of each of the two data sets. Compare your results here to the educated guess that you made in part (a).

Working with Large Data Sets

In each of Exercises 3.82–3.90, use the technology of your choice to determine and interpret the range and sample standard deviation for those data sets to which those concepts apply. If those concepts don't apply, explain why. Note: If an exercise contains more than one data set, perform the aforementioned tasks for each data set.

3.82 Japanese Vehicle Exports. The Japan Automobile Manufacturer's Association provides data on exported vehicles in *Motor Vehicle Statistics of Japan*. In 2010, cars, trucks, and buses constituted 88.3%, 9.3%, and 2.4% of vehicle exports, respectively. A random sample of last year's exports yielded the vehicle-type data on the WeissStats site.

3.83 U.S. Hospitals. The American Hospital Association conducts annual surveys of hospitals in the United States and publishes its findings in *AHA Hospital Statistics*. Data on hospital type for U.S. registered hospitals can be found on the WeissStats site. For convenience, we use the following abbreviations:

- NPC: Nongovernment not-for-profit community hospitals
- IOC: Investor-owned (for-profit) community hospitals
- SLC: State and local government community hospitals
- FGH: Federal government hospitals
- NFP: Nonfederal psychiatric hospitals
- NLT: Nonfederal long-term-care hospitals
- HUI: Hospital units of institutions

3.84 Marital Status and Drinking. Research by W. Clark and L. Midanik (*Alcohol Consumption and Related Problems: Alcohol and Health Monograph 1*. DHHS Pub. No. (ADM) 82–1190) examined, among other issues, alcohol consumption patterns of U.S. adults by marital status. Data for marital status and number of drinks per month, based on the researcher's survey results, are provided on the WeissStats site.

3.85 Ballot Preferences. In Issue 338 of the *Amstat News*, then-president of the American Statistical Association, F. Scheuren reported the results of a survey on how members would prefer to receive ballots in annual elections. On the WeissStats site, you will find data for preference and highest degree obtained for the 566 respondents.

3.86 The Great White Shark. In an article titled "Great White, Deep Trouble" (*National Geographic*, Vol. 197(4), pp. 2–29), Peter Benchley—the author of *JAWS*—discussed various aspects of the Great White Shark (*Carcharodon carcharias*). Data on the number of pups borne in a lifetime by each of 80 Great White Shark females are provided on the WeissStats site.

3.87 The Beatles. In the article, "Length of The Beatles' Songs" (*Chance*, Vol. 25, No. 1, pp. 30–33), T. Koyama discusses aspects and interpretations of the lengths of songs by The Beatles. Data on the length, in seconds, of 229 Beatles' songs are presented on the WeissStats site.

3.88 High School Completion. As reported by the U.S. Census Bureau in *Educational Attainment in the United States*, the percentage of adults in each state and the District of Columbia who have completed high school is provided on the WeissStats site.

3.89 Bachelor's Completion. As reported by the U.S. Census Bureau in *Educational Attainment in the United States*, the percentage of adults in each state and the District of Columbia who have completed a bachelor's degree is provided on the WeissStats site.

3.90 Body Temperature. A study by researchers at the University of Maryland addressed the question of whether the mean body temperature of humans is 98.6°F. The results of the study by P. Mackowiak et al. appeared in the article "A Critical Appraisal of 98.6°F, the Upper Limit of the Normal Body Temperature, and Other Legacies of Carl Reinhold August Wunderlich" (*Journal of the American Medical Association*, Vol. 268, pp. 1578–1580). Among other data, the researchers obtained the body temperatures of 93 healthy humans, as provided on the WeissStats site.

In each of Exercises 3.91–3.92,
a. *use the technology of your choice to determine the range and sample standard deviation of each of the two data sets.*
b. *compare the two data sets by using your results from part (a).*

3.91 Treating Psychotic Illness. L. Petersen et al. evaluated the effects of integrated treatment for patients with a first episode of psychotic illness in the paper "A Randomised Multicentre Trial of Integrated Versus Standard Treatment for Patients With a First Episode of Psychotic Illness" (*British Medical Journal*, Vol. 331, (7517):602).

Part of the study included a questionnaire that was designed to measure client satisfaction for both the integrated treatment and a standard treatment. The data on the WeissStats site are based on the results of the client questionnaire.

3.92 The Etruscans. Anthropologists are still trying to unravel the mystery of the origins of the Etruscan empire, a highly advanced Italic civilization formed around the eighth century B.C. in central Italy. Were they native to the Italian peninsula or, as many aspects of their civilization suggest, did they migrate from the East by land or sea? The maximum head breadth, in millimeters, of 70 modern Italian male skulls and that of 84 preserved Etruscan male skulls were analyzed to help researchers decide whether the Etruscans were native to Italy. The resulting data can be found on the WeissStats site. [SOURCE: N. Barnicot and D. Brothwell, "The Evaluation of Metrical Data in the Comparison of Ancient and Modern Bones." In *Medical Biology and Etruscan Origins*, G. Wolstenholme and C. O'Connor, eds., Little, Brown & Co., 1959]

Extending the Concepts and Skills

3.93 Outliers. In Exercise 3.54 on pages 104–105, we discussed *outliers,* or observations that fall well outside the overall pattern of the data. The following table contains two data sets. Data Set II was obtained by removing the outliers from Data Set I.

Data Set I					Data Set II			
0	12	14	15	23	10	14	15	17
0	14	15	16	24	12	14	15	
10	14	15	17		14	15	16	

a. Compute the sample standard deviation of each of the two data sets.
b. Compute the range of each of the two data sets.
c. What effect do outliers have on variation? Explain your answer.

Grouped-Data Formulas. When data are grouped in a frequency distribution, we use the following formulas to obtain the sample mean and sample standard deviation.

Grouped-Data Formulas

$$\bar{x} = \frac{\Sigma x_i f_i}{n} \quad \text{and} \quad s = \sqrt{\frac{\Sigma (x_i - \bar{x})^2 f_i}{n - 1}},$$

where x_i denotes either class mark or midpoint, f_i denotes class frequency, and $n \ (= \Sigma f_i)$ denotes sample size. The sample standard deviation can also be obtained by using the computing formula

$$s = \sqrt{\frac{\Sigma x_i^2 f_i - (\Sigma x_i f_i)^2 / n}{n - 1}}.$$

In general, these formulas yield only approximations to the actual sample mean and sample standard deviation. We ask you to apply the grouped-data formulas in Exercises 3.94 and 3.95.

3.94 Weekly Salaries. In the following table, we repeat the salary data in Data Set II from Example 3.1.

300	300	940	450	400
400	300	300	1050	300

a. Use Definitions 3.4 and 3.6 on pages 99 and 108, respectively, to obtain the sample mean and sample standard deviation of this (ungrouped) data set.
b. A frequency distribution for Data Set II, using single-value grouping, is presented in the first two columns of the following table. The third column of the table is for the xf-values, that is, class mark or midpoint (which here is the same as the class) times class frequency. Complete the missing entries in the table and then use the grouped-data formula to obtain the sample mean.

Salary x	Frequency f	Salary · Frequency xf
300	5	1500
400	2	
450	1	
940	1	
1050	1	

c. Compare the answers that you obtained for the sample mean in parts (a) and (b). Explain why the grouped-data formula always yields the actual sample mean when the data are grouped by using single-value grouping. (*Hint:* What does xf represent for each class?)
d. Construct a table similar to the one in part (b) but with columns for x, f, $x - \bar{x}$, $(x - \bar{x})^2$, and $(x - \bar{x})^2 f$. Use the table and the grouped-data formula to obtain the sample standard deviation.
e. Compare your answers for the sample standard deviation in parts (a) and (d). Explain why the grouped-data formula always yields the actual sample standard deviation when the data are grouped by using single-value grouping.

3.95 Days to Maturity. The first two columns of the following table provide a frequency distribution, using limit grouping, for the days to maturity of 40 short-term investments, as found in *BARRON'S*. The third column shows the class marks.

Days to maturity	Frequency f	Class mark x
30–39	3	34.5
40–49	1	44.5
50–59	8	54.5
60–69	10	64.5
70–79	7	74.5
80–89	7	84.5
90–99	4	94.5

a. Use the grouped-data formulas to estimate the sample mean and sample standard deviation of the days-to-maturity data. Round your final answers to one decimal place.
b. The following table gives the raw days-to-maturity data.

70	64	99	55	64	89	87	65
62	38	67	70	60	69	78	39
75	56	71	51	99	68	95	86
57	53	47	50	55	81	80	98
51	36	63	66	85	79	83	70

Using Definitions 3.4 and 3.6 on pages 99 and 108, respectively, gives the true sample mean and sample standard deviation of the days-to-maturity data as 68.3 and 16.7, respectively, rounded to one decimal place. Compare these actual values of $\bar{x}$ and s to the estimates from part (a). Explain why the grouped-data formulas generally yield only approximations to the sample mean and sample standard deviation for non–single-value grouping.

3.3 Chebyshev's Rule and the Empirical Rule*

The three-standard-deviations rule (Key Fact 3.2 on page 111) states that almost all the observations in a data set lie within three standard deviations to either side of the mean. That statement is somewhat vague—what does "almost all" mean?

Chebyshev's Rule

A more precise version of the three-standard-deviations rule can be obtained from **Chebyshev's rule**.

KEY FACT 3.3

Chebyshev's Rule

For any quantitative data set and any real number k greater than or equal to 1, at least $1 - 1/k^2$ of the observations lie within k standard deviations to either side of the mean, that is, between $\bar{x} - k \cdot s$ and $\bar{x} + k \cdot s$.

Two specific cases of Chebyshev's rule are applied frequently, namely, when $k = 2$ and $k = 3$. These specific cases can be stated as follows.

- $k = 2$: At least 75% of the observations in any data set lie within two standard deviations to either side of the mean, that is, between $\bar{x} - 2s$ and $\bar{x} + 2s$.
- $k = 3$: At least 89% of the observations in any data set lie within three standard deviations to either side of the mean, that is, between $\bar{x} - 3s$ and $\bar{x} + 3s$.

We note that the second bulleted item (Chebyshev's rule with $k = 3$) provides a more precise version of the three-standard-deviations rule.

Let us illustrate how to get these two specific cases of Chebyshev's rule by considering the specific case $k = 2$. Then

$$1 - \frac{1}{k^2} = 1 - \frac{1}{2^2} = 1 - \frac{1}{4} = \frac{3}{4} = 0.75,$$

or 75%. Thus, by Chebyshev's rule with $k = 2$, at least 75% of the observations lie within two standard deviations to either side of the mean.

Figure 3.7 provides a graphic for our two specific cases of Chebyshev's rule.

FIGURE 3.7

Chebyshev's rule with $k = 2$ and $k = 3$

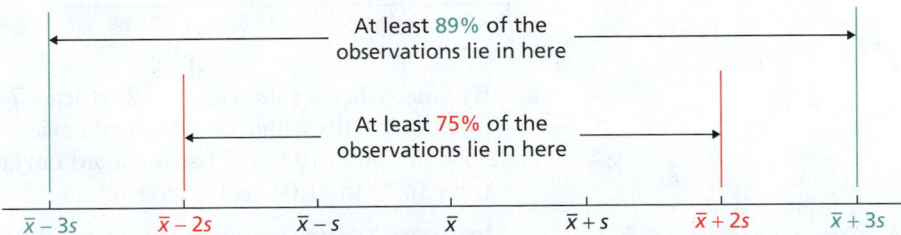

At least 89% of the
observations lie in here

At least 75% of the
observations lie in here

$\bar{x} - 3s$ $\bar{x} - 2s$ $\bar{x} - s$ $\bar{x}$ $\bar{x} + s$ $\bar{x} + 2s$ $\bar{x} + 3s$

EXAMPLE 3.15   Chebyshev's Rule

Consider again the data set portrayed in Fig. 3.5 on page 111.

a.   Compare the percentage of the observations that actually lie within two standard deviations to either side of the mean with that given by Chebyshev's rule with $k = 2$.

b.   Repeat part (a) with $k = 3$.

Solution

a.   Chebyshev's rule with $k = 2$ says that at least 75% of the observations lie within two standard deviations to either side of the mean. However, as we see from

Fig. 3.5, nine of the ten, or 90% of the observations actually lie within two standard deviations to either side of the mean.

b. Chebyshev's rule with $k = 3$ says that at least 89% of the observations lie within three standard deviations to either side of the mean. However, as we see from Fig. 3.5, all (100%) of the observations actually lie within three standard deviations to either side of the mean.

Interpretation Chebyshev's rule gives only a minimum for the percentage of observations that lie within a specified number of standard deviations to either side of the mean; the actual percentage will usually be higher.

The power of Chebyshev's rule is its generality—it holds for any data set. Moreover, it can be shown that, in general, Chebyshev's rule cannot be improved upon.

Chebyshev's rule also permits us to make pertinent statements about a data set when we know only its mean and standard deviation, which is frequently all that we do know. The next example illustrates this point.

You try it!

Exercise 3.133
on page 123

■■ ■■   **EXAMPLE 3.16**   Chebyshev's Rule

Forearm Length In 1903, K. Pearson and A. Lee published the paper "On the Laws of Inheritance in Man. I. Inheritance of Physical Characters" (*Biometrika*, Vol. 2, pp. 357–462). The article examined data on forearm length, in inches, for a sample of 140 men. The mean and standard deviation of the forearm lengths are 18.8 in. and 1.12 in., respectively.

a. Apply Chebyshev's rule with $k = 2$ to make pertinent statements about the forearm lengths of the men in the sample.

b. Repeat part (a) with $k = 3$.

Solution Based on the fact that $\bar{x} = 18.8$ and $s = 1.12$, we constructed Fig. 3.8. Notice, for instance, that $\bar{x} - 2s = 18.8 - 2 \cdot 1.12 = 16.56$.

FIGURE 3.8

The mean and one, two, and three standard deviations to either side of the mean for the forearm-length data

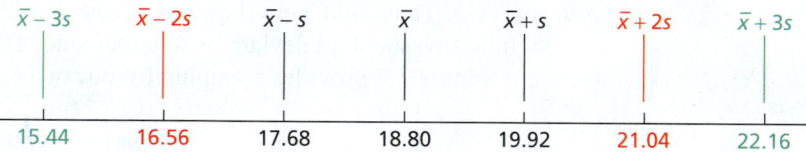

a. By Chebyshev's rule with $k = 2$, at least 75% of the men in the sample have forearm lengths within two standard deviations to either side of the mean. Now, 75% of 140 is 105, and two standard deviations to either side of the mean is from 16.56 to 21.04, as we see from Fig. 3.8.

Interpretation At least 105 of the 140 men in the sample have forearm lengths between 16.56 in. and 21.04 in.

b. By Chebyshev's rule with $k = 3$, at least 89% of the men in the sample have forearm lengths within three standard deviations to either side of the mean. Now, 89% of 140 is 124.6, and three standard deviations to either side of the mean is from 15.44 to 22.16, as we see from Fig. 3.8.

Interpretation At least 125 (124.6 rounded up) of the 140 men in the sample have forearm lengths between 15.44 in. and 22.16 in. *Question:* Why did we round 124.6 up to the nearest integer?

You try it!

Exercise 3.135
on page 123

The Empirical Rule

For data sets with approximately bell-shaped distributions, we can improve on the estimates given by Chebyshev's rule by applying the **empirical rule.**

KEY FACT 3.4

Empirical Rule

For any quantitative data set with roughly a bell-shaped distribution, the following properties hold.

Property 1: Approximately 68% of the observations lie within one standard deviation to either side of the mean, that is, between $\bar{x} - s$ and $\bar{x} + s$.

Property 2: Approximately 95% of the observations lie within two standard deviations to either side of the mean, that is, between $\bar{x} - 2s$ and $\bar{x} + 2s$.

Property 3: Approximately 99.7% of the observations lie within three standard deviations to either side of the mean, that is, between $\bar{x} - 3s$ and $\bar{x} + 3s$.

These three properties are illustrated together in Fig. 3.9.

FIGURE 3.9

The empirical rule

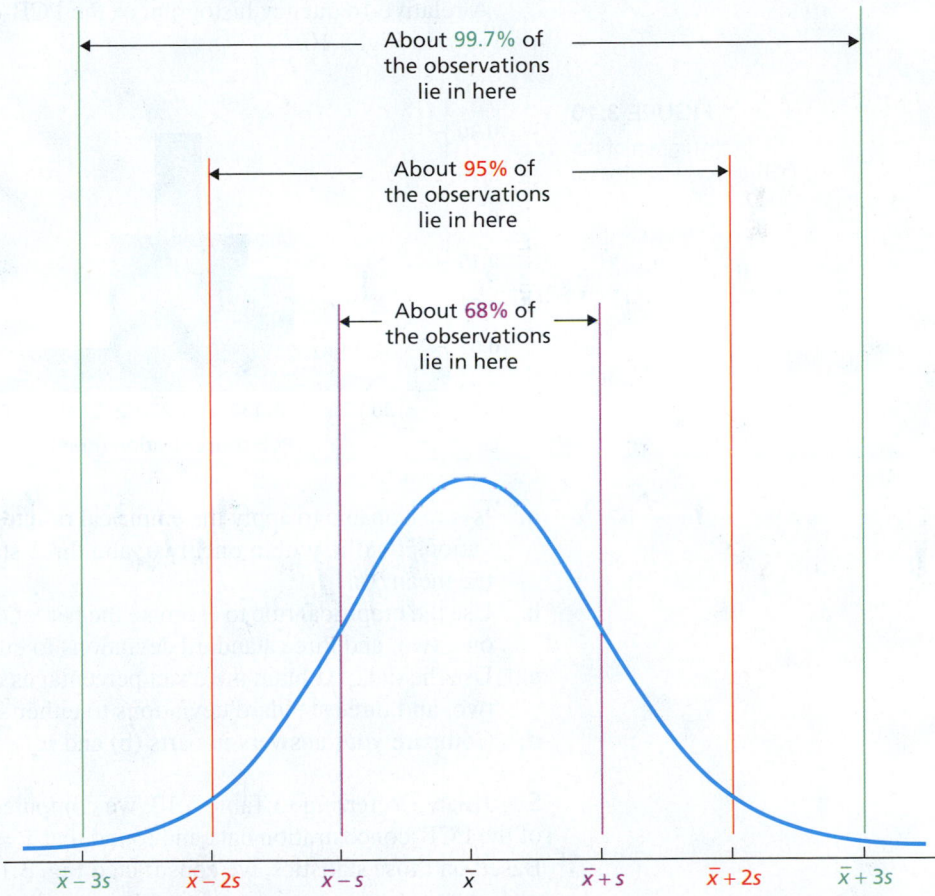

Note the following:

- The empirical rule is also known as the **68-95-99.7 rule.**
- The percentage approximations given by the empirical rule can be somewhat rough.
- The percentage approximations given by the empirical rule tend to be accurate when the distribution of the data set is close to bell shaped.

■ ■ ■ EXAMPLE 3.17   **The Empirical Rule**

PCBs and Pelicans Polychlorinated biphenyls (PCBs), industrial pollutants, are known to be carcinogens and a great danger to natural ecosystems. As a result of several studies, PCB production was banned in the United States in 1979 and by the Stockholm Convention on Persistent Organic Pollutants in 2001. One study, published in 1972 by R. Risebrough, is titled "Effects of Environmental Pollutants Upon

Animals Other Than Man" (*Proceedings of the 6th Berkeley Symposium on Mathematics and Statistics, VI*, University of California Press, pp. 443–463). In that study, 60 Anacapa pelican eggs were collected and measured for their shell thickness, in millimeters (mm), and concentration of PCBs, in parts per million (ppm). The PCB concentrations are presented in Table 3.12.

TABLE 3.12

PCB concentrations, in parts per million, of 60 pelican eggs

139	166	175	260	204	138	316	396	46	218
173	220	147	216	216	177	246	296	188	89
198	122	250	256	261	132	212	171	164	199
214	177	205	208	320	191	305	230	204	143
175	119	216	185	236	356	289	324	109	265
193	203	214	150	229	236	144	232	87	237

A relative-frequency histogram of the PCB-concentration data in Table 3.12 is provided in Fig. 3.10.

FIGURE 3.10

Histogram of the PCB-concentration data

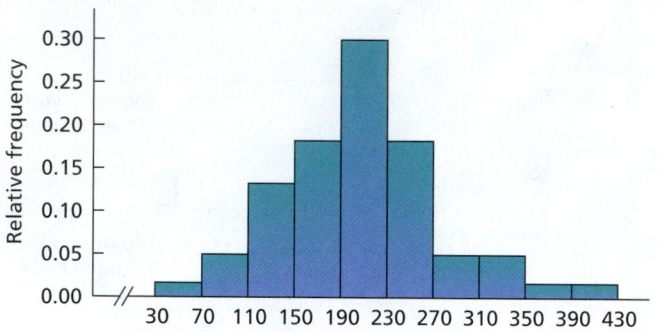

a. Is it reasonable to apply the empirical rule to estimate the percentages of observations that lie within one, two, and three standard deviations to either side of the mean?

b. Use the empirical rule to estimate the percentages of observations that lie within one, two, and three standard deviations to either side of the mean.

c. Use the data to obtain the exact percentages of observations that lie within one, two, and three standard deviations to either side of the mean.

d. Compare your answers in parts (b) and (c).

Solution Referring to Table 3.12, we computed the mean and standard deviation of the PCB-concentration data and found that $\bar{x} = 206.45$ ppm and $s = 66.42$ ppm. Based on those statistics, we constructed Fig. 3.11.

FIGURE 3.11

The mean and one, two, and three standard deviations to either side of the mean for the PCB-concentration data

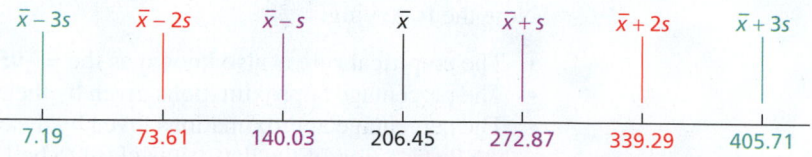

a. Yes, it is reasonable to apply the empirical rule to estimate the percentages of observations that lie within one, two, and three standard deviations to either side of the mean because the distribution of the sample of PCB concentrations is roughly bell shaped, as seen in Fig. 3.10.

b. Using the empirical rule, we estimate that 68% of the observations lie within one standard deviation to either side of the mean, 95% lie within two standard deviations to either side of the mean, and 99.7% lie within three standard deviations to either side of the mean.

c. Referring to Table 3.12 and Fig. 3.11, we found that 43 of the 60 observations, or 71.7%, lie within one standard deviation to either side of the mean; 57 of the 60 observations, or 95%, lie within two standard deviations to either side of the mean; and 60 of the 60 observations, or 100%, lie within three standard deviations to either side of the mean.

You try it!

Exercise 3.139 on page 124

d. Comparing the percentages obtained in parts (b) and (c) [68%, 95%, and 99.7% versus 71.7%, 95%, and 100%], we see that, in this case, the empirical rule provides excellent estimates of the actual percentages. This result is not surprising because the distribution of PCB concentrations is quite close to bell shaped, as seen in Fig. 3.10. ■

Provided that a data set has a distribution that is roughly bell shaped, the empirical rule also permits us to make pertinent statements about the data set when we know only its mean and standard deviation, which is frequently all that we know. The next example illustrates this point.

EXAMPLE 3.18   The Empirical Rule

Forearm Length From Example 3.16 (page 118), recall that the mean and standard deviation of forearm lengths for a sample of 140 men are 18.8 in. and 1.12 in., respectively.

In Example 3.16, we used Chebyshev's rule to make pertinent statements about the forearm lengths of the men in the sample. Presuming that the distribution of forearm lengths of the men in the sample is roughly bell shaped (which, in fact, it actually is), we can use the empirical rule to get more accurate estimates.

a. Apply Property 1 of the empirical rule to make pertinent statements about the forearm lengths of the men in the sample.
b. Repeat part (a) for Property 2 of the empirical rule.
c. Repeat part (a) for Property 3 of the empirical rule.

Solution We refer to Fig. 3.8 on page 118.

a. By Property 1 of the empirical rule, approximately 68% of the men in the sample have forearm lengths within one standard deviation to either side of the mean. Now, 68% of 140 is 95.2, and one standard deviation to either side of the mean is from 17.68 to 19.92, as we see from Fig. 3.8.

Interpretation Approximately 95 (95.2 rounded) of the 140 men in the sample have forearm lengths between 17.68 in. and 19.92 in.

b. By Property 2 of the empirical rule, approximately 95% of the men in the sample have forearm lengths within two standard deviations to either side of the mean. Now, 95% of 140 is 133, and two standard deviations to either side of the mean is from 16.56 to 21.04, as we see from Fig. 3.8.

Interpretation Approximately 133 of the 140 men in the sample have forearm lengths between 16.56 in. and 21.04 in.

c. By Property 3 of the empirical rule, approximately 99.7% of the men in the sample have forearm lengths within three standard deviations to either side of the mean. Now, 99.7% of 140 is 139.58, and three standard deviations to either side of the mean is from 15.44 to 22.16, as we see from Fig. 3.8.

Interpretation Approximately 140 (139.58 rounded), that is, approximately all of the 140 men in the sample have forearm lengths between 15.44 in. and 22.16 in.

You try it!

Exercise 3.141 on page 124

You should compare the results in parts (b) and (c) in this example to that of parts (a) and (b) of Example 3.16. ■

Exercises 3.3

Understanding the Concepts and Skills

3.96 Discuss the pros and cons of Chebyshev's rule.

3.97 If the condition for using the empirical rule is met, why should that rule be used instead of Chebyshev's rule?

3.98 Apply Chebyshev's rule with $k = 3$ to verify that at least 89% of the observations in any data set lie within three standard deviations to either side of the mean, that is, between $\bar{x} - 3s$ and $\bar{x} + 3s$.

3.99 What does Chebyshev's rule say about the percentage of observations in any data set that lie within
a. four standard deviations to either side of the mean?
b. 2.5 standard deviations to either side of the mean?

3.100 What does Chebyshev's rule say about the percentage of observations in any data set that lie within
a. 1.25 standard deviations to either side of the mean?
b. 3.5 standard deviations to either side of the mean?

3.101 Apply Chebyshev's rule with $k = 5$ and interpret your answer.

3.102 Consider the following data set.

30	16	22	23	18	18
20	24	19	13	9	28

a. Draw a graph similar to Fig. 3.5 on page 111.
b. Compare the percentage of the observations that actually lie within two standard deviations to either side of the mean with that given by Chebyshev's rule with $k = 2$.
c. Repeat part (b) with $k = 3$.

3.103 Consider the following data set.

82	85	65	91	81
78	94	84	86	84

a. Draw a graph similar to Fig. 3.5 on page 111.
b. Compare the percentage of the observations that actually lie within two standard deviations to either side of the mean with that given by Chebyshev's rule with $k = 2$.
c. Repeat part (b) with $k = 3$.

3.104 What condition on a data set is required to apply the empirical rule?

3.105 Each of the following smooth curves represents the shape of a data set. In each case, decide whether application of the empirical rule to the data set is appropriate. Explain your answers.

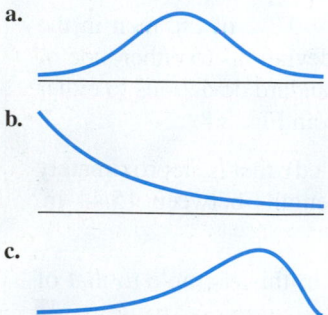

3.106 What is another name for the empirical rule? Why is that name appropriate?

3.107 In this exercise, you will compare Chebyshev's rule and the empirical rule.
a. Compare the estimates given by the two rules for the percentage of observations that lie within two standard deviations to either side of the mean. Comment on the differences.
b. Compare the estimates given by the two rules for the percentage of observations that lie within three standard deviations to either side of the mean. Comment on the differences.

*Apply Chebyshev's rule to solve Exercises **3.108–3.119**.*

3.108 A quantitative data set has mean 10 and standard deviation 3. Fill in the following blanks:
a. At least 75% of the observations lie between _____ and _____.
b. At least _____ % of the observations lie between 1 and 19.

3.109 A quantitative data set has mean 25 and standard deviation 5. Fill in the following blanks:
a. At least 89% of the observations lie between _____ and _____.
b. At least _____ % of the observations lie between 15 and 35.

3.110 A quantitative data set has mean 15 and standard deviation 2. At least what percentage of the observations lie between 7 and 23?

3.111 A quantitative data set has mean 30 and standard deviation 4. At least what percentage of the observations lie between 10 and 50?

3.112 A quantitative data set has size 40. At least how many observations lie within two standard deviations to either side of the mean?

3.113 A quantitative data set has size 80. At least how many observations lie within two standard deviations to either side of the mean?

3.114 A quantitative data set has size 60. At least how many observations lie within three standard deviations to either side of the mean?

3.115 A quantitative data set has size 50. At least how many observations lie within three standard deviations to either side of the mean?

3.116 A quantitative data set of size 80 has mean 30 and standard deviation 5. At least how many observations lie between 20 and 40?

3.117 A quantitative data set of size 60 has mean 100 and standard deviation 16. At least how many observations lie between 68 and 132?

3.118 A quantitative data set of size 200 has mean 20 and standard deviation 4. At least how many observations lie between 8 and 32?

3.119 A quantitative data set of size 150 has mean 35 and standard deviation 4. At least how many observations lie between 23 and 47?

*In each of Exercises **3.120–3.131**, the quantitative data set under consideration has roughly a bell-shaped distribution. Apply the empirical rule to solve each exercise.*

3.120 The data set has mean 10 and standard deviation 3. Fill in the following blanks:
a. Approximately 68% of the observations lie between _____ and _____.
b. Approximately 95% of the observations lie between _____ and _____.
c. Approximately 99.7% of the observations lie between _____ and _____.

3.121 The data set has mean 25 and standard deviation 5. Fill in the following blanks:

a. Approximately 68% of the observations lie between _____ and _____.

b. Approximately 95% of the observations lie between _____ and _____.

c. Approximately 99.7% of the observations lie between _____ and _____.

3.122 The data set has mean 15 and standard deviation 2. Approximately what percentage of the observations lie between 13 and 17?

3.123 The data set has mean 30 and standard deviation 4. Approximately what percentage of the observations lie between 22 and 38?

3.124 The data set has size 40. Approximately how many observations lie within two standard deviations to either side of the mean?

3.125 The data set has size 80. Approximately how many observations lie within two standard deviations to either side of the mean?

3.126 The data set has size 60. Approximately how many observations lie within three standard deviations to either side of the mean?

3.127 The data set has size 50. Approximately how many observations lie within one standard deviation to either side of the mean?

3.128 The data set has 80 observations and has mean 30 and standard deviation 5. Approximately how many observations lie between 20 and 40?

3.129 The data set has 250 observations and has mean 100 and standard deviation 16. Approximately how many observations lie between 52 and 148?

3.130 The data set has 200 observations and has mean 20 and standard deviation 4. Approximately how many observations lie between 16 and 24?

3.131 The data set has 150 observations and has mean 35 and standard deviation 4. Approximately how many observations lie between 31 and 39?

Applying the Concepts and Skills

3.132 Exam Scores. Consider the following sample of exam scores, arranged in increasing order.

28	57	58	64	69	74
79	80	83	85	85	87
87	89	89	90	92	93
94	94	95	96	96	97
97	97	97	98	100	100

The sample mean and sample standard deviation of these exam scores are 85 and 16.1, respectively. Modeling your solutions after those in Example 3.15 on pages 117–118, solve the following problems.

a. Compare the percentage of the observations that actually lie within two standard deviations to either side of the mean with that given by Chebyshev's rule with $k = 2$.

b. Repeat part (a) with $k = 3$.

c. Interpret your results from parts (a) and (b).

3.133 Near-Earth Objects. Objects such as asteroids and comets that come into proximity with the Earth are called near-Earth

objects (NEOs). The National Aeronautics and Space Administration (NASA) tracks and catalogues all NEOs that are at least 1 kilometer wide. Data on NEOs can be found on the NASA website. The following table gives the relative velocities in kilometers per second (km/s), arranged in increasing order, for the NEO close approaches to the earth during June 2013.

4	5	5	5	5	6	6
8	8	8	9	9	9	9
9	9	9	10	11	11	12
13	13	14	14	17	17	19
20	20	21	28	30		

The sample mean and sample standard deviation of these velocities are 11.9 km/s and 6.5 km/s, respectively. Modeling your solutions after those in Example 3.15 on pages 117–118, solve the following problems.

a. Compare the percentage of the observations that actually lie within two standard deviations to either side of the mean with that given by Chebyshev's rule with $k = 2$.

b. Repeat part (a) with $k = 3$.

c. Interpret your results from parts (a) and (b).

In each of Exercises 3.134–3.137,

a. construct a graph similar to Fig. 3.8 on page 118.

b. apply Chebyshev's rule with $k = 2$ to make pertinent statements about the observations in the sample.

c. repeat part (b) with $k = 3$.

Model your solutions after those in Example 3.16 on page 118.

3.134 Early-Onset Dementia. Dementia is the loss of intellectual and social abilities severe enough to interfere with judgment, behavior, and daily functioning. Alzheimer's disease is the most common type of dementia. In the article "Living with Early Onset Dementia: Exploring the Experience and Developing Evidence-Based Guidelines for Practice" (*Alzheimer's Care Quarterly*, Vol. 5, Issue 2, pp. 111–122), P. Harris and J. Keady explored the experience and struggles of people diagnosed with dementia and their families. The ages of a simple random sample of 21 people with early-onset dementia have a mean of 52.5 years and a standard deviation of 6.5 years.

3.135 Iron Deficiency? Iron is essential to most life forms and to normal human physiology. It is an integral part of many proteins and enzymes that maintain good health. Recommendations for iron are provided in *Dietary Reference Intakes*, developed by the Institute of Medicine of the National Academy of Sciences. The recommended dietary allowance (RDA) of iron for adult females under the age of 51 is 18 milligrams (mg) per day. The iron intakes during a 24-hour period for a random sample of 45 adult females under the age of 51 have a mean of 14.7 mg and a standard deviation of 3.1 mg.

3.136 Body Mass Index. Body mass index (BMI) is a measure of body fat based on height and weight. According to the document *Dietary Guidelines for Americans* published by the U.S. Department of Agriculture and the U.S. Department of Health and Human Services, for adults, a BMI of greater than 25 indicates an above healthy weight (i.e., overweight or obese). The BMIs of 75 randomly selected U.S. adults have a mean of 26.0 and a standard deviation of 5.0.

3.137 Body Temperature. A study by researchers at the University of Maryland addressed the question of whether the mean body temperature of humans is 98.6°F. The results of the study by P. Mackowiak et al. appeared in the article "A Critical Appraisal of 98.6°F, the Upper Limit of the Normal Body Temperature, and Other Legacies of Carl Reinhold August Wunderlich" (*Journal of the American Medical Association*, Vol. 268, pp. 1578–1580). Among other data, the researchers obtained the body temperatures of 93 healthy humans. The temperatures had a mean of 98.1°F and a standard deviation of 0.65°F.

3.138 Clocking the Cheetah. The cheetah (*Acinonyx jubatus*) is the fastest land mammal and is highly specialized to run down prey. The cheetah often exceeds speeds of 60 miles per hour (mph) and, according to the online document "Cheetah Conservation in Southern Africa" (*Trade & Environment Database (TED) Case Studies*, Vol. 8, No. 2) by J. Urbaniak, the cheetah is capable of speeds up to 72 mph. The following table gives the top speeds, in miles per hour, arranged in increasing order, for a sample of 35 cheetahs.

52.4	52.6	54.7	54.8	55.4	55.5	55.9
56.5	57.3	57.5	57.6	57.8	57.8	58.1
58.3	58.7	59.0	59.2	59.6	59.7	59.8
60.1	60.2	60.6	60.7	60.9	61.3	61.6
62.3	62.6	63.4	65.0	65.2	66.0	75.3

The sample mean and sample standard deviation of these speeds are 59.53 mph and 4.27 mph, respectively. A histogram of the speeds is bell shaped. Modeling your solutions after those in Example 3.17 on pages 119–121, solve the following problems.

a. Is it reasonable to apply the empirical rule to estimate the percentages of observations that lie within one, two, and three standard deviations to either side of the mean?

b. Use the empirical rule to estimate the percentages of observations that lie within one, two, and three standard deviations to either side of the mean.

c. Use the data to obtain the exact percentages of observations that lie within one, two, and three standard deviations to either side of the mean.

d. Compare your answers in parts (b) and (c).

3.139 Malnutrition and Poverty. R. Reifen et al. studied various nutritional measures of Ethiopian school children and published their findings in the paper "Ethiopian-Born and Native Israeli School Children Have Different Growth Patterns" (*Nutrition*, Vol. 19, pp. 427–431). The study, conducted in Azezo, North West Ethiopia, found that malnutrition is prevalent in primary and secondary school children because of economic poverty. The weights, in kilograms (kg), of 60 randomly selected male Ethiopian-born school children aged 12–15 years old are presented in increasing order in the following table.

36.3	37.7	38.0	38.8	38.9	39.0
39.3	40.9	41.1	41.3	41.5	41.8
42.0	42.0	42.1	42.5	42.5	42.8
42.9	43.3	43.4	43.5	44.0	44.4
44.7	44.8	45.2	45.2	45.2	45.4
45.5	45.7	45.9	45.9	46.2	46.3
46.5	46.6	46.8	47.2	47.4	47.5
47.8	47.9	48.1	48.2	48.3	48.4
48.5	48.6	48.9	49.1	49.2	49.5
50.9	51.4	51.8	52.8	53.8	56.6

The sample mean and sample standard deviation of these weights are 45.30 kg and 4.16 kg, respectively. A histogram of the weights

is shown in Exercise 2.142 on page 81. Modeling your solutions after those in Example 3.17 on pages 119–121, solve the following problems.

a. Is it reasonable to apply the empirical rule to estimate the percentages of observations that lie within one, two, and three standard deviations to either side of the mean?

b. Use the empirical rule to estimate the percentages of observations that lie within one, two, and three standard deviations to either side of the mean.

c. Use the data to obtain the exact percentages of observations that lie within one, two, and three standard deviations to either side of the mean.

d. Compare your answers in parts (b) and (c).

In each of Exercises 3.140–3.143, the quantitative data set under consideration has roughly a bell-shaped distribution. For each exercise,

a. *construct a graph similar to Fig. 3.8 on page 118.*

b. *apply Property 1 of the empirical rule to make pertinent statements about the observations in the sample.*

c. *repeat part (b) for Property 2 of the empirical rule.*

d. *repeat part (b) for Property 3 of the empirical rule.*

Model your solutions after those in Example 3.18 on page 121.

3.140 Giant Tarantulas. One of the larger species of tarantulas is the *Grammostola mollicoma*, whose common name is the Brazilian giant tawny red. A tarantula has two body parts. The anterior part of the body is covered above by a shell, or carapace. F. Costa and F. Perez–Miles discussed the carapace length of the adult male *G. mollicoma* in the article "Reproductive Biology of Uruguayan Theraphosids" (*The Journal of Arachnology*, Vol. 30, No. 3, pp. 571–587). The carapace lengths of a random sample of 50 adult male *G. mollicoma* have a mean of 18.14 mm and a standard deviation of 1.76 mm.

3.141 Student-to-Faculty Ratio. A sample of fifth-grade classes was studied in the journal article "Predicting Feelings of School Safety for Lower, Middle, and Upper School Students" (*Applied Psychology in Criminal Justice*, Vol. 7, Issue 2, pp. 59–76) by R. Bachman et al. One of the variables collected was the class size in terms of student-to-faculty ratio. The student-to-faculty ratios of the 81 fifth-grade classes sampled have a mean of 15.83 and a standard deviation of 1.74.

3.142 PGA Driving Distances. The PGA TOUR provides various statistics on performance of players in the Professional Golfers' Association of America. For the week ending September 9, 2013, the year-to-date leader for longest average drive was Bubba Watson. For his 773 drives, he averaged 298.5 yards. Assume a standard deviation of 19.7 yards.

3.143 Brain Weights. In 1905, R. Pearl published the article "Biometrical Studies on Man. I. Variation and Correlation in Brain Weight" (*Biometrika*, Vol. 4, pp. 13–104) in which he studied the brain weights of Swedish males. The brain weights of a random sample of 225 Swedish men have a mean of 1.40 kg and standard deviation 0.11 kg.

Extending the Concepts and Skills

3.144 What does Chebyshev's rule say about the percentage of observations that lie within one standard deviation to either side of the mean? Discuss your answer.

3.145 How many standard deviations to either side of the mean must we go to ensure that, for any data set, at least 99% of the observations lie within?

3.146 How many standard deviations to either side of the mean must we go to ensure that, for any data set, at least 95% of the observations lie within?

3.147 A data set consists of $2m^2 - 1$ zeros, one $-m$, and one m.
a. Compute $\bar{x}$ and s for this data set.
b. How many standard deviations from the mean is the observation m?
c. Assuming that $m \geq 4$, what percentage of the observations lie within three standard deviations to either side of the mean?

3.4 The Five-Number Summary; Boxplots

So far, we have focused on the mean and standard deviation to measure center and variation. We now examine several descriptive measures based on percentiles.

Unlike the mean and standard deviation, descriptive measures based on percentiles are *resistant*—they are not sensitive to the influence of a few extreme observations. For this reason, descriptive measures based on percentiles are often preferred over those based on the mean and standard deviation.

Quartiles

As you learned in Section 3.1, the median of a data set divides the data into two equal parts: the bottom 50% and the top 50%. More generally, for each p between 0 and 100, the **pth percentile** is the number that divides the bottom $p\%$ of the data from the top $(100 - p)\%$.[†] We use the letter P, subscripted with the percent in question, to denote a percentile.

For example, the 30th percentile is denoted P_{30} and is the number that divides the bottom 30% of the data from the top 70%; the 2.5th percentile is denoted $P_{2.5}$ and is the number that divides the bottom 2.5% of the data from the top 97.5%. Note that the median is also the 50th percentile.

Certain percentiles are particularly important: the 10th, 20th, ..., 90th percentiles are called the **deciles** and divide a data set into tenths (10 equal parts); the 20th, 40th, 60th, and 80th percentiles are called the **quintiles** and divide a data set into fifths (five equal parts).

The most commonly used percentiles other than the median are the **quartiles,** which are the 25th, 50th, and 75th percentiles, and divide a data set into quarters (four equal parts). Because of their importance, we use a special notation for the three quartiles, namely, Q_1, Q_2, and Q_3.

Hence, roughly speaking, the **first quartile, Q_1,** is the number that divides the bottom 25% of the data from the top 75%; the **second quartile, Q_2,** is the median, which, as you know, is the number that divides the bottom 50% of the data from the top 50%; and the **third quartile, Q_3,** is the number that divides the bottom 75% of the data from the top 25%. Figure 3.12 depicts the quartiles for uniform, bell-shaped, right-skewed, and left-skewed distributions.

FIGURE 3.12

Quartiles for (a) uniform, (b) bell-shaped, (c) right-skewed, and (d) left-skewed distributions

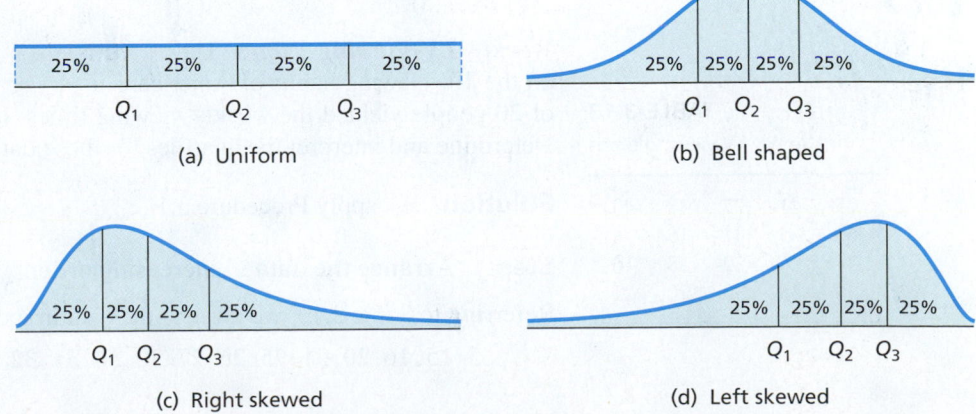

(a) Uniform

(b) Bell shaped

(c) Right skewed

(d) Left skewed

[†] This is a rather rough definition of percentiles. More precise definitions exist but, for our purposes, this intuitive definition will suffice for most percentiles.

DEFINITION 3.7

Quartiles

First, arrange the data in increasing order. Next, determine the median. Then, divide the (ordered) data set into two halves, a bottom half and a top half; if the number of observations is odd, include the median in both halves.

- The **first quartile** (Q_1) is the median of the bottom half of the data set.
- The **second quartile** (Q_2) is the median of the entire data set.
- The **third quartile** (Q_3) is the median of the top half of the data set.

? What Does It Mean?

The quartiles divide a data set into quarters (four equal parts).

Note: Not all statisticians define quartiles in exactly the same way.[†] Our method for computing quartiles is consistent with the one used by Professor John Tukey for the construction of boxplots (which will be discussed shortly). Other definitions may lead to different values, but, in practice, the differences tend to be small with large data sets.

As the definition of quartiles is somewhat complex, it is helpful to have a step-by-step procedure for actually determining the quartiles of a data set. To that end, we present the following procedure.

PROCEDURE 3.1

To Determine the Quartiles

Step 1   Arrange the data in increasing order.

Step 2   Find the median of the entire data set. This value is the second quartile, Q_2.

Step 3   Divide the ordered data set into two halves, a bottom half and a top half; if the number of observations is odd, include the median in both halves.

Step 4   Find the median of the bottom half of the data set. This value is the first quartile, Q_1.

Step 5   Find the median of the top half of the data set. This value is the third quartile, Q_3.

Step 6   Summarize the results.

EXAMPLE 3.19   **Quartiles**

TABLE 3.13

Weekly TV-viewing times

25	41	27	32	43
66	35	31	15	5
34	26	32	38	16
30	38	30	20	21

Weekly TV-Viewing Times The A. C. Nielsen Company publishes information on the TV-viewing habits of Americans in *Nielsen Report on Television*. A sample of 20 people yielded the weekly viewing times, in hours, displayed in Table 3.13. Determine and interpret the quartiles for these data.

Solution   We apply Procedure 3.1.

Step 1 **Arrange the data in increasing order.**

Referring to Table 3.13, we arrange the data in increasing order and get

5 15 16 20 21 25 26 27 30 **30 31** 32 32 34 35 38 38 41 43 66

[†]For a detailed discussion of the different methods for computing quartiles, see the online article "Quartiles in Elementary Statistics" by E. Langford (*Journal of Statistics Education*, Vol 14, No. 3, www.amstat.org/publications/jse/v14n3/langford.html).

Step 2 Find the median of the entire data set. This value is the second quartile, Q_2.

The number of observations is 20 and, consequently, the median is at position $(20 + 1)/2 = 10.5$, halfway between the tenth and eleventh observations (shown in boldface) in the ordered list. Thus, the median of the entire data set is $(30 + 31)/2 = 30.5$. So, $Q_2 = 30.5$.

Step 3 Divide the ordered data set into two halves, a bottom half and a top half; if the number of observations is odd, include the median in both halves.

Referring to the ordered list in Step 1, we see that the bottom half and top half of the data set are as follows:

$$\underbrace{5 \ 15 \ 16 \ 20 \ \mathbf{21} \ \mathbf{25} \ 26 \ 27 \ 30 \ 30}_{\text{Bottom half}} \qquad \underbrace{31 \ 32 \ 32 \ 34 \ \mathbf{35} \ \mathbf{38} \ 38 \ 41 \ 43 \ 66}_{\text{Top half}}$$

Step 4 Find the median of the bottom half of the data set. This value is the first quartile, Q_1.

Referring to Step 3, we see that the bottom half of the data set has 10 observations, so its median is at position $(10 + 1)/2 = 5.5$, halfway between the fifth and sixth observations (shown in boldface) in the ordered list. Thus, the median of this data set—and hence the first quartile—is $(21 + 25)/2 = 23$; that is, $Q_1 = 23$.

Step 5 Find the median of the top half of the data set. This value is the third quartile, Q_3.

Referring again to Step 3, we see that the top half of the data set has 10 observations, so its median is at position $(10 + 1)/2 = 5.5$, halfway between the fifth and sixth observations (shown in boldface) in the ordered list. Thus, the median of this data set—and hence the third quartile—is $(35 + 38)/2 = 36.5$; that is, $Q_3 = 36.5$.

Step 6 Summarize the results.

In summary, then, the three quartiles for the TV-viewing times in Table 3.13 are $Q_1 = 23$ hours, $Q_2 = 30.5$ hours, and $Q_3 = 36.5$ hours.

Interpretation We see that 25% of the TV-viewing times are less than 23 hours, 25% are between 23 hours and 30.5 hours, 25% are between 30.5 hours and 36.5 hours, and 25% are greater than 36.5 hours. ■

StatCrunch

Report 3.6

You try it!

Exercise 3.167(a) on page 136

In Example 3.19, the number of observations is 20, which is even. To illustrate how to find quartiles when the number of observations is odd, we consider the following example.

■■■ **EXAMPLE 3.20** Quartiles

TABLE 3.14
Maximum wind speeds, in miles per hour, for tropical cyclones

60	70	85	65	100
60	110	45	80	40
105	80	115	90	50
45	90	115	50	

Cyclone Wind Speeds From *Tropical Cyclone Reports*, published by the National Hurricane Center, we obtained the data shown in Table 3.14 on maximum wind speeds, in miles per hour (mph), for one year's tropical cyclones in the Atlantic Basin. Determine and interpret the quartiles for these data.

Solution We apply Procedure 3.1.

Step 1 Arrange the data in increasing order.

40 45 45 50 50 60 60 65 70 **80** 80 85 90 90 100 105 110 115 115

Step 2 **Find the median of the entire data set. This value is the second quartile, Q_2.**

The number of observations is 19 and, consequently, the median is at position $(19 + 1)/2 = 10$, the tenth observation (shown in boldface) in the ordered list. Thus, the median of the entire data set is 80. So, $Q_2 = 80$.

Step 3 **Divide the ordered data set into two halves, a bottom half and a top half; if the number of observations is odd, include the median in both halves.**

Because the number of observations is 19, which is an odd number, we include the median in both the bottom and top halves of the data set. So, we see by referring to the ordered list in Step 1 that the bottom half and top half of the data set are as follows:

$$\underbrace{40\ \ 45\ \ 45\ \ 50\ \ \mathbf{50}\ \ \mathbf{60}\ \ 60\ \ 65\ \ 70\ \ 80}_{\text{Bottom half}}\qquad\underbrace{80\ \ 80\ \ 85\ \ 90\ \ \mathbf{90}\ \ \mathbf{100}\ \ 105\ \ 110\ \ 115\ \ 115}_{\text{Top half}}$$

Step 4 **Find the median of the bottom half of the data set. This value is the first quartile, Q_1.**

Referring to Step 3, we see that the bottom half of the data set has 10 observations, so its median is at position $(10 + 1)/2 = 5.5$, halfway between the fifth and sixth observations (shown in boldface) in the ordered list. Thus, the median of this data set—and hence the first quartile—is $(50 + 60)/2 = 55$; that is, $Q_1 = 55$.

Step 5 **Find the median of the top half of the data set. This value is the third quartile, Q_3.**

Referring again to Step 3, we see that the top half of the data set has 10 observations, so its median is at position $(10 + 1)/2 = 5.5$, halfway between the fifth and sixth observations (shown in boldface) in the ordered list. Thus, the median of this data set—and hence the third quartile—is $(90 + 100)/2 = 95$; that is, $Q_3 = 95$.

Step 6 **Summarize the results.**

You try it!

Exercise 3.169(a) on page 137

In summary, the three quartiles for the maximum wind speeds in Table 3.14 on page 127 are $Q_1 = 55$ mph, $Q_2 = 80$ mph, and $Q_3 = 95$ mph.

Interpretation We see that roughly 25% of the maximum wind speeds are less than 55 mph, roughly 25% are between 55 mph and 80 mph, roughly 25% are between 80 mph and 95 mph, and roughly 25% are greater than 95 mph. ∎

The Interquartile Range

Next, we discuss the *interquartile range*. Because quartiles are used to define the interquartile range, it is the preferred measure of variation when the median is used as the measure of center. Like the median, the interquartile range is a resistant measure.

DEFINITION 3.8

? What Does It Mean?

Roughly speaking, the IQR gives the range of the middle 50% of the observations.

Interquartile Range

The **interquartile range**, or **IQR**, is the difference between the first and third quartiles; that is, $\text{IQR} = Q_3 - Q_1$.

In Example 3.21, we show how to obtain the interquartile range for the data on TV-viewing times.

■■■   **EXAMPLE 3.21**   **The Interquartile Range**

Weekly TV-Viewing Times Find the IQR for the TV-viewing-time data given in Table 3.13 on page 126.

Solution As we discovered in Example 3.19 (pages 126–127), the first and third quartiles are 23 and 36.5, respectively. Therefore,

$$\text{IQR} = Q_3 - Q_1 = 36.5 - 23 = 13.5 \text{ hours.}$$

Interpretation The middle 50% of the TV-viewing times are spread out over a 13.5-hour interval, roughly.

Exercise 3.167(b)
on page 136

The Five-Number Summary

From the three quartiles, we can obtain a measure of center (the median, Q_2) and measures of variation of the two middle quarters of the data, $Q_2 - Q_1$ for the second quarter and $Q_3 - Q_2$ for the third quarter. But the three quartiles don't tell us anything about the variation of the first and fourth quarters.

To gain that information, we need only include the minimum and maximum observations as well. Then the variation of the first quarter can be measured as the difference between the minimum and the first quartile, $Q_1 - \text{Min}$, and the variation of the fourth quarter can be measured as the difference between the third quartile and the maximum, $\text{Max} - Q_3$.

Thus the minimum, maximum, and quartiles together provide, among other things, information on center and variation.

DEFINITION 3.9

? **What Does It Mean?**

The five-number summary of a data set consists of the minimum, maximum, and quartiles, in increasing order.

Five-Number Summary

The **five-number summary** of a data set is Min, Q_1, Q_2, Q_3, Max.

In Example 3.22, we show how to obtain and interpret the five-number summary of a set of data.

■■■   **EXAMPLE 3.22**   **The Five-Number Summary**

Weekly TV-Viewing Times Find and interpret the five-number summary for the TV-viewing-time data given in Table 3.13 on page 126.

Solution From the ordered list of the entire data set (see page 126), $\text{Min} = 5$ and $\text{Max} = 66$. Furthermore, as we showed earlier, $Q_1 = 23$, $Q_2 = 30.5$, and $Q_3 = 36.5$. Consequently, the five-number summary of the data on TV-viewing times is 5, 23, 30.5, 36.5, and 66 hours. The variations of the four quarters of the TV-viewing-time data are therefore 18, 7.5, 6, and 29.5 hours, respectively.

StatCrunch

Report 3.7

Exercise 3.167(c)
on page 136

Interpretation There is less variation in the middle two quarters of the TV-viewing times than in the first and fourth quarters, and the fourth quarter has the greatest variation of all.

Outliers

In data analysis, the identification of **outliers**—observations that fall well outside the overall pattern of the data—is important. An outlier requires special attention. It may be the result of a measurement or recording error, an observation from a different population, or an unusual extreme observation. Note that an extreme observation need not be an outlier; it may instead be an indication of skewness.

As an example of an outlier, consider the data set consisting of the individual wealths (in dollars) of all U.S. residents. For this data set, the wealth of Bill Gates is an outlier—in this case, an unusual extreme observation.

Whenever you observe an outlier, try to determine its cause. If an outlier is caused by a measurement or recording error, or if for some other reason it clearly does not belong in the data set, the outlier can simply be removed. However, if no explanation for the outlier is apparent, the decision whether to retain it in the data set can be a difficult judgment call.

We can use quartiles and the IQR to identify potential outliers, that is, as a diagnostic tool for spotting observations that may be outliers. To do so, we first define the *lower limit* and the *upper limit* of a data set.

DEFINITION 3.10

What Does It Mean?

The lower limit is the number that lies 1.5 IQRs below the first quartile; the upper limit is the number that lies 1.5 IQRs above the third quartile.

Lower and Upper Limits

The **lower limit** and **upper limit** of a data set are

$$\text{Lower limit} = Q_1 - 1.5 \cdot \text{IQR};$$
$$\text{Upper limit} = Q_3 + 1.5 \cdot \text{IQR}.$$

Note: The lower limit is also called the **lower fence** and the upper limit is also called the **upper fence.**

Observations that lie below the lower limit or above the upper limit are **potential outliers.** To determine whether a potential outlier is truly an outlier, you should perform further data analyses by constructing a histogram, stem-and-leaf diagram, and other appropriate graphics that we present later.

EXAMPLE 3.23 Outliers

Weekly TV-Viewing Times For the TV-viewing-time data in Table 3.13 on page 126,

a. obtain the lower and upper limits.
b. determine potential outliers, if any.

Solution

a. As before, $Q_1 = 23$, $Q_3 = 36.5$, and IQR $= 13.5$. Therefore

$$\text{Lower limit} = Q_1 - 1.5 \cdot \text{IQR} = 23 - 1.5 \cdot 13.5 = 2.75 \text{ hours};$$
$$\text{Upper limit} = Q_3 + 1.5 \cdot \text{IQR} = 36.5 + 1.5 \cdot 13.5 = 56.75 \text{ hours}.$$

These limits are shown in Fig. 3.13.

FIGURE 3.13
Lower and upper limits for TV-viewing times

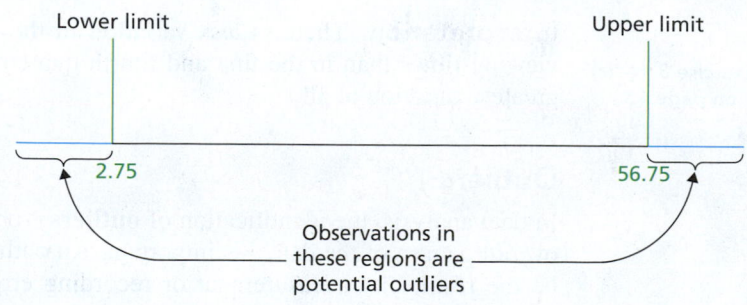

Observations in these regions are potential outliers

Exercise 3.167(d)
on page 136

b. The ordered list of the entire data set on page 126 reveals one observation, 66, that lies outside the lower and upper limits—specifically, above the upper limit. Consequently, 66 is a potential outlier. A histogram and a stem-and-leaf diagram both indicate that the observation of 66 hours is truly an outlier.

Interpretation The weekly viewing time of 66 hours lies outside the overall pattern of the other 19 viewing times in the data set.

Boxplots

A **boxplot,** also called a **box-and-whisker diagram,** is based on the five-number summary and can be used to provide a graphical display of the center and variation of a data set. These diagrams, like stem-and-leaf diagrams, were invented by Professor John Tukey.[†]

To construct a boxplot, we also need the concept of *adjacent values.* The **adjacent values** of a data set are the most extreme observations that still lie within the lower and upper limits; they are the most extreme observations that are not potential outliers. Note that, if a data set has no potential outliers, the adjacent values are just the minimum and maximum observations.

PROCEDURE 3.2   **To Construct a Boxplot**

Step 1   Determine the quartiles.

Step 2   Determine potential outliers and the adjacent values.

Step 3   Draw a horizontal axis on which the numbers obtained in Steps 1 and 2 can be located. Above this axis, mark the quartiles and the adjacent values with vertical lines.

Step 4   Connect the quartiles to make a box, and then connect the box to the adjacent values with lines.

Step 5   Plot each potential outlier with an asterisk.

Note:
- In a boxplot, the two lines emanating from the box are called **whiskers.**
- Boxplots are frequently drawn vertically instead of horizontally.
- Symbols other than an asterisk are often used to plot potential outliers.

EXAMPLE 3.24   **Boxplots**

Weekly TV-Viewing Times The weekly TV-viewing times for a sample of 20 people are given in Table 3.13 on page 126. Construct a boxplot for these data.

Solution We apply Procedure 3.2. For easy reference, we repeat here the ordered list of the TV-viewing times.

5 15 16 20 21 25 26 27 30 30 31 32 32 34 35 38 38 41 43 66

Step 1   Determine the quartiles.

In Example 3.19 (pages 126–127), we found the quartiles for the TV-viewing times to be $Q_1 = 23$, $Q_2 = 30.5$, and $Q_3 = 36.5$.

[†]Several types of boxplots are in common use. Here we discuss a type that displays any potential outliers, sometimes called a **modified boxplot.**

Step 2 Determine potential outliers and the adjacent values.

As we found in Example 3.23(b) on pages 130–131, the TV-viewing times contain one potential outlier, 66. Therefore, from the ordered list of the data, we see that the adjacent values are 5 and 43.

Step 3 Draw a horizontal axis on which the numbers obtained in Steps 1 and 2 can be located. Above this axis, mark the quartiles and the adjacent values with vertical lines.

See Fig. 3.14(a).

Step 4 Connect the quartiles to make a box, and then connect the box to the adjacent values with lines.

See Fig. 3.14(b).

Step 5 Plot each potential outlier with an asterisk.

As we noted in Step 2, this data set contains one potential outlier—namely, 66. It is plotted with an asterisk in Fig. 3.14(c).

FIGURE 3.14   Constructing a boxplot for the TV-viewing times

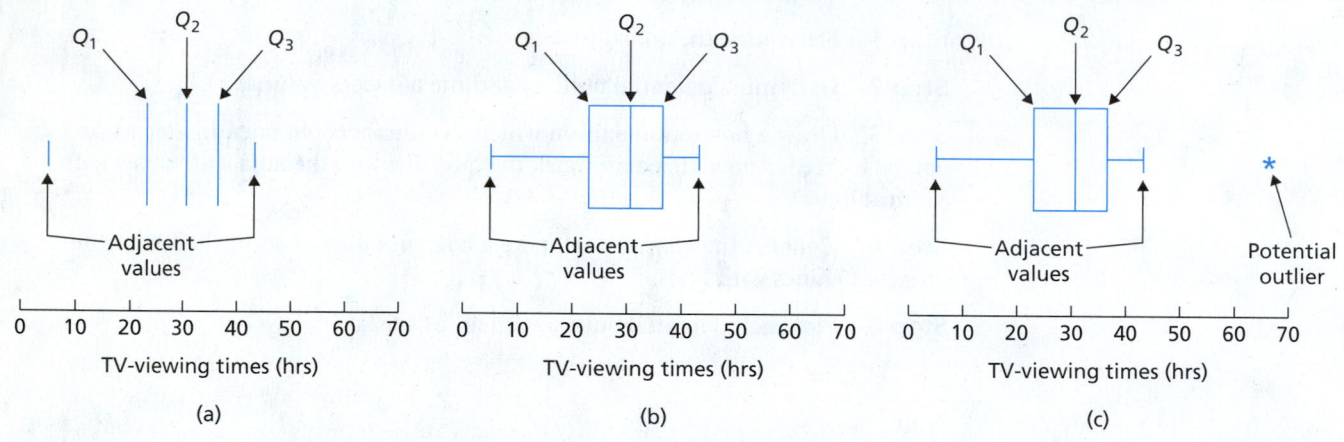

Figure 3.14(c) is a boxplot for the TV-viewing-times data. Because the ends of the combined box are at the quartiles, the width of that box equals the interquartile range, IQR. Notice also that the left whisker represents the spread of the first quarter of the data, the two individual boxes represent the spreads of the second and third quarters, and the right whisker and asterisk represent the spread of the fourth quarter.

Interpretation   There is less variation in the middle two quarters of the TV-viewing times than in the first and fourth quarters, and the fourth quarter has the greatest variation of all.

StatCrunch

Report 3.8

Exercise 3.167(e)
on page 136

You try it!

Other Uses of Boxplots

Boxplots are especially suited for comparing two or more data sets. In doing so, the same scale should be used for all the boxplots.

EXAMPLE 3.25   **Comparing Data Sets by Using Boxplots**

Skinfold Thickness   A study titled "Body Composition of Elite Class Distance Runners" was conducted by M. Pollock et al. to determine whether elite distance runners are actually thinner than other people. Their results were published in

The Marathon: Physiological, Medical, Epidemiological, and Psychological Studies (P. Milvey (ed.), New York: New York Academy of Sciences, p. 366). The researchers measured skinfold thickness, an indirect indicator of body fat, of samples of runners and nonrunners in the same age group. The sample data, in millimeters (mm), presented in Table 3.15 are based on their results. Use boxplots to compare these two data sets, paying special attention to center and variation.

TABLE 3.15

Skinfold thickness (mm) for samples of elite runners and others

Runners			Others			
7.3	6.7	8.7	24.0	19.9	7.5	18.4
3.0	5.1	8.8	28.0	29.4	20.3	19.0
7.8	3.8	6.2	9.3	18.1	22.8	24.2
5.4	6.4	6.3	9.6	19.4	16.3	16.3
3.7	7.5	4.6	12.4	5.2	12.2	15.6

Solution Figure 3.15 displays boxplots for the two data sets, using the same scale.

FIGURE 3.15

Boxplots of the data sets in Table 3.15

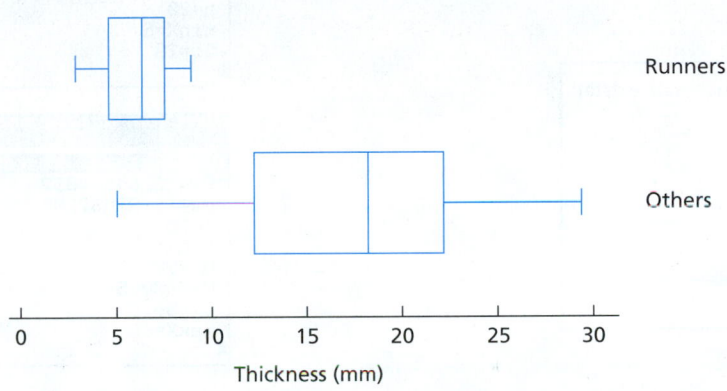

From Fig. 3.15, it is apparent that, on average, the elite runners sampled have smaller skinfold thickness than the other people sampled. Furthermore, there is much less variation in skinfold thickness among the elite runners sampled than among the other people sampled. By the way, when you study inferential statistics, you will be able to decide whether these descriptive properties of the samples can be extended to the populations from which the samples were drawn.

You try it!

Exercise 3.179 on page 138

It is also instructive to examine boxplots for right-skewed, symmetric, and left-skewed distributions. Figure 3.16 shows such distributions and their boxplots. Pay particular attention to how box width and whisker length relate to skewness and symmetry.

FIGURE 3.16

Boxplots for (a) right-skewed, (b) symmetric, and (c) left-skewed distributions

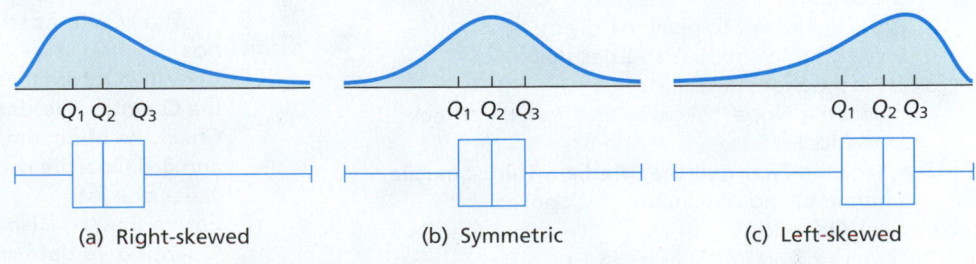

(a) Right-skewed   (b) Symmetric   (c) Left-skewed

THE TECHNOLOGY CENTER

Most statistical technologies have programs that automatically determine five-number summaries and produce boxplots. In this subsection, we present output and step-by-step instructions for such programs. We begin with the five-number summary.

EXAMPLE 3.26   **Using Technology to Obtain a Five-Number Summary**

Weekly TV-Viewing Times   Use Minitab, Excel, or the TI-83/84 Plus to obtain the five-number summary for the TV-viewing times given in Table 3.13 on page 126.

Solution   We applied the five-number-summary programs to the data, resulting in Output 3.3. Steps for generating that output are presented in Instructions 3.3.

OUTPUT 3.3   Five-number summary for the TV-viewing times

MINITAB

Descriptive Statistics: TIMES

Variable	Minimum	Q1	Median	Q3	Maximum
TIMES	5.00	22.00	30.50	37.25	66.00

EXCEL

Descriptive statistics (Quantitative data):

Statistic	TIMES
Minimum	5.0000
Maximum	66.0000
1st Quartile	24.0000
Median	30.5000
3rd Quartile	35.7500

TI-83/84 PLUS

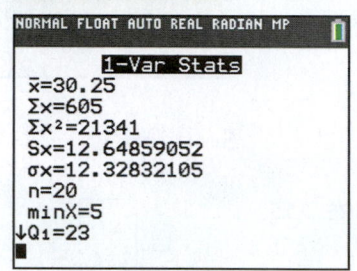

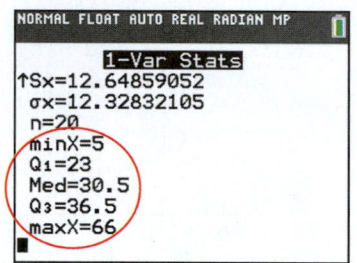

As we noted previously, not all statisticians, statistical software packages, or statistical calculators define quartiles in exactly the same way. We can see this fact by referring to the outputs in Output 3.3.

INSTRUCTIONS 3.3   Steps for generating Output 3.3

MINITAB

1 Store the data from Table 3.13 in a column named TIMES
2 Choose **Stat ➤ Basic Statistics ➤ Display Descriptive Statistics...**
3 Press the F3 key to reset the dialog box
4 Specify TIMES in the **Variables** text box
5 Click the **Statistics...** button
6 Select the **None** option button from the **Check statistics** list
7 Check the **First quartile, Median, Third quartile, Minimum,** and **Maximum** check boxes
8 Click **OK** twice

EXCEL

1 Store the data from Table 3.13 in a column named TIMES
2 Choose **XLSTAT ➤ Describing data ➤ Descriptive statistics**

3 Click the reset button in the lower left corner of the dialog box
4 Click in the **Quantitative data** selection box and then select the column of the worksheet that contains the TIMES data
5 Click the **Options** tab and uncheck the **Charts** check box
6 Click the **Outputs** tab and click the **None** button below the **Quantitative data** check box list
7 Check the **Minimum, Maximum, 1st Quartile, Median,** and **3rd Quartile** check boxes in the **Quantitative data** check box list
8 Ensure that the **Display vertically** check box under the **Quantitative data** check box list is checked
9 Click **OK**
10 Click the **Continue** button in the **XLSTAT – Selections** dialog box

(continued)

Note to Minitab users: Versions 15 and earlier of Minitab do not provide a **Check statistics** list. Users of such versions must manually check and/or uncheck check boxes.

Next we explain how to use Minitab, Excel, and the TI-83/84 Plus to obtain a boxplot for the TV-viewing-times data.

Note to Minitab and Excel users: Both of these statistical software packages default to displaying boxplots vertically. However, both have an option for horizontal display. For consistency with the text, we will employ that option.

EXAMPLE 3.27   **Using Technology to Obtain a Boxplot**

Weekly TV-Viewing Times Use Minitab, Excel, or the TI-83/84 Plus to obtain a boxplot for the TV-viewing times given in Table 3.13 on page 126.

Solution We applied the boxplot programs to the data, resulting in Output 3.4. Steps for generating that output are presented in Instructions 3.4 (next page).

OUTPUT 3.4   Boxplot for the TV-viewing times

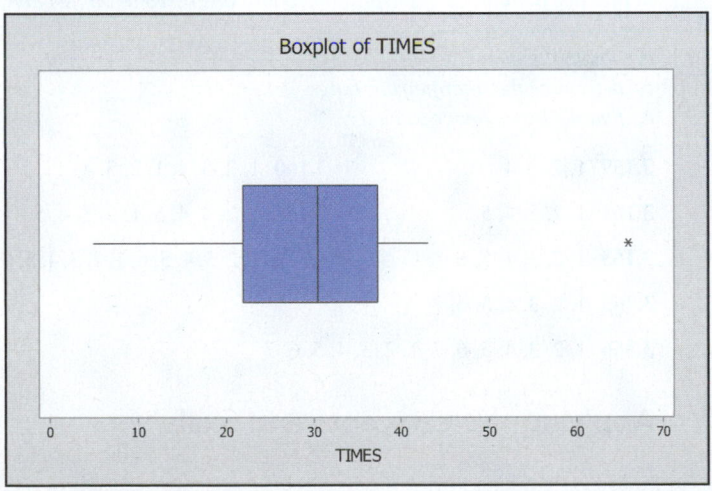

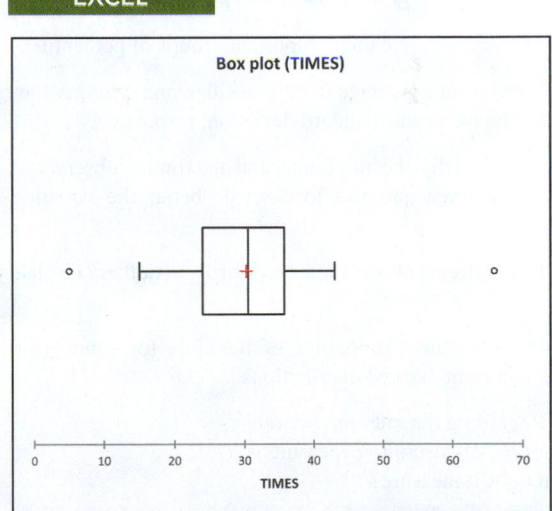

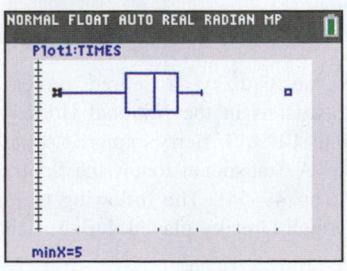

Observe that the Excel output shows two outliers, whereas the Minitab and TI-83/84 Plus outputs show only one outlier. This discrepancy is due to the fact that these statistical technologies use different methods for computing quartiles. Compare the boxplots in Output 3.4 to the one in Fig. 3.14(c) on page 132.

INSTRUCTIONS 3.4 Steps for generating Output 3.4

MINITAB

1 Store the data from Table 3.13 in a column named TIMES
2 Choose **Graph ➤ Boxplot...**
3 Click **OK**
4 Press the F3 key to reset the dialog box
5 Specify TIMES in the **Graph variables** text box
6 Click the **Scale...** button
7 Within the **Axes and Ticks** tab, check the **Transpose value and category scales** check box
8 Click **OK** twice

5 Click the **Options** tab
6 Uncheck the **Descriptive statistics** check box
7 Click the **Charts (1)** tab
8 Click the **Options** tab next to the **Chart types** tab
9 Select the **Horizontal** option button
10 Uncheck the **Minimum/Maximum** check box
11 Check the **Outliers** check box
12 Click **OK**
13 Click the **Continue** button in the **XLSTAT – Selections** dialog box

EXCEL

1 Store the data from Table 3.13 in a column named TIMES
2 Choose **XLSTAT ➤ Visualizing data ➤ Univariate plots**
3 Click the reset button in the lower left corner of the dialog box
4 Click in the **Quantitative data** selection box and then select the column of the worksheet that contains the TIMES data

TI-83/84 PLUS

1 Store the data from Table 3.13 in a list named TIMES
2 Ensure that all stat plots and all **Y =** functions are off
3 Press **2ND ➤ STAT PLOT** and then press **ENTER** twice
4 Arrow to the fourth graph icon and press **ENTER**
5 Press the down-arrow key
6 Press **2ND ➤ LIST**
7 Arrow down to TIMES and press **ENTER** twice
8 Type 1 for **Freq** and press **ENTER**
9 Press **ZOOM,** then **9,** and then **TRACE**

Exercises 3.4

Understanding the Concepts and Skills

3.148 Identify by name three important groups of percentiles.

3.149 Identify an advantage that the median and interquartile range have over the mean and standard deviation, respectively.

3.150 Explain why the minimum and maximum observations are added to the three quartiles to describe better the variation in a data set.

3.151 Is an extreme observation necessarily an outlier? Explain your answer.

3.152 State pertinent properties of boxplots for symmetric, left-skewed, and right-skewed distributions.

3.153 Regarding the interquartile range,
a. what type of descriptive measure is it?
b. what does it measure?

3.154 Identify a use of the lower and upper limits.

3.155 When are the adjacent values just the minimum and maximum observations?

3.156 Which measure of variation is preferred when
a. the mean is used as a measure of center?
b. the median is used as a measure of center?

3.157 *Fill in the blanks:* Roughly, when arranged in increasing order, the middle 50% of a data set are found between _____ and _____ .

3.158 *Fill in the blank:* Roughly, when arranged in increasing order, the uppermost 25% of a data set are greater than or equal to _____ .

In Exercises 3.159–3.166, we have provided simple data sets for you to practice finding the descriptive measures discussed in this section. For each data set,
a. *obtain the quartiles.*
b. *determine the interquartile range.*
c. *find the five-number summary.*

3.159 1, 2, 3, 4

3.160 1, 2, 3, 4, 1, 2, 3, 4

3.161 1, 2, 3, 4, 5

3.162 1, 2, 3, 4, 5, 1, 2, 3, 4, 5

3.163 1, 2, 3, 4, 5, 6

3.164 1, 2, 3, 4, 5, 6, 1, 2, 3, 4, 5, 6

3.165 1, 2, 3, 4, 5, 6, 7

3.166 1, 2, 3, 4, 5, 6, 7, 1, 2, 3, 4, 5, 6, 7

Applying the Concepts and Skills

In Exercises 3.167–3.176,
a. *obtain and interpret the quartiles.*
b. *determine and interpret the interquartile range.*
c. *find and interpret the five-number summary.*
d. *identify potential outliers, if any.*
e. *construct and interpret a boxplot.*

3.167 **The Great Gretzky.** Wayne Gretzky, a retired professional hockey player, played 20 seasons in the National Hockey League (NHL), from 1980 through 1999. S. Berry explored some of Gretzky's accomplishments in "A Statistician Reads the Sports Pages" (*Chance*, Vol. 16, No. 1, pp. 49–54). The following table shows the number of games in which Gretzky played during each of his 20 seasons in the NHL.

79	80	80	80	74
80	80	79	64	78
73	78	74	45	81
48	80	82	82	70

3.168 Parenting Grandparents. In the article "Grandchildren Raised by Grandparents, a Troubling Trend" (*California Agriculture*, Vol. 55, No. 2, pp. 10–17), M. Blackburn considered the rates of children (under 18 years of age) living in California with grandparents as their primary caretakers. A sample of 14 California counties yielded the following percentages of children under 18 living with grandparents.

5.9	4.0	5.7	5.1	4.1	4.4	6.5
4.4	5.8	5.1	6.1	4.5	4.9	4.9

3.169 Hospital Stays. The U.S. National Center for Health Statistics compiles data on the length of stay by patients in short-term hospitals and publishes its findings in *Vital and Health Statistics*. A random sample of 21 patients yielded the following data on length of stay, in days.

4	4	12	18	9	6	12
3	6	15	7	3	55	1
10	13	5	7	1	23	9

3.170 Miles Driven. The U.S. Federal Highway Administration conducts studies on motor vehicle travel by type of vehicle. Results are published annually in *Highway Statistics*. A sample of 15 cars yields the following data on number of miles driven, in thousands, for last year.

13.2	13.3	11.9	15.7	11.3
12.2	16.7	10.7	3.3	13.6
14.8	9.6	11.6	8.7	15.0

3.171 Hurricanes. An article by D. Schaefer et al. (*Journal of Tropical Ecology*, Vol. 16, pp. 189–207) reported on a long-term study of the effects of hurricanes on tropical streams of the Luquillo Experimental Forest in Puerto Rico. The study shows that Hurricane Hugo had a significant impact on stream water chemistry. The following table shows a sample of 10 ammonia fluxes in the first year after Hugo. Data are in kilograms per hectare per year.

96	66	147	147	175
116	57	154	88	154

3.172 Sky Guide. The publication *California Wild: Natural Sciences for Thinking Animals* has a monthly feature called the "Sky Guide" that keeps track of the sunrise and sunset for the first day of each month in San Francisco. Over several issues, B. Quock from the Morrison Planetarium recorded the following sunrise times from July 1 of one year through June 1 of the next year. The times are given in minutes past midnight.

352	374	400	426	396	427
445	434	400	354	374	349

3.173 Capital Spending. An issue of *Brokerage Report* discussed the capital spending of telecommunications companies in the United States and Canada. The capital spending, in thousands of dollars, for each of 27 telecommunications companies is shown in the following table.

9,310	2,515	3,027	1,300	1,800	70	3,634
656	664	5,947	649	682	1,433	389
17,341	5,299	195	8,543	4,200	7,886	11,189
1,006	1,403	1,982	21	125	2,205	

3.174 Medieval Cremation Burials. In the article "Material Culture as Memory: Combs and Cremations in Early Medieval Britain" (*Early Medieval Europe*, Vol. 12, Issue 2, pp. 89–128), H. Williams discussed the frequency of cremation burials found in 17 archaeological sites in eastern England. Here are the data.

83	64	46	48	523	35	34	265	2484
46	385	21	86	429	51	258	119	

3.175 Pizza Diameters. In the article, "Assessing Claims Made by a Pizza Chain" by P. Dunn, (*Journal of Statistics Education*, Vol. 20, No. 1, pp. 1–19), a sample of pizzas from Domino's was collected. The intention of the study was to test the claim of a local pizza shop that its pizzas are bigger than Domino's. The following table gives the diameters, in centimeters (cm), of 41 large Supreme pizzas from Domino's.

29.40	27.45	26.38	26.50	26.55	26.64	26.87
26.61	29.33	26.67	26.09	29.32	27.12	26.83
27.04	29.17	27.14	27.04	29.19	26.26	26.16
26.70	28.78	29.33	27.27	28.90	29.14	29.23
28.32	27.12	27.05	28.44	26.97	27.52	26.79
26.54	26.70	28.79	28.97	28.84	26.71	

3.176 Water-Park Attendance. Water parks are a huge summer attraction for vacationers in the United States. The *Global Attractions Attendance Report*, published by the Themed Entertainment Association, provides the attendance report for theme parks and water parks around the world. The following table provides the total yearly attendance for the top 20 water parks in the United States, in thousands, during one year.

367	374	395	398	400
432	461	471	500	500
535	559	643	644	723
982	1223	1500	1891	2058

3.177 Nicotine Patches. In the paper "The Smoking Cessation Efficacy of Varying Doses of Nicotine Patch Delivery Systems 4 to 5 Years Post-Quit Day" (*Preventative Medicine*, 28, pp. 113–118), D. Daughton et al. discussed the long-term effectiveness of transdermal nicotine patches on participants who had previously smoked at least 20 cigarettes per day. A sample of 15 participants in the Transdermal Nicotine Study Group (TNSG) reported that they now smoke the following number of cigarettes per day.

10	9	10	8	7
6	10	9	10	8
9	10	8	8	10

a. Determine the quartiles for these data.

b. Remark on the usefulness of quartiles with respect to this data set.

3.178 Starting Salaries. The National Association of Colleges and Employers (NACE) conducts surveys of salary offers to new college graduates and publishes the results in *Salary Survey*. The following diagram provides boxplots for the starting annual salaries, in thousands of dollars, obtained from samples of 35 business graduates (top boxplot) and 32 education graduates (bottom boxplot). Use the boxplots to compare the starting salaries of the sampled business graduates and education graduates, paying special attention to center and variation.

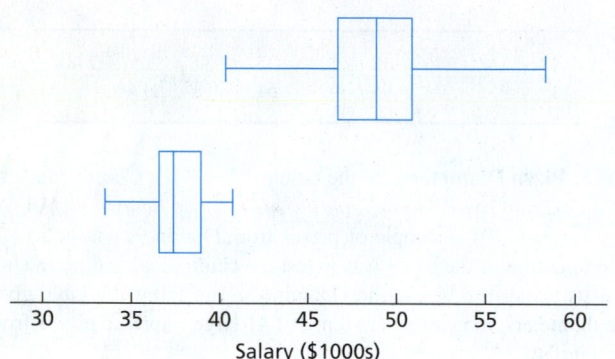

3.179 Obesity. Researchers in obesity wanted to compare the effectiveness of dieting with exercise against dieting without exercise. Seventy-three patients were randomly divided into two groups. Group 1, composed of 37 patients, was put on a program of dieting with exercise. Group 2, composed of 36 patients, dieted only. The results for weight loss, in pounds, after 2 months are summarized in the following boxplots. The top boxplot is for Group 1 and the bottom boxplot is for Group 2. Use the boxplots to compare the weight losses for the two groups, paying special attention to center and variation.

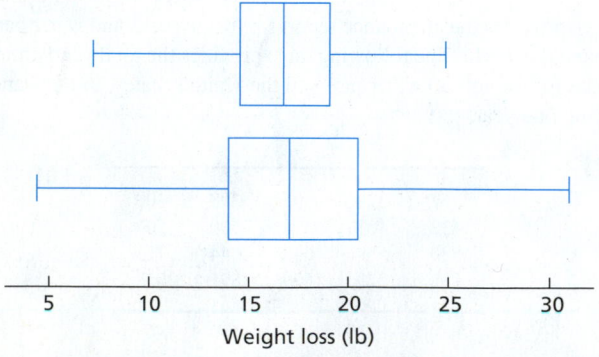

3.180 Cuckoo Care. Many species of cuckoos are brood parasites. The females lay their eggs in the nests of smaller bird species, who then raise the young cuckoos at the expense of their own young. Data on the lengths, in millimeters (mm), of cuckoo eggs found in the nests of three bird species—the Tree Pipit, Hedge Sparrow, and Pied Wagtail—were collected by the late O. M. Latter in 1902 and used by L. H. C. Tippett in his text *The Methods of Statistics* (New York: Wiley, 1952, p. 176). Use the following boxplots to compare the

lengths of cuckoo eggs found in the nests of the three bird species, paying special attention to center and variation.

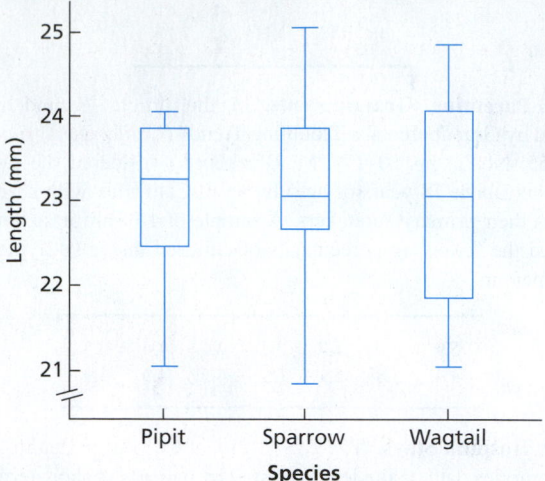

3.181 Sickle Cell Disease. A study published by E. Anionwu et al. in the *British Medical Journal* (Vol. 282, pp. 283–286) examined the steady-state hemoglobin levels of patients with three different types of sickle cell disease: HB SC, HB SS, and HB ST. Use the following boxplots to compare the hemoglobin levels for the three groups of patients, paying special attention to center and variation.

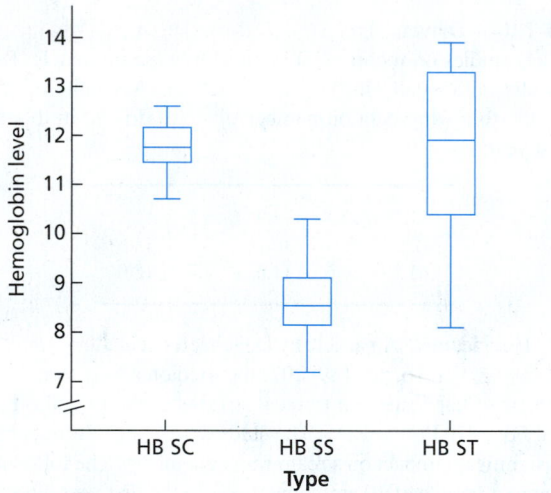

Working with Large Data Sets

In Exercises 3.182–3.187, use the technology of your choice to

a. *obtain and interpret the quartiles.*

b. *determine and interpret the interquartile range.*

c. *find and interpret the five-number summary.*

d. *identify potential outliers, if any.*

e. *obtain and interpret a boxplot.*

3.182 Women Students. The U.S. Department of Education sponsors a report on educational institutions, including colleges and universities, titled *Digest of Education Statistics*. Among many of the statistics provided are the numbers of men and women enrolled in 2-year and 4-year degree-granting institutions. During one year, the percentage of full-time enrolled students that were women, for each of the 50 states and the District of Columbia, is as presented on the WeissStats site.

3.183 The Great White Shark. In an article titled "Great White, Deep Trouble" (*National Geographic*, Vol. 197(4), pp. 2–29), Peter Benchley—the author of *JAWS*—discussed various aspects of the Great White Shark (*Carcharodon carcharias*). Data on the number of pups borne in a lifetime by each of 80 Great White Shark females are provided on the WeissStats site.

3.184 The Beatles. In the article, "Length of The Beatles' Songs" (*Chance*, Vol. 25, No. 1, pp. 30–33), T. Koyama discusses aspects and interpretations of the lengths of songs by The Beatles. Data on the length, in seconds, of 229 Beatles' songs are presented on the WeissStats site.

3.185 High School Completion. As reported by the U.S. Census Bureau in *Educational Attainment in the United States*, the percentage of adults in each state and the District of Columbia who have completed high school is provided on the WeissStats site.

3.186 Bachelor's Completion. As reported by the U.S. Census Bureau in *Educational Attainment in the United States*, the percentage of adults in each state and the District of Columbia who have completed a bachelor's degree is provided on the WeissStats site.

3.187 Body Temperature. A study by researchers at the University of Maryland addressed the question of whether the mean body temperature of humans is 98.6°F. The results of the study by P. Mackowiak et al. appeared in the article "A Critical Appraisal of 98.6°F, the Upper Limit of the Normal Body Temperature, and Other Legacies of Carl Reinhold August Wunderlich" (*Journal of the American Medical Association*, Vol. 268, pp. 1578–1580). Among other data, the researchers obtained the body temperatures of 93 healthy humans, as provided on the WeissStats site.

In each of Exercises 3.188–3.191,
a. use the technology of your choice to obtain boxplots for the data sets, using the same scale.
b. compare the data sets by using your results from part (a), paying special attention to center and variation.

3.188 Treating Psychotic Illness. L. Petersen et al. evaluated the effects of integrated treatment for patients with a first episode of psychotic illness in the paper "A Randomised Multicentre Trial of Integrated Versus Standard Treatment for Patients with a First Episode of Psychotic Illness" (*British Medical Journal*, Vol. 331, (7517):602).

Part of the study included a questionnaire that was designed to measure client satisfaction for both the integrated treatment and a standard treatment. The data on the WeissStats site are based on the results of the client questionnaire.

3.189 The Etruscans. Anthropologists are still trying to unravel the mystery of the origins of the Etruscan empire, a highly advanced Italic civilization formed around the eighth century B.C. in central Italy. Were they native to the Italian peninsula or, as many aspects of their civilization suggest, did they migrate from the East by land or sea? The maximum head breadth, in millimeters, of 70 modern Italian male skulls and that of 84 preserved Etruscan male skulls were analyzed to help researchers decide whether the Etruscans were native to Italy. The resulting data can be found on the WeissStats site. [SOURCE: N. Barnicot and D. Brothwell, "The Evaluation of Metrical Data in the Comparison of Ancient and Modern Bones." In *Medical Biology and Etruscan Origins*, G. Wolstenholme and C. O'Connor, eds., Little, Brown & Co., 1959]

3.190 Magazine Ads. Advertising researchers F. Shuptrine and D. McVicker wanted to determine whether there were significant differences in the readability of magazine advertisements. Thirty magazines were classified based on their educational level—high, mid, or low—and then three magazines were randomly selected from each level. From each magazine, six advertisements were randomly chosen and examined for readability. In this particular case, readability was characterized by the numbers of words, sentences, and words of three syllables or more in each ad. The researchers published their findings in the article "Readability Levels of Magazine Ads" (*Journal of Advertising Research*, Vol. 21, No. 5, pp. 45–51). The number of words of three syllables or more in each ad are provided on the WeissStats site.

3.191 Prolonging Life. Vitamin C (ascorbate) boosts the human immune system and is effective in preventing a variety of illnesses. In a study by E. Cameron and L. Pauling titled "Supplemental Ascorbate in the Supportive Treatment of Cancer: Reevaluation of Prolongation of Survival Times in Terminal Human Cancer" (*Proceedings of the National Academy of Science*, Vol. 75, No. 9, pp. 4538–4542), patients in advanced stages of cancer were given a vitamin C supplement. Patients were grouped according to the organ affected by cancer: stomach, bronchus, colon, ovary, or breast. The study yielded the survival times, in days, given on the WeissStats site.

3.5 Descriptive Measures for Populations; Use of Samples

In this section, we discuss several descriptive measures for *population data*—the data obtained by observing the values of a variable for an entire population. Although, in reality, we often don't have access to population data, it is nonetheless helpful to become familiar with the notation and formulas used for descriptive measures of such data.

The Population Mean

Recall that, for a variable x and a sample of size n from a population, the sample mean is

$$\bar{x} = \frac{\sum x_i}{n}.$$

TABLE 3.16

Notation used for a sample and for the population

	Size	Mean
Sample	n	$\bar{x}$
Population	N	μ

First, we sum the observations of the variable for the sample, and then we divide by the size of the sample.

We can find the mean of a finite population similarly: first, we sum all possible observations of the variable for the entire population, and then we divide by the size of the population. However, to distinguish the *population mean* from a sample mean, we use the Greek letter μ (pronounced "mew") to denote the population mean. We also use the uppercase English letter N to represent the size of the population. Table 3.16 summarizes the notation that is used for both a sample and the population.

DEFINITION 3.11

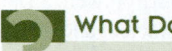

What Does It Mean?

A population mean (mean of a variable) is the arithmetic average (mean) of population data.

Population Mean (Mean of a Variable)

For a variable x, the mean of all possible observations for the entire population is called the **population mean** or **mean of the variable x.** It is denoted μ_x or, when no confusion will arise, simply μ. For a finite population,

$$\mu = \frac{\Sigma x_i}{N},$$

where N is the population size.

Note: For a particular variable on a particular population:

- There is only one population mean—namely, the mean of all possible observations of the variable for the entire population.
- There are many sample means—one for each possible sample of the population.

 **EXAMPLE 3.28** The Population Mean

Augusta National Golf Course Located in Augusta, Georgia, the Augusta National Golf Club was founded by Bobby Jones and Clifford Roberts. The course itself was designed by Jones and Alister MacKenzie. It opened for play in January of 1933 and has hosted the Masters Tournament since 1934.

Prior to becoming a golf course, Augusta National was a plant nursery. The name of each hole derives from the tree or shrub associated with it. Table 3.17 provides data for the holes. Find the population mean length of the holes at Augusta National Golf Club.

TABLE 3.17

Holes at the Augusta National Golf Club

Hole	Name	Length (yd)	Par
1	Tea olive	445	4
2	Pink dogwood	575	5
3	Flowering peach	350	4
4	Flowering crab apple	240	3
5	Magnolia	455	4
6	Juniper	180	3
7	Pampas	450	4
8	Yellow jasmine	570	5
9	Carolina cherry	460	4
10	Camellia	495	4
11	White dogwood	505	4
12	Golden bell	155	3
13	Azalea	510	5
14	Chinese fir	440	4
15	Firethorn	530	5
16	Redbud	170	3
17	Nandina	440	4
18	Holly	465	4

You try it!

Exercise 3.209(a) on page 147

Solution   Here the variable is length and the population consists of the holes at the Augusta National Golf Club. The sum of the lengths in the third column of Table 3.17 is 7435 yd. Because there are 18 holes, $N = 18$. Therefore, by Definition 3.11,

$$\mu = \frac{\Sigma x_i}{N} = \frac{7435}{18} = 413.1 \text{ yd (rounded)}.$$

Interpretation   The population mean length of the holes at the Augusta National Golf Club is 413.1 yd.

Using a Sample Mean to Estimate a Population Mean

In inferential studies, we analyze sample data. Nonetheless, the objective is to describe the entire population. We use samples because they are usually more practical, as illustrated in the next example.

■■■   **EXAMPLE 3.29**   A Use of a Sample Mean

Estimating Mean Household Income   The U.S. Census Bureau reports the mean (annual) income of U.S. households in the publication *Current Population Survey*. To obtain the population data—the incomes of all U.S. households—would be extremely expensive and time consuming. It is also unnecessary because accurate estimates of the mean income of all U.S. households can be obtained from the mean income of a sample of such households. The Census Bureau samples only 57,000 households from a total of more than 100 million.

Here are the basic elements for this problem, also summarized in Fig. 3.17:

- *Variable:* income
- *Population:* all U.S. households
- *Population data:* incomes of all U.S. households
- *Population mean:* mean income, μ, of all U.S. households
- *Sample:* 57,000 U.S. households sampled by the Census Bureau
- *Sample data:* incomes of the 57,000 U.S. households sampled
- *Sample mean:* mean income, $\bar{x}$, of the 57,000 U.S. households sampled

FIGURE 3.17

Population and sample for incomes of U.S. households

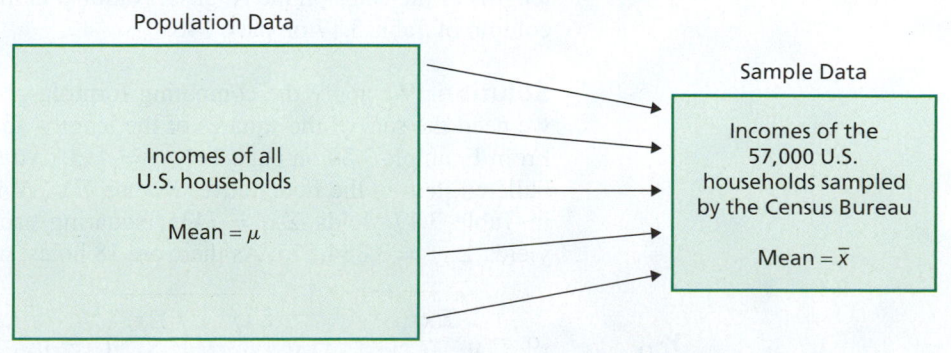

The Census Bureau uses the sample mean income, $\bar{x}$, of the 57,000 U.S. households sampled to estimate the population mean income, μ, of all U.S. households.

The Population Standard Deviation

Recall that, for a variable x and a sample of size n from a population, the sample standard deviation is

$$s = \sqrt{\frac{\Sigma(x_i - \bar{x})^2}{n - 1}}.$$

The standard deviation of a finite population is obtained in a similar, but slightly different, way. To distinguish the *population standard deviation* from a sample standard deviation, we use the Greek letter σ (pronounced "sigma") to denote the population standard deviation.

DEFINITION 3.12

? What Does It Mean?

Roughly speaking, the population standard deviation indicates how far, on average, the observations in the population are from the mean of the population.

Population Standard Deviation (Standard Deviation of a Variable)

For a variable x, the standard deviation of all possible observations for the entire population is called the **population standard deviation** or **standard deviation of the variable x.** It is denoted σ_x or, when no confusion will arise, simply σ. For a finite population, the defining formula is

$$\sigma = \sqrt{\frac{\Sigma(x_i - \mu)^2}{N}},$$

where N is the population size.

The population standard deviation can also be found from the computing formula

$$\sigma = \sqrt{\frac{\Sigma x_i^2}{N} - \mu^2}.$$

Note:

- The rounding rule on page 110 says not to perform any rounding until a computation is complete. Thus, in computing a population standard deviation by hand, you should replace μ by $\Sigma x_i/N$ in the formulas given in Definition 3.12, unless μ is unrounded.
- Just as s^2 is called a sample variance, σ^2 is called the **population variance** (or **variance of the variable**).

EXAMPLE 3.30   ### The Population Standard Deviation

Augusta National Golf Course   Calculate the population standard deviation of the lengths of the holes on the Augusta National Golf Course, as presented in the third column of Table 3.17 on page 140.

Solution   We apply the computing formula given in Definition 3.12. To do so, we need the sum of the squares of the lengths and the population mean length, μ. From Example 3.28 on page 141, $\mu = 413.1$ yd (rounded); so, in view of the first bulleted item in the note above, we use $\Sigma x_i/N$ instead of μ. Adding the lengths in Table 3.17 yields $\Sigma x_i = 7435$; squaring each length and adding the results yields $\Sigma x_i^2 = 3,384,575$. As there are 18 holes, we have that

You try it!

Exercise 3.209(b)
on page 147

$$\sigma = \sqrt{\frac{\Sigma x_i^2}{N} - \mu^2} = \sqrt{\frac{\Sigma x_i^2}{N} - \left(\frac{\Sigma x_i}{N}\right)^2} = \sqrt{\frac{3,384,575}{18} - \left(\frac{7,435}{18}\right)^2} = 132.0 \text{ yd.}$$

Interpretation   The population standard deviation of the lengths of the holes at Augusta National is 132.0 yd. Roughly speaking, the lengths of the holes fall, on average, 132.0 yd from their mean length of 413.1 yd.

Using a Sample Standard Deviation to Estimate a Population Standard Deviation

We have shown that a sample mean can be used to estimate a population mean. Likewise, a sample standard deviation can be used to estimate a population standard deviation, as illustrated in the next example.

■■■   **EXAMPLE 3.31**   **A Use of a Sample Standard Deviation**

Xenical Capsules   Xenical is a trade name used by Roche for Orlistat (tetrahydrolipstatin). This drug is used to treat obesity in people with risk factors such as diabetes, high blood pressure, and high cholesterol or triglycerides. Xenical works in the intestines, where it blocks about one-third of the fat in the food a person eats from being digested.

A standard prescription of Xenical is given in 120-milligram (mg) capsules. Although the capsule weights can vary somewhat from 120 mg and also from each other, keeping the variation small is important for various medical reasons.

To evaluate the variation in capsule weights, we need to know the population standard deviation, σ, of capsule weights. Because, in this case, σ cannot be determined exactly (do you know why?), the standard deviation of the weights of a sample of capsules must be used to estimate σ. Suppose that it is decided to use a sample of 10 capsules.

Here are the basic elements for this problem, also summarized in Fig. 3.18:

- *Variable:* weight
- *Population:* all ("120-mg" Xenical) capsules
- *Population data:* weights of all capsules
- *Population standard deviation:* standard deviation, σ, of the weights of all capsules
- *Sample:* the 10 capsules sampled
- *Sample data:* weights of the 10 capsules sampled
- *Sample standard deviation:* standard deviation, s, of the weights of the 10 capsules sampled

FIGURE 3.18

Population and sample for capsule weights

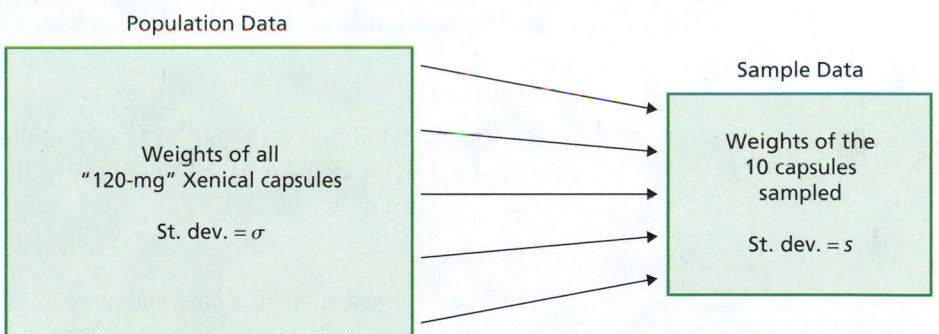

The sample standard deviation, s, of the weights of the 10 capsules sampled can be used to estimate the population standard deviation, σ, of the weights of all capsules. We discuss this type of inference in Chapter 11. ■

Parameter and Statistic

The following terminology helps us distinguish between descriptive measures for populations and samples.

DEFINITION 3.13   **Parameter and Statistic**

Parameter: A descriptive measure for a population

Statistic: A descriptive measure for a sample

Thus, for example, μ and σ are parameters, whereas $\bar{x}$ and s are statistics.

Standardized Variables

From any variable x, we can form a new variable z, defined as follows.

DEFINITION 3.14

> **Standardized Variable**
>
> For a variable x, the variable
>
> $$z = \frac{x - \mu}{\sigma}$$
>
> is called the **standardized version** of x or the **standardized variable** corresponding to the variable x.

What Does It Mean?

The standardized version of a variable x is obtained by first subtracting from x its mean and then dividing by its standard deviation.

A standardized variable always has mean 0 and standard deviation 1. For this and other reasons, standardized variables play an important role in many aspects of statistical theory and practice. We present a few applications of standardized variables in this section; several others appear throughout the rest of the book.

■■■ EXAMPLE 3.32   Standardized Variables

TABLE 3.18

Possible observations of x and z

x	−1	3	3	3	5	5
z	−2	0	0	0	1	1

Understanding the Basics Let's consider a simple variable x—namely, one with possible observations shown in the first row of Table 3.18.

a. Determine the standardized version of x.
b. Find the observed value of z corresponding to an observed value of x of 5.
c. Calculate all possible observations of z.
d. Find the mean and standard deviation of z using Definitions 3.11 and 3.12. Was it necessary to do these calculations to obtain the mean and standard deviation?
e. Show dotplots of the distributions of both x and z. Interpret the results.

Solution

a. Using Definitions 3.11 and 3.12, we find that the mean and standard deviation of x are $\mu = 3$ and $\sigma = 2$. Consequently, the standardized version of x is

$$z = \frac{x - 3}{2}.$$

b. The observed value of z corresponding to an observed value of x of 5 is

$$z = \frac{x - 3}{2} = \frac{5 - 3}{2} = 1.$$

c. Applying the formula $z = (x - 3)/2$ to each of the possible observations of the variable x shown in the first row of Table 3.18, we obtain the possible observations of the standardized variable z shown in the second row of Table 3.18.

d. From the second row of Table 3.18,

$$\mu_z = \frac{\Sigma z_i}{N} = \frac{0}{6} = 0$$

and

$$\sigma_z = \sqrt{\frac{\Sigma(z_i - \mu_z)^2}{N}} = \sqrt{\frac{6}{6}} = 1.$$

The results of these two computations illustrate that the mean of a standardized variable is always 0 and its standard deviation is always 1. We didn't need to perform these calculations.

e. Figures 3.19(a) and 3.19(b) show dotplots of the distributions of x and z, respectively.

FIGURE 3.19

Dotplots of the distributions of x and
its standardized version z

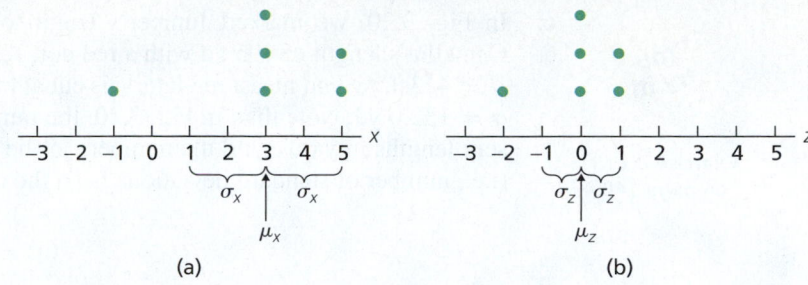

(a)   (b)

Interpretation   The two dotplots in Fig. 3.19 show how standardizing shifts
a distribution so the new mean is 0 and changes the scale so the new standard
deviation is 1.

z-Scores

An important concept associated with standardized variables is that of the *z-score*, or
standard score.

DEFINITION 3.15

? What Does It Mean?

The z-score of an
observation tells us the number
of standard deviations that the
observation is from the mean,
that is, how far the observation
is from the mean in units of
standard deviation.

z-Score

For an observed value of a variable x, the corresponding value of the standardized variable z is called the **z-score** of the observation. The term **standard score** is often used instead of *z-score*.

A negative z-score indicates that the observation is below (less than) the mean,
whereas a positive z-score indicates that the observation is above (greater than)
the mean. Example 3.33 illustrates calculation and interpretation of z-scores.

EXAMPLE 3.33   *z-Scores*

Augusta National Golf Course   The lengths of the holes on the Augusta National
Golf Course are presented in the third column of Table 3.17 on page 140. We determined earlier that the mean and standard deviation of the lengths are 413.1 yd
and 132.0 yd, respectively. So, in this case, the standardized variable is

$$z = \frac{x - 413.1}{132.0}.$$

a.   Find and interpret the z-score of Juniper's length of 180 yd.
b.   Find and interpret the z-score of Camellia's length of 495 yd.
c.   Construct a graph showing the results obtained in parts (a) and (b).

Solution

a.   The z-score of Juniper's length of 180 yd is

$$z = \frac{x - 413.1}{132.0} = \frac{180 - 413.1}{132.0} = -1.77.$$

Interpretation   Juniper's length is 1.77 standard deviations below the mean.

b.   The z-score of Camellia's length of 495 yd is

$$z = \frac{x - 413.1}{132.0} = \frac{495 - 413.1}{132.0} = 0.62.$$

Interpretation   Camellia's length is 0.62 standard deviation above the mean.

Exercise 3.213
on page 148

c. In Fig. 3.20, we marked Juniper's length of 180 yd with a green dot and Camellia's length of 495 yd with a red dot. Additionally, we located the mean, $\mu = 413.1$ yd, and measured intervals equal in length to the standard deviation, $\sigma = 132.0$ yd. Note that, in Fig. 3.20, the numbers in the row labeled x represent lengths in yards, and the numbers in the row labeled z represent z-scores (i.e., number of standard deviations from the mean). ∎

FIGURE 3.20   Graph showing Juniper's length (green dot) and Camellia's length (red dot)

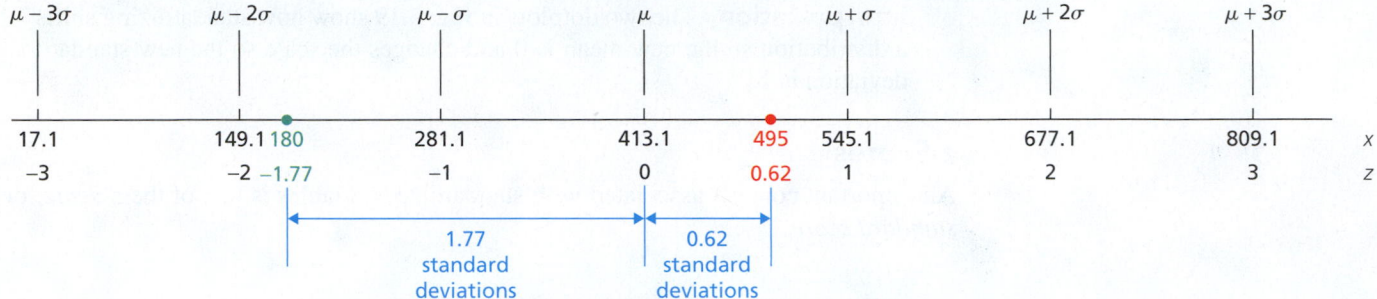

The z-Score as a Measure of Relative Standing

The three-standard-deviations rule (Key Fact 3.2 on page 111) states that almost all the observations in any data set lie within three standard deviations to either side of the mean. Thus, for any variable, almost all possible observations have z-scores between -3 and 3.

The z-score of an observation, therefore, can be used as a rough measure of its relative standing among all the observations comprising a data set. For instance, a z-score of 3 or more indicates that the observation is larger than most of the other observations; a z-score of -3 or less indicates that the observation is smaller than most of the other observations; and a z-score near 0 indicates that the observation is located near the mean.

The use of z-scores as a measure of relative standing can be refined and made more precise by applying Chebyshev's rule, as you are asked to explore in Exercises 3.222 and 3.223. Moreover, if the distribution of the variable under consideration is roughly bell shaped, then, as you will see in Chapter 6, the use of z-scores as a measure of relative standing can be improved even further.

Percentiles usually give a more exact method of measuring relative standing than do z-scores. However, if only the mean and standard deviation of a variable are known, z-scores provide a feasible alternative to percentiles for measuring relative standing.

Other Descriptive Measures for Populations

Up to this point, we have concentrated on the mean and standard deviation in our discussion of descriptive measures for populations. The reason is that many of the classical inference procedures for center and variation concern those two parameters.

However, modern statistical analyses also rely heavily on descriptive measures based on percentiles. Quartiles, the IQR, and other descriptive measures based on percentiles are defined in the same way for (finite) populations as they are for samples. For simplicity and with one exception, we use the same notation for descriptive measures based on percentiles whether we are considering a sample or a population. The exception is that we use M to denote a sample median and η (eta) to denote a population median.

Exercises 3.5

Understanding the Concepts and Skills

3.192 Identify each quantity as a parameter or a statistic.
a. μ **b.** s **c.** $\bar{x}$ **d.** σ

3.193 Although, in practice, sample data are generally analyzed in inferential studies, what is the ultimate objective of such studies?

3.194 Microwave Popcorn. For a given brand of microwave popcorn, what property is desirable for the population standard deviation of the cooking time? Explain your answer.

3.195 Fill in the following blanks.
a. A standardized variable always has mean _____ and standard deviation _____ .
b. The z-score corresponding to an observed value of a variable tells you _____ .
c. A positive z-score indicates that the observation is _____ the mean, whereas a negative z-score indicates that the observation is _____ the mean.

3.196 Identify the statistic that is used to estimate
a. a population mean.
b. a population standard deviation.

3.197 Augusta National Golf Course. Earlier in this section, we found that the population mean length of the holes at the Augusta National Golf Club is 413.1 yd. In this context, is the number 413.1 a parameter or a statistic? Explain your answer.

3.198 Augusta National Golf Course. Earlier in this section, we found that the population standard deviation of the lengths of the holes at the Augusta National Golf Club is 132.0 yd. In this context, is the number 132.0 a parameter or a statistic? Explain your answer.

3.199 Heights of Basketball Players. In Section 3.2, we analyzed the heights of the starting five players on each of two men's college basketball teams. The heights, in inches, of the players on Team II are 67, 72, 76, 76, and 84. Regarding the five players as a population, solve the following problems.
a. Compute the population mean height, μ.
b. Compute the population standard deviation of the heights, σ.

3.200 Heights of Basketball Players. This exercise requires that you have first done Exercise 3.199. In Example 3.12 on page 109, we found that, considering the five starting players on Team II a sample of all male starting college basketball players, the mean and standard deviation of the heights are 75 inches and 6.2 inches, respectively. Explain why, numerically, the sample mean of 75 inches is the same as the population mean found in Exercise 3.199(a) but that the sample standard deviation of 6.2 inches differs from the population standard deviation found in Exercise 3.199(b).

In Exercises 3.201–3.206, we have provided simple data sets for you to practice the basics of finding a
a. population mean.
b. population standard deviation.

3.201 4, 0, 5   **3.202** 3, 5, 7

3.203 1, 2, 4, 4   **3.204** 2, 5, 0, −1

3.205 1, 9, 8, 4, 3   **3.206** 4, 2, 0, 2, 2

Applying the Concepts and Skills

3.207 Age of U.S. Residents. The U.S. Census Bureau collects information about the ages of people in the United States. Results are published in *Current Population Reports*.
a. Identify the variable and population under consideration.
b. A sample of six U.S. residents yielded the following data on ages (in years). Determine the median of these age data. Decide whether this descriptive measure is a parameter or a statistic, and use statistical notation to express the result.

29	54	9	45	51	7

c. By consulting the most recent census data, we found that the median age of all U.S. residents is 37.2 years. Decide whether that descriptive measure is a parameter or a statistic, and use statistical notation to express the result.

3.208 Chinchillidae Range. The family of mammals called Chinchillidae contains the chinchilla and viscachas, which are large South American rodents. The article, "Species Richness and Distribution of Neotropical Rodents, with Conservation Implications" (*Mammalia*, Vol. 77, Issue 1, pp. 1–19) by G. Amori et al., reports the range sizes of each species of Chinchilladae. There are six species in the Chinchillidae family and the size of each species' range, in square kilometers (km^2), is given in the following table.

370,954.56	762,822.31	11,213.32
2,070.38	11,943.02	1,988,086.68

a. Obtain and interpret the population mean range of Chinchilladae species.
b. Obtain and interpret the population standard deviation of the ranges of Chinchilladae species.

3.209 Atlantic Basin Hurricanes. The *Tropical Cyclone Report*, a publication of the National Hurricane Center, contains comprehensive information on each tropical cyclone, including synoptic history, meteorological statistics, casualties, and damages. A hurricane is a tropical cyclone with winds that have reached a constant speed of 74 miles per hour or more. During one year, there were 10 Atlantic basin hurricanes. Their maximum wind speeds, in miles per hour (mph), were as shown in the following table.

85	100	110	80	105
80	115	90	90	115

Consider these storms a population of interest. Obtain the following parameters for the maximum wind speeds. Use the appropriate mathematical notation for the parameters to express your answers.
a. Mean   **b.** Standard deviation
c. Median   **d.** Mode   **e.** IQR

3.210 Dallas Mavericks. From the ESPN website, in the *Dallas Mavericks Roster*, we obtained the following ages, in years, for the players on that basketball team for the 2013–2014 season.

25	24	31	36	23	32	23
25	27	30	28	38	23	20
21	35	24	25	23	35	25

Obtain the following parameters for these ages. Use the appropriate mathematical notation for the parameters to express your answers.
a. Mean   b. Standard deviation
c. Median   d. Mode(s)   e. IQR

3.211 STD Surveillance. The Centers for Disease Control and Prevention compiles reported cases and rates of diseases in United States cities and outlying areas. In a document titled *Sexually Transmitted Disease Surveillance*, the number of reported cases of all stages of syphilis is provided for cities, including Orlando, Florida, and Cincinnati, Ohio. Following is the number of reported cases of syphilis for those two cities for the years 2007–2011.

| Orlando | 583 | 460 | 408 | 391 | 485 |
| Cincinnati | 77 | 105 | 227 | 484 | 436 |

a. Obtain the individual population means of the number of cases for both cities.
b. Without doing any calculations, decide for which city the population standard deviation of the number of cases is smaller. Explain your answer.
c. Obtain the individual population standard deviations of the number of cases for both cities.
d. Are your answers to parts (b) and (c) consistent? Why or why not?

3.212 Dart Doubles. The top two players in the 2001–2002 Professional Darts Corporation World Championship were Phil Taylor and Peter Manley. Taylor and Manley dominated the competition with a record number of doubles. A *double* is a throw that lands in either the outer ring of the dartboard or the outer ring of the bull's-eye. The following table provides the number of doubles thrown by each of the two players during the five rounds of competition, as found in *Chance* (Vol. 15, No. 3, pp. 48–55).

| Taylor | 21 | 18 | 18 | 19 | 13 |
| Manley | 5 | 24 | 20 | 26 | 14 |

a. Obtain the individual population means of the number of doubles.
b. Without doing any calculations, decide for which player the standard deviation of the number of doubles is smaller. Explain your answer.
c. Obtain the individual population standard deviations of the number of doubles.
d. Are your answers to parts (b) and (c) consistent? Why or why not?

3.213 Doing Time. According to *Compendium of Federal Justice Statistics*, published by the Bureau of Justice Statistics, the mean time served to first release by Federal prisoners is 32.9 months. Assume the standard deviation of the times served is 17.9 months. Let x denote time served to first release by a Federal prisoner.
a. Find the standardized version of x.
b. Find the mean and standard deviation of the standardized variable.
c. Determine the z-scores for prison times served of 81.3 months and 20.8 months. Round your answers to two decimal places.
d. Interpret your answers in part (c).
e. Construct a graph similar to Fig. 3.20 on page 146 that depicts your results from parts (b) and (c).

3.214 Gestation Periods of Humans. Gestation periods of humans have a mean of 266 days and a standard deviation of 16 days. Let y denote the variable "gestation period" for humans.
a. Find the standardized variable corresponding to y.
b. What are the mean and standard deviation of the standardized variable?
c. Obtain the z-scores for gestation periods of 227 days and 315 days. Round your answers to two decimal places.
d. Interpret your answers in part (c).
e. Construct a graph similar to Fig. 3.20 on page 146 that shows your results from parts (b) and (c).

3.215 Frog Thumb Length. W. Duellman and J. Kohler explore a new species of frog in the article "New Species of Marsupial Frog (Hylidae: Hemiphractinae: *Gastrotheca*) from the Yungas of Bolivia" (*Journal of Herpetology*, Vol. 39, No. 1, pp. 91–100). These two museum researchers collected information on the lengths and widths of different body parts for the male and female *Gastrotheca piperata*. Thumb length for the female *Gastrotheca piperata* has a mean of 6.71 mm and a standard deviation of 0.67 mm. Let x denote thumb length for a female specimen.
a. Find the standardized version of x.
b. Determine and interpret the z-scores for thumb lengths of 5.2 mm and 8.1 mm. Round your answers to two decimal places.

3.216 Low-Birth-Weight Hospital Stays. Data on low-birth-weight babies were collected over a 2-year period by 14 participating centers of the National Institute of Child Health and Human Development Neonatal Research Network. Results were reported by J. Lemons et al. in the on-line paper "Very Low Birth Weight Outcomes of the National Institute of Child Health and Human Development Neonatal Research Network" (*Pediatrics*, Vol. 107, No. 1, p. e1). For the 1084 surviving babies whose birth weights were 751–1000 grams, the average length of stay in the hospital was 86 days, although one center had an average of 66 days and another had an average of 108 days.
a. Can the mean lengths of stay be considered population means? Explain your answer.
b. Assuming that the population standard deviation is 12 days, determine the z-score for a baby's length of stay of 86 days at the center where the mean was 66 days.
c. Assuming that the population standard deviation is 12 days, determine the z-score for a baby's length of stay of 86 days at the center where the mean was 108 days.
d. What can you conclude from parts (b) and (c) about an infant with a length of stay equal to the mean at all centers if that infant was born at a center with a mean of 66 days? mean of 108 days?

3.217 Low Gas Mileage. Suppose you buy a new car whose advertised mileage is 25 miles per gallon (mpg). After driving your car for several months, you find that its mileage is 21.4 mpg. You telephone the manufacturer and learn that the standard deviation of gas mileages for all cars of the model you bought is 1.15 mpg.
a. Find the z-score for the gas mileage of your car, assuming the advertised claim is correct.
b. Does it appear that your car is getting unusually low gas mileage?

3.218 Exam Scores. Suppose that you take an exam with 400 possible points and are told that the mean score is 280 and that the standard deviation is 20. You are also told that you got 350. Did you do well on the exam? Explain your answer.

Extending the Concepts and Skills

Population and Sample Standard Deviations. In Exercises 3.219–3.221, you examine the numerical relationship between the population standard deviation and the sample standard deviation computed from the same data. This relationship is helpful when the computer or statistical calculator being used has a built-in program for sample standard deviation but not for population standard deviation.

3.219 Consider the following three data sets.

Data Set 1		Data Set 2				Data Set 3				
2	4	7	5	5	3	4	7	8	9	7
7	3	9	8	6		4	5	3	4	5

a. Assuming that each of these data sets is sample data, compute the standard deviations. (Round your final answers to two decimal places.)
b. Assuming that each of these data sets is population data, compute the standard deviations. (Round your final answers to two decimal places.)
c. Using your results from parts (a) and (b), make an educated guess about the answer to the following question: If both s and σ are computed for the same data set, will they tend to be closer together if the data set is large or if it is small?

3.220 Consider a data set with m observations. If the data are sample data, you compute the sample standard deviation, s, whereas if the data are population data, you compute the population standard deviation, σ.
a. Derive a mathematical formula that gives σ in terms of s when both are computed for the same data set. (*Hint:* First note that, numerically, the values of $\bar{x}$ and μ are identical. Consider the ratio of the defining formula for σ to the defining formula for s.)
b. Refer to the three data sets in Exercise 3.219. Verify that your formula in part (a) works for each of the three data sets.
c. Suppose that a data set consists of 15 observations. You compute the sample standard deviation of the data and obtain $s = 38.6$. Then you realize that the data are actually population data and that you should have obtained the population standard deviation instead. Use your formula from part (a) to obtain σ.

3.221 Dallas Mavericks. From the ESPN website, in the *Dallas Mavericks Roster*, we obtained the following weights, in pounds, for the players on that basketball team for the 2013–2014 season.

185	270	211	220	235	250
215	200	185	192	240	188
215	176	195	228	189	191
255	245	210			

Use the technology of your choice to determine
a. the population mean weight.
b. the population standard deviation of the weights. *Note:* Depending on the technology that you're using, you may need to refer to the formula derived in Exercise 3.220(a).

Estimating Relative Standing. In Key Fact 3.3, we stated Chebyshev's rule: For any quantitative data set and any real number k greater than or equal to 1, at least $1 - 1/k^2$ of the observations lie within k standard deviations to either side of the mean, that is, between $\bar{x} - k \cdot s$ and $\bar{x} + k \cdot s$. You can use z-scores and Chebyshev's rule to estimate the relative standing of an observation.

To see how, let us consider again the lengths of the holes at Augusta National Golf Club, shown in the third column of Table 3.17 on page 140. Earlier, we found that the population mean and standard deviation of these lengths are 413.1 yd and 132.0 yd, respectively. We note, for instance, that the z-score for Pink dogwood's length of 575 yd is $(575 - 413.1)/132.0$, or 1.23. Applying Chebyshev's rule

to that z-score, we conclude that at least $1 - 1/1.23^2$, or 33.9%, of the lengths lie within 1.23 standard deviations to either side of the mean. Therefore, Pink dogwood's length, which is 1.23 standard deviations above the mean, is greater than at least 33.9% of the other holes' lengths.

3.222 Stewed Tomatoes. A company produces cans of stewed tomatoes with an advertised weight of 14 oz. The standard deviation of the weights is known to be 0.4 oz. A quality-control engineer selects a can of stewed tomatoes at random and finds its net weight to be 17.28 oz.
a. Estimate the relative standing of that can of stewed tomatoes, assuming the true mean weight is 14 oz. Use the z-score and Chebyshev's rule.
b. Does the quality-control engineer have reason to suspect that the true mean weight of all cans of stewed tomatoes being produced is not 14 oz? Explain your answer.

3.223 Buying a Home. Suppose that you are thinking of buying a resale home in a large tract. The owner is asking $205,500. Your realtor obtains the sale prices of comparable homes in the area that have sold recently. The mean of the prices is $220,258 and the standard deviation is $5,237. Does it appear that the home you are contemplating buying is a bargain? Explain your answer using the z-score and Chebyshev's rule.

Comparing Relative Standing. If two distributions have the same shape or, more generally, if they differ only by center and variation, then z-scores can be used to compare the relative standings of two observations from those distributions. The two observations can be of the same variable from different populations or they can be of different variables from the same population. Consider Exercise 3.224.

3.224 SAT Scores. Each year, thousands of high school students bound for college take the Scholastic Assessment Test (SAT). This test measures the verbal and mathematical abilities of prospective college students. Student scores are reported on a scale that ranges from a low of 200 to a high of 800. Summary results for the scores are published by the College Entrance Examination Board in *College Bound Seniors*. In one high school graduating class, the mean SAT math score is 528 with a standard deviation of 105; the mean SAT verbal score is 475 with a standard deviation of 98. A student in the graduating class scored 740 on the SAT math and 715 on the SAT verbal.
a. Under what conditions would it be reasonable to use z-scores to compare the standings of the student on the two tests relative to the other students in the graduating class?
b. Assuming that a comparison using z-scores is legitimate, relative to the other students in the graduating class, on which test did the student do better?

3.225 Copperhead and Tiger Snakes. S. Fearn et al. compare two types of snakes in the article "Body Size and Trophic Divergence of Two Large Sympatric Elapic Snakes in Tasmania" (*Australian Journal of Zoology*, Vol. 60, No. 3, pp. 159–165). Tiger snakes and lowland copperheads are both large snakes confined to the cooler parts of Tasmania. The weights of the male lowland copperhead in Tasmania have a mean of 812.07 g and a standard deviation of 330.24 g; the weights of the male tiger snake in Tasmania have a mean of 743.65 g and a standard deviation of 336.36 g.
a. Determine the z-scores for both a male lowland copperhead snake and a male tiger snake whose weights are 850 g.
b. Under what conditions would it be reasonable to use z-scores to compare the relative standings of the weights of the two snakes?
c. Assuming that a comparison using z-scores is legitimate, relative to the other snakes of its type, which snake is heavier?

CHAPTER IN REVIEW

You Should Be Able to

1. use and understand the formulas in this chapter.

2. explain the purpose of a measure of center.

3. obtain and interpret the mean, the median, and the mode(s) of a data set.

4. choose an appropriate measure of center for a data set.

5. use and understand summation notation.

6. define, compute, and interpret a sample mean.

7. explain the purpose of a measure of variation.

8. define, compute, and interpret the range of a data set.

9. define, compute, and interpret a sample standard deviation.

*10. state and apply Chebyshev's rule.

*11. state and apply the empirical rule.

12. define percentiles, deciles, and quartiles.

13. obtain and interpret the quartiles, IQR, and five-number summary of a data set.

14. obtain the lower and upper limits of a data set and identify potential outliers.

15. construct and interpret a boxplot.

16. use boxplots to compare two or more data sets.

17. state pertinent properties of boxplots for right-skewed, symmetric, and left-skewed distributions.

18. define and interpret the population mean (mean of a variable).

19. define and interpret the population standard deviation (standard deviation of a variable).

20. compute the population mean and population standard deviation of a finite population.

21. distinguish between a parameter and a statistic.

22. understand how and why statistics are used to estimate parameters.

23. define and obtain standardized variables.

24. obtain and interpret z-scores.

Key Terms

adjacent values, *131*
box-and-whisker diagram, *131*
boxplot, *131*
Chebyshev's rule,* *117*
deciles, *125*
descriptive measures, *93*
deviations from the mean, *106*
empirical rule,* *119*
first quartile (Q_1), *126*
five-number summary, *129*
indices, *98*
interquartile range (IQR), *128*
lower fence, *130*
lower limit, *130*
mean, *94*
mean of a variable (μ), *140*
measures of center, *94*
measures of central tendency, *94*
measures of spread, *105*

measures of variation, *105*
median, *95*
mode, *96*
outliers, *104, 116, 129*
parameter, *143*
percentiles, *125*
population mean (μ), *140*
population standard deviation (σ), *142*
population variance (σ^2), *142*
potential outlier, *130*
pth percentile, *125*
quartiles, *125, 126*
quintiles, *125*
range, *106*
resistant measure, *97*
sample mean ($\bar{x}$), *99*
sample size (n), *99*
sample standard deviation (s), *108*
sample variance (s^2), *108*

second quartile (Q_2), *126*
standard deviation, *106*
standard deviation of a
    variable (σ), *142*
standard score, *145*
standardized variable, *144*
standardized version, *144*
statistic, *143*
subscripts, *98*
sum of squared deviations, *107*
summation notation, *98*
third quartile (Q_3), *126*
trimmed mean, *97, 104*
upper fence, *130*
upper limit, *130*
variance of a variable (σ^2), *142*
whiskers, *131*
z-score, *145*

REVIEW PROBLEMS

Understanding the Concepts and Skills

1. Define
a. descriptive measures.
b. measures of center.
c. measures of variation.

2. Identify the two most commonly used measures of center for quantitative data. Explain the relative advantages and disadvantages of each.

3. Among the measures of center discussed, which is the only one appropriate for qualitative data?

4. Identify the most appropriate measure of variation corresponding to each of the following measures of center.
a. Mean   **b.** Median

5. Specify the mathematical symbol used for each of the following descriptive measures.
a. Sample mean
b. Sample standard deviation
c. Population mean
d. Population standard deviation

6. Data Set A has more variation than Data Set B. Decide which of the following statements are necessarily true.
a. Data Set A has a larger mean than Data Set B.
b. Data Set A has a larger standard deviation than Data Set B.

7. Complete the statement: Almost all the observations in any data set lie within _____ standard deviations to either side of the mean.

***8.** What does Chebyshev's rule say about the percentage of observations in any data set that lie within
a. six standard deviations to either side of the mean?
b. 1.5 standard deviations to either side of the mean?

***9.** A quantitative data set of size 87 has mean 80 and standard deviation 10. At least how many observations lie between 60 and 100?

***10.** A data set with roughly a bell-shaped distribution has mean 45 and standard deviation 12. Approximately what percentage of the observations lie between 33 and 57?

***11.** A data set of size 152 with roughly a bell-shaped distribution has mean 25 and standard deviation 4. Approximately how many observations lie between 17 and 33?

12. Regarding the five-number summary:
a. Identify its components.
b. How can it be employed to describe center and variation?
c. What graphical display is based on it?

13. Regarding outliers:
a. What is an outlier?
b. Explain how you can identify potential outliers, using only the first and third quartiles.

14. Regarding z-scores:
a. How is a z-score obtained?
b. What is the interpretation of a z-score?
c. An observation has a z-score of 2.9. Roughly speaking, what is the relative standing of the observation?

Applying the Concepts and Skills

15. Party Time. An integral part of doing business in the dot-com culture of the late 1990s was frequenting the party circuit centered in San Francisco. Here high-tech companies threw as many as five parties a night to recruit or retain talented workers in a highly competitive job market. With as many as 700 guests at a single party, the food and booze flowed, with an average alcohol cost per guest of $15–$18 and an average food bill of $75–$150. A sample of guests at a dot-com party yielded the following data on number of alcoholic drinks consumed per person. [SOURCE: *USA TODAY* Online]

4	4	1	0	5
1	1	2	4	3
1	5	3	0	2
2	2	1	2	4

a. Find the mean, median, and mode of these data.
b. Which measure of center do you think is best here? Explain your answer.

16. Duration of Marriages. The National Center for Health Statistics publishes information on the duration of marriages in *Vital Statistics of the United States*. Which measure of center is more appropriate for data on the duration of marriages, the mean or the median? Explain your answer.

17. Causes of Death. The U.S. National Center for Health Statistics collects data on causes of death and publishes its findings in *National Vital Statistics Reports*. Which of the three main measures of center is appropriate for causes of death? Explain your answer.

18. Fossil Argonauts. In the article "Fossil Argonauts (Mollusca: Cephalopoda: Octopodida) from Late Miocene Siltstones of the Los Angeles Basin, California" (*Journal of Paleontology*, Vol. 79, No. 3, pp. 520–531), paleontologists L. Saul and C. Stadum discussed fossilized Argonaut egg cases from the late Miocene period found in California. A sample of 10 fossilized egg cases yielded the following data on height, in millimeters. Obtain the mean, median, and mode(s) of these data.

37.5	31.5	27.4	21.0	32.0
33.0	33.0	38.0	17.4	34.5

19. Road Patrol. In the paper "Injuries and Risk Factors in a 100-Mile (161-km) Infantry Road March" (*Preventative Medicine*, Vol. 28, pp. 167–173), K. Reynolds et al. reported on a study commissioned by the U.S. Army. The purpose of the study was to improve medical planning and identify risk factors during multiple-day road patrols by examining the acute effects of long-distance marches by light-infantry soldiers. Each soldier carried a standard U.S. Army rucksack, Meal-Ready-to-Eat packages, and other field equipment. A sample of 10 participating soldiers revealed the following data on total load mass, in kilograms.

48	50	45	49	44
47	37	54	40	43

a. Obtain the sample mean of these 10 load masses.
b. Obtain the range of the load masses.
c. Obtain the sample standard deviation of the load masses.

20. Millionaires. Dr. Thomas Stanley of Georgia State University has collected information on millionaires, including their ages, since 1973. A sample of 36 millionaires has a mean age of 58.5 years and a standard deviation of 13.4 years.
a. Complete the following graph.

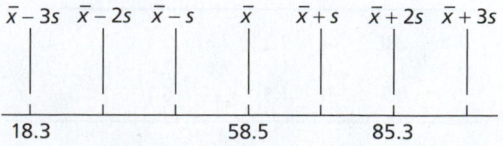

b. Fill in the blanks: Almost all the 36 millionaires are between _____ and _____ years old.

***21. Caffeinated Beverages and Diabetes.** The objective of the article, "Caffeinated and Caffeine-free Beverages and Risk of Type 2 Diabetes" (*American Journal of Clinical Nutrition*, Vol. 97, No. 1, pp. 155–166) by S. Bhupathiraju et al., was to examine the association between caffeinated beverages and type 2 diabetes risk. The mean and standard deviation of the body mass index (BMI) for a sample of 10,215 women who drink at least one caffeinated carbonated beverage a day are 25.7 and 5.3, respectively.

a. At least how many women in the sample have a BMI of between 15.1 and 36.3?

b. *Fill in the blanks:* At least 89% of the women in the sample have BMIs between _____ and _____.

*22. Prices of New Mobile Homes. The U.S. Census Bureau publishes annual price figures for new mobile homes in *Manufactured Housing Statistics*. The prices of a sample of 250 new mobile homes have roughly a bell-shaped distribution with mean $63.3 thousand and standard deviation $7.9 thousand.

a. *Fill in the blanks:* Approximately 95% of the mobile homes in the sample have prices between $ _____ thousand and $ _____ thousand.

b. *Fill in the blank:* Approximately _____% of the mobile homes in the sample have prices between $39.6 thousand and $87.0 thousand.

c. Approximately how many mobile homes in the sample have prices between $55.4 thousand and $71.2 thousand?

23. Millionaires. Refer to Problem 20. The ages of the 36 millionaires sampled are arranged in increasing order in the following table.

31	38	39	39	42	42	45	47	48
48	48	52	52	53	54	55	57	59
60	61	64	64	66	66	67	68	68
69	71	71	74	75	77	79	79	79

a. Determine the quartiles for the data.
b. Obtain and interpret the interquartile range.
c. Find and interpret the five-number summary.
d. Calculate the lower and upper limits.
e. Identify potential outliers, if any.
f. Construct and interpret a boxplot.

24. Oxygen Distribution. In the article "Distribution of Oxygen in Surface Sediments from Central Sagami Bay, Japan: In Situ Measurements by Microelectrodes and Planar Optodes" (*Deep Sea Research Part I: Oceanographic Research Papers*, Vol. 52, Issue 10, pp. 1974–1987), R. Glud et al. explored the distributions of oxygen in surface sediments from central Sagami Bay. The oxygen distribution gives important information on the general biogeochemistry of marine sediments. Measurements were performed at 16 sites. A sample of 22 depths yielded the following data, in millimoles per square meter per day (mmol m^{-2} d^{-1}), on diffusive oxygen uptake (DOU).

1.8	2.0	1.8	2.3	3.8	3.4	2.7	1.1
3.3	1.2	3.6	1.9	7.6	2.0	1.5	2.0
1.1	0.7	1.0	1.8	1.8	6.7		

a. Obtain the five-number summary for these data.
b. Identify potential outliers, if any.
c. Construct a boxplot.

25. A Better Golf Tee? An independent golf equipment testing facility compared the difference in the performance of golf balls hit off a regular 2-3/4″ wooden tee to those hit off a 3″ Stinger Competition golf tee. A Callaway Great Big Bertha driver with 10 degrees of loft was used for the test, and a robot swung the club head at approximately 95 miles per hour. Boxplots of the distances traveled, in yards, are shown in the following figure. Use the boxplots to compare the driving distances for the two golf tees, paying special attention to center and variation.

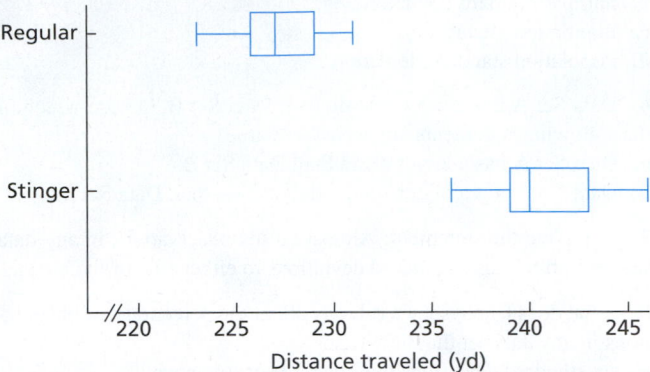

26. Power 90 Extreme. Beachbody, LLC, provides fitness programs, including home workout videos and nutrition. P90X, or Power 90 Extreme, is a home exercise program that consists of an intense series of workout DVDs. It is a 90-day program that uses the term "muscle confusion" to refer to their training methods. There are a total of 12 workout DVDs included with the program, each targeting a different muscle group. According to their official website, the lengths, in minutes, of the 12 DVDs are as follows.

53	58	60	92	59	58
57	57	55	51	43	16

a. Obtain and interpret the population mean length of the P90X DVDs.
b. Obtain and interpret the population standard deviation of the lengths of the P90X DVDs.

27. UC Enrollment. According to the *Statistical Summary of Students and Staff*, prepared by the Department of Information Resources and Communications, Office of the President, University of California, the Fall 2012 enrollment figures for undergraduates at the University of California campuses were as follows.

Campus	Enrollment (1000s)
Berkeley	25.8
Davis	25.8
Irvine	22.1
Los Angeles	27.7
Merced	5.4
Riverside	18.6
San Diego	22.7
Santa Barbara	19.0
Santa Cruz	16.0

a. Compute the population mean enrollment, μ, of the UC campuses. (Round your answer to two decimal places.)
b. Compute σ. (Round your answer to two decimal places.)
c. Letting x denote enrollment, specify the standardized variable, z, corresponding to x.
d. Without performing any calculations, give the mean and standard deviation of z. Explain your answers.

e. Construct dotplots for the distributions of both x and z. Interpret your graphs.

f. Obtain and interpret the z-scores for the enrollments at the Los Angeles and Riverside campuses.

28. Gasoline Prices. The U.S. Energy Information Administration reports weekly figures on retail gasoline prices in *Weekly Retail Gasoline and Diesel Prices*. Every Monday, retail prices for all three grades of gasoline are collected by telephone from a sample of approximately 800 retail gasoline outlets out of a total of more than 100,000 retail gasoline outlets. For the 800 stations sampled on September 16, 2013, the mean price per gallon for unleaded regular gasoline was $3.547.

a. Is the mean price given here a sample mean or a population mean? Explain your answer.

b. What letter or symbol would you use to designate the mean of $3.547?

c. Is the mean price given here a statistic or a parameter? Explain your answer.

Working with Large Data Sets

29. U.S. Divisions and Regions. The U.S. Census Bureau classifies the states in the United States by region and division. The data giving the region and division of each state are presented on the WeissStats site. Use the technology of your choice to determine the mode(s) of the

a. regions.

b. divisions.

In Problems 30–32, use the technology of your choice to

a. *obtain the mean, median, and mode(s) of the data. Determine which of these measures of center is best, and explain your answer.*

b. *determine the range and sample standard deviation of the data.*

c. *find the five-number summary and interquartile range of the data.*

d. *identify potential outliers, if any.*

e. *obtain and interpret a boxplot.*

30. Agricultural Exports. The U.S. Department of Agriculture collects data pertaining to the value of agricultural exports and publishes its findings in *U.S. Agricultural Trade Update*. For one year, the values of these exports, by state, are provided on the WeissStats site. Data are in millions of dollars.

31. Life Expectancy. From The World Bank, in the document *Life Expectancy at Birth*, we obtained data on the expectation of life (in years) at birth for people in various countries. Those data are presented on the WeissStats site.

32. High and Low Temperatures. The U.S. National Oceanic and Atmospheric Administration publishes temperature data in *Climatography of the United States*. According to that document, the annual average maximum and minimum temperatures for selected cities in the United States are as provided on the WeissStats site. [*Note:* Do parts (a)–(e) for both the maximum and minimum temperatures.]

33. Vegetarians and Omnivores. Philosophical and health issues are prompting an increasing number of Taiwanese to switch to a vegetarian lifestyle. In the paper "LDL of Taiwanese Vegetarians Are Less Oxidizable than Those of Omnivores" (*Journal of Nutrition*, Vol. 130, pp. 1591–1596), S. Lu et al. compared the daily intake of nutrients by vegetarians and omnivores living in Taiwan. Among the nutrients considered was protein. Too little protein stunts growth and interferes with all bodily functions; too much protein puts a strain on the kidneys, can cause diarrhea and dehydration, and can leach calcium from bones and teeth. The data on the WeissStats site, based on the results of the aforementioned study, give the daily protein intake, in grams, by samples of 51 female vegetarians and 53 female omnivores.

a. Apply the technology of your choice to obtain boxplots, using the same scale, for the protein-intake data in the two samples.

b. Use the boxplots obtained in part (a) to compare the protein intakes of the females in the two samples, paying special attention to center and variation.

FOCUSING ON DATA ANALYSIS

UWEC UNDERGRADUATES

Recall from Chapter 1 (see page 34) that the Focus database and Focus sample contain information on the undergraduate students at the University of Wisconsin - Eau Claire (UWEC). Now would be a good time for you to review the discussion about these data sets.

a. Open the Focus sample (FocusSample) in the statistical software package of your choice and then obtain the mean and standard deviation of the ages of the sample of 200 UWEC undergraduate students. Are these descriptive measures parameters or statistics? Explain your answer.

b. If your statistical software package will accommodate the entire Focus database (Focus), open that worksheet and then obtain the mean and standard deviation of the ages of all UWEC undergraduate students. (*Answers:* 20.75 years and 1.87 years) Are these descriptive measures parameters or statistics? Explain your answer.

c. Compare your means and standard deviations from parts (a) and (b). What do these results illustrate?

d. If you used a different simple random sample of 200 UWEC undergraduate students than the one in the Focus sample, would you expect the mean and standard deviation of the ages to be the same as that in part (a)? Explain your answer.

e. Open the Focus sample and then obtain the mode of the classifications (class levels) of the sample of 200 UWEC undergraduate students.

f. If your statistical software package will accommodate the entire Focus database, open that worksheet and then obtain the mode of the classifications of all UWEC undergraduate students. (*Answer:* Senior)

g. From parts (e) and (f), you found that the mode of the classifications is the same for both the population and sample of UWEC undergraduate students. Would this necessarily always be the case? Explain your answer.

h. Open the Focus sample and then obtain the five-number summary of the ACT math scores, individually for males and females. Use those statistics to compare the two samples of scores, paying particular attention to center and variation.

i. Open the Focus sample and then obtain the five-number summary of the ACT English scores, individually for males and females. Use those statistics to compare the two samples of scores, paying particular attention to center and variation.

j. Open the Focus sample and then obtain boxplots of the cumulative GPAs, individually for males and females. Use

those statistics to compare the two samples of cumulative GPAs, paying particular attention to center and variation.

k. Open the Focus sample and then obtain boxplots of the cumulative GPAs, individually for each classification (class level). Use those statistics to compare the four samples of cumulative GPAs, paying particular attention to center and variation.

CASE STUDY DISCUSSION

THE BEATLES' SONG LENGTH

At the beginning of this chapter (page 94), we presented some information on the 12 studio albums by The Beatles, specifically, names, release dates, and number of songs per album. The following table gives the lengths, in seconds, of the songs, by album: Please Please Me (PPM), With The Beatles (WTB), A Hard Day's Night (HDN), Beatles for Sale (BFS), Help! (HLP), Rubber Soul (RBS), Revolver (RVR), Sgt. Pepper's Lonely Hearts Club Band (SPL), The Beatles (TBS), Yellow Submarine (YSB), Abbey Road (ABR), and Let It Be (LIB).

Note that the albums are ordered by release date (PPM was the first album released, WTB was the second album released, etc.). We have arranged the song-length data for each album in increasing order.

a. Determine the median song length of each album.

b. Construct a graph with album name on the horizontal axis and median song length on the vertical axis. Interpret your

graph, keeping in mind that the albums are ordered by release date.

c. Determine the range of song lengths for each album.

d. Construct a graph with album name on the horizontal axis and range of song lengths on the vertical axis. Interpret your graph.

e. Find the quartiles of song length for each album.

f. Determine the IQR of song lengths for each album.

g. Construct a graph with album name on the horizontal axis and IQR of song lengths on the vertical axis. Interpret your graph.

h. Find the lower and upper limits of song length for each album. Use them to identify potential outliers.

i. For each album, construct a boxplot of the song lengths, and interpret your results in terms of the variation of the song lengths.

PPM	WTB	HDN	BFS	HLP	RBS	RVR	SPL	TBS		YSB	ABR	LIB
110	108	109	106	114	125	121	80	52	183	130	23	41
112	119	119	121	123	136	121	124	100	188	158	66	49
119	124	131	122	124	138	127	157	105	191	194	72	145
124	128	132	123	125	139	129	157	123	193	207	91	152
125	129	135	132	128	143	135	158	124	195	227	96	172
125	133	140	135	137	147	145	163	137	212	388	117	187
142	136	140	143	138	147	149	164	138	230		125	213
144	147	144	150	148	150	150	166	144	236		146	217
146	149	152	151	149	153	157	169	148	240		165	217
147	152	153	153	153	162	158	210	160	255		171	227
153	156	158	153	155	163	159	217	162	269		183	232
158	167	158	153	156	164	177	307	163	285		185	241
175	167	166	163	173	170	181	333	163	493		207	
177	182		175	190	202	181		167			213	
								174			242	
								174			260	
								182			467	

BIOGRAPHY

JOHN TUKEY: A PIONEER OF EDA

John Wilder Tukey was born on June 16, 1915, in New Bedford, Massachusetts. After earning bachelor's and master's degrees in chemistry from Brown University in 1936 and 1937, respectively, he enrolled in the mathematics program at Princeton

University, where he received a master's degree in 1938 and a doctorate in 1939.

After graduating, Tukey was appointed Henry B. Fine Instructor in Mathematics at Princeton; 10 years later he was advanced to a full professorship. In 1965, Princeton established a department of statistics, and Tukey was named its first chairperson. In addition to his position at Princeton, he was a member of the Technical Staff at AT&T Bell Laboratories, where he served as Associate Executive Director, Research in the Information Sciences Division, from 1945 until his retirement in 1985.

Tukey was among the leaders in the field of exploratory data analysis (EDA), which provides techniques such as stem-and-leaf diagrams for effectively investigating data. He also made fundamental contributions to the areas of robust estimation and time series analysis. Tukey wrote numerous books and more than 350 technical papers on mathematics, statistics, and other scientific subjects. In addition, he coined the word *bit,* a contraction of *binary digit* (a unit of information, often as processed by a computer).

Tukey's participation in educational, public, and government service was most impressive. He was appointed to serve on the President's Science Advisory Committee by President Eisenhower; was chairperson of the committee that prepared "Restoring the Quality of our Environment" in 1965; helped develop the National Assessment of Educational Progress; and was a member of the Special Advisory Panel on the 1990 Census of the U.S. Department of Commerce, Bureau of the Census—to name only a few of his involvements.

Among many honors, Tukey received the National Medal of Science, the IEEE Medal of Honor, Princeton University's James Madison Medal, and Foreign Member, The Royal Society (London). He was the first recipient of the Samuel S. Wilks Award of the American Statistical Association. Until his death, Tukey remained on the faculty at Princeton as Donner Professor of Science, Emeritus; Professor of Statistics, Emeritus; and Senior Research Statistician. Tukey died on July 26, 2000, after a short illness. He was 85 years old.

Probability Concepts

CHAPTER OBJECTIVES

Until now, we have concentrated on descriptive statistics—methods for organizing and summarizing data. Another important aspect of this text is to present the fundamentals of inferential statistics—methods of drawing conclusions about a population based on information from a sample of the population.

Because inferential statistics involves using information from part of a population (a sample) to draw conclusions about the entire population, we can never be certain that our conclusions are correct; that is, uncertainty is inherent in inferential statistics. Consequently, you need to become familiar with uncertainty before you can understand, develop, and apply the methods of inferential statistics.

The science of uncertainty is called **probability theory.** It enables you to evaluate and control the likelihood that a statistical inference is correct. More generally, probability theory provides the mathematical basis for inferential statistics. This chapter begins your study of probability.

CASE STUDY

Texas Hold'em

Texas hold'em or, more simply, hold'em, is now considered the most popular poker game. The Texas State Legislature officially recognizes Robstown, Texas, as the game's birthplace and dates the game back to the early 1900s.

Three reasons for the current popularity of Texas hold'em can be attributed to (1) the emergence of Internet poker sites, (2) the hole cam (a camera that allows people watching television to see the hole cards of the

players), and (3) Tennessee accountant and then-amateur poker-player Chris Moneymaker's first-place win of $2.5 million in the 2003 World Series of Poker after winning his seat to the tournament through a $39 PokerStars satellite tournament.

Following are the details of Texas hold'em.

- Each player is dealt two cards face down, called "hole cards," and then there is a betting round.
- Next, three cards are dealt face up in the center of the table. These three cards are termed "the flop" and are *community cards*, meaning that they can be used by all the players; again there is a betting round.
- Next, an additional community card is dealt face up, called "the turn," and once again there is a betting round.

- Finally, a fifth community card is dealt face up, called "the river," and then there is a final betting round.

A player can use any five cards from the seven cards consisting of his two hole cards and the five community cards to constitute his or her hand. The player with the best hand (using the same hand-ranking as in five-card draw) wins the pot, that is, all the money that has been bet on the hand.

There is one other way that a player can win the pot. Namely, if during any one of the four betting rounds all players but one have folded (i.e., thrown their hole cards face down in the center of the table), then the remaining player is awarded the pot.

The best possible starting hand (hole cards) is two aces. What are the chances of being dealt those hole cards? After studying probability, you will be able to answer that question and similar ones. You will be asked to do so when you revisit Texas hold'em at the end of this chapter.

4.1 Probability Basics

Although most applications of probability theory to statistical inference involve large populations, we will explain the fundamental concepts of probability in this chapter with examples that involve relatively small populations or games of chance.

The Equal-Likelihood Model

We discussed an important aspect of probability when we examined probability sampling in Chapter 1. The following example returns to the illustration of simple random sampling from Example 1.7 on page 11.

EXAMPLE 4.1   Introducing Probability

TABLE 4.1

Five top Oklahoma state officials

| Governor (G) |
| Lieutenant Governor (L) |
| Secretary of State (S) |
| Attorney General (A) |
| Treasurer (T) |

Oklahoma State Officials As reported by the *World Almanac*, the top five state officials of Oklahoma are as shown in Table 4.1. Suppose that we take a simple random sample without replacement of two officials from the five officials.

a. Find the probability that we obtain the governor and treasurer.
b. Find the probability that the attorney general is included in the sample.

Solution For convenience, we use the letters in parentheses after the titles in Table 4.1 to represent the officials. As we saw in Example 1.7, there are 10 possible samples of two officials from the population of five officials. They are listed in Table 4.2. If we take a simple random sample of size 2, each of the possible samples of two officials is equally likely to be the one selected.

TABLE 4.2

The 10 possible samples of two officials

| G, L | G, S | G, A | G, T | L, S |
| L, A | L, T | S, A | S, T | A, T |

a. Because there are 10 possible samples, the probability is $\frac{1}{10}$, or 0.1, of selecting the governor and treasurer (G, T). Another way of looking at this result is that 1 out of 10, or 10%, of the samples include both the governor and the treasurer; hence the probability of obtaining such a sample is 10%, or 0.1. The same goes for any other two particular officials.

b. Table 4.2 shows that the attorney general (A) is included in 4 of the 10 possible samples of size 2. As each of the 10 possible samples is equally likely to be the one selected, the probability is $\frac{4}{10}$, or 0.4, that the attorney general is included in the sample. Another way of looking at this result is that 4 out of 10, or 40%, of the samples include the attorney general; hence the probability of obtaining such a sample is 40%, or 0.4.

You try it!

Exercise 4.13 on page 161

The essential idea in Example 4.1 is that when outcomes are equally likely, probabilities are nothing more than percentages (relative frequencies).

DEFINITION 4.1

Probability for Equally Likely Outcomes (f/N Rule)

Suppose an experiment has N possible outcomes, all equally likely. An event that can occur in f ways has probability f/N of occurring:

Number of ways event can occur ↙

$$\text{Probability of an event} = \frac{f}{N}.$$

↖ Total number of possible outcomes

? What Does It Mean?

For an experiment with equally likely outcomes, probabilities are identical to relative frequencies (or percentages).

In stating Definition 4.1, we used the terms *experiment* and *event* in their intuitive sense. Basically, by an **experiment,** we mean an action whose outcome cannot be predicted with certainty. By an **event,** we mean some specified result that may or may not occur when an experiment is performed.

For instance, in Example 4.1, the experiment consists of taking a random sample of size 2 from the five officials. It has 10 possible outcomes ($N = 10$), all equally likely. In part (b), the event is that the sample obtained includes the attorney general, which can occur in four ways ($f = 4$); hence its probability equals

$$\frac{f}{N} = \frac{4}{10} = 0.4,$$

as we noted in Example 4.1(b).

■■■ **EXAMPLE 4.2**   **Probability for Equally Likely Outcomes**

Family Income   The U.S. Census Bureau compiles data on family income and publishes its findings in *Income, Poverty and Health Insurance Coverage in the United States*. Table 4.3 gives a frequency distribution of annual income for U.S. families.

A U.S. family is selected **at random,** meaning that each family is equally likely to be the one obtained (simple random sample of size 1). Determine the probability that the family selected has an annual income of

a. between $50,000 and $74,999, inclusive (i.e., greater than or equal to $50,000 but less than or equal to $74,999).

b. between $15,000 and $49,999, inclusive.

c. under $25,000.

TABLE 4.3

Frequency distribution of annual income for U.S. families

Income	Frequency (1000s)
Under $15,000	6,827
$15,000–$24,999	7,194
$25,000–$34,999	7,863
$35,000–$49,999	10,898
$50,000–$74,999	15,260
$75,000–$99,999	10,668
$100,000 and over	20,157
	78,867

Solution   The second column of Table 4.3 shows that there are 78,867 thousand U.S. families; so $N = 78{,}867$ thousand.

a. The event in question is that the family selected makes between $50,000 and $74,999. Table 4.3 shows that the number of such families is 15,260 thousand, so $f = 15{,}260$ thousand. Applying the f/N rule, we find that the probability that the family selected makes between $50,000 and $74,999 is

$$\frac{f}{N} = \frac{15{,}260}{78{,}867} = 0.193.$$

Interpretation   19.3% of families in the United States have annual incomes between $50,000 and $74,999, inclusive.

b. The event in question is that the family selected makes between $15,000 and $49,999. Table 4.3 reveals that the number of such families is $7{,}194 + 7{,}863 + 10{,}898$, or 25,955 thousand. Consequently, $f = 25{,}955$ thousand, and the

required probability is

$$\frac{f}{N} = \frac{25,955}{78,867} = 0.329.$$

Interpretation   32.9% of families in the United States make between $15,000 and $49,999, inclusive.

c.   Proceeding as in parts (a) and (b), we find that the probability that the family selected makes under $25,000 is

$$\frac{f}{N} = \frac{6,827 + 7,194}{78,867} = 0.178.$$

Interpretation   17.8% of families in the United States make under $25,000.

You try it!

Exercise 4.19
on page 162

EXAMPLE 4.3   **Probability for Equally Likely Outcomes**

Dice   When two balanced dice are rolled, 36 equally likely outcomes are possible, as depicted in Fig. 4.1. Find the probability that

a.   the sum of the dice is 11.
b.   doubles are rolled; that is, both dice come up the same number.

FIGURE 4.1
Possible outcomes for
rolling a pair of dice

Solution   For this experiment, $N = 36$.

a.   The sum of the dice can be 11 in two ways, as is apparent from Fig. 4.1. Hence the probability that the sum of the dice is 11 equals $f/N = 2/36 = 0.056$.

Interpretation   There is a 5.6% chance of a sum of 11 when two balanced dice are rolled.

You try it!

Exercise 4.25
on page 163

b.   Figure 4.1 also shows that doubles can be rolled in six ways. Consequently, the probability of rolling doubles equals $f/N = 6/36 = 0.167$.

Interpretation   There is a 16.7% chance of doubles when two balanced dice are rolled.

The Meaning of Probability

Essentially, probability is a generalization of the concept of percentage. When we select a member at random from a finite population, as we did in Example 4.2, probability is nothing more than percentage. In general, however, how do we interpret probability? For instance, what do we mean when we say that

- the probability is 0.314 that the gestation period of a woman will exceed 9 months or
- the probability is 0.667 that the favorite in a horse race will finish in the money (first, second, or third place) or

- the probability is 0.40 that a traffic fatality will involve an intoxicated or alcohol-impaired driver or nonoccupant?

Some probabilities are easy to interpret: A probability near 0 indicates that the event in question is very unlikely to occur when the experiment is performed, whereas a probability near 1 (100%) suggests that the event is quite likely to occur. More generally, the **frequentist interpretation of probability** construes the probability of an event to be the proportion of times it occurs in a large number of repetitions of the experiment.

Consider, for instance, the simple experiment of tossing a balanced coin once. Because the coin is balanced, we reason that there is a 50–50 chance the coin will land with heads facing up. Consequently, we attribute a probability of 0.5 to that event. The frequentist interpretation is that in a large number of tosses, the coin will land with heads facing up about half the time.

We used a computer to perform two simulations of tossing a balanced coin 100 times. The results are displayed in Fig. 4.2. Each graph shows the number of tosses of the coin versus the proportion of heads. Both graphs seem to corroborate the frequentist interpretation.

FIGURE 4.2

Two computer simulations of tossing a balanced coin 100 times

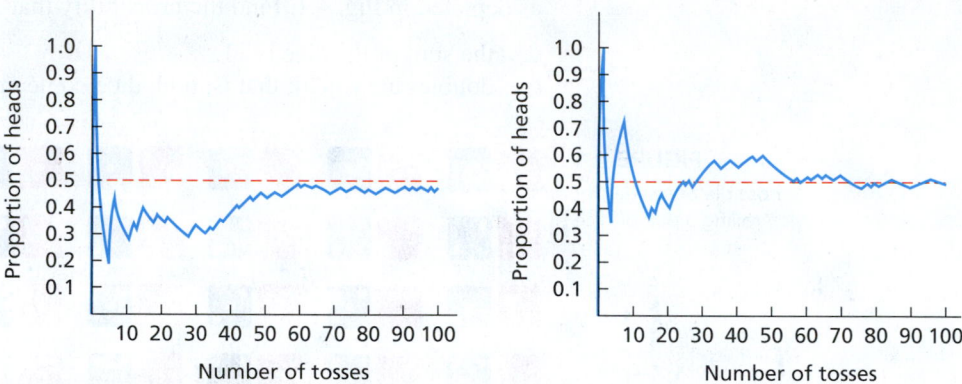

Although the frequentist interpretation is helpful for understanding the meaning of probability, it cannot be used as a definition of probability. One common way to define probabilities is to specify a **probability model**—a mathematical description of the experiment based on certain primary aspects and assumptions.

The **equal-likelihood model** discussed earlier in this section is an example of a probability model. Its primary aspect and assumption are that all possible outcomes are equally likely to occur. We discuss other probability models later in this and subsequent chapters.

Basic Properties of Probabilities

Some basic properties of probabilities are as follows.

KEY FACT 4.1

Basic Properties of Probabilities

Property 1: The probability of an event is always between 0 and 1, inclusive.

Property 2: The probability of an event that cannot occur is 0. (An event that cannot occur is called an **impossible event.**)

Property 3: The probability of an event that must occur is 1. (An event that must occur is called a **certain event.**)

Property 1 indicates that numbers such as 5 or −0.23 could not possibly be probabilities. Example 4.4 illustrates Properties 2 and 3.

EXAMPLE 4.4 Basic Properties of Probabilities

Dice Let's return to Example 4.3, in which two balanced dice are rolled. Determine the probability that

a. the sum of the dice is 1.
b. the sum of the dice is 12 or less.

Solution

a. Figure 4.1 on page 159 shows that the sum of the dice must be 2 or more. Thus the probability that the sum of the dice is 1 equals $f/N = 0/36 = 0$.

Interpretation Getting a sum of 1 when two balanced dice are rolled is impossible and hence has probability 0.

b. From Fig. 4.1, the sum of the dice must be 12 or less. Thus the probability of that event equals $f/N = 36/36 = 1$.

Interpretation Getting a sum of 12 or less when two balanced dice are rolled is certain and hence has probability 1.

Exercise 4.27
on page 163

Exercises 4.1

Understanding the Concepts and Skills

4.1 Roughly speaking, what is an experiment? an event?

4.2 Concerning the equal-likelihood model of probability,
a. what is it?
b. how is the probability of an event found?

4.3 What is the difference between selecting a member at random from a finite population and taking a simple random sample of size 1?

4.4 If a member is selected at random from a finite population, probabilities are identical to _____.

4.5 An experiment has 20 possible outcomes, all equally likely. An event can occur in five ways. The probability that the event occurs is _____.

4.6 An experiment has 40 possible outcomes, all equally likely. An event can occur in 25 ways. The probability that the event occurs is _____.

4.7 State the frequentist interpretation of probability.

4.8 Interpret each of the following probability statements, using the frequentist interpretation of probability.
a. The probability is 0.487 that a newborn baby will be a girl.
b. The probability of a single ticket winning a prize in the Powerball lottery is 0.031.

4.9 Interpret each of the following probability statements, using the frequentist interpretation of probability.
a. The probability of being dealt a pocket pair in Texas hold'em is 0.059.
b. If a balanced dime is tossed three times, the probability that it will come up heads all three times is 0.125.

4.10 Which of the following numbers could not possibly be a probability? Justify your answer.
a. 0.462   b. −0.201   c. 1

4.11 Which of the following numbers could not possibly be a probability? Justify your answer.
a. 5/6   b. 3.5   c. 0

Applying the Concepts and Skills

4.12 Oklahoma State Officials. Refer to Table 4.1 on page 157.
a. List the possible samples without replacement of size 3 that can be obtained from the population of five officials. (*Hint:* There are 10 possible samples.)
If a simple random sample without replacement of three officials is taken from the five officials, determine the probability that
b. the governor, attorney general, and treasurer are obtained.
c. the governor and treasurer are included in the sample.
d. the governor is included in the sample.

4.13 Oklahoma State Officials. Refer to Table 4.1 on page 157.
a. List the possible samples without replacement of size 4 that can be obtained from the population of five officials. (*Hint:* There are five possible samples.)
If a simple random sample without replacement of four officials is taken from the five officials, determine the probability that
b. the governor, secretary of state, attorney general, and treasurer are obtained.
c. the governor and treasurer are included in the sample.
d. the governor is included in the sample.

4.14 Playing Cards. An ordinary deck of playing cards has 52 cards. There are four suits—spades, hearts, diamonds, and clubs—with 13 cards in each suit. Spades and clubs are black; hearts and diamonds are red. If one of these cards is selected at random, what is the probability that it is
a. a spade?   b. red?   c. not a club?

4.15 Poker Chips. A bowl contains 12 poker chips—3 red, 4 white, and 5 blue. If one of these poker chips is selected at random from the bowl, what is the probability that its color is
a. red?   b. red or white?   c. not white?

*In Exercises **4.16–4.26**, express your probability answers as a decimal rounded to three places.*

4.16 Preeclampsia. Preeclampsia is a medical condition characterized by high blood pressure and protein in the urine of a pregnant woman. It is a serious condition that can be life threatening to the mother and child. In the article "Women's Experiences of Preeclampsia: Australian Action on Preeclampsia Survey of Women and Their Confidants" (*Journal of Pregnancy*, Vol. 2011, Issue 1, Article ID 375653), C. East et al. examined the experiences of 68 women with preeclampsia. The following table provides a frequency distribution of instances of prenatal or infant death for infants of the women with preeclampsia.

Prenatal/Infant Death	Frequency
Stillborn	9
Death within one week	4
Death one week to six weeks	3
Death six weeks to six months	1
Death after six months	1
Did not die	50

Suppose that one of these women with preeclampsia is randomly selected. Find the probability that the child of the woman selected
a. died.
b. died one week to six months after birth.
c. lived at least six weeks.

4.17 Russian Presidential Election. According to the Central Election Commission of the Russian Federation, a frequency distribution for the March 4, 2012 Russian presidential election is as follows.

Candidate	Votes
Vladimir Putin	45,513,001
Gennady Zyuganov	12,288,624
Mikhail Prokhorov	5,680,558
Vladimir Zhirinovsky	4,448,959
Sergey Mironov	2,755,642

Find the probability that a randomly selected voter voted for
a. Putin.
b. either Zhirinovsky or Mironov.
c. someone other than Putin.

4.18 Cardiovascular Hospitalizations. From the Florida State Center for Health Statistics report *Women and Cardiovascular Disease Hospitalization*, we obtained the following table showing the number of female hospitalizations for cardiovascular disease, by age group, during one year.

Age group (yr)	Number
0–19	810
20–39	5,029
40–49	10,977
50–59	20,983
60–69	36,884
70–79	65,017
80 and over	69,167

One of these case records is selected at random. Find the probability that the woman was

a. in her 50s.
b. less than 50 years old.
c. between 40 and 69 years old, inclusive.
d. 70 years old or older.

4.19 Housing Units. The U.S. Census Bureau publishes data on housing units in *American Housing Survey for the United States*. The following table provides a frequency distribution for the number of rooms in U.S. housing units. The frequencies are in thousands.

Rooms	No. of units
1	601
2	1,404
3	11,433
4	23,636
5	30,440
6	27,779
7	17,868
8+	19,257

A U.S. housing unit is selected at random. Find the probability that the housing unit obtained has
a. four rooms.
b. more than four rooms.
c. one or two rooms.
d. fewer than one room.
e. one or more rooms.

4.20 Murder Victims. As reported by the Federal Bureau of Investigation in *Crime in the United States*, the age distribution of murder victims between 20 and 59 years old is as shown in the following table.

Age (yr)	Frequency
20–24	2329
25–29	1892
30–34	1414
35–39	1046
40–44	900
45–49	854
50–54	673
55–59	500

A murder case in which the person murdered was between 20 and 59 years old is selected at random. Find the probability that the murder victim was
a. between 40 and 44 years old, inclusive.
b. at least 25 years old, that is, 25 years old or older.
c. between 45 and 59 years old, inclusive.
d. under 30 or over 54.

4.21 Occupations in Seoul. The population of Seoul was studied in an article by B. Lee and J. McDonald, "Determinants of Commuting Time and Distance for Seoul Residents: The Impact of Family Status on the Commuting of Women" (*Urban Studies*, Vol. 40, No. 7, pp. 1283–1302). The authors examined the different occupations for males and females in Seoul. The table at the top of the next page is a frequency distribution of occupation type for males taking part in a survey. (*Note:* M = manufacturing, N = nonmanufacturing.)

If one of these males is selected at random, find the probability that his occupation is
a. service.
b. administrative.
c. manufacturing.
d. not manufacturing.

Occupation	Frequency
Administrative/M	2,197
Administrative/N	6,450
Technical/M	2,166
Technical/N	6,677
Clerk/M	1,640
Clerk/N	4,538
Production workers/M	5,721
Production workers/N	10,266
Service	9,274
Agriculture	159

4.22 Nobel Laureates. From Wikipedia and the article "Which Country Has the Best Brains?" from *BBC News Magazine*, we obtained a frequency distribution of the number of Nobel Prize winners, by country.

Country	Winners
United States	338
United Kingdom	119
Germany	103
France	59
Sweden	29
Switzerland	25
Other countries	166

Suppose that a recipient of a Nobel Prize is selected at random. Find the probability that the Nobel Laureate is from
a. Sweden.   **b.** either France or Germany.
c. any country other than the United States.

4.23 Graduate Science Students. According to *Survey of Graduate Science Engineering Students and Postdoctorates*, published by the U.S. National Science Foundation, the distribution of graduate science students in doctorate-granting institutions is as follows. Frequencies are in thousands. *Note:* Earth sciences include atmospheric and ocean sciences as well.

Field	Frequency
Agricultural sciences	15.1
Biological sciences	69.7
Computer sciences	46.1
Earth sciences	14.9
Mathematical sciences	22.0
Physical sciences	38.7
Psychology	45.6
Social sciences	97.2

A graduate science student who is attending a doctorate-granting institution is selected at random. Determine the probability that the field of the student obtained is
a. psychology.
b. physical or social science.
c. not computer science.

4.24 Family Size. A *family* is defined to be a group of two or more persons related by birth, marriage, or adoption and residing together in a household. According to *Current Population Survey*, published by the U.S. Census Bureau, the size distribution of U.S. families is as follows. Frequencies are in thousands.

Size	Frequency
2	37,148
3	17,810
4	15,054
5	6,660
6	2,438
7+	1,395

A U.S. family is selected at random. Find the probability that the family obtained has
a. two persons.
b. more than three persons.
c. between one and three persons, inclusive.
d. one person.
e. one or more persons.

4.25 Dice. Two balanced dice are rolled. Refer to Fig. 4.1 on page 159 and determine the probability that the sum of the dice is
a. 6.   **b.** even.
c. 7 or 11.   **d.** 2, 3, or 12.

4.26 Coin Tossing. A balanced dime is tossed three times. The possible outcomes can be represented as follows.

HHH	HTH	THH	TTH
HHT	HTT	THT	TTT

Here, for example, HHT means that the first two tosses come up heads and the third tails. Find the probability that
a. exactly two of the three tosses come up heads.
b. the last two tosses come up tails.
c. all three tosses come up the same.
d. the second toss comes up heads.

4.27 Housing Units. Refer to Exercise 4.19.
a. Which, if any, of the events in parts (a)–(e) are certain? impossible?
b. Determine the probability of each event identified in part (a).

4.28 Family Size. Refer to Exercise 4.24.
a. Which, if any, of the events in parts (a)–(e) are certain? impossible?
b. Determine the probability of each event identified in part (a).

4.29 Gender and Handedness. This problem requires that you first obtain the gender and handedness of each student in your class. Subsequently, determine the probability that a randomly selected student in your class is
a. female.
b. left-handed.
c. female and left-handed.
d. neither female nor left-handed.

4.30 The probability is 0.314 that the gestation period of a woman will exceed 9 months. In 4000 human gestation periods, roughly how many will exceed 9 months?

4.31 The probability is 0.667 that the favorite in a horse race will finish in the money (first, second, or third place). In 500 horse races, roughly how many times will the favorite finish in the money?

4.32 U.S. Governors. In 2013, according to the National Governors Association, 30 of the 50 state governors were Republicans. Suppose that on each day of 2013, one U.S. state governor was randomly selected to read the invocation on a popular radio program. On approximately how many of those days should we expect that a Republican was chosen?

Extending the Concepts and Skills

4.33 Explain what is wrong with the following argument: When two balanced dice are rolled, the sum of the dice can be 2, 3, 4, 5, 6, 7, 8, 9, 10, 11, or 12, giving 11 possibilities. Therefore the probability is $\frac{1}{11}$ that the sum is 12.

4.34 Bilingual and Trilingual. At a certain university in the United States, 62% of the students are at least bilingual—speaking English and at least one other language. Of these students, 80% speak Spanish and, of the 80% who speak Spanish, 10% also speak French. Determine the probability that a randomly selected student at this university

a. does not speak Spanish.　　**b.** speaks Spanish and French.

4.35 Consider the random experiment of tossing a coin once. There are two possible outcomes for this experiment, namely, a head (H) or a tail (T).

a. Repeat the random experiment five times—that is, toss a coin five times—and record the information required in the following table. (The third and fourth columns are for running totals and running proportions, respectively.)

Toss	Outcome	Number of heads	Proportion of heads
1			
2			
3			
4			
5			

b. Based on your five tosses, what estimate would you give for the probability of a head when this coin is tossed once? Explain your answer.

c. Now toss the coin five more times and continue recording in the table so that you now have entries for tosses 1–10. Based on your 10 tosses, what estimate would you give for the probability of a head when this coin is tossed once? Explain your answer.

d. Now toss the coin 10 more times and continue recording in the table so that you now have entries for tosses 1–20. Based on your 20 tosses, what estimate would you give for the probability of a head when this coin is tossed once? Explain your answer.

e. In view of your results in parts (b)–(d), explain why the frequentist interpretation cannot be used as the definition of probability.

Odds. Closely related to probabilities are *odds*. Newspapers, magazines, and other popular publications often express likelihood in terms of odds instead of probabilities, and odds are used much more than probabilities in gambling contexts. If the probability that an event occurs is p, the odds that the event occurs are p to $1 - p$. This fact is also expressed by saying that the odds are p to $1 - p$ *in favor of the*

event or that the odds are $1 - p$ to p *against the event*. Conversely, if the odds in favor of an event are a to b (or, equivalently, the odds against it are b to a), the probability the event occurs is $a/(a + b)$. For example, if an event has probability 0.75 of occurring, the odds that the event occurs are 0.75 to 0.25, or 3 to 1; if the odds against an event are 3 to 2, the probability that the event occurs is $2/(2 + 3)$, or 0.4. We examine odds in Exercises 4.36–4.40.

4.36 Roulette. An American roulette wheel contains 38 numbers, of which 18 are red, 18 are black, and 2 are green. When the roulette wheel is spun, the ball is equally likely to land on any of the 38 numbers. For a bet on red, the house pays even odds (i.e., 1 to 1). What should the odds actually be to make the bet fair?

4.37 Cyber Affair. As found in *USA TODAY*, results of a survey by International Communications Research revealed that roughly 75% of adult women believe that a romantic relationship over the Internet while in an exclusive relationship in the real world is cheating. What are the odds against randomly selecting an adult female Internet user who believes that having a "cyber affair" is cheating?

4.38 Belmont Stakes. The Belmont Stakes is the third leg, after the Kentucky Derby and Preakness Stakes, of the Triple Crown of thoroughbred horseracing. The morning-line betting odds of the two favorites, Orb and Revolutionary, for the 2013 Belmont Stakes were 7 to 2 (against) and 5 to 1 (against), respectively. Based on the morning-line betting odds, determine the probability that the winner of the race would be

a. Orb.　　　　　　　　　**b.** Revolutionary.

4.39 Cursing Your Computer. A study was conducted by the firm Coleman & Associates, Inc. to determine who curses at their computer. The results, which appeared in *USA TODAY*, indicated that 46% of people age 18–34 years have cursed at their computer. What are the odds against a randomly selected 18- to 34-year-old having cursed at his or her computer?

4.40 Lightning Casualties. An issue of *Travel + Leisure Golf* magazine reported several facts about lightning. Here are three of them.

- The odds of an individual being struck by lightning in a year in the United States are about 280,000 to 1 (against).
- The odds of an individual being struck by lightning in a year in Florida—the state with the most golf courses—are about 80,000 to 1 (against).
- About 5% of all lightning fatalities occur on golf courses.

Based on these data, answer the following questions.

a. What is the probability of a person being struck by lightning in a year in the United States? Express your answer as a decimal rounded to eight places.

b. What is the probability of a person being struck by lightning in a year in Florida? Express your answer as a decimal rounded to seven decimal places.

c. If a person dies from being hit by lightning, what are the odds that the fatality did not occur on a golf course?

4.2　Events

Before continuing, we need to discuss events in greater detail. In Section 4.1, we used the word *event* intuitively. More precisely, an **event** is a collection of outcomes, as illustrated in Example 4.5.

◼◼◼ **EXAMPLE 4.5** **Introducing Events**

Playing Cards A deck of playing cards contains 52 cards, as displayed in Fig. 4.3. When we perform the experiment of randomly selecting one card from the deck, we will get one of these 52 cards. The collection of all 52 cards—the possible outcomes—is called the **sample space** for this experiment.

FIGURE 4.3
A deck of playing cards

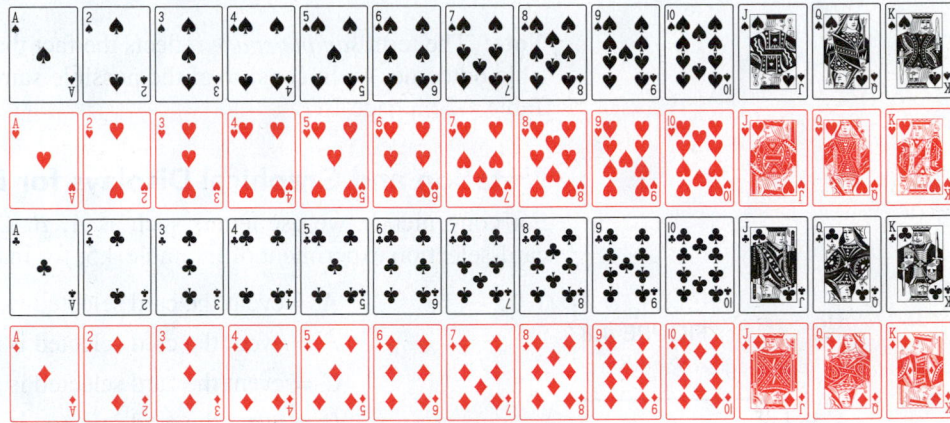

Many different events can be associated with this card-selection experiment. Let's consider four:

a. The event that the card selected is the king of hearts.
b. The event that the card selected is a king.
c. The event that the card selected is a heart.
d. The event that the card selected is a face card.

FIGURE 4.4
The event the king of hearts is selected

List the outcomes constituting each of these four events.

Solution

a. The event that the card selected is the king of hearts consists of the single outcome "king of hearts," as pictured in Fig. 4.4.

FIGURE 4.5
The event a king is selected

b. The event that the card selected is a king consists of the four outcomes "king of spades," "king of hearts," "king of clubs," and "king of diamonds," as depicted in Fig. 4.5.

c. The event that the card selected is a heart consists of the 13 outcomes "ace of hearts," "two of hearts,". . . , "king of hearts," as shown in Fig. 4.6.

FIGURE 4.6
The event a heart is selected

d. The event that the card selected is a face card consists of 12 outcomes, namely, the 12 face cards shown in Fig. 4.7.

FIGURE 4.7
The event a face card is selected

You try it!

Exercise 4.53 on page 170

When the experiment of randomly selecting a card from the deck is performed, a specified event **occurs** if that event contains the card selected. For instance, if the card selected turns out to be the king of spades, the second and fourth events (Figs. 4.5 and 4.7) occur, whereas the first and third events (Figs. 4.4 and 4.6) do not. ◼

DEFINITION 4.2

Sample Space and Event

Sample space: The collection of all possible outcomes for an experiment.

Event: A collection of outcomes for the experiment, that is, any subset of the sample space. An event **occurs** if and only if the outcome of the experiment is a member of the event.

Note: The term *sample space* reflects the fact that, in statistics, the collection of possible outcomes often consists of the possible samples of a given size, as illustrated in Table 4.2 on page 157.

Notation and Graphical Displays for Events

For convenience, we use letters such as A, B, C, D, ... to represent events. In the card-selection experiment of Example 4.5, for instance, we might let

$A =$ event the card selected is the king of hearts,

$B =$ event the card selected is a king,

$C =$ event the card selected is a heart, and

$D =$ event the card selected is a face card.

FIGURE 4.8

Venn diagram for event E

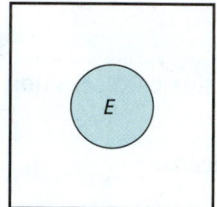

Venn diagrams, named after English logician John Venn (1834–1923), are one of the best ways to portray events and relationships among events visually. The sample space is depicted as a rectangle, and the various events are drawn as disks (or other geometric shapes) inside the rectangle. In the simplest case, only one event is displayed, as shown in Fig. 4.8, with the colored portion representing the event.

Relationships Among Events

Each event E has a corresponding event defined by the condition that "E does not occur." That event is called the **complement** of E, denoted **(not E).** Event (not E) consists of all outcomes not in E, as shown in the Venn diagram in Fig. 4.9(a).

FIGURE 4.9

Venn diagrams for (a) event (not E),
(b) event (A & B), and (c) event (A or B)

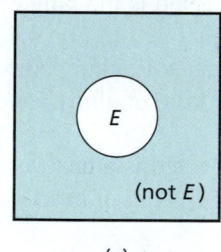 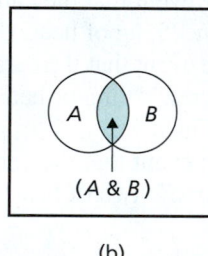 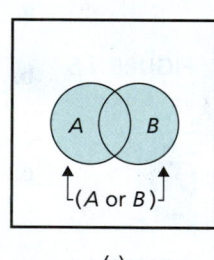

    (a)   (b)   (c)

What Does It Mean?

Event (not E) consists of all outcomes not in event E; event (A & B) consists of all outcomes common to event A and event B; event (A or B) consists of all outcomes either in event A or in event B or both.

With any two events, say, A and B, we can associate two new events. One new event is defined by the condition that "both event A and event B occur" and is denoted **(A & B).** Event (A & B) consists of all outcomes common to both event A and event B, as illustrated in Fig. 4.9(b).

The other new event associated with A and B is defined by the condition that "either event A or event B or both occur" or, equivalently, that "at least one of events A and B occurs." That event is denoted **(A or B)** and consists of all outcomes in either event A or event B or both, as Fig. 4.9(c) shows.

DEFINITION 4.3

Relationships Among Events

(not E):   The event "E does not occur"

(A & B):   The event "both A and B occur"

(A or B):   The event "either A or B or both occur"

Note: Because the event "both A and B occur" is the same as the event "both B and A occur," event (A & B) is the same as event (B & A). Similarly, event (A or B) is the same as event (B or A).

EXAMPLE 4.6   **Relationships Among Events**

Playing Cards For the experiment of randomly selecting one card from a deck of 52, let

A = event the card selected is the king of hearts,

B = event the card selected is a king,

C = event the card selected is a heart, and

D = event the card selected is a face card.

We showed the outcomes for each of those four events in Figs. 4.4–4.7, respectively, in Example 4.5 on page 165. Determine the following events.

a. (not D)   **b.** (B & C)   **c.** (B or C)   **d.** (C & D)

Solution

a. (not D) is the event D does not occur—the event that a face card is not selected. Event (not D) consists of the 40 cards in the deck that are not face cards, as depicted in Fig. 4.10.

FIGURE 4.10

Event (not D)

FIGURE 4.11

Event (B & C)

b. (B & C) is the event both B and C occur—the event that the card selected is both a king and a heart. Consequently, (B & C) is the event that the card selected is the king of hearts and consists of the single outcome shown in Fig. 4.11.

Note: Event (B & C) is the same as event A, so we can write A = (B & C).

c. (B or C) is the event either B or C or both occur—the event that the card selected is either a king or a heart or both. Event (B or C) consists of 16 outcomes—namely, the 4 kings and the 12 non-king hearts—as illustrated in Fig. 4.12.

FIGURE 4.12

Event (B or C)

FIGURE 4.13

Event (C & D)

Note: Event (*B* or *C*) can occur in 16, not 17, ways because the outcome "king of hearts" is common to both event *B* and event *C*.

d. (*C* & *D*) is the event both *C* and *D* occur—the event that the card selected is both a heart and a face card. For that event to occur, the card selected must be the jack, queen, or king of hearts. Thus event (*C* & *D*) consists of the three outcomes displayed in Fig. 4.13. These three outcomes are those common to events *C* and *D*. ■

You try it!

Exercise 4.57
on page 171

In the previous example, we described events by listing their outcomes. Sometimes, describing events verbally is more appropriate, as in the next example.

EXAMPLE 4.7 Relationships Among Events

TABLE 4.4

Frequency distribution for students' ages

Age (yr)	Frequency
17	1
18	1
19	9
20	7
21	7
22	5
23	3
24	4
26	1
35	1
36	1

Student Ages A frequency distribution for the ages of the 40 students in Professor Weiss's introductory statistics class is presented in Table 4.4. One student is selected at random. Let

A = event the student selected is under 21,

B = event the student selected is over 30,

C = event the student selected is in his or her 20s, and

D = event the student selected is over 18.

Determine the following events.

a. (not *D*)   **b.** (*A* & *D*)   **c.** (*A* or *D*)   **d.** (*B* or *C*)

Solution

a. (not *D*) is the event *D* does not occur—the event that the student selected is not over 18, that is, is 18 or under. From Table 4.4, (not *D*) comprises the two students in the class who are 18 or under.

b. (*A* & *D*) is the event both *A* and *D* occur—the event that the student selected is both under 21 and over 18, that is, is either 19 or 20. Event (*A* & *D*) comprises the 16 students in the class who are 19 or 20.

c. (*A* or *D*) is the event either *A* or *D* or both occur—the event that the student selected is either under 21 or over 18 or both. But every student in the class is either under 21 or over 18. Consequently, event (*A* or *D*) comprises all 40 students in the class and is certain to occur.

d. (*B* or *C*) is the event either *B* or *C* or both occur—the event that the student selected is either over 30 or in his or her 20s. Table 4.4 shows that (*B* or *C*) comprises the 29 students in the class who are 20 or over. ■

You try it!

Exercise 4.61
on page 171

At Least, At Most, and Inclusive

Events are sometimes described in words by using phrases such as **at least, at most,** and **inclusive**. For instance, consider the experiment of randomly selecting a U.S. housing unit. The event that the housing unit selected has *at most* four rooms means that it has four or fewer rooms; the event that the housing unit selected has *at least* two rooms means that it has two or more rooms; and the event that the housing unit selected has between three and five rooms, *inclusive*, means that it has at least three rooms but at most five rooms (i.e., three, four, or five rooms).

More generally, for any numbers *x* and *y*, the phrase "at least *x*" means "greater than or equal to *x*," the phrase "at most *x*" means "less than or equal to *x*," and the phrase "between *x* and *y*, inclusive," means "greater than or equal to *x* but less than or equal to *y*."

Mutually Exclusive Events

Next, we introduce the concept of *mutually exclusive events*.

DEFINITION 4.4

Mutually Exclusive Events

Two or more events are **mutually exclusive events** if no two of them have outcomes in common.

What Does It Mean?

Events are mutually exclusive if no two of them can occur simultaneously or, equivalently, if at most one of the events can occur when the experiment is performed.

The Venn diagrams shown in Fig. 4.14 portray the difference between two events that are mutually exclusive and two events that are not mutually exclusive. In Fig. 4.15, we show one case of three mutually exclusive events and two cases of three events that are not mutually exclusive.

FIGURE 4.14
(a) Two mutually exclusive events;
(b) two non–mutually exclusive events

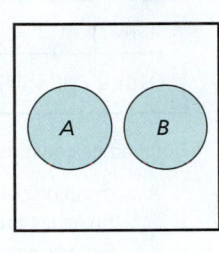
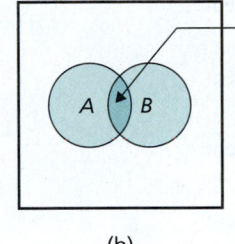

(a)   (b)

FIGURE 4.15
(a) Three mutually exclusive events;
(b) three non–mutually exclusive events;
(c) three non–mutually exclusive events

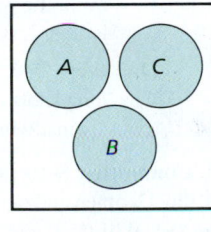
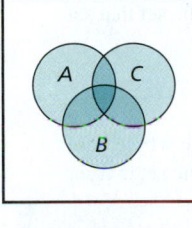
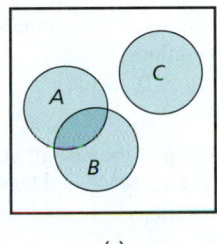

(a)   (b)   (c)

EXAMPLE 4.8   **Mutually Exclusive Events**

Playing Cards For the experiment of randomly selecting one card from a deck of 52, let

C = event the card selected is a heart,

D = event the card selected is a face card,

E = event the card selected is an ace,

F = event the card selected is an 8, and

G = event the card selected is a 10 or a jack.

Which of the following collections of events are mutually exclusive?

a. C and D **b.** C and E **c.** D and E
d. D, E, and F **e.** D, E, F, and G

Solution

a. Event C and event D are not mutually exclusive because they have the common outcomes "king of hearts," "queen of hearts," and "jack of hearts." Both events occur if the card selected is the king, queen, or jack of hearts.

b. Event C and event E are not mutually exclusive because they have the common outcome "ace of hearts." Both events occur if the card selected is the ace of hearts.

You try it!

Exercise 4.67
on page 172

c. Event D and event E are mutually exclusive because they have no common outcomes. They cannot both occur when the experiment is performed because selecting a card that is both a face card and an ace is impossible.

d. Events D, E, and F are mutually exclusive because no two of them can occur simultaneously.

e. Events D, E, F, and G are not mutually exclusive because event D and event G both occur if the card selected is a jack.

Exercises 4.2

Understanding the Concepts and Skills

4.41 What type of graphical displays are useful for portraying events and relationships among events?

4.42 Construct a Venn diagram representing the event
a. (not E).   **b.** (A or B).

4.43 Construct a Venn diagram representing the event
a. (A & B).   **b.** (A & B & C).

4.44 Construct a Venn diagram representing the event
a. (A or B or C).   **b.** ((not A) & B).

4.45 Construct a Venn diagram representing the event
a. (A & (not B)).   **b.** ((A or B) & (not (A & B))).

4.46 Consider the set consisting of the first 12 positive whole numbers (i.e., 1–12). Determine explicitly the numbers in the set that satisfy each of the following conditions:
a. at least 9   **b.** at most 10
c. between 5 and 8, inclusive

4.47 Consider the set consisting of the first 10 positive whole numbers (i.e., 1–10). Determine explicitly the numbers in the set that satisfy each of the following conditions:
a. at least 6
b. at most 3
c. between 2 and 5, inclusive

4.48 What does it mean for two events to be mutually exclusive?

4.49 What does it mean for three events to be mutually exclusive?

4.50 Answer *true* or *false* to the following statement and justify your answer: If event A and event B are not mutually exclusive, neither are events A, B, and C for every event C.

4.51 Answer *true* or *false* to the following statement and justify your answer: If event A and event B are mutually exclusive, so are events A, B, and C for every event C.

4.52 Draw a Venn diagram portraying four mutually exclusive events.

Applying the Concepts and Skills

4.53 Dice. When one die is rolled, the following six outcomes are possible:

List the outcomes constituting
A = event the die comes up even,
B = event the die comes up 4 or more,
C = event the die comes up at most 2, and
D = event the die comes up 3.

4.54 Horse Racing. In a horse race, the odds against winning are as shown in the following table. For example, the odds against winning are 8 to 1 for horse #1.

Horse	#1	#2	#3	#4	#5	#6	#7	#8
Odds	8	15	2	3	30	5	10	5

List the outcomes constituting

A = event one of the top two favorites wins (the top two favorites are the two horses with the lowest odds against winning),

B = event the winning horse's number is above 5,

C = event the winning horse's number is at most 3, that is, 3 or less, and

D = event one of the two long shots wins (the two long shots are the two horses with the highest odds against winning).

4.55 Committee Selection. A committee consists of five executives, three women and two men. Their names are Maria (M), John (J), Susan (S), Will (W), and Holly (H). The committee needs to select a chairperson and a secretary. It decides to make the selection randomly by drawing straws. The person getting the longest straw will be appointed chairperson, and the one getting the shortest straw will be appointed secretary. The possible outcomes can be represented in the following manner.

MS	SM	HM	JM	WM
MH	SH	HS	JS	WS
MJ	SJ	HJ	JH	WH
MW	SW	HW	JW	WJ

Here, for example, MS represents the outcome that Maria is appointed chairperson and Susan is appointed secretary. List the outcomes constituting each of the following four events.

A = event a male is appointed chairperson,
B = event Holly is appointed chairperson,
C = event Will is appointed secretary,
D = event only females are appointed.

4.56 Coin Tossing. When a dime is tossed four times, there are the following 16 possible outcomes.

HHHH	HTHH	THHH	TTHH
HHHT	HTHT	THHT	TTHT
HHTH	HTTH	THTH	TTTH
HHTT	HTTT	THTT	TTTT

Here, for example, HTTH represents the outcome that the first toss is heads, the next two tosses are tails, and the fourth toss is heads. List the outcomes constituting each of the following four events.

A = event exactly two heads are tossed,
B = event the first two tosses are tails,
C = event the first toss is heads,
D = event all four tosses come up the same.

4.57 Dice. Refer to Exercise 4.53. For each of the following events, list the outcomes that constitute the event and describe the event in words.
a. (not A)   **b.** (A & B)   **c.** (B or C)

4.58 Horse Racing. Refer to Exercise 4.54. For each of the following events, list the outcomes that constitute the event and describe the event in words.
a. (not C)   **b.** (C & D)   **c.** (A or C)

4.59 Committee Selection. Refer to Exercise 4.55. For each of the following events, list the outcomes that constitute the event, and describe the event in words.
a. (not A)   **b.** (B & D)   **c.** (B or C)

4.60 Coin Tossing. Refer to Exercise 4.56. For each of the following events, list the outcomes that constitute the event, and describe the event in words.
a. (not B)   **b.** (A & B)   **c.** (C or D)

4.61 Diabetes Prevalence. In a report titled *Behavioral Risk Factor Surveillance System Summary Prevalence Report*, the Centers for Disease Control and Prevention discusses the prevalence of diabetes in the United States. The following table provides a diabetes-prevalence frequency distribution for the 50 U.S. states based on the question "Have you ever been told by a doctor that you have diabetes?"

Diabetes (%)	Frequency
5–under 6	1
6–under 7	5
7–under 8	13
8–under 9	10
9–under 10	9
10–under 11	8
11–under 12	2
12–under 13	1
13–under 14	1

For a randomly selected state, let

A = event that the state has a diabetes prevalence percentage of at least 8%,

B = event that the state has a diabetes prevalence percentage of less than 7%,

C = event that the state has a diabetes prevalence percentage of at least 6% but less than 13%, and

D = event that the state has a diabetes prevalence percentage of less than 9%.

Describe each of the following events in words and determine the number of outcomes (states) that constitute each event.
a. (not C)   **b.** (A & B)   **c.** (C or D)   **d.** (C & B)

4.62 Family Planning. The following table provides a frequency distribution for the ages of adult women seeking pregnancy tests at public health facilities in Missouri during a 3-month period. It appeared in the article "Factors Affecting Contraceptive Use in Women

Seeking Pregnancy Tests" (*Family Planning Perspectives*, Vol. 32, No. 3, pp. 124–131) by M. Sable et al.

Age (yr)	Frequency
18–19	89
20–24	130
25–29	66
30–39	26

For one of these woman selected at random, let

A = event the woman is at least 25 years old,
B = event the woman is at most 29 years old,
C = event that the woman is between 18 and 29 years old, and
D = event that the woman is at least 20 years old.

Describe the following events in words, and determine the number of outcomes (women) that constitute each event.
a. (not D)   **b.** (B & D)   **c.** (C or A)   **d.** (A & B)

4.63 World Series. The World Series in baseball is won by the first team to win four games (ignoring the 1903 and 1919–1921 World Series, when it was a best of nine). Thus it takes at least four games and no more than seven games to establish a winner. From the document *World Series History* on the Baseball Almanac website, as of November 2013, the lengths of the World Series are as given in the following table.

Games required	Frequency
4	21
5	24
6	24
7	36

For one of these series selected at random, let

A = event that the World Series was decided in four games.
B = event that the World Series was decided in less than six games.
C = event that the World Series was decided in seven games.

Describe the following events in words, and determine the number of outcomes (World Series) that constitute each event.
a. (not A)   **b.** (A & B)
c. (A or C)   **d.** (A & C)

4.64 NBA Finals. The NBA Finals of basketball is played in a best of seven series. The number of games necessary to decide a winner can range from four to seven. Wikipedia lists the NBA Finals winners and number of games played per series. The following table provides the number of series that were decided in 4, 5, 6, or 7 games for the NBA Finals from 1950–2012.

Games required	Frequency
4	8
5	15
6	24
7	16

For one of these series selected at random, let

A = event that the NBA Finals was decided in five games.
B = event that the NBA Finals was decided in at most five games.
C = event that the NBA Finals was decided in at least five games.

Describe the following events in words, and determine the number of outcomes (NBA Finals) that constitute each event.

a. (*B* & *C*)   **b.** (*B* or *C*)   **c.** (not *A*)

4.65 Housing Units. The U.S. Census Bureau publishes data on housing units in *American Housing Survey for the United States*. The following table provides a frequency distribution for the number of rooms in U.S. housing units. The frequencies are in thousands.

Rooms	No. of units
1	601
2	1,404
3	11,433
4	23,636
5	30,440
6	27,779
7	17,868
8+	19,257

For a U.S. housing unit selected at random, let

A = event the unit has at most four rooms,
B = event the unit has at least two rooms,
C = event the unit has between five and seven rooms, inclusive, and
D = event the unit has more than seven rooms.

Describe each of the following events in words, and determine the number of outcomes (housing units) that constitute each event.

a. (not *A*)   **b.** (*A* & *B*)   **c.** (*C* or *D*)

4.66 Protecting the Environment. A survey was conducted in Canada to ascertain public opinion about a major national park region in the Banff-Bow Valley. One question asked the amount that respondents would be willing to contribute per year to protect the environment in the Banff-Bow Valley region. The following frequency distribution was found in an article by J. Ritchie et al. titled "Public Reactions to Policy Recommendations from the Banff-Bow Valley Study" (*Journal of Sustainable Tourism*, Vol. 10, No. 4, pp. 295–308).

Contribution ($)	Frequency
0	85
1–50	116
51–100	59
101–200	29
201–300	5
301–500	7
501–1000	3

For a respondent selected at random, let

A = event that the respondent would be willing to contribute at least $101,
B = event that the respondent would not be willing to contribute more than $50,
C = event that the respondent would be willing to contribute between $1 and $200, and
D = event that the respondent would be willing to contribute at least $1.

Describe the following events in words, and determine the number of outcomes (respondents) that make up each event.

a. (not *D*)   **b.** (*A* & *B*)   **c.** (*C* or *A*)   **d.** (*B* & *D*)

4.67 Dice. Refer to Exercise 4.53.

a. Are events *A* and *B* mutually exclusive?
b. Are events *B* and *C* mutually exclusive?
c. Are events *A*, *C*, and *D* mutually exclusive?

d. Are there three mutually exclusive events among *A*, *B*, *C*, and *D*? four?

4.68 Horse Racing. Each part of this exercise contains events from Exercise 4.54. In each case, decide whether the events are mutually exclusive.

a. *A* and *B*   **b.** *B* and *C*   **c.** *A*, *B*, and *C*
d. *A*, *B*, and *D*   **e.** *A*, *B*, *C*, and *D*

4.69 Housing Units. Refer to Exercise 4.65. Among the events *A*, *B*, *C*, and *D*, identify the collections of events that are mutually exclusive.

4.70 Protecting the Environment. Refer to Exercise 4.66. Among the events *A*, *B*, *C*, and *D*, identify the collections of events that are mutually exclusive.

4.71 Die and Coin. Consider the following random experiment: First, roll a die and observe the number of dots facing up; then, toss a coin the number of times that the die shows and observe the total number of heads. Thus, if the die shows three dots facing up and the coin (which is then tossed three times) comes up heads exactly twice, then the outcome of the experiment can be represented as (3, 2).

a. Determine a sample space for this experiment.
b. Determine the event that the total number of heads is even.

4.72 Jurors. From 10 men and 8 women in a pool of potential jurors, 12 are chosen at random to constitute a jury. Suppose that you observe the number of men who are chosen for the jury. Let *A* be the event that at least half of the 12 jurors are men, and let *B* be the event that at least half of the 8 women are on the jury.

a. Determine the sample space for this experiment.
b. Find (*A* or *B*), (*A* & *B*), and (*A* & (not *B*)), listing all the outcomes for each of those three events.
c. Are events *A* and *B* mutually exclusive? Are events *A* and (not *B*)? Are events (not *A*) and (not *B*)? Explain.

4.73 Let *A* and *B* be events of a sample space.

a. Suppose that *A* and (not *B*) are mutually exclusive. Explain why *B* occurs whenever *A* occurs.
b. Suppose that *B* occurs whenever *A* occurs. Explain why *A* and (not *B*) are mutually exclusive.

Extending the Concepts and Skills

4.74 Construct a Venn diagram that portrays four events, *A*, *B*, *C*, and *D* that have the following properties: Events *A*, *B*, and *C* are mutually exclusive; events *A*, *B*, and *D* are mutually exclusive; no other three of the four events are mutually exclusive.

4.75 Suppose that *A*, *B*, and *C* are three events that cannot all occur simultaneously. Does this condition necessarily imply that *A*, *B*, and *C* are mutually exclusive? Justify your answer and illustrate it with a Venn diagram.

4.76 Let *A*, *B*, and *C* be events of a sample space. Complete the following table.

Event	Description
(*A* & *B*)	Both *A* and *B* occur
	At least one of *A* and *B* occurs
(*A* & (not *B*))	
	Neither *A* nor *B* occur
(*A* or *B* or *C*)	
	All three of *A*, *B*, and *C* occur
	Exactly one of *A*, *B*, and *C* occurs
	Exactly two of *A*, *B*, and *C* occur
	At most one of *A*, *B*, and *C* occurs

4.3 Some Rules of Probability

In this section, we discuss several rules of probability, after we introduce an additional notation used in probability.

EXAMPLE 4.9 Probability Notation

Dice When a balanced die is rolled once, six equally likely outcomes are possible, as shown in Fig. 4.16. Use probability notation to express the probability that the die comes up an even number.

FIGURE 4.16

Sample space for rolling a die once

What Does It Mean?

Keep in mind that A refers to the event that the die comes up even, whereas $P(A)$ refers to the probability of that event occurring.

Solution The event that the die comes up an even number can occur in three ways—namely, if 2, 4, or 6 is rolled. Because $f/N = 3/6 = 0.5$, *the probability that the die comes up even is 0.5.* We want to express the italicized phrase using probability notation.

Let A denote the event that the die comes up even. We use the notation $P(A)$ to represent the probability that event A occurs. Hence we can rewrite the italicized statement simply as $P(A) = 0.5$, which is read "the probability of A is 0.5." ■

DEFINITION 4.5

Probability Notation

If E is an event, then **$P(E)$** represents the probability that event E occurs. It is read "the probability of E."

The Special Addition Rule

FIGURE 4.17

Two mutually exclusive events

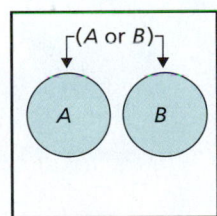

The first rule of probability that we present is the **special addition rule,** which states that, for mutually exclusive events, the probability that one or another of the events occurs equals the sum of the individual probabilities.

We use the Venn diagram in Fig. 4.17, which shows two mutually exclusive events A and B, to illustrate the special addition rule. If you think of the colored regions as probabilities, the colored disk on the left is $P(A)$, the colored disk on the right is $P(B)$, and the total colored region is $P(A \text{ or } B)$. Because events A and B are mutually exclusive, the total colored region equals the sum of the two colored disks; that is, $P(A \text{ or } B) = P(A) + P(B)$.

FORMULA 4.1

What Does It Mean?

For mutually exclusive events, the probability that at least one occurs equals the sum of their individual probabilities.

The Special Addition Rule

If event A and event B are mutually exclusive, then

$$P(A \text{ or } B) = P(A) + P(B).$$

More generally, if events A, B, C, ... are mutually exclusive, then

$$P(A \text{ or } B \text{ or } C \text{ or } \cdots) = P(A) + P(B) + P(C) + \cdots.$$

Example 4.10 illustrates use of the special addition rule.

EXAMPLE 4.10 The Special Addition Rule

Size of Farms The first two columns of Table 4.5 show a relative-frequency distribution for the size of farms (adjusted for coverage) in the United States. The U.S. Department of Agriculture compiled this information and published it in *Census of Agriculture*.

TABLE 4.5

Size of farms in the United States

Size (acres)	Relative frequency	Event
Under 10	0.106	A
10–49	0.281	B
50–179	0.300	C
180–499	0.167	D
500–999	0.068	E
1000–1999	0.042	F
2000 & over	0.036	G

You try it!

Exercise 4.87 on page 177

In the third column of Table 4.5, we introduce events that correspond to the size classes. For example, if a farm is selected at random, D denotes the event that the farm has between 180 and 499 acres, inclusive. The probabilities of the events in the third column of Table 4.5 equal the relative frequencies in the second column. For instance, the probability is 0.167 that a randomly selected farm has between 180 and 499 acres, inclusive: $P(D) = 0.167$.

Use Table 4.5 and the special addition rule to determine the probability that a randomly selected farm has between 180 and 1999 acres, inclusive.

Solution The event that the farm selected has between 180 and 1999 acres, inclusive, can be expressed as $(D \text{ or } E \text{ or } F)$. Because events D, E, and F are mutually exclusive, the special addition rule gives

$$P(D \text{ or } E \text{ or } F) = P(D) + P(E) + P(F)$$
$$= 0.167 + 0.068 + 0.042 = 0.277.$$

The probability that a randomly selected U.S. farm has between 180 and 1999 acres, inclusive, is 0.277.

Interpretation 27.7% of U.S. farms have between 180 and 1999 acres, inclusive.

FIGURE 4.18

An event and its complement

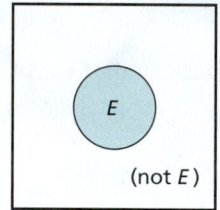

The Complementation Rule

The second rule of probability that we discuss is the **complementation rule.** It states that the probability an event occurs equals 1 minus the probability the event does not occur.

We use the Venn diagram in Fig. 4.18, which shows an event E and its complement (not E), to illustrate the complementation rule. If you think of the regions as probabilities, the entire region enclosed by the rectangle is the probability of the sample space, or 1. Furthermore, the colored region is $P(E)$ and the uncolored region is $P(\text{not } E)$. Thus, $P(E) + P(\text{not } E) = 1$ or, equivalently, $P(E) = 1 - P(\text{not } E)$.

FORMULA 4.2

What Does It Mean?

The probability that an event occurs equals 1 minus the probability that it does not occur.

The Complementation Rule

For any event E,

$$P(E) = 1 - P(\text{not } E).$$

The complementation rule is useful because sometimes computing the probability that an event does not occur is easier than computing the probability that it does occur. In such cases, we can subtract the former from 1 to find the latter.

EXAMPLE 4.11 The Complementation Rule

Size of Farms We saw that the first two columns of Table 4.5 provide a relative-frequency distribution for the size of U.S. farms. Find the probability that a randomly selected farm has

a. less than 2000 acres.　　　　**b.** 50 acres or more.

Solution

a. Let

$$J = \text{event the farm selected has less than 2000 acres.}$$

To determine $P(J)$, we apply the complementation rule because $P(\text{not } J)$ is easier to compute than $P(J)$. Note that (not J) is the event the farm obtained has 2000 or more acres, which is event G in Table 4.5. Thus $P(\text{not } J) = P(G) = 0.036$. Applying the complementation rule yields

$$P(J) = 1 - P(\text{not } J) = 1 - 0.036 = 0.964.$$

The probability that a randomly selected U.S. farm has less than 2000 acres is 0.964.

Interpretation   96.4% of U.S. farms have less than 2000 acres.

b. Let

$$K = \text{event the farm selected has 50 acres or more.}$$

We apply the complementation rule to find $P(K)$. Now, (not K) is the event the farm obtained has less than 50 acres. From Table 4.5, event (not K) is the same as event (A or B). Because events A and B are mutually exclusive, the special addition rule implies that

$$P(\text{not } K) = P(A \text{ or } B) = P(A) + P(B) = 0.106 + 0.281 = 0.387.$$

Using this result and the complementation rule, we conclude that

$$P(K) = 1 - P(\text{not } K) = 1 - 0.387 = 0.613.$$

The probability that a randomly selected U.S. farm has 50 acres or more is 0.613.

Interpretation   61.3% of U.S. farms have at least 50 acres.

You try it!

Exercise 4.93
on page 178

The General Addition Rule

FIGURE 4.19
Non–mutually exclusive events

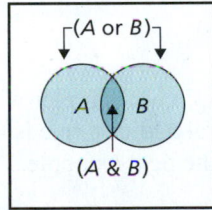

The special addition rule concerns mutually exclusive events. For events that are not mutually exclusive, we must use the **general addition rule.** To introduce it, we use the Venn diagram shown in Fig. 4.19.

If you think of the colored regions as probabilities, the colored disk on the left is $P(A)$, the colored disk on the right is $P(B)$, and the total colored region is $P(A \text{ or } B)$. To obtain the total colored region, $P(A \text{ or } B)$, we first sum the two colored disks, $P(A)$ and $P(B)$. When we do so, however, we count the common colored region, $P(A \& B)$, twice. Thus, we must subtract $P(A \& B)$ from the sum. So, we see that $P(A \text{ or } B) = P(A) + P(B) - P(A \& B)$.

FORMULA 4.3

? **What Does It Mean?**

For any two events, the probability that at least one occurs equals the sum of their individual probabilities less the probability that both occur.

The General Addition Rule

If A and B are any two events, then

$$P(A \text{ or } B) = P(A) + P(B) - P(A \& B).$$

In the next example, we consider a situation in which a required probability can be computed both with and without use of the general addition rule.

■■■ **EXAMPLE 4.12**   **The General Addition Rule**

Playing Cards   Consider again the experiment of selecting one card at random from a deck of 52 playing cards. Find the probability that the card selected is either a spade or a face card

a. without using the general addition rule.
b. by using the general addition rule.

Solution

a. Let

$$E = \text{event the card selected is either a spade or a face card.}$$

Event E consists of 22 cards—namely, the 13 spades plus the other nine face cards that are not spades—as shown in Fig. 4.20. So, by the f/N rule,

$$P(E) = \frac{f}{N} = \frac{22}{52} = 0.423.$$

The probability that a randomly selected card is either a spade or a face card is 0.423.

FIGURE 4.20
Event E

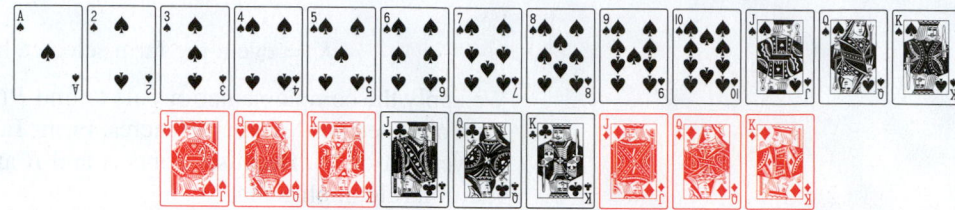

b. To determine $P(E)$ by using the general addition rule, we first note that we can write $E = (C \text{ or } D)$, where

$$C = \text{event the card selected is a spade, and}$$
$$D = \text{event the card selected is a face card.}$$

Event C consists of the 13 spades, and event D consists of the 12 face cards. In addition, event $(C \& D)$ consists of the three spades that are face cards—the jack, queen, and king of spades. Applying the general addition rule gives

You try it!

Exercise 4.95
on page 178

$$P(E) = P(C \text{ or } D) = P(C) + P(D) - P(C \& D)$$
$$= \frac{13}{52} + \frac{12}{52} - \frac{3}{52} = 0.250 + 0.231 - 0.058 = 0.423,$$

which agrees with the answer found in part (a).

Computing the probability in the previous example was simpler without using the general addition rule. Frequently, however, the general addition rule is the easier or the only way to compute a probability, as illustrated in the next example.

EXAMPLE 4.13   **The General Addition Rule**

Characteristics of People Arrested Data on people who have been arrested are published by the Federal Bureau of Investigation in *Uniform Crime Reports*. Records for one year show that 73.9% of the people arrested were male, 12.0% were under 18 years of age, and 8.5% were males under 18 years of age. If a person arrested that year is selected at random, what is the probability that that person is either male or under 18?

Solution   Let

$$M = \text{event the person obtained is male, and}$$
$$E = \text{event the person obtained is under 18.}$$

We can represent the event that the selected person is either male or under 18 as $(M \text{ or } E)$. To find the probability of that event, we apply the general addition rule to the data provided:

$$P(M \text{ or } E) = P(M) + P(E) - P(M \& E)$$
$$= 0.739 + 0.120 - 0.085 = 0.774.$$

You try it!

The probability that the person obtained is either male or under 18 is 0.774.

Exercise 4.99
on page 179

Interpretation   77.4% of those arrested during the year in question were either male or under 18 years of age (or both).

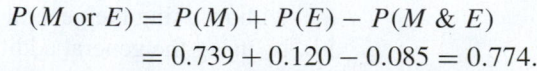

Note the following:

- The general addition rule is consistent with the special addition rule—if two events are mutually exclusive, both rules yield the same result.
- There are also general addition rules for more than two events. For instance, the general addition rule for three events is

$$P(A \text{ or } B \text{ or } C) = P(A) + P(B) + P(C) - P(A \& B) - P(A \& C) - P(B \& C) + P(A \& B \& C).$$

Exercises 4.3

Understanding the Concepts and Skills

4.77 Playing Cards. An ordinary deck of playing cards has 52 cards. There are four suits—spades, hearts, diamonds, and clubs—with 13 cards in each suit. Spades and clubs are black; hearts and diamonds are red. One of these cards is selected at random. Let R denote the event that a red card is chosen. Find the probability that a red card is chosen, and express your answer in probability notation.

4.78 Poker Chips. A bowl contains 12 poker chips—3 red, 4 white, and 5 blue. One of these poker chips is selected at random from the bowl. Let B denote the event that the chip selected is blue. Find the probability that a blue chip is selected, and express your answer in probability notation.

4.79 Suppose that A and B are mutually exclusive events such that $P(A) = 0.25$ and $P(B) = 0.40$. Determine $P(A \text{ or } B)$.

4.80 Suppose that C and D are mutually exclusive events such that $P(C) = 0.14$ and $P(D) = 0.32$. Determine $P(C \text{ or } D)$.

4.81 Let E be an event with probability 0.35. Find the probability of (not E).

4.82 Let F be an event with probability 0.72. Find the probability of (not F).

4.83 Suppose that C and D are events such that $P(C) = 0.35$, $P(D) = 0.40$, and $P(C \& D) = 0.30$. Determine $P(C \text{ or } D)$.

4.84 Suppose that A and B are events such that $P(A) = 0.84$, $P(B) = 0.46$, and $P(A \& B) = 0.38$. Determine $P(A \text{ or } B)$.

4.85 Suppose that A and B are events such that $P(A) = 1/3$, $P(A \text{ or } B) = 1/2$, and $P(A \& B) = 1/10$.
a. Are events A and B mutually exclusive? Explain your answer.
b. Find $P(B)$.

4.86 Suppose that A and B are events such that $P(A) = 1/4$, $P(B) = 1/3$, and $P(A \text{ or } B) = 1/2$.
a. Are events A and B mutually exclusive? Explain your answer.
b. Find $P(A \& B)$.

Applying the Concepts and Skills

4.87 Ages of Senators. According to the *Congressional Directory*, the official directory of the U.S. Congress, prepared by the Joint Committee on Printing, the age distribution for senators in the U.S. Congress as of Fall 2013, is as shown in the following table.

Age (yr)	No. of senators
Under 50	11
50–59	30
60–69	37
70–79	20
80 and over	2

Suppose that a U.S. senator is selected at random. Let

A = event the senator is under 50,
B = event the senator is in his or her 50s,
C = event the senator is in his or her 60s, and
S = event the senator is under 70.

a. Use the table and the f/N rule to find $P(S)$.
b. Express event S in terms of events A, B, and C.
c. Determine $P(A)$, $P(B)$, and $P(C)$.
d. Compute $P(S)$, using the special addition rule and your answers from parts (b) and (c). Compare your answer with that in part (a).

4.88 Sales Tax Receipts. The State of Texas maintains records pertaining to the economic development of corporations in the state. From the *Economic Development Corporation Report*, published by the Texas Comptroller of Public Accounts, we obtained the following frequency distribution summarizing the sales tax receipts from the state's Type 4A development corporations during one fiscal year.

Receipts	Frequency
$0–24,999	25
$25,000–49,999	23
$50,000–74,999	21
$75,000–99,999	11
$100,000–199,999	34
$200,000–499,999	44
$500,000–999,999	17
$1,000,000 & over	32

Suppose that one of these Type 4A development corporations is selected at random. Let

A = event the receipts are less than $25,000,
B = event the receipts are between $25,000 and $49,999,
C = event the receipts are between $500,000 and $999,999,
D = event the receipts are at least $1,000,000, and
R = event the receipts are either less than $50,000 or at least $500,000.

a. Use the table and the f/N rule to find $P(R)$.
b. Express event R in terms of events A, B, C, and D.
c. Determine $P(A)$, $P(B)$, $P(C)$, and $P(D)$.
d. Compute $P(R)$ by using the special addition rule and your answers from parts (b) and (c). Compare your answer with that in part (a).

4.89 Twelfth-Grade Smokers. The National Institute on Drug Abuse issued the report *Monitoring the Future*, which addressed the issue of drinking, cigarette, and smokeless tobacco use for eighth, tenth, and twelfth graders. During one year, 12,900 twelfth graders were asked the question, "How frequently have you smoked cigarettes during the past 30 days?" Based on their responses, we constructed the following percentage distribution for all twelfth graders.

Cigarettes per day	Percentage	Event
None	73.3	A
Some, but less than 1	9.8	B
1–5	7.8	C
6–14	5.3	D
15–25	2.8	E
26–34	0.7	F
35 or more	0.3	G

Find the probability that, within the last 30 days, a randomly selected twelfth grader

a. smoked.

b. smoked at least one cigarette per day.

c. smoked between 6 and 34 cigarettes per day, inclusive.

4.90 Oil Spills. The U.S. Coast Guard maintains a database of the number, source, and location of oil spills in U.S. navigable and territorial waters. The following is a probability distribution for location of oil spill events. [SOURCE: *Statistical Abstract of the United States*]

Location	Probability
Atlantic Ocean	0.008
Pacific Ocean	0.037
Gulf of Mexico	0.233
Great Lakes	0.020
Other lakes	0.002
Rivers and canals	0.366
Bays and sounds	0.146
Harbors	0.161
Other	0.027

Apply the special addition rule to find the percentage of oil spills in U.S. navigable and territorial waters that

a. occur in an ocean.

b. occur in a lake or harbor.

c. do not occur in a lake, ocean, river, or canal.

4.91 Internet Access. From the document "Computer and Internet Use in the United States: Population Characteristics" (*Current Population Survey*) by T. File, we obtained the following percentage distribution of household income for U.S. households with Internet access.

Household income	Percentage	Event
Less than $25,000	17.1	A
$25,000–$49,999	24.0	B
$50,000–$99,999	35.0	C
$100,000–$149,999	14.1	D
$150,000 or more	9.8	E

Suppose that a U.S. household with Internet access is selected at random. Let A denote the event that the household has an income under $25,000, B denote the event that the household has an income between $25,000 and $49,999, and so on (see the third column of the table). Apply the special addition rule to find the probability that the household obtained has an income

a. under $50,000.

b. $25,000 or above.

c. between $25,000 and $149,999, inclusive.

d. Interpret each of your answers in parts (a)–(c) in terms of percentages.

4.92 Religion in America. According to the *Religious Landscape Survey*, sponsored by the Pew Forum on Religion and Public Life, a distribution of religious affiliation among U.S. adults is as shown in the following table.

Affiliation	Relative frequency
Protestant	0.513
Catholic	0.239
Jewish	0.017
Mormon	0.017
Other	0.214

Find the probability that the religious affiliation of a randomly selected U.S. adult is

a. Catholic or Protestant.

b. not Jewish.

c. not Catholic, Protestant, or Jewish.

4.93 Ages of Senators. Refer to Exercise 4.87. Use the complementation rule to find the probability that a randomly selected U.S. senator is

a. 50 years old or older.   **b.** under 70 years old.

4.94 Sales Tax Receipts. Refer to Exercise 4.88. Use the complementation rule to find the probability that one of these Type 4A development corporations selected at random has receipts

a. of at least $25,000.   **b.** less than $500,000.

4.95 Student Debt. The Association of American Universities published a report titled "Looking More Closely at Student Debt." This report explores the issue about the cost of a college education and its impact on student loan debt. Using information from a credit reporting company, the following table provides a percentage distribution for the loan balance of outstanding student loans from individuals with graduate, professional, and undergraduate degree debt.

Loan balance	Percentage
$1–$10,000	43.1
$10,001–$25,000	29.2
$25,001–$50,000	16.5
$50,001–$75,000	5.8
$75,001–$100,000	2.3
$100,001 or more	3.1

Suppose that one of these individuals is selected at random.

a. Without using the general addition rule, determine the probability that the individual obtained has a loan balance either between $10,001 and $100,000, inclusive, or at most $75,000.

b. Obtain the probability in part (a) by using the general addition rule.

c. Which method did you find easier?

4.96 Naturalization. The U.S. Bureau of Citizenship and Immigration Services collects and reports information about naturalized persons in *Statistical Yearbook*. Following is an age distribution for persons naturalized during one year.

Age (yr)	Frequency	Age (yr)	Frequency
18–19	5,958	45–49	42,820
20–24	50,905	50–54	32,574
25–29	58,829	55–59	25,534
30–34	64,735	60–64	18,767
35–39	69,844	65–74	25,528
40–44	57,834	75 & over	9,872

Suppose that one of these naturalized persons is selected at random.

a. Without using the general addition rule, determine the probability that the age of the person obtained is either between 30 and 64, inclusive, or at least 50.

b. Find the probability in part (a), using the general addition rule.

c. Which method did you find easier?

4.97 Craps. In the game of *craps*, a player rolls two balanced dice. Thirty-six equally likely outcomes are possible, as shown in Fig. 4.1 on page 159. Let

A = event the sum of the dice is 7,
B = event the sum of the dice is 11,
C = event the sum of the dice is 2,
D = event the sum of the dice is 3,
E = event the sum of the dice is 12,
F = event the sum of the dice is 8, and
G = event doubles are rolled.

a. Compute the probability of each of the seven events.

b. The player wins on the first roll if the sum of the dice is 7 or 11. Find the probability of that event by using the special addition rule and your answers from part (a).

c. The player loses on the first roll if the sum of the dice is 2, 3, or 12. Determine the probability of that event by using the special addition rule and your answers from part (a).

d. Compute the probability that either the sum of the dice is 8 or doubles are rolled, without using the general addition rule.

e. Compute the probability that either the sum of the dice is 8 or doubles are rolled by using the general addition rule, and compare your answer to the one you obtained in part (d).

4.98 Gender and Divorce. According to *America's Families and Living Arrangements*, published by the U.S. Census Bureau, 51.5% of U.S. adults are female, 10.4% of U.S. adults are divorced, and 6.0% of U.S. adults are divorced females. For a U.S. adult selected at random, let

F = event the person is female, and
D = event the person is divorced.

a. Obtain $P(F)$, $P(D)$, and $P(F \& D)$.

b. Determine $P(F \text{ or } D)$, and interpret your answer in terms of percentages.

c. Find the probability that a randomly selected adult is male.

4.99 School Enrollment. The National Center for Education Statistics publishes information about school enrollment in *Digest of Education Statistics*. According to that document, 85.3% of students attend public schools, 27.9% of students attend college, and 20.1% of students attend public colleges. What percentage of students attend either public school or college?

Extending the Concepts and Skills

4.100 Suppose that A and B are mutually exclusive events.

a. Use the special addition rule to express $P(A \text{ or } B)$ in terms of $P(A)$ and $P(B)$.

b. Show that the general addition rule gives the same answer as that in part (a).

4.101 Newspaper Subscription. A certain city has three major newspapers, the *Times*, the *Herald*, and the *Examiner*. Circulation information indicates that 47.0% of households get the *Times*, 33.4% get the *Herald*, 34.6% get the *Examiner*, 11.9% get the *Times* and the *Herald*, 15.1% get the *Times* and the *Examiner*, 10.4% get the *Herald* and the *Examiner*, and 4.8% get all three. If a household in this city is selected at random, determine the probability that it gets at least one of the three major newspapers.

4.102 General Addition Rule Extended. The general addition rule for two events is presented in Formula 4.3 on page 175 and that for three events is displayed on page 177.

a. Verify the general addition rule for three events.

b. Write the general addition rule for four events and explain your reasoning.

| 4.4 | # Contingency Tables; Joint and Marginal Probabilities* |

In Section 2.2, we discussed how to group data from one variable of a population into a frequency distribution. Data from one variable of a population are called **univariate data.**

We often need to group and analyze data from two variables of a population. Data from two variables of a population are called **bivariate data,** and a frequency distribution for bivariate data is called a **contingency table** or **two-way table.**

■■ ■■ **EXAMPLE 4.14**   **Introducing Contingency Tables**

Age and Rank of Faculty Data about two variables—age and rank—of the faculty members at a university yielded the contingency table shown in Table 4.6. Discuss and interpret the numbers in the table.

---

*Sections 4.4–4.6 are recommended for classes that will study the binomial distribution (Section 5.3).

TABLE 4.6

Contingency table for age and
rank of faculty members

	Rank				
	Full professor R_1	Associate professor R_2	Assistant professor R_3	Instructor R_4	**Total**
Under 30 A_1	2	3	57	6	68
30–39 A_2	52	170	163	17	402
40–49 A_3	156	125	61	6	348
50–59 A_4	145	68	36	4	253
60 & over A_5	75	15	3	0	93
Total	430	381	320	33	1164

(row label: Age (yr))

Solution The small boxes inside the rectangle formed by the heavy lines are called **cells.** The upper left cell indicates that two faculty members are full professors under the age of 30 years. The cell diagonally below and to the right of the upper left cell shows that 170 faculty members are associate professors in their 30s.

The first row total reveals that 68 $(2 + 3 + 57 + 6)$ of the faculty members are under the age of 30 years. Similarly, the third column total shows that 320 of the faculty members are assistant professors. The number 1164 in the lower right corner gives the total number of faculty. That total can be found by summing the row totals, the column totals, or the frequencies in the 20 cells of the contingency table.

You try it!

Exercise 4.111
on page 183

Joint and Marginal Probabilities

We now use the age and rank data from Table 4.6 to introduce the concepts of *joint probabilities* and *marginal probabilities*.

EXAMPLE 4.15   Joint and Marginal Probabilities

Age and Rank of Faculty Refer to Example 4.14. Suppose that a faculty member is selected at random.

a. Identify the events represented by the subscripted letters that label the rows and columns of the contingency table shown in Table 4.6.
b. Identify the events represented by the cells of the contingency table.
c. Determine the probabilities of the events discussed in parts (a) and (b).
d. Summarize the results of part (c) in a table.
e. Discuss the relationship among the probabilities in the table obtained in part (d).

Solution

a. The subscripted letter A_1 that labels the first row of Table 4.6 represents the event that the selected faculty member is under 30 years of age:

$$A_1 = \text{event the faculty member is under 30.}$$

Similarly, the subscripted letter R_2 that labels the second column represents the event that the selected faculty member is an associate professor:

$$R_2 = \text{event the faculty member is an associate professor.}$$

Likewise, we can identify the remaining seven events represented by the subscripted letters that label the rows and columns. Note that the events A_1, A_2, A_3, A_4, and A_5 are mutually exclusive, as are the events R_1, R_2, R_3, and R_4.

b. In addition to considering events A_1 through A_5 and R_1 through R_4 separately, we can also consider them jointly. For instance, the event that the selected faculty member is under 30 (event A_1) *and* is also an associate professor (event R_2) can be expressed as (A_1 & R_2):

$$(A_1 \text{ \& } R_2) = \text{event the faculty member is an associate professor under 30.}$$

The joint event (A_1 & R_2) is represented by the cell in the first row and second column of Table 4.6. That joint event is one of 20 different joint events—one for each cell of the contingency table—associated with this random experiment.

Thinking of a contingency table as a Venn diagram can be useful. The Venn diagram corresponding to Table 4.6 is shown in Fig. 4.21. This figure makes it clear that the 20 joint events (A_1 & R_1), (A_1 & R_2), ..., (A_5 & R_4) are mutually exclusive.

FIGURE 4.21

Venn diagram corresponding to Table 4.6

	R_1	R_2	R_3	R_4
A_1	(A_1 & R_1)	(A_1 & R_2)	(A_1 & R_3)	(A_1 & R_4)
A_2	(A_2 & R_1)	(A_2 & R_2)	(A_2 & R_3)	(A_2 & R_4)
A_3	(A_3 & R_1)	(A_3 & R_2)	(A_3 & R_3)	(A_3 & R_4)
A_4	(A_4 & R_1)	(A_4 & R_2)	(A_4 & R_3)	(A_4 & R_4)
A_5	(A_5 & R_1)	(A_5 & R_2)	(A_5 & R_3)	(A_5 & R_4)

c. To find the probabilities of the events discussed in parts (a) and (b), we begin by observing that the total number of faculty members is 1164, or, $N = 1164$. The probability that the selected faculty member is under 30 (event A_1) is found by first noting from Table 4.6 that $f = 68$ and then applying the f/N rule:

$$P(A_1) = \frac{f}{N} = \frac{68}{1164} = 0.058.$$

Similarly, we can find the probability that the selected faculty member is an associate professor (event R_2):

$$P(R_2) = \frac{f}{N} = \frac{381}{1164} = 0.327.$$

Likewise, we can determine the probabilities of the remaining seven of the nine events represented by the subscripted letters. These nine probabilities are often called **marginal probabilities** because they correspond to events represented in the margin of the contingency table.

We can also find probabilities for joint events, so-called **joint probabilities.** For instance, the probability that the selected faculty member is an associate professor under 30 [event (A_1 & R_2)] is

$$P(A_1 \text{ \& } R_2) = \frac{f}{N} = \frac{3}{1164} = 0.003.$$

Similarly, we can find the probabilities of the remaining 19 joint events.

d. By referring to part (c), we can replace the joint frequency distribution in Table 4.6 with the **joint probability distribution** in Table 4.7, where probabilities are displayed instead of frequencies.

TABLE 4.7

Joint probability distribution corresponding to Table 4.6

			Rank			
		Full professor R_1	Associate professor R_2	Assistant professor R_3	Instructor R_4	$P(A_i)$
Age (yr)	Under 30 A_1	0.002	0.003	0.049	0.005	0.058
	30–39 A_2	0.045	0.146	0.140	0.015	0.345
	40–49 A_3	0.134	0.107	0.052	0.005	0.299
	50–59 A_4	0.125	0.058	0.031	0.003	0.217
	60 & over A_5	0.064	0.013	0.003	0.000	0.080
	$P(R_j)$	0.369	0.327	0.275	0.028	1.000

Note that in Table 4.7 the joint probabilities are displayed in the cells and the marginal probabilities in the margin. Also observe that the row and column labels "Total" in Table 4.6 have been changed in Table 4.7 to $P(R_j)$ and $P(A_i)$, respectively. The reason is that in Table 4.7 the last row gives the probabilities of events R_1 through R_4, and the last column gives the probabilities of events A_1 through A_5.

e. The sum of the joint probabilities in a row or column of a joint probability distribution equals the marginal probability in that row or column, with any observed discrepancy being due to roundoff error. For example, for the A_4 row of Table 4.7, the sum of the joint probabilities is

$$0.125 + 0.058 + 0.031 + 0.003 = 0.217,$$

which equals the marginal probability at the end of the A_4 row.

You try it!

Exercise 4.117 on page 184

Exercises 4.4

Understanding the Concepts and Skills

4.103 Identify three ways in which the total number of observations of bivariate data can be obtained from the frequencies in a contingency table.

4.104 Suppose that bivariate data are to be grouped into a contingency table. Determine the number of cells that the contingency table will have if the number of possible values for the two variables are
a. two and three.
b. four and three.
c. m and n.

4.105 Fill in the blanks.
a. Data from one variable of a population are called _____ data.
b. Data from two variables of a population are called _____ data.

4.106 Give an example of
a. univariate data.   **b.** bivariate data.

In each of Exercises 4.107–4.110,
a. fill in the missing entries in the contingency table.
b. determine $P(C_1)$, $P(R_2)$, and $P(C_1 \& R_2)$.
c. construct the corresponding joint probability distribution.

4.107

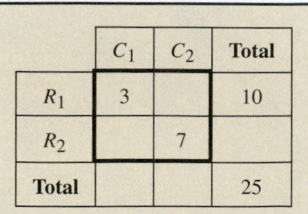

	C_1	C_2	Total
R_1	3		10
R_2		7	
Total			25

4.108

	C_1	C_2	Total
R_1			18
R_2	20		32
Total		15	

4.109

	C_1	C_2	C_3	Total
R_1	12	4		
R_2	5			28
Total		14	19	

4.110

	C_1	C_2	Total
R_1	3		
R_2	2	4	
R_3		1	
Total		8	20

4.111 New England Patriots. From the National Football League (NFL) Web site, in the *New England Patriots Roster*, we obtained information on the weights and years of experience for the players on that team, as of September 26, 2013. The following contingency table provides a cross-classification of those data.

Weight (lb)	Years of experience				
	Rookie Y_1	1–5 Y_2	6–10 Y_3	Over 10 Y_4	Total
Under 200 W_1	3	5	0	0	8
200–300 W_2	11	21	7	2	41
Over 300 W_3	4	4	5	0	13
Total	18	30	12	2	62

a. How many cells are in this contingency table?
b. How many players are on the New England Patriots roster as of September 26, 2013?
c. How many players are rookies?
d. How many players weigh between 200 and 300 lb?
e. How many players are rookies who weigh between 200 and 300 lb?

4.112 Motor Vehicle Use. The Federal Highway Administration compiles information on motor vehicle use around the globe and publishes its findings in *Highway Statistics*. Following is a contingency table for the number of motor vehicles in use in North American countries, by country and type of vehicle, during one year. Frequencies are in thousands.

Vehicle type	Country			
	U.S. C_1	Canada C_2	Mexico C_3	Total
Automobiles V_1	129,728	13,138	8,607	151,473
Motorcycles V_2	3,871	320	270	4,461
Trucks V_3	75,940	6,933	4,287	87,160
Total	209,539	20,391	13,164	243,094

a. How many cells are in this contingency table?
b. How many vehicles are Canadian?
c. How many vehicles are motorcycles?
d. How many vehicles are Canadian motorcycles?
e. How many vehicles are either Canadian or motorcycles?
f. How many automobiles are Mexican?
g. How many vehicles are not automobiles?

4.113 Naval Classifications. The U.S. Department of Defense collects various data on military personnel. The following contingency table provides a joint frequency distribution for marital status and pay grade of active Naval military during one year. Pay grades are divided into three groups: Enlisted, Officer, and Warrant Officer, and are used to determine a military member's entitlements.

Marital status	Pay grade			
	Enlisted G_1	Officer G_2	Warrant G_3	Total
Single w/o children M_1	118,116	13,915	75	132,106
Single with children M_2	14,784			16,365
Joint service marriage M_3	14,722	2,681		
Civilian marriage M_4	124,976	32,031	1,423	158,430
Total	272,598		1,664	324,371

For the active Naval military personnel during the year in question,
a. fill in the five missing entries.
b. how many are officers?
c. how many are enlisted and have a civilian marriage?
d. how many are either enlisted or have a civilian marriage?
e. how many are neither enlisted nor have a civilian marriage?
f. how many do not have a joint service marriage?

4.114 Farms. The U.S. Department of Agriculture publishes information about U.S. farms in *Census of Agriculture*. A joint frequency distribution for number of farms, by acreage and tenure of operator, is provided in the following contingency table. Frequencies are in thousands.

	Tenure of operator			
Acreage	Full owner T_1	Part owner T_2	Tenant T_3	Total
Under 50 A_1		70	44	
50–under 180 A_2	492	130	38	660
180–under 500 A_3	198			368
500–under 1000 A_4	51	84	14	149
1000 & over A_5	41	114	17	172
Total	1521	541		

a. Fill in the six missing entries.
b. How many cells does this contingency table have?
c. How many farms have under 50 acres?
d. How many farms are tenant operated?
e. How many farms are operated by part owners and have between 500 and 1000 acres?
f. How many farms are not full-owner operated?
g. How many tenant-operated farms have 180 acres or more?

4.115 Field Trips. P. Li et al. analyzed existing problems in teaching geography in rural counties in the article "Geography Education in Rural Tennessee Counties" (*Geography*, Vol. 88, No. 1, pp. 63–74). Fifty-one high-school teachers from the Upper Cumberland Region of Tennessee were surveyed. The following contingency table cross-classifies these teachers by highest degree obtained and whether they offered field trips.

	Field trips		
Degree	Yes F_1	No F_2	Total
Bachelor's D_1	14	14	28
Master's D_2	18	5	23
Total	32	19	51

a. How many of these teachers offered field trips?
b. How many of these teachers have master's degrees?
c. How many teachers with only bachelor's degrees offered field trips?

d. Describe the events D_1 and (D_2 & F_2) in words.
e. Compute the probability of each event in part (d).

4.116 Fear of Gangs. In the article "Growing Pains and Fear of Gangs" (*Applied Psychology in Criminal Justice*, Vol. 5, No. 2, pp. 139–164), B. Brown and W. Benedict examined the relationship between worry about a gang attack and actually being a victim of a gang attack. Interviews of a sample of high school students yielded the following contingency table.

	Worry		
Victim	Yes W_1	No W_2	Total
Yes V_1	18	21	39
No V_2	22	152	174
Total	40	173	213

a. How many of these students were victims of a gang attack?
b. How many of these students did not worry about gang attacks?
c. How many of these students who worried about gang attacks were actual victims of such an attack?
d. Describe the events W_1 and (W_2 & V_1) in words.
e. Compute the probability of each event in part (d).

4.117 New England Patriots. Refer to Exercise 4.111.
a. For a randomly selected player on the New England Patriots, describe the events Y_3, W_2, and (W_1 & Y_2) in words.
b. Compute the probability of each event in part (a). Interpret your answers in terms of percentages.
c. Construct a joint probability distribution similar to that shown in Table 4.7 on page 182.
d. Verify that the sum of each row and column of joint probabilities equals the marginal probability in that row or column. (*Note:* Rounding may cause slight deviations.)

4.118 Motor Vehicle Use. Refer to Exercise 4.112.
a. For a randomly selected vehicle, describe the events C_1, V_3, and (C_1 & V_3) in words.
b. Compute the probability of each event in part (a).
c. Compute $P(C_1$ or $V_3)$, using the contingency table and the f/N rule.
d. Compute $P(C_1$ or $V_3)$, using the general addition rule and your answers from part (b).
e. Construct a joint probability distribution.

4.119 Naval Classifications. Refer to Exercise 4.113. An active Naval military personnel during the year in question is selected at random.
a. Use the letters in the margins of the contingency table to represent each of the following three events: The person obtained is (i) single with children, (ii) a warrant officer, and (iii) single without children and enlisted.
b. Compute the probability of each event in part (a).
c. Construct a **joint percentage distribution,** a table similar to a joint probability distribution except with percentages instead of probabilities.

4.120 Farms. Refer to Exercise 4.114. A U.S. farm is selected at random.
a. Use the letters in the margins of the contingency table to represent each of the following three events: The farm obtained (i) has between 180 and 500 acres, (ii) is part-owner operated, and (iii) is full-owner operated and has at least 1000 acres.
b. Compute the probability of each event in part (a).
c. Construct a **joint percentage distribution**, a table similar to a joint probability distribution except with percentages instead of probabilities.

	C_1	$\cdots$	C_n	$P(R_i)$
R_1	$P(R_1 \& C_1)$	$\cdots$	$P(R_1 \& C_n)$	$P(R_1)$
.	.	$\cdots$	.	.
.	.	$\cdots$	.	.
.	.	$\cdots$	.	.
R_m	$P(R_m \& C_1)$	$\cdots$	$P(R_m \& C_n)$	$P(R_m)$
$P(C_j)$	$P(C_1)$	$\cdots$	$P(C_n)$	1

Extending the Concepts and Skills

4.121 Explain why the joint events in a contingency table are mutually exclusive.

4.122 What does the general addition rule (Formula 4.3 on page 175) mean in the context of the probabilities in a joint probability distribution?

4.123 In this exercise, you are asked to verify that the sum of the joint probabilities in a row or column of a joint probability distribution equals the marginal probability in that row or column. Consider the following joint probability distribution.

a. Explain why
$$R_1 = ((R_1 \& C_1) \text{ or } \cdots \text{ or } (R_1 \& C_n)).$$
b. Why are the events $(R_1 \& C_1), \ldots, (R_1 \& C_n)$ mutually exclusive?
c. Explain why parts (a) and (b) imply that
$$P(R_1) = P(R_1 \& C_1) + \cdots + P(R_1 \& C_n).$$

This equation shows that the first row of joint probabilities sums to the marginal probability at the end of that row. A similar argument applies to any other row or column.

4.5 Conditional Probability*

In this section, we introduce the concept of *conditional probability.*

DEFINITION 4.6

What Does It Mean?

 A conditional probability of an event is the probability that the event occurs under the assumption that another event occurs.

Conditional Probability

The probability that event B occurs given that event A occurs is called a **conditional probability.** It is denoted $P(B\,|\,A)$, which is read "the probability of B given A." We call A the **given event.**

In the next example, we illustrate the calculation of conditional probabilities with the simple experiment of rolling a balanced die once.

EXAMPLE 4.16 **Conditional Probability**

Rolling a Die When a balanced die is rolled once, six equally likely outcomes are possible, as displayed in Fig. 4.22.

FIGURE 4.22
Sample space for rolling a die once

Let

F = event a 5 is rolled, and
O = event the die comes up odd.

Determine the following probabilities:

a. $P(F)$, the probability that a 5 is rolled.
b. $P(F\,|\,O)$, the conditional probability that a 5 is rolled, given that the die comes up odd.

c. $P\big(O\,|\,(\text{not }F)\big)$, the conditional probability that the die comes up odd, given that a 5 is not rolled.

Solution

a. From Fig. 4.22, we see that six outcomes are possible. Also, event F can occur in only one way: if the die comes up 5. Thus the probability that a 5 is rolled is

$$P(F) = \frac{f}{N} = \frac{1}{6} = 0.167.$$

Interpretation   There is a 16.7% chance of rolling a 5.

FIGURE 4.23

Event O

b. Given that the die comes up odd, that is, that event O occurs, there are no longer six possible outcomes. There are only three, as Fig. 4.23 shows. Therefore the conditional probability that a 5 is rolled, given that the die comes up odd, is

$$P(F\,|\,O) = \frac{f}{N} = \frac{1}{3} = 0.333.$$

Comparison of this probability with the one obtained in part (a) shows that $P(F\,|\,O) \neq P(F)$; that is, the conditional probability that a 5 is rolled, given that the die comes up odd, is not the same as the (unconditional) probability that a 5 is rolled.

Interpretation   Given that the die comes up odd, there is a 33.3% chance of rolling a 5, compared with a 16.7% (unconditional) chance of rolling a 5. Knowing that the die comes up odd affects the chance of rolling a 5.

FIGURE 4.24

Event (not F)

You try it!

Exercise 4.135
on page 189

c. Given that a 5 is not rolled, that is, that event (not F) occurs, the possible outcomes are the five shown in Fig. 4.24. Under these circumstances, event O (odd) can occur in two ways: if a 1 or a 3 is rolled. So the conditional probability that the die comes up odd, given that a 5 is not rolled, is

$$P\big(O\,|\,(\text{not }F)\big) = \frac{f}{N} = \frac{2}{5} = 0.4.$$

Compare this probability with the (unconditional) probability that the die comes up odd, which is 0.5.

Conditional probability is often used to analyze bivariate data. In Section 4.4, we discussed contingency tables as a method for tabulating such data. We show next how to obtain conditional probabilities for bivariate data directly from a contingency table.

EXAMPLE 4.17   Conditional Probability

Age and Rank of Faculty   Table 4.8 repeats the contingency table for age and rank of faculty members at a university. Suppose that a faculty member is selected at random.

a. Determine the (unconditional) probability that the selected faculty member is in his or her 50s.
b. Determine the (conditional) probability that the selected faculty member is in his or her 50s given that an assistant professor is selected.

Solution

a. We are to determine the probability of event A_4. From Table 4.8, $N = 1164$, the total number of faculty members. Also, because 253 of the faculty members are in their 50s, we have $f = 253$. Therefore

$$P(A_4) = \frac{f}{N} = \frac{253}{1164} = 0.217.$$

Interpretation   21.7% of the faculty are in their 50s.

TABLE 4.8

Contingency table for age and rank of faculty members

	Rank				
Age (yr)	Full professor R_1	Associate professor R_2	Assistant professor R_3	Instructor R_4	**Total**
Under 30 A_1	2	3	57	6	68
30–39 A_2	52	170	163	17	402
40–49 A_3	156	125	61	6	348
50–59 A_4	145	68	36	4	253
60 & over A_5	75	15	3	0	93
Total	430	381	320	33	1164

You try it!

Exercise 4.139 on page 190

b. We are to find the probability of event A_4, given that an assistant professor is selected (event R_3); in other words, we want to determine $P(A_4 \mid R_3)$. To do so, we restrict our attention to the assistant professor column of Table 4.8. We have $N = 320$, the total number of assistant professors. Also, because 36 of the assistant professors are in their 50s, we have $f = 36$. Consequently,

$$P(A_4 \mid R_3) = \frac{f}{N} = \frac{36}{320} = 0.113.$$

Interpretation 11.3% of the assistant professors are in their 50s.

The Conditional Probability Rule

In the previous two examples, we computed conditional probabilities *directly*, meaning that we first obtained the new sample space determined by the given event and then, using the new sample space, we calculated probabilities in the usual manner.

Sometimes we cannot determine conditional probabilities directly but must instead compute them in terms of unconditional probabilities. We obtain a formula for doing so in the next example.

EXAMPLE 4.18 **Introducing the Conditional Probability Rule**

Age and Rank of Faculty In Example 4.17(b), we used a direct computation to determine the conditional probability that a faculty member is in his or her 50s (event A_4), given that an assistant professor is selected (event R_3). To do that, we restricted our attention to the R_3 column of Table 4.8 and obtained

$$P(A_4 \mid R_3) = \frac{36}{320} = 0.113.$$

Now express the conditional probability $P(A_4 \mid R_3)$ in terms of unconditional probabilities.

Solution First, we note that the number 36 in the numerator of the preceding fraction is the number of assistant professors in their 50s, that is, the number of ways event (R_3 & A_4) can occur. Next, we observe that the number 320 in the denominator is the total number of assistant professors, that is, the number of ways event R_3 can occur. Thus the numbers 36 and 320 are those used to compute the unconditional probabilities of events (R_3 & A_4) and R_3, respectively:

$$P(R_3 \text{ \& } A_4) = \frac{36}{1164} = 0.031 \quad \text{and} \quad P(R_3) = \frac{320}{1164} = 0.275.$$

From the previous three probabilities,

$$P(A_4 \mid R_3) = \frac{36}{320} = \frac{36/1164}{320/1164} = \frac{P(R_3 \text{ \& } A_4)}{P(R_3)}.$$

In other words, we can express the conditional probability $P(A_4 \mid R_3)$ in terms of the unconditional probabilities $P(R_3$ & $A_4)$ and $P(R_3)$ by using the formula

$$P(A_4 \mid R_3) = \frac{P(R_3 \text{ \& } A_4)}{P(R_3)}.$$

The general form of this formula is called the **conditional probability rule.** ◼

FORMULA 4.4

The Conditional Probability Rule

What Does It Mean?

If A and B are any two events with $P(A) > 0$, then

The conditional probability of one event given another equals the probability that both events occur divided by the probability of the given event.

$$P(B \mid A) = \frac{P(A \text{ \& } B)}{P(A)}.$$

For the faculty-member example, we can find conditional probabilities either directly or by applying the conditional probability rule. Using the conditional probability rule, however, is sometimes the only way to find conditional probabilities.

◼◻◻◼ EXAMPLE 4.19 The Conditional Probability Rule

Marital Status and Gender From *America's Families and Living Arrangements*, a publication of the U.S. Census Bureau, we obtained a joint probability distribution for the marital status of U.S. adults by gender, as shown in Table 4.9. We used "Single" to mean "Never married."

TABLE 4.9

Joint probability distribution of marital status and gender

		Marital status				
		Single M_1	Married M_2	Widowed M_3	Divorced M_4	$P(S_i)$
Gender	Male S_1	0.147	0.281	0.013	0.044	0.485
	Female S_2	0.121	0.284	0.050	0.060	0.515
	$P(M_j)$	0.268	0.565	0.063	0.104	1.000

Suppose a U.S. adult is selected at random.

a. Determine the probability that the adult selected is divorced, given that the adult selected is a male.

b. Determine the probability that the adult selected is a male, given that the adult selected is divorced.

Solution Unlike our previous work with contingency tables, we do not have frequency data here; rather, we have only probability (relative-frequency) data. Hence we cannot compute conditional probabilities directly; we must instead use the conditional probability rule.

a. We want $P(M_4 \mid S_1)$. Using the conditional probability rule and Table 4.9, we get

$$P(M_4 \mid S_1) = \frac{P(S_1 \ \& \ M_4)}{P(S_1)} = \frac{0.044}{0.485} = 0.091.$$

Interpretation In the United States, 9.1% of adult males are divorced.

b. We want $P(S_1 \mid M_4)$. Using the conditional probability rule and Table 4.9, we get

$$P(S_1 \mid M_4) = \frac{P(M_4 \ \& \ S_1)}{P(M_4)} = \frac{0.044}{0.104} = 0.423.$$

Interpretation In the United States, 42.3% of divorced adults are males. ■

You try it!

Exercise 4.143
on page 191

Exercises 4.5

Understanding the Concepts and Skills

4.124 Regarding conditional probability:
a. What is it?
b. Which event is the "given event"?

4.125 Give an example of the conditional probability of an event being the same as the unconditional probability of the event. (*Hint:* Consider the experiment of tossing a coin twice.)

4.126 Suppose that A and B are events such that $P(A) = 0.75$ and $P(A \ \& \ B) = 0.25$. Determine $P(B \mid A)$.

4.127 Suppose that C and D are events such that $P(C) = 0.5$ and $P(C \ \& \ D) = 0.2$. Determine $P(D \mid C)$.

4.128 Suppose that C and D are events such that $P(C) = 4/5$ and $P(C \ \& \ D) = 3/10$. Find $P(D \mid C)$.

4.129 Suppose that A and B are events such that $P(A) = 2/7$ and $P(A \ \& \ B) = 3/25$. Determine $P(B \mid A)$.

In each of Exercises 4.130–4.133, we have presented either a contingency table or a joint probability distribution. In each case, determine both $P(C_1 \mid R_2)$ and $P(R_2 \mid C_1)$.

4.130

	C_1	C_2	Total
R_1	3	7	10
R_2	8	7	15
Total	11	14	25

4.131

	C_1	C_2	Total
R_1	15	3	18
R_2	20	12	32
Total	35	15	50

4.132

	C_1	C_2	C_3	$P(R_j)$
R_1	0.24	0.08	0.12	0.44
R_2	0.10	0.20	0.26	0.56
$P(C_i)$	0.34	0.28	0.38	1.00

4.133

	C_1	C_2	$P(R_j)$
R_1	0.15	0.15	0.30
R_2	0.10	0.20	0.30
R_3	0.35	0.05	0.40
$P(C_i)$	0.60	0.40	1.00

Applying the Concepts and Skills

4.134 Coin Tossing. A balanced dime is tossed twice. The four possible equally likely outcomes are HH, HT, TH, TT. Let

A = event the first toss is heads,
B = event the second toss is heads, and
C = event at least one toss is heads.

Determine the following probabilities and express your results in words. Compute the conditional probabilities directly; do not use the conditional probability rule.
a. $P(B)$ **b.** $P(B \mid A)$ **c.** $P(B \mid C)$
d. $P(C)$ **e.** $P(C \mid A)$ **f.** $P\big(C \mid (\text{not } B)\big)$

4.135 Playing Cards. One card is selected at random from an ordinary deck of 52 playing cards. Let

A = event a face card is selected,
B = event a king is selected, and
C = event a heart is selected.

Find the following probabilities and express your results in words. Compute the conditional probabilities directly; do not use the conditional probability rule.

a. $P(B)$ **b.** $P(B \mid A)$
c. $P(B \mid C)$ **d.** $P(B \mid (\text{not } A))$
e. $P(A)$ **f.** $P(A \mid B)$
g. $P(A \mid C)$ **h.** $P(A \mid (\text{not } B))$

4.136 State Populations. From Infoplease, we obtained data on state population from which we constructed the following frequency distribution.

Population size (millions)	Frequency
Under 1	6
1–under 2	8
2–under 3	6
3–under 5	8
5–under 10	15
10 & over	7

Compute the following conditional probabilities directly; that is, do not use the conditional probability rule. For a state selected at random, find the probability that the population of the state obtained is

a. between 2 million and 3 million.
b. between 2 million and 3 million given that it is at least 1 million.
c. less than 5 million given that it is at least 1 million.
d. Interpret your answers in parts (a)–(c) in terms of percentages.

4.137 Housing Units. The U.S. Census Bureau publishes data on housing units in *American Housing Survey for the United States*. The following table provides a frequency distribution for the number of rooms in U.S. housing units. The frequencies are in thousands.

Rooms	No. of units
1	601
2	1,404
3	11,433
4	23,636
5	30,440
6	27,779
7	17,868
8+	19,257

Compute the following conditional probabilities directly; that is, do not use the conditional probability rule. For a U.S. housing unit selected at random, determine

a. the probability that the unit has exactly four rooms.
b. the conditional probability that the unit has exactly four rooms, given that it has at least two rooms.
c. the conditional probability that the unit has at most four rooms, given that it has at least two rooms.
d. Interpret your answers in parts (a)–(c) in terms of percentages.

4.138 Protective Orders. In the article "Judicial Dispositions of Ex-Parte and Domestic Violence Protection Order Hearings: A Comparative Analysis of Victim Requests and Court Authorized Relief" (*Journal of Family Violence*, Vol. 20, No. 3, pp. 161–170), D. Yearwood looked at the discrepancies between what a victim of domestic violence requests and what the courts reward. The following contingency table cross-classifies, by race and gender, a sample of

407 domestic violence protective orders from the North Carolina Criminal Justice Analysis Center.

		Race			
		White R_1	Black R_2	Other R_3	Total
Gender	Male G_1	30	26	0	56
	Female G_2	210	121	20	351
	Total	240	147	20	407

Compute the following conditional probabilities directly; that is, do not use the conditional probability rule. One of these protective orders is selected at random. Find the probability that the order was filed by

a. a Black.   **b.** a white female.
c. a male, given that the filer was white.
d. a male, given that the filer was Black.

4.139 New England Patriots. From the National Football League (NFL) Web site, in the *New England Patriots Roster*, we obtained information on the weights and years of experience for the players on that team, as of September 26, 2013. The following contingency table provides a cross-classification of those data.

		Years of experience				
		Rookie Y_1	1–5 Y_2	6–10 Y_3	Over 10 Y_4	Total
Weight (lb)	Under 200 W_1	3	5	0	0	8
	200–300 W_2	11	21	7	2	41
	Over 300 W_3	4	4	5	0	13
	Total	18	30	12	2	62

Compute the following conditional probabilities directly; that is, do not use the conditional probability rule. A player on the New England Patriots is selected at random. Find the probability that the player selected

a. is a rookie.   **b.** weighs under 200 pounds.
c. is a rookie, given that he weighs under 200 pounds.
d. weighs under 200 pounds, given that he is a rookie.
e. Interpret your answers in parts (a)–(d) in terms of percentages.

4.140 Acclaim or Attack. In the article "A Functional Analysis Comparison of Web-Only Advertisements and Traditional Television Advertisements from the 2004 and 2008 Presidential Campaigns" (*Journalism & Mass Communication Quarterly*, Vol. 90, No. 1, pp. 23–38), author C. Roberts explores whether a difference exists in purpose and content of web-only ads and television ads created by candidates in presidential campaigns. The following contingency table cross-classifies a sample of such advertisements by purpose and source. The purpose of the ad is classified as "acclaim" if the candidate acclaims himself/herself with positive statements about his/her credentials to be a better candidate. The purpose of the ad is classified as "attack" if the candidate attacks the opponent's credentials or defends with refutations against the opponent's attack.

Purpose

Source		Acclaim P_1	Attack P_2	Total
	Web-only ad S_1	20	42	62
	Television ad S_2	332	176	508
	Total	352	218	570

a. Find $P(S_2)$.   **b.** Find $P(S_2 \ \& \ P_1)$.
c. Obtain $P(P_1 \mid S_2)$ directly from the table.
d. Obtain $P(P_1 \mid S_2)$ by using the conditional probability rule and your answers from parts (a) and (b).

4.141 Self-Concept and Sightedness. Self-concept can be defined as the general view of oneself in terms of personal value and capabilities. A study of whether visual impairment affects self-concept was reported in the article ''An Exploration into Self Concept: A Comparative Analysis between the Adolescents Who Are Sighted and Blind in India'' (*British Journal of Visual Impairment*, Vol. 30, No. 1, pp. 31–41) by S. Halder and P. Datta. The following contingency table cross-classifies the Indian adolescents in the study by self-concept and sightedness.

Self-concept

Sightedness		High S_1	Moderate S_2	Low S_3	Total
	Sighted V_1	13	73	14	100
	Blind V_2	3	40	17	60
	Total	16	113	31	160

a. Find $P(V_2)$.   **b.** Find $P(V_2 \ \& \ S_3)$.
c. Obtain $P(S_3 \mid V_2)$ directly from the table.
d. Obtain $P(S_3 \mid V_2)$ by using the conditional probability rule and your answers from parts (a) and (b).

4.142 Living Arrangements. As reported by the U.S. Census Bureau in *America's Families and Living Arrangements*, the living arrangements by age of U.S. citizens 15 years of age and older are as shown in the following joint probability distribution.

Living arrangement

Age (yr)		Alone L_1	With spouse L_2	With others L_3	$P(A_i)$
	15–24 A_1	0.006	0.012	0.157	0.175
	25–44 A_2	0.030	0.184	0.123	0.337
	45–64 A_3	0.047	0.216	0.067	0.330
	Over 64 A_4	0.046	0.088	0.024	0.158
	$P(L_j)$	0.129	0.500	0.371	1.000

A U.S. citizen 15 years of age or older is selected at random. Determine the probability that the person selected
a. lives with spouse.   **b.** is over 64.
c. lives with spouse and is over 64.
d. lives with spouse, given that the person is over 64.
e. is over 64, given that the person lives with spouse.
f. Interpret your answers in parts (a)–(e) in terms of percentages.

4.143 Smartphone Ownership. From the document *Smartphone Ownership* by the Pew Research Center and the document *Current Population Survey* by the U.S. Census Bureau, we constructed the following joint probability distribution for educational attainment and smartphone ownership for U.S. adults.

Owner

Education		Yes S_1	No S_2	$P(E_i)$
	Not HS grad E_1	0.046	0.083	0.129
	HS grad E_2	0.144	0.168	0.312
	Some college E_3	0.101	0.067	0.168
	College grad E_4	0.274	0.117	0.391
	$P(S_j)$	0.564	0.436	1.000

A U.S. adult is selected at random. Determine the probability that the person selected
a. owns a smartphone.   **b.** is a college grad.
c. owns a smartphone and is a college grad.
d. owns a smartphone, given that the person is a college grad.
e. is a college grad, given that the person owns a smartphone.
f. Interpret your answers in parts (a)–(e) in terms of percentages.

4.144 HPV Vaccine. In the article "Correlates for Completion of 3-dose Regimen of HPV Vaccine in Female Members of a Managed Care Organization" (*Mayo Clinical Proceedings*, Vol. 84. pp. 864–870), C. Chao et al. examined characteristics that may influence whether young female patients complete a three-injection sequence of the Gardasil quadrivalent human papillomavirus vaccine (HPV4). HPV is a virus that has been linked to the development of cervical cancer. The following joint probability distribution summarizes completion of treatment versus practice type.

Owner

Practice		Yes C_1	No C_2	$P(T_i)$
	Pediatric T_1	0.115	0.250	0.365
	Family T_2	0.075	0.183	0.258
	OB/GYN T_3	0.142	0.235	0.377
	$P(C_j)$	0.332	0.668	1.000

For a patient in the study selected at random, find the following probabilities and interpret your results in terms of percentages.

a. $P(C_1)$ b. $P(T_2)$ c. $P(C_1 \& T_2)$

d. $P(C_1 \mid T_2)$ e. $P(T_2 \mid C_1)$

4.145 Dentist Visits. The National Center for Health Statistics publishes information about visits to the dentist in *National Health Interview Survey*. Here is a joint probability distribution for the length of time (in years) since last visit to a dentist or other dental health professional, by age, for U.S. adults during one year.

	Age (yr)				
Time (yr)	18–44 A_1	45–64 A_2	65–74 A_3	75+ A_4	$P(T_i)$
Less than 0.5 T_1	0.221	0.154	0.038	0.028	0.441
0.5–under 1 T_2	0.101	0.052	0.011	0.011	0.175
1–under 2 T_3	0.076	0.036	0.008	0.006	0.126
2–under 5 T_4	0.070	0.034	0.010	0.008	0.122
5 or more T_5	0.058	0.038	0.018	0.022	0.136
$P(A_j)$	0.526	0.314	0.085	0.075	1.000

For a U.S. adult selected at random, determine the following probabilities and interpret your results in terms of percentages.

a. $P(T_1)$ b. $P(\text{not } A_2)$ c. $P(A_4 \& T_5)$

d. $P(T_4 \mid A_1)$ e. $P(A_1 \mid T_4)$

4.146 Scientists and Engineers. The U.S. National Science Foundation collects data on science and engineering (S&E) degrees awarded and publishes the results in *Science and Engineering Degrees*. During one year, 75.2% of S&E degrees awarded were for Bachelor's degrees and 37.8% of S&E degrees were Bachelor's degrees awarded to women. What percentage of S&E Bachelor's degrees were awarded to women?

4.147 Property Crime. As reported by the Federal Bureau of Investigation in *Crime in the United States*, 5.1% of property crimes are committed in rural areas and 1.6% of property crimes are burglaries committed in rural areas. What percentage of property crimes committed in rural areas are burglaries?

4.148 National Guard. The U.S. Department of Defense collects data on active members in the National Guard. Results are published in *Selected Manpower Statistics*. According to that document, 77.0% of members are in the Army National Guard and 10.2% of members are Black and in the Army National Guard. What percentage of members in the Army National Guard are Black?

4.149 Physician Characteristics. The Association of American Medical Colleges (AAMC) compiles data on active U.S. physicians and publishes its results in the *Physician Specialty Data Book*. From that document, 13.6% of active U.S. physicians specialize in internal medicine and 8.67% of active U.S. physicians are under age 55 and specialize in internal medicine. What percentage of active U.S. physicians who specialize in internal medicine are under age 55?

4.150 Dice. Two balanced dice are thrown, one red and one black. What is the probability that the red die comes up 1, given that the

a. black die comes up 3?

b. sum of the dice is 4?

c. sum of the dice is 9?

4.151 Royal Offspring. A king and queen have two children. Assuming that a child of the king and queen is equally likely to be a boy or a girl, what is the probability that both children are boys, given that

a. the first child born is a boy?

b. at least one child is a boy?

Extending the Concepts and Skills

4.152 New England Patriots. Refer to Exercise 4.139.

a. Construct a joint probability distribution.

b. Determine the probability distribution of weight for rookies; that is, construct a table showing the conditional probabilities that a rookie weighs under 200 pounds, between 200 and 300 pounds, and over 300 pounds.

c. Determine the probability distribution of years of experience for players who weigh under 200 pounds.

d. The probability distributions in parts (b) and (c) are examples of a **conditional probability distribution.** Determine two other conditional probability distributions for the data on weight and years of experience for the New England Patriots.

Correlation of Events. One important application of conditional probability is to the concept of the **correlation of events.**

- Event B is said to be **positively correlated** with event A if $P(B \mid A) > P(B)$.
- Event B is said to be **negatively correlated** with event A if $P(B \mid A) < P(B)$.
- Event B is said to be **independent** of event A if $P(B \mid A) = P(B)$.

You are asked to examine correlation of events in Exercises 4.153 and 4.154.

4.153 Let A and B be events, each with positive probability.

a. State in words what it means for event B to be positively correlated with event A; negatively correlated with event A; independent of event A.

b. Show that event B is positively correlated with event A if and only if event A is positively correlated with event B.

c. Show that event B is negatively correlated with event A if and only if event A is negatively correlated with event B.

d. Show that event B is independent of event A if and only if event A is independent of event B.

4.154 Drugs and Car Accidents. Suppose that it has been determined that "one-fourth of drivers at fault in a car accident use a certain drug."

a. Explain in words what it means to say that being the driver at fault in a car accident is positively correlated with use of the drug.

b. Under what condition on the percentage of drivers involved in car accidents who use the drug does the statement in quotes imply that being the driver at fault in a car accident is positively correlated with use of the drug? negatively correlated with use of the drug? independent of use of the drug? Explain your answers.

c. Suppose that, in fact, being the driver at fault in a car accident is positively correlated with use of the drug. Can you deduce that a cause-and-effect relationship exists between use of the drug and being the driver at fault in a car accident? Explain your answer.

4.6   The Multiplication Rule; Independence*

The conditional probability rule is used to compute conditional probabilities in terms of unconditional probabilities. That is,

$$P(B \mid A) = \frac{P(A \& B)}{P(A)}.$$

Multiplying both sides of this equation by $P(A)$, we obtain a formula for computing joint probabilities in terms of marginal and conditional probabilities. It is called the **general multiplication rule,** and we express it as the following formula.

FORMULA 4.5

? **What Does It Mean?**

For any two events, the probability that both occur equals the probability that a specified one occurs times the conditional probability of the other event, given the specified event.

The General Multiplication Rule

If A and B are any two events, then

$$P(A \& B) = P(A) \cdot P(B \mid A).$$

The conditional probability rule and the general multiplication rule are simply variations of each other. On one hand, when the joint and marginal probabilities are known or can be easily determined directly, we use the conditional probability rule to obtain conditional probabilities. On the other hand, when the marginal and conditional probabilities are known or can be easily determined directly, we use the general multiplication rule to obtain joint probabilities.

EXAMPLE 4.20   ### The General Multiplication Rule

U.S. Congress   The U.S. Congress, Joint Committee on Printing, provides information on the composition of the Congress in the *Congressional Directory*. For the 113th Congress, 18.7% of the members are senators and 53% of the senators are Democrats. What is the probability that a randomly selected member of the 113th Congress is a Democratic senator?

Solution   Let

$$D = \text{event the member selected is a Democrat, and}$$
$$S = \text{event the member selected is a senator.}$$

The event that the member selected is a Democratic senator can be expressed as $(S \& D)$. We want to determine the probability of that event.

Because 18.7% of members are senators, $P(S) = 0.187$; and because 53% of senators are Democrats, $P(D \mid S) = 0.530$. Applying the general multiplication rule, we get

$$P(S \& D) = P(S) \cdot P(D \mid S) = 0.187 \cdot 0.530 = 0.099.$$

The probability that a randomly selected member of the 113th Congress is a Democratic senator is 0.099.

Exercise 4.179 on page 198

Interpretation   9.9% of members of the 113th Congress are Democratic senators.

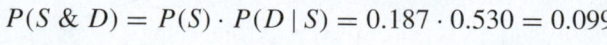

Another application of the general multiplication rule relates to sampling two or more members from a population. Example 4.21 provides an illustration.

■■ ■ **EXAMPLE 4.21** **The General Multiplication Rule**

TABLE 4.10

Frequency distribution of males and females in Professor Weiss's introductory statistics class

Gender	Frequency
Male	17
Female	23
	40

Gender of Students In Professor Weiss's introductory statistics class, the number of males and females are as shown in the frequency distribution presented in Table 4.10. Two students are selected at random from the class. The first student selected is not returned to the class for possible reselection; that is, the sampling is without replacement. Find the probability that the first student selected is female and the second is male.

Solution Let

$$F1 = \text{event the first student obtained is female, and}$$
$$M2 = \text{event the second student obtained is male.}$$

We want to determine $P(F1 \,\&\, M2)$. Using the general multiplication rule, we write

$$P(F1 \,\&\, M2) = P(F1) \cdot P(M2 \,|\, F1).$$

Computing the two probabilities on the right side of this equation is easy. To find $P(F1)$—the probability that the first student selected is female—we note from Table 4.10 that 23 of the 40 students are female, so

$$P(F1) = \frac{f}{N} = \frac{23}{40}.$$

Next, we find $P(M2 \,|\, F1)$—the conditional probability that the second student selected is male, given that the first one selected is female. Given that the first student selected is female, of the 39 students remaining in the class 17 are male, so

$$P(M2 \,|\, F1) = \frac{f}{N} = \frac{17}{39}.$$

Applying the general multiplication rule, we conclude that

$$P(F1 \,\&\, M2) = P(F1) \cdot P(M2 \,|\, F1) = \frac{23}{40} \cdot \frac{17}{39} = 0.251.$$

You try it!

Exercise 4.183 on page 198

Interpretation When two students are randomly selected from the class, the probability is 0.251 that the first student selected is female and the second student selected is male.

You will find that drawing a **tree diagram** is often helpful when you are applying the general multiplication rule. An appropriate tree diagram for Example 4.21 is shown in Fig. 4.25.

FIGURE 4.25

Tree diagram for student-selection problem

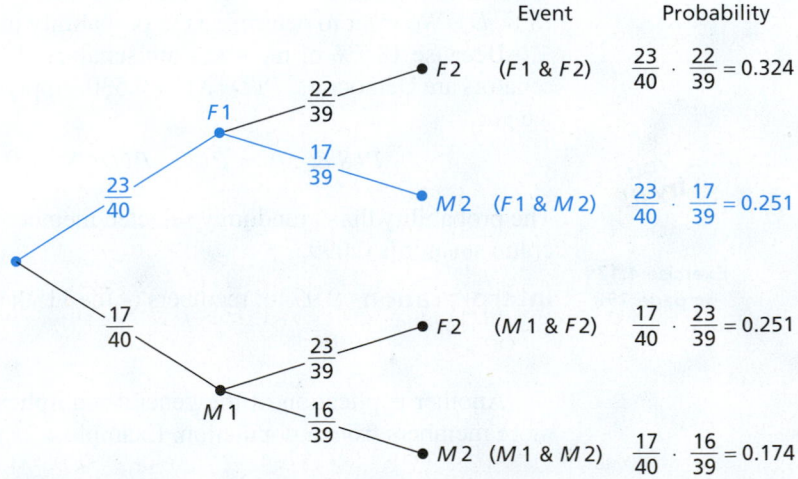

Each branch of the tree corresponds to one possibility for selecting two students at random from the class. For instance, the second branch of the tree, shown in color, corresponds to event ($F1$ & $M2$)—the event that the first student selected is female (event $F1$) and the second is male (event $M2$).

Starting from the left on that branch, the number $\frac{23}{40}$ is the probability that the first student selected is female, $P(F1)$; the number $\frac{17}{39}$ is the conditional probability that the second student selected is male, given that the first student selected is female, $P(M2\,|\,F1)$. The product of those two probabilities is, by the general multiplication rule, the probability that the first student selected is female and the second is male, $P(F1\,\&\,M2)$. The second entry in the Probability column of Fig. 4.25 shows that this probability is 0.251, as we discovered at the end of Example 4.21.

Note: The general multiplication rule can be extended to more than two events. Exercises 4.200–4.202 discuss and apply this extension.

Independence

One of the most important concepts in probability is that of **statistical independence** of events. For two events, statistical independence or, more simply, **independence,** is defined as follows.

DEFINITION 4.7

? **What Does It Mean?**

One event is independent of another event if knowing whether the latter event occurs does not affect the probability of the former event.

Independent Events

Event B is said to be **independent** of event A if $P(B\,|\,A) = P(B)$.

In the next example, we illustrate how to determine whether one event is independent of another event by returning to the experiment of randomly selecting a card from a deck.

EXAMPLE 4.22   **Independent Events**

Playing Cards   Consider again the experiment of randomly selecting one card from a deck of 52 playing cards. Let

$$F = \text{event a face card is selected,}$$
$$K = \text{event a king is selected, and}$$
$$H = \text{event a heart is selected.}$$

a. Determine whether event K is independent of event F.
b. Determine whether event K is independent of event H.

Solution   First we note that the unconditional probability that event K occurs is

$$P(K) = \frac{f}{N} = \frac{4}{52} = \frac{1}{13} = 0.077.$$

a. To determine whether event K is independent of event F, we must compute $P(K\,|\,F)$ and compare it to $P(K)$. If those two probabilities are equal, event K is independent of event F; otherwise, event K is not independent of event F. Now, given that event F occurs, 12 outcomes are possible (four jacks, four queens, and four kings), and event K can occur in 4 ways out of those 12 possibilities. Hence

$$P(K\,|\,F) = \frac{f}{N} = \frac{4}{12} = 0.333,$$

which does not equal $P(K)$; event K is not independent of event F.

Interpretation Event K (king) is not independent of event F (face card) because the percentage of kings among the face cards (33.3%) is not the same as the percentage of kings among all the cards (7.7%).

b. We need to compute $P(K \mid H)$ and compare it to $P(K)$. Given that event H occurs, 13 outcomes are possible (the 13 hearts), and event K can occur in 1 way out of those 13 possibilities. Therefore

$$P(K \mid H) = \frac{f}{N} = \frac{1}{13} = 0.077,$$

which equals $P(K)$; event K is independent of event H.

Interpretation Event K (king) is independent of event H (heart) because the percentage of kings among the hearts is the same as the percentage of kings among all the cards, namely, 7.7%.

You try it!

Exercise 4.185
on page 198

If event B is independent of event A, then event A is also independent of event B. In such cases, we often say that event A and event B are independent, or that A and B are **independent events.** If two events are not independent, we say that they are **dependent events.** In Example 4.22, F and K are dependent events, whereas K and H are independent events.

The Special Multiplication Rule

Recall that the general multiplication rule states that, for any two events A and B,

$$P(A \text{ \& } B) = P(A) \cdot P(B \mid A).$$

If A and B are independent events, $P(B \mid A) = P(B)$. Thus, for the special case of independent events, we can replace the term $P(B \mid A)$ in the general multiplication rule by the term $P(B)$. Doing so yields the **special multiplication rule,** which we express as the following formula.

FORMULA 4.6

? What Does It Mean?

Two events are independent if and only if the probability that both occur equals the product of their individual probabilities.

The Special Multiplication Rule (for Two Independent Events)

If A and B are independent events, then

$$P(A \text{ \& } B) = P(A) \cdot P(B),$$

and conversely, if $P(A \text{ \& } B) = P(A) \cdot P(B)$, then A and B are independent events.

We can decide whether event A and event B are independent by using either of two methods. As we saw in Example 4.22, we can determine whether $P(B \mid A) = P(B)$. Alternatively, we can use the special multiplication rule, that is, determine whether $P(A \text{ \& } B) = P(A) \cdot P(B)$.

The definition of independence for three or more events is more complicated than that for two events. Nevertheless, the special multiplication rule still holds, as expressed in the following formula.

FORMULA 4.7

? What Does It Mean?

For independent events, the probability that they all occur equals the product of their individual probabilities.

The Special Multiplication Rule

If events $A, B, C, \ldots$ are independent, then

$$P(A \text{ \& } B \text{ \& } C \text{ \& } \cdots) = P(A) \cdot P(B) \cdot P(C) \cdots.$$

We can use the special multiplication rule to compute joint probabilities when we know or can reasonably assume that two or more events are independent, as shown in the next example.

EXAMPLE 4.23   The Special Multiplication Rule

Roulette An American roulette wheel contains 38 numbers, of which 18 are red, 18 are black, and 2 are green. When the roulette wheel is spun, the ball is equally likely to land on any of the 38 numbers. In three plays at a roulette wheel, what is the probability that the ball will land on green the first time and on black the second and third times?

Solution First, we can reasonably assume that outcomes on successive plays at the wheel are independent. Now, we let

$G1$ = event the ball lands on green the first time,

$B2$ = event the ball lands on black the second time, and

$B3$ = event the ball lands on black the third time.

We want to determine $P(G1 \ \& \ B2 \ \& \ B3)$.

Because outcomes on successive plays at the wheel are independent, we know that event $G1$, event $B2$, and $B3$ are independent. Applying the special multiplication rule, we conclude that

You try it!

Exercise 4.193 on page 199

$$P(G1 \ \& \ B2 \ \& \ B3) = P(G1) \cdot P(B2) \cdot P(B3) = \frac{2}{38} \cdot \frac{18}{38} \cdot \frac{18}{38} = 0.012.$$

Interpretation In three plays at a roulette wheel, there is a 1.2% chance that the ball will land on green the first time and on black the second and third times.

Mutually Exclusive Versus Independent Events

The terms *mutually exclusive* and *independent* refer to different concepts. Mutually exclusive events are those that cannot occur simultaneously; independent events are those for which the occurrence of some does not affect the probabilities of the others occurring.

In fact, if two or more events are mutually exclusive, the occurrence of one precludes the occurrence of the others. Two or more (nonimpossible) events cannot be both mutually exclusive and independent.

Exercises 4.6

Understanding the Concepts and Skills

4.155 Regarding the general multiplication rule and the conditional probability rule:
a. State these two rules.
b. Explain the relationship between them.
c. Why are two different variations of essentially the same rule emphasized?

4.156 Suppose that A and B are two events.
a. What does it mean for event B to be independent of event A?
b. If event A and event B are independent, how can their joint probability be obtained from their marginal probabilities?

4.157 Suppose $P(A) = 0.6$ and $P(B \mid A) = 0.4$. Find $P(A \ \& \ B)$.

4.158 Suppose $P(C) = 0.3$ and $P(D \mid C) = 0.8$. Find $P(C \ \& \ D)$.

4.159 Suppose $P(C) = 1/4$ and $P(D \mid C) = 4/5$. Find $P(C \ \& \ D)$.

4.160 Suppose $P(A) = 7/8$ and $P(B \mid A) = 5/7$. Find $P(A \ \& \ B)$.

4.161 Suppose that C and D are independent events such that $P(C) = 0.7$ and $P(D) = 0.6$. Find $P(C \& D)$.

4.162 Suppose that A and B are independent events such that $P(A) = 0.3$ and $P(B) = 0.2$. Find $P(A \ \& \ B)$.

4.163 Suppose that A and B are independent events such that $P(A) = 5/8$ and $P(B) = 4/7$. Find $P(A \ \& \ B)$.

4.164 Suppose that C and D are independent events such that $P(C) = 3/4$ and $P(D) = 2/5$. Find $P(C \ \& \ D)$.

In each of Exercises 4.165–4.176, decide whether or not the two events in question are independent or whether it is not possible to tell. Justify your answers.

4.165 $P(B) = 0.8$ and $P(B \mid A) = 0.6$

4.166 $P(D) = 0.6$ and $P(D \mid C) = 0.6$

4.167 $P(D) = 3/4$ and $P(D \mid C) = 3/4$

4.168 $P(B) = 5/8$ and $P(B \mid A) = 1/3$

4.169 $P(B) = 0.3$ and $P(A \mid B) = 0.3$

4.170 $P(D) = 2/5$ and $P(C \mid D) = 2/5$

4.171 $P(D) = 3/5$ and $P(C \mid D) = 2/5$

4.172 $P(B) = 0.7$ and $P(A \mid B) = 0.6$

4.173 $P(C) = 0.8$, $P(D) = 0.5$, and $P(C \& D) = 0.4$

4.174 $P(C) = 0.3$, $P(D) = 0.6$, and $P(C \& D) = 0.2$

4.175 $P(A) = 5/7$, $P(B) = 1/4$, and $P(A \& B) = 1/7$

4.176 $P(A) = 2/5$, $P(B) = 3/4$, and $P(A \& B) = 3/10$

4.177 Suppose that A, B, and C are independent events such that $P(A) = 0.8$, $P(B) = 0.5$, and $P(C) = 0.3$. Find $P(A \& B \& C)$.

4.178 Suppose that C, D, and E are independent events such that $P(C) = 7/8$, $P(D) = 5/7$, and $P(E) = 2/3$. Find $P(C \& D \& E)$.

Applying the Concepts and Skills

4.179 Holiday Depression. According to the Opinion Research Corporation, 44% of U.S. women suffer from holiday depression, and, from the U.S. Census Bureau's *Current Population Reports*, 52% of U.S. adults are women. Find the probability that a randomly selected U.S. adult is a woman who suffers from holiday depression. Interpret your answer in terms of percentages.

4.180 Addiction Medicine. The report "Addiction Medicine: Closing the Gap between Science and Practice" from the National Center on Addiction and Substance Abuse at Columbia University reveals that addiction treatment is neglected by the medical system in the United States. It shows, in particular, that 16% of Americans have the disease of addiction and that 90% of them receive no form of treatment. Determine the probability that a randomly selected American has the disease of addiction and receives no form of treatment. Interpret your answer in terms of percentages.

4.181 ESP Experiment. A person has agreed to participate in an extrasensory perception (ESP) experiment. He is asked to randomly pick two numbers between 1 and 6. The second number must be different from the first. Let

H = event the first number picked is a 3, and

K = event the second number picked exceeds 4.

Determine

a. $P(H)$.   **b.** $P(K \mid H)$.   **c.** $P(H \& K)$.

Find the probability that both numbers picked are

d. less than 3.   **e.** greater than 3.

4.182 Cards. Cards numbered 1, 2, 3, ..., 10 are placed in a box. The box is shaken, and a blindfolded person selects two successive cards without replacement.

a. What is the probability that the first card selected is numbered 6?

b. Given that the first card is numbered 6, what is the probability that the second is numbered 9?

c. Find the probability of selecting first a 6 and then a 9.

d. What is the probability that both cards selected are numbered over 5?

4.183 Class Levels. A frequency distribution for the class level of students in Professor Weiss's introductory statistics course is as follows.

Class	Frequency
Freshman	6
Sophomore	15
Junior	12
Senior	7

Two students are randomly selected without replacement. Determine the probability that

a. the first student obtained is a junior and the second a senior.

b. both students obtained are sophomores.

c. Draw a tree diagram for this problem similar to Fig. 4.25 on page 194.

d. What is the probability that one of the students obtained is a freshman and the other a sophomore?

4.184 Governors. The National Governors Association publishes data on U.S. governors in *Governors' Political Affiliations & Terms of Office*. Based on that document, we obtained the following frequency distribution for U.S. governors, as of 2013.

Party	Frequency
Democratic	20
Republican	30

Two U.S. governors are selected at random without replacement.

a. Find the probability that the first is a Republican and the second a Democrat.

b. Find the probability that both are Republicans.

c. Draw a tree diagram for this problem similar to the one shown in Fig. 4.25 on page 194.

d. What is the probability that the two governors selected have the same political-party affiliation?

e. What is the probability that the two governors selected have different political-party affiliations?

4.185 Medical School Faculty. The Association of American Medical Colleges (AAMC) compiles data on medical school faculty and publishes the results in *AAMC Faculty Roster*. The following contingency table cross-classifies medical school faculty by the characteristics gender and rank.

		Gender		
		Male G_1	Female G_2	Total
Rank	Professor R_1	21,224	3,194	24,418
	Associate professor R_2	16,332	5,400	21,732
	Assistant professor R_3	25,888	14,491	40,379
	Instructor R_4	5,775	5,185	10,960
	Other R_5	781	723	1,504
	Total	70,000	28,993	98,993

a. Find $P(R_3)$.

b. Find $P(R_3 \mid G_1)$.

c. Are events G_1 and R_3 independent? Explain your answer.

d. For a medical school faculty member, is the event that the person is female independent of the event that the person is an associate professor? Explain your answer.

4.186 Injured Americans. The National Center for Health Statistics compiles data on injuries and publishes the information in

Vital and Health Statistics. A contingency table for injuries in the United States, by circumstance and gender, is as follows. Frequencies are in millions.

	Circumstance			
	Work C_1	Home C_2	Other C_3	**Total**
Male S_1	8.0	9.8	17.8	35.6
Female S_2	1.3	11.6	12.9	25.8
Total	9.3	21.4	30.7	61.4

(Gender on left axis)

a. Find $P(C_1)$.
b. Find $P(C_1 \mid S_2)$.
c. Are events C_1 and S_2 independent? Explain your answer.
d. Is the event that an injured person is male independent of the event that an injured person was hurt at home? Explain your answer.

4.187 U.S. Congress. The U.S. Congress, Joint Committee on Printing, provides information on the composition of Congress in *Congressional Directory*. Here is a joint probability distribution for the members of the 113th Congress by legislative group and political party. The "other" category includes Independents and vacancies. (Rep = Representative.)

	Group		
	Rep C_1	Senator C_2	$P(P_i)$
Democratic P_1	0.374	0.097	0.471
Republican P_2	0.434	0.086	0.520
Other P_3	0.006	0.004	0.009
$P(C_j)$	0.813	0.187	1.000

(Party on left axis)

a. Determine $P(P_1)$, $P(C_2)$, and $P(P_1 \& C_2)$.
b. Use the special multiplication rule to determine whether events P_1 and C_2 are independent.

4.188 Doctoral Degrees. The U.S. National Science Foundation publishes data on doctoral degrees conferred in the document *Science and Engineering Doctorate Awards*. The following joint probability distribution summarizes the doctoral degrees conferred in the social sciences and psychology by gender during one year.

	Field		
	Social science F_1	Psychology F_2	$P(G_i)$
Male G_1	0.300	0.119	0.419
Female G_2	0.282	0.298	0.581
$P(F_j)$	0.582	0.418	1.000

(Gender on left axis)

a. Find $P(G_1)$, $P(F_1)$, and $P(G_1 \& F_1)$.
b. Use the special multiplication rule to determine whether events G_1 and F_1 are independent.

4.189 Coin Tossing. When a balanced dime is tossed three times, eight equally likely outcomes are possible:

HHH	HTH	THH	TTH
HHT	HTT	THT	TTT

Let

A = event the first toss is heads,
B = event the third toss is tails, and
C = event the total number of heads is 1.

a. Compute $P(A)$, $P(B)$, and $P(C)$.
b. Compute $P(B \mid A)$.
c. Are A and B independent events? Explain your answer.
d. Compute $P(C \mid A)$.
e. Are A and C independent events? Explain your answer.

4.190 Dice. When two balanced dice are rolled, 36 equally likely outcomes are possible, as depicted in Fig. 4.1 on page 159. Let

A = event the red die comes up even,
B = event the black die comes up odd,
C = event the sum of the dice is 10, and
D = event the sum of the dice is even.

a. Compute $P(A)$, $P(B)$, $P(C)$, and $P(D)$.
b. Compute $P(B \mid A)$.
c. Are events A and B independent? Why or why not?
d. Compute $P(C \mid A)$.
e. Are events A and C independent? Why or why not?
f. Compute $P(D \mid A)$.
g. Are events A and D independent? Why or why not?

4.191 Drawing Cards. Two cards are drawn at random from an ordinary deck of 52 cards. Determine the probability that both cards are aces if
a. the first card is replaced before the second card is drawn.
b. the first card is not replaced before the second card is drawn.

4.192 Yahtzee. In the game of Yahtzee, five balanced dice are rolled.
a. What is the probability of rolling all 2s?
b. What is the probability that all the dice come up the same number?

4.193 The *Challenger* Disaster. In a letter to the editor that appeared in the February 23, 1987, issue of *U.S. News and World Report*, a reader discussed the issue of space shuttle safety. Each "criticality 1" item must have 99.99% reliability, according to NASA standards, meaning that the probability of failure for such an item is 0.0001. Mission 25, the mission in which the *Challenger* exploded, had 748 "criticality 1" items. Determine the probability that
a. none of the "criticality 1" items would fail.
b. at least one "criticality 1" item would fail.
c. Interpret your answer in part (b) in words.

4.194 Bar Dice. It is not uncommon after a round of golf to find a foursome in the clubhouse shaking the bar dice to see who buys the refreshments. In the first two rounds, each person gets to shake

the five dice once. The person with the most dice with the highest number is eliminated from the competition to see who pays. So, for instance, four 3s beats three 5s, but four 6s beats four 3s. The 1s on the dice are wild, that is, they can be used as any number.
a. What is the probability of getting five 6s? (Remember 1s are wild.)
b. What is the probability of getting no 6s and no 1s?

4.195 Traffic Fatalities. According to *Accident Facts*, published by the National Safety Council, a probability distribution of age group for drivers at fault in fatal crashes is as follows.

Age (yr)	Probability
16–24	0.255
25–34	0.238
35–64	0.393
65 & over	0.114

Of three fatal automobile crashes, find the probability that
a. the drivers at fault in the first, second, and third crashes are in the age groups 16–24, 25–34, and 35–64, respectively.
b. two of the drivers at fault are between 16 and 24 years old, and one of the drivers at fault is 65 years old or older.

4.196 Death Penalty. One of the more contentious issues is whether there should be the death penalty for a person convicted of murder. According to estimates by Gallup, the percentages of U.S. adults that approve of, disapprove of, and have no opinion about the death penalty are 61%, 35%, and 4%, respectively. Suppose that three U.S. adults are randomly selected and asked their position on the death penalty.
a. Find the probability that the first two approve and the third disapproves.
b. Find the probability that exactly two of the three approve and one disapproves.
c. In doing your calculations in parts (a) and (b), are you assuming sampling with replacement or sampling without replacement? Does it make a difference which of those two types of sampling is used? Explain your answers.

4.197 Nuts and Bolts. A hardware manufacturer produces nuts and bolts. Each bolt produced is attached to a nut to make a single unit. It is known that 2% of the nuts produced and 3% of the bolts produced are defective in some way. A nut–bolt unit is considered defective if either the nut or the bolt has a defect.
a. Determine the percentage of defective nut–bolt units.
b. What assumptions are you making in solving part (a)?

4.198 Activity Limitations. The National Center for Health Statistics compiles information on activity limitations. Results are published in *Vital and Health Statistics*. The data show that 13.6% of males and 14.4% of females have an activity limitation. Are gender and activity limitation statistically independent? Explain your answer.

4.199 Scholarships. Marilyn vos Savant writes the column "Ask Marilyn," which appears in *Parade Magazine*. One reader submitted the following question: My daughter is one of five finalists for a grant that awards two scholarships and is also one of five finalists for a separate grant that awards one scholarship. All things being equal, determine the chance that my daughter wins at least one scholarship. *Hint:* First compute the probability that the daughter does not win a scholarship from either grant.

Extending the Concepts and Skills

4.200 General Multiplication Rule Extended. For three events, say, A, B, and C, the general multiplication rule is

$$P(A \& B \& C) = P(A) \cdot P(B \mid A) \cdot P(C \mid (A \& B)).$$

a. Suppose that three cards are randomly selected without replacement from an ordinary deck of 52 cards. Find the probability that all three cards are hearts; the first two cards are hearts and the third is a spade.
b. State the general multiplication rule for four events.

4.201 Gender of Students. In Example 4.21 on page 194, we discussed randomly selecting, without replacement, two students from Professor Weiss's introductory statistics class. Suppose now that three students are selected without replacement. What is the probability that the first two students chosen are female and the third is male? (*Hint:* Refer to Exercise 4.200.)

4.202 Calculus Pretest. Students are given three chances to pass a basic skills exam for permission to enroll in Calculus I. Sixty percent of the students pass on the first try; of those that fail on the first try, 54% pass on the second try; and of those remaining, 48% pass on the third try.
a. What is the probability that a student passes on the second try?
b. What is the probability that a student passes on the third try?
c. What percentage of students pass?

4.203 In this exercise, you examine further the concepts of independent events and mutually exclusive events.
a. If two events are mutually exclusive, determine their joint probability.
b. If two nonimpossible (i.e., positive probability) events are independent, explain why their joint probability is not 0.
c. Give an example of two events that are neither mutually exclusive nor independent.

4.204 Independence Extended. Three events A, B, and C are said to be independent if

$$P(A \& B) = P(A) \cdot P(B),$$
$$P(A \& C) = P(A) \cdot P(C),$$
$$P(B \& C) = P(B) \cdot P(C), \text{ and}$$
$$P(A \& B \& C) = P(A) \cdot P(B) \cdot P(C).$$

What is required for four events to be independent? Explain your definition in words.

4.205 Dice. When two balanced dice are rolled, 36 equally likely outcomes are possible, as illustrated in Fig. 4.1 on page 159. Let

A = event the red die comes up even,
B = event the black die comes up even,
C = event the sum of the dice is even,
D = event the red die comes up 1, 2, or 3,
E = event the red die comes up 3, 4, or 5, and
F = event the sum of the dice is 5.

Apply the definition of independence for three events stated in Exercise 4.204 to solve each problem.
a. Are A, B, and C independent events?
b. Show that $P(D \& E \& F) = P(D) \cdot P(E) \cdot P(F)$ but that D, E, and F are not independent events.

4.206 Coin Tossing. When a balanced coin is tossed four times, 16 equally likely outcomes are possible, as shown in the following table.

HHHH	THHH	THHT	THTT
HHHT	HHTT	THTH	TTHT
HHTH	HTHT	TTHH	TTTH
HTHH	HTTH	HTTT	TTTT

Let

A = event the first toss is heads,

B = event the second toss is tails, and

C = event the last two tosses are heads.

Apply the definition of independence for three events stated in Exercise 4.204 to show that A, B, and C are independent events.

4.7 Bayes's Rule*

In this section, we discuss *Bayes's rule*, which was developed by Thomas Bayes, an eighteenth-century clergyman. One of the primary uses of Bayes's rule is to revise probabilities in accordance with newly acquired information. Such revised probabilities are actually conditional probabilities, and so, in some sense, we have already examined much of the material in this section. However, as you will see, application of Bayes's rule involves some new concepts and the use of some new techniques.

The Rule of Total Probability

In preparation for discussion of Bayes's rule, we need to study another rule of probability called the *rule of total probability*. First, we consider the concept of *exhaustive events*. Events $A_1, A_2, \ldots, A_k$ are said to be **exhaustive events** if one or more of them must occur.

For instance, the National Governors Association classifies governors as Democrat, Republican, or Independent. Suppose that a governor is selected at random; let E_1, E_2, and E_3 denote the events that the governor selected is a Democrat, a Republican, and an Independent, respectively. Then events E_1, E_2, and E_3 are exhaustive because at least one of them must occur when a governor is selected—the governor selected must be a Democrat, a Republican, or an Independent.

The events E_1, E_2, and E_3 are not only exhaustive, but they are also mutually exclusive; a governor cannot have more than one political party affiliation at the same time. In general, if events are both exhaustive and mutually exclusive, exactly one of them must occur. This statement is true because at least one of the events must occur (the events are exhaustive) and at most one of the events can occur (the events are mutually exclusive).

An event and its complement are always mutually exclusive and exhaustive. Figure 4.26(a) portrays three events, A_1, A_2, and A_3, that are both mutually exclusive and exhaustive. Note that the three events do not overlap, indicating that they are mutually exclusive; furthermore, they fill out the entire region enclosed by the heavy rectangle (the sample space), indicating that they are exhaustive.

FIGURE 4.26

(a) Three mutually exclusive and exhaustive events; (b) an event B and three mutually exclusive and exhaustive events

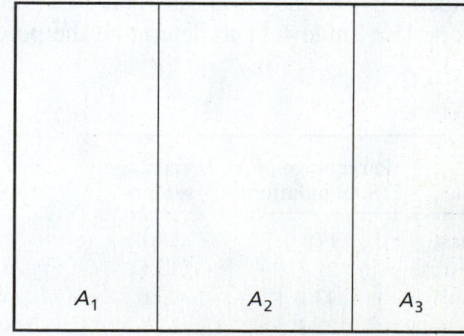

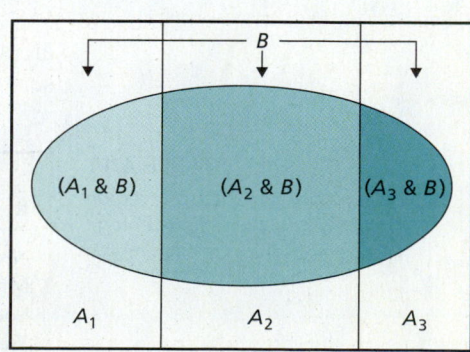

(a)

(b)

Now consider, say, three mutually exclusive and exhaustive events, A_1, A_2, and A_3, and any event B, as shown in Fig. 4.26(b). Note that event B comprises the mutually exclusive events (A_1 & B), (A_2 & B), and (A_3 & B), which are shown in color. This condition means that event B must occur in conjunction with exactly one of the events, A_1, A_2, or A_3.

If you think of the colored regions in Fig. 4.26(b) as probabilities, the total colored region is $P(B)$ and the three colored subregions are, from left to right, $P(A_1$ & $B)$, $P(A_2$ & $B)$, and $P(A_3$ & $B)$. Because events (A_1 & B), (A_2 & B), and (A_3 & B) are mutually exclusive, the total colored region equals the sum of the three colored subregions; in other words,

$$P(B) = P(A_1 \text{ \& } B) + P(A_2 \text{ \& } B) + P(A_3 \text{ \& } B).$$

Applying the general multiplication rule (Formula 4.5 on page 193) to each term on the right side of this equation, we obtain

$$P(B) = P(A_1) \cdot P(B \mid A_1) + P(A_2) \cdot P(B \mid A_2) + P(A_3) \cdot P(B \mid A_3).$$

This formula holds in general and is called the **rule of total probability,** which we express as Formula 4.8. It is also referred to as the **stratified sampling theorem** because of its importance in stratified sampling.

FORMULA 4.8

What Does It Mean?

Let $A_1, A_2, \ldots, A_k$ be mutually exclusive and exhaustive events. Then the probability of an event B can be obtained by multiplying the probability of each A_j by the conditional probability of B given A_j and then summing those products.

The Rule of Total Probability

Suppose that events $A_1, A_2, \ldots, A_k$ are mutually exclusive and exhaustive; that is, exactly one of the events must occur. Then for any event B,

$$P(B) = \sum_{j=1}^{k} P(A_j) \cdot P(B \mid A_j).$$

We apply the rule of total probability in the next example.

EXAMPLE 4.24   The Rule of Total Probability

U.S. Demographics   The U.S. Census Bureau presents data on age of residents and region of residence in *Demographic Profiles*. The first two columns of Table 4.11 give a percentage distribution for region of residence; the third column shows the percentage of seniors (age 65 years or over) in each region. For instance, 17.9% of U.S. residents live in the Northeast, and 14.1% of those who live in the Northeast are seniors. Use Table 4.11 to determine the percentage of U.S. residents that are seniors.

TABLE 4.11

Percentage distribution for region of residence and percentage of seniors in each region

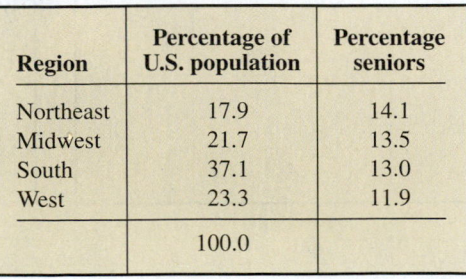

Region	Percentage of U.S. population	Percentage seniors
Northeast	17.9	14.1
Midwest	21.7	13.5
South	37.1	13.0
West	23.3	11.9
	100.0	

Solution To solve this problem, we first translate the information displayed in Table 4.11 into the language of probability. Suppose that a U.S. resident is selected at random. Let

$$S = \text{event the resident selected is a senior,}$$

and

$$R_1 = \text{event the resident selected lives in the Northeast,}$$
$$R_2 = \text{event the resident selected lives in the Midwest,}$$
$$R_3 = \text{event the resident selected lives in the South, and}$$
$$R_4 = \text{event the resident selected lives in the West.}$$

TABLE 4.12

Probabilities derived from Table 4.11

$P(R_1) = 0.179$ $P(S \mid R_1) = 0.141$
$P(R_2) = 0.217$ $P(S \mid R_2) = 0.135$
$P(R_3) = 0.371$ $P(S \mid R_3) = 0.130$
$P(R_4) = 0.233$ $P(S \mid R_4) = 0.119$

The percentages shown in the second and third columns of Table 4.11 translate into the probabilities displayed in Table 4.12.

The problem is to determine the percentage of U.S. residents that are seniors, or, in terms of probability, $P(S)$. Because a U.S. resident must reside in exactly one of the four regions, events R_1, R_2, R_3, and R_4 are mutually exclusive and exhaustive. Therefore, by the rule of total probability applied to the event S and from Table 4.12, we have

$$P(S) = \sum_{j=1}^{4} P(R_j) \cdot P(S \mid R_j)$$

$$= 0.179 \cdot 0.141 + 0.217 \cdot 0.135 + 0.371 \cdot 0.130 + 0.233 \cdot 0.119$$

$$= 0.130.$$

A tree diagram for this calculation is shown in Fig. 4.27, where J represents the event that the resident selected is not a senior. We obtain $P(S)$ from the tree diagram by first multiplying the two probabilities on each branch of the tree that ends with S (the colored branches) and then summing all those products.

FIGURE 4.27

Tree diagram for calculating $P(S)$, using the rule of total probability

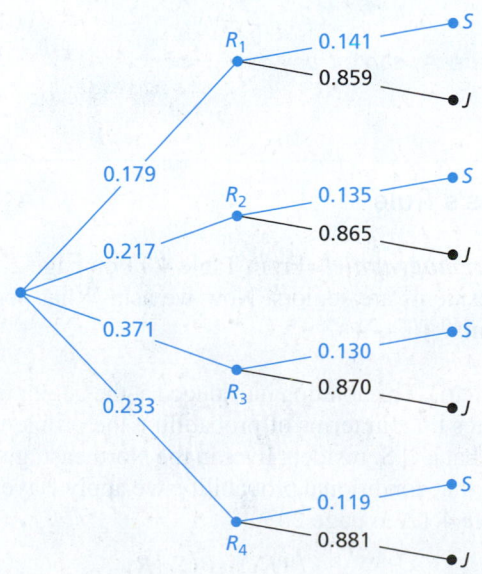

You try it!

**Exercise 4.219(a)
on page 206**

In any case, we see that $P(S) = 0.130$; the probability is 0.130 that a randomly selected U.S. resident is a senior.

Interpretation 13.0% of U.S. residents are seniors.

Bayes's Rule

Using the rule of total probability, we can derive Bayes's rule. For simplicity, let's consider three events, A_1, A_2, and A_3, that are mutually exclusive and exhaustive and let B be any event. For Bayes's rule, we assume that the probabilities $P(A_1)$, $P(A_2)$, $P(A_3)$, $P(B \mid A_1)$, $P(B \mid A_2)$, and $P(B \mid A_3)$ are known. The problem is to use those six probabilities to determine the conditional probabilities $P(A_1 \mid B)$, $P(A_2 \mid B)$, and $P(A_3 \mid B)$.

We now show how to express $P(A_2 \mid B)$ in terms of the six known probabilities; $P(A_1 \mid B)$ and $P(A_3 \mid B)$ are handled similarly. First, we apply the conditional probability rule (Formula 4.4 on page 188), to write

$$P(A_2 \mid B) = \frac{P(B \ \& \ A_2)}{P(B)} = \frac{P(A_2 \ \& \ B)}{P(B)}. \tag{4.1}$$

Next, to the fraction on the right in Equation (4.1), we apply the general multiplication rule (Formula 4.5 on page 193) to the numerator, giving

$$P(A_2 \ \& \ B) = P(A_2) \cdot P(B \mid A_2),$$

and the rule of total probability to the denominator, giving

$$P(B) = P(A_1) \cdot P(B \mid A_1) + P(A_2) \cdot P(B \mid A_2) + P(A_3) \cdot P(B \mid A_3).$$

Substituting these results into the right-hand fraction in Equation (4.1) gives

$$P(A_2 \mid B) = \frac{P(A_2) \cdot P(B \mid A_2)}{P(A_1) \cdot P(B \mid A_1) + P(A_2) \cdot P(B \mid A_2) + P(A_3) \cdot P(B \mid A_3)}.$$

This formula holds in general and is called **Bayes's rule.**

FORMULA 4.9

Bayes's Rule

Suppose that events A_1, A_2, ..., A_k are mutually exclusive and exhaustive. Then for any event B,

$$P(A_i \mid B) = \frac{P(A_i) \cdot P(B \mid A_i)}{\sum_{j=1}^{k} P(A_j) \cdot P(B \mid A_j)},$$

where A_i can be any one of events A_1, A_2, ..., A_k.

■■■ **EXAMPLE 4.25**   Bayes's Rule

U.S. Demographics   From Table 4.11 on page 202, we know that 14.1% of Northeast residents are seniors. Now we ask: What percentage of seniors are Northeast residents?

Solution   The notation introduced at the beginning of the solution to Example 4.24 indicates that, in terms of probability, the problem is to find $P(R_1 \mid S)$—the probability that a U.S. resident lives in the Northeast, given that the resident is a senior. To obtain that conditional probability, we apply Bayes's rule to the probabilities shown in Table 4.12 on page 203:

$$\begin{aligned}
P(R_1 \mid S) &= \frac{P(R_1) \cdot P(S \mid R_1)}{\sum_{j=1}^{4} P(R_j) \cdot P(S \mid R_j)} \\
&= \frac{0.179 \cdot 0.141}{0.179 \cdot 0.141 + 0.217 \cdot 0.135 + 0.371 \cdot 0.130 + 0.233 \cdot 0.119} \\
&= 0.193.
\end{aligned}$$

You try it!

Exercise 4.219(b)
on page 206

Interpretation   19.3% of seniors are Northeast residents. ■

EXAMPLE 4.26   Bayes's Rule

Smoking and Lung Disease According to the Arizona Chapter of the American Lung Association, 7.0% of the population has lung disease. Of those having lung disease, 90.0% are smokers; of those not having lung disease, 25.3% are smokers. Determine the probability that a randomly selected smoker has lung disease.

Solution Suppose that a person is selected at random. Let

$$S = \text{event the person selected is a smoker,}$$

and

$$L_1 = \text{event the person selected has no lung disease, and}$$
$$L_2 = \text{event the person selected has lung disease.}$$

Note that events L_1 and L_2 are complementary, which implies that they are mutually exclusive and exhaustive.

The data given in the statement of the problem indicate that $P(L_2) = 0.070$, $P(S \mid L_2) = 0.900$, and $P(S \mid L_1) = 0.253$. Also, $L_1 = (\text{not } L_2)$, so we can conclude that $P(L_1) = P(\text{not } L_2) = 1 - P(L_2) = 1 - 0.070 = 0.930$. We summarize this information in Table 4.13.

The problem is to determine the probability that a randomly selected smoker has lung disease, $P(L_2 \mid S)$. Applying Bayes's rule to the probability data in Table 4.13, we obtain

$$P(L_2 \mid S) = \frac{P(L_2) \cdot P(S \mid L_2)}{P(L_1) \cdot P(S \mid L_1) + P(L_2) \cdot P(S \mid L_2)}$$

$$= \frac{0.070 \cdot 0.900}{0.930 \cdot 0.253 + 0.070 \cdot 0.900} = 0.211.$$

The probability is 0.211 that a randomly selected smoker has lung disease.

Interpretation 21.1% of smokers have lung disease.

TABLE 4.13
Known probability information

$P(L_1) = 0.930$	$P(S \mid L_1) = 0.253$
$P(L_2) = 0.070$	$P(S \mid L_2) = 0.900$

You try it!

Exercise 4.223 on page 207

Example 4.26 shows that the rate of lung disease among smokers (21.1%) is more than three times the rate among the general population (7.0%). Using arguments similar to those in Example 4.26, we can show that the probability is 0.010 that a randomly selected nonsmoker has lung disease; in other words, 1.0% of nonsmokers have lung disease.

Hence the rate of lung disease among smokers (21.1%) is more than 20 times that among nonsmokers (1.0%). Because this study is observational, however, we cannot conclude that smoking causes lung disease; we can only infer that a strong positive association exists between smoking and lung disease.

Prior and Posterior Probabilities

Two important terms associated with Bayes's rule are *prior probability* and *posterior probability*. In Example 4.26, we saw that the probability is 0.070 that a randomly selected person has lung disease: $P(L_2) = 0.070$. This probability does not take into consideration whether the person is a smoker. It is therefore called a **prior probability** because it represents the probability that the person selected has lung disease *before* knowing whether the person is a smoker.

Now suppose that the person selected is found to be a smoker. Using this additional information, we can revise the probability that the person has lung disease. We do so by determining the conditional probability that the person selected has lung disease, given that the person selected is a smoker: $P(L_2 \mid S) = 0.211$ (from Example 4.26). This revised probability is called a **posterior probability** because it represents the probability that the person selected has lung disease *after* we learn that the person is a smoker.

Exercises 4.7

Understanding the Concepts and Skills

4.207 What does it mean for four events to be exhaustive?

4.208 What does it mean for four events to be mutually exclusive?

4.209 Are exhaustive events necessarily mutually exclusive? Are mutually exclusive events necessarily exhaustive? Explain your answers.

4.210 Explain why an event and its complement are always mutually exclusive and exhaustive.

In each of Exercises 4.211–4.214, the events A_1, A_2, ... are mutually exclusive and exhaustive. We have provided $P(A_1)$, $P(A_2)$, ... and $P(B \mid A_1)$, $P(B \mid A_2)$, ... In each case,
a. use the rule of total probability to find $P(B)$.
b. apply Bayes's rule to find $P(A_1 \mid B)$.

4.211

j	$P(A_j)$	$P(B \mid A_j)$
1	0.4	0.8
2	0.6	0.7

4.212

j	$P(A_j)$	$P(B \mid A_j)$
1	0.3	0.2
2	0.7	0.6

4.213

j	$P(A_j)$	$P(B \mid A_j)$
1	0.5	0.4
2	0.2	0.8
3	0.3	0.6

4.214

j	$P(A_j)$	$P(B \mid A_j)$
1	0.2	0.7
2	0.3	0.4
3	0.1	0.8
4	0.4	0.3

4.215 U.S. Demographics. Refer to Example 4.24 on pages 202–203. In probability notation, the percentage of Midwest residents can be expressed as $P(R_2)$. Do the same for the percentage of
a. Southern residents.
b. Southern residents who are seniors.
c. seniors who are Southern residents.

Applying the Concepts and Skills

4.216 Playing Golf. From the National Golf Foundation and the U.S. Census Bureau, we obtained the following statistics for Americans: 15.1% of males and 4.3% of females play golf, respectively; 49.2% and 50.8% of the population are male and female, respectively. Find the probability that a randomly selected American
a. plays golf.
b. plays golf, given that the person is a male.
c. is a female, given that the person plays golf.
d. Interpret your answers in parts (a)–(c) in terms of percentages.

4.217 Belief in Extraterrestrial Aliens. According to an Opinion Dynamics Poll published in *USA TODAY*, roughly 54% of U.S. men and 33% of U.S. women believe in extraterrestrial aliens. Of U.S. adults, roughly 48% are men and 52% women.
a. What percentage of U.S. adults believe in such aliens?
b. What percentage of U.S. women believe in such aliens?
c. What percentage of U.S. adults that believe in such aliens are women?

4.218 Moviegoers. A *moviegoer* is defined as a person who goes to the movies at least once a month. From the article *Hollywood Demographics* on the Indiewire website, we obtained the following data on moviegoers in the United States and Canada. The first two columns provide an age distribution; the third column gives the percentage of people in each age group who are moviegoers.

Age group (years)	Percentage of population	Percentage moviegoers
2–11	14	11
12–17	8	13
18–24	10	12
25–39	21	11
40–49	14	10
50–59	14	9
60 & over	19	7

A person from among those living in either the United States or Canada is selected at random.
a. Find the probability that the person is a moviegoer.
b. Find the probability that the person is between 18 and 24 years old, given that he or she is a moviegoer.
c. Interpret your answers in parts (a) and (b) in terms of percentages.

4.219 Education and Astrology. The following table provides statistics found in the document *Science and Engineering Indicators*, issued by the National Science Foundation, for a sample of 1564 adults. The first two columns of the table present an educational-level distribution for the adults; the third column gives the percentage of the adults in each educational-level category who read an astrology report every day.

Educational level	Percentage of adults	Percentage astrology
Less than high school	7.4	9.0
High school graduate	53.3	7.0
Baccalaureate or higher	39.3	4.0

For one of these adults selected at random, determine the probability that he or she
a. reads an astrology report every day.
b. is not a high school graduate, given that he or she reads an astrology report every day.
c. holds a baccalaureate degree or higher, given that he or she reads an astrology report every day.

4.220 Chronic Illness and Participation. The aim of a study in the article, "Separate and Joint Effects of Physical and Mental Health on Participation of People with Somatic Chronic Illness" (*Journal of Nursing and Healthcare of Chronic Illness*, Vol. 3, Issue 1, pp. 61–72), by D. Jansen and M. Rijken, is to examine the extent to which people with a chronic illness participate in activities such as jobs, volunteer work, or other social activities. In the study, a sample of 1456 chronic illness patients resulted in the statistics provided in the following table. The first two columns of the table present a percentage distribution for the degree of urbanism of the respondent's home address; the third column gives the percentage of the respondents in each degree of urbanism category who participated in social activities such as sports, church, or clubs.

Degree of urbanization	Percentage of respondents	Percentage social participation
Extreme	10.1	49
Strong	29.9	53
Moderate	21.3	59
Weak	30.8	64
None	7.9	65

For one of these respondents selected at random, determine the probability that he or she
a. participates in social activities.
b. lives in an extremely urbanized environment, given that he or she participates in social activities.
c. does not live in an extremely urbanized environment, given that he or she participates in social activities.
d. Interpret your answers in parts (a)–(c) in terms of percentages.

4.221 Obesity and Age. A person is said to be *overweight* if his or her body mass index (BMI) is between 25 and 29, inclusive; a person is said to be obese if his or her BMI is 30 or greater. From the document *Utah Behavioral Risk Factor Surveillance System (BRFSS) Local Health District Report*, issued by the Utah Department of Health, we obtained the following table. The first two columns of the table provide an age distribution for adults living in Utah. The third column gives the percentage of adults in each age group who are either obese or overweight.

Age (yr)	Percentage of adults	Percentage obese or overweight
18–34	42.5	41.1
35–49	28.5	57.9
50–64	16.4	68.2
65 & over	12.6	55.3

a. What percentage of Utah adults are overweight or obese?
b. Of those Utah adults who are between 35 and 49 years old, inclusive, what percentage are overweight or obese?
c. Of those Utah adults who are overweight or obese, what percentage are between 35 and 49 years old, inclusive?
d. Interpret your answers to parts (a)–(c) in terms of percentages.

4.222 Corporations and Elections. Based on research by the Associated Press and the National Constitution Center, 85% of Democrats, 81% of Republicans, and 78% of Independents support limiting corporate influence in political elections. Estimates from Gallup show that 32% of U.S. adults are Democrats, 28% are Republicans, and 40% are Independents. What percentage of U.S. adults who support limiting corporate influence in political elections are Republicans?

4.223 Textbook Revision. Textbook publishers must estimate the sales of new (first-edition) books. The records of one major publishing company indicate that 10% of all new books sell more than projected, 30% sell close to the number projected, and 60% sell less than projected. Of those that sell more than projected, 70% are revised for

a second edition, as are 50% of those that sell close to the number projected and 20% of those that sell less than projected.
a. What percentage of books published by this publishing company go to a second edition?
b. What percentage of books published by this publishing company that go to a second edition sold less than projected in their first edition?

Extending the Concepts and Skills

4.224 Broken Eggs. At a grocery store, eggs come in cartons that hold a dozen eggs. Experience indicates that 78.5% of the cartons have no broken eggs, 19.2% have one broken egg, 2.2% have two broken eggs, and 0.1% have three broken eggs, and that the percentage of cartons with four or more broken eggs is negligible. An egg selected at random from a carton is found to be broken. What is the probability that this egg is the only broken one in the carton?

4.225 Pap Tests. Medical tests are frequently used to decide whether a person has a particular illness. The *sensitivity* of a test is the probability that a person having the illness will test positive; the *specificity* of a test is the probability that a person not having the illness will test negative. According to the National Cancer Institute and the *Surveillance Epidemiology and End Results Fact Sheet*, roughly 7.9 per 100,000 women have cancer of the cervix. A Pap test, or Papanicolaou test, is a screening test used to detect precancerous and cancerous cells in the cervix. The specificity of the Pap test is 98% and the sensitivity of the Pap test is 51% according to the *Alliance for Cervical Cancer Protection*.
a. Interpret the sensitivity and specificity of this test in terms of percentages.
b. Determine the probability that a randomly selected female tests positive.
c. Determine the probability that a female testing positive actually has cervical cancer.
d. Interpret your answers from parts (b) and (c) in terms of percentages.

4.226 Monty Hall Problem. Several years ago, in a column published by Marilyn vos Savant in *Parade Magazine*, an interesting probability problem was posed. That problem is now referred to as the Monty Hall Problem because of its origins from the television show *Let's Make a Deal*. Following is a version of the Monty Hall Problem. On a game show, there are three doors, behind each of which is one prize. Two of the prizes are worthless and one is valuable. A contestant selects one of the doors, following which, the game-show host—who knows where the valuable prize lies—opens one of the remaining two doors to reveal a worthless prize. The host then offers the contestant the opportunity to change his or her selection. Should the contestant switch? Verify your answer.

4.227 Red and Black. You have two cards. One is red on both sides, and the other is red on one side and black on the other. After shuffling the cards behind your back, you select one of them at random and place it on your desk with your hand covering it. Upon lifting your hand, you observe that the face showing is red.
a. What is the probability that the other side is red?
b. Provide an intuitive explanation for the result in part (a).

4.228 Smoking and Lung Disease. Refer to Example 4.26 on page 205.
a. Determine the probability that a randomly selected nonsmoker has lung disease.
b. Use the probability obtained in part (a) and the result of Example 4.26 to compare the rates of lung disease for smokers and nonsmokers.

4.8   Counting Rules*

We often need to determine the number of ways something can happen—the number of possible outcomes for an experiment, the number of ways an event can occur, the number of ways a certain task can be performed, and so on. Sometimes, we can list the possibilities and count them, but, usually, doing so is impractical.

Therefore we need to develop techniques that do not rely on a direct listing for determining the number of ways something can happen. Such techniques are called **counting rules.** In this section, we examine some widely used counting rules.

The Basic Counting Rule

The **basic counting rule** (**BCR**), which we introduce next, is fundamental to all the counting techniques we discuss.

EXAMPLE 4.27   **Introducing the Basic Counting Rule**

Home Models and Elevations   Robson Communities, Inc. builds new-home communities in several parts of the western United States. In one subdivision, it offers four models—the Shalimar, Palacia, Valencia, and Monterey—each in three different elevations, designated A, B, and C. How many choices are there for the selection of a home, including both model and elevation?

Solution   We first use a tree diagram (see Fig. 4.28) to obtain systematically a direct listing of the possibilities. We use S for Shalimar, P for Palacia, V for Valencia, and M for Monterey.

FIGURE 4.28

Tree diagram for model and elevation possibilities

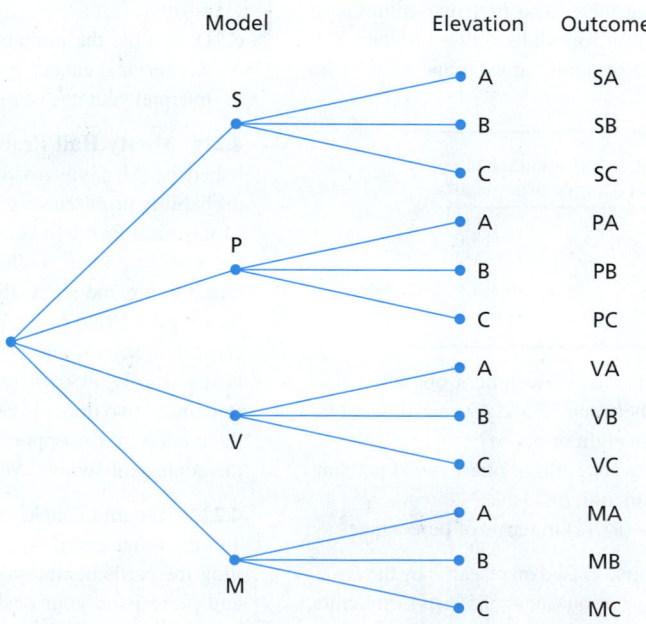

Each branch of the tree corresponds to one possibility for model and elevation. For instance, the first branch of the tree, ending in SA, corresponds to the Shalimar model with the A elevation. We can find the total number of possibilities by counting the number of branches, which is 12.

The tree-diagram approach also provides a clue for finding the number of possibilities without resorting to a direct listing. Specifically, there are four possibilities for model, indicated by the four subbranches emanating from the starting point of the tree; corresponding to each possibility for model are three possibilities for elevation, indicated by the three subbranches emanating from the end of each model subbranch. Consequently, there are

$$\underbrace{3+3+3+3}_{4 \text{ times}} = 4 \cdot 3 = 12$$

You try it!

Exercise 4.245(a)–(b) on page 216

possibilities altogether. Thus we can obtain the total number of possibilities by multiplying the number of possibilities for the model by the number of possibilities for the elevation.

The same multiplication principle applies regardless of the number of actions. We state this principle more precisely in the following key fact.

KEY FACT 4.2

What Does It Mean?

The total number of ways that several actions can occur equals the product of the individual number of ways for each action.

The Basic Counting Rule (BCR)[†]

Suppose that r actions are to be performed in a definite order. Further suppose that there are m_1 possibilities for the first action and that corresponding to each of these possibilities are m_2 possibilities for the second action, and so on. Then there are $m_1 \cdot m_2 \cdots m_r$ possibilities altogether for the r actions.

In Example 4.27 there are two actions ($r = 2$)—selecting a model and selecting an elevation. Because there are four possibilities for model, $m_1 = 4$, and because corresponding to each model are three possibilities for elevation, $m_2 = 3$. Therefore, by the BCR, the total number of possibilities, including both model and elevation, again is

$$m_1 \cdot m_2 = 4 \cdot 3 = 12.$$

Because the number of possibilities in the model/elevation problem is small, determining the number by a direct listing is relatively simple. It is even easier, however, to find the number by applying the BCR. Moreover, in problems having a large number of possibilities, the BCR is the only practical way to proceed.

EXAMPLE 4.28   **The Basic Counting Rule**

License Plates   The license plates of a state consist of three letters followed by three digits.

a. How many different license plates are possible?
b. How many possibilities are there for license plates on which no letter or digit is repeated?

Solution   For both parts (a) and (b), we apply the BCR with six actions ($r = 6$).

a. There are 26 possibilities for the first letter, 26 for the second, and 26 for the third; there are 10 possibilities for the first digit, 10 for the second, and 10 for the third. Applying the BCR gives

$$m_1 \cdot m_2 \cdot m_3 \cdot m_4 \cdot m_5 \cdot m_6 = 26 \cdot 26 \cdot 26 \cdot 10 \cdot 10 \cdot 10 = 17,576,000$$

possibilities for different license plates. Obviously, finding the number of possibilities by a direct listing would be impractical—the tree diagram would have 17,576,000 branches!

b. Again, there are 26 possibilities for the first letter. However, for each possibility for the first letter, there are 25 corresponding possibilities for the second letter because the second letter cannot be the same as the first, and for each possibility for the first two letters, there are 24 corresponding possibilities for the third letter because the third letter cannot be the same as either the first or the second.

[†]The basic counting rule is also known as the *basic principle of counting,* the *fundamental counting rule,* and the *multiplication rule.*

Exercise 4.245(c)
on page 216

Similarly, there are 10 possibilities for the first digit, 9 for the second, and 8 for the third. So by the BCR, there are

$$m_1 \cdot m_2 \cdot m_3 \cdot m_4 \cdot m_5 \cdot m_6 = 26 \cdot 25 \cdot 24 \cdot 10 \cdot 9 \cdot 8 = 11{,}232{,}000$$

possibilities for license plates on which no letter or digit is repeated.   ■

Factorials

Before continuing our presentation of counting rules, we need to discuss *factorials*.

DEFINITION 4.8

Factorials

The product of the first k positive integers (counting numbers) is called **k factorial** and is denoted **$k!$**. In symbols,

$$k! = k(k-1)\cdots 2 \cdot 1.$$

We also define $0! = 1$.

■ ■ ■   **EXAMPLE 4.29**   Factorials

Determine 3!, 4!, and 5!.

Exercise 4.235
on page 215

Solution   Applying Definition 4.8 gives $3! = 3 \cdot 2 \cdot 1 = 6$, $4! = 4 \cdot 3 \cdot 2 \cdot 1 = 24$, and $5! = 5 \cdot 4 \cdot 3 \cdot 2 \cdot 1 = 120$.   ■

Notice that $6! = 6 \cdot 5!$, $6! = 6 \cdot 5 \cdot 4!$, $6! = 6 \cdot 5 \cdot 4 \cdot 3!$, and so on. In general, if $j \le k$, then $k! = k(k-1)\cdots(k-j+1)(k-j)!$.

Permutations

A **permutation** of r objects from a collection of m objects is any *ordered* arrangement of r of the m objects. The number of possible permutations of r objects that can be formed from a collection of m objects is denoted $_m P_r$ (read "m permute r").[†]

■ ■ ■   **EXAMPLE 4.30**   Introducing Permutations

Arrangement of Letters   Consider the collection of objects consisting of the five letters a, b, c, d, e.

a.   List all possible permutations of three letters from this collection of five letters.
b.   Use part (a) to determine the number of possible permutations of three letters that can be formed from the collection of five letters; that is, find $_5 P_3$.
c.   Use the BCR to determine the number of possible permutations of three letters that can be formed from the collection of five letters; that is, find $_5 P_3$ by using the BCR.

Solution

a.   The list of all possible permutations (ordered arrangements) of three letters from the five letters is shown in Table 4.14.

TABLE 4.14

Possible permutations of three letters from the collection of five letters

abc	abd	abe	acd	ace	ade	bcd	bce	bde	cde
acb	adb	aeb	adc	aec	aed	bdc	bec	bed	ced
bac	bad	bae	cad	cae	dae	cbd	cbe	dbe	dce
bca	bda	bea	cda	cea	dea	cdb	ceb	deb	dec
cab	dab	eab	dac	eac	ead	dbc	ebc	ebd	ecd
cba	dba	eba	dca	eca	eda	dcb	ecb	edb	edc

[†]Other notations used for the number of possible permutations include P_r^m and $(m)_r$.

b. Table 4.14 indicates that there are 60 possible permutations of three letters from the collection of five letters; in other words, $_5P_3 = 60$.

c. There are five possibilities for the first letter, four possibilities for the second letter, and three possibilities for the third letter. Hence, by the BCR, there are

$$m_1 \cdot m_2 \cdot m_3 = 5 \cdot 4 \cdot 3 = 60$$

possibilities altogether, again giving $_5P_3 = 60$.

We can make two observations from Example 4.30. First, listing all possible permutations is generally tedious or impractical. Second, listing all possible permutations is not necessary in order to determine how many there are—we can use the BCR to count them.

Part (c) of Example 4.30 reveals that we can use the BCR to obtain a general formula for $_mP_r$. Specifically, $_mP_r = m(m-1) \cdots (m - r + 1)$. Multiplying and dividing the right side of this formula by $(m - r)!$, we get the equivalent expression $_mP_r = m!/(m - r)!$. This formula is called the **permutations rule**.

FORMULA 4.10

You try it!

Exercise 4.237 on page 215

The Permutations Rule

The number of possible permutations of r objects from a collection of m objects is given by the formula

$$_mP_r = \frac{m!}{(m - r)!}.$$

EXAMPLE 4.31 | **The Permutations Rule**

Exacta Wagering In an exacta wager at the race track, a bettor picks the two horses that he or she thinks will finish first and second in a specified order. For a race with 12 entrants, determine the number of possible exacta wagers.

Solution Selecting two horses from the 12 horses for an exacta wager is equivalent to specifying a permutation of two objects from a collection of 12 objects. The first object is the horse selected to finish in first place, and the second object is the horse selected to finish in second place.

Thus the number of possible exacta wagers is $_{12}P_2$—the number of possible permutations of two objects from a collection of 12 objects. Applying the permutations rule, with $m = 12$ and $r = 2$, we obtain

$$_{12}P_2 = \frac{12!}{(12-2)!} = \frac{12!}{10!} = \frac{12 \cdot 11 \cdot \cancel{10!}}{\cancel{10!}} = 12 \cdot 11 = 132.$$

You try it!

Exercise 4.253 on page 216

Interpretation In a 12-horse race, there are 132 possible exacta wagers.

EXAMPLE 4.32 | **The Permutations Rule**

Arranging Books on a Shelf A student has 10 books to arrange on a shelf of a bookcase. In how many ways can the 10 books be arranged?

Solution Any particular arrangement of the 10 books on the shelf is a permutation of 10 objects from a collection of 10 objects. Hence we need to determine $_{10}P_{10}$, the number of possible permutations of 10 objects from a collection of 10 objects, more commonly expressed as the number of possible permutations of 10 objects among themselves. Applying the permutations rule, we get

$$_{10}P_{10} = \frac{10!}{(10-10)!} = \frac{10!}{0!} = \frac{10!}{1} = 10! = 3{,}628{,}800.$$

Interpretation There are 3,628,800 ways to arrange 10 books on a shelf. ■

Let's generalize Example 4.32 to find the number of possible permutations of m objects among themselves. Using the permutations rule, we conclude that

$$_{m}P_{m} = \frac{m!}{(m-m)!} = \frac{m!}{0!} = \frac{m!}{1} = m!,$$

which is called the **special permutations rule.**

FORMULA 4.11

The Special Permutations Rule

The number of possible permutations of m objects among themselves is $m!$.

Combinations

A **combination** of r objects from a collection of m objects is any *unordered* arrangement of r of the m objects—in other words, any subset of r objects from the collection of m objects. Note that order matters in permutations but not in combinations.

The number of possible combinations of r objects that can be formed from a collection of m objects is denoted $_{m}C_{r}$ (read "m choose r").[†]

■■■ **EXAMPLE 4.33** Introducing Combinations

Arrangement of Letters Consider the collection of objects consisting of the five letters a, b, c, d, e.

a. List all possible combinations of three letters from this collection of five letters.
b. Use part (a) to determine the number of possible combinations of three letters that can be formed from the collection of five letters; that is, find $_{5}C_{3}$.

Solution

TABLE 4.15

Combinations

$\{a,b,c\}$	$\{a,b,d\}$	$\{a,b,e\}$	$\{a,c,d\}$
$\{a,c,e\}$	$\{a,d,e\}$	$\{b,c,d\}$	$\{b,c,e\}$
$\{b,d,e\}$	$\{c,d,e\}$		

a. The list of all possible combinations (unordered arrangements) of three letters from the five letters is shown in Table 4.15.
b. Table 4.15 reveals that there are 10 possible combinations of three letters from the collection of five letters; in other words, $_{5}C_{3} = 10$. ■

In the previous example, we found the number of possible combinations by a direct listing. Let's find a simpler method.

Look back at the first combination in Table 4.15, $\{a, b, c\}$. By the special permutations rule, there are $3! = 6$ permutations of these three letters among themselves; they are abc, acb, bac, bca, cab, and cba. These six permutations are the ones displayed in the first column of Table 4.14 on page 210. Similarly, there are $3! = 6$ permutations of the three letters appearing as the second combination in Table 4.15, $\{a, b, d\}$. These six permutations are the ones displayed in the second column of Table 4.14. The same comments apply to the other eight combinations in Table 4.15.

[†]Other notations used for the number of possible combinations include C_{r}^{m} and $\binom{m}{r}$.

Thus, for each combination of three letters from the collection of five letters, there are 3! corresponding permutations of three letters from the collection of five letters. Because any such permutation is accounted for in this way, there must be 3! times as many permutations as combinations. Equivalently, the number of possible combinations of three letters from the collection of five letters must equal the number of possible permutations of three letters from the collection of five letters divided by 3!. Thus

$$_5C_3 = \frac{_5P_3}{3!} = \frac{5!/(5-3)!}{3!} = \frac{5!}{3!\,(5-3)!} = \frac{5 \cdot 4 \cdot \cancel{3!}}{\cancel{3!}\,2!} = \frac{5 \cdot 4}{2} = 10,$$

which is the number we obtained in Example 4.33 by a direct listing. The same type of argument holds in general and yields the **combinations rule.**

FORMULA 4.12

Exercise 4.239
on page 215

> ### The Combinations Rule
>
> The number of possible combinations of r objects from a collection of m objects is given by the formula
>
> $$_mC_r = \frac{m!}{r!\,(m-r)!}.$$

■■■■   **EXAMPLE 4.34**   **The Combinations Rule**

CD-Club Introductory Offer   To recruit new members, a compact-disc (CD) club advertises a special introductory offer: A new member agrees to buy 1 CD at regular club prices and receives free any 4 CDs of his or her choice from a collection of 69 CDs. How many possibilities does a new member have for the selection of the 4 free CDs?

Solution   Any particular selection of 4 CDs from 69 CDs is a combination of 4 objects from a collection of 69 objects. By the combinations rule, the number of possible selections is

$$_{69}C_4 = \frac{69!}{4!\,(69-4)!} = \frac{69!}{4!\,65!} = \frac{69 \cdot 68 \cdot 67 \cdot 66 \cdot \cancel{65!}}{4!\,\cancel{65!}} = 864{,}501.$$

Exercise 4.257
on page 216

Interpretation   There are 864,501 possibilities for the selection of 4 CDs from a collection of 69 CDs. ■

■■■■   **EXAMPLE 4.35**   **The Combinations Rule**

Sampling Students   An economics professor is using a new method to teach a junior-level course with an enrollment of 42 students. The professor wants to conduct in-depth interviews with the students to get feedback on the new teaching method but does not want to interview all 42 of them. The professor decides to interview a sample of 5 students from the class. How many different samples are possible?

Solution   A sample of 5 students from the class of 42 students can be considered a combination of 5 objects from a collection of 42 objects. By the combinations rule, the number of possible samples is

$$_{42}C_5 = \frac{42!}{5!\,(42-5)!} = \frac{42!}{5!\,37!} = 850{,}668.$$

Interpretation   There are 850,668 possible samples of 5 students from a class of 42 students. ■

Example 4.35 shows how to determine the number of possible samples of a specified size from a finite population. This method is so important that we record it as the following formula.

FORMULA 4.13

Number of Possible Samples

The number of possible samples of size n from a population of size N is $_NC_n$.

Applications to Probability

Suppose that an experiment has N equally likely possible outcomes. Then, according to the f/N rule, the probability that a specified event occurs equals the number of ways, f, that the event can occur divided by the total number of possible outcomes, N.

In the probability problems that we have considered so far, determining f and N has been easy, but that isn't always the case. We must often use counting rules to obtain the number of possible outcomes and the number of ways that the specified event can occur.

EXAMPLE 4.36

Applying Counting Rules to Probability

Quality Assurance The quality assurance engineer of a television manufacturer inspects TVs in lots of 100. He selects 5 of the 100 TVs at random and inspects them thoroughly. Assuming that 6 of the 100 TVs in the current lot are defective, find the probability that exactly 2 of the 5 TVs selected by the engineer are defective.

Solution Because the engineer makes his selection at random, each of the possible outcomes is equally likely. We can therefore apply the f/N rule to find the probability.

First, we determine the number of possible outcomes for the experiment. It is the number of ways that 5 TVs can be selected from the 100 TVs—the number of possible combinations of 5 objects from a collection of 100 objects. Applying the combinations rule yields

$$_{100}C_5 = \frac{100!}{5!\,(100-5)!} = \frac{100!}{5!\,95!} = 75{,}287{,}520,$$

or $N = 75{,}287{,}520$.

Next, we determine the number of ways the specified event can occur, that is, the number of outcomes in which exactly 2 of the 5 TVs selected are defective. To do so, we think of the 100 TVs as partitioned into two groups—namely, the defective TVs and the nondefective TVs, as shown in the top part of Fig. 4.29.

FIGURE 4.29

Calculating the number of outcomes in which exactly 2 of the 5 TVs selected are defective

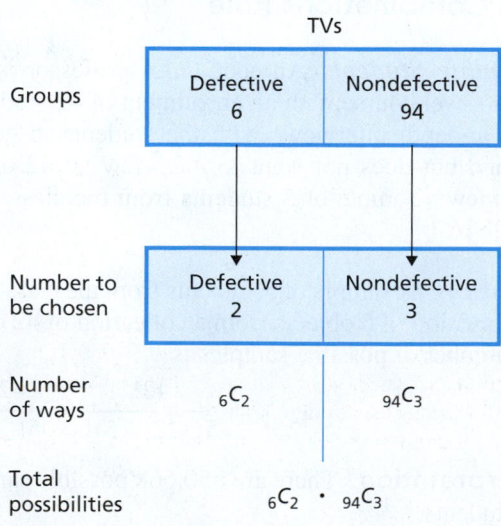

There are 6 TVs in the defective group and 2 are to be selected, which can be done in

$$_6C_2 = \frac{6!}{2!\,(6-2)!} = \frac{6!}{2!\,4!} = 15$$

ways. There are 94 TVs in the nondefective group and 3 are to be selected, which can be done in

$$_{94}C_3 = \frac{94!}{3!\,(94-3)!} = \frac{94!}{3!\,91!} = 134{,}044$$

ways. Consequently, by the BCR, there are a total of

$$_6C_2 \cdot {_{94}C_3} = 15 \cdot 134{,}044 = 2{,}010{,}660$$

outcomes in which exactly 2 of the 5 TVs selected are defective, so $f = 2{,}010{,}660$. Figure 4.29 summarizes these calculations.

Applying the f/N rule, we now conclude that the probability that exactly 2 of the 5 TVs selected are defective is

$$\frac{f}{N} = \frac{2{,}010{,}660}{75{,}287{,}520} = 0.027.$$

Exercise 4.263 on page 217

Interpretation There is a 2.7% chance that exactly 2 of the 5 TVs selected by the engineer will be defective.

Exercises 4.8

Understanding the Concepts and Skills

4.229 What are counting rules? Why are they important?

4.230 Why is the basic counting rule (BCR) often referred to as the multiplication rule?

4.231 Regarding permutations and combinations,
a. what is a permutation?
b. what is a combination?
c. what is the major distinction between the two?

4.232 Identify the notation used for each of the following concepts.
a. k factorial
b. The number of possible permutations of r objects that can be formed from a collection of m objects
c. The number of possible combinations of r objects that can be formed from a collection of m objects

4.233 Three actions are to be performed in a definite order. There are four possibilities for the first action; corresponding to each of these possibilities are five possibilities for the second action; and corresponding to each of these possibilities are two possibilities for the third action. How many possibilities are there altogether for the three actions?

4.234 Four actions are to be performed in a definite order. There are five possibilities for the first action; corresponding to each of these possibilities are two possibilities for the second action; corresponding to each of these possibilities are two possibilities for the third action; and corresponding to each of these possibilities are four possibilities for the fourth action. How many possibilities are there altogether for the four actions?

4.235 Find $0!$, $5!$, and $6!$.

4.236 Find $3!$, $7!$, and $2!$.

4.237 Determine the value of each quantity.
a. $_4P_3$ b. $_{15}P_4$ c. $_6P_2$ d. $_{10}P_0$ e. $_8P_8$

4.238 Determine the value of each quantity.
a. $_7P_3$ b. $_5P_2$ c. $_8P_4$ d. $_6P_0$ e. $_9P_9$

4.239 Determine the value of each quantity.
a. $_4C_3$ b. $_{15}C_4$ c. $_6C_2$ d. $_{10}C_0$ e. $_8C_8$

4.240 Determine the value of each quantity.
a. $_7C_3$ b. $_5C_2$ c. $_8C_4$ d. $_6C_0$ e. $_9C_9$

4.241 *Fill in the blank:* The number of possible permutations of m objects among themselves is _____.

4.242 *Fill in the blank:* The number of possible samples of size n from a population of size N is _____.

4.243 How many possible samples of size 5 are possible from a population of size 70?

Applying the Concepts and Skills

4.244 Home Models and Elevations. Refer to Example 4.27 on pages 208–209. Suppose that the developer discontinues the Shalimar model but provides an additional elevation choice, D, for each of the remaining three model choices.
a. Draw a tree diagram similar to the one shown in Fig. 4.28 depicting the possible choices for the selection of a home, including both model and elevation.

b. Use the tree diagram in part (a) to determine the total number of choices for the selection of a home, including both model and elevation.

c. Use the BCR to determine the total number of choices for the selection of a home, including both model and elevation.

4.245 Home Models and Elevations. Refer to Example 4.27 on pages 208–209. Suppose that the developer provides an additional model choice, called the Nanaimo.

a. Draw a tree diagram similar to the one shown in Fig. 4.28 depicting the possible choices for the selection of a home, including both model and elevation.

b. Use the tree diagram in part (a) to determine the total number of choices for the selection of a home, including both model and elevation.

c. Use the BCR to determine the total number of choices for the selection of a home, including both model and elevation.

4.246 Tesla Model S. There are many choices to make when buying a new car. The options for a Tesla Model S can be found on the Tesla Motors website. For 2013, choices are available, among others, for roof (3), exterior color (9), seat fabric/color (4), interior trim (5), and wheels (5). How many possibilities are there altogether, taking into account choices for the five aforementioned items?

4.247 Zip Codes. The author spoke with a representative of the U.S. Postal Service and obtained the following information about zip codes. A five-digit zip code consists of five digits, of which the first three give the sectional center and the last two the post office or delivery area. In addition to the five-digit zip code, there is a trailing *plus four zip code*. The first two digits of the plus four zip code give the sector or several blocks and the last two the segment or side of the street. For the five-digit zip code, the first four digits can be any of the digits 0–9 and the fifth any of the digits 1–8. For the plus four zip code, the first three digits can be any of the digits 0–9 and the fourth any of the digits 1–9.

a. How many possible five-digit zip codes are there?

b. How many possible plus four zip codes are there?

c. How many possibilities are there in all, including both the five-digit zip code and the plus four zip code?

4.248 Computerized Testing. A statistics professor needs to construct a five-question quiz, one question for each of five topics. The computerized testing system she uses provides eight choices for the question on the first topic, nine choices for the question on the second topic, seven choices for the question on the third topic, eight choices for the question on the fourth topic, and six choices for the question on the fifth topic. How many possibilities are there for the five-question quiz?

4.249 Telephone Numbers. In the United States, telephone numbers consist of a three-digit area code followed by a seven-digit local number. Suppose neither the first digit of an area code nor the first digit of a local number can be a zero but that all other choices are acceptable.

a. How many different area codes are possible?

b. For a given area code, how many local telephone numbers are possible?

c. How many telephone numbers are possible?

4.250 *i* Dolls. An advertisement for *i* Dolls states: "Choose from 69 billion combinations to create a one-of-a-kind doll." The ad goes on to say that there are 39 choices for hairstyle, 19 for eye color, 8 for hair color, 6 for face shape, 24 for lip color, 5 for freckle pattern, 5 for line of clothing, 6 for blush color, and 5 for skin tone. Exactly how many possibilities are there for these options?

4.251 Mutual Fund Investing. Investment firms usually have a large selection of mutual funds from which an investor can choose. One such firm has 30 mutual funds. Suppose that you plan to invest in four of these mutual funds, one during each quarter of next year. In how many different ways can you make these four investments?

4.252 Testing for ESP. An extrasensory perception (ESP) experiment is conducted by a psychologist. For part of the experiment, the psychologist takes 10 cards numbered 1–10 and shuffles them. Then she looks at the cards one at a time. While she looks at each card, the subject writes down the number he thinks is on the card.

a. How many possibilities are there for the order in which the subject writes down the numbers?

b. If the subject has no ESP and is just guessing each time, what is the probability that he writes down the numbers in the correct order, that is, in the order that the cards are actually arranged?

c. Based on your result from part (b), what would you conclude if the subject writes down the numbers in the correct order? Explain your answer.

4.253 Los Angeles Dodgers. From the official website of the 2013 Los Angeles Dodgers major league baseball team, we found that there were five active players on roster available to play outfield. Assuming that these five players could play any outfield position, how many possible assignments could manager Don Mattingly have made for the three outfield positions?

4.254 A Movie Festival. At a movie festival, a team of judges is to pick the first, second, and third place winners from the 18 films entered. How many possibilities are there?

4.255 Assigning Sales Territories. The sales manager of a clothing company needs to assign seven salespeople to seven different territories. How many possibilities are there for the assignments?

4.256 Five-Card Stud. A hand of five-card stud poker consists of an ordered arrangement of five cards from an ordinary deck of 52 playing cards.

a. How many five-card stud poker hands are possible?

b. How many different hands consisting of three kings and two queens are possible?

c. The hand in part (b) is an example of a full house: three cards of one denomination and two of another. How many different full houses are possible?

d. Calculate the probability of being dealt a full house.

4.257 IRS Audits. The Internal Revenue Service (IRS) decides that it will audit the returns of 3 people from a group of 18. Use combination notation to express the number of possibilities and then evaluate that expression.

4.258 A Lottery. At a lottery, 100 tickets were sold and three prizes are to be given. How many possible outcomes are there if

a. the three prizes are equivalent?

b. there is a first, second, and third prize?

4.259 Shake. Ten people attend a party. If each pair of people shakes hands, how many handshakes will occur?

4.260 Championship Series. Professional sports leagues commonly end their seasons with a championship series between two teams. The series ends when one team has won four games and so must last at least four games and at most seven games. How many different sequences of game winners are there in which the series ends in

a. 4 games?   b. 5 games?   c. 6 games?   d. 7 games?

e. Assuming that the two teams are evenly matched, determine the probability of each of the outcomes in parts (a)–(d).

4.261 Five-Card Draw. A hand of five-card draw poker consists of an unordered arrangement of five cards from an ordinary deck of 52 playing cards.

a. How many five-card draw poker hands are possible?

b. How many different hands consisting of three kings and two queens are possible?

c. The hand in part (b) is an example of a full house: three cards of one denomination and two of another. How many different full houses are possible?

d. Calculate the probability of being dealt a full house.

e. Compare your answers in parts (a)–(d) to those in Exercise 4.256.

4.262 Senate Committees. The U.S. Senate consists of 100 senators, 2 from each state. A committee consisting of 5 senators is to be formed.

a. How many different committees are possible?

b. How many are possible if no state may have more than 1 senator on the committee?

c. If the committee is selected at random from all 100 senators, what is the probability that no state will have both of its senators on the committee?

4.263 Venus Throw. The Dr. Fisher's Casebook feature "Codex magistri Piscatori," appearing in an issue of *Significance Magazine* (Vol. 10, Issue 2, p. 28), showed an image, including the probability, of a "Venus throw," one in which the outcome of tossing four balanced dice results in four different numbers. Determine that probability.

4.264 Which Key? Suppose that you have a key ring with eight keys on it, one of which is your house key. Further suppose that you get home after dark and can't see the keys on the key ring. You randomly try one key at a time, being careful not to mix the keys that you've already tried with the ones you haven't. What is the probability that you get the right key

a. on the first try?   b. on the eighth try?

c. on or before the fifth try?

4.265 Quality Assurance. Refer to Example 4.36, which starts on page 214. Determine the probability that the number of defective TVs obtained by the engineer is

a. exactly one.   b. at most one.   c. at least one.

4.266 The Birthday Problem. A biology class has 38 students. Find the probability that at least 2 students in the class have the same birthday. For simplicity, assume that there are always 365 days in a year and that birth rates are constant throughout the year. (*Hint:* First, determine the probability that no 2 students have the same birthday and then apply the complementation rule.)

4.267 Mega Millions. *Mega Millions* is a multi-state jackpot draw game with a jackpot starting at $15 million and growing until someone wins. In order to play, the player selects five white numbers from the numbers 1–75 and one Mega Ball number from the numbers 1–15. *Mega Millions* has nine winning ball combinations summarized by the following table

Match	Prize
Five white balls and Mega Ball	Jackpot
Five white balls	$1,000,000
Four white balls and Mega Ball	$5,000
Four white balls	$500
Three white balls and Mega Ball	$50
Two white balls and Mega Ball	$5
Three white balls	$5
One white ball and Mega Ball	$2
Mega Ball	$1

If you buy one *Mega Millions* ticket, determine the probability (to three significant digits) that you win

a. the jackpot.

b. exactly $5,000.

c. at least $5,000.

d. a prize.

4.268 True–False Tests. A student takes a true–false test consisting of 15 questions. Assume that the student guesses at each question and find the probability that

a. the student gets at least 1 question correct.

b. the student gets a 60% or better on the exam.

Extending the Concepts and Skills

4.269 Florida Battleground State. From the *Washington Post* website, we found final results of the 2012 presidential election. According to that site, Barack Obama received about 50% of the popular vote in the battleground state of Florida. Suppose that 10 Floridians who voted in 2012 are selected at random. Determine the approximate probability that

a. exactly 5 voted for Obama.

b. 8 or more voted for Obama.

c. Even presuming that exactly 50% of the voters in Florida voted for Obama, why would the probabilities in parts (a) and (b) still be only approximately correct?

4.270 Sampling Without Replacement. A simple random sample of size n is to be taken without replacement from a population of size N.

a. Determine the probability that any particular sample of size n is the one selected.

b. Determine the probability that any specified member of the population is included in the sample.

c. Determine the probability that any k specified members of the population are included in the sample.

4.271 The Birthday Problem. Refer to Exercise 4.266, but now assume that the class consists of N students.

a. Determine the probability that at least 2 of the students have the same birthday.

b. If you have access to a computer or a programmable calculator, use it and your answer from part (a) to construct a table giving the probability that at least 2 of the students in the class have the same birthday for $N = 2, 3, \ldots, 70$.

CHAPTER IN REVIEW

You Should Be Able to

1. use and understand the formulas in this chapter.

2. compute probabilities for experiments having equally likely outcomes.

3. interpret probabilities, using the frequentist interpretation of probability.

4. state and understand the basic properties of probability.

5. construct and interpret Venn diagrams.

6. find and describe (not *E*), (*A* & *B*), and (*A* or *B*).

7. determine whether two or more events are mutually exclusive.

8. understand and use probability notation.

9. state and apply the special addition rule.

10. state and apply the complementation rule.

11. state and apply the general addition rule.

*12. read and interpret contingency tables.

*13. construct a joint probability distribution.

*14. compute conditional probabilities both directly and by using the conditional probability rule.

*15. state and apply the general multiplication rule.

*16. state and apply the special multiplication rule.

*17. determine whether two events are independent.

*18. understand the difference between mutually exclusive events and independent events.

*19. determine whether two or more events are exhaustive.

*20. state and apply the rule of total probability.

*21. state and apply Bayes's rule.

*22. state and apply the basic counting rule (BCR).

*23. state and apply the permutations and combinations rules.

*24. apply counting rules to solve probability problems where appropriate.

Key Terms

(*A* & *B*), *166*
(*A* or *B*), *166*
at least, *168*
at most, *168*
at random, *158*
basic counting rule (BCR),* *209*
Bayes's rule,* *204*
bivariate data,* *179*
cells,* *180*
certain event, *160*
combination,* *212*
combinations rule,* *213*
complement, *166*
complementation rule, *174*
conditional probability,* *185*
conditional probability rule,* *188*
contingency table,* *179*
counting rules,* *208*
dependent events,* *196*
equal-likelihood model, *160*

event, *158, 166*
exhaustive events,* *201*
experiment, *158*
f/N rule, *158*
factorials,* *210*
frequentist interpretation of
    probability, *160*
general addition rule, *175*
general multiplication rule,* *193*
given event,* *185*
impossible event, *160*
inclusive, *168*
independence,* *195*
independent events,* *195, 196*
joint probabilities,* *181*
joint probability distribution,* *182*
marginal probabilities,* *181*
mutually exclusive events, *169*
(not *E*), *166*
occurs, *166*

$P(B \mid A),$* *185*
$P(E),$ *173*
permutation,* *210*
permutations rule,* *211*
posterior probability,* *205*
prior probability,* *205*
probability model, *160*
probability theory, *156*
rule of total probability,* *202*
sample space, *166*
special addition rule, *173*
special multiplication rule,* *196*
special permutations rule,* *212*
statistical independence,* *195*
stratified sampling theorem,* *202*
tree diagram,* *194*
two-way table,* *179*
univariate data,* *179*
Venn diagrams, *166*

REVIEW PROBLEMS

Understanding the Concepts and Skills

1. Why is probability theory important to statistics?

2. Regarding the equal-likelihood model,
a. what is it?
b. how are probabilities computed?

3. What meaning is given to the probability of an event by the frequentist interpretation of probability?

4. Decide which of these numbers could not possibly be probabilities. Explain your answers.
a. 0.047   **b.** −0.047   **c.** 3.5   **d.** 1/3.5

5. Identify a commonly used graphical technique for portraying events and relationships among events.

6. What does it mean for two or more events to be mutually exclusive?

7. Suppose that E is an event. Use probability notation to represent
a. the probability that event E occurs.
b. the probability that event E occurs is 0.436.

8. Answer true or false to each statement and explain your answers.
a. For any two events, the probability that one or the other of the events occurs equals the sum of the two individual probabilities.
b. For any event, the probability that it occurs equals 1 minus the probability that it does not occur.

9. Identify one reason why the complementation rule is useful.

*10. Fill in the blanks.
a. Data obtained by observing values of one variable of a population are called _____ data.
b. Data obtained by observing values of two variables of a population are called _____ data.
c. A frequency distribution for bivariate data is called a _____.

*11. The sum of the joint probabilities in a row or column of a joint probability distribution equals the _____ probability in that row or column.

*12. Let A and B be events.
a. Use probability notation to represent the conditional probability that event B occurs, given that event A has occurred.
b. In part (a), which is the given event, A or B?

*13. Identify two possible ways in which conditional probabilities can be computed.

*14. What is the relationship between the joint probability and marginal probabilities of two independent events?

*15. If two or more events have the property that at least one of them must occur when the experiment is performed, the events are said to be _____.

*16. State the basic counting rule (BCR).

17. A, B, and C are mutually exclusive events such that $P(A) = 0.2$, $P(B) = 0.6$, and $P(C) = 0.1$. Find $P(A \text{ or } B \text{ or } C)$.

18. E is an event and $P(\text{not } E) = 0.4$. Find $P(E)$.

19. A and B are events such that $P(A) = 0.2$, $P(B) = 0.6$, and $P(A \& B) = 0.1$. Find $P(A \text{ or } B)$.

*20. A and B are events such that $P(A) = 0.2$, $P(B) = 0.6$, and $P(A \& B) = 0.1$. Find $P(A \mid B)$ and $P(B \mid A)$.

*21. A and B are events such that $P(A) = 0.4$, $P(B) = 0.5$, and $P(A \& B) = 0.2$. Answer each question and explain your reasoning.
a. Are A and B mutually exclusive?
b. Are A and B independent?

*22. A and B are events such that $P(A) = 0.4$ and $P(B \mid A) = 0.7$. Find $P(A \& B)$.

*23. A and B are independent events such that $P(A) = 0.3$ and $P(B) = 0.6$. Find $P(A \& B)$.

*24. A_1, A_2, and A_3 are mutually exclusive and exhaustive events such that $P(A_1) = 0.2$, $P(A_2) = 0.5$, and $P(A_3) = 0.3$. B is an event such that $P(B \mid A_1) = 0.7$, $P(B \mid A_2) = 0.8$, and $P(B \mid A_3) = 0.4$.
a. Find $P(A_2 \mid B)$.   **b.** Find $P(B)$.

*25. Four actions are to be performed in a definite order. There are three possibilities for the first action; corresponding to each of these possibilities are four possibilities for the second action; corresponding to each of these possibilities are two possibilities for the third action; and corresponding to each of these possibilities are three possibilities for the fourth action. How many possibilities are there altogether for the four actions?

*26. Determine 0!, 1!, 3!, and 4!.

*27. How many samples of size 3 are possible from a population of size 10?

*28. For the first four letters in the English alphabet,
a. list the possible permutations of three letters from the four.
b. list the possible combinations of three letters from the four.
c. Use parts (a) and (b) to obtain $_4P_3$ and $_4C_3$.
d. Use the permutations and combinations rules to obtain $_4P_3$ and $_4C_3$. Compare your answers in parts (c) and (d).

Applying the Concepts and Skills

29. TV Location. The Television Bureau of Advertising publishes a report titled *TV Basics* for the purpose of providing information to help advertisers make the most effective and efficient use of local and national spot television advertisements. The following table gives a percentage distribution for the location of a television set in the home.

Location	Percentage
Living or Sitting Room	27
Master Bedroom	24
Child or other bedroom	19
Den or Office	14
Kitchen or Dining Room	5
Basement or Garage	5
Other room or storage	6

Suppose that a home television set is randomly selected. What is the probability that it is
a. located in a bedroom?
b. not located in a bedroom?
c. located in the kitchen, dining room, basement, or garage?

30. Adjusted Gross Incomes. The Internal Revenue Service compiles data on income tax returns and summarizes its findings in *Statistics of Income*. The first two columns of Table 4.16 show a frequency distribution (number of returns) for adjusted gross income (AGI) from federal individual income tax returns, where K = thousand.

TABLE 4.16
Adjusted gross incomes

Adjusted gross income	Frequency (1000s)	Event	Probability
Under $10K	26,268	A	
$10K–under $20K	22,778	B	
$20K–under $30K	18,610	C	
$30K–under $40K	14,554	D	
$40K–under $50K	11,087	E	
$50K–under $100K	30,926	F	
$100K & over	18,227	G	
	142,450		

A federal individual income tax return is selected at random.

a. Determine $P(A)$, the probability that the return selected shows an AGI under $10K.

b. Find the probability that the return selected shows an AGI between $30K and $100K (i.e., at least $30K but less than $100K).

c. Compute the probability of each of the seven events in the third column of Table 4.16, and record those probabilities in the fourth column.

31. Adjusted Gross Incomes. Refer to Problem 30. A federal individual income tax return is selected at random. Let

H = event the return shows an AGI between $20K and $100K,

I = event the return shows an AGI of less than $50K,

J = event the return shows an AGI of less than $100K, and

K = event the return shows an AGI of at least $50K.

Describe each of the following events in words and determine the number of outcomes (returns) that constitute each event.

a. (not J)   **b.** (H & I)

c. (H or K)   **d.** (H & K)

32. Adjusted Gross Incomes. For the following groups of events from Problem 31, determine which are mutually exclusive.

a. H and I **b.** I and K

c. H and (not J)   **d.** H, (not J), and K

33. Adjusted Gross Incomes. Refer to Problems 30 and 31.

a. Use the second column of Table 4.16 and the f/N rule to compute the probability of each of the events H, I, J, and K.

b. Express each of the events H, I, J, and K in terms of the mutually exclusive events displayed in the third column of Table 4.16.

c. Compute the probability of each of the events H, I, J, and K, using your answers from part (b), the special addition rule, and the fourth column of Table 4.16, which you completed in Problem 30(c).

34. Adjusted Gross Incomes. Consider the events (not J), (H & I), (H or K), and (H & K) discussed in Problem 31.

a. Find the probability of each of those four events, using the f/N rule and your answers from Problem 31.

b. Compute $P(J)$, using the complementation rule and your answer for $P(\text{not } J)$ from part (a).

c. In Problem 33(a), you determined that $P(H) = 0.528$ and $P(K) = 0.345$; and, in part (a) of this problem, you found that $P(H \& K) = 0.217$. Using those probabilities and the general addition rule, find $P(H \text{ or } K)$.

d. Compare the answers that you obtained for $P(H \text{ or } K)$ in parts (a) and (c).

*35. School Enrollment. The National Center for Education Statistics publishes information about school enrollment in the *Digest of Education Statistics*. Table 4.17 provides a contingency table for enrollment in public and private schools by level. Frequencies are in thousands of students.

a. How many cells are in this contingency table?

b. How many students are in high school?

c. How many students attend public schools?

d. How many students attend private colleges?

TABLE 4.17

Enrollment by level and type

		Type		
		Public T_1	Private T_2	Total
Level	Elementary L_1	36,161	4,073	40,234
	High school L_2	14,877	1,103	15,980
	College L_3	15,909	6,059	21,968
	Total	66,947	11,235	78,182

*36. School Enrollment. Refer to the information given in Problem 35. A student is selected at random.

a. Describe the events L_3, T_1, and (T_1 & L_3) in words.

b. Find the probability of each event in part (a), and interpret your answers in terms of percentages.

c. Construct a joint probability distribution corresponding to Table 4.17.

d. Compute $P(T_1 \text{ or } L_3)$, using Table 4.17 and the f/N rule.

e. Compute $P(T_1 \text{ or } L_3)$, using the general addition rule and your answers from part (b).

f. Compare your answers from parts (d) and (e). Explain any discrepancy.

*37. School Enrollment. Refer to the information given in Problem 35. A student is selected at random.

a. Find $P(L_3 \mid T_1)$ directly, using Table 4.17 and the f/N rule. Interpret the probability you obtain in terms of percentages.

b. Use the conditional probability rule and your answers from Problem 36(b) to find $P(L_3 \mid T_1)$.

c. Compare your answers from parts (a) and (b). Explain any discrepancy.

*38. School Enrollment. Refer to the information given in Problem 35. A student is selected at random.

a. Use Table 4.17 to find $P(T_2)$ and $P(T_2 \mid L_2)$.

b. Are events L_2 and T_2 independent? Explain your answer in terms of percentages.

c. Are events L_2 and T_2 mutually exclusive?

*39. Public Programs. During one year, the College of Public Programs at a major university awarded the following number of master's degrees.

Type of degree	Frequency
Master of arts	3
Master of public administration	28
Master of science	19

Two students who received such master's degrees are selected at random without replacement. Determine the probability that

a. the first student selected received a master of arts and the second a master of science.

b. both students selected received a master of public administration.

c. Construct a tree diagram for this problem similar to the one shown in Fig. 4.25 on page 194.

d. Find the probability that the two students selected received the same degree.

***40. Divorced Birds.** Research by B. Hatchwell et al. on divorce rates among the long-tailed tit (*Aegithalos caudatus*) appeared in *Science News* (Vol. 157, No. 20, p. 317). Tracking birds in York-shire from one breeding season to the next, the researchers noted that 63% of pairs divorced and that "...compared with moms whose off-spring had died, nearly twice the percentage of females that raised their youngsters to the fledgling stage moved out of the family flock and took mates elsewhere the next season—81% versus 43%." For the females in this study, find

a. the percentage whose offspring died. (*Hint:* You will need to use the rule of total probability and the complementation rule.)

b. the percentage that divorced and whose offspring died.

c. the percentage whose offspring died among those that divorced.

***41. Color Blindness.** According to M. Neitz and J. Neitz of the Medical College of Wisconsin Eye Institute, 9% of men are color blind. For four randomly selected men, determine the probability that

a. none are color blind.

b. the first three are not color blind and the fourth is color blind.

c. exactly one of the four is color blind.

***42. Smartphone Ownership.** The Pew Internet & American Life Project compiles data on smartphone ownership and publishes the results in the document *Smartphone Ownership*. The first two columns of the following table provide a percentage distribution for age of U.S. adults. The third column of the table shows the percentage of people in each age group who own a smartphone.

Age group (years)	Percentage of adults	Percentage who own a smartphone
18–24	12.5	79
25–34	18.0	81
35–44	16.7	69
45–54	17.5	55
55–64	16.4	39
65+	18.9	18

Suppose that a U.S. adult is selected at random. Determine the probability that the person selected

a. owns a smartphone, given that he or she is between 18 and 24 years old.

b. owns a smartphone

c. is between 18 and 24 years old, given that he or she owns a smart-phone.

d. Interpret your answers in parts (a)–(c) in terms of percentages.

***43. Quinella and Trifecta Wagering.** In Example 4.31 on page 211, we considered exacta wagering in horse racing. Two sim-ilar wagers are the quinella and the trifecta. In a quinella wager, the bettor picks the two horses that he or she believes will finish first and second, but not in a specified order. In a trifecta wager, the bettor picks the three horses he or she thinks will finish first, second, and third in a specified order. For a 12-horse race,

a. how many different quinella wagers are there?

b. how many different trifecta wagers are there?

c. Repeat parts (a) and (b) for an 8-horse race.

***44. Bridge.** A bridge hand consists of an unordered arrangement of 13 cards dealt at random from an ordinary deck of 52 playing cards.

a. How many possible bridge hands are there?

b. Find the probability of being dealt a bridge hand that contains exactly two of the four aces.

c. Find the probability of being dealt an 8-4-1 distribution, that is, eight cards of one suit, four of another, and one of another.

d. Determine the probability of being dealt a 5-5-2-1 distribution.

e. Determine the probability of being dealt a hand void in a speci-fied suit.

***45. Sweet Sixteen.** In the NCAA basketball tournament, 64 teams compete in 63 games during six rounds of single-elimination bracket competition. During the "Sweet Sixteen" competition (the third round of the tournament), 16 teams compete in eight games. If you were to choose in advance of the tournament the 8 teams that would win in the "Sweet Sixteen" competition and thus play in the fourth round of competition, how many different possibilities would you have?

***46. TVs and DVDs.** According to Nielsen Media Research, 98.6% of (U.S.) households own a TV and 85.2% of TV households own a DVD player.

a. Under what condition can you use the information provided to determine the percentage of households that own a DVD player? Explain your reasoning.

b. Assuming that the condition you stated in part (a) actually holds, determine the percentage of households that own a DVD player.

c. Assuming that the condition you stated in part (a) does not hold, what other piece of information would you need in order to find the percentage of households that own a DVD player?

FOCUSING ON DATA ANALYSIS

UWEC UNDERGRADUATES

Recall from Chapter 1 (see page 34) that the Focus database and Focus sample contain information on the undergraduate stu-dents at the University of Wisconsin-Eau Claire (UWEC). Now would be a good time for you to review the discussion about these data sets.

The following problems are designed for use with the entire Focus database (Focus). If your statistical software package won't accommodate the entire Focus database, use the Focus sam-ple (FocusSample) instead. Of course, in that case, your results will apply to the 200 UWEC undergraduate students in the Focus sample rather than to all UWEC undergraduate students.

a. Obtain a relative-frequency distribution for the classification (class-level) data.

b. Using your answer from part (a), determine the probability that a randomly selected UWEC undergraduate student is a freshman.

c. Consider the experiment of selecting a UWEC undergradu-ate student at random and observing the classification of the student obtained. Simulate that experiment 1000 times. (*Hint:* The simulation is equivalent to taking a random sample of size 1000 with replacement.)

d. Referring to the simulation performed in part (c), in approximately what percentage of the 1000 experiments would you

expect a freshman to be selected? Compare that percentage with the actual percentage of the 1000 experiments in which a freshman was selected.

e. Repeat parts (b)–(d) for sophomores; for juniors; for seniors.

CASE STUDY DISCUSSION

TEXAS HOLD'EM

At the beginning of this chapter on pages 156–157, we discussed Texas hold'em and described the basic rules of the game. Here we examine some of the simplest probabilities associated with the game.

Recall that, to begin, each player is dealt 2 cards face down, called "hole cards," from an ordinary deck of 52 playing cards, as pictured in Fig. 4.3 on page 165. The best possible starting hand is two aces, referred to as "pocket aces."

a. The probability that you are dealt pocket aces is 1/221, or 0.00452 to three significant digits. If you studied either Sections 4.5 and 4.6 or Section 4.8, verify that probability.

b. Using the result from part (a), obtain the probability that you are dealt "pocket kings."

c. Using the result from part (a) and your analysis in part (b), find the probability that you are dealt a "pocket pair," that is, two cards of the same denomination.

Next recall that, after receiving your hole cards, there is a betting round. Subsequently, 3 cards, called "the flop," are dealt face up in the center of the table. To do the remaining problems, you need to have studied either Sections 4.5 and 4.6 or Section 4.8. Assuming that you are dealt a pocket pair, determine the probability that the flop

*d. contains at least 1 card of your denomination. (*Hint:* Complementation rule.)

*e. gives you "trips," that is, contains exactly 1 card of your denomination and 2 other unpaired cards.

*f. gives you "quads," that is, contains 2 cards of your denomination.

*g. gives you a "boat," that is, contains 1 card of your denomination and 2 cards of another denomination.

BIOGRAPHY

ANDREI KOLMOGOROV: FATHER OF MODERN PROBABILITY THEORY

Andrei Nikolaevich Kolmogorov was born on April 25, 1903, in Tambov, Russia. At the age of 17, Kolmogorov entered Moscow State University, from which he graduated in 1925. His contributions to the world of mathematics, many of which appear in his numerous articles and books, encompass a formidable range of subjects.

Kolmogorov revolutionized probability theory with the introduction of the modern axiomatic approach to probability and by proving many of the fundamental theorems that are a consequence of that approach. He also developed two systems of partial differential equations that bear his name. Those systems extended the development of probability theory and allowed its broader application to the fields of physics, chemistry, biology, and civil engineering.

In 1938, Kolmogorov published an extensive article entitled "Mathematics," which appeared in the first edition of the *Bolshaya Sovyetskaya Entsiklopediya* (Great Soviet Encyclopedia). In this article he discussed the development of mathematics from

ancient to modern times and interpreted it in terms of dialectical materialism, the philosophy originated by Karl Marx and Friedrich Engels.

Kolmogorov became a member of the faculty at Moscow State University in 1925 at the age of 22. In 1931, he was promoted to professor; in 1933, he was appointed a director of the Institute of Mathematics of the university; and in 1937, he became Head of the University.

In addition to his work in higher mathematics, Kolmogorov was interested in the mathematical education of schoolchildren. He was chairman of the Commission for Mathematical Education under the Presidium of the Academy of Sciences of the U.S.S.R. During his tenure as chairman, he was instrumental in the development of a new mathematics training program that was introduced into Soviet schools.

Kolmogorov remained on the faculty at Moscow State University until his death in Moscow on October 20, 1987.

Discrete Random Variables*

CHAPTER OBJECTIVES

In Chapters 2 and 3, we examined, among other things, variables and their distributions. Most of the variables we discussed in those chapters were variables of finite populations. However, many variables are not of that type: the number of people waiting in line at a bank, the lifetime of an automobile tire, and the weight of a newborn baby, to name just three.

Probability theory enables us to extend concepts that apply to variables of finite populations—concepts such as relative-frequency distribution, mean, and standard deviation—to other types of variables. In doing so, we are led to the notion of a *random variable* and its *probability distribution*.

In this chapter, we discuss the fundamentals of discrete random variables and probability distributions and examine the concepts of the mean and standard deviation of a discrete random variable. In addition, we describe in detail two of the most important discrete random variables: the binomial and the Poisson.

Note: Those readers studying the normal approximation to the binomial distribution (Section 6.5) should cover Sections 5.1–5.3.

CASE STUDY

Aces Wild on the Sixth at Oak Hill

A most amazing event occurred during the second round of the 1989 U.S. Open at Oak Hill in Pittsford, New York. Four golfers—Doug Weaver, Mark Wiebe, Jerry Pate, and Nick Price—made holes in one on the sixth hole. What are the chances for the occurrence of such a remarkable event?

On the next day, a *Boston Globe* article (copyright ©1989) discussed the event in detail. To quote the article, "...for perspective, consider this: This is the 89th U.S. Open, and through the thousands and thousands and thousands of rounds played in the previous 88, there had been only 17 holes in one. Yet on this dark Friday morning, there were four holes in one on the same hole in less than two hours. Four times into a cup $4\frac{1}{2}$ inches in diameter from 160 yards away."

The article also reported odds estimates obtained from several different sources. These estimates varied considerably, from about 1 in 10 million to 1 in 1,890,000,000,000,000 to 1 in 8.7 million to 1 in 332,000.

After you have completed this chapter, you will be able to compute the odds for yourself.

5.1   Discrete Random Variables and Probability Distributions*

In this section, we introduce discrete random variables and probability distributions. As you will discover, these concepts are natural extensions of the ideas of variables and relative-frequency distributions.

EXAMPLE 5.1   Introducing Random Variables

TABLE 5.1

Frequency and relative-frequency distributions for number of siblings for students in introductory statistics

Siblings x	Frequency f	Relative frequency
0	8	0.200
1	17	0.425
2	11	0.275
3	3	0.075
4	1	0.025
	40	1.000

Number of Siblings Professor Weiss asked his introductory statistics students to state how many siblings they have. Table 5.1 presents frequency and relative-frequency distributions for that information. The table shows, for instance, that 11 of the 40 students, or 27.5%, have two siblings. Discuss the "number of siblings" in the context of randomness.

Solution Because the "number of siblings" varies from student to student, it is a variable. Suppose now that a student is selected at random. Then the "number of siblings" of the student obtained is called a *random variable* because its value depends on chance—namely, on which student is selected.

Keeping the previous example in mind, we now present the definition of a random variable.

DEFINITION 5.1

Random Variable

A **random variable** is a quantitative variable whose value depends on chance.

Example 5.1 shows how random variables arise naturally as (quantitative) variables of finite populations in the context of randomness. Specifically, as you learned in Chapter 2, a variable is a characteristic that varies from one member of a population to another. When one or more members are selected at random from the population, the variable, in that context, is called a random variable.

However, there are random variables that are not quantitative variables of finite populations in the context of randomness. Four examples of such random variables are

- the sum of the dice when a pair of fair dice are rolled,
- the number of puppies in a litter,
- the return on an investment, and
- the lifetime of a flashlight battery.

As you also learned in Chapter 2, a *discrete variable* is a variable whose possible values can be listed, even though the list may continue indefinitely. This property holds, for instance, if either the variable has only a finite number of possible values or its possible values are some collection of whole numbers. The variable "number of siblings" in Example 5.1 is therefore a discrete variable. We use the adjective *discrete* for random variables in the same way that we do for variables—hence the term *discrete random variable*.

What Does It Mean?

A discrete random variable usually involves a count of something.

DEFINITION 5.2

Discrete Random Variable

A **discrete random variable** is a random variable whose possible values can be listed. In particular, a random variable with only a finite number of possible values is a discrete random variable.

Random-Variable Notation

Recall that we use lowercase letters such as x, y, and z to denote variables. To represent random variables, however, we usually use uppercase letters. For instance, we could use x to denote the variable "number of siblings," but we would generally use X to denote the random variable "number of siblings."

Random-variable notation is a useful shorthand for discussing and analyzing random variables. For example, let X denote the number of siblings of a randomly selected student. Then we can represent the event that the selected student has two siblings by $\{X = 2\}$, read "X equals two," and the probability of that event as $P(X = 2)$, read "the probability that X equals two."

More generally, if X is any discrete random variable and x is any number, then we use the following notation:

- $\{X = x\}$ denotes the event that the random variable X equals x.
- $P(X = x)$ denotes the probability that the random variable X equals x.

Probability Distributions and Histograms

Recall that the relative-frequency distribution or relative-frequency histogram of a discrete variable gives the possible values of the variable and the proportion of times each value occurs. Using the language of probability, we can extend the notions of relative-frequency distribution and relative-frequency histogram—concepts applying to variables of finite populations—to any discrete random variable. In doing so, we use the terms *probability distribution* and *probability histogram*.

DEFINITION 5.3

? What Does It Mean?

The probability distribution and probability histogram of a discrete random variable show its possible values and their likelihood.

Probability Distribution and Probability Histogram

Probability distribution: A listing of the possible values and corresponding probabilities of a discrete random variable, or a formula for the probabilities.

Probability histogram: A graph of the probability distribution that displays the possible values of a discrete random variable on the horizontal axis and the probabilities of those values on the vertical axis. The probability of each value is represented by a vertical bar whose height equals the probability.

■■■   **EXAMPLE 5.2**   **Probability Distributions and Histograms**

Number of Siblings   Refer to Example 5.1, and let X denote the number of siblings of a randomly selected student.

a. Determine the probability distribution of the random variable X.
b. Construct a probability histogram for the random variable X.

TABLE 5.2

Probability distribution of the random variable X, the number of siblings of a randomly selected student

Siblings x	Probability $P(X = x)$
0	0.200
1	0.425
2	0.275
3	0.075
4	0.025
	1.000

Solution

a. We want to determine the probability of each of the possible values of the random variable X. To obtain, for instance, $P(X = 2)$, the probability that the student selected has two siblings, we apply the f/N rule. From Table 5.1, we find that

$$P(X = 2) = \frac{f}{N} = \frac{11}{40} = 0.275.$$

The other probabilities are found in the same way. Table 5.2 displays these probabilities and provides the probability distribution of the random variable X.

b. To construct a probability histogram for X, we plot its possible values on the horizontal axis and display the corresponding probabilities as vertical bars. Referring to Table 5.2, we get the probability histogram of the random variable X, as shown in Fig. 5.1.

FIGURE 5.1

Probability histogram for the random variable X, the number of siblings of a randomly selected student

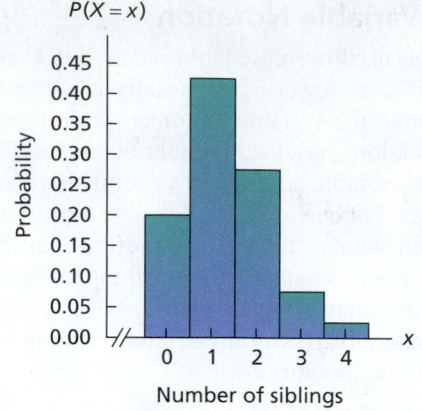

You try it!

Exercise 5.11 on page 230

The probability histogram provides a quick and easy way to visualize how the probabilities are distributed.

The variable "number of siblings" is a variable of a finite population, so its probabilities are identical to its relative frequencies. As a consequence, its probability distribution, given in the first and second columns of Table 5.2, is the same as its relative-frequency distribution, shown in the first and third columns of Table 5.1. Apart from labeling, the variable's probability histogram is identical to its relative-frequency histogram. These statements hold for any variable of a finite population.

Note also that the probabilities in the second column of Table 5.2 sum to 1, which is always the case for discrete random variables.

KEY FACT 5.1

 **What Does It Mean?**

The sum of the probabilities of the possible values of a discrete random variable equals 1.

Sum of the Probabilities of a Discrete Random Variable

For any discrete random variable X, we have $\Sigma P(X = x) = 1$.[†]

Examples 5.3 and 5.4 provide additional illustrations of random-variable notation and probability distributions.

EXAMPLE 5.3   Random Variables and Probability Distributions

TABLE 5.3

Frequency distribution for enrollment by grade level in U.S. public elementary schools

Grade level y	Frequency f
0	4,819
1	3,708
2	3,699
3	3,708
4	3,647
5	3,629
6	3,614
7	3,653
8	3,692
	34,169

Elementary-School Enrollment The National Center for Education Statistics compiles enrollment data on U.S. public schools and publishes the results in the *Digest of Education Statistics*. Table 5.3 displays a frequency distribution for the enrollment by grade level in public elementary schools, where 0 = pre-kindergarten and kindergarten, 1 = first grade, and so on. Frequencies are in thousands of students.

Let Y denote the grade level of a randomly selected elementary-school student. Then Y is a discrete random variable whose possible values are 0, 1, 2, ..., 8.

a. Use random-variable notation to represent the event that the selected student is in the fifth grade.

b. Determine $P(Y = 5)$ and express the result in terms of percentages.

c. Determine the probability distribution of Y.

Solution

a. The event that the selected student is in the fifth grade can be represented as $\{Y = 5\}$.

[†]The sum $\Sigma P(X = x)$ represents adding the individual probabilities, $P(X = x)$, for all possible values, x, of the random variable X.

TABLE 5.4

Probability distribution of the random variable Y, the grade level of a randomly selected elementary-school student

Grade level y	Probability $P(Y = y)$
0	0.141
1	0.109
2	0.108
3	0.109
4	0.107
5	0.106
6	0.106
7	0.107
8	0.108
	1.001

b. $P(Y = 5)$ is the probability that the selected student is in the fifth grade. Using Table 5.3 and the f/N rule, we get

$$P(Y = 5) = \frac{f}{N} = \frac{3{,}629}{34{,}169} = 0.106.$$

Interpretation 10.6% of elementary-school students in the United States are in the fifth grade.

c. The probability distribution of Y is obtained by computing $P(Y = y)$ for $y = 0, 1, 2, \ldots, 8$. We have already done that for $y = 5$. The other probabilities are computed similarly and are displayed in Table 5.4.

Note: In Table 5.4, the sum of the probabilities is given as 1.001. Key Fact 5.1, however, states that the sum of the probabilities must be exactly 1. Our computation is off slightly because we rounded the probabilities for Y. ■

Once we have the probability distribution of a discrete random variable, we can easily determine any probability involving that random variable. The basic tool for accomplishing this is the special addition rule, Formula 4.1 on page 173.[†] We illustrate this technique in part (e) of the next example. Before reading the example, you might find it helpful to review the discussion of the phrases "at least," "at most," and "inclusive," as presented on page 168.

EXAMPLE 5.4 Random Variables and Probability Distributions

TABLE 5.5

Possible outcomes

HHH	HTH	THH	TTH
HHT	HTT	THT	TTT

Coin Tossing When a balanced dime is tossed three times, eight equally likely outcomes are possible, as shown in Table 5.5. Here, for instance, HHT means that the first two tosses are heads and the third is tails. Let X denote the total number of heads obtained in the three tosses. Then X is a discrete random variable whose possible values are 0, 1, 2, and 3.

a. Use random-variable notation to represent the event that exactly two heads are tossed.
b. Determine $P(X = 2)$.
c. Find the probability distribution of X.
d. Use random-variable notation to represent the event that at most two heads are tossed.
e. Find $P(X \leq 2)$.

Solution

a. The event that exactly two heads are tossed can be represented as $\{X = 2\}$.
b. $P(X = 2)$ is the probability that exactly two heads are tossed. Table 5.5 shows that there are three ways to get exactly two heads and that there are eight possible (equally likely) outcomes altogether. So, by the f/N rule,

$$P(X = 2) = \frac{f}{N} = \frac{3}{8} = 0.375.$$

The probability that exactly two heads are tossed is 0.375.

Interpretation There is a 37.5% chance of obtaining exactly two heads in three tosses of a balanced dime.

[†] Specifically, to find the probability that a discrete random variable takes a value in some set of real numbers, we simply sum the individual probabilities of that random variable over the values in the set. In symbols, if X is a discrete random variable and A is a set of real numbers, then

$$P(X \in A) = \sum_{x \in A} P(X = x),$$

where the sum on the right represents adding the individual probabilities, $P(X = x)$, for all possible values, x, of the random variable X that belong to the set A.

TABLE 5.6

Probability distribution of the random variable X, the number of heads obtained in three tosses of a balanced dime

No. of heads x	Probability $P(X = x)$
0	0.125
1	0.375
2	0.375
3	0.125
	1.000

You try it!

Exercise 5.15 on page 230

c. The remaining probabilities for X are computed as in part (b) and are shown in Table 5.6.

d. The event that at most two heads are tossed can be represented as $\{X \le 2\}$, read as "X is less than or equal to two."

e. $P(X \le 2)$ is the probability that at most two heads are tossed. The event that at most two heads are tossed can be expressed as

$$\{X \le 2\} = (\{X = 0\} \text{ or } \{X = 1\} \text{ or } \{X = 2\}).$$

Because the three events on the right are mutually exclusive, we use the special addition rule and Table 5.6 to conclude that

$$P(X \le 2) = P(X = 0) + P(X = 1) + P(X = 2)$$
$$= 0.125 + 0.375 + 0.375 = 0.875.$$

The probability that at most two heads are tossed is 0.875.

Interpretation There is an 87.5% chance of obtaining two or fewer heads in three tosses of a balanced dime.

Interpretation of Probability Distributions

Recall that the frequentist interpretation of probability construes the probability of an event to be the proportion of times it occurs in a large number of independent repetitions of the experiment. Using that interpretation, we clarify the meaning of a probability distribution.

EXAMPLE 5.5 Interpreting a Probability Distribution

Coin Tossing Suppose we repeat the experiment of observing the number of heads, X, obtained in three tosses of a balanced dime a large number of times. Then the proportion of those times in which, say, no heads are obtained ($X = 0$) should approximately equal the probability of that event [$P(X = 0)$]. The same statement holds for the other three possible values of the random variable X. Use simulation to verify these facts.

TABLE 5.7

Frequencies and proportions for the numbers of heads obtained in three tosses of a balanced dime for 1000 observations

No. of heads x	Frequency f	Proportion $f/1000$
0	136	0.136
1	377	0.377
2	368	0.368
3	119	0.119
	1000	1.000

Solution Simulating a random variable means that we use a computer or statistical calculator to generate observations of the random variable. In this instance, we used a computer to simulate 1000 observations of the random variable X, the number of heads obtained in three tosses of a balanced dime.

Table 5.7 shows the frequencies and proportions for the numbers of heads obtained in the 1000 observations. For example, 136 of the 1000 observations resulted in no heads out of three tosses, which gives a proportion of 0.136.

As expected, the proportions in the third column of Table 5.7 are fairly close to the true probabilities in the second column of Table 5.6. This result is more easily seen if we compare the proportion histogram to the probability histogram of the random variable X, as shown in Fig. 5.2.

If we simulated, say, 10,000 observations instead of 1000, the proportions that would appear in the third column of Table 5.7 would most likely be even closer to the true probabilities listed in the second column of Table 5.6.

KEY FACT 5.2

Interpretation of a Probability Distribution

In a large number of independent observations of a random variable X, the proportion of times each possible value occurs will approximate the probability distribution of X; or, equivalently, the proportion histogram will approximate the probability histogram for X.

FIGURE 5.2

(a) Histogram of proportions for the numbers of heads obtained in three tosses of a balanced dime for 1000 observations; (b) probability histogram for the number of heads obtained in three tosses of a balanced dime

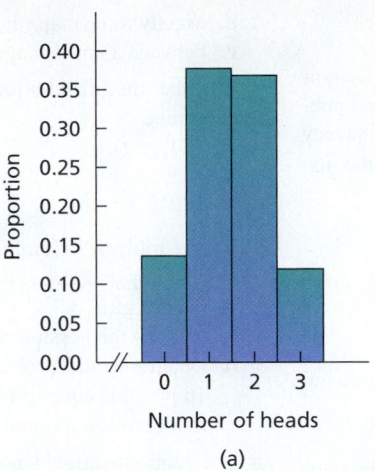

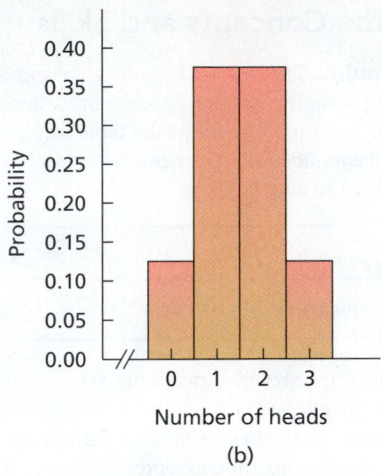

Exercises 5.1

Understanding the Concepts and Skills

5.1 Fill in the blanks.
a. A relative-frequency distribution is to a variable as a _____ distribution is to a random variable.
b. A relative-frequency histogram is to a variable as a _____ histogram is to a random variable.

5.2 Provide an example (other than one discussed in the text) of a random variable that does not arise from a quantitative variable of a finite population in the context of randomness.

5.3 Let X denote the number of siblings of a randomly selected student. Explain the difference between $\{X = 3\}$ and $P(X = 3)$.

5.4 Fill in the blank. For a discrete random variable, the sum of the probabilities of its possible values equals _____.

5.5 Suppose that you make a large number of independent observations of a random variable and then construct a table giving the possible values of the random variable and the proportion of times each value occurs. What will this table resemble?

5.6 What rule of probability permits you to obtain any probability for a discrete random variable by simply knowing its probability distribution?

5.7 A variable x of a finite population has the following frequency distribution:

x	1	2	3
f	6	8	6

Suppose a member is selected at random from the population and let X denote the value of the variable x for the member obtained.
a. Determine the probability distribution of the random variable X.
b. Use random-variable notation to describe the events that X takes on the value 2, a value of at most 2, and a value greater than 2.
c. Find $P(X = 2)$, $P(X \leq 2)$, and $P(X > 2)$. Interpret your results.
d. Construct a probability histogram for the random variable X.

5.8 A variable y of a finite population has the following frequency distribution:

y	1	2	3	4
f	2	2	10	6

Suppose a member is selected at random from the population and let Y denote the value of the variable y for the member obtained.
a. Determine the probability distribution of the random variable Y.
b. Use random-variable notation to describe the events that Y takes on the value 3, a value less than 3, and a value of at least 3.
c. Find $P(Y = 3)$, $P(Y < 3)$, and $P(Y \geq 3)$. Interpret your results.
d. Construct a probability histogram for the random variable Y.

5.9 A variable y of a finite population has the following frequency distribution:

y	0	1	4	6
f	18	14	8	10

Suppose a member is selected at random from the population and let Y denote the value of the variable y for the member obtained.
a. Determine the probability distribution of the random variable Y.
b. Use random-variable notation to describe the events that Y takes on the value 3, a value less than 3, and a value of at least 3.
c. Find $P(Y = 3)$, $P(Y < 3)$, and $P(Y \geq 3)$. Interpret your results.
d. Construct a probability histogram for the random variable Y.

5.10 A variable x of a finite population has the following frequency distribution:

x	5	7	8
f	25	40	60

Suppose a member is selected at random from the population and let X denote the value of the variable x for the member obtained.
a. Determine the probability distribution of the random variable X.
b. Use random-variable notation to describe the events that X takes on the value 6, a value of at most 6, and a value greater than 6.
c. Find $P(X = 6)$, $P(X \leq 6)$, and $P(X > 6)$. Interpret your results.
d. Construct a probability histogram for the random variable X.

Applying the Concepts and Skills

5.11 Space Shuttles. The National Aeronautics and Space Administration (NASA) compiles data on space-shuttle launches and publishes them on its website. The following table displays a frequency distribution for the number of crew members on each shuttle mission from April 12, 1981 to July 8, 2011.

Crew size	2	4	5	6	7	8
Frequency	4	3	36	28	63	1

Let X denote the crew size of a randomly selected shuttle mission between the aforementioned dates.
a. What are the possible values of the random variable X?
b. Use random-variable notation to represent the event that the shuttle mission obtained has a crew size of 7.
c. Find $P(X = 4)$; interpret in terms of percentages.
d. Obtain the probability distribution of X.
e. Construct a probability histogram for X.

5.12 Persons per Housing Unit. From the document *American Housing Survey for the United States*, published by the U.S. Census Bureau, we obtained the following frequency distribution for the number of persons per occupied housing unit, where we have used "7" in place of "7 or more." Frequencies are in millions of housing units.

Persons	1	2	3	4	5	6	7
Frequency	27.9	34.4	17.0	15.5	6.8	2.3	1.4

For a randomly selected housing unit, let Y denote the number of persons living in that unit.
a. Identify the possible values of the random variable Y.
b. Use random-variable notation to represent the event that a housing unit has exactly three persons living in it.
c. Determine $P(Y = 3)$; interpret in terms of percentages.
d. Determine the probability distribution of Y.
e. Construct a probability histogram for Y.

5.13 Major Hurricanes. The *Atlantic Hurricane Database* extends back to 1851, recording among other things the number of major hurricanes striking the U.S. Atlantic and Gulf Coast per year. A major hurricane is a hurricane measuring at least a Category 3 on the Saffir-Simpson hurricane wind scale (i.e., with winds of at least 110 mph). As published by the National Oceanic & Atmospheric Administration and the Atlantic Oceanographic & Meteorological Laboratory, the following table provides a probability distribution for the number of major hurricanes, Y, for a randomly selected year between 1851 and 2012.

y	$P(Y = y)$	y	$P(Y = y)$
0	0.185	5	0.056
1	0.296	6	0.037
2	0.266	7	0.012
3	0.093	8	0.006
4	0.049		

Use random-variable notation to represent each of the following events. The year had
a. at least one major hurricane.

b. exactly three major hurricanes.
c. between 2 and 4 major hurricanes, inclusive.

Use the special addition rule and the probability distribution to determine
d. $P(Y \geq 1)$.
e. $P(Y = 3)$.
f. $P(2 \leq Y \leq 4)$.

5.14 Children's Gender. A certain couple is equally likely to have either a boy or a girl. If the family has four children, let X denote the number of girls.
a. Identify the possible values of the random variable X.
b. Determine the probability distribution of X. (*Hint:* There are 16 possible equally likely outcomes. One is GBBB, meaning the first born is a girl and the next three born are boys.)

Use random-variable notation to represent each of the following events. Also use the special addition rule and the probability distribution obtained in part (b) to determine each event's probability. The couple has
c. exactly two girls.
d. at least two girls.
e. at most two girls.
f. between one and three girls, inclusive.
g. children all of the same gender.

5.15 Dice. When two balanced dice are rolled, 36 equally likely outcomes are possible, as depicted in Fig. 4.1 on page 159. Let Y denote the sum of the dice.
a. What are the possible values of the random variable Y?
b. Use random-variable notation to represent the event that the sum of the dice is 7.
c. Find $P(Y = 7)$.
d. Find the probability distribution of Y. Leave your probabilities in fraction form.
e. Construct a probability histogram for Y.

In the game of craps, a first roll of a sum of 7 or 11 wins, whereas a first roll of a sum of 2, 3, or 12 loses. To win with any other first sum, that sum must be repeated before a sum of 7 is rolled. Determine the probability of
f. a win on the first roll.
g. a loss on the first roll.

5.16 World Series. The World Series in baseball is won by the first team to win four games (ignoring the 1903 and 1919–1921 World Series, when it was a best of nine). Thus it takes at least four games and no more than seven games to establish a winner. From the document *World Series History* on the Baseball Almanac website, as of November 2013, the lengths of the World Series are as given in the following table.

Number of games	Frequency	Relative frequency
4	21	0.200
5	24	0.229
6	24	0.229
7	36	0.343

a. If X denotes the number of games that it takes to complete a World Series, identify the possible values of the random variable X.
b. Do the first and third columns of the table provide a probability distribution for X? Explain your answer.
c. Historically, what is the most likely number of games it takes to complete a series?

d. Historically, for a randomly chosen series, what is the probability that it ends in five games?

e. Historically, for a randomly chosen series, what is the probability that it ends in five or more games?

f. The data in the table exhibit a statistical oddity. If the two teams in a series are evenly matched and one team is ahead three games to two, either team has the same chance of winning game number six. Thus there should be about an equal number of six- and seven-game series. If the teams are not evenly matched, the series should tend to be shorter, ending in six or fewer games, not seven games. Can you explain why the series tend to last longer than expected?

5.17 Archery. An archer shoots an arrow into a square target 6 feet on a side whose center we call the origin. The outcome of this random experiment is the point in the target hit by the arrow. The archer scores 10 points if she hits the bull's eye—a disk of radius 1 foot centered at the origin; she scores 5 points if she hits the ring with inner radius 1 foot and outer radius 2 feet centered at the origin; and she scores 0 points otherwise. Assume that the archer will actually hit the target and is equally likely to hit any portion of the target. For one arrow shot, let S be the score.

a. Obtain and interpret the probability distribution of the random variable S. (*Hint:* The area of a square is the square of its side length; the area of a disk is the square of its radius times π.)

b. Use the special addition rule and the probability distribution obtained in part (a) to determine and interpret the probability of each of the following events: $\{S = 5\}$; $\{S > 0\}$; $\{S \leq 7\}$; $\{5 < S \leq 15\}$; $\{S < 15\}$; and $\{S < 0\}$.

5.18 Solar Eclipses. The *World Almanac* provides information on past and projected total solar eclipses from 1955 to 2015. Unlike total lunar eclipses, observing a total solar eclipse from Earth is rare because it can be seen along only a very narrow path and for only a short period of time.

a. Let X denote the duration, in minutes, of a total solar eclipse. Is X a discrete random variable? Explain your answer.

b. Let Y denote the duration, to the nearest minute, of a total solar eclipse. Is Y a discrete random variable? Explain your answer.

5.19 Black Bear Litters. In the article "Reproductive Ecology and Cub Survival of Florida Black Bears" (*Journal of Wildlife Management*, Vol. 71, Issue 3, pp. 720–727), E. Garrison et al. investigated cub survival rates of the Florida black bear. Through the observation of female Florida black bears and their litters, the following estimate is given for the probability distribution of the number of young per litter for female Florida black bears.

Litter size	1	2	3	4
Probability	0.212	0.606	0.152	0.030

a. Explain why the probabilities in the table provide only an estimate of the true probabilities.

b. Estimate the probability that a Florida black bear has a litter of either two or three cubs.

5.20 All-Numeric Passwords. The technology consultancy Data-Genetics published the online document *PIN analysis*. In addition to analyzing PIN numbers, password trends were examined. Seven million all-numeric passwords were collected and yielded the following estimate of the probability distribution of the number of digits used in an all-numeric password.

Digits	Probability	Digits	Probability
4	0.492	9	0.029
5	0.092	10	0.015
6	0.178	11	0.004
7	0.073	12	0.003
8	0.113	13	0.001

a. Explain why the probabilities in the table provide only an estimate of the true probabilities.

b. Estimate the probability that an all-numeric password contains at least six digits.

Extending the Concepts and Skills

5.21 Suppose that $P(Z > 1.96) = 0.025$. Find $P(Z \leq 1.96)$. (*Hint:* Use the complementation rule.)

5.22 Suppose that T and Z are random variables.

a. If $P(T > 2.02) = 0.05$ and $P(T < -2.02) = 0.05$, determine $P(-2.02 \leq T \leq 2.02)$.

b. Suppose that $P(-1.64 \leq Z \leq 1.64) = 0.90$ and also suppose that $P(Z > 1.64) = P(Z < -1.64)$. Find $P(Z > 1.64)$.

5.23 Let $c > 0$ and $0 \leq \alpha \leq 1$. Also let X, Y, and T be random variables.

a. If $P(X > c) = \alpha$, determine $P(X \leq c)$ in terms of α.

b. Suppose that $P(Y > c) = \alpha/2$ and $P(Y < -c) = P(Y > c)$. Determine $P(-c \leq Y \leq c)$ in terms of α.

c. Suppose that $P(-c \leq T \leq c) = 1 - \alpha$ and also suppose that $P(T < -c) = P(T > c)$. Find $P(T > c)$ in terms of α.

5.24 Simulation. Refer to the probability distribution displayed in Table 5.6 on page 228.

a. Use the technology of your choice to repeat the simulation done in Example 5.5 on page 228.

b. Obtain the proportions for the number of heads in three tosses and compare them to the probability distribution in Table 5.6.

c. Obtain a histogram of the proportions and compare it to the probability histogram in Fig. 5.2(b) on page 229.

d. What do parts (b) and (c) illustrate?

5.2

The Mean and Standard Deviation of a Discrete Random Variable*

In this section, we introduce the mean and standard deviation of a discrete random variable. As you will see, the mean and standard deviation of a discrete random variable are analogous to the population mean and population standard deviation, respectively.

Mean of a Discrete Random Variable

Recall that, for a variable x, the mean of all possible observations for the entire population is called the *population mean* or *mean of the variable* x. In Section 3.5, we gave a formula for the mean of a variable x:

$$\mu = \frac{\Sigma x_i}{N}.$$

Although this formula applies only to variables of finite populations, we can use it and the language of probability to extend the concept of the mean to any discrete variable. We show how to do so in Example 5.6.

EXAMPLE 5.6 Introducing the Mean of a Discrete Random Variable

Student Ages Consider a population of eight students whose ages, in years, are those given in Table 5.8. Let X denote the age of a randomly selected student. From a relative-frequency distribution of the age data in Table 5.8, we get the probability distribution of the random variable X shown in Table 5.9. Express the mean age of the students in terms of the probability distribution of X.

TABLE 5.8

Ages of eight students

19	20	20	19
21	27	20	21

Solution Referring first to Table 5.8 and then to Table 5.9, we get

$$\mu = \frac{\Sigma x_i}{N} = \frac{19 + 20 + 20 + 19 + 21 + 27 + 20 + 21}{8}$$

$$= \frac{\overbrace{19 + 19}^{2} + \overbrace{20 + 20 + 20}^{3} + \overbrace{21 + 21}^{2} + \overbrace{27}^{1}}{8}$$

$$= \frac{19 \cdot 2 + 20 \cdot 3 + 21 \cdot 2 + 27 \cdot 1}{8} = 19 \cdot \frac{2}{8} + 20 \cdot \frac{3}{8} + 21 \cdot \frac{2}{8} + 27 \cdot \frac{1}{8}$$

$$= 19 \cdot P(X = 19) + 20 \cdot P(X = 20) + 21 \cdot P(X = 21) + 27 \cdot P(X = 27)$$

$$= \Sigma x P(X = x).$$

TABLE 5.9

Probability distribution of X, the age of a randomly selected student

Age x	Probability $P(X = x)$	
19	0.250	← 2/8
20	0.375	← 3/8
21	0.250	← 2/8
27	0.125	← 1/8

The previous example shows that we can express the mean of a variable of a finite population in terms of the probability distribution of the corresponding random variable: $\mu = \Sigma x P(X = x)$. Because the expression on the right of this equation is meaningful for any discrete random variable, we can define the *mean of a discrete random variable* as follows.

DEFINITION 5.4

? **What Does It Mean?**

To obtain the mean of a discrete random variable, multiply each possible value by its probability and then add those products.

Mean of a Discrete Random Variable

The **mean of a discrete random variable** X is denoted μ_X or, when no confusion will arise, simply μ. It is defined by

$$\mu = \Sigma x P(X = x).$$

The terms **expected value** and **expectation** are commonly used in place of the term *mean*.

EXAMPLE 5.7 The Mean of a Discrete Random Variable

Busy Tellers Prescott National Bank has six tellers available to serve customers. The number of tellers busy with customers at, say, 1:00 P.M. varies from day to day

TABLE 5.10

Table for computing the mean of the random variable X, the number of tellers busy with customers

x	P(X = x)	xP(X = x)
0	0.029	0.000
1	0.049	0.049
2	0.078	0.156
3	0.155	0.465
4	0.212	0.848
5	0.262	1.310
6	0.215	1.290
		4.118

You try it!

Exercise 5.35(a) on page 235

and depends on chance; hence it is a random variable, say, X. Past records indicate that the probability distribution of X is as shown in the first two columns of Table 5.10. Find the mean of the random variable X.

Solution The third column of Table 5.10 provides the products of x with $P(X = x)$, which, in view of Definition 5.4, are required to determine the mean of X. Summing that column gives

$$\mu = \Sigma x P(X = x) = 4.118.$$

Interpretation The mean number of tellers busy with customers is 4.118.

Interpretation of the Mean of a Random Variable

Recall that the mean of a variable of a finite population is the arithmetic average of all possible observations. A similar interpretation holds for the mean of a random variable.

For instance, in the previous example, the random variable X is the number of tellers busy with customers at 1:00 P.M., and the mean is 4.118. Of course, there never will be a day when 4.118 tellers are busy with customers at 1:00 P.M. Over many days, however, the average number of busy tellers at 1:00 P.M. will be about 4.118.

This interpretation holds in all cases. It is commonly known as the **law of averages** and in mathematical circles as the **law of large numbers.**

KEY FACT 5.3

? What Does It Mean?

The mean of a random variable can be considered the long-run-average value of the random variable in repeated independent observations.

Interpretation of the Mean of a Random Variable

In a large number of independent observations of a random variable X, the average value of those observations will approximately equal the mean, μ, of X. The larger the number of observations, the closer the average tends to be to μ.

We used a computer to simulate the number of busy tellers at 1:00 P.M. on 100 randomly selected days; that is, we obtained 100 independent observations of the random variable X. The data are displayed in Table 5.11.

TABLE 5.11

One hundred observations of the random variable X, the number of tellers busy with customers

5	3	5	3	4	3	4	3	6	5	6	4	5	4	3	5	4	5	6	3
4	1	6	5	3	6	3	5	5	4	6	4	1	6	5	3	3	6	4	5
3	4	2	5	5	6	5	4	6	2	4	5	4	6	4	5	5	3	4	6
1	5	4	6	4	4	4	5	6	2	5	4	5	1	3	3	6	4	6	4
5	6	5	5	3	2	4	6	6	1	5	1	3	6	5	3	5	4	3	6

The average value of the 100 observations in Table 5.11 is 4.25. This value is quite close to the mean, $\mu = 4.118$, of the random variable X. If we made, say, 1000 observations instead of 100, the average value of those 1000 observations would most likely be even closer to 4.118.

Figure 5.3(a) on the following page shows a plot of the average number of busy tellers versus the number of observations for the data in Table 5.11. The dashed line is at $\mu = 4.118$. Figure 5.3(b) depicts a plot for a different simulation of the number of busy tellers at 1:00 P.M. on 100 randomly selected days. Both plots suggest that, as the number of observations increases, the average number of busy tellers approaches the mean, $\mu = 4.118$, of the random variable X.

Standard Deviation of a Discrete Random Variable

Similar reasoning also lets us extend the concept of population standard deviation (standard deviation of a variable) to any discrete variable.

FIGURE 5.3

Graphs showing the average number of busy tellers versus the number of observations for two simulations of 100 observations each

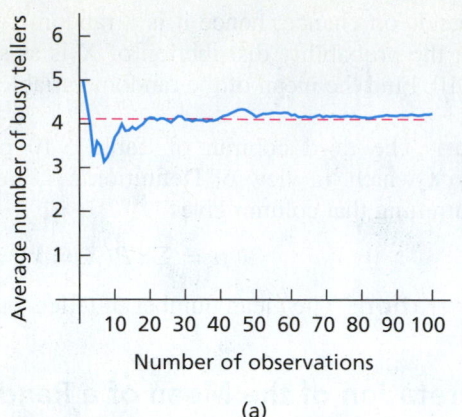

(a)

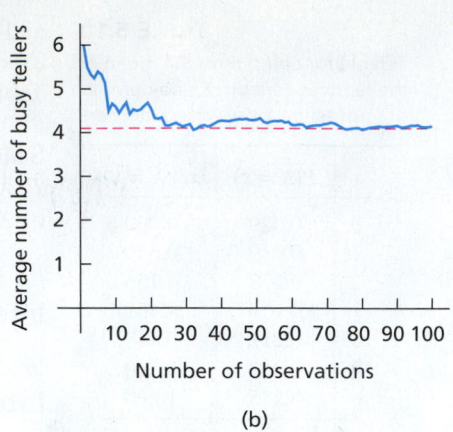

(b)

DEFINITION 5.5

What Does It Mean?

Roughly speaking, the standard deviation of a random variable X indicates how far, on average, an observed value of X is from its mean. In particular, the smaller the standard deviation of X, the more likely it is that an observed value of X will be close to its mean.

Standard Deviation of a Discrete Random Variable

The **standard deviation of a discrete random variable X** is denoted σ_X or, when no confusion will arise, simply σ. It is defined as

$$\sigma = \sqrt{\Sigma(x - \mu)^2 P(X = x)}.$$

The standard deviation of a discrete random variable can also be obtained from the computing formula

$$\sigma = \sqrt{\Sigma x^2 P(X = x) - \mu^2}.$$

Note: The square of the standard deviation, σ^2, is called the **variance** of X.

EXAMPLE 5.8 **The Standard Deviation of a Discrete Random Variable**

Busy Tellers Recall Example 5.7, where X denotes the number of tellers busy with customers at 1:00 P.M. Find the standard deviation of X.

Solution We apply the computing formula given in Definition 5.5. To use that formula, we need the mean of X, which we found in Example 5.7 to be 4.118, and columns for x^2 and $x^2 P(X = x)$, which are presented in the last two columns of Table 5.12.

TABLE 5.12

Table for computing the standard deviation of the random variable X, the number of tellers busy with customers

x	$P(X = x)$	x^2	$x^2 P(X = x)$
0	0.029	0	0.000
1	0.049	1	0.049
2	0.078	4	0.312
3	0.155	9	1.395
4	0.212	16	3.392
5	0.262	25	6.550
6	0.215	36	7.740
			19.438

You try it!

Exercise 5.35(b) on page 235

From the final column of Table 5.12, $\Sigma x^2 P(X = x) = 19.438$. Thus

$$\sigma = \sqrt{\Sigma x^2 P(X = x) - \mu^2} = \sqrt{19.438 - (4.118)^2} = 1.6.$$

Interpretation Roughly speaking, on average, the number of busy tellers is 1.6 from the mean of 4.118 busy tellers.

Exercises 5.2

Understanding the Concepts and Skills

5.25 What concept does the mean of a discrete random variable generalize?

5.26 Comparing Investments. Suppose that the random variables X and Y represent the amount of return on two different investments. Further suppose that the mean of X equals the mean of Y but that the standard deviation of X is greater than the standard deviation of Y.
a. On average, is there a difference between the returns of the two investments? Explain your answer.
b. Which investment is more conservative? Why?

In Exercises 5.27–5.30, we have provided the probability distributions of the random variables considered in Exercises 5.7–5.10 of Section 5.1. For each exercise, do the following tasks.
a. Find the mean of the random variable.
b. Obtain the standard deviation of the random variable by using one of the formulas given in Definition 5.5.

5.27

x	1	2	3
$P(X = x)$	0.3	0.4	0.3

5.28

y	1	2	3	4
$P(Y = y)$	0.1	0.1	0.5	0.3

5.29

y	0	1	4	6
$P(Y = y)$	0.36	0.28	0.16	0.20

5.30

x	5	7	8
$P(X = x)$	0.20	0.32	0.48

Applying the Concepts and Skills

In Exercises 5.31–5.35, we have provided the probability distributions of the random variables considered in Exercises 5.11–5.15 of Section 5.1. For each exercise, do the following tasks.
a. Find and interpret the mean of the random variable.
b. Obtain the standard deviation of the random variable by using one of the formulas given in Definition 5.5.
c. Construct a probability histogram for the random variable; locate the mean; and show one-, two-, and three-standard-deviation intervals.

5.31 Space Shuttles. The random variable X is the crew size of a randomly selected shuttle mission between April 12, 1981 and July 8, 2011. Its probability distribution is as follows.

x	2	4	5	6	7	8
$P(X = x)$	0.030	0.022	0.267	0.207	0.467	0.007

5.32 Persons per Housing Unit. The random variable Y is the number of persons living in a randomly selected occupied housing unit. Its probability distribution is as follows.

y	1	2	3	4	5	6	7
$P(Y = y)$	0.265	0.327	0.161	0.147	0.065	0.022	0.013

5.33 Major Hurricanes. The random variable Y is the number of major hurricanes for a randomly selected year between 1851 and 2012. Its probability distribution is as follows.

y	$P(Y = y)$	y	$P(Y = y)$
0	0.185	5	0.056
1	0.296	6	0.037
2	0.266	7	0.012
3	0.093	8	0.006
4	0.049		

5.34 Children's Gender. The random variable X is the number of girls of four children born to a couple that is equally likely to have either a boy or a girl. Its probability distribution is as follows.

x	0	1	2	3	4
$P(X = x)$	0.0625	0.2500	0.3750	0.2500	0.0625

5.35 Dice. The random variable Y is the sum of the dice when two balanced dice are rolled. Its probability distribution is as follows.

y	2	3	4	5	6	7	8	9	10	11	12
$P(Y = y)$	$\frac{1}{36}$	$\frac{1}{18}$	$\frac{1}{12}$	$\frac{1}{9}$	$\frac{5}{36}$	$\frac{1}{6}$	$\frac{5}{36}$	$\frac{1}{9}$	$\frac{1}{12}$	$\frac{1}{18}$	$\frac{1}{36}$

5.36 World Series. The World Series in baseball is won by the first team to win four games (ignoring the 1903 and 1919–1921 World Series, when it was a best of nine). From the document *World Series History* on the Baseball Almanac website, as of November 2013, the lengths of the World Series are as given in the following table.

Number of games	Frequency	Relative frequency
4	21	0.200
5	24	0.229
6	24	0.229
7	36	0.343

Let X denote the number of games that it takes to complete a World Series, and let Y denote the number of games that it took to complete a randomly selected World Series from among those considered in the table.
a. Determine the mean and standard deviation of the random variable Y. Interpret your results.
b. Provide an estimate for the mean and standard deviation of the random variable X. Explain your reasoning.

5.37 Archery. An archer shoots an arrow into a square target 6 feet on a side whose center we call the origin. The outcome of this random experiment is the point in the target hit by the arrow. The archer scores 10 points if she hits the bull's eye—a disk of radius 1 foot centered at the origin; she scores 5 points if she hits the ring with inner radius 1 foot and outer radius 2 feet centered at the origin; and she scores 0 points otherwise. Assume that the archer will actually hit the target and is equally likely to hit any portion of the target. For one arrow shot, let S be the score. A probability distribution for the random variable S is as follows.

s	0	5	10
$P(S = s)$	0.651	0.262	0.087

a. On average, how many points will the archer score per arrow shot?
b. Obtain and interpret the standard deviation of the score per arrow shot.

5.38 All-Numeric Passwords. The technology consultancy Data-Genetics published the online document *PIN analysis*. In addition to analyzing PIN numbers, passwords trends were examined. Seven million all-numeric passwords were collected and yielded the following estimate of the probability distribution of the number of digits used in an all-numeric password.

Digits	Probability	Digits	Probability
4	0.492	9	0.029
5	0.092	10	0.015
6	0.178	11	0.004
7	0.073	12	0.003
8	0.113	13	0.001

On average, approximately how many digits would you expect an all-numeric password to have? Explain your answer.

Expected Value. *As noted in Definition 5.4 on page 232, the mean of a random variable is also called its expected value. This terminology is especially useful in gambling, decision theory, and the insurance industry, as illustrated in Exercises 5.39–5.42.*

5.39 Roulette. An American roulette wheel contains 38 numbers: 18 are red, 18 are black, and 2 are green. When the roulette wheel is spun, the ball is equally likely to land on any of the 38 numbers. Suppose that you bet $1 on red. If the ball lands on a red number, you win $1; otherwise you lose your $1. Let X be the amount you win on your $1 bet. Then X is a random variable whose probability distribution is as follows.

x	1	−1
$P(X = x)$	0.474	0.526

a. Verify that the probability distribution is correct.
b. Find the expected value of the random variable X.
c. On average, how much will you lose per play?
d. Approximately how much would you expect to lose if you bet $1 on red 100 times? 1000 times?
e. Is roulette a profitable game to play? Explain.

5.40 Evaluating Investments. An investor plans to put $50,000 in one of four investments. The return on each investment depends on whether next year's economy is strong or weak. The following table summarizes the possible payoffs, in dollars, for the four investments.

	Next year's economy	
Investment	Strong	Weak
Certificate of deposit	6,000	6,000
Office complex	15,000	5,000
Land speculation	33,000	−17,000
Technical school	5,500	10,000

Let V, W, X, and Y denote the payoffs for the certificate of deposit, office complex, land speculation, and technical school, respectively. Then V, W, X, and Y are random variables. Assume that next year's economy has a 40% chance of being strong and a 60% chance of being weak.

a. Find the probability distribution of each random variable V, W, X, and Y.
b. Determine the expected value of each random variable.
c. Which investment has the best expected payoff? the worst?
d. Which investment would you select? Explain.

5.41 Homeowner's Policy. An insurance company wants to design a homeowner's policy for mid-priced homes. From data compiled by the company, it is known that the annual claim amount, X, in thousands of dollars, per homeowner is a random variable with the following probability distribution.

x	0	10	50	100	200
$P(X = x)$	0.95	0.045	0.004	0.0009	0.0001

a. Determine the expected annual claim amount per homeowner.
b. How much should the insurance company charge for the annual premium if it wants to average a net profit of $50 per policy?

5.42 Expected Utility. One method for deciding among various investments involves the concept of **expected utility**. Economists describe the importance of various levels of wealth by using **utility functions**. For instance, in most cases, a single dollar is more important (has greater utility) for someone with little wealth than for someone with great wealth. Consider two investments, say, Investment A and Investment B. Measured in thousands of dollars, suppose that Investment A yields 0, 1, and 4 with probabilities 0.1, 0.5, and 0.4, respectively, and that Investment B yields 0, 1, and 16 with probabilities 0.5, 0.3, and 0.2, respectively. Let Y denote the yield of an investment. For the two investments, determine and compare
a. the mean of Y, the expected yield.
b. the mean of $\sqrt{Y}$, the expected utility, using the utility function $u(y) = \sqrt{y}$. Interpret the utility function u.
c. the mean of $Y^{3/2}$, the expected utility, using the utility function $v(y) = y^{3/2}$. Interpret the utility function v.

5.43 Equipment Breakdowns. A factory manager collected data on the number of equipment breakdowns per day. From those data, she derived the probability distribution shown in the following table, where W denotes the number of breakdowns on a given day.

w	0	1	2
$P(W = w)$	0.80	0.15	0.05

a. Determine μ_W and σ_W. Round your answer for the standard deviation to three decimal places.

b. On average, how many breakdowns occur per day?

c. About how many breakdowns are expected during a 1-year period, assuming 250 work days per year?

Extending the Concepts and Skills

5.44 Simulation. Let X be the value of a randomly selected decimal digit, that is, a whole number between 0 and 9, inclusive.

a. Use simulation to estimate the mean of X. Explain your reasoning.

b. Obtain the exact mean of X by applying Definition 5.4 on page 232. Compare your result with that in part (a).

5.45 Queuing Simulation. Benny's Barber Shop in Cleveland has five chairs for waiting customers. The number of customers waiting is a random variable Y with the following probability distribution.

y	0	1	2	3	4	5
$P(Y = y)$	0.424	0.161	0.134	0.111	0.093	0.077

a. Compute and interpret the mean of the random variable Y.

b. In a large number of independent observations, how many customers will be waiting, on average?

c. Use the technology of your choice to simulate 500 observations of the number of customers waiting.

d. Obtain the mean of the observations in part (c) and compare it to μ_Y.

e. What does part (d) illustrate?

5.46 Mean as Center of Gravity. Let X be a discrete random variable with a finite number of possible values, say, x_1, $x_2, \ldots, x_m$. For convenience, set $p_k = P(X = x_k)$, for $k = 1$, $2, \ldots, m$. Think of a horizontal axis as a seesaw and each p_k as a mass placed at point x_k on the seesaw. The *center of gravity* of these masses is defined to be the point c on the horizontal axis at which a fulcrum could be placed to balance the seesaw.

Relative to the center of gravity, the torque acting on the seesaw by the mass p_k is proportional to the product of that mass with the signed distance of the point x_k from c, that is, to $(x_k - c) \cdot p_k$. Show that the center of gravity equals the mean of the random variable X. (*Hint:* To balance, the total torque acting on the seesaw must be 0.)

Properties of the Mean and Standard Deviation. In Exercises 5.47 and 5.48, you will develop some important properties of the mean and standard deviation of a random variable. Two of them relate the mean and standard deviation of the sum of two random variables to the individual means and standard deviations, respectively; two others relate the mean and standard deviation of a constant times a random variable to the constant and the mean and standard deviation of the random variable, respectively.

In developing these properties, you will need to use the concept of independent random variables. Two discrete random variables, X and Y, are said to be *independent random variables* if

$$P(\{X = x\} \& \{Y = y\}) = P(X = x) \cdot P(Y = y)$$

for all x and y—that is, if the joint probability distribution of X and Y equals the product of their marginal probability distributions. This condition is equivalent to requiring that events $\{X = x\}$ and $\{Y = y\}$ are independent for all x and y. A similar definition holds for independence of more than two discrete random variables.

5.47 Equipment Breakdowns. Refer to Exercise 5.43. Assume that the number of breakdowns on different days are independent of one another. Let X and Y denote the number of breakdowns on each of two consecutive days.

a. Complete the preceding joint probability distribution table. *Hint:* To obtain the joint probability in the first row, third column, use the definition of independence for discrete random variables and the table in Exercise 5.43:

$$P(\{X = 0\} \& \{Y = 2\}) = P(X = 0) \cdot P(Y = 2)$$
$$= 0.80 \cdot 0.05 = 0.04.$$

b. Use the joint probability distribution you obtained in part (a) to determine the probability distribution of the random variable $X + Y$, the total number of breakdowns in two days; that is, complete the following table.

u	0	1	2	3	4
$P(X + Y = u)$					

c. Use part (b) to find μ_{X+Y} and σ^2_{X+Y}.

d. Use part (c) to verify that the following equations hold for this example:

$$\mu_{X+Y} = \mu_X + \mu_Y \quad \text{and} \quad \sigma^2_{X+Y} = \sigma^2_X + \sigma^2_Y.$$

(*Note:* The mean and variance of X and Y are the same as that of W in Exercise 5.43.)

e. The equations in part (d) hold in general: If X and Y are any two random variables,

$$\mu_{X+Y} = \mu_X + \mu_Y.$$

In addition, if X and Y are independent,

$$\sigma^2_{X+Y} = \sigma^2_X + \sigma^2_Y.$$

Interpret these two equations in words.

5.48 Equipment Breakdowns. The factory manager in Exercise 5.43 estimates that each breakdown costs the company $500 in repairs and loss of production. If W is the number of breakdowns in a day, then $\$500W$ is the cost of breakdowns for that day.

a. Refer to the probability distribution shown in Exercise 5.43 and determine the probability distribution of the random variable $500W$.

b. Determine the mean daily breakdown cost, μ_{500W}, by using your answer from part (a).

c. What is the relationship between μ_{500W} and μ_W? (*Note:* From Exercise 5.43, $\mu_W = 0.25$.)

d. Find σ_{500W} by using your answer from part (a).

e. What is the apparent relationship between σ_{500W} and σ_W? (*Note:* From Exercise 5.43, $\sigma_W = 0.536$.)

f. The results in parts (c) and (e) hold in general: If W is any random variable and c is a constant,

$$\mu_{cW} = c\mu_W \quad \text{and} \quad \sigma_{cW} = |c|\sigma_W.$$

Interpret these two equations in words.

5.3 The Binomial Distribution*

Many applications of probability and statistics concern the repetition of an experiment. We call each repetition a **trial,** and we are particularly interested in cases in which the experiment (each trial) has only two possible outcomes. Here are three examples.

- Testing the effectiveness of a drug: Several patients take the drug (the trials), and for each patient, the drug is either effective or not effective (the two possible outcomes).
- Weekly sales of a car salesperson: The salesperson has several customers during the week (the trials), and for each customer, the salesperson either makes a sale or does not make a sale (the two possible outcomes).
- Taste tests for colas: A number of people taste two different colas (the trials), and for each person, the preference is either for the first cola or for the second cola (the two possible outcomes).

To analyze repeated trials of an experiment that has two possible outcomes, we require knowledge of factorials, binomial coefficients, Bernoulli trials, and the binomial distribution. We begin with factorials.

Factorials

Factorials are defined as follows.

DEFINITION 5.6

? **What Does It Mean?**

The factorial of a counting number is obtained by successively multiplying it by the next-smaller counting number until reaching 1.

Factorials

The product of the first k positive integers (counting numbers) is called **k factorial** and is denoted **$k!$**. In symbols,

$$k! = k(k-1)\cdots 2 \cdot 1.$$

We also define $0! = 1$.

We illustrate the calculation of factorials in the next example.

■■■ **EXAMPLE 5.9** Factorials

You try it!

Exercise 5.57 on page 248

Doing the Calculations Determine 3!, 4!, and 5!.

Solution Applying Definition 5.6 gives $3! = 3 \cdot 2 \cdot 1 = 6$, $4! = 4 \cdot 3 \cdot 2 \cdot 1 = 24$, and $5! = 5 \cdot 4 \cdot 3 \cdot 2 \cdot 1 = 120$. ■

Notice that $6! = 6 \cdot 5!$, $6! = 6 \cdot 5 \cdot 4!$, $6! = 6 \cdot 5 \cdot 4 \cdot 3!$, and so on. In general, if $j \le k$, then $k! = k(k-1)\cdots(k-j+1)(k-j)!$.

Binomial Coefficients

You may have already encountered *binomial coefficients* in algebra when you studied the binomial expansion, the expansion of $(a+b)^n$.

DEFINITION 5.7

Binomial Coefficients

If n is a positive integer and x is a nonnegative integer less than or equal to n, then the **binomial coefficient** $\binom{n}{x}$ is defined as

$$\binom{n}{x} = \frac{n!}{x!\,(n-x)!}.^{\dagger}$$

[†] If you have read Section 4.8, you will note that the binomial coefficient $\binom{n}{x}$ equals the number of possible combinations of x objects from a collection of n objects.

■■■ **EXAMPLE 5.10**   Binomial Coefficients

Doing the Calculations Determine the value of each binomial coefficient.

a. $\binom{6}{1}$ b. $\binom{5}{3}$ c. $\binom{7}{3}$ d. $\binom{4}{4}$

Solution We apply Definition 5.7.

You try it!

a. $\binom{6}{1} = \dfrac{6!}{1!\,(6-1)!} = \dfrac{6!}{1!\,5!} = \dfrac{6 \cdot 5!}{1!\,5!} = \dfrac{6}{1} = 6$

b. $\binom{5}{3} = \dfrac{5!}{3!\,(5-3)!} = \dfrac{5!}{3!\,2!} = \dfrac{5 \cdot 4 \cdot 3!}{3!\,2!} = \dfrac{5 \cdot 4}{2} = 10$

c. $\binom{7}{3} = \dfrac{7!}{3!\,(7-3)!} = \dfrac{7!}{3!\,4!} = \dfrac{7 \cdot 6 \cdot 5 \cdot 4!}{3!\,4!} = \dfrac{7 \cdot 6 \cdot 5}{6} = 35$

d. $\binom{4}{4} = \dfrac{4!}{4!\,(4-4)!} = \dfrac{4!}{4!\,0!} = \dfrac{4!}{4!\,0!} = \dfrac{1}{1} = 1$

Exercise 5.59
on page 248

■

Bernoulli Trials

Next we define *Bernoulli trials* and some related concepts.

DEFINITION 5.8

What Does It Mean?

Bernoulli trials are identical and independent repetitions of an experiment with two possible outcomes.

Bernoulli Trials

Repeated trials of an experiment are called **Bernoulli trials** if the following three conditions are satisfied:

1. The experiment (each trial) has two possible outcomes, denoted generically *s*, for **success**, and *f*, for **failure.**
2. The trials are independent, meaning that the outcome on one trial in no way affects the outcome on other trials.
3. The probability of a success, called the **success probability** and denoted *p*, remains the same from trial to trial.

Introducing the Binomial Distribution

The **binomial distribution** is the probability distribution for the number of successes in a sequence of Bernoulli trials.

■■■ **EXAMPLE 5.11**   Introducing the Binomial Distribution

Mortality Mortality tables enable actuaries to obtain the probability that a person at any particular age will live a specified number of years. Insurance companies and others use such probabilities to determine life-insurance premiums, retirement pensions, and annuity payments.

According to tables provided by the National Center for Health Statistics in *Vital Statistics of the United States*, a person of age 20 years has about an 80% chance of being alive at age 65 years. Suppose three people of age 20 years are selected at random.

a. Formulate the process of observing which people are alive at age 65 as a sequence of three Bernoulli trials.
b. Obtain the possible outcomes of the three Bernoulli trials.
c. Determine the probability of each outcome in part (b).
d. Find the probability that exactly two of the three people will be alive at age 65.
e. Obtain the probability distribution of the number of people of the three who are alive at age 65.

Solution

a. Each trial consists of observing whether a person currently of age 20 is alive at age 65 and has two possible outcomes: alive or dead. The trials are independent. If we let a success, s, correspond to being alive at age 65, the success probability is 0.8 (80%); that is, $p = 0.8$.

TABLE 5.13

Possible outcomes

sss	ssf	sfs	sff
fss	fsf	ffs	fff

b. The possible outcomes of the three Bernoulli trials are shown in Table 5.13 (s = success = alive, f = failure = dead). For instance, ssf represents the outcome that at age 65 the first two people are alive and the third is not.

c. As Table 5.13 indicates, eight outcomes are possible. However, because these eight outcomes are not equally likely, we cannot use the f/N rule to determine their probabilities; instead, we must proceed as follows. First of all, by part (a), the success probability equals 0.8, or

$$P(s) = p = 0.8.$$

Therefore the failure probability is

$$P(f) = 1 - p = 1 - 0.8 = 0.2.$$

Now, because the trials are independent, we can apply the special multiplication rule (Formula 4.7 on page 196) to obtain the probability of each outcome. For instance, the probability of the outcome ssf is

$$P(ssf) = P(s) \cdot P(s) \cdot P(f) = 0.8 \cdot 0.8 \cdot 0.2 = 0.128.$$

All eight possible outcomes and their probabilities are shown in Table 5.14. Note that outcomes containing the same number of successes have the same probability. For instance, the three outcomes containing exactly two successes—ssf, sfs, and fss—have the same probability: 0.128. Each probability is the product of two success probabilities of 0.8 and one failure probability of 0.2.

A tree diagram is useful for organizing and summarizing the possible outcomes of this experiment and their probabilities. See Fig. 5.4.

TABLE 5.14

Outcomes and probabilities for observing whether each of three people is alive at age 65

Outcome	Probability
sss	$(0.8)(0.8)(0.8) = 0.512$
ssf	$(0.8)(0.8)(0.2) = 0.128$
sfs	$(0.8)(0.2)(0.8) = 0.128$
sff	$(0.8)(0.2)(0.2) = 0.032$
fss	$(0.2)(0.8)(0.8) = 0.128$
fsf	$(0.2)(0.8)(0.2) = 0.032$
ffs	$(0.2)(0.2)(0.8) = 0.032$
fff	$(0.2)(0.2)(0.2) = 0.008$

FIGURE 5.4

Tree diagram corresponding to Table 5.14

First person	Second person	Third person	Outcome	Probability
	s 0.8	s	sss	$(0.8)(0.8)(0.8) = 0.512$
s 0.8		0.2 f	ssf	$(0.8)(0.8)(0.2) = 0.128$
	0.2 f 0.8	s	sfs	$(0.8)(0.2)(0.8) = 0.128$
0.8		0.2 f	sff	$(0.8)(0.2)(0.2) = 0.032$
	s 0.8	s	fss	$(0.2)(0.8)(0.8) = 0.128$
0.2 f 0.8		0.2 f	fsf	$(0.2)(0.8)(0.2) = 0.032$
	0.2 f 0.8	s	ffs	$(0.2)(0.2)(0.8) = 0.032$
		0.2 f	fff	$(0.2)(0.2)(0.2) = 0.008$

d. Table 5.14 shows that the event that exactly two of the three people are alive at age 65 consists of the outcomes ssf, sfs, and fss. So, by the special addition rule (Formula 4.1 on page 173),

$$P(\text{Exactly two will be alive}) = P(ssf) + P(sfs) + P(fss)$$

$$= \underbrace{0.128 + 0.128 + 0.128}_{3 \text{ times}} = 3 \cdot 0.128 = 0.384.$$

The probability that exactly two of the three people will be alive at age 65 is 0.384.

You try it!

Exercise 5.65 on page 248

FIGURE 5.5

Probability histogram for the random variable X, the number of people of three who are alive at age 65

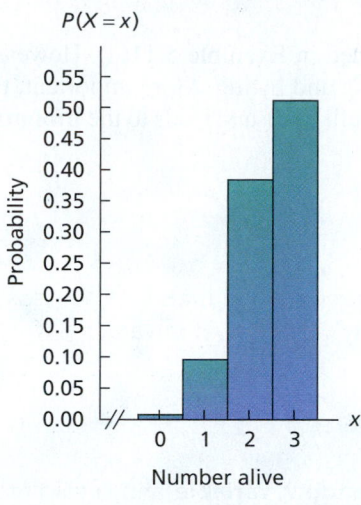

$P(X = x)$

e. Let X denote the number of people of the three who are alive at age 65. In part (d), we found $P(X = 2)$. We can proceed in the same way to find the remaining three probabilities: $P(X = 0)$, $P(X = 1)$, and $P(X = 3)$. The results are given in Table 5.15 and also in the probability histogram in Fig. 5.5. Note for future reference that this probability distribution is left skewed.

TABLE 5.15

Probability distribution of the random variable X, the number of people of three who are alive at age 65

Number alive x	Probability $P(X = x)$
0	0.008
1	0.096
2	0.384
3	0.512

The Binomial Probability Formula

We obtained the probability distribution in Table 5.15 by using a tabulation method (Table 5.14), which required much work. In most practical applications, the amount of work required would be even more and often would be prohibitive because the number of trials is generally much larger than three. For instance, with twenty, rather than three, 20-year-olds, there would be over 1 million possible outcomes. The tabulation method certainly would not be feasible in that case.

The good news is that a relatively simple formula will give us binomial probabilities. Before we develop that formula, we need the following fact.

KEY FACT 5.4

What Does It Mean?

There are $\binom{n}{x}$ ways of getting exactly x successes in n Bernoulli trials.

Number of Outcomes Containing a Specified Number of Successes

In n Bernoulli trials, the number of outcomes that contain exactly x successes equals the binomial coefficient $\binom{n}{x}$.

We won't stop to prove Key Fact 5.4, but let's check it against the results in Example 5.11. For instance, in Table 5.13, we saw that there are three outcomes in which exactly two of the three people are alive at age 65. Key Fact 5.4 gives us that information more easily:

$$\begin{bmatrix} \text{Number of outcomes} \\ \text{comprising the event} \\ \text{exactly two alive} \end{bmatrix} = \binom{3}{2} = \frac{3!}{2!\,(3-2)!} = \frac{3!}{2!\,1!} = 3.$$

We can now develop a probability formula for the number of successes in Bernoulli trials. We illustrate how that formula is derived by referring to Example 5.11. For instance, to determine the probability that exactly two of the three people will be alive at age 65, $P(X = 2)$, we reason as follows:

1. Any particular outcome in which exactly two of the three people are alive at age 65 (e.g., sfs) has probability

$$\underset{\underset{\text{Probability}}{\uparrow}}{\overset{\overset{\text{Two alive}}{\downarrow}}{(0.8)^2}} \cdot \underset{\underset{\text{Probability}}{\uparrow}}{\overset{\overset{\text{One dead}}{\downarrow}}{(0.2)^1}} = 0.64 \cdot 0.2 = 0.128.$$

Probability alive   Probability dead

2. By Key Fact 5.4, the number of outcomes in which exactly two of the three people are alive at age 65 is

Number of trials

$$\underset{\underset{\text{Number alive}}{\uparrow}}{\binom{3}{2}} = \frac{3!}{2!\,(3-2)!} = 3.$$

3. By the special addition rule, the probability that exactly two of the three people will be alive at age 65 is

$$P(X = 2) = \binom{3}{2} \cdot (0.8)^2(0.2)^1 = 3 \cdot 0.128 = 0.384.$$

Of course, this result is the same as that obtained in Example 5.11(d). However, this time we found the probability without tabulating and listing. More important, the reasoning we used applies to any sequence of Bernoulli trials and leads to the **binomial probability formula.**

FORMULA 5.1

Binomial Probability Formula

Let X denote the total number of successes in n Bernoulli trials with success probability p. Then the probability distribution of the random variable X is given by

$$P(X = x) = \binom{n}{x} p^x(1 - p)^{n-x}, \qquad x = 0, 1, 2, \ldots, n.$$

The random variable X is called a **binomial random variable** and is said to have the **binomial distribution** with parameters n and p.

APPLET

Applet 5.1

To determine a binomial probability formula in specific problems, having a well-organized strategy, such as the one presented in Procedure 5.1, is useful.

PROCEDURE 5.1

To Find a Binomial Probability Formula

Assumptions

1. n trials are to be performed.
2. Two outcomes, success or failure, are possible for each trial.
3. The trials are independent.
4. The success probability, p, remains the same from trial to trial.

Step 1 Identify a success.

Step 2 Determine p, the success probability.

Step 3 Determine n, the number of trials.

Step 4 The binomial probability formula for the number of successes, X, is

$$P(X = x) = \binom{n}{x} p^x(1 - p)^{n-x}.$$

In the following example, we illustrate this procedure by applying it to the random variable considered in Example 5.11.

EXAMPLE 5.12 **Obtaining Binomial Probabilities**

Mortality According to tables provided by the National Center for Health Statistics in *Vital Statistics of the United States*, there is roughly an 80% chance that a person of age 20 years will be alive at age 65 years. Suppose that three people of age 20 years are selected at random. Find the probability that the number alive at age 65 years will be

a. exactly two.   **b.** at most one.   **c.** at least one.

d. Determine the probability distribution of the number alive at age 65.

Solution   Let X denote the number of people of the three who are alive at age 65. To solve parts (a)–(d), we first apply Procedure 5.1.

Step 1   Identify a success.

A success is that a person currently of age 20 will be alive at age 65.

Step 2   Determine p, the success probability.

The probability that a person currently of age 20 will be alive at age 65 is 80%, so $p = 0.8$.

Step 3   Determine n, the number of trials.

The number of trials is the number of people in the study, which is three, so $n = 3$.

Step 4   The binomial probability formula for the number of successes, X, is

$$P(X = x) = \binom{n}{x} p^x (1 - p)^{n-x}.$$

Because $n = 3$ and $p = 0.8$, the formula becomes

$$P(X = x) = \binom{3}{x}(0.8)^x (0.2)^{3-x}.$$

We see that X is a binomial random variable and has the binomial distribution with parameters $n = 3$ and $p = 0.8$. Now we can solve parts (a)–(d) relatively easily.

a.   Applying the binomial probability formula with $x = 2$ yields

$$P(X = 2) = \binom{3}{2}(0.8)^2(0.2)^{3-2} = \frac{3!}{2!\,(3-2)!}(0.8)^2(0.2)^1 = 0.384.$$

Interpretation   Chances are 38.4% that exactly two of the three people will be alive at age 65.

b.   The probability that at most one person will be alive at age 65 is

$$P(X \le 1) = P(X = 0) + P(X = 1)$$

$$= \binom{3}{0}(0.8)^0(0.2)^{3-0} + \binom{3}{1}(0.8)^1(0.2)^{3-1}$$

$$= 0.008 + 0.096 = 0.104.$$

Interpretation   Chances are 10.4% that one or fewer of the three people will be alive at age 65.

c.   The probability that at least one person will be alive at age 65 is $P(X \ge 1)$, which we can obtain by first using the fact that

$$P(X \ge 1) = P(X = 1) + P(X = 2) + P(X = 3)$$

and then applying the binomial probability formula to calculate each of the three individual probabilities. However, using the complementation rule is easier:

$$P(X \ge 1) = 1 - P(X < 1) = 1 - P(X = 0)$$

$$= 1 - \binom{3}{0}(0.8)^0(0.2)^{3-0} = 1 - 0.008 = 0.992.$$

Interpretation   Chances are 99.2% that one or more of the three people will be alive at age 65.

d.   To obtain the probability distribution of the random variable X, we need to use the binomial probability formula to compute $P(X = x)$ for $x = 0, 1, 2$, and 3.

Report 5.1

**Exercise 5.77(a)–(e)
on page 249**

We have already done so for $x = 0$, 1, and 2 in parts (a) and (b). For $x = 3$, we have

$$P(X = 3) = \binom{3}{3}(0.8)^3(0.2)^{3-3} = (0.8)^3 = 0.512.$$

Thus the probability distribution of X is as shown in Table 5.15 on page 241. This time, however, we computed the probabilities quickly and easily by using the binomial probability formula. ■

Note: The probability $P(X \le 1)$ required in part (b) of the previous example is a *cumulative probability*. In general, a **cumulative probability** is the probability that a random variable is less than or equal to a specified number, that is, a probability of the form $P(X \le x)$. The concept of cumulative probability applies to any random variable, not just binomial random variables.

We can express the probability that a random variable lies between two specified numbers—say, a and b—in terms of cumulative probabilities:

$$P(a < X \le b) = P(X \le b) - P(X \le a).$$

Binomial Probability Tables

Because of the importance of the binomial distribution, tables of binomial probabilities have been extensively compiled. Table XI in Appendix A gives a binomial probability table for values of n between 1 and 7. As you can see, the table displays the number of trials, n, in the far left column; the number of successes, x, in the next column to the right; and the success probability, p, across the top row.

To illustrate a use of Table XI, we determine the probability required in part (a) of the preceding example. The number of trials is three ($n = 3$), and the success probability is 0.8 ($p = 0.8$). To find the required probability, $P(X = 2)$, we first go down the leftmost column, labeled n, to "3." Next we concentrate on the row for x labeled "2." Then, going across that row to the column labeled "0.8," we reach 0.384. This number is the required probability, that is, $P(X = 2) = 0.384$.

Binomial probability tables eliminate most of the computations required in working with the binomial distribution. Such tables are of limited usefulness, however, because they contain only a relatively small number of different values of n and p. For instance, Table XI has only 11 different values of p and stops at $n = 7$. A more extensive (but still of limited usefulness) binomial probability table is available on the WeissStats site.

Consequently, if we want to determine a binomial probability whose n or p parameter is not included in the table, we must use either the binomial probability formula or statistical software. The latter method is discussed at the end of this section.

Shape of a Binomial Distribution

Figure 5.5 on page 241 shows that, for three people currently 20 years old, the probability distribution of the number who will be alive at age 65 is left skewed. The reason is that the success probability, $p = 0.8$, exceeds 0.5.

Applet 5.2

More generally, *a binomial distribution is right skewed if $p < 0.5$, is symmetric if $p = 0.5$, and is left skewed if $p > 0.5$*. Figure 5.6 illustrates these facts for three different binomial distributions with $n = 6$.

Mean and Standard Deviation of a Binomial Random Variable

In Section 5.2, we discussed the mean and standard deviation of a discrete random variable. We presented formulas to compute these parameters in Definition 5.4 on page 232 and Definition 5.5 on page 234.

Because these formulas apply to any discrete random variable, they work for a binomial random variable. Hence we can determine the mean and standard deviation of a binomial random variable by first using the binomial probability formula to obtain its probability distribution and then applying Definitions 5.4 and 5.5.

FIGURE 5.6

Probability histograms for three different binomial distributions with parameter $n = 6$

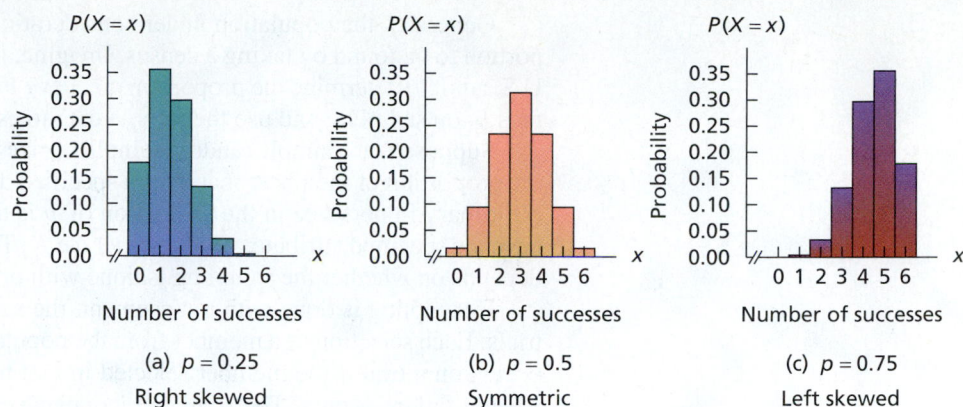

(a) $p = 0.25$
Right skewed

(b) $p = 0.5$
Symmetric

(c) $p = 0.75$
Left skewed

You try it!

Exercise 5.77(f)–(g) on page 249

However, there is an easier way. If we substitute the binomial probability formula into the formulas for the mean and standard deviation of a discrete random variable and then simplify mathematically, we obtain the following formulas.

FORMULA 5.2

? What Does It Mean?

The mean of a binomial random variable equals the product of the number of trials and success probability; its standard deviation equals the square root of the product of the number of trials, success probability, and failure probability.

Mean and Standard Deviation of a Binomial Random Variable

The mean and standard deviation of a binomial random variable with parameters n and p are

$$\mu = np \quad \text{and} \quad \sigma = \sqrt{np(1 - p)},$$

respectively.

In the next example, we apply the two formulas in Formula 5.2 to determine the mean and standard deviation of the binomial random variable considered in the mortality illustration.

■ ■ ■   EXAMPLE 5.13   Mean and Standard Deviation of a Binomial Random Variable

Mortality For three randomly selected 20-year-olds, let X denote the number who are still alive at age 65. Find the mean and standard deviation of X.

Solution As we stated in the previous example, X is a binomial random variable with parameters $n = 3$ and $p = 0.8$. Applying Formula 5.2 gives

$$\mu = np = 3 \cdot 0.8 = 2.4$$

and

$$\sigma = \sqrt{np(1 - p)} = \sqrt{3 \cdot 0.8 \cdot 0.2} = 0.7.$$

Interpretation On average, 2.4 of every three 20-year-olds will still be alive at age 65. And, roughly speaking, on average, the number out of three given 20-year-olds who will still be alive at age 65 will differ from the mean number of 2.4 by 0.7.

You try it!

Exercise 5.77(h)–(i) on page 249

Binomial Approximation to the Hypergeometric Distribution

We often want to determine the proportion (percentage) of members of a finite population that have a specified attribute. For instance, we might be interested in the proportion of U.S. adults that have Internet access. Here the population consists of all U.S. adults, and the attribute is "has Internet access." Or we might want to know the proportion of U.S. businesses that are minority owned. In this case, the population consists of all U.S. businesses, and the attribute is "minority owned."

Generally, the population under consideration is too large for the population proportion to be found by taking a census. Imagine, for instance, trying to interview every U.S. adult to determine the proportion that have Internet access. So, in practice, we rely mostly on sampling and use the sample data to estimate the population proportion.

Suppose that a simple random sample of size n is taken from a population in which the proportion of members that have a specified attribute is p. Then a random variable of primary importance in the estimation of p is the number of members sampled that have the specified attribute, which we denote X. The exact probability distribution of X depends on whether the sampling is done with or without replacement.

If sampling is done with replacement, the sampling process constitutes Bernoulli trials: Each selection of a member from the population corresponds to a trial. A success occurs on a trial if the member selected in that trial has the specified attribute; otherwise, a failure occurs. The trials are independent because the sampling is done with replacement. The success probability remains the same from trial to trial—it always equals the proportion of the population that has the specified attribute. Therefore the random variable X has the binomial distribution with parameters n (the sample size) and p (the population proportion).

In reality, however, sampling is ordinarily done without replacement. Under these circumstances, the sampling process does not constitute Bernoulli trials because the trials are not independent and the success probability varies from trial to trial. In other words, the random variable X does not have a binomial distribution. Its distribution is important, however, and is referred to as a **hypergeometric distribution.**

We won't present the hypergeometric probability formula here because, in practice, a hypergeometric distribution can usually be approximated by a binomial distribution. The reason is that, if the sample size does not exceed 5% of the population size, there is little difference between sampling with and without replacement. We summarize the previous discussion as follows.

KEY FACT 5.5

? **What Does It Mean?**

When a simple random sample is taken from a finite population, you can use a binomial distribution for the number of members obtained having a specified attribute, regardless of whether the sampling is with or without replacement, provided that, in the latter case, the sample size is small relative to the population size.

Sampling and the Binomial Distribution

Suppose that a simple random sample of size n is taken from a finite population in which the proportion of members that have a specified attribute is p. Then the number of members sampled that have the specified attribute

- has exactly a binomial distribution with parameters n and p if the sampling is done with replacement and
- has approximately a binomial distribution with parameters n and p if the sampling is done without replacement and the sample size does not exceed 5% of the population size.

For example, according to the U.S. Census Bureau publication *Current Population Survey*, 86.8% of U.S. adults have completed high school. Suppose that eight U.S. adults are to be randomly selected without replacement. Let X denote the number of those sampled that have completed high school. Then, because the sample size does not exceed 5% of the population size, the random variable X has approximately a binomial distribution with parameters $n = 8$ and $p = 0.868$.

Other Discrete Probability Distributions

The binomial distribution is the most important and most widely used discrete probability distribution. Other common discrete probability distributions are the Poisson, hypergeometric, and geometric distributions, which you are asked to consider in the exercises. We discuss the Poisson distribution in detail in Section 5.4.

 **THE TECHNOLOGY CENTER**

Almost all statistical technologies include programs that determine binomial probabilities. In this subsection, we present output and step-by-step instructions for such programs.

EXAMPLE 5.14   Using Technology to Obtain Binomial Probabilities

Mortality Consider once again the mortality illustration discussed in Example 5.12 on pages 242–243. Use Minitab, Excel, or the TI-83/84 Plus to determine the probability that exactly two of the three people will be alive at age 65.

Solution Recall that, of three randomly selected people of age 20 years, the number, X, who are alive at age 65 years has a binomial distribution with parameters $n = 3$ and $p = 0.8$. We want the probability that exactly two of the three people will be alive at age 65 years—that is, $P(X = 2)$.

We applied the binomial probability programs, resulting in Output 5.1. Steps for generating that output are presented in Instructions 5.1. As shown in Output 5.1, the required probability is 0.384. *Note to Excel users:* The output that you get may differ somewhat from that shown in Output 5.1 depending on the version of Excel that you use.

OUTPUT 5.1   Probability that exactly two of the three people will be alive at age 65

MINITAB

Probability Density Function

Binomial with n = 3 and p = 0.8

```
x   P( X = x )
2    0.384
```

EXCEL

Function Arguments

BINOM.DIST

Number_s	2		= 2
Trials	3		= 3
Probability_s	0.8		= 0.8
Cumulative	FALSE		= FALSE

= 0.384

Returns the individual term binomial distribution probability.

                Cumulative   is a logical value: for the cumulative distribution function, use TRUE; for the probability mass function, use FALSE.

Formula result = 0.384

Help on this function   OK   Cancel

TI-83/84 PLUS

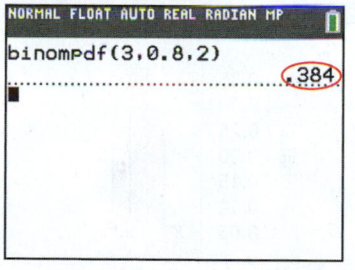

```
NORMAL FLOAT AUTO REAL RADIAN MP
binompdf(3,0.8,2)
                            .384
■
```

INSTRUCTIONS 5.1   Steps for generating Output 5.1

MINITAB

1 Choose **Calc ➤ Probability Distributions ➤ Binomial...**
2 Press the F3 key to reset the dialog box
3 Select the **Probability** option button
4 Click in the **Number of trials** text box and type 3
5 Click in the **Event probability** text box and type 0.8
6 Select the **Input constant** option button
7 Click in the **Input constant** text box and type 2
8 Click **OK**

EXCEL

1 Click **fₓ** (Insert Function)
2 Select **Statistical** from the **Or select a category** drop down list box
3 Select **BINOM.DIST** from the **Select a function** list
4 Click **OK**

5 Type 2 in the **Number_s** text box
6 Click in the **Trials** text box and type 3
7 Click in the **Probability_s** text box and type 0.8
8 Click in the **Cumulative** text box and type FALSE

TI-83/84 PLUS

FOR THE TI-84 PLUS *C*
1 Press **2nd ➤ DISTR**
2 Arrow down to **binompdf(** and press **ENTER**
3 Type 3 for **trials** and press **ENTER**
4 Type 0.8 for **p** and press **ENTER**
5 Type 2 for **x value** and press **ENTER** three times

FOR THE TI-83/84 PLUS
1 Press **2nd ➤ DISTR**
2 Arrow down to **binompdf(** and press **ENTER**
3 Type 3,0.8,2) and press **ENTER**

You can also obtain cumulative probabilities for a binomial distribution by using Minitab, Excel, or the TI-83/84 Plus. To do so, modify Instructions 5.1 as follows:

- For Minitab, in step 3, select the **Cumulative probability** option button instead of the **Probability** option button.
- For Excel, in step 8, type <u>TRUE</u> instead of <u>FALSE</u>.
- For the TI-83/84 Plus, in step 2, arrow down to **binomcdf(** instead of **binompdf(**.

Exercises 5.3

Understanding the Concepts and Skills

5.49 In probability and statistics, what is each repetition of an experiment called?

5.50 Under what three conditions are repeated trials of an experiment called Bernoulli trials?

5.51 Explain the significance of binomial coefficients with respect to Bernoulli trials.

5.52 Discuss the pros and cons of binomial probability tables.

5.53 What is the binomial distribution?

5.54 Suppose that a simple random sample is taken from a finite population in which each member is classified as either having or not having a specified attribute. Fill in the following blanks.
a. If sampling is with replacement, the probability distribution of the number of members sampled that have the specified attribute is a _____ distribution.
b. If sampling is without replacement, the probability distribution of the number of members sampled that have the specified attribute is a _____ distribution.
c. If sampling is without replacement and the sample size does not exceed _____% of the population size, the probability distribution of the number of members sampled that have the specified attribute can be approximated by a _____ distribution.

5.55 Give two examples of Bernoulli trials other than those presented in the text.

5.56 What does the "bi" in "binomial" signify?

5.57 Compute 3!, 7!, 8!, and 9!.

5.58 Find 1!, 2!, 4!, and 6!.

5.59 Evaluate the following binomial coefficients.
a. $\binom{5}{2}$ b. $\binom{7}{4}$ c. $\binom{10}{3}$ d. $\binom{12}{5}$

5.60 Evaluate the following binomial coefficients.
a. $\binom{3}{2}$ b. $\binom{6}{0}$ c. $\binom{6}{6}$ d. $\binom{7}{3}$

5.61 Evaluate the following binomial coefficients.
a. $\binom{4}{1}$ b. $\binom{6}{2}$ c. $\binom{8}{3}$ d. $\binom{9}{6}$

5.62 Determine the value of each binomial coefficient.
a. $\binom{5}{3}$ b. $\binom{10}{0}$ c. $\binom{10}{10}$ d. $\binom{9}{5}$

5.63 For each of the following probability histograms of binomial distributions, specify whether the success probability is less than, equal to, or greater than 0.5. Explain your answers.

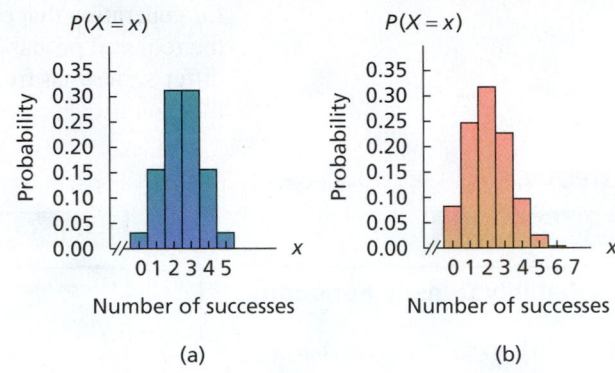

(a)   (b)

5.64 For each of the following probability histograms of binomial distributions, specify whether the success probability is less than, equal to, or greater than 0.5. Explain your answers.

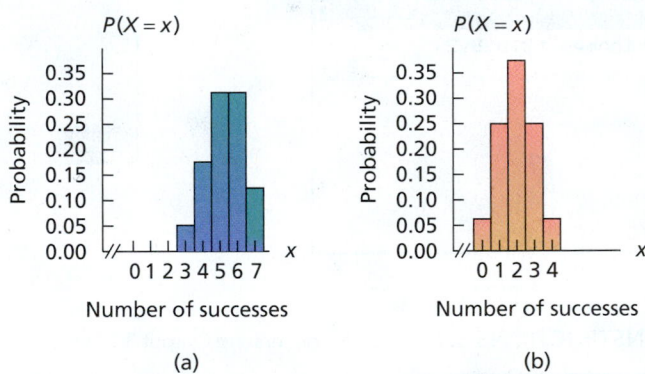

(a)   (b)

5.65 Pinworm Infestation. Pinworm infestation, which is commonly found in children, can be treated with the drug pyrantel pamoate. According to the *Merck Manual*, the treatment is effective in 90% of cases. Suppose that three children with pinworm infestation are given pyrantel pamoate.
a. Considering a success in a given case to be "a cure," formulate the process of observing which children are cured and which children are not cured as a sequence of three Bernoulli trials.
b. Construct a table similar to Table 5.14 on page 240 for the three cases. Display the probabilities to three decimal places.
c. Draw a tree diagram for this problem similar to the one shown in Fig. 5.4 on page 240.
d. List the outcomes in which exactly two of the three children are cured.
e. Find the probability of each outcome in part (d). Why are those probabilities all the same?

f. Use parts (d) and (e) to determine the probability that exactly two of the three children will be cured.

g. Without using the binomial probability formula, obtain the probability distribution of the random variable X, the number of children out of three who are cured.

5.66 Psychiatric Disorders. The National Institute of Mental Health reports that there is a 20% chance of an adult American suffering from a psychiatric disorder. Four randomly selected adult Americans are examined for psychiatric disorders.

a. If you let a success correspond to an adult American having a psychiatric disorder, what is the success probability, p? (*Note:* The use of the word *success* in Bernoulli trials need not reflect its usually positive connotation.)

b. Construct a table similar to Table 5.14 on page 240 for the four people examined. Display the probabilities to four decimal places.

c. Draw a tree diagram for this problem similar to the one shown in Fig. 5.4 on page 240.

d. List the outcomes in which exactly three of the four people examined have a psychiatric disorder.

e. Find the probability of each outcome in part (d). Why are those probabilities all the same?

f. Use parts (d) and (e) to determine the probability that exactly three of the four people examined have a psychiatric disorder.

g. Without using the binomial probability formula, obtain the probability distribution of the random variable Y, the number of adults out of four who have a psychiatric disorder.

In each of Exercises 5.67–5.72, we have provided the number of trials and success probability for Bernoulli trials. Let X denote the total number of successes. Determine the required probabilities by using

a. *the binomial probability formula, Formula 5.1 on page 242. Round your probability answers to three decimal places.*

b. *Table XI in Appendix A. Compare your answer here to that in part (a).*

5.67 $n = 4$, $p = 0.3$, $P(X = 2)$

5.68 $n = 5$, $p = 0.6$, $P(X = 3)$

5.69 $n = 6$, $p = 0.5$, $P(X = 4)$

5.70 $n = 3$, $p = 0.4$, $P(X = 1)$

5.71 $n = 5$, $p = 3/4$, $P(X = 4)$

5.72 $n = 4$, $p = 1/4$, $P(X = 2)$

Applying the Concepts and Skills

5.73 Pinworm Infestation. Use Procedure 5.1 on page 242 to solve part (g) of Exercise 5.65.

5.74 Psychiatric Disorders. Use Procedure 5.1 on page 242 to solve part (g) of Exercise 5.66.

5.75 Tossing a Coin. If we repeatedly toss a balanced coin, then, in the long run, it will come up heads about half the time. But what is the probability that such a coin will come up heads exactly half the time in 10 tosses?

5.76 Rolling a Die. If we repeatedly roll a balanced die, then, in the long run, it will come up "4" about one-sixth of the time. But what is the probability that such a die will come up "4" exactly once in six rolls?

5.77 Horse Racing. According to the *Daily Racing Form*, the probability is about 0.67 that the favorite in a horse race will finish in the money (first, second, or third place). In the next five races, what is the probability that the favorite finishes in the money

a. exactly twice?

b. exactly four times?

c. at least four times?

d. between two and four times, inclusive?

e. Determine the probability distribution of the random variable X, the number of times the favorite finishes in the money in the next five races.

f. Identify the probability distribution of X as right skewed, symmetric, or left skewed without consulting its probability distribution or drawing its probability histogram.

g. Draw a probability histogram for X.

h. Use your answer from part (e) and Definitions 5.4 and 5.5 on pages 232 and 234, respectively, to obtain the mean and standard deviation of the random variable X.

i. Use Formula 5.2 on page 245 to obtain the mean and standard deviation of the random variable X.

j. Interpret your answer for the mean in words.

5.78 Gestation Periods. The probability is 0.314 that the gestation period of a woman will exceed 9 months. In six human births, what is the probability that the number in which the gestation period exceeds 9 months is

a. exactly three?

b. exactly five?

c. at least five?

d. between three and five, inclusive?

e. Determine the probability distribution of the random variable X, the number of six human births in which the gestation period exceeds 9 months.

f. Identify the probability distribution of X as right skewed, symmetric, or left skewed without consulting its probability distribution or drawing its probability histogram.

g. Draw a probability histogram for X.

h. Use your answer from part (e) and Definitions 5.4 and 5.5 on pages 232 and 234, respectively, to obtain the mean and standard deviation of the random variable X.

i. Use Formula 5.2 on page 245 to obtain the mean and standard deviation of the random variable X.

j. Interpret your answer for the mean in words.

5.79 Traffic Fatalities and Intoxication. The National Safety Council publishes information about automobile accidents in *Accident Facts*. According to that document, the probability is 0.40 that a traffic fatality will involve an intoxicated or alcohol-impaired driver or nonoccupant. In eight traffic fatalities, find the probability that the number, Y, that involve an intoxicated or alcohol-impaired driver or nonoccupant is

a. exactly three; at least three; at most three.

b. between two and four, inclusive.

c. Find and interpret the mean of the random variable Y.

d. Obtain the standard deviation of Y.

5.80 Multiple-Choice Exams. A student takes a multiple-choice exam with 10 questions, each with four possible selections for the answer. A passing grade is 60% or better. Suppose that the student was unable to find time to study for the exam and just guesses at each question. Find the probability that the student

a. gets at least one question correct.

b. passes the exam.

c. receives an "A" on the exam (90% or better).

d. How many questions would you expect the student to get correct?

e. Obtain the standard deviation of the number of questions that the student gets correct.

5.81 Love Stinks? J. Fetto, in the article "Love Stinks" (*American Demographics*, Vol. 25, No. 1, pp. 10–11), reports that Americans split with their significant other for many reasons—including indiscretion, infidelity, and simply "growing apart." According to the article, 35% of American adults have experienced a breakup at least once during the last 10 years. Of nine randomly selected American adults, find the probability that the number, X, who have experienced a breakup at least once during the last 10 years is

a. exactly five; at most five; at least five.

b. at least one; at most one.

c. between six and eight, inclusive.

d. Determine the probability distribution of the random variable X.

e. Strictly speaking, why is the probability distribution that you obtained in part (d) only approximately correct? What is the exact distribution called?

5.82 Carbon Tax. A poll commissioned by Friends of the Earth and conducted by the Mellman Group found that 72% of American voters are in favor of a carbon tax. Suppose that six voters in the United States are randomly sampled and asked whether they favor a carbon tax. Determine the probability that the number answering in the affirmative is

a. exactly two. b. exactly four. c. at least two.

d. Determine the probability distribution of the number of American voters in a sample of six who favor a carbon tax.

e. Strictly speaking, why is the probability distribution that you obtained in part (d) only approximately correct? What is the exact distribution called?

5.83 Video Games. A pathological video game user (PVGU) is a video game user that averages 31 or more hours a week of gameplay. According to the article "Pathological Video Game Use among Youths: A Two-Year Longitudinal Study" (*Pediatrics*, Vol. 127, No. 2, pp. 319–329) by D. Gentile et al., in 2011, about 9% of children in grades 3–8 were PVGUs. Suppose that, today, seven youths in grades 3–8 are randomly selected.

a. Assuming that the percentage of PVGUs in grades 3–8 is the same today as it was in 2011, determine the probability distribution for the number, X, who are PVGUs.

b. Determine and interpret the mean of X.

c. If, in fact, exactly three of the seven youths selected are PVGUs, would you be inclined to conclude that the percentage of PVGUs in grades 3–8 has increased from the 2011 percentage? Explain your reasoning. *Hint:* First consider the probability $P(X \geq 3)$.

d. If, in fact, exactly two of the seven youths selected are PVGUs, would you be inclined to conclude that the percentage of PVGUs in grades 3–8 has increased from the 2011 percentage? Explain your reasoning.

5.84 Recidivism. In the *Scientific American* article "Reducing Crime: Rehabilitation is Making a Comeback," R. Doyle examined rehabilitation of felons. One aspect of the article discussed recidivism of juvenile prisoners between 14 and 17 years old, indicating that 82% of those released in 1994 were rearrested within 3 years. Suppose that, today, six newly released juvenile prisoners between 14 and 17 years old are selected at random.

a. Assuming that the recidivism rate is the same today as it was in 1994, determine the probability distribution for the number, Y, who are rearrested within 3 years.

b. Determine and interpret the mean of Y.

c. If, in fact, exactly two of the six newly released juvenile prisoners are rearrested within 3 years, would you be inclined to conclude that the recidivism rate today has decreased from the 82% rate in 1994? Explain your reasoning. *Hint:* First consider the probability $P(Y \leq 2)$.

d. If, in fact, exactly four of the six newly released juvenile prisoners are rearrested within 3 years, would you be inclined to conclude that the recidivism rate today has decreased from the 82% rate in 1994? Explain your reasoning.

Extending the Concepts and Skills

5.85 Roulette. A success, s, in Bernoulli trials is often derived from a collection of outcomes. For example, an American roulette wheel consists of 38 numbers, of which 18 are red, 18 are black, and 2 are green. When the roulette wheel is spun, the ball is equally likely to land on any one of the 38 numbers. If you are interested in which number the ball lands on, each play at the roulette wheel has 38 possible outcomes. Suppose, however, that you are betting on red. Then you are interested only in whether the ball lands on a red number. From this point of view, each play at the wheel has only two possible outcomes—either the ball lands on a red number or it doesn't. Hence, successive bets on red constitute a sequence of Bernoulli trials with success probability $\frac{18}{38}$. In four plays at a roulette wheel, what is the probability that the ball lands on red

a. exactly twice? b. at least once?

5.86 Sampling and the Binomial Distribution. Refer to the discussion on the binomial approximation to the hypergeometric distribution that begins on page 245.

a. If sampling is with replacement, explain why the trials are independent and the success probability remains the same from trial to trial—always the proportion of the population that has the specified attribute.

b. If sampling is without replacement, explain why the trials are not independent and the success probability varies from trial to trial.

5.87 Sampling and the Binomial Distribution. Following is a gender frequency distribution for students in Professor Weiss's introductory statistics class.

Gender	Frequency
Male	17
Female	23

Two students are selected at random. Find the probability that both students are male if the selection is done

a. with replacement.

b. without replacement.

c. Compare the answers obtained in parts (a) and (b).

Suppose that Professor Weiss's class had 10 times the students, but in the same proportions, that is, 170 males and 230 females.

d. Repeat parts (a)–(c), using this hypothetical distribution of students.

e. In which case is there less difference between sampling without and with replacement? Explain why this is so.

5.88 The Hypergeometric Distribution. In this exercise, we discuss the *hypergeometric distribution* in more detail. When sampling is done without replacement from a finite population, the hypergeometric distribution is the exact probability distribution for the number of members sampled that have a specified attribute. The hypergeometric probability formula is

$$P(X = x) = \frac{\binom{Np}{x}\binom{N(1-p)}{n-x}}{\binom{N}{n}},$$

where X denotes the number of members sampled that have the specified attribute, N is the population size, n is the sample size, and p is the population proportion.

To illustrate, suppose that a customer purchases 4 fuses from a shipment of 250, of which 94% are not defective. Let a success correspond to a fuse that is not defective.

a. Determine N, n, and p.

b. Apply the hypergeometric probability formula to determine the probability distribution of the number of nondefective fuses that the customer gets.

Key Fact 5.5 shows that a hypergeometric distribution can be approximated by a binomial distribution, provided the sample size does not exceed 5% of the population size. In particular, you can use the binomial probability formula

$$P(X = x) = \binom{n}{x}p^x(1-p)^{n-x},$$

with $n = 4$ and $p = 0.94$, to approximate the probability distribution of the number of nondefective fuses that the customer gets.

c. Obtain the binomial distribution with parameters $n = 4$ and $p = 0.94$.

d. Compare the hypergeometric distribution that you obtained in part (b) with the binomial distribution that you obtained in part (c).

5.89 The Geometric Distribution. In this exercise, we discuss the *geometric distribution,* the probability distribution for the number of trials until the first success in Bernoulli trials. The geometric probability formula is

$$P(X = x) = p(1-p)^{x-1},$$

where X denotes the number of trials until the first success and p the success probability. Using the geometric probability formula and Definition 5.4 on page 232, we can show that the mean of the random variable X is $1/p$.

To illustrate, again consider the *Mega Millions* lottery as described in Exercise 4.267 on page 217. Suppose that you buy one *Mega Millions* ticket per week. Let X denote the number of weeks until you win a prize.

a. Find and interpret the probability formula for the random variable X. (*Note:* The appropriate success probability was obtained in Exercise 4.267(d).)

b. Compute the probability that the number of weeks until you win a prize is exactly 3; at most 3; at least 3.

c. On average, how long will it be until you win a prize?

5.90 The Poisson Distribution. Another important discrete probability distribution is the *Poisson distribution,* named in honor of the French mathematician and physicist Simeon Poisson (1781–1840). This probability distribution is often used to model the frequency with which a specified event occurs during a particular period of time. The Poisson probability formula is

$$P(X = x) = e^{-\lambda}\frac{\lambda^x}{x!},$$

where X is the number of times the event occurs and λ is a parameter equal to the mean of X. The number e is the base of natural logarithms and is approximately equal to 2.7183.

To illustrate, consider the following problem: Desert Samaritan Hospital, located in Mesa, Arizona, keeps records of emergency room traffic. Those records reveal that the number of patients who arrive between 6:00 P.M. and 7:00 P.M. has a Poisson distribution with parameter $\lambda = 6.9$. Determine the probability that, on a given day, the number of patients who arrive at the emergency room between 6:00 P.M. and 7:00 P.M. will be

a. exactly 4.

b. at most 2.

c. between 4 and 10, inclusive.

5.4 The Poisson Distribution*

Another important discrete probability distribution is the *Poisson distribution,* named in honor of the French mathematician and physicist Simeon D. Poisson (1781–1840). The Poisson distribution is often used to model the frequency with which a specified event occurs during a particular period of time. For instance, we might apply the Poisson distribution when analyzing

- the number of patients who arrive at an emergency room between 6:00 P.M. and 7:00 P.M.,
- the number of telephone calls received per day at a switchboard, or
- the number of alpha particles emitted per minute by a radioactive substance.

In addition, we might use the Poisson distribution to describe the probability distribution of the number of misprints in a book, the number of earthquakes occurring during a 1-year period of time, or the number of bacterial colonies appearing on a petri dish smeared with a bacterial suspension.

The Poisson Probability Formula

Any particular Poisson distribution is identified by one parameter, usually denoted λ (the Greek letter lambda). Here is the **Poisson probability formula.**

FORMULA 5.3

Poisson Probability Formula

Probabilities for a random variable X that has a Poisson distribution are given by the formula

$$P(X = x) = e^{-\lambda}\frac{\lambda^x}{x!}, \qquad x = 0, 1, 2, \ldots,$$

where λ is a positive real number and $e \approx 2.718$. (Most calculators have an e key.) The random variable X is called a **Poisson random variable** and is said to have the **Poisson distribution** with parameter λ.

Note: A Poisson random variable has infinitely many possible values—namely, all whole numbers. Consequently, we cannot display all the probabilities for a Poisson random variable in a probability distribution table.

EXAMPLE 5.15   ## The Poisson Distribution

Emergency Room Traffic   Desert Samaritan Hospital keeps records of emergency room (ER) traffic. Those records indicate that the number of patients arriving between 6:00 P.M. and 7:00 P.M. has a Poisson distribution with parameter $\lambda = 6.9$. Determine the probability that, on a given day, the number of patients who arrive at the emergency room between 6:00 P.M. and 7:00 P.M. will be

a. exactly 4.
b. at most 2.
c. between 4 and 10, inclusive.
d. Obtain a table of probabilities for the random variable X, the number of patients arriving between 6:00 P.M. and 7:00 P.M. Stop when the probabilities become zero to three decimal places.
e. Use part (d) to construct a (partial) probability histogram for X.
f. Identify the shape of the probability distribution of X.

Solution   The random variable X—the number of patients arriving between 6:00 P.M. and 7:00 P.M.—has a Poisson distribution with parameter $\lambda = 6.9$. Thus, by Formula 5.3, the probabilities for X are given by the Poisson probability formula,

$$P(X = x) = e^{-6.9}\frac{(6.9)^x}{x!}.$$

Using this formula, we can now solve parts (a)–(f).

a. Applying the Poisson probability formula with $x = 4$ gives

$$P(X = 4) = e^{-6.9}\frac{(6.9)^4}{4!} = e^{-6.9}\cdot\frac{2266.7121}{24} = 0.095.$$

Interpretation   Chances are 9.5% that exactly 4 patients will arrive at the ER between 6:00 P.M. and 7:00 P.M.

b. The probability of at most 2 arrivals is

$$P(X \leq 2) = P(X = 0) + P(X = 1) + P(X = 2)$$

$$= e^{-6.9}\frac{(6.9)^0}{0!} + e^{-6.9}\frac{(6.9)^1}{1!} + e^{-6.9}\frac{(6.9)^2}{2!}$$

$$= e^{-6.9}\left(\frac{6.9^0}{0!} + \frac{6.9^1}{1!} + \frac{6.9^2}{2!}\right)$$

$$= e^{-6.9}(1 + 6.9 + 23.805) = e^{-6.9}\cdot 31.705 = 0.032.$$

Interpretation   Chances are only 3.2% that 2 or fewer patients will arrive at the ER between 6:00 P.M. and 7:00 P.M.

c. The probability of between 4 and 10 arrivals, inclusive, is

$$P(4 \leq X \leq 10) = P(X = 4) + P(X = 5) + \cdots + P(X = 10)$$

$$= e^{-6.9}\left(\frac{6.9^4}{4!} + \frac{6.9^5}{5!} + \cdots + \frac{6.9^{10}}{10!}\right) = 0.821.$$

Interpretation Chances are 82.1% that between 4 and 10 patients, inclusive, will arrive at the ER between 6:00 P.M. and 7:00 P.M.

d. We use the method of part (a) to generate Table 5.16, a partial probability distribution of the random variable X.

e. Figure 5.7, a partial probability histogram for X, is based on Table 5.16.

f. Figure 5.7 shows that the probability distribution is right skewed.

TABLE 5.16

Partial probability distribution of the random variable X, the number of patients arriving at the emergency room between 6:00 P.M. and 7:00 P.M.

Number arriving x	Probability $P(X = x)$	Number arriving x	Probability $P(X = x)$
0	0.001	10	0.068
1	0.007	11	0.043
2	0.024	12	0.025
3	0.055	13	0.013
4	0.095	14	0.006
5	0.131	15	0.003
6	0.151	16	0.001
7	0.149	17	0.001
8	0.128	18	0.000
9	0.098		

FIGURE 5.7

Partial probability histogram for the random variable X, the number of patients arriving at the emergency room between 6:00 P.M. and 7:00 P.M.

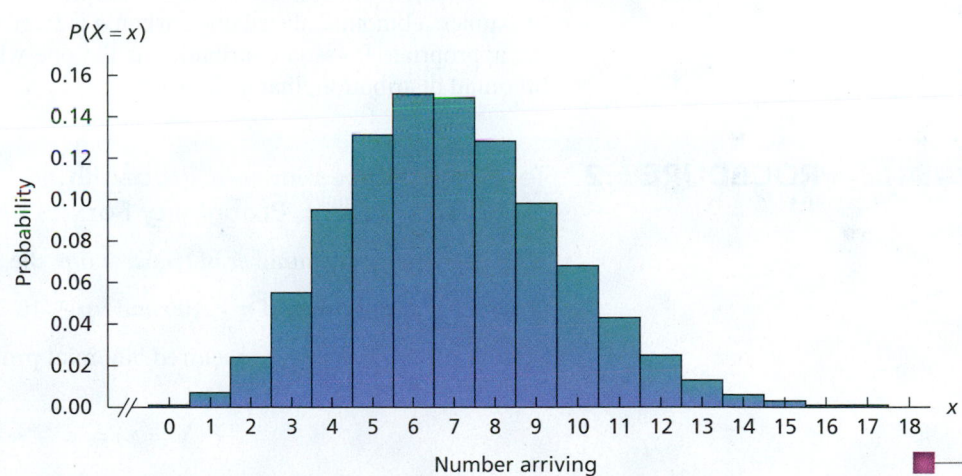

 StatCrunch

 You try it!

Report 5.2

Exercise 5.101(a)–(e)
on page 257

Shape of a Poisson Distribution

In the previous example, we found that the probability distribution is right skewed. As a matter of fact, *all Poisson distributions are right skewed.*

Mean and Standard Deviation of a Poisson Random Variable

If we substitute the Poisson probability formula into the formulas for the mean and standard deviation of a discrete random variable and then simplify mathematically, we obtain the following formulas.

What Does It Mean?

The mean and standard deviation of a Poisson random variable are its parameter and square root of its parameter, respectively.

FORMULA 5.4

Mean and Standard Deviation of a Poisson Random Variable

The mean and standard deviation of a Poisson random variable with parameter λ are

$$\mu = \lambda \quad \text{and} \quad \sigma = \sqrt{\lambda},$$

respectively.

■■■   **EXAMPLE 5.16**   **Mean and Standard Deviation of a Poisson Random Variable**

Emergency Room Traffic Let X denote the number of patients arriving at the emergency room of Desert Samaritan Hospital between 6:00 P.M. and 7:00 P.M.

a. Determine and interpret the mean of the random variable X.
b. Determine the standard deviation of X.

Exercise 5.103(d)–(e)
on page 257

Solution As we know, X has the Poisson distribution with parameter $\lambda = 6.9$. So we apply Formula 5.4 to determine the mean and standard deviation of X.

a. The mean of X is $\mu = \lambda = 6.9$.

    Interpretation On average, 6.9 patients arrive at the emergency room between 6:00 P.M. and 7:00 P.M.

b. The standard deviation of X is $\sigma = \sqrt{\lambda} = \sqrt{6.9} = 2.6$. ■

Poisson Approximation to the Binomial Distribution

Recall that the binomial probability formula is

$$P(X = x) = \binom{n}{x} p^x (1 - p)^{n-x}.$$

We use this formula to obtain probabilities for the number of successes, X, in n Bernoulli trials with success probability p.

Because of computational difficulties, the binomial probability formula can be difficult or impractical to use when n is large. We can use a Poisson distribution to approximate a binomial distribution when n is large and p is small. As you might expect, the appropriate Poisson distribution is the one whose mean is the same as that of the binomial distribution; that is, $\lambda = np$.

■■■   **PROCEDURE 5.2**   **To Approximate Binomial Probabilities by Using a Poisson Probability Formula**

Step 1 Find n, the number of trials, and p, the success probability.

Step 2 Continue only if $n \geq 100$ and $np \leq 10$.

Step 3 Approximate the required binomial probabilities by using the Poisson probability formula

$$P(X = x) = e^{-np} \frac{(np)^x}{x!}.$$

■■■   **EXAMPLE 5.17**   **Poisson Approximation to the Binomial**

IMR in Finland The infant mortality rate (IMR) is the number of deaths of children under 1 year old per 1000 live births during a calendar year. From the *World Factbook*, the Central Intelligence Agency's most popular publication, we found that the IMR in Finland is 3.4. Use the Poisson approximation to determine the probability that, of 500 randomly selected live births in Finland, there are

a. no infant deaths.   **b.** at most three infant deaths.

Solution Let X denote the number of infant deaths out of 500 live births in Finland. We use Procedure 5.2 to approximate the required probabilities for X.

Step 1 Find n, the number of trials, and p, the success probability.

We have $n = 500$ (number of live births) and $p = \frac{3.4}{1000} = 0.0034$ (probability of an infant death).

Step 2 Continue only if $n \geq 100$ and $np \leq 10$.

We have $n = 500$ and $np = 500 \cdot 0.0034 = 1.7$. So $n \geq 100$ and $np \leq 10$.

Step 3 Approximate the required binomial probabilities by using the Poisson probability formula

$$P(X = x) = e^{-np}\frac{(np)^x}{x!}.$$

Because $np = 1.7$, the appropriate Poisson probability formula is

$$P(X = x) = e^{-1.7}\frac{(1.7)^x}{x!}.$$

a. The approximate probability of no infant deaths in 500 live births is

$$P(X = 0) = e^{-1.7}\frac{(1.7)^0}{0!} = 0.183.$$

Interpretation Chances are about 18.3% that there will be no infant deaths in 500 live births.

b. The approximate probability of at most three infant deaths in 500 live births is

$$P(X \leq 3) = P(X = 0) + P(X = 1) + P(X = 2) + P(X = 3)$$

$$= e^{-1.7}\left(\frac{1.7^0}{0!} + \frac{1.7^1}{1!} + \frac{1.7^2}{2!} + \frac{1.7^3}{3!}\right) = 0.907.$$

Interpretation Chances are about 90.7% that there will be three or fewer infant deaths in 500 live births.

You try it!

Exercise 5.107 on page 257

Let's use the previous example to illustrate the accuracy of the Poisson approximation. Table 5.17 shows both the binomial distribution with parameters $n = 500$ and $p = 0.0034$ and the Poisson distribution with parameter $\lambda = np = 500 \cdot 0.0034 = 1.7$. We rounded to four decimal places and did not list probabilities that are zero to four decimal places. In any case, notice how well the Poisson distribution approximates the binomial distribution.

TABLE 5.17

Comparison of the binomial distribution with parameters $n = 500$ and $p = 0.0034$ to the Poisson distribution with parameter $\lambda = 1.7$

x	0	1	2	3	4	5	6	7	8	9
Binomial probability	0.1822	0.3107	0.2645	0.1498	0.0635	0.0215	0.0060	0.0015	0.0003	0.0001
Poisson probability	0.1827	0.3106	0.2640	0.1496	0.0636	0.0216	0.0061	0.0015	0.0003	0.0001

THE TECHNOLOGY CENTER

Most statistical technologies include programs that determine Poisson probabilities. In this subsection, we present output and step-by-step instructions for such programs.

EXAMPLE 5.18   Using Technology to Obtain Poisson Probabilities

Emergency Room Traffic Consider again the illustration of emergency room traffic discussed in Example 5.15, which begins on page 252. Use Minitab, Excel, or the TI-83/84 Plus to determine the probability that exactly four patients will arrive at the emergency room between 6:00 P.M. and 7:00 P.M.

Solution Recall that the number of patients, X, that arrive at the ER between 6:00 P.M. and 7:00 P.M. has a Poisson distribution with parameter $\lambda = 6.9$. We want the probability of exactly four arrivals, that is, $P(X = 4)$.

We applied the Poisson probability programs, resulting in Output 5.2. Steps for generating that output are presented in Instructions 5.2. As shown in Output 5.2, the required probability is 0.095.

OUTPUT 5.2   Probability that exactly four patients will arrive at the emergency room between 6:00 P.M. and 7:00 P.M.

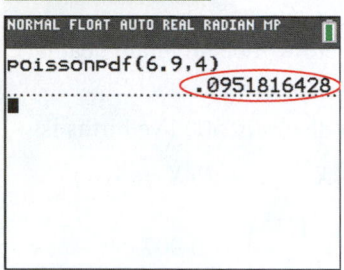

MINITAB

Probability Density Function

Poisson with mean = 6.9

x   P(X = x)
4   0.0951816

TI-83/84 PLUS

NORMAL FLOAT AUTO REAL RADIAN MP
poissonpdf(6.9,4)
                    .0951816428

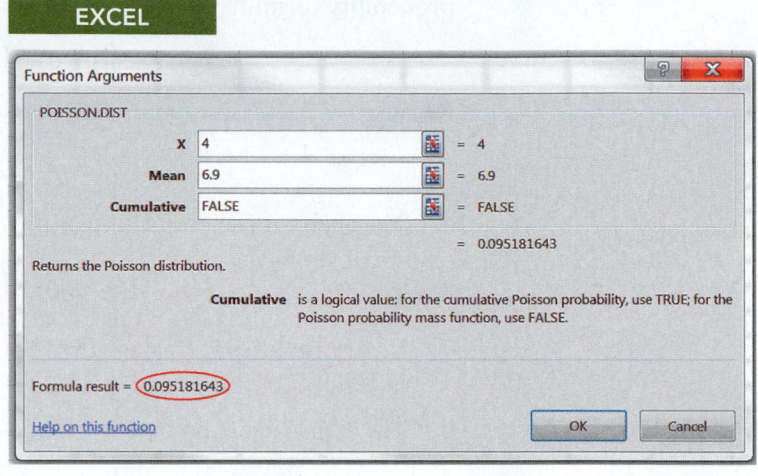

EXCEL

INSTRUCTIONS 5.2   Steps for generating Output 5.2

MINITAB

1 Choose **Calc ➤ Probability Distributions ➤ Poisson...**
2 Press the F3 key to reset the dialog box
3 Select the **Probability** option button
4 Click in the **Mean** text box and type 6.9
5 Select the **Input constant** option button
6 Click in the **Input constant** text box and type 4
7 Click **OK**

EXCEL

1 Click **f_x** (Insert Function)
2 Select **Statistical** from the **Or select a category** drop down list box
3 Select **POISSON.DIST** from the **Select a function** list
4 Click **OK**

5 Type 4 in the **X** text box
6 Click in the **Mean** text box and type 6.9
7 Click in the **Cumulative** text box and type FALSE

TI-83/84 PLUS

FOR THE TI-84 PLUS C

1 Press **2nd ➤ DISTR**
2 Arrow down to **poissonpdf(** and press **ENTER**
3 Type 6.9 for λ and press **ENTER**
4 Type 4 for **x value** and press **ENTER** three times

FOR THE TI-83/84 PLUS

1 Press **2nd ➤ DISTR**
2 Arrow down to **poissonpdf(** and press **ENTER**
3 Type 6.9,4) and press **ENTER**

You can also obtain cumulative probabilities for a Poisson distribution by using Minitab, Excel, or the TI-83/84 Plus. To do so, modify Instructions 5.2 as follows:

• For Minitab, in step 3, select the **Cumulative probability** option button instead of the **Probability** option button.
• For Excel, in step 7, type TRUE instead of FALSE.
• For the TI-83/84 Plus, in step 2, arrow down to **poissoncdf(** instead of **poissonpdf(**.

Exercises 5.4

Understanding the Concepts and Skills

5.91 Identify two uses of Poisson distributions.

5.92 Why can't all the probabilities for a Poisson random variable be displayed in a probability distribution table?

5.93 For a Poisson random variable, what is the relationship between its parameter and its mean?

5.94 What conditions should be satisfied in order to approximate binomial probabilities by Poisson probabilities?

5.95 Suppose that we plan to approximate a binomial distribution with parameters n and p by a Poisson distribution. What is the parameter for the Poisson distribution?

In each of Exercises 5.96–5.99, we have provided the parameter of a Poisson random variable, X. For each exercise,
a. determine the required probabilities. Round your probability answers to three decimal places.
b. find the mean and standard deviation of X.

5.96 $\lambda = 3$; $P(X = 2)$, $P(X \leq 3)$, $P(X > 0)$. (*Hint:* For the third probability, use the complementation rule.)

5.97 $\lambda = 5$; $P(X = 5)$, $P(X < 2)$, $P(X \geq 3)$. (*Hint:* For the third probability, use the complementation rule.)

5.98 $\lambda = 6.3$; $P(X = 7)$, $P(5 \leq X \leq 8)$, $P(X \geq 2)$.

5.99 $\lambda = 4.7$; $P(X = 3)$, $P(5 \leq X \leq 7)$, $P(X > 2)$.

Applying the Concepts and Skills

5.100 Amusement Ride Safety. Approximately 297 million guests visit the 400 American amusement parks annually and take 1.7 billion safe rides. The National Safety Council publishes a report titled *Fixed-Site Amusement Ride Injury Survey* for the International Association of Amusement Parks and Attractions. The number of injuries per million patron-rides, X, has a Poisson distribution with parameter 0.8. In one million patron-rides, what is the probability that there are
a. no injuries?
b. more than two injuries?
c. Construct a table of probabilities for the random variable X. Compute the probabilities until they are zero to three decimal places.
d. Draw a histogram of the probabilities in part (c).

5.101 Polonium. In the 1910 article "The Probability Variations in the Distribution of α Particles" (*Philosophical Magazine*, Series 6, No. 20, pp. 698–707), E. Rutherford and H. Geiger described the results of experiments with polonium. The experiments indicate that the number of α (alpha) particles that reach a small screen during an 8-minute interval has a Poisson distribution with parameter $\lambda = 3.87$. Determine the probability that, during an 8-minute interval, the number, Y, of α particles that reach the screen is
a. exactly four.   **b.** at most one.
c. between two and five, inclusive.
d. Construct a table of probabilities for the random variable Y. Compute the probabilities until they are zero to three decimal places.
e. Draw a histogram of the probabilities in part (d).
f. On average, how many alpha particles reach the screen during an 8-minute interval?

5.102 Wasps. M. Goodisman et al. studied patterns in queen and worker wasps and published their findings in the article "Mating and Reproduction in the Wasp *Vespula germanica*" (*Behavioral Ecology and Sociobiology*, Vol. 51, No. 6, pp. 497–502). The number of male mates of a queen wasp has a Poisson distribution with parameter $\lambda = 2.7$. Find the probability that the number, Y, of male mates of a queen wasp is
a. exactly two.   **b.** at most two.
c. between one and three, inclusive.
d. On average, how many male mates does a queen wasp have?
e. Construct a table of probabilities for the random variable Y. Compute the probabilities until they are zero to three decimal places.
f. Draw a histogram of the probabilities in part (e).

5.103 Wars. In the paper "The Distribution of Wars in Time" (*Journal of the Royal Statistical Society*, Vol. 107, No. 3/4, pp. 242–250),

L. F. Richardson analyzed the distribution of wars in time. From the data, we determined that the number of wars that begin during a given calendar year has roughly a Poisson distribution with parameter $\lambda = 0.7$. If a calendar year is selected at random, find the probability that the number, X, of wars that begin during that calendar year will be
a. zero.   **b.** at most two.
c. between one and three, inclusive.
d. Find and interpret the mean of the random variable X.
e. Determine the standard deviation of X.

5.104 Motel Reservations. M. Driscoll and N. Weiss discussed the modeling and solution of problems concerning motel reservation networks in "An Application of Queuing Theory to Reservation Networks" (*TIMS*, Vol. 22, No. 5, pp. 540–546). They defined a Type 1 call to be a call from a motel's computer terminal to the national reservation center. For a certain motel, the number, X, of Type 1 calls per hour has a Poisson distribution with parameter $\lambda = 1.7$. Determine the probability that the number of Type 1 calls made from this motel during a period of 1 hour will be
a. exactly one.   **b.** at most two.
c. at least two. (*Hint:* Use the complementation rule.)
d. Find and interpret the mean of the random variable X.
e. Determine the standard deviation of X.

5.105 Cherry Pies. At one time, a well-known restaurant chain sold cherry pies. Professor D. Lund of the University of Wisconsin - Eau Claire enlisted the help of one of his classes to gather data on the number of cherries per pie. The data obtained by the students are presented in the following table.

0	1	2	1	0	2	3
0	0	2	1	4	1	2
1	1	0	1	3	2	0
0	0	0	1	0	2	1
1	0	1	2	2	1	4

a. For the student data, find the mean number of cherries per pie.
b. For the student data, construct a relative-frequency distribution for the number of cherries per pie.
c. Assuming that, for cherry pies sold by the restaurant, the number of cherries per pie has a Poisson distribution with the mean from part (a), obtain the probability distribution of the number of cherries per pie.
d. Compare the relative frequencies in part (b) to the probabilities in part (c). What conclusions can you draw?

5.106 Motor-Vehicle Deaths. According to *Injury Facts*, a publication of the National Safety Council, in the United States, the lifetime probability of dying in a motor-vehicle accident is 1 in 108. Use the Poisson distribution to determine the approximate probability that, of 200 randomly selected deaths in the United States,
a. none are due to motor-vehicle accidents.
b. three or more are due to motor-vehicle accidents.

5.107 Prisoners. From the U.S. Census Bureau and the article "U.S. Prison Population Falls for Third Year" by B. Montopoli of CBS News, we found that about 1 in every 200 Americans are in prison. Use this information and the Poisson distribution to determine the approximate probability that at most three people in a random sample of 500 Americans are currently in prison.

5.108 The *Challenger* Disaster. In a letter to the editor that appeared in the February 23, 1987, issue of *U.S. News and World Report*, a reader discussed the issue of space-shuttle safety. Each "criticality 1" item must have a 99.99% reliability, by NASA standards, which means that the probability of failure for a "criticality 1" item is only 0.0001. Mission 25, the mission in which the *Challenger* exploded on takeoff, had 748 "criticality 1" items. Use the Poisson approximation to the binomial distribution to determine the approximate probability that

a. none of the "criticality 1" items would fail.

b. at least one "criticality 1" item would fail.

5.109 Fragile X Syndrome. The second-leading genetic cause of mental retardation is Fragile X Syndrome, named for the fragile appearance of the tip of the X chromosome in affected individuals. One in 1500 males is affected worldwide, with no ethnic bias.

a. In a sample of 10,000 males, how many would you expect to have Fragile X Syndrome?

b. For a sample of 10,000 males, use the Poisson approximation to the binomial distribution to determine the probability that more than 7 of the males have Fragile X Syndrome; that at most 10 of the males have Fragile X Syndrome.

5.110 Holes in One. Refer to the case study on page 223. According to the experts, the odds against a PGA golfer making a hole in one are 3708 to 1; that is, the probability is $\frac{1}{3709}$. Use the Poisson approximation to the binomial distribution to determine the probability that at least 4 of the 155 golfers playing the second round would get a hole in one on the sixth hole.

5.111 A Yellow Lobster! As reported by the Associated Press, a veteran lobsterman recently hauled up a yellow lobster less than a quarter mile south of Prince Point in Harpswell Cove, Maine. Yellow lobsters are considerably rarer than blue lobsters and, according to B. Ballenger's *The Lobster Almanac* (Darby, PA: Diane Publishing Company, 1998), roughly 1 in every 30 million lobsters hatched is yellow. Apply the Poisson approximation to the binomial distribution to answer the following questions:

a. Of 100 million lobsters hatched, what is the probability that between 3 and 5, inclusive, are yellow?

b. Roughly how many lobsters must be hatched in order to be at least 90% sure that at least one is yellow?

Extending the Concepts and Skills

5.112 With regard to the use of a Poisson distribution to approximate binomial probabilities, on page 254 we stated that "As you might expect, the appropriate Poisson distribution is the one whose mean is the same as that of the binomial distribution...." Explain why you might expect this result.

5.113 Roughly speaking, you can use the Poisson probability formula to approximate binomial probabilities when n is large and p is small (i.e., near 0). Explain how to use the Poisson probability formula to approximate binomial probabilities when n is large and p is large (i.e., near 1).

CHAPTER IN REVIEW

You Should Be Able to

1. use and understand the formulas in this chapter.

2. determine the probability distribution of a discrete random variable.

3. construct a probability histogram.

4. describe events by using random-variable notation, when appropriate.

5. use the frequentist interpretation of probability to understand the meaning of the probability distribution of a random variable.

6. find and interpret the mean and standard deviation of a discrete random variable.

7. compute factorials and binomial coefficients.

8. define and apply the concept of Bernoulli trials.

9. assign probabilities to the outcomes in a sequence of Bernoulli trials.

10. obtain binomial probabilities.

11. compute the mean and standard deviation of a binomial random variable.

12. obtain Poisson probabilities.

13. compute the mean and standard deviation of a Poisson random variable.

14. use the Poisson distribution to approximate binomial probabilities, when appropriate.

Key Terms

Bernoulli trials, *239*
binomial coefficients, *238*
binomial distribution, *239, 242*
binomial probability formula, *242*
binomial random variable, *242*
cumulative probability, *244*
discrete random variable, *224*
expectation, *232*
expected value, *232*
factorials, *238*

failure, *239*
hypergeometric distribution, *246*
law of averages, *233*
law of large numbers, *233*
mean of a discrete random
    variable, *232*
Poisson distribution, *252*
Poisson probability formula, *252*
Poisson random variable, *252*
probability distribution, *225*

probability histogram, *225*
random variable, *224*
standard deviation of a discrete random
    variable, *234*
success, *239*
success probability, *239*
trial, *238*
variance of a discrete random
    variable, *234*

REVIEW PROBLEMS

Understanding the Concepts and Skills

1. Fill in the blanks.

a. A _____ is a quantitative variable whose value depends on chance.

b. A discrete random variable is a random variable whose possible values _____.

2. What does the probability distribution of a discrete random variable tell you?

3. How do you graphically portray the probability distribution of a discrete random variable?

4. If you sum the probabilities of the possible values of a discrete random variable, the result always equals _____.

5. A random variable X equals 2 with probability 0.386.

a. Use probability notation to express that fact.

b. If you make repeated independent observations of the random variable X, in approximately what percentage of those observations will you observe the value 2?

c. Roughly how many times would you expect to observe the value 2 in 50 observations? 500 observations?

6. A random variable X has mean 3.6. If you make a large number of repeated independent observations of the random variable X, the average value of those observations will be approximately _____.

7. Two random variables, X and Y, have standard deviations 2.4 and 3.6, respectively. Which one is more likely to take a value close to its mean? Explain your answer.

8. Determine 0!, 3!, 4!, and 7!.

9. Determine the value of each binomial coefficient.

a. $\binom{8}{3}$ **b.** $\binom{8}{5}$ **c.** $\binom{6}{6}$ **d.** $\binom{10}{2}$ **e.** $\binom{40}{4}$ **f.** $\binom{100}{0}$

10. List the three requirements for repeated trials of an experiment to constitute Bernoulli trials.

11. What is the relationship between Bernoulli trials and the binomial distribution?

12. In 10 Bernoulli trials, how many outcomes contain exactly three successes?

13. Craps. The game of craps is played by rolling two balanced dice. A first roll of a sum of 7 or 11 wins; and a first roll of a sum of 2, 3, or 12 loses. To win with any other first sum, that sum must be repeated before a sum of 7 is thrown. It can be shown that the probability is 0.493 that a player wins a game of craps. Suppose we consider a win by a player to be a success, s.

a. Identify the success probability, p.

b. Construct a table showing the possible win–lose results and their probabilities for three games of craps. Round each probability to three decimal places.

c. Draw a tree diagram for part (b).

d. List the outcomes in which the player wins exactly two out of three times.

e. Determine the probability of each of the outcomes in part (d). Explain why those probabilities are equal.

f. Find the probability that the player wins exactly two out of three times.

g. Without using the binomial probability formula, obtain the probability distribution of the random variable Y, the number of times out of three that the player wins.

h. Identify the probability distribution in part (g).

14. Following are two probability histograms of binomial distributions. For each, specify whether the success probability is less than, equal to, or greater than 0.5.

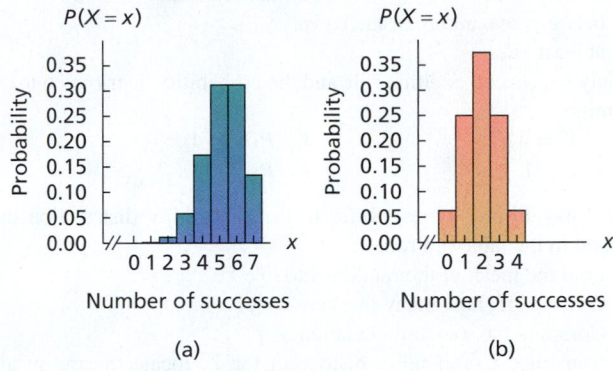

(a)　　　　　　　(b)

15. Explain how the special formulas for the mean and standard deviation of a binomial or Poisson random variable are derived.

16. Suppose that a simple random sample of size n is taken from a finite population in which the proportion of members having a specified attribute is p. Let X be the number of members sampled that have the specified attribute.

a. If the sampling is done with replacement, identify the probability distribution of X.

b. If the sampling is done without replacement, identify the probability distribution of X.

c. Under what conditions is it acceptable to approximate the probability distribution in part (b) by the probability distribution in part (a)? Why is it acceptable?

Applying the Concepts and Skills

17. ASU Enrollment Summary. According to the *Arizona State University Enrollment Summary*, a frequency distribution for the number of undergraduate students attending Arizona State University (ASU) in the Fall 2012 semester, by class level, is as shown in the following table. Here, 1 = freshman, 2 = sophomore, 3 = junior, and 4 = senior.

Class level	1	2	3	4
No. of students	9,652	11,115	17,302	21,114

Let X denote the class level of a randomly selected ASU undergraduate.

a. What are the possible values of the random variable X?

b. Use random-variable notation to represent the event that the student selected is a junior (class-level 3).

c. Determine $P(X = 3)$, and interpret your answer in terms of percentages.

d. Determine the probability distribution of the random variable X.

e. Construct a probability histogram for the random variable X.

18. Busy Phone Lines. An accounting office has six incoming telephone lines. The probability distribution of the number of busy lines, Y, is as follows.

y	0	1	2	3	4	5	6
$P(Y = y)$	0.052	0.154	0.232	0.240	0.174	0.105	0.043

Use random-variable notation to express each of the following events. The number of busy lines is

a. exactly four.   **b.** at least four.

c. between two and four, inclusive.

d. at least one.

Apply the special addition rule and the probability distribution to determine

e. $P(Y = 4)$.   **f.** $P(Y \geq 4)$.

g. $P(2 \leq Y \leq 4)$.   **h.** $P(Y \geq 1)$.

19. Busy Phone Lines. Refer to the probability distribution displayed in the table in Problem 18.

a. Find the mean of the random variable Y.

b. On average, how many lines are busy?

c. Compute the standard deviation of Y.

d. Construct a probability histogram for Y; locate the mean; and show one-, two-, and three-standard-deviation intervals.

20. Craps. Use the binomial probability formula to solve Problem 13(g). Compare your results.

21. Penalty Kicks. In the game of soccer, a penalty kick is a direct free kick, taken from 12 yards out from the goal on the penalty mark. According to the article "Penalty Kicks in Soccer: An Empirical Analysis of Shooting Strategies and Goalkeeper's Preferences" (*Soccer & Society*, Vol. 10, No. 2, pp. 183–191) by M. Bar-Eli and O. Azar, 85% of penalty kicks placed by professional soccer players are successful. In 15 penalty kicks by professional soccer players, what is the probability that

a. all are successful?

b. at least 13 are successful?

c. Find and interpret the mean and standard deviation of the number of successful penalty kicks out of 15.

22. Pets. According to *JAVMA News*, a publication of the American Veterinary Medical Association, roughly 60% of U.S. households own one or more pets. Four U.S. households are selected at random. Use Table XI in Appendix A to solve the following problems.

a. Find the probability that, of the four households sampled, the number that own one or more pets is exactly three; at least three; at most three.

b. Find the probability distribution of the random variable X, the number of U.S. households in a random sample of four that own one or more pets.

c. Without referring to the probability distribution obtained in part (b) or constructing a probability histogram, decide whether the probability distribution is right skewed, symmetric, or left skewed. Explain your answer.

23. Pets. Refer to Problem 22.

a. Draw a probability histogram for the random variable X.

b. The selection of the four households was done without replacement. Strictly speaking, then, why is the probability distribution that you obtained in Problem 22(b) only approximately correct? What is the exact distribution called?

c. Do you think the probability distribution that you obtained is a reasonable approximation to the actual one? Explain your answer.

24. Wrong Number. A classic study by F. Thorndike on the number of calls to a wrong number appeared in the paper "Applications of Poisson's Probability Summation" (*Bell Systems Technical*

Journal, Vol. 5, pp. 604–624). The study examined the number of calls to a wrong number from coin-box telephones in a large transportation terminal. According to the paper, the number of calls to a wrong number, X, in a 1-minute period has a Poisson distribution with parameter $\lambda = 1.75$. Determine the probability that during a 1-minute period the number of calls to a wrong number will be

a. exactly two.

b. between four and six, inclusive.

c. at least one.

25. Wrong Number. Refer to Problem 24.

a. Obtain a table of probabilities for the random variable X, stopping when the probabilities become zero to three decimal places.

b. Use part (a) to construct a partial probability histogram for X.

c. Identify the shape of the probability distribution of X. Is this shape typical of Poisson distributions?

d. Find and interpret the mean of the random variable X.

e. Find the standard deviation of the random variable X.

26. Meteoroids. In the article "Interstellar Pelting" (*Scientific American*, Vol. 288, No. 5, pp. 28–30), G. Musser explained that information on extrasolar planets can be discerned from foreign material and dust found in our solar system. Studies show that 1 in every 100 meteoroids entering Earth's atmosphere is actually alien matter from outside our solar system.

a. Of 300 meteoroids entering the Earth's atmosphere, how many would you expect to be alien matter from outside our solar system? Justify your answer.

b. Apply the Poisson approximation to the binomial distribution to determine the probability that, of 300 meteoroids entering the Earth's atmosphere, between 2 and 4, inclusive, are alien matter from outside our solar system.

c. Apply the Poisson approximation to the binomial distribution to determine the probability that, of 300 meteoroids entering the Earth's atmosphere, at least 1 is alien matter from outside our solar system.

27. Emphysema. The respiratory disease emphysema, which is most commonly caused by smoking, causes damage to the air sacs in the lungs. According to the National Center for Health Statistics report *Data from the National Health Interview Survey*, 1.5% of the adult American population suffer from emphysema. Of 100 randomly selected adult Americans, let X denote the number who have emphysema.

a. What are the parameters for the appropriate binomial distribution?

b. What is the parameter for the approximating Poisson distribution?

c. Compute the individual probabilities for the binomial distribution in part (a). Obtain the probabilities until they are zero to four decimal places.

d. Compute the individual probabilities for the Poisson distribution in part (b). Obtain the probabilities until they are zero to four decimal places.

e. Compare the probabilities that you obtained in parts (c) and (d).

28. Emphysema. Refer to Problem 27. Use both the binomial probabilities and Poisson probabilities that you obtained in parts (c) and (d), respectively, to find the probability that, in a sample of 100 randomly selected Americans, the number who suffer from emphysema is

a. exactly three.

b. between two and five, inclusive.

c. less than 4% of those surveyed.

d. more than two.

Compare your two answers in each case.

FOCUSING ON DATA ANALYSIS

UWEC UNDERGRADUATES

Recall from Chapter 1 (see page 34) that the Focus database and Focus sample contain information on the undergraduate students at the University of Wisconsin - Eau Claire (UWEC). Now would be a good time for you to review the discussion about these data sets.

The following problems are designed for use with the entire Focus database (Focus). If your statistical software package won't accommodate the entire Focus database, use the Focus sample (FocusSample) instead. Of course, in that case, your results will apply to the 200 UWEC undergraduate students in the Focus sample rather than to all UWEC undergraduate students.

a. Let X denote the age of a randomly selected undergraduate student at UWEC. Obtain the probability distribution of the random variable X. Display the probabilities to six decimal places.

b. Obtain a probability histogram or similar graphic for the random variable X.

c. Determine the mean and standard deviation of the random variable X.

d. Simulate 100 observations of the random variable X.

e. Roughly, what would you expect the average value of the 100 observations obtained in part (d) to be? Explain your reasoning.

f. In actuality, what is the average value of the 100 observations obtained in part (d)? Compare this value to the value you expected, as answered in part (e).

g. Consider the experiment of randomly selecting 10 UWEC undergraduates with replacement and observing the number of those selected who are 21 years old. Simulate that experiment 1000 times. (*Hint:* Simulate an appropriate binomial distribution.)

h. Referring to the simulation in part (g), in approximately what percentage of the 1000 experiments would you expect exactly 3 of the 10 students selected to be 21 years old? Compare that percentage to the actual percentage of the 1000 experiments in which exactly 3 of the 10 students selected are 21 years old.

CASE STUDY DISCUSSION

ACES WILD ON THE SIXTH AT OAK HILL

As we reported at the beginning of this chapter, on June 16, 1989, during the second round of the 1989 U.S. Open, four golfers—Doug Weaver, Mark Wiebe, Jerry Pate, and Nick Price—made holes in one on the sixth hole at Oak Hill in Pittsford, New York. Now that you have studied the material in this chapter, you can determine for yourself the likelihood of such an event.

According to the experts, the odds against a professional golfer making a hole in one are 3708 to 1; in other words, the probability is $\frac{1}{3709}$ that a professional golfer will make a hole in one. One hundred fifty-five golfers participated in the second round.

a. Determine the probability that at least 4 of the 155 golfers would get a hole in one on the sixth hole. Discuss your result.

b. What assumptions did you make in solving part (a)? Do those assumptions seem reasonable to you? Explain your answer.

BIOGRAPHY

JAMES BERNOULLI: PAVING THE WAY FOR PROBABILITY THEORY

James Bernoulli was born on December 27, 1654, in Basle, Switzerland. He was the first of the Bernoulli family of mathematicians; his younger brother John and various nephews and grandnephews were also renowned mathematicians. His father, Nicolaus Bernoulli (1623–1708), planned the ministry as James's career. James rebelled, however; to him, mathematics was much more interesting.

Although Bernoulli was schooled in theology, he studied mathematics on his own. He was especially fascinated with calculus. In a 1690 issue of the journal *Acta eruditorum*, Bernoulli used the word *integral* to describe the inverse of differential. The results of his studies of calculus and the catenary (the curve formed by a cord freely suspended between two fixed points) were soon applied to the building of suspension bridges.

Some of Bernoulli's most important work was published posthumously in *Ars Conjectandi* (The Art of Conjecturing)

in 1713. This book contains his theory of permutations and combinations, the Bernoulli numbers, and his writings on probability, which include the weak law of large numbers for Bernoulli trials. *Ars Conjectandi* has been regarded as the beginning of the theory of probability.

Both James and his brother John were highly accomplished mathematicians. Rather than collaborating in their work, however, they were most often competing. James would publish a question inviting solutions in a professional journal. John would reply in the same journal with a solution, only to find that an ensuing issue would contain another article by James, telling him that he was wrong. In their later years, they communicated only in this manner.

Bernoulli began lecturing in natural philosophy and mechanics at the University of Basle in 1682 and became a Professor of Mathematics there in 1687. He remained at the university until his death of a "slow fever" on August 10, 1705.

CHAPTER
6

The Normal Distribution

CHAPTER OBJECTIVES

In this chapter, we discuss the most important distribution in statistics—the *normal distribution.* As you will see, its importance lies in the fact that it appears again and again in both theory and practice.

A variable is said to be *normally distributed* or to have a *normal distribution* if its distribution has the shape of a normal curve, a special type of bell-shaped curve. In Section 6.1, we first briefly discuss *density curves.* Then we introduce normally distributed variables, show that percentages (or probabilities) for such a variable are equal to areas under its associated normal curve, and explain how all normal distributions can be converted to a single normal distribution—the *standard normal distribution.*

In Section 6.2, we demonstrate how to determine areas under the *standard normal curve,* the normal curve corresponding to a variable that has the standard normal distribution. Then, in Section 6.3, we describe an efficient procedure for finding percentages (or probabilities) for any normally distributed variable from areas under the standard normal curve.

We present a method for graphically assessing whether a variable is normally distributed —*the normal probability plot*—in Section 6.4. Finally, in Section 6.5, we show how to approximate binomial probabilities with areas under a suitable normal curve.

CASE STUDY

Chest Sizes of Scottish Militiamen

In 1817, an article entitled "Statement of the Sizes of Men in Different Counties of Scotland, Taken from the Local Militia" appeared in the *Edinburgh Medical and Surgical Journal* (Vol. 13, pp. 260–264). Included in the article were data on chest circumference for 5732 Scottish militiamen. The data were collected by an army contractor who was responsible for providing clothing for the militia. A frequency distribution for the chest circumferences, in inches, is given in the following table.

Chest size (in.)	Frequency	Chest size (in.)	Frequency
33	3	41	935
34	19	42	646
35	81	43	313
36	189	44	168
37	409	45	50
38	753	46	18
39	1062	47	3
40	1082	48	1

In his book *Lettres à S.A.R. le Duc Régnant de Saxe-Cobourg et Gotha sur la théorie des probabilités appliquée aux sciences morales et politiques* (Brussels: Hayez, 1846), Adolphe Quetelet discussed a procedure for fitting a normal curve (a special type of bell-shaped curve) to the data on chest circumference.

At the end of this chapter, you will be asked to fit a normal curve to the data, using a technique different from the one used by Quetelet.

6.1 Introducing Normally Distributed Variables

Before beginning our discussion of the normal distribution, we briefly discuss *density curves*. From Section 2.4, we know that an important aspect of the distribution of a variable is its shape and that we can frequently identify the shape of a distribution with a smooth curve. Such curves are called **density curves.**

Theoretically, a density curve represents the distribution of a continuous variable. However, as we have seen, a density curve can often be used to approximate the distribution of a discrete variable.

Two basic properties of every density curve are presented in Key Fact 6.1 and illustrated visually in Fig. 6.1.

KEY FACT 6.1

Basic Properties of Density Curves

Property 1: A density curve is always on or above the horizontal axis.

Property 2: The total area under a density curve (and above the horizontal axis) equals 1.

FIGURE 6.1

Properties 1 and 2 of Key Fact 6.1

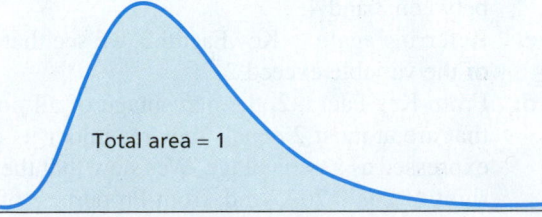

Total area = 1

One of the most important uses of the density curve of a variable relies on the fact that percentages for the variable are equal to areas under its density curve. More precisely, we have Key Fact 6.2 and an illustration of it in Fig. 6.2.

KEY FACT 6.2

Variables and Their Density Curves

For a variable with a density curve, the percentage of all possible observations of the variable that lie within any specified range equals (at least approximately) the corresponding area under the density curve, expressed as a percentage.

FIGURE 6.2   Illustration of Key Fact 6.2

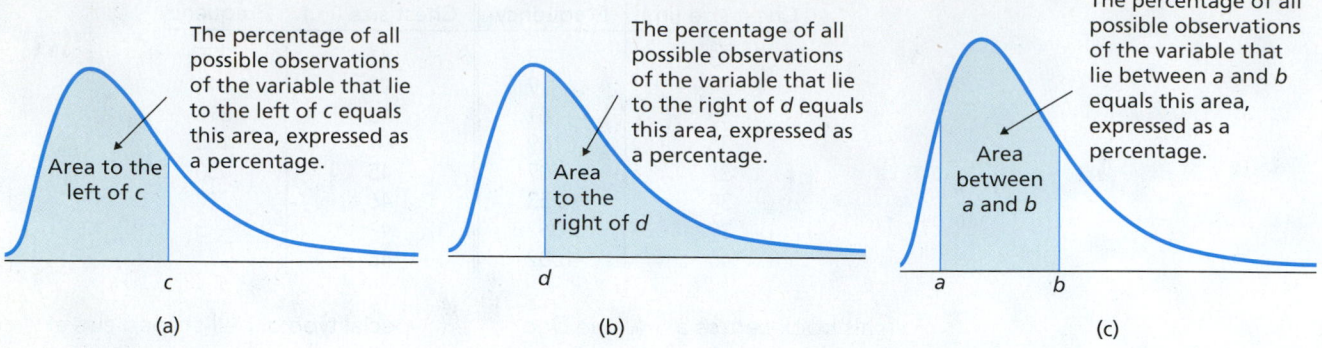

Figure 6.2 illustrates Key Fact 6.2 in the cases where (a) the specified range is to the left of a number, (b) the specified range is to the right of a number, and (c) the specified ranged is between two numbers. A similar illustration would be for the case where, for instance, the specified range is either to the left of one number or to the right of another number.

EXAMPLE 6.1   Density Curves

A certain variable has a density curve.

a.   In terms of its density curve, what percentage of all possible observations of the variable lie between 3 and 4?

b.   The area under the density curve between 3 and 4 is 0.387. What percentage of all possible observations of the variable lie between 3 and 4?

c.   The area under the density curve to the right of 2 is 0.762. What percentage of all possible observations of the variable exceed 2?

d.   What percentage of all possible observations of the variable are at most 2 (i.e., 2 or less)?

Solution

a.   Referring to Key Fact 6.2, we see that the percentage of all possible observations of the variable that lie between 3 and 4 equals the area under its density curve between 3 and 4, expressed as a decimal.

b.   In view of the solution to part (a) and the information provided in the statement of part (b), we see that 38.7% of all possible observations of the variable lie between 3 and 4.

c.   Referring again to Key Fact 6.2, we see that 76.2% of all possible observations of the variable exceed 2.

d.   From Key Fact 6.2, the percentage of all possible observations of the variable that are at most 2 equals the area under its density curve at or to the left of 2, expressed as a percentage. We know that the area under the density curve to the right of 2 is 0.762. And, from Property 2 of Key Fact 6.1, the total area under the density curve is 1. So, the area under the density curve at or to the left of 2 is $1 - 0.762 = 0.238$. Therefore, 23.8% of all possible observations of the variable are at most 2.

You try it!

Exercise 6.7
on page 271

EXAMPLE 6.2   Density Curves

Crystal size   The crystals of a certain mineral are cubes. The length of the sides is a variable that has the density curve shown in Fig. 6.3(a), where length is measured in millimeters (mm). Using the facts that the equation of this density curve is $y = x/2$

for $0 < x < 2$ and that the area of a triangle equals one-half its base times its height, we see that the area under this density curve to the left of any number x between 0 and 2 equals $x^2/4$, as shown in Fig. 6.3(b).

FIGURE 6.3

(a) Density curve for crystal side length, in mm (b) Area under the density curve to the left of x

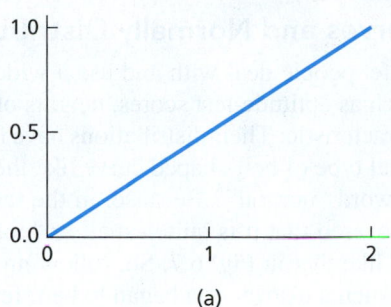
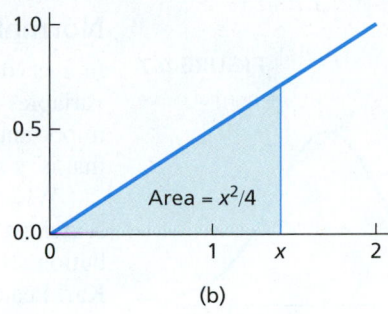

(a)   (b)

Determine the percentage of these crystals that have side lengths

a. less than 0.5 mm.
b. between 1 mm and 1.5 mm.
c. at least 0.25 mm.

Solution   We apply Key Fact 6.2.

a. The percentage of these crystals that have side lengths less than 0.5 mm equals the area under the density curve to the left of 0.5, expressed as a percentage. The required area is $(0.5)^2/4$, or 0.0625, as shown in Fig. 6.4.

Interpretation   Only 6.25% of these crystals have side lengths less than 0.5 mm.

FIGURE 6.4   Area under the density curve to the left of 0.5

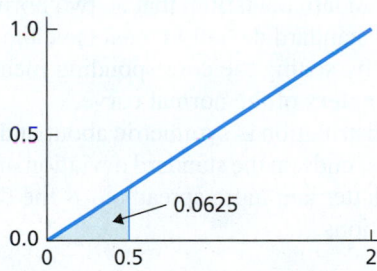

FIGURE 6.5   Area under the density curve between 1 and 1.5

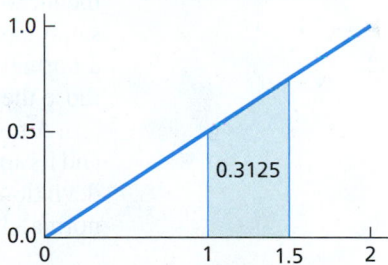

FIGURE 6.6

Area under the density curve at or to the right of 0.25

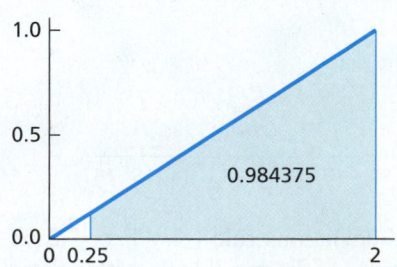

You try it!

Exercise 6.31 on page 271

b. The percentage of these crystals that have side lengths between 1 mm and 1.5 mm equals the area under the density curve between 1 and 1.5, expressed as a percentage. The required area is equal to the area to the left of 1.5 less the area to the left of 1. Consequently, the required area is $(1.5)^2/4 - 1^2/4 = 0.3125$, as shown in Fig. 6.5.

Interpretation   31.25% of these crystals have side lengths between 1 mm and 1.5 mm.

c. The percentage of these crystals that have side lengths at least 0.25 mm equals the area under the density curve at or to the right of 0.25, expressed as a percentage. In view of Property 2 of Key Fact 6.1, the area under the density curve at or to the right of 0.25 equals 1 minus the area under the density curve to the left of 0.25. The area under the density curve to the left of 0.25 equals $(0.25)^2/4$, or 0.015625. So, the area under the density curve to the right of 0.25 is $1 - (0.25)^2/4$, or 0.984375, as shown in Fig. 6.6.

Interpretation   About 98.4% of these crystals have side lengths at least 0.25 mm.

In this chapter, we concentrate on the most important density curve—the *normal density curve* or, simply, the *normal curve*. Later, we discuss other important density curves such as *t*-curves, χ^2-curves, and *F*-curves.

Normal Curves and Normally Distributed Variables

FIGURE 6.7

A normal curve

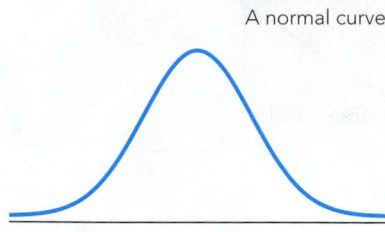

In everyday life, people deal with and use a wide variety of variables. Some of these variables—such as aptitude-test scores, heights of women, and wheat yield—share an important characteristic: Their distributions have roughly the shape of a **normal curve,** that is, a special type of bell-shaped curve like the one shown in Fig. 6.7.

Why the word "normal"? Because, in the last half of the nineteenth century, researchers discovered that it is quite usual, or "normal," for a variable to have a distribution shaped like that in Fig. 6.7. So, following the lead of noted British statistician Karl Pearson, such a distribution began to be referred to as a *normal distribution*.

DEFINITION 6.1

Normally Distributed Variable

A variable is said to be a **normally distributed variable** or to have a **normal distribution** if its distribution has the shape of a normal curve.

Here is some important terminology associated with normal distributions.

- If a variable of a population is normally distributed and is the only variable under consideration, common practice is to say that the **population is normally distributed** or that it is a **normally distributed population.**
- In practice, a distribution is unlikely to have exactly the shape of a normal curve. If a variable's distribution is shaped roughly like a normal curve, we say that the variable is an **approximately normally distributed variable** or that it has **approximately a normal distribution.**

A normal distribution (and hence a normal curve) is completely determined by the mean and standard deviation; that is, two normally distributed variables having the same mean and standard deviation must have the same distribution. We often identify a normal curve by stating the corresponding mean and standard deviation and calling those the **parameters** of the normal curve.[†]

A normal distribution is symmetric about and centered at the mean of the variable, and its spread depends on the standard deviation of the variable—the larger the standard deviation, the flatter and more spread out is the distribution. Figure 6.8 displays three normal distributions.

FIGURE 6.8

Three normal distributions

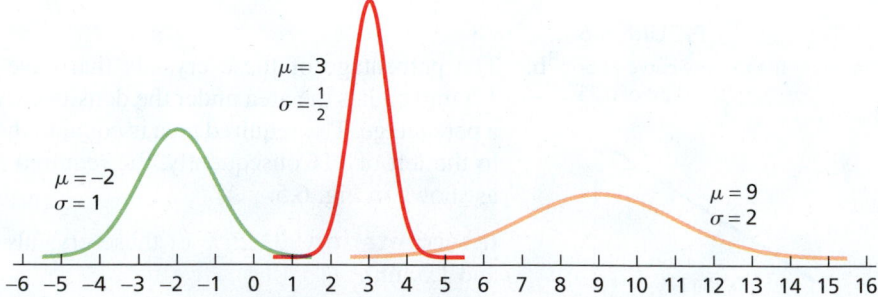

When applied to a variable, the three-standard-deviations rule (Key Fact 3.2 on page 111) states that almost all the possible observations of the variable lie within three standard deviations to either side of the mean. This rule is illustrated by the three normal distributions in Fig. 6.8: Each normal curve is close to the horizontal axis outside the range of three standard deviations to either side of the mean.

[†]The equation of the normal curve with parameters μ and σ is $y = e^{-(x-\mu)^2/2\sigma^2}/(\sqrt{2\pi}\sigma)$, where $e \approx 2.718$ and $\pi \approx 3.142$.

For instance, the third normal distribution in Fig. 6.8 has mean $\mu = 9$ and standard deviation $\sigma = 2$. Three standard deviations to the left of the mean is

$$\mu - 3\sigma = 9 - 3 \cdot 2 = 3,$$

and three standard deviations to the right of the mean is

$$\mu + 3\sigma = 9 + 3 \cdot 2 = 15.$$

As shown in Fig. 6.8, the corresponding normal curve is close to the horizontal axis outside the range from 3 to 15.

In summary, the normal curve associated with a normal distribution is

- bell shaped,
- centered at μ, and
- close to the horizontal axis outside the range from $\mu - 3\sigma$ to $\mu + 3\sigma$,

as depicted in Figs. 6.8 and 6.9. This information helps us sketch a normal distribution.

You try it!

Exercise 6.23
on page 271

FIGURE 6.9

Graph of generic normal distribution

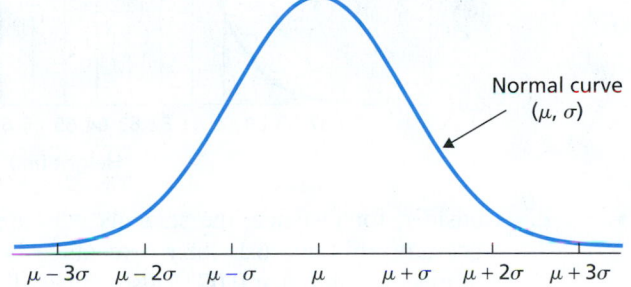

Normal curve (μ, σ)

$\mu - 3\sigma$ $\mu - 2\sigma$ $\mu - \sigma$ μ $\mu + \sigma$ $\mu + 2\sigma$ $\mu + 3\sigma$

Example 6.3 illustrates a normally distributed variable and discusses some additional properties of such variables.

EXAMPLE 6.3   A Normally Distributed Variable

TABLE 6.1

Frequency and relative-frequency distributions for heights

Height (in.)	Frequency f	Relative freq.
56–under 57	3	0.0009
57–under 58	6	0.0018
58–under 59	26	0.0080
59–under 60	74	0.0227
60–under 61	147	0.0450
61–under 62	247	0.0757
62–under 63	382	0.1170
63–under 64	483	0.1480
64–under 65	559	0.1713
65–under 66	514	0.1575
66–under 67	359	0.1100
67–under 68	240	0.0735
68–under 69	122	0.0374
69–under 70	65	0.0199
70–under 71	24	0.0074
71–under 72	7	0.0021
72–under 73	5	0.0015
73–under 74	1	0.0003
	3264	1.0000

Heights of Female College Students A midwestern college has an enrollment of 3264 female students. Records show that the mean height of these students is 64.4 inches and that the standard deviation is 2.4 inches. Here the variable is height, and the population consists of the 3264 female students attending the college. Frequency and relative-frequency distributions for these heights appear in Table 6.1. The table shows, for instance, that 7.35% (0.0735) of the students are between 67 and 68 inches tall.

a. Show that the variable "height" is approximately normally distributed for this population.

b. Identify the normal curve associated with the variable "height" for this population.

c. Discuss the relationship between the percentage of female students whose heights lie within a specified range and the corresponding area under the associated normal curve.

Solution

a. Figure 6.10 on the following page displays a relative-frequency histogram for the heights of the female students. It shows that the distribution of heights has roughly the shape of a normal curve and, consequently, that the variable "height" is approximately normally distributed for this population.

b. The associated normal curve is the one whose parameters are the same as the mean and standard deviation of the variable, which are 64.4 and 2.4, respectively. Thus the required normal curve has parameters $\mu = 64.4$ and $\sigma = 2.4$. It is superimposed on the histogram in Fig. 6.10.

FIGURE 6.10
Relative-frequency histogram
for heights with superimposed
normal curve

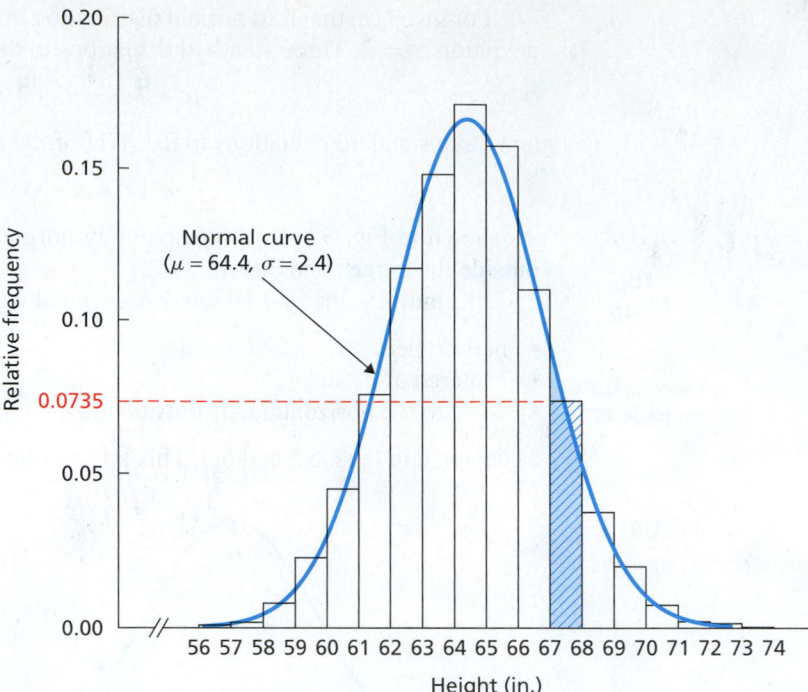

c. Consider, for instance, the students who are between 67 and 68 inches tall. According to Table 6.1, their percentage is 7.35%, or 0.0735 (to four decimal places). Note that 0.0735 also equals the area of the cross-hatched bar in Fig. 6.10 because the bar has height 0.0735 and width 1. Now look at the area under the curve between 67 and 68, shaded in Fig. 6.10. This area approximates the area of the cross-hatched bar. Thus we can approximate the percentage of students between 67 and 68 inches tall by the area under the normal curve between 67 and 68. This result holds in general.

Interpretation The percentage of female students whose heights lie within any specified range can be approximated by the corresponding area under the normal curve associated with the variable "height" for this population of female students.

You try it!

Exercise 6.35 on page 272

The interpretation just given is not surprising. In fact, it simply provides an illustration of Key Fact 6.2 on page 263. However, for emphasis, we present Key Fact 6.3, which is a special case of Key Fact 6.2 when applied to normally distributed variables.

KEY FACT 6.3

Normally Distributed Variables and Normal-Curve Areas

For a normally distributed variable, the percentage of all possible observations that lie within any specified range equals the corresponding area under its associated normal curve, expressed as a percentage. This result holds approximately for a variable that is approximately normally distributed.

Note: For brevity, we often paraphrase the content of Key Fact 6.3 with the statement "percentages for a normally distributed variable are equal to areas under its associated normal curve."

Standardizing a Normally Distributed Variable

Now the question is: How do we find areas under a normal curve? Conceptually, we need a table of areas for each normal curve. This, of course, is impossible because there are infinitely many different normal curves—one for each choice of μ and σ. The way out of this difficulty is standardizing, which transforms every normal distribution into one particular normal distribution, the *standard normal distribution*.

DEFINITION 6.2

FIGURE 6.11
Standard normal distribution

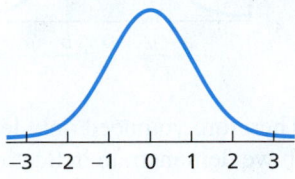

Standard Normal Distribution; Standard Normal Curve

A normally distributed variable having mean 0 and standard deviation 1 is said to have the **standard normal distribution.** Its associated normal curve is called the **standard normal curve,** which is shown in Fig. 6.11.

Recall from Chapter 3 (page 144) that we standardize a variable x by subtracting its mean and then dividing by its standard deviation. The resulting variable, $z = (x - \mu)/\sigma$, is called the standardized version of x or the standardized variable corresponding to x.

The standardized version of any variable has mean 0 and standard deviation 1. A normally distributed variable furthermore has a normally distributed standardized version.

KEY FACT 6.4

? **What Does It Mean?**

Subtracting from a normally distributed variable its mean and then dividing by its standard deviation results in a variable with the standard normal distribution.

Standardized Normally Distributed Variable

The standardized version of a normally distributed variable x,

$$z = \frac{x - \mu}{\sigma},$$

has the standard normal distribution.

We can interpret Key Fact 6.4 in several ways. Theoretically, it says that standardizing converts all normal distributions to the standard normal distribution, as depicted in Fig. 6.12.

FIGURE 6.12
Standardizing normal distributions

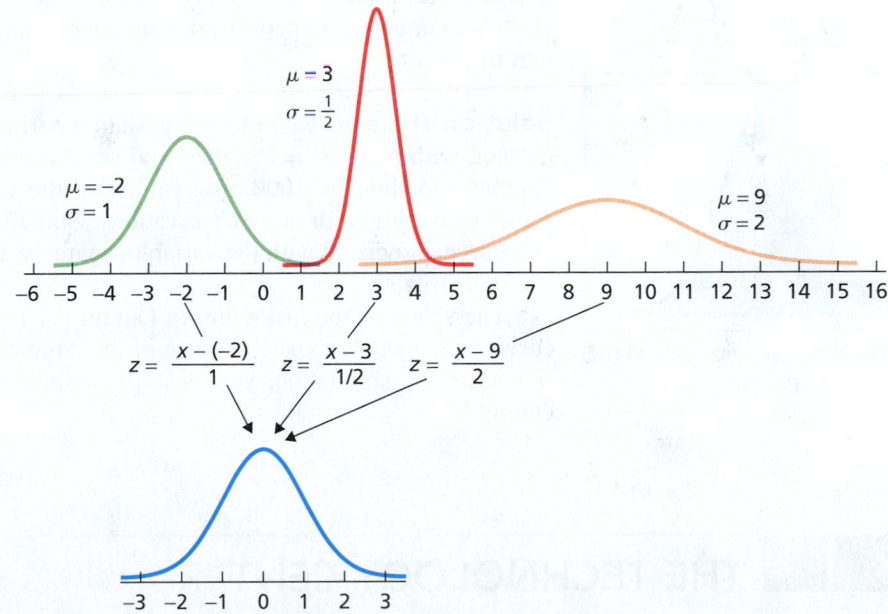

We need a more practical interpretation of Key Fact 6.4. Let x be a normally distributed variable with mean μ and standard deviation σ, and let a and b be real numbers with $a < b$. The percentage of all possible observations of x that lie between a and b is the same as the percentage of all possible observations of z that lie between $(a - \mu)/\sigma$ and $(b - \mu)/\sigma$. In light of Key Fact 6.4, this latter percentage equals the area under the standard normal curve between $(a - \mu)/\sigma$ and $(b - \mu)/\sigma$. We summarize these ideas graphically in Fig. 6.13 on the next page.

Consequently, for a normally distributed variable, we can find the percentage of all possible observations that lie within any specified range by

1. expressing the range in terms of z-scores, and
2. determining the corresponding area under the standard normal curve.

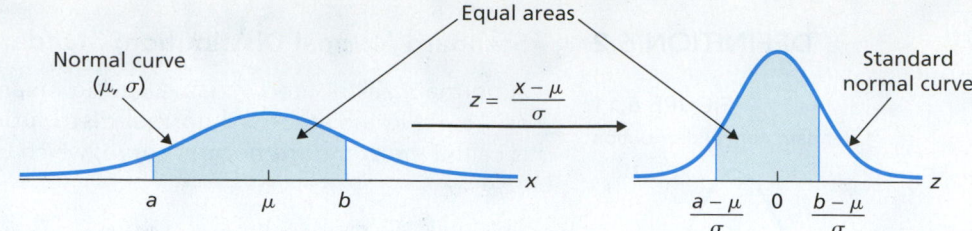

FIGURE 6.13

Finding percentages for a normally distributed variable from areas under the standard normal curve

You already know how to convert to z-scores. Therefore you need only learn how to find areas under the standard normal curve, which we demonstrate in Section 6.2.

Simulating a Normal Distribution

For understanding and for research, simulating a variable is often useful. Doing so involves use of a computer or statistical calculator to generate observations of the variable.

When we simulate a normally distributed variable, a histogram of the observations will have roughly the same shape (and center) as that of the normal curve associated with the variable. The shape of the histogram will tend to look more like that of the normal curve when the number of observations is large. We illustrate the simulation of a normally distributed variable in the next example.

EXAMPLE 6.4 **Simulating a Normally Distributed Variable**

OUTPUT 6.1

Histogram of 1000 simulated human gestation periods with superimposed normal curve

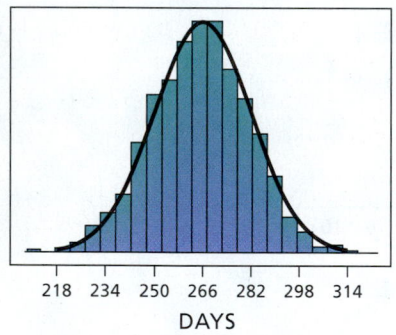

DAYS

Report 6.1

Gestation Periods of Humans Gestation periods of humans are normally distributed with a mean of 266 days and a standard deviation of 16 days. Simulate 1000 human gestation periods, obtain a histogram of the simulated data, and interpret the results.

Solution Here the variable x is gestation period. For humans, it is normally distributed with mean $\mu = 266$ days and standard deviation $\sigma = 16$ days. We used a computer to simulate 1000 observations of the variable x for humans. Output 6.1 shows a histogram for those observations. Note that we have superimposed the normal curve associated with the variable—namely, the one with parameters $\mu = 266$ and $\sigma = 16$.

The shape of the histogram in Output 6.1 is quite close to that of the normal curve, as we would expect because of the large number of simulated observations. If you do the simulation, your histogram should be similar to the one shown in Output 6.1.

THE TECHNOLOGY CENTER

Most statistical software packages and some graphing calculators have built-in procedures to simulate observations of normally distributed variables. In Example 6.4, we used Minitab, but Excel and the TI-83/84 Plus can also be used to conduct that simulation and obtain the histogram.

Exercises 6.1

Understanding the Concepts and Skills

6.1 What is a density curve?

6.2 State the two basic properties of every density curve.

6.3 For a variable with a density curve, what is the relationship between the percentage of all possible observations of the variable that lie within any specified range and the corresponding area under its density curve?

In each of Exercises 6.4–6.11, assume that the variable under consideration has a density curve. Note that the answers required here may be only approximately correct.

6.4 The percentage of all possible observations of the variable that lie between 7 and 12 equals the area under its density curve between _____ and _____, expressed as a percentage.

6.5 The percentage of all possible observations of the variable that lie to the right of 4 equals the area under its density curve to the right of _____, expressed as a percentage.

6.6 The area under the density curve that lies to the left of 10 is 0.654. What percentage of all possible observations of the variable are
a. less than 10?   **b.** at least 10?

6.7 The area under the density curve that lies to the right of 15 is 0.324. What percentage of all possible observations of the variable
a. exceed 15?   **b.** are at most 15?

6.8 The area under the density curve that lies between 30 and 40 is 0.832. What percentage of all possible observations of the variable are either less than 30 or greater than 40?

6.9 The area under the density curve that lies between 15 and 20 is 0.414. What percentage of all possible observations of the variable are either less than 15 or greater than 20?

6.10 Given that 33.6% of all possible observations of the variable exceed 8, determine the area under the density curve that lies to the
a. right of 8.   **b.** left of 8.

6.11 Given that 28.4% of all possible observations of the variable are less than 11, determine the area under the density curve that lies to the
a. left of 11.   **b.** right of 11.

6.12 A curve has area 0.425 to the left of 4 and area 0.585 to the right of 4. Could this curve be a density curve for some variable? Explain your answer.

6.13 A curve has area 0.613 to the left of 65 and area 0.287 to the right of 65. Could this curve be a density curve for some variable? Explain your answer.

6.14 Explain in your own words why a density curve has the two properties listed in Key Fact 6.1 on page 263.

6.15 A variable is approximately normally distributed. If you draw a histogram of the distribution of the variable, roughly what shape will it have?

6.16 Precisely what is meant by the statement that a population is normally distributed?

6.17 Two normally distributed variables have the same means and the same standard deviations. What can you say about their distributions? Explain your answer.

6.18 Which normal distribution has a wider spread: the one with mean 1 and standard deviation 2 or the one with mean 2 and standard deviation 1? Explain your answer.

6.19 Consider two normal distributions, one with mean −4 and standard deviation 3, and the other with mean 6 and standard deviation 3. Answer true or false to each statement and explain your answers.
a. The two normal distributions have the same spread.
b. The two normal distributions are centered at the same place.

6.20 Consider two normal distributions, one with mean −4 and standard deviation 3, and the other with mean −4 and standard deviation 6. Answer true or false to each statement and explain your answers.
a. The two normal distributions have the same spread.
b. The two normal distributions are centered at the same place.

6.21 True or false: The mean of a normal distribution has no effect on its spread. Explain your answer.

6.22 What are the parameters for a normal curve?

6.23 Sketch the normal distribution with
a. $\mu = 3$ and $\sigma = 3$.   **b.** $\mu = 1$ and $\sigma = 3$.
c. $\mu = 3$ and $\sigma = 1$.

6.24 Sketch the normal distribution with
a. $\mu = -2$ and $\sigma = 2$.   **b.** $\mu = -2$ and $\sigma = 1/2$.
c. $\mu = 0$ and $\sigma = 2$.

6.25 For a normally distributed variable, what is the relationship between the percentage of all possible observations that lie between 2 and 3 and the area under the associated normal curve between 2 and 3? What if the variable is only approximately normally distributed?

6.26 For a normally distributed variable, what is the relationship between the percentage of all possible observations that lie to the right of 7 and the area under the associated normal curve to the right of 7? What if the variable is only approximately normally distributed?

6.27 The area under a particular normal curve to the left of 105 is 0.6227. A normally distributed variable has the same mean and standard deviation as the parameters for this normal curve. What percentage of all possible observations of the variable lie to the left of 105? Explain your answer.

6.28 The area under a particular normal curve between 10 and 15 is 0.6874. A normally distributed variable has the same mean and standard deviation as the parameters for this normal curve. What percentage of all possible observations of the variable lie between 10 and 15? Explain your answer.

6.29 A variable has the density curve whose equation is $y = 2x$ for $0 < x < 1$, and $y = 0$ otherwise.
a. Graph the density curve of this variable.
b. Show that the area under this density curve to the left of any number x between 0 and 1 equals x^2.

What percentage of all possible observations of the variable
c. lie between 1/2 and 3/4?
d. are at least 1/4?

6.30 A variable has the density curve whose equation is $y = 1$ for $0 < x < 1$, and $y = 0$ otherwise.
a. Graph the density curve of this variable.
b. Show that the area under this density curve to the left of any number x between 0 and 1 equals x.

What percentage of all possible observations of the variable
c. lie between 1/4 and 3/4?
d. exceed 7/8?

Applying the Concepts and Skills

6.31 Waiting for the Train. A commuter train arrives punctually at a station every half hour. Each morning, a commuter named John leaves his house and casually strolls to the train station. The time, in minutes, that John waits for the train is a variable with density curve $y = 1/30$ for $0 < x < 30$, and $y = 0$ otherwise.
a. Graph the density curve of this variable.
b. Show that the area under this density curve to the left of any number x between 0 and 30 equals $x/30$.

What percentage of the time does John wait for the train

c. less than 5 minutes?

d. between 10 and 15 minutes?

e. at least 20 minutes?

6.32 Bacteria on a Petri Dish. A petri dish is a small, shallow dish of thin glass or plastic, used especially for cultures in bacteriology. A 2-inch-radius petri dish, containing nutrients upon which bacteria can multiply, is smeared with a uniform suspension of bacteria. Subsequently, spots indicating colonies of bacteria appear. The distance of the center of the first spot to appear from the center of the petri dish is a variable with density curve $y = x/2$ for $0 < x < 2$, and $y = 0$ otherwise.

a. Graph the density curve of this variable.

b. Show that the area under this density curve to the left of any number x between 0 and 2 equals $x^2/4$.

What percentage of the time is the distance of the center of the first spot to appear from the center of the petri dish

c. at most 1 inch?

d. between 0.25 inch and 1.5 inches?

e. more than 0.75 inch?

6.33 Fire Loss. The loss, in millions of dollars, due to a fire in a commercial building is a variable with density curve $y = 1 - x/2$ for $0 < x < 2$, and $y = 0$ otherwise. Using the fact that the area of a triangle equals one-half its base times its height, we find that the area under this density curve to the left of any number x between 0 and 2 equals $x - x^2/4$.

a. Graph the density curve of this variable.

b. What percentage of losses exceed $1.5 million?

6.34 Emergency Room Traffic. Desert Samaritan Hospital in Mesa, Arizona, keeps records of its emergency-room traffic. Beginning at 6:00 P.M. on any given day, the elapsed time, in hours, until the first patient arrives is a variable with density curve $y = 6.9e^{-6.9x}$ for $x > 0$, and $y = 0$ otherwise. Here e is Euler's number, which is approximately 2.71828. Most calculators have an e-key. Using calculus, it can be shown that the area under this density curve to the left of any number x greater than 0 equals $1 - e^{-6.9x}$.

a. Graph the density curve of this variable.

b. What percentage of the time does the first patient arrive between 6:15 P.M. and 6:30 P.M.?

6.35 Female College Students. Refer to Example 6.3 on page 267.

a. Use the relative-frequency distribution in Table 6.1 to obtain the percentage of female students who are between 60 and 65 inches tall.

b. Use your answer from part (a) to estimate the area under the normal curve having parameters $\mu = 64.4$ and $\sigma = 2.4$ that lies between 60 and 65. Why do you get only an estimate of the true area?

6.36 Female College Students. Refer to Example 6.3 on page 267.

a. The area under the normal curve with parameters $\mu = 64.4$ and $\sigma = 2.4$ that lies to the left of 61 is 0.0783. Use this information to estimate the percentage of female students who are shorter than 61 inches.

b. Use the relative-frequency distribution in Table 6.1 to obtain the actual percentage of female students who are shorter than 61 inches.

c. Compare your answers from parts (a) and (b).

6.37 Giant Tarantulas. One of the larger species of tarantulas is the *Grammostola mollicoma*, whose common name is the Brazilian giant tawny red. A tarantula has two body parts. The anterior part of the body is covered above by a shell, or carapace. From a recent article by F. Costa and F. Perez–Miles titled "Reproductive Biology of Uruguayan Theraphosids" (*The Journal of Arachnology*, Vol. 30, No. 3, pp. 571–587), we find that the carapace length of the adult male *G. mollicoma* is normally distributed with a mean of 18.14 mm and a standard deviation of 1.76 mm. Let x denote carapace length for the adult male *G. mollicoma*.

a. Sketch the distribution of the variable x.

b. Obtain the standardized version, z, of x.

c. Identify and sketch the distribution of z.

d. The percentage of adult male *G. mollicoma* that have carapace length between 16 mm and 17 mm is equal to the area under the standard normal curve between _____ and _____.

e. The percentage of adult male *G. mollicoma* that have carapace length exceeding 19 mm is equal to the area under the standard normal curve that lies to the _____ of _____.

6.38 Serum Cholesterol Levels. According to the *National Health and Nutrition Examination Survey*, published by the National Center for Health Statistics, the serum (noncellular portion of blood) total cholesterol level of U.S. females 20 years old or older is normally distributed with a mean of 206 mg/dL (milligrams per deciliter) and a standard deviation of 44.7 mg/dL. Let x denote serum total cholesterol level for U.S. females 20 years old or older.

a. Sketch the distribution of the variable x.

b. Obtain the standardized version, z, of x.

c. Identify and sketch the distribution of z.

d. The percentage of U.S. females 20 years old or older who have a serum total cholesterol level between 150 mg/dL and 250 mg/dL is equal to the area under the standard normal curve between _____ and _____.

e. The percentage of U.S. females 20 years old or older who have a serum total cholesterol level below 220 mg/dL is equal to the area under the standard normal curve that lies to the _____ of _____.

6.39 New York City 10-km Run. As reported in *Runner's World* magazine, the times of the finishers in the New York City 10-km run are normally distributed with mean 61 minutes and standard deviation 9 minutes. Let x denote finishing time for finishers in this race.

a. Sketch the distribution of the variable x.

b. Obtain the standardized version, z, of x.

c. Identify and sketch the distribution of z.

d. The percentage of finishers with times between 50 and 70 minutes is equal to the area under the standard normal curve between _____ and _____.

e. The percentage of finishers with times less than 75 minutes is equal to the area under the standard normal curve that lies to the _____ of _____.

6.40 Green Sea Urchins. From the paper "Effects of Chronic Nitrate Exposure on Gonad Growth in Green Sea Urchin *Strongylocentrotus droebachiensis*" (*Aquaculture*, Vol. 242, No. 1–4, pp. 357–363) by S. Siikavuopio et al., we found that weights of adult green sea urchins are normally distributed with mean 52.0 g and standard deviation 17.2 g. Let x denote weight of adult green sea urchins.

a. Sketch the distribution of the variable x.

b. Obtain the standardized version, z, of x.

c. Identify and sketch the distribution of z.

d. The percentage of adult green sea urchins with weights between 50 g and 60 g is equal to the area under the standard normal curve between _____ and _____.

e. The percentage of adult green sea urchins with weights above 40 g is equal to the area under the standard normal curve that lies to the _____ of _____.

6.41 Ages of Mothers. From the document *National Vital Statistics Reports*, a publication of the National Center for Health Statistics, we obtained the following frequency distribution for the ages of women who became mothers during one year.

Age (yr)	Frequency
10–under 15	7,315
15–under 20	425,493
20–under 25	1,022,106
25–under 30	1,060,391
30–under 35	951,219
35–under 40	453,927
40–under 45	95,788
45–under 50	54,872

a. Obtain a relative-frequency histogram of these age data.
b. Based on your histogram, do you think that the ages of women who became mothers that year are approximately normally distributed? Explain your answer.

6.42 Birth Rates. The National Center for Health Statistics publishes information about birth rates (per 1000 population) in the document *National Vital Statistics Report*. The following table provides a frequency distribution for birth rates during one year for the 50 states and the District of Columbia.

Rate	Frequency	Rate	Frequency
9–under 10	3	14–under 15	7
10–under 11	3	15–under 16	1
11–under 12	7	16–under 17	0
12–under 13	19	17–under 18	0
13–under 14	10	18–under 19	1

a. Obtain a frequency histogram of these birth-rate data.
b. Based on your histogram, do you think that birth rates for the 50 states and the District of Columbia are approximately normally distributed? Explain your answer.

6.43 Cloudiness in Breslau. In the paper "Cloudiness: Note on a Novel Case of Frequency" (*Proceedings of the Royal Society of London*, Vol. 62, pp. 287–290), K. Pearson examined data on daily degree of cloudiness, on a scale of 0 to 10, at Breslau (Wroclaw), Poland, during the decade 1876–1885. A frequency distribution of the data is presented in the following table.

Degree	Frequency	Degree	Frequency
0	751	6	21
1	179	7	71
2	107	8	194
3	69	9	117
4	46	10	2089
5	9		

a. Draw a frequency histogram of these degree-of-cloudiness data.
b. Based on your histogram, do you think that degree of cloudiness in Breslau during the decade in question is approximately normally distributed? Explain your answer.

6.44 Wrong Number. A classic study by F. Thorndike on the number of calls to a wrong number appeared in the paper "Applications of Poisson's Probability Summation" (*Bell Systems Technical Journal*, Vol. 5, pp. 604–624). The study examined the number of calls to a wrong number from coin-box telephones in a large transportation terminal. Based on the results of that paper, we obtained the following percent distribution for the number of wrong numbers during a 1-minute period.

Wrong	0	1	2	3	4	5	6	7	8
Percent	17.2	30.5	26.6	15.1	7.3	2.4	0.7	0.1	0.1

a. Construct a relative-frequency histogram of these wrong-number data.
b. Based on your histogram, do you think that the number of wrong numbers from these coin-box telephones is approximately normally distributed? Explain your answer.

Working with Large Data Sets

6.45 SAT Scores. Each year, thousands of high school students bound for college take the Scholastic Assessment Test (SAT). This test measures the verbal and mathematical abilities of prospective college students. Student scores are reported on a scale that ranges from a low of 200 to a high of 800. Summary results for the scores are published by the College Entrance Examination Board in *College Bound Seniors*. In one high school graduating class, the SAT scores are as provided on the WeissStats site. Use the technology of your choice to answer the following questions.
a. Do the SAT verbal scores for this class appear to be approximately normally distributed? Explain your answer.
b. Do the SAT math scores for this class appear to be approximately normally distributed? Explain your answer.

6.46 Fertility Rates. From the U.S. Census Bureau, in the document *International Data Base*, we obtained data on the total fertility rates for women in various countries. Those data are presented on the WeissStats site. The total fertility rate gives the average number of children that would be born if all women in a given country lived to the end of their childbearing years and, at each year of age, they experienced the birth rates occurring in the specified year. Use the technology of your choice to decide whether total fertility rates for countries appear to be approximately normally distributed. Explain your answer.

Extending the Concepts and Skills

6.47 "Chips Ahoy! 1,000 Chips Challenge." Students in an introductory statistics course at the U.S. Air Force Academy participated in Nabisco's "Chips Ahoy! 1,000 Chips Challenge" by confirming that there were at least 1000 chips in every 18-ounce bag of cookies that they examined. As part of their assignment, they concluded that the number of chips per bag is approximately normally distributed. Could the number of chips per bag be exactly normally distributed? Explain your answer. [SOURCE: B. Warner and J. Rutledge, "Checking the Chips Ahoy! Guarantee," *Chance*, Vol. 12(1), pp. 10–14]

6.48 Gestation Periods of Humans. Refer to the simulation of human gestation periods discussed in Example 6.4 on page 270.
a. Sketch the normal curve for human gestation periods.
b. Simulate 1000 human gestation periods. (*Note:* Users of the TI-83/84 Plus should simulate 500 human gestation periods.)
c. Approximately what values would you expect for the sample mean and sample standard deviation of the 1000 observations? Explain your answers.
d. Obtain the sample mean and sample standard deviation of the 1000 observations, and compare your answers to your estimates in part (c).

e. Roughly what would you expect a histogram of the 1000 observations to look like? Explain your answer.

f. Obtain a histogram of the 1000 observations, and compare your result to your expectation in part (e).

6.49 Delaying Adulthood. In the paper, "Delayed Metamorphosis of a Tropical Reef Fish (*Acanthurus triostegus*): A Field Experiment" (*Marine Ecology Progress Series*, Vol. 176, pp. 25–38), M. McCormick studied larval duration of the convict surgeonfish, a common tropical reef fish. This fish has been found to delay metamorphosis into adulthood by extending its larval phase, a delay that often leads to enhanced survivorship in the species by increasing the chances of finding suitable habitat. Duration of the larval phase for convict surgeonfish is normally distributed with mean 53 days and standard deviation 3.4 days. Let x denote larval-phase duration for convict surgeonfish.

a. Sketch the normal curve for the variable x.

b. Simulate 1500 observations of x. (*Note:* Users of the TI-83/84 Plus should simulate 750 observations.)

c. Approximately what values would you expect for the sample mean and sample standard deviation of the 1500 observations? Explain your answers.

d. Obtain the sample mean and sample standard deviation of the 1500 observations, and compare your answers to your estimates in part (c).

e. Roughly what would you expect a histogram of the 1500 observations to look like? Explain your answer.

f. Obtain a histogram of the 1500 observations, and compare your result to your expectation in part (e).

6.2   Areas under the Standard Normal Curve

In Section 6.1, we demonstrated, among other things, that we can obtain the percentage of all possible observations of a normally distributed variable that lie within any specified range by (1) expressing the range in terms of z-scores and (2) determining the corresponding area under the standard normal curve.

You already know how to convert to z-scores. In this section, you will discover how to implement the second step—determining areas under the standard normal curve.

Basic Properties of the Standard Normal Curve

We first need to discuss some of the basic properties of the standard normal curve. Recall that this curve is the one associated with the standard normal distribution, which has mean 0 and standard deviation 1. Figure 6.14 again shows the standard normal distribution and the standard normal curve.

In Section 6.1, we showed that a normal curve is bell shaped, is centered at μ, and is close to the horizontal axis outside the range from $\mu - 3\sigma$ to $\mu + 3\sigma$. Applied to the standard normal curve, these characteristics mean that it is bell shaped, is centered at 0, and is close to the horizontal axis outside the range from −3 to 3. Thus the standard normal curve is symmetric about 0. All of these properties are reflected in Fig. 6.14.

Another property of the standard normal curve is that the total area under it is 1. This property is shared by all density curves, as noted in Key Fact 6.1 on page 263.

FIGURE 6.14

Standard normal distribution and standard normal curve

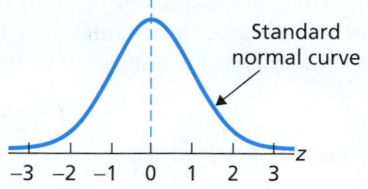

Standard normal curve

KEY FACT 6.5

Basic Properties of the Standard Normal Curve

Property 1: The total area under the standard normal curve is 1.

Property 2: The standard normal curve extends indefinitely in both directions, approaching, but never touching, the horizontal axis as it does so.

Property 3: The standard normal curve is symmetric about 0; that is, the part of the curve to the left of the dashed line in Fig. 6.14 is the mirror image of the part of the curve to the right of it.

Property 4: Almost all the area under the standard normal curve lies between −3 and 3.

Because the standard normal curve is the associated normal curve for a standardized normally distributed variable, we labeled the horizontal axis in Fig. 6.14 with the letter z and refer to numbers on that axis as z-scores. For these reasons, the standard normal curve is sometimes called the **z-curve**.

Using the Standard Normal Table (Table II)

Areas under the standard normal curve are so important that we have tables of those areas. Table II, located inside the back cover of this book and in Appendix A, is such a table.

A typical four-decimal-place number in the body of Table II gives the area under the standard normal curve that lies to the left of a specified z-score. The left page of Table II is for negative z-scores, and the right page is for positive z-scores.

■■■■ **EXAMPLE 6.5**   **Finding the Area to the Left of a Specified z-Score**

Determine the area under the standard normal curve that lies to the left of 1.23, as shown in Fig. 6.15(a).

FIGURE 6.15

Finding the area under the standard normal curve to the left of 1.23

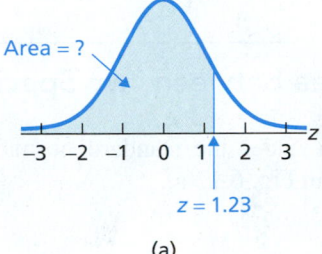
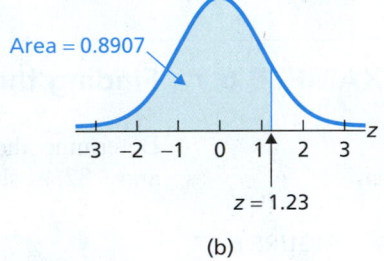

(a)   (b)

Solution   We use the right page of Table II because 1.23 is positive. For ease of reference, a portion of that page is provided in Table 6.2.

TABLE 6.2

Areas under the standard normal curve

z	0.00	0.01	0.02	0.03	0.04	0.05	0.06	0.07	0.08	0.09
.	.	.	.	.	.	.	.	.	.	.
.	.	.	.	.	.	.	.	.	.	.
.	.	.	.	.	.	.	.	.	.	.
1.0	0.8413	0.8438	0.8461	0.8485	0.8508	0.8531	0.8554	0.8577	0.8599	0.8621
1.1	0.8643	0.8665	0.8686	0.8708	0.8729	0.8749	0.8770	0.8790	0.8810	0.8830
1.2	0.8849	0.8869	0.8888	0.8907	0.8925	0.8944	0.8962	0.8980	0.8997	0.9015
1.3	0.9032	0.9049	0.9066	0.9082	0.9099	0.9115	0.9131	0.9147	0.9162	0.9177
1.4	0.9192	0.9207	0.9222	0.9236	0.9251	0.9265	0.9279	0.9292	0.9306	0.9319
.	.	.	.	.	.	.	.	.	.	.
.	.	.	.	.	.	.	.	.	.	.
.	.	.	.	.	.	.	.	.	.	.

Second decimal place in z

You try it!

Exercise 6.59 on page 279

First, we go down the left-hand column, labeled z, to "1.2." Then, going across that row to the column labeled "0.03," we reach 0.8907. This number is the area under the standard normal curve that lies to the left of 1.23, as shown in Fig. 6.15(b).

■

We can also use Table II to find the area to the right of a specified z-score and to find the area between two specified z-scores.

■■■■ **EXAMPLE 6.6**   **Finding the Area to the Right of a Specified z-Score**

Determine the area under the standard normal curve that lies to the right of 0.76, as shown in Fig. 6.16(a) on the following page.

Solution   Because the total area under the standard normal curve is 1 (Property 1 of Key Fact 6.5), the area to the right of 0.76 equals 1 minus the area to the left of 0.76. We find this latter area as in the previous example, by first going down the

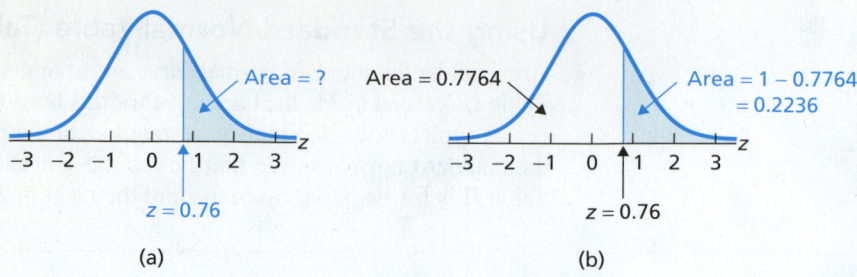

FIGURE 6.16
Finding the area under the standard
normal curve to the right of 0.76

You try it!

Exercise 6.61
on page 279

z column to "0.7." Then, going across that row to the column labeled "0.06," we reach 0.7764, which is the area under the standard normal curve that lies to the left of 0.76. Thus, the area under the standard normal curve that lies to the right of 0.76 is $1 - 0.7764 = 0.2236$, as shown in Fig. 6.16(b).

◼◼◼ EXAMPLE 6.7 Finding the Area between Two Specified z-Scores

Determine the area under the standard normal curve that lies between -0.68 and 1.82, as shown in Fig. 6.17(a).

FIGURE 6.17
Finding the area under the standard
normal curve that lies between
–0.68 and 1.82

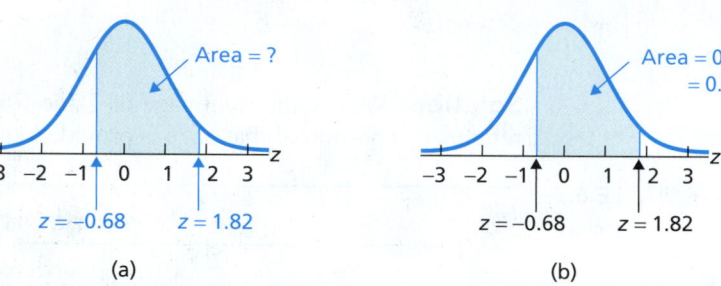

You try it!

Exercise 6.63
on page 279

Solution The area under the standard normal curve that lies between -0.68 and 1.82 equals the area to the left of 1.82 minus the area to the left of -0.68. Table II shows that these latter two areas are 0.9656 and 0.2483, respectively. So the area we seek is $0.9656 - 0.2483 = 0.7173$, as shown in Fig 6.17(b).

The discussion presented in Examples 6.5–6.7 is summarized by the three graphs in Fig. 6.18.

FIGURE 6.18
Using Table II to find the area
under the standard normal curve that
lies (a) to the left of a specified z-score,
(b) to the right of a specified z-score,
and (c) between two specified z-scores

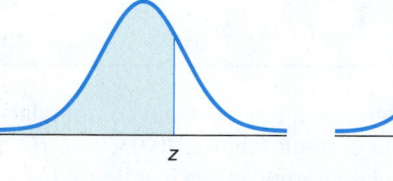 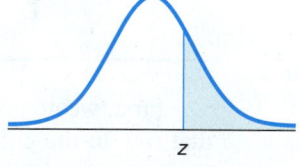 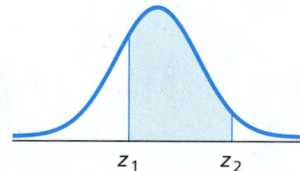

(a) Shaded area:
Area to left of z

(b) Shaded area:
$1 -$ (Area to left of z)

(c) Shaded area:
(Area to left of z_2)
$-$ (Area to left of z_1)

A Note Concerning Table II

The first area given in Table II, 0.0000, is for $z = -3.90$. This entry does not mean that the area under the standard normal curve that lies to the left of -3.90 is exactly 0, but only that it is 0 to four decimal places (the area is 0.0000481 to seven decimal places). Indeed, because the standard normal curve extends indefinitely to the left without ever touching the axis, the area to the left of any z-score is greater than 0.

Similarly, the last area given in Table II, 1.0000, is for $z = 3.90$. This entry does not mean that the area under the standard normal curve that lies to the left of 3.90 is exactly 1, but only that it is 1 to four decimal places (the area is 0.9999519 to seven decimal places). Indeed, the area to the left of any z-score is less than 1.

Finding the z-Score for a Specified Area

So far, we have used Table II to find areas. Now we show how to use Table II to find the z-score(s) corresponding to a specified area under the standard normal curve.

■■■   **EXAMPLE 6.8**   Finding the z-Score Having a Specified Area to Its Left

Determine the z-score having an area of 0.04 to its left under the standard normal curve, as shown in Fig. 6.19(a).

FIGURE 6.19

Finding the z-score having an area of 0.04 to its left

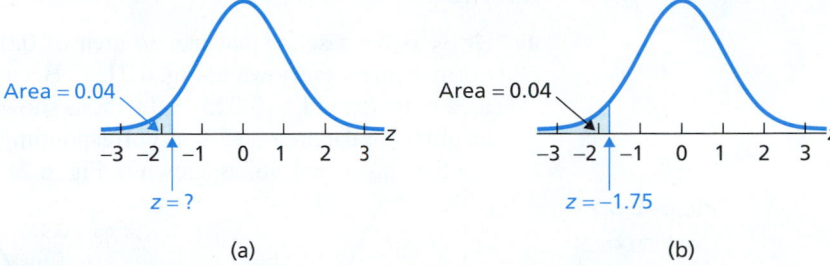

(a)   (b)

Solution   Use Table II, a portion of which is given in Table 6.3.

TABLE 6.3

Areas under the standard normal curve

				Second decimal place in z						
0.09	0.08	0.07	0.06	0.05	0.04	0.03	0.02	0.01	0.00	z
.	.	.	.	.	.	.	.	.	.	.
.	.	.	.	.	.	.	.	.	.	.
.	.	.	.	.	.	.	.	.	.	.
0.0233	0.0239	0.0244	0.0250	0.0256	0.0262	0.0268	0.0274	0.0281	0.0287	−1.9
0.0294	0.0301	0.0307	0.0314	0.0322	0.0329	0.0336	0.0344	0.0351	0.0359	−1.8
0.0367	0.0375	0.0384	0.0392	0.0401	0.0409	0.0418	0.0427	0.0436	0.0446	−1.7
0.0455	0.0465	0.0475	0.0485	0.0495	0.0505	0.0516	0.0526	0.0537	0.0548	−1.6
0.0559	0.0571	0.0582	0.0594	0.0606	0.0618	0.0630	0.0643	0.0655	0.0668	−1.5
.	.	.	.	.	.	.	.	.	.	.
.	.	.	.	.	.	.	.	.	.	.
.	.	.	.	.	.	.	.	.	.	.

Exercise 6.73 on page 280

Search the body of the table for the area 0.04. There is no such area in the table, so use the area closest to 0.04, which is 0.0401. The z-score corresponding to that area is −1.75. Thus the z-score having area 0.04 to its left under the standard normal curve is roughly −1.75, as shown in Fig. 6.19(b).   ■

The previous example shows that, when no area entry in Table II equals the one desired, we take the z-score corresponding to the closest area entry as an approximation of the required z-score. Two other cases are possible.

If an area entry in Table II equals the one desired, we of course use its corresponding z-score. If two area entries are equally closest to the one desired, we take the mean of the two corresponding z-scores as an approximation of the required z-score. Both of these cases are illustrated in the next example.

Finding the z-score that has a specified area to its right is often necessary. We have to make this determination so frequently that we use a special notation, z_α.

FIGURE 6.20

The z_α notation

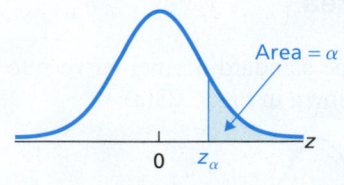

DEFINITION 6.3   **The z_α Notation**

The symbol **z_α** is used to denote the z-score that has an area of α (alpha) to its right under the standard normal curve, as illustrated in Fig. 6.20. Read "z_α" as "z sub α" or more simply as "$z\alpha$."

In the following two examples, we illustrate the z_α notation in a couple of different ways.

EXAMPLE 6.9 **Finding z_α**

Use Table II to find

a. $z_{0.025}$.

b. $z_{0.05}$.

Solution

a. $z_{0.025}$ is the z-score that has an area of 0.025 to its right under the standard normal curve, as shown in Fig. 6.21(a). Because the area to its right is 0.025, the area to its left is $1 - 0.025 = 0.975$, as shown in Fig. 6.21(b). Table II contains an entry for the area 0.975; its corresponding z-score is 1.96. Consequently, we see that $z_{0.025} = 1.96$, as shown in Fig. 6.21(b).

FIGURE 6.21

Finding $z_{0.025}$

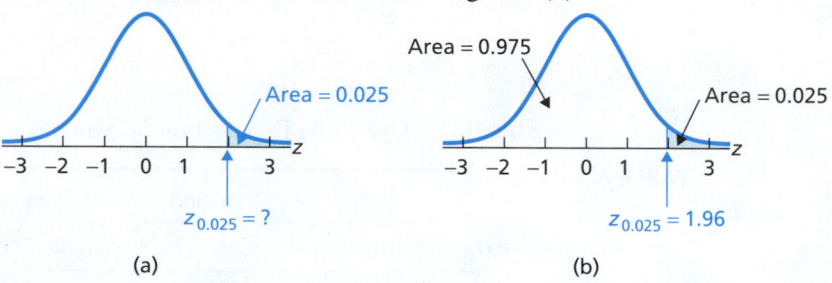

(a)　　　　　　　　　　　　(b)

b. $z_{0.05}$ is the z-score that has an area of 0.05 to its right under the standard normal curve, as shown in Fig. 6.22(a). Because the area to its right is 0.05, the area to its left is $1 - 0.05 = 0.95$, as shown in Fig. 6.22(b). Table II does not contain an entry for the area 0.95 and has two area entries equally closest to 0.95—namely, 0.9495 and 0.9505. The z-scores corresponding to those two areas are 1.64 and 1.65, respectively. So our approximation of $z_{0.05}$ is the mean of 1.64 and 1.65; that is, $z_{0.05} = 1.645$, as shown in Fig. 6.22(b).

FIGURE 6.22

Finding $z_{0.05}$

You try it!

Exercise 6.79
on page 280

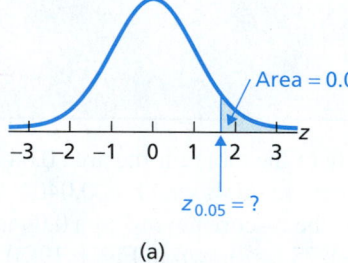

(a)

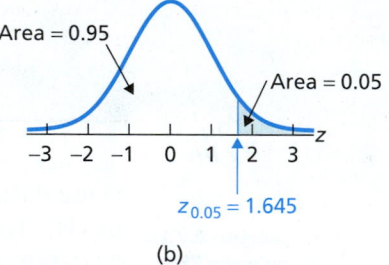

(b)

The next example shows how to find the two z-scores that divide the area under the standard normal curve into three specified areas.

EXAMPLE 6.10 **Finding the z-Scores for a Specified Area**

Find the two z-scores that divide the area under the standard normal curve into a middle 0.95 area and two outside 0.025 areas, as shown in Fig. 6.23(a).

FIGURE 6.23

Finding the two z-scores that divide the area under the standard normal curve into a middle 0.95 area and two outside 0.025 areas

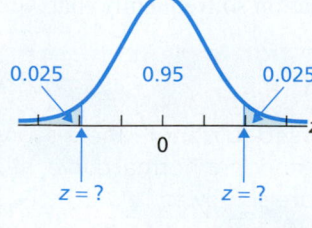

(a)

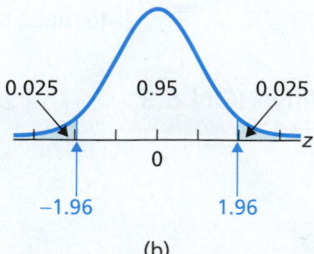

(b)

You try it!

Exercise 6.81
on page 280

Solution The area of the shaded region on the right in Fig. 6.23(a) is 0.025. In Example 6.9(a), we found that the corresponding z-score, $z_{0.025}$, is 1.96. Because the standard normal curve is symmetric about 0, the z-score on the left is -1.96. Therefore the two required z-scores are ± 1.96, as shown in Fig. 6.23(b). ∎

Note: We could also solve the previous example by first using Table II to find the z-score on the left in Fig. 6.23(a), which is -1.96, and then applying the symmetry property to obtain the z-score on the right, which is 1.96. Can you think of a third way to solve the problem?

Exercises 6.2

Understanding the Concepts and Skills

6.50 With which normal distribution is the standard normal curve associated?

6.51 Without consulting Table II, explain why the area under the standard normal curve that lies to the right of 0 is 0.5.

6.52 According to Table II, the area under the standard normal curve that lies to the left of -2.08 is 0.0188. Without further consulting Table II, determine the area under the standard normal curve that lies to the right of 2.08. Explain your reasoning.

6.53 According to Table II, the area under the standard normal curve that lies to the left of 0.43 is 0.6664. Without further consulting Table II, determine the area under the standard normal curve that lies to the right of 0.43. Explain your reasoning.

6.54 According to Table II, the area under the standard normal curve that lies to the left of 1.96 is 0.975. Without further consulting Table II, determine the area under the standard normal curve that lies to the left of -1.96. Explain your reasoning.

6.55 Property 4 of Key Fact 6.5 states that most of the area under the standard normal curve lies between -3 and 3. Use Table II to determine precisely the percentage of the area under the standard normal curve that lies between -3 and 3.

6.56 Why is the standard normal curve sometimes referred to as the z-curve?

6.57 Explain how Table II is used to determine the area under the standard normal curve that lies
a. to the left of a specified z-score.
b. to the right of a specified z-score.
c. between two specified z-scores.

6.58 The area under the standard normal curve that lies to the left of a z-score is always strictly between _____ and _____.

Applying the Concepts and Skills

Use Table II to obtain the areas under the standard normal curve required in Exercises 6.59–6.66. Sketch a standard normal curve and shade the area of interest in each problem.

6.59 Determine the area under the standard normal curve that lies to the left of
a. 2.24.   **b.** -1.56.   **c.** 0.   **d.** -4.

6.60 Determine the area under the standard normal curve that lies to the left of
a. -0.87.   **b.** 3.56.   **c.** 5.12.

6.61 Find the area under the standard normal curve that lies to the right of
a. -1.07.   **b.** 0.6.   **c.** 0.   **d.** 4.2.

6.62 Find the area under the standard normal curve that lies to the right of
a. 2.02.   **b.** -0.56.   **c.** -4.

6.63 Determine the area under the standard normal curve that lies between
a. -2.18 and 1.44.   **b.** -2 and -1.5.
c. 0.59 and 1.51.   **d.** 1.1 and 4.2.

6.64 Determine the area under the standard normal curve that lies between
a. -0.88 and 2.24.   **b.** -2.5 and -2.
c. 1.48 and 2.72.   **d.** -5.1 and 1.

6.65 Find the area under the standard normal curve that lies
a. either to the left of -2.12 or to the right of 1.67.
b. either to the left of 0.63 or to the right of 1.54.

6.66 Find the area under the standard normal curve that lies
a. either to the left of -1 or to the right of 2.
b. either to the left of -2.51 or to the right of -1.

6.67 Use Table II to obtain each shaded area under the standard normal curve.

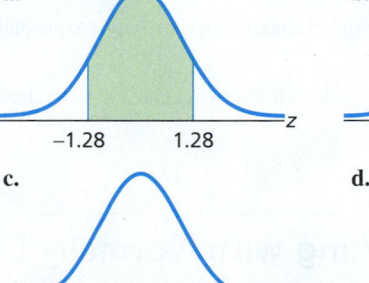
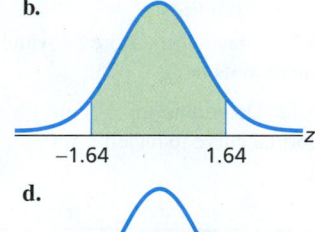
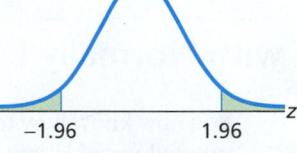
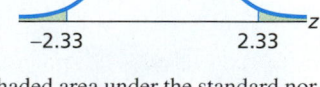

6.68 Use Table II to obtain each shaded area under the standard normal curve.

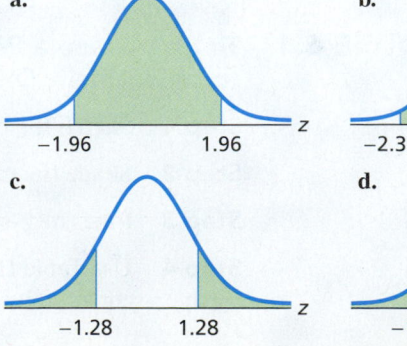
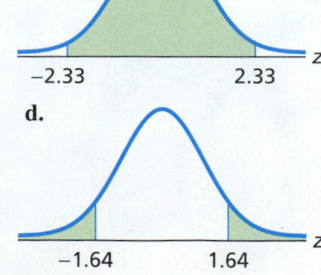

6.69 In each part, find the area under the standard normal curve that lies between the specified z-scores, sketch a standard normal curve, and shade the area of interest.
a. -1 and 1 **b.** -2 and 2 **c.** -3 and 3

6.70 The total area under the following standard normal curve is divided into eight regions.

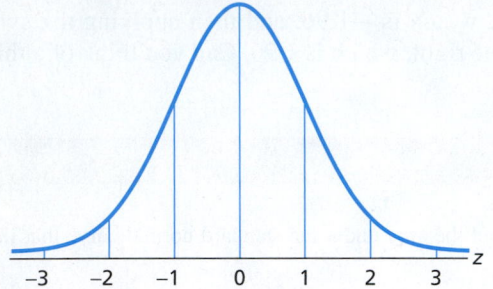

a. Determine the area of each region.
b. Complete the following table.

Region	Area	Percentage of total area
$-\infty$ to -3	0.0013	0.13
-3 to -2		
-2 to -1		
-1 to 0		
0 to 1	0.3413	34.13
1 to 2		
2 to 3		
3 to ∞		
	1.0000	100.00

In Exercises 6.71–6.82, use Table II to obtain the required z-scores. Illustrate your work with graphs.

6.71 Obtain the z-score for which the area under the standard normal curve to its left is 0.025.

6.72 Determine the z-score for which the area under the standard normal curve to its left is 0.01.

6.73 Find the z-score that has an area of 0.75 to its left under the standard normal curve.

6.74 Obtain the z-score that has area 0.80 to its left under the standard normal curve.

6.75 Obtain the z-score that has an area of 0.95 to its right.

6.76 Obtain the z-score that has area 0.70 to its right.

6.77 Determine $z_{0.33}$.

6.78 Determine $z_{0.015}$.

6.79 Find the following z-scores.
a. $z_{0.03}$ **b.** $z_{0.005}$

6.80 Obtain the following z-scores.
a. $z_{0.20}$ **b.** $z_{0.06}$

6.81 Determine the two z-scores that divide the area under the standard normal curve into a middle 0.90 area and two outside 0.05 areas.

6.82 Determine the two z-scores that divide the area under the standard normal curve into a middle 0.99 area and two outside 0.005 areas.

6.83 Complete the following table.

$z_{0.10}$	$z_{0.05}$	$z_{0.025}$	$z_{0.01}$	$z_{0.005}$
1.28				

Extending the Concepts and Skills

6.84 In this section, we mentioned that the total area under any curve representing the distribution of a variable equals 1. Explain why.

6.85 Let $0 < \alpha < 1$. Determine the
a. z-score having an area of α to its right in terms of z_{α}.
b. z-score having an area of α to its left in terms of z_{α}.
c. two z-scores that divide the area under the curve into a middle $1 - \alpha$ area and two outside areas of $\alpha/2$.
d. Draw graphs to illustrate your results in parts (a)–(c).

6.3   Working with Normally Distributed Variables

You now know how to find the percentage of all possible observations of a normally distributed variable that lie within any specified range: First express the range in terms of z-scores, and then determine the corresponding area under the standard normal curve. More formally, use Procedure 6.1.

■■■   PROCEDURE 6.1   To Determine a Percentage or Probability for a Normally Distributed Variable

Step 1   Sketch the normal curve associated with the variable.

Step 2   Shade the region of interest and mark its delimiting x-value(s).

Step 3   Find the z-score(s) for the delimiting x-value(s) found in Step 2.

Step 4   Use Table II to find the area under the standard normal curve delimited by the z-score(s) found in Step 3.

FIGURE 6.24

Graphical portrayal of Procedure 6.1

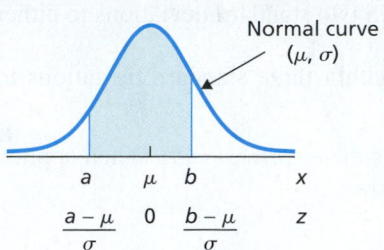

The steps in Procedure 6.1 are illustrated in Fig. 6.24, with the specified range lying between two numbers, a and b. If the specified range is to the left (or right) of a number, it is represented similarly. However, there will be only one x-value, and the shaded region will be the area under the normal curve that lies to the left (or right) of that x-value.

Note: When computing z-scores in Step 3 of Procedure 6.1, round to two decimal places, the precision provided in Table II.

EXAMPLE 6.11   Percentages for a Normally Distributed Variable

Intelligence Quotients Intelligence quotients (IQs) measured on the Stanford Revision of the Binet–Simon Intelligence Scale are normally distributed with a mean of 100 and a standard deviation of 16. Determine the percentage of people who have IQs between 115 and 140.

Solution Here the variable is IQ, and the population consists of all people. Because IQs are normally distributed, we can determine the required percentage by applying Procedure 6.1.

Step 1 Sketch the normal curve associated with the variable.

Here $\mu = 100$ and $\sigma = 16$. The normal curve associated with the variable is shown in Fig. 6.25. Note that the tick marks are 16 units apart; that is, the distance between successive tick marks is equal to the standard deviation.

FIGURE 6.25

Determination of the percentage of people having IQs between 115 and 140

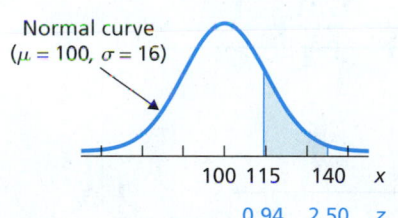

Step 2 Shade the region of interest and mark its delimiting x-values.

Figure 6.25 shows the required shaded region and its delimiting x-values, which are 115 and 140.

Step 3 Find the z-scores for the delimiting x-values found in Step 2.

We need to compute the z-scores for the x-values 115 and 140:

$$x = 115 \quad \longrightarrow \quad z = \frac{115 - \mu}{\sigma} = \frac{115 - 100}{16} = 0.94,$$

and

$$x = 140 \quad \longrightarrow \quad z = \frac{140 - \mu}{\sigma} = \frac{140 - 100}{16} = 2.50.$$

These z-scores are marked beneath the x-values in Fig. 6.25.

Step 4 Use Table II to find the area under the standard normal curve delimited by the z-scores found in Step 3.

We need to find the area under the standard normal curve that lies between 0.94 and 2.50. The area to the left of 0.94 is 0.8264, and the area to the left of 2.50 is 0.9938. The required area, shaded in Fig. 6.25, is therefore $0.9938 - 0.8264 = 0.1674$.

StatCrunch

Report 6.2

You try it!

Exercise 6.99(a)–(b) on page 288

Interpretation 16.74% of all people have IQs between 115 and 140. Equivalently, the probability is 0.1674 that a randomly selected person will have an IQ between 115 and 140.

Visualizing a Normal Distribution

We now present a rule that helps us "visualize" a normally distributed variable. Experience has shown that, for any data set whose distribution is roughly bell-shaped, the following properties hold:

- Approximately 68% of the observations lie within one standard deviation to either side of the mean.
- Approximately 95% of the observations lie within two standard deviations to either side of the mean.
- Approximately 99.7% of the observations lie within three standard deviations to either side of the mean.

These three properties are referred to collectively as the *empirical rule*. When applied to variables, the empirical rule can be stated as follows.

KEY FACT 6.6

Empirical Rule for Variables

For any variable whose distribution is bell-shaped (in particular, for any normally distributed variable), the following three properties hold.

Property 1: Approximately 68% of all possible observations lie within one standard deviation to either side of the mean, that is, between $\mu - \sigma$ and $\mu + \sigma$.

Property 2: Approximately 95% of all possible observations lie within two standard deviations to either side of the mean, that is, between $\mu - 2\sigma$ and $\mu + 2\sigma$.

Property 3: Approximately 99.7% of all possible observations lie within three standard deviations to either side of the mean, that is, between $\mu - 3\sigma$ and $\mu + 3\sigma$.

These three properties are illustrated together in Fig. 6.26.

FIGURE 6.26

The empirical rule for variables

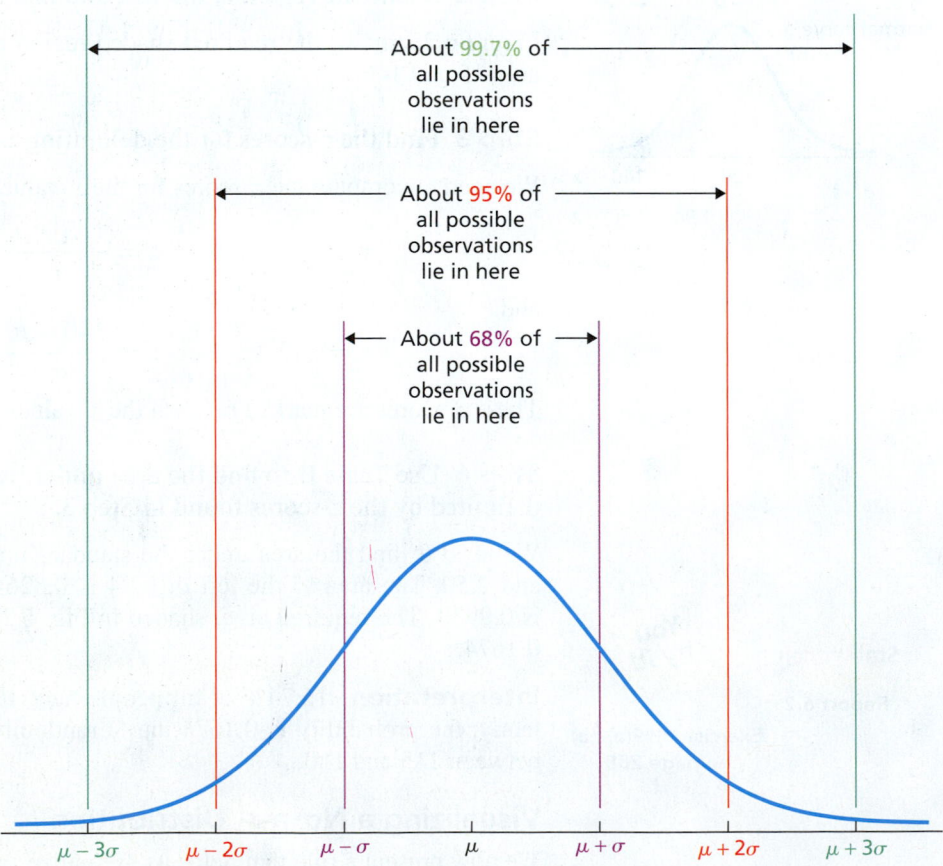

About 99.7% of all possible observations lie in here

About 95% of all possible observations lie in here

About 68% of all possible observations lie in here

$\mu - 3\sigma$ $\mu - 2\sigma$ $\mu - \sigma$ μ $\mu + \sigma$ $\mu + 2\sigma$ $\mu + 3\sigma$

Note: For obvious reasons, the empirical rule is also known as the **68-95-99.7 rule.**

The percentage approximations given by the empirical rule can, in general, be somewhat rough. However, for a normally distributed variable, the percentages, although still approximate, are exact to the specified number of decimal places. To see this fact, we use Table II as follows.

Recall that the z-score of an observation tells us how many standard deviations the observation is from the mean. Thus, for instance, the percentage of all possible observations that lie within one standard deviation to either side of the mean equals the percentage of all observations whose z-scores lie between −1 and 1. For a normally distributed variable, that percentage is the same as the area under the standard normal curve between −1 and 1, which, by Table II, is 0.6826, or 68% to the nearest percentage point.

Proceeding similarly, we can show that, for a normally distributed variable, the other two percentages in the empirical rule are exact to the specified number of decimal places.

EXAMPLE 6.12   The Empirical Rule

Intelligence Quotients   Apply the empirical rule to IQs.

Solution   Recall that IQs (measured on the Stanford Revision of the Binet–Simon Intelligence Scale) are normally distributed with a mean of 100 and a standard deviation of 16. In particular, we have $\mu = 100$ and $\sigma = 16$.

Property 1 of the empirical rule implies that approximately 68% of all people have IQs within one standard deviation to either side of the mean. One standard deviation below the mean is $\mu - \sigma = 100 - 16 = 84$; one standard deviation above the mean is $\mu + \sigma = 100 + 16 = 116$.

Interpretation   Approximately 68% of all people have IQs between 84 and 116, as illustrated in Fig. 6.27(a).

FIGURE 6.27

Graphical display of the empirical rule for IQs

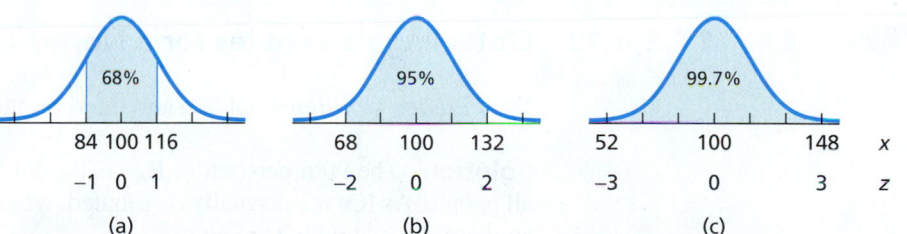

(a) (b) (c)

Property 2 of the empirical rule implies that approximately 95% of all people have IQs within two standard deviations to either side of the mean; that is, from $\mu - 2\sigma = 100 - 2 \cdot 16 = 68$ to $\mu + 2\sigma = 100 + 2 \cdot 16 = 132$.

Interpretation   Approximately 95% of all people have IQs between 68 and 132, as illustrated in Fig. 6.27(b).

You try it!

Exercise 6.105 on page 289

Property 3 of the empirical rule implies that approximately 99.7% of all people have IQs within three standard deviations to either side of the mean; that is, from $\mu - 3\sigma = 100 - 3 \cdot 16 = 52$ to $\mu + 3\sigma = 100 + 3 \cdot 16 = 148$.

Interpretation   Approximately 99.7% of all people have IQs between 52 and 148, as illustrated in Fig. 6.27(c).

As illustrated in the preceding example, the empirical rule allows us to quickly and easily get useful information about any variable whose distribution is (roughly) bell-shaped. Specifically, it allows us to determine, at least approximately, the percentage of all possible observations of the variable that lie within one, two, or three standard deviations to either side of the mean.

For normally distributed variables, however, we can obtain percentages for any number of standard deviations by using Table II. For instance, applying that table, we find that, for any normally distributed variable, approximately 87% of all possible observations lie within 1.5 standard deviations to either side of the mean.

Finding the Observations for a Specified Percentage

Procedure 6.1 shows how to determine the percentage of all possible observations of a normally distributed variable that lie within any specified range. Frequently, however, we want to carry out the reverse procedure, that is, to find the observations corresponding to a specified percentage. Procedure 6.2 allows us to do that.

◼◻◼◼ PROCEDURE 6.2 **To Determine the Observations Corresponding to a Specified Percentage or Probability for a Normally Distributed Variable**

Step 1 Sketch the normal curve associated with the variable.

Step 2 Shade the region of interest.

Step 3 Use Table II to determine the z-score(s) delimiting the region found in Step 2.

Step 4 Find the x-value(s) having the z-score(s) found in Step 3.

Note: To find each x-value in Step 4 from its z-score in Step 3, use the formula

$$x = \mu + z \cdot \sigma,$$

where μ and σ are the mean and standard deviation, respectively, of the variable under consideration.

Among other things, we can use Procedure 6.2 to obtain quartiles, deciles, or any other percentile for a normally distributed variable. Example 6.13 shows how to find percentiles by this method.

◼◻◼ EXAMPLE 6.13 **Obtaining Percentiles for a Normally Distributed Variable**

Intelligence Quotients Obtain and interpret the 90th percentile for IQs.

Solution The 90th percentile, P_{90}, is the IQ that is higher than those of 90% of all people. As IQs are normally distributed, we can determine the 90th percentile by applying Procedure 6.2.

Step 1 Sketch the normal curve associated with the variable.

Here $\mu = 100$ and $\sigma = 16$. The normal curve associated with IQs is shown in Fig. 6.28.

Step 2 Shade the region of interest.

See the shaded region in Fig. 6.28.

Step 3 Use Table II to determine the z-score delimiting the region found in Step 2.

The z-score corresponding to P_{90} is the one having an area of 0.90 to its left under the standard normal curve. From Table II, that z-score is 1.28, approximately, as shown in Fig. 6.28.

Step 4 Find the x-value having the z-score found in Step 3.

We must find the x-value having the z-score 1.28—the IQ that is 1.28 standard deviations above the mean. It is $100 + 1.28 \cdot 16 = 100 + 20.48 = 120.48$.

Interpretation The 90th percentile for IQs is 120.48. Thus, 90% of people have IQs below 120.48 and 10% have IQs above 120.48.

FIGURE 6.28
Finding the 90th percentile for IQs

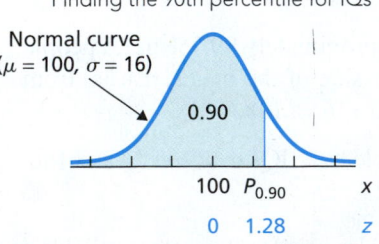

Report 6.3

Exercise 6.99(c)–(d) on page 288

THE TECHNOLOGY CENTER

Most statistical technologies have programs that carry out the procedures discussed in this section—namely, to obtain for a normally distributed variable

- the percentage of all possible observations that lie within any specified range and
- the observations corresponding to a specified percentage.

In this subsection, we present output and step-by-step instructions for such programs.

Minitab, Excel, and the TI-83/84 Plus each have a program for determining the area under the associated normal curve of a normally distributed variable that lies to the left of a specified value. Such an area corresponds to a **cumulative probability,** the probability that the variable will be less than or equal to the specified value.

EXAMPLE 6.14   Using Technology to Obtain Normal Percentages

Intelligence Quotients Recall that IQs are normally distributed with mean 100 and standard deviation 16. Use Minitab, Excel, or the TI-83/84 Plus to find the percentage of people who have IQs between 115 and 140.

Solution We applied the cumulative-normal programs, resulting in Output 6.2. Steps for generating that output are presented in Instructions 6.1 on the next page. *Note to Excel users:* The output that you get may differ somewhat from that shown in Output 6.2 depending on the version of Excel that you use.

OUTPUT 6.2   The percentage of people with IQs between 115 and 140

MINITAB

Cumulative Distribution Function

Normal with mean = 100 and standard deviation = 16

x	$P(X \le x)$
115	0.825749
140	0.993790

TI-83/84 PLUS

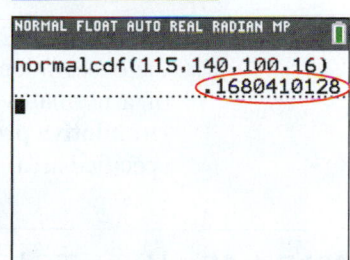

EXCEL

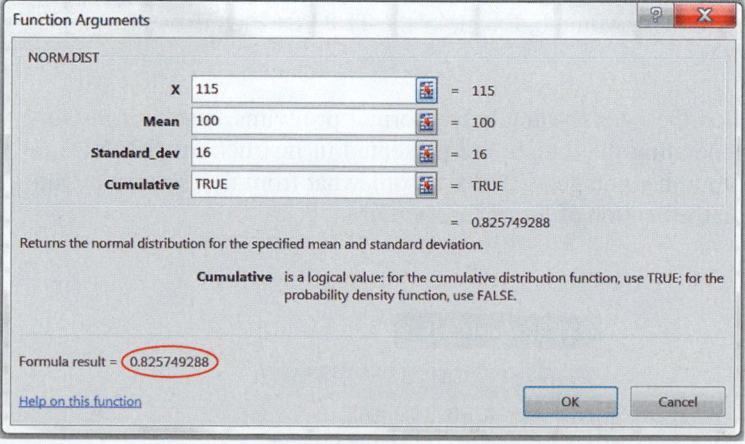

Note: Replacing 115 by 140 in the X text box yields 0.993790335.

We get the required percentage from Output 6.2 as follows.

- *Minitab:* Subtract the two cumulative probabilities: $0.993790 - 0.825749 = 0.168041$, or 16.80%.

- *Excel:* Subtract the two cumulative probabilities, the one circled in red from the one in the note: 0.993790355 − 0.825749288 = 0.168041047, or 16.80%.
- *TI-83/84 Plus:* Direct from the output: 0.1680410128, or 16.80%.

Note that the percentages obtained by the three technologies differ slightly from the percentage of 16.74% that we found in Example 6.11. The differences reflect the fact that the technologies retain more accuracy than we can get from Table II. ∎

INSTRUCTIONS 6.1   Steps for generating Output 6.2

MINITAB

1. Store the delimiting IQs, 115 and 140, in a column named IQ
2. Choose **Calc ➤ Probability Distributions ➤ Normal...**
3. Press the F3 key to reset the dialog box
4. Click in the **Mean** text box and type 100
5. Click in the **Standard deviation** text box and type 16
6. Click in the **Input column** text box and specify IQ
7. Click **OK**

EXCEL

1. Click **f_x** (Insert Function)
2. Select **Statistical** from the **Or select a category** drop down list box
3. Select **NORM.DIST** from the **Select a function** list
4. Click **OK**
5. Type 115 in the **X** text box
6. Click in the **Mean** text box and type 100

7. Click in the **Standard_dev** text box and type 16
8. Click in the **Cumulative** text box and type TRUE
9. To obtain the cumulative probability for 140, replace 115 by 140 in the **X** text box

TI-83/84 PLUS

FOR THE TI-84 PLUS C
1. Press **2nd ➤ DISTR**
2. Arrow down to **normalcdf(** and press **ENTER**
3. Type 115 for **lower** and press **ENTER**
4. Type 140 for **upper** and press **ENTER**
5. Type 100 for μ and press **ENTER**
6. Type 16 for σ and press **ENTER** three times

FOR THE TI-83/84 PLUS
1. Press **2nd ➤ DISTR**
2. Arrow down to **normalcdf(** and press **ENTER**
3. Type 115,140,100,16) and press **ENTER**

Minitab, Excel, and the TI-83/84 Plus also each have a program for determining the observation that has a specified area to its left under the associated normal curve of a normally distributed variable. Such an observation corresponds to an **inverse cumulative probability,** the observation whose cumulative probability is equal to the specified area.

EXAMPLE 6.15   Using Technology to Obtain Normal Percentiles

Intelligence Quotients   Use Minitab, Excel, or the TI-83/84 Plus to determine the 90th percentile for IQs.

Solution   We applied the inverse-cumulative-normal programs, resulting in Output 6.3. Steps for generating that output are presented in Instructions 6.2. *Note to Excel users:* The output that you get may differ somewhat from that shown in Output 6.3 depending on the version of Excel that you use.

OUTPUT 6.3   The 90th percentile for IQs

MINITAB

Inverse Cumulative Distribution Function

Normal with mean = 100 and standard deviation = 16

P(X ≤ x)   x
0.9   120.505

TI-83/84 PLUS

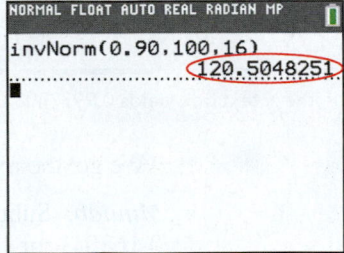

OUTPUT 6.3 (cont.)
The 90th percentile for IQs

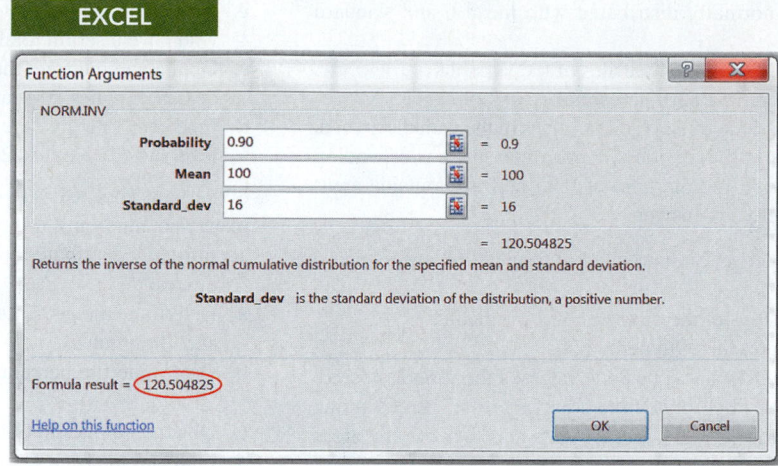

As shown in Output 6.3, the 90th percentile for IQs is 120.505. Note that this value differs slightly from the value of 120.48 that we obtained in Example 6.13. The difference reflects the fact that the three technologies retain more accuracy than we can get from Table II.

INSTRUCTIONS 6.2   Steps for generating Output 6.3

MINITAB

1 Choose **Calc ➤ Probability Distributions ➤ Normal...**
2 Press the F3 key to reset the dialog box
3 Select the **Inverse cumulative probability** option button
4 Click in the **Mean** text box and type 100
5 Click in the **Standard deviation** text box and type 16
6 Select the **Input constant** option button
7 Click in the **Input constant** text box and type 0.90
8 Click **OK**

EXCEL

1 Click f_x (Insert Function)
2 Select **Statistical** from the **Or select a category** drop down list box
3 Select **NORM.INV** from the **Select a function** list

4 Click **OK**
5 Type 0.90 in the **Probability** text box
6 Click in the **Mean** text box and type 100
7 Click in the **Standard_dev** text box and type 16

TI-83/84 PLUS

FOR THE TI-84 PLUS C
1 Press **2nd ➤ DISTR**
2 Arrow down to **invNorm(** and press **ENTER**
3 Type 0.90 for **area** and press **ENTER**
4 Type 100 for μ and press **ENTER**
5 Type 16 for σ and press **ENTER** three times

FOR THE TI-83/84 PLUS
1 Press **2nd ➤ DISTR**
2 Arrow down to **invNorm(** and press **ENTER**
3 Type 0.90,100,16) and press **ENTER**

Exercises 6.3

Understanding the Concepts and Skills

6.86 Briefly, for a normally distributed variable, how do you obtain the percentage of all possible observations that lie within a specified range?

6.87 Explain why the percentage of all possible observations of a normally distributed variable that lie within two standard deviations to either side of the mean equals the area under the standard normal curve between −2 and 2.

6.88 State the empirical rule as specialized to variables.

6.89 A variable is normally distributed with mean 6 and standard deviation 2. Find the percentage of all possible values of the variable that
a. lie between 1 and 7.   **b.** exceed 5.
c. are less than 4.

6.90 A variable is normally distributed with mean 68 and standard deviation 10. Find the percentage of all possible values of the variable that
a. lie between 73 and 80.   **b.** are at least 75.
c. are at most 90.

6.91 A variable is normally distributed with mean 10 and standard deviation 3. Find the percentage of all possible values of the variable that
a. lie between 6 and 7.   **b.** are at least 10.
c. are at most 17.5.

6.92 A variable is normally distributed with mean 0 and standard deviation 4. Find the percentage of all possible values of the variable that
a. lie between −8 and 8.   **b.** exceed −1.5.
c. are less than 2.75.

6.93 A variable is normally distributed with mean 6 and standard deviation 2.

a. Determine and interpret the quartiles of the variable.

b. Obtain and interpret the 85th percentile.

c. Find the value that 65% of all possible values of the variable exceed.

d. Find the two values that divide the area under the corresponding normal curve into a middle area of 0.95 and two outside areas of 0.025. Interpret your answer.

6.94 A variable is normally distributed with mean 68 and standard deviation 10.

a. Determine and interpret the quartiles of the variable.

b. Obtain and interpret the 99th percentile.

c. Find the value that 85% of all possible values of the variable exceed.

d. Find the two values that divide the area under the corresponding normal curve into a middle area of 0.90 and two outside areas of 0.05. Interpret your answer.

6.95 A variable is normally distributed with mean 10 and standard deviation 3.

a. Determine and interpret the quartiles of the variable.

b. Obtain and interpret the seventh decile.

c. Find the value that 35% of all possible values of the variable exceed.

d. Find the two values that divide the area under the corresponding normal curve into a middle area of 0.99 and two outside areas of 0.005. Interpret your answer.

6.96 A variable is normally distributed with mean 0 and standard deviation 4.

a. Determine and interpret the quartiles of the variable.

b. Obtain and interpret the second decile.

c. Find the value that 15% of all possible values of the variable exceed.

d. Find the two values that divide the area under the corresponding normal curve into a middle area of 0.80 and two outside areas of 0.10. Interpret your answer.

Applying the Concepts and Skills

6.97 Giant Tarantulas. One of the larger species of tarantulas is the *Grammostola mollicoma*, whose common name is the Brazilian giant tawny red. A tarantula has two body parts. The anterior part of the body is covered above by a shell, or carapace. From a recent article by F. Costa and F. Perez–Miles titled "Reproductive Biology of Uruguayan Theraphosids" (*The Journal of Arachnology*, Vol. 30, No. 3, pp. 571–587), we find that the carapace length of the adult male *G. mollicoma* is normally distributed with mean 18.14 mm and standard deviation 1.76 mm.

a. Find the percentage of adult male *G. mollicoma* that have carapace length between 16 mm and 17 mm.

b. Find the percentage of adult male *G. mollicoma* that have carapace length exceeding 19 mm.

c. Determine and interpret the quartiles for carapace length of the adult male *G. mollicoma*.

d. Obtain and interpret the 95th percentile for carapace length of the adult male *G. mollicoma*.

6.98 Serum Cholesterol Levels. According to the *National Health and Nutrition Examination Survey*, published by the National Center for Health Statistics, the serum (noncellular portion of blood) total cholesterol level of U.S. females 20 years old or older is normally distributed with a mean of 206 mg/dL (milligrams per deciliter) and a standard deviation of 44.7 mg/dL.

a. Determine the percentage of U.S. females 20 years old or older who have a serum total cholesterol level between 150 mg/dL and 250 mg/dL.

b. Determine the percentage of U.S. females 20 years old or older who have a serum total cholesterol level below 220 mg/dL.

c. Obtain and interpret the quartiles for serum total cholesterol level of U.S. females 20 years old or older.

d. Find and interpret the fourth decile for serum total cholesterol level of U.S. females 20 years old or older.

6.99 New York City 10-km Run. As reported in *Runner's World* magazine, the times of the finishers in the New York City 10-km run are normally distributed with mean 61 minutes and standard deviation 9 minutes.

a. Determine the percentage of finishers who have times between 50 and 70 minutes.

b. Determine the percentage of finishers who have times less than 75 minutes.

c. Obtain and interpret the 40th percentile for the finishing times.

d. Find and interpret the 8th decile for the finishing times.

6.100 Green Sea Urchins. From the paper "Effects of Chronic Nitrate Exposure on Gonad Growth in Green Sea Urchin *Strongylocentrotus droebachiensis*" (*Aquaculture*, Vol. 242, No. 1–4, pp. 357–363) by S. Siikavuopio et al., we found that weights of adult green sea urchins are normally distributed with mean 52.0 g and standard deviation 17.2 g.

a. Find the percentage of adult green sea urchins with weights between 50 g and 60 g.

b. Obtain the percentage of adult green sea urchins with weights above 40 g.

c. Determine and interpret the 90th percentile for the weights.

d. Find and interpret the 6th decile for the weights.

6.101 Arterial Cord pH. Umbilical cord blood analysis immediately after delivery is one way to measure the health of an infant after birth. Researchers G. Natalucci et al. used it as a predictor of brain maturation of preterm infants in the article "Functional Brain Maturation Assessed During Early Life Correlates with Anatomical Brain Maturation at Term-Equivalent Age in Preterm Infants" (*Pediatric Research*, Vol. 74. No. 1, pp. 68–74). Based on this study, we will assume that, for preterm infants, the pH level of the arterial cord (one vessel in the umbilical cord) is normally distributed with mean 7.32 and standard deviation 0.1. Find the percentage of preterm infants who have arterial cord pH levels

a. between 7.0 and 7.5.   **b.** over 7.4.

6.102 Elephant Pregnancies. G. Wittemeyer et al. studied demographic data on African elephants living in Kenya in the article "Comparative Demography of an At-Risk African Elephant Population" (*PLOS ONE*, Vol. 8. No. 1). Based on this study, we will assume that the time between pregnancies of the African elephant in Kenya, for elephants that have more than one calf, is normally distributed with mean 4.01 years and standard deviation 0.94 years. Determine the percentage of such times that are

a. less than 2 years.   **b.** between 3 and 5 years.

6.103 Gibbon Song Duration. A preliminary behavioral study of the Jingdong black gibbon, a primate endemic to the Wuliang Mountains in China, found that the mean song bout duration in the wet season is 12.59 minutes with a standard deviation of 5.31 minutes. [SOURCE: L. Sheeran et al., "Preliminary Report on the Behavior of the Jingdong Black Gibbon (*Hylobates concolor jingdongensis*)," *Tropical Biodiversity*, Vol. 5(2), pp. 113–125] Assuming that song bout is normally distributed, determine the percentage of song bouts that have durations within

a. one standard deviation to either side of the mean.

b. two standard deviations to either side of the mean.

c. three standard deviations to either side of the mean.

6.104 Friendship Motivation. In the article "Assessing Friendship Motivation During Preadolescence and Early Adolescence" (*Journal of Early Adolescence*, Vol. 25, No. 3, pp. 367–385), J. Richard and B. Schneider described the properties of the Friendship Motivation Scale for Children (FMSC), a scale designed to assess children's desire for friendships. Two interesting conclusions are that friends generally report similar levels of the FMSC and girls tend to score higher on the FMSC than boys. Boys in the seventh grade scored a mean of 9.32 with a standard deviation of 1.71, and girls in the seventh grade scored a mean of 10.04 with a standard deviation of 1.83. Assuming that FMSC scores are normally distributed, determine the percentage of seventh-grade boys who have FMSC scores within

a. one standard deviation to either side of the mean.
b. two standard deviations to either side of the mean.
c. three standard deviations to either side of the mean.
d. Repeat parts (a)–(c) for seventh-grade girls.

6.105 Brain Weights. In 1905, R. Pearl published the article "Biometrical Studies on Man. I. Variation and Correlation in Brain Weight" (*Biometrika*, Vol. 4, pp. 13–104). According to the study, brain weights of Swedish men are normally distributed with mean 1.40 kg and standard deviation 0.11 kg. Apply the empirical rule to fill in the blanks.

a. Approximately 68% of Swedish men have brain weights between _____ and _____.
b. Approximately 95% of Swedish men have brain weights between _____ and _____.
c. Approximately 99.7% of Swedish men have brain weights between _____ and _____.
d. Draw graphs similar to those in Fig. 6.27 on page 283 to portray your results.

6.106 Children Watching TV. The A. C. Nielsen Company reported in the *Nielsen Report on Television* that the mean weekly television viewing time for children aged 2–6 years is 24.85 hours. Assume that the weekly television viewing times of such children are normally distributed with a standard deviation of 6.23 hours and apply the empirical rule to fill in the blanks.

a. Approximately 68% of all such children watch between _____ and _____ hours of TV per week.
b. Approximately 95% of all such children watch between _____ and _____ hours of TV per week.
c. Approximately 99.7% of all such children watch between _____ and _____ hours of TV per week.
d. Draw graphs similar to those in Fig. 6.27 on page 283 to portray your results.

6.107 Heights of Female Students. Refer to Example 6.3 on pages 267–268. The heights of the 3264 female students attending a midwestern college are approximately normally distributed with mean 64.4 inches and standard deviation 2.4 inches. Thus we can use the normal distribution with $\mu = 64.4$ and $\sigma = 2.4$ to approximate the percentage of these students having heights within any specified range. In each part, (i) obtain the exact percentage from Table 6.1, (ii) use the normal distribution to approximate the percentage, and (iii) compare your answers.

a. The percentage of female students with heights between 62 and 63 inches.
b. The percentage of female students with heights between 65 and 70 inches.

6.108 Women's Shoes. Research reveals that foot length of women is normally distributed with mean 9.58 inches and standard deviation 0.51 inch. This distribution is useful to shoe manufacturers, shoe stores, and related merchants because it permits them to make

informed decisions about shoe production, inventory, and so forth. Along these lines, the following table provides a foot-length-to-shoe-size conversion, obtained from *Payless ShoeSource*.

Length (in.)	Size (U.S.)	Length (in.)	Size (U.S.)
8	3	10	9
$8\frac{3}{16}$	$3\frac{1}{2}$	$10\frac{3}{16}$	$9\frac{1}{2}$
$8\frac{5}{16}$	4	$10\frac{5}{16}$	10
$8\frac{8}{16}$	$4\frac{1}{2}$	$10\frac{8}{16}$	$10\frac{1}{2}$
$8\frac{11}{16}$	5	$10\frac{11}{16}$	11
$8\frac{13}{16}$	$5\frac{1}{2}$	$10\frac{13}{16}$	$11\frac{1}{2}$
9	6	11	12
$9\frac{3}{16}$	$6\frac{1}{2}$	$11\frac{3}{16}$	$12\frac{1}{2}$
$9\frac{5}{16}$	7	$11\frac{5}{16}$	13
$9\frac{8}{16}$	$7\frac{1}{2}$	$11\frac{8}{16}$	$13\frac{1}{2}$
$9\frac{11}{16}$	8	$11\frac{11}{16}$	14
$9\frac{13}{16}$	$8\frac{1}{2}$		

a. Sketch the distribution of women's foot length.
b. What percentage of women have foot lengths between 9 and 10 inches?
c. What percentage of women have foot lengths that exceed 11 inches?
d. Shoe manufacturers suggest that if a foot length is between two sizes, wear the larger size. Referring to the preceding table, determine the percentage of women who wear size 8 shoes; size $11\frac{1}{2}$ shoes.
e. If an owner of a chain of shoe stores intends to purchase 10,000 pairs of women's shoes, roughly how many should he purchase of size 8? of size $11\frac{1}{2}$? Explain your reasoning.

6.109 College-Math Success. Researchers S. Lesik and M. Mitchell explore the difficulty of predicting success in college-level mathematics in the article "The Investigation of Multiple Paths to Success in College-Level Mathematics" (*Journal of Applied Research in Higher Education*, Vol. 5, Issue 1, pp. 48–57). One of the variables explored as an indicator of success was the length of time since a college freshman has taken a mathematics course. The article reports that the mean length of time is 0.18 years with a standard deviation of 0.624 years. For college freshmen, let x represent the time, in years, since taking a math course.

a. What percentage of times are at least 0 years?
b. Assuming that x is approximately normally distributed, use normal-curve areas to determine the approximate percentage of times that are at least 0 years.
c. Based on your results from parts (a) and (b), do you think that the length of time since taking a math course for college freshmen is approximately a normally distributed variable? Explain your answer.

6.110 Tipping. In the article "Are Christian/Religious People Poor Tippers?" (*Journal of Applied Social Psychology*, Vol. 43, Issue 5, pp. 928–935), M. Lynn and B. Katz report that, for customers who receive bad service, the mean percentage tip is 8.56% of the bill with a standard deviation of 5.37%. Let x represent the percentage of the bill that a customer tips when he or she receives bad service.

a. What percentage of tips for bad service are at least 0%?
b. Assuming that x is approximately normally distributed, use normal-curve areas to determine the approximate percentage of tips for bad service that are at least 0%.

c. Based on your results from parts (a) and (b), do you think that the percentage of the bill that a customer tips when he or she receives bad service is approximately a normally distributed variable? Explain your answer.

Extending the Concepts and Skills

6.111 Booted Eagles. The rare booted eagle of western Europe was the focus of a study by S. Suarez et al. to identify optimal nesting habitat for this raptor. According to their paper "Nesting Habitat Selection by Booted Eagles (*Hieraaetus pennatus*) and Implications for Management" (*Journal of Applied Ecology*, Vol. 37, pp. 215–223), the distances of such nests to the nearest marshland are normally distributed with mean 4.66 km and standard deviation 0.75 km. Let Y be the distance of a randomly selected nest to the nearest marshland. Determine and interpret

a. $P(Y > 5)$. **b.** $P(3 \leq Y \leq 6)$.

6.112 Emergency Room Traffic. Desert Samaritan Hospital in Mesa, Arizona, keeps records of emergency room traffic. Those records reveal that the times between arriving patients have a mean of 8.7 minutes with a standard deviation of 8.7 minutes. Based solely on the values of these two parameters, explain why it is unreasonable to assume that the times between arriving patients is normally distributed or even approximately so.

6.113 Let $0 < \alpha < 1$. For a normally distributed variable, show that $100(1 - \alpha)\%$ of all possible observations lie within $z_{\alpha/2}$ standard deviations to either side of the mean, that is, between $\mu - z_{\alpha/2} \cdot \sigma$ and $\mu + z_{\alpha/2} \cdot \sigma$.

6.114 Express the quartiles, Q_1, Q_2, and Q_3, of a normally distributed variable in terms of its mean, μ, and standard deviation, σ.

6.115 Express the kth percentile, P_k, of a normally distributed variable in terms of its mean, μ, and standard deviation, σ.

6.4 Assessing Normality; Normal Probability Plots

You have now seen how to work with normally distributed variables. For instance, you know how to determine the percentage of all possible observations that lie within any specified range and how to obtain the observations corresponding to a specified percentage.

Another problem involves deciding whether a variable is normally distributed, or at least approximately so, based on a sample of observations. Such decisions often play a major role in subsequent analyses—from percentage or percentile calculations to statistical inferences.

From Key Fact 2.1 on page 80, if a simple random sample is taken from a population, the distribution of the observed values of a variable will approximate the distribution of the variable—and the larger the sample, the better the approximation tends to be. We can use this fact to help decide whether a variable is normally distributed.

If a variable is normally distributed, then, for a large sample, a histogram of the observations should be roughly bell shaped; for a very large sample, even moderate departures from a bell shape cast doubt on the normality of the variable. However, for a relatively small sample, ascertaining a clear shape in a histogram and, in particular, whether it is bell shaped is often difficult. These comments also hold for stem-and-leaf diagrams and dotplots.

Thus, for relatively small samples, a more sensitive graphical technique than the ones we have presented so far is required for assessing normality. *Normal probability plots* provide such a technique.

The idea behind a normal probability plot is simple: Compare the observed values of the variable to the observations expected for a normally distributed variable. More precisely, a **normal probability plot** is a plot of the observed values of the variable versus the **normal scores**—the observations expected for a variable having the standard normal distribution. If the variable is normally distributed, the normal probability plot should be roughly linear (i.e., fall roughly in a straight line) and vice versa.

When you use a normal probability plot to assess the normality of a variable, you must remember two things: (1) that the decision of whether a normal probability plot is roughly linear is a subjective one, and (2) that you are using a sample of observations of the variable to make a judgment about all possible observations of the variable. Keep these considerations in mind when using the following guidelines.

KEY FACT 6.7

What Does It Mean?

Roughly speaking, a normal probability plot that falls nearly in a straight line indicates a normal variable, and one that does not indicates a nonnormal variable.

Guidelines for Assessing Normality Using a Normal Probability Plot

To assess the normality of a variable using sample data, construct a normal probability plot.

- If the plot is roughly linear, you can assume that the variable is approximately normally distributed.
- If the plot is not roughly linear, you can assume that the variable is not approximately normally distributed.

These guidelines should be interpreted loosely for small samples but usually interpreted strictly for large samples.

In practice, normal probability plots are generated by computer. However, to better understand these plots, constructing a few by hand is helpful. Table III in Appendix A gives the normal scores for sample sizes from 5 to 30. In the next example, we explain how to use Table III to obtain a normal probability plot.

■■■ **EXAMPLE 6.16**   **Normal Probability Plots**

TABLE 6.4

Adjusted gross incomes ($1000s)

9.7	93.1	33.0	21.2
81.4	51.1	43.5	10.6
12.8	7.8	18.1	12.7

TABLE 6.5

Ordered data and normal scores

Adjusted gross income	Normal score
7.8	−1.64
9.7	−1.11
10.6	−0.79
12.7	−0.53
12.8	−0.31
18.1	−0.10
21.2	0.10
33.0	0.31
43.5	0.53
51.1	0.79
81.4	1.11
93.1	1.64

Adjusted Gross Incomes The Internal Revenue Service publishes data on federal individual income tax returns in *Statistics of Income, Individual Income Tax Returns*. A simple random sample of 12 returns from last year revealed the adjusted gross incomes, in thousands of dollars, shown in Table 6.4. Construct a normal probability plot for these data, and use the plot to assess the normality of adjusted gross incomes.

Solution Here the variable is adjusted gross income, and the population consists of all of last year's federal individual income tax returns. To construct a normal probability plot, we first arrange the data in increasing order and obtain the normal scores from Table III. The ordered data are shown in the first column of Table 6.5; the normal scores, from the $n = 12$ column of Table III, are shown in the second column of Table 6.5.

Next, we plot the points in Table 6.5, using the horizontal axis for the adjusted gross incomes and the vertical axis for the normal scores. For instance, the first point plotted has a horizontal coordinate of 7.8 and a vertical coordinate of −1.64. Figure 6.29 shows all 12 points from Table 6.5. This graph is the normal probability plot for the sample of adjusted gross incomes. Note that the normal probability plot in Fig. 6.29 is curved, not linear.

FIGURE 6.29 Normal probability plot for the sample of adjusted gross incomes

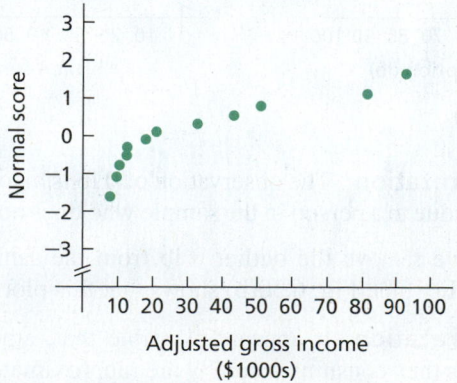

Adjusted gross income ($1000s)

StatCrunch

Report 6.4

Exercise 6.127(a), (c)
on page 295

Interpretation In light of Key Fact 6.7, last year's adjusted gross incomes apparently are not (approximately) normally distributed.

Note: If two or more observations in a sample are equal, you can think of them as slightly different from one another for purposes of obtaining their normal scores.

In some books and statistical technologies, you may encounter one or more of the following differences in normal probability plots:

- The vertical axis is used for the data and the horizontal axis for the normal scores.
- A probability or percent scale is used instead of normal scores.
- An averaging process is used to assign equal normal scores to equal observations.
- The method used for computing normal scores differs from the one used to obtain Table III.

Detecting Outliers with Normal Probability Plots

Recall that outliers are observations that fall well outside the overall pattern of the data. We can also use normal probability plots to detect outliers.

■■■   **EXAMPLE 6.17**   **Using Normal Probability Plots to Detect Outliers**

TABLE 6.6

Sample of last year's chicken consumption (lb)

57	69	63	49	63	61
72	65	91	59	0	82
60	75	55	80	73	

Chicken Consumption The U.S. Department of Agriculture publishes data on U.S. chicken consumption in *Food Consumption, Prices, and Expenditures*. The annual chicken consumption, in pounds, for 17 randomly selected people is displayed in Table 6.6. A normal probability plot for these observations is presented in Fig. 6.30(a). Use the plot to discuss the distribution of chicken consumption and to detect any outliers.

Solution Figure 6.30(a) reveals that the normal probability plot falls roughly in a straight line, except for the point corresponding to 0 lb, which falls well outside the overall pattern of the plot.

FIGURE 6.30 Normal probability plots for chicken consumption: (a) original data; (b) data with outlier removed

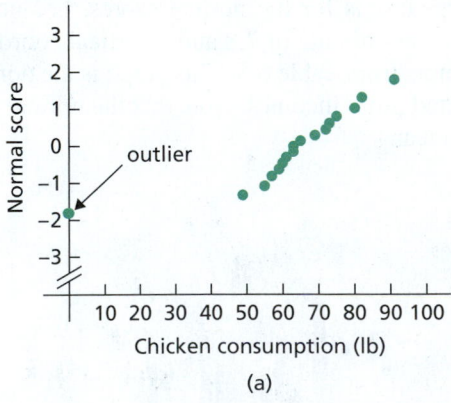

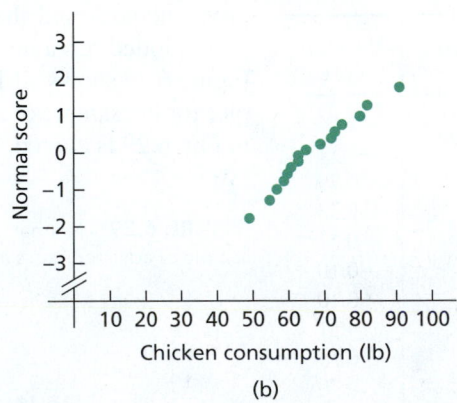

Interpretation The observation of 0 lb is an outlier, which might be a recording error or due to a person in the sample who does not eat chicken, such as a vegetarian.

You
try it!

Exercise 6.127(b)
on page 295

If we remove the outlier 0 lb from the sample data and draw a new normal probability plot, Fig. 6.30(b) shows that this plot is quite linear.

Interpretation It appears plausible that, among people who eat chicken, the amounts they consume annually are (approximately) normally distributed.   ■

Although the visual assessment of normality that we studied in this section is subjective, it is sufficient for most statistical analyses.

THE TECHNOLOGY CENTER

Most statistical technologies have programs that automatically construct normal probability plots. In this subsection, we present output and step-by-step instructions for such programs.

EXAMPLE 6.18   **Using Technology to Obtain Normal Probability Plots**

Adjusted Gross Incomes   Use Minitab, Excel, or the TI-83/84 Plus to obtain a normal probability plot for the adjusted gross incomes in Table 6.4 on page 291.

Solution   We applied the normal-probability-plot programs to the data, resulting in Output 6.4. Steps for generating that output are presented in Instructions 6.3 on the following page.

OUTPUT 6.4   Normal probability plots for the sample of adjusted gross incomes

MINITAB

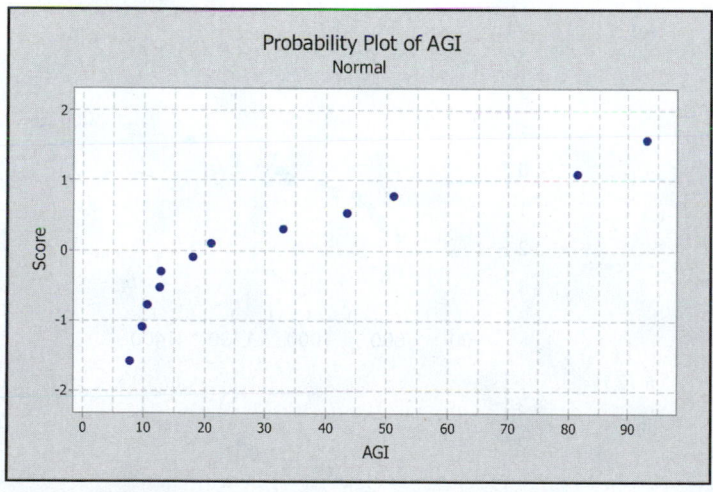

EXCEL

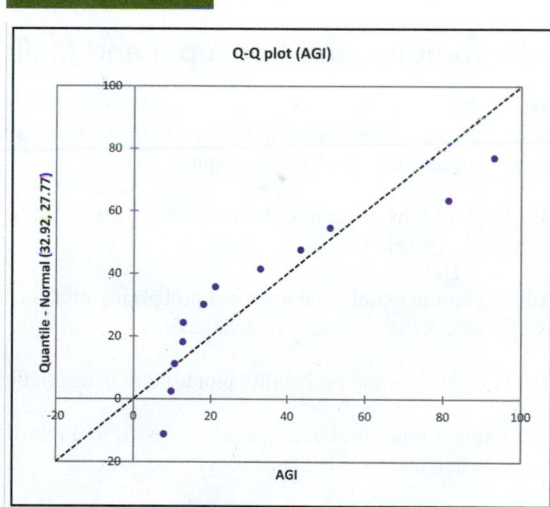

TI-83/84 PLUS

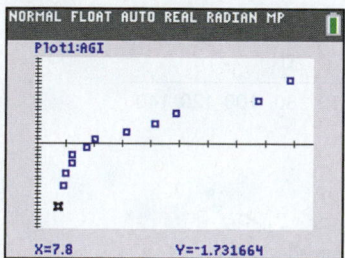

INSTRUCTIONS 6.3   Steps for generating Output 6.4

MINITAB

1 Store the data from Table 6.4 in a column named AGI
2 Choose **Graph ➤ Probability Plot...**
3 Click **OK**
4 Press the F3 key to reset the dialog box
5 Specify AGI in the **Graph variables** text box
6 Click the **Distribution...** button
7 Click the **Data Display** tab, select the **Symbols only** option button from the **Data Display** list, and click **OK**
8 Click the **Scale...** button
9 Click the **Y-Scale Type** tab, select the **Score** option button from the **Y-Scale Type** list, and click **OK**
10 Click **OK**

EXCEL

1 Store the data from Table 6.4 in a column named AGI
2 Choose **XLSTAT ➤ Visualizing data ➤ Univariate plots**
3 Click the reset button in the lower left corner of the dialog box

4 Click in the **Quantitative data** selection box and then select the column of the worksheet that contains the AGI data
5 Click the **Options** tab and uncheck the **Descriptive statistics** check box
6 Click the **Charts (1)** tab, uncheck the **Box plots** check box, and check the **Normal Q-Q plots** check box
7 Click **OK**
8 Click the **Continue** button in the XLSTAT – Selections dialog box

TI-83/84 PLUS

1 Store the data from Table 6.4 in a list named AGI
2 Ensure that all stat plots and all **Y =** functions are off
3 Press **2nd ➤ STAT PLOT** and then press **ENTER** twice
4 Arrow to the sixth graph icon and press **ENTER**
5 Press the down-arrow key
6 Press **2nd ➤ LIST**
7 Arrow down to AGI and press **ENTER**
8 Press **ZOOM**, then **9**, and then **TRACE**

Exercises 6.4

Understanding the Concepts and Skills

6.116 Under what circumstances is using a normal probability plot to assess the normality of a variable usually better than using a histogram, stem-and-leaf diagram, or dotplot?

6.117 Explain why assessing the normality of a variable is often important.

6.118 Explain in detail what a normal probability plot is and how it is used to assess the normality of a variable.

6.119 How is a normal probability plot used to detect outliers?

6.120 Explain how to obtain normal scores from Table III in Appendix A when a sample contains equal observations.

In each of Exercises 6.121–6.126, we have provided a normal probability plot of data from a sample of a population. In each case, assess the normality of the variable under consideration.

6.121

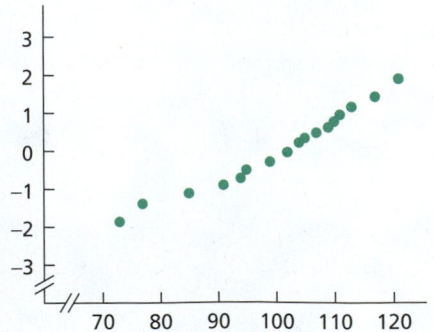

6.122

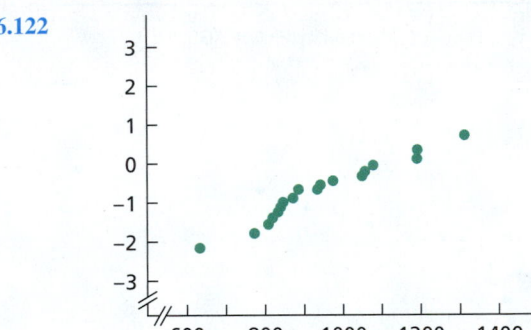

6.123

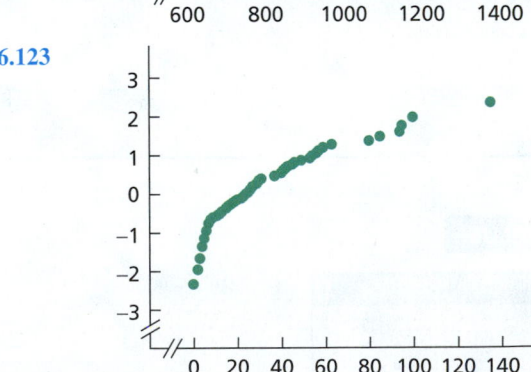

6.124

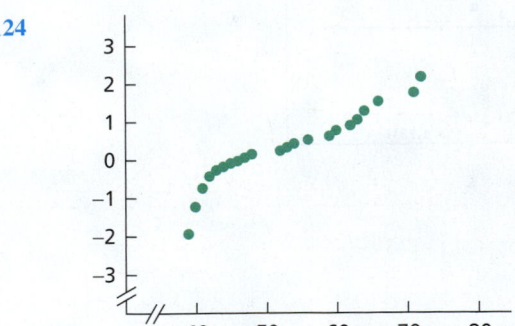

6.125

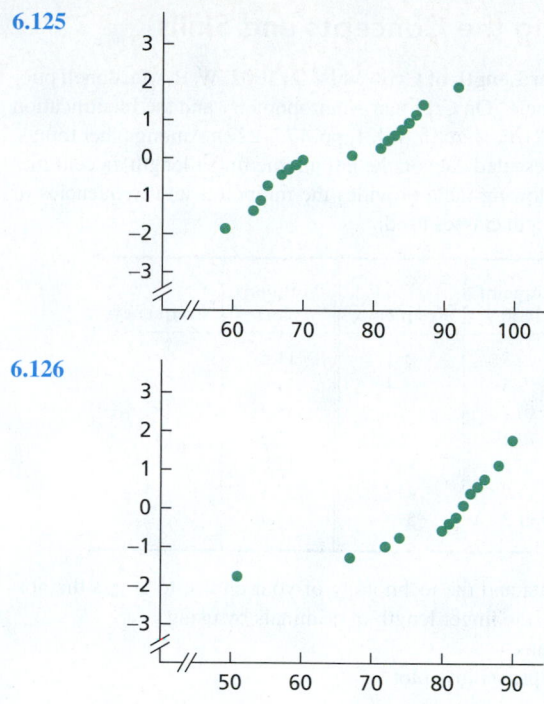

6.126

Applying the Concepts and Skills

In Exercises 6.127–6.130,
a. *use Table III in Appendix A to construct a normal probability plot of the given data.*
b. *use part (a) to identify any outliers.*
c. *use part (a) to assess the normality of the variable under consideration.*

6.127 Exam Scores. A sample of the final exam scores in a large introductory statistics course is as follows.

88	67	64	76	86
85	82	39	75	34
90	63	89	90	84
81	96	100	70	96

6.128 Cellular Bills. CTIA–The Wireless Association collects data on cell phones and publishes the results in *Semi-annual Wireless Survey*. A sample of 15 monthly cell-phone bills gave the following data (to the nearest dollar).

25	95	75	15	55
55	15	45	45	35
45	55	20	65	60

6.129 Thoroughbred Racing. The following table displays finishing times, in seconds, for the winners of fourteen 1-mile thoroughbred horse races, as found in *Thoroughbred Times*.

94.15	93.37	103.02	95.57	97.73	101.09	99.38
97.19	96.63	101.05	97.91	98.44	97.47	95.10

6.130 Beverage Expenditures. The Bureau of Labor Statistics publishes information on average annual expenditures by consumers in the *Consumer Expenditure Survey*. In 2010, the mean amount spent by consumers on nonalcoholic beverages was $333. A random sample

of 12 consumers yielded the following data, in dollars, on last year's expenditures on nonalcoholic beverages.

472	287	295	376
370	392	351	384
370	360	305	369

In Exercises 6.131–6.134,
a. *obtain a normal probability plot of the given data.*
b. *use part (a) to identify any outliers.*
c. *use part (a) to assess the normality of the variable under consideration.*

6.131 Shoe and Apparel E-Tailers. In the special report "Mousetrap: The Most-Visited Shoe and Apparel E-tailers" (*Footwear News*, Vol. 58, No. 3, p. 18), we found the following data on the average time, in minutes, spent per user per month from January to June of one year for a sample of 15 shoe and apparel retail Web sites.

13.3	9.0	11.1	9.1	8.4
15.6	8.1	8.3	13.0	17.1
16.3	13.5	8.0	15.1	5.8

6.132 Hotels and Motels. The following table provides the daily charges, in dollars, for a sample of 15 hotels and motels operating in South Carolina. The data were found in the report *South Carolina Statistical Abstract*, sponsored by the South Carolina Budget and Control Board.

81.05	69.63	74.25	53.39	57.48
47.87	61.07	51.40	50.37	106.43
47.72	58.07	56.21	130.17	95.23

6.133 Oxygen Distribution. In the article "Distribution of Oxygen in Surface Sediments from Central Sagami Bay, Japan: In Situ Measurements by Microelectrodes and Planar Optodes" (*Deep Sea Research Part I: Oceanographic Research Papers*, Vol. 52, Issue 10, pp. 1974–1987), R. Glud et al. explored the distributions of oxygen in surface sediments from central Sagami Bay. The oxygen distribution gives important information on the general biogeochemistry of marine sediments. Measurements were performed at 16 sites. A sample of 22 depths yielded the following data, in millimoles per square meter per day (mmol m^{-2} d^{-1}), on diffusive oxygen uptake (DOU).

1.8	2.0	1.8	2.3	3.8	3.4	2.7	1.1
3.3	1.2	3.6	1.9	7.6	2.0	1.5	2.0
1.1	0.7	1.0	1.8	1.8	6.7		

6.134 Medieval Cremation Burials. In the article "Material Culture as Memory: Combs and Cremations in Early Medieval Britain" (*Early Medieval Europe*, Vol. 12, Issue 2, pp. 89–128), H. Williams discussed the frequency of cremation burials found in 17 archaeological sites in eastern England. Here are the data.

83	64	46	48	523	35	34	265	2484
46	385	21	86	429	51	258	119	

Working with Large Data Sets

6.135 Body Temperature. A study by researchers at the University of Maryland addressed the question of whether the mean body temperature of humans is 98.6°F. The results of the study by P. Mackowiak et al. appeared in the article "A Critical Appraisal of 98.6°F, the Upper Limit of the Normal Body Temperature, and Other Legacies of Carl Reinhold August Wunderlich" (*Journal of the American Medical Association*, Vol. 268, pp. 1578–1580). Among other data, the researchers obtained the body temperatures of 93 healthy humans, as provided on the WeissStats site. Use the technology of your choice to do the following.

a. Obtain a histogram of the data and use it to assess the (approximate) normality of the variable under consideration.

b. Obtain a normal probability plot of the data and use it to assess the (approximate) normality of the variable under consideration.

c. Compare your results in parts (a) and (b).

6.136 Vegetarians and Omnivores. Philosophical and health issues are prompting an increasing number of Taiwanese to switch to a vegetarian lifestyle. In the paper "LDL of Taiwanese Vegetarians Are Less Oxidizable than Those of Omnivores" (*Journal of Nutrition*, Vol. 130, pp. 1591–1596), S. Lu et al. compared the daily intake of nutrients by vegetarians and omnivores living in Taiwan. Among the nutrients considered was protein. Too little protein stunts growth and interferes with all bodily functions; too much protein puts a strain on the kidneys, can cause diarrhea and dehydration, and can leach calcium from bones and teeth. The daily protein intakes, in grams, for 51 female vegetarians and 53 female omnivores are provided on the WeissStats site. Use the technology of your choice to do the following for each of the two sets of sample data.

a. Obtain a histogram of the data and use it to assess the (approximate) normality of the variable under consideration.

b. Obtain a normal probability plot of the data and use it to assess the (approximate) normality of the variable under consideration.

c. Compare your results in parts (a) and (b).

6.137 "Chips Ahoy! 1,000 Chips Challenge." Students in an introductory statistics course at the U.S. Air Force Academy participated in Nabisco's "Chips Ahoy! 1,000 Chips Challenge" by confirming that there were at least 1000 chips in every 18-ounce bag of cookies that they examined. As part of their assignment, they concluded that the number of chips per bag is approximately normally distributed. Their conclusion was based on the data provided on the WeissStats site, which gives the number of chips per bag for 42 bags. Do you agree with the conclusion of the students? Explain your answer. [SOURCE: B. Warner and J. Rutledge, "Checking the Chips Ahoy! Guarantee," *Chance*, Vol. 12(1), pp. 10–14]

Extending the Concepts and Skills

6.138 Finger Length of Criminals. In 1902, W. R. Macdonell published the article "On Criminal Anthropometry and the Identification of Criminals" (*Biometrika*, Vol. 1, pp. 177–227). Among other things, the author presented data on the left middle finger length, in centimeters. The following table provides the midpoints and frequencies of the finger-length classes used.

Midpoint (cm)	Frequency	Midpoint (cm)	Frequency
9.5	1	11.6	691
9.8	4	11.9	509
10.1	24	12.2	306
10.4	67	12.5	131
10.7	193	12.8	63
11.0	417	13.1	16
11.3	575	13.4	3

Use these data and the technology of your choice to assess the normality of middle finger length of criminals by using

a. a histogram.

b. a normal probability plot.

6.139 Gestation Periods of Humans. For humans, gestation periods are normally distributed with a mean of 266 days and a standard deviation of 16 days.

a. Use the technology of your choice to simulate four random samples of 50 human gestation periods each.

b. Obtain a normal probability plot of each sample in part (a).

c. Are the normal probability plots in part (b) what you expected? Explain your answer.

6.140 Emergency Room Traffic. Desert Samaritan Hospital in Mesa, Arizona, keeps records of emergency room traffic. Those records reveal that the times between arriving patients have a special type of reverse-J-shaped distribution called an *exponential distribution*. The records also show that the mean time between arriving patients is 8.7 minutes.

a. Use the technology of your choice to simulate four random samples of 75 interarrival times each.

b. Obtain a normal probability plot of each sample in part (a).

c. Are the normal probability plots in part (b) what you expected? Explain your answer.

6.5 Normal Approximation to the Binomial Distribution*†

In this section, we demonstrate the approximation of binomial probabilities by using areas under a suitable normal curve. The development of the mathematical theory for doing so is credited to Abraham de Moivre (1667–1754) and Pierre-Simon Laplace (1749–1827). For more information on de Moivre and Laplace, see the biographies at the end of Chapters 12 and 7, respectively.

First, we need to review briefly the binomial distribution, which we discussed in detail in Section 5.3. Suppose that n identical independent success–failure experiments are performed, with the probability of success on any given trial being p. Let X denote the total number of successes in the n trials. Then, the probability distribution of the

---

†Coverage of the binomial distribution (Section 5.3) is prerequisite to this section.

random variable X is given by the binomial probability formula,

$$P(X = x) = \binom{n}{x} p^x (1 - p)^{n-x}, \qquad x = 0, 1, 2, \ldots, n.$$

We say that X has the binomial distribution with parameters n and p.

You might be wondering why we would use normal-curve areas to approximate binomial probabilities when we can obtain them exactly with the binomial probability formula. Example 6.19 provides the reason.

EXAMPLE 6.19   **The Need to Approximate Binomial Probabilities**

Mortality Mortality tables enable actuaries to obtain the probability that a person at any particular age will live a specified number of years. Insurance companies and others use such probabilities to determine life-insurance premiums, retirement pensions, and annuity payments.

According to tables provided by the National Center for Health Statistics in *Vital Statistics of the United States*, a person of age 20 years has about an 80% chance of being alive at age 65 years. In Example 5.12 on pages 242–244, we used the binomial probability formula to determine probabilities for the number of 20-year-olds out of three who will be alive at age 65.

For most real-world problems, the number of people under investigation is much larger than three. Although in principle we can use the binomial probability formula to determine probabilities regardless of number, in practice we do not. Suppose, for instance, that 500 people of age 20 years are selected at random. Find the probability that

a. exactly 390 of them will be alive at age 65.
b. between 375 and 425 of them, inclusive, will be alive at age 65.

Solution Let X denote the number of people of the 500 who are alive at age 65. Then X has the binomial distribution with parameters $n = 500$ (the 500 people) and $p = 0.8$ (the probability a person of age 20 will be alive at age 65). In principle, we can determine probabilities for X exactly by using the binomial probability formula,

$$P(X = x) = \binom{500}{x} (0.8)^x (0.2)^{500-x}.$$

Let's use that formula for parts (a) and (b).

a. The "answer" is

$$P(X = 390) = \binom{500}{390} (0.8)^{390} (0.2)^{110}.$$

However, obtaining the numerical value of the expression on the right-hand side is not easy, even with a calculator. Such computations often lead to roundoff errors and to numbers so large or so small that they are outside the range of the calculator. Fortunately, we can sidestep the calculations altogether by using normal-curve areas.

b. The "answer" is

$$P(375 \leq X \leq 425) = P(X = 375) + P(X = 376) + \cdots + P(X = 425)$$

$$= \binom{500}{375} (0.8)^{375} (0.2)^{125} + \binom{500}{376} (0.8)^{376} (0.2)^{124}$$

$$+ \cdots + \binom{500}{425} (0.8)^{425} (0.2)^{75}.$$

Here we have the same computational difficulties as we did in part (a), except that we must evaluate 51 complex expressions instead of 1. Again, the binomial probability formula is too difficult to use, and we will need to use normal-curve areas.

The previous example makes clear that using the binomial probability formula when the number of trials, n, is very large is impractical. Under certain conditions on n and p, the distribution of a binomial random variable is roughly bell shaped. In such cases, we can approximate probabilities for the random variable by areas under a suitable normal curve, as shown in the next example.

EXAMPLE 6.20   Approximating Binomial Probabilities, Using Normal-Curve Areas

True–False Exams   A student is taking a true–false exam with 10 questions. Assume that the student guesses at all 10 questions.

a.   Determine the probability that the student gets either 7 or 8 answers correct.
b.   Approximate the probability obtained in part (a) by an area under a suitable normal curve.

Solution   Let X denote the number of correct answers by the student. Then X has the binomial distribution with parameters $n = 10$ (the 10 questions) and $p = 0.5$ (the probability of a correct guess).

a.   Probabilities for X are given by the binomial probability formula

$$P(X = x) = \binom{10}{x}(0.5)^x(1 - 0.5)^{10-x}.$$

Using this formula, we get the probability distribution of X, as shown in Table 6.7. According to that table, the probability the student gets either 7 or 8 answers correct is

$$P(X = 7 \text{ or } 8) = P(X = 7) + P(X = 8) = 0.1172 + 0.0439 = 0.1611.$$

b.   Referring to Table 6.7, we drew the probability histogram of X in Fig. 6.31. Because the probability histogram is bell shaped, probabilities for X can be approximated by areas under a normal curve. The appropriate normal curve is the one whose parameters are the same as the mean and standard deviation of X, which, by Formula 5.2 on page 245, are

$$\mu = np = 10 \cdot 0.5 = 5$$

and

$$\sigma = \sqrt{np(1 - p)} = \sqrt{10 \cdot 0.5 \cdot (1 - 0.5)} = 1.58.$$

Therefore, the required normal curve has parameters $\mu = 5$ and $\sigma = 1.58$; it is superimposed on the probability histogram in Fig. 6.31.

TABLE 6.7

Probability distribution of the number of correct answers of 10 by the student

Number correct x	Probability $P(X = x)$
0	0.0010
1	0.0098
2	0.0439
3	0.1172
4	0.2051
5	0.2461
6	0.2051
7	0.1172
8	0.0439
9	0.0098
10	0.0010

FIGURE 6.31

Probability histogram for X with superimposed normal curve

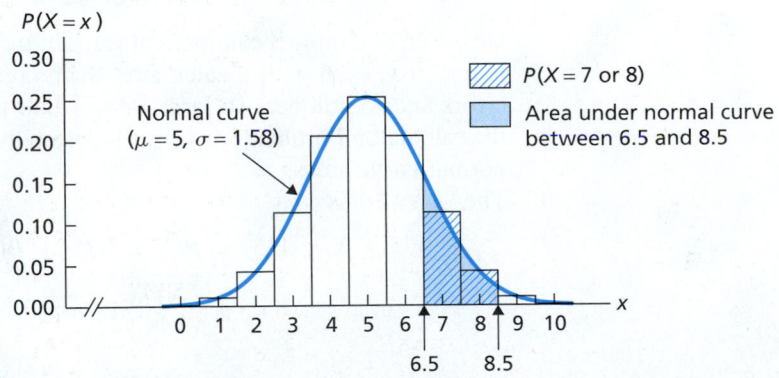

The probability $P(X = 7 \text{ or } 8)$ equals the area of the corresponding bars of the histogram, cross-hatched in Fig. 6.31. Note that the cross-hatched area approximately equals the area under the normal curve between 6.5 and 8.5, shaded in Fig. 6.31.

Figure 6.31 makes clear why we consider the area under the normal curve between 6.5 and 8.5 instead of between 7 and 8. This adjustment is called the *correction for continuity*. It is required because we are approximating the distribution of a discrete variable by that of a continuous variable.

Figure 6.31 shows that $P(X = 7 \text{ or } 8)$ roughly equals the area under the normal curve with parameters $\mu = 5$ and $\sigma = 1.58$ that lies between 6.5 and 8.5. To compute this area, we convert to z-scores and then find the corresponding area under the standard normal curve in the usual way, as shown in Fig. 6.32.

The last line in Fig. 6.32 shows that the area under the normal curve between 6.5 and 8.5 is 0.1579. This area is close to $P(X = 7 \text{ or } 8)$, which, as we found in part (a), is 0.1611.

What Does It Mean?

The normal-curve area provides an excellent approximation of the exact probability.

FIGURE 6.32

Determination of the area under the normal curve with parameters $\mu = 5$ and $\sigma = 1.58$ that lies between 6.5 and 8.5

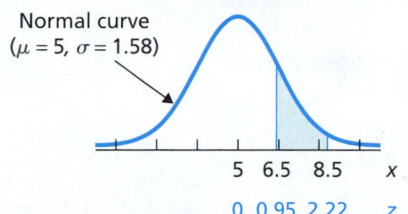

Normal curve
($\mu = 5$, $\sigma = 1.58$)

	5	6.5	8.5	x
	0	0.95	2.22	z

z-score computations: Area to the left of z:

$x = 6.5 \longrightarrow z = \dfrac{6.5 - 5}{1.58} = 0.95$ 0.8289

$x = 8.5 \longrightarrow z = \dfrac{8.5 - 5}{1.58} = 2.22$ 0.9868

Shaded area = 0.9868 − 0.8289 = 0.1579

As indicated by the previous example, we can use normal-curve areas to approximate probabilities for binomial random variables that have bell-shaped distributions. Whether a particular binomial random variable has a bell-shaped distribution depends on its parameters, n and p. Figure 6.33 on the next page shows nine different binomial distributions.

As illustrated in Figs. 6.33(a) and 6.33(c), a binomial distribution with $p \neq 0.5$ is always skewed. For small n, such a distribution is too skewed to allow a normal approximation but, for large n, is sufficiently bell shaped to permit it. In contrast, Fig. 6.33(b) illustrates that a binomial distribution with $p = 0.5$ is always symmetric. Nonetheless, such a distribution will not be sufficiently bell shaped to permit a normal approximation if n is too small.

The customary rule of thumb for using the normal approximation is that *both np and $n(1 - p)$ are 5 or greater*. This restriction indicates that the farther the success probability is from 0.5, the larger the number of trials must be to use the normal approximation.

Procedure for Using the Normal Approximation to the Binomial Distribution

We can now write a general step-by-step method for approximating binomial probabilities by areas under a normal curve.

PROCEDURE 6.3   **To Approximate Binomial Probabilities by Normal-Curve Areas**

Step 1   Find n, the number of trials, and p, the success probability.

Step 2   Continue only if both np and $n(1 - p)$ are 5 or greater.

Step 3   Find μ and σ, using the formulas $\mu = np$ and $\sigma = \sqrt{np(1 - p)}$.

Step 4   Make the correction for continuity, and find the required area under the normal curve with parameters μ and σ.

FIGURE 6.33 Nine different binomial distributions

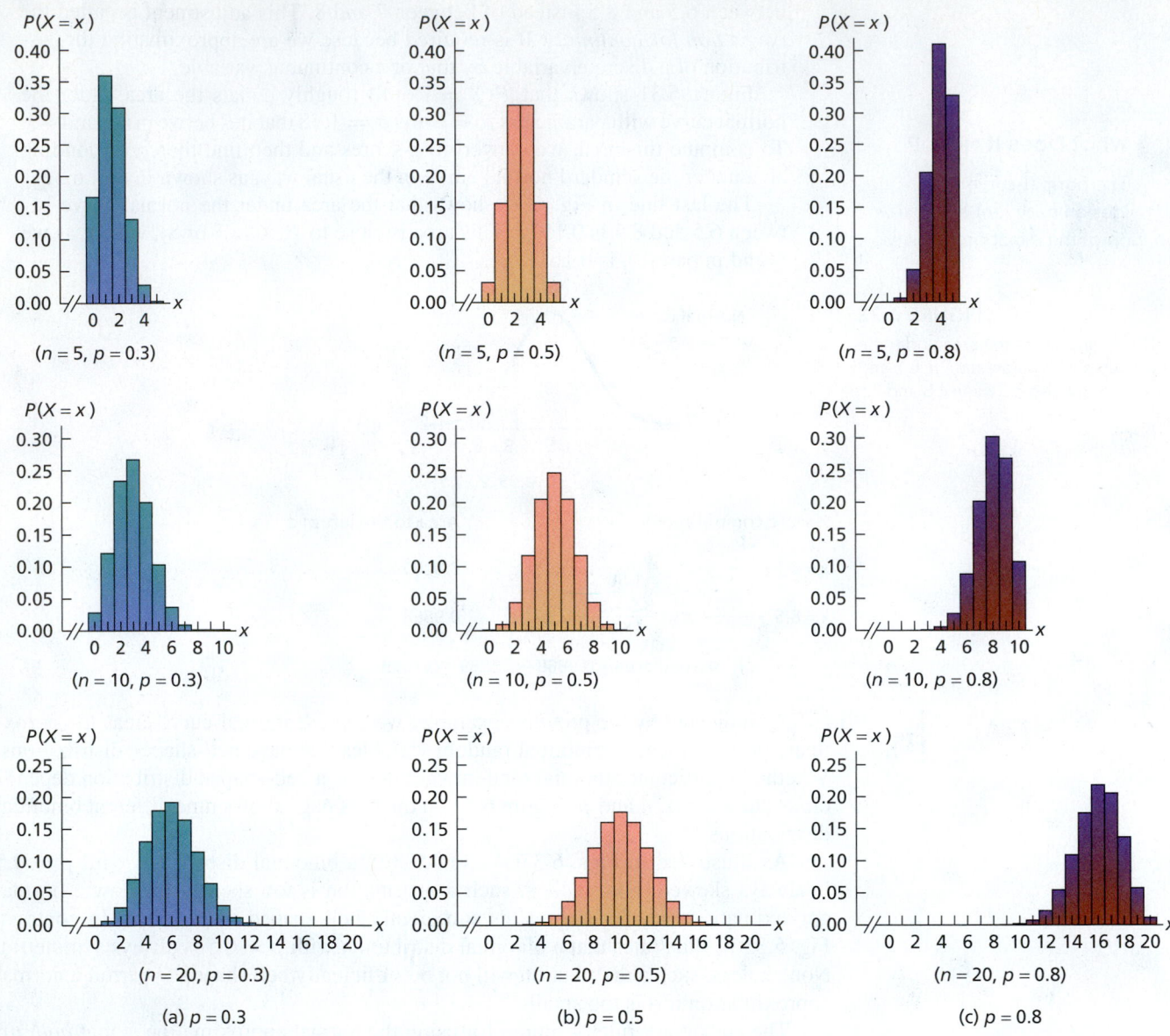

Step 4 of Procedure 6.3 requires the **correction for continuity,** as illustrated in Example 6.20. For instance, when using normal-curve areas to approximate the probability that an observed value of a binomial random variable will be between two whole numbers, inclusive, we subtract 0.5 from the smaller whole number and add 0.5 to the larger whole number before finding the area under the normal curve.

In general, we always make the correction factor (add or subtract 0.5) that leads us to the original whole numbers. For example, if we want to approximate $P(X < 16)$, the whole numbers in question are $0, 1, 2, \ldots, 15$; thus, we would find the area under the normal curve that lies between -0.5 and 15.5. Similarly, if we want to approximate $P(12 < X \leq 16)$, the whole numbers in question are 13, 14, 15, and 16; hence, we would find the area under the normal curve that lies between 12.5 and 16.5.

■■■ **EXAMPLE 6.21** **Normal Approximation to the Binomial**

Mortality The probability is 0.80 that a person of age 20 years will be alive at age 65 years. Suppose that 500 people of age 20 are selected at random. Determine the probability that

a. exactly 390 of them will be alive at age 65.

b. between 375 and 425 of them, inclusive, will be alive at age 65.

Solution We will approximate the probabilities in parts (a) and (b) by using Procedure 6.3.

Step 1 Find n, the number of trials, and p, the success probability.

We have $n = 500$ and $p = 0.8$.

Step 2 Continue only if both np and $n(1 - p)$ are 5 or greater.

From the values for n and p noted in Step 1,

$$np = 500 \cdot 0.8 = 400 \quad \text{and} \quad n(1 - p) = 500 \cdot 0.2 = 100.$$

Both np and $n(1 - p)$ are greater than 5, so we can continue.

Step 3 Find μ and σ, using the formulas $\mu = np$ and $\sigma = \sqrt{np(1 - p)}$.

We get $\mu = 500 \cdot 0.8 = 400$ and $\sigma = \sqrt{500 \cdot 0.8 \cdot 0.2} = 8.94$.

Step 4 Make the correction for continuity, and find the required area under the normal curve with parameters μ and σ.

a. To make the correction for continuity, we subtract 0.5 from 390 and add 0.5 to 390. Thus we need to find the area under the normal curve with parameters $\mu = 400$ and $\sigma = 8.94$ that lies between 389.5 and 390.5. This area, 0.0236, is found in Fig. 6.34. So, $P(X = 390) = 0.0236$, approximately.

FIGURE 6.34

Determination of the area under the normal curve with parameters $\mu = 400$ and $\sigma = 8.94$ that lies between 389.5 and 390.5

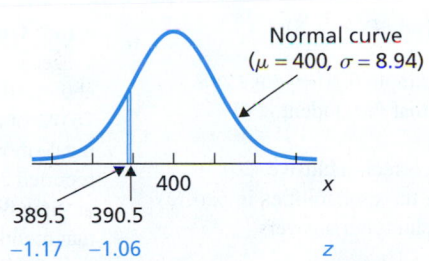

z-score computations:

$x = 389.5 \longrightarrow z = \dfrac{389.5 - 400}{8.94} = -1.17$ Area to the left of z: 0.1210

$x = 390.5 \longrightarrow z = \dfrac{390.5 - 400}{8.94} = -1.06$ 0.1446

Shaded area = 0.1446 − 0.1210 = 0.0236

Interpretation The probability is about 0.0236 that exactly 390 of the 500 people selected will be alive at age 65.

b. To make the correction for continuity, we subtract 0.5 from 375 and add 0.5 to 425. Thus we need to determine the area under the normal curve with parameters $\mu = 400$ and $\sigma = 8.94$ that lies between 374.5 and 425.5. As in part (a), we convert to z-scores, and then find the corresponding area under the standard normal curve. This area is 0.9956. So, $P(375 \le X \le 425) = 0.9956$, approximately.

Interpretation The probability is approximately 0.9956 that between 375 and 425 of the 500 people selected will be alive at age 65.

You try it!

Exercise 6.167 on page 302

Exercises 6.5

Understanding the Concepts and Skills

6.141 Why should you sometimes use normal-curve areas to approximate binomial probabilities even though you have a formula for computing them exactly?

6.142 The rule of thumb for using the normal approximation to the binomial is that both np and $n(1 - p)$ are 5 or greater. Why is this restriction necessary?

In Exercises 6.143–6.160, X denotes a binomial random variable with parameters n and p. For each exercise, indicate which area under the appropriate normal curve would be determined to approximate the specified binomial probability.

6.143 $P(X = 8)$

6.144 $P(X = 6)$

6.145 $P(X < 7)$

6.146 $P(X \leq 4)$

6.147 $P(X \leq 7)$

6.148 $P(X < 4)$

6.149 $P(7 < X \leq 10)$

6.150 $P(4 \leq X < 8)$

6.151 $P(7 \leq X < 10)$

6.152 $P(4 < X \leq 8)$

6.153 $P(7 \leq X \leq 10)$

6.154 $P(4 < X < 8)$

6.155 $P(7 < X < 10)$

6.156 $P(4 \leq X \leq 8)$

6.157 $P(X > 10)$

6.158 $P(X \geq 8)$

6.159 $P(X \geq 10)$

6.160 $P(X > 8)$

6.161 **True–False Exams.** Refer to Example 6.20 on page 298.
a. Use Table 6.7 to find the probability that the student gets
   i. either four or five answers correct.
   ii. between three and seven answers correct, inclusive.
b. Apply Procedure 6.3 to approximate the probabilities in part (a) by areas under a normal curve. Compare your answers.

6.162 **True–False Exams.** Refer to Example 6.20 on page 298.
a. Use Table 6.7 to find the probability that the student gets
   i. at most five answers correct.
   ii. at least six answers correct.
b. Apply Procedure 6.3 to approximate the probabilities in part (a) by areas under a normal curve. Compare your answers.

6.163 **True–False Exams.** If, in Example 6.20, the true–false exam had 25 questions instead of 10, which normal curve would you use to approximate probabilities for the number of correct guesses?

6.164 **True–False Exams.** If, in Example 6.20, the true–false exam had 30 questions instead of 10, which normal curve would you use to approximate probabilities for the number of correct guesses?

Applying the Concepts and Skills

In Exercises 6.165–6.172, apply Procedure 6.3 on page 299 to approximate the required binomial probabilities.

6.165 **Stalked College Women.** In the article "Violence against Women: How Safe Are Our Institutions of Higher Learning?" (*Family & Intimate Partner Violence Quarterly*, Vol. 05, No. 04, pp. 341–349), researcher D. Thomas asserts that the population of students

attending colleges and universities, and women in particular, are at higher risk of being victims of sexual assault than are their non–college-bound peers. According to research, 13% of college women have been stalked. If 100 college women are selected at random, what is the probability that the number who have been stalked is
a. less than 10?
b. between 15 and 20, inclusive?
c. at least 15?

6.166 **Naturalization.** The Office of Immigration Statistics collects and reports information about naturalized persons in the document *U.S. Naturalizations*. During one year, there were 694,193 persons who became naturalized citizens of the United States and, of those, 36% were originally from Asia. If 200 people who became naturalized citizens of the United States that year are selected at random, what is the probability that the number who were originally from Asia is
a. fewer than 65?
b. between 70 and 80, inclusive?
c. less than 60 or more than 80?

6.167 **High School Graduates.** According to the document *Current Population Survey*, published by the U.S. Census Bureau, 30.4% of U.S. adults 25 years old or older have a high school degree as their highest educational level. If 100 such adults are selected at random, determine the probability that the number who have a high school degree as their highest educational level is
a. exactly 32.
b. between 30 and 35, inclusive.
c. at least 25.

6.168 **On-Time Airlines.** The Office of Aviation Enforcement and Proceedings (OAEP) publishes important consumer information about airlines in *Air Travel Consumer Report*. During one month, 76.0% of all flights arriving at U.S. airports arrived "on time," meaning no more than 15 minutes late at the arrival gate. The Boston airport reported 8960 arrivals during that month. If the overall percentage of on-time flights applies to Boston, what is the probability that, during that month, the number of on-time flights to Boston
a. exceeded 6800?
b. was between 6750 and 6850, inclusive?

6.169 **Airline Reservations.** As reported by a spokesperson for Southwest Airlines, the no-show rate for reservations is 16%. In other words, the probability is 0.16 that a person making a reservation will not take the flight. For the next flight, 42 people have reservations. What is the probability that
a. exactly 5 do not take the flight?
b. between 9 and 12, inclusive, do not take the flight?
c. at least 1 does not take the flight?
d. at most 2 do not take the flight?
e. Comment on the accuracy of the normal approximation in this case.

6.170 **Lightning-Induced Fatalities.** As reported in an issue of *Weatherwise*, according to the National Oceanic and Atmospheric Administration, people at ballparks and playgrounds are in more danger of being struck by lightning than are those on golf courses. Of lightning-induced fatalities, 3.9% occur on golf courses. What is the probability that, of 250 randomly selected lightning-induced fatalities, the number occurring on golf courses is
a. exactly 4?
b. between 4 and 10, inclusive?

c. at least 10?

d. Comment on the accuracy of the normal approximation in this case.

6.171 More Adults Exercising. In the article "More US Adults Getting Some Exercise" appearing on the YAHOO News website, R. Rettner reports that about 34% of U.S. adults are physically inactive, meaning that they do not engage in any physical activity lasting at least 10 minutes a day. If 250 U.S. adults are selected at random, find the probability that the number who are physically inactive

a. is exactly 34% of those sampled.

b. exceeds 34% of those sampled.

c. is less than 34% of those sampled.

6.172 Children's Sleep Problems. Families and pediatricians often have difficulty managing sleep problems of young children. In the article "An Individualized and Comprehensive Approach to Treating Sleep Problems in Young Children" (*Journal of Applied Behavioral Analysis*, Vol. 46, No. 1, pp. 161–180), researchers C. Jin et al. estimate that sleep problems affect 70% of children with an autism spectrum disorder. If 50 children with an autism spectrum disorder are selected at random, what is the probability that the number who have sleep problems

a. is exactly 70% of those sampled?

b. at most 70% of those sampled?

c. at least 70% of those sampled?

Extending the Concepts and Skills

6.173 Roulette. An American roulette wheel consists of 38 numbers, of which 18 are red, 18 are black, and 2 are green. When the roulette wheel is spun, the ball is equally likely to land on each of the 38 numbers. A gambler is playing roulette and bets $10 on red each time. If the ball lands on a red number, the gambler wins $10; otherwise, the gambler loses $10. What is the probability that the gambler will be ahead after

a. 100 bets?   **b.** 1000 bets?   **c.** 5000 bets?

(*Hint:* The gambler will be ahead after a series of bets if and only if he or she has won more than half the bets.)

6.174 Flashlight Battery Lifetimes. A brand of flashlight battery has normally distributed lifetimes with a mean of 30 hours and a standard deviation of 5 hours. A supermarket purchases 500 of these batteries from the manufacturer. What is the probability that at least 80% of them will last longer than 25 hours?

6.175 Fragile X Syndrome. The second-leading genetic cause of mental retardation is Fragile X Syndrome, named for the fragile appearance of the tip of the X chromosome in affected individuals. One in 1500 males is affected world-wide, with no ethnic bias.

a. In a sample of 10,000 males, how many would you expect to have Fragile X Syndrome?

b. For a sample of 10,000 males, use the normal approximation to the binomial distribution to determine the probability that more than 7 of the males have Fragile X Syndrome; that at most 10 of the males have Fragile X Syndrome.

c. The probabilities in part (b) were obtained in Exercise 5.109 on page 258 by using the Poisson approximation to the binomial distribution. Which estimates of the true binomial probabilities would you expect to be better, the ones using the normal approximation or those using the Poisson approximation? Explain your answer.

CHAPTER IN REVIEW

You Should Be Able to

1. use and understand the formulas in this chapter.

2. define, state the basic properties of, and apply density curves.

3. explain what it means for a variable to be normally distributed or approximately normally distributed.

4. explain the meaning of the parameters for a normal curve.

5. identify the basic properties of and sketch a normal curve.

6. identify the standard normal distribution and the standard normal curve.

7. use Table II to determine areas under the standard normal curve.

8. use Table II to determine the z-score(s) corresponding to a specified area under the standard normal curve.

9. use and understand the z_α notation.

10. determine a percentage or probability for a normally distributed variable.

11. state and apply the empirical rule for variables.

12. determine the observations corresponding to a specified percentage or probability for a normally distributed variable.

13. explain how to assess the normality of a variable with a normal probability plot.

14. construct a normal probability plot with the aid of Table III.

15. use a normal probability plot to detect outliers.

*16. approximate binomial probabilities by normal-curve areas, when appropriate.

Key Terms

68-95-99.7 rule, *282*

approximately normally distributed variable, *266*

correction for continuity,* *300*

cumulative probability, *285*

density curves, *263*

empirical rule, *282*

inverse cumulative probability, *286*

normal curve, *266*

normal distribution, *266*

normal probability plot, *290*

normal scores, *290*

normally distributed population, *266*

normally distributed variable, *266*

parameters, *266*

standard normal curve, *269*

standard normal distribution, *269*

standardized normally distributed variable, *269*

z_α, *277*

z-curve, *274*

REVIEW PROBLEMS

Understanding the Concepts and Skills

1. What is a density curve, and why are such curves important?

In each of Problems 2–4, assume that the variable under consideration has a density curve. Note that the answers required here may be only approximately correct.

2. The percentage of all possible observations of a variable that lie between 25 and 50 equals the area under its density curve between _____ and _____, expressed as a percentage.

3. The area under a density curve that lies to the left of 60 is 0.364. What percentage of all possible observations of the variable are
a. less than 60?　　　**b.** at least 60?

4. The area under a density curve that lies between 5 and 6 is 0.728. What percentage of all possible observations of the variable are either less than 5 or greater than 6?

5. A variable has the density curve with equation $y = 1 - x/2$ for $0 < x < 2$, and $y = 0$ otherwise.
a. Graph the density curve of this variable.
b. Show that the area under this density curve to the left of any number x between 0 and 2 equals $x - x^2/4$. *Hint:* First find the area to the right of x and then apply Property 2 of Key Fact 6.1 on page 263.

What percentage of all possible observations of the variable
c. lie between 1/2 and 1?
d. are at least 1.5?

6. State two of the main reasons for studying the normal distribution.

7. Define
a. normally distributed variable.
b. normally distributed population.
c. parameters for a normal curve.

8. Answer true or false to each statement. Give reasons for your answers.
a. Two variables that have the same mean and standard deviation have the same distribution.
b. Two normally distributed variables that have the same mean and standard deviation have the same distribution.

9. Explain the relationship between percentages for a normally distributed variable and areas under the corresponding normal curve.

10. Identify the distribution of the standardized version of a normally distributed variable.

11. Answer true or false to each statement. Explain your answers.
a. Two normal distributions that have the same mean are centered at the same place, regardless of the relationship between their standard deviations.
b. Two normal distributions that have the same standard deviation have the same spread, regardless of the relationship between their means.

12. Consider the normal curves that have the parameters $\mu = 1.5$ and $\sigma = 3$; $\mu = 1.5$ and $\sigma = 6.2$; $\mu = -2.7$ and $\sigma = 3$; $\mu = 0$ and $\sigma = 1$.
a. Which curve has the largest spread?

b. Which curves are centered at the same place?
c. Which curves have the same spread?
d. Which curve is centered farthest to the left?
e. Which curve is the standard normal curve?

13. What key fact permits you to determine percentages for a normally distributed variable by first converting to z-scores and then determining the corresponding area under the standard normal curve?

14. Explain how to use Table II to determine the area under the standard normal curve that lies
a. to the left of a specified z-score.
b. to the right of a specified z-score.
c. between two specified z-scores.

15. Explain how to use Table II to determine the z-score that has a specified area to its
a. left under the standard normal curve.
b. right under the standard normal curve.

16. What does the symbol z_α signify?

17. State the empirical rule for variables.

18. Roughly speaking, what are the normal scores corresponding to a sample of observations?

19. If you observe the values of a normally distributed variable for a sample, a normal probability plot should be roughly _____.

20. Sketch the normal curve having the parameters
a. $\mu = -1$ and $\sigma = 2$.　　　**b.** $\mu = 3$ and $\sigma = 2$.
c. $\mu = -1$ and $\sigma = 0.5$.

21. According to Table II, the area under the standard normal curve that lies to the left of 1.05 is 0.8531. Without further reference to Table II, determine the area under the standard normal curve that lies
a. to the right of 1.05.　　　**b.** to the left of -1.05.
c. between -1.05 and 1.05.

22. Determine and sketch the area under the standard normal curve that lies
a. to the left of -3.02.　　　**b.** to the right of 0.61.
c. between 1.11 and 2.75.　　**d.** between -2.06 and 5.02.
e. between -4.11 and -1.5.
f. either to the left of 1 or to the right of 3.

23. For the standard normal curve, find the z-score(s)
a. that has area 0.30 to its left.
b. that has area 0.10 to its right.
c. $z_{0.025}$, $z_{0.05}$, $z_{0.01}$, and $z_{0.005}$.
d. that divide the area under the curve into a middle 0.99 area and two outside 0.005 areas.

Applying the Concepts and Skills

24. Dispensing Coffee. A coffee machine is supposed to dispense 6 fluid ounces (fl oz) of coffee into a paper cup. In reality, the amounts dispensed vary from cup to cup. In fact, the amount dispensed, in fl oz, is a variable with density curve $y = 2$ for $5.75 < x < 6.25$, and $y = 0$ otherwise.
a. Graph the density curve of this variable.
b. Show that the area under this density curve to the left of any number x between 5.75 and 6.25 equals $2x - 11.5$.

What percentage of cups dispensed by this machine contain

c. less than 6 fl oz?

d. between 5.9 and 6.1 fl oz?

e. at least 5.8 fl oz?

25. Forearm Length. In 1903, K. Pearson and A. Lee published a paper entitled "On the Laws of Inheritance in Man. I. Inheritance of Physical Characters" (*Biometrika*, Vol. 2, pp. 357–462). From information presented in that paper, forearm length of men, measured from the elbow to the middle fingertip, is (roughly) normally distributed with a mean of 18.8 inches and a standard deviation of 1.1 inches. Let x denote forearm length, in inches, for men.

a. Sketch the distribution of the variable x.

b. Obtain the standardized version, z, of x.

c. Identify and sketch the distribution of z.

d. The area under the normal curve with parameters 18.8 and 1.1 that lies between 17 and 20 is 0.8115. Determine the probability that a randomly selected man will have a forearm length between 17 inches and 20 inches.

e. The percentage of men who have forearm length less than 16 inches equals the area under the standard normal curve that lies to the _____ of _____.

26. Birth Weights. The *WONDER database*, maintained by the Centers for Disease Control and Prevention, provides a single point of access to a wide variety of reports and numeric public health data. From that database, we obtained the following data for one year's birth weights of male babies who weighed under 5000 grams (about 11 pounds).

Weight (g)	Frequency
0–under 500	2,025
500–under 1000	8,400
1000–under 1500	10,215
1500–under 2000	19,919
2000–under 2500	67,068
2500–under 3000	274,913
3000–under 3500	709,110
3500–under 4000	609,719
4000–under 4500	191,826
4500–under 5000	31,942

a. Obtain a relative-frequency histogram of these weight data.

b. Based on your histogram, do you think that, for the year in question, the birth weights of male babies who weighed under 5000 grams are approximately normally distributed? Explain your answer.

27. Lower Limb Surgery. The study "Intrathecal Sufentanil versus Fentanyl for Lower Limb Surgeries – A Randomized Controlled Trial" (*Journal of Anaesthesiology Clinical Pharmacology*, Vol. 27, Issue 1, pp. 67–73) by P. Motiani et al. compares two different agents, intrathecal sufentail and fentanyl, used in enhancing the anesthesiology of patients receiving major lower limb surgery. One variable compared between the two agents was the amount of blood loss during the surgery. Based on the study, we will assume that, using fentanyl, the amount of blood loss during major lower limb surgery is normally distributed with mean 283.3 ml and standard deviation 83.3 ml. Find the percentage of patients whose amount of blood loss during major lower limb surgery using fentanyl is

a. less than 304 ml.

b. between 221 and 429 ml.

c. more than 450 ml.

28. Verbal GRE Scores. The Graduate Record Examination (GRE) is a standardized test that students usually take before entering graduate school. According to the document *GRE Guide to the Use of Scores*, a publication of the Educational Testing Service, the scores on the verbal portion of the GRE have mean 150 points and standard deviation 8.75 points. Assuming that these scores are (approximately) normally distributed,

a. obtain and interpret the quartiles.

b. find and interpret the 99th percentile.

29. Verbal GRE Scores. Refer to Problem 28, and fill in the following blanks.

a. Approximately 68% of students who took the verbal portion of the GRE scored between _____ and _____.

b. Approximately 95% of students who took the verbal portion of the GRE scored between _____ and _____.

c. Approximately 99.7% of students who took the verbal portion of the GRE scored between _____ and _____.

30. Northwest Atlantic Cod. Researchers M. Kroll et al. studied the influence of paternity on rates of mortality and development in eggs and larvae of Northwest Atlantic cod in the article, "Paternal Effects on Early Life History Traits in Northwest Atlantic Cod, *Gadus morhua*" (*Journal of Applied Ichthyology*, Vol. 29, Issue 3, pp. 623–629). A sample of adult male Northwest Atlantic cod was collected in Nova Scotia, Canada. The following table gives the lengths, in centimeters (cm), of the cod sampled.

47.6	65.1	54.6	48.5	58.6	67.9
48.1	62.5	52.4	55.4	57.1	55.8

a. Using Table III in Appendix A, construct a normal probability plot of the given data.

b. Use part (a) to identify any outliers.

c. Use part (a) to assess the normality of the variable under consideration.

31. Mortgage Industry Employees. In an issue of *National Mortgage News*, a special report was published on publicly traded mortgage industry companies. A sample of 25 mortgage industry companies had the following numbers of employees.

260	20,800	1,801	2,073	3,596
3,223	2,128	1,796	17,540	15
29,272	6,929	2,468	7,000	6,600
2,458	3,216	209	726	9,200
650	4,800	19,400	24,886	3,082

a. Obtain a normal probability plot of the data.

b. Use part (a) to identify any outliers.

c. Use part (a) to assess the normality of the variable under consideration.

*32. Diarrhea Vaccine. Acute rotavirus diarrhea is the leading cause of death among children under age 5, killing an estimated 4.5 million annually in developing countries. Scientists from Finland and Belgium claim that a new oral vaccine is 80% effective against rotavirus diarrhea. Assuming that the claim is correct, use the normal approximation to the binomial distribution to find the probability that, out of 1500 cases, the vaccine will be effective in

a. exactly 1225 cases.   **b.** at least 1175 cases.

c. between 1150 and 1250 cases, inclusive.

FOCUSING ON DATA ANALYSIS

UWEC UNDERGRADUATES

Recall from Chapter 1 (see page 34) that the Focus database and Focus sample contain information on the undergraduate students at the University of Wisconsin - Eau Claire (UWEC). Now would be a good time for you to review the discussion about these data sets.

Begin by opening the Focus sample (FocusSample) in the statistical software package of your choice.

a. Obtain a normal probability plot of the sample data for each of the following variables: high school percentile, cumulative GPA, age, total earned credits, ACT English score, ACT math score, and ACT composite score.

b. Based on your results from part (a), which of the variables considered there appear to be approximately normally distributed?

c. Based on your results from part (a), which of the variables considered there appear to be far from normally distributed?

If your statistical software package will accommodate the entire Focus database (Focus), open that worksheet.

d. Obtain a histogram for each of the following variables: high school percentile, cumulative GPA, age, total earned credits, ACT English score, ACT math score, and ACT composite score.

e. In view of the histograms that you obtained in part (d), comment on your answers in parts (b) and (c).

CASE STUDY DISCUSSION

CHEST SIZES OF SCOTTISH MILITIAMEN

On page 263, we presented a frequency distribution for data on chest circumference, in inches, for 5732 Scottish militiamen. As mentioned there, Adolphe Quetelet used a procedure for fitting a normal curve to the data based on the binomial distribution. Here you are to accomplish that task by using techniques that you studied in this chapter.

a. Construct a relative-frequency histogram for the chest circumference data, using classes based on a single value.

b. The population mean and population standard deviation of the chest circumferences are 39.85 and 2.07, respectively. Identify the normal curve that should be used for the chest circumferences.

c. Use the table on page 263 to find the percentage of militiamen in the survey with chest circumference between 36 and 41 inches, inclusive. *Note:* As the circumferences were rounded to the nearest inch, you are actually finding the percentage of militiamen in the survey with chest circumference between 35.5 and 41.5 inches.

d. Use the normal curve you identified in part (b) to obtain an approximation to the percentage of militiamen in the survey with chest circumference between 35.5 and 41.5 inches. Compare your answer to the exact percentage found in part (c).

BIOGRAPHY

CARL FRIEDRICH GAUSS: CHILD PRODIGY

Carl Friedrich Gauss was born on April 30, 1777, in Brunswick, Germany, the only son in a poor, semiliterate peasant family; he taught himself to calculate before he could talk. At the age of 3, he pointed out an error in his father's calculations of wages. In addition to his arithmetic experimentation, he taught himself to read. At the age of 8, Gauss instantly solved the summing of all numbers from 1 to 100. His father was persuaded to allow him to stay in school and to study after school instead of working to help support the family.

Impressed by Gauss's brilliance, the Duke of Brunswick supported him monetarily from the ages of 14 to 30. This patronage permitted Gauss to pursue his studies exclusively. He conceived most of his mathematical discoveries by the time he was 17. Gauss was granted a doctorate in absentia from the university at Helmstedt; his doctoral thesis developed the concept of complex numbers and proved the fundamental theorem of algebra, which had previously been only partially established. Shortly thereafter, Gauss published his theory of numbers, which is considered one of the most brilliant achievements in mathematics.

Gauss made important discoveries in mathematics, physics, astronomy, and statistics. Two of his major contributions to statistics were the development of the least-squares method and fundamental work with the normal distribution, often called the *Gaussian distribution* in his honor.

In 1807, Gauss accepted the directorship of the observatory at the University of Göttingen, which ended his dependence on the Duke of Brunswick. He remained there the rest of his life. In 1833, Gauss and a colleague, Wilhelm Weber, invented a working electric telegraph, 5 years before Samuel Morse. Gauss died in Göttingen in 1855.

The Sampling Distribution of the Sample Mean

CHAPTER OBJECTIVES

In the preceding chapters, you have studied sampling, descriptive statistics, probability, and the normal distribution. Now you will learn how these seemingly diverse topics can be integrated to lay the groundwork for inferential statistics.

In Section 7.1, we introduce the concepts of *sampling error* and *sampling distribution* and explain the essential role these concepts play in the design of inferential studies. The *sampling distribution* of a statistic is the distribution of the statistic, that is, the distribution of all possible observations of the statistic for samples of a given size from a population. In this chapter, we concentrate on the sampling distribution of the sample mean.

In Sections 7.2 and 7.3, we provide the required background for applying the sampling distribution of the sample mean. Specifically, in Section 7.2, we present formulas for the mean and standard deviation of the sample mean. Then, in Section 7.3, we indicate that, under certain general conditions, the sampling distribution of the sample mean is a normal distribution, or at least approximately so.

We apply this momentous fact in Chapters 8 and 9 to develop two important statistical-inference procedures: using the mean, $\bar{x}$, of a sample from a population to estimate and to draw conclusions about the mean, μ, of the entire population.

CASE STUDY

The Chesapeake and Ohio Freight Study

Can relatively small samples actually provide results that are nearly as accurate as those obtained from a census? Statisticians have proven that such is the case, but a real study with sample and census results can be enlightening.

When a freight shipment travels over several railroads, the revenue from the freight charge is appropriately divided among those railroads. A *waybill*, which accompanies each freight shipment, provides information on the goods, route, and total charges. From the waybill, the amount due each railroad can be calculated.

Calculating these allocations for a large number of shipments is time consuming and costly. If the division of total revenue to the railroads could be done accurately on the basis of a sample—as statisticians contend—considerable savings could be realized in accounting and clerical costs.

To convince themselves of the validity of the sampling approach, officials of the Chesapeake and

Ohio Railroad Company (C&O) undertook a study of freight shipments that had traveled over its Pere Marquette district and another railroad during a 6-month period. The total number of waybills for that period (22,984) and the total freight revenue were known.

The study used statistical theory to determine the smallest number of waybills needed to estimate, with a prescribed accuracy, the total freight revenue due C&O. In all, 2072 of the 22,984 waybills, roughly 9%, were sampled. For each waybill in the sample, the amount of freight revenue due C&O was calculated and, from those amounts, the total revenue due C&O was estimated to be $64,568.

How close was the estimate of $64,568, based on a sample of only 2072 waybills, to the total revenue actually due C&O for the 22,984 waybills? Take a guess! We'll discuss the answer at the end of this chapter.

7.1 Sampling Error; the Need for Sampling Distributions

We have already seen that using a sample to acquire information about a population is often preferable to conducting a census. Generally, sampling is less costly and can be done more quickly than a census; it is often the only practical way to gather information.

However, because a sample provides data for only a portion of an entire population, we cannot expect the sample to yield perfectly accurate information about the population. Thus we should anticipate that a certain amount of error—called *sampling error*—will result simply because we are sampling.

DEFINITION 7.1

Sampling Error

Sampling error is the error resulting from using a sample to estimate a population characteristic.

EXAMPLE 7.1   **Sampling Error and the Need for Sampling Distributions**

Income Tax   The Internal Revenue Service (IRS) publishes annual figures on individual income tax returns in *Statistics of Income, Individual Income Tax Returns*. For the year 2010, the IRS reported that the mean tax of individual income tax returns was $11,266. In actuality, the IRS reported the mean tax of a sample of 308,946 individual income tax returns from a total of more than 130 million such returns.

  a.   Identify the population under consideration.
  b.   Identify the variable under consideration.
  c.   Is the mean tax reported by the IRS a sample mean or the population mean?
  d.   Should we expect the mean tax, $\bar{x}$, of the 308,946 returns sampled by the IRS to be exactly the same as the mean tax, μ, of all individual income tax returns for 2010?
  e.   How can we answer questions about sampling error? For instance, is the sample mean tax, $\bar{x}$, reported by the IRS likely to be within $100 of the population mean tax, μ?

Solution

  a.   The population consists of all individual income tax returns for the year 2010.
  b.   The variable is "tax" (amount of income tax).

c. The mean tax reported is a sample mean, namely, the mean tax, $\bar{x}$, of the 308,946 returns sampled. It is not the population mean tax, μ, of all individual income tax returns for 2010.

d. We certainly cannot expect the mean tax, $\bar{x}$, of the 308,946 returns sampled by the IRS to be exactly the same as the mean tax, μ, of all individual income tax returns for 2010—some sampling error is to be anticipated.

e. To answer questions about sampling error, we need to know the distribution of all possible sample mean tax amounts (i.e., all possible $\bar{x}$-values) that could be obtained by sampling 308,946 individual income tax returns. That distribution is called the *sampling distribution of the sample mean.*

The distribution of a statistic (i.e., of all possible observations of the statistic for samples of a given size) is called the **sampling distribution** of the statistic. In this chapter, we concentrate on the *sampling distribution of the sample mean*, that is, of the statistic $\bar{x}$.

DEFINITION 7.2

? What Does It Mean?

The sampling distribution of the sample mean is the distribution of all possible sample means for samples of a given size.

Sampling Distribution of the Sample Mean

For a variable x and a given sample size, the distribution of the variable $\bar{x}$ is called the **sampling distribution of the sample mean.**

In statistics, the following terms and phrases are synonymous.

• Sampling distribution of the sample mean
• Distribution of the variable $\bar{x}$
• Distribution of all possible sample means of a given sample size

We, therefore, use these three terms interchangeably.

Introducing the sampling distribution of the sample mean with an example that is both realistic and concrete is difficult because even for moderately large populations the number of possible samples is enormous, thus prohibiting an actual listing of the possibilities. For example, the number of possible samples of size 50 from a population of size 10,000 is about 3×10^{135}, a 3 followed by 135 zeros.

Consequently, we use an unrealistically small population to introduce the sampling distribution of the mean. Note that the emphasis here is not on learning to do some particular task, but more on understanding the concept of sampling distributions.

EXAMPLE 7.2   Sampling Distribution of the Sample Mean

Heights of Starting Players Suppose that the population of interest consists of the five starting players on a men's basketball team, who we will call A, B, C, D, and E. Further suppose that the variable of interest is height, in inches. Table 7.1 lists the players and their heights.

TABLE 7.1

Heights, in inches, of the five starting players

Player	A	B	C	D	E
Height	76	78	79	81	86

a. Obtain the sampling distribution of the sample mean for samples of size 2.

b. Make some observations about sampling error when the mean height of a random sample of two players is used to estimate the population mean height.[†]

c. Find the probability that, for a random sample of size 2, the sampling error made in estimating the population mean by the sample mean will be 1 inch or less; that is, determine the probability that $\bar{x}$ will be within 1 inch of μ.

[†]As we mentioned in Section 1.2, the statistical-inference techniques considered in this book are intended for use only with simple random sampling. Therefore, unless otherwise specified, when we say *random sample,* we mean *simple random sample.* Furthermore, we assume that sampling is without replacement unless explicitly stated otherwise.

TABLE 7.2

Possible samples and sample means for samples of size 2

Sample	Heights	$\bar{x}$
A, B	76, 78	77.0
A, C	76, 79	77.5
A, D	76, 81	78.5
A, E	76, 86	81.0
B, C	78, 79	78.5
B, D	78, 81	79.5
B, E	78, 86	82.0
C, D	79, 81	80.0
C, E	79, 86	82.5
D, E	81, 86	83.5

Solution For future reference, we first compute the population mean height:

$$\mu = \frac{\Sigma x_i}{N} = \frac{76 + 78 + 79 + 81 + 86}{5} = 80 \text{ inches.}$$

a. The population is so small that we can list the possible samples of size 2. The first column of Table 7.2 gives the 10 possible samples, the second column the corresponding heights (values of the variable "height"), and the third column the sample means. Figure 7.1 is a dotplot for the distribution of the sample means (the sampling distribution of the sample mean for samples of size 2).

b. From Table 7.2 or Fig. 7.1, we see that the mean height of the two players selected isn't likely to equal the population mean of 80 inches. In fact, only 1 of the 10 samples has a mean of 80 inches, the eighth sample in Table 7.2. The chances are, therefore, only $\frac{1}{10}$, or 10%, that $\bar{x}$ will equal μ; some sampling error is likely.

c. Figure 7.1 shows that 3 of the 10 samples have means within 1 inch of the population mean of 80 inches (i.e., between 79 and 81 inches, inclusive). So the probability is $\frac{3}{10}$, or 0.3, that the sampling error made in estimating μ by $\bar{x}$ will be 1 inch or less.

FIGURE 7.1

Dotplot for the sampling distribution of the sample mean for samples of size 2 ($n = 2$)

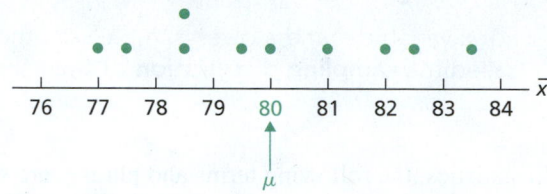

Exercise 7.11 on page 312

Interpretation There is a 30% chance that the mean height of the two players selected will be within 1 inch of the population mean.

In the previous example, we determined the sampling distribution of the sample mean for samples of size 2. If we consider samples of another size, we obtain a different sampling distribution of the sample mean, as demonstrated in the next example.

EXAMPLE 7.3 Sampling Distribution of the Sample Mean

Heights of Starting Players Refer to Table 7.1 on the preceding page, which gives the heights of the five starting players on a men's basketball team.

TABLE 7.3

Possible samples and sample means for samples of size 4

Sample	Heights	$\bar{x}$
A, B, C, D	76, 78, 79, 81	78.50
A, B, C, E	76, 78, 79, 86	79.75
A, B, D, E	76, 78, 81, 86	80.25
A, C, D, E	76, 79, 81, 86	80.50
B, C, D, E	78, 79, 81, 86	81.00

a. Obtain the sampling distribution of the sample mean for samples of size 4.

b. Make some observations about sampling error when the mean height of a random sample of four players is used to estimate the population mean height.

c. Find the probability that, for a random sample of size 4, the sampling error made in estimating the population mean by the sample mean will be 1 inch or less; that is, determine the probability that $\bar{x}$ will be within 1 inch of μ.

Solution

a. There are five possible samples of size 4. The first column of Table 7.3 gives the possible samples, the second column the corresponding heights (values of the variable "height"), and the third column the sample means. Figure 7.2 is a dotplot for the distribution of the sample means.

FIGURE 7.2

Dotplot for the sampling distribution of the sample mean for samples of size 4 ($n = 4$)

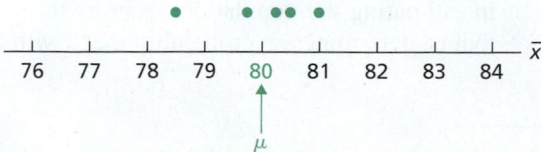

b. From Table 7.3 or Fig. 7.2, we see that none of the samples of size 4 has a mean equal to the population mean of 80 inches. Thus, some sampling error is certain.

Exercise 7.13
on page 312

c. Figure 7.2 shows that four of the five samples have means within 1 inch of the population mean of 80 inches. So the probability is $\frac{4}{5}$, or 0.8, that the sampling error made in estimating μ by $\bar{x}$ will be 1 inch or less.

Interpretation There is an 80% chance that the mean height of the four players selected will be within 1 inch of the population mean.

Sample Size and Sampling Error

We continue our look at the sampling distributions of the sample mean for the heights of the five starting players on a basketball team. In Figs. 7.1 and 7.2, we drew dotplots for the sampling distributions of the sample mean for samples of sizes 2 and 4, respectively. Those two dotplots and dotplots for samples of sizes 1, 3, and 5 are displayed in Fig. 7.3.

FIGURE 7.3

Dotplots for the sampling distributions of the sample mean for the heights of the five starting players for samples of sizes 1, 2, 3, 4, and 5

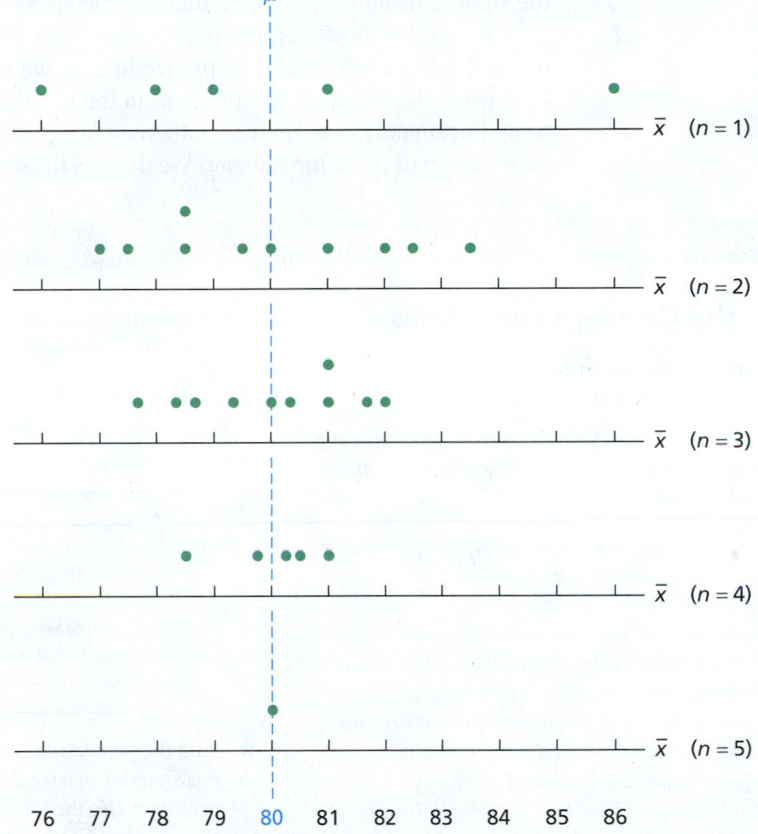

Figure 7.3 vividly illustrates that the possible sample means cluster more closely around the population mean as the sample size increases. This result suggests that sampling error tends to be smaller for large samples than for small samples.

For example, for samples of size 1, Fig. 7.3 reveals that 2 of 5 (40%) of the possible sample means lie within 1 inch of μ. Likewise, for samples of sizes 2, 3, 4, and 5, respectively, 3 of 10 (30%), 5 of 10 (50%), 4 of 5 (80%), and 1 of 1 (100%) of the possible sample means lie within 1 inch of μ. The first four columns of Table 7.4 summarize these results. The last two columns of that table provide other sampling-error results, easily obtained from Fig. 7.3.

TABLE 7.4

Sample size and sampling error illustrations for the heights of the basketball players ("No." is an abbreviation of "Number")

Sample size n	No. possible samples	No. within 1″ of μ	% within 1″ of μ	No. within 0.5″ of μ	% within 0.5″ of μ
1	5	2	40%	0	0%
2	10	3	30%	2	20%
3	10	5	50%	2	20%
4	5	4	80%	3	60%
5	1	1	100%	1	100%

More generally, we can make the following qualitative statement.

Sample Size and Sampling Error

The larger the sample size, the smaller the sampling error tends to be in estimating a population mean, μ, by a sample mean, $\bar{x}$.

What We Do in Practice

We used the heights of a population of five basketball players to illustrate and explain the importance of the sampling distribution of the sample mean. For that small population with known population data, we easily determined the sampling distribution of the sample mean for any particular sample size by listing all possible sample means.

In practice, however, the populations with which we work are large and the population data are unknown, so proceeding as we did in the basketball-player example isn't possible. What do we do, then, in the usual case of a large and unknown population? Fortunately, we can use mathematical relationships to approximate the sampling distribution of the sample mean. We discuss those relationships in Sections 7.2 and 7.3.

Exercises 7.1

Understanding the Concepts and Skills

7.1 Why is sampling often preferable to conducting a census for the purpose of obtaining information about a population?

7.2 Why should you generally expect some error when estimating a parameter (e.g., a population mean) by a statistic (e.g., a sample mean)? What is this kind of error called?

In Exercises 7.3–7.10, we have given population data for a variable. For each exercise, do the following tasks.
a. *Find the mean, μ, of the variable.*
b. *For each of the possible sample sizes, construct a table similar to Table 7.2 on page 310 and draw a dotplot for the sampling distribution of the sample mean similar to Fig. 7.1 on page 310.*
c. *Construct a graph similar to Fig. 7.3 and interpret your results.*
d. *For each of the possible sample sizes, find the probability that the sample mean will equal the population mean.*
e. *For each of the possible sample sizes, find the probability that the sampling error made in estimating the population mean by the sample mean will be 0.5 or less (in magnitude), that is, that the absolute value of the difference between the sample mean and the population mean is at most 0.5.*

7.3 Population data: 1, 2, 3.

7.4 Population data: 2, 5, 8.

7.5 Population data: 1, 2, 3, 4.

7.6 Population data: 3, 4, 7, 8.

7.7 Population data: 1, 2, 3, 4, 5.

7.8 Population data: 2, 3, 5, 7, 8.

7.9 Population data: 1, 2, 3, 4, 5, 6.

7.10 Population data: 2, 3, 5, 5, 7, 8.

Applying the Concepts and Skills

Exercises 7.11–7.23 are intended solely to provide concrete illustrations of the sampling distribution of the sample mean. For that reason,
the populations considered are unrealistically small. In each exercise, assume that sampling is without replacement.

7.11 **NBA Champs.** The winner of the 2012–2013 National Basketball Association (NBA) championship was the Miami Heat. One possible starting lineup for that team is as follows.

Player	Position	Height (in.)
Chris Bosh (B)	Center	83
Dwyane Wade (W)	Guard	76
LeBron James (J)	Forward	80
Mario Chalmers (C)	Guard	74
Udonis Haslem (H)	Forward	80

a. Find the population mean height of the five players.
b. For samples of size 2, construct a table similar to Table 7.2 on page 310. Use the letter in parentheses after each player's name to represent each player.
c. Draw a dotplot for the sampling distribution of the sample mean for samples of size 2.
d. For a random sample of size 2, what is the chance that the sample mean will equal the population mean?
e. For a random sample of size 2, obtain the probability that the sampling error made in estimating the population mean by the sample mean will be 1 inch or less; that is, determine the probability that $\bar{x}$ will be within 1 inch of μ. Interpret your result in terms of percentages.

7.12 **NBA Champs.** Repeat parts (b)–(e) of Exercise 7.11 for samples of size 1.

7.13 **NBA Champs.** Repeat parts (b)–(e) of Exercise 7.11 for samples of size 3.

7.14 **NBA Champs.** Repeat parts (b)–(e) of Exercise 7.11 for samples of size 4.

7.15 **NBA Champs.** Repeat parts (b)–(e) of Exercise 7.11 for samples of size 5.

7.16 NBA Champs. This exercise requires that you have done Exercises 7.11–7.15.

a. Draw a graph similar to that shown in Fig. 7.3 on page 311 for sample sizes of 1, 2, 3, 4, and 5.

b. What does your graph in part (a) illustrate about the impact of increasing sample size on sampling error?

c. Construct a table similar to Table 7.4 on page 311 for some values of your choice.

7.17 America's Richest. Each year, *Forbes* magazine publishes a list of the richest people in the United States. As of September 16, 2013, the six richest Americans and their wealth (to the nearest billion dollars) are as shown in the following table. Consider these six people a population of interest.

Person	Wealth ($ billions)
Bill Gates (G)	72
Warren Buffett (B)	59
Larry Ellison (E)	41
Charles Koch (C)	36
David Koch (D)	36
Christy Walton (W)	35

a. Calculate the mean wealth, μ, of the six people.

b. For samples of size 2, construct a table similar to Table 7.2 on page 310. (There are 15 possible samples of size 2.)

c. Draw a dotplot for the sampling distribution of the sample mean for samples of size 2.

d. For a random sample of size 2, what is the chance that the sample mean will equal the population mean?

e. For a random sample of size 2, determine the probability that the mean wealth of the two people obtained will be within 3 (i.e., $3 billion) of the population mean. Interpret your result in terms of percentages.

7.18 America's Richest. Repeat parts (b)–(e) of Exercise 7.17 for samples of size 1.

7.19 America's Richest. Repeat parts (b)–(e) of Exercise 7.17 for samples of size 3. (There are 20 possible samples.)

7.20 America's Richest. Repeat parts (b)–(e) of Exercise 7.17 for samples of size 4. (There are 15 possible samples.)

7.21 America's Richest. Repeat parts (b)–(e) of Exercise 7.17 for samples of size 5. (There are six possible samples.)

7.22 America's Richest. Repeat parts (b)–(e) of Exercise 7.17 for samples of size 6. What is the relationship between the only possible sample here and the population?

7.23 America's Richest. Explain what the dotplots in part (c) of Exercises 7.17–7.22 illustrate about the impact of increasing sample size on sampling error.

Extending the Concepts and Skills

7.24 Suppose that a sample is to be taken without replacement from a finite population of size N. If the sample size is the same as the population size,

a. how many possible samples are there?

b. what are the possible sample means?

c. what is the relationship between the only possible sample and the population?

7.25 Suppose that a random sample of size 1 is to be taken from a finite population of size N.

a. How many possible samples are there?

b. Identify the relationship between the possible sample means and the possible observations of the variable under consideration.

c. What is the difference between taking a random sample of size 1 from a population and selecting a member at random from the population?

7.2   The Mean and Standard Deviation of the Sample Mean

In Section 7.1, we discussed the sampling distribution of the sample mean—the distribution of all possible sample means for any specified sample size or, equivalently, the distribution of the variable $\bar{x}$. We use that distribution to make inferences about the population mean based on a sample mean.

As we said earlier, we generally do not know the sampling distribution of the sample mean exactly. Fortunately, however, we can often approximate that sampling distribution by a normal distribution; that is, under certain conditions, the variable $\bar{x}$ is approximately normally distributed.

Recall that a variable is normally distributed if its distribution has the shape of a normal curve and that a normal distribution is determined by the mean and standard deviation. Hence a first step in learning how to approximate the sampling distribution of the sample mean by a normal distribution is to obtain the mean and standard deviation of the sample mean, that is, of the variable $\bar{x}$. We describe how to do that in this section.

To begin, let's review the notation used for the mean and standard deviation of a variable. Recall that the mean of a variable is denoted μ, subscripted if necessary with the letter representing the variable. So the mean of x is written as μ_x, the mean of y as μ_y, and so on. In particular, then, the mean of $\bar{x}$ is written as $\mu_{\bar{x}}$; similarly, the standard deviation of $\bar{x}$ is written as $\sigma_{\bar{x}}$.

The Mean of the Sample Mean

There is a simple relationship between the mean of the variable $\bar{x}$ and the mean of the variable under consideration: They are equal, or $\mu_{\bar{x}} = \mu$. In other words, for any particular sample size, the mean of all possible sample means equals the population mean. This equality holds regardless of the size of the sample. In Example 7.4, we illustrate the relationship $\mu_{\bar{x}} = \mu$ by returning to the heights of the basketball players considered in Section 7.1.

EXAMPLE 7.4 **Mean of the Sample Mean**

TABLE 7.5

Heights, in inches, of the five starting players

Player	A	B	C	D	E
Height	76	78	79	81	86

Heights of Starting Players The heights, in inches, of the five starting players on a men's basketball team are repeated in Table 7.5. Here the population is the five players and the variable is height.

a. Determine the population mean, μ.
b. Obtain the mean, $\mu_{\bar{x}}$, of the variable $\bar{x}$ for samples of size 2. Verify that the relation $\mu_{\bar{x}} = \mu$ holds.
c. Repeat part (b) for samples of size 4.

Solution

a. To determine the population mean (the mean of the variable "height"), we apply Definition 3.11 on page 140 to the heights in Table 7.5:

$$\mu = \frac{\Sigma x_i}{N} = \frac{76 + 78 + 79 + 81 + 86}{5} = 80 \text{ inches.}$$

Thus the mean height of the five players is 80 inches.

b. To obtain the mean of the variable $\bar{x}$ for samples of size 2, we again apply Definition 3.11, but this time to $\bar{x}$. Referring to the third column of Table 7.2 on page 310, we get

$$\mu_{\bar{x}} = \frac{77.0 + 77.5 + \cdots + 83.5}{10} = 80 \text{ inches.}$$

By part (a), $\mu = 80$ inches. So, for samples of size 2, $\mu_{\bar{x}} = \mu$.

Interpretation For samples of size 2, the mean of all possible sample means equals the population mean.

APPLET

Applet 7.1

You try it!

Exercise 7.41 on page 317

c. Proceeding as in part (b), but this time referring to the third column of Table 7.3 on page 310, we obtain the mean of the variable $\bar{x}$ for samples of size 4:

$$\mu_{\bar{x}} = \frac{78.50 + 79.75 + 80.25 + 80.50 + 81.00}{5} = 80 \text{ inches,}$$

which again is the same as μ.

Interpretation For samples of size 4, the mean of all possible sample means equals the population mean.

For emphasis, we restate the relationship $\mu_{\bar{x}} = \mu$ in Formula 7.1.

What Does It Mean?

For any sample size, the mean of all possible sample means equals the population mean.

FORMULA 7.1 **Mean of the Sample Mean**

For samples of size *n*, the mean of the variable $\bar{x}$ equals the mean of the variable under consideration. In symbols,

$$\mu_{\bar{x}} = \mu.$$

The Standard Deviation of the Sample Mean

Next, we investigate the standard deviation of the variable $\bar{x}$ to discover any relationship it has to the standard deviation of the variable under consideration. We begin by returning to the basketball players.

EXAMPLE 7.5 ### Standard Deviation of the Sample Mean

Heights of Starting Players Refer back to Table 7.5.

a. Determine the population standard deviation, σ.
b. Obtain the standard deviation, $\sigma_{\bar{x}}$, of the variable $\bar{x}$ for samples of size 2. Indicate any apparent relationship between $\sigma_{\bar{x}}$ and σ.
c. Repeat part (b) for samples of sizes 1, 3, 4, and 5.
d. Summarize and discuss the results obtained in parts (a)–(c).

Solution

a. To determine the population standard deviation (the standard deviation of the variable "height"), we apply Definition 3.12 on page 142 to the heights in Table 7.5. Recalling that $\mu = 80$ inches, we have

$$\sigma = \sqrt{\frac{\Sigma(x_i - \mu)^2}{N}}$$

$$= \sqrt{\frac{(76 - 80)^2 + (78 - 80)^2 + (79 - 80)^2 + (81 - 80)^2 + (86 - 80)^2}{5}}$$

$$= \sqrt{\frac{16 + 4 + 1 + 1 + 36}{5}} = \sqrt{11.6} = 3.41 \text{ inches.}$$

Thus the standard deviation of the heights of the five players is 3.41 inches.

b. To obtain the standard deviation of the variable $\bar{x}$ for samples of size 2, we again apply Definition 3.12, but this time to $\bar{x}$. Referring to the third column of Table 7.2 on page 310 and recalling that $\mu_{\bar{x}} = \mu = 80$ inches, we have

$$\sigma_{\bar{x}} = \sqrt{\frac{(77.0 - 80)^2 + (77.5 - 80)^2 + \cdots + (83.5 - 80)^2}{10}}$$

$$= \sqrt{\frac{9.00 + 6.25 + \cdots + 12.25}{10}} = \sqrt{4.35} = 2.09 \text{ inches,}$$

to two decimal places. Note that this result is not the same as the population standard deviation, which is $\sigma = 3.41$ inches. Also note that $\sigma_{\bar{x}}$ is smaller than σ.

c. Using the same procedure as in part (b), we compute $\sigma_{\bar{x}}$ for samples of sizes 1, 3, 4, and 5 and summarize the results in Table 7.6.

d. Table 7.6 suggests that the standard deviation of $\bar{x}$ gets smaller as the sample size gets larger. We could have predicted this result from the dotplots shown in Fig. 7.3 on page 311 and the fact that the standard deviation of a variable measures the variation of its possible values.

TABLE 7.6

The standard deviation of $\bar{x}$ for sample sizes 1, 2, 3, 4, and 5

Sample size n	Standard deviation of $\bar{x}$ $\sigma_{\bar{x}}$
1	3.41
2	2.09
3	1.39
4	0.85
5	0.00

Example 7.5 provides evidence that the standard deviation of $\bar{x}$ gets smaller as the sample size gets larger; that is, the variation of all possible sample means decreases as the sample size increases. The question now is whether there is a formula that relates the standard deviation of $\bar{x}$ to the sample size and standard deviation of the population. The answer is yes! In fact, two different formulas express the precise relationship.

When sampling is done without replacement from a finite population, as in Example 7.5, the appropriate formula is

$$\sigma_{\bar{x}} = \sqrt{\frac{N - n}{N - 1}} \cdot \frac{\sigma}{\sqrt{n}}, \tag{7.1}$$

where, as usual, n denotes the sample size and N the population size. When sampling is done with replacement from a finite population or when it is done from an infinite population, the appropriate formula is

$$\sigma_{\bar{x}} = \frac{\sigma}{\sqrt{n}}. \tag{7.2}$$

APPLET

Applet 7.2

When the sample size is small relative to the population size, there is little difference between sampling with and without replacement.[†] So, in such cases, the two formulas for $\sigma_{\bar{x}}$ yield almost the same numbers. In most practical applications, the sample size is small relative to the population size, so in this book, we use the second formula only (with the understanding that the equality may be approximate).

FORMULA 7.2

? What Does It Mean?

For each sample size, the standard deviation of all possible sample means equals the population standard deviation divided by the square root of the sample size.

Standard Deviation of the Sample Mean

For samples of size n, the standard deviation of the variable $\bar{x}$ equals the standard deviation of the variable under consideration divided by the square root of the sample size. In symbols,

$$\sigma_{\bar{x}} = \frac{\sigma}{\sqrt{n}}.$$

Note: In the formula for the standard deviation of $\bar{x}$, the sample size, n, appears in the denominator. This explains mathematically why the standard deviation of $\bar{x}$ decreases as the sample size increases.

Applying the Formulas

We have shown that simple formulas relate the mean and standard deviation of $\bar{x}$ to the mean and standard deviation of the population, namely, $\mu_{\bar{x}} = \mu$ and $\sigma_{\bar{x}} = \sigma/\sqrt{n}$ (at least approximately). We apply those formulas next.

■ ■ ■ **EXAMPLE 7.6**   **Mean and Standard Deviation of the Sample Mean**

Living Space of Homes As reported by the U.S. Census Bureau in *Current Housing Reports*, the mean living space for single-family detached homes is 1742 sq. ft. Assume a standard deviation of 568 sq. ft.

a. For samples of 25 single-family detached homes, determine the mean and standard deviation of the variable $\bar{x}$.

b. Repeat part (a) for a sample of size 500.

Solution Here the variable is living space, and the population consists of all single-family detached homes in the United States. From the given information, we know that $\mu = 1742$ sq. ft. and $\sigma = 568$ sq. ft.

a. We use Formula 7.1 (page 314) and Formula 7.2 to get

$$\mu_{\bar{x}} = \mu = 1742 \quad \text{and} \quad \sigma_{\bar{x}} = \frac{\sigma}{\sqrt{n}} = \frac{568}{\sqrt{25}} = 113.6.$$

b. We again use Formula 7.1 and Formula 7.2 to get

$$\mu_{\bar{x}} = \mu = 1742 \quad \text{and} \quad \sigma_{\bar{x}} = \frac{\sigma}{\sqrt{n}} = \frac{568}{\sqrt{500}} = 25.4.$$

You try it!

Exercise 7.47 on page 318

Interpretation For samples of 25 single-family detached homes, the mean and standard deviation of all possible sample mean living spaces are 1742 sq. ft. and 113.6 sq. ft., respectively. For samples of 500, these numbers are 1742 sq. ft. and 25.4 sq. ft., respectively.

■

[†]As a rule of thumb, we say that the sample size is small relative to the population size if the size of the sample does not exceed 5% of the size of the population ($n \leq 0.05N$).

Sample Size and Sampling Error (Revisited)

Key Fact 7.1 states that the possible sample means cluster more closely around the population mean as the sample size increases, and therefore the larger the sample size, the smaller the sampling error tends to be in estimating a population mean by a sample mean. Here is why that key fact is true.

- The larger the sample size, the smaller is the standard deviation of $\bar{x}$.
- The smaller the standard deviation of $\bar{x}$, the more closely the possible values of $\bar{x}$ (the possible sample means) cluster around the mean of $\bar{x}$.
- The mean of $\bar{x}$ equals the population mean.

Because the standard deviation of $\bar{x}$ determines the amount of sampling error to be expected when a population mean is estimated by a sample mean, it is often referred to as the **standard error of the sample mean.** In general, the standard deviation of a statistic used to estimate a parameter is called the **standard error (SE)** of the statistic.

Exercises 7.2

Understanding the Concepts and Skills

7.26 Although, in general, you cannot know the sampling distribution of the sample mean exactly, by what distribution can you often approximate it?

7.27 Why is obtaining the mean and standard deviation of $\bar{x}$ a first step in approximating the sampling distribution of the sample mean by a normal distribution?

7.28 Does the sample size have an effect on the mean of all possible sample means? Explain your answer.

7.29 Does the sample size have an effect on the standard deviation of all possible sample means? Explain your answer.

7.30 Explain why increasing the sample size tends to result in a smaller sampling error when a sample mean is used to estimate a population mean.

7.31 What is another name for the standard deviation of the variable $\bar{x}$? What is the reason for that name?

7.32 You have seen that the larger the sample size, the smaller the sampling error tends to be in estimating a population mean by a sample mean. This fact is reflected mathematically by the formula for the standard deviation of the sample mean: $\sigma_{\bar{x}} = \sigma/\sqrt{n}$. For a fixed sample size, explain what this formula implies about the relationship between the population standard deviation and sampling error.

Exercises 7.33–7.40 require that you have done Exercises 7.3–7.10, respectively.

7.33 Refer to Exercise 7.3 on page 312.
a. Use your answers from Exercise 7.3(b) to determine the mean, $\mu_{\bar{x}}$, of the variable $\bar{x}$ for each of the possible sample sizes.
b. For each of the possible sample sizes, determine the mean, $\mu_{\bar{x}}$, of the variable $\bar{x}$, using only your answer from Exercise 7.3(a).

7.34 Refer to Exercise 7.4 on page 312.
a. Use your answers from Exercise 7.4(b) to determine the mean, $\mu_{\bar{x}}$, of the variable $\bar{x}$ for each of the possible sample sizes.
b. For each of the possible sample sizes, determine the mean, $\mu_{\bar{x}}$, of the variable $\bar{x}$, using only your answer from Exercise 7.4(a).

7.35 Refer to Exercise 7.5 on page 312.
a. Use your answers from Exercise 7.5(b) to determine the mean, $\mu_{\bar{x}}$, of the variable $\bar{x}$ for each of the possible sample sizes.
b. For each of the possible sample sizes, determine the mean, $\mu_{\bar{x}}$, of the variable $\bar{x}$, using only your answer from Exercise 7.5(a).

7.36 Refer to Exercise 7.6 on page 312.
a. Use your answers from Exercise 7.6(b) to determine the mean, $\mu_{\bar{x}}$, of the variable $\bar{x}$ for each of the possible sample sizes.
b. For each of the possible sample sizes, determine the mean, $\mu_{\bar{x}}$, of the variable $\bar{x}$, using only your answer from Exercise 7.6(a).

7.37 Refer to Exercise 7.7 on page 312.
a. Use your answers from Exercise 7.7(b) to determine the mean, $\mu_{\bar{x}}$, of the variable $\bar{x}$ for each of the possible sample sizes.
b. For each of the possible sample sizes, determine the mean, $\mu_{\bar{x}}$, of the variable $\bar{x}$, using only your answer from Exercise 7.7(a).

7.38 Refer to Exercise 7.8 on page 312.
a. Use your answers from Exercise 7.8(b) to determine the mean, $\mu_{\bar{x}}$, of the variable $\bar{x}$ for each of the possible sample sizes.
b. For each of the possible sample sizes, determine the mean, $\mu_{\bar{x}}$, of the variable $\bar{x}$, using only your answer from Exercise 7.8(a).

7.39 Refer to Exercise 7.9 on page 312.
a. Use your answers from Exercise 7.9(b) to determine the mean, $\mu_{\bar{x}}$, of the variable $\bar{x}$ for each of the possible sample sizes.
b. For each of the possible sample sizes, determine the mean, $\mu_{\bar{x}}$, of the variable $\bar{x}$, using only your answer from Exercise 7.9(a).

7.40 Refer to Exercise 7.10 on page 312.
a. Use your answers from Exercise 7.10(b) to determine the mean, $\mu_{\bar{x}}$, of the variable $\bar{x}$ for each of the possible sample sizes.
b. For each of the possible sample sizes, determine the mean, $\mu_{\bar{x}}$, of the variable $\bar{x}$, using only your answer from Exercise 7.10(a).

Applying the Concepts and Skills

Exercises 7.41–7.45 require that you have done Exercises 7.11–7.15, respectively.

7.41 **NBA Champs.** The winner of the 2012–2013 National Basketball Association (NBA) championship was the Miami Heat. One possible starting lineup for that team is as follows.

Player	Position	Height (in.)
Chris Bosh (B)	Center	83
Dwyane Wade (W)	Guard	76
LeBron James (J)	Forward	80
Mario Chalmers (C)	Guard	74
Udonis Haslem (H)	Forward	80

a. Determine the population mean height, μ, of the five players.

b. Consider samples of size 2 without replacement. Use your answer to Exercise 7.11(b) on page 312 and Definition 3.11 on page 140 to find the mean, $\mu_{\bar{x}}$, of the variable $\bar{x}$.

c. Find $\mu_{\bar{x}}$, using only the result of part (a).

7.42 NBA Champs. Repeat parts (b) and (c) of Exercise 7.41 for samples of size 1. For part (b), use your answer to Exercise 7.12(b).

7.43 NBA Champs. Repeat parts (b) and (c) of Exercise 7.41 for samples of size 3. For part (b), use your answer to Exercise 7.13(b).

7.44 NBA Champs. Repeat parts (b) and (c) of Exercise 7.41 for samples of size 4. For part (b), use your answer to Exercise 7.14(b).

7.45 NBA Champs. Repeat parts (b) and (c) of Exercise 7.41 for samples of size 5. For part (b), use your answer to Exercise 7.15(b).

7.46 Young Adults at Risk. Research by R. Pyhala et al. shows that young adults who were born prematurely with very low birth weights (below 1500 grams) have higher blood pressure than those born at term. The study can be found in the article, "Blood Pressure Responses to Physiological Stress in Young Adults with Very Low Birth Weight" (*Pediatrics*, Vol. 123, No. 2, pp. 731–734). The researchers found that systolic blood pressures of young adults who were born prematurely with very low birth weights have mean 120.7 mm Hg and standard deviation 13.8 mm Hg.

a. Identify the population and variable.

b. For samples of 30 young adults who were born prematurely with very low birth weights, find the mean and standard deviation of all possible sample mean systolic blood pressures. Interpret your results in words.

c. Repeat part (b) for samples of size 90.

7.47 Baby Weight. The paper "Are Babies Normal?" by T. Clemons and M. Pagano (*The American Statistician*, Vol. 53, No. 4, pp. 298–302) focused on birth weights of babies. According to the article, the mean birth weight is 3369 grams (7 pounds, 6.5 ounces) with a standard deviation of 581 grams.

a. Identify the population and variable.

b. For samples of size 200, find the mean and standard deviation of all possible sample mean weights.

c. Repeat part (b) for samples of size 400.

7.48 Menopause in Mexico. In the article "Age at Menopause in Puebla, Mexico" (*Human Biology*, Vol. 75, No. 2, pp. 205–206), authors L. Sievert and S. Hautaniemi compared the age of menopause for different populations. Menopause, the last menstrual period, is a universal phenomenon among females. According to the article, the mean age of menopause, surgical or natural, in Puebla, Mexico is 44.8 years with a standard deviation of 5.87 years. Let $\bar{x}$ denote the mean age of menopause for a sample of females in Puebla, Mexico.

a. For samples of size 40, find the mean and standard deviation of $\bar{x}$. Interpret your results in words.

b. Repeat part (a) with $n = 120$.

7.49 Mobile Homes. According to the U.S. Census Bureau publication *Manufactured Housing Statistics*, the mean price of new mobile homes is $65,100. Assume a standard deviation of $7200. Let $\bar{x}$ denote the mean price of a sample of new mobile homes.

a. For samples of size 50, find the mean and standard deviation of $\bar{x}$. Interpret your results in words.

b. Repeat part (a) with $n = 100$.

7.50 Undergraduate Binge Drinking. Alcohol consumption on college and university campuses has gained attention because undergraduate students drink significantly more than young adults who are not students. Researchers I. Balodis et al. studied binge drinking in undergraduates in the article "Binge Drinking in Undergraduates: Relationships with Gender, Drinking Behaviors, Impulsivity, and the Perceived Effects of Alcohol" (*Behavioural Pharmacology*, Vol. 20, No. 5, pp. 518–526). The researchers found that students who are binge drinkers drink many times a month with the span of each outing having a mean of 4.9 hours and a standard deviation of 1.1 hours.

a. For samples of size 40, find the mean and standard deviation of all possible sample mean spans of binge drinking episodes. Interpret your results in words.

b. Repeat part (a) with $n = 120$.

7.51 Earthquakes. According to *The Earth: Structure, Composition and Evolution* (The Open University, S237), for earthquakes with a magnitude of 7.5 or greater on the Richter scale, the time between successive earthquakes has a mean of 437 days and a standard deviation of 399 days. Suppose that you observe a sample of four times between successive earthquakes that have a magnitude of 7.5 or greater on the Richter scale.

a. On average, what would you expect to be the mean of the four times?

b. How much variation would you expect from your answer in part (a)? (*Hint:* Use the three-standard-deviations rule.)

Working with Large Data Sets

7.52 Bachelor's Completion. As reported by the U.S. Census Bureau in *Educational Attainment in the United States*, the percentage of adults in each state who have completed a bachelor's degree is provided on the WeissStats site. Use the technology of your choice to solve the following problems.

a. Obtain the standard deviation of the variable "percentage of adults who have completed a bachelor's degree" for the population of 50 states.

b. Consider simple random samples without replacement from the population of 50 states. Strictly speaking, which is the correct formula for obtaining the standard deviation of the sample mean—Equation (7.1) or Equation (7.2)? Explain your answer.

c. Referring to part (b), obtain $\sigma_{\bar{x}}$ for simple random samples of size 30 by using both formulas. Why does Equation (7.2) provide such a poor estimate of the true value given by Equation (7.1)?

d. Referring to part (b), obtain $\sigma_{\bar{x}}$ for simple random samples of size 2 by using both formulas. Why does Equation (7.2) provide a somewhat reasonable estimate of the true value given by Equation (7.1)?

7.53 SAT Scores. Each year, thousands of high school students bound for college take the Scholastic Assessment Test (SAT). This test measures the verbal and mathematical abilities of prospective college students. Student scores are reported on a scale that ranges from a low of 200 to a high of 800. Summary results for the scores are published by the College Entrance Examination Board in *College Bound Seniors*. The SAT math scores for one high school graduating class are as provided on the WeissStats site. Use the technology of your choice to solve the following problems.

a. Obtain the standard deviation of the variable "SAT math score" for this population of students.

b. For simple random samples without replacement of sizes 1–487, construct a table to compare the true values of $\sigma_{\bar{x}}$—obtained by using Equation (7.1)—with the values of $\sigma_{\bar{x}}$ obtained by using Equation (7.2). Explain why the results found by using Equation (7.2) are sometimes reasonably accurate and sometimes not.

Extending the Concepts and Skills

7.54 Unbiased and Biased Estimators. A statistic is said to be an **unbiased estimator** of a parameter if the mean of all its possible values equals the parameter; otherwise, it is said to be a **biased estimator.** An unbiased estimator yields, on average, the correct value of the parameter, whereas a biased estimator does not.
a. Is the sample mean an unbiased estimator of the population mean? Explain your answer.
b. Is the sample median an unbiased estimator of the population median? (*Hint:* Refer to Example 7.2 on pages 309–310. Consider samples of size 2.)

For Exercises 7.55–7.57, refer to Equations (7.1) and (7.2) on pages 315 and 316, respectively.

7.55 Suppose that a simple random sample is taken without replacement from a finite population of size N.
a. Show mathematically that Equations (7.1) and (7.2) are identical for samples of size 1.
b. Explain in words why part (a) is true.
c. Without doing any computations, determine $\sigma_{\bar{x}}$ for samples of size N without replacement. Explain your reasoning.
d. Use Equation (7.1) to verify your answer in part (c).

7.56 Heights of Starting Players. In Example 7.5, we used the definition of the standard deviation of a variable (Definition 3.12 on page 142) to obtain the standard deviation of the heights of the five starting players on a men's basketball team and also the standard deviation of $\bar{x}$ for samples of sizes 1, 2, 3, 4, and 5. The results are summarized in Table 7.6 on page 315. Because the sampling is without replacement from a finite population, Equation (7.1) can also be used to obtain $\sigma_{\bar{x}}$.
a. Apply Equation (7.1) to compute $\sigma_{\bar{x}}$ for samples of sizes 1, 2, 3, 4, and 5. Compare your answers with those in Table 7.6.
b. Use the simpler formula, Equation (7.2), to compute $\sigma_{\bar{x}}$ for samples of sizes 1, 2, 3, 4, and 5. Compare your answers with those in Table 7.6. Why does Equation (7.2) generally yield such poor approximations to the true values?
c. What percentages of the population size are samples of sizes 1, 2, 3, 4, and 5?

7.57 Finite-Population Correction Factor. Consider simple random samples of size n without replacement from a population of size N.
a. Show that if $n \leq 0.05N$, then
$$0.97 \leq \sqrt{\frac{N-n}{N-1}} \leq 1.$$
b. Use part (a) to explain why there is little difference in the values provided by Equations (7.1) and (7.2) when the sample size is small relative to the population size—that is, when the size of the sample does not exceed 5% of the size of the population.
c. Explain why the finite-population correction factor can be ignored and the simpler formula, Equation (7.2), can be used when the sample size is small relative to the population size.

d. The term $\sqrt{(N-n)/(N-1)}$ is known as the **finite-population correction factor.** Can you explain why?

7.58 Class Project Simulation. This exercise can be done individually or, better yet, as a class project.
a. Use a random-number table or random-number generator to obtain a sample (with replacement) of four digits between 0 and 9. Do so a total of 50 times and compute the mean of each sample.
b. Theoretically, what are the mean and standard deviation of all possible sample means for samples of size 4?
c. Roughly what would you expect the mean and standard deviation of the 50 sample means you obtained in part (a) to be? Explain your answers.
d. Determine the mean and standard deviation of the 50 sample means you obtained in part (a).
e. Compare your answers in parts (c) and (d). Why are they different?

7.59 Gestation Periods of Humans. For humans, gestation periods are normally distributed with a mean of 266 days and a standard deviation of 16 days. Suppose that you observe the gestation periods for a sample of nine humans.
a. Theoretically, what are the mean and standard deviation of all possible sample means?
b. Use the technology of your choice to simulate 2000 samples of nine human gestation periods each.
c. Determine the mean of each of the 2000 samples you obtained in part (b).
d. Roughly what would you expect the mean and standard deviation of the 2000 sample means you obtained in part (c) to be? Explain your answers.
e. Determine the mean and standard deviation of the 2000 sample means you obtained in part (c).
f. Compare your answers in parts (d) and (e). Why are they different?

7.60 Emergency Room Traffic. Desert Samaritan Hospital in Mesa, Arizona, keeps records of emergency room traffic. Those records reveal that the times between arriving patients have a special type of reverse-J-shaped distribution called an *exponential distribution.* They also indicate that the mean time between arriving patients is 8.7 minutes, as is the standard deviation. Suppose that you observe a sample of 10 interarrival times.
a. Theoretically, what are the mean and standard deviation of all possible sample means?
b. Use the technology of your choice to simulate 1000 samples of 10 interarrival times each.
c. Determine the mean of each of the 1000 samples you obtained in part (b).
d. Roughly what would you expect the mean and standard deviation of the 1000 sample means you obtained in part (c) to be? Explain your answers.
e. Determine the mean and standard deviation of the 1000 sample means you obtained in part (c).
f. Compare your answers in parts (d) and (e). Why are they different?

7.3   The Sampling Distribution of the Sample Mean

In Section 7.2, we took the first step in describing the sampling distribution of the sample mean, that is, the distribution of the variable $\bar{x}$. There, we showed that the mean and standard deviation of $\bar{x}$ can be expressed in terms of the sample size and the population mean and standard deviation: $\mu_{\bar{x}} = \mu$ and $\sigma_{\bar{x}} = \sigma/\sqrt{n}$.

In this section, we take the final step in describing the sampling distribution of the sample mean. In doing so, we distinguish between the case in which the variable under consideration is normally distributed and the case in which it may not be so.

Sampling Distribution of the Sample Mean for Normally Distributed Variables

Although it is by no means obvious, if the variable under consideration is normally distributed, so is the variable $\bar{x}$. The proof of this fact requires advanced mathematics, but we can make it plausible by simulation, as shown next.

◼◼ ◼ **EXAMPLE 7.7**

OUTPUT 7.1
Histogram of the sample means
for 1000 samples of four IQs
with superimposed normal curve

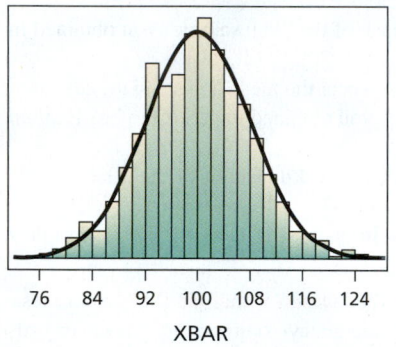

76 84 92 100 108 116 124
XBAR

Sampling Distribution of the Sample Mean for a Normally Distributed Variable

Intelligence Quotients Intelligence quotients (IQs) measured on the Stanford Revision of the Binet–Simon Intelligence Scale are normally distributed with mean 100 and standard deviation 16. For a sample size of 4, use simulation to make plausible the fact that $\bar{x}$ is normally distributed.

Solution First, we apply Formula 7.1 (page 314) and Formula 7.2 (page 316) to conclude that $\mu_{\bar{x}} = \mu = 100$ and $\sigma_{\bar{x}} = \sigma/\sqrt{n} = 16/\sqrt{4} = 8$; that is, the variable $\bar{x}$ has mean 100 and standard deviation 8.

We simulated 1000 samples of four IQs each, determined the sample mean of each of the 1000 samples, and obtained a histogram (Output 7.1) of the 1000 sample means. We also superimposed on the histogram the normal distribution with mean 100 and standard deviation 8. The histogram is shaped roughly like a normal curve (with parameters 100 and 8).

Interpretation The histogram in Output 7.1 suggests that $\bar{x}$ is normally distributed, that is, that the possible sample mean IQs for samples of four people have a normal distribution. ◼

KEY FACT 7.2

? What Does It Mean?

For a normally distributed variable, the possible sample means for samples of a given size are also normally distributed.

Sampling Distribution of the Sample Mean for a Normally Distributed Variable

Suppose that a variable x of a population is normally distributed with mean μ and standard deviation σ. Then, for samples of size n, the variable $\bar{x}$ is also normally distributed and has mean μ and standard deviation $\sigma/\sqrt{n}$.

We illustrate Key Fact 7.2 in the next example.

◼◼ ◼ **EXAMPLE 7.8**

Sampling Distribution of the Sample Mean for a Normally Distributed Variable

Intelligence Quotients Consider again the variable IQ, which is normally distributed with mean 100 and standard deviation 16. Obtain the sampling distribution of the sample mean for samples of size

a. 4.   **b.** 16.

Solution The normal distribution for IQs is shown in Fig. 7.4(a). Because IQs are normally distributed, Key Fact 7.2 implies that, for any particular sample size n, the variable $\bar{x}$ is also normally distributed and has mean 100 and standard deviation $16/\sqrt{n}$.

FIGURE 7.4

(a) Normal distribution for IQs;
(b) sampling distribution of the sample
mean for n = 4; (c) sampling distribution
of the sample mean for n = 16

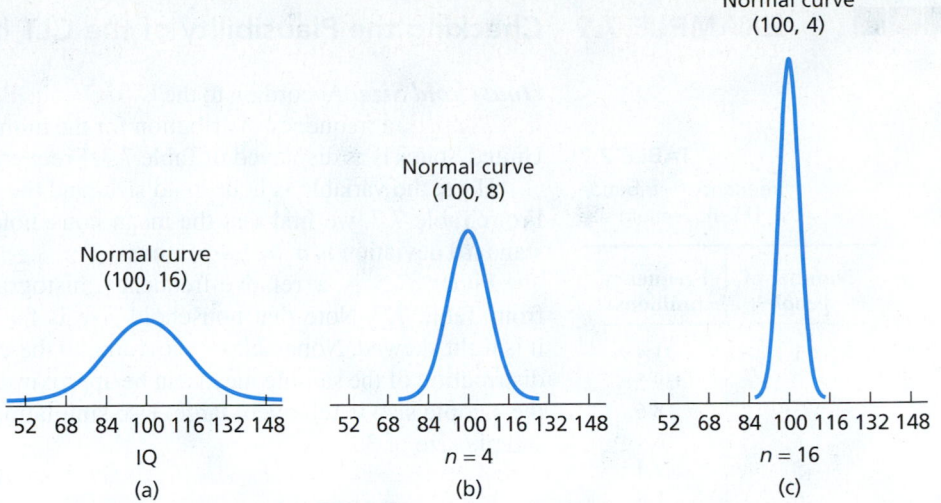

a. For samples of size 4, we have $16/\sqrt{n} = 16/\sqrt{4} = 8$, and therefore the sampling distribution of the sample mean is a normal distribution with mean 100 and standard deviation 8. Figure 7.4(b) shows this normal distribution.

Interpretation The possible sample mean IQs for samples of four people have a normal distribution with mean 100 and standard deviation 8.

You try it!

Exercise 7.67 on page 324

b. For samples of size 16, we have $16/\sqrt{n} = 16/\sqrt{16} = 4$, and therefore the sampling distribution of the sample mean is a normal distribution with mean 100 and standard deviation 4. Figure 7.4(c) shows this normal distribution.

Interpretation The possible sample mean IQs for samples of 16 people have a normal distribution with mean 100 and standard deviation 4.

The normal curves in Figs. 7.4(b) and 7.4(c) are drawn to scale so that you can visualize two important things that you already know: both curves are centered at the population mean ($\mu_{\bar{x}} = \mu$), and the spread decreases as the sample size increases ($\sigma_{\bar{x}} = \sigma/\sqrt{n}$).

Figure 7.4 also illustrates something else that you already know: The possible sample means cluster more closely around the population mean as the sample size increases, and therefore the larger the sample size, the smaller the sampling error tends to be in estimating a population mean by a sample mean.

Central Limit Theorem

According to Key Fact 7.2, if the variable x is normally distributed, so is the variable $\bar{x}$. That key fact also holds approximately if x is not normally distributed, provided only that the sample size is relatively large. This extraordinary fact, one of the most important theorems in statistics, is called the **central limit theorem.**

KEY FACT 7.3

? **What Does It Mean?**

For a large sample size, the possible sample means are approximately normally distributed, regardless of the distribution of the variable under consideration.

The Central Limit Theorem (CLT)

For a relatively large sample size, the variable $\bar{x}$ is approximately normally distributed, regardless of the distribution of the variable under consideration. The approximation becomes better with increasing sample size.

Roughly speaking, the farther the variable under consideration is from being normally distributed, the larger the sample size must be for a normal distribution to provide an adequate approximation to the distribution of $\bar{x}$. Usually, however, a sample size of 30 or more ($n \geq 30$) is large enough.

The proof of the central limit theorem is difficult, but we can make it plausible by simulation, as shown in the next example.

EXAMPLE 7.9   **Checking the Plausibility of the CLT by Simulation**

Household Size   According to the U.S. Census Bureau publication *Current Population Reports*, a frequency distribution for the number of people per household in the United States is as displayed in Table 7.7. Frequencies are in millions of households.

Here, the variable is household size, and the population is all U.S. households. From Table 7.7, we find that the mean household size is $\mu = 2.5$ persons and the standard deviation is $\sigma = 1.4$ persons.

Figure 7.5 is a relative-frequency histogram for household size, obtained from Table 7.7. Note that household size is far from being normally distributed; it is right skewed. Nonetheless, according to the central limit theorem, the sampling distribution of the sample mean can be approximated by a normal distribution when the sample size is relatively large. Use simulation to make that fact plausible for a sample size of 30.

TABLE 7.7

Frequency distribution for U.S. household size

Number of people	Frequency (millions)
1	31.4
2	39.5
3	18.6
4	16.1
5	7.4
6	2.8
7	1.7

FIGURE 7.5

Relative-frequency histogram for household size

OUTPUT 7.2

Histogram of the sample means for 1000 samples of 30 household sizes with superimposed normal curve

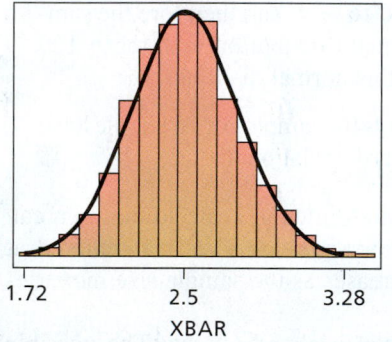

Solution   First, we apply Formula 7.1 (page 314) and Formula 7.2 (page 316) to conclude that, for samples of size 30,

$$\mu_{\bar{x}} = \mu = 2.5 \quad \text{and} \quad \sigma_{\bar{x}} = \sigma/\sqrt{n} = 1.4/\sqrt{30} = 0.26.$$

Thus the variable $\bar{x}$ has a mean of 2.5 and a standard deviation of 0.26.

We simulated 1000 samples of 30 households each, determined the sample mean of each of the 1000 samples, and obtained a histogram (Output 7.2) of the 1000 sample means. We also superimposed on the histogram the normal distribution with mean 2.5 and standard deviation 0.26. The histogram is shaped roughly like a normal curve (with parameters 2.5 and 0.26).

Interpretation   The histogram in Output 7.2 suggests that $\bar{x}$ is approximately normally distributed, as guaranteed by the central limit theorem. Thus, for samples of 30 households, the possible sample mean household sizes have approximately a normal distribution.

The Sampling Distribution of the Sample Mean

We now summarize the facts that we have learned about the sampling distribution of the sample mean.

KEY FACT 7.4

? **What Does It Mean?**

   If a variable is normally distributed or the sample size is large, then the possible sample means have, at least approximately, a normal distribution with mean μ and standard deviation $\sigma/\sqrt{n}$.

Sampling Distribution of the Sample Mean

Suppose that a variable x of a population has mean μ and standard deviation σ. Then, for samples of size n,

- the mean of $\bar{x}$ equals the population mean, or $\mu_{\bar{x}} = \mu$;
- the standard deviation of $\bar{x}$ equals the population standard deviation divided by the square root of the sample size, or $\sigma_{\bar{x}} = \sigma/\sqrt{n}$;
- if x is normally distributed, so is $\bar{x}$, regardless of sample size; and
- if the sample size is large, $\bar{x}$ is approximately normally distributed, regardless of the distribution of x.

FIGURE 7.6   Sampling distributions of the sample mean for (a) normal, (b) reverse-J-shaped, and (c) uniform variables

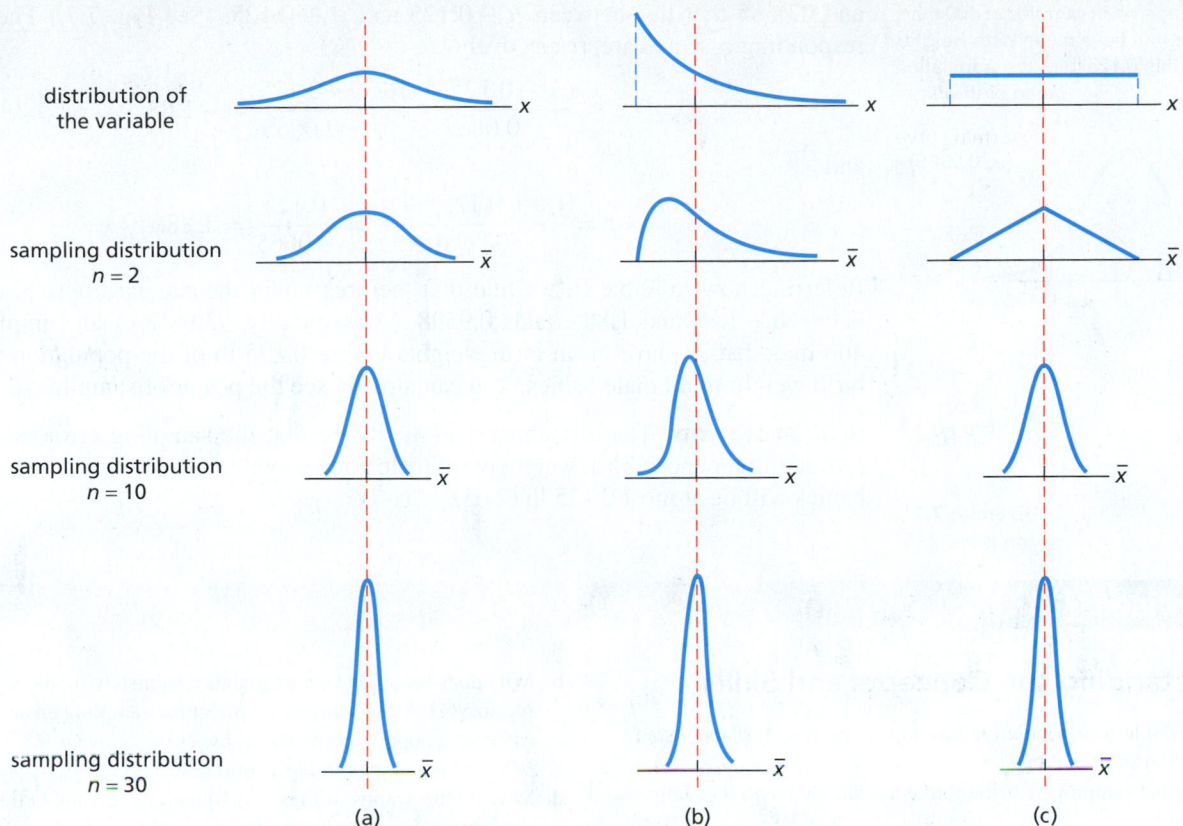

APPLET

Applet 7.3

From Key Fact 7.4, we know that, if the variable under consideration is normally distributed, so is the variable $\bar{x}$, regardless of sample size, as illustrated by Fig. 7.6(a).

In addition, we know that, if the sample size is large, the variable $\bar{x}$ is approximately normally distributed, regardless of the distribution of the variable under consideration. Figures 7.6(b) and 7.6(c) illustrate this fact for two nonnormal variables, one having a reverse-J-shaped distribution and the other having a uniform distribution.

In each of these latter two cases, for samples of size 2, the variable $\bar{x}$ is far from being normally distributed; for samples of size 10, it is already somewhat normally distributed; and for samples of size 30, it is very close to being normally distributed.

Figure 7.6 further illustrates that the mean of each sampling distribution equals the population mean (see the dashed red lines) and that the standard error of the sample mean decreases with increasing sample size.

■■■   **EXAMPLE 7.10**   **Sampling Distribution of the Sample Mean**

Birth Weight   The National Center for Health Statistics publishes information about birth weights in *Vital Statistics of the United States*. According to that document, birth weights of male babies have a standard deviation of 1.33 lb. Determine the percentage of all samples of 400 male babies that have mean birth weights within 0.125 lb (2 oz) of the population mean birth weight of all male babies. Interpret your answer in terms of sampling error.

Solution   Let μ denote the population mean birth weight of all male babies. From Key Fact 7.4, for samples of size 400, the sample mean birth weight, $\bar{x}$, is approximately normally distributed with

$$\mu_{\bar{x}} = \mu \quad \text{and} \quad \sigma_{\bar{x}} = \frac{\sigma}{\sqrt{n}} = \frac{1.33}{\sqrt{400}} = 0.0665.$$

Thus, the percentage of all samples of 400 male babies that have mean birth weights within 0.125 lb of the population mean birth weight of all male babies is

FIGURE 7.7

Percentage of all samples of 400 male babies that have mean birth weights within 0.125 lb of the population mean birth weight

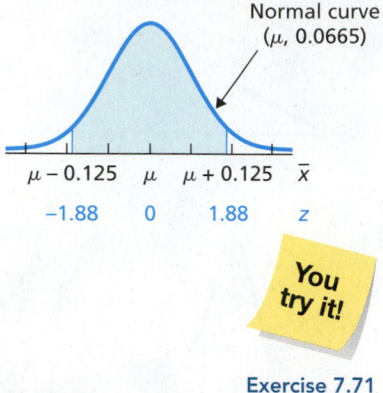

You try it!

Exercise 7.71 on page 325

(approximately) equal to the area under the normal curve with parameters μ and 0.0665 that lies between $\mu - 0.125$ and $\mu + 0.125$. (See Fig. 7.7.) The corresponding z-scores are, respectively,

$$z = \frac{(\mu - 0.125) - \mu}{0.0665} = \frac{-0.125}{0.0665} = -1.88$$

and

$$z = \frac{(\mu + 0.125) - \mu}{0.0665} = \frac{0.125}{0.0665} = 1.88.$$

Referring now to Table II, we find that the area under the standard normal curve between -1.88 and 1.88 equals 0.9398. Consequently, 93.98% of all samples of 400 male babies have mean birth weights within 0.125 lb of the population mean birth weight of all male babies. You can already see the power of sampling.

Interpretation There is about a 94% chance that the sampling error made in estimating the mean birth weight of all male babies by that of a sample of 400 male babies will be at most 0.125 lb (2 oz).

Exercises 7.3

Understanding the Concepts and Skills

7.61 A variable of a population has a mean of $\mu = 100$ and a standard deviation of $\sigma = 28$.
a. Identify the sampling distribution of the sample mean for samples of size 49.
b. In answering part (a), what assumptions did you make about the distribution of the variable?
c. Can you answer part (a) if the sample size is 16 instead of 49? Why or why not?

7.62 A variable of a population has a mean of $\mu = 35$ and a standard deviation of $\sigma = 42$.
a. If the variable is normally distributed, identify the sampling distribution of the sample mean for samples of size 9.
b. Can you answer part (a) if the distribution of the variable under consideration is unknown? Explain your answer.
c. Can you answer part (a) if the distribution of the variable under consideration is unknown but the sample size is 36 instead of 9? Why or why not?

7.63 A variable of a population is normally distributed with mean μ and standard deviation σ.
a. Identify the distribution of $\bar{x}$.
b. Does your answer to part (a) depend on the sample size? Explain your answer.
c. Identify the mean and the standard deviation of $\bar{x}$.
d. Does your answer to part (c) depend on the assumption that the variable under consideration is normally distributed? Why or why not?

7.64 A variable of a population has mean μ and standard deviation σ. For a large sample size n, answer the following questions.
a. Identify the distribution of $\bar{x}$.
b. Does your answer to part (a) depend on n being large? Explain your answer.
c. Identify the mean and the standard deviation of $\bar{x}$.
d. Does your answer to part (c) depend on the sample size being large? Why or why not?

7.65 Refer to Fig. 7.6 on page 323.
a. Why are the four graphs in Fig. 7.6(a) all centered at the same place?

b. Why does the spread of the graphs diminish with increasing sample size? How does this result affect the sampling error when you estimate a population mean, μ, by a sample mean, $\bar{x}$?
c. Why are the graphs in Fig. 7.6(a) bell shaped?
d. Why do the graphs in Figs. 7.6(b) and (c) become bell shaped as the sample size increases?

7.66 According to the central limit theorem, for a relatively large sample size, the variable $\bar{x}$ is approximately normally distributed.
a. What rule of thumb is used for deciding whether the sample size is relatively large?
b. Roughly speaking, what property of the distribution of the variable under consideration determines how large the sample size must be for a normal distribution to provide an adequate approximation to the distribution of $\bar{x}$?

Applying the Concepts and Skills

7.67 Brain Weights. In 1905, R. Pearl published the article "Biometrical Studies on Man. I. Variation and Correlation in Brain Weight" (*Biometrika*, Vol. 4, pp. 13–104). According to the study, brain weights of Swedish men are normally distributed with a mean of 1.40 kg and a standard deviation of 0.11 kg.
a. Determine the sampling distribution of the sample mean for samples of size 3. Interpret your answer in terms of the distribution of all possible sample mean brain weights for samples of three Swedish men.
b. Repeat part (a) for samples of size 12.
c. Construct graphs similar to those shown in Fig. 7.4 on page 321.
d. Determine the percentage of all samples of three Swedish men that have mean brain weights within 0.1 kg of the population mean brain weight of 1.40 kg. Interpret your answer in terms of sampling error.
e. Repeat part (d) for samples of size 12.

7.68 New York City 10-km Run. As reported by *Runner's World* magazine, the times of the finishers in the New York City 10-km run are normally distributed with a mean of 61 minutes and a standard deviation of 9 minutes. Do the following for the variable "finishing time" of finishers in the New York City 10-km run.

a. Find the sampling distribution of the sample mean for samples of size 4.

b. Repeat part (a) for samples of size 9.

c. Construct graphs similar to those shown in Fig. 7.4 on page 321.

d. Obtain the percentage of all samples of four finishers that have mean finishing times within 5 minutes of the population mean finishing time of 61 minutes. Interpret your answer in terms of sampling error.

e. Repeat part (d) for samples of size 9.

7.69 Teacher Salaries. Data on salaries in the public school system are published annually in *Ranking of the States and Estimates of School Statistics* by the National Education Association. The mean annual salary of (public) classroom teachers is $55.4 thousand. Assume a standard deviation of $9.2 thousand. Do the following tasks for the variable "annual salary" of classroom teachers.

a. Determine the sampling distribution of the sample mean for samples of size 64. Interpret your answer in terms of the distribution of all possible sample mean salaries for samples of 64 classroom teachers.

b. Repeat part (a) for samples of size 256.

c. Do you need to assume that classroom teacher salaries are normally distributed to answer parts (a) and (b)? Explain your answer.

d. What is the probability that the sampling error made in estimating the population mean salary of all classroom teachers by the mean salary of a sample of 64 classroom teachers will be at most $1000?

e. Repeat part (d) for samples of size 256.

7.70 Loan Amounts. B. Ciochetti et al. studied mortgage loans in the article "A Proportional Hazards Model of Commercial Mortgage Default with Originator Bias" (*Journal of Real Estate and Economics*, Vol. 27, No. 1, pp. 5–23). According to the article, the loan amounts of loans originated by a large insurance-company lender have a mean of $6.74 million with a standard deviation of $15.37 million. The variable "loan amount" is known to have a right-skewed distribution.

a. Using units of millions of dollars, determine the sampling distribution of the sample mean for samples of size 200. Interpret your result.

b. Repeat part (a) for samples of size 600.

c. Why can you still answer parts (a) and (b) when the distribution of loan amounts is not normal, but rather right skewed?

d. What is the probability that the sampling error made in estimating the population mean loan amount by the mean loan amount of a simple random sample of 200 loans will be at most $1 million?

e. Repeat part (d) for samples of size 600.

7.71 Nurses and Hospital Stays. In the article "A Multifactorial Intervention Program Reduces the Duration of Delirium, Length of Hospitalization, and Mortality in Delirious Patients" (*Journal of the American Geriatrics Society*, Vol. 53, No. 4, pp. 622–628), M. Lundstrom et al. investigated whether education programs for nurses improve the outcomes for their older patients. The standard deviation of the lengths of hospital stay on the intervention ward is 8.3 days.

a. For the variable "length of hospital stay," determine the sampling distribution of the sample mean for samples of 80 patients on the intervention ward.

b. The distribution of the length of hospital stay is right skewed. Does this invalidate your result in part (a)? Explain your answer.

c. Obtain the probability that the sampling error made in estimating the population mean length of stay on the intervention ward by the mean length of stay of a sample of 80 patients will be at most 2 days.

7.72 Women at Work. In the article "Job Mobility and Wage Growth" (*Monthly Labor Review*, Vol. 128, No. 2, pp. 33–39),

A. Light examined data on employment and answered questions regarding why workers separate from their employers. According to the article, the standard deviation of the length of time that women with one job are employed during the first 8 years of their career is 92 weeks. Length of time employed during the first 8 years of career is a left-skewed variable. For that variable, do the following tasks.

a. Determine the sampling distribution of the sample mean for simple random samples of 50 women with one job. Explain your reasoning.

b. Obtain the probability that the sampling error made in estimating the mean length of time employed by all women with one job by that of a random sample of 50 such women will be at most 20 weeks.

7.73 Ethanol Railroad Tariffs. An ethanol railroad tariff is a fee charged for shipments of ethanol on public railroads. The Agricultural Marketing Service publishes tariff rates for railroad-car shipments of ethanol in the *Biofuel Transportation Database*. Assuming that the standard deviation of such tariff rates is $1150, determine the probability that the mean tariff rate of 500 randomly selected railroad-car shipments of ethanol will be within $100 of the mean tariff rate of all railroad-car shipments of ethanol. Interpret your answer in terms of sampling error.

7.74 Taller Young Women. In the document *Anthropometric Reference Data for Children and Adults*, C. Fryer et al. present data from the *National Health and Nutrition Examination Survey* on a variety of human body measurements. A half-century ago, the mean height of (U.S.) women in their 20s was 62.6 inches. Assume that the heights of today's women in their 20s are approximately normally distributed with a standard deviation of 2.88 inches. If the mean height today is the same as that of a half-century ago, what percentage of all samples of 25 of today's women in their 20s have mean heights of at least 64.24 inches?

7.75 Poverty and Dietary Calcium. Calcium is the most abundant mineral in the human body and has several important functions. Recommendations for calcium are provided in *Dietary Reference Intakes*, developed by the Institute of Medicine of the National Academy of Sciences. The recommended adequate intake (RAI) of calcium for adults (ages 19–50) is 1000 milligrams (mg) per day. If adults with incomes below the poverty level have a mean calcium intake equal to the RAI, what percentage of all samples of 18 such adults have mean calcium intakes of at most 947.4 mg? Assume that $\sigma = 188$ mg. State any assumptions that you are making in solving this problem.

7.76 Early-Onset Dementia. Dementia is the loss of the intellectual and social abilities severe enough to interfere with judgment, behavior, and daily functioning. Alzheimer's disease is the most common type of dementia. In the article "Living with Early Onset Dementia: Exploring the Experience and Developing Evidence-Based Guidelines for Practice" (*Alzheimer's Care Quarterly*, Vol. 5, Issue 2, pp. 111–122), P. Harris and J. Keady explored the experience and struggles of people diagnosed with dementia and their families. If the mean age at diagnosis of all people with early-onset dementia is 55 years, find the probability that a random sample of 21 such people will have a mean age at diagnosis less than 52.5 years. Assume that the population standard deviation is 6.8 years. State any assumptions that you are making in solving this problem.

7.77 Worker Fatigue. A study by M. Chen et al. titled "Heat Stress Evaluation and Worker Fatigue in a Steel Plant" (*American Industrial Hygiene Association*, Vol. 64, pp. 352–359) assessed fatigue in steel-plant workers due to heat stress. If the mean post-work heart rate for casting workers equals the normal resting heart rate of 72 beats per minute (bpm), find the probability that a random sample of 29 casting workers will have a mean post-work heart rate exceeding 78.3 bpm. Assume that the population standard deviation of post-work heart rates for casting workers is 11.2 bpm. State any assumptions that you are making in solving this problem.

Extending the Concepts and Skills

Use the empirical rule for variables to answer the questions posed in parts (a)–(c) of Exercises 7.78 and 7.79.

7.78 A variable of a population is normally distributed with mean μ and standard deviation σ. For samples of size n, fill in the blanks. Justify your answers.

a. Approximately 68% of all possible samples have means that lie within _____ of the population mean, μ.

b. Approximately 95% of all possible samples have means that lie within _____ of the population mean, μ.

c. Approximately 99.7% of all possible samples have means that lie within _____ of the population mean, μ.

d. $100(1 - \alpha)\%$ of all possible samples have means that lie within _____ of the population mean, μ. (*Hint:* Draw a graph for the distribution of $\bar{x}$, and determine the z-scores dividing the area under the normal curve into a middle $1 - \alpha$ area and two outside areas of $\alpha/2$.)

7.79 A variable of a population has mean μ and standard deviation σ. For a large sample size n, fill in the blanks. Justify your answers.

a. Approximately _____% of all possible samples have means within $\sigma/\sqrt{n}$ of the population mean, μ.

b. Approximately _____% of all possible samples have means within $2\sigma/\sqrt{n}$ of the population mean, μ.

c. Approximately _____% of all possible samples have means within $3\sigma/\sqrt{n}$ of the population mean, μ.

d. Approximately _____% of all possible samples have means within $z_{\alpha/2}$ of the population mean, μ.

7.80 Testing for Content Accuracy. A brand of water-softener salt comes in packages marked "net weight 40 lb." The company that packages the salt claims that the bags contain an average of 40 lb of salt and that the standard deviation of the weights is 1.5 lb. Assume that the weights are normally distributed.

a. Obtain the probability that the weight of one randomly selected bag of water-softener salt will be 39 lb or less, if the company's claim is true.

b. Determine the probability that the mean weight of 10 randomly selected bags of water-softener salt will be 39 lb or less, if the company's claim is true.

c. If you bought one bag of water-softener salt and it weighed 39 lb, would you consider this evidence that the company's claim is incorrect? Explain your answer.

d. If you bought 10 bags of water-softener salt and their mean weight was 39 lb, would you consider this evidence that the company's claim is incorrect? Explain your answer.

7.81 Household Size. In Example 7.9 on page 322, we conducted a simulation to check the plausibility of the central limit theorem. The variable under consideration there is household size, and the population consists of all U.S. households. A frequency distribution for household size of U.S. households is presented in Table 7.7.

a. Suppose that you simulate 1000 samples of four households each, determine the sample mean of each of the 1000 samples, and obtain a histogram of the 1000 sample means. Would you expect the histogram to be bell shaped? Explain your answer.

b. Carry out the tasks in part (a) and note the shape of the histogram.

c. Repeat parts (a) and (b) for samples of size 10.

d. Repeat parts (a) and (b) for samples of size 100.

7.82 Gestation Periods of Humans. For humans, gestation periods are normally distributed with a mean of 266 days and a standard deviation of 16 days.

a. Use the technology of your choice to simulate 2000 samples of nine human gestation periods each.

b. Find the sample mean of each of the 2000 samples.

c. Obtain the mean, the standard deviation, and a histogram of the 2000 sample means.

d. Theoretically, what are the mean, standard deviation, and distribution of all possible sample means for samples of size 9?

e. Compare your results from parts (c) and (d).

7.83 Emergency Room Traffic. A variable is said to have an *exponential distribution* or to be *exponentially distributed* if its distribution has the shape of an exponential curve, that is, a curve of the form $y = e^{-x/\mu}/\mu$ for $x > 0$, where μ is the mean of the variable. The standard deviation of such a variable also equals μ. At the emergency room at Desert Samaritan Hospital in Mesa, Arizona, the time from the arrival of one patient to the next, called an interarrival time, has an exponential distribution with a mean of 8.7 minutes.

a. Sketch the exponential curve for the distribution of the variable "interarrival time." Note that this variable is far from being normally distributed. What shape does its distribution have?

b. Use the technology of your choice to simulate 1000 samples of four interarrival times each.

c. Find the sample mean of each of the 1000 samples.

d. Determine the mean and standard deviation of the 1000 sample means.

e. Theoretically, what are the mean and the standard deviation of all possible sample means for samples of size 4? Compare your answers to those you obtained in part (d).

f. Obtain a histogram of the 1000 sample means. Is the histogram bell shaped? Would you necessarily expect it to be?

g. Repeat parts (b)–(f) for a sample size of 40.

CHAPTER IN REVIEW

You Should Be Able to

1. use and understand the formulas in this chapter.

2. define sampling error, and explain the need for sampling distributions.

3. find the mean and standard deviation of the variable $\bar{x}$, given the mean and standard deviation of the population and the sample size.

4. state and apply the central limit theorem.

5. determine the sampling distribution of the sample mean when the variable under consideration is normally distributed.

6. determine the sampling distribution of the sample mean when the sample size is relatively large.

Key Terms

central limit theorem, *321*
sampling distribution, *309*

sampling distribution of the sample
mean, *309*
sampling error, *308*

standard error (SE), *317*
standard error of the sample
mean, *317*

REVIEW PROBLEMS

Understanding the Concepts and Skills

1. Define sampling error.

2. What is the sampling distribution of a statistic? Why is it important?

3. Provide two synonyms for "the distribution of all possible sample means for samples of a given size."

4. Relative to the population mean, what happens to the possible sample means for samples of the same size as the sample size increases? Explain the relevance of this property in estimating a population mean by a sample mean.

5. **Officer Salaries.** The following table gives the monthly salaries (in \$1000s) of the six officers of a company.

Officer	A	B	C	D	E	F
Salary	8	12	16	20	24	28

a. Calculate the population mean monthly salary, μ.

There are 15 possible samples of size 4 from the population of six officers. They are listed in the first column of the following table.

Sample	Salaries	$\bar{x}$
A, B, C, D	8, 12, 16, 20	14
A, B, C, E	8, 12, 16, 24	15
A, B, C, F	8, 12, 16, 28	16
A, B, D, E	8, 12, 20, 24	16
A, B, D, F	8, 12, 20, 28	17
A, B, E, F	8, 12, 24, 28	18
A, C, D, E		
A, C, D, F		
A, C, E, F		
A, D, E, F		
B, C, D, E		
B, C, D, F		
B, C, E, F		
B, D, E, F		
C, D, E, F		

b. Complete the second and third columns of the table.
c. Complete the dotplot for the sampling distribution of the sample mean for samples of size 4. Locate the population mean on the graph.

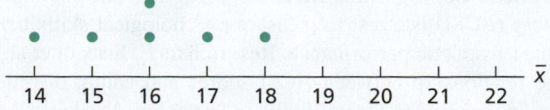

d. Obtain the probability that the mean salary of a random sample of four officers will be within 1 (i.e., \$1000) of the population mean.

6. **Officer Salaries.** Refer to Problem 5.
a. Use the answer you obtained in Problem 5(b) and Definition 3.11 on page 140 to find the mean of the variable $\bar{x}$. Interpret your answer.
b. Can you obtain the mean of the variable $\bar{x}$ without doing the calculation in part (a)? Explain your answer.

7. The following graph shows the curve for a normally distributed variable. Superimposed are the curves for the sampling distributions of the sample mean for two different sample sizes.

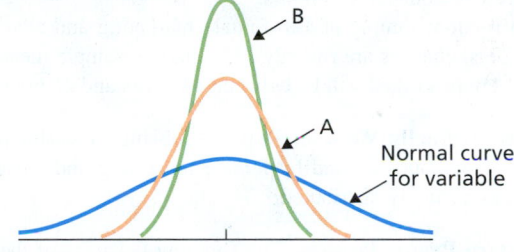

a. Explain why all three curves are centered at the same place.
b. Which curve corresponds to the larger sample size? Explain your answer.
c. Why is the spread of each curve different?
d. Which of the two sampling-distribution curves corresponds to the sample size that will tend to produce less sampling error? Explain your answer.
e. Why are the two sampling-distribution curves normal curves?

Applying the Concepts and Skills

8. **Income Tax and the IRS.** In 2010, the Internal Revenue Service (IRS) sampled 308,946 tax returns to obtain estimates of various parameters. Data were published in *Statistics of Income, Individual Income Tax Returns*. According to that document, the mean income tax per return for the returns sampled was \$11,266.
a. Explain the meaning of sampling error in this context.
b. If, in reality, the population mean income tax per return in 2010 was \$11,354, how much sampling error was made in estimating that parameter by the sample mean of \$11,266?
c. If the IRS had sampled 400,000 returns instead of 308,946, would the sampling error necessarily have been smaller? Explain your answer.
d. In future surveys, how can the IRS increase the likelihood of small sampling error?

9. **New Car Passion.** Edmunds.com publishes information on new car prices in *Car Shopping Trends Report*. During a recent year, Americans spent an average of \$30,803 for a new car. Assume a standard deviation of \$10,200.
a. Identify the population and variable under consideration.
b. For samples of 50 new car sales during the year in question, determine the mean and standard deviation of all possible sample mean prices.

c. Repeat part (b) for samples of size 100.

d. For samples of size 1000, answer the following question without doing any computations: Will the standard deviation of all possible sample mean prices be larger than, smaller than, or the same as that in part (c)? Explain your answer.

10. Hours Actually Worked. In the article "How Hours of Work Affect Occupational Earnings" (*Monthly Labor Review*, Vol. 121), D. Hecker discussed the number of hours actually worked as opposed to the number of hours paid for. The study examines both full-time men and full-time women in 87 different occupations. According to the article, the mean number of hours (actually) worked by female marketing and advertising managers is $\mu = 45$ hours. Assuming a standard deviation of $\sigma = 7$ hours, decide whether each of the following statements is true or false or whether the information is insufficient to decide. Give a reason for each of your answers.

a. For a random sample of 196 female marketing and advertising managers, chances are roughly 95% that the sample mean number of hours worked will be between 31 hours and 59 hours.

b. Approximately 95% of all possible observations of the number of hours worked by female marketing and advertising managers lie between 31 hours and 59 hours.

c. For a random sample of 196 female marketing and advertising managers, chances are roughly 95% that the sample mean number of hours worked will be between 44 hours and 46 hours.

11. Hours Actually Worked. Repeat Problem 10, assuming that the number of hours worked by female marketing and advertising managers is normally distributed.

12. Western Pygmy-Possum. The foraging behavior of the western pygmy-possum was investigated in the article "Strategies of a Small Nectarivorous Marsupial, the Western Pygmy-Possum, in Response to Seasonal Variation in Food Availability" (*Journal of Mammalogy*, Vol. 96, No. 6, pp. 1525–1535) by D. Morrant and S. Petit. The weights of adult male pygmy-possums in Australia are normally distributed with a mean of 8.5 g and a standard deviation of 0.3 g.

a. Sketch the normal curve for the pygmy-possum weights.

b. Find the sampling distribution of the sample mean for samples of size 4. Draw a graph of the normal curve associated with $\bar{x}$.

c. Repeat part (b) for samples of size 9.

13. Western Pygmy-Possum. Refer to Problem 12.

a. Find the percentage of all samples of four pygmy-possums that have mean weights within 0.225 g of the population mean weight of 8.5 g.

b. Obtain the probability that the mean weight of four randomly selected pygmy-possums will be within 0.225 g of the population mean weight of 8.5 g.

c. Interpret the probability you obtained in part (b) in terms of sampling error.

d. Repeat parts (a)–(c) for samples of size 9.

14. Blood Glucose Level. In the article "Drinking Glucose Improves Listening Span in Students Who Miss Breakfast" (*Educational Research*, Vol. 43, No. 2, pp. 201–207), authors N. Morris and P. Sarll explored the relationship between students who skip breakfast and their performance on a number of cognitive tasks. According to their findings, blood glucose levels in the morning, after a 9-hour fast, have a mean of 4.60 mmol/L with a standard deviation of 0.16 mmol/L. (*Note:* mmol/L is an abbreviation of millimoles/liter, which is the world standard unit for measuring glucose in blood.)

a. Determine the sampling distribution of the sample mean for samples of size 60.

b. Repeat part (a) for samples of size 120.

c. Must you assume that the blood glucose levels are normally distributed to answer parts (a) and (b)? Explain your answer.

15. Life Insurance in Force. The American Council of Life Insurers provides information about life insurance in force per covered family in the *Life Insurers Fact Book*. Assume that the standard deviation of life insurance in force is $50,900.

a. Determine the probability that the sampling error made in estimating the population mean life insurance in force by that of a sample of 500 covered families will be $2000 or less.

b. Must you assume that life-insurance amounts are normally distributed in order to answer part (a)? What if the sample size is 20 instead of 500?

c. Repeat part (a) for a sample size of 5000.

16. Paint Durability. A paint manufacturer in Pittsburgh claims that his paint will last an average of 5 years. Assuming that paint life is normally distributed and has a standard deviation of 0.5 year, answer the following questions:

a. Suppose that you paint one house with the paint and that the paint lasts 4.5 years. Would you consider that evidence against the manufacturer's claim? (*Hint:* Assuming that the manufacturer's claim is correct, determine the probability that the paint life for a randomly selected house painted with the paint is 4.5 years or less.)

b. Suppose that you paint 10 houses with the paint and that the paint lasts an average of 4.5 years for the 10 houses. Would you consider that evidence against the manufacturer's claim?

c. Repeat part (b) if the paint lasts an average of 4.9 years for the 10 houses painted.

17. Cloudiness in Breslau. In the paper "Cloudiness: Note on a Novel Case of Frequency" (*Proceedings of the Royal Society of London*, Vol. 62, pp. 287–290), K. Pearson examined data on daily degree of cloudiness, on a scale of 0 to 10, at Breslau (Wroclaw), Poland, during the decade 1876–1885. A frequency distribution of the data is presented in the following table. From the table, we find that the mean degree of cloudiness is 6.83 with a standard deviation of 4.28.

Degree	Frequency	Degree	Frequency
0	751	6	21
1	179	7	71
2	107	8	194
3	69	9	117
4	46	10	2089
5	9		

a. Consider simple random samples of 100 days during the decade in question. Approximately what percentage of such samples have a mean degree of cloudiness exceeding 7.5?

b. Would it be reasonable to use a normal distribution to obtain the percentage required in part (a) for samples of size 5? Explain your answer.

Extending the Concepts and Skills

18. Athletic Coping Skills Inventory. The Athletic Coping Skills Inventory (ACSI) is a test to measure psychological skills believed to influence athletic performance. Researchers E. Estanol et al. studied the relationship between ACSI scores and eating disorders in dancers in the article "Mental Skills as Protective Attributes Against Eating Disorder Risk in Dancers (*Journal of Applied Sport Psychology*, Vol. 25, No. 2, pp. 209–222). The study found that dancers'

ACSI scores are (approximately) normally distributed with mean 49.34 points and standard deviation 10.14 points.

a. Use the technology of your choice to simulate 1000 samples of four dancers' ACSI scores each.

b. Find the sample mean of each of the 1000 samples obtained in part (a).

c. Obtain the mean, the standard deviation, and a histogram of the 1000 sample means.

d. Theoretically, what are the mean, standard deviation, and distribution of all possible sample means for samples of size 4?

e. Compare your answers from parts (c) and (d).

19. Random Numbers. A variable is said to be *uniformly distributed* or to have a *uniform distribution* with parameters a and b if its distribution has the shape of the horizontal line segment with equation $y = 1/(b - a)$, for $a < x < b$. The mean and standard deviation of such a variable are $(a + b)/2$ and $(b - a)/\sqrt{12}$, respectively. The basic random-number generator on a computer or calculator, which returns a number between 0 and 1, simulates a variable having a uniform distribution with parameters 0 and 1.

a. Sketch the distribution of a uniformly distributed variable with parameters 0 and 1. Observe from your sketch that such a variable is far from being normally distributed.

b. Use the technology of your choice to simulate 2000 samples of two random numbers between 0 and 1.

c. Find the sample mean of each of the 2000 samples obtained in part (b).

d. Determine the mean and standard deviation of the 2000 sample means.

e. Theoretically, what are the mean and the standard deviation of all possible sample means for samples of size 2? Compare your answers to those you obtained in part (d).

f. Obtain a histogram of the 2000 sample means. Is the histogram bell shaped? Would you expect it to be?

g. Repeat parts (b)–(f) for a sample size of 35.

FOCUSING ON DATA ANALYSIS

UWEC UNDERGRADUATES

Recall from Chapter 1 (see page 34) that the Focus database and Focus sample contain information on the undergraduate students at the University of Wisconsin - Eau Claire (UWEC). Now would be a good time for you to review the discussion about these data sets.

Suppose that you want to conduct extensive interviews with a simple random sample of 25 UWEC undergraduate students. Use the technology of your choice to obtain such a sample and the corresponding data for the 13 variables in the Focus database (Focus).

Note: If your statistical software package will not accommodate the entire Focus database, use the Focus sample (FocusSample) instead. Of course, in that case, your simple random sample of 25 UWEC undergraduate students will come from the 200 UWEC undergraduate students in the Focus sample rather than from all UWEC undergraduate students in the Focus database.

CASE STUDY DISCUSSION

THE CHESAPEAKE AND OHIO FREIGHT STUDY

At the beginning of this chapter, we discussed a freight study commissioned by the Chesapeake and Ohio Railroad Company (C&O). A sample of 2072 waybills from a population of 22,984 waybills was used to estimate the total revenue due C&O. The estimate arrived at was $64,568.

Because all 22,984 waybills were available, a census could be taken to determine exactly the total revenue due C&O and thereby reveal the accuracy of the estimate obtained by sampling. The exact amount due C&O was found to be $64,651.

a. What percentage of the waybills constituted the sample?

b. What percentage error was made by using the sample to estimate the total revenue due C&O?

c. At the time of the study, the cost of a census was approximately $5000, whereas the cost for the sample estimate was only $1000. Knowing this information and your answers to parts (a) and (b), do you think that sampling was preferable to a census? Explain your answer.

d. In the study, the $83 error was against C&O. Could the error have been in C&O's favor?

BIOGRAPHY

PIERRE-SIMON LAPLACE: THE NEWTON OF FRANCE

Pierre-Simon Laplace was born on March 23, 1749, at Beaumount-en-Auge, Normandy, France, the son of a peasant farmer. His early schooling was at the military academy at Beaumount, where he developed his mathematical abilities. At the age of 18, he went to Paris. Within 2 years he was recommended for a professorship at the École Militaire by the French mathematician and philosopher Jean d'Alembert. (It is said that Laplace examined and passed Napoleon Bonaparte there in 1785.) In 1773 Laplace was granted membership in the Academy of Sciences.

Laplace held various positions in public life: He was president of the Bureau des Longitudes, professor at the École Normale, Minister of the Interior under Napoleon for six weeks (at which time he was replaced by Napoleon's brother), and Chancellor of the Senate; he was also made a marquis.

Laplace's professional interests were also varied. He published several volumes on celestial mechanics (which the Scottish geologist and mathematician John Playfair said were "the highest point to which man has yet ascended in the scale of intellectual attainment"), a book entitled *Théorie analytique des probabilités* (Analytic Theory of Probability), and other works on physics and mathematics. Laplace's primary contribution to the field of probability and statistics was the remarkable and all-important central limit theorem, which appeared in an 1809 publication and was read to the Academy of Sciences on April 9, 1810.

Astronomy was Laplace's major area of interest; approximately half of his publications were concerned with the solar system and its gravitational interactions. These interactions were so complex that even Sir Isaac Newton had concluded "divine intervention was periodically required to preserve the system in equilibrium." Laplace, however, proved that planets' average angular velocities are invariable and periodic, and thus made the most important advance in physical astronomy since Newton.

When Laplace died in Paris on March 5, 1827, he was eulogized by the famous French mathematician and physicist Simeon Poisson as "the Newton of France."

CHAPTER

8

Confidence Intervals for One Population Mean

CHAPTER OBJECTIVES

In this chapter, you begin your study of inferential statistics by examining methods for estimating the mean of a population. As you might suspect, the statistic used to estimate the population mean, μ, is the sample mean, $\bar{x}$. Because of sampling error, you cannot expect $\bar{x}$ to equal μ exactly. Thus, providing information about the accuracy of the estimate is important, which leads to a discussion of confidence intervals, the main topic of this chapter.

In Section 8.1, we provide the intuitive foundation for confidence intervals. Then, in Section 8.2, we present confidence intervals for one population mean when the population standard deviation, σ, is known. Although, in practice, σ is usually unknown, we first consider, for pedagogical reasons, the case where σ is known.

In Section 8.3, we discuss confidence intervals for one population when the population standard deviation is unknown. As a prerequisite to that topic, we introduce and describe one of the most important distributions in inferential statistics—*Student's t*.

CASE STUDY

Bank Robberies: A Statistical Analysis

In the article "Robbing Banks" (*Significance*, Vol. 9, Issue 3, pp. 17–21) B. Reilly et al. studied several aspects of bank robberies. As these researchers state, "Robbing a bank is the staple crime of thrillers, movies and newspapers. But . . . bank robbery is not all it is cracked up to be."

The researchers concentrated on the factors that determine the amount of proceeds from bank robberies, and thus were able to work out both the economics of attempting one and of preventing one. In particular, the researchers revealed that the return on an average bank robbery per person per raid is modest indeed—so modest that it is not worthwhile for the banks to spend too much money on such preventative measures as fast-rising screens at tellers' windows.

The researchers obtained exclusive data from the British Bankers' Association. In one aspect of their study, they analyzed the data from a sample of 364 bank raids over a several-year period in the United Kingdom. The following table repeats a portion of Table 1 on page 19 of the article.

Variable	Mean	Std. dev.
Amount stolen (pounds sterling)	20,330.5	53,510.2
Number of bank staff present	5.417	4.336
Number of customers present	2.000	3.684
Number of bank raiders	1.637	0.971
Travel time, in minutes from bank to nearest police station	4.557	4.028

After studying point estimates and confidence intervals in this chapter, you will be asked to use these summary statistics to estimate the (population) means of the variables in the table.

8.1   Estimating a Population Mean

A common problem in statistics is to obtain information about the mean, μ, of a population. For example, we might want to know

- the mean age of people in the civilian labor force,
- the mean cost of a wedding,
- the mean gas mileage of a new-model car, or
- the mean starting salary of liberal-arts graduates.

If the population is small, we can ordinarily determine μ exactly by first taking a census and then computing μ from the population data. If the population is large, however, as it often is in practice, taking a census is generally impractical, extremely expensive, or impossible. Nonetheless, we can usually obtain sufficiently accurate information about μ by taking a sample from the population.

Point Estimate

One way to obtain information about a population mean μ without taking a census is to estimate it by a sample mean $\bar{x}$, as illustrated in the next example.

EXAMPLE 8.1   Point Estimate of a Population Mean

Prices of New Mobile Homes The U.S. Census Bureau publishes annual price figures for new mobile homes in *Manufactured Housing Statistics*. The figures are obtained from sampling, not from a census. A simple random sample of 36 new mobile homes yielded the prices, in thousands of dollars, shown in Table 8.1. Use the data to estimate the population mean price, μ, of all new mobile homes.

TABLE 8.1
Prices ($1000s) of 36 randomly selected new mobile homes

67.8	68.4	59.2	56.9	63.9	62.2	55.6	72.9	62.6
67.1	73.4	63.7	57.7	66.7	61.7	55.5	49.3	72.9
49.9	56.5	71.2	59.1	64.3	64.0	55.9	51.3	53.7
56.0	76.7	76.8	60.6	74.5	57.9	70.4	63.8	77.9

Solution We estimate the population mean price, μ, of all new mobile homes by the sample mean price, $\bar{x}$, of the 36 new mobile homes sampled. From Table 8.1,

$$\bar{x} = \frac{\Sigma x_i}{n} = \frac{2278}{36} = 63.28.$$

You try it!

Interpretation Based on the sample data, we estimate the mean price, μ, of all new mobile homes to be approximately $63.28 thousand, that is, $63,280.

Exercise 8.17 on page 337

An estimate of this kind is called a **point estimate** for μ because it consists of a single number, or point.

As indicated in the following definition, the term point estimate applies to the use of a statistic to estimate any parameter, not just a population mean.

DEFINITION 8.1

Point Estimate

A **point estimate** of a parameter is the value of a statistic used to estimate the parameter.

In the previous example, the parameter is the mean price, μ, of all new mobile homes, which is unknown. The point estimate of that parameter is the mean price, $\bar{x}$, of the 36 mobile homes sampled, which is $63,280.

In Section 7.2, we learned that the mean of the sample mean equals the population mean ($\mu_{\bar{x}} = \mu$). In other words, on average, the sample mean equals the population mean. For this reason, the sample mean is called an *unbiased estimator* of the population mean.

More generally, a statistic is called an **unbiased estimator** of a parameter if the mean of all its possible values equals the parameter; otherwise, the statistic is called a **biased estimator** of the parameter. Ideally, we want our statistic to be unbiased and have small standard error. In that case, chances are good that our point estimate (the value of the statistic) will be close to the parameter.

Confidence-Interval Estimate

As you learned in Chapter 7, a sample mean is usually not equal to the population mean; generally, there is sampling error. Therefore, we should accompany any point estimate of μ with information that indicates the accuracy of that estimate. This information is called a *confidence-interval estimate* for μ, which we introduce in the next example.

EXAMPLE 8.2   **Introducing Confidence Intervals**

Prices of New Mobile Homes Consider again the problem of estimating the (population) mean price, μ, of all new mobile homes by using the sample data in Table 8.1. Let's assume that the population standard deviation of all such prices is $7.2 thousand, that is, $7200.[†]

a. Identify the distribution of the variable $\bar{x}$, that is, the sampling distribution of the sample mean for samples of size 36.
b. Use part (a) to show that approximately 95% of all samples of 36 new mobile homes have the property that the interval from $\bar{x} - 2.4$ to $\bar{x} + 2.4$ contains μ.
c. Use part (b) and the sample data in Table 8.1 to find a 95% *confidence interval* for μ, that is, an interval of numbers that we can be 95% confident contains μ.

FIGURE 8.1

Normal probability plot of the price data in Table 8.1

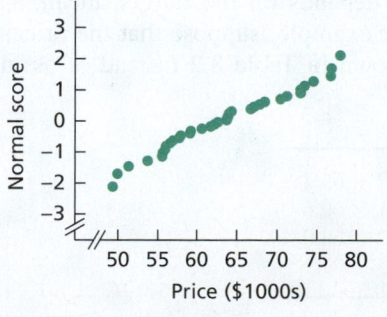

Solution

a. Figure 8.1 is a normal probability plot of the price data in Table 8.1. The plot shows we can reasonably presume that prices of new mobile homes are normally distributed. Because $n = 36$, $\sigma = 7.2$, and prices of new mobile homes are normally distributed, Key Fact 7.2 on page 320 implies that

- $\mu_{\bar{x}} = \mu$ (which we don't know),
- $\sigma_{\bar{x}} = \sigma/\sqrt{n} = 7.2/\sqrt{36} = 1.2$, and
- $\bar{x}$ is normally distributed.

In other words, for samples of size 36, the variable $\bar{x}$ is normally distributed with mean μ and standard deviation 1.2.

[†]We might know the population standard deviation from previous research or from a preliminary study of prices. We examine the more usual case where σ is unknown in Section 8.3.

b. Property 2 of the empirical rule (Key Fact 6.6 on page 282) implies that, for a normally distributed variable, approximately 95% of all possible observations lie within two standard deviations to either side of the mean. Applying this rule to the variable $\bar{x}$ and referring to part (a), we see that approximately 95% of all samples of 36 new mobile homes have mean prices within 2.4 (i.e., $2 \cdot 1.2$) of μ. Equivalently, approximately 95% of all samples of 36 new mobile homes have the property that the interval from $\bar{x} - 2.4$ to $\bar{x} + 2.4$ contains μ.

c. Because we are taking a simple random sample, each possible sample of size 36 is equally likely to be the one obtained. From part (b), (approximately) 95% of all such samples have the property that the interval from $\bar{x} - 2.4$ to $\bar{x} + 2.4$ contains μ. Hence, chances are 95% that the sample we obtain has that property. Consequently, we can be 95% confident that the sample of 36 new mobile homes whose prices are shown in Table 8.1 has the property that the interval from $\bar{x} - 2.4$ to $\bar{x} + 2.4$ contains μ. For that sample, $\bar{x} = 63.28$, so

$$\bar{x} - 2.4 = 63.28 - 2.4 = 60.88 \quad \text{and} \quad \bar{x} + 2.4 = 63.28 + 2.4 = 65.68.$$

Thus our 95% confidence interval is from 60.88 to 65.68.

Interpretation We can be 95% confident that the mean price, μ, of all new mobile homes is somewhere between \$60,880 and \$65,680.

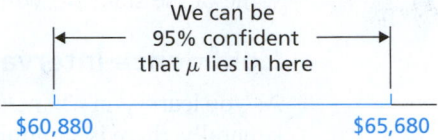

We can be
95% confident
that μ lies in here

\$60,880   \$65,680

You try it!

Exercise 8.19
on page 337

Note: Although this or any other 95% confidence interval may or may not contain μ, we can be 95% confident that it does because the method that we used to construct the confidence interval gives correct results 95% of the time. ■

With the previous example in mind, we now define confidence-interval estimate and related terms. As indicated, the terms apply to estimating any parameter, not just a population mean.

DEFINITION 8.2

? What Does It Mean?

A confidence-interval estimate for a parameter provides a range of numbers along with a percentage confidence that the parameter lies in that range.

Confidence-Interval Estimate

Confidence interval (CI): An interval of numbers obtained from a point estimate of a parameter.

Confidence level: The confidence we have that the parameter lies in the confidence interval (i.e., that the confidence interval contains the parameter).

Confidence-interval estimate: The confidence level and confidence interval.

A confidence interval for a population mean depends on the sample mean, $\bar{x}$, which in turn depends on the sample selected. For example, suppose that the prices of the 36 new mobile homes sampled were as shown in Table 8.2 instead of as in Table 8.1.

TABLE 8.2

Prices (\$1000s) of another sample of 36 randomly selected new mobile homes

73.0	72.1	61.2	53.0	75.5	63.8	56.0	75.7	65.7
53.2	66.6	65.3	68.9	58.4	69.1	65.8	64.1	60.6
66.5	64.7	62.5	61.3	62.1	68.0	79.2	69.2	68.0
60.2	72.1	54.9	66.1	64.1	72.0	68.8	64.3	77.9

Then we would have $\bar{x} = 65.83$ so that

$$\bar{x} - 2.4 = 65.83 - 2.4 = 63.43 \quad \text{and} \quad \bar{x} + 2.4 = 65.83 + 2.4 = 68.23.$$

In this case, the 95% confidence interval for μ would be from 63.43 to 68.23. We could be 95% confident that the mean price, μ, of all new mobile homes is somewhere between \$63,430 and \$68,230.

Interpreting Confidence Intervals

The next example stresses the importance of interpreting a confidence interval correctly. It also illustrates that the population mean, μ, may or may not lie in the confidence interval obtained.

EXAMPLE 8.3   **Interpreting Confidence Intervals**

Prices of New Mobile Homes Consider again the prices of new mobile homes. As demonstrated in part (b) of Example 8.2, (approximately) 95% of all samples of 36 new mobile homes have the property that the interval from $\bar{x} - 2.4$ to $\bar{x} + 2.4$ contains μ. In other words, if 36 new mobile homes are selected at random and their mean price, $\bar{x}$, is computed, the interval from

$$\bar{x} - 2.4 \quad \text{to} \quad \bar{x} + 2.4 \tag{8.1}$$

will be a 95% confidence interval for the mean price of all new mobile homes. Illustrate that the mean price, μ, of all new mobile homes may or may not lie in the 95% confidence interval obtained.

Solution We used a computer to simulate 25 samples of 36 new mobile home prices each. For the simulation, we assumed that $\mu = 65$ (i.e., \$65 thousand) and $\sigma = 7.2$ (i.e., \$7.2 thousand). In reality, we don't know μ; we are assuming a value for μ to illustrate a point.

For each of the 25 samples of 36 new mobile home prices, we did three things: computed the sample mean price, $\bar{x}$; used Equation (8.1) to obtain the 95% confidence interval; and noted whether the population mean, $\mu = 65$, actually lies in the confidence interval.

Figure 8.2 on the next page summarizes our results. For each sample, we have drawn a graph on the right-hand side of Fig. 8.2. The dot represents the sample mean, $\bar{x}$, in thousands of dollars, and the horizontal line represents the corresponding 95% confidence interval. Note that the population mean, μ, lies in the confidence interval only when the horizontal line crosses the dashed line.

Figure 8.2 reveals that μ lies in the 95% confidence interval in 24 of the 25 samples, that is, in 96% of the samples. If, instead of 25 samples, we simulated 1000, we would probably find that the percentage of those 1000 samples for which μ lies in the 95% confidence interval would be even closer to 95%. Hence we can be 95% confident that any computed 95% confidence interval will contain μ.

Margin of Error

In Example 8.2(c), we found a 95% confidence interval for the mean price, μ, of all new mobile homes. Looking back at the construction of that confidence interval on page 334, we see that the endpoints of the confidence interval are 60.88 and 65.68 (in thousands of dollars). These two numbers were obtained, respectively, by subtracting 2.4 from and adding 2.4 to the sample mean of 63.28. In other words, the endpoints of the confidence interval can be expressed as 63.28 ± 2.4.

The number 2.4 is called the **margin of error** because it indicates how accurate our guess (in this case, $\bar{x}$) is as an estimate for the value of the unknown parameter (in this case, μ). Here, we can be 95% confident that the mean price, μ, of all new mobile homes is within \$2.4 thousand of the sample mean price of \$63.28 thousand.

Using this terminology, we can express the (endpoints of the) confidence interval as follows:

point estimate $\pm$ margin of error.

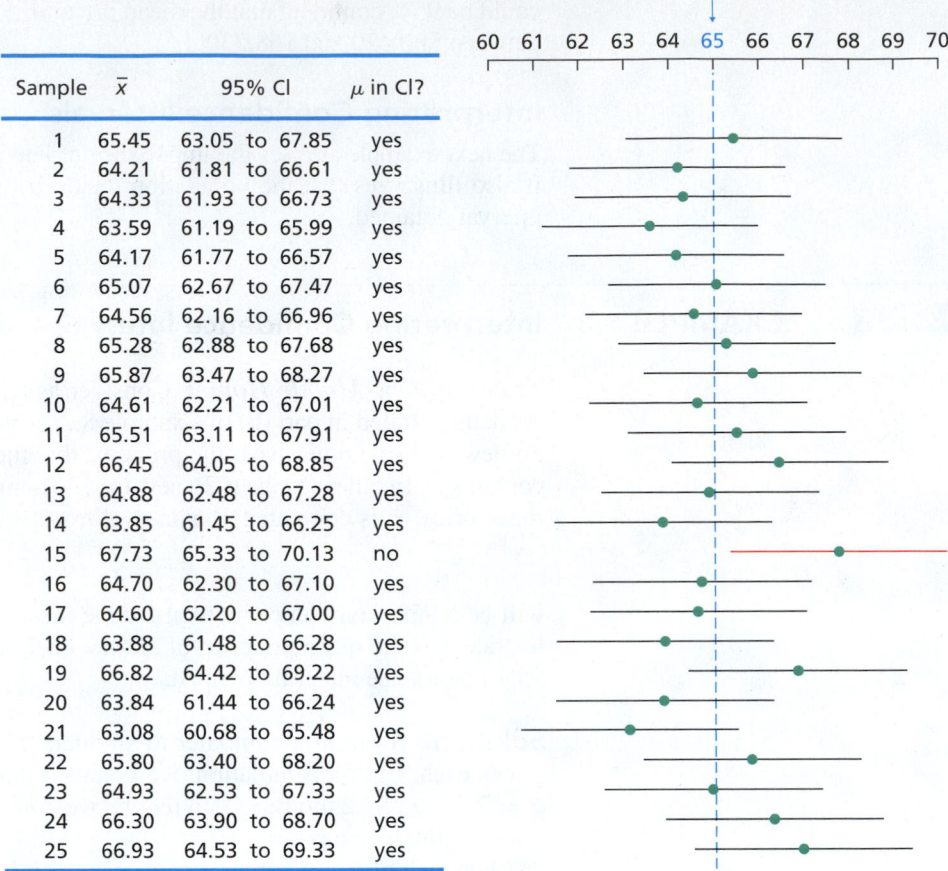

FIGURE 8.2

Twenty-five confidence intervals for the mean price of all new mobile homes, each based on a sample of 36 new mobile homes

Sample	$\bar{x}$	95% CI	μ in CI?
1	65.45	63.05 to 67.85	yes
2	64.21	61.81 to 66.61	yes
3	64.33	61.93 to 66.73	yes
4	63.59	61.19 to 65.99	yes
5	64.17	61.77 to 66.57	yes
6	65.07	62.67 to 67.47	yes
7	64.56	62.16 to 66.96	yes
8	65.28	62.88 to 67.68	yes
9	65.87	63.47 to 68.27	yes
10	64.61	62.21 to 67.01	yes
11	65.51	63.11 to 67.91	yes
12	66.45	64.05 to 68.85	yes
13	64.88	62.48 to 67.28	yes
14	63.85	61.45 to 66.25	yes
15	67.73	65.33 to 70.13	no
16	64.70	62.30 to 67.10	yes
17	64.60	62.20 to 67.00	yes
18	63.88	61.48 to 66.28	yes
19	66.82	64.42 to 69.22	yes
20	63.84	61.44 to 66.24	yes
21	63.08	60.68 to 65.48	yes
22	65.80	63.40 to 68.20	yes
23	64.93	62.53 to 67.33	yes
24	66.30	63.90 to 68.70	yes
25	66.93	64.53 to 69.33	yes

This expression will be the form of most of the confidence intervals that we encounter in our study of statistics. Observe that the margin of error is half the length of the confidence interval or, equivalently, the length of the confidence interval is twice the margin of error.

By the way, it is interesting to note that margin of error is analogous to tolerance in manufacturing and production processes.

Exercises 8.1

Understanding the Concepts and Skills

8.1 The value of a statistic used to estimate a parameter is called a _____ of the parameter.

8.2 What is a confidence-interval estimate of a parameter? Why is such an estimate superior to a point estimate?

8.3 When estimating an unknown parameter, what does the margin of error indicate?

8.4 Express the form of most of the confidence intervals that you will encounter in your study of statistics in terms of "point estimate" and "margin of error."

8.5 Suppose that you take 1000 simple random samples from a population and that, for each sample, you obtain a 95% confidence interval for an unknown parameter. Approximately how many of those confidence intervals will contain the value of the unknown parameter?

8.6 Suppose that you take 500 simple random samples from a population and that, for each sample, you obtain a 90% confidence interval for an unknown parameter. Approximately how many of those confidence intervals will not contain the value of the unknown parameter?

8.7 A simple random sample is taken from a population and yields the following data for a variable of the population:

9	32	37	32	20
24	10	15	21	30

Find a point estimate for the population mean (i.e., the mean of the variable).

8.8 A simple random sample is taken from a population and yields the following data for a variable of the population:

20	2	6	2	12	6
9	8	16	8	21	

Find a point estimate for the population mean (i.e., the mean of the variable).

8.9 Refer to Exercise 8.7 and find a point estimate for the population standard deviation (i.e., the standard deviation of the variable).

8.10 Refer to Exercise 8.8 and find a point estimate for the population standard deviation (i.e., the standard deviation of the variable).

In each of Exercises 8.11–8.16, we provide a sample mean, sample size, and population standard deviation. In each case, perform the following tasks.
a. Find a 95% confidence interval for the population mean. (Note: You may want to review Example 8.2, which begins on page 333.)
b. Identify and interpret the margin of error.
c. Express the endpoints of the confidence interval in terms of the point estimate and the margin of error.

8.11 $\bar{x} = 20$, $n = 36$, $\sigma = 3$ **8.12** $\bar{x} = 25$, $n = 36$, $\sigma = 3$

8.13 $\bar{x} = 30$, $n = 25$, $\sigma = 4$ **8.14** $\bar{x} = 35$, $n = 25$, $\sigma = 4$

8.15 $\bar{x} = 50$, $n = 16$, $\sigma = 5$ **8.16** $\bar{x} = 55$, $n = 16$, $\sigma = 5$

Applying the Concepts and Skills

8.17 Wedding Costs. According to *Bride's Magazine*, getting married these days can be expensive when the costs of the reception, engagement ring, bridal gown, pictures—just to name a few—are included. A simple random sample of 20 recent U.S. weddings yielded the following data on wedding costs, in dollars.

19,496	23,789	18,312	14,554	18,460
27,806	21,203	29,288	34,081	27,896
30,098	13,360	33,178	42,646	24,053
32,269	40,406	35,050	21,083	19,510

a. Use the data to obtain a point estimate for the population mean wedding cost, μ, of all recent U.S. weddings. (*Note:* The sum of the data is $526,538.)
b. Is your point estimate in part (a) likely to equal μ exactly? Explain your answer.

8.18 Cottonmouth Litter Size. In the article "The Eastern Cottonmouth (*Agkistrodon piscivorus*) at the Northern Edge of Its Range" (*Journal of Herpetology*, Vol. 29, No. 3, pp. 391–398), C. Blem and L. Blem examined the reproductive characteristics of the eastern cottonmouth, a once widely distributed snake whose numbers have decreased recently due to encroachment by humans. A simple random sample of 44 female cottonmouths yielded the following data on number of young per litter.

5	12	7	7	6	8	12	9	7
4	9	6	12	7	5	6	10	3
10	8	8	12	5	6	10	11	3
8	4	5	7	6	11	7	6	8
8	14	8	7	11	7	5	4	

a. Use the data to obtain a point estimate for the mean number of young per litter, μ, of all female eastern cottonmouths. (*Note:* $\Sigma x_i = 334$.)

b. Is your point estimate in part (a) likely to equal μ exactly? Explain your answer.

For Exercises 8.19–8.24, you may want to review Example 8.2, which begins on page 333.

8.19 Wedding Costs. Refer to Exercise 8.17. Assume that recent wedding costs in the United States are normally distributed with a standard deviation of $8100.
a. Determine a 95% confidence interval for the mean cost, μ, of all recent U.S. weddings.
b. Interpret your result in part (a).
c. Does the mean cost of all recent U.S. weddings lie in the confidence interval you obtained in part (a)? Explain your answer.

8.20 Cottonmouth Litter Size. Refer to Exercise 8.18. Assume that $\sigma = 2.4$.
a. Obtain a 95% confidence interval for the mean number of young per litter of all female eastern cottonmouths.
b. Interpret your result in part (a).
c. Does the mean number of young per litter of all female eastern cottonmouths lie in the confidence interval you obtained in part (a)? Explain your answer.

8.21 Fuel Tank Capacity. *Consumer Reports* provides information on new automobile models—including price, mileage ratings, engine size, body size, and indicators of features. A simple random sample of 35 new models yielded the following data on fuel tank capacity, in gallons.

17.2	23.1	17.5	15.7	19.8	16.9	15.3
18.5	18.5	25.5	18.0	17.5	14.5	20.0
17.0	20.0	24.0	26.0	18.1	21.0	19.3
20.0	20.0	12.5	13.2	15.9	14.5	22.2
21.1	14.4	25.0	26.4	16.9	16.4	23.0

a. Determine a point estimate for the mean fuel tank capacity of all new automobile models. Interpret your answer in words. (*Note:* $\Sigma x_i = 664.9$ gallons.)
b. Determine a 95% confidence interval for the mean fuel tank capacity of all new automobile models. Assume $\sigma = 3.50$ gallons.
c. How would you decide whether fuel tank capacities for new automobile models are approximately normally distributed?
d. Must fuel tank capacities for new automobile models be exactly normally distributed for the confidence interval that you obtained in part (b) to be approximately correct? Explain your answer.

8.22 Home Improvements. The *American Express Retail Index* provides information on budget amounts for home improvements. The following table displays the budgets, in dollars, of 45 randomly sampled home improvement jobs in the United States.

3179	1032	1822	4093	2285	1478	955	2773	514
3915	4800	3843	5265	2467	2353	4200	3146	551
2659	4660	3570	1598	2605	3643	2816	3125	3104
4503	2911	3605	2948	1421	1910	5145	4557	2026
2750	2069	3056	2550	631	4550	5069	2124	1573

a. Determine a point estimate for the population mean budget, μ, for such home improvement jobs. Interpret your answer in words. (*Note:* The sum of the data is $129,849.)
b. Obtain a 95% confidence interval for the population mean budget, μ, for such home improvement jobs and interpret your result in words. Assume that the population standard deviation of budgets for home improvement jobs is $1350.

c. How would you decide whether budgets for such home improvement jobs are approximately normally distributed?

d. Must the budgets for such home improvement jobs be exactly normally distributed for the confidence interval that you obtained in part (b) to be approximately correct? Explain your answer.

8.23 Giant Tarantulas. A tarantula has two body parts. The anterior part of the body is covered above by a shell, or carapace. In the paper "Reproductive Biology of Uruguayan Theraphosids" (*The Journal of Arachnology*, Vol. 30, No. 3, pp. 571–587), F. Costa and F. Perez–Miles discussed a large species of tarantula whose common name is the Brazilian giant tawny red. A simple random sample of 15 of these adult male tarantulas provided the following data on carapace length, in millimeters (mm).

15.7	18.3	19.7	17.6	19.0
19.2	19.8	18.1	18.0	20.9
16.4	16.8	18.9	18.5	19.5

a. Obtain a normal probability plot of the data.

b. Based on your result from part (a), is it reasonable to presume that carapace length of adult male Brazilian giant tawny red tarantulas is normally distributed? Explain your answer.

c. Find and interpret a 95% confidence interval for the mean carapace length of all adult male Brazilian giant tawny red tarantulas. The population standard deviation is 1.76 mm.

d. In Exercise 6.97, we noted that the mean carapace length of all adult male Brazilian giant tawny red tarantulas is 18.14 mm. Does your confidence interval in part (c) contain the population mean? Would it necessarily have to? Explain your answers.

8.24 Serum Cholesterol Levels. Information on serum total cholesterol level is published by the Centers for Disease Control and

Prevention in *National Health and Nutrition Examination Survey*. A simple random sample of 12 U.S. females 20 years old or older provided the following data on serum total cholesterol level, in milligrams per deciliter (mg/dL).

260	289	190	214	110	241
169	173	191	178	129	185

a. Obtain a normal probability plot of the data.

b. Based on your result from part (a), is it reasonable to presume that serum total cholesterol level of U.S. females 20 years old or older is normally distributed? Explain your answer.

c. Find and interpret a 95% confidence interval for the mean serum total cholesterol level of U.S. females 20 years old or older. The population standard deviation is 44.7 mg/dL.

d. In Exercise 6.98, we noted that the mean serum total cholesterol level of U.S. females 20 years old or older is 206 mg/dL. Does your confidence interval in part (c) contain the population mean? Would it necessarily have to? Explain your answers.

Extending the Concepts and Skills

8.25 New Mobile Homes. Refer to Examples 8.1 and 8.2. Use the data in Table 8.1 on page 332 to obtain a 99.7% confidence interval for the mean price of all new mobile homes. (*Hint:* Proceed as in Example 8.2, but use Property 3 of the empirical rule on page 282 instead of Property 2.)

8.26 New Mobile Homes. Refer to Examples 8.1 and 8.2. Use the data in Table 8.1 on page 332 to obtain a 68% confidence interval for the mean price of all new mobile homes. (*Hint:* Proceed as in Example 8.2, but use Property 1 of the empirical rule on page 282 instead of Property 2.)

8.2 ## Confidence Intervals for One Population Mean When σ Is Known

In Section 8.1, we showed how to find a 95% confidence interval for a population mean, that is, a confidence interval at a confidence level of 95%. In this section, we generalize the arguments used there to obtain a confidence interval for a population mean at any prescribed confidence level.

To begin, we introduce some general notation used with confidence intervals. Frequently, we want to write the confidence level in the form $1 - \alpha$, where α is a number between 0 and 1; that is, if the confidence level is expressed as a decimal, α is the number that must be subtracted from 1 to get the confidence level. To find α, we simply subtract the confidence level from 1. If the confidence level is 95%, then $\alpha = 1 - 0.95 = 0.05$; if the confidence level is 90%, then $\alpha = 1 - 0.90 = 0.10$; and so on.

Next, recall from Section 6.2 that the symbol z_α denotes the z-score that has area α to its right under the standard normal curve. So, for example, $z_{0.05}$ denotes the z-score that has area 0.05 to its right, and $z_{\alpha/2}$ denotes the z-score that has area $\alpha/2$ to its right. Note that, for any normally distributed variable, $100(1 - \alpha)\%$ of all possible observations lie within $z_{\alpha/2}$ standard deviations to either side of the mean. You should draw a graph to verify that result.

Obtaining Confidence Intervals for a Population Mean When σ Is Known

We now develop a step-by-step procedure to obtain a confidence interval for a population mean when the population standard deviation is known. In doing so, we assume that the variable under consideration is normally distributed. Because of the central limit theorem, however, the procedure will also work to obtain an approximately correct confidence interval when the sample size is large, regardless of the distribution of the variable.

The basis of our confidence-interval procedure is stated in Key Fact 7.2: If x is a normally distributed variable with mean μ and standard deviation σ, then, for samples of size n, the variable $\bar{x}$ is also normally distributed and has mean μ and standard deviation $\sigma/\sqrt{n}$. As in Section 8.1, we can use Property 2 of the empirical rule to conclude that approximately 95% of all samples of size n have means within $2 \cdot \sigma/\sqrt{n}$ of μ, as depicted in Fig. 8.3(a).

FIGURE 8.3

(a) Approximately 95% of all samples have means within 2 standard deviations of μ; (b) $100(1 - \alpha)$% of all samples have means within $z_{\alpha/2}$ standard deviations of μ

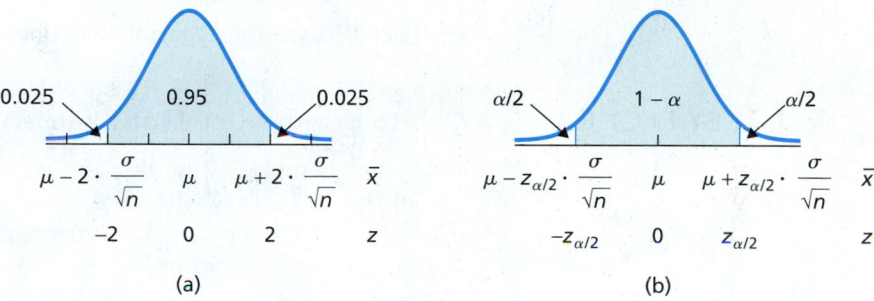

More generally (and more precisely), we can say that $100(1 - \alpha)$% of all samples of size n have means within $z_{\alpha/2} \cdot \sigma/\sqrt{n}$ of μ, as depicted in Fig. 8.3(b). Equivalently, we can say that $100(1 - \alpha)$% of all samples of size n have the property that the interval from

$$\bar{x} - z_{\alpha/2} \cdot \frac{\sigma}{\sqrt{n}} \quad \text{to} \quad \bar{x} + z_{\alpha/2} \cdot \frac{\sigma}{\sqrt{n}}$$

contains μ. Consequently, we have Procedure 8.1, called the **one-mean z-interval procedure**, or, when no confusion can arise, simply the **z-interval procedure**.[†]

PROCEDURE 8.1   **One-Mean z-Interval Procedure**

Purpose To find a confidence interval for a population mean, μ

Assumptions
1. Simple random sample
2. Normal population or large sample
3. σ known

Step 1 For a confidence level of $1 - \alpha$, use Table II to find $z_{\alpha/2}$.

Step 2 The confidence interval for μ is from

$$\bar{x} - z_{\alpha/2} \cdot \frac{\sigma}{\sqrt{n}} \quad \text{to} \quad \bar{x} + z_{\alpha/2} \cdot \frac{\sigma}{\sqrt{n}},$$

where $z_{\alpha/2}$ is found in Step 1, n is the sample size, and $\bar{x}$ is computed from the sample data.

Step 3 Interpret the confidence interval.

Note: The confidence interval is exact for normal populations and is approximately correct for large samples from nonnormal populations.

[†]The one-mean z-interval procedure is also known as the **one-sample z-interval procedure** and the **one-variable z-interval procedure.** We prefer "one-mean" because it makes clear the parameter being estimated.

Note: By saying that the confidence interval is *exact,* we mean that the true confidence level equals $1 - \alpha$; by saying that the confidence interval is *approximately correct,* we mean that the true confidence level only approximately equals $1 - \alpha$.

Before applying Procedure 8.1, we need to make several comments about it and the assumptions for its use.

- We use the term **normal population** as an abbreviation for "the variable under consideration is normally distributed."
- The z-interval procedure works reasonably well even when the variable is not normally distributed and the sample size is small or moderate, provided the variable is not too far from being normally distributed. Thus we say that the z-interval procedure is **robust** to moderate violations of the normality assumption.[†]
- Watch for outliers because their presence calls into question the normality assumption. Moreover, even for large samples, outliers can sometimes unduly affect a z-interval because the sample mean is not resistant to outliers.

Key Fact 8.1 lists some general guidelines for use of the z-interval procedure.

KEY FACT 8.1

When to Use the One-Mean z-Interval Procedure[‡]

- For small samples—say, of size less than 15—the z-interval procedure should be used only when the variable under consideration is normally distributed or very close to being so.
- For samples of moderate size—say, between 15 and 30—the z-interval procedure can be used unless the data contain outliers or the variable under consideration is far from being normally distributed.
- For large samples—say, of size 30 or more—the z-interval procedure can be used essentially without restriction. However, if outliers are present and their removal is not justified, you should compare the confidence intervals obtained with and without the outliers to see what effect the outliers have. If the effect is substantial, use a different procedure or take another sample, if possible.
- If outliers are present but their removal is justified and results in a data set for which the z-interval procedure is appropriate (as previously stated), the procedure can be used.

Key Fact 8.1 makes it clear that you should conduct preliminary data analyses before applying the z-interval procedure. More generally, the following fundamental principle of data analysis is relevant to all inferential procedures.

KEY FACT 8.2

? What Does It Mean?

Always look at the sample data (by constructing a histogram, normal probability plot, boxplot, etc.) prior to performing a statistical-inference procedure to help check whether the procedure is appropriate.

A Fundamental Principle of Data Analysis

Before performing a statistical-inference procedure, examine the sample data. If any of the conditions required for using the procedure appear to be violated, do not apply the procedure. Instead use a different, more appropriate procedure, if one exists.

Even for small samples, where graphical displays must be interpreted carefully, it is far better to examine the data than not to. Remember, though, to proceed cautiously

---

[†]A statistical procedure that works reasonably well even when one of its assumptions is violated (or moderately violated) is called a **robust procedure** relative to that assumption.

[‡]Statisticians also consider skewness. Roughly speaking, the more skewed the distribution of the variable under consideration, the larger is the sample size required for the validity of the z-interval procedure. See, for instance, the paper "How Large Does n Have to Be for Z and t Intervals?" by D. Boos and J. Hughes-Oliver (*The American Statistician*, Vol. 54, No. 2, pp. 121–128).

TABLE 8.3

Some important values of $z_{\alpha/2}$

Confidence level	α	$z_{\alpha/2}$
90%	0.10	1.645
95%	0.05	1.960
99%	0.01	2.575

when conducting graphical analyses of small samples, especially very small samples—say, of size 10 or less.

In Step 1 of Procedure 8.1, we need to find $z_{\alpha/2}$. Using the standard normal table, Table II, we obtained the values of $z_{\alpha/2}$ corresponding to the three most commonly used confidence levels, as shown in Table 8.3. You may find this table handy in constructing confidence intervals that have one of the three most commonly used levels.

Note that, for a 95% confidence interval, we use $z_{0.025} = 1.96$, which is more accurate than the approximate value of 2 given by Property 2 of the empirical rule.

EXAMPLE 8.4   The One-Mean z-Interval Procedure

TABLE 8.4

Ages, in years, of 50 randomly selected people in the civilian labor force

16	37	52	65	36
50	47	51	34	45
40	37	61	46	62
39	40	19	33	59
24	47	45	48	26
60	42	46	33	20
24	31	38	22	61
30	34	70	34	58
61	39	49	41	21
32	60	45	32	27

The Civilian Labor Force   The Bureau of Labor Statistics collects information on the ages of people in the civilian labor force and publishes the results in *Current Population Survey*. Fifty people in the civilian labor force are randomly selected; their ages are displayed in Table 8.4. Find a 95% confidence interval for the mean age, μ, of all people in the civilian labor force. Assume that the population standard deviation of the ages is 12.1 years.

Solution   We note that the population standard deviation is known. Because the sample size is 50, which is large, we need only check for outliers in the age data before applying Procedure 8.1. (See the third bulleted item in Key Fact 8.1.)

To check for outliers, we constructed a boxplot of the age data, as shown in Fig. 8.4. The boxplot indicates no outliers, so we proceed to apply Procedure 8.1 to find the required confidence interval.

Step 1   For a confidence level of $1 - \alpha$, use Table II to find $z_{\alpha/2}$.

We want a 95% confidence interval, so $\alpha = 1 - 0.95 = 0.05$. From Table II or Table 8.3, $z_{\alpha/2} = z_{0.05/2} = z_{0.025} = 1.96$.

Step 2   The confidence interval for μ is from

$$\bar{x} - z_{\alpha/2} \cdot \frac{\sigma}{\sqrt{n}} \quad \text{to} \quad \bar{x} + z_{\alpha/2} \cdot \frac{\sigma}{\sqrt{n}}.$$

We know $\sigma = 12.1$, $n = 50$, and, from Step 1, $z_{\alpha/2} = 1.96$. To compute $\bar{x}$ for the data in Table 8.4, we apply the usual formula:

$$\bar{x} = \frac{\Sigma x_i}{n} = \frac{2069}{50} = 41.4,$$

to one decimal place. Consequently, a 95% confidence interval for μ is from

$$41.4 - 1.96 \cdot \frac{12.1}{\sqrt{50}} \quad \text{to} \quad 41.4 + 1.96 \cdot \frac{12.1}{\sqrt{50}},$$

or 38.0 to 44.8.

FIGURE 8.4

Boxplot of age data

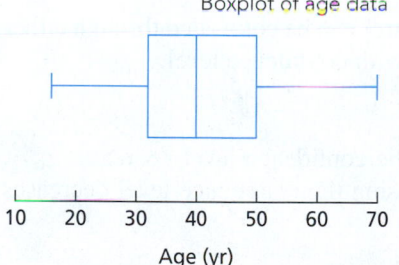

10   20   30   40   50   60   70

Age (yr)

StatCrunch

Report 8.1

You try it!

Exercise 8.69 on page 348

Step 3   Interpret the confidence interval.

Interpretation   We can be 95% confident that the mean age, μ, of all people in the civilian labor force is somewhere between 38.0 years and 44.8 years.

Margin of Error Revisited

At the end of Section 8.1 (see page 335), we introduced the *margin of error*, which indicates the accuracy of our guess (point estimate) for the value of the unknown parameter under consideration. We noted that most confidence intervals that we encounter in our study of statistics will have endpoints of the form

$$\text{point estimate} \pm \text{margin of error}.$$

For the one-mean z-interval procedure for a population mean (Procedure 8.1 on page 339), the point estimate is the sample mean, $\bar{x}$. Referring now to Step 2 of Procedure 8.1, we see that the margin of error for a one-mean z-interval is $z_{\alpha/2} \cdot \sigma/\sqrt{n}$, which we denote by the letter E. Formula 8.1 summarizes our discussion.

FORMULA 8.1

Margin of Error for the Estimate of μ

The **margin of error** for the estimate of μ is $z_{\alpha/2} \cdot \sigma/\sqrt{n}$, which is denoted by the letter E. Thus,

$$E = z_{\alpha/2} \cdot \sigma/\sqrt{n}.$$

See Fig. 8.5.

FIGURE 8.5

Margin of error, E

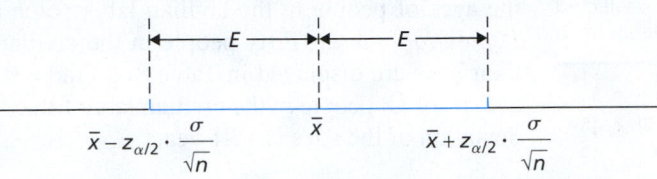

What Does It Mean?

The margin of error for the estimate of a population mean indicates the accuracy with which a sample mean estimates the unknown population mean.

In Fig. 8.5, the blue line represents the confidence interval. We can see from Fig. 8.5 that the margin of error equals half the length of the confidence interval or, equivalently, the length of the confidence interval equals twice the margin of error. So, we can use either the length of the confidence interval or the margin of error to measure the accuracy of our point estimate.

The margin of error indicates the accuracy of our confidence-interval estimate, in this case, the accuracy with which a sample mean estimates the unknown population mean. A small margin of error indicates good accuracy, whereas, a large margin of error indicates poor accuracy.

As we will now see, the size of the margin of error can be controlled through either the confidence level or the sample size. We begin with confidence level.

Confidence and Accuracy

Table 8.3 on page 341 suggests that decreasing the confidence level decreases $z_{\alpha/2}$. Referring now to Formula 8.1, we see that decreasing the confidence level decreases the margin of error. Here is an example.

EXAMPLE 8.5   Confidence and Accuracy

The Civilian Labor Force In Example 8.4 on page 341, we applied the one-mean z-interval procedure to the ages of a sample of 50 people in the civilian labor force to obtain a 95% confidence interval for the mean age, μ, of all people in the civilian labor force. The confidence interval is from 38.0 years to 44.8 years.

a.   Determine the margin of error for the 95% confidence interval.
b.   Find a 90% confidence interval for μ based on the same data.
c.   Compare the 90% and 95% confidence intervals.
d.   Compare the margins of error for the 90% and 95% confidence intervals.

Solution   Recall that $n = 50$, $\bar{x} = 41.4$ years, and $\sigma = 12.1$ years.

a.   We can determine the margin of error, E, by applying Formula 8.1. For a 95% confidence interval, $z_{\alpha/2} = z_{0.025} = 1.96$. Consequently,

$$E = z_{\alpha/2} \cdot \frac{\sigma}{\sqrt{n}} = 1.96 \cdot \frac{12.1}{\sqrt{50}} = 3.4 \text{ years.}$$

Note: Alternatively, we could determine E by taking half the length of the confidence interval: $E = (44.8 - 38.0)/2 = 3.4$ years.

b. For a 90% confidence interval, $z_{\alpha/2} = z_{0.05} = 1.645$. Thus, by Procedure 8.1, the resulting confidence interval, using the same sample data (Table 8.4), is from

$$41.4 - 1.645 \cdot \frac{12.1}{\sqrt{50}} \quad \text{to} \quad 41.4 + 1.645 \cdot \frac{12.1}{\sqrt{50}},$$

or 38.6 years to 44.2 years.

c. Figure 8.6 shows both the 90% and 95% confidence intervals.

FIGURE 8.6

90% and 95% confidence intervals for μ, using the data in Table 8.4

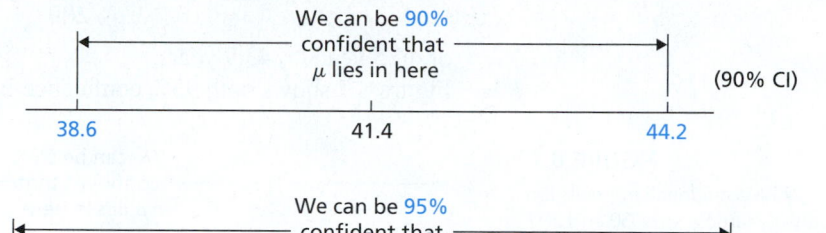

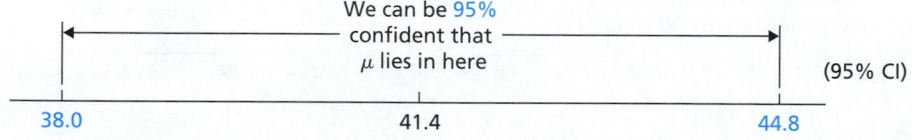

d. Because, as we see from Fig. 8.6, the 90% confidence interval is shorter than the 95% confidence interval, we can conclude that the margin of error for the former is less than that for the latter. More precisely, by applying Formula 8.1 (or taking half the length of the confidence interval found in part (b) above), we find that the margin of error for the 90% confidence interval is 2.8 years. This margin of error is indeed smaller than that for the 95% confidence interval, which, by part (a), is 3.4 years.

You try it!

Exercises 8.77 & 8.79 on page 349

Interpretation　Decreasing the confidence level decreases the margin of error.

KEY FACT 8.3　**Confidence and Accuracy**

For a fixed sample size, decreasing the confidence level decreases the margin of error and, hence, improves the accuracy of a confidence-interval estimate.

Sample Size and Accuracy

Next we consider how the size of the margin of error can be controlled through the sample size. Because the sample size, n, appears in the denominator of the expression for the margin of error, E, in Formula 8.1, it follows that increasing the sample size decreases the margin of error. This fact, of course, makes sense, because we expect more accurate information from larger samples. Here is an example.

EXAMPLE 8.6　**Sample Size and Accuracy**

The Civilian Labor Force　In Example 8.4 on page 341, we applied the one-mean z-interval procedure to the ages of a sample of 50 people in the civilian labor force to obtain a 95% confidence interval for the mean age, μ, of all people in the civilian labor force. The confidence interval is from 38.0 years to 44.8 years.

a. Determine the margin of error for the 95% confidence interval.
b. A random sample of 200 people in the civilian labor force gave a mean age of 42.2 years. Find a 95% confidence interval for μ based on that data.
c. Compare the two 95% confidence intervals.
d. Compare the margins of error for the two 95% confidence intervals.

Solution Recall that $\sigma = 12.1$ years and, moreover, for a 95% confidence interval, $z_{\alpha/2} = z_{0.025} = 1.96$.

a. In Example 8.5(a), we determined that the margin of error, E, for the sample of size 50 is 3.4 years.

b. For the sample of size 200, we have $n = 200$ and $\bar{x} = 42.2$ years. Therefore, by Procedure 8.1, the confidence interval is from

$$42.2 - 1.96 \cdot \frac{12.1}{\sqrt{200}} \quad \text{to} \quad 42.2 + 1.96 \cdot \frac{12.1}{\sqrt{200}},$$

or 40.5 years to 43.9 years.

c. Figure 8.7 shows both 95% confidence intervals.

FIGURE 8.7

95% confidence intervals for μ with sample sizes 50 and 200

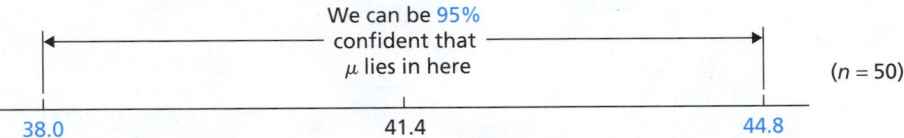

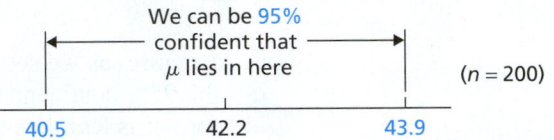

d. Because, as we see from Fig. 8.7, the confidence interval with $n = 200$ is shorter than the confidence interval with $n = 50$, we can conclude that the margin of error for the former is less than that for the latter. More precisely, by applying Formula 8.1 (or taking half the length of the confidence interval in part (b) above), we find that the margin of error for the confidence interval when $n = 200$ is 1.7 years. This margin of error is indeed smaller than that when $n = 50$, which, by part (a), is 3.4 years.

You try it!

Exercise 8.81 on page 349

Interpretation Increasing the sample size decreases the margin of error.

KEY FACT 8.4

Sample Size and Accuracy

For a fixed confidence level, increasing the sample size decreases the margin of error and, hence, improves the accuracy of a confidence-interval estimate.

Determining the Required Sample Size

If the margin of error and confidence level are specified in advance, then we must determine the sample size needed to meet those specifications. To find the formula for the required sample size, we solve the margin-of-error formula, $E = z_{\alpha/2} \cdot \sigma/\sqrt{n}$, for n. The result is given in Formula 8.2.

FORMULA 8.2

Sample Size for Estimating μ

The sample size required for a $(1 - \alpha)$-level confidence interval for μ with a specified margin of error, E, is given by the formula

$$n = \left(\frac{z_{\alpha/2} \cdot \sigma}{E}\right)^2,$$

rounded up to the nearest whole number.

EXAMPLE 8.7   **Sample Size for Estimating μ**

The Civilian Labor Force Consider again the problem of estimating the mean age, μ, of all people in the civilian labor force.

a. Determine the sample size needed in order to be 95% confident that μ is within 0.5 year of the point estimate, $\bar{x}$. Recall that $\sigma = 12.1$ years.
b. Find a 95% confidence interval for μ if a sample of the size determined in part (a) has a mean age of 43.8 years.

Solution

a. To find the sample size, we use Formula 8.2. We know that $\sigma = 12.1$ and we want a margin of error of 0.5, that is, $E = 0.5$. The confidence level is 0.95, which means that $\alpha = 0.05$ and $z_{\alpha/2} = z_{0.025} = 1.96$. Thus by Formula 8.2,

$$n = \left(\frac{z_{\alpha/2} \cdot \sigma}{E} \right)^2 = \left(\frac{1.96 \cdot 12.1}{0.5} \right)^2 = 2249.79,$$

which, rounded up to the nearest whole number, is 2250.

Interpretation If 2250 people in the civilian labor force are randomly selected, we can be 95% confident that the mean age of all people in the civilian labor force is within 0.5 year of the mean age of the people in the sample.

b. Applying Procedure 8.1 with $\alpha = 0.05$, $\sigma = 12.1$, $\bar{x} = 43.8$, and $n = 2250$, we get the confidence interval

$$43.8 - 1.96 \cdot \frac{12.1}{\sqrt{2250}} \quad \text{to} \quad 43.8 + 1.96 \cdot \frac{12.1}{\sqrt{2250}},$$

or 43.3 to 44.3.

Exercise 8.89
on page 350

Interpretation We can be 95% confident that the mean age, μ, of all people in the civilian labor force is somewhere between 43.3 years and 44.3 years. ■

Note: The sample size of 2250 was determined in part (a) of Example 8.7 to guarantee a margin of error of 0.5 year for a 95% confidence interval. Therefore, instead of applying Procedure 8.1 to find the confidence interval required in part (b) of Example 8.7, we could have simply computed

$$\text{point estimate} \pm \text{margin of error} = \bar{x} \pm E = 43.8 \pm 0.5.$$

Doing so would give the same confidence interval, 43.3 to 44.3, but with much less work. The simpler method might have yielded a somewhat wider confidence interval because the sample size is rounded up. Hence, this simpler method gives, at worst, a slightly conservative estimate, so it is acceptable in practice.

Two additional noteworthy items are the following:

- The formula for finding the required sample size, Formula 8.2, involves the population standard deviation, σ, which is usually unknown. In such cases, we can take a preliminary large sample, say, of size 30 or more, and use the sample standard deviation, s, in place of σ in Formula 8.2.
- Ideally, we want both a high confidence level and a small margin of error. Accomplishing these specifications generally takes a large sample size. However, available resources (e.g., money or personnel) often place a restriction on the size of the sample that can be used, requiring us to perhaps lower our confidence level or increase our margin of error. Exercises 8.95 and 8.96 explore such situations.

THE TECHNOLOGY CENTER

Most statistical technologies have programs that automatically perform the one-mean z-interval procedure. In this subsection, we present output and step-by-step instructions for such programs.

EXAMPLE 8.8   Using Technology to Obtain a One-Mean z-Interval

The Civilian Labor Force Table 8.4 on page 341 displays the ages of 50 randomly selected people in the civilian labor force. Use Minitab, Excel, or the TI-83/84 Plus to determine a 95% confidence interval for the mean age, μ, of all people in the civilian labor force. Assume that the population standard deviation of the ages is 12.1 years.

Solution We applied the one-mean z-interval programs to the data, resulting in Output 8.1. Steps for generating that output are presented in Instructions 8.1. *Note to Excel users:* For brevity, we have presented only the essential portions of the actual output.

OUTPUT 8.1   One-mean z-interval on the sample of ages

MINITAB

```
One-Sample Z: AGE

The assumed standard deviation = 12.1


Variable    N    Mean   StDev   SE Mean      95% CI
AGE        50   41.38   13.56     1.71    (38.03, 44.73)
```

EXCEL

Summary statistics:			
Variable	Observations	Mean	Std. deviation
AGE	50	41.3800	13.5645
95% confidence interval on the mean:			
(38.0261, 44.7339)			

TI-83/84 PLUS

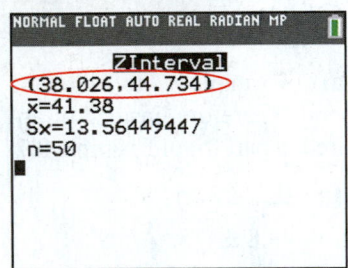

As shown in Output 8.1, the required 95% confidence interval is from 38.03 to 44.73. We can be 95% confident that the mean age of all people in the civilian labor force is somewhere between 38.0 years and 44.7 years. Compare this confidence interval to the one obtained in Example 8.4. Can you explain the slight discrepancy?

INSTRUCTIONS 8.1   Steps for generating Output 8.1

MINITAB

1 Store the data from Table 8.4 in a column named AGE
2 Choose **Stat ➤ Basic Statistics ➤ 1-Sample Z...**
3 Press the F3 key to reset the dialog box
4 Click in the text box directly below the **One or more samples, each in a column** drop-down list box and specify AGE
5 Click in the **Known standard deviation** text box and type 12.1
6 Click the **Options...** button
7 Type 95 in the **Confidence level** text box
8 Click **OK** twice

EXCEL

1 Store the data from Table 8.4 in a column named AGE
2 Choose **XLSTAT ➤ Parametric tests ➤ One-sample t-test and z-test**
3 Click the reset button in the lower left corner of the dialog box
4 Click in the **Data** selection box and then select the column of the worksheet that contains the AGE data
5 Check the **z test** check box and uncheck the **Student's t test** check box

(continued)

EXCEL	TI-83/84 PLUS
6 Click the **Options** tab 7 Type <u>5</u> in the **Significance level (%)** text box 8 In the **Variance for the z-test** list, select the **User defined** option button 9 Type <u>146.41</u> in the **Variance** text box 10 Click **OK** 11 Click the **Continue** button in the **XLSTAT – Selections** dialog box	1 Store the data from Table 8.4 in a list named AGE 2 Press **STAT**, arrow over to **TESTS**, and press **7** 3 Highlight **Data** and press **ENTER** 4 Press the down-arrow key, type <u>12.1</u> for σ, and press **ENTER** 5 Press **2nd ➤ LIST** 6 Arrow down to AGE and press **ENTER** twice 7 Type <u>1</u> for **Freq** and then press **ENTER** 8 Type <u>.95</u> for **C-Level** and press **ENTER** twice

Notes to Excel users:

- Step 7 of the Excel instructions states to type 5 in the **Significance level (%)** text box. Indeed, for this procedure, XLSTAT uses α (i.e., 1 minus the confidence level), expressed as a percentage, instead of the confidence level. So, for instance, type 5 for a 95% confidence interval and type 10 for a 90% confidence interval.
- Step 9 of the Excel instructions states to type 146.41 in the **Variance** text box. Note that 146.41 is the assumed population variance of the ages of all people in the civilian labor force; it is the square of 12.1, the assumed population standard deviation of all people in the civilian labor force. In general, the **Variance** text box requires the assumed population variance (square of the assumed population standard deviation) of the variable under consideration.

Exercises 8.2

Understanding the Concepts and Skills

8.27 Find the confidence level and α for
a. a 90% confidence interval.
b. a 99% confidence interval.

8.28 Find the confidence level and α for
a. an 85% confidence interval.
b. a 95% confidence interval.

8.29 What is meant by saying that a $1 - \alpha$ confidence interval is
a. exact?   **b.** approximately correct?

8.30 In developing Procedure 8.1, we assumed that the variable under consideration is normally distributed.
a. Explain why we needed that assumption.
b. Explain why the procedure yields an approximately correct confidence interval for large samples, regardless of the distribution of the variable under consideration.

8.31 For what is *normal population* an abbreviation?

8.32 Refer to Procedure 8.1.
a. Explain in detail the assumptions required for using the z-interval procedure.
b. How important is the normality assumption? Explain your answer.

8.33 What is meant by saying that a statistical procedure is robust?

In each of Exercises 8.34–8.39, assume that the population standard deviation is known and decide whether use of the z-interval procedure to obtain a confidence interval for the population mean is reasonable. Explain your answers.

8.34 The variable under consideration is very close to being normally distributed, and the sample size is 10.

8.35 The variable under consideration is very close to being normally distributed, and the sample size is 75.

8.36 The sample data contain outliers, and the sample size is 20.

8.37 The sample data contain no outliers, the variable under consideration is roughly normally distributed, and the sample size is 20.

8.38 The distribution of the variable under consideration is highly skewed, and the sample size is 20.

8.39 The sample data contain no outliers, the sample size is 250, and the variable under consideration is far from being normally distributed.

8.40 Suppose that you have obtained data by taking a random sample from a population. Before performing a statistical inference, what should you do?

8.41 Suppose that you have obtained data by taking a random sample from a population and that you intend to find a confidence interval for the population mean, μ. Which confidence level, 95% or 99%, will result in the confidence interval giving a more accurate estimate of μ?

8.42 Suppose that you will be taking a random sample from a population and that you intend to find a 95% confidence interval for the population mean, μ. Which sample size, 25 or 50, will result in the confidence interval giving a more accurate estimate of μ?

8.43 Discuss the relationship between the margin of error and the standard error of the mean.

8.44 Explain why the margin of error determines the accuracy with which a sample mean estimates a population mean.

In each of Exercises 8.45–8.48, explain the effect on the margin of error and hence the effect on the accuracy of estimating a population mean by a sample mean.

8.45 Increasing the sample size while keeping the same confidence level

8.46 Decreasing the confidence level while keeping the same sample size

8.47 Increasing the confidence level while keeping the same sample size

8.48 Decreasing the sample size while keeping the same confidence level

8.49 A confidence interval for a population mean has a margin of error of 3.4.
a. Determine the length of the confidence interval.
b. If the sample mean is 52.8, obtain the confidence interval.
c. Construct a graph that illustrates your results.

8.50 A confidence interval for a population mean has a margin of error of 0.047.
a. Determine the length of the confidence interval.
b. If the sample mean is 0.205, obtain the confidence interval.
c. Construct a graph that illustrates your results.

8.51 A confidence interval for a population mean has length 20.
a. Determine the margin of error.
b. If the sample mean is 60, obtain the confidence interval.
c. Construct a graph that illustrates your results.

8.52 A confidence interval for a population mean has a length of 162.6.
a. Determine the margin of error.
b. If the sample mean is 643.1, determine the confidence interval.
c. Construct a graph that illustrates your results.

In each of Exercises 8.53–8.60, answer true or false to each statement concerning a confidence interval for a population mean. Give reasons for your answers.

8.53 The length of the confidence interval can be determined if you know only the margin of error.

8.54 The margin of error can be determined if you know only the length of the confidence interval.

8.55 The confidence interval can be obtained if you know only the margin of error.

8.56 The confidence interval can be obtained if you know only the margin of error and the sample mean.

8.57 The margin of error can be determined if you know only the confidence level.

8.58 The confidence level can be determined if you know only the margin of error.

8.59 The margin of error can be determined if you know only the confidence level, population standard deviation, and sample size.

8.60 The confidence level can be determined if you know only the margin of error, population standard deviation, and sample size.

8.61 Formula 8.2 on page 344 provides a method for computing the sample size required to obtain a confidence interval with a specified confidence level and margin of error. The number resulting from the formula should be rounded up to the nearest whole number.
a. Why do we want a whole number?
b. Why do we round up instead of down?

8.62 The margin of error is also called the maximum error of the estimate. Explain why.

In each of Exercises 8.63–8.68, we provide a sample mean, sample size, population standard deviation, and confidence level. In each case, perform the following tasks:
a. *Use the one-mean z-interval procedure to find a confidence interval for the mean of the population from which the sample was drawn.*
b. *Obtain the margin of error by taking half the length of the confidence interval.*
c. *Obtain the margin of error by using Formula 8.1 on page 342.*

8.63 $\bar{x} = 20$, $n = 36$, $\sigma = 3$, confidence level $= 95\%$

8.64 $\bar{x} = 25$, $n = 36$, $\sigma = 3$, confidence level $= 95\%$

8.65 $\bar{x} = 30$, $n = 25$, $\sigma = 4$, confidence level $= 90\%$

8.66 $\bar{x} = 35$, $n = 25$, $\sigma = 4$, confidence level $= 90\%$

8.67 $\bar{x} = 50$, $n = 16$, $\sigma = 5$, confidence level $= 99\%$

8.68 $\bar{x} = 55$, $n = 16$, $\sigma = 5$, confidence level $= 99\%$

Applying the Concepts and Skills

Preliminary data analyses indicate that you can reasonably apply the z-interval procedure (Procedure 8.1 on page 339) in Exercises 8.69–8.74.

8.69 **Venture-Capital Investments.** Data on investments in the high-tech industry by venture capitalists are compiled by VentureOne Corporation and published in *America's Network Telecom Investor Supplement*. A random sample of 18 venture-capital investments in the fiber optics business sector yielded the following data, in millions of dollars.

5.60	6.27	5.96	10.51	2.04	5.48
5.74	5.58	4.13	8.63	5.95	6.67
4.21	7.71	9.21	4.98	8.64	6.66

Determine and interpret a 95% confidence interval for the mean amount, μ, of all venture-capital investments in the fiber optics business sector. Assume that the population standard deviation is $2.04 million. (*Note:* The sum of the data is $113.97 million.)

8.70 **Poverty and Dietary Calcium.** Calcium is the most abundant mineral in the human body and has several important functions. Most body calcium is stored in the bones and teeth, where it functions to support their structure. Recommendations for calcium are provided in *Dietary Reference Intakes*, developed by the Institute of Medicine of the National Academy of Sciences. The recommended adequate intake (RAI) of calcium for adults (ages 19–50) is 1000 milligrams (mg) per day. A simple random sample of 18 adults with incomes below the poverty level gave the following daily calcium intakes.

886	633	943	847	934	841
1193	820	774	834	1050	1058
1192	975	1313	872	1079	809

Determine and interpret a 95% confidence interval for the mean calcium intake, μ, of all adults with incomes below the poverty level. Assume that the population standard deviation is 188 mg. (*Note:* The sum of the data is 17,053 mg.)

8.71 Toxic Mushrooms? Cadmium, a heavy metal, is toxic to animals. Mushrooms, however, are able to absorb and accumulate cadmium at high concentrations. The Czech and Slovak governments have set a safety limit for cadmium in dry vegetables at 0.5 part per million (ppm). M. Melgar et al. measured the cadmium levels in a random sample of the edible mushroom *Boletus pinicola* and published the results in the paper "Influence of Some Factors in Toxicity and Accumulation of Cd from Edible Wild Macrofungi in NW Spain" (*Journal of Environmental Science and Health*, Vol. B33(4), pp. 439–455). Here are the data obtained by the researchers.

0.24	0.59	0.62	0.16	0.77	1.33
0.92	0.19	0.33	0.25	0.59	0.32

Find and interpret a 99% confidence interval for the mean cadmium level of all *Boletus pinicola* mushrooms. Assume a population standard deviation of cadmium levels in *Boletus pinicola* mushrooms of 0.37 ppm. (*Note:* The sum of the data is 6.31 ppm.)

8.72 Smelling Out the Enemy. Snakes deposit chemical trails as they travel through their habitats. These trails are often detected and recognized by lizards, which are potential prey. The ability to recognize their predators via tongue flicks can often mean life or death for lizards. Scientists from the University of Antwerp were interested in quantifying the responses of juveniles of the common lizard (*Lacerta vivipara*) to natural predator cues to determine whether the behavior is learned or congenital. Seventeen juvenile common lizards were exposed to the chemical cues of the viper snake. Their responses, in number of tongue flicks per 20 minutes, are presented in the following table. [SOURCE: Van Damme et al., "Responses of Naïve Lizards to Predator Chemical Cues," *Journal of Herpetology*, Vol. 29(1), pp. 38–43]

425	510	629	236	654	200
276	501	811	332	424	674
676	694	710	662	633	

Find and interpret a 90% confidence interval for the mean number of tongue flicks per 20 minutes for all juvenile common lizards. Assume a population standard deviation of 190.0.

8.73 Political Prisoners. A. Ehlers et al. studied various characteristics of political prisoners from the former East Germany and presented their findings in the paper "Posttraumatic Stress Disorder (PTSD) Following Political Imprisonment: The Role of Mental Defeat, Alienation, and Perceived Permanent Change" (*Journal of Abnormal Psychology*, Vol. 109, pp. 45–55). According to the article, the mean duration of imprisonment for 32 patients with chronic PTSD was 33.4 months. Assuming that $\sigma = 42$ months, determine a 95% confidence interval for the mean duration of imprisonment, μ, of all East German political prisoners with chronic PTSD. Interpret your answer in words.

8.74 Keep on Rolling. The Rolling Stones, a rock group formed in the 1960s, have toured extensively in support of new albums. *Pollstar* has collected data on the earnings from the Stones's North American tours. For 30 randomly selected Rolling Stones concerts, the mean gross earnings is $2.27 million. Assuming a population standard deviation gross earnings of $0.5 million, obtain a 99% confidence interval for the mean gross earnings of all Rolling Stones concerts. Interpret your answer in words.

8.75 Venture-Capital Investments. Refer to Exercise 8.69.
a. Find a 99% confidence interval for μ.

b. Why is the confidence interval you found in part (a) longer than the one in Exercise 8.69?
c. Draw a graph similar to that shown in Fig. 8.6 on page 343 to display both confidence intervals.
d. Which confidence interval yields a more accurate estimate of μ? Explain your answer.

8.76 Poverty and Dietary Calcium. Refer to Exercise 8.70.
a. Find a 90% confidence interval for μ.
b. Why is the confidence interval you found in part (a) shorter than the one in Exercise 8.70?
c. Draw a graph similar to that shown in Fig. 8.6 on page 343 to display both confidence intervals.
d. Which confidence interval yields a more accurate estimate of μ? Explain your answer.

8.77 Medical Marijuana. An issue with legalization of medical marijuana is "diversion," the process in which medical marijuana prescribed for one person is given, traded, or sold to someone who is not registered for medical marijuana use. Researchers S. Sautel et al. study the issue of diversion in the article "Medical Marijuana Use Among Adolescents in Substance Abuse Treatment" (*Journal of the American Academy of Child and Adolescent Psychiatry*, Vol. 51, No. 7, pp. 694–702). The mean number of days that 120 adolescents in substance abuse treatment used medical marijuana in the last 6 months was 102.72. Assume the population standard deviation is 32 days.
a. Find a 95% confidence interval for the mean number of days, μ, of diverted medical marijuana use in the last 6 months of all adolescents in substance abuse treatment.
b. Repeat part (a) at a 90% confidence level.
c. Draw a graph similar to Fig. 8.6 on page 343 to display both confidence intervals.
d. Which confidence interval yields a more accurate estimate of μ? Explain your answer.

8.78 American Alligators. Multi-sensor data loggers were attached to free-ranging American alligators in a study conducted by Y. Watanabe for the article "Behavior of American Alligators Monitored by Multi-Sensor Data Loggers" (*Aquatic Biology*, Vol. 18, pp. 1–8). The mean duration for a sample of 68 dives was 338.0 seconds. Assume the population standard deviation is 100 seconds.
a. Find a 95% confidence interval for the mean duration, μ, of an American-alligator dive.
b. Repeat part (a) at a 99% confidence level.
c. Draw a graph similar to Fig. 8.6 on page 343 to display both confidence intervals.
d. Which confidence interval yields a more accurate estimate of μ? Explain your answer.

8.79 Medical Marijuana. Refer to Exercise 8.77.
a. Determine the margin of error for the 95% confidence interval.
b. Determine the margin of error for the 90% confidence interval.
c. Compare the margins of error found in parts (a) and (b).
d. What principle is being illustrated?

8.80 American Alligators. Refer to Exercise 8.78.
a. Determine the margin of error for the 95% confidence interval.
b. Determine the margin of error for the 99% confidence interval.
c. Compare the margins of error found in parts (a) and (b).
d. What principle is being illustrated?

8.81 Medical Marijuana. Refer to Exercise 8.77.
a. The mean number of days that 30 adolescents in substance abuse treatment used medical marijuana in the last 6 months was 105.43. Find a 95% confidence interval for μ based on that data.

b. Compare the 95% confidence intervals obtained here and in Exercise 8.77(a) by drawing a graph similar to Fig. 8.7 on page 344.

c. Compare the margins of error for the two 95% confidence intervals.

d. What principle is being illustrated?

8.82 American Alligators. Refer to Exercise 8.78.

a. The mean duration for a sample of 612 dives was 322 seconds. Find a 99% confidence interval for μ based on that data.

b. Compare the 99% confidence intervals obtained here and in Exercise 8.78(b) by drawing a graph similar to Fig. 8.7 on page 344.

c. Compare the margins of error for the two 99% confidence intervals.

d. What principle is being illustrated?

8.83 Prices of New Mobile Homes. Recall that a simple random sample of 36 new mobile homes yielded the prices, in thousands of dollars, shown in Table 8.1 on page 332. We found the mean of those prices to be $63.28 thousand.

a. Use this information and Procedure 8.1 on page 339 to find a 95% confidence interval for the mean price of all new mobile homes. Recall that $\sigma = \$7.2$ thousand.

b. Compare your 95% confidence interval in part (a) to the one found in Example 8.2(c) on page 334 and explain any discrepancy that you observe.

8.84 Body Fat. J. McWhorter et al. of the College of Health Sciences at the University of Nevada, Las Vegas, studied physical therapy students during their graduate-school years. The researchers were interested in the fact that, although graduate physical-therapy students are taught the principles of fitness, some have difficulty finding the time to implement those principles. In the study, published as "An Evaluation of Physical Fitness Parameters for Graduate Students" (*Journal of American College Health*, Vol. 51, No. 1, pp. 32–37), a sample of 27 female graduate physical-therapy students had a mean of 22.46 percent body fat.

a. Assuming that percent body fat of female graduate physical-therapy students is normally distributed with standard deviation 4.10 percent body fat, determine a 95% confidence interval for the mean percent body fat of all female graduate physical-therapy students.

b. Obtain the margin of error, E, for the confidence interval you found in part (a).

c. Explain the meaning of E in this context in terms of the accuracy of the estimate.

d. Determine the sample size required to have a margin of error of 1.55 percent body fat with a 99% confidence level.

8.85 Pulmonary Hypertension. In the paper "Persistent Pulmonary Hypertension of the Neonate and Asymmetric Growth Restriction" (*Obstetrics & Gynecology*, Vol. 91, No. 3, pp. 336–341), M. Williams et al. reported on a study of characteristics of neonates. Infants treated for pulmonary hypertension, called the PH group, were compared with those not so treated, called the control group. One of the characteristics measured was head circumference. The mean head circumference of the 10 infants in the PH group was 34.2 centimeters (cm).

a. Assuming that head circumferences for infants treated for pulmonary hypertension are normally distributed with standard deviation 2.1 cm, determine a 90% confidence interval for the mean head circumference of all such infants.

b. Obtain the margin of error, E, for the confidence interval you found in part (a).

c. Explain the meaning of E in this context in terms of the accuracy of the estimate.

d. Determine the sample size required to have a margin of error of 0.5 cm with a 95% confidence level.

8.86 Fuel Expenditures. In estimating the mean monthly fuel expenditure, μ, per household vehicle, the Energy Information Administration takes a sample of size 6841. Assuming that $\sigma = \$20.65$, determine the margin of error in estimating μ at the 95% level of confidence.

8.87 Venture-Capital Investments. In Exercise 8.69, you found a 95% confidence interval for the mean amount of all venture-capital investments in the fiber optics business sector to be from $5.389 million to $7.274 million. Obtain the margin of error by

a. taking half the length of the confidence interval.

b. using Formula 8.1 on page 342. (Recall that $n = 18$ and that $\sigma = \$2.04$ million.)

8.88 Smelling Out the Enemy. In Exercise 8.72, you found a 90% confidence interval for the mean number of tongue flicks per 20 minutes for all juvenile common lizards to be from 456.4 to 608.0. Obtain the margin of error by

a. taking half the length of the confidence interval.

b. using Formula 8.1 on page 342. (Recall that $n = 17$ and that $\sigma = 190.0$.)

8.89 Political Prisoners. In Exercise 8.73, you found a 95% confidence interval of 18.8 months to 48.0 months for the mean duration of imprisonment, μ, of all East German political prisoners with chronic PTSD.

a. Determine the margin of error, E.

b. Explain the meaning of E in this context in terms of the accuracy of the estimate.

c. Find the sample size required to have a margin of error of 12 months and a 99% confidence level. (Recall that $\sigma = 42$ months.)

d. Find a 99% confidence interval for the mean duration of imprisonment, μ, if a sample of the size determined in part (c) has a mean of 36.2 months.

8.90 Keep on Rolling. In Exercise 8.74, you found a 99% confidence interval of $2.03 million to $2.51 million for the mean gross earnings of all Rolling Stones concerts.

a. Determine the margin of error, E.

b. Explain the meaning of E in this context in terms of the accuracy of the estimate.

c. Find the sample size required to have a margin of error of $0.1 million and a 95% confidence level. (Recall that $\sigma = \$0.5$ million.)

d. Obtain a 95% confidence interval for the mean gross earnings if a sample of the size determined in part (c) has a mean of $2.35 million.

8.91 LEDs and CFLs. Light-emitting diodes (LEDs) and compact fluorescent lights (CFLs) are lightbulbs that are supposed to last up to fifty times longer than old fashioned incandescent lightbulbs and also use less energy. *Consumer Reports* sampled eighteen different 60-watt LED and CFL lightbulbs. The following table lists their brightness, in lumens. Use the technology of your choice to decide whether applying the z-interval procedure to these data is reasonable. Explain your answer.

900	690	910	745	765	820
700	360	925	805	850	730
865	680	750	735	865	635

8.92 Long Drives. The Professional Golfer's Association of America (PGA) organizes golf tournaments for professional golfers. The following table lists the longest drives, in yards, recorded during a

PGA tournament for a random sample of 26 golfers. Use the technology of your choice to decide whether applying the z-interval procedure to these data is reasonable. Explain your answer.

395	400	377	367	386	407	383
396	371	376	373	384	369	391
386	393	374	366	388	371	416
375	450	370	379	381		

8.93 Doing Time. The U.S. Department of Justice, Office of Justice Programs, Bureau of Justice Statistics provides information on prison sentences in the document *National Corrections Reporting Program*. A random sample of 20 maximum sentences for murder yielded the data, in months, presented on the WeissStats site. Use the technology of your choice to do the following.
a. Find a 95% confidence interval for the mean maximum sentence of all murders. Assume a population standard deviation of 30 months.
b. Obtain a normal probability plot, boxplot, histogram, and stem-and-leaf diagram of the data.
c. Remove the outliers (if any) from the data, and then repeat part (a).
d. Comment on the advisability of using the z-interval procedure on these data.

8.94 Ages of Diabetics. According to the document *All About Diabetes*, found on the website of the American Diabetes Association, "...diabetes is a disease in which the body does not produce or properly use insulin, a hormone that is needed to convert sugar, starches, and other food into energy needed for daily life." A random sample of 15 diabetics yielded the data on ages, in years, presented on the WeissStats site. Use the technology of your choice to do the following.
a. Find a 95% confidence interval for the mean age, μ, of all people with diabetes. Assume that $\sigma = 21.2$ years.
b. Obtain a normal probability plot, boxplot, histogram, and stem-and-leaf diagram of the data.
c. Remove the outliers (if any) from the data, and then repeat part (a).
d. Comment on the advisability of using the z-interval procedure on these data.

8.95 Civilian Labor Force. Consider again the problem of estimating the mean age, μ, of all people in the civilian labor force. In Example 8.7 on page 345, we found that a sample size of 2250 is required to have a margin of error of 0.5 year and a 95% confidence level. Suppose that, due to financial constraints, the largest sample size possible is 900. Determine the smallest margin of error, given that the confidence level is to be kept at 95%. Recall that $\sigma = 12.1$ years.

8.96 Civilian Labor Force. Consider again the problem of estimating the mean age, μ, of all people in the civilian labor force. In Example 8.7 on page 345, we found that a sample size of 2250 is required to have a margin of error of 0.5 year and a 95% confidence level. Suppose that, due to financial constraints, the largest sample size possible is 900. Determine the greatest confidence level, given that the margin of error is to be kept at 0.5 year. Recall that $\sigma = 12.1$ years.

8.97 Millionaires. Professor Thomas Stanley of Georgia State University has surveyed millionaires since 1973. Among other information, Professor Stanley obtains estimates for the mean age, μ, of all U.S. millionaires. Suppose that one year's study involved a simple random sample of 36 U.S. millionaires whose mean age was 58.53 years with a sample standard deviation of 13.36 years.
a. If, for next year's study, a confidence interval for μ is to have a margin of error of 2 years and a confidence level of 95%, determine the required sample size.

b. Why did you use the sample standard deviation, $s = 13.36$, in place of σ in your solution to part (a)? Why is it permissible to do so?

8.98 Corporate Farms. The U.S. Census Bureau estimates the mean value of the land and buildings per corporate farm. Those estimates are published in the *Census of Agriculture*. Suppose that an estimate, $\bar{x}$, is obtained and that the margin of error is $1000. Does this result imply that the true mean, μ, is within $1000 of the estimate? Explain your answer.

Working with Large Data Sets

8.99 Body Temperature. A study by researchers at the University of Maryland addressed the question of whether the mean body temperature of humans is 98.6°F. The results of the study by P. Mackowiak et al. appeared in the article "A Critical Appraisal of 98.6°F, the Upper Limit of the Normal Body Temperature, and Other Legacies of Carl Reinhold August Wunderlich" (*Journal of the American Medical Association*, Vol. 268, pp. 1578–1580). Among other data, the researchers obtained the body temperatures of 93 healthy humans, as provided on the WeissStats site. Use the technology of your choice to do the following.
a. Obtain a normal probability plot, boxplot, histogram, and stem-and-leaf diagram of the data.
b. Based on your results from part (a), can you reasonably apply the z-interval procedure to the data? Explain your reasoning.
c. Find and interpret a 99% confidence interval for the mean body temperature of all healthy humans. Assume that $\sigma = 0.63°F$. Does the result surprise you? Why?

8.100 Malnutrition and Poverty. R. Reifen et al. studied various nutritional measures of Ethiopian school children and published their findings in the paper "Ethiopian-Born and Native Israeli School Children Have Different Growth Patterns" (*Nutrition*, Vol. 19, pp. 427–431). The study, conducted in Azezo, North West Ethiopia, found that malnutrition is prevalent in primary and secondary school children because of economic poverty. The weights, in kilograms (kg), of 60 randomly selected male Ethiopian-born school children of ages 12–15 years are presented on the WeissStats site. Use the technology of your choice to do the following.
a. Obtain a normal probability plot, boxplot, histogram, and stem-and-leaf diagram of the data.
b. Based on your results from part (a), can you reasonably apply the z-interval procedure to the data? Explain your reasoning.
c. Find and interpret a 95% confidence interval for the mean weight of all male Ethiopian-born school children of ages 12–15 years. Assume that the population standard deviation is 4.5 kg.

8.101 Clocking the Cheetah. The cheetah (*Acinonyx jubatus*) is the fastest land mammal and is highly specialized to run down prey. The cheetah often exceeds speeds of 60 mph and, according to the online document "Cheetah Conservation in Southern Africa" (*Trade & Environment Database (TED) Case Studies*, Vol. 8, No. 2) by J. Urbaniak, the cheetah is capable of speeds up to 72 mph. The WeissStats site contains the top speeds, in miles per hour, for a sample of 35 cheetahs. Use the technology of your choice to do the following tasks.
a. Find a 95% confidence interval for the mean top speed, μ, of all cheetahs. Assume that the population standard deviation of top speeds is 3.2 mph.
b. Obtain a normal probability plot, boxplot, histogram, and stem-and-leaf diagram of the data.
c. Remove the outliers (if any) from the data, and then repeat part (a).
d. Comment on the advisability of using the z-interval procedure on these data.

Extending the Concepts and Skills

8.102 Class Project: Gestation Periods of Humans. This exercise can be done individually or, better yet, as a class project. Gestation periods of humans are normally distributed with a mean of 266 days and a standard deviation of 16 days.

a. Simulate 100 samples of nine human gestation periods each.
b. For each sample in part (a), obtain a 95% confidence interval for the population mean gestation period.
c. For the 100 confidence intervals that you obtained in part (b), roughly how many would you expect to contain the population mean gestation period of 266 days?
d. For the 100 confidence intervals that you obtained in part (b), determine the number that contain the population mean gestation period of 266 days.
e. Compare your answers from parts (c) and (d), and comment on any observed difference.

8.103 Suppose that a simple random sample is taken from a normal population having a standard deviation of 10 for the purpose of obtaining a 95% confidence interval for the mean of the population.

a. If the sample size is 4, obtain the margin of error.
b. Repeat part (a) for a sample size of 16.
c. Can you guess the margin of error for a sample size of 64? Explain your reasoning.

8.104 For a fixed confidence level, show that (approximately) quadrupling the sample size is necessary to halve the margin of error. (*Hint:* Use Formula 8.2.)

*Another type of confidence interval is called a **one-sided confidence interval**. A one-sided confidence interval provides either a lower confidence bound or an upper confidence bound for the parameter in question. You are asked to examine one-sided confidence intervals in Exercises 8.105–8.107.*

8.105 One-Sided One-Mean z-Intervals. Presuming that the assumptions for a one-mean z-interval are satisfied, we have the following formulas for $(1 - \alpha)$-level confidence bounds for a population mean μ:

- Lower confidence bound: $\bar{x} - z_\alpha \cdot \sigma/\sqrt{n}$
- Upper confidence bound: $\bar{x} + z_\alpha \cdot \sigma/\sqrt{n}$

Interpret the preceding formulas for lower and upper confidence bounds in words.

8.106 Poverty and Dietary Calcium. Refer to Exercise 8.70.

a. Determine and interpret a 95% upper confidence bound for the mean calcium intake of all people with incomes below the poverty level.
b. Compare your one-sided confidence interval in part (a) to the (two-sided) confidence interval found in Exercise 8.70.

8.107 Toxic Mushrooms? Refer to Exercise 8.71.

a. Determine and interpret a 99% lower confidence bound for the mean cadmium level of all *Boletus pinicola* mushrooms.
b. Compare your one-sided confidence interval in part (a) to the (two-sided) confidence interval found in Exercise 8.71.

8.3 Confidence Intervals for One Population Mean When σ Is Unknown

In Section 8.2, you learned how to determine a confidence interval for a population mean, μ, when the population standard deviation, σ, is known. The basis of the procedure is in Key Fact 7.2: If x is a normally distributed variable with mean μ and standard deviation σ, then, for samples of size n, the variable $\bar{x}$ is also normally distributed and has mean μ and standard deviation $\sigma/\sqrt{n}$. Equivalently, the **standardized version of $\bar{x}$**,

$$z = \frac{\bar{x} - \mu}{\sigma/\sqrt{n}}, \tag{8.2}$$

has the standard normal distribution.

What if, as is usual in practice, the population standard deviation is unknown? Then we cannot base our confidence-interval procedure on the standardized version of $\bar{x}$. The best we can do is estimate the population standard deviation, σ, by the sample standard deviation, s; in other words, we replace σ by s in Equation (8.2) and base our confidence-interval procedure on the resulting variable

$$t = \frac{\bar{x} - \mu}{s/\sqrt{n}} \tag{8.3}$$

called the **studentized version of $\bar{x}$.**

Unlike the standardized version, the studentized version of $\bar{x}$ does not have a normal distribution. To get an idea of how their distributions differ, we used statistical software to simulate each variable for samples of size 4, assuming that $\mu = 15$ and $\sigma = 0.8$. (Any sample size, population mean, and population standard deviation will do.)

1. We simulated 5000 samples of size 4 each.
2. For each of the 5000 samples, we obtained the sample mean and sample standard deviation.

3. For each of the 5000 samples, we determined the observed values of the standardized and studentized versions of $\bar{x}$.

4. We obtained histograms of the 5000 observed values of the standardized version of $\bar{x}$ and the 5000 observed values of the studentized version of $\bar{x}$, as shown in Output 8.2.

OUTPUT 8.2
Histograms of z (standardized version of $\bar{x}$) and t (studentized version of $\bar{x}$) for 5000 samples of size 4

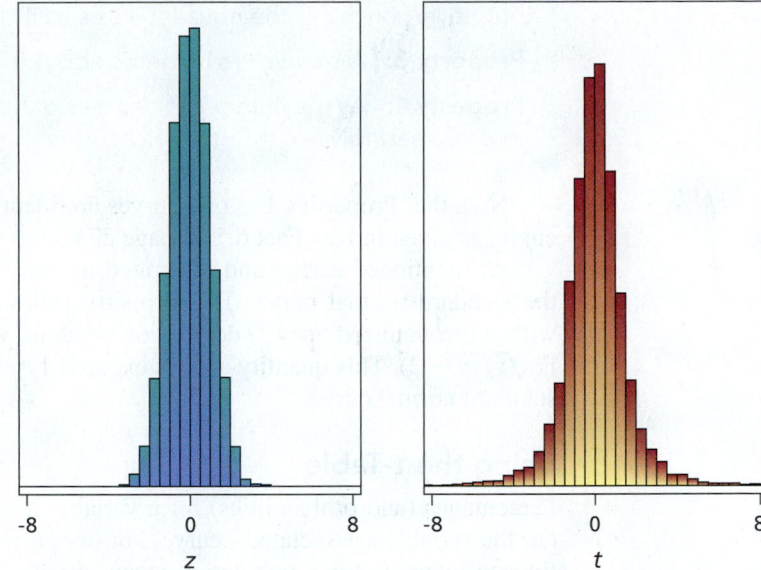

The two histograms suggest that the distributions of both the standardized version of $\bar{x}$—the variable z in Equation (8.2)—and the studentized version of $\bar{x}$—the variable t in Equation (8.3)—are bell shaped and symmetric about 0. However, there is an important difference in the distributions: The studentized version has more spread than the standardized version. This difference is not surprising because the variation in the possible values of the standardized version is due solely to the variation of sample means, whereas that of the studentized version is due to the variation of both sample means and sample standard deviations.

As you know, the standardized version of $\bar{x}$ has the standard normal distribution. In 1908, William Gosset determined the distribution of the studentized version of $\bar{x}$, a distribution now called **Student's *t*-distribution** or, simply, the ***t*-distribution.** (The biography on page 366 has more on Gosset and the Student's *t*-distribution.)

What Does It Mean?

For a normally distributed variable, the studentized version of the sample mean has the *t*-distribution with degrees of freedom 1 less than the sample size.

t-Distributions and *t*-Curves

There is a different *t*-distribution for each sample size. We identify a particular *t*-distribution by its number of **degrees of freedom (df)**. For the studentized version of $\bar{x}$, the number of degrees of freedom is 1 less than the sample size, which we indicate symbolically by **df = *n* − 1.**

KEY FACT 8.5

FIGURE 8.8
Standard normal curve and two *t*-curves

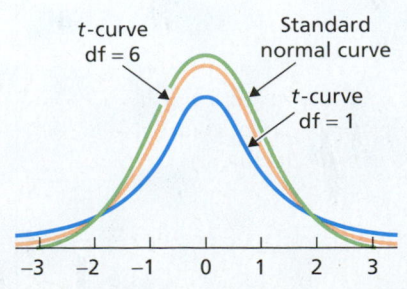

Studentized Version of the Sample Mean

Suppose that a variable x of a population is normally distributed with mean μ. Then, for samples of size n, the variable

$$t = \frac{\bar{x} - \mu}{s/\sqrt{n}}$$

has the *t*-distribution with $n - 1$ degrees of freedom.

A variable with a *t*-distribution has an associated curve, called a ***t*-curve.** In this book, you need to understand the basic properties of a *t*-curve, but not its equation.

Although there is a different *t*-curve for each number of degrees of freedom, all *t*-curves are similar and resemble the standard normal curve, as illustrated in Fig. 8.8. That figure also illustrates the basic properties of *t*-curves, listed in Key Fact 8.6.

KEY FACT 8.6

> ## Basic Properties of t-Curves
>
> **Property 1:** The total area under a t-curve equals 1.
>
> **Property 2:** A t-curve extends indefinitely in both directions, approaching, but never touching, the horizontal axis as it does so.
>
> **Property 3:** A t-curve is symmetric about 0.
>
> **Property 4:** As the number of degrees of freedom becomes larger, t-curves look increasingly like the standard normal curve.

Note that Properties 1–3 of t-curves are identical to those of the standard normal curve, as given in Key Fact 6.5 on page 274.

As mentioned earlier and illustrated in Fig. 8.8, t-curves have more spread than the standard normal curve. This property follows from the fact that, for a t-curve with ν (pronounced "new") degrees of freedom, where $\nu > 2$, the standard deviation is $\sqrt{\nu/(\nu - 2)}$. This quantity always exceeds 1, which is the standard deviation of the standard normal curve.

Using the t-Table

Percentages (and probabilities) for a variable having a t-distribution equal areas under the variable's associated t-curve. For our purposes, one of which is obtaining confidence intervals for a population mean, we don't need a complete t-table for each t-curve; only certain areas will be important. Table IV, which appears in Appendix A and in abridged form inside the back cover, is sufficient for our purposes.

The two outside columns of Table IV, labeled df, display the number of degrees of freedom. As expected, the symbol t_α denotes the t-value having area α to its right under a t-curve. Thus the column headed $t_{0.10}$, for example, contains t-values having area 0.10 to their right.

EXAMPLE 8.9

Finding the t-Value Having a Specified Area to Its Right

For a t-curve with 13 degrees of freedom, determine $t_{0.05}$; that is, find the t-value having area 0.05 to its right, as shown in Fig. 8.9(a).

FIGURE 8.9

Finding the t-value having area 0.05 to its right

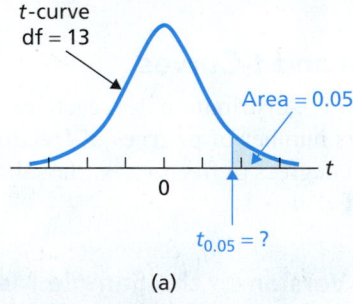

(a)

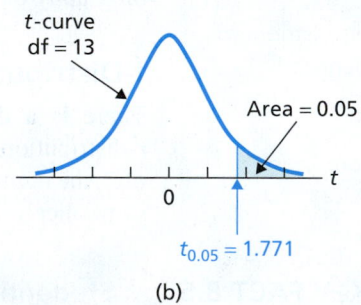

(b)

Solution To find the t-value in question, we use Table IV, a portion of which is given in Table 8.5.

TABLE 8.5

Values of t_α

df	$t_{0.10}$	$t_{0.05}$	$t_{0.025}$	$t_{0.01}$	$t_{0.005}$	df
.	.	.	.	.	.	.
.	.	.	.	.	.	.
.	.	.	.	.	.	.
12	1.356	1.782	2.179	2.681	3.055	12
13	1.350	1.771	2.160	2.650	3.012	13
14	1.345	1.761	2.145	2.624	2.977	14
15	1.341	1.753	2.131	2.602	2.947	15
.	.	.	.	.	.	.
.	.	.	.	.	.	.

Exercise 8.117
on page 359

The number of degrees of freedom is 13, so we first go down the outside columns, labeled df, to "13." Then, going across that row to the column labeled $t_{0.05}$, we reach 1.771. This number is the t-value having area 0.05 to its right, as shown in Fig. 8.9(b). In other words, for a t-curve with df $= 13$, $t_{0.05} = 1.771$.   ■

Note that Table IV in Appendix A contains degrees of freedom from 1 to 75, but then has only selected degrees of freedom. If the number of degrees of freedom you seek is not in Table IV, you could find a more detailed t-table, use technology, or use linear interpolation and Table IV. A less exact option is to use the degrees of freedom in Table IV closest to the one required.

As we noted earlier, t-curves look increasingly like the standard normal curve as the number of degrees of freedom gets larger. For degrees of freedom greater than 2000, a t-curve and the standard normal curve are virtually indistinguishable. Consequently, we stopped the t-table at df $= 2000$ and supplied the corresponding values of z_α beneath. These values can be used not only for the standard normal distribution, but also for any t-distribution having degrees of freedom greater than 2000.

The values of z_α given at the bottom of Table IV are accurate to three decimal places. Because of that fact, some of these values of z_α differ slightly from those that you get by using Table 8.3 on page 341 and, more generally, from those that you get by applying the method that you learned for using Table II.

Obtaining Confidence Intervals for a Population Mean When σ Is Unknown

Having discussed t-distributions and t-curves, we can now develop a procedure for obtaining a confidence interval for a population mean when the population standard deviation is unknown. We proceed in essentially the same way as we did when the population standard deviation is known, except now we invoke a t-distribution instead of the standard normal distribution.

Hence we use $t_{\alpha/2}$ instead of $z_{\alpha/2}$ in the formula for the confidence interval. As a result, we have Procedure 8.2, which we call the **one-mean t-interval procedure** or, when no confusion can arise, simply the **t-interval procedure.**[†]

■ ■ ■ ■ **PROCEDURE 8.2**   **One-Mean t-Interval Procedure**

Purpose To find a confidence interval for a population mean, μ

Assumptions
1. Simple random sample
2. Normal population or large sample
3. σ unknown

Step 1 For a confidence level of $1 - \alpha$, use Table IV to find $t_{\alpha/2}$ with df $= n - 1$, where n is the sample size.

Step 2 The confidence interval for μ is from

$$\bar{x} - t_{\alpha/2} \cdot \frac{s}{\sqrt{n}} \quad \text{to} \quad \bar{x} + t_{\alpha/2} \cdot \frac{s}{\sqrt{n}},$$

where $t_{\alpha/2}$ is found in Step 1 and $\bar{x}$ and s are computed from the sample data.

Step 3 Interpret the confidence interval.

Note: The confidence interval is exact for normal populations and is approximately correct for large samples from nonnormal populations.

Properties and guidelines for use of the t-interval procedure are the same as those for the z-interval procedure, as given in Key Fact 8.1 on page 340. In particular, the

---

[†]The one-mean t-interval procedure is also known as the **one-sample t-interval procedure** and the **one-variable t-interval procedure.** We prefer "one-mean" because it makes clear the parameter being estimated.

APPLET

Applet 8.1

t-interval procedure is robust to moderate violations of the normality assumption but, even for large samples, can sometimes be unduly affected by outliers because the sample mean and sample standard deviation are not resistant to outliers.

EXAMPLE 8.10   The One-Mean *t*-Interval Procedure

TABLE 8.6

Losses ($) for a sample of 25 pickpocket offenses

447	207	627	430	883
313	844	253	397	214
217	768	1064	26	587
833	277	805	653	549
649	554	570	223	443

FIGURE 8.10

Normal probability plot of the loss data in Table 8.6

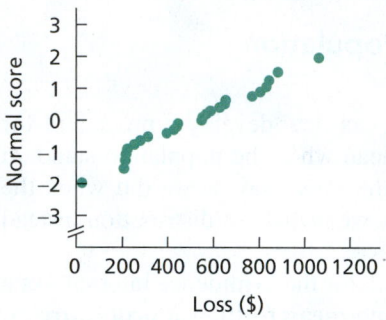

Pickpocket Offenses   The Federal Bureau of Investigation (FBI) compiles data on robbery and property crimes and publishes the information in *Population-at-Risk Rates and Selected Crime Indicators*. A simple random sample of pickpocket offenses yielded the losses, in dollars, shown in Table 8.6. Use the data to find a 95% confidence interval for the mean loss, μ, of all pickpocket offenses.

Solution   Because the sample size, $n = 25$, is moderate, we first need to consider questions of normality and outliers. (See the second bulleted item in Key Fact 8.1 on page 340.) To do that, we constructed the normal probability plot in Fig. 8.10. The plot reveals no outliers and falls roughly in a straight line. So, we can apply Procedure 8.2 to find the confidence interval.

Step 1   For a confidence level of $1 - \alpha$, use Table IV to find $t_{\alpha/2}$ with df $= n - 1$, where n is the sample size.

We want a 95% confidence interval, so $\alpha = 1 - 0.95 = 0.05$. For $n = 25$, we have df $= 25 - 1 = 24$. From Table IV, $t_{\alpha/2} = t_{0.05/2} = t_{0.025} = 2.064$.

Step 2   The confidence interval for μ is from

$$\bar{x} - t_{\alpha/2} \cdot \frac{s}{\sqrt{n}} \quad \text{to} \quad \bar{x} + t_{\alpha/2} \cdot \frac{s}{\sqrt{n}}.$$

From Step 1, $t_{\alpha/2} = 2.064$. Applying the usual formulas for $\bar{x}$ and s to the data in Table 8.6 gives $\bar{x} = 513.32$ and $s = 262.23$. So a 95% confidence interval for μ is from

$$513.32 - 2.064 \cdot \frac{262.23}{\sqrt{25}} \quad \text{to} \quad 513.32 + 2.064 \cdot \frac{262.23}{\sqrt{25}},$$

or 405.07 to 621.57.

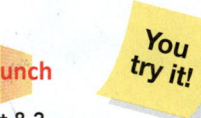

StatCrunch

Report 8.2

Exercise 8.129
on page 360

You try it!

Step 3   Interpret the confidence interval.

Interpretation   We can be 95% confident that the mean loss of all pickpocket offenses is somewhere between $405.07 and $621.57.

EXAMPLE 8.11   The One-Mean *t*-Interval Procedure

TABLE 8.7

Sample of year's chicken consumption (lb)

57	69	63	49	63	61
72	65	91	59	0	82
60	75	55	80	73	

Chicken Consumption   The U.S. Department of Agriculture publishes data on chicken consumption in *Food Consumption, Prices, and Expenditures*. Table 8.7 shows a year's chicken consumption, in pounds, for 17 randomly selected people. Find a 90% confidence interval for the year's mean chicken consumption, μ.

Solution   A normal probability plot of the data, shown in Fig. 8.11(a), reveals an outlier (0 lb). Because the sample size is only moderate, applying Procedure 8.2 here is inappropriate.

The outlier of 0 lb might be a recording error or it might reflect a person in the sample who does not eat chicken (e.g., a vegetarian). If we remove the outlier from the data, the normal probability plot for the abridged data shows no outliers and is roughly linear, as seen in Fig. 8.11(b).

Thus, if we are willing to take as our population only people who eat chicken, we can use Procedure 8.2 to obtain a confidence interval. Doing so yields a 90% confidence interval of 62.3 to 72.0.

FIGURE 8.11 Normal probability plots for chicken consumption: (a) original data and (b) data with outlier removed

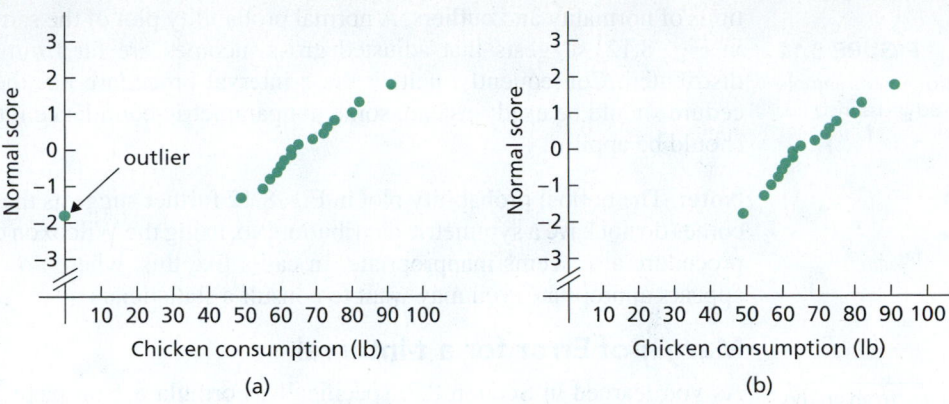

(a)   (b)

Interpretation   We can be 90% confident that the year's mean chicken consumption, among people who eat chicken, is somewhere between 62.3 lb and 72.0 lb. ■

By restricting our population of interest to only those people who eat chicken, we were justified in removing the outlier of 0 lb. Generally, an outlier should not be removed without careful consideration. *Simply removing an outlier because it is an outlier is unacceptable statistical practice.*

In Example 8.11, if we had been careless in our analysis by blindly finding a confidence interval without first examining the data, our result would have been invalid and misleading.

What If the Assumptions Are Not Satisfied?

Suppose you want to obtain a confidence interval for a population mean based on a small sample, but preliminary data analyses indicate either the presence of outliers or that the variable under consideration is far from normally distributed. As neither the z-interval procedure nor the t-interval procedure is appropriate, what can you do?

Under certain conditions, you can use a *nonparametric method.*[†] For example, if the variable under consideration has a symmetric distribution, you can use a nonparametric method called the *Wilcoxon confidence-interval procedure* to find a confidence interval for the population mean.

Most nonparametric methods do not require even approximate normality, are resistant to outliers and other extreme values, and can be applied regardless of sample size. However, parametric methods, such as the z-interval and t-interval procedures, tend to give more accurate results than nonparametric methods when the normality assumption and other requirements for their use are met.

We do not cover the Wilcoxon confidence-interval procedure in this book. We do discuss several other nonparametric procedures, however, beginning in Chapter 9 with the Wilcoxon signed-rank test.

> **? What Does It Mean?**
>
> Performing preliminary data analyses to check assumptions before applying inferential procedures is essential.

■■■ **EXAMPLE 8.12**   **Choosing a Confidence-Interval Procedure**

TABLE 8.8

Adjusted gross incomes ($1000)

9.7	93.1	33.0	21.2
81.4	51.1	43.5	10.6
12.8	7.8	18.1	12.7

Adjusted Gross Incomes   The Internal Revenue Service (IRS) publishes data on federal individual income tax returns in *Statistics of Income, Individual Income Tax Returns*. A sample of 12 returns from a recent year revealed the adjusted gross incomes, in thousands of dollars, shown in Table 8.8. Which procedure should be used to obtain a confidence interval for the mean adjusted gross income, μ, of all the year's individual income tax returns?

[†]Recall that descriptive measures for a population, such as μ and σ, are called parameters. Technically, inferential methods concerned with parameters are called **parametric methods;** those that are not are called **nonparametric methods.** However, common practice is to refer to most methods that can be applied without assuming normality (regardless of sample size) as nonparametric. Thus the term *nonparametric method* as used in contemporary statistics is somewhat of a misnomer.

FIGURE 8.12

Normal probability plot for the sample of adjusted gross incomes

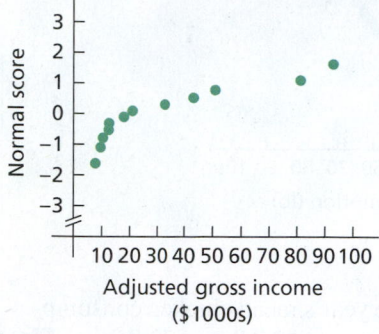

Solution Because the sample size is small ($n = 12$), we must first consider questions of normality and outliers. A normal probability plot of the sample data, shown in Fig. 8.12, suggests that adjusted gross incomes are far from being normally distributed. Consequently, neither the z-interval procedure nor the t-interval procedure should be used; instead, some nonparametric confidence interval procedure should be applied.

Note: The normal probability plot in Fig. 8.12 further suggests that adjusted gross incomes do not have a symmetric distribution; so, using the Wilcoxon confidence-interval procedure also seems inappropriate. In cases like this, where no common procedure appears appropriate, you may want to consult a statistician.

Margin of Error for a *t*-Interval

As you learned in Section 8.2, specifically, Formula 8.1 on page 342, the margin of error for the estimate of a population mean is $z_{\alpha/2} \cdot \sigma/\sqrt{n}$. This margin of error is for the case when the population standard deviation, σ, is known.

When σ is unknown, the margin of error for the estimate of a population mean is $t_{\alpha/2} \cdot s/\sqrt{n}$, as we see from Step 2 of Procedure 8.2 on page 355. The margin of error in this case has the same basic properties regarding confidence level and sample size (Key Facts 8.3 and 8.4 on pages 343 and 344, respectively) as in the case when σ is known.

▪▪▪ THE TECHNOLOGY CENTER

Most statistical technologies have programs that automatically perform the one-mean t-interval procedure. In this subsection, we present output and step-by-step instructions for such programs.

EXAMPLE 8.13   Using Technology to Obtain a One-Mean *t*-Interval

Pickpocket Offenses The losses, in dollars, of 25 randomly selected pickpocket offenses are displayed in Table 8.6 on page 356. Use Minitab, Excel, or the TI-83/84 Plus to find a 95% confidence interval for the mean loss, μ, of all pickpocket offenses.

Solution We applied the one-mean t-interval programs to the data, resulting in Output 8.3. Steps for generating that output are presented in Instructions 8.2. *Note to Excel users:* For brevity, we have presented only the essential portions of the actual output.

OUTPUT 8.3   One-mean t-interval on the sample of losses

MINITAB

One-Sample T: LOSS

Variable	N	Mean	StDev	SE Mean	95% CI
LOSS	25	513.3	262.2	52.4	(405.1, 621.6)

EXCEL

Summary statistics:

Variable	Observations	Mean	Std. deviation
LOSS	25	513.3200	262.2309

95% confidence interval on the mean:

(405.0764, 621.5636)

TI-83/84 PLUS

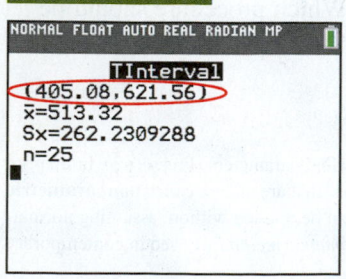

As shown in Output 8.3, the required 95% confidence interval is from 405.1 to 621.6. We can be 95% confident that the mean loss of all pickpocket offenses is somewhere between $405.1 and $621.6.

INSTRUCTIONS 8.2   Steps for generating Output 8.3

MINITAB

1. Store the data from Table 8.6 in a column named LOSS
2. Choose **Stat ➤ Basic Statistics ➤ 1-Sample t...**
3. Press the F3 key to reset the dialog box
4. Click in the text box directly below the **One or more samples, each in a column** drop-down list box and specify LOSS
5. Click the **Options...** button
6. Type 95 in the **Confidence level** text box
7. Click **OK** twice

EXCEL

1. Store the data from Table 8.6 in a column named LOSS
2. Choose **XLSTAT ➤ Parametric tests ➤ One-sample t-test and z-test**
3. Click the reset button in the lower left corner of the dialog box

4. Click in the **Data** selection box and then select the column of the worksheet that contains the LOSS data
5. Click the **Options** tab
6. Type 5 in the **Significance level (%)** text box
7. Click **OK**
8. Click the **Continue** button in the **XLSTAT – Selections** dialog box

TI-83/84 PLUS

1. Store the data from Table 8.6 in a list named LOSS
2. Press **STAT**, arrow over to **TESTS**, and press **8**
3. Highlight **Data** and press **ENTER**
4. Press the down-arrow key
5. Press **2nd ➤ LIST**
6. Arrow down to LOSS and press **ENTER** twice
7. Type 1 for **Freq** and then press **ENTER**
8. Type .95 for **C-Level** and press **ENTER** twice

Notes to Excel users: Step 6 of the Excel instructions states to type 5 in the **Significance level (%)** text box. Indeed, for this procedure, XLSTAT uses α (i.e., 1 minus the confidence level), expressed as a percentage, instead of the confidence level. So, for instance, type 5 for a 95% confidence interval and type 10 for a 90% confidence interval.

Exercises 8.3

Understanding the Concepts and Skills

8.108 Why do you need to consider the studentized version of $\bar{x}$ to develop a confidence-interval procedure for a population mean when the population standard deviation is unknown?

8.109 A variable has a mean of 100 and a standard deviation of 16. Four observations of this variable have a mean of 108 and a sample standard deviation of 12. Determine the observed value of the
a. standardized version of $\bar{x}$.
b. studentized version of $\bar{x}$.

8.110 A variable of a population has a normal distribution. Suppose that you want to find a confidence interval for the population mean.
a. If you know the population standard deviation, which procedure would you use?
b. If you do not know the population standard deviation, which procedure would you use?

8.111 Green Sea Urchins. From the paper "Effects of Chronic Nitrate Exposure on Gonad Growth in Green Sea Urchin *Strongylocentrotus droebachiensis*" (*Aquaculture*, Vol. 242, No. 1–4, pp. 357–363) by S. Siikavuopio et al., the weights, x, of adult green sea urchins are normally distributed with mean 52.0 g and standard deviation 17.2 g. For samples of 12 such weights, identify the distribution of each of the following variables.
a. $\dfrac{\bar{x} - 52.0}{17.2/\sqrt{12}}$
b. $\dfrac{\bar{x} - 52.0}{s/\sqrt{12}}$

8.112 Batting Averages. An issue of *Scientific American* revealed that batting averages, x, of major-league baseball players are normally distributed and have a mean of 0.270 and a standard deviation of 0.031. For samples of 20 batting averages, identify the distribution of each variable.
a. $\dfrac{\bar{x} - 0.270}{0.031/\sqrt{20}}$
b. $\dfrac{\bar{x} - 0.270}{s/\sqrt{20}}$

8.113 Explain why there is more variation in the possible values of the studentized version of $\bar{x}$ than in the possible values of the standardized version of $\bar{x}$.

8.114 Two t-curves have degrees of freedom 12 and 20, respectively. Which one more closely resembles the standard normal curve? Explain your answer.

8.115 For a t-curve with df $= 6$, use Table IV to find each t-value.
a. $t_{0.10}$
b. $t_{0.025}$
c. $t_{0.01}$

8.116 For a t-curve with df $= 17$, use Table IV to find each t-value.
a. $t_{0.05}$
b. $t_{0.025}$
c. $t_{0.005}$

8.117 For a t-curve with df $= 21$, find each t-value, and illustrate your results graphically.
a. The t-value having area 0.10 to its right
b. $t_{0.01}$
c. The t-value having area 0.025 to its left (*Hint:* A t-curve is symmetric about 0.)
d. The two t-values that divide the area under the curve into a middle 0.90 area and two outside areas of 0.05

8.118 For a t-curve with df $= 8$, find each t-value, and illustrate your results graphically.
a. The t-value having area 0.05 to its right
b. $t_{0.10}$

c. The *t*-value having area 0.01 to its left (*Hint:* A *t*-curve is symmetric about 0.)

d. The two *t*-values that divide the area under the curve into a middle 0.95 area and two outside 0.025 areas

8.119 A simple random sample of size 100 is taken from a population with unknown standard deviation. A normal probability plot of the data displays significant curvature but no outliers. Can you reasonably apply the *t*-interval procedure? Explain your answer.

8.120 A simple random sample of size 17 is taken from a population with unknown standard deviation. A normal probability plot of the data reveals an outlier but is otherwise roughly linear. Can you reasonably apply the *t*-interval procedure? Explain your answer.

8.121 Identify the formula for the margin of error for the estimate of a population mean when the population standard deviation is unknown.

8.122 For the one-mean *t*-interval procedure, express the formula for the endpoints of a confidence interval in the form

$$\text{point estimate} \pm \text{margin of error.}$$

In each of Exercises 8.123–8.128, we provide a sample mean, sample size, sample standard deviation, and confidence level. In each exercise,

a. *use the one-mean t-interval procedure to find a confidence interval for the mean of the population from which the sample was drawn.*

b. *obtain the margin of error by taking half the length of the confidence interval.*

c. *obtain the margin of error by using the formula $t_{\alpha/2} \cdot s/\sqrt{n}$.*

8.123 $\bar{x} = 20$, $n = 36$, $s = 3$, confidence level $= 95\%$

8.124 $\bar{x} = 25$, $n = 36$, $s = 3$, confidence level $= 95\%$

8.125 $\bar{x} = 30$, $n = 25$, $s = 4$, confidence level $= 90\%$

8.126 $\bar{x} = 35$, $n = 25$, $s = 4$, confidence level $= 90\%$

8.127 $\bar{x} = 50$, $n = 16$, $s = 5$, confidence level $= 99\%$

8.128 $\bar{x} = 55$, $n = 16$, $s = 5$, confidence level $= 99\%$

Applying the Concepts and Skills

Preliminary data analyses indicate that you can reasonably apply the t-interval procedure (Procedure 8.2 on page 355) in Exercises 8.129–8.134.

8.129 **Northeast Commutes.** According to Scarborough Research, more than 85% of working adults commute by car. Of all U.S. cities, Washington, D.C., and New York City have the longest commute times. A sample of 30 commuters in the Washington, D.C., area yielded the following commute times, in minutes.

24	28	31	29	54	28
27	38	24	14	46	38
31	16	21	11	21	15
30	29	17	23	27	18
29	44	19	35	34	38

Find and interpret a 90% confidence interval for the mean commute time of all commuters in Washington, D.C. (*Note:* $\bar{x} = 27.97$ minutes and $s = 10.04$ minutes.)

8.130 **Digital Viewing Times.** According to eMarketer, the average time spent per day with digital media in 2010 was 3 hours and 14 minutes. For last year, a random sample of 20 American adults spent the following number of hours per day with digital media.

1.4	5.0	5.8	4.1	5.6
2.1	6.5	2.3	8.0	9.5
7.3	5.9	9.4	4.3	2.9
2.8	5.1	4.5	4.1	6.6

Find and interpret a 90% confidence interval for last year's mean time spent per day with digital media by American adults. (*Note:* $\bar{x} = 5.16$ hr and $s = 2.30$ hr.)

8.131 **Sleep.** In 1908, W. S. Gosset published the article "The Probable Error of a Mean" (*Biometrika*, Vol. 6, pp. 1–25). In this pioneering paper, written under the pseudonym "Student," Gosset introduced what later became known as Student's *t*-distribution. Gosset used the following data set, which gives the additional sleep in hours obtained by a sample of 10 patients using laevohysocyamine hydrobromide.

1.9	0.8	1.1	0.1	−0.1
4.4	5.5	1.6	4.6	3.4

a. Obtain and interpret a 95% confidence interval for the additional sleep that would be obtained on average for all people using laevohysocyamine hydrobromide. (*Note:* $\bar{x} = 2.33$ hr; $s = 2.002$ hr.)

b. Was the drug effective in increasing sleep? Explain your answer.

8.132 **Family Fun?** Taking the family to an amusement park has become increasingly costly according to the industry publication *Amusement Business*, which provides figures on the cost for a family of four to spend the day at one of America's amusement parks. A random sample of 25 families of four that attended amusement parks yielded the following costs, rounded to the nearest dollar.

156	212	218	189	172
221	175	208	152	184
209	195	207	179	181
202	166	213	221	237
130	217	161	208	220

Obtain and interpret a 95% confidence interval for the mean cost of a family of four to spend the day at an American amusement park. (*Note:* $\bar{x} = \$193.32$; $s = \$26.73$.)

8.133 **"Chips Ahoy! 1,000 Chips Challenge."** As reported by B. Warner and J. Rutledge in the paper "Checking the Chips Ahoy! Guarantee" (*Chance*, Vol. 12, Issue 1, pp. 10–14), a random sample of forty-two 18-ounce bags of Chips Ahoy! cookies yielded a mean of 1261.6 chips per bag with a standard deviation of 117.6 chips per bag.

a. Determine a 95% confidence interval for the mean number of chips per bag for all 18-ounce bags of Chips Ahoy! cookies, and interpret your result in words.

b. Can you conclude that the average 18-ounce bag of Chips Ahoy! cookies contains at least 1000 chocolate chips? Explain your answer.

8.134 **Bottlenose Dolphins.** The webpage "Bottlenose Dolphin" produced by the National Geographic Society provides information about the bottlenose dolphin. A random sample of 50 adult bottlenose dolphins have a mean length of 12.04 ft with a standard deviation of 1.03 ft. Find and interpret a 90% confidence interval for the mean length of all adult bottlenose dolphins.

In each of Exercises 8.135–8.138, use the technology of your choice to decide whether applying the t-interval procedure to obtain a confidence interval for the population mean in question appears reasonable. Explain your answers.

8.135 Military Assistance Loans. The annual update of *U.S. Overseas Loans and Grants*, informally known as the "Greenbook," contains data on U.S. government monetary economic and military assistance loans. The following table shows military assistance loans, in thousands of dollars, to a sample of 10 countries, as reported by the U.S. Agency for International Development.

102	280	33	1643	177
69	180	89	205	695

8.136 Positively Selected Genes. R. Nielsen et al. compared 13,731 annotated genes from humans with their chimpanzee orthologs to identify genes that show evidence of positive selection. The researchers published their findings in "A Scan for Positively Selected Genes in the Genomes of Humans and Chimpanzees" (*PLOS Biology*, Vol. 3, Issue 6, pp. 976–985). A simple random sample of 14 tissue types yielded the following number of genes.

66	47	43	101	201	83	93
82	120	64	244	51	70	14

8.137 Big Bucks. In the article "The $350,000 Club" (*The Business Journal*, Vol. 24, Issue 14, pp. 80–82), J. Trunelle et al. examined Arizona public-company executives with salaries and bonuses totaling over $350,000. The following data provide the salaries, to the nearest thousand dollars, of a random sample of 20 such executives.

516	574	560	623	600
770	680	672	745	450
450	545	630	650	461
836	404	428	620	604

8.138 Shoe and Apparel E-Tailers. In the special report "Mousetrap: The Most-Visited Shoe and Apparel E-tailers" (*Footwear News*, Vol. 58, No. 3, p. 18), we found the following data on the average time, in minutes, spent per user per month from January to June of one year for a sample of 15 shoe and apparel retail websites.

13.3	9.0	11.1	9.1	8.4
15.6	8.1	8.3	13.0	17.1
16.3	13.5	8.0	15.1	5.8

Working with Large Data Sets

8.139 The Coruro's Burrow. The subterranean coruro (*Spalacopus cyanus*) is a social rodent that lives in large colonies in underground burrows that can reach lengths of up to 600 meters. Zoologists S. Begall and M. Gallardo studied the characteristics of the burrow systems of the subterranean coruro in central Chile and published their findings in the paper "*Spalacopus cyanus* (Rodentia: Octodontidae): An Extremist in Tunnel Constructing and Food Storing among Subterranean Mammals" (*Journal of Zoology*, Vol. 251, pp. 53–60). A sample of 51 burrows had the depths, in centimeters (cm), presented on the WeissStats site. Use the technology of your choice to do the following.

a. Obtain a normal probability plot, boxplot, histogram, and stem-and-leaf diagram of the data.

b. Based on your results from part (a), can you reasonably apply the *t*-interval procedure to the data? Explain your reasoning.

c. Find and interpret a 90% confidence interval for the mean depth of all subterranean coruro burrows.

8.140 Forearm Length. In 1903, K. Pearson and A. Lee published the paper "On the Laws of Inheritance in Man. I. Inheritance of Physical Characters" (*Biometrika*, Vol. 2, pp. 357–462). The article examined and presented data on forearm length, in inches, for a sample of 140 men, which we have provided on the WeissStats site. Use the technology of your choice to do the following.

a. Obtain a normal probability plot, boxplot, and histogram of the data.

b. Is it reasonable to apply the *t*-interval procedure to the data? Explain your answer.

c. If you answered "yes" to part (b), find a 95% confidence interval for the mean forearm length of men. Interpret your result.

8.141 Blood Cholesterol and Heart Disease. Numerous studies have shown that high blood cholesterol leads to artery clogging and subsequent heart disease. One such study by D. Scott et al. was published in the paper "Plasma Lipids as Collateral Risk Factors in Coronary Artery Disease: A Study of 371 Males With Chest Pain" (*Journal of Chronic Diseases*, Vol. 31, pp. 337–345). The research compared the plasma cholesterol concentrations of independent random samples of patients with and without evidence of heart disease. Evidence of heart disease was based on the degree of narrowing in the arteries. The data on plasma cholesterol concentrations, in milligrams/deciliter (mg/dL), are provided on the WeissStats site. Use the technology of your choice to do the following.

a. Obtain a normal probability plot, boxplot, and histogram of the data for patients without evidence of heart disease.

b. Is it reasonable to apply the *t*-interval procedure to those data? Explain your answer.

c. If you answered "yes" to part (b), determine a 95% confidence interval for the mean plasma cholesterol concentration of all males without evidence of heart disease. Interpret your result.

d. Repeat parts (a)–(c) for males with evidence of heart disease.

Extending the Concepts and Skills

8.142 Bicycle Commuting Times. A city planner working on bikeways designs a questionnaire to obtain information about local bicycle commuters. One of the questions asks how long it takes the rider to pedal from home to his or her destination. A sample of local bicycle commuters yields the following times, in minutes.

22	19	24	31	29	29
21	15	27	23	37	31
30	26	16	26	12	
23	48	22	29	28	

a. Find a 90% confidence interval for the mean commuting time of all local bicycle commuters in the city. (*Note:* The sample mean and sample standard deviation of the data are 25.82 minutes and 7.71 minutes, respectively.)

b. Interpret your result in part (a).

c. Graphical analyses of the data indicate that the time of 48 minutes may be an outlier. Remove this potential outlier and repeat part (a). (*Note:* The sample mean and sample standard deviation of the abridged data are 24.76 and 6.05, respectively.)

d. Should you have used the procedure that you did in part (a)? Explain your answer.

8.143 Table IV in Appendix A contains degrees of freedom from 1 to 75 consecutively but then contains only selected degrees of freedom.
a. Why couldn't we provide entries for all possible degrees of freedom?
b. Why did we construct the table so that consecutive entries appear for smaller degrees of freedom but that only selected entries occur for larger degrees of freedom?
c. If you had only Table IV, what value would you use for $t_{0.05}$ with df = 87? with df = 125? with df = 650? with df = 3000? Explain your answers.

8.144 Let $0 < \alpha < 1$. For a t-curve, determine
a. the t-value having area α to its right in terms of t_α.
b. the t-value having area α to its left in terms of t_α.
c. the two t-values that divide the area under the curve into a middle $1 - \alpha$ area and two outside $\alpha/2$ areas.
d. Draw graphs to illustrate your results in parts (a)–(c).

8.145 Batting Averages. An issue of *Scientific American* revealed that the batting averages of major-league baseball players are normally distributed with mean .270 and standard deviation .031.
a. Simulate 2000 samples of five batting averages each.
b. Determine the sample mean and sample standard deviation of each of the 2000 samples.
c. For each of the 2000 samples, determine the observed value of the standardized version of $\bar{x}$.
d. Obtain a histogram of the 2000 observations in part (c).
e. Theoretically, what is the distribution of the standardized version of $\bar{x}$?
f. Compare your results from parts (d) and (e).
g. For each of the 2000 samples, determine the observed value of the studentized version of $\bar{x}$.
h. Obtain a histogram of the 2000 observations in part (g).
i. Theoretically, what is the distribution of the studentized version of $\bar{x}$?
j. Compare your results from parts (h) and (i).
k. Compare your histograms from parts (d) and (h). How and why do they differ?

*Another type of confidence interval is called a **one-sided confidence interval**. A one-sided confidence interval provides either a lower confidence bound or an upper confidence bound for the parameter in question. You are asked to examine one-sided confidence intervals in Exercises 8.146–8.150.*

8.146 One-Sided One-Mean t-Intervals. Presuming that the assumptions for a one-mean t-interval are satisfied, we have the following formulas for $(1 - \alpha)$-level confidence bounds for a population mean μ:

- Lower confidence bound: $\bar{x} - t_\alpha \cdot s/\sqrt{n}$
- Upper confidence bound: $\bar{x} + t_\alpha \cdot s/\sqrt{n}$

Interpret the preceding formulas for lower and upper confidence bounds in words.

8.147 Northeast Commutes. Refer to Exercise 8.129.
a. Determine and interpret a 90% upper confidence bound for the mean commute time of all commuters in Washington, DC.
b. Compare your one-sided confidence interval in part (a) to the (two-sided) confidence interval found in Exercise 8.129.

8.148 Digital Viewing Times. Refer to Exercise 8.130.
a. Find and interpret a 90% lower confidence bound for last year's mean time spent per day with digital media by American adults.
b. Compare your one-sided confidence interval in part (a) to the (two-sided) confidence interval found in Exercise 8.130.

8.149 M&Ms. In the article "Sweetening Statistics—What M&M's Can Teach Us" (Minitab Inc., August 2008), M. Paret and E. Martz discussed several statistical analyses that they performed on bags of M&Ms. The authors took a random sample of 30 small bags of peanut M&Ms and obtained the following weights, in grams (g).

55.02	50.76	52.08	57.03	52.13	53.51
51.31	51.46	46.35	55.29	45.52	54.10
55.29	50.34	47.18	53.79	50.68	51.52
50.45	51.75	53.61	51.97	51.91	54.32
48.04	53.34	53.50	55.98	49.06	53.92

a. Determine a 95% lower confidence bound for the mean weight of all small bags of peanut M&Ms. (*Note:* The sample mean and sample standard deviation of the data are 52.040 g and 2.807 g, respectively.)
b. Interpret your result in part (a).
c. According to the package, each small bag of peanut M&Ms should weigh 49.3 g. Comment on this specification in view of your answer to part (b).

8.150 Christmas Spending. In a national poll of 1039 U.S. adults, conducted November 7–10, 2013, Gallup asked "Roughly how much money do you think you personally will spend on Christmas gifts this year?". The data provided on the WeissStats site are based on the results of the poll.
a. Determine a 95% upper confidence bound for the mean amount spent on Christmas gifts in 2013. (*Note:* The sample mean and sample standard deviation of the data are $704.00 and $477.98, respectively.)
b. Interpret your result in part (a).
c. In 2012, the mean amount spent on Christmas gifts was $770. Comment on this information in view of your answer to part (b).

8.151 Bootstrap Confidence Intervals. With the advent of high-speed computing, new procedures have been developed that permit statistical inferences to be performed under less restrictive conditions than those of classical procedures. **Bootstrap confidence intervals** constitute one such collection of new procedures. To obtain a bootstrap confidence interval for one population mean, proceed as follows.
1. Take a random sample of size n (the sample size) with replacement from the original sample.
2. Compute the mean of the new sample.
3. Repeat steps 1 and 2 a large number (hundreds or thousands) of times.
4. The distribution of the resulting sample means provides an estimate of the sampling distribution of the sample mean. This estimate is called a **bootstrap distribution.**
5. The (estimated) endpoints of a 95% confidence interval for the population mean are the 2.5th and 97.5th percentiles of the bootstrap distribution (i.e., $P_{2.5}$ and $P_{97.5}$).

Refer to Example 8.10 on page 356. Use the technology of your choice to find a 95% bootstrap confidence interval and compare your result with that found by using the one-mean t-interval procedure. Discuss any discrepancy that you encounter.

CHAPTER IN REVIEW

You Should Be Able to

1. use and understand the formulas in this chapter.

2. obtain a point estimate for a population mean.

3. find and interpret a confidence interval for a population mean when the population standard deviation is known.

4. compute and interpret the margin of error for the estimate of μ.

5. understand the relationship between sample size, standard deviation, confidence level, and margin of error for a confidence interval for μ.

6. determine the sample size required for a specified confidence level and margin of error for the estimate of μ.

7. understand the difference between the standardized and studentized versions of $\bar{x}$.

8. state the basic properties of t-curves.

9. use Table IV to find $t_{\alpha/2}$ for df $= n - 1$ and selected values of α.

10. find and interpret a confidence interval for a population mean when the population standard deviation is unknown.

11. decide whether it is appropriate to use the z-interval procedure, the t-interval procedure, or neither.

Key Terms

biased estimator, *333*
confidence interval (CI), *334*
confidence-interval estimate, *334*
confidence level, *334*
degrees of freedom (df), *353*
margin of error (E), *335, 342*
nonparametric methods, *357*
normal population, *340*

one-mean t-interval procedure, *355*
one-mean z-interval procedure, *339*
parametric methods, *357*
point estimate, *333*
robust procedures, *340*
standardized version of $\bar{x}$, *352*
studentized version of $\bar{x}$, *352*
Student's t-distribution, *353*

t_α, *354*
t-curve, *353*
t-distribution, *353*
t-interval procedure, *355*
unbiased estimator, *333*
z_α, *338*
z-interval procedure, *339*

REVIEW PROBLEMS

Understanding the Concepts and Skills

1. Explain the difference between a point estimate of a parameter and a confidence-interval estimate of a parameter.

2. Answer true or false to the following statement, and give a reason for your answer: If a 95% confidence interval for a population mean, μ, is from 33.8 to 39.0, the mean of the population must lie somewhere between 33.8 and 39.0.

3. Must the variable under consideration be normally distributed for you to use the z-interval procedure or t-interval procedure? Explain your answer.

4. If you obtained one thousand 95% confidence intervals for a population mean, μ, roughly how many of the intervals would actually contain μ?

5. Suppose that you have obtained a sample with the intent of performing a particular statistical-inference procedure. What should you do before applying the procedure to the sample data? Why?

6. Suppose that you intend to find a 95% confidence interval for a population mean by applying the one-mean z-interval procedure to a sample of size 100.
a. What would happen to the accuracy of the estimate if you used a sample of size 50 instead but kept the same confidence level of 0.95?

b. What would happen to the accuracy of the estimate if you changed the confidence level to 0.90 but kept the same sample size of 100?

7. A confidence interval for a population mean has a margin of error of 10.7.
a. Obtain the length of the confidence interval.
b. If the mean of the sample is 75.2, determine the confidence interval.
c. Express the confidence interval in the form "point estimate ± margin of error."

8. Suppose that you plan to apply the one-mean z-interval procedure to obtain a 90% confidence interval for a population mean, μ. You know that $\sigma = 12$ and that you are going to use a sample of size 9.
a. What will be your margin of error?
b. What else do you need to know in order to obtain the confidence interval?

9. A variable of a population has a mean of 266 and a standard deviation of 16. Ten observations of this variable have a mean of 262.1 and a sample standard deviation of 20.4. Obtain the observed value of the
a. standardized version of $\bar{x}$.
b. studentized version of $\bar{x}$.

10. Baby Weight. The paper "Are Babies Normal?" by T. Clemons and M. Pagano (*The American Statistician*, Vol. 53, No. 4, pp. 298–302) focused on birth weights of babies. According to the article, for babies born within the "normal" gestational range of

37–43 weeks, birth weights are normally distributed with a mean of 3432 grams (7 pounds 9 ounces) and a standard deviation of 482 grams (1 pound 1 ounce). For samples of 15 such birth weights, identify the distribution of each variable.

a. $\dfrac{\bar{x} - 3432}{482/\sqrt{15}}$

b. $\dfrac{\bar{x} - 3432}{s/\sqrt{15}}$

11. The following figure shows the standard normal curve and two *t*-curves. Which of the two *t*-curves has the larger degrees of freedom? Explain your answer.

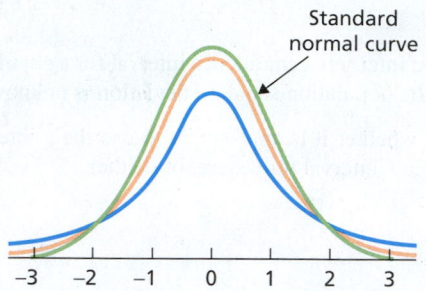

Standard normal curve

In each of Problems 12–17, we have provided a scenario for a confidence interval. Decide, in each case, whether the appropriate method for obtaining the confidence interval is the z-interval procedure, the t-interval procedure, or neither.

12. A random sample of size 17 is taken from a population. A normal probability plot of the sample data is found to be very close to linear (straight line). The population standard deviation is unknown.

13. A random sample of size 50 is taken from a population. A boxplot of the sample data reveals no outliers. The population standard deviation is known.

14. A random sample of size 25 is taken from a population. A normal probability plot of the sample data shows three outliers but is otherwise roughly linear. Checking reveals that the outliers are due to recording errors and are really not outliers. The population standard deviation is known.

15. A random sample of size 20 is taken from a population. A normal probability plot of the sample data shows three outliers but is otherwise roughly linear. Removal of the outliers is questionable. The population standard deviation is unknown.

16. A random sample of size 128 is taken from a population. A normal probability plot of the sample data shows no outliers but has significant curvature. The population standard deviation is known.

17. A random sample of size 13 is taken from a population. A normal probability plot of the sample data shows no outliers but has significant curvature. The population standard deviation is unknown.

18. For a *t*-curve with df = 18, obtain the *t*-value and illustrate your results graphically.
a. The *t*-value having area 0.025 to its right
b. $t_{0.05}$
c. The *t*-value having area 0.10 to its left
d. The two *t*-values that divide the area under the curve into a middle 0.99 area and two outside 0.005 areas

Applying the Concepts and Skills

19. Millionaires. Dr. Thomas Stanley of Georgia State University has surveyed millionaires since 1973. Among other information, Stanley obtains estimates for the mean age, μ, of all U.S. millionaires.

Suppose that 36 randomly selected U.S. millionaires are the following ages, in years.

31	45	79	64	48	38	39	68	52
59	68	79	42	79	53	74	66	66
71	61	52	47	39	54	67	55	71
77	64	60	75	42	69	48	57	48

Determine a 95% confidence interval for the mean age, μ, of all U.S. millionaires. Assume that the standard deviation of ages of all U.S. millionaires is 13.0 years. (*Note:* The mean of the data is 58.53 years.)

20. Millionaires. From Problem 19, we know that "a 95% confidence interval for the mean age of all U.S. millionaires is from 54.3 years to 62.8 years." Decide which of the following sentences provide a correct interpretation of the statement in quotes. Justify your answers.
a. Ninety-five percent of all U.S. millionaires are between the ages of 54.3 years and 62.8 years.
b. There is a 95% chance that the mean age of all U.S. millionaires is between 54.3 years and 62.8 years.
c. We can be 95% confident that the mean age of all U.S. millionaires is between 54.3 years and 62.8 years.
d. The probability is 0.95 that the mean age of all U.S. millionaires is between 54.3 years and 62.8 years.

21. Prison Sentences. Researchers M. Dhami et al. discussed how people adjust to prison life in the article "Adaption to Imprisonment" (*Criminal Justice and Behavior*, Vol. 34, No. 8, pp. 1085–1100). A sample of 712 federally sentenced adult male prisoners had an average sentence of 9.15 years. Assume that, for federally sentenced adult male prisoners, the population standard deviation of sentence length is 17.2 years.
a. Find and interpret a 90% confidence interval for the mean sentence length, μ, of all federally sentenced adult male prisoners.
b. Under what conditions can you freely apply the procedure that you used in part (a)?

22. Prison Sentences. Refer to Problem 21.
a. Find the margin of error, E.
b. Explain the meaning of E as far as the accuracy of the estimate is concerned.
c. Determine the sample size required to have a margin of error of 0.5 year and a 90% confidence level.
d. Find a 90% confidence interval for μ if a sample of the size determined in part (c) yields a mean of 10.1 years.

23. Children of Diabetic Mothers. The paper "Correlations between the Intrauterine Metabolic Environment and Blood Pressure in Adolescent Offspring of Diabetic Mothers" (*Journal of Pediatrics*, Vol. 136, Issue 5, pp. 587–592) by N. Cho et al. presented findings of research on children of diabetic mothers. Past studies showed that maternal diabetes results in obesity, blood pressure, and glucose tolerance complications in the offspring. Following are the arterial blood pressures, in millimeters of mercury (mm Hg), for a random sample of 16 children of diabetic mothers.

81.6	84.1	87.6	82.8	82.0	88.9	86.7	96.4
84.6	101.9	90.8	94.0	69.4	78.9	75.2	91.0

a. Apply the *t*-interval procedure to these data to find a 95% confidence interval for the mean arterial blood pressure of all children of diabetic mothers. Interpret your result in words. (*Note:* $\bar{x} = 85.99$ mm Hg and $s = 8.08$ mm Hg.)

b. Obtain a normal probability plot, a boxplot, a histogram, and a stem-and-leaf diagram of the data.

c. Based on your graphs from part (b), is it reasonable to apply the t-interval procedure as you did in part (a)? Explain your answer.

24. Diamond Pricing. In a Singapore edition of *Business Times*, diamond pricing was explored. The price of a diamond is based on the diamond's weight, color, and clarity. A simple random sample of 18 one-half-carat diamonds had the following prices, in dollars.

1676	1442	1995	1718	1826	2071	1947	1983	2146
1995	1876	2032	1988	2071	2234	2108	1941	2316

a. Apply the t-interval procedure to these data to find a 90% confidence interval for the mean price of all one-half-carat diamonds. Interpret your result. (*Note:* $\bar{x} = \$1964.7$ and $s = \$206.5$.)

b. Obtain a normal probability plot, a boxplot, a histogram, and a stem-and-leaf diagram of the data.

c. Based on your graphs from part (b), is it reasonable to apply the t-interval procedure as you did in part (a)? Explain your answer.

25. Wildfires. Wildfires are uncontrolled fires that usually spread quickly and are common in wilderness areas that have long and dry summers. The *National Interagency Fire Center* reports statistics on wildfires on their website www.nifc.gov. The following data lists the size, in thousands of acres, of a sample of 14 large (at least 100,000 acres) wildfires.

341	414	907	163	161	220	113
500	316	102	131	125	517	240

Use the technology of your choice to decide whether applying the one-mean t-interval procedure to these data is reasonable. Justify your answer.

Working with Large Data Sets

26. Delaying Adulthood. The convict surgeonfish is a common tropical reef fish that has been found to delay metamorphosis into adult by extending its larval phase. This delay often leads to enhanced survivorship in the species by increasing the chances of finding suitable habitat. In the paper "Delayed Metamorphosis of a Tropical Reef Fish (*Acanthurus triostegus*): A Field Experiment" (*Marine Ecology Progress Series*, Vol. 176, pp. 25–38), M. McCormick published data that he obtained on the larval duration, in days, of 90 convict surgeonfish. The data are contained on the WeissStats site.

a. Import the data into the technology of your choice.

b. Use the technology of your choice to obtain a normal probability plot, boxplot, and histogram of the data.

c. Is it reasonable to apply the t-interval procedure to the data? Explain your answer.

d. If you answered "yes" to part (c), obtain a 99% confidence interval for the mean larval duration of convict surgeonfish. Interpret your result.

27. Fuel Economy. The U.S. Department of Energy collects fuel-economy information on new motor vehicles and publishes its findings in *Fuel Economy Guide*. The data included are the result of vehicle testing done at the Environmental Protection Agency's National Vehicle and Fuel Emissions Laboratory in Ann Arbor, Michigan, and by vehicle manufacturers themselves with oversight by the Environmental Protection Agency. On the WeissStats site, we provide the highway mileages, in miles per gallon (mpg), for one year's cars. Use the technology of your choice to do the following.

a. Obtain a random sample of 35 of the mileages.

b. Use your data from part (b) and the t-interval procedure to find a 95% confidence interval for the mean highway gas mileage of all cars of the year in question.

c. Does the mean highway gas mileage of all cars of the year in question lie in the confidence interval that you found in part (c)? Would it necessarily have to? Explain your answers.

28. Old Faithful Geyser. In the online article "Old Faithful at Yellowstone, a Bimodal Distribution," D. Howell examined various aspects of the Old Faithful Geyser at Yellowstone National Park. Despite its name, there is considerable variation in both the length of the eruptions and in the time interval between eruptions. The times between eruptions, in minutes, for 500 recent observations are provided on the WeissStats site.

a. Identify the population and variable under consideration.

b. Use the technology of your choice to determine and interpret a 99% confidence interval for the mean time between eruptions.

c. Discuss the relevance of your confidence interval for future eruptions, say, 5 years from now.

29. Booted Eagles. The rare booted eagle of western Europe was the focus of a study by S. Suarez et al. to identify optimal nesting habitat for this raptor. According to their paper "Nesting Habitat Selection by Booted Eagles (*Hieraaetus pennatus*) and Implications for Management" (*Journal of Applied Ecology*, Vol. 37, pp. 215–223), the distances of such nests to the nearest marshland are normally distributed with mean 4.66 km and standard deviation 0.75 km.

a. Simulate 3000 samples of four distances each.

b. Determine the sample mean and sample standard deviation of each of the 3000 samples.

c. For each of the 3000 samples, determine the observed value of the standardized version of $\bar{x}$.

d. Obtain a histogram of the 3000 observations in part (c).

e. Theoretically, what is the distribution of the standardized version of $\bar{x}$?

f. Compare your results from parts (d) and (e).

g. For each of the 3000 samples, determine the observed value of the studentized version of $\bar{x}$.

h. Obtain a histogram of the 3000 observations in part (g).

i. Theoretically, what is the distribution of the studentized version of $\bar{x}$?

j. Compare your results from parts (h) and (i).

k. Compare your histograms from parts (d) and (h). How and why do they differ?

FOCUSING ON DATA ANALYSIS

UWEC UNDERGRADUATES

Recall from Chapter 1 (see page 34) that the Focus database and Focus sample contain information on the undergraduate students at the University of Wisconsin - Eau Claire (UWEC). Now would be a good time for you to review the discussion about these data sets.

a. Open the Focus sample (FocusSample) in the statistical software package of your choice and then obtain and interpret a 95% confidence interval for the mean high school percentile of all UWEC undergraduate students. Interpret your result.

b. In practice, the (population) mean of the variable under consideration is unknown. However, in this case, we actually do

have the population data, namely, in the Focus database (Focus). If your statistical software package will accommodate the entire Focus database, open that worksheet and then obtain the mean high school percentile of all UWEC undergraduate students. (*Answer:* 74.0)

c. Does your confidence interval in part (a) contain the population mean found in part (b)? Would it necessarily have to? Explain your answers.

d. Repeat parts (a)–(c) for the variables cumulative GPA, age, total earned credits, ACT English score, ACT math score, and ACT composite score. (*Note:* The means of these variables are 3.055, 20.7, 70.2, 23.0, 23.5, and 23.6, respectively.)

CASE STUDY DISCUSSION

BANK ROBBERIES: A STATISTICAL ANALYSIS

At the beginning of this chapter, on page 332, we presented summary statistics for data on bank robberies for five variables: amount stolen, number of bank staff present, number of customers present, number of bank raiders, and travel time from the bank to the nearest police station. These summary statistics were obtained by three researchers for data from a sample of 364 bank raids over a several-year period in the United Kingdom. For all bank raids in the United Kingdom during the years in question:

a. Identify and interpret a point estimate for the mean of each of the five aforementioned variables.

b. Find and interpret a 95% confidence interval for the mean amount stolen.

c. Find and interpret a 95% confidence interval for the mean number of bank staff present at the time of robberies.

d. Determine and interpret a 95% confidence interval for the mean number of customers present at the time of robberies.

e. Determine and interpret a 95% confidence interval for the mean number of bank raiders.

f. Obtain and interpret a 95% confidence interval for the mean travel time from the nearest police station to the bank outlet.

BIOGRAPHY

WILLIAM GOSSET: THE "STUDENT" IN STUDENT'S *t*-DISTRIBUTION

William Sealy Gosset was born in Canterbury, England, on June 13, 1876, the eldest son of Colonel Frederic Gosset and Agnes Sealy. He studied mathematics and chemistry at Winchester College and New College, Oxford, receiving a first-class degree in natural sciences in 1899.

After graduation Gosset began work with Arthur Guinness and Sons, a brewery in Dublin, Ireland. He saw the need for accurate statistical analyses of various brewing processes ranging from barley production to yeast fermentation, and pressed the firm to solicit mathematical advice. In 1906, the brewery sent him to work under Karl Pearson (see the biography in Chapter 13) at University College in London.

During the next few years, Gosset developed what has come to be known as Student's *t*-distribution. This distribution has proved to be fundamental in statistical analyses involving normal

distributions. In particular, Student's *t*-distribution is used in performing inferences for a population mean when the population being sampled is (approximately) normally distributed and the population standard deviation is unknown. Although the statistical theory for large samples had been completed in the early 1800s, no small-sample theory was available before Gosset's work.

Because Guinness's brewery prohibited its employees from publishing any of their research, Gosset published his contributions to statistical theory under the pseudonym "Student"—consequently the name "Student" in Student's *t*-distribution.

Gosset remained with Guinness his entire working life. In 1935, he moved to London to take charge of a new brewery. His tenure there was short lived; he died in Beaconsfield, England, on October 16, 1937.

Hypothesis Tests for One Population Mean

CHAPTER OBJECTIVES

In Chapter 8, we examined methods for obtaining confidence intervals for one population mean. We know that a confidence interval for a population mean, μ, is based on a sample mean, $\bar{x}$. Now we show how that statistic can be used to make decisions about hypothesized values of a population mean.

For example, suppose that we want to decide whether the mean prison sentence, μ, of all people imprisoned last year for drug offenses exceeds the year 2000 mean of 75.5 months. To make that decision, we can take a random sample of people imprisoned last year for drug offenses, compute their sample mean sentence, $\bar{x}$, and then apply a statistical-inference technique called a *hypothesis test*.

In this chapter, we describe hypothesis tests for one population mean. In doing so, we consider three different procedures. The first two are called the *one-mean z-test* and the *one-mean t-test,* which are the hypothesis-test analogues of the one-mean z-interval and one-mean *t*-interval confidence-interval procedures, respectively, discussed in Chapter 8. The third is a nonparametric method called the *Wilcoxon signed-rank test,* which applies when the variable under consideration has a symmetric distribution.

We also examine two different approaches to hypothesis testing—namely, the critical-value approach and the *P*-value approach.

CASE STUDY

Gender and Sense of Direction

Many of you have been there, a classic scene: mom yelling at dad to turn left, while dad decides to do just the opposite. Well, who made the right call? More generally, who has a better sense of direction, women or men?

Dr. J. Sholl et al. considered these and related questions in the paper "The Relation of Sex and Sense of Direction to Spatial Orientation in an Unfamiliar Environment" (*Journal of Environmental Psychology*, Vol. 20, pp. 17–28).

In their study, the spatial orientation skills of 30 male students and 30 female students from Boston College were challenged in Houghton Garden Park, a wooded park near campus in Newton, Massachusetts. Before driving to the park, the participants were asked to rate their own sense of direction as either good or poor.

In the park, students were instructed to point to predesignated landmarks and also to the direction of south. Pointing was carried out by students moving a pointer attached to a 360° protractor; the angle of the pointing response was then recorded to the nearest degree. For the female students who had rated their sense of direction to be good, the following table displays the pointing errors (in degrees) when they attempted to point south.

14	122	128	109	12
91	8	78	31	36
27	68	20	69	18

Based on these data, can you conclude that, in general, women who consider themselves to have a good sense of direction really do better, on average, than they would by randomly guessing at the direction of south? To answer that question, you need to conduct a hypothesis test, which you will do after you study hypothesis testing in this chapter.

9.1 The Nature of Hypothesis Testing

We often use inferential statistics to make decisions or judgments about the value of a parameter, such as a population mean. For example, we might need to decide whether the mean weight, μ, of all bags of pretzels packaged by a particular company differs from the advertised weight of 454 grams (g), or we might want to determine whether the mean age, μ, of all cars in use has increased from the year 2000 mean of 9.0 years.

One of the most commonly used methods for making such decisions or judgments is to perform a *hypothesis test*. A **hypothesis** is a statement that something is true. For example, the statement "the mean weight of all bags of pretzels packaged differs from the advertised weight of 454 g" is a hypothesis.

Typically, a hypothesis test involves two hypotheses: the *null hypothesis* and the *alternative hypothesis* (or *research hypothesis*), which we define as follows.

DEFINITION 9.1

What Does It Mean?

Originally, the word *null* in *null hypothesis* stood for "no difference" or "the difference is null." Over the years, however, *null hypothesis* has come to mean simply a hypothesis to be tested.

Null and Alternative Hypotheses; Hypothesis Test

Null hypothesis: A hypothesis to be tested. We use the symbol H_0 to represent the null hypothesis.

Alternative hypothesis: A hypothesis to be considered as an alternative to the null hypothesis. We use the symbol H_a to represent the alternative hypothesis.

Hypothesis test: The problem in a hypothesis test is to decide whether the null hypothesis should be rejected in favor of the alternative hypothesis.

For instance, in the pretzel packaging example, the null hypothesis might be "the mean weight of all bags of pretzels packaged equals the advertised weight of 454 g," and the alternative hypothesis might be "the mean weight of all bags of pretzels packaged differs from the advertised weight of 454 g."

Choosing the Hypotheses

The first step in setting up a hypothesis test is to decide on the null hypothesis and the alternative hypothesis. The following are some guidelines for choosing these two hypotheses. Although the guidelines refer specifically to hypothesis tests for one population mean, μ, they apply to any hypothesis test concerning one parameter.

Null Hypothesis

In this book, the null hypothesis for a hypothesis test concerning a population mean, μ, always specifies a single value for that parameter. Hence we can express the null hypothesis as

$$H_0: \mu = \mu_0,$$

where μ_0 is some number.

Alternative Hypothesis

The choice of the alternative hypothesis depends on and should reflect the purpose of the hypothesis test. Three choices are possible for the alternative hypothesis.

- If the primary concern is deciding whether a population mean, μ, is *different from* a specified value μ_0, we express the alternative hypothesis as

$$H_a: \mu \neq \mu_0.$$

A hypothesis test whose alternative hypothesis has this form is called a **two-tailed test.**

- If the primary concern is deciding whether a population mean, μ, is *less than* a specified value μ_0, we express the alternative hypothesis as

$$H_a: \mu < \mu_0.$$

A hypothesis test whose alternative hypothesis has this form is called a **left-tailed test.**

- If the primary concern is deciding whether a population mean, μ, is *greater than* a specified value μ_0, we express the alternative hypothesis as

$$H_a: \mu > \mu_0.$$

A hypothesis test whose alternative hypothesis has this form is called a **right-tailed test.**

A hypothesis test is called a **one-tailed test** if it is either left tailed or right tailed.

■■ ■■ **EXAMPLE 9.1**   **Choosing the Null and Alternative Hypotheses**

Quality Assurance A snack-food company produces a 454-g bag of pretzels. Although the actual net weights deviate slightly from 454 g and vary from one bag to another, the company insists that the mean net weight of the bags be 454 g.

As part of its program, the quality assurance department periodically performs a hypothesis test to decide whether the packaging machine is working properly, that is, to decide whether the mean net weight of all bags packaged is 454 g.

a.   Determine the null hypothesis for the hypothesis test.
b.   Determine the alternative hypothesis for the hypothesis test.
c.   Classify the hypothesis test as two tailed, left tailed, or right tailed.

Solution   Let μ denote the mean net weight of all bags packaged.

a.   The null hypothesis is that the packaging machine is working properly, that is, that the mean net weight, μ, of all bags packaged *equals* 454 g. In symbols, $H_0: \mu = 454$ g.
b.   The alternative hypothesis is that the packaging machine is not working properly, that is, that the mean net weight, μ, of all bags packaged is *different from* 454 g. In symbols, $H_a: \mu \neq 454$ g.
c.   This hypothesis test is two tailed because a does-not-equal sign ($\neq$) appears in the alternative hypothesis.   ■

■■ ■■ **EXAMPLE 9.2**   **Choosing the Null and Alternative Hypotheses**

Taller Young Women In the document *Anthropometric Reference Data for Children and Adults*, C. Fryer et al. present data from the *National Health and Nutrition Examination Survey* on a variety of human body measurements, such as weight, height, and size. Anthropometry is a key component of nutritional status assessment in children and adults.

A half-century ago, the average (U.S.) woman in her 20s was 62.6 inches tall. Suppose that we want to perform a hypothesis test to decide whether today's women in their 20s are, on average, taller than such women were a half-century ago.

a. Determine the null hypothesis for the hypothesis test.
b. Determine the alternative hypothesis for the hypothesis test.
c. Classify the hypothesis test as two tailed, left tailed, or right tailed.

Solution Let μ denote the mean height of today's women in their 20s.

a. The null hypothesis is that the mean height of today's women in their 20s *equals* the mean height of women in their 20s a half-century ago; that is, $H_0: \mu = 62.6$ inches.
b. The alternative hypothesis is that the mean height of today's women in their 20s is *greater than* the mean height of women in their 20s a half-century ago; that is, $H_a: \mu > 62.6$ inches.
c. This hypothesis test is right tailed because a greater-than sign ($>$) appears in the alternative hypothesis.

EXAMPLE 9.3 **Choosing the Null and Alternative Hypotheses**

Poverty and Dietary Calcium Calcium is the most abundant mineral in the human body and has several important functions. Most body calcium is stored in the bones and teeth, where it functions to support their structure. Recommendations for calcium are provided in *Dietary Reference Intakes*, developed by the Institute of Medicine of the National Academy of Sciences. The recommended adequate intake (RAI) of calcium for adults (ages 19–50 years) is 1000 milligrams (mg) per day.

Suppose that we want to perform a hypothesis test to decide whether the average adult with an income below the poverty level gets less than the RAI of 1000 mg.

a. Determine the null hypothesis for the hypothesis test.
b. Determine the alternative hypothesis for the hypothesis test.
c. Classify the hypothesis test as two tailed, left tailed, or right tailed.

Solution Let μ denote the mean calcium intake (per day) of all adults with incomes below the poverty level.

You try it!

Exercise 9.15
on page 375

a. The null hypothesis is that the mean calcium intake of all adults with incomes below the poverty level *equals* the RAI of 1000 mg per day; that is, $H_0: \mu = 1000$ mg.
b. The alternative hypothesis is that the mean calcium intake of all adults with incomes below the poverty level is *less than* the RAI of 1000 mg per day; that is, $H_a: \mu < 1000$ mg.
c. This hypothesis test is left tailed because a less-than sign ($<$) appears in the alternative hypothesis.

The Logic of Hypothesis Testing

After we have chosen the null and alternative hypotheses, we must decide whether to reject the null hypothesis in favor of the alternative hypothesis. The procedure for deciding is roughly as follows.

Basic Logic of Hypothesis Testing

Take a random sample from the population. If the sample data are consistent with the null hypothesis, do not reject the null hypothesis; if the sample data are inconsistent with the null hypothesis and supportive of the alternative hypothesis, reject the null hypothesis in favor of the alternative hypothesis.

In practice, of course, we must have a precise criterion for deciding whether to reject the null hypothesis. We discuss such criteria in Sections 9.2 and 9.3. At this point, we simply note that a precise criterion involves a **test statistic,** a statistic calculated from the data that is used as a basis for deciding whether the null hypothesis should be rejected.

Type I and Type II Errors

Any decision we make based on a hypothesis test may be incorrect because we have used partial information obtained from a sample to draw conclusions about the entire population. There are two types of incorrect decisions—*Type I error* and *Type II error*, as indicated in Table 9.1 and Definition 9.2.

TABLE 9.1

Correct and incorrect decisions for a hypothesis test

<table>
<tr><td rowspan="2"></td><td rowspan="2"></td><td colspan="2">H_0 is:</td></tr>
<tr><td>True</td><td>False</td></tr>
<tr><td rowspan="2">Decision:</td><td>Do not reject H_0</td><td>Correct decision</td><td>Type II error</td></tr>
<tr><td>Reject H_0</td><td>Type I error</td><td>Correct decision</td></tr>
</table>

DEFINITION 9.2

Type I and Type II Errors

Type I error: Rejecting the null hypothesis when it is in fact true.

Type II error: Not rejecting the null hypothesis when it is in fact false.

In understanding and applying the concepts of Type I and Type II errors, you may also find it useful to refer to one or both of the flowcharts in Fig. 9.1 on the next page. The flowchart in Fig. 9.1(a) starts by asking whether the null hypothesis is in fact true, whereas the flowchart in Fig. 9.1(b) starts by asking whether the null hypothesis was rejected.

EXAMPLE 9.4 Type I and Type II Errors

Quality Assurance Consider again the pretzel-packaging hypothesis test. The null and alternative hypotheses are, respectively,

H_0: $\mu = 454$ g (the packaging machine is working properly)

H_a: $\mu \neq 454$ g (the packaging machine is not working properly),

where μ is the mean net weight of all bags of pretzels packaged. Explain what each of the following would mean.

a. Type I error   **b.** Type II error   **c.** Correct decision

Now suppose that the results of carrying out the hypothesis test lead to rejection of the null hypothesis $\mu = 454$ g, that is, to the conclusion that $\mu \neq 454$ g. Classify that conclusion by error type or as a correct decision if

d. the mean net weight, μ, is in fact 454 g.
e. the mean net weight, μ, is in fact not 454 g.

Solution

a. A Type I error occurs when a true null hypothesis is rejected. In this case, a Type I error would occur if in fact $\mu = 454$ g but the results of the sampling lead to the conclusion that $\mu \neq 454$ g.

FIGURE 9.1 Flowcharts for Type I and Type II errors

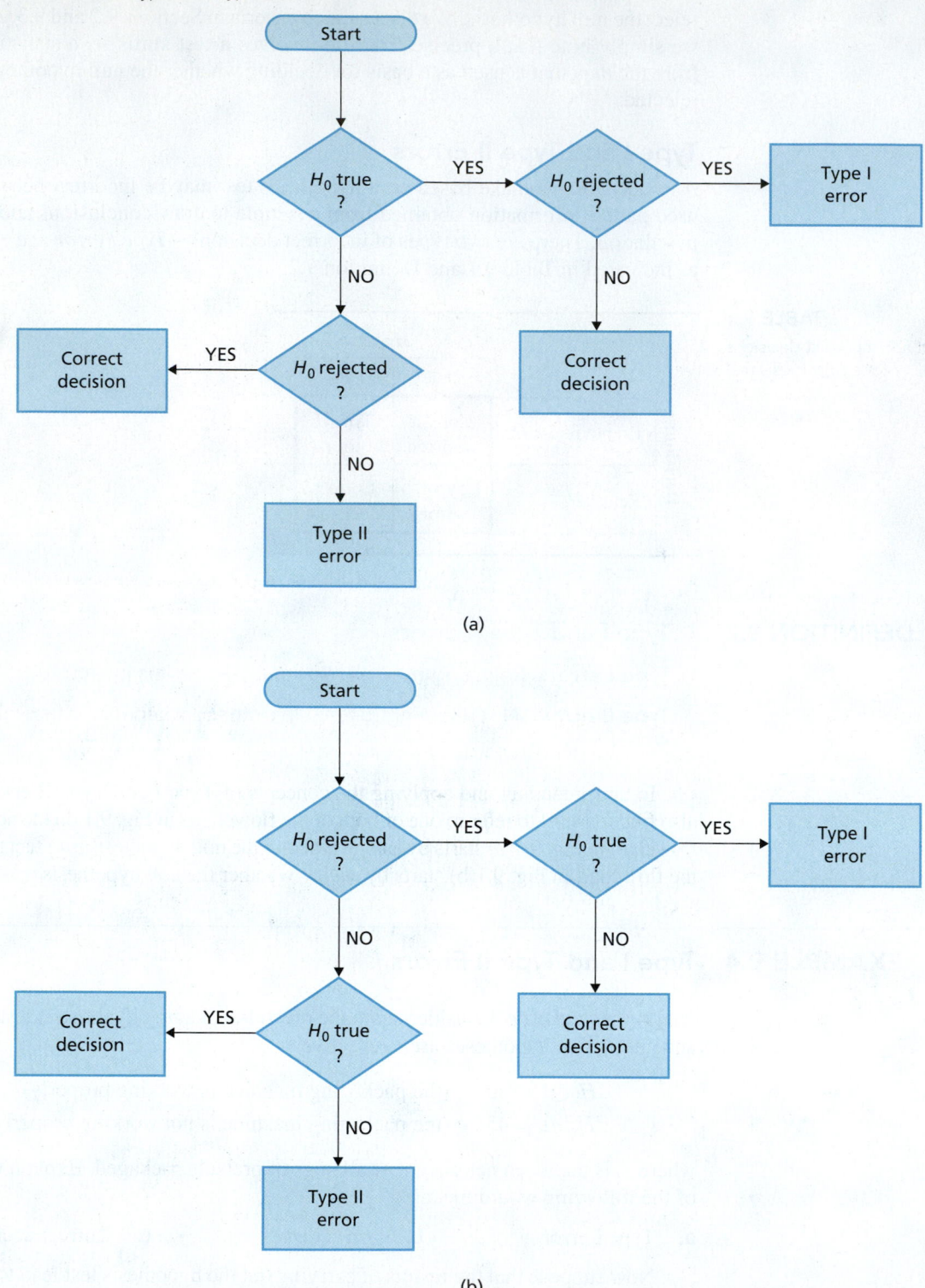

(a)

(b)

Interpretation A Type I error occurs if we conclude that the packaging machine is not working properly when in fact it is working properly.

b. A Type II error occurs when a false null hypothesis is not rejected. In this case, a Type II error would occur if in fact $\mu \neq 454$ g but the results of the sampling fail to lead to that conclusion.

Interpretation A Type II error occurs if we fail to conclude that the packaging machine is not working properly when in fact it is not working properly.

c. A correct decision can occur in either of two ways.

- A true null hypothesis is not rejected. That would happen if in fact $\mu = 454$ g and the results of the sampling do not lead to the rejection of that fact.
- A false null hypothesis is rejected. That would happen if in fact $\mu \neq 454$ g and the results of the sampling lead to that conclusion.

Interpretation   A correct decision occurs if either we fail to conclude that the packaging machine is not working properly when in fact it is working properly, or we conclude that the packaging machine is not working properly when in fact it is not working properly.

You try it!

Exercise 9.23
on page 375

d. If in fact $\mu = 454$ g, the null hypothesis is true. Consequently, by rejecting the null hypothesis $\mu = 454$ g, we have made a Type I error—we have rejected a true null hypothesis.

e. If in fact $\mu \neq 454$ g, the null hypothesis is false. Consequently, by rejecting the null hypothesis $\mu = 454$ g, we have made a correct decision—we have rejected a false null hypothesis.

Probabilities of Type I and Type II Errors

Part of evaluating the effectiveness of a hypothesis test involves analyzing the chances of making an incorrect decision. A Type I error occurs if a true null hypothesis is rejected. The probability of that happening, the **Type I error probability,** commonly called the **significance level** of the hypothesis test, is denoted α (the lowercase Greek letter alpha).

DEFINITION 9.3

> ### Significance Level
>
> The probability of making a Type I error, that is, of rejecting a true null hypothesis, is called the **significance level, α,** of a hypothesis test.

A Type II error occurs if a false null hypothesis is not rejected. The probability of that happening, the **Type II error probability,** is denoted β (the lowercase Greek letter beta). Calculation of Type II error probabilities is examined in Section 9.7.

Ideally, both Type I and Type II errors should have small probabilities. Then the chance of making an incorrect decision would be small, regardless of whether the null hypothesis is true or false. As we soon demonstrate, we can design a hypothesis test to have any specified significance level. So, for instance, if not rejecting a true null hypothesis is important, we should specify a small value for α. However, in making our choice for α, we must keep Key Fact 9.1 in mind.

KEY FACT 9.1

> ### Relation between Type I and Type II Error Probabilities
>
> For a fixed sample size, the smaller we specify the significance level, α, the larger will be the probability, β, of not rejecting a false null hypothesis.

Consequently, we must always assess the risks involved in committing both types of errors and use that assessment as a method for balancing the Type I and Type II error probabilities.

Possible Conclusions for a Hypothesis Test

The significance level, α, is the probability of making a Type I error, that is, of rejecting a true null hypothesis. Therefore, if the hypothesis test is conducted at a small significance level (e.g., $\alpha = 0.05$), the chance of rejecting a true null hypothesis will be small. In this text, we generally specify a small significance level. Thus, if we do reject the null hypothesis, we can be reasonably confident that the null hypothesis is false.

In other words, if we do reject the null hypothesis, we conclude that the data provide sufficient evidence to support the alternative hypothesis.

However, we usually do not know the probability, β, of making a Type II error, that is, of not rejecting a false null hypothesis. Consequently, if we do not reject the null hypothesis, we simply reserve judgment about which hypothesis is true. In other words, if we do not reject the null hypothesis, we conclude only that the data do not provide sufficient evidence to support the alternative hypothesis; we do not conclude that the data provide sufficient evidence to support the null hypothesis.

KEY FACT 9.2

Possible Conclusions for a Hypothesis Test

Suppose that a hypothesis test is conducted at a small significance level.

- If the null hypothesis is rejected, we conclude that the data provide sufficient evidence to support the alternative hypothesis.
- If the null hypothesis is not rejected, we conclude that the data do not provide sufficient evidence to support the alternative hypothesis.

Note: Another way of viewing the use of a small significance level is as follows: The null hypothesis gets the benefit of the doubt; the alternative hypothesis has the burden of proof.

When the null hypothesis is rejected in a hypothesis test performed at the significance level α, we frequently express that fact with the phrase "the test results are **statistically significant** at the α level." Similarly, when the null hypothesis is not rejected in a hypothesis test performed at the significance level α, we often express that fact with the phrase "the test results are **not statistically significant** at the α level."

Exercises 9.1

Understanding the Concepts and Skills

9.1 Explain the meaning of the term *hypothesis* as used in inferential statistics.

9.2 Regarding the term *null hypothesis:*
a. Originally, what did the word *null* in *null hypothesis* stand for?
b. What has the term *null hypothesis* come to mean now?

9.3 What role does the decision criterion play in a hypothesis test?

9.4 Suppose that you want to perform a hypothesis test for a population mean, μ.
a. Express the null hypothesis both in words and in symbolic form.
b. Express each of the three possible alternative hypotheses in words and in symbolic form.

9.5 Suppose that you are considering a hypothesis test for a population mean, μ. In each part, express the alternative hypothesis symbolically and identify the hypothesis test as two tailed, left tailed, or right tailed.
a. You want to decide whether the population mean is different from a specified value μ_0.
b. You want to decide whether the population mean is less than a specified value μ_0.
c. You want to decide whether the population mean is greater than a specified value μ_0.

9.6 Suppose that, in a hypothesis test, the null hypothesis is in fact true.
a. Is it possible to make a Type I error? Explain your answer.
b. Is it possible to make a Type II error? Explain your answer.

9.7 Suppose that, in a hypothesis test, the null hypothesis is in fact false.
a. Is it possible to make a Type I error? Explain your answer.
b. Is it possible to make a Type II error? Explain your answer.

9.8 What is the relation between the significance level of a hypothesis test and the probability of making a Type I error?

9.9 Answer true or false and explain your answer: If it is important not to reject a true null hypothesis, the hypothesis test should be performed at a small significance level.

9.10 Answer true or false and explain your answer: For a fixed sample size, decreasing the significance level of a hypothesis test results in an increase in the probability of making a Type II error.

9.11 Identify the two types of incorrect decisions in a hypothesis test. For each incorrect decision, what symbol is used to represent the probability of making that type of error?

9.12 Suppose that a hypothesis test is performed at a small significance level. State the appropriate conclusion in each case by referring to Key Fact 9.2.
a. The null hypothesis is rejected.
b. The null hypothesis is not rejected.

9.13 Approving Nuclear Reactors. Suppose that you are performing a statistical test to decide whether a nuclear reactor should be approved for use. Further suppose that failing to reject the null hypothesis corresponds to approval. What property would you want the Type II error probability, β, to have?

9.14 Guilty or Innocent? In the U.S. court system, a defendant is assumed innocent until proven guilty. Suppose that you regard a court trial as a hypothesis test with null and alternative hypotheses

H_0: Defendant is innocent

H_a: Defendant is guilty.

a. Explain the meaning of a Type I error.

b. Explain the meaning of a Type II error.

c. If you were the defendant, would you want α to be large or small? Explain your answer.

d. If you were the prosecuting attorney, would you want β to be large or small? Explain your answer.

e. What are the consequences to the court system if you make $\alpha = 0$? $\beta = 0$?

Applying the Concepts and Skills

In Exercises 9.15–9.22, hypothesis tests are proposed. For each hypothesis test,

a. *determine the null hypothesis.*

b. *determine the alternative hypothesis.*

c. *classify the hypothesis test as two tailed, left tailed, or right tailed.*

9.15 Toxic Mushrooms? Cadmium, a heavy metal, is toxic to animals. Mushrooms, however, are able to absorb and accumulate cadmium at high concentrations. The Czech and Slovak governments have set a safety limit for cadmium in dry vegetables at 0.5 part per million (ppm). M. Melgar et al. measured the cadmium levels in a random sample of the edible mushroom *Boletus pinicola* and published the results in the paper "Influence of Some Factors in Toxicity and Accumulation of Cd from Edible Wild Macrofungi in NW Spain" (*Journal of Environmental Science and Health*, Vol. B33(4), pp. 439–455). A hypothesis test is to be performed to decide whether the mean cadmium level in *Boletus pinicola* mushrooms is greater than the government's recommended limit.

9.16 Grey-Seal Nursing. Grey seals are one of several types of earless seals. The length of time that a female grey seal nurses her pup is studied by S. Twiss et al. in the article "Variation in Female Grey Seal (*Halichoerus grypus*) Reproductive Performance Correlates to Proactive-Reactive Behavioural Types" (*PLOS ONE* 7(11): e49598. doi:10.1371/journal.pone.0049598). The average lactation (nursing) period of all earless seals is 23 days. A hypothesis test is to be performed to decide whether the mean lactation period of grey seals differs from 23 days.

9.17 Iron Deficiency? Iron is essential to most life forms and to normal human physiology. It is an integral part of many proteins and enzymes that maintain good health. Recommendations for iron are provided in *Dietary Reference Intakes*, developed by the Institute of Medicine of the National Academy of Sciences. The recommended dietary allowance (RDA) of iron for adult females under the age of 51 years is 18 milligrams (mg) per day. A hypothesis test is to be performed to decide whether adult females under the age of 51 years are, on average, getting less than the RDA of 18 mg of iron.

9.18 Early-Onset Dementia. Dementia is the loss of the intellectual and social abilities severe enough to interfere with judgment, behavior, and daily functioning. Alzheimer's disease is the most common type of dementia. In the article "Living with Early Onset Dementia: Exploring the Experience and Developing Evidence-Based Guidelines for Practice" (*Alzheimer's Care Quarterly*, Vol. 5, Issue 2, pp. 111–122), P. Harris and J. Keady explored the experience and struggles of people diagnosed with dementia and their families. A hypothesis test is to be performed to decide whether the mean age at diagnosis of all people with early-onset dementia is less than 55 years old.

9.19 Serving Time. According to the Bureau of Crime Statistics and Research of Australia, as reported on *Lawlink*, the mean length of imprisonment for motor-vehicle-theft offenders in Australia is 16.7 months. You want to perform a hypothesis test to decide whether the mean length of imprisonment for motor-vehicle-theft offenders in Sydney differs from the national mean in Australia.

9.20 Worker Fatigue. A study by M. Chen et al. titled "Heat Stress Evaluation and Worker Fatigue in a Steel Plant" (*American Industrial Hygiene Association*, Vol. 64, pp. 352–359) assessed fatigue in steel-plant workers due to heat stress. Among other things, the researchers monitored the heart rates of a random sample of 29 casting workers. A hypothesis test is to be conducted to decide whether the mean post-work heart rate of casting workers exceeds the normal resting heart rate of 72 beats per minute (bpm).

9.21 Body Temperature. A study by researchers at the University of Maryland addressed the question of whether the mean body temperature of humans is 98.6°F. The results of the study by P. Mackowiak et al. appeared in the article "A Critical Appraisal of 98.6°F, the Upper Limit of the Normal Body Temperature, and Other Legacies of Carl Reinhold August Wunderlich" (*Journal of the American Medical Association*, Vol. 268, pp. 1578–1580). Among other data, the researchers obtained the body temperatures of 93 healthy humans. Suppose that you want to use those data to decide whether the mean body temperature of healthy humans differs from 98.6°F.

9.22 Teacher Salaries. Data on salaries in the public school system are published annually in *Ranking of the States and Estimates of School Statistics* by the National Education Association. The mean annual salary of (public) classroom teachers is $55.4 thousand. A hypothesis test is to be performed to decide whether the mean annual salary of classroom teachers in Ohio is greater than the national mean.

9.23 Toxic Mushrooms? Refer to Exercise 9.15. Explain what each of the following would mean.

a. Type I error   **b.** Type II error   **c.** Correct decision

Now suppose that the results of carrying out the hypothesis test lead to nonrejection of the null hypothesis. Classify that conclusion by error type or as a correct decision if in fact the mean cadmium level in *Boletus pinicola* mushrooms

d. equals the safety limit of 0.5 ppm.

e. exceeds the safety limit of 0.5 ppm.

9.24 Grey-Seal Nursing. Refer to Exercise 9.16. Explain what each of the following would mean.

a. Type I error   **b.** Type II error   **c.** Correct decision

Now suppose that the results of carrying out the hypothesis test lead to rejection of the null hypothesis. Classify that conclusion by error type or as a correct decision if in fact the mean lactation period of grey seals

d. equals 23 days.   **e.** differs from 23 days.

9.25 Iron Deficiency? Refer to Exercise 9.17. Explain what each of the following would mean.

a. Type I error   **b.** Type II error   **c.** Correct decision

Now suppose that the results of carrying out the hypothesis test lead to rejection of the null hypothesis. Classify that conclusion by error type or as a correct decision if in fact the mean iron intake of all adult females under the age of 51 years

d. equals the RDA of 18 mg per day.

e. is less than the RDA of 18 mg per day.

9.26 Early-Onset Dementia. Refer to Exercise 9.18. Explain what each of the following would mean.

a. Type I error   **b.** Type II error   **c.** Correct decision

Now suppose that the results of carrying out the hypothesis test lead to nonrejection of the null hypothesis. Classify that conclusion by error type or as a correct decision if in fact the mean age at diagnosis of all people with early-onset dementia

d. is 55 years old.   **e.** is less than 55 years old.

9.27 Serving Time. Refer to Exercise 9.19. Explain what each of the following would mean.

a. Type I error   **b.** Type II error   **c.** Correct decision

Now suppose that the results of carrying out the hypothesis test lead to nonrejection of the null hypothesis. Classify that conclusion by error type or as a correct decision if in fact the mean length of imprisonment for motor-vehicle-theft offenders in Sydney

d. equals the national mean of 16.7 months.
e. differs from the national mean of 16.7 months.

9.28 Worker Fatigue. Refer to Exercise 9.20. Explain what each of the following would mean.

a. Type I error   **b.** Type II error   **c.** Correct decision

Now suppose that the results of carrying out the hypothesis test lead to rejection of the null hypothesis. Classify that conclusion by error type or as a correct decision if in fact the mean post-work heart rate of casting workers

d. equals the normal resting heart rate of 72 bpm.
e. exceeds the normal resting heart rate of 72 bpm.

9.29 Body Temperature. Refer to Exercise 9.21. Explain what each of the following would mean.

a. Type I error   **b.** Type II error   **c.** Correct decision

Now suppose that the results of carrying out the hypothesis test lead to rejection of the null hypothesis. Classify that conclusion by error type or as a correct decision if in fact the mean body temperature of all healthy humans

d. is 98.6°F.
e. is not 98.6°F.

9.30 Teacher Salaries. Refer to Exercise 9.22. Explain what each of the following would mean.

a. Type I error   **b.** Type II error   **c.** Correct decision

Now suppose that the results of carrying out the hypothesis test lead to nonrejection of the null hypothesis. Classify that conclusion by error type or as a correct decision if in fact the mean salary of classroom teachers in Ohio

d. equals the national mean of $55.4 thousand.
e. exceeds the national mean of $55.4 thousand.

9.2   Critical-Value Approach to Hypothesis Testing[†]

With the **critical-value approach to hypothesis testing,** we choose a "cutoff point" (or cutoff points) based on the significance level of the hypothesis test. The criterion for deciding whether to reject the null hypothesis involves a comparison of the value of the test statistic to the cutoff point(s). Our next example introduces these ideas.

EXAMPLE 9.5   The Critical-Value Approach

Golf Driving Distances Jack tells Jean that his average drive of a golf ball is 275 yards. Jean is skeptical and asks for substantiation. To that end, Jack hits 25 drives. The results, in yards, are shown in Table 9.2.

The (sample) mean of Jack's 25 drives is only 264.4 yards. Jack still maintains that, on average, he drives a golf ball 275 yards and that his (relatively) poor performance can reasonably be attributed to chance.

At the 5% significance level, do the data provide sufficient evidence to conclude that Jack's mean driving distance is less than 275 yards? We use the following steps to answer the question.

TABLE 9.2

Distances (yards) of 25 drives by Jack

266	254	248	249	297
261	293	261	266	279
222	212	282	281	265
240	284	253	274	243
272	279	261	273	295

a. State the null and alternative hypotheses.
b. Discuss the logic of this hypothesis test.
c. Obtain a precise criterion for deciding whether to reject the null hypothesis in favor of the alternative hypothesis.
d. Apply the criterion in part (c) to the sample data and state the conclusion.

For our analysis, we assume that Jack's driving distances are normally distributed (which can be shown to be reasonable) and that the population standard deviation of all such driving distances is 20 yards.[‡]

[†]Those concentrating on the *P*-value approach to hypothesis testing can skip this section if so desired.

[‡]We are assuming that the population standard deviation is known, for simplicity. The more usual case in which the population standard deviation is unknown is discussed in Section 9.5.

Solution

a. Let μ denote the population mean of (all) Jack's driving distances. The null hypothesis is Jack's claim of an overall driving-distance average of 275 yards. The alternative hypothesis is Jean's suspicion that Jack's overall driving-distance average is less than 275 yards. Hence, the null and alternative hypotheses are, respectively,

$$H_0: \mu = 275 \text{ yards (Jack's claim)}$$
$$H_a: \mu < 275 \text{ yards (Jean's suspicion)}.$$

Note that this hypothesis test is left tailed.

b. Basically, the logic of this hypothesis test is as follows: If the null hypothesis is true, then the mean distance, $\bar{x}$, of the sample of Jack's 25 drives should approximately equal 275 yards. We say "approximately equal" because we cannot expect a sample mean to exactly equal the population mean; some sampling error is anticipated. However, if the sample mean driving distance is "too much smaller" than 275 yards, we would be inclined to reject the null hypothesis in favor of the alternative hypothesis.

c. We use our knowledge of the sampling distribution of the sample mean and the specified significance level to decide how much smaller is "too much smaller." Assuming that the null hypothesis is true, Key Fact 7.2 on page 320 shows that, for samples of size 25, the sample mean driving distance, $\bar{x}$, is normally distributed with mean and standard deviation

$$\mu_{\bar{x}} = \mu = 275 \text{ yards} \quad \text{and} \quad \sigma_{\bar{x}} = \frac{\sigma}{\sqrt{n}} = \frac{20}{\sqrt{25}} = 4 \text{ yards},$$

respectively. Therefore, from Key Fact 6.4 on page 269, the standardized version of $\bar{x}$,

$$z = \frac{\bar{x} - \mu_{\bar{x}}}{\sigma_{\bar{x}}} = \frac{\bar{x} - \mu}{\sigma/\sqrt{n}} = \frac{\bar{x} - 275}{4},$$

has the standard normal distribution. We use this variable, $z = (\bar{x} - 275)/4$, as our test statistic.

Because the hypothesis test is left tailed and we want a 5% significance level (i.e., $\alpha = 0.05$), we choose the cutoff point to be the z-score with area 0.05 to its left under the standard normal curve. From Table II, we find that z-score to be -1.645.

Consequently, "too much smaller" is a sample mean driving distance with a z-score of -1.645 or less. Figure 9.2 displays our criterion for deciding whether to reject the null hypothesis.

FIGURE 9.2

Criterion for deciding whether to reject the null hypothesis

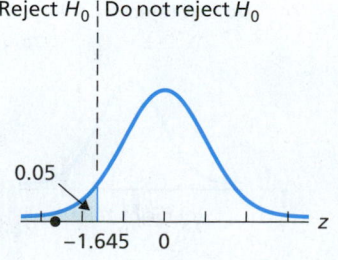

d. Now we compute the value of the test statistic and compare it to our cutoff point of -1.645. As we noted, the sample mean driving distance of Jack's 25 drives is 264.4 yards. Hence, the value of the test statistic is

$$z = \frac{\bar{x} - 275}{4} = \frac{264.4 - 275}{4} = -2.65.$$

This value of z is marked with a dot in Fig. 9.2. We see that the value of the test statistic, -2.65, is less than the cutoff point of -1.645 and, hence, we reject H_0.

Interpretation At the 5% significance level, the data provide sufficient evidence to conclude that Jack's mean driving distance is less than his claimed 275 yards. ∎

Note: The curve in Fig. 9.2—which is the standard normal curve—is the normal curve for the test statistic $z = (\bar{x} - 275)/4$, provided that the null hypothesis is true. We see then from Fig. 9.2 that the probability of rejecting the null hypothesis if it is in fact true (i.e., the probability of making a Type I error) is 0.05. In other words, the significance level of the hypothesis test is indeed 0.05 (5%), as required.

Terminology of the Critical-Value Approach

Referring to the preceding example, we present some important terminology that is used with the critical-value approach to hypothesis testing. The set of values for the test statistic that leads us to reject the null hypothesis is called the **rejection region.** In this case, the rejection region consists of all z-scores that lie to the left of -1.645—that part of the horizontal axis under the shaded area in Fig. 9.2.

The set of values for the test statistic that leads us not to reject the null hypothesis is called the **nonrejection region.** Here, the nonrejection region consists of all z-scores that lie to the right of -1.645—that part of the horizontal axis under the unshaded area in Fig. 9.2.

The value of the test statistic that separates the rejection and nonrejection regions (i.e., the cutoff point) is called the **critical value.** In this case, the critical value is $z = -1.645$.

We summarize the preceding discussion in Fig. 9.3, and, with that discussion in mind, we present Definition 9.4. Before doing so, however, we note the following:

- The rejection region pictured in Fig. 9.3 is typical of that for a left-tailed test. Soon we will discuss the form of the rejection regions for a two-tailed test and a right-tailed test.
- The terminology introduced so far in this section (and most of that which will be presented later) applies to any hypothesis test, not just to hypothesis tests for a population mean.

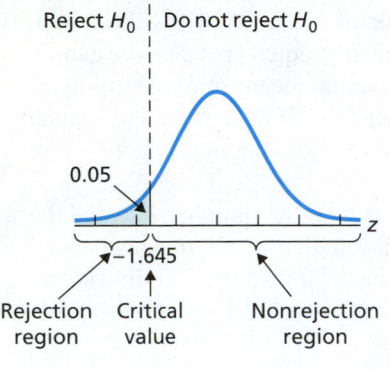

FIGURE 9.3

Rejection region, nonrejection region, and critical value for the golf-driving-distances hypothesis test

DEFINITION 9.4

? What Does It Mean?

If the value of the test statistic falls in the rejection region, reject the null hypothesis; otherwise, do not reject the null hypothesis.

Rejection Region, Nonrejection Region, and Critical Values

Rejection region: The set of values for the test statistic that leads to rejection of the null hypothesis.

Nonrejection region: The set of values for the test statistic that leads to non-rejection of the null hypothesis.

Critical value(s): The value or values of the test statistic that separate the rejection and nonrejection regions. A critical value is considered part of the rejection region.

For a two-tailed test, as in Example 9.1 on page 369, the null hypothesis is rejected when the test statistic is either too small or too large. Thus the rejection region for such a test consists of two parts: one on the left and one on the right, as shown in Fig. 9.4(a).

FIGURE 9.4

Graphical display of rejection regions for two-tailed, left-tailed, and right-tailed tests

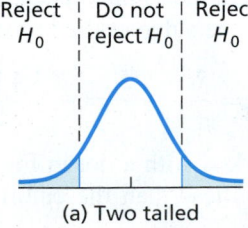

(a) Two tailed

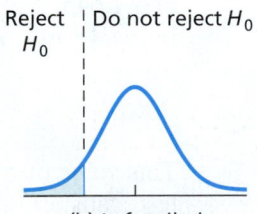

(b) Left tailed

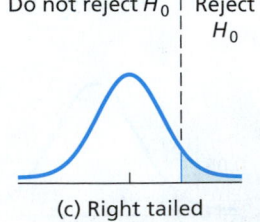

(c) Right tailed

For a left-tailed test, as in Example 9.3 on page 370, the null hypothesis is rejected only when the test statistic is too small. Thus the rejection region for such a test consists of only one part, which is on the left, as shown in Fig. 9.4(b).

For a right-tailed test, as in Example 9.2 on pages 369–370, the null hypothesis is rejected only when the test statistic is too large. Thus the rejection region for such a test consists of only one part, which is on the right, as shown in Fig. 9.4(c).

Table 9.3 and Fig. 9.4 summarize our discussion. Figure 9.4 shows why the term *tailed* is used: The rejection region is in both tails for a two-tailed test, in the left tail for a left-tailed test, and in the right tail for a right-tailed test.

You try it!

Exercise 9.35 on page 381

TABLE 9.3

Rejection regions for two-tailed, left-tailed, and right-tailed tests

	Two-tailed test	Left-tailed test	Right-tailed test
Sign in H_a	$\neq$	$<$	$>$
Rejection region	Both sides	Left side	Right side

Obtaining Critical Values

Recall that the significance level of a hypothesis test is the probability of rejecting a true null hypothesis. With the critical-value approach, we reject the null hypothesis if and only if the test statistic falls in the rejection region. Therefore, we have Key Fact 9.3.

KEY FACT 9.3

Obtaining Critical Values

Suppose that a hypothesis test is to be performed at the significance level α. Then the critical value(s) must be chosen so that, if the null hypothesis is true, the probability is α that the test statistic will fall in the rejection region.

Obtaining Critical Values for a One-Mean z-Test

The first hypothesis-testing procedure that we discuss is called the **one-mean z-test.** This procedure is used to perform a hypothesis test for one population mean when the population standard deviation is known and the variable under consideration is normally distributed. Keep in mind, however, that because of the central limit theorem, the one-mean z-test will work reasonably well when the sample size is large, regardless of the distribution of the variable.

As you have seen, the null hypothesis for a hypothesis test concerning one population mean, μ, has the form H_0: $\mu = \mu_0$, where μ_0 is some number. Referring to part (c) of the solution to Example 9.5 (page 377), we see that the test statistic for a one-mean z-test is

$$z = \frac{\bar{x} - \mu_0}{\sigma/\sqrt{n}},$$

which, by the way, tells you how many standard deviations the observed sample mean, $\bar{x}$, is from μ_0 (the value specified for the population mean in the null hypothesis).

The basis of the hypothesis-testing procedure is in Key Fact 7.2: If x is a normally distributed variable with mean μ and standard deviation σ, then, for samples of size n, the variable $\bar{x}$ is also normally distributed and has mean μ and standard deviation $\sigma/\sqrt{n}$. This fact and Key Fact 6.4 (page 269) applied to $\bar{x}$ imply that, if the null hypothesis is true, the test statistic z has the standard normal distribution.

Consequently, in view of Key Fact 9.3, for a specified significance level α, we need to choose the critical value(s) so that the area under the standard normal curve that lies above the rejection region equals α.

EXAMPLE 9.6   **Obtaining the Critical Values for a One-Mean z-Test**

Determine the critical value(s) for a one-mean z-test at the 5% significance level ($\alpha = 0.05$) if the test is

a. two tailed.   **b.** left tailed.   **c.** right tailed.

Solution Because $\alpha = 0.05$, we need to choose the critical value(s) so that the area under the standard normal curve that lies above the rejection region equals 0.05.

a. For a two-tailed test, the rejection region is on both the left and right. So the critical values are the two z-scores that divide the area under the standard normal curve into a middle 0.95 area and two outside areas of 0.025. In other words, the critical values are $\pm z_{0.025}$. From Table II in Appendix A, $\pm z_{0.025} = \pm 1.96$, as shown in Fig. 9.5(a).

FIGURE 9.5

Critical value(s) for a one-mean z-test at the 5% significance level if the test is (a) two tailed, (b) left tailed, or (c) right tailed

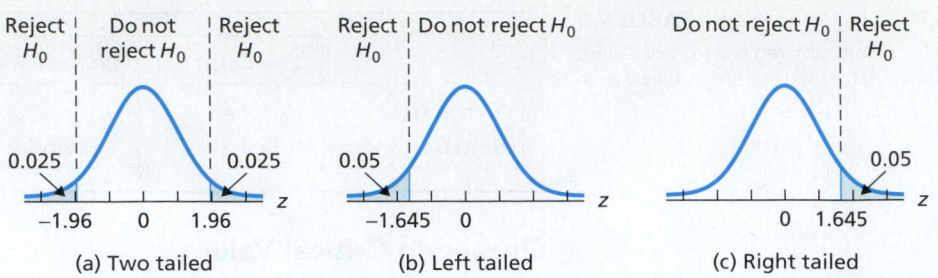

(a) Two tailed (b) Left tailed (c) Right tailed

b. For a left-tailed test, the rejection region is on the left. So the critical value is the z-score with area 0.05 to its left under the standard normal curve, which is $-z_{0.05}$. From Table II, $-z_{0.05} = -1.645$, as shown in Fig. 9.5(b).

c. For a right-tailed test, the rejection region is on the right. So the critical value is the z-score with area 0.05 to its right under the standard normal curve, which is $z_{0.05}$. From Table II, $z_{0.05} = 1.645$, as shown in Fig. 9.5(c).

By reasoning as we did in the previous example, we can obtain the critical value(s) for any specified significance level α. As shown in Fig. 9.6, for a two-tailed test, the critical values are $\pm z_{\alpha/2}$; for a left-tailed test, the critical value is $-z_\alpha$; and for a right-tailed test, the critical value is z_α.

FIGURE 9.6

Critical value(s) for a one-mean z-test at the significance level α if the test is (a) two tailed, (b) left tailed, or (c) right tailed

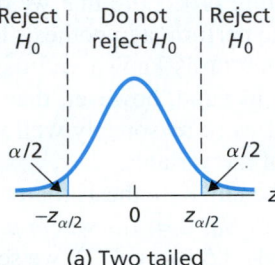 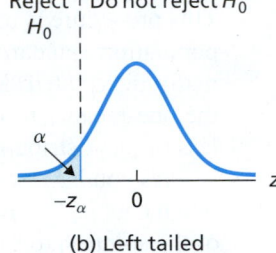 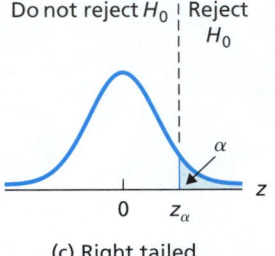

(a) Two tailed (b) Left tailed (c) Right tailed

You try it!

Exercise 9.41 on page 381

The most commonly used significance levels are 0.10, 0.05, and 0.01. If we consider both one-tailed and two-tailed tests, these three significance levels give rise to five "tail areas." Using the standard-normal table, Table II, we obtained the value of z_α corresponding to each of those five tail areas as shown in Table 9.4.

TABLE 9.4

Some important values of z_α

$z_{0.10}$	$z_{0.05}$	$z_{0.025}$	$z_{0.01}$	$z_{0.005}$
1.28	1.645	1.96	2.33	2.575

Alternatively, we can find these five values of z_α at the bottom of the t-table, Table IV, where they are displayed to three decimal places. Can you explain the slight discrepancy between the values given for $z_{0.005}$ in the two tables?

Steps in the Critical-Value Approach to Hypothesis Testing

We have now covered all the concepts required for the critical-value approach to hypothesis testing. The general steps involved in that approach are presented in Table 9.5.

TABLE 9.5

General steps for the critical-value approach to hypothesis testing

CRITICAL-VALUE APPROACH TO HYPOTHESIS TESTING
Step 1 State the null and alternative hypotheses.
Step 2 Decide on the significance level, α.
Step 3 Compute the value of the test statistic.
Step 4 Determine the critical value(s).
Step 5 If the value of the test statistic falls in the rejection region, reject H_0; otherwise, do not reject H_0.
Step 6 Interpret the result of the hypothesis test.

Throughout the text, we present dedicated step-by-step procedures for specific hypothesis-testing procedures. Those using the critical-value approach, however, are all based on the steps shown in Table 9.5.

Exercises 9.2

Understanding the Concepts and Skills

In each of Exercises 9.31–9.34, define the term given.

9.31 test statistic

9.32 rejection region

9.33 nonrejection region

9.34 critical values

Exercises 9.35–9.40 contain graphs portraying the decision criterion for a one-mean z-test. The curve in each graph is the normal curve for the test statistic under the assumption that the null hypothesis is true. For each exercise, determine the

a. rejection region. **b.** nonrejection region.
c. critical value(s). **d.** significance level.
e. Construct a graph similar to that in Fig. 9.3 on page 378 that depicts your results from parts (a)–(d).
f. Identify the hypothesis test as two tailed, left tailed, or right tailed.

9.35

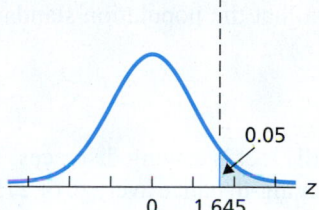

9.36

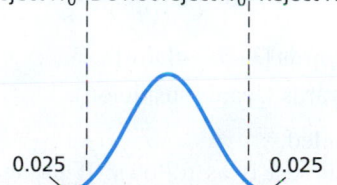

9.37

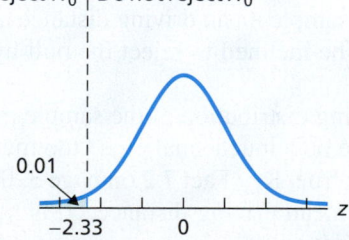

9.38

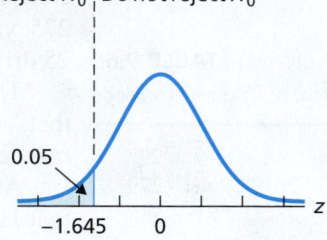

9.39

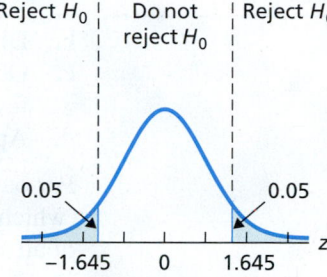

9.40

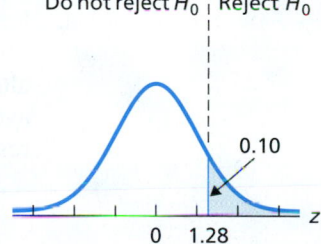

In each of Exercises 9.41–9.46, determine the critical value(s) for a one-mean z-test. For each exercise, draw a graph that illustrates your answer.

9.41 A two-tailed test with $\alpha = 0.10$.

9.42 A right-tailed test with $\alpha = 0.05$.

9.43 A left-tailed test with $\alpha = 0.01$.

9.44 A left-tailed test with $\alpha = 0.05$.

9.45 A right-tailed test with $\alpha = 0.01$.

9.46 A two-tailed test with $\alpha = 0.05$.

9.3 *P*-Value Approach to Hypothesis Testing[†]

Roughly speaking, with the ***P*-value approach to hypothesis testing,** we first evaluate how likely observation of the value obtained for the test statistic would be if the null hypothesis is true. The criterion for deciding whether to reject the null hypothesis involves a comparison of that likelihood with the specified significance level of the hypothesis test. Our next example introduces these ideas.

[†]Those concentrating on the critical-value approach to hypothesis testing can skip this section if so desired. Note, however, that this section is prerequisite to the (optional) technology materials that appear in The Technology Center sections.

   **EXAMPLE 9.7**   The *P*-Value Approach

Golf Driving Distances Jack tells Jean that his average drive of a golf ball is 275 yards. Jean is skeptical and asks for substantiation. To that end, Jack hits 25 drives. The results, in yards, are shown in Table 9.6.

TABLE 9.6

Distances (yards) of 25 drives by Jack

266	254	248	249	297
261	293	261	266	279
222	212	282	281	265
240	284	253	274	243
272	279	261	273	295

The (sample) mean of Jack's 25 drives is only 264.4 yards. Jack still maintains that, on average, he drives a golf ball 275 yards and that his (relatively) poor performance can reasonably be attributed to chance.

At the 5% significance level, do the data provide sufficient evidence to conclude that Jack's mean driving distance is less than 275 yards? We use the following steps to answer the question.

a. State the null and alternative hypotheses.
b. Discuss the logic of this hypothesis test.
c. Obtain a precise criterion for deciding whether to reject the null hypothesis in favor of the alternative hypothesis.
d. Apply the criterion in part (c) to the sample data and state the conclusion.

For our analysis, we assume that Jack's driving distances are normally distributed (which can be shown to be reasonable) and that the population standard deviation of all such driving distances is 20 yards.[†]

Solution

a. Let μ denote the population mean of (all) Jack's driving distances. The null hypothesis is Jack's claim of an overall driving-distance average of 275 yards. The alternative hypothesis is Jean's suspicion that Jack's overall driving-distance average is less than 275 yards. Hence, the null and alternative hypotheses are, respectively,

$$H_0: \mu = 275 \text{ yards (Jack's claim)}$$
$$H_a: \mu < 275 \text{ yards (Jean's suspicion).}$$

Note that this hypothesis test is left tailed.

b. Basically, the logic of this hypothesis test is as follows: If the null hypothesis is true, then the mean distance, $\bar{x}$, of the sample of Jack's 25 drives should approximately equal 275 yards. We say "approximately equal" because we cannot expect a sample mean to exactly equal the population mean; some sampling error is anticipated. However, if the sample mean driving distance is "too much smaller" than 275 yards, we would be inclined to reject the null hypothesis in favor of the alternative hypothesis.

c. We use our knowledge of the sampling distribution of the sample mean and the specified significance level to decide how much smaller is "too much smaller." Assuming that the null hypothesis is true, Key Fact 7.2 on page 320 shows that, for samples of size 25, the sample mean driving distance, $\bar{x}$, is normally distributed with mean and standard deviation

$$\mu_{\bar{x}} = \mu = 275 \text{ yards} \quad \text{and} \quad \sigma_{\bar{x}} = \frac{\sigma}{\sqrt{n}} = \frac{20}{\sqrt{25}} = 4 \text{ yards,}$$

respectively. Thus, from Key Fact 6.4 on page 269, the standardized version of $\bar{x}$,

$$z = \frac{\bar{x} - \mu_{\bar{x}}}{\sigma_{\bar{x}}} = \frac{\bar{x} - \mu}{\sigma/\sqrt{n}} = \frac{\bar{x} - 275}{4},$$

has the standard normal distribution. We use this variable, $z = (\bar{x} - 275)/4$, as our test statistic.

Because the hypothesis test is left tailed, we compute the probability of observing a value of the test statistic z that is as small as or smaller than the value

---

[†]We are assuming that the population standard deviation is known, for simplicity. The more usual case in which the population standard deviation is unknown is discussed in Section 9.5.

actually observed. This probability is called the *P-value* of the hypothesis test and is denoted by the letter *P*.

　　Our criterion for deciding whether to reject the null hypothesis is then as follows: If the *P*-value is less than or equal to the specified significance level, we reject the null hypothesis; otherwise, we do not reject the null hypothesis.

d. Now we obtain the *P*-value and compare it to the specified significance level of 0.05. As we have noted, the sample mean driving distance of Jack's 25 drives is 264.4 yards. Hence, the value of the test statistic is

$$z = \frac{\bar{x} - 275}{4} = \frac{264.4 - 275}{4} = -2.65.$$

Consequently, the *P*-value is the probability of observing a value of z of -2.65 or smaller if the null hypothesis is true. That probability equals the area under the standard normal curve to the left of -2.65, the shaded region in Fig. 9.7. From Table II, we find that area to be 0.0040. Because the *P*-value, 0.0040, is less than the specified significance level of 0.05, we reject H_0.

FIGURE 9.7

P-value for golf-driving-distances hypothesis test

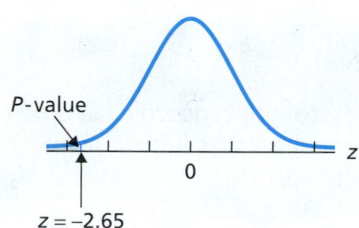

Interpretation　At the 5% significance level, the data provide sufficient evidence to conclude that Jack's mean driving distance is less than his claimed 275 yards. ∎

Note: The *P*-value will be less than or equal to 0.05 whenever the value of the test statistic z has area 0.05 or less to its left under the standard normal curve, which is exactly 5% of the time if the null hypothesis is true. Thus, we see that, by using the decision criterion "reject the null hypothesis if $P \le 0.05$; otherwise, do not reject the null hypothesis," the probability of rejecting the null hypothesis if it is in fact true (i.e., the probability of making a Type I error) is 0.05. In other words, the significance level of the hypothesis test is indeed 0.05 (5%), as required.

　　Let us emphasize the meaning of the *P*-value, 0.0040, obtained in the preceding example. Specifically, if the null hypothesis is true, we would observe a value of the test statistic z of -2.65 or less only 4 times in 1000. In other words, if the null hypothesis is true, a random sample of 25 of Jack's drives would have a mean distance of 264.4 yards or less only 0.4% of the time. The sample data provide very strong evidence against the null hypothesis (Jack's claim) and in favor of the alternative hypothesis (Jean's suspicion).

Terminology of the *P*-Value Approach

We introduced the *P*-value in the context of the preceding example. More generally, we define the *P*-value as follows.

DEFINITION 9.5

What Does It Mean?

Small *P*-values provide evidence against the null hypothesis; larger *P*-values do not.

P-Value

The **P-value** of a hypothesis test is the probability of getting sample data at least as inconsistent with the null hypothesis (and supportive of the alternative hypothesis) as the sample data actually obtained.[†] We use the letter **P** to denote the *P*-value.

Note: The smaller (closer to 0) the *P*-value, the stronger is the evidence against the null hypothesis and, hence, in favor of the alternative hypothesis. Stated simply, *an outcome that would rarely occur if the null hypothesis were true provides evidence against the null hypothesis and, hence, in favor of the alternative hypothesis.*

　　As illustrated in the solution to part (c) of Example 9.7 (golf driving distances), with the *P*-value approach to hypothesis testing, we use the following criterion to decide whether to reject the null hypothesis.

[†]Alternatively, we can define the *P*-value to be the percentage of samples that are at least as inconsistent with the null hypothesis (and supportive of the alternative hypothesis) as the sample actually obtained.

KEY FACT 9.4

Decision Criterion for a Hypothesis Test Using the *P*-Value

If the *P*-value is less than or equal to the specified significance level, reject the null hypothesis; otherwise, do not reject the null hypothesis. In other words, if $P \leq \alpha$, reject H_0; otherwise, do not reject H_0.

The *P*-value of a hypothesis test is also referred to as the **observed significance level.** To understand why, suppose that the *P*-value of a hypothesis test is $P = 0.07$. Then, for instance, we see from Key Fact 9.4 that we can reject the null hypothesis at the 10% significance level (because $P \leq 0.10$), but we cannot reject the null hypothesis at the 5% significance level (because $P > 0.05$). In fact, here, the null hypothesis can be rejected at any significance level of at least 0.07 and cannot be rejected at any significance level less than 0.07.

More generally, we have the following fact.

KEY FACT 9.5

P-Value as the Observed Significance Level

The *P*-value of a hypothesis test equals the smallest significance level at which the null hypothesis can be rejected, that is, the smallest significance level for which the observed sample data results in rejection of H_0.

Determining *P*-Values

We defined the *P*-value of a hypothesis test in Definition 9.5 (preceding page). To actually determine a *P*-value, however, we rely on the value of the test statistic, as follows.

KEY FACT 9.6

Determining a *P*-Value

To determine the *P*-value of a hypothesis test, we assume that the null hypothesis is true and compute the probability of observing a value of the test statistic as extreme as or more extreme than that observed. By *extreme* we mean "far from what we would expect to observe if the null hypothesis is true."

Determining the *P*-Value for a One-Mean *z*-Test

The first hypothesis-testing procedure that we discuss is called the **one-mean *z*-test.** This procedure is used to perform a hypothesis test for one population mean when the population standard deviation is known and the variable under consideration is normally distributed. Keep in mind, however, that because of the central limit theorem, the one-mean *z*-test will work reasonably well when the sample size is large, regardless of the distribution of the variable.

As you have seen, the null hypothesis for a hypothesis test concerning one population mean, μ, has the form $H_0: \mu = \mu_0$, where μ_0 is some number. Referring to part (c) of the solution to Example 9.7, we see that the test statistic for a one-mean *z*-test is

$$z = \frac{\bar{x} - \mu_0}{\sigma/\sqrt{n}},$$

which, by the way, tells you how many standard deviations the observed sample mean, $\bar{x}$, is from μ_0 (the value specified for the population mean in the null hypothesis).

The basis of the hypothesis-testing procedure is in Key Fact 7.2: If x is a normally distributed variable with mean μ and standard deviation σ, then, for samples of size n, the variable $\bar{x}$ is also normally distributed and has mean μ and standard deviation $\sigma/\sqrt{n}$. This fact and Key Fact 6.4 (page 269) applied to $\bar{x}$ imply that, if the null hypothesis is true, the test statistic z has the standard normal distribution and hence that its probabilities equal areas under the standard normal curve.

Therefore, in view of Key Fact 9.6, if we let z_0 denote the observed value of the test statistic z, we determine the *P*-value as follows:

- *Two-tailed test:* The *P*-value equals the probability of observing a value of the test statistic z that is at least as large in magnitude as the value actually observed, which is the area under the standard normal curve that lies outside the interval from $-|z_0|$ to $|z_0|$, as illustrated in Fig. 9.8(a).
- *Left-tailed test:* The *P*-value equals the probability of observing a value of the test statistic z that is as small as or smaller than the value actually observed, which is the area under the standard normal curve that lies to the left of z_0, as illustrated in Fig. 9.8(b).
- *Right-tailed test:* The *P*-value equals the probability of observing a value of the test statistic z that is as large as or larger than the value actually observed, which is the area under the standard normal curve that lies to the right of z_0, as illustrated in Fig. 9.8(c).

FIGURE 9.8

P-value for a one-mean *z*-test if the test is (a) two tailed, (b) left tailed, or (c) right tailed

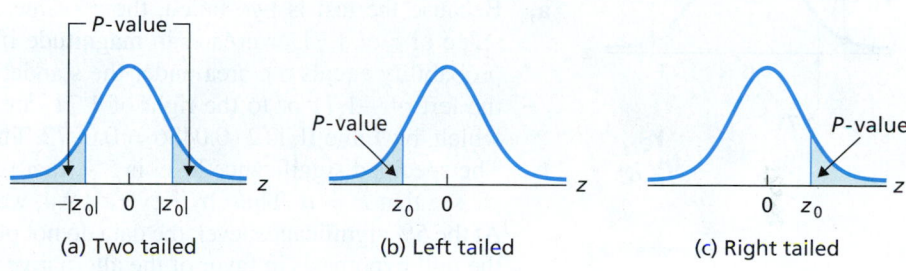

(a) Two tailed   (b) Left tailed   (c) Right tailed

EXAMPLE 9.8   **Determining the *P*-Value for a One-Mean *z*-Test**

The value of the test statistic for a left-tailed one-mean *z*-test is $z = -1.19$.

a. Determine the *P*-value.
b. At the 5% significance level, do the data provide sufficient evidence to reject the null hypothesis in favor of the alternative hypothesis?

FIGURE 9.9

Value of the test statistic and the *P*-value

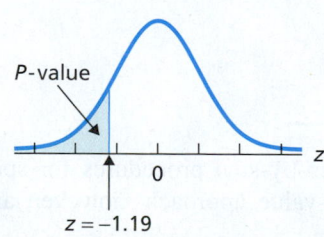

Solution

a. Because the test is left tailed, the *P*-value is the probability of observing a value of z of -1.19 or less if the null hypothesis is true. That probability equals the area under the standard normal curve to the left of -1.19, the shaded area shown in Fig. 9.9, which, by Table II, is 0.1170. Therefore, $P = 0.1170$.
b. The specified significance level is 5%, that is, $\alpha = 0.05$. Hence, from part (a), we see that $P > \alpha$. Thus, by Key Fact 9.4, we do not reject the null hypothesis. At the 5% significance level, the data do not provide sufficient evidence to reject the null hypothesis in favor of the alternative hypothesis.

EXAMPLE 9.9   **Determining the *P*-Value for a One-Mean *z*-Test**

The value of the test statistic for a right-tailed one-mean *z*-test is $z = 2.85$.

FIGURE 9.10

Value of the test statistic and the *P*-value

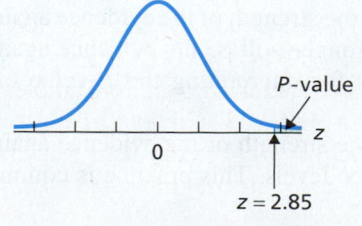

a. Determine the *P*-value.
b. At the 1% significance level, do the data provide sufficient evidence to reject the null hypothesis in favor of the alternative hypothesis?

Solution

a. Because the test is right tailed, the *P*-value is the probability of observing a value of z of 2.85 or greater if the null hypothesis is true. That probability equals the area under the standard normal curve to the right of 2.85, the shaded area shown in Fig. 9.10, which, by Table II, is $1 - 0.9978 = 0.0022$. Therefore, $P = 0.0022$.

b. The specified significance level is 1%, that is, $\alpha = 0.01$. Hence, from part (a), we see that $P \leq \alpha$. Thus, by Key Fact 9.4, we reject the null hypothesis. At the 1% significance level, the data provide sufficient evidence to reject the null hypothesis in favor of the alternative hypothesis.

■

■■ ■ EXAMPLE 9.10 Determining the *P*-Value for a One-Mean *z*-Test

FIGURE 9.11

Value of the test statistic and the *P*-value

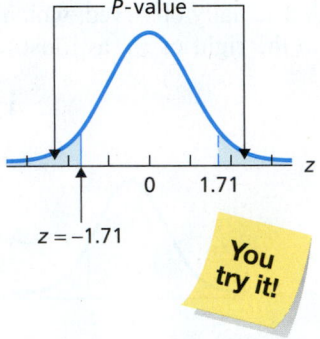

Exercise 9.63
on page 387

The value of the test statistic for a two-tailed one-mean *z*-test is $z = -1.71$.

a. Determine the *P*-value.
b. At the 5% significance level, do the data provide sufficient evidence to reject the null hypothesis in favor of the alternative hypothesis?

Solution

a. Because the test is two tailed, the *P*-value is the probability of observing a value of z of 1.71 or greater in magnitude if the null hypothesis is true. That probability equals the area under the standard normal curve that lies either to the left of -1.71 or to the right of 1.71, the shaded area shown in Fig. 9.11, which, by Table II, is $2 \cdot 0.0436 = 0.0872$. Therefore, $P = 0.0872$.

b. The specified significance level is 5%, that is, $\alpha = 0.05$. Hence, from part (a), we see that $P > \alpha$. Thus, by Key Fact 9.4, we do not reject the null hypothesis. At the 5% significance level, the data do not provide sufficient evidence to reject the null hypothesis in favor of the alternative hypothesis.

■

Steps in the *P*-Value Approach to Hypothesis Testing

We have now covered all the concepts required for the *P*-value approach to hypothesis testing. The general steps involved in that approach are presented in Table 9.7.

TABLE 9.7

General steps for the *P*-value approach to hypothesis testing

***P*-VALUE APPROACH TO HYPOTHESIS TESTING**
Step 1 State the null and alternative hypotheses.
Step 2 Decide on the significance level, α.
Step 3 Compute the value of the test statistic.
Step 4 Determine the *P*-value, *P*.
Step 5 If $P \leq \alpha$, reject H_0; otherwise, do not reject H_0.
Step 6 Interpret the result of the hypothesis test.

Throughout the text, we present dedicated step-by-step procedures for specific hypothesis-testing procedures. Those using the *P*-value approach, however, are all based on the steps shown in Table 9.7.

Using the *P*-Value to Assess the Evidence Against the Null Hypothesis

TABLE 9.8

Guidelines for using the *P*-value to assess the evidence against the null hypothesis

P-value	Evidence against H_0
$P > 0.10$	Weak or none
$0.05 < P \leq 0.10$	Moderate
$0.01 < P \leq 0.05$	Strong
$P \leq 0.01$	Very strong

Key Fact 9.5 asserts that the *P*-value is the smallest significance level at which the null hypothesis can be rejected. Consequently, knowing the *P*-value allows us to assess significance at any level we desire. For instance, if the *P*-value of a hypothesis test is 0.03, the null hypothesis can be rejected at any significance level larger than or equal to 0.03, and it cannot be rejected at any significance level smaller than 0.03.

Knowing the *P*-value also allows us to evaluate the strength of the evidence against the null hypothesis: the smaller the *P*-value, the stronger will be the evidence against the null hypothesis. Table 9.8 presents guidelines for interpreting the *P*-value of a hypothesis test.

Note that we can use the *P*-value to evaluate the strength of the evidence against the null hypothesis without reference to significance levels. This practice is common among researchers.

> **Hypothesis Tests Without Significance Levels:** Many researchers do not explicitly refer to significance levels. Instead, they simply obtain the *P*-value and use it (or let the reader use it) to assess the strength of the evidence against the null hypothesis.

Exercises 9.3

Understanding the Concepts and Skills

9.47 State two reasons why including the *P*-value is prudent when you are reporting the results of a hypothesis test.

9.48 What is the *P*-value of a hypothesis test? When does it provide evidence against the null hypothesis?

9.49 Explain how the *P*-value is obtained for a one-mean *z*-test in case the hypothesis test is
a. left tailed.   **b.** right tailed.   **c.** two tailed.

9.50 True or false: The *P*-value is the smallest significance level for which the observed sample data result in rejection of the null hypothesis.

9.51 The *P*-value for a hypothesis test is 0.06. For each of the following significance levels, decide whether the null hypothesis should be rejected.
a. $\alpha = 0.05$ **b.** $\alpha = 0.10$ **c.** $\alpha = 0.06$

9.52 The *P*-value for a hypothesis test is 0.083. For each of the following significance levels, decide whether the null hypothesis should be rejected.
a. $\alpha = 0.05$ **b.** $\alpha = 0.10$ **c.** $\alpha = 0.06$

9.53 Which provides stronger evidence against the null hypothesis, a *P*-value of 0.02 or a *P*-value of 0.03? Explain your answer.

9.54 Which provides stronger evidence against the null hypothesis, a *P*-value of 0.06 or a *P*-value of 0.04? Explain your answer.

In each of Exercises 9.55–9.62, we have given the P-value for a hypothesis test. For each exercise, refer to Table 9.8 to determine the strength of the evidence against the null hypothesis.

9.55 $P = 0.06$ **9.56** $P = 0.35$

9.57 $P = 0.027$ **9.58** $P = 0.004$

9.59 $P = 0.184$ **9.60** $P = 0.086$

9.61 $P = 0.001$ **9.62** $P = 0.012$

In Exercises 9.63–9.68, we have given the value obtained for the test statistic, z, in a one-mean z-test. We have also specified whether the test is two tailed, left tailed, or right tailed. Determine the P-value in each case and decide whether, at the 5% significance level, the data provide sufficient evidence to reject the null hypothesis in favor of the alternative hypothesis.

9.63 Right-tailed test:
a. $z = 2.03$ **b.** $z = -0.31$

9.64 Left-tailed test:
a. $z = -1.84$ **b.** $z = 1.25$

9.65 Left-tailed test:
a. $z = -0.74$ **b.** $z = 1.16$

9.66 Two-tailed test:
a. $z = 3.08$ **b.** $z = -2.42$

9.67 Two-tailed test:
a. $z = -1.66$ **b.** $z = 0.52$

9.68 Right-tailed test:
a. $z = 1.24$ **b.** $z = -0.69$

Extending the Concepts and Skills

9.69 Consider a one-mean *z*-test. Denote z_0 as the observed value of the test statistic *z*. If the test is right tailed, then the *P*-value can be expressed as $P(z \geq z_0)$. Determine the corresponding expression for the *P*-value if the test is
a. left tailed.   **b.** two tailed.

9.70 The symbol $\Phi(z)$ is often used to denote the area under the standard normal curve that lies to the left of a specified value of *z*. Consider a one-mean *z*-test. Denote z_0 as the observed value of the test statistic *z*. Express the *P*-value of the hypothesis test in terms of Φ if the test is
a. left tailed.   **b.** right tailed.   **c.** two tailed.

9.71 **Obtaining the *P*-value.** Let *x* denote the test statistic for a hypothesis test and x_0 its observed value. Then the *P*-value of the hypothesis test equals
a. $P(x \geq x_0)$ for a right-tailed test,
b. $P(x \leq x_0)$ for a left-tailed test,
c. $2 \cdot \min\{P(x \leq x_0), P(x \geq x_0)\}$ for a two-tailed test,
where the probabilities are computed under the assumption that the null hypothesis is true. Suppose that you are considering a one-mean *z*-test. Verify that the probability expressions in parts (a)–(c) are equivalent to those obtained in Exercise 9.69.

9.4   Hypothesis Tests for One Population Mean When σ Is Known

As we mentioned earlier, the first hypothesis-testing procedure that we discuss is used to perform a hypothesis test for one population mean when the population standard deviation is known. We call this hypothesis-testing procedure the **one-mean *z*-test** or, when no confusion can arise, simply the **z-test.**[†]

[†] The one-mean *z*-test is also known as the **one-sample *z*-test** and the **one-variable *z*-test**. We prefer "one-mean" because it makes clear the parameter being tested.

Procedure 9.1 provides a step-by-step method for performing a one-mean z-test. As you can see, Procedure 9.1 includes options for either the critical-value approach (keep left) or the P-value approach (keep right). The bases for these approaches were discussed in Sections 9.2 and 9.3, respectively.

■ ■ ■ ■   **PROCEDURE 9.1**   One-Mean z-Test

Purpose   To perform a hypothesis test for a population mean, μ

Assumptions
1. Simple random sample
2. Normal population or large sample
3. σ known

Step 1   The null hypothesis is $H_0: \mu = \mu_0$, and the alternative hypothesis is

$$H_a: \mu \neq \mu_0 \quad \text{or} \quad H_a: \mu < \mu_0 \quad \text{or} \quad H_a: \mu > \mu_0$$
$$\text{(Two tailed)} \qquad \text{(Left tailed)} \qquad \text{(Right tailed)}.$$

Step 2   Decide on the significance level, α.

Step 3   Compute the value of the test statistic

$$z = \frac{\bar{x} - \mu_0}{\sigma/\sqrt{n}}$$

and denote that value z_0.

CRITICAL-VALUE APPROACH	OR	P-VALUE APPROACH

Step 4   The critical value(s) are

$$\pm z_{\alpha/2} \quad \text{or} \quad -z_\alpha \quad \text{or} \quad z_\alpha$$
$$\text{(Two tailed)} \qquad \text{(Left tailed)} \qquad \text{(Right tailed)}.$$

Use Table II to find the critical value(s).

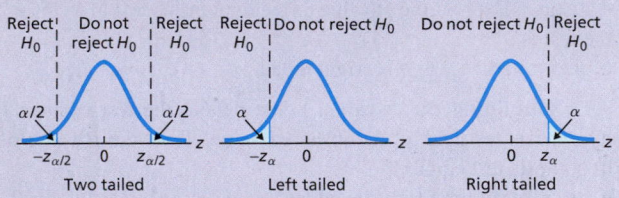

Step 5   If the value of the test statistic falls in the rejection region, reject H_0; otherwise, do not reject H_0.

Step 4   Use Table II to obtain the P-value.

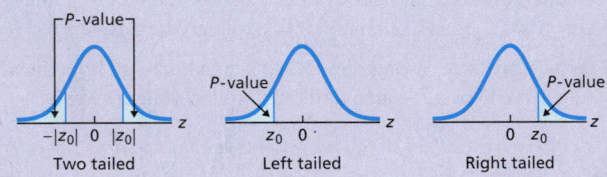

Step 5   If $P \leq \alpha$, reject H_0; otherwise, do not reject H_0.

Step 6   Interpret the results of the hypothesis test.

Note: The hypothesis test is exact for normal populations and is approximately correct for large samples from nonnormal populations.

Note: By saying that the hypothesis test is *exact*, we mean that the true significance level equals α; by saying that it is *approximately correct*, we mean that the true significance level only approximately equals α.

Properties and guidelines for use of the one-mean z-test are similar to those for the one-mean z-interval procedure. In particular, the one-mean z-test is robust to moderate violations of the normality assumption but, even for large samples, can sometimes be unduly affected by outliers because the sample mean is not resistant to outliers. Key Fact 9.7 lists some general guidelines for use of the one-mean z-test.

KEY FACT 9.7

When to Use the One-Mean z-Test[†]

- For small samples—say, of size less than 15—the z-test should be used only when the variable under consideration is normally distributed or very close to being so.
- For samples of moderate size—say, between 15 and 30—the z-test can be used unless the data contain outliers or the variable under consideration is far from being normally distributed.
- For large samples—say, of size 30 or more—the z-test can be used essentially without restriction. However, if outliers are present and their removal is not justified, you should perform the hypothesis test once with the outliers and once without them to see what effect the outliers have. If the conclusion is affected, use a different procedure or take another sample, if possible.
- If outliers are present but their removal is justified and results in a data set for which the z-test is appropriate (as previously stated), the procedure can be used.

Applying the One-Mean z-Test

Examples 9.11–9.13 illustrate use of the z-test, Procedure 9.1.

EXAMPLE 9.11

The One-Mean z-Test

Taller Young Women In the document *Anthropometric Reference Data for Children and Adults*, C. Fryer et al. present data from the *National Health and Nutrition Examination Survey* on a variety of human body measurements, such as weight, height, and size. Anthropometry is a key component of nutritional status assessment in children and adults.

A half-century ago, the average (U.S.) woman in her 20s was 62.6 inches tall. The heights, in inches, of a random sample of 25 of today's women in their 20s is presented in Table 9.9.

At the 1% significance level, do the data provide sufficient evidence to conclude that the mean height of today's women in their 20s is greater than the mean height of women in their 20s a half-century ago? Assume that the population standard deviation of heights for today's women in their 20s is 2.88 inches.

TABLE 9.9

Heights, in inches, of a random sample of 25 of today's women in their 20s

61.8	66.1	60.7	66.6	61.1
66.9	66.5	61.9	64.9	67.4
61.5	59.3	64.5	68.6	63.7
66.7	69.4	62.1	62.3	68.2
65.0	62.3	65.0	62.0	61.4

Solution We note that the population standard deviation is known. Because the sample size, $n = 25$, is moderate, we first need to consider questions of normality and outliers. (See the second bulleted item in Key Fact 9.7.) Hence we constructed a normal probability plot for the data, shown in Fig. 9.12. The plot reveals no outliers and falls roughly in a straight line. Thus, we can apply Procedure 9.1 to perform the required hypothesis test.

Step 1 State the null and alternative hypotheses.

Let μ denote the mean height of today's women in their 20s. We obtained the null and alternative hypotheses in Example 9.2. They are

$$H_0: \mu = 62.6 \text{ inches (mean height has not increased)}$$

$$H_a: \mu > 62.6 \text{ inches (mean height has increased)}$$

FIGURE 9.12

Normal probability plot of the heights in Table 9.9

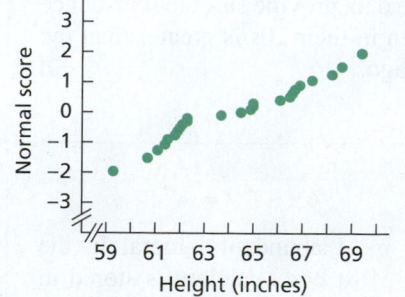

Height (inches)

_____

[†]We can refine these guidelines further by considering the impact of skewness. Roughly speaking, the more skewed the distribution of the variable under consideration, the larger is the sample size required to use the z-test.

Note that the hypothesis test is right tailed because a greater-than sign (>) appears in the alternative hypothesis.

Step 2 Decide on the significance level, α.

We are to perform the test at the 1% significance level, or $\alpha = 0.01$.

Step 3 Compute the value of the test statistic

$$z = \frac{\bar{x} - \mu_0}{\sigma/\sqrt{n}}.$$

We have $\mu_0 = 62.6$, $\sigma = 2.88$, and $n = 25$. The mean of the sample data in Table 9.9 is $\bar{x} = 64.24$. Thus the value of the test statistic is

$$z = \frac{64.24 - 62.6}{2.88/\sqrt{25}} = 2.85.$$

| **CRITICAL-VALUE APPROACH** | **OR** | **P-VALUE APPROACH** |

Step 4 The critical value for a right-tailed test is z_α. Use Table II to find the critical value.

Because $\alpha = 0.01$, the critical value is $z_{0.01}$. From Table II (or Table 9.4 on page 380), $z_{0.01} = 2.33$, as shown in Fig. 9.13A.

FIGURE 9.13A

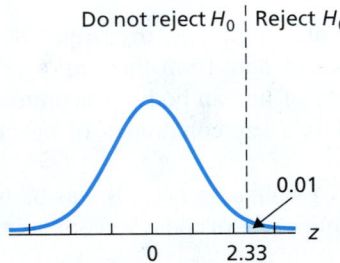

Do not reject H_0 | Reject H_0

0.01

0   2.33

z

Step 5 If the value of the test statistic falls in the rejection region, reject H_0; otherwise, do not reject H_0.

The value of the test statistic found in Step 3 is $z = 2.85$. Figure 9.13A reveals that this value falls in the rejection region, so we reject H_0. The test results are statistically significant at the 1% level.

Step 4 Use Table II to obtain the P-value.

From Step 3, the value of the test statistic is $z = 2.85$. The test is right tailed, so the P-value is the probability of observing a value of z of 2.85 or greater if the null hypothesis is true. That probability equals the shaded area in Fig. 9.13B, which, by Table II, is 0.0022. Hence $P = 0.0022$.

FIGURE 9.13B

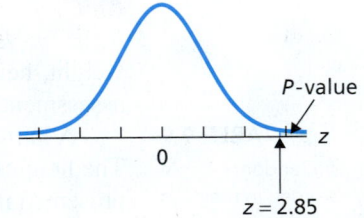

P-value

0

z = 2.85

z

Step 5 If $P \le \alpha$, reject H_0; otherwise, do not reject H_0.

From Step 4, $P = 0.0022$. Because the P-value is less than the specified significance level of 0.01, we reject H_0. The test results are statistically significant at the 1% level and (see Table 9.8 on page 386) provide very strong evidence against the null hypothesis.

Step 6 Interpret the results of the hypothesis test.

Interpretation At the 1% significance level, the data provide sufficient evidence to conclude that the mean height of today's women in their 20s is greater than the mean height of women in their 20s a half-century ago.

■■■■ **EXAMPLE 9.12** **The One-Mean z-Test**

Poverty and Dietary Calcium Calcium is the most abundant mineral in the human body and has several important functions. Most body calcium is stored in the bones and teeth, where it functions to support their structure. Recommendations for calcium are provided in *Dietary Reference Intakes*, developed by the Institute of Medicine of the National Academy of Sciences. The recommended adequate intake (RAI) of calcium for adults (ages 19–50 years) is 1000 milligrams (mg) per day.

A simple random sample of 18 adults with incomes below the poverty level gives the daily calcium intakes shown in Table 9.10. At the 5% significance level, do the data provide sufficient evidence to conclude that the mean calcium intake of all adults with incomes below the poverty level is less than the RAI of 1000 mg? Assume that $\sigma = 188$ mg.

Solution   Because the sample size, $n = 18$, is moderate, we first need to consider questions of normality and outliers. (See the second bulleted item in Key Fact 9.7 on page 389.) Hence we constructed a normal probability plot for the data, shown in Fig. 9.14. The plot reveals no outliers and falls roughly in a straight line. Thus, we can apply Procedure 9.1 to perform the required hypothesis test.

TABLE 9.10

Daily calcium intake (mg) for 18 adults with incomes below the poverty level

886	633	943	847	934	841
1193	820	774	834	1050	1058
1192	975	1313	872	1079	809

Step 1 **State the null and alternative hypotheses.**

Let μ denote the mean calcium intake (per day) of all adults with incomes below the poverty level. The null and alternative hypotheses, which we obtained in Example 9.3, are, respectively,

$$H_0: \mu = 1000 \text{ mg (mean calcium intake is not less than the RAI)}$$

$$H_a: \mu < 1000 \text{ mg (mean calcium intake is less than the RAI)}.$$

Note that the hypothesis test is left tailed because a less-than sign ($<$) appears in the alternative hypothesis.

FIGURE 9.14

Normal probability plot of the calcium-intake data in Table 9.10

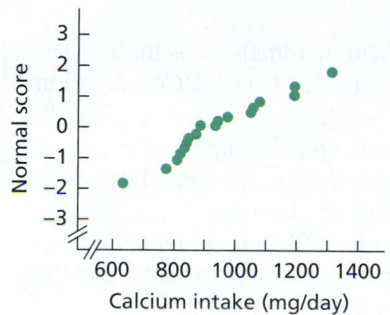

Step 2 **Decide on the significance level, α.**

We are to perform the test at the 5% significance level, or $\alpha = 0.05$.

Step 3 **Compute the value of the test statistic**

$$z = \frac{\bar{x} - \mu_0}{\sigma/\sqrt{n}}.$$

We have $\mu_0 = 1000$, $\sigma = 188$, and $n = 18$. From the data in Table 9.10, we find that $\bar{x} = 947.4$. Thus the value of the test statistic is

$$z = \frac{947.4 - 1000}{188/\sqrt{18}} = -1.19.$$

CRITICAL-VALUE APPROACH	OR	**P-VALUE APPROACH**

Step 4 **The critical value for a left-tailed test is $-z_\alpha$. Use Table II to find the critical value.**

Because $\alpha = 0.05$, the critical value is $-z_{0.05}$. From Table II (or Table 9.4 on page 380), $z_{0.05} = 1.645$. Hence the critical value is $-z_{0.05} = -1.645$, as shown in Fig. 9.15A.

FIGURE 9.15A

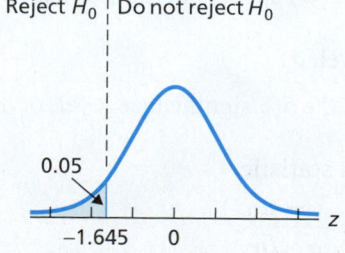

Step 4 **Use Table II to obtain the P-value.**

From Step 3, the value of the test statistic is $z = -1.19$. The test is left tailed, so the P-value is the probability of observing a value of z of -1.19 or less if the null hypothesis is true. That probability equals the shaded area in Fig. 9.15B, which, by Table II, is 0.1170. Hence $P = 0.1170$.

FIGURE 9.15B

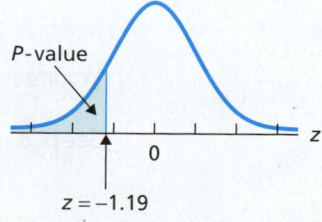

CRITICAL-VALUE APPROACH	OR	P-VALUE APPROACH

Step 5 If the value of the test statistic falls in the rejection region, reject H_0; otherwise, do not reject H_0.

The value of the test statistic from Step 3 is $z = -1.19$. Figure 9.15A reveals that this value does not fall in the rejection region, so we do not reject H_0. The test results are not statistically significant at the 5% level.

Step 5 If $P \leq \alpha$, reject H_0; otherwise, do not reject H_0.

From Step 4, $P = 0.1170$. Because the P-value exceeds the specified significance level of 0.05, we do not reject H_0. The test results are not statistically significant at the 5% level and (see Table 9.8 on page 386) provide at most weak evidence against the null hypothesis.

Step 6 Interpret the results of the hypothesis test.

Report 9.1

Interpretation At the 5% significance level, the data do not provide sufficient evidence to conclude that the mean calcium intake of all adults with incomes below the poverty level is less than the RAI of 1000 mg per day.

■ ■ ■ ■ **EXAMPLE 9.13**   **The One-Mean z-Test**

Clocking the Cheetah   The cheetah is the fastest land mammal and is highly specialized to run down prey. According to the online document "Cheetah Conservation in Southern Africa" (*Trade & Environment Database (TED) Case Studies*, Vol. 8, No. 2) by J. Urbaniak, the cheetah is capable of speeds up to 72 mph.

One common estimate of mean top speed for cheetahs is 60 mph. Table 9.11 gives the top speeds, in miles per hour, for a sample of 35 cheetahs.

At the 5% significance level, do the data provide sufficient evidence to conclude that the mean top speed of all cheetahs differs from 60 mph? Assume that the population standard deviation of top speeds is 3.2 mph.

TABLE 9.11

Top speeds, in miles per hour, for a sample of 35 cheetahs

57.3	57.5	59.0	56.5	61.3
57.6	59.2	65.0	60.1	59.7
62.6	52.6	60.7	62.3	65.2
54.8	55.4	55.5	57.8	58.7
57.8	60.9	75.3	60.6	58.1
55.9	61.6	59.6	59.8	63.4
54.7	60.2	52.4	58.3	66.0

Solution   We note that the population standard deviation is known. Because the sample size is 35, which is large, we need only check for outliers in the speed data before applying Procedure 9.1. (See the third bulleted item in Key Fact 9.7 on page 389.)

To check for outliers, we constructed a boxplot of the speed data, as shown in Fig. 9.16(a). The boxplot indicates that the top speed of 75.3 mph (third entry of the fifth row of Table 9.11) is a potential outlier. To decide whether 75.3 mph is in fact an outlier, we also constructed a normal probability plot, histogram, and stem-and-leaf diagram of the speed data, as shown in Figs. 9.16(b)–(d). *Note:* For simplicity in constructing the stem-and-leaf diagram, we first removed the decimal parts of the speeds.

From the four graphs in Fig. 9.16, we see that the top speed of 75.3 mph is indeed an outlier. Thus, as suggested in the third bulleted item in Key Fact 9.7, we apply Procedure 9.1 first to the full data set in Table 9.11 and then to that data set with the outlier removed.

Step 1   State the null and alternative hypotheses.

The null and alternative hypotheses are, respectively,

H_0: $\mu = 60$ mph (mean top speed of cheetahs is 60 mph)

H_a: $\mu \neq 60$ mph (mean top speed of cheetahs is not 60 mph),

where μ denotes the mean top speed of all cheetahs. Note that the hypothesis test is two tailed because a does-not-equal sign ($\neq$) appears in the alternative hypothesis.

Step 2   Decide on the significance level, α.

We are to perform the hypothesis test at the 5% significance level, or $\alpha = 0.05$.

Step 3   Compute the value of the test statistic

$$z = \frac{\bar{x} - \mu_0}{\sigma/\sqrt{n}}.$$

FIGURE 9.16

Graphs of top-speed data in Table 9.11:
(a) boxplot, (b) normal probability plot,
(c) histogram, (d) stem-and-leaf diagram

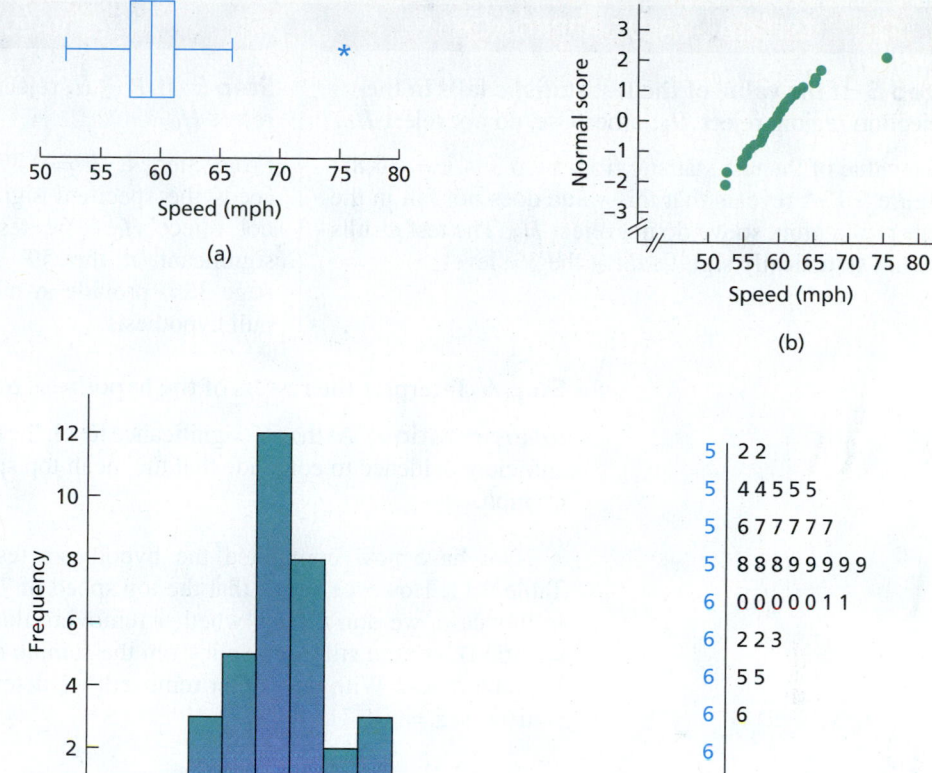

(a)

(b)

(c)

5	2 2
5	4 4 5 5 5
5	6 7 7 7 7 7
5	8 8 8 9 9 9 9 9
6	0 0 0 0 0 1 1
6	2 2 3
6	5 5
6	6
6	
7	
7	
7	5

(d)

We have $\mu_0 = 60$, $\sigma = 3.2$, and $n = 35$. From the data in Table 9.11, we find that $\bar{x} = 59.526$. Thus the value of the test statistic is

$$z = \frac{59.526 - 60}{3.2/\sqrt{35}} = -0.88.$$

CRITICAL-VALUE APPROACH	OR	P-VALUE APPROACH

Step 4 The critical values for a two-tailed test are $\pm z_{\alpha/2}$. Use Table II to find the critical values.

Because $\alpha = 0.05$, we find from Table II (or Table 9.4) the critical values of $\pm z_{0.05/2} = \pm z_{0.025} = \pm 1.96$, as shown in Fig. 9.17A.

FIGURE 9.17A

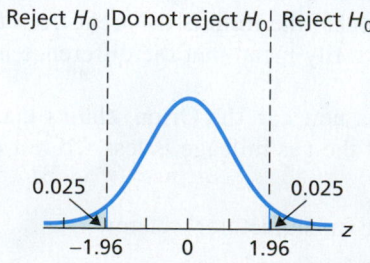

Step 4 Use Table II to obtain the P-value.

From Step 3, the value of the test statistic is $z = -0.88$. The test is two tailed, so the P-value is the probability of observing a value of z of 0.88 or greater in magnitude if the null hypothesis is true. That probability equals the shaded area in Fig. 9.17B, which, by Table II, is $2 \cdot 0.1894$ or 0.3788. Hence $P = 0.3788$.

FIGURE 9.17B

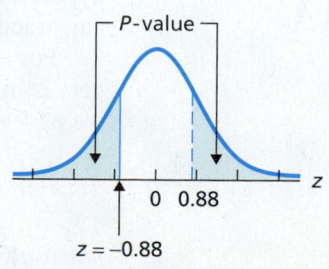

CRITICAL-VALUE APPROACH	OR	P-VALUE APPROACH

Step 5 If the value of the test statistic falls in the rejection region, reject H_0; otherwise, do not reject H_0.

The value of the test statistic from Step 3 is $z = -0.88$. Figure 9.17A reveals that this value does not fall in the rejection region, so we do not reject H_0. The test results are not statistically significant at the 5% level.

Step 5 If $P \leq \alpha$, reject H_0; otherwise, do not reject H_0.

From Step 4, $P = 0.3788$. Because the P-value exceeds the specified significance level of 0.05, we do not reject H_0. The test results are not statistically significant at the 5% level and (see Table 9.8 on page 386) provide at most weak evidence against the null hypothesis.

Step 6 Interpret the results of the hypothesis test.

Interpretation At the 5% significance level, the (unabridged) data do not provide sufficient evidence to conclude that the mean top speed of all cheetahs differs from 60 mph.

We have now completed the hypothesis test, using all 35 top speeds in Table 9.11. However, recall that the top speed of 75.3 mph is an outlier. Although in this case, we don't know whether removing this outlier is justified (a common situation), we can still remove it from the sample data and assess the effect on the hypothesis test. With the outlier removed, we determined that the value of the test statistic is $z = -1.71$.

CRITICAL-VALUE APPROACH	OR	P-VALUE APPROACH

We see from Fig. 9.17A that the value of the test statistic, $z = -1.71$, for the abridged data does not fall in the rejection region (although it is much closer to the rejection region than the value of the test statistic for the unabridged data, $z = -0.88$). Hence we do not reject H_0. The test results are not statistically significant at the 5% level.

For the abridged data, the P-value is the probability of observing a value of z of 1.71 or greater in magnitude if the null hypothesis is true. Referring to Table II, we find that probability to be $2 \cdot 0.0436$, or 0.0872. Hence $P = 0.0872$.

Because the P-value exceeds the specified significance level of 0.05, we do not reject H_0. The test results are not statistically significant at the 5% level but, as we see from Table 9.8 on page 386, the abridged data do provide moderate evidence against the null hypothesis.

Exercise 9.83 on page 397

Interpretation At the 5% significance level, the (abridged) data do not provide sufficient evidence to conclude that the mean top speed of all cheetahs differs from 60 mph. Thus, we see that removing the outlier does not affect the conclusion of this hypothesis test.

Statistical Significance Versus Practical Significance

Recall that the results of a hypothesis test are *statistically significant* if the null hypothesis is rejected at the chosen level of α. Statistical significance means that the data provide sufficient evidence to conclude that the truth is different from the stated null hypothesis. However, it does not necessarily mean that the difference is important in any practical sense.

For example, the manufacturer of a new car, the Orion, claims that a typical car gets 26 miles per gallon. We think that the gas mileage is less. To test our suspicion, we perform the hypothesis test

$$H_0: \mu = 26 \text{ mpg (manufacturer's claim)}$$
$$H_a: \mu < 26 \text{ mpg (our suspicion)},$$

where μ is the mean gas mileage of all Orions.

We take a random sample of 1000 Orions and find that their mean gas mileage is 25.9 mpg. Assuming $\sigma = 1.4$ mpg, the value of the test statistic for a z-test is $z = -2.26$. This result is statistically significant at the 5% level. Thus, at the 5% significance level, we reject the manufacturer's claim.

Because the sample size, 1000, is so large, the sample mean, $\bar{x} = 25.9$ mpg, is probably nearly the same as the population mean. As a result, we rejected the manufacturer's claim because μ is about 25.9 mpg instead of 26 mpg. From a practical point of view, however, the difference between 25.9 mpg and 26 mpg is not important.

The Relation between Hypothesis Tests and Confidence Intervals

Hypothesis tests and confidence intervals are closely related. Consider, for example, a two-tailed hypothesis test for a population mean at the significance level α. In this case, the null hypothesis will be rejected if and only if the value μ_0 given for the mean in the null hypothesis lies outside the $(1 - \alpha)$-level confidence interval for μ. You can examine the relation between hypothesis tests and confidence intervals in greater detail in Exercises 9.95–9.97.

THE TECHNOLOGY CENTER

Most statistical technologies have programs that automatically perform a one-mean z-test. In this subsection, we present output and step-by-step instructions for such programs.

EXAMPLE 9.14   Using Technology to Conduct a One-Mean z-Test

Poverty and Dietary Calcium Table 9.10 on page 391 shows the daily calcium intakes for a simple random sample of 18 adults with incomes below the poverty level. Use Minitab, Excel, or the TI-83/84 Plus to decide, at the 5% significance level, whether the data provide sufficient evidence to conclude that the mean calcium intake of all adults with incomes below the poverty level is less than the RAI of 1000 mg per day. Assume that $\sigma = 188$ mg.

Solution Let μ denote the mean calcium intake (per day) of all adults with incomes below the poverty level. We want to perform the hypothesis test

$H_0: \mu = 1000$ mg (mean calcium intake is not less than the RAI)

$H_a: \mu < 1000$ mg (mean calcium intake is less than the RAI)

at the 5% significance level ($\alpha = 0.05$). Note that the hypothesis test is left tailed.

We applied the one-mean z-test programs to the data, resulting in Output 9.1. Steps for generating that output are presented in Instructions 9.1 on the following page. *Note to Excel users:* For brevity, we have presented only the essential portions of the actual output.

OUTPUT 9.1   One-mean z-test on the sample of calcium intakes

MINITAB

```
One-Sample Z: CALCIUM

Test of μ = 1000 vs < 1000
The assumed standard deviation = 188


Variable   N    Mean   StDev   SE Mean   95% Upper Bound      Z       P
CALCIUM    18   947.4  172.0   44.3              1020.3     -1.19   0.118
```

TI-83/84 PLUS

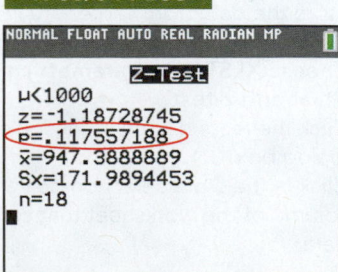

OUTPUT 9.1 (cont.)
One-mean z-test on the sample
of calcium intakes

EXCEL

Theoretical mean: 1000			
Significance level (%): 5			
Summary statistics:			
Variable	Observations	Mean	Std. deviation
CALCIUM	18	947.3889	171.9894
One-sample z-test / Lower-tailed test:			
Difference	-52.6111		
z (Observed value)	-1.1873		
z (Critical value)	-1.6449		
p-value (one-tailed)	0.1176		
alpha	0.05		

As shown in Output 9.1, the *P*-value for the hypothesis test is 0.118. Because the *P*-value exceeds the specified significance level of 0.05, we do not reject H_0. At the 5% significance level, the data do not provide sufficient evidence to conclude that the mean calcium intake of all adults with incomes below the poverty level is less than the RAI of 1000 mg per day.

INSTRUCTIONS 9.1   Steps for generating Output 9.1

MINITAB

1 Store the data from Table 9.10 in a column named CALCIUM
2 Choose **Stat ➤ Basic Statistics ➤ 1-Sample Z...**
3 Press the F3 key to reset the dialog box
4 Click in the text box directly below the **One or more samples, each in a column** drop-down list box and specify CALCIUM
5 Click in the **Known standard deviation** text box and type 188
6 Check the **Perform hypothesis test** check box
7 Click in the **Hypothesized mean** text box and type 1000
8 Click the **Options...** button
9 Click the arrow button at the right of the **Alternative hypothesis** drop-down list box and select **Mean < hypothesized mean**
10 Click **OK** twice

EXCEL

1 Store the data from Table 9.10 in a column named CALCIUM
2 Choose **XLSTAT ➤ Parametric tests ➤ One-sample t-test and z-test**
3 Click the reset button in the lower left corner of the dialog box
4 Click in the **Data** selection box and then select the column of the worksheet that contains the CALCIUM data

5 Check the **z test** check box and uncheck the **Student's t test** check box
6 Click the **Options** tab
7 Click the arrow button at the right of the **Alternative hypothesis** drop-down list box and select **Mean 1 < Theoretical mean**
8 Type 1000 in the **Theoretical mean** text box
9 Type 5 in the **Significance level (%)** text box
10 In the **Variance for the z-test** list, select the **User defined** option button
11 Type 35344 (188^2) in the **Variance** text box
12 Click **OK**
13 Click the **Continue** button in the **XLSTAT – Selections** dialog box

TI-83/84 PLUS

1 Store the data from Table 9.10 in a list named CALCI
2 Press **STAT**, arrow over to **TESTS**, and press **1**
3 Highlight **Data** and press **ENTER**
4 Press the down-arrow key, type 1000 for μ_0, and press **ENTER**
5 Type 188 for σ and press **ENTER**
6 Press **2nd ➤ LIST**
7 Arrow down to CALCI and press **ENTER** twice
8 Type 1 for **Freq** and then press **ENTER**
9 Highlight $< \mu_0$ and press **ENTER**
10 Arrow down to **Calculate** and press **ENTER**

Exercises 9.4

Understanding the Concepts and Skills

9.72 Explain why considering outliers is important when you are conducting a one-mean z-test.

In each of Exercises 9.73–9.76, we have provided a scenario for a hypothesis test for a population mean. Decide whether the z-test is an appropriate method for conducting the hypothesis test. Assume that the population standard deviation is known in each case.

9.73 Preliminary data analyses reveal that the sample data contain no outliers but that the distribution of the variable under consideration is probably highly skewed. The sample size is 24.

9.74 A normal probability plot of the sample data shows no outliers and is quite linear. The sample size is 12.

9.75 Preliminary data analyses reveal that the sample data contain no outliers but that the distribution of the variable under consideration is probably mildly skewed. The sample size is 70.

9.76 Preliminary data analyses reveal that the sample data contain an outlier. It is determined that the outlier is a legitimate observation and should not be removed. The sample size is 17.

In each of Exercises 9.77–9.82, we have provided a sample mean, sample size, and population standard deviation. In each case, use the one-mean z-test to perform the required hypothesis test at the 5% significance level.

9.77 $\bar{x} = 20$, $n = 32$, $\sigma = 4$, H_0: $\mu = 22$, H_a: $\mu < 22$

9.78 $\bar{x} = 21$, $n = 32$, $\sigma = 4$, H_0: $\mu = 22$, H_a: $\mu < 22$

9.79 $\bar{x} = 24$, $n = 15$, $\sigma = 4$, H_0: $\mu = 22$, H_a: $\mu > 22$

9.80 $\bar{x} = 23$, $n = 15$, $\sigma = 4$, H_0: $\mu = 22$, H_a: $\mu > 22$

9.81 $\bar{x} = 23$, $n = 24$, $\sigma = 4$, H_0: $\mu = 22$, H_a: $\mu \neq 22$

9.82 $\bar{x} = 20$, $n = 24$, $\sigma = 4$, H_0: $\mu = 22$, H_a: $\mu \neq 22$

Applying the Concepts and Skills

Preliminary data analyses indicate that applying the z-test (Procedure 9.1 on page 388) in Exercises 9.83–9.88 is reasonable.

9.83 **Toxic Mushrooms?** Cadmium, a heavy metal, is toxic to animals. Mushrooms, however, are able to absorb and accumulate cadmium at high concentrations. The Czech and Slovak governments have set a safety limit for cadmium in dry vegetables at 0.5 part per million (ppm). M. Melgar et al. measured the cadmium levels in a random sample of the edible mushroom *Boletus pinicola* and published the results in the paper "Influence of Some Factors in Toxicity and Accumulation of Cd from Edible Wild Macrofungi in NW Spain" (*Journal of Environmental Science and Health*, Vol. B33(4), pp. 439–455). Here are the data.

0.24	0.59	0.62	0.16	0.77	1.33
0.92	0.19	0.33	0.25	0.59	0.32

At the 5% significance level, do the data provide sufficient evidence to conclude that the mean cadmium level in *Boletus pinicola* mushrooms is greater than the government's recommended limit of 0.5 ppm? Assume that the population standard deviation of cadmium

levels in *Boletus pinicola* mushrooms is 0.37 ppm. (*Note:* The sum of the data is 6.31 ppm.)

9.84 **Grey-Seal Nursing.** The average lactation (nursing) period of all earless seals is 23 days. Grey seals are one of several types of earless seals. The length of time that a female grey seal nurses her pup is studied by S. Twiss et al. in the article "Variation in Female Grey Seal (*Halichoerus grypus*) Reproductive Performance Correlates to Proactive-Reactive Behavioural Types" (*PLOS ONE* 7(11): e49598. doi:10.1371/journal.pone.0049598). A sample of 14 female grey seals had the following lactation periods, in days.

20.2	20.9	20.6	23.6	19.6	15.9	19.8
15.4	21.4	19.5	17.4	21.9	22.3	16.4

At the 5% significance level, do the data provide sufficient evidence to conclude that the mean lactation period of grey seals differs from 23 days? Assume that the population standard deviation is 3.0 days. (*Note:* The sum of the data is 274.9 days.)

9.85 **Iron Deficiency?** Iron is essential to most life forms and to normal human physiology. It is an integral part of many proteins and enzymes that maintain good health. Recommendations for iron are provided in *Dietary Reference Intakes*, developed by the Institute of Medicine of the National Academy of Sciences. The recommended dietary allowance (RDA) of iron for adult females under the age of 51 is 18 milligrams (mg) per day. The following iron intakes, in milligrams, were obtained during a 24-hour period for 45 randomly selected adult females under the age of 51.

15.0	18.1	14.4	14.6	10.9	18.1	18.2	18.3	15.0
16.0	12.6	16.6	20.7	19.8	11.6	12.8	15.6	11.0
15.3	9.4	19.5	18.3	14.5	16.6	11.5	16.4	12.5
14.6	11.9	12.5	18.6	13.1	12.1	10.7	17.3	12.4
17.0	6.3	16.8	12.5	16.3	14.7	12.7	16.3	11.5

At the 1% significance level, do the data suggest that adult females under the age of 51 are, on average, getting less than the RDA of 18 mg of iron? Assume that the population standard deviation is 4.2 mg. (*Note:* $\bar{x} = 14.68$ mg.)

9.86 **Early-Onset Dementia.** Dementia is the loss of the intellectual and social abilities severe enough to interfere with judgment, behavior, and daily functioning. Alzheimer's disease is the most common type of dementia. In the article "Living with Early Onset Dementia: Exploring the Experience and Developing Evidence-Based Guidelines for Practice" (*Alzheimer's Care Quarterly*, Vol. 5, Issue 2, pp. 111–122), P. Harris and J. Keady explored the experience and struggles of people diagnosed with dementia and their families. A simple random sample of 21 people with early-onset dementia gave the following data on age at diagnosis, in years.

60	58	52	58	59	58	51
61	54	59	55	53	44	46
47	42	56	57	49	41	43

At the 1% significance level, do the data provide sufficient evidence to conclude that the mean age at diagnosis of all people with early-onset dementia is less than 55 years old? Assume that the population standard deviation is 6.8 years. (*Note:* $\bar{x} = 52.5$ years.)

9.87 Serving Time. According to the Bureau of Crime Statistics and Research of Australia, as reported on *Lawlink*, the mean length of imprisonment for motor-vehicle-theft offenders in Australia is 16.7 months. One hundred randomly selected motor-vehicle-theft offenders in Sydney, Australia, had a mean length of imprisonment of 17.8 months. At the 5% significance level, do the data provide sufficient evidence to conclude that the mean length of imprisonment for motor-vehicle-theft offenders in Sydney differs from the national mean in Australia? Assume that the population standard deviation of the lengths of imprisonment for motor-vehicle-theft offenders in Sydney is 6.0 months.

9.88 Worker Fatigue. A study by M. Chen et al. titled "Heat Stress Evaluation and Worker Fatigue in a Steel Plant" (*American Industrial Hygiene Association*, Vol. 64, pp. 352–359) assessed fatigue in steel-plant workers due to heat stress. A random sample of 29 casting workers had a mean post-work heart rate of 78.3 beats per minute (bpm). At the 5% significance level, do the data provide sufficient evidence to conclude that the mean post-work heart rate for casting workers exceeds the normal resting heart rate of 72 bpm? Assume that the population standard deviation of post-work heart rates for casting workers is 11.2 bpm.

9.89 Job Gains and Losses. In the article "Business Employment Dynamics: New Data on Gross Job Gains and Losses" (*Monthly Labor Review*, Vol. 127, Issue 4, pp. 29–42), J. Spletzer et al. examined gross job gains and losses as a percentage of the average of previous and current employment figures. A simple random sample of 20 quarters provided the net percentage gains (losses are negative gains) for jobs as presented on the WeissStats site. Use the technology of your choice to do the following.

a. Decide whether, on average, the net percentage gain for jobs exceeds 0.2. Assume a population standard deviation of 0.42. Apply the one-mean z-test with a 5% significance level.
b. Obtain a normal probability plot, boxplot, histogram, and stem-and-leaf diagram of the data.
c. Remove the outliers (if any) from the data and then repeat part (a).
d. Comment on the advisability of using the z-test here.

9.90 Hotels and Motels. The daily charges, in dollars, for a sample of 15 hotels and motels operating in South Carolina are provided on the WeissStats site. The data were found in the report *South Carolina Statistical Abstract*, sponsored by the South Carolina Budget and Control Board.

a. Use the one-mean z-test to decide, at the 5% significance level, whether the data provide sufficient evidence to conclude that the mean daily charge for hotels and motels operating in South Carolina is less than $75. Assume a population standard deviation of $22.40.
b. Obtain a normal probability plot, boxplot, histogram, and stem-and-leaf diagram of the data.
c. Remove the outliers (if any) from the data and then repeat part (a).
d. Comment on the advisability of using the z-test here.

Working with Large Data Sets

9.91 Body Temperature. A study by researchers at the University of Maryland addressed the question of whether the mean body temperature of humans is 98.6°F. The results of the study by P. Mackowiak et al. appeared in the article "A Critical Appraisal of 98.6°F, the Upper Limit of the Normal Body Temperature, and Other Legacies of Carl Reinhold August Wunderlich" (*Journal of the American Medical Association*, Vol. 268, pp. 1578–1580). Among other data, the researchers obtained the body temperatures of 93 healthy humans,

which we provide on the WeissStats site. Use the technology of your choice to do the following.

a. Obtain a normal probability plot, boxplot, histogram, and stem-and-leaf diagram of the data.
b. Based on your results from part (a), can you reasonably apply the one-mean z-test to the data? Explain your reasoning.
c. At the 1% significance level, do the data provide sufficient evidence to conclude that the mean body temperature of healthy humans differs from 98.6°F? Assume that $\sigma = 0.63°F$.

9.92 Teacher Salaries. Data on salaries in the public school system are published annually in *Ranking of the States and Estimates of School Statistics* by the National Education Association. The mean annual salary of (public) classroom teachers is $55.4 thousand. A random sample of 90 classroom teachers in Ohio yielded the annual salaries, in thousands of dollars, presented on the WeissStats site. Use the technology of your choice to perform the following tasks.

a. Obtain a normal probability plot, boxplot, histogram, and stem-and-leaf diagram of the data.
b. Based on your results from part (a), can you reasonably apply the one-mean z-test to the data? Explain your reasoning.
c. At the 5% significance level, do the data provide sufficient evidence to conclude that the mean annual salary of classroom teachers in Ohio is greater than the national mean? Assume that the standard deviation of annual salaries for all classroom teachers in Ohio is $9.2 thousand.

9.93 Cell Phones. The number of cell phone users has increased dramatically since 1987. According to the *Semi-annual Wireless Survey*, published by the Cellular Telecommunications & Internet Association, the mean local monthly bill for cell phone users in the United States was $48.16 in 2009. Last year's local monthly bills, in dollars, for a random sample of 75 cell phone users are given on the WeissStats site. Use the technology of your choice to do the following.

a. Obtain a normal probability plot, boxplot, histogram, and stem-and-leaf diagram of the data.
b. At the 5% significance level, do the data provide sufficient evidence to conclude that last year's mean local monthly bill for cell phone users decreased from the 2009 mean of $48.16? Assume that the population standard deviation of last year's local monthly bills for cell phone users is $25.
c. Remove the two outliers from the data and repeat parts (a) and (b).
d. State your conclusions regarding the hypothesis test.

Extending the Concepts and Skills

9.94 Class Project: Quality Assurance. This exercise can be done individually or, better yet, as a class project. For the pretzel-packaging hypothesis test in Example 9.1 on page 369, the null and alternative hypotheses are, respectively,

$$H_0: \mu = 454 \text{ g (machine is working properly)}$$
$$H_a: \mu \neq 454 \text{ g (machine is not working properly)},$$

where μ is the mean net weight of all bags of pretzels packaged. The net weights are normally distributed with a standard deviation of 7.8 g.

a. Assuming that the null hypothesis is true, simulate 100 samples of 25 net weights each.
b. Suppose that the hypothesis test is performed at the 5% significance level. Of the 100 samples obtained in part (a), roughly how many would you expect to lead to rejection of the null hypothesis? Explain your answer.

c. Of the 100 samples obtained in part (a), determine the number that lead to rejection of the null hypothesis.

d. Compare your answers from parts (b) and (c), and comment on any observed difference.

9.95 Two-Tailed Hypothesis Tests and CIs. As we mentioned on page 395, the following relationship holds between hypothesis tests and confidence intervals for one-mean z-procedures: For a two-tailed hypothesis test at the significance level α, the null hypothesis H_0: $\mu = \mu_0$ will be rejected in favor of the alternative hypothesis H_a: $\mu \neq \mu_0$ if and only if μ_0 lies outside the $(1 - \alpha)$-level confidence interval for μ. In each case, illustrate the preceding relationship by obtaining the appropriate one-mean z-interval (Procedure 8.1 on page 339) and comparing the result to the conclusion of the hypothesis test in the specified exercise.

a. Exercise 9.84   **b.** Exercise 9.87

9.96 Left-Tailed Hypothesis Tests and CIs. In Exercise 8.105 on page 352, we introduced one-sided one-mean z-intervals. The following relationship holds between hypothesis tests and confidence intervals for one-mean z-procedures: For a left-tailed hypothesis test at the significance level α, the null hypothesis H_0: $\mu = \mu_0$ will be rejected in favor of the alternative hypothesis H_a: $\mu < \mu_0$ if and only if μ_0 is greater than or equal to the $(1 - \alpha)$-level upper confidence bound for μ. In each case, illustrate the preceding relationship by obtaining the appropriate upper confidence bound and comparing the result to the conclusion of the hypothesis test in the specified exercise.

a. Exercise 9.85   **b.** Exercise 9.86

9.97 Right-Tailed Hypothesis Tests and CIs. In Exercise 8.105 on page 352, we introduced one-sided one-mean z-intervals. The following relationship holds between hypothesis tests and confidence intervals for one-mean z-procedures: For a right-tailed hypothesis test at the significance level α, the null hypothesis H_0: $\mu = \mu_0$ will be rejected in favor of the alternative hypothesis H_a: $\mu > \mu_0$ if and only if μ_0 is less than or equal to the $(1 - \alpha)$-level lower confidence bound for μ. In each case, illustrate the preceding relationship by obtaining the appropriate lower confidence bound and comparing the result to the conclusion of the hypothesis test in the specified exercise.

a. Exercise 9.83   **b.** Exercise 9.88

9.5	Hypothesis Tests for One Population Mean When σ Is Unknown

In Section 9.4, you learned how to perform a hypothesis test for one population mean when the population standard deviation, σ, is known. However, as we have mentioned, the population standard deviation is usually not known.

To develop a hypothesis-testing procedure for a population mean when σ is unknown, we begin by recalling Key Fact 8.5: If a variable x of a population is normally distributed with mean μ, then, for samples of size n, the studentized version of $\bar{x}$,

$$t = \frac{\bar{x} - \mu}{s/\sqrt{n}},$$

has the t-distribution with $n - 1$ degrees of freedom.

Because of Key Fact 8.5, we can perform a hypothesis test for a population mean when the population standard deviation is unknown by proceeding in essentially the same way as when it is known. The only difference is that we invoke a t-distribution instead of the standard normal distribution. Specifically, for a test with null hypothesis H_0: $\mu = \mu_0$, we employ the variable

$$t = \frac{\bar{x} - \mu_0}{s/\sqrt{n}}$$

as our test statistic and use the t-table, Table IV, to obtain the critical value(s) or P-value. We call this hypothesis-testing procedure the **one-mean t-test** or, when no confusion can arise, simply the **t-test.**[†]

P-Values for a t-Test[‡]

Before presenting a step-by-step procedure for conducting a (one-mean) t-test, we need to discuss P-values for such a test. P-values for a t-test are obtained in a manner similar to that for a z-test.

As we know, if the null hypothesis is true, the test statistic for a t-test has the t-distribution with $n - 1$ degrees of freedom, so its probabilities equal areas under the t-curve with df $= n - 1$. Therefore, if we let t_0 denote the observed value of the test statistic t, we determine the P-value as follows.

[†] The one-mean t-test is also known as the **one-sample t-test** and the **one-variable t-test.** We prefer "one-mean" because it makes clear the parameter being tested.

[‡] Those concentrating on the critical-value approach to hypothesis testing can skip to the subsection on the "The One-Mean t-Test," beginning on page 401.

- *Two-tailed test:* The *P*-value equals the probability of observing a value of the test statistic *t* that is at least as large in magnitude as the value actually observed, which is the area under the *t*-curve that lies outside the interval from $-|t_0|$ to $|t_0|$, as shown in Fig. 9.18(a).
- *Left-tailed test:* The *P*-value equals the probability of observing a value of the test statistic *t* that is as small as or smaller than the value actually observed, which is the area under the *t*-curve that lies to the left of t_0, as shown in Fig. 9.18(b).
- *Right-tailed test:* The *P*-value equals the probability of observing a value of the test statistic *t* that is as large as or larger than the value actually observed, which is the area under the *t*-curve that lies to the right of t_0, as shown in Fig. 9.18(c).

FIGURE 9.18

P-value for a *t*-test if the test is (a) two tailed, (b) left tailed, or (c) right tailed

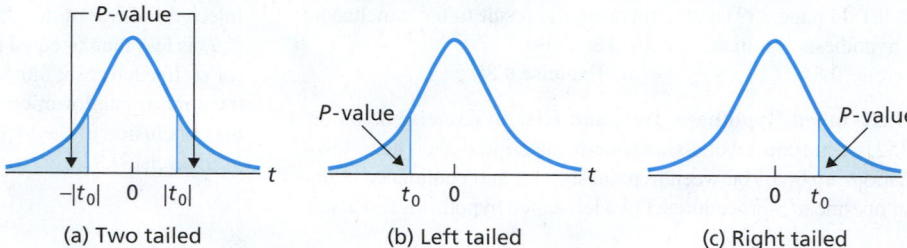

(a) Two tailed (b) Left tailed (c) Right tailed

Estimating the *P*-Value of a *t*-Test

To obtain the exact *P*-value of a *t*-test, we need statistical software or a statistical calculator. However, we can use *t*-tables, such as Table IV, to estimate the *P*-value of a *t*-test, and an estimate of the *P*-value is usually sufficient for deciding whether to reject the null hypothesis.

For instance, consider a right-tailed *t*-test with $n = 15$, $\alpha = 0.05$, and a value of the test statistic of $t = 3.458$. For df $= 15 - 1 = 14$, the *t*-value 3.458 is larger than any *t*-value in Table IV, the largest one being $t_{0.005} = 2.977$ (which means that the area under the *t*-curve that lies to the right of 2.977 equals 0.005). This fact, in turn, implies that the area to the right of 3.458 is less than 0.005; in other words, $P < 0.005$. Because the *P*-value is less than the designated significance level of 0.05, we reject H_0.

Example 9.15 provides two more illustrations of how Table IV can be used to estimate the *P*-value of a *t*-test.

■■■ **EXAMPLE 9.15** **Using Table IV to Estimate the *P*-Value of a *t*-Test**

Use Table IV to estimate the *P*-value of each one-mean *t*-test.

a. Left-tailed test, $n = 12$, and $t = -1.938$
b. Two-tailed test, $n = 25$, and $t = -0.895$

Solution

a. Because the test is left tailed, the *P*-value is the area under the *t*-curve with df $= 12 - 1 = 11$ that lies to the left of -1.938, as shown in Fig. 9.19(a).

FIGURE 9.19

Estimating the *P*-value of a left-tailed *t*-test with a sample size of 12 and test statistic $t = -1.938$

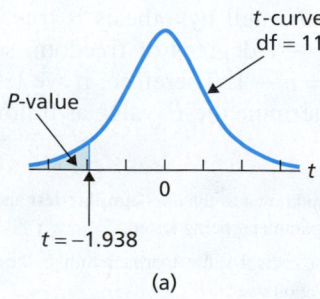

(a)

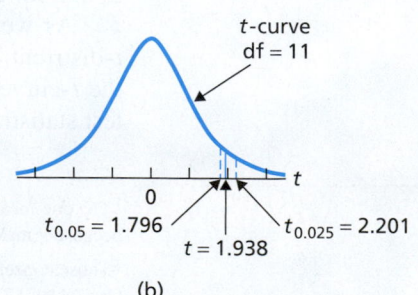

(b)

A t-curve is symmetric about 0, so the area to the left of -1.938 equals the area to the right of 1.938, which we can estimate by using Table IV. In the df $= 11$ row of Table IV, the two t-values that straddle 1.938 are $t_{0.05} = 1.796$ and $t_{0.025} = 2.201$. Therefore the area under the t-curve that lies to the right of 1.938 is between 0.025 and 0.05, as shown in Fig. 9.19(b).

Consequently, the area under the t-curve that lies to the left of -1.938 is also between 0.025 and 0.05, so $0.025 < P < 0.05$. Hence we can reject H_0 at any significance level of 0.05 or larger, and we cannot reject H_0 at any significance level of 0.025 or smaller. For significance levels between 0.025 and 0.05, Table IV is not sufficiently detailed to help us to decide whether to reject H_0.

b. Because the test is two tailed, the P-value is the area under the t-curve with df $= 25 - 1 = 24$ that lies either to the left of -0.895 or to the right of 0.895, as shown in Fig. 9.20(a).

FIGURE 9.20

Estimating the P-value of a two-tailed t-test with a sample size of 25 and test statistic $t = -0.895$

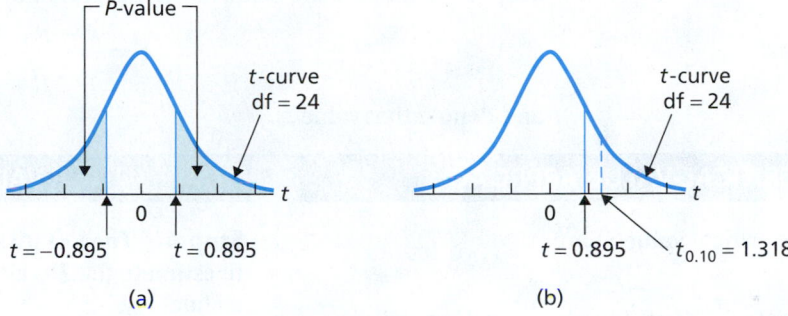

(a)   (b)

Because a t-curve is symmetric about 0, the areas to the left of -0.895 and to the right of 0.895 are equal. In the df $= 24$ row of Table IV, 0.895 is smaller than any other t-value, the smallest being $t_{0.10} = 1.318$. The area under the t-curve that lies to the right of 0.895, therefore, is greater than 0.10, as shown in Fig. 9.20(b).

Consequently, the area under the t-curve that lies either to the left of -0.895 or to the right of 0.895 is greater than 0.20, so $P > 0.20$. Hence we cannot reject H_0 at any significance level of 0.20 or smaller. For significance levels larger than 0.20, Table IV is not sufficiently detailed to help us to decide whether to reject H_0.

You try it!

Exercise 9.101
on page 405

The One-Mean t-Test

We now present, on the next page, Procedure 9.2, a step-by-step method for performing a one-mean t-test. As you can see, Procedure 9.2 includes both the critical-value approach for a one-mean t-test and the P-value approach for a one-mean t-test.

APPLET

Applet 9.1

Properties and guidelines for use of the t-test are the same as those for the z-test, as given in Key Fact 9.7 on page 389. In particular, the t-test is robust to moderate violations of the normality assumption but, even for large samples, can sometimes be unduly affected by outliers because the sample mean and sample standard deviation are not resistant to outliers.

EXAMPLE 9.16   The One-Mean t-Test

TABLE 9.12

pH levels for 15 lakes

7.2	7.3	6.1	6.9	6.6
7.3	6.3	5.5	6.3	6.5
5.7	6.9	6.7	7.9	5.8

Acid Rain and Lake Acidity Acid rain from the burning of fossil fuels has caused many of the lakes around the world to become acidic. The biology in these lakes often collapses because of the rapid and unfavorable changes in water chemistry. A lake is classified as nonacidic if it has a pH greater than 6.

A. Marchetto and A. Lami measured the pH of high mountain lakes in the Southern Alps and reported their findings in the paper "Reconstruction of pH by Chrysophycean Scales in Some Lakes of the Southern Alps" (*Hydrobiologia*, Vol. 274, pp. 83–90). Table 9.12 shows the pH levels obtained by the researchers

■■■■ **PROCEDURE 9.2**   **One-Mean *t*-Test**

Purpose To perform a hypothesis test for a population mean, μ

Assumptions
1.   Simple random sample
2.   Normal population or large sample
3. σ unknown

Step 1   The null hypothesis is $H_0: \mu = \mu_0$, and the alternative hypothesis is

$$H_a: \mu \neq \mu_0 \quad \text{or} \quad H_a: \mu < \mu_0 \quad \text{or} \quad H_a: \mu > \mu_0$$
$$\text{(Two tailed)} \qquad \text{(Left tailed)} \qquad \text{(Right tailed)}.$$

Step 2   Decide on the significance level, α.

Step 3   Compute the value of the test statistic

$$t = \frac{\bar{x} - \mu_0}{s/\sqrt{n}}$$

and denote that value t_0.

CRITICAL-VALUE APPROACH	OR	P-VALUE APPROACH

Step 4   The critical value(s) are

$$\pm t_{\alpha/2} \quad \text{or} \quad -t_\alpha \quad \text{or} \quad t_\alpha$$
$$\text{(Two tailed)} \qquad \text{(Left tailed)} \qquad \text{(Right tailed)}$$

with df $= n - 1$. Use Table IV to find the critical value(s).

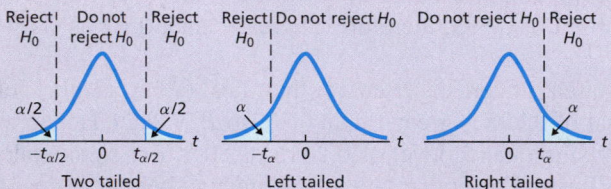

Step 5   If the value of the test statistic falls in the rejection region, reject H_0; otherwise, do not reject H_0.

Step 4   The *t*-statistic has df $= n - 1$. Use Table IV to estimate the *P*-value, or obtain it exactly by using technology.

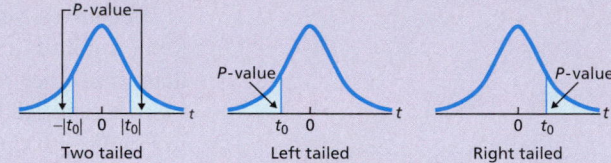

Step 5   If $P \leq \alpha$, reject H_0; otherwise, do not reject H_0.

Step 6   Interpret the results of the hypothesis test.

Note: The hypothesis test is exact for normal populations and is approximately correct for large samples from nonnormal populations.

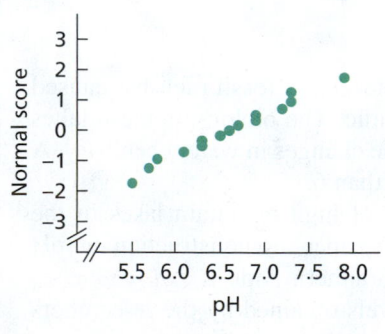

FIGURE 9.21

Normal probability plot of pH levels in Table 9.12

for 15 lakes. At the 5% significance level, do the data provide sufficient evidence to conclude that, on average, high mountain lakes in the Southern Alps are nonacidic?

Solution   Figure 9.21, a normal probability plot of the data in Table 9.12, reveals no outliers and is quite linear. Consequently, we can apply Procedure 9.2 to conduct the required hypothesis test.

Step 1   **State the null and alternative hypotheses.**

Let μ denote the mean pH level of all high mountain lakes in the Southern Alps. Then the null and alternative hypotheses are, respectively,

$$H_0: \mu = 6 \text{ (on average, the lakes are acidic)}$$
$$H_a: \mu > 6 \text{ (on average, the lakes are nonacidic)}.$$

Note that the hypothesis test is right tailed.

Step 2   Decide on the significance level, α.

We are to perform the test at the 5% significance level, so $\alpha = 0.05$.

Step 3   Compute the value of the test statistic

$$t = \frac{\bar{x} - \mu_0}{s/\sqrt{n}}.$$

We have $\mu_0 = 6$ and $n = 15$ and calculate the mean and standard deviation of the sample data in Table 9.12 as 6.6 and 0.672, respectively. Hence the value of the test statistic is

$$t = \frac{6.6 - 6}{0.672/\sqrt{15}} = 3.458.$$

CRITICAL-VALUE APPROACH	OR	P-VALUE APPROACH

Step 4   The critical value for a right-tailed test is t_α with df $= n - 1$. Use Table IV to find the critical value.

We have $n = 15$ and $\alpha = 0.05$. Table IV shows that for df $= 15 - 1 = 14$, $t_{0.05} = 1.761$. See Fig. 9.22A.

FIGURE 9.22A

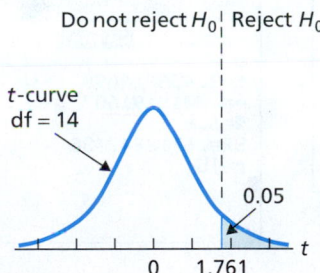

Step 5   If the value of the test statistic falls in the rejection region, reject H_0; otherwise, do not reject H_0.

The value of the test statistic, found in Step 3, is $t = 3.458$. Figure 9.22A reveals that it falls in the rejection region. Consequently, we reject H_0. The test results are statistically significant at the 5% level.

Step 4   The t-statistic has df $= n - 1$. Use Table IV to estimate the P-value, or obtain it exactly by using technology.

From Step 3, the value of the test statistic is $t = 3.458$. The test is right tailed, so the P-value is the probability of observing a value of t of 3.458 or greater if the null hypothesis is true. That probability equals the shaded area in Fig. 9.22B.

FIGURE 9.22B

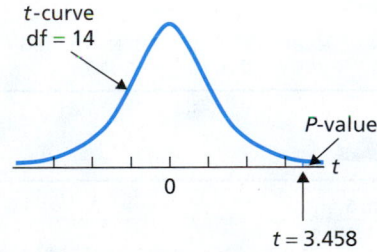

We have $n = 15$, and so df $= 15 - 1 = 14$. From Fig. 9.22B and Table IV, $P < 0.005$. (Using technology, we obtain $P = 0.00192$.)

Step 5   If $P \leq \alpha$, reject H_0; otherwise, do not reject H_0.

From Step 4, $P < 0.005$. Because the P-value is less than the specified significance level of 0.05, we reject H_0. The test results are statistically significant at the 5% level and (see Table 9.8 on page 386) provide very strong evidence against the null hypothesis.

StatCrunch

You try it!

Report 9.2

Exercise 9.113 on page 405

Step 6   Interpret the results of the hypothesis test.

Interpretation   At the 5% significance level, the data provide sufficient evidence to conclude that, on average, high mountain lakes in the Southern Alps are nonacidic.

THE TECHNOLOGY CENTER

Most statistical technologies have programs that automatically perform a one-mean *t*-test. In this subsection, we present output and step-by-step instructions for such programs.

EXAMPLE 9.17   Using Technology to Conduct a One-Mean *t*-Test

Acid Rain and Lake Acidity   Table 9.12 on page 401 gives the pH levels of a sample of 15 lakes in the Southern Alps. Use Minitab, Excel, or the TI-83/84 Plus to decide, at the 5% significance level, whether the data provide sufficient evidence to conclude that, on average, high mountain lakes in the Southern Alps are nonacidic.

Solution   Let μ denote the mean pH level of all high mountain lakes in the Southern Alps. We want to perform the hypothesis test

$$H_0: \mu = 6 \text{ (on average, the lakes are acidic)}$$
$$H_a: \mu > 6 \text{ (on average, the lakes are nonacidic)}$$

at the 5% significance level. Note that the hypothesis test is right tailed.

We applied the one-mean *t*-test programs to the data, resulting in Output 9.2. Steps for generating that output are presented in Instructions 9.2. *Note to Excel users:* For brevity, we have presented only the essential portions of the actual output.

OUTPUT 9.2   One-mean *t*-test on the sample of pH levels

MINITAB

One-Sample T: PH

Test of μ = 6 vs > 6

Variable	N	Mean	StDev	SE Mean	95% Lower Bound	T	P
PH	15	6.600	0.672	0.173	6.294	3.46	0.002

TI-83/84 PLUS

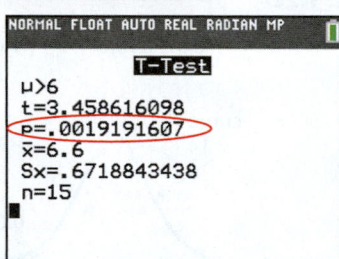

EXCEL

Theoretical mean: 6			
Significance level (%): 5			
Summary statistics:			
Variable	Observations	Mean	Std. deviation
PH	15	6.6000	0.6719
One-sample t-test / Upper-tailed test:			
Difference	0.6000		
t (Observed value)	3.4586		
t (Critical value)	1.7613		
DF	14		
p-value (one-tailed)	0.0019		
alpha	0.05		

As shown in Output 9.2, the *P*-value for the hypothesis test is 0.002. The *P*-value is less than the specified significance level of 0.05, so we reject H_0. At the 5% significance level, the data provide sufficient evidence to conclude that, on average, high mountain lakes in the Southern Alps are nonacidic.

INSTRUCTIONS 9.2   Steps for generating Output 9.2

MINITAB

1 Store the data from Table 9.12 in a column named PH
2 Choose **Stat ➤ Basic Statistics ➤ 1-Sample t...**
3 Press the F3 key to reset the dialog box
4 Click in the text box directly below the **One or more samples, each in a column** drop-down list box and specify PH
5 Check the **Perform hypothesis test** check box
6 Click in the **Hypothesized mean** text box and type 6
7 Click the **Options...** button
8 Click the arrow button at the right of the **Alternative hypothesis** drop-down list box and select **Mean > hypothesized mean**
9 Click **OK** twice

EXCEL

1 Store the data from Table 9.12 in a column named PH
2 Choose **XLSTAT ➤ Parametric tests ➤ One-sample t-test and z-test**
3 Click the reset button in the lower left corner of the dialog box
4 Click in the **Data** selection box and then select the column of the worksheet that contains the PH data

5 Click the **Options** tab
6 Click the arrow button at the right of the **Alternative hypothesis** drop-down list box and select **Mean 1 > Theoretical mean**
7 Type 6 in the **Theoretical mean** text box
8 Type 5 in the **Significance level (%)** text box
9 Click **OK**
10 Click the **Continue** button in the **XLSTAT – Selections** dialog box

TI-83/84 PLUS

1 Store the data from Table 9.12 in a list named PH
2 Press **STAT**, arrow over to **TESTS**, and press **2**
3 Highlight **Data** and press **ENTER**
4 Press the down-arrow key, type 6 for μ_0, and press **ENTER**
5 Press **2nd ➤ LIST**
6 Arrow down to PH and press **ENTER** twice
7 Type 1 for **Freq** and then press **ENTER**
8 Highlight > μ_0 and press **ENTER**
9 Arrow down to **Calculate** and press **ENTER**

Exercises 9.5

Understanding the Concepts and Skills

9.98 What is the difference in assumptions between the one-mean t-test and the one-mean z-test?

9.99 Suppose that you want to perform a hypothesis test for a population mean based on a small sample but that preliminary data analyses indicate either the presence of outliers or that the variable under consideration is far from normally distributed.
a. Is either the z-test or t-test appropriate?
b. If not, what type of procedure might be appropriate?

9.100 Fill in the following blanks.
a. The t-test is _____ to moderate violations of the normality assumption.
b. The t-test can sometimes be unduly affected by outliers because the sample mean and sample standard deviation are not _____ to outliers.

Exercises 9.101–9.106 pertain to P-values for a one-mean t-test. For each exercise, do the following tasks.
a. Use Table IV in Appendix A to estimate the P-value.
b. Based on your estimate in part (a), state at which significance levels the null hypothesis can be rejected, at which significance levels it cannot be rejected, and at which significance levels it is not possible to decide.

9.101 Right-tailed test, $n = 20$, and $t = 2.235$

9.102 Right-tailed test, $n = 11$, and $t = 1.246$

9.103 Left-tailed test, $n = 10$, and $t = -3.381$

9.104 Left-tailed test, $n = 30$, and $t = -1.572$

9.105 Two-tailed test, $n = 17$, and $t = -2.733$

9.106 Two-tailed test, $n = 8$, and $t = 3.725$

In each of Exercises 9.107–9.112, we have provided a sample mean, sample standard deviation, and sample size. In each case, use the one-mean t-test to perform the required hypothesis test at the 5% significance level.

9.107 $\bar{x} = 20$, $s = 4$, $n = 32$, H_0: $\mu = 22$, H_a: $\mu < 22$

9.108 $\bar{x} = 21$, $s = 4$, $n = 32$, H_0: $\mu = 22$, H_a: $\mu < 22$

9.109 $\bar{x} = 24$, $s = 4$, $n = 15$, H_0: $\mu = 22$, H_a: $\mu > 22$

9.110 $\bar{x} = 23$, $s = 4$, $n = 15$, H_0: $\mu = 22$, H_a: $\mu > 22$

9.111 $\bar{x} = 23$, $s = 4$, $n = 24$, H_0: $\mu = 22$, H_a: $\mu \neq 22$

9.112 $\bar{x} = 20$, $s = 4$, $n = 24$, H_0: $\mu = 22$, H_a: $\mu \neq 22$

Applying the Concepts and Skills

Preliminary data analyses indicate that you can reasonably use a t-test to conduct each of the hypothesis tests required in Exercises 9.113–9.118.

9.113 TV Viewing. According to *Communications Industry Forecast & Report*, published by Veronis Suhler Stevenson, the average person watched 4.55 hours of television per day in 2005. A random sample of 20 people gave the following number of hours of television watched per day for last year.

1.0	4.6	5.4	3.7	5.2
1.7	6.1	1.9	7.6	9.1
6.9	5.5	9.0	3.9	2.5
2.4	4.7	4.1	3.7	6.2

At the 10% significance level, do the data provide sufficient evidence to conclude that the amount of television watched per day last year by the average person differed from that in 2005? (*Note:* $\bar{x} = 4.760$ hours and $s = 2.297$ hours.)

9.114 Golf Robots. Serious golfers and golf equipment companies sometimes use golf equipment testing labs to obtain precise information about particular club heads, club shafts, and golf balls. One golfer requested information about the Jazz Fat Cat 5-iron from Golf Laboratories, Inc. The company tested the club by using a robot to hit a Titleist NXT Tour ball six times with a head velocity of 85 miles per hour. The golfer wanted a club that, on average, would hit the ball more than 180 yards at that club speed. The total yards each ball traveled was as follows.

180	187	181	182	185	181

a. At the 5% significance level, do the data provide sufficient evidence to conclude that the club does what the golfer wants? (*Note:* The sample mean and sample standard deviation of the data are 182.7 yards and 2.7 yards, respectively.)
b. Repeat part (a) for a test at the 1% significance level.

9.115 Death Rolls. Alligators perform a spinning maneuver, referred to as a "death roll," to subdue their prey. Videos were taken of juvenile alligators performing this maneuver in a study for the article "Death Roll of the Alligator, Mechanics of Twist and Feeding in Water" (*Journal of Experimental Biology*, Vol. 210, pp. 2811–2818) by F. Fish et al. One of the variables measured was the degree of the angle between the body and head of the alligator while performing the roll. A sample of 20 rolls yielded the following data, in degrees.

58.6	58.7	57.3	54.5	52.9
59.5	29.4	43.4	31.8	52.3
42.7	34.8	39.2	61.3	60.4
51.5	42.8	57.5	43.6	47.6

At the 5% significance level, do the data provide sufficient evidence to conclude that, on average, the angle between the body and head of an alligator during a death roll is greater than 45°? (*Note:* $\bar{x} = 49.0$ and $s = 10.0$)

9.116 Apparel and Services. According to the document *Consumer Expenditures*, a publication of the Bureau of Labor Statistics, the average consumer unit spent $1736 on apparel and services in 2012. That same year, 25 consumer units in the Northeast had the following annual expenditures, in dollars, on apparel and services.

1279	1457	2020	1682	1273
2223	2233	2192	1611	1734
2688	2029	2166	1860	2444
1844	1765	2267	1522	2012
1990	1751	2113	2202	1712

At the 5% significance level, do the data provide sufficient evidence to conclude that the 2012 mean annual expenditure on apparel and services for consumer units in the Northeast differed from the national mean of $1736? (*Note:* The sample mean and sample standard deviation of the data are $1922.76 and $350.90, respectively.)

9.117 Ankle Brachial Index. The ankle brachial index (ABI) compares the blood pressure of a patient's arm to the blood pressure of the patient's leg. The ABI can be an indicator of different diseases, including arterial diseases. A healthy (or normal) ABI is 0.9 or greater. In a study by M. McDermott et al. titled "Sex Differences in Peripheral Arterial Disease: Leg Symptoms and Physical Functioning" (*Journal of the American Geriatrics Society*, Vol. 51, No. 2,

pp. 222–228), the researchers obtained the ABI of 187 women with peripheral arterial disease. The results were a mean ABI of 0.64 with a standard deviation of 0.15. At the 1% significance level, do the data provide sufficient evidence to conclude that, on average, women with peripheral arterial disease have an unhealthy ABI?

9.118 Dirt Bikes. Dirt bikes are simpler and lighter motorcycles that are designed for off-road events. Specifications for dirt bikes can be found through Motorcycle USA on their website www.motorcycle-usa.com. A random sample of 30 dirt bikes have a mean fuel capacity of 1.91 gallons with a standard deviation of 0.74 gallons. At the 10% significance level, do the data provide sufficient evidence to conclude that the mean fuel tank capacity of all dirt bikes is less than 2 gallons?

In each of Exercises 9.119–9.122, use the technology of your choice to decide whether applying the t-test to perform a hypothesis test for the population mean in question appears reasonable. Explain your answers.

9.119 Cardiovascular Hospitalizations. From the Florida State Center for Health Statistics report, *Women and Cardiovascular Disease Hospitalizations*, we found that, for cardiovascular hospitalizations, the mean age of women is 71.9 years. At one hospital, a random sample of 20 of its female cardiovascular patients had the following ages, in years.

75.9	83.7	87.3	74.5	82.5
78.2	76.1	52.8	56.4	53.8
88.2	78.9	81.7	54.4	52.7
58.9	97.6	65.8	86.4	72.4

9.120 Medieval Cremation Burials. In the article "Material Culture as Memory: Combs and Cremations in Early Medieval Britain" (*Early Medieval Europe*, Vol. 12, Issue 2, pp. 89–128), H. Williams discussed the frequency of cremation burials found in 17 archaeological sites in eastern England. Here are the data.

83	64	46	48	523	35	34	265	2484
46	385	21	86	429	51	258	119	

9.121 Capital Spending. An issue of *Brokerage Report* discussed the capital spending of telecommunications companies in the United States and Canada. The capital spending, in thousands of dollars, for each of 27 telecommunications companies is shown in the following table.

9,310	2,515	3,027	1,300	1,800	70	3,634
656	664	5,947	649	682	1,433	389
17,341	5,299	195	8,543	4,200	7,886	11,189
1,006	1,403	1,982	21	125	2,205	

9.122 Dating Artifacts. In the paper "Reassessment of TL Age Estimates of Burnt Flint from the Paleolithic Site of Tabun Cave, Israel" (*Journal of Human Evolution*, Vol. 45, Issue 5, pp. 401–409), N. Mercier and H. Valladas discussed the re-dating of artifacts and human remains found at Tabun Cave by using new methodological improvements. A random sample of 18 excavated pieces yielded the following new thermoluminescence (TL) ages, in thousands of years.

195	243	215	282	361	222
237	266	244	251	282	290
276	248	357	301	224	191

Working with Large Data Sets

9.123 Stressed-Out Bus Drivers. Previous studies have shown that urban bus drivers have an extremely stressful job, and a large proportion of drivers retire prematurely with disabilities due to occupational stress. In the paper, "Hassles on the Job: A Study of a Job Intervention With Urban Bus Drivers" (*Journal of Organizational Behavior*, Vol. 20, pp. 199–208), G. Evans et al. examined the effects of an intervention program to improve the conditions of urban bus drivers. Among other variables, the researchers monitored diastolic blood pressure of bus drivers in downtown Stockholm, Sweden. The data, in millimeters of mercury (mm Hg), on the WeissStats site are based on the blood pressures obtained prior to intervention for the 41 bus drivers in the study. Use the technology of your choice to do the following.

a. Obtain a normal probability plot, boxplot, histogram, and stem-and-leaf diagram of the data.

b. Based on your results from part (a), can you reasonably apply the one-mean t-test to the data? Explain your reasoning.

c. At the 10% significance level, do the data provide sufficient evidence to conclude that the mean diastolic blood pressure of bus drivers in Stockholm exceeds the normal diastolic blood pressure of 80 mm Hg?

9.124 How Far People Drive. In 2011, the average car in the United States was driven 13.5 thousand miles, as reported by the Federal Highway Administration in *Highway Statistics*. On the WeissStats site, we provide last year's distance driven, in thousands of miles, by each of 500 randomly selected cars. Use the technology of your choice to do the following.

a. Obtain a normal probability plot and histogram of the data.

b. Based on your results from part (a), can you reasonably apply the one-mean t-test to the data? Explain your reasoning.

c. At the 5% significance level, do the data provide sufficient evidence to conclude that the mean distance driven last year differs from that in 2011?

9.125 Fair Market Rent. According to the document *Out of Reach*, published by the National Low Income Housing Coalition, the fair market rent (FMR) for a two-bedroom unit in the United States is $949. A sample of 100 randomly selected two-bedroom units yielded the data on monthly rents, in dollars, given on the WeissStats site. Use the technology of your choice to do the following.

a. At the 5% significance level, do the data provide sufficient evidence to conclude that the mean monthly rent for two-bedroom units is greater than the FMR of $949? Apply the one-mean t-test.

b. Remove the outlier from the data and repeat the hypothesis test in part (a).

c. Comment on the effect that removing the outlier has on the hypothesis test.

d. State your conclusion regarding the hypothesis test and explain your answer.

Extending the Concepts and Skills

9.126 Two-Tailed Hypothesis Tests and CIs. The following relationship holds between hypothesis tests and confidence intervals for one-mean t-procedures: For a two-tailed hypothesis test at the significance level α, the null hypothesis H_0: $\mu = \mu_0$ will be rejected in favor of the alternative hypothesis H_a: $\mu > \mu_0$ if and only if μ_0 lies outside the $(1 - \alpha)$-level confidence interval for μ. In each case, illustrate the preceding relationship by obtaining the appropriate one-mean t-interval (Procedure 8.2 on page 355) and comparing the result to the conclusion of the hypothesis test in the specified exercise.

a. Exercise 9.113   b. Exercise 9.116

9.127 Left-Tailed Hypothesis Tests and CIs. In Exercise 8.146 on page 362, we introduced one-sided one-mean t-intervals. The following relationship holds between hypothesis tests and confidence intervals for one-mean t-procedures: For a left-tailed hypothesis test at the significance level α, the null hypothesis H_0: $\mu = \mu_0$ will be rejected in favor of the alternative hypothesis H_a: $\mu < \mu_0$ if and only if μ_0 is greater than or equal to the $(1 - \alpha)$-level upper confidence bound for μ. In each case, illustrate the preceding relationship by obtaining the appropriate upper confidence bound and comparing the result to the conclusion of the hypothesis test in the specified exercise.

a. Exercise 9.117   b. Exercise 9.118

9.128 Right-Tailed Hypothesis Tests and CIs. In Exercise 8.146 on page 362, we introduced one-sided one-mean t-intervals. The following relationship holds between hypothesis tests and confidence intervals for one-mean t-procedures: For a right-tailed hypothesis test at the significance level α, the null hypothesis H_0: $\mu = \mu_0$ will be rejected in favor of the alternative hypothesis H_a: $\mu > \mu_0$ if and only if μ_0 is less than or equal to the $(1 - \alpha)$-level lower confidence bound for μ. In each case, illustrate the preceding relationship by obtaining the appropriate lower confidence bound and comparing the result to the conclusion of the hypothesis test in the specified exercise.

a. Exercise 9.114 (both parts)

b. Exercise 9.115

9.6 The Wilcoxon Signed-Rank Test*

Up to this point, we have presented two methods for performing a hypothesis test for a population mean. If the population standard deviation is known, we can use the z-test; if it is unknown, we can use the t-test.

Both procedures require another assumption for their use: The variable under consideration should be approximately normally distributed, or the sample size should be relatively large. For small samples, both procedures should be avoided in the presence of outliers.

In this section, we describe a third method for performing a hypothesis test for a population mean—the **Wilcoxon signed-rank test.**[†] This test, which is sometimes

---

[†]The Wilcoxon signed-rank text is also known as the **one-sample Wilcoxon signed-rank test** and the **one-variable Wilcoxon signed-rank test.**

more appropriate than either the z-test or the t-test, is an example of a *nonparametric method*.

What Is a Nonparametric Method?

Recall that descriptive measures for population data, such as μ and σ, are called parameters. Technically, inferential methods concerned with parameters are called **parametric methods;** those that are not are called **nonparametric methods.** However, common statistical practice is to refer to most methods that can be applied without assuming normality as nonparametric. Thus the term *nonparametric method* as used in contemporary statistics is a misnomer.

Nonparametric methods have both advantages and disadvantages. On one hand, they usually entail fewer and simpler computations than parametric methods and are resistant to outliers and other extreme values. On the other hand, they are not as powerful as parametric methods, such as the z-test and t-test, when the requirements for use of parametric methods are met.[†]

The Logic Behind the Wilcoxon Signed-Rank Test

The Wilcoxon signed-rank test is based on the assumption that the variable under consideration has a *symmetric distribution*—one that can be divided into two pieces that are mirror images of each other—but does not require that its distribution be normal or have any other specific shape. Thus, for instance, the Wilcoxon signed-rank test applies to a variable that has a normal, triangular, uniform, or symmetric bimodal distribution but not to one that has a right-skewed or left-skewed distribution. The next example explains the reasoning behind this test.

■ ■ ■ **EXAMPLE 9.18** **Introducing the Wilcoxon Signed-Rank Test**

Weekly Food Costs The U.S. Department of Agriculture publishes information about food costs in *Agricultural Research Service*. According to that document, a typical U.S. family of three spends about $157 per week on food. Ten randomly selected Kansas families of three have the weekly food costs shown in Table 9.13. Do the data provide sufficient evidence to conclude that the mean weekly food cost for Kansas families of three is less than the national mean of $157?

TABLE 9.13

Sample of weekly food costs ($)

143	169	149	135	161
138	152	150	141	159

Solution Let μ denote the mean weekly food cost for all Kansas families of three. We want to perform the hypothesis test

H_0: $\mu = \$157$ (mean weekly food cost is not less than $157)

H_a: $\mu < \$157$ (mean weekly food cost is less than $157).

As we said, a condition for the use of the Wilcoxon signed-rank test is that the variable under consideration have a symmetric distribution. If the weekly food costs for Kansas families of three have a symmetric distribution, a graphic of the sample data should be roughly symmetric.

FIGURE 9.23

Stem-and-leaf diagram of sample data in Table 9.13

13	5 8
14	1 3
14	9
15	0 2
15	9
16	1
16	9

Figure 9.23 shows a stem-and-leaf diagram of the sample data in Table 9.13. The diagram is roughly symmetric and so does not reveal any obvious violations of the symmetry condition.[‡] We therefore apply the Wilcoxon signed-rank test to carry out the hypothesis test.

To begin, we rank the data in Table 9.13 according to distance and direction from the null hypothesis mean, $\mu_0 = \$157$. The steps for doing so are presented in Table 9.14.

[†]A precise definition of *power* is presented in Section 9.7.

[‡]For ease in explaining the Wilcoxon signed-rank test, we have chosen an example in which the sample size is very small. This selection, however, makes it difficult to effectively check the symmetry condition. In general, we must proceed cautiously when dealing with very small samples.

TABLE 9.14

Steps for ranking the data in Table 9.13 according to distance and direction from the null hypothesis mean

Cost ($) x	Difference $D = x - 157$	$\|D\|$	Rank of $\|D\|$	Signed rank R
143	−14	14	7	− 7
138	−19	19	9	− 9
169	12	12	6	6
152	−5	5	3	− 3
149	−8	8	5	− 5
150	−7	7	4	− 4
135	−22	22	10	−10
141	−16	16	8	− 8
161	4	4	2	2
159	2	2	1	1

Step 1 *Subtract μ_0 from x.*

Step 2 *Make each difference positive by taking absolute values.*

Step 3 *Rank the absolute differences in order from smallest (1) to largest (10).*

Step 4 *Give each rank the same sign as the sign in the Difference column.*

The absolute differences, $|D|$, displayed in the third column, identify how far each observation is from 157. The ranks of those absolute differences, displayed in the fourth column, show which observations are closer to 157 and which are farther away. The signed ranks, R, displayed in the last column, indicate in addition whether an observation is greater than 157 (+) or less than 157 (−). Figure 9.24 depicts the information for the second and third rows of Table 9.14.

FIGURE 9.24

Meaning of signed ranks for the observations 138 and 169

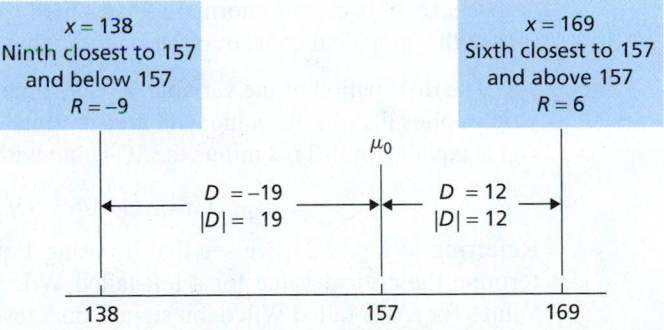

The reasoning behind the Wilcoxon signed-rank test is as follows: If the null hypothesis, $\mu = \$157$, is true, then, because the distribution of weekly food costs is symmetric, we expect the sum of the positive ranks and the sum of the negative ranks to be roughly the same in magnitude. For the sample size of 10, the sum of all the ranks must be $1 + 2 + \cdots + 10 = 55$, and half of 55 is 27.5.

Thus, if the null hypothesis is true, we expect the sum of the positive ranks (and the sum of the negative ranks) to be roughly 27.5. If the sum of the positive ranks is too much smaller than 27.5, we conclude that the null hypothesis is false and, therefore, that the mean weekly food cost is less than $157. From the last column of Table 9.14, the sum of the positive ranks, which we call W, equals $6 + 2 + 1 = 9$. This value is much smaller than 27.5 (the value we would expect if the mean is $157).

The question now is, can the difference between the observed and expected values of W be reasonably attributed to sampling error, or does it indicate that the mean weekly food cost for Kansas families of three is actually less than $157? We answer that question and complete the hypothesis test after we discuss some prerequisite material.

Using the Wilcoxon Signed-Rank Table[†]

Table V in Appendix A gives values of W_α for a Wilcoxon signed-rank test.[‡] The two outside columns of Table V give the sample size, n. As expected, the symbol W_α denotes the W-value with area (percentage, probability) α to its right. Thus the column headed $W_{0.10}$ contains W-values with area 0.10 to their right, the column headed $W_{0.05}$ contains W-values with area 0.05 to their right, and so on.

We can express the critical value(s) for a Wilcoxon signed-rank test at the significance level α as follows:

- For a two-tailed test, the critical values are the W-values with area $\alpha/2$ to its left (or, equivalently, area $1 - \alpha/2$ to its right) and area $\alpha/2$ to its right, which are $W_{1-\alpha/2}$ and $W_{\alpha/2}$, respectively. See Fig. 9.25(a).
- For a left-tailed test, the critical value is the W-value with area α to its left or, equivalently, area $1 - \alpha$ to its right, which is $W_{1-\alpha}$. See Fig. 9.25(b).
- For a right-tailed test, the critical value is the W-value with area α to its right, which is W_α. See Fig. 9.25(c).

FIGURE 9.25

Critical value(s) for a Wilcoxon signed-rank test at the significance level α if the test is (a) two tailed, (b) left tailed, or (c) right tailed

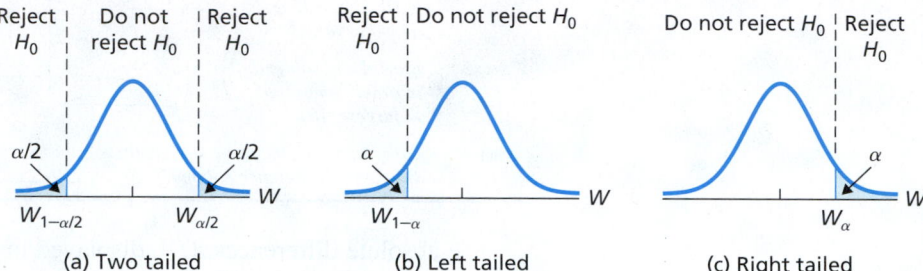

(a) Two tailed   (b) Left tailed   (c) Right tailed

Note the following:

- A critical value from Table V is to be included as part of the rejection region.
- Although the variable W is discrete, we drew the "histograms" in Fig. 9.25 in the shape of a normal curve. This approach is not only convenient, it is also acceptable because W is close to normally distributed except for very small sample sizes. We use this graphical convention throughout this section.

The distribution of the variable W is symmetric about $n(n + 1)/4$. This characteristic implies that the W-value with area A to its left (or, equivalently, area $1 - A$ to its right) equals $n(n + 1)/2$ minus the W-value with area A to its right. In symbols,

$$W_{1-A} = n(n + 1)/2 - W_A. \qquad (9.1)$$

Referring to Fig. 9.25, we see that by using Equation (9.1) and Table V, we can determine the critical value for a left-tailed Wilcoxon signed-rank test and the critical values for a two-tailed Wilcoxon signed-rank test. The next example illustrates the use of Table V to determine critical values for a Wilcoxon signed-rank test.

EXAMPLE 9.19   Using the Wilcoxon Signed-Rank Table

In each case, use Table V to determine the critical value(s) for a Wilcoxon signed-rank test. Sketch graphs to illustrate your results.

a. Sample size = 12; significance level = 0.01; right tailed
b. Sample size = 14; significance level = 0.10; left tailed
c. Sample size = 10; significance level = 0.05; two tailed

[†]We can use the Wilcoxon signed-rank table to estimate the P-value of a Wilcoxon signed-rank test. Because doing so can be awkward or tedious, however, using statistical software is preferable. Thus, those concentrating on the P-value approach to hypothesis testing can skip to the subsection "Performing the Wilcoxon Signed-Rank Test" on the next page.

[‡]Actually, the α-levels in Table V are only approximate but are used in practice.

Solution   In solving these problems, it helps to refer to Fig. 9.25.

a.   The critical value for a right-tailed test at the 1% significance level is $W_{0.01}$. To find the critical value, we use Table V. First we go down the outside columns, labeled n, to "12." Then, going across that row to the column labeled $W_{0.01}$, we reach 68, the required critical value. See Fig. 9.26(a).

FIGURE 9.26

Critical value(s) for a Wilcoxon signed-rank test: (a) right tailed, $\alpha = 0.01$, $n = 12$; (b) left tailed, $\alpha = 0.10$, $n = 14$; (c) two tailed, $\alpha = 0.05$, $n = 10$

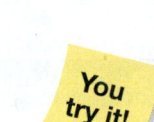

b.   The critical value for a left-tailed test at the 10% significance level is $W_{1-0.10}$. To find the critical value, we use Table V and Equation (9.1). First we go down the outside columns, labeled n, to "14." Then, going across that row to the column labeled $W_{0.10}$, we reach 74; thus $W_{0.10} = 74$. Now we apply Equation (9.1) and the result just obtained to get

$$W_{1-0.10} = 14(14 + 1)/2 - W_{0.10} = 105 - 74 = 31,$$

which is the required critical value. See Fig. 9.26(b).

c.   The critical values for a two-tailed test at the 5% significance level are $W_{1-0.05/2}$ and $W_{0.05/2}$, that is, $W_{1-0.025}$ and $W_{0.025}$. First we use Table V to find $W_{0.025}$. We go down the outside columns, labeled n, to "10." Then, going across that row to the column labeled $W_{0.025}$, we reach 47; thus $W_{0.025} = 47$. Now we apply Equation (9.1) and the result just obtained to get $W_{1-0.025}$:

$$W_{1-0.025} = 10(10 + 1)/2 - W_{0.025} = 55 - 47 = 8.$$

See Fig. 9.26(c).

You try it!

Exercise 9.139 on page 418

Performing the Wilcoxon Signed-Rank Test

Procedure 9.3 on the next page provides a step-by-step method for performing a Wilcoxon signed-rank test by using either the critical-value approach or the P-value approach. Note that we often use the phrase **symmetric population** to indicate that the variable under consideration has a symmetric distribution.

EXAMPLE 9.20   The Wilcoxon Signed-Rank Test

Weekly Food Costs   Let's complete the hypothesis test of Example 9.18. A random sample of 10 Kansas families of three yielded the data on weekly food costs shown in Table 9.13 on page 408. At the 5% significance level, do the data provide sufficient evidence to conclude that the mean weekly food cost for Kansas families of three is less than the national mean of $157?

Solution   We apply Procedure 9.3.

Step 1   **State the null and alternative hypotheses.**

Let μ denote the mean weekly food cost for all Kansas families of three. Then the null and alternative hypotheses are, respectively,

H_0: $\mu = \$157$ (mean weekly food cost is not less than $157)

H_a: $\mu < \$157$ (mean weekly food cost is less than $157).

Note that the hypothesis test is left tailed.

■ ■□ ■ **PROCEDURE 9.3** Wilcoxon Signed-Rank Test

Purpose To perform a hypothesis test for a population mean, μ

Assumptions
1. Simple random sample
2. Symmetric population

Step 1 The null hypothesis is $H_0: \mu = \mu_0$, and the alternative hypothesis is

$$H_a: \mu \neq \mu_0 \quad \text{or} \quad H_a: \mu < \mu_0 \quad \text{or} \quad H_a: \mu > \mu_0$$
$$\text{(Two tailed)} \qquad \text{(Left tailed)} \qquad \text{(Right tailed)}.$$

Step 2 Decide on the significance level, α.

Step 3 Compute the value of the test statistic

$$W = \text{sum of the positive ranks}$$

and denote that value W_0. To do so, construct a work table of the following form.

Observation x	Difference $D = x - \mu_0$	$\lvert D \rvert$	Rank of $\lvert D \rvert$	Signed rank R
.	.	.	.	.
.	.	.	.	.
.	.	.	.	.

CRITICAL-VALUE APPROACH	OR	P-VALUE APPROACH

Step 4 The critical value(s) are

$$W_{1-\alpha/2} \text{ and } W_{\alpha/2} \quad \text{or} \quad W_{1-\alpha} \quad \text{or} \quad W_\alpha$$
$$\text{(Two tailed)} \qquad \text{(Left tailed)} \qquad \text{(Right tailed)}.$$

Use Table V to find the critical value(s). For a left-tailed or two-tailed test, you will also need the relation $W_{1-A} = n(n+1)/2 - W_A$.

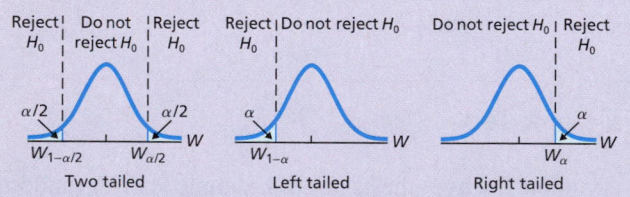

Step 5 If the value of the test statistic falls in the rejection region, reject H_0; otherwise, do not reject H_0.

Step 4 Obtain the *P*-value by using technology.

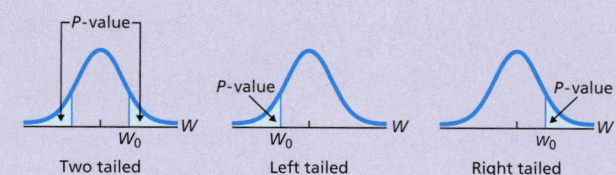

Step 5 If $P \leq \alpha$, reject H_0; otherwise, do not reject H_0.

Step 6 Interpret the results of the hypothesis test.

Step 2 Decide on the significance level, α.

The test is to be performed at the 5% significance level, or $\alpha = 0.05$.

Step 3 Compute the value of the test statistic

$$W = \text{sum of the positive ranks.}$$

The last column of Table 9.14 on page 409 shows that the sum of the positive ranks equals

$$W = 6 + 2 + 1 = 9.$$

| CRITICAL-VALUE APPROACH | OR | P-VALUE APPROACH |

Step 4 The critical value for a left-tailed test is $W_{1-\alpha}$. Use Table V and the relation $W_{1-A} = n(n+1)/2 - W_A$ to find the critical value.

From Table 9.13 on page 408, we see that the sample size is 10. The critical value for a left-tailed test at the 5% significance level is $W_{1-0.05}$. To find the critical value, first we go down the outside columns of Table V, labeled n, to "10." Then, going across that row to the column labeled $W_{0.05}$, we reach 44; thus $W_{0.05} = 44$. Now we apply the aforementioned relation and the result just obtained to get

$$W_{1-0.05} = 10(10+1)/2 - W_{0.05} = 55 - 44 = 11,$$

which is the required critical value. See Fig. 9.27A.

FIGURE 9.27A

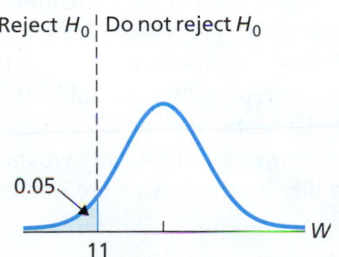

Step 5 If the value of the test statistic falls in the rejection region, reject H_0; otherwise, do not reject H_0.

The value of the test statistic is $W = 9$, as found in Step 3, which falls in the rejection region shown in Fig. 9.27A. Thus we reject H_0. The test results are statistically significant at the 5% level.

Step 4 Obtain the P-value by using technology.

Using technology, we find that the P-value for the hypothesis test is $P = 0.03$, as shown in Fig. 9.27B.

FIGURE 9.27B

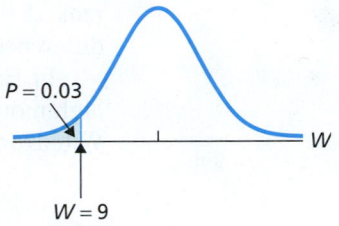

Step 5 If $P \leq \alpha$, reject H_0; otherwise, do not reject H_0.

From Step 4, $P = 0.03$. Because the P-value is less than the specified significance level of 0.05, we reject H_0. The test results are statistically significant at the 5% level and (see Table 9.8 on page 386) provide strong evidence against the null hypothesis.

Step 6 Interpret the results of the hypothesis test.

Report 9.3

Interpretation At the 5% significance level, the data provide sufficient evidence to conclude that the mean weekly food cost for Kansas families of three is less than the national mean of $157.

As mentioned earlier, one advantage of nonparametric methods is that they are resistant to outliers. We can illustrate that advantage for the Wilcoxon signed-rank test by referring to Example 9.20.

The stem-and-leaf diagram depicted in Fig. 9.23 on page 408 shows that the sample data presented in Table 9.13 contain no outliers. The smallest observation, and also the farthest from the null hypothesis mean of 157, is 135. Replacing 135 by, say, 85, introduces an outlier but has no effect on the value of the test statistic and hence none on the hypothesis test itself. (Why is that so?)

> **Note:** The following points may be relevant when performing a Wilcoxon signed-rank test:
>
> - If an observation equals μ_0 (the value for the mean in the null hypothesis), that observation should be removed and the sample size reduced by 1.
> - If two or more absolute differences are tied, each should be assigned the mean of the ranks they would have had if there were no ties.

To illustrate the second bulleted item, suppose that two absolute differences are tied for second place. Then each should be assigned rank $(2 + 3)/2 = 2.5$, and rank 4 should be assigned to the next-largest absolute difference, which really is fourth. Similarly, if three absolute differences are tied for fifth place, each should be assigned rank $(5 + 6 + 7)/3 = 6$, and rank 8 should be assigned to the next-largest absolute difference.

In Example 9.16, we used the one-mean t-test to decide whether, on average, high mountain lakes in the Southern Alps are nonacidic. Now we do so by using the Wilcoxon signed-rank test.

EXAMPLE 9.21 The Wilcoxon Signed-Rank Test

Acid Rain and Lake Acidity A lake is classified as nonacidic if it has a pH greater than 6. A. Marchetto and A. Lami measured the pH of high mountain lakes in the Southern Alps and reported their findings in the paper "Reconstruction of pH by Chrysophycean Scales in Some Lakes of the Southern Alps" (*Hydrobiologia*, Vol. 274, pp 83–90). Table 9.12, which we repeat here as Table 9.15, shows the pH levels obtained by the researchers for 15 lakes.

TABLE 9.15
pH levels for 15 lakes

7.2	7.3	6.1	6.9	6.6
7.3	6.3	5.5	6.3	6.5
5.7	6.9	6.7	7.9	5.8

At the 5% significance level, do the data provide sufficient evidence to conclude that, on average, high mountain lakes in the Southern Alps are nonacidic? Use the Wilcoxon signed-rank test.

Solution Figure 9.28 shows a stem-and-leaf diagram of the sample data in Table 9.15. The diagram is relatively symmetric. Hence, we can reasonably apply Procedure 9.3 to carry out the required hypothesis test.

FIGURE 9.28
Stem-and-leaf diagram of pH levels in Table 9.15

```
5 | 5 7 8
6 | 1 3 3
6 | 5 6 7 9 9
7 | 2 3 3
7 | 9
```

Step 1 State the null and alternative hypotheses.

Let μ denote the mean pH level of all high mountain lakes in the Southern Alps. Then the null and alternative hypotheses are, respectively,

$$H_0: \mu = 6 \text{ (on average, the lakes are acidic)}$$

$$H_a: \mu > 6 \text{ (on average, the lakes are nonacidic)}.$$

Note that the hypothesis test is right tailed.

Step 2 Decide on the significance level, α.

We are to perform the test at the 5% significance level, so $\alpha = 0.05$.

Step 3 Compute the value of the test statistic

$$W = \text{sum of the positive ranks.}$$

To do so, first construct a worktable to obtain the signed ranks.

We construct the following work table. Note that, in several instances, we applied the aforementioned method to deal with tied absolute differences.

pH x	Difference $D = x - 6$	$\lvert D \rvert$	Rank of $\lvert D \rvert$	Signed rank R
7.2	1.2	1.2	12	12
7.3	1.3	1.3	13.5	13.5
6.1	0.1	0.1	1	1
6.9	0.9	0.9	10.5	10.5
6.6	0.6	0.6	8	8
7.3	1.3	1.3	13.5	13.5
6.3	0.3	0.3	4	4
5.5	−0.5	0.5	6.5	−6.5
6.3	0.3	0.3	4	4
6.5	0.5	0.5	6.5	6.5
5.7	−0.3	0.3	4	−4
6.9	0.9	0.9	10.5	10.5
6.7	0.7	0.7	9	9
7.9	1.9	1.9	15	15
5.8	−0.2	0.2	2	−2

Referring to the last column of the work table, we find that the value of the test statistic is

$$W = 12 + 13.5 + 1 + \cdots + 9 + 15 = 107.5.$$

| CRITICAL-VALUE APPROACH | OR | P-VALUE APPROACH |

CRITICAL-VALUE APPROACH

Step 4 The critical value for a right-tailed test is W_α. Use Table V to find the critical value.

From Table 9.15, we see that the sample size is 15. The critical value for a right-tailed test at the 5% significance level is $W_{0.05}$. To find the critical value, first we go down the outside columns of Table V, labeled n, to "15." Then, going across that row to the column labeled $W_{0.05}$, we reach 90, the required critical value. See Fig. 9.29A.

FIGURE 9.29A

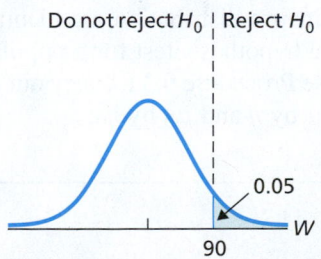

Do not reject H_0 | Reject H_0

0.05

90

Step 5 If the value of the test statistic falls in the rejection region, reject H_0; otherwise, do not reject H_0.

The value of the test statistic is $W = 107.5$, as found in Step 3, which falls in the rejection region shown in Fig. 9.29A. Thus we reject H_0. The test results are statistically significant at the 5% level.

P-VALUE APPROACH

Step 4 Obtain the P-value by using technology.

Using technology, we find that the P-value for the hypothesis test is $P = 0.004$, as shown in Fig. 9.29B.

FIGURE 9.29B

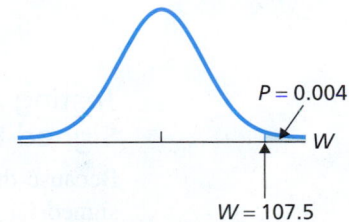

$P = 0.004$

W

$W = 107.5$

Step 5 If $P \leq \alpha$, reject H_0; otherwise, do not reject H_0.

From Step 4, $P = 0.004$. Because the P-value is less than the specified significance level of 0.05, we reject H_0. The test results are statistically significant at the 5% level and (see Table 9.8 on page 386) provide very strong evidence against the null hypothesis.

You try it!

Exercise 9.149 on page 419

Step 6 Interpret the results of the hypothesis test.

Interpretation At the 5% significance level, the data provide sufficient evidence to conclude that, on average, high mountain lakes in the Southern Alps are non-acidic.

We note that both the one-mean t-test of Example 9.16 and the Wilcoxon signed-rank test of Example 9.21 reject the null hypothesis that high mountain lakes in the Southern Alps are, on average, acidic in favor of the alternative hypothesis that they are, on average, nonacidic. Furthermore, with both tests, the data provide very strong evidence against that null hypothesis (and, hence, in favor of the alternative hypothesis). Indeed, as we have seen, $P = 0.002$ for the one-mean t-test, and $P = 0.004$ for the Wilcoxon signed-rank test.

Comparing the Wilcoxon Signed-Rank Test and the t-Test

As you learned in Section 9.5, a t-test can be used to conduct a hypothesis test for a population mean when the variable under consideration is normally distributed. Because normally distributed variables have symmetric distributions, we can also use the Wilcoxon signed-rank test to perform such a hypothesis test.

For a normally distributed variable, the t-test is more powerful than the Wilcoxon signed-rank test because it is designed expressly for such variables; surprisingly, though, the t-test is not much more powerful than the Wilcoxon signed-rank test. However, if the variable under consideration has a symmetric distribution but is not normally distributed, the Wilcoxon signed-rank test is usually more powerful than the t-test and is often considerably more powerful.

KEY FACT 9.8

Wilcoxon Signed-Rank Test Versus the t-Test

Suppose that you want to perform a hypothesis test for a population mean. When deciding between the t-test and the Wilcoxon signed-rank test, follow these guidelines:

- If you are reasonably sure that the variable under consideration is normally distributed, use the t-test.
- If you are not reasonably sure that the variable under consideration is normally distributed but are reasonably sure that it has a symmetric distribution, use the Wilcoxon signed-rank test.

Testing a Population Median with the Wilcoxon Signed-Rank Procedure

Because the mean and median of a symmetric distribution are identical, a Wilcoxon signed-rank test can be used to perform a hypothesis test for a population median, η, as well as for a population mean, μ. To use Procedure 9.3 to carry out a hypothesis test for a population median, simply replace μ by η and μ_0 by η_0.

THE TECHNOLOGY CENTER

Some statistical technologies have programs that automatically perform a Wilcoxon signed-rank test, but others do not. In this subsection, we present output and step-by-step instructions for such programs.

As you will see, different programs may report slightly different P-values for a Wilcoxon signed-rank test. These differences are due to the fact that different programs may use different methods for obtaining or approximating such P-values.

Note to Excel users: The Excel program that we use to perform a (one-sample) Wilcoxon signed-rank test is actually designed for a two-sample test. Nonetheless, as we show, it is possible to use that program to perform a (one-sample) Wilcoxon signed-rank test.

Note to TI-83/84 Plus users: At the time of this writing, the TI-83/84 Plus does not have a built-in program for conducting a Wilcoxon signed-rank test. However, we have written a TI program called WILCOX for performing that test. It is located in the TI Programs section on the WeissStats site. Your instructor can show you how to download the program to your calculator. *Warning:* Any data that you may have previously stored in Lists 1–6 will be erased during program execution, so copy those data to other lists prior to program execution if you want to retain them.

As we said earlier, a Wilcoxon signed-rank test can be used to perform a hypothesis test for a population median, η, as well as for a population mean, μ. Many statistical technologies present the output of that procedure in terms of the median, but that output can also be interpreted in terms of the mean.

EXAMPLE 9.22   Using Technology to Conduct a Wilcoxon Signed-Rank Test

Weekly Food Costs Table 9.13 on page 408 gives the weekly food costs for 10 Kansas families of three. Use Minitab, Excel, or the TI-83/84 Plus to decide, at the 5% significance level, whether the data provide sufficient evidence to conclude that the mean weekly food cost for Kansas families of three is less than the national mean of $157.

Solution Let μ denote the mean weekly food cost for all Kansas families of three. We want to perform the hypothesis test

$$H_0: \mu = \$157 \text{ (mean weekly food cost is not less than \$157)}$$
$$H_a: \mu < \$157 \text{ (mean weekly food cost is less than \$157)}$$

at the 5% significance level. Note that the hypothesis test is left tailed.

We applied the Wilcoxon signed-rank test programs to the data, resulting in Output 9.3. Steps for generating that output are presented in Instructions 9.3. *Note to Excel users:* For brevity, we have presented only the essential portions of the actual output.

OUTPUT 9.3   Wilcoxon signed-rank test output on the sample of weekly food costs

MINITAB

Wilcoxon Signed Rank Test: COST

Test of median = 157.0 versus median < 157.0

	N	N for Test	Wilcoxon Statistic	P	Estimated Median
COST	10	10	9.0	0.033	149.5

EXCEL

Hypothesized difference (D): 0			
Significance level (%): 5			
p-value: Exact p-value			
Summary statistics:			
Variable	Observations	Mean	Std. deviation
COST	10	149.7000	10.8837
Wilcoxon signed-rank test / Lower-tailed test:			
V	9		
Expected value	27.5000		
Variance (V)	96.2500		
p-value (one-tailed)	0.0322		
alpha	0.05		

TI-83/84 PLUS

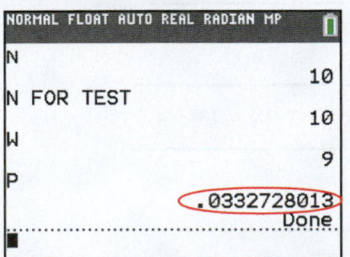

```
NORMAL FLOAT AUTO REAL RADIAN MP
N
                            10
N FOR TEST
                            10
W
                             9
P
                   .0332728013
                          Done
```

As shown in Output 9.3, the *P*-value for the hypothesis test is 0.03. Because the *P*-value is less than the specified significance level of 0.05, we reject H_0. At the 5% significance level, the data provide sufficient evidence to conclude that the mean weekly food cost for Kansas families of three is less than the national mean of $157.

INSTRUCTIONS 9.3   Steps for generating Output 9.3

MINITAB

1 Store the data from Table 9.13 in a column named COST
2 Choose **Stat ➤ Nonparametrics ➤ 1-Sample Wilcoxon...**
3 Press the F3 key to reset the dialog box
4 Specify COST in the **Variables** text box
5 Select the **Test median** option button
6 Click in the **Test median** text box and type 157
7 Click the arrow button at the right of the **Alternative** drop-down list box and select **less than**
8 Click **OK**

EXCEL

1 Store the data from Table 9.13 in a column named COST
2 Store the null hypothesis mean, 157, repeated 10 times (the sample size), in a column named MU_0
3 Choose **XLSTAT ➤ Nonparametric tests ➤ Comparison of two samples (Wilcoxon, Mann-Whitney, …)**
4 Click the reset button in the lower left corner of the dialog box

5 Click in the **Sample 1** selection box and then select the column of the worksheet that contains the COST data
6 Click in the **Sample 2** selection box and then select the column of the worksheet that contains the MU_0 data
7 Uncheck the **Sign test** check box
8 Click the **Options** tab
9 Click the arrow button at the right of the **Alternative hypothesis** drop-down list box and select **Sample 1 – Sample 2 < D**
10 Type 5 in the **Significance level (%)** text box
11 Check the **Exact p-value** check box
12 Click **OK**
13 Click the **Continue** button in the **XLSTAT – Selections** dialog box

TI-83/84 PLUS

1 Store the data from Table 9.13 in a list named COST
2 Press **PRGM**
3 Arrow down to WILCOX and press **ENTER** twice
4 Press **2ND ➤ LIST**, arrow down to COST, and press **ENTER** twice
5 Type 157 for **MU0** and press **ENTER**
6 Type –1 for **TYPE** and press **ENTER**

Exercises 9.6

Understanding the Concepts and Skills

9.129 Technically, what is a *nonparametric method?* In current statistical practice, how is that term used?

9.130 What distributional assumption must be met in order to use the Wilcoxon signed-rank test?

9.131 We mentioned that if, in a Wilcoxon signed-rank test, an observation equals μ_0 (the value given for the mean in the null hypothesis), that observation should be removed and the sample size reduced by 1. Why does that need to be done?

In each of Exercises 9.132–9.137, suppose that you want to perform a hypothesis test for a population mean. Assume that the population standard deviation is unknown and that the sample size is relatively small. In each exercise, we have given the distribution shape of the variable under consideration. Decide whether you would use the t-test, the Wilcoxon signed-rank test, or neither. Explain your answers.

9.132 Uniform

9.133 Normal

9.134 Reverse J shaped

9.135 Triangular

9.136 Symmetric bimodal

9.137 Left skewed

9.138 The Wilcoxon signed-rank test can be used to perform a hypothesis test for a population median, η, as well as for a population mean, μ. Why is that so?

Exercises 9.139–9.142 pertain to critical values for a Wilcoxon signed-rank test. Use Table V in Appendix A to determine the critical value(s) in each case. For a left-tailed or two-tailed test, you will also need the relation $W_{1-A} = n(n + 1)/2 - W_A$.

9.139 Sample size = 8; Significance level = 0.05
a. Right tailed   b. Left tailed   c. Two tailed

9.140 Sample size = 10; Significance level = 0.01
a. Right tailed   b. Left tailed   c. Two tailed

9.141 Sample size = 19; Significance level = 0.10
a. Right tailed   b. Left tailed   c. Two tailed

9.142 Sample size = 15; Significance level = 0.05
a. Right tailed   b. Left tailed   c. Two tailed

In each of Exercises 9.143–9.148, we have provided a null hypothesis and alternative hypothesis and a sample from the population under consideration. In each case, use the Wilcoxon signed-rank test to perform the required hypothesis test at the 10% significance level.

9.143 $H_0: \mu = 5$, $H_a: \mu > 5$

12	7	11	9	3	2	8	6

9.144 $H_0: \mu = 10$, $H_a: \mu < 10$

7	6	5	12	15	14	13	4

9.145 $H_0: \mu = 6$, $H_a: \mu \neq 6$

6	4	8	4	1	1	4	7

9.146 $H_0: \mu = 3$, $H_a: \mu \neq 3$

6	6	3	3	2	5	4	7	4

9.147 H_0: $\mu = 12$, H_a: $\mu < 12$

16	11	10	14	13	15	5	8	11

9.148 H_0: $\mu = 8$, H_a: $\mu > 8$

8	10	11	11	5	9	9	12

Applying the Concepts and Skills

In each of Exercises 9.149–9.154, use the Wilcoxon signed-rank test to perform the required hypothesis test.

9.149 Global Warming? During the late 1800s, Lake Wingra in Madison, Wisconsin, was frozen over an average of 124.9 days per year. A random sample of eight recent years provided the following data on numbers of days that the lake was frozen over.

103	80	79	135	134	77	80	111

At the 5% significance level, do the data provide sufficient evidence to conclude that the average number of ice days is less now than in the late 1800s?

9.150 Happy-Life Years. In the article, "Apparent Quality-of-Life in Nations: How Long and Happy People Live" (*Social Indicators Research*, Vol. 71, pp. 61–86) R. Veenhoven discussed how the quality of life in nations can be measured by how long and happy people live. In the 1990s, the median number of happy-life years across nations was 46.7. A random sample of eight nations for this year provided the following data on number of happy-life years.

30.3	47.0	56.4	30.5	39.6	47.9	29.7	52.5

At the 5% significance level, do the data provide sufficient evidence to conclude that the median number of happy-life years has changed from that in the 1990s?

9.151 How Old People Are. In 2010, the median age of U.S. residents was 37.2 years, as reported by the U.S. Census Bureau in *Current Population Reports*. A random sample of 10 U.S. residents taken this year yielded the following ages, in years.

44	64	16	59	38
47	51	41	13	28

At the 1% significance level, do the data provide sufficient evidence to conclude that the median age of today's U.S. residents has increased from the 2010 median age of 37.2 years?

9.152 Beverage Expenditures. The Bureau of Labor Statistics publishes information on average annual expenditures by consumers in *Consumer Expenditures*. In 2012, the mean amount spent per consumer unit on nonalcoholic beverages was $370. A random sample of 12 consumer units yielded the following data, in dollars, on last year's expenditures on nonalcoholic beverages.

511	326	334	415	409	431
390	423	409	399	344	408

At the 5% significance level, do the data provide sufficient evidence to conclude that last year's mean amount spent by consumers on nonalcoholic beverages has increased from the 2012 mean of $370?

9.153 Pricing Mustangs. According to the *Kelley Blue Book*, the fair purchase price from dealers for a 2-year-old Ford Mustang coupe is about $18,000. A random sample of 10 purchase prices from private parties yielded the following data, in dollars.

16,594	16,106	16,102	15,914	15,713
15,613	14,614	13,514	15,614	15,714

At the 1% significance level, do the data provide sufficient evidence to conclude that the mean purchase price from private parties for 2-year-old Ford Mustang coupes is less than the fair purchase price from dealers?

9.154 Birth Weights. According to the article "Baby Birth Weight Statistics" by V. Iannelli, which appears on the About.com Pediatrics website, in 2005, the median birth weight of U.S. babies was 3389 g, or about 7.5 lb. A random sample of this year's births provided the following weights, in pounds.

8.7	7.5	5.4	13.9	7.9	5.8	9.3
8.9	8.3	9.3	5.7	6.1	11.7	7.3

Can we conclude that this year's median birth weight differs from that in 2005? Use a significance level of 0.05.

9.155 Death Rolls. Alligators perform a spinning maneuver, referred to as a "death roll", to subdue their prey. Videos were taken of juvenile alligators performing this maneuver in a study for the article "Death Roll of the Alligator, Mechanics of Twist and Feeding in Water" (*Journal of Experimental Biology*, Vol. 210, pp. 2811–2818) by F. Fish et al. One of the variables measured was the degree of the angle between the body and head of the alligator while performing the roll. A sample of 20 rolls yielded the following data, in degrees.

58.6	58.7	57.3	54.5	52.9
59.5	29.4	43.4	31.8	52.3
42.7	34.8	39.2	61.3	60.4
51.5	42.8	57.5	43.6	47.6

a. Do the data provide sufficient evidence to conclude that, on average, the angle between the body and head of an alligator during a death roll is greater than 45°? Perform a Wilcoxon signed-rank test at the 5% significance level.

b. The hypothesis test considered in part (a) was done in Exercise 9.115 with a t-test. The assumption in that exercise is that the angle between the body and head of an alligator during a death roll is (approximately) normally distributed. If that is the case, why is it permissible to perform a Wilcoxon signed-rank test for the mean angle between the body and head of an alligator during a death roll?

9.156 Ethical Food Choice Motives. In the paper "Measurement of Ethical Food Choice Motives" (*Appetite*, Vol. 34, pp. 55–59), research psychologists M. Lindeman and M. Väänänen of the University of Helsinki published a study on the factors that most influence peoples' choice of food. One of the questions asked of the participants was how important, on a scale of 1 to 4 (1 = not at all important, 4 = very important), is ecological welfare in food choice motive, where ecological welfare includes animal welfare and environmental protection. Following are the responses of a random sample of 18 Helsinkians.

3	2	2	3	3	3	2	2	3
3	3	1	3	4	2	1	3	1

At the 5% significance level, do the data provide sufficient evidence to conclude that, on average, Helsinkians respond with an ecological welfare food choice motive greater than 2?
a. Use the Wilcoxon signed-rank test.
b. Use the t-test.
c. Compare the results of the two tests.

9.157 Checking Advertised Contents. A manufacturer of liquid soap produces a bottle with an advertised content of 310 milliliters (mL). Sixteen bottles are randomly selected and found to have the following contents, in mL.

297	318	306	300	311	303	291	298
322	307	312	300	315	296	309	311

A normal probability plot of the data indicates that you can assume the contents are normally distributed. Let μ denote the mean content of all bottles produced. To decide whether the mean content is less than advertised, perform the hypothesis test

$$H_0: \mu = 310 \text{ mL}$$
$$H_a: \mu < 310 \text{ mL}$$

at the 5% significance level.
a. Use the t-test.
b. Use the Wilcoxon signed-rank test.
c. If the mean content is in fact less than 310 mL, how do you explain the discrepancy between the two tests?

9.158 Education of Jail Inmates. Thirty years ago, the Bureau of Justice Statistics reported in *Profile of Jail Inmates* that the median educational attainment of jail inmates was 10.2 years. Ten current inmates are randomly selected and found to have the following educational attainments, in years.

14	10	5	6	8
10	10	8	9	9

Assume that educational attainments of current jail inmates have a symmetric, nonnormal distribution. At the 10% significance level, do the data provide sufficient evidence to conclude that this year's median educational attainment has changed from what it was 30 years ago?
a. Use the t-test.
b. Use the Wilcoxon signed-rank test.
c. If this year's median educational attainment has in fact changed from what it was 30 years ago, how do you explain the discrepancy between the two tests?

In each of Exercises **9.159** *and* **9.160**, *use the technology of your choice to decide whether applying the Wilcoxon signed-rank test is reasonable. Explain your answers.*

9.159 Head Injury Criterion. The Head Injury Criterion (HIC) is a measure of the likelihood of an injury arising from an accident such as a vehicle crash. At an HIC of 1000, one in six people will suffer a life-threatening injury to the brain. The Insurance Institute for Highway Safety performs safety rating tests on vehicles. One of the

variables measured is the HIC. The following data provide the HIC levels for a sample of small SUVs.

493	127	95	101	283	82
147	358	81	158	102	196

9.160 Asian Elephants. In the paper "A Survey of African and Asian Elephant Diets and Measured Body Dimensions Compared to Their Estimated Nutrient Requirements" (*Proceedings of the American Zoo and Aquarium Association Nutrition Advisory Group*, 4:13–27), K. Ange et al. studied nutrient levels of African and Asian elephants at a sample of zoos in the United States and Europe. The following table gives the number of Asian elephants at each of the fifteen zoos sampled.

0	2	4	0	3	2	0	0
0	1	0	7	8	2	0	

Working with Large Data Sets

9.161 Delaying Adulthood. The convict surgeonfish is a common tropical reef fish that has been found to delay metamorphosis into adult by extending its larval phase. This delay often leads to enhanced survivorship in the species by increasing the chances of finding suitable habitat. In the paper "Delayed Metamorphosis of a Tropical Reef Fish (*Acanthurus triostegus*): A Field Experiment" (*Marine Ecology Progress Series*, Vol. 176, pp. 25–38), M. McCormick published data that he obtained on the larval duration, in days, of 90 convict surgeonfish. The data are given on the WeissStats site. At the 5% significance level, do the data provide sufficient evidence to conclude that the mean larval duration of convict surgeonfish exceeds 52 days?
a. Employ the Wilcoxon signed-rank test.
b. Employ the t-test.
c. Compare your results from parts (a) and (b).

9.162 Easy Hole at the British Open? The Old Course at St. Andrews in Scotland is home of the British Open, one of the major tournaments in professional golf. The *Hole O'Cross Out*, known by both European and American professional golfers as one of the friendliest holes at St. Andrews, is the fifth hole, a 514-yard, par 5 hole with an open fairway and a large green. As one reporter put it, "If players think before they drive, they will easily walk away with birdies and pars." The scores on the *Hole O'Cross Out* posted by a sample of 156 golf professionals are presented on the WeissStats site. Use those data and the technology of your choice to decide whether, on average, professional golfers score better than par on the *Hole O'Cross Out*. Perform the required hypothesis test at the 0.01 level of significance.
a. Employ the Wilcoxon signed-rank test.
b. Employ the t-test.
c. Compare your results from parts (a) and (b).

In Exercises **9.163**–**9.165**, *we have repeated the contexts of Exercises 9.123–9.125 from Section 9.5. For each exercise, use the technology of your choice to do the following.*
a. *Apply the Wilcoxon signed-rank test to perform the required hypothesis test.*
b. *Compare your result in part (a) to that obtained in the corresponding exercise in Section 9.5, where the t-test was used.*

9.163 Stressed-Out Bus Drivers. In the paper "Hassles on the Job: A Study of a Job Intervention With Urban Bus Drivers" (*Journal of Organizational Behavior*, Vol. 20, pp. 199–208), G. Evans et al.

examined the effects of an intervention program to improve the conditions of urban bus drivers. Among other variables, the researchers monitored diastolic blood pressure of bus drivers in downtown Stockholm, Sweden. The data, in millimeters of mercury (mm Hg), on the WeissStats site are based on the blood pressures obtained prior to intervention for the 41 bus drivers in the study. At the 10% significance level, do the data provide sufficient evidence to conclude that the mean diastolic blood pressure of bus drivers in Stockholm exceeds the normal diastolic blood pressure of 80 mm Hg?

9.164 How Far People Drive. In 2011, the average car in the United States was driven 13.5 thousand miles, as reported by the Federal Highway Administration in *Highway Statistics*. On the WeissStats site, we provide last year's distance driven, in thousands of miles, by each of 500 randomly selected cars. At the 5% significance level, do the data provide sufficient evidence to conclude that the mean distance driven last year differs from that in 2011?

9.165 Fair Market Rent. According to the document *Out of Reach*, published by the National Low Income Housing Coalition, the fair market rent (FMR) for a two-bedroom unit in the United States is $949. A sample of 100 randomly selected two-bedroom units yielded the data on monthly rents, in dollars, given on the WeissStats site. At the 5% significance level, do the data provide sufficient evidence to conclude that the mean monthly rent for two-bedroom units is greater than the FMR of $949? Perform the required hypothesis test both with and without the outlier.

Extending the Concepts and Skills

Normal Approximation for W. The Wilcoxon signed-rank table, Table V, stops at $n = 20$. For larger samples, a normal approximation can be used. In fact, the normal approximation works well even for sample sizes as small as 10.

Normal Approximation for W

Suppose that the variable under consideration has a symmetric distribution. Then, for samples of size n,

- $\mu_W = n(n + 1)/4$,
- $\sigma_W = \sqrt{n(n + 1)(2n + 1)/24}$, and
- W is approximately normally distributed for $n \geq 10$.

Thus, for samples of size 10 or more, the standardized variable

$$z = \frac{W - n(n + 1)/4}{\sqrt{n(n + 1)(2n + 1)/24}}$$

has approximately the standard normal distribution.

9.166 Large-Sample Wilcoxon Signed-Rank Test. Formulate a hypothesis-testing procedure for a Wilcoxon signed-rank test that uses the test statistic z given in the preceding box. *Note:* Using a continuity correction provides even more accurate results.

9.167 Birth Weights. Refer to Exercise 9.154.
a. Use the procedure you formulated in Exercise 9.166 to perform the hypothesis test in Exercise 9.154.
b. Compare your result in part (a) to the one you obtained in Exercise 9.154, where the normal approximation was not used.

9.168 The Distribution of W. In this exercise, you are to obtain the distribution of the variable W for samples of size 3 so that you can see how the Wilcoxon signed-rank table is constructed.

a. The rows of the following table give all possible signs for the signed ranks in a Wilcoxon signed-rank test with $n = 3$. For instance, the first row covers the possibility that all three observations are greater than μ_0 and thus have positive sign ranks. Fill in the empty column with values of W. (*Hint:* The first entry is 6, and the last is 0.)

Rank			
1	2	3	W
+	+	+	
+	+	−	
+	−	+	
+	−	−	
−	+	+	
−	+	−	
−	−	+	
−	−	−	

b. If the null hypothesis $H_0: \mu = \mu_0$ is true, what percentages of samples will match any particular row of the table? (*Hint:* The answer is the same for all rows.)
c. Use the answer from part (b) to obtain the distribution of W for samples of size 3.
d. Draw a relative-frequency histogram of the distribution obtained in part (c).
e. Use your histogram from part (d) to find $W_{0.125}$ for a sample size of 3.

One-Median Sign Test. Recall that the Wilcoxon signed-rank test, which can be used to perform a hypothesis test for a population median, η, requires that the variable under consideration has a symmetric distribution. If that is not the case, the **one-median sign test** (or simply the **sign test**) can be used instead. The one-median sign test is also known as the **one-sample sign test** and the **one-variable sign test**. Technically, like the Wilcoxon signed-rank test, use of the sign test requires that the variable under consideration has a continuous distribution. In practice, however, that restriction is usually ignored.

If the null hypothesis $H_0: \eta = \eta_0$ is true, the probability is 0.5 of an observation exceeding η_0. Therefore, in a simple random sample of size n, the number of observations, s, that exceed η_0 has a binomial distribution with parameters n and 0.5.

To perform a sign test, first assign a "+" sign to each observation in the sample that exceeds η_0 and then obtain the number of "+" signs, which we denote s_0. The P-value for the hypothesis test can be found by applying Exercise 9.71 on page 387 and obtaining the required binomial probability.

9.169 Assuming that the null hypothesis $H_0: \eta = \eta_0$ is true, answer the following questions.
a. Why is the probability of an observation exceeding η_0 equal to 0.5?
b. In a simple random sample of size n, why does the number of observations that exceed η_0 have a binomial distribution with parameters n and 0.5?

9.170 The sign test can be used whether or not the variable under consideration has a symmetric distribution. If the distribution is in fact symmetric, the Wilcoxon signed-rank test is preferable. Why do you think that is so?

9.171 What advantage does the sign test have over the Wilcoxon signed-rank test?

9.172 Explain how to proceed with a sign test if one or more of the observations equals η_0, the value specified in the null hypothesis for the population median.

In Exercises 9.173–9.178,
a. *apply the sign test to the specified exercise.*
b. *compare your result in part (a) to that obtained by using the Wilcoxon signed-rank test earlier in this exercise section.*

9.173 Exercise 9.149   **9.174** Exercise 9.150

9.175 Exercise 9.151   **9.176** Exercise 9.152

9.177 Exercise 9.153   **9.178** Exercise 9.154

9.7   Type II Error Probabilities; Power*

As you learned in Section 9.1, hypothesis tests do not always yield correct conclusions; they have built-in margins of error. An important part of planning a study is to consider both types of errors that can be made and their effects.

Recall that two types of errors are possible with hypothesis tests. One is a Type I error: rejecting a true null hypothesis. The other is a Type II error: not rejecting a false null hypothesis. Also recall that the probability of making a Type I error is called the significance level of the hypothesis test and is denoted α, and that the probability of making a Type II error is denoted β.

In this section, we show how to compute Type II error probabilities. We also investigate the concept of the *power of a hypothesis test.* Although the discussion is limited to the one-mean z-test, the ideas apply to any hypothesis test.

Computing Type II Error Probabilities

The probability of making a Type II error depends on the sample size, the significance level, and the true value of the parameter under consideration.

■■ ■ **EXAMPLE 9.23**   **Computing Type II Error Probabilities**

Questioning Gas Mileage Claims The manufacturer of a new model car, the Orion, claims that a typical car gets 26 miles per gallon (mpg). A consumer advocacy group is skeptical of this claim and thinks that the mean gas mileage, μ, of all Orions may be less than 26 mpg. The group plans to perform the hypothesis test

$$H_0\text{: } \mu = 26 \text{ mpg (manufacturer's claim)}$$
$$H_a\text{: } \mu < 26 \text{ mpg (consumer group's conjecture)},$$

at the 5% significance level, using a sample of 30 Orions. Find the probability of making a Type II error if the true mean gas mileage of all Orions is

a.   25.8 mpg.   **b.**   25.0 mpg.

Assume that gas mileages of Orions are normally distributed with a standard deviation of 1.4 mpg.

Solution   The inference under consideration is a left-tailed hypothesis test for a population mean at the 5% significance level. The test statistic is

$$z = \frac{\bar{x} - \mu_0}{\sigma/\sqrt{n}} = \frac{\bar{x} - 26}{1.4/\sqrt{30}}.$$

We first express the decision criterion of whether or not to reject the null hypothesis in terms of the value of the test statistic, z.

- *Critical-value approach:* The critical value is $-z_\alpha = -z_{0.05} = -1.645$. Consequently, the value of the test statistic falls in the rejection region, and hence we reject the null hypothesis, if and only if $z \leq -1.645$.
- *P-value approach:* We reject the null hypothesis if and only if $P \leq \alpha = 0.05$, which happens if and only if the area under the standard normal curve to the left of the value of the test statistic is at most 0.05, that is, $z \leq -z_{0.05}$. Referring to Table II, we see that we reject the null hypothesis if and only if $z \leq -1.645$.

Therefore, a decision criterion for the hypothesis test is: If $z \leq -1.645$, reject H_0; if $z > -1.645$, do not reject H_0. Note that $-1.645 = -z_{0.05} = -z_\alpha$.

Computing Type II error probabilities is somewhat simpler if the decision criterion is expressed in terms of $\bar{x}$ instead of z. To do that here, we must find the sample mean that is 1.645 standard deviations below the null hypothesis population mean of 26:

$$\bar{x} = 26 - 1.645 \cdot \frac{1.4}{\sqrt{30}} = 25.6.$$

We call this value the $\bar{x}$ **critical value**. The decision criterion can thus be expressed in terms of $\bar{x}$ as: If $\bar{x} \leq 25.6$ mpg, reject H_0; if $\bar{x} > 25.6$ mpg, do not reject H_0. See Fig. 9.30. Note that the curve in Fig. 9.30 is the normal curve for $\bar{x}$ under the assumption that the null hypothesis is true, which is the normal curve with parameters 26 and $1.4/\sqrt{30} = 0.26$.

a. If $\mu = 25.8$ mpg, then

- $\mu_{\bar{x}} = \mu = 25.8$,
- $\sigma_{\bar{x}} = \sigma/\sqrt{n} = 1.4/\sqrt{30} = 0.26$, and
- $\bar{x}$ is normally distributed.

Thus, the variable $\bar{x}$ is normally distributed with a mean of 25.8 mpg and a standard deviation of 0.26 mpg. The normal curve for $\bar{x}$ is shown in Fig. 9.31.

FIGURE 9.30

Decision criterion for the gas mileage illustration ($\alpha = 0.05$, $n = 30$)

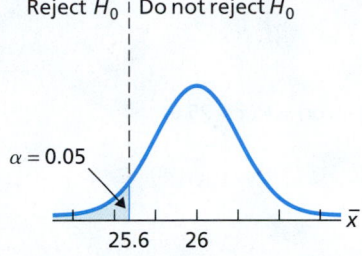

FIGURE 9.31

Determining the probability of a Type II error if $\mu = 25.8$ mpg

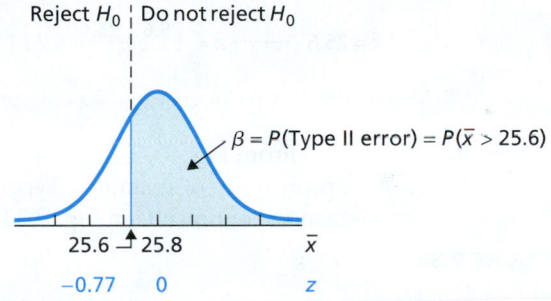

z-score computation: Area to the left of z:

$$\bar{x} = 25.6 \longrightarrow z = \frac{25.6 - 25.8}{0.26} = -0.77 \quad\quad 0.2206$$

Shaded area $= 1 - 0.2206 = 0.7794$

A Type II error occurs if we do not reject H_0, that is, if $\bar{x} > 25.6$ mpg. The probability of this happening equals the percentage of all samples whose means exceed 25.6 mpg, which we obtain in Fig. 9.31 by applying Procedure 6.1 on page 280. Thus, if the true mean gas mileage of all Orions is 25.8 mpg, the probability of making a Type II error is 0.7794; that is, $\beta = 0.7794$. We summarize our results in Fig. 9.32.

FIGURE 9.32

Decision criterion and Type II error probability when true mean is 25.8 mpg ($\alpha = 0.05$, $n = 30$)

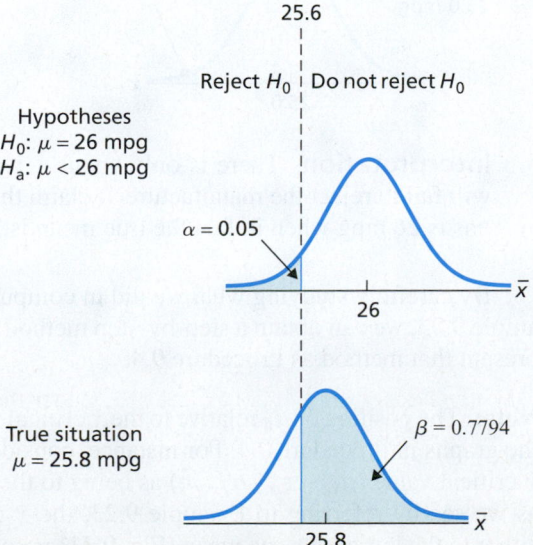

Interpretation   There is roughly a 78% chance that the consumer group will fail to reject the manufacturer's claim that the mean gas mileage of all Orions is 26 mpg when in fact the true mean is 25.8 mpg.

Although this result is a rather high chance of error, we probably would not expect the hypothesis test to detect such a small difference in mean gas mileage (25.8 mpg as opposed to 26 mpg) with a sample size of only 30.

b. We proceed as we did in part (a), but this time we assume that $\mu = 25.0$ mpg. Figure 9.33 shows the required computations.

FIGURE 9.33

Determining the probability of a Type II error if $\mu = 25.0$ mpg

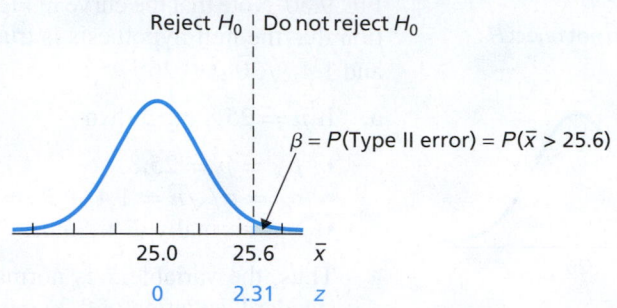

z-score computation:

$$\bar{x} = 25.6 \longrightarrow z = \frac{25.6 - 25.0}{0.26} = 2.31$$

Area to the left of z:

0.9896

Shaded area $= 1 - 0.9896 = 0.0104$

From Fig. 9.33, if the true mean gas mileage of all Orions is 25.0 mpg, the probability of making a Type II error is 0.0104; that is, $\beta = 0.0104$. We summarize our results in Fig. 9.34.

FIGURE 9.34

Decision criterion and Type II error probability when true mean is 25.0 mpg ($\alpha = 0.05$, $n = 30$)

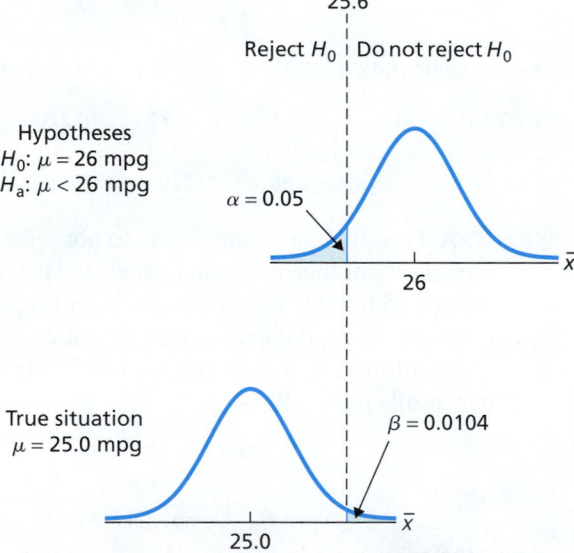

Interpretation   There is only about a 1% chance that the consumer group will fail to reject the manufacturer's claim that the mean gas mileage of all Orions is 26 mpg when in fact the true mean is 25.0 mpg.

By carefully studying what we did in computing Type II error probabilities in Example 9.23, we can obtain a step-by-step method for calculating such probabilities. We present that method as Procedure 9.4.

Note: The position of μ_a relative to the $\bar{x}$ critical value(s) may differ from that shown in the graphs in Procedure 9.4. For instance, consider the middle graph, which shows the $\bar{x}$ critical value ($\mu_0 - z_\alpha \cdot \sigma/\sqrt{n}$) as being to the right of the true mean, μ_a. However, as we see by referring to Example 9.23, the $\bar{x}$ critical value, which is 25.6, can fall either to the left of the true mean (Fig. 9.31) or to the right of the true mean (Fig. 9.33).

▪ ▪ ▪ PROCEDURE 9.4   Type II Error Probabilities for a One-Mean z-Test

Purpose To find Type II error probabilities for a one-mean z-test when the true population mean equals μ_a

Assumptions
1. Simple random sample
2. Normal population or large sample
3. σ known

Step 1 The $\bar{x}$ critical value(s) are

$$\mu_0 \pm z_{\alpha/2} \cdot \sigma/\sqrt{n} \quad \text{or} \quad \mu_0 - z_{\alpha} \cdot \sigma/\sqrt{n} \quad \text{or} \quad \mu_0 + z_{\alpha} \cdot \sigma/\sqrt{n}$$
$$\text{(Two tailed)} \qquad\qquad \text{(Left tailed)} \qquad\qquad \text{(Right tailed)} \; .$$

where μ_0 is the null-hypothesis value of the mean, α is the significance level, σ is the population standard deviation, and n is the sample size.

Step 2 The probability of a Type II error, β, equals the area under the normal curve with parameters μ_a and $\sigma/\sqrt{n}$ that lies

between the two		to the right of the		to the left of the
$\bar{x}$ critical values	or	$\bar{x}$ critical value	or	$\bar{x}$ critical value .
(Two tailed)		(Left tailed)		(Right tailed)

Use Procedure 6.1 on page 280 to obtain the required area.

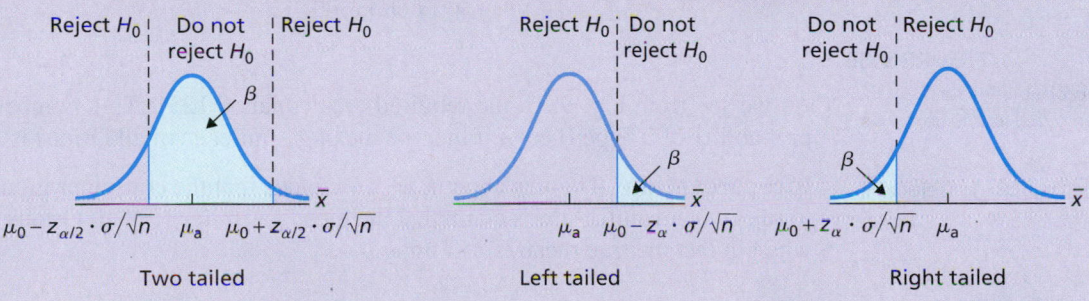

Two tailed	Left tailed	Right tailed

Let us now return to the gas-mileage example in order to illustrate the use of Procedure 9.4.

▪ ▪ ▪ EXAMPLE 9.24   Type II Error Probabilities

Questioning Gas Mileage Claims Refer to Example 9.23, starting on page 422. The null and alternative hypotheses are

$$H_0: \mu = 26 \text{ mpg (manufacturer's claim)}$$
$$H_a: \mu < 26 \text{ mpg (consumer group's conjecture)},$$

where μ is the mean gas mileage of all Orions. The hypothesis test is to be performed at the 5% significance level with a sample size of 30. Find the probability of making a Type II error if the true mean gas mileage of all Orions is 25.3 mpg. Assume that gas mileages of Orions are normally distributed with a standard deviation of 1.4 mpg.

Solution We apply Procedure 9.4. Note that the hypothesis test is left tailed and that $\mu_0 = 26$, $\alpha = 0.05$, $n = 30$, $\sigma = 1.4$, and $\mu_a = 25.3$.

Step 1 The $\bar{x}$ critical value for a left-tailed test is $\mu_0 - z_{\alpha} \cdot \sigma/\sqrt{n}$.

We have $z_{\alpha} = z_{0.05} = 1.645$. Therefore, the $\bar{x}$ critical value is

$$\mu_0 - z_{\alpha} \cdot \frac{\sigma}{\sqrt{n}} = 26 - 1.645 \cdot \frac{1.4}{\sqrt{30}} = 25.6,$$

as we also found in Example 9.23.

Step 2 The probability of a Type II error, β, equals the area under the normal curve with parameters μ_a and $\sigma/\sqrt{n}$ that lies to the right of the $\bar{x}$ critical value. Use Procedure 6.1 on page 280 to obtain the required area.

We have $\mu_a = 25.3$, $\sigma/\sqrt{n} = 1.4/\sqrt{30} = 0.26$, and, from Step 1, the $\bar{x}$ critical value is 25.6. Therefore, we need to find the area under the normal curve with parameters 25.3 and 0.26 that lies to the right of 25.6. That area is obtained in Fig. 9.35 by applying Procedure 6.1.

FIGURE 9.35

Determining the probability of a Type II error if $\mu = 25.3$ mpg

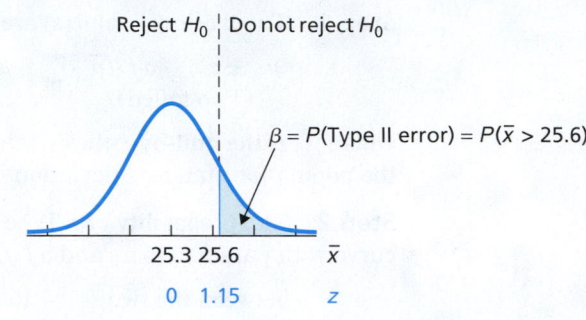

z-score computation:

$$\bar{x} = 25.6 \longrightarrow z = \frac{25.6 - 25.3}{0.26} = 1.15$$

Area to the left of z:

0.8749

Shaded area = $1 - 0.8749 = 0.1251$

You try it!

Exercise 9.191(b)
on page 432

FIGURE 9.36

Type II error probabilities for $\mu = 25.8$, 25.6, 25.3, and 25.0 ($\alpha = 0.05$, $n = 30$)

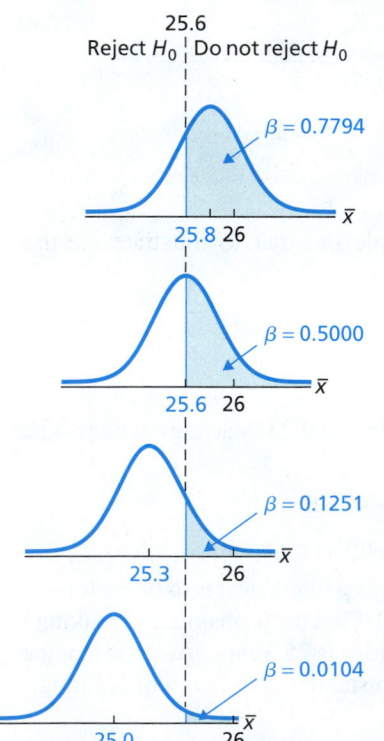

As we see from Fig. 9.35, the required area equals 0.1251. This number is β, the probability of a Type II error if the true mean gas mileage of all Orions is 25.3 mpg.

Interpretation   There is about a 12.5% chance that the consumer group will fail to reject the manufacturer's claim that the mean gas mileage of all Orions is 26 mpg when in fact the true mean is 25.3 mpg.

Combining figures such as Figs. 9.31, 9.33, and 9.35 gives a better understanding of Type II error probabilities. In Fig. 9.36, we combine those three figures with one other. The Type II error probabilities for the additional value of μ was obtained by applying Procedure 9.4, although, in this case, it could be obtained simply by inspection (Why is that?).

Figure 9.36 shows clearly that the farther the true mean is from the null hypothesis mean of 26 mpg, the smaller will be the probability of a Type II error. This result is hardly surprising: We would expect that a false null hypothesis is more likely to be detected when the true mean is far from the null hypothesis mean than when the true mean is close to the null hypothesis mean.

Power and Power Curves

In modern statistical practice, analysts generally use the probability of not making a Type II error, called the **power,** to appraise the performance of a hypothesis test. Once we know the Type II error probability, β, obtaining the power is simple—we just subtract β from 1.

DEFINITION 9.6

Power

The **power** of a hypothesis test is the probability of not making a Type II error, that is, the probability of rejecting a false null hypothesis. We have

$$\text{Power} = 1 - P(\text{Type II error}) = 1 - \beta.$$

What Does It Mean?

The power of a hypothesis test is between 0 and 1 and measures the ability of the hypothesis test to detect a false null hypothesis. If the power is near 0, the hypothesis test is not very good at detecting a false null hypothesis; if the power is near 1, the hypothesis test is extremely good at detecting a false null hypothesis.

Note: Of course, it is also possible to calculate power directly without first obtaining the probability of a Type II error. For instance, we can do so for a one-mean z-test by applying Procedure 9.4 (page 425) but, in Step 2, obtain the unshaded area under the normal curve with parameters μ_a and $\sigma/\sqrt{n}$ instead of the shaded area.

In reality, the true value of the parameter in question will be unknown. Consequently, constructing a table of powers for various values of the parameter consistent with the alternative hypothesis is helpful in evaluating the overall effectiveness of a hypothesis test.

Even more helpful is a visual display of the effectiveness of the hypothesis test, obtained by plotting points of power against various values of the parameter and then connecting the points with a smooth curve. The resulting curve is called a **power curve**. In general, the closer a power curve is to 1 (i.e., the horizontal line 1 unit above the horizontal axis), the better the hypothesis test is at detecting a false null hypothesis.

As we mentioned earlier, in this section, we will limit our discussion to the one-mean z-test, although the ideas apply to any hypothesis test. Procedure 9.5 provides a step-by-step method for obtaining a power curve for a one-mean z-test.

■ ■ ■ ■ PROCEDURE 9.5   **Power Curve for a One-Mean z-Test**

Purpose To construct a power curve for a one-mean z-test

Assumptions
1. Simple random sample
2. Normal population or large sample
3. σ known

Step 1 Decide on equidistant values of μ_a to be used in plotting the power curve. Choose values of μ_a

on both sides of μ_0		to the left of μ_0		to the right of μ_0
(Two tailed)	or	(Left tailed)	or	(Right tailed)

Step 2 Construct a work table of the form

True mean μ	P (Type II error) β	Power $1 - \beta$
.	.	.
.	.	.
.	.	.

Fill in the first column of the work table with the values of μ_a chosen in Step 1.

Step 3 For each value of μ_a chosen in Step 1, use Procedure 9.4 (page 425) to obtain the probability of a Type II error. Fill in the second column of the work table from Step 2.

Step 4 Use the results of Step 3 and the relation Power $= 1 - \beta$ to fill in the third column of the work table from Step 2.

Step 5 Plot the values in the first and third columns of the work table from Step 2 as points on a graph with μ on the horizontal axis and power on the vertical axis. Include on the graph a hollow point for the null-hypothesis mean (μ_0) and the significance level (α).

Step 6 Connect the points in the graph from Step 5 with a smooth curve.

■■■   **EXAMPLE 9.25**   Power Curves

Questioning Gas Mileage Claims Refer again to Example 9.23, starting on page 422. The null and alternative hypotheses are

$$H_0: \mu = 26 \text{ mpg (manufacturer's claim)}$$
$$H_a: \mu < 26 \text{ mpg (consumer group's conjecture)},$$

where μ is the mean gas mileage of all Orions. The hypothesis test is to be performed at the 5% significance level with a sample size of 30. Construct a power curve.

Solution   We apply Procedure 9.5. Note that the hypothesis test is left tailed.

Step 1   Decide on equidistant values of μ_a to be used in plotting the power curve. For a left-tailed test, choose values of μ_a to the left of μ_0.

We have $\mu_0 = 26$. Let us choose values of μ_a from 24.8 to 25.9 in increments of one-tenth.

Step 2   Construct a work table for the true mean (μ), the Type II error probability (β), and the power ($1 - \beta$). Fill in the first column of the work table with the values of μ_a chosen in Step 1.

See the first column of the following table.

True mean μ	P (Type II error) β	Power $1 - \beta$
24.8	0.0010	0.9990
24.9	0.0036	0.9964
25.0	0.0104	0.9896
25.1	0.0274	0.9726
25.2	0.0618	0.9382
25.3	0.1251	0.8749
25.4	0.2206	0.7794
25.5	0.3520	0.6480
25.6	0.5000	0.5000
25.7	0.6480	0.3520
25.8	0.7794	0.2206
25.9	0.8749	0.1251
↑	↑	↑
From Step 1	From Step 3	From Step 4

Step 3   For each value of μ_a chosen in Step 1, use Procedure 9.4 (page 425) to obtain the probability of a Type II error. Fill in the second column of the work table from Step 2.

In Example 9.24 (pages 425–426), we applied Procedure 9.4 to obtain the probability of a Type II error when $\mu_a = 25.3$ (i.e., when the true mean is 25.3); we found that $\beta = 0.1251$. Applying Procedure 9.4 to the other chosen values of μ_a, we obtained the other required β-values, as shown in the second column of the work table from Step 2.

Step 4   Use the results of Step 3 and the relation Power = $1 - \beta$ to fill in the third column of the work table from Step 2.

Applying the relation Power = $1 - \beta$ to the Type II error probabilities in the second column of the work table from Step 2, we get the powers in the third column of the work table from Step 2.

Step 5 **Plot the values in the first and third columns of the work table from Step 2 as points on a graph with μ on the horizontal axis and power on the vertical axis. Include on the graph a hollow point for the null-hypothesis mean (μ_0) and the significance level (α).**

Referring to values in the first and third columns of the work table from Step 2, we get the solid green points in Fig. 9.37. Recalling that $\mu_0 = 26$ and $\alpha = 0.05$, we also get the hollow green point in Fig. 9.37.

FIGURE 9.37

Power curve for gas mileage illustration ($\alpha = 0.05$, $n = 30$)

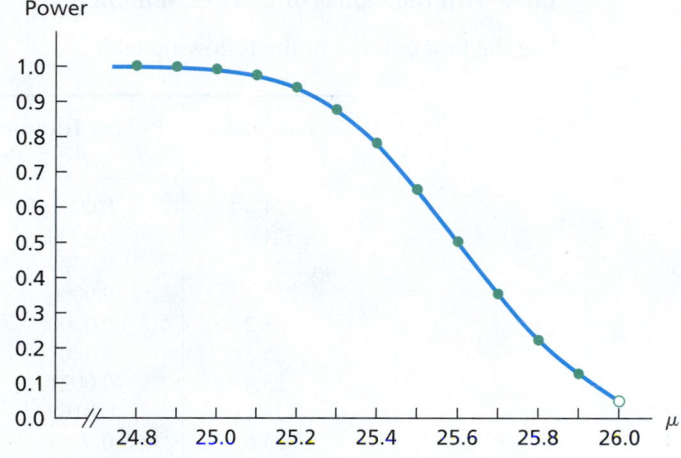

Applet 9.2

Exercise 9.191(c) on page 432

Step 6 **Connect the points in the graph from Step 5 with a smooth curve.**

We connected the points in the graph from Step 5 with a smooth curve, as shown in Fig. 9.37. That figure provides a power curve for the gas mileage illustration when $\alpha = 0.05$ and $n = 30$.

Sample Size and Power

Ideally, both Type I and Type II errors should have small probabilities. In terms of significance level and power, then, we want to specify a small significance level (close to 0) and yet have large power (close to 1).

Key Fact 9.1 (page 373) implies that the smaller we specify the significance level, the smaller will be the power. However, by using a large sample, we can have both a small significance level and large power, as shown in the next example.

EXAMPLE 9.26 **Sample Size and Power**

Questioning Gas Mileage Claims Consider again the hypothesis test for the gas-mileage illustration of Example 9.23, starting on page 422. The null and alternative hypotheses are

$$H_0: \mu = 26 \text{ mpg (manufacturer's claim)}$$
$$H_a: \mu < 26 \text{ mpg (consumer group's conjecture)},$$

where μ is the mean gas mileage of all Orions. In Example 9.25, we determined a power curve when $\alpha = 0.05$ and $n = 30$. See Fig. 9.37.

Now suppose that we keep the significance level at 0.05 but increase the sample size from 30 to 100. Construct a power curve for $n = 100$ and compare it to the power curve for $n = 30$.

Solution To construct a power curve for $n = 100$, we apply Procedure 9.5. Note that the hypothesis test is left tailed.

Step 1 **Decide on equidistant values of μ_a to be used in plotting the power curve. For a left-tailed test, choose values of μ_a to the left of μ_0.**

We have $\mu_0 = 26$. Because we want to compare the power curve for $n = 100$ to that for $n = 30$, let us choose the same values of μ_a as we used in Example 9.25; namely, from 24.8 to 25.9 in increments of one-tenth.

Step 2 **Construct a work table for the true mean (μ), the Type II error probability (β), and the power $(1 - \beta)$. Fill in the first column of the work table with the values of μ_a chosen in Step 1.**

See the first column of the following table.

True mean μ	P (Type II error) β	Power $1 - \beta$
24.8	0.0000	1.0000
24.9	0.0000	1.0000
25.0	0.0000	1.0000
25.1	0.0000	1.0000
25.2	0.0000	1.0000
25.3	0.0002	0.9998
25.4	0.0021	0.9979
25.5	0.0162	0.9838
25.6	0.0764	0.9236
25.7	0.2389	0.7611
25.8	0.5000	0.5000
25.9	0.7611	0.2389

↑	↑	↑
From Step 1	From Step 3	From Step 4

Step 3 **For each value of μ_a chosen in Step 1, use Procedure 9.4 (page 425) to obtain the probability of a Type II error. Fill in the second column of the work table from Step 2.**

Recall that we are assuming that gas mileages of Orions are normally distributed with a standard deviation of 1.4 mpg. Step 1 of Procedure 9.4 requires the $\bar{x}$ critical value, which, for this left-tailed test, is

$$\mu_0 - z_\alpha \cdot \frac{\sigma}{\sqrt{n}} = 26 - 1.645 \cdot \frac{1.4}{\sqrt{100}} = 25.8.$$

Now we can apply Step 2 of Procedure 9.4, as we have done previously, to obtain the probability of a Type II error for each value of μ_a in the first column of the work table from Step 2 here. The result is shown in the second column of the work table from Step 2. Observe that we have displayed the Type II error probabilities to four decimal places. So, for instance, when the true mean is 24.8, the probability of a Type II error is 0 to four decimal places.

Step 4 **Use the results of Step 3 and the relation Power $= 1 - \beta$ to fill in the third column of the work table from Step 2.**

Applying the relation Power $= 1 - \beta$ to the Type II error probabilities in the second column of the work table from Step 2, we get the powers in the third column of the work table from Step 2. Observe that we have displayed the powers to four decimal places. So, for instance, when the true mean is 24.8, the power is 1 to four decimal places.

Step 5 **Plot the values in the first and third columns of the work table from Step 2 as points on a graph with μ on the horizontal axis and power on the vertical axis. Include on the graph a hollow point for the null-hypothesis mean (μ_0) and the significance level (α).**

Referring to the values in the first and third columns of the work table from Step 2, we get the solid green points in the top graph of Fig. 9.38. Recalling that $\mu_0 = 26$ and $\alpha = 0.05$, we also get the hollow green point in the top graph of Fig. 9.38.

FIGURE 9.38

Power curves for the gas mileage illustration when $n = 30$ and $n = 100$ ($\alpha = 0.05$)

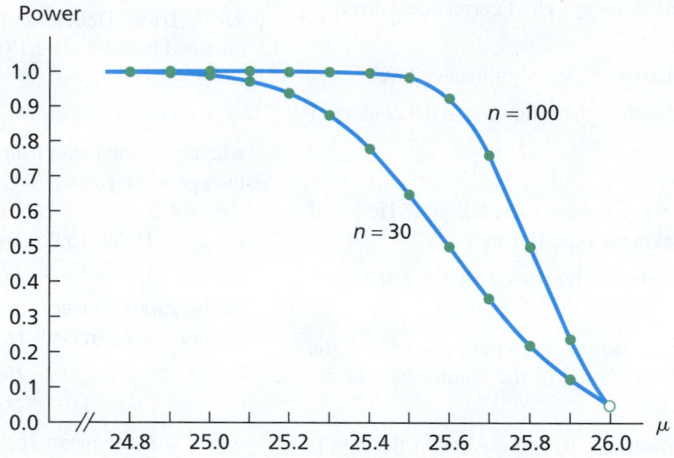

Step 6 Connect the points in the graph from Step 5 with a smooth curve.

You try it!

Exercise 9.197 on page 432

We connected the points in the top graph from Step 5 with a smooth curve, as shown in Fig. 9.38. That graph provides a power curve for the gas mileage illustration when $\alpha = 0.05$ and $n = 100$. For comparative purposes, Fig. 9.38 includes the graph of the power curve for $n = 30$ from Fig. 9.37 on page 429.

Interpretation Comparing the two power curves in Fig. 9.38, we see that each power is greater when $n = 100$ than when $n = 30$.

In the preceding example, we found that increasing the sample size without changing the significance level increased the power. This relationship is true in general.

KEY FACT 9.9

What Does It Mean?

By using a sufficiently large sample size, we can obtain a hypothesis test with as much power as we want.

Sample Size and Power

For a fixed significance level, increasing the sample size increases the power.

In practice, larger sample sizes tend to increase the cost of a study. Consequently, we must balance, among other things, the cost of a large sample against the cost of possible errors.

As we have indicated, power is a useful way to evaluate the overall effectiveness of a hypothesis-testing procedure. Additionally, power can be used to compare different procedures. For example, a researcher might decide between two hypothesis-testing procedures on the basis of which test is more powerful for the situation under consideration.

THE TECHNOLOGY CENTER

As we have shown, obtaining Type II error probabilities or powers is computationally intensive. Moreover, determining those quantities by hand can result in substantial roundoff error. Therefore, in practice, Type II error probabilities and powers are almost always determined by computer.

Exercises 9.7

Understanding the Concepts and Skills

9.179 Why don't hypothesis tests always yield correct decisions?

9.180 Define each term.
a. Type I error **b.** Type II error **c.** Significance level

9.181 Explain the meaning of each of the following in the context of hypothesis testing.
a. α **b.** β **c.** $1 - \beta$

9.182 What does the power of a hypothesis test tell you? How is it related to the probability of making a Type II error?

9.183 Why is it useful to obtain the power curve for a hypothesis test?

9.184 What happens to the power of a hypothesis test if the sample size is increased without changing the significance level? Explain your answer.

9.185 What happens to the power of a hypothesis test if the significance level is decreased without changing the sample size? Explain your answer.

9.186 Suppose that you must choose between two procedures for performing a hypothesis test—say, Procedure A and Procedure B. Further suppose that, for the same sample size and significance level, Procedure A has less power than Procedure B. Which procedure would you choose? Explain your answer.

9.187 Explain why Type II errors and powers are almost always obtained by computer.

In each of Exercises 9.188–9.190, we have specified the "tailed-ness" of a hypothesis test for a population mean with null hypothesis H_0: $\mu = \mu_0$. For each exercise,
a. draw the ideal power curve.
b. explain what your curve in part (a) portrays.

9.188 right-tailed   **9.189** left-tailed   **9.190** two-tailed

Applying the Concepts and Skills

In Exercises 9.191–9.196, we have given a hypothesis testing situation and (i) the population standard deviation, σ, (ii) a significance level, (iii) a sample size, and (iv) some values of μ_a. For each exercise,
a. determine the probability of a Type I error.
b. apply Procedure 9.4 on page 425 to determine the probability of a Type II error for the first given value of μ_a.
c. apply Procedure 9.5 on page 427 to construct a power curve based on the given values of μ_a.

9.191 **Toxic Mushrooms?** The null and alternative hypotheses obtained in Exercise 9.15 on page 375 are, respectively,

$$H_0\text{: } \mu = 0.5 \text{ ppm}$$
$$H_a\text{: } \mu > 0.5 \text{ ppm,}$$

where μ is the mean cadmium level in *Boletus pinicola* mushrooms.
i. $\sigma = 0.37$ ii. $\alpha = 0.05$ iii. $n = 12$
iv. $\mu_a = 0.55, 0.60, 0.65, 0.70, 0.75, 0.80, 0.85$

9.192 **Grey-Seal Nursing.** The null and alternative hypotheses obtained in Exercise 9.16 on page 375 are, respectively,

$$H_0\text{: } \mu = 23 \text{ days}$$
$$H_a\text{: } \mu \neq 23 \text{ days,}$$

where μ is the mean lactation period of grey seals.

i. $\sigma = 3.0$ ii. $\alpha = 0.05$ iii. $n = 14$
iv. $\mu_a = 20.5, 21.0, 21.5, 22.0, 22.5, 23.5, 24.0, 24.5, 25.0, 25.5$

9.193 **Iron Deficiency?** The null and alternative hypotheses obtained in Exercise 9.17 on page 375 are, respectively,

$$H_0\text{: } \mu = 18 \text{ mg}$$
$$H_a\text{: } \mu < 18 \text{ mg,}$$

where μ is the mean iron intake (per day) of all adult females under the age of 51 years.
i. $\sigma = 4.2$ ii. $\alpha = 0.01$ iii. $n = 45$
iv. $\mu_a = 15.50, 15.75, 16.00, 16.25, 16.50, 16.75,$
    $17.00, 17.25, 17.50, 17.75$

9.194 **Early-Onset Dementia.** The null and alternative hypotheses obtained in Exercise 9.18 on page 375 are, respectively,

$$H_0\text{: } \mu = 55 \text{ years old}$$
$$H_a\text{: } \mu < 55 \text{ years old,}$$

where μ is the mean age at diagnosis of all people with early-onset dementia.
i. $\sigma = 6.8$ ii. $\alpha = 0.01$ iii. $n = 21$
iv. $\mu_a = 47, 48, 49, 50, 51, 52, 53, 54$

9.195 **Serving Time.** The null and alternative hypotheses obtained in Exercise 9.19 on page 375 are, respectively,

$$H_0\text{: } \mu = 16.7 \text{ months}$$
$$H_a\text{: } \mu \neq 16.7 \text{ months,}$$

where μ is the mean length of imprisonment for motor-vehicle-theft offenders in Sydney, Australia.
i. $\sigma = 6.0$ ii. $\alpha = 0.05$ iii. $n = 100$
iv. $\mu_a = 14.0, 14.5, 15.0, 15.5, 16.0, 16.5, 17.0,$
    $17.5, 18.0, 18.5, 19.0$

9.196 **Worker Fatigue.** The null and alternative hypotheses obtained in Exercise 9.20 on page 375 are, respectively,

$$H_0\text{: } \mu = 72 \text{ bpm}$$
$$H_a\text{: } \mu > 72 \text{ bpm,}$$

where μ is the mean post-work heart rate of all casting workers.
i. $\sigma = 11.2$ ii. $\alpha = 0.05$ iii. $n = 29$
iv. $\mu_a = 73, 74, 75, 76, 77, 78, 79, 80$

9.197 **Toxic Mushrooms?** Repeat parts (a)–(c) of Exercise 9.191 for a sample size of 20. Compare your power curves for the two sample sizes, and explain the principle being illustrated.

9.198 **Grey-Seal Nursing.** Repeat parts (a)–(c) of Exercise 9.192 for a sample size of 30. Compare your power curves for the two sample sizes, and explain the principle being illustrated.

9.199 **Serving Time.** Repeat parts (a)–(c) of Exercise 9.195 for a sample size of 40. Compare your power curves for the two sample sizes, and explain the principle being illustrated.

9.200 **Early-Onset Dementia.** Repeat parts (a)–(c) of Exercise 9.194 for a sample size of 15. Compare your power curves for the two sample sizes, and explain the principle being illustrated.

Extending the Concepts and Skills

9.201 **Class Project: Questioning Gas Mileage.** This exercise can be done individually or, better yet, as a class project. Refer to the gas mileage hypothesis test of Example 9.23 on page 422. Recall that the null and alternative hypotheses are

$$H_0\text{: } \mu = 26 \text{ mpg (manufacturer's claim)}$$
$$H_a\text{: } \mu < 26 \text{ mpg (consumer group's conjecture),}$$

where μ is the mean gas mileage of all Orions. Also recall that the mileages are normally distributed with a standard deviation of 1.4 mpg. Figure 9.30 on page 423 portrays the decision criterion for a test at the 5% significance level with a sample size of 30. Suppose that, in reality, the mean gas mileage of all Orions is 25.4 mpg.
a. Determine the probability of making a Type II error.
b. Simulate 100 samples of 30 gas mileages each.
c. Determine the mean of each sample in part (b).

d. For the 100 samples obtained in part (b), about how many would you expect to lead to nonrejection of the null hypothesis? Explain your answer.
e. For the 100 samples obtained in part (b), determine the number that lead to nonrejection of the null hypothesis.
f. Compare your answers from parts (d) and (e), and comment on any observed difference.

CHAPTER IN REVIEW

You Should Be Able to

1. use and understand the formulas in this chapter.

2. define and apply the terms that are associated with hypothesis testing.

3. choose the null and alternative hypotheses for a hypothesis test.

4. explain the basic logic behind hypothesis testing.

5. define and apply the concepts of Type I and Type II errors.

6. understand the relation between Type I and Type II error probabilities.

7. state and interpret the possible conclusions for a hypothesis test.

8. understand and apply the critical-value approach to hypothesis testing and/or the P-value approach to hypothesis testing.

9. perform a hypothesis test for one population mean when the population standard deviation is known.

10. perform a hypothesis test for one population mean when the population standard deviation is unknown.

*11. perform a hypothesis test for one population mean when the variable under consideration has a symmetric distribution.

*12. compute Type II error probabilities for a one-mean z-test.

*13. calculate the power of a hypothesis test.

*14. draw a power curve.

*15. understand the relationship between sample size, significance level, and power.

**16. decide which procedure should be used to perform a hypothesis test for one population mean.

Key Terms

alternative hypothesis, *368*
critical-value approach to hypothesis
   testing, *376*
critical values, *378*
hypothesis, *368*
hypothesis test, *368*
left-tailed test, *369*
nonparametric methods,* *408*
nonrejection region, *378*
not statistically significant, *374*
null hypothesis, *368*
observed significance level, *384*
one-mean t-test, *402*

one-mean z-test, *388*
one-tailed test, *369*
P-value (P), *383*
P-value approach to hypothesis
   testing, *381*
parametric methods,* *408*
power,* *426*
power curve,* *427*
rejection region, *378*
right-tailed test, *369*
significance level (α), *373*
statistically significant, *374*
symmetric population,* *411*

t-test, *399*
test statistic, *371*
two-tailed test, *369*
Type I error, *371*
Type I error probability (α), *373*
Type II error, *371*
Type II error probability (β), *373*
W_α,* *410*
Wilcoxon signed-rank test,* *412*
$\bar{x}$ critical value,* *423*
z-test, *387*

REVIEW PROBLEMS

Understanding the Concepts and Skills

1. Explain the meaning of each term.
a. null hypothesis
b. alternative hypothesis
c. test statistic
d. significance level

2. The following statement appeared on a box of Tide laundry detergent: "Individual packages of Tide may weigh slightly more or less

than the marked weight due to normal variations incurred with high speed packaging machines, but each day's production of Tide will average slightly above the marked weight."
a. Explain in statistical terms what the statement means.
b. Describe in words a hypothesis test for checking the statement.
c. Suppose that the marked weight is 76 ounces. State in words the null and alternative hypotheses for the hypothesis test. Express those hypotheses in statistical terminology.

3. Regarding a hypothesis test:
a. What is the procedure, generally, for deciding whether the null hypothesis should be rejected?
b. How can the procedure identified in part (a) be made objective and precise?

4. There are three possible alternative hypotheses in a hypothesis test for a population mean. Identify them, and explain when each is used.

5. Two types of incorrect decisions can be made in a hypothesis test: a Type I error and a Type II error.
a. Explain the meaning of each type of error.
b. Identify the letter used to represent the probability of each type of error.
c. If the null hypothesis is in fact true, only one type of error is possible. Which type is that? Explain your answer.
d. If you fail to reject the null hypothesis, only one type of error is possible. Which type is that? Explain your answer.

6. For a fixed sample size, what happens to the probability of a Type II error if the significance level is decreased from 0.05 to 0.01?

Problems 7–12 pertain to the critical-value approach to hypothesis testing.

7. Explain the meaning of each term.
a. rejection region   **b.** nonrejection region
c. critical value(s)

8. True or false: A critical value is considered part of the rejection region.

9. Suppose that you want to conduct a left-tailed hypothesis test at the 5% significance level. How must the critical value be chosen?

10. Determine the critical value(s) for a one-mean z-test at the 1% significance level if the test is
a. right tailed.   **b.** left tailed.   **c.** two tailed.

11. The following graph portrays the decision criterion for a one-mean z-test, using the critical-value approach to hypothesis testing. The curve in the graph is the normal curve for the test statistic under the assumption that the null hypothesis is true.

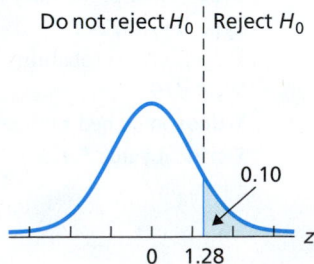

Determine the
a. rejection region.   **b.** nonrejection region.
c. critical value(s).   **d.** significance level.
e. Draw a graph that depicts the answers that you obtained in parts (a)–(d).
f. Classify the hypothesis test as two tailed, left tailed, or right tailed.

12. State the general steps of the critical-value approach to hypothesis testing.

Problems 13–20 pertain to the P-value approach to hypothesis testing.

13. Define the P-value of a hypothesis test.

14. True or false: A P-value of 0.02 provides more evidence against the null hypothesis than a P-value of 0.03. Explain your answer.

15. State the decision criterion for a hypothesis test, using the P-value.

16. Explain why the P-value of a hypothesis test is also referred to as the observed significance level.

17. How is the P-value of a hypothesis test actually determined?

18. In each part, we have given the value obtained for the test statistic, z, in a one-mean z-test. We have also specified whether the test is two tailed, left tailed, or right tailed. Determine the P-value in each case and decide whether, at the 5% significance level, the data provide sufficient evidence to reject the null hypothesis in favor of the alternative hypothesis.
a. $z = -1.25$; left-tailed test
b. $z = 2.36$; right-tailed test
c. $z = 1.83$; two-tailed test

19. State the general steps of the P-value approach to hypothesis testing.

20. Assess the evidence against the null hypothesis if the P-value of the hypothesis test is 0.062.

21. What is meant when we say that a hypothesis test is
a. exact?   **b.** approximately correct?

22. Discuss the difference between statistical significance and practical significance.

23. In each part, we have identified a hypothesis-testing procedure for one population mean. State the assumptions required and the test statistic used in each case.
a. one-mean t-test   **b.** one-mean z-test
*c. Wilcoxon signed-rank test

*24. Identify two advantages of nonparametric methods over parametric methods. When is a parametric procedure preferred? Explain your answer.

*25. Regarding the power of a hypothesis test:
a. What does it represent?
b. What happens to the power of a hypothesis test if the significance level is kept at 0.01 while the sample size is increased from 50 to 100?

Applying the Concepts and Skills

26. Cheese Consumption. The U.S. Department of Agriculture reports in *Food Consumption, Prices, and Expenditures* that the average American consumed 33 lb of cheese in 2010. Suppose that you want to decide whether last year's mean cheese consumption is greater than the 2010 mean.
a. Identify the null hypothesis.
b. Identify the alternative hypothesis.
c. Classify the hypothesis test as two tailed, left tailed, or right tailed.

27. Cheese Consumption. The null and alternative hypotheses for the hypothesis test in Problem 26 are, respectively,

$$H_0: \mu = 33 \text{ lb (mean has not increased)}$$
$$H_a: \mu > 33 \text{ lb (mean has increased)},$$

where μ is last year's mean cheese consumption for all Americans. Explain what each of the following would mean.
a. Type I error   **b.** Type II error   **c.** Correct decision

Now suppose that the results of carrying out the hypothesis test lead to rejection of the null hypothesis. Classify that decision by error type or as a correct decision if in fact last year's mean cheese consumption

d. has not increased from the 2010 mean of 33 lb.

e. has increased from the 2010 mean of 33 lb.

28. Cheese Consumption. Refer to Problem 26. The following table provides last year's cheese consumption, in pounds, for 35 randomly selected Americans.

46	29	33	38	42	40	34
33	32	36	28	47	26	42
36	32	45	24	39	28	33
44	33	26	37	27	31	36
37	37	36	22	44	36	29

a. At the 10% significance level, do the data provide sufficient evidence to conclude that last year's mean cheese consumption for all Americans has increased over the 2010 mean? Assume that $\sigma = 6.9$ lb. Use a z-test. (*Note:* The sum of the data is 1218 lb.)

b. Given the conclusion in part (a), if an error has been made, what type must it be? Explain your answer.

29. Purse Snatching. The Federal Bureau of Investigation (FBI) compiles information on robbery and property crimes by type and selected characteristic and publishes its findings in *Uniform Crime Reports*. According to that document, the mean value lost to purse snatching was $468 in 2012. For last year, 12 randomly selected purse-snatching offenses yielded the following values lost, to the nearest dollar.

415	539	365	479	375	303
572	487	550	481	371	523

Use a t-test to decide, at the 5% significance level, whether last year's mean value lost to purse snatching has decreased from the 2012 mean. The mean and standard deviation of the data are $455.0 and $86.8, respectively.

*30. Purse Snatching. Refer to Problem 29.

a. Perform the required hypothesis test, using the Wilcoxon signed-rank test.

b. In performing the hypothesis test in part (a), what assumption did you make about the distribution of last year's values lost to purse snatching?

c. In Problem 29, we used the t-test to perform the hypothesis test. The assumption in that problem is that last year's values lost to purse snatching are normally distributed. If that assumption is true, why is it permissible to perform a Wilcoxon signed-rank test for the mean value lost?

*31. Purse Snatching. Refer to Problems 29 and 30. If in fact last year's values lost to purse snatching are normally distributed, which is the preferred procedure for performing the hypothesis test—the t-test or the Wilcoxon signed-rank test? Explain your answer.

32. Betting the Spreads. College basketball, and particularly the NCAA basketball tournament, is a popular venue for gambling, from novices in office betting pools to high rollers. To encourage uniform betting across teams, Las Vegas oddsmakers assign a point spread to each game. The *point spread* is the oddsmakers' prediction for the number of points by which the favored team will win. If you bet on the favorite, you win the bet provided the favorite wins by more than the point spread; otherwise, you lose the bet. Is the point spread a

good measure of the relative ability of the two teams? H. Stern and B. Mock addressed this question in the paper "College Basketball Upsets: Will a 16-Seed Ever Beat a 1-Seed?" (*Chance*, Vol. 11(1), pp. 27–31). They obtained the difference between the actual margin of victory and the point spread, called the *point-spread error*, for 2109 college basketball games. The mean point-spread error was found to be -0.2 point with a standard deviation of 10.9 points. For a particular game, a point-spread error of 0 indicates that the point spread was a perfect estimate of the two teams' relative abilities.

a. If, on average, the oddsmakers are estimating correctly, what is the (population) mean point-spread error?

b. Use the data to decide, at the 5% significance level, whether the (population) mean point-spread error differs from 0.

c. Interpret your answer in part (b).

*33. Cheese Consumption. Refer to Problem 26. Suppose that you decide to use a z-test with a significance level of 0.10 and a sample size of 35. Assume that $\sigma = 6.9$ lb.

a. Determine the probability of a Type I error.

b. If last year's mean cheese consumption was 36.5 lb, identify the distribution of the variable $\bar{x}$, that is, the sampling distribution of the mean for samples of size 35.

c. Use part (b) to determine the probability, β, of a Type II error if in fact last year's mean cheese consumption was 36.5 lb.

d. Repeat parts (b) and (c) if in fact last year's mean cheese consumption was 33.5 lb, 34.0 lb, 34.5 lb, 35.0 lb, 35.5 lb, 36.0 lb, and 37.0 lb.

e. Use your answers from parts (c) and (d) to construct a table of selected Type II error probabilities and powers.

f. Use your answer from part (e) to construct the power curve.

Using a sample size of 60 instead of 35, repeat

g. part (b).   **h.** part (c).   **i.** part (d).

j. part (e).   **k.** part (f).

l. Compare your power curves for the two sample sizes and explain the principle being illustrated.

Problems 34 and 35 each include a normal probability plot and either a frequency histogram or a stem-and-leaf diagram for a set of sample data. The intent is to use the sample data to perform a hypothesis test for the mean of the population from which the data were obtained. In each case, consult the graphs provided to decide whether to use the z-test, the t-test, or neither. Explain your answer.

34. The normal probability plot and histogram of the data are depicted in Fig. 9.49 on the next page; σ is known.

35. The normal probability plot and stem-and-leaf diagram of the data are depicted in Fig. 9.50 on the next page; σ is unknown.

*36. Refer to Problems 34 and 35.

a. In each case, consult the appropriate graphs to decide whether using the Wilcoxon signed-rank test is reasonable for performing a hypothesis test for the mean of the population from which the data were obtained. Give reasons for your answers.

b. For each case where using either the z-test or the t-test is reasonable and where using the Wilcoxon signed-rank test is also appropriate, decide which test is preferable. Give reasons for your answers.

*37. Nursing-Home Costs. According to the document *Market Survey of Long-Term Care Costs*, published by the MetLife Mature Market Institute, the average cost of a private room in a nursing home was $239 per day in 2011. A random sample of 11 nursing homes yielded

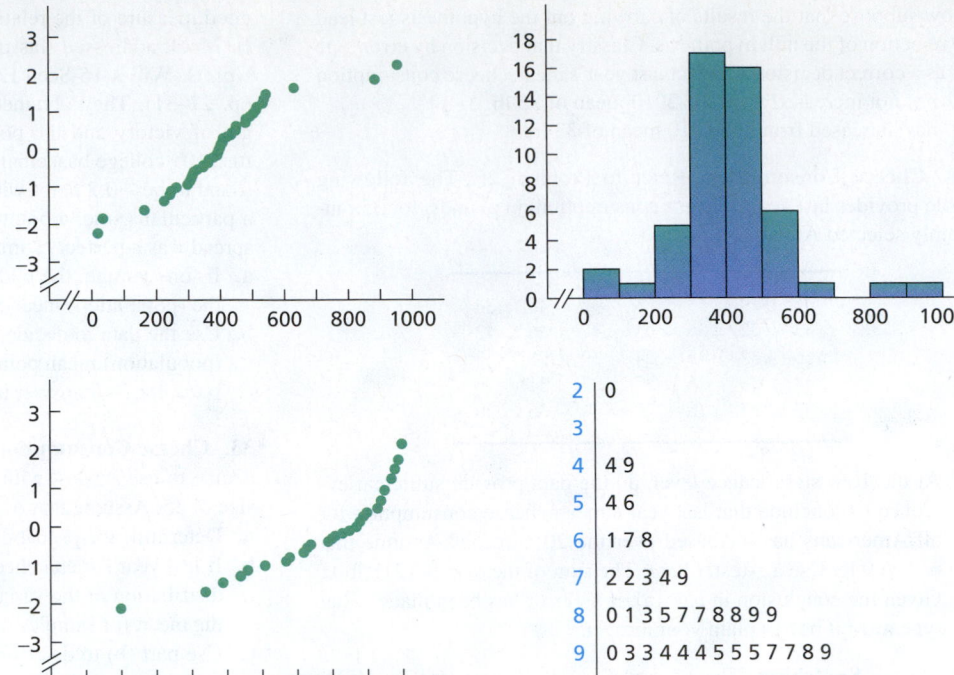

FIGURE 9.49

Normal probability plot and histogram for Problem 34

FIGURE 9.50

Normal probability plot and stem-and-leaf diagram for Problem 35

the following daily costs, in dollars, for a private room in a nursing home for this year.

265	283	266	366	243	334
213	266	140	276	292	

Use the technology of your choice to perform the following tasks.

a. Apply the *t*-test to decide at the 5% significance level whether this year's average cost for a private room in a nursing home exceeds that in 2011.

b. Repeat part (a) by using the Wilcoxon signed-rank test.

c. Obtain a normal probability plot, a boxplot, a stem-and-leaf diagram, and a histogram of the sample data.

d. Discuss the discrepancy in results between the *t*-test and the Wilcoxon signed-rank test.

Working with Large Data Sets

38. Beef Consumption. According to *Food Consumption, Prices, and Expenditures*, published by the U.S. Department of Agriculture, the mean consumption of beef per person in 2011 was 57.5 lb. A sample of 40 people taken this year yielded the data, in pounds, on last year's beef consumption given on the WeissStats site. Use the technology of your choice to do the following.

a. Obtain a normal probability plot, a boxplot, a histogram, and a stem-and-leaf diagram of the data on beef consumptions.

b. Decide, at the 5% significance level, whether last year's mean beef consumption is less than the 2011 mean of 57.5 lb. Apply the one-mean *t*-test.

c. The sample data contain four potential outliers: 0, 0, 0, and 13. Remove those four observations, repeat the hypothesis test in part (b), and compare your result with that obtained in part (b).

d. Assuming that the four potential outliers are not recording errors, comment on the advisability of removing them from the sample data before performing the hypothesis test.

e. What action would you take regarding this hypothesis test?

*39. Beef Consumption. Use the technology of your choice to do the following.

a. Repeat parts (b) and (c) of Problem 38 by using the Wilcoxon signed-rank test.

b. Compare your results from part (a) with those in Problem 38.

c. Discuss the reasonableness of using the Wilcoxon signed-rank test here.

40. Body Mass Index. Body mass index (BMI) is a measure of body fat based on height and weight. According to *Dietary Guidelines for Americans*, published by the U.S. Department of Agriculture and the U.S. Department of Health and Human Services, for adults, a BMI of greater than 25 indicates an above healthy weight (i.e., overweight or obese). The BMIs of 75 randomly selected U.S. adults provided the data on the WeissStats site. Use the technology of your choice to do the following.

a. Obtain a normal probability plot, a boxplot, and a histogram of the data.

b. Based on your graphs from part (a), is it reasonable to apply the one-mean *z*-test to the data? Explain your answer.

c. At the 5% significance level, do the data provide sufficient evidence to conclude that the average U.S. adult has an above healthy weight? Apply the one-mean *z*-test, assuming a standard deviation of 5.0 for the BMIs of all U.S. adults.

41. Beer Drinking. According to the *Beer Institute Annual Report*, the mean annual consumption of beer per person in the United States is 28.2 gallons (roughly 300 twelve-ounce bottles). A random sample of 300 Missouri residents yielded the annual beer consumptions provided on the WeissStats site. Use the technology of your choice to do the following.

a. Obtain a histogram of the data.

b. Does your histogram in part (a) indicate any outliers?

c. At the 1% significance level, do the data provide sufficient evidence to conclude that the mean annual consumption of beer per person in Missouri differs from the national mean? (*Note:* See the third bulleted item in Key Fact 9.7 on page 389.)

FOCUSING ON DATA ANALYSIS

UWEC UNDERGRADUATES

Recall from Chapter 1 (see page 34) that the Focus database and Focus sample contain information on the undergraduate students at the University of Wisconsin—Eau Claire (UWEC). Now would be a good time for you to review the discussion about these data sets.

According to *ACT High School Profile Report*, published by ACT, Inc., the national means for ACT composite, English, and math scores are 21.1, 20.6, and 21.0, respectively. You will use these national means in the following problems.

a. Apply the one-mean *t*-test to the ACT composite score data in the Focus sample (FocusSample) to decide, at the 5% significance level, whether the mean ACT composite score of UWEC undergraduates exceeds the national mean of 21.1 points.

b. In practice, the population mean of the variable under consideration is unknown. However, in this case, we actually do have the population data, namely, in the Focus database (Focus). If your statistical software package will accommodate the entire Focus database, open that worksheet and then obtain the mean ACT composite score of all UWEC undergraduate students. (*Answer:* 23.6)

c. Was the decision concerning the hypothesis test in part (a) correct? Would it necessarily have to be? Explain your answers.

d. Repeat parts (a)–(c) for ACT English scores. (*Note:* The mean ACT English score of all UWEC undergraduate students is 23.0.)

e. Repeat parts (a)–(c) for ACT math scores. (*Note:* The mean ACT math score of all UWEC undergraduate students is 23.5.)

CASE STUDY DISCUSSION

GENDER AND SENSE OF DIRECTION

At the beginning of this chapter, we discussed research by J. Sholl et al. on the relationship between gender and sense of direction. Recall that, in their study, the spatial orientation skills of 30 male and 30 female students were challenged in a wooded park near the Boston College campus in Newton, Massachusetts. The participants were asked to rate their own sense of direction as either good or poor.

In the park, students were instructed to point to predesignated landmarks and also to the direction of south. For the female students who had rated their sense of direction to be good, the table on page 368 provides the pointing errors (in degrees) when they attempted to point south.

a. If, on average, women who consider themselves to have a good sense of direction do no better than they would by just randomly guessing at the direction of south, what would their mean pointing error be?

b. At the 1% significance level, do the data provide sufficient evidence to conclude that women who consider themselves to have a good sense of direction really do better, on average, than they would by just randomly guessing at the direction of south? Use a one-mean *t*-test.

c. Obtain a normal probability plot, boxplot, and stem-and-leaf diagram of the data. Based on these plots, is use of the *t*-test reasonable? Explain your answer.

d. Use the technology of your choice to perform the data analyses in parts (b) and (c).

*e. Solve part (b) by using the Wilcoxon signed-rank test.

*f. Based on the plots you obtained in part (c), is use of the Wilcoxon signed-rank test reasonable? Explain your answer.

*g. Use the technology of your choice to perform the required Wilcoxon signed-rank test of part (e).

BIOGRAPHY

JERZY NEYMAN: A PRINCIPAL FOUNDER OF MODERN STATISTICAL THEORY

Jerzy Neyman was born on April 16, 1894, in Bendery, Russia. His father, Czeslaw, was a member of the Polish nobility, a lawyer, a judge, and an amateur archaeologist. Because Russian authorities prohibited the family from living in Poland, Jerzy Neyman grew up in various cities in Russia. He entered the university in Kharkov in 1912. At Kharkov he was at first interested in physics, but, because of his clumsiness in the laboratory, he decided to pursue mathematics.

After World War I, when Russia was at war with Poland over borders, Neyman was jailed as an enemy alien. In 1921, as a result of a prisoner exchange, he went to Poland for the first time. In 1924, he received his doctorate from the University of Warsaw. Between 1924 and 1934, Neyman worked with Karl Pearson (see Biography in Chapter 13) and his son Egon Pearson and held a position at the University of Kraków. In 1934, Neyman took a position in Karl Pearson's statistical laboratory at University College in London. He stayed in England, where he worked with Egon

Pearson until 1938, at which time he accepted an offer to join the faculty at the University of California at Berkeley.

When the United States entered World War II, Neyman set aside development of a statistics program and did war work. After the war ended, Neyman organized a symposium to celebrate its end and "the return to theoretical research." That symposium, held in August 1945, and succeeding ones, held every 5 years until 1970, were instrumental in establishing Berkeley as a preeminent statistical center.

Neyman was a principal founder of the theory of modern statistics. His work on hypothesis testing, confidence intervals, and survey sampling transformed both the theory and the practice of statistics. His achievements were acknowledged by the granting of many honors and awards, including election to the U.S. National Academy of Sciences and receiving the Guy Medal in Gold of the Royal Statistical Society and the U.S. National Medal of Science.

Neyman remained active until his death of heart failure on August 5, 1981, at the age of 87, in Oakland, California.

Inferences for Two Population Means

CHAPTER OBJECTIVES

In Chapters 8 and 9, you learned how to obtain confidence intervals and perform hypothesis tests for one population mean. Frequently, however, inferential statistics is used to compare the means of two or more populations.

For example, we might want to perform a hypothesis test to decide whether the mean age of buyers of new domestic cars is greater than the mean age of buyers of new imported cars, or we might want to find a confidence interval for the difference between the two mean ages.

Broadly speaking, in this chapter we examine two types of inferential procedures for comparing the means of two populations. The first type applies when the samples from the two populations are *independent*, meaning that the sample selected from one of the populations has no effect or bearing on the sample selected from the other population.

The second type of inferential procedure for comparing the means of two populations applies when the samples from the two populations are *paired*. A paired sample may be appropriate when there is a natural pairing of the members of the two populations such as husband and wife.

CASE STUDY

Dexamethasone Therapy and IQ

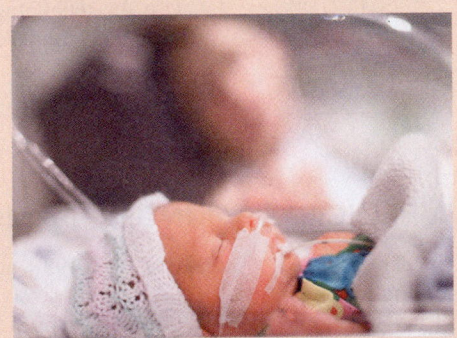

In the paper "Outcomes at School Age after Postnatal Dexamethasone Therapy for Lung Disease of Prematurity" (*New England Journal of Medicine*, Vol. 350, No. 13, pp. 1304–1313), T. Yeh et al. studied several characteristics at school age of children who as premature infants participated in a double-blind, placebo-controlled trial of early

postnatal dexamethasone therapy for the prevention of chronic lung disease of prematurity.

The study included 146 Chinese children of which 72 had been in the dexamethasone group and 74 had been in the control group. All the children in both groups had severe respiratory distress syndrome (RDS) as infants, due to insufficiently developed lungs. Infants in the dexamethasone group were given 0.25 mg of dexamethasone per kilogram of body weight intravenously every 12 hours for 1 week, and subsequently the dose was tapered off.

At school age, the researchers evaluated the children's growth, neurologic and motor function, cognition, and school performance. Their conclusions were that "Early postnatal dexamethasone therapy should not be recommended for the routine prevention or treatment of chronic lung disease, because it leads to substantial adverse effects on neuromotor and cognitive function at school age."

In this case study, we will consider only a few of the many characteristics measured by the researchers, namely, those related to cognition.

The researchers employed the Wechsler Intelligence Scale for Children, Third Edition (WISC-III). Following are the data obtained by the researchers. The researchers noted that the WISC-III scores found here were lower than those that have been reported in other studies and stated that "We did not have an established standard for Chinese children; the racial, ethnic, or cultural bias of the tests might explain the low scores...."

Main scale	Dexamethasone group (n = 72)		Control group (n = 74)	
	Mean	Std. dev.	Mean	Std. dev.
Full IQ	78.2	15.0	84.4	12.6
Verbal IQ	84.1	13.2	88.4	11.8
Performance IQ	76.5	14.6	84.5	12.7
Subscale				
Verbal comprehension	84.9	13.1	87.6	12.9
Perceptual organization	78.0	16.5	85.4	12.4
Freedom from distractibility	86.8	14.8	96.8	13.4
Processing speed	83.2	16.8	91.0	15.9

At the end of this chapter, after you have studied inferences for two population means, you will be asked to analyze and interpret these data in terms of the effects of dexamethasone in this context.

10.1   The Sampling Distribution of the Difference between Two Sample Means for Independent Samples

In this section, we lay the groundwork for making statistical inferences to compare the means of two populations. The methods that we first consider require not only that the samples selected from the two populations be simple random samples, but also that they be **independent samples.** That is, the sample selected from one of the populations has no effect or bearing on the sample selected from the other population.

With **independent simple random samples,** each possible pair of samples (one from one population and one from the other) is equally likely to be the pair of samples selected. Example 10.1 provides an unrealistically simple illustration of independent samples, but it will help you understand the concept.

EXAMPLE 10.1   **Introducing Independent Random Samples**

Males and Females Let's consider two small populations, one consisting of three men and the other of four women, as shown in the following figure.

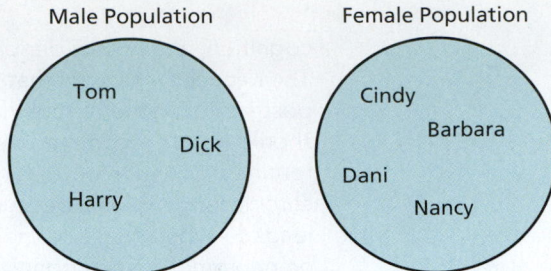

Suppose that we take a sample of size 2 from the male population and a sample of size 3 from the female population.

a. List the possible pairs of independent samples.
b. If the samples are selected at random, determine the chance of obtaining any particular pair of independent samples.

Solution For convenience, we use the first letter of each name as an abbreviation for the actual name.

a. In Table 10.1, the possible samples of size 2 from the male population are listed on the left; the possible samples of size 3 from the female population are listed on the right. To obtain the possible pairs of independent samples, we list each possible male sample of size 2 with each possible female sample of size 3, as shown in Table 10.2. There are 12 possible pairs of independent samples of two men and three women.

TABLE 10.1

Possible samples of size 2 from the male population and possible samples of size 3 from the female population

Male sample of size 2	Female sample of size 3
T, D	C, B, D
T, H	C, B, N
D, H	C, D, N
	B, D, N

TABLE 10.2

Possible pairs of independent samples of two men and three women

Male sample of size 2	Female sample of size 3
T, D	C, B, D
T, D	C, B, N
T, D	C, D, N
T, D	B, D, N
T, H	C, B, D
T, H	C, B, N
T, H	C, D, N
T, H	B, D, N
D, H	C, B, D
D, H	C, B, N
D, H	C, D, N
D, H	B, D, N

b. For independent simple random samples, each of the 12 possible pairs of samples shown in Table 10.2 is equally likely to be the pair selected. Therefore the chance of obtaining any particular pair of independent samples is $\frac{1}{12}$. ■

The previous example provides a concrete illustration of independent samples and emphasizes that, for independent simple random samples of any given sizes, each possible pair of independent samples is equally likely to be the one selected. In practice, we neither obtain the number of possible pairs of independent samples nor explicitly compute the chance of selecting a particular pair of independent samples. But these concepts underlie the methods we do use.

Note: Recall that, when we say *random sample,* we mean *simple random sample* unless specifically stated otherwise. Likewise, when we say *independent random samples,* we mean *independent simple random samples,* unless specifically stated otherwise.

Comparing Two Population Means, Using Independent Samples

We can now examine the process for comparing the means of two populations based on independent samples.

■■■■   **EXAMPLE 10.2**   ### Comparing Two Population Means, Using Independent Samples

Faculty Salaries The American Association of University Professors (AAUP) conducts salary studies of college professors and publishes its findings in *AAUP Annual Report on the Economic Status of the Profession.* Suppose that we want to decide whether the mean salaries of college faculty in private and public institutions are different.

a. Pose the problem as a hypothesis test.
b. Explain the basic idea for carrying out the hypothesis test.
c. Suppose that 35 faculty members from private institutions and 30 faculty members from public institutions are randomly and independently selected and that their salaries are as shown in Table 10.3, in thousands of dollars rounded to the nearest hundred. Discuss the use of these data to make a decision concerning the hypothesis test.

TABLE 10.3
Annual salaries ($1000s) for 35 faculty members in private institutions and 30 faculty members in public institutions

Sample 1 (private institutions)							Sample 2 (public institutions)					
97.3	85.9	118.8	93.9	66.6	109.2	64.9	59.9	115.7	126.1	50.3	133.1	89.3
83.1	100.6	99.3	94.9	94.4	139.3	108.8	82.5	67.1	60.7	79.9	50.1	81.7
158.1	142.4	85.0	108.2	116.3	141.5	51.4	83.9	102.5	109.9	105.1	67.9	107.5
125.6	70.6	74.6	69.9	115.4	84.6	92.0	54.9	41.5	59.5	65.9	76.9	66.9
97.2	55.1	126.6	116.7	76.0	109.6	63.0	85.9	113.9	70.3	90.1	99.7	96.7

Solution

a. We first note that we have one variable (salary) and two populations (all faculty in private institutions and all faculty in public institutions). Let the two populations in question be designated Populations 1 and 2, respectively:

> Population 1: All faculty in private institutions
> Population 2: All faculty in public institutions.

Next, we denote the means of the variable "salary" for the two populations by μ_1 and μ_2, respectively:

> μ_1 = mean salary of all faculty in private institutions;
> μ_2 = mean salary of all faculty in public institutions.

Then, we can state the hypothesis test we want to perform as

> H_0: $\mu_1 = \mu_2$ (mean salaries are the same)
> H_a: $\mu_1 \neq \mu_2$ (mean salaries are different).

b. Roughly speaking, we can carry out the hypothesis test as follows.

1. Independently and randomly take a sample of faculty members from private institutions (Population 1) and a sample of faculty members from public institutions (Population 2).

2. Compute the mean salary, $\bar{x}_1$, of the sample from private institutions and the mean salary, $\bar{x}_2$, of the sample from public institutions.

3. Reject the null hypothesis if the sample means, $\bar{x}_1$ and $\bar{x}_2$, differ by too much; otherwise, do not reject the null hypothesis.

This process is depicted in Fig. 10.1.

FIGURE 10.1

Process for comparing two population means, using independent samples

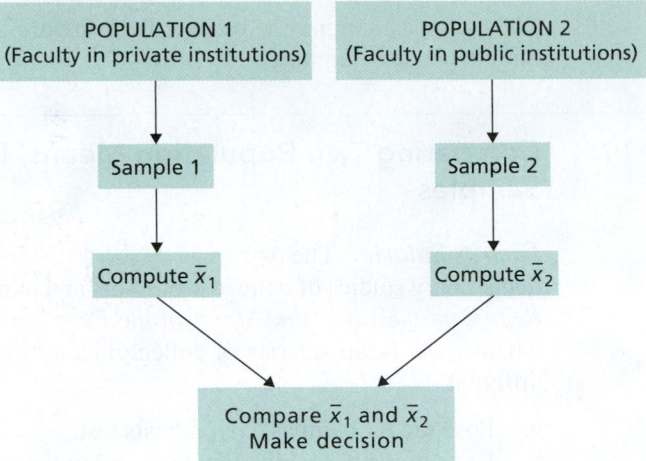

c. The means of the two samples in Table 10.3 are, respectively,

$$\bar{x}_1 = \frac{\sum x_i}{n_1} = \frac{3436.8}{35} = 98.19 \quad \text{and} \quad \bar{x}_2 = \frac{\sum x_i}{n_2} = \frac{2495.4}{30} = 83.18.$$

The question now is, can the difference of 15.01 ($15,010) between these two sample means reasonably be attributed to sampling error, or is the difference large enough to indicate that the two populations have different means? To answer that question, we need to know the distribution of the difference between two sample means—the *sampling distribution of the difference between two sample means*. We examine that sampling distribution in this section and complete the hypothesis test in the next section.

Choosing the Hypotheses

As illustrated in the preceding example, for a variable of two populations, say, Population 1 and Population 2, we denote the means of that variable on those two populations by μ_1 and μ_2, respectively. With this notation in mind, we now present some guidelines for choosing the null and alternative hypotheses in a hypothesis test to compare two population means.

Null Hypothesis

In this book, the null hypothesis for a hypothesis test to compare two population means is that the two population means are *equal*. Hence, we can express the null hypothesis as

$$H_0: \mu_1 = \mu_2.$$

Alternative Hypothesis

The choice of the alternative hypothesis depends on and should reflect the purpose of the hypothesis test. Three choices are possible for the alternative hypothesis.

*   If the primary concern is deciding whether the two population means are *different*, we express the alternative hypothesis as

$$H_a: \mu_1 \neq \mu_2,$$

which is a two-tailed test.

- If the primary concern is deciding whether the mean of Population 1 is *less than* the mean of Population 2, we express the alternative hypothesis as

$$H_a: \mu_1 < \mu_2,$$

which is a left-tailed test.

- If the primary concern is deciding whether the mean of Population 1 is *greater than* the mean of Population 2, we express the alternative hypothesis as

$$H_a: \mu_1 > \mu_2,$$

which is a right-tailed test.

You try it!

Exercise 10.23 on page 445

Comparing Two Population Means with Confidence Intervals

We can also compare two population means by finding a confidence interval for the difference between them. One important aspect of that inference is the interpretation of the confidence interval.

To interpret confidence intervals for the difference, $\mu_1 - \mu_2$, between the two population means, considering three cases is helpful.

Case 1: The endpoints of the confidence interval are both positive numbers.

To illustrate, suppose that a 95% confidence interval for $\mu_1 - \mu_2$ is from 3 to 5. Then we can be 95% confident that $\mu_1 - \mu_2$ lies somewhere between 3 and 5. Equivalently, we can be 95% confident that μ_1 is somewhere between 3 and 5 greater than μ_2.

Case 2: The endpoints of the confidence interval are both negative numbers.

To illustrate, suppose that a 95% confidence interval for $\mu_1 - \mu_2$ is from -5 to -3. Then we can be 95% confident that $\mu_1 - \mu_2$ lies somewhere between -5 and -3. Equivalently, we can be 95% confident that μ_1 is somewhere between 3 and 5 less than μ_2.

Case 3: One endpoint of the confidence interval is negative and the other is positive.

To illustrate, suppose that a 95% confidence interval for $\mu_1 - \mu_2$ is from -3 to 5. Then we can be 95% confident that $\mu_1 - \mu_2$ lies somewhere between -3 and 5. Equivalently, we can be 95% confident that μ_1 is somewhere between 3 less than and 5 more than μ_2.

You try it!

Exercise 10.13 on page 445

We present real examples throughout the chapter to further help you understand how to interpret confidence intervals for the difference between two population means. For instance, in the next section, we find and interpret a 95% confidence interval for the difference between the mean salaries of faculty in private and public institutions.

The Sampling Distribution of the Difference between Two Sample Means for Independent Samples

We need to discuss the notation used for parameters and statistics when we are analyzing two populations. Let's call the two populations Population 1 and Population 2. Then, as indicated previously, we use a subscript 1 when referring to parameters or statistics for Population 1 and a subscript 2 when referring to them for Population 2. See Table 10.4.

TABLE 10.4

Notation for parameters and statistics when considering two populations

	Population 1	Population 2
Population mean	μ_1	μ_2
Population standard deviation	σ_1	σ_2
Sample mean	$\bar{x}_1$	$\bar{x}_2$
Sample standard deviation	s_1	s_2
Sample size	n_1	n_2

Armed with this notation, we describe in Key Fact 10.1 the **sampling distribution of the difference between two sample means.** Understanding Key Fact 10.1 is aided by recalling Key Fact 7.2 on page 320.

KEY FACT 10.1

The Sampling Distribution of the Difference between Two Sample Means for Independent Samples

Suppose that x is a normally distributed variable on each of two populations. Then, for independent samples of sizes n_1 and n_2 from the two populations,

- $\mu_{\bar{x}_1 - \bar{x}_2} = \mu_1 - \mu_2$,
- $\sigma_{\bar{x}_1 - \bar{x}_2} = \sqrt{(\sigma_1^2/n_1) + (\sigma_2^2/n_2)}$, and
- $\bar{x}_1 - \bar{x}_2$ is normally distributed.

In words, the first bulleted item says that the mean of all possible differences between the two sample means equals the difference between the two population means (i.e., the difference between sample means is an unbiased estimator of the difference between population means). The second bulleted item indicates that the standard deviation of all possible differences between the two sample means equals the square root of the sum of the population variances each divided by the corresponding sample size.

The formulas for the mean and standard deviation of $\bar{x}_1 - \bar{x}_2$ given in the first and second bulleted items, respectively, hold regardless of the distributions of the variable on the two populations. The assumption that the variable is normally distributed on each of the two populations is needed only to conclude that $\bar{x}_1 - \bar{x}_2$ is normally distributed (third bulleted item) and, because of the central limit theorem, that too holds approximately for large samples, regardless of distribution type.

Under the conditions of Key Fact 10.1, the standardized version of $\bar{x}_1 - \bar{x}_2$,

$$z = \frac{(\bar{x}_1 - \bar{x}_2) - (\mu_1 - \mu_2)}{\sqrt{(\sigma_1^2/n_1) + (\sigma_2^2/n_2)}},$$

has the standard normal distribution. Using this fact, we can develop hypothesis-testing and confidence-interval procedures for comparing two population means when the population standard deviations are known.[†] However, because population standard deviations are usually unknown, we won't discuss those procedures. Instead, in Sections 10.2 and 10.3, we concentrate on the more usual situation where the population standard deviations are unknown.

Exercises 10.1

Understanding the Concepts and Skills

10.1 Give an example of interest to you for comparing two population means. Identify the variable under consideration and the two populations.

10.2 Define the phrase *independent samples*.

10.3 Consider the quantities $\mu_1, \sigma_1, \bar{x}_1, s_1, \mu_2, \sigma_2, \bar{x}_2,$ and s_2.
a. Which quantities represent parameters and which represent statistics?
b. Which quantities are fixed numbers and which are variables?

10.4 Discuss the basic strategy for performing a hypothesis test to compare the means of two populations, based on independent samples.

In each of Exercises 10.5–10.10, we have stated the purpose for performing a hypothesis test to compare the means, μ_1 and μ_2, of two populations. For each exercise,
a. *determine the null and alternative hypotheses. Note: Always place the mean of Population 1 on the left.*
b. *classify the hypothesis test as two tailed, left tailed, or right tailed.*

[†]We call these procedures the **two-means z-test** and the **two-means z-interval procedure,** respectively. The two-means z-test is also known as the **two-sample z-test** and the **two-variable z-test.** Likewise, the two-means z-interval procedure is also known as the **two-sample z-interval procedure** and the **two-variable z-interval procedure.**

10.5 The primary concern is deciding whether the mean of Population 1 differs from the mean of Population 2.

10.6 The primary concern is deciding whether the mean of Population 2 is less than the mean of Population 1.

10.7 The primary concern is deciding whether the mean of Population 1 is greater than the mean of Population 2.

10.8 The primary concern is deciding whether the mean of Population 2 is greater than the mean of Population 1.

10.9 The primary concern is deciding whether the mean of Population 1 is less than the mean of Population 2.

10.10 The primary concern is deciding whether the mean of Population 2 differs from the mean of Population 1.

10.11 Why do you need to know the sampling distribution of the difference between two sample means in order to perform a hypothesis test to compare two population means?

10.12 Identify the assumption for using the two-means z-test and the two-means z-interval procedure that renders those procedures generally impractical.

In each of Exercises 10.13–10.18, we have presented a confidence interval (CI) for the difference, $\mu_1 - \mu_2$, between two population means. Interpret each confidence interval.

10.13 95% CI is from 15 to 20.

10.14 95% CI is from -20 to -15.

10.15 90% CI is from -10 to -5.

10.16 90% CI is from 5 to 10.

10.17 99% CI is from -20 to 15.

10.18 99% CI is from -10 to 5.

10.19 A variable of two populations has a mean of 40 and a standard deviation of 12 for one of the populations and a mean of 40 and a standard deviation of 6 for the other population.
a. For independent samples of sizes 9 and 4, respectively, find the mean and standard deviation of $\bar{x}_1 - \bar{x}_2$.
b. Must the variable under consideration be normally distributed on each of the two populations for you to answer part (a)? Explain your answer.
c. Can you conclude that the variable $\bar{x}_1 - \bar{x}_2$ is normally distributed? Explain your answer.

10.20 A variable of two populations has a mean of 7.9 and a standard deviation of 5.4 for one of the populations and a mean of 7.1 and a standard deviation of 4.6 for the other population.
a. For independent samples of sizes 3 and 6, respectively, find the mean and standard deviation of $\bar{x}_1 - \bar{x}_2$.
b. Must the variable under consideration be normally distributed on each of the two populations for you to answer part (a)? Explain your answer.
c. Can you conclude that the variable $\bar{x}_1 - \bar{x}_2$ is normally distributed? Explain your answer.

10.21 A variable of two populations has a mean of 40 and a standard deviation of 12 for one of the populations and a mean of 40 and a standard deviation of 6 for the other population. Moreover, the variable is normally distributed on each of the two populations.
a. For independent samples of sizes 9 and 4, respectively, determine the mean and standard deviation of $\bar{x}_1 - \bar{x}_2$.
b. Can you conclude that the variable $\bar{x}_1 - \bar{x}_2$ is normally distributed? Explain your answer.

c. Determine the percentage of all pairs of independent samples of sizes 9 and 4, respectively, from the two populations with the property that the difference $\bar{x}_1 - \bar{x}_2$ between the sample means is between -10 and 10.

10.22 A variable of two populations has a mean of 7.9 and a standard deviation of 5.4 for one of the populations and a mean of 7.1 and a standard deviation of 4.6 for the other population. Moreover, the variable is normally distributed on each of the two populations.
a. For independent samples of sizes 3 and 6, respectively, determine the mean and standard deviation of $\bar{x}_1 - \bar{x}_2$.
b. Can you conclude that the variable $\bar{x}_1 - \bar{x}_2$ is normally distributed? Explain your answer.
c. Determine the percentage of all pairs of independent samples of sizes 4 and 16, respectively, from the two populations with the property that the difference $\bar{x}_1 - \bar{x}_2$ between the sample means is between -3 and 4.

Applying the Concepts and Skills

10.23 Faculty Salaries. Suppose, in Example 10.2 on page 441, you want to decide whether the mean salary of faculty in private institutions is greater than the mean salary of faculty in public institutions. State the null and alternative hypotheses for that hypothesis test.

10.24 Faculty Salaries. Suppose, in Example 10.2 on page 441, you want to decide whether the mean salary of faculty in private institutions is less than the mean salary of faculty in public institutions. State the null and alternative hypotheses for that hypothesis test.

In Exercises 10.25–10.30, hypothesis tests are proposed. For each hypothesis test,
a. identify the variable.
b. identify the two populations.
c. determine the null and alternative hypotheses.
d. classify the hypothesis test as two tailed, left tailed, or right tailed.

10.25 Children of Diabetic Mothers. Samples of adolescent offspring of diabetic mothers (ODM) and nondiabetic mothers (ONM) were taken by N. Cho et al. and evaluated for potential differences in vital measurements, including blood pressure and glucose tolerance. The study was published in the paper "Correlations Between the Intrauterine Metabolic Environment and Blood Pressure in Adolescent Offspring of Diabetic Mothers" (*Journal of Pediatrics*, Vol. 136, Issue 5, pp. 587–592). A hypothesis test is to be performed to decide whether the mean systolic blood pressure of ODM adolescents exceeds that of ONM adolescents.

10.26 Teaching Duties. Contingent faculty members in higher education are non-tenure track faculty, adjuncts, postdocs, lecturers, or instructors. R. Bowden and L. Gonzalez researched whether contingent faculty members are different from tenure-track faculty members with regards to teaching, research, and service in the article "The Rise of Contingent Faculty: Its Impact on the Professoriate and Higher Education" (*Journal of Applied Research in Higher Education*, Vol. 4, No. 1, pp. 5–22). A hypothesis test was conducted to decide whether the mean number of classes taught for credit per semester was less for contingent faculty than for tenure-track faculty.

10.27 Driving Distances. Data on household vehicle miles of travel (VMT) are compiled annually by the Federal Highway Administration and are published in *National Household Travel Survey,*

Summary of Travel Trends. A hypothesis test is to be performed to decide whether a difference exists in last year's mean VMT for households in the Midwest and South.

10.28 Presidential-Election Commercials. Television commercials are becoming increasingly important and prevalent in presidential elections. A study by D. Lowry and M. Naser examined whether rhetoric in presidential TV commercials is different between the winners and losers. The researchers published their results in the article "From Eisenhower to Obama: Lexical Characteristics of Winning versus Losing Presidential Campaign Commercials" (*Journalism & Mass Communication Quarterly*, Vol. 87, Nos. 3/4, pp. 530–547). A hypothesis test was conducted to see whether, on average, the loser's TV-commercial rhetoric was more frequent on terms that were self-related (I/me/my words) than that of the winner's commercials.

10.29 Neurosurgery Operative Times. An Arizona State University professor, R. Jacobowitz, Ph.D., in consultation with G. Vishteh, M.D., and other neurosurgeons obtained data on operative times, in minutes, for both a dynamic system (Z-plate) and a static system (ALPS plate). They wanted to perform a hypothesis test to decide whether the mean operative time is less with the dynamic system than with the static system.

10.30 Wing Length. D. Cristol et al. published results of their studies of two subspecies of dark-eyed juncos in the paper "Migratory Dark-Eyed Juncos, *Junco hyemalis*, Have Better Spatial Memory and Denser Hippocampal Neurons Than Nonmigratory Conspecifics" (*Animal Behaviour*, Vol. 66, Issue 2, pp. 317–328). One of the subspecies migrates each year, and the other does not migrate. A hypothesis test is to be performed to decide whether the mean wing lengths for the two subspecies (migratory and nonmigratory) are different.

Extending the Concepts and Skills

10.31 Simulation. To obtain the sampling distribution of the difference between two sample means for independent samples, as stated in Key Fact 10.1 on page 444, we need to know that, for independent observations, the difference of two normally distributed variables is also a normally distributed variable. In this exercise, you are to perform a computer simulation to make that fact plausible.

a. Simulate 2000 observations from a normally distributed variable with a mean of 100 and a standard deviation of 16.
b. Repeat part (a) for a normally distributed variable with a mean of 120 and a standard deviation of 12.
c. Determine the difference between each pair of observations in parts (a) and (b).
d. Obtain a histogram of the 2000 differences found in part (c). Why is the histogram bell shaped?

10.32 Simulation. In this exercise, you are to perform a computer simulation to illustrate the sampling distribution of the difference between two sample means for independent samples, Key Fact 10.1 on page 444.

a. Simulate 1000 samples of size 12 from a normally distributed variable with a mean of 640 and a standard deviation of 70. Obtain the sample mean of each of the 1000 samples.
b. Simulate 1000 samples of size 15 from a normally distributed variable with a mean of 715 and a standard deviation of 150. Obtain the sample mean of each of the 1000 samples.
c. Obtain the difference, $\bar{x}_1 - \bar{x}_2$, for each of the 1000 pairs of sample means obtained in parts (a) and (b).
d. Obtain the mean, the standard deviation, and a histogram of the 1000 differences found in part (c).
e. Theoretically, what are the mean, standard deviation, and distribution of all possible differences, $\bar{x}_1 - \bar{x}_2$?
f. Compare your answers from parts (d) and (e).

10.2   Inferences for Two Population Means, Using Independent Samples: Standard Deviations Assumed Equal[†]

In Section 10.1, we laid the groundwork for developing inferential methods to compare the means of two populations based on independent samples. In this section, we develop such methods when the two populations have equal standard deviations; in Section 10.3, we develop such methods without that requirement.

Hypothesis Tests for the Means of Two Populations with Equal Standard Deviations, Using Independent Samples

We now develop a procedure for performing a hypothesis test based on independent samples to compare the means of two populations with equal but unknown standard deviations. We must first find a test statistic for this test. In doing so, we assume that the variable under consideration is normally distributed on each population.

Let's use σ to denote the common standard deviation of the two populations. We know from Key Fact 10.1 on page 444 that, for independent samples, the standardized version of $\bar{x}_1 - \bar{x}_2$,

$$z = \frac{(\bar{x}_1 - \bar{x}_2) - (\mu_1 - \mu_2)}{\sqrt{(\sigma_1^2/n_1) + (\sigma_2^2/n_2)}},$$

[†]We recommend covering the pooled *t*-procedures discussed in this section because they provide valuable motivation for one-way ANOVA.

has the standard normal distribution. Replacing σ_1 and σ_2 with their common value σ and using some algebra, we obtain the variable

$$z = \frac{(\bar{x}_1 - \bar{x}_2) - (\mu_1 - \mu_2)}{\sigma\sqrt{(1/n_1) + (1/n_2)}}. \tag{10.1}$$

However, we cannot use this variable as a basis for the required test statistic because σ is unknown.

Consequently, we need to use sample information to estimate σ, the unknown population standard deviation. We do so by first estimating the unknown population variance, σ^2. The best way to do that is to regard the sample variances, s_1^2 and s_2^2, as two estimates of σ^2 and then **pool** those estimates by weighting them according to sample size (actually by degrees of freedom). Thus our estimate of σ^2 is

$$s_p^2 = \frac{(n_1 - 1)s_1^2 + (n_2 - 1)s_2^2}{n_1 + n_2 - 2},$$

and hence that of σ is

$$s_p = \sqrt{\frac{(n_1 - 1)s_1^2 + (n_2 - 1)s_2^2}{n_1 + n_2 - 2}}.$$

The subscript "p" stands for "pooled," and the quantity s_p is called the **pooled sample standard deviation.**

Replacing σ in Equation (10.1) with its estimate, s_p, we get the variable

$$\frac{(\bar{x}_1 - \bar{x}_2) - (\mu_1 - \mu_2)}{s_p\sqrt{(1/n_1) + (1/n_2)}},$$

which we can use as the required test statistic. Although the variable in Equation (10.1) has the standard normal distribution, this one has a t-distribution, with which you are already familiar.

KEY FACT 10.2

Distribution of the Pooled t-Statistic

Suppose that x is a normally distributed variable on each of two populations and that the population standard deviations are equal. Then, for independent samples of sizes n_1 and n_2 from the two populations, the variable

$$t = \frac{(\bar{x}_1 - \bar{x}_2) - (\mu_1 - \mu_2)}{s_p\sqrt{(1/n_1) + (1/n_2)}}$$

has the t-distribution with df $= n_1 + n_2 - 2$.

In light of Key Fact 10.2, for a hypothesis test that has null hypothesis $H_0: \mu_1 = \mu_2$ (population means are equal), we can use the variable

$$t = \frac{\bar{x}_1 - \bar{x}_2}{s_p\sqrt{(1/n_1) + (1/n_2)}}$$

as the test statistic and obtain the critical value(s) or P-value from the t-table, Table IV in Appendix A. We call this hypothesis-testing procedure the **pooled t-test.**[†] Procedure 10.1 provides a step-by-step method for performing a pooled t-test by using either the critical-value approach or the P-value approach.

[†]The pooled t-test is also known as the **two-sample t-test with equal variances assumed**, the **pooled two-variable t-test**, and the **pooled independent samples t-test**.

■■■■ PROCEDURE 10.1   Pooled *t*-Test

Purpose To perform a hypothesis test to compare two population means, μ_1 and μ_2

Assumptions
1. Simple random samples
2. Independent samples
3. Normal populations or large samples
4. Equal population standard deviations

Step 1   The null hypothesis is $H_0: \mu_1 = \mu_2$, and the alternative hypothesis is

$$H_a: \mu_1 \neq \mu_2 \quad \text{or} \quad H_a: \mu_1 < \mu_2 \quad \text{or} \quad H_a: \mu_1 > \mu_2$$
$$\text{(Two tailed)} \qquad \text{(Left tailed)} \qquad \text{(Right tailed)}$$

Step 2   Decide on the significance level, α.

Step 3   Compute the value of the test statistic

$$t = \frac{\bar{x}_1 - \bar{x}_2}{s_p\sqrt{(1/n_1) + (1/n_2)}},$$

where

$$s_p = \sqrt{\frac{(n_1 - 1)s_1^2 + (n_2 - 1)s_2^2}{n_1 + n_2 - 2}}.$$

Denote the value of the test statistic t_0.

CRITICAL-VALUE APPROACH	OR	*P*-VALUE APPROACH

Step 4   The critical value(s) are

$$\pm t_{\alpha/2} \quad \text{or} \quad -t_\alpha \quad \text{or} \quad t_\alpha$$
$$\text{(Two tailed)} \qquad \text{(Left tailed)} \qquad \text{(Right tailed)}$$

with df $= n_1 + n_2 - 2$. Use Table IV to find the critical value(s).

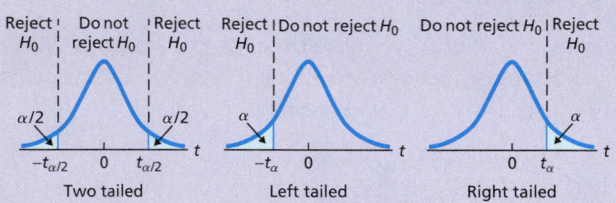

Step 5   If the value of the test statistic falls in the rejection region, reject H_0; otherwise, do not reject H_0.

Step 4   The *t*-statistic has df $= n_1 + n_2 - 2$. Use Table IV to estimate the *P*-value, or obtain it exactly by using technology.

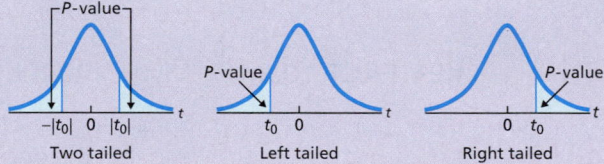

Step 5   If $P \leq \alpha$, reject H_0; otherwise, do not reject H_0.

Step 6   Interpret the results of the hypothesis test.

Note: The hypothesis test is exact for normal populations and is approximately correct for large samples from nonnormal populations.

Regarding Assumptions 1 and 2, we note that the pooled *t*-test can also be used as a method for comparing two means with a designed experiment. Additionally, the pooled *t*-test is robust to moderate violations of Assumption 3 (normal populations) but, even for large samples, can sometimes be unduly affected by outliers because the sample mean and sample standard deviation are not resistant to outliers. The pooled *t*-test is also robust to moderate violations of Assumption 4 (equal population standard deviations) provided the sample sizes are roughly equal. We will say more about the robustness of the pooled t-test at the end of Section 10.3.

How can the conditions of normality and equal population standard deviations (Assumptions 3 and 4, respectively) be checked? As before, normality can be checked by using normal probability plots.

Checking equal population standard deviations can be difficult, especially when the sample sizes are small. As a rough rule of thumb, you can consider the condition of equal population standard deviations met if the ratio of the larger to the smaller sample standard deviation is less than 2. Comparing stem-and-leaf diagrams, histograms, or boxplots of the two samples is also helpful; be sure to use the same scales for each pair of graphs.[†]

■■■■ **EXAMPLE 10.3**   **The Pooled t-Test**

Faculty Salaries Let's return to the salary problem of Example 10.2, in which we want to perform a hypothesis test to decide whether the mean salaries of faculty in private institutions and public institutions are different.

Independent simple random samples of 35 faculty members in private institutions and 30 faculty members in public institutions yielded the data in Table 10.5. At the 5% significance level, do the data provide sufficient evidence to conclude that mean salaries for faculty in private and public institutions differ?

TABLE 10.5

Annual salaries ($1000s) for 35 faculty members in private institutions and 30 faculty members in public institutions

Sample 1 (private institutions)							Sample 2 (public institutions)					
97.3	85.9	118.8	93.9	66.6	109.2	64.9	59.9	115.7	126.1	50.3	133.1	89.3
83.1	100.6	99.3	94.9	94.4	139.3	108.8	82.5	67.1	60.7	79.9	50.1	81.7
158.1	142.4	85.0	108.2	116.3	141.5	51.4	83.9	102.5	109.9	105.1	67.9	107.5
125.6	70.6	74.6	69.9	115.4	84.6	92.0	54.9	41.5	59.5	65.9	76.9	66.9
97.2	55.1	126.6	116.7	76.0	109.6	63.0	85.9	113.9	70.3	90.1	99.7	96.7

Solution First, we find the required summary statistics for the two samples, as shown in Table 10.6. Next, we check the four conditions required for using the pooled t-test, as listed in Procedure 10.1.

TABLE 10.6

Summary statistics for the samples in Table 10.5

Private institutions	Public institutions
$\bar{x}_1 = 98.19$	$\bar{x}_2 = 83.18$
$s_1 = 26.21$	$s_2 = 23.95$
$n_1 = 35$	$n_2 = 30$

- The samples are given as simple random samples and, therefore, Assumption 1 is satisfied.
- The samples are given as independent samples and, therefore, Assumption 2 is satisfied.
- The sample sizes are 35 and 30, both of which are large; furthermore, Figs. 10.2 and 10.3, both on the next page, suggest no outliers for either sample. So, we can consider Assumption 3 satisfied.
- According to Table 10.6, the sample standard deviations are 26.21 and 23.95. These statistics are certainly close enough for us to consider Assumption 4 satisfied, as we also see from the boxplots in Fig. 10.3.

[†]The assumption of equal population standard deviations is sometimes checked by performing a formal hypothesis test, called the two-standard-deviations F-test. We don't recommend that strategy because, although the pooled t-test is robust to moderate violations of normality, the two-standard-deviations F-test is extremely nonrobust to such violations. As the noted statistician George E. P. Box remarked, "To make a preliminary test on variances [standard deviations] is rather like putting to sea in a rowing boat to find out whether conditions are sufficiently calm for an ocean liner to leave port!"

FIGURE 10.2

Normal probability plots of the sample data for faculty in (a) private institutions and (b) public institutions

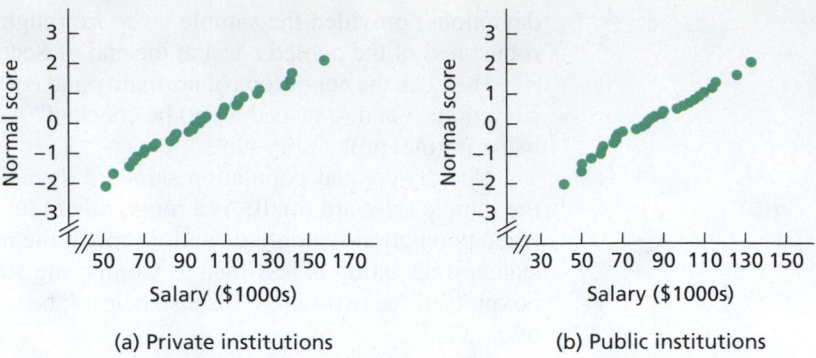

(a) Private institutions

(b) Public institutions

FIGURE 10.3

Boxplots of the salary data for faculty in private institutions and public institutions

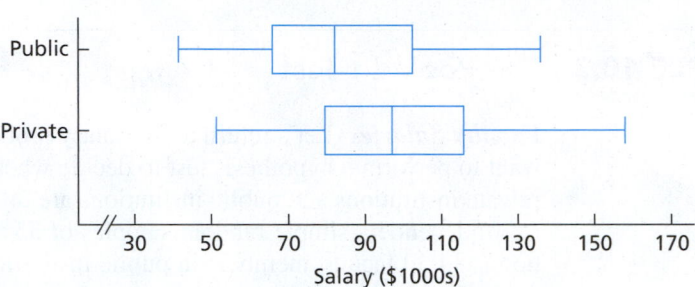

The preceding items suggest that the pooled t-test can be used to carry out the hypothesis test. We apply Procedure 10.1.

Step 1 State the null and alternative hypotheses.

The null and alternative hypotheses are, respectively,

$$H_0: \mu_1 = \mu_2 \text{ (mean salaries are the same)}$$
$$H_a: \mu_1 \neq \mu_2 \text{ (mean salaries are different)},$$

where μ_1 and μ_2 are the mean salaries of all faculty in private and public institutions, respectively. Note that the hypothesis test is two tailed.

Step 2 Decide on the significance level, α.

The test is to be performed at the 5% significance level, or $\alpha = 0.05$.

Step 3 Compute the value of the test statistic

$$t = \frac{\bar{x}_1 - \bar{x}_2}{s_p\sqrt{(1/n_1) + (1/n_2)}},$$

where

$$s_p = \sqrt{\frac{(n_1 - 1)s_1^2 + (n_2 - 1)s_2^2}{n_1 + n_2 - 2}}.$$

To find the pooled sample standard deviation, s_p, we refer to Table 10.6:

$$s_p = \sqrt{\frac{(35 - 1)\cdot(26.21)^2 + (30 - 1)\cdot(23.95)^2}{35 + 30 - 2}} = 25.19.$$

Referring again to Table 10.6, we calculate the value of the test statistic:

$$t = \frac{\bar{x}_1 - \bar{x}_2}{s_p\sqrt{(1/n_1) + (1/n_2)}} = \frac{98.19 - 83.18}{25.19\sqrt{(1/35) + (1/30)}} = 2.395.$$

CRITICAL-VALUE APPROACH	OR	P-VALUE APPROACH

Step 4 The critical values for a two-tailed test are $\pm t_{\alpha/2}$ with df $= n_1 + n_2 - 2$. Use Table IV to find the critical values.

From Table 10.6, $n_1 = 35$ and $n_2 = 30$ and, therefore, df $= 35 + 30 - 2 = 63$. Also, from Step 2, we have $\alpha = 0.05$. In Table IV with df $= 63$, we find that the critical values are $\pm t_{\alpha/2} = \pm t_{0.025} = \pm 1.998$, as shown in Fig. 10.4A.

Step 4 The t-statistic has df $= n_1 + n_2 - 2$. Use Table IV to estimate the P-value, or obtain it exactly by using technology.

From Step 3, the value of the test statistic is $t = 2.395$. The test is two tailed, so the P-value is the probability of observing a value of t of 2.395 or greater in magnitude if the null hypothesis is true. That probability equals the shaded area in Fig. 10.4B.

FIGURE 10.4A

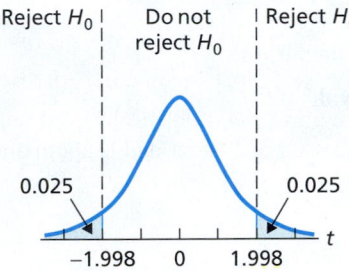

FIGURE 10.4B

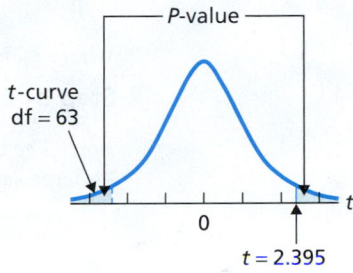

Step 5 If the value of the test statistic falls in the rejection region, reject H_0; otherwise, do not reject H_0.

From Step 3, the value of the test statistic is $t = 2.395$, which falls in the rejection region (see Fig. 10.4A). Thus we reject H_0. The test results are statistically significant at the 5% level.

From Table 10.6, $n_1 = 35$ and $n_2 = 30$ and, therefore, df $= 35 + 30 - 2 = 63$. Referring to Fig. 10.4B and to Table IV with df $= 63$, we find that $0.01 < P < 0.02$. (Using technology, we obtain $P = 0.0196$.)

Step 5 If $P \leq \alpha$, reject H_0; otherwise, do not reject H_0.

From Step 4, $0.01 < P < 0.02$. Because the P-value is less than the specified significance level of 0.05, we reject H_0. The test results are statistically significant at the 5% level and (see Table 9.8 on page 386) provide strong evidence against the null hypothesis.

StatCrunch

Report 10.1

You try it!

Exercise 10.45 on page 455

Step 6 Interpret the results of the hypothesis test.

Interpretation At the 5% significance level, the data provide sufficient evidence to conclude that a difference exists between the mean salaries of faculty in private and public institutions.

Confidence Intervals for the Difference between the Means of Two Populations with Equal Standard Deviations

We can also use Key Fact 10.2 on page 447 to derive a confidence-interval procedure, Procedure 10.2, for the difference between two population means, which we call the **pooled t-interval procedure.**[†]

[†]The pooled t-interval procedure is also known as the **two-sample t-interval procedure with equal variances assumed,** the **pooled two-variable t-interval procedure,** and the **pooled independent samples t-interval procedure.**

■■■ **PROCEDURE 10.2** Pooled *t*-Interval Procedure

Purpose To find a confidence interval for the difference between two population means, μ_1 and μ_2.

Assumptions
1.   Simple random samples
2.   Independent samples
3.   Normal populations or large samples
4.   Equal population standard deviations

Step 1   **For a confidence level of $1 - \alpha$, use Table IV to find $t_{\alpha/2}$ with df $= n_1 + n_2 - 2$.**

Step 2   **The endpoints of the confidence interval for $\mu_1 - \mu_2$ are**

$$(\bar{x}_1 - \bar{x}_2) \pm t_{\alpha/2} \cdot s_p\sqrt{(1/n_1) + (1/n_2)},$$

where s_p is the pooled sample standard deviation.

Step 3   **Interpret the confidence interval.**

Note: The confidence interval is exact for normal populations and is approximately correct for large samples from nonnormal populations.

■■■   **EXAMPLE 10.4**   **The Pooled *t*-Interval Procedure**

Faculty Salaries Obtain a 95% confidence interval for the difference, $\mu_1 - \mu_2$, between the mean salaries of faculty in private and public institutions.

Solution   We apply Procedure 10.2.

Step 1 **For a confidence level of $1 - \alpha$, use Table IV to find $t_{\alpha/2}$ with df $= n_1 + n_2 - 2$.**

For a 95% confidence interval, $\alpha = 0.05$. From Table 10.6, $n_1 = 35$ and $n_2 = 30$, so df $= n_1 + n_2 - 2 = 35 + 30 - 2 = 63$. In Table IV, we find that with df $= 63$, $t_{\alpha/2} = t_{0.05/2} = t_{0.025} = 1.998$.

Step 2 **The endpoints of the confidence interval for $\mu_1 - \mu_2$ are**

$$(\bar{x}_1 - \bar{x}_2) \pm t_{\alpha/2} \cdot s_p\sqrt{(1/n_1) + (1/n_2)},$$

where s_p is the pooled sample standard deviation.

From Step 1, $t_{\alpha/2} = 1.998$. Also, $n_1 = 35$, $n_2 = 30$, and, from Example 10.3, we know that $\bar{x}_1 = 98.19$, $\bar{x}_2 = 83.18$, and $s_p = 25.19$. Hence the endpoints of the confidence interval for $\mu_1 - \mu_2$ are

$$(98.19 - 83.18) \pm 1.998 \cdot 25.19\sqrt{(1/35) + (1/30)},$$

or 15.01 ± 12.52. Thus the 95% confidence interval is from 2.49 to 27.53.

Step 3 **Interpret the confidence interval.**

Interpretation   We can be 95% confident that the difference between the mean salaries of faculty in private institutions and public institutions is somewhere between $2,490 and $27,530. In other words (see page 443), we can be 95% confident that the mean salary of faculty in private institutions exceeds that of faculty in public institutions by somewhere between $2,490 and $27,530.

StatCrunch

Report 10.2

Exercise 10.51 on page 456

The Relation between Hypothesis Tests and Confidence Intervals

Hypothesis tests and confidence intervals are closely related. Consider, for example, a two-tailed hypothesis test for comparing two population means at the significance level α. In this case, the null hypothesis will be rejected if and only if the $(1 - \alpha)$-level confidence interval for $\mu_1 - \mu_2$ does not contain 0. You are asked to examine the relation between hypothesis tests and confidence intervals in greater detail in Exercises 10.63–10.65.

THE TECHNOLOGY CENTER

Most statistical technologies have programs that automatically perform pooled t-procedures. In this subsection, we present output and step-by-step instructions for such programs.

EXAMPLE 10.5   Using Technology to Conduct Pooled t-Procedures

Faculty Salaries Table 10.5 on page 449 shows the annual salaries, in thousands of dollars, for independent samples of 35 faculty members in private institutions and 30 faculty members in public institutions. Use Minitab, Excel, or the TI-83/84 Plus to perform the hypothesis test in Example 10.3 and obtain the confidence interval required in Example 10.4.

Solution Let μ_1 and μ_2 denote the mean salaries of all faculty in private and public institutions, respectively. The task in Example 10.3 is to perform the hypothesis test

$$H_0: \mu_1 = \mu_2 \text{ (mean salaries are the same)}$$
$$H_a: \mu_1 \neq \mu_2 \text{ (mean salaries are different)}$$

at the 5% significance level; the task in Example 10.4 is to obtain a 95% confidence interval for $\mu_1 - \mu_2$.

We applied the pooled t-procedures programs to the data, resulting in Output 10.1 on this and the next page. Steps for generating that output are presented in Instructions 10.1 on the next page. *Note to Excel users:* For brevity, we have presented only the essential portions of the actual output.

As shown in Output 10.1, the P-value for the hypothesis test is about 0.02. Because the P-value is less than the specified significance level of 0.05, we reject H_0. Output 10.1 also shows that a 95% confidence interval for the difference between the means is from 2.49 to 27.54.

OUTPUT 10.1 Pooled t-procedures on the salary data

MINITAB

```
Two-Sample T-Test and CI: PRIVATE, PUBLIC

Two-sample T for PRIVATE vs PUBLIC

          N   Mean   StDev   SE Mean
PRIVATE   35  98.2   26.2    4.4
PUBLIC    30  83.2   24.0    4.4


Difference = μ (PRIVATE) - μ (PUBLIC)
Estimate for difference:   15.01
95% CI for difference:  (2.49, 27.54)
T-Test of difference = 0 (vs ≠): T-Value = 2.40  P-Value = 0.020  DF = 63
Both use Pooled StDev = 25.1948
```

OUTPUT 10.1 (cont.)   Pooled *t*-procedures on the salary data

EXCEL

Hypothesized difference (D): 0
Significance level (%): 5
Population variances for the t-test: Assume equality

Summary statistics:

Variable	Observations	Mean	Std. deviation
PRIVATE	35	98.1943	26.2078
PUBLIC	30	83.1800	23.9528

t-test for two independent samples / Two-tailed test:

95% confidence interval on the difference between the means:

(2.4874, 27.5412)

Difference	15.0143
t (Observed value)	2.3951
|t| (Critical value)	1.9983
DF	63
p-value (Two-tailed)	0.0196
alpha	0.05

TI-83/84 PLUS

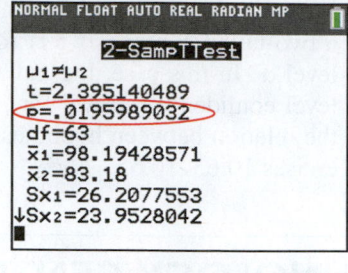

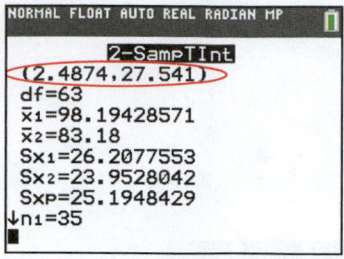

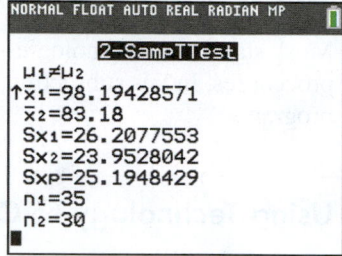

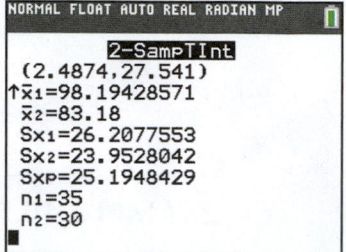

Using **2-SampTTest**   Using **2-SampTInt**

INSTRUCTIONS 10.1   Steps for generating Output 10.1

MINITAB

1 Store the two samples of salary data from Table 10.5 in columns named PRIVATE and PUBLIC
2 Choose **Stat ➤ Basic Statistics ➤ 2-Sample t...**
3 Press the F3 key to reset the dialog box
4 Select **Each sample is in its own column** from the drop-down list box
5 Click in the **Sample 1** text box and specify PRIVATE
6 Click in the **Sample 2** text box and specify PUBLIC
7 Click the **Options...** button
8 Click in the **Confidence level** text box and type 95
9 Click the arrow button at the right of the **Alternative hypothesis** drop-down list box and select **Difference ≠ hypothesized difference**
10 Check the **Assume equal variances** check box
11 Click **OK** twice

EXCEL

1 Store the two samples of salary data from Table 10.5 in columns named PRIVATE and PUBLIC
2 Choose **XLSTAT ➤ Parametric tests ➤ Two-sample t-test and z-test**
3 Click the reset button in the lower left corner of the dialog box
4 Click in the **Sample 1** selection box and then select the column of the worksheet that contains the PRIVATE data
5 Click in the **Sample 2** selection box and then select the column of the worksheet that contains the PUBLIC data
6 Click the **Options** tab
7 Click the arrow button at the right of the **Alternative hypothesis** drop-down list box and select **Mean 1 – Mean 2 ≠ D**
8 Type 5 in the **Significance level (%)** text box
9 Click **OK**

10 Click the **Continue** button in the **XLSTAT – Selections** dialog box

TI-83/84 PLUS

Store the two samples of salary data from Table 10.5 in lists named PRIV and PUBL.

FOR THE HYPOTHESIS TEST:
1 Press **STAT**, arrow over to **TESTS**, and press **4**
2 Highlight **Data** and press **ENTER**
3 Press the down-arrow key
4 Press **2nd ➤ LIST**, arrow down to PRIV, and press **ENTER** twice
5 Press **2nd ➤ LIST**, arrow down to PUBL, and press **ENTER** twice
6 Type 1 for **Freq1**, press **ENTER**, type 1 for **Freq2**, and press **ENTER**
7 Highlight ≠ *μ*2 and press **ENTER**
8 Press the down-arrow key, highlight **Yes**, and press **ENTER**
9 Arrow down to **Calculate** and press **ENTER**

FOR THE CI:
1 Press **STAT**, arrow over to **TESTS**, and press **0**
2 Highlight **Data** and press **ENTER**
3 Press the down-arrow key
4 Press **2nd ➤ LIST**, arrow down to PRIV, and press **ENTER** twice
5 Press **2nd ➤ LIST**, arrow down to PUBL, and press **ENTER** twice
6 Type 1 for **Freq1**, press **ENTER**, type 1 for **Freq2**, and press **ENTER**
7 Type .95 for **C-Level** and press **ENTER**
8 Highlight **Yes**, and press **ENTER**
9 Press the down-arrow key and press **ENTER**

Note to Minitab and Excel users: Although Minitab and Excel simultaneously perform a hypothesis test and obtain a confidence interval, the type of confidence interval found depends on the type of hypothesis test. Specifically, Minitab and Excel compute a two-sided confidence interval for a two-tailed test and a one-sided confidence interval for a one-tailed test. To perform a one-tailed hypothesis test and obtain a two-sided confidence interval, apply Minitab's or Excel's pooled t-procedure twice: once for the one-tailed hypothesis test and once for the confidence interval specifying a two-tailed hypothesis test.

Exercises 10.2

Understanding the Concepts and Skills

10.33 Regarding the four conditions required for using the pooled t-procedures:
a. what are they?　　　　**b.** how important is each condition?

10.34 Explain why s_p is called the pooled sample standard deviation.

In each of Exercises 10.35–10.38, we have provided summary statistics for independent simple random samples from two populations. Preliminary data analyses indicate that the variable under consideration is normally distributed on each population. Decide, in each case, whether use of the pooled t-test and pooled t-interval procedure is reasonable. Explain your answer.

10.35 $\bar{x}_1 = 468.3$, $s_1 = 38.2$, $n_1 = 6$,
$\bar{x}_2 = 394.6$, $s_2 = 84.7$, $n_2 = 14$

10.36 $\bar{x}_1 = 115.1$, $s_1 = 79.4$, $n_1 = 51$,
$\bar{x}_2 = 24.3$, $s_2 = 10.5$, $n_2 = 19$

10.37 $\bar{x}_1 = 118$, $s_1 = 12.04$, $n_1 = 99$,
$\bar{x}_2 = 110$, $s_2 = 11.25$, $n_2 = 80$

10.38 $\bar{x}_1 = 39.04$, $s_1 = 18.82$, $n_1 = 51$,
$\bar{x}_2 = 49.92$, $s_2 = 18.97$, $n_2 = 53$

In each of Exercises 10.39–10.44, we have provided summary statistics for independent simple random samples from two populations. In each case, use the pooled t-test and the pooled t-interval procedure to conduct the required hypothesis test and obtain the specified confidence interval.

10.39 $\bar{x}_1 = 10$, $s_1 = 2.1$, $n_1 = 15$, $\bar{x}_2 = 12$, $s_2 = 2.3$, $n_2 = 15$
a. Two-tailed test, $\alpha = 0.05$
b. 95% confidence interval

10.40 $\bar{x}_1 = 10$, $s_1 = 4$, $n_1 = 15$, $\bar{x}_2 = 12$, $s_2 = 5$, $n_2 = 15$
a. Two-tailed test, $\alpha = 0.05$
b. 95% confidence interval

10.41 $\bar{x}_1 = 20$, $s_1 = 4$, $n_1 = 10$, $\bar{x}_2 = 18$, $s_2 = 5$, $n_2 = 15$
a. Right-tailed test, $\alpha = 0.05$
b. 90% confidence interval

10.42 $\bar{x}_1 = 20$, $s_1 = 4$, $n_1 = 10$, $\bar{x}_2 = 23$, $s_2 = 5$, $n_2 = 15$
a. Left-tailed test, $\alpha = 0.05$
b. 90% confidence interval

10.43 $\bar{x}_1 = 20$, $s_1 = 4$, $n_1 = 20$, $\bar{x}_2 = 24$, $s_2 = 5$, $n_2 = 15$
a. Left-tailed test, $\alpha = 0.05$
b. 90% confidence interval

10.44 $\bar{x}_1 = 20$, $s_1 = 4$, $n_1 = 30$, $\bar{x}_2 = 18$, $s_2 = 5$, $n_2 = 40$
a. Right-tailed test, $\alpha = 0.05$
b. 90% confidence interval

Applying the Concepts and Skills

Preliminary data analyses indicate that you can reasonably consider the assumptions for using pooled t-procedures satisfied in Exercises 10.45–10.50. For each exercise, perform the required hypothesis test by using either the critical-value approach or the P-value approach.

10.45 **Doing Time.** The Federal Bureau of Prisons publishes data in *Prison Statistics* on the times served by prisoners released from federal institutions for the first time. Independent random samples of released prisoners in the fraud and firearms offense categories yielded the following information on time served, in months.

Fraud		Firearms	
3.6	17.9	25.5	23.8
5.3	5.9	10.4	17.9
10.7	7.0	18.4	21.9
8.5	13.9	19.6	13.3
11.8	16.6	20.9	16.1

At the 5% significance level, do the data provide sufficient evidence to conclude that the mean time served for fraud is less than that for firearms offenses? (*Note:* $\bar{x}_1 = 10.12$, $s_1 = 4.90$, $\bar{x}_2 = 18.78$, and $s_2 = 4.64$.)

10.46 **Gender and Direction.** In the paper "The Relation of Sex and Sense of Direction to Spatial Orientation in an Unfamiliar Environment" (*Journal of Environmental Psychology*, Vol. 20, pp. 17–28), J. Sholl et al. published the results of examining the sense of direction of 30 male and 30 female students. After being taken to an unfamiliar wooded park, the students were given some spatial orientation tests, including pointing to south, which tested their absolute frame of reference. The students pointed by moving a pointer attached to a 360° protractor. Following are the absolute pointing errors, in degrees, of the participants.

Male					Female				
13	130	39	33	10	14	8	20	3	138
13	68	18	3	11	122	78	69	111	3
38	23	60	5	9	128	31	18	35	111
59	5	86	22	70	109	36	27	32	35
58	3	167	15	30	12	27	8	3	80
8	20	67	26	19	91	68	66	176	15

At the 1% significance level, do the data provide sufficient evidence to conclude that, on average, males have a better sense of direction and, in particular, a better frame of reference than females? (*Note:* $\bar{x}_1 = 37.6$, $s_1 = 38.5$, $\bar{x}_2 = 55.8$, and $s_2 = 48.3$.)

10.47 Fortified Juice and PTH. V. Tangpricha et al. did a study to determine whether fortifying orange juice with Vitamin D would result in changes in the blood levels of five biochemical variables. One of those variables was the concentration of parathyroid hormone (PTH), measured in picograms/milliliter (pg/mL). The researchers published their results in the paper "Fortification of Orange Juice with Vitamin D: A Novel Approach for Enhancing Vitamin D Nutritional Health" (*American Journal of Clinical Nutrition*, Vol. 77, pp. 1478–1483). Concentration levels were recorded at the beginning of the experiment and again at the end of 12 weeks. The following data, based on the results of the study, provide the decrease (negative values indicate increase) in PTH levels, in pg/mL, for those drinking the fortified juice and for those drinking the unfortified juice.

Fortified				Unfortified		
−7.7	11.2	65.8	−45.6	65.1	0.0	40.0
−4.8	26.4	55.9	−15.5	−48.8	15.0	8.8
34.4	−5.0	−2.2		13.5	−6.1	29.4
−20.1	−40.2	73.5		−20.5	−48.4	−28.7

At the 5% significance level, do the data provide sufficient evidence to conclude that drinking fortified orange juice reduces PTH level more than drinking unfortified orange juice? (*Note:* The mean and standard deviation for the data on fortified juice are 9.0 pg/mL and 37.4 pg/mL, respectively, and for the data on unfortified juice, they are 1.6 pg/mL and 34.6 pg/mL, respectively.)

10.48 Driving Distances. Data on household vehicle miles of travel (VMT) are compiled annually by the Federal Highway Administration and are published in *National Household Travel Survey, Summary of Travel Trends*. Independent random samples of 15 midwestern households and 14 southern households provided the following data on last year's VMT, in thousands of miles.

Midwest			South		
16.2	12.9	17.3	22.2	19.2	9.3
14.6	18.6	10.8	24.6	20.2	15.8
11.2	16.6	16.6	18.0	12.2	20.1
24.4	20.3	20.9	16.0	17.5	18.2
9.6	15.1	18.3	22.8	11.5	

At the 5% significance level, does there appear to be a difference in last year's mean VMT for midwestern and southern households? (*Note:* $\bar{x}_1 = 16.23$, $s_1 = 4.06$, $\bar{x}_2 = 17.69$, and $s_2 = 4.42$.)

10.49 Nigerian Spleen Length. O. Ehimwenma and M. Tagbo, researchers in Nigeria, were interested in how characteristics of the spleen of residents in their tropical environment compare to those found elsewhere in the world. They published their findings in the article "Determination of Normal Dimensions of the Spleen by Ultrasound in an Endemic Tropical Environment" (*Nigerian Medical Journal*, Vol. 52, No. 3, pp. 198–203). The researchers randomly sampled 91 males and 109 females in Nigeria. The mean and standard deviation of the spleen lengths for the males were 11.1 cm and 0.9 cm, respectively, and those for the females were 10.1 cm and 0.7 cm, respectively. At the 1% significance level, do the data provide sufficient evidence to conclude that a difference exists in mean spleen lengths of male and female Nigerians?

10.50 Recess and Wasted Food. E. Bergman et al. conducted a study to determine, among other things, the impact that scheduling recess before or after the lunch period has on wasted food for students in grades three through five. Results were published in the online article "The Relationship of Meal and Recess Schedules to Plate Waste in Elementary Schools" (*Journal of Child Nutrition and Management*, Vol. 28, Issue 2). Summary statistics for the amount of food wasted, in grams, by randomly selected students are presented in the following table.

Lunch before recess	Lunch after recess
$\bar{x}_1 = 223.1$	$\bar{x}_2 = 156.6$
$s_1 = 122.9$	$s_2 = 108.1$
$n_1 = 889$	$n_2 = 1119$

At the 1% significance level, do the data provide sufficient evidence to conclude that, in grades three through five, the mean amount of food wasted for lunches before recess exceeds that for lunches after recess?

In Exercises 10.51–10.56, apply Procedure 10.2 on page 452 to obtain the required confidence interval. Interpret your result in each case.

10.51 Doing Time. Refer to Exercise 10.45 and obtain a 90% confidence interval for the difference between the mean times served by prisoners in the fraud and firearms offense categories.

10.52 Gender and Direction. Refer to Exercise 10.46 and obtain a 98% confidence interval for the difference between the mean absolute pointing errors for males and females.

10.53 Fortified Juice and PTH. Refer to Exercise 10.47 and find a 90% confidence interval for the difference between the mean reductions in PTH levels for fortified and unfortified orange juice.

10.54 Driving Distances. Refer to Exercise 10.48 and determine a 95% confidence interval for the difference between last year's mean VMTs by midwestern and southern households.

10.55 Nigerian Spleen Length. Refer to Exercise 10.49 and determine a 99% confidence interval for the difference between mean spleen lengths of Nigerian males and females.

10.56 Recess and Wasted Food. Refer to Exercise 10.50 and find a 98% confidence interval for the difference between the mean amount of food wasted for lunches before recess and that for lunches after recess.

Working with Large Data Sets

10.57 Vegetarians and Omnivores. Philosophical and health issues are prompting an increasing number of Taiwanese to switch to a vegetarian lifestyle. In the paper "LDL of Taiwanese Vegetarians Are Less Oxidizable than Those of Omnivores" (*Journal of Nutrition*, Vol. 130, pp. 1591–1596), S. Lu et al. compared the daily intake of nutrients by vegetarians and omnivores living in Taiwan. Among the nutrients considered was protein. Too little protein stunts growth and interferes with all bodily functions; too much protein puts a strain on the kidneys, can cause diarrhea and dehydration, and can leach calcium from bones and teeth. Independent random samples of 51 female vegetarians and 53 female omnivores yielded the data, in grams, on daily protein intake presented on the WeissStats site. Use the technology of your choice to do the following.

a. Obtain normal probability plots, boxplots, and the standard deviations for the two samples.

b. Do the data provide sufficient evidence to conclude that the mean daily protein intakes of female vegetarians and female omnivores differ? Perform the required hypothesis test at the 1% significance level.

c. Find a 99% confidence interval for the difference between the mean daily protein intakes of female vegetarians and female omnivores.

d. Are your procedures in parts (b) and (c) justified? Explain your answer.

10.58 Children of Diabetic Mothers. The paper "Correlations Between the Intrauterine Metabolic Environment and Blood Pressure in Adolescent Offspring of Diabetic Mothers" (*Journal of Pediatrics*, Vol. 136, Issue 5, pp. 587–592) by N. Cho et al. presented findings of research on children of diabetic mothers. Past studies have shown that maternal diabetes results in obesity, blood pressure, and glucose-tolerance complications in the offspring. The WeissStats site provides data on systolic blood pressure, in mm Hg, from independent random samples of 99 adolescent offspring of diabetic mothers (ODM) and 80 adolescent offspring of nondiabetic mothers (ONM).

a. Obtain normal probability plots, boxplots, and the standard deviations for the two samples.

b. At the 5% significance level, do the data provide sufficient evidence to conclude that the mean systolic blood pressure of ODM children exceeds that of ONM children?

c. Determine a 95% confidence interval for the difference between the mean systolic blood pressures of ODM and ONM children.

d. Are your procedures in parts (b) and (c) justified? Explain your answer.

10.59 A Better Golf Tee? An independent golf equipment testing facility compared the difference in the performance of golf balls hit off a regular 2-3/4″ wooden tee to those hit off a 3″ Stinger Competition golf tee. A Callaway Great Big Bertha driver with 10 degrees of loft was used for the test, and a robot swung the club head at approximately 95 miles per hour. Data on total distance traveled (in yards) with each type of tee, based on the test results, are provided on the WeissStats site.

a. Obtain normal probability plots, boxplots, and the standard deviations for the two samples.

b. At the 1% significance level, do the data provide sufficient evidence to conclude that, on average, the Stinger tee improves total distance traveled?

c. Find a 99% confidence interval for the difference between the mean total distance traveled with the regular and Stinger tees.

d. Are your procedures in parts (b) and (c) justified? Why or why not?

Extending the Concepts and Skills

10.60 In this section, we introduced the pooled t-test, which provides a method for comparing two population means. In deriving the pooled t-test, we stated that the variable

$$z = \frac{(\bar{x}_1 - \bar{x}_2) - (\mu_1 - \mu_2)}{\sigma\sqrt{(1/n_1) + (1/n_2)}}$$

cannot be used as a basis for the required test statistic because σ is unknown. Why can't that variable be used as a basis for the required test statistic?

10.61 The formula for the pooled variance, s_p^2, is given on page 447. Show that, if the sample sizes, n_1 and n_2, are equal, then s_p^2 is the mean of s_1^2 and s_2^2.

10.62 Simulation. In this exercise, you are to perform a computer simulation to illustrate the distribution of the pooled t-statistic, given in Key Fact 10.2 on page 447.

a. Simulate 1000 random samples of size 4 from a normally distributed variable with a mean of 100 and a standard deviation of 16. Then obtain the sample mean and sample standard deviation of each of the 1000 samples.

b. Simulate 1000 random samples of size 3 from a normally distributed variable with a mean of 110 and a standard deviation of 16. Then obtain the sample mean and sample standard deviation of each of the 1000 samples.

c. Determine the value of the pooled t-statistic for each of the 1000 pairs of samples obtained in parts (a) and (b).

d. Obtain a histogram of the 1000 values found in part (c).

e. Theoretically, what is the distribution of all possible values of the pooled t-statistic?

f. Compare your results from parts (d) and (e).

10.63 Two-Tailed Hypothesis Tests and CIs. As we mentioned on page 453, the following relationship holds between hypothesis tests and confidence intervals: For a two-tailed hypothesis test at the significance level α, the null hypothesis H_0: $\mu_1 = \mu_2$ will be rejected in favor of the alternative hypothesis H_a: $\mu_1 \neq \mu_2$ if and only if the $(1 - \alpha)$-level confidence interval for $\mu_1 - \mu_2$ does not contain 0. In each case, illustrate the preceding relationship by comparing the results of the hypothesis test and confidence interval in the specified exercises.

a. Exercises 10.48 and 10.54

b. Exercises 10.49 and 10.55

10.64 Left-Tailed Hypothesis Tests and CIs. If the assumptions for a pooled t-interval are satisfied, the formula for a $(1 - \alpha)$-level upper confidence bound for the difference, $\mu_1 - \mu_2$, between two population means is

$$(\bar{x}_1 - \bar{x}_2) + t_\alpha \cdot s_p\sqrt{(1/n_1) + (1/n_2)}.$$

For a left-tailed hypothesis test at the significance level α, the null hypothesis H_0: $\mu_1 = \mu_2$ will be rejected in favor of the alternative hypothesis H_a: $\mu_1 < \mu_2$ if and only if the $(1 - \alpha)$-level upper confidence bound for $\mu_1 - \mu_2$ is less than or equal to 0. In each case, illustrate the preceding relationship by obtaining the appropriate upper confidence bound and comparing the result to the conclusion of the hypothesis test in the specified exercise.

a. Exercise 10.45

b. Exercise 10.46

10.65 Right-Tailed Hypothesis Tests and CIs. If the assumptions for a pooled t-interval are satisfied, the formula for a $(1 - \alpha)$-level lower confidence bound for the difference, $\mu_1 - \mu_2$, between two population means is

$$(\bar{x}_1 - \bar{x}_2) - t_\alpha \cdot s_p\sqrt{(1/n_1) + (1/n_2)}.$$

For a right-tailed hypothesis test at the significance level α, the null hypothesis H_0: $\mu_1 = \mu_2$ will be rejected in favor of the alternative hypothesis H_a: $\mu_1 > \mu_2$ if and only if the $(1 - \alpha)$-level lower confidence bound for $\mu_1 - \mu_2$ is greater than or equal to 0. In each case, illustrate the preceding relationship by obtaining the appropriate lower confidence bound and comparing the result to the conclusion of the hypothesis test in the specified exercise.

a. Exercise 10.47

b. Exercise 10.50

10.66 Permutation Tests. With the advent of high-speed computing, new procedures have been developed that permit statistical inferences to be performed under less restrictive conditions than those of classical procedures. **Permutation tests** constitute one such collection of new procedures. To perform a permutation test to compare two population means using independent samples, proceed as follows.

1. Combine the two samples.

2. Randomly select n_1 members from the combined sample. Now treat these n_1 members as the first sample and the remaining n_2 members as the second sample.

3. Compute the difference between the means of the two new samples.
4. Repeat steps 2 and 3 a large number (hundreds or thousands) of times.
5. The distribution of the resulting differences between sample means provides an estimate of the sampling distribution of the sample-mean differences when the null hypothesis of equal population means is true. This estimate is called a **permutation distribution.**

6. The (estimated) *P*-value of the hypothesis test equals the proportion of values of the permutation distribution that are as extreme as or more extreme than the difference between the two observed sample means.

Refer to Example 10.3 on page 449. Use the technology of your choice to conduct a permutation test and compare your results with those found by using the pooled *t*-test. Discuss any discrepancy that you encounter.

10.3 Inferences for Two Population Means, Using Independent Samples: Standard Deviations Not Assumed Equal

In Section 10.2, we examined methods based on independent samples for performing inferences to compare the means of two populations. The methods discussed, called pooled *t*-procedures, require that the standard deviations of the two populations be equal.

In this section, we develop inferential procedures based on independent samples to compare the means of two populations that do not require the population standard deviations to be equal, even though they may be. As before, we assume that the population standard deviations are unknown, because that is usually the case in practice.

For our derivation, we also assume that the variable under consideration is normally distributed on each population. However, like the pooled *t*-procedures, the resulting inferential procedures are approximately correct for large samples, regardless of distribution type.

Hypothesis Tests for the Means of Two Populations, Using Independent Samples

We begin by finding a test statistic. We know from Key Fact 10.1 on page 444 that, for independent samples, the standardized version of $\bar{x}_1 - \bar{x}_2$,

$$z = \frac{(\bar{x}_1 - \bar{x}_2) - (\mu_1 - \mu_2)}{\sqrt{(\sigma_1^2/n_1) + (\sigma_2^2/n_2)}},$$

has the standard normal distribution. We are assuming that the population standard deviations, σ_1 and σ_2, are unknown, so we cannot use this variable as a basis for the required test statistic. We therefore replace σ_1 and σ_2 with their sample estimates, s_1 and s_2, and obtain the variable

$$\frac{(\bar{x}_1 - \bar{x}_2) - (\mu_1 - \mu_2)}{\sqrt{(s_1^2/n_1) + (s_2^2/n_2)}},$$

which we can use as a basis for the required test statistic. This variable does not have the standard normal distribution, but it does have roughly a *t*-distribution.

KEY FACT 10.3

Distribution of the Nonpooled *t*-Statistic

Suppose that x is a normally distributed variable on each of two populations. Then, for independent samples of sizes n_1 and n_2 from the two populations, the variable

$$t = \frac{(\bar{x}_1 - \bar{x}_2) - (\mu_1 - \mu_2)}{\sqrt{(s_1^2/n_1) + (s_2^2/n_2)}}$$

has approximately a *t*-distribution. The degrees of freedom used is obtained from the sample data. It is denoted Δ and given by

$$\Delta = \frac{[(s_1^2/n_1) + (s_2^2/n_2)]^2}{\dfrac{(s_1^2/n_1)^2}{n_1 - 1} + \dfrac{(s_2^2/n_2)^2}{n_2 - 1}},$$

rounded down to the nearest integer.

In light of Key Fact 10.3, for a hypothesis test with null hypothesis H_0: $\mu_1 = \mu_2$, we can use the variable

$$t = \frac{\bar{x}_1 - \bar{x}_2}{\sqrt{(s_1^2/n_1) + (s_2^2/n_2)}}$$

as the test statistic and obtain the critical value(s) or P-value from the t-table, Table IV. We call this hypothesis-testing procedure the **nonpooled t-test**.[†] Procedure 10.3 provides a step-by-step method for performing a nonpooled t-test by using either the critical-value approach or the P-value approach.

PROCEDURE 10.3 Nonpooled t-Test

Purpose To perform a hypothesis test to compare two population means, μ_1 and μ_2

Assumptions
1. Simple random samples
2. Independent samples
3. Normal populations or large samples

Step 1 The null hypothesis is H_0: $\mu_1 = \mu_2$, and the alternative hypothesis is

| H_a: $\mu_1 \neq \mu_2$ (Two tailed) | or | H_a: $\mu_1 < \mu_2$ (Left tailed) | or | H_a: $\mu_1 > \mu_2$ (Right tailed) |

Step 2 Decide on the significance level, α.

Step 3 Compute the value of the test statistic

$$t = \frac{\bar{x}_1 - \bar{x}_2}{\sqrt{(s_1^2/n_1) + (s_2^2/n_2)}}.$$

Denote the value of the test statistic t_0.

| CRITICAL-VALUE APPROACH | OR | P-VALUE APPROACH |

CRITICAL-VALUE APPROACH

Step 4 The critical value(s) are

$$\pm t_{\alpha/2} \quad \text{or} \quad -t_{\alpha} \quad \text{or} \quad t_{\alpha}$$
(Two tailed)   (Left tailed)   (Right tailed)

with df $= \Delta$, where

$$\Delta = \frac{\left[(s_1^2/n_1) + (s_2^2/n_2)\right]^2}{\dfrac{(s_1^2/n_1)^2}{n_1 - 1} + \dfrac{(s_2^2/n_2)^2}{n_2 - 1}}$$

rounded down to the nearest integer. Use Table IV to find the critical value(s).

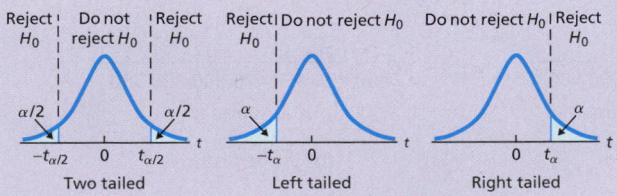

Step 5 If the value of the test statistic falls in the rejection region, reject H_0; otherwise, do not reject H_0.

P-VALUE APPROACH

Step 4 The t-statistic has df $= \Delta$, where

$$\Delta = \frac{\left[(s_1^2/n_1) + (s_2^2/n_2)\right]^2}{\dfrac{(s_1^2/n_1)^2}{n_1 - 1} + \dfrac{(s_2^2/n_2)^2}{n_2 - 1}}$$

rounded down to the nearest integer. Use Table IV to estimate the P-value, or obtain it exactly by using technology.

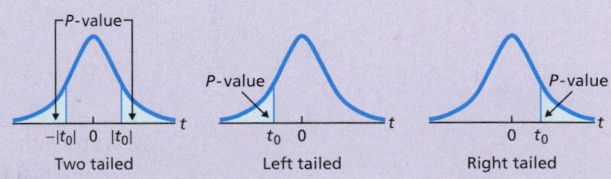

Step 5 If $P \leq \alpha$, reject H_0; otherwise, do not reject H_0.

Step 6 Interpret the results of the hypothesis test.

[†]The nonpooled t-test is also known as the **two-sample t-test** (with equal variances not assumed), the (nonpooled) **two-variable t-test**, and the (nonpooled) **independent samples t-test**.

Regarding Assumptions 1 and 2, we note that the nonpooled t-test can also be used as a method for comparing two means with a designed experiment. In addition, the nonpooled t-test is robust to moderate violations of Assumption 3 (normal populations), but even for large samples, it can sometimes be unduly affected by outliers because the sample mean and sample standard deviation are not resistant to outliers.

■■■ **EXAMPLE 10.6** **The Nonpooled t-Test**

Neurosurgery Operative Times Several neurosurgeons wanted to determine whether a dynamic system (Z-plate) reduced the operative time relative to a static system (ALPS plate). R. Jacobowitz, Ph.D., an Arizona State University professor, along with G. Vishteh, M.D., and other neurosurgeons obtained the data displayed in Table 10.7 on operative times, in minutes, for the two systems. At the 5% significance level, do the data provide sufficient evidence to conclude that the mean operative time is less with the dynamic system than with the static system?

TABLE 10.7

Operative times, in minutes, for dynamic and static systems

Dynamic							Static		
370	360	510	445	295	315	490	430	445	455
345	450	505	335	280	325	500	455	490	535

Solution First, we find the required summary statistics for the two samples, as shown in Table 10.8. Because the two sample standard deviations are considerably different, as seen in Table 10.8 or Fig. 10.6, the pooled t-test is inappropriate here.

TABLE 10.8

Summary statistics for the samples in Table 10.7

Dynamic	Static
$\bar{x}_1 = 394.6$	$\bar{x}_2 = 468.3$
$s_1 = 84.7$	$s_2 = 38.2$
$n_1 = 14$	$n_2 = 6$

Next, we check the three conditions required for using the nonpooled t-test. These data were obtained from a randomized comparative experiment, a type of designed experiment. Therefore, we can consider Assumptions 1 and 2 satisfied.

To check Assumption 3, we refer to the normal probability plots and boxplots in Figs. 10.5 and 10.6, respectively. These graphs reveal no outliers and, given that the nonpooled t-test is robust to moderate violations of normality, show that we can consider Assumption 3 satisfied.

FIGURE 10.5

Normal probability plots of the sample data for the (a) dynamic system and (b) static system

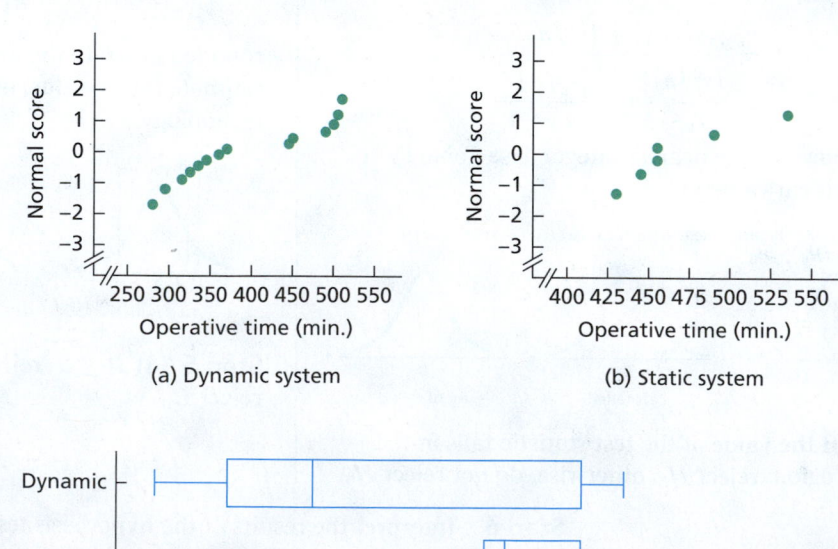

FIGURE 10.6

Boxplots of the operative times for the dynamic and static systems

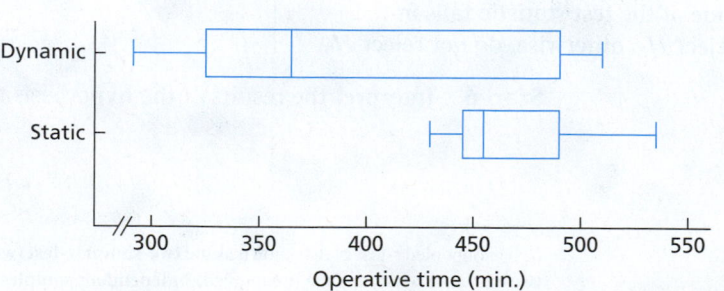

The preceding two paragraphs suggest that the nonpooled t-test can be used to carry out the hypothesis test. We apply Procedure 10.3.

Step 1 **State the null and alternative hypotheses.**

Let μ_1 and μ_2 denote the mean operative times for the dynamic and static systems, respectively. Then the null and alternative hypotheses are, respectively,

H_0: $\mu_1 = \mu_2$ (mean dynamic time is not less than mean static time)

H_a: $\mu_1 < \mu_2$ (mean dynamic time is less than mean static time).

Note that the hypothesis test is left tailed.

Step 2 **Decide on the significance level, α.**

The test is to be performed at the 5% significance level, or $\alpha = 0.05$.

Step 3 **Compute the value of the test statistic**

$$t = \frac{\bar{x}_1 - \bar{x}_2}{\sqrt{(s_1^2/n_1) + (s_2^2/n_2)}}.$$

Referring to Table 10.8, we get

$$t = \frac{394.6 - 468.3}{\sqrt{(84.7^2/14) + (38.2^2/6)}} = -2.681.$$

CRITICAL-VALUE APPROACH	OR	P-VALUE APPROACH

Step 4 **The critical value for a left-tailed test is $-t_\alpha$ with df $= \Delta$. Use Table IV to find the critical value.**

From Step 2, $\alpha = 0.05$. Also, from Table 10.8, we see that

$$\text{df} = \Delta = \frac{\left[(84.7^2/14) + (38.2^2/6)\right]^2}{\dfrac{(84.7^2/14)^2}{14 - 1} + \dfrac{(38.2^2/6)^2}{6 - 1}},$$

which equals 17 when rounded down. From Table IV with df $= 17$, we determine that the critical value is $-t_\alpha = -t_{0.05} = -1.740$, as shown in Fig. 10.7A.

FIGURE 10.7A

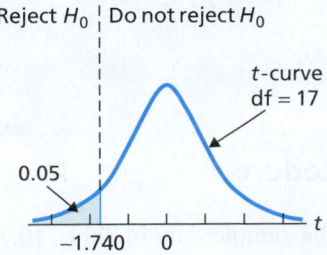

Step 4 **The t-statistic has df $= \Delta$. Use Table IV to estimate the P-value, or obtain it exactly by using technology.**

From Step 3, the value of the test statistic is $t = -2.681$. The test is left tailed, so the P-value is the probability of observing a value of t of -2.681 or less if the null hypothesis is true. That probability equals the shaded area shown in Fig. 10.7B.

FIGURE 10.7B

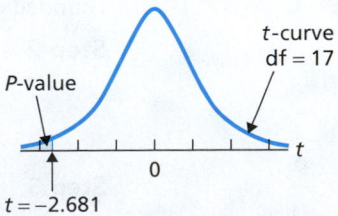

From Table 10.8, we find that

$$\text{df} = \Delta = \frac{\left[(84.7^2/14) + (38.2^2/6)\right]^2}{\dfrac{(84.7^2/14)^2}{14 - 1} + \dfrac{(38.2^2/6)^2}{6 - 1}},$$

which equals 17 when rounded down. Referring to Fig. 10.7B and Table IV with df $= 17$, we determine that $0.005 < P < 0.01$. (Using technology, we find that $P = 0.00789$.)

CRITICAL-VALUE APPROACH	OR	P-VALUE APPROACH

Step 5 If the value of the test statistic falls in the rejection region, reject H_0; otherwise, do not reject H_0.

From Step 3, the value of the test statistic is $t = -2.681$, which, as we see from Fig. 10.7A, falls in the rejection region. Thus we reject H_0. The test results are statistically significant at the 5% level.

Step 5 If $P \leq \alpha$, reject H_0; otherwise, do not reject H_0.

From Step 4, $0.005 < P < 0.01$. Because the P-value is less than the specified significance level of 0.05, we reject H_0. The test results are statistically significant at the 5% level and (see Table 9.8 on page 386) provide very strong evidence against the null hypothesis.

Step 6 Interpret the results of the hypothesis test.

StatCrunch

Report 10.3

Exercise 10.79
on page 467

Interpretation At the 5% significance level, the data provide sufficient evidence to conclude that the mean operative time is less with the dynamic system than with the static system.

Confidence Intervals for the Difference between the Means of Two Populations, Using Independent Samples

Key Fact 10.3 on page 458 can also be used to derive a confidence-interval procedure for the difference between two means. We call this procedure the **nonpooled t-interval procedure.**[†]

PROCEDURE 10.4   Nonpooled t-Interval Procedure

Purpose To find a confidence interval for the difference between two population means, μ_1 and μ_2

Assumptions
1. Simple random samples
2. Independent samples
3. Normal populations or large samples

Step 1 For a confidence level of $1 - \alpha$, use Table IV to find $t_{\alpha/2}$ with df $= \Delta$, where

$$\Delta = \frac{\left[(s_1^2/n_1) + (s_2^2/n_2)\right]^2}{\dfrac{(s_1^2/n_1)^2}{n_1 - 1} + \dfrac{(s_2^2/n_2)^2}{n_2 - 1}}$$

rounded down to the nearest integer.

Step 2 The endpoints of the confidence interval for $\mu_1 - \mu_2$ are

$$(\bar{x}_1 - \bar{x}_2) \pm t_{\alpha/2} \cdot \sqrt{(s_1^2/n_1) + (s_2^2/n_2)}.$$

Step 3 Interpret the confidence interval.

EXAMPLE 10.7   The Nonpooled t-Interval Procedure

Neurosurgery Operative Times Use the sample data in Table 10.7 on page 460 to obtain a 90% confidence interval for the difference, $\mu_1 - \mu_2$, between the mean operative times of the dynamic and static systems.

[†]The nonpooled t-interval procedure is also known as the **two-sample t-interval procedure** (with equal variances not assumed), the (nonpooled) **two-variable t-interval procedure,** and the (nonpooled) **independent samples** t-interval procedure.

Solution   We apply Procedure 10.4.

Step 1   For a confidence level of $1 - \alpha$, use Table IV to find $t_{\alpha/2}$ with df = Δ.

For a 90% confidence interval, $\alpha = 0.10$. From Example 10.6, df = 17. In Table IV, with df = 17, $t_{\alpha/2} = t_{0.10/2} = t_{0.05} = 1.740$.

Step 2   The endpoints of the confidence interval for $\mu_1 - \mu_2$ are

$$(\bar{x}_1 - \bar{x}_2) \pm t_{\alpha/2} \cdot \sqrt{(s_1^2/n_1) + (s_2^2/n_2)}.$$

From Step 1, $t_{\alpha/2} = 1.740$. Referring to Table 10.8 on page 460, we conclude that the endpoints of the confidence interval for $\mu_1 - \mu_2$ are

$$(394.6 - 468.3) \pm 1.740 \cdot \sqrt{(84.7^2/14) + (38.2^2/6)}$$

or -121.5 to -25.9.

Step 3   Interpret the confidence interval.

Interpretation   We can be 90% confident that the difference between the mean operative times of the dynamic and static systems is somewhere between -121.5 minutes and -25.9 minutes. In other words (see page 443), we can be 90% confident that the dynamic system, relative to the static system, reduces the mean operative time by somewhere between 25.9 minutes and 121.5 minutes.   ■

StatCrunch

Report 10.4

You try it!

Exercise 10.85 on page 468

Pooled Versus Nonpooled *t*-Procedures

Suppose that we want to perform a hypothesis test based on independent simple random samples to compare the means of two populations. Further suppose that either the variable under consideration is normally distributed on each of the two populations or the sample sizes are large. Then two tests are candidates for the job: the pooled *t*-test and the nonpooled *t*-test.

In theory, the pooled *t*-test requires that the population standard deviations be equal, but what if they are not? The answer depends on several factors. If the population standard deviations are not too unequal and the sample sizes are nearly the same, using the pooled *t*-test will not cause serious difficulties. If the population standard deviations are quite different, however, using the pooled *t*-test can result in a significantly larger Type I error probability than the specified one.

In contrast, the nonpooled *t*-test applies whether or not the population standard deviations are equal. Then why use the pooled *t*-test at all? The reason is that, if the population standard deviations are equal or nearly so, then, on average, the pooled *t*-test is slightly more powerful; that is, the probability of making a Type II error is somewhat smaller. Similar remarks apply to the pooled *t*-interval and nonpooled *t*-interval procedures.

KEY FACT 10.4

Choosing between a Pooled and a Nonpooled *t*-Procedure

Suppose you want to use independent simple random samples to compare the means of two populations. To decide between a pooled *t*-procedure and a nonpooled *t*-procedure, follow these guidelines: If you are reasonably sure that the populations have nearly equal standard deviations, use a pooled *t*-procedure; otherwise, use a nonpooled *t*-procedure.

THE TECHNOLOGY CENTER

Most statistical technologies have programs that automatically perform nonpooled *t*-procedures. In this subsection, we present output and step-by-step instructions for such programs.

EXAMPLE 10.8 Using Technology to Conduct Nonpooled *t*-Procedures

Neurosurgery Operative Times Table 10.7 on page 460 displays samples of neurosurgery operative times, in minutes, for dynamic and static systems. Use Minitab, Excel, or the TI-83/84 Plus to perform the hypothesis test in Example 10.6 and obtain the confidence interval required in Example 10.7.

Solution Let μ_1 and μ_2 denote, respectively, the mean operative times of the dynamic and static systems. The task in Example 10.6 is to perform the hypothesis test

$$H_0: \mu_1 = \mu_2 \text{ (mean dynamic time is not less than mean static time)}$$
$$H_a: \mu_1 < \mu_2 \text{ (mean dynamic time is less than mean static time)}$$

at the 5% significance level; the task in Example 10.7 is to obtain a 90% confidence interval for $\mu_1 - \mu_2$.

We applied the nonpooled *t*-procedures programs to the data, resulting in Output 10.2. Steps for generating that output are presented in Instructions 10.2. *Note to Excel users:* For brevity, we have presented only the essential portions of the actual output.

As shown in Output 10.2, the *P*-value for the hypothesis test is about 0.008. Because the *P*-value is less than the specified significance level of 0.05, we reject H_0. Output 10.2 also shows that a 90% confidence interval for the difference between the means is from -121 to -26.

OUTPUT 10.2 Nonpooled *t*-procedures on the operative-time data

MINITAB

```
[FOR THE HYPOTHESIS TEST]
Two-Sample T-Test and CI: DYNAMIC, STATIC

Two-sample T for DYNAMIC vs STATIC

                        SE
        N   Mean  StDev  Mean
DYNAMIC  14  394.6  84.7   23
STATIC   6  468.3  38.2   16


Difference = μ (DYNAMIC) - μ (STATIC)
Estimate for difference:  -73.7
95% upper bound for difference:  -25.9
T-Test of difference = 0 (vs <): T-Value = -2.68  P-Value = 0.008  DF = 17


[FOR THE CONFIDENCE INTERVAL]
Two-Sample T-Test and CI: DYNAMIC, STATIC

Two-sample T for DYNAMIC vs STATIC

                        SE
        N   Mean  StDev  Mean
DYNAMIC  14  394.6  84.7   23
STATIC   6  468.3  38.2   16


Difference = μ (DYNAMIC) - μ (STATIC)
Estimate for difference:  -73.7
90% CI for difference:  (-121.5, -25.9)
T-Test of difference = 0 (vs ≠): T-Value = -2.68  P-Value = 0.016  DF = 17
```

OUTPUT 10.2 (cont.)   Nonpooled *t*-procedures on the operative-time data

EXCEL				

[FOR THE HYPOTHESIS TEST]			
Hypothesized difference (D): 0			
Significance level (%): 5			
Population variances for the t-test:			
Summary statistics:			
Variable	Observations	Mean	Std. deviation
DYNAMIC	14	394.6429	84.7500
STATIC	6	468.3333	38.1663
t-test for two independent samples / Lower-tailed test:			
Difference	-73.6905		
t (Observed value)	-2.6804		
t (Critical value)	-1.7349		
DF	18		
p-value (one-tailed)	0.0077		
alpha	0.05		
[FOR THE CONFIDENCE INTERVAL]			
t-test for two independent samples / Two-tailed test:			
90% confidence interval on the difference between the means:			
(-121.3879 , -25.9931)			

TI-83/84 PLUS

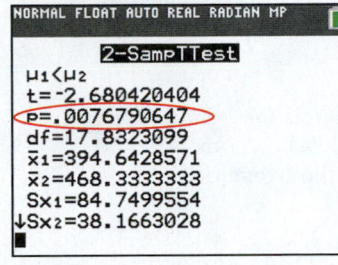

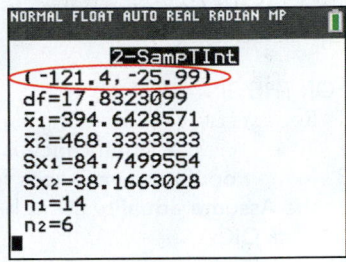

Using **2-SampTInt**

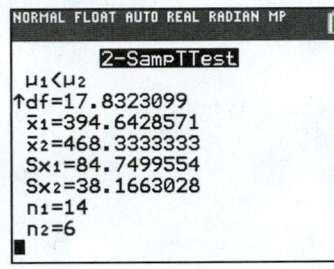

Using **2-SampTTest**

Note: For nonpooled *t*-procedures, discrepancies may occur among results provided by statistical technologies because some round the number of degrees of freedom and others do not.

INSTRUCTIONS 10.2   Steps for generating Output 10.2

MINITAB	EXCEL

MINITAB

Store the two samples of operative-time data from Table 10.7 in columns named DYNAMIC and STATIC.

FOR THE HYPOTHESIS TEST:
1 Choose **Stat ➤ Basic Statistics ➤ 2-Sample t...**
2 Press the F3 key to reset the dialog box
3 Select **Each sample is in its own column** from the drop-down list box
4 Click in the **Sample 1** text box and specify DYNAMIC
5 Click in the **Sample 2** text box and specify STATIC
6 Click the **Options...** button
7 Click the arrow button at the right of the **Alternative hypothesis** drop-down list box and select **Difference < hypothesized difference**
8 Click **OK** twice

FOR THE CI:
1 Repeat steps 1–6 from the hypothesis-test instructions
2 Click in the **Confidence level** text box and type 90
3 Click **OK** twice

EXCEL

Store the two samples of operative-time data from Table 10.7 in columns named DYNAMIC and STATIC.

FOR THE HYPOTHESIS TEST:
1 Choose **XLSTAT ➤ Parametric tests ➤ Two-sample t-test and z-test**
2 Click the reset button in the lower left corner of the dialog box
3 Click in the **Sample 1** selection box and then select the column of the worksheet that contains the DYNAMIC data
4 Click in the **Sample 2** selection box and then select the column of the worksheet that contains the STATIC data
5 Click the **Options** tab
6 Click the arrow button at the right of the **Alternative hypothesis** drop-down list box and select **Mean 1 – Mean 2 < D**
7 Type 5 in the **Significance level (%)** text box
8 In the **Population variances for the t-test** list, uncheck the **Assume equality** check box

(continued)

Note to Minitab and Excel users: As we have previously noted, Minitab and Excel compute a two-sided confidence interval for a two-tailed test and a one-sided confidence interval for a one-tailed test. To perform a one-tailed hypothesis test and obtain a two-sided confidence interval, apply Minitab's or Excel's nonpooled *t*-procedure twice: once for the one-tailed hypothesis test and once for the confidence interval specifying a two-tailed hypothesis test.

Exercises 10.3

Understanding the Concepts and Skills

In each of Exercises 10.67–10.70, suppose that you know that a variable is normally distributed on each of two populations. Further suppose that you want to perform a hypothesis test based on independent random samples to compare the two population means. For each exercise, decide whether you would use the pooled or nonpooled t-test, and give a reason for your answer.

10.67 You know that the population standard deviations are equal.

10.68 You know that the population standard deviations are not equal.

10.69 The sample standard deviations are 23.6 and 59.2.

10.70 The sample standard deviations are 23.6 and 25.2, and each sample size is 25.

10.71 Each pair of graphs in Fig. 10.8 shows the distributions of a variable on two populations. Suppose that, in each case, you want to perform a small-sample hypothesis test based on independent

FIGURE 10.8
Figure for Exercise 10.71

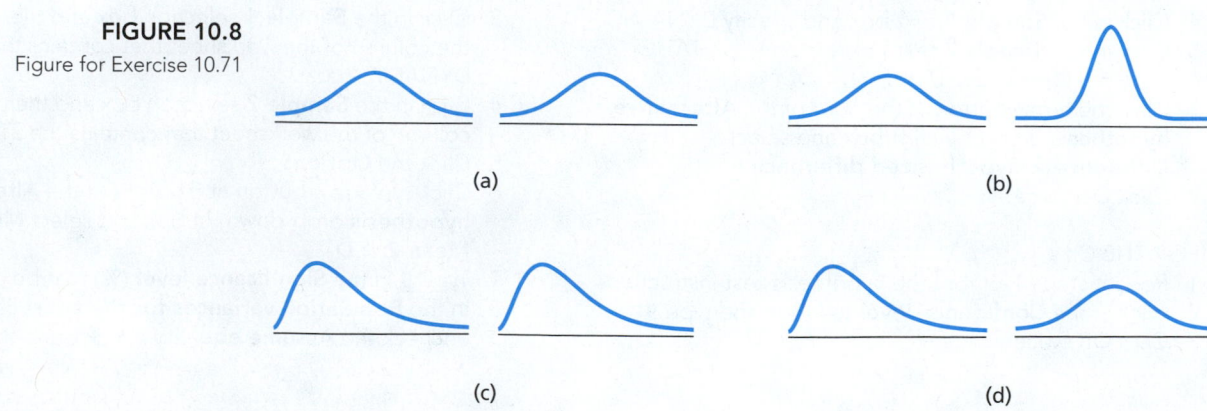

(a)   (b)

(c)   (d)

simple random samples to compare the means of the two populations. In each case, decide whether the pooled t-test, nonpooled t-test, or neither should be used. Explain your answers.

10.72 Discuss the relative advantages and disadvantages of using pooled and nonpooled t-procedures.

In each of Exercises 10.73–10.78, we have provided summary statistics for independent simple random samples from two populations. In each case, use the nonpooled t-test and the nonpooled t-interval procedure to conduct the required hypothesis test and obtain the specified confidence interval.

10.73 $\bar{x}_1 = 10$, $s_1 = 2$, $n_1 = 15$, $\bar{x}_2 = 12$, $s_2 = 5$, $n_2 = 15$
a. Two-tailed test, $\alpha = 0.05$ **b.** 95% confidence interval

10.74 $\bar{x}_1 = 15$, $s_1 = 2$, $n_1 = 15$, $\bar{x}_2 = 12$, $s_2 = 5$, $n_2 = 15$
a. Two-tailed test, $\alpha = 0.05$ **b.** 95% confidence interval

10.75 $\bar{x}_1 = 20$, $s_1 = 4$, $n_1 = 10$, $\bar{x}_2 = 18$, $s_2 = 5$, $n_2 = 15$
a. Right-tailed test, $\alpha = 0.05$ **b.** 90% confidence interval

10.76 $\bar{x}_1 = 20$, $s_1 = 4$, $n_1 = 10$, $\bar{x}_2 = 23$, $s_2 = 5$, $n_2 = 15$
a. Left-tailed test, $\alpha = 0.05$ **b.** 90% confidence interval

10.77 $\bar{x}_1 = 20$, $s_1 = 6$, $n_1 = 20$, $\bar{x}_2 = 24$, $s_2 = 2$, $n_2 = 15$
a. Left-tailed test, $\alpha = 0.05$ **b.** 90% confidence interval

10.78 $\bar{x}_1 = 20$, $s_1 = 2$, $n_1 = 30$, $\bar{x}_2 = 18$, $s_2 = 5$, $n_2 = 40$
a. Right-tailed test, $\alpha = 0.05$ **b.** 90% confidence interval

Applying the Concepts and Skills

Preliminary data analyses indicate that you can reasonably use non-pooled t-procedures in Exercises 10.79–10.84. For each exercise, apply a nonpooled t-test to perform the required hypothesis test, using either the critical-value approach or the P-value approach.

10.79 Political Prisoners. According to the American Psychiatric Association, posttraumatic stress disorder (PTSD) is a common psychological consequence of traumatic events that involve a threat to life or physical integrity. During the Cold War, some 200,000 people in East Germany were imprisoned for political reasons. Many were subjected to physical and psychological torture during their imprisonment, resulting in PTSD. A. Ehlers et al. studied various characteristics of political prisoners from the former East Germany and presented their findings in the paper "Posttraumatic Stress Disorder (PTSD) Following Political Imprisonment: The Role of Mental Defeat, Alienation, and Perceived Permanent Change" (*Journal of Abnormal Psychology*, Vol. 109, pp. 45–55). The researchers randomly and independently selected 32 former prisoners diagnosed with chronic PTSD and 20 former prisoners that were diagnosed with PTSD after release from prison but had since recovered (remitted). The ages, in years, at arrest yielded the following summary statistics.

Chronic	Remitted
$\bar{x}_1 = 25.8$	$\bar{x}_2 = 22.1$
$s_1 = 9.2$	$s_2 = 5.7$
$n_1 = 32$	$n_2 = 20$

At the 10% significance level, is there sufficient evidence to conclude that a difference exists in the mean age at arrest of East German prisoners with chronic PTSD and remitted PTSD?

10.80 Phyllodes Tumors. Phyllodes tumors of the breast are rare tumors that represent less than one percent of growths in the breast. Researchers I. Youn et al. presented characteristics of phyllodes tumors in the article "Phyllodes Tumors of the Breast: Ultrasonographic Findings and Diagnostic Performance of Ultrasound-Guided Core Needle Biopsy" (*Ultrasound in Medicine and Biology*, Vol. 39, No. 6, pp. 987–992). The following table provides summary statistics for the sizes, in millimeters (mm), of independent samples of benign and malignant phyllodes tumors.

Malignant	Benign
$\bar{x}_1 = 48.5$	$\bar{x}_2 = 33.7$
$s_1 = 38.5$	$s_2 = 26.6$
$n_1 = 52$	$n_2 = 116$

At the 1% significance level, do the data provide sufficient evidence to conclude that, on average, malignant phyllodes tumors are larger than benign phyllodes tumors?

10.81 Acute Postoperative Days. Refer to Example 10.6 (page 460). The researchers also obtained the following data on the number of acute postoperative days in the hospital using the dynamic and static systems.

Dynamic							Static		
7	5	8	8	6	7	7	6	18	9
9	10	7	7	7	7	8	7	14	9

At the 5% significance level, do the data provide sufficient evidence to conclude that the mean number of acute postoperative days in the hospital is smaller with the dynamic system than with the static system? (*Note:* $\bar{x}_1 = 7.36$, $s_1 = 1.22$, $\bar{x}_2 = 10.50$, and $s_2 = 4.59$.)

10.82 Stressed-Out Bus Drivers. An intervention program designed by the Stockholm Transit District was implemented to improve the work conditions of the city's bus drivers. Improvements were evaluated by G. Evans et al., who collected physiological and psychological data for bus drivers who drove on the improved routes (intervention) and for drivers who were assigned the normal routes (control). Their findings were published in the article "Hassles on the Job: A Study of a Job Intervention with Urban Bus Drivers" (*Journal of Organizational Behavior*, Vol. 20, pp. 199–208). Following are data, based on the results of the study, for the heart rates, in beats per minute, of the intervention and control drivers.

Intervention		Control						
68	66	74	52	67	63	77	57	80
74	58	77	53	76	54	73	54	
69	63	60	77	63	60	68	64	
68	73	66	71	66	55	71	84	
64	76	63	73	59	68	64	82	

a. At the 5% significance level, do the data provide sufficient evidence to conclude that the intervention program reduces mean heart rate of urban bus drivers in Stockholm? (*Note:* $\bar{x}_1 = 67.90$, $s_1 = 5.49$, $\bar{x}_2 = 66.81$, and $s_2 = 9.04$.)
b. Can you provide an explanation for the somewhat surprising results of the study?
c. Is the study a designed experiment or an observational study? Explain your answer.

10.83 Schizophrenia and Dopamine. Previous research has suggested that changes in the activity of dopamine, a neurotransmitter in the brain, may be a causative factor for schizophrenia. In the paper "Schizophrenia: Dopamine β-Hydroxylase Activity and Treatment Response" (*Science*, Vol. 216, pp. 1423–1425), D. Sternberg et al. published the results of their study in which they examined 25 schizophrenic patients who had been classified as either psychotic or not psychotic by hospital staff. The activity of dopamine was measured in each patient by using the enzyme dopamine β-hydroxylase to assess differences in dopamine activity between the two groups. The following are the data, in nanomoles per milliliter-hour per milligram (nmol/mL-hr/mg).

Psychotic		Not psychotic		
0.0150	0.0222	0.0104	0.0230	0.0145
0.0204	0.0275	0.0200	0.0116	0.0180
0.0306	0.0270	0.0210	0.0252	0.0154
0.0320	0.0226	0.0105	0.0130	0.0170
0.0208	0.0245	0.0112	0.0200	0.0156

At the 1% significance level, do the data suggest that dopamine activity is higher, on average, in psychotic patients? (*Note:* $\bar{x}_1 = 0.02426$, $s_1 = 0.00514$, $\bar{x}_2 = 0.01643$, and $s_2 = 0.00470$.)

10.84 Wing Length. D. Cristol et al. published results of their studies of two subspecies of dark-eyed juncos in the article "Migratory Dark-Eyed Juncos, *Junco Hyemalis*, Have Better Spatial Memory and Denser Hippocampal Neurons than Nonmigratory Conspecifics" (*Animal Behaviour*, Vol. 66, pp. 317–328). One of the subspecies migrates each year, and the other does not migrate. Several physical characteristics of 14 birds of each subspecies were measured, one of which was wing length. The following data, based on results obtained by the researchers, provide the wing lengths, in millimeters (mm), for the samples of two subspecies.

Migratory			Nonmigratory		
84.5	81.0	82.6	82.1	82.4	83.9
82.8	84.5	81.2	87.1	84.6	85.1
80.5	82.1	82.3	86.3	86.6	83.9
80.1	83.4	81.7	84.2	84.3	86.2
83.0	79.7		87.8	84.1	

a. At the 1% significance level, do the data provide sufficient evidence to conclude that the mean wing lengths for the two subspecies are different? (*Note:* The mean and standard deviation for the migratory-bird data are 82.1 mm and 1.501 mm, respectively, and that for the nonmigratory-bird data are 84.9 mm and 1.698 mm, respectively.)

b. Would it be reasonable to use a pooled *t*-test here? Explain your answer.

c. If your answer to part (b) was *yes*, then perform a pooled *t*-test to answer the question in part (a) and compare your results to that found in part (a) by using a nonpooled *t*-test.

In Exercises 10.85–10.90, apply Procedure 10.4 on page 462 to obtain the required confidence interval. Interpret your result in each case.

10.85 Political Prisoners. Refer to Exercise 10.79 and obtain a 90% confidence interval for the difference, $\mu_1 - \mu_2$, between the mean ages at arrest of East German prisoners with chronic PTSD and remitted PTSD.

10.86 Phyllodes Tumors. Refer to Exercise 10.80 and determine a 98% confidence interval for the difference, $\mu_1 - \mu_2$, between the mean sizes of malignant and benign phyllodes tumors.

10.87 Acute Postoperative Days. Refer to Exercise 10.81 and find a 90% confidence interval for the difference between the mean numbers of acute postoperative days in the hospital with the dynamic and static systems.

10.88 Stressed-Out Bus Drivers. Refer to Exercise 10.82 and find a 90% confidence interval for the difference between the mean heart rates of urban bus drivers in Stockholm in the two environments.

10.89 Schizophrenia and Dopamine. Refer to Exercise 10.83 and determine a 98% confidence interval for the difference between the mean dopamine activities of psychotic and nonpsychotic patients.

10.90 Wing Length. Refer to Exercise 10.84 and find a 99% confidence interval for the difference between the mean wing lengths of the two subspecies.

10.91 Sleep Apnea. In the article "Sleep Apnea in Adults With Traumatic Brain Injury: A Preliminary Investigation" (*Archives of Physical Medicine and Rehabilitation*, Vol. 82, Issue 3, pp. 316–321), J. Webster et al. investigated sleep-related breathing disorders in adults with traumatic brain injuries (TBI). The respiratory disturbance index (RDI), which is the number of apneic and hypopneic episodes per hour of sleep, was used as a measure of severity of sleep apnea. An RDI of 5 or more indicates sleep-related breathing disturbances. The RDIs for the females and males in the study are as follows.

Female				Male						
0.1	0.5	0.3	2.3	2.6	19.3	1.4	1.0	0.0	39.2	4.1
2.0	1.4	0.0		0.0	2.1	1.1	5.6	5.0	7.0	2.3
				4.3	7.5	16.5	7.8	3.3	8.9	7.3

Use the technology of your choice to answer the following questions. Explain your answers.
a. If you had to choose between the use of pooled *t*-procedures and nonpooled *t*-procedures here, which would you choose?
b. Is it reasonable to use the type of procedure that you selected in part (a)?

10.92 Mandate Perceptions. L. Grossback et al. examined mandate perceptions and their causes in the paper "Comparing Competing Theories on the Causes of Mandate Perceptions" (*American Journal of Political Science*, Vol. 49, Issue 2, pp. 406–419). Following are data on the percentage of members in each chamber of Congress who reacted to mandates in various years.

House				Senate				
30.3	41.1	15.6	10.1	21	38	40	39	27
23.9	15.2	11.7		27	17	25	25	

Use the technology of your choice to answer the following questions. Explain your answers.
a. If you had to choose between the use of pooled *t*-procedures and nonpooled *t*-procedures here, which would you choose?
b. Is it reasonable to use the type of procedure that you selected in part (a)?

10.93 Mutual Funds. A *mutual fund* is a professionally managed investment that can be sold to the general public. Mutual funds can be specialized into different categories such as healthcare-related or technology-related mutual funds. The following table lists the 3-month rates of return, in percent, for samples of health-care related and technology-related mutual funds, as reported by Morningstar, an independent investment research company.

Healthcare					Technology				
9.7	9.5	7.6	6.8	6.7	6.5	6.4	6.4	6.2	4.9
6.7	6.6	4.6	3.9	3.9	4.6	4.6	4.2	3.6	3.2
3.7	3.3	3.0	0.8	−9.9	2.9	2.8	2.7	2.5	

Use the technology of your choice to answer the following questions. Explain your answers.

a. If you had to choose between the use of pooled t-procedures and nonpooled t-procedures here, which would you choose?

b. Is it reasonable to use the type of procedure that you selected in part (a)?

10.94 Acute Postoperative Days. In Exercise 10.81, you conducted a nonpooled t-test to decide whether the mean number of acute postoperative days spent in the hospital is smaller with the dynamic system than with the static system. Use the technology of your choice to perform the following tasks.

a. Using a pooled t-test, repeat that hypothesis test.

b. Compare your results from the pooled and nonpooled t-tests.

c. Which test do you think is more appropriate, the pooled or nonpooled t-test? Explain your answer.

10.95 Neurosurgery Operative Times. In Example 10.6 on pages 460–462, we conducted a nonpooled t-test, at the 5% significance level, to decide whether the mean operative time is less with the dynamic system than with the static system. Use the technology of your choice to perform the following tasks.

a. Using a pooled t-test, repeat that hypothesis test.

b. Compare your results from the pooled and nonpooled t-tests.

c. Repeat both tests, using a 1% significance level, and compare your results.

d. Which test do you think is more appropriate, the pooled or nonpooled t-test? Explain your answer.

Working with Large Data Sets

10.96 Treating Psychotic Illness. L. Petersen et al. evaluated the effects of integrated treatment for patients with a first episode of psychotic illness in the paper "A Randomised Multicentre Trial of Integrated Versus Standard Treatment for Patients With a First Episode of Psychotic Illness" (*British Medical Journal*, Vol. 331, (7517):602). Part of the study included a questionnaire that was designed to measure client satisfaction for both the integrated treatment and a standard treatment. The data on the WeissStats site are based on the results of the client questionnaire. Use the technology of your choice to do the following.

a. Obtain normal probability plots, boxplots, and the standard deviations for the two samples.

b. Based on your results from part (a), which would you be inclined to use to compare the population means: a pooled or a nonpooled t-procedure? Explain your answer.

c. Do the data provide sufficient evidence to conclude that, on average, clients preferred the integrated treatment? Perform the required hypothesis test at the 1% significance level by using both the pooled t-test and the nonpooled t-test. Compare your results.

d. Find a 98% confidence interval for the difference between mean client satisfaction scores for the two treatments. Obtain the required confidence interval by using both the pooled t-interval procedure and the nonpooled t-interval procedure. Compare your results.

10.97 A Better Golf Tee? An independent golf equipment testing facility compared the difference in the performance of golf balls hit off a regular 2-3/4″ wooden tee to those hit off a 3″ Stinger Competition golf tee. A Callaway Great Big Bertha driver with 10 degrees of loft was used for the test and a robot swung the club head at approximately 95 miles per hour. Data on ball velocity (in miles per hour) with each type of tee, based on the test results, are provided on the WeissStats site. Use the technology of your choice to do the following.

a. Obtain normal probability plots, boxplots, and the standard deviations for the two samples.

b. Based on your results from part (a), which would you be inclined to use to compare the population means: a pooled or a nonpooled t-procedure? Explain your answer.

c. At the 5% significance level, do the data provide sufficient evidence to conclude that, on average, ball velocity is less with the regular tee than with the Stinger tee? Perform the required hypothesis test by using both the pooled t-test and the nonpooled t-test, and compare results.

d. Find a 90% confidence interval for the difference between the mean ball velocities with the regular and Stinger tees. Obtain the required confidence interval by using both the pooled t-interval procedure and the nonpooled t-interval procedure. Compare your results.

10.98 The Etruscans. Anthropologists are still trying to unravel the mystery of the origins of the Etruscan empire, a highly advanced Italic civilization formed around the eighth century B.C. in central Italy. Were they native to the Italian peninsula or, as many aspects of their civilization suggest, did they migrate from the East by land or sea? The maximum head breadth, in millimeters, of 70 modern Italian male skulls and 84 preserved Etruscan male skulls was analyzed to help researchers decide whether the Etruscans were native to Italy. The resulting data can be found on the WeissStats site. [SOURCE: N. Barnicot and D. Brothwell, "The Evaluation of Metrical Data in the Comparison of Ancient and Modern Bones." In *Medical Biology and Etruscan Origins*, G. Wolstenholme and C. O'Connor, eds., Little, Brown & Co., 1959]

a. Obtain normal probability plots, boxplots, and the standard deviations for the two samples.

b. Based on your results from part (a), which would you be inclined to use to compare the population means: a pooled or a nonpooled t-procedure? Explain your answer.

c. Do the data provide sufficient evidence to conclude that a difference exists between the mean maximum head breadths of modern Italian males and Etruscan males? Perform the required hypothesis test at the 5% significance level by using both the pooled t-test and the nonpooled t-test. Compare your results.

d. Find a 95% confidence interval for the difference between the mean maximum head breadths of modern Italian males and Etruscan males. Obtain the required confidence interval by using both the pooled t-interval procedure and the nonpooled t-interval procedure. Compare your results.

Extending the Concepts and Skills

10.99 Suppose that the sample sizes, n_1 and n_2, are equal for independent simple random samples from two populations.

a. Show that the values of the pooled and nonpooled t-statistics will be identical. (*Hint:* Refer to Exercise 10.61 on page 457.)

b. Explain why part (a) does not imply that the two t-tests are equivalent (i.e., will necessarily lead to the same conclusion) when the sample sizes are equal.

10.100 Tukey's Quick Test. In this exercise, we examine an alternative method, conceived by the late Professor John Tukey, for performing a two-tailed hypothesis test for two population means based on independent random samples. To apply this procedure, one of the samples must contain the largest observation (high group) and the other sample must contain the smallest observation (low group). Here are the steps for performing Tukey's quick test.

Step 1 Count the number of observations in the high group that are greater than or equal to the largest observation in the low group. Count ties as 1/2.

Step 2 Count the number of observations in the low group that are less than or equal to the smallest observation in the high group. Count ties as 1/2.

Step 3 Add the two counts obtained in Steps 1 and 2, and denote the sum c.

Step 4 Reject the null hypothesis at the 5% significance level if and only if $c \geq 7$; reject it at the 1% significance level if and only if $c \geq 10$; and reject it at the 0.1% significance level if and only if $c \geq 13$.

a. Can Tukey's quick test be applied to Exercise 10.48 on page 456? Explain your answer.
b. If your answer to part (a) was *yes*, apply Tukey's quick test and compare your result to that found in Exercise 10.48, where a *t*-test was used.
c. Can Tukey's quick test be applied to Exercise 10.84? Explain your answer.
d. If your answer to part (c) was *yes*, apply Tukey's quick test and compare your result to that found in Exercise 10.84, where a *t*-test was used.

For more details about Tukey's quick test, see J. Tukey, "A Quick, Compact, Two-Sample Test to Duckworth's Specifications" (*Technometrics*, Vol. 1, No. 1, pp. 31–48).

10.101 Two-Tailed Hypothesis Tests and CIs. As we mentioned on page 453, the following relationship holds between hypothesis tests and confidence intervals: For a two-tailed hypothesis test at the significance level α, the null hypothesis H_0: $\mu_1 = \mu_2$ will be rejected in favor of the alternative hypothesis H_a: $\mu_1 \neq \mu_2$ if and only if the $(1 - \alpha)$-level confidence interval for $\mu_1 - \mu_2$ does not contain 0. In each case, illustrate the preceding relationship by comparing the results of the hypothesis test and confidence interval in the specified exercises.
a. Exercises 10.79 and 10.85   **b.** Exercises 10.84 and 10.90

10.102 Left-Tailed Hypothesis Tests and CIs. If the assumptions for a nonpooled *t*-interval are satisfied, the formula for a $(1 - \alpha)$-level upper confidence bound for the difference, $\mu_1 - \mu_2$, between two population means is

$$(\bar{x}_1 - \bar{x}_2) + t_\alpha \cdot \sqrt{(s_1^2/n_1) + (s_2^2/n_2)}.$$

For a left-tailed hypothesis test at the significance level α, the null hypothesis H_0: $\mu_1 = \mu_2$ will be rejected in favor of the alternative hypothesis H_a: $\mu_1 < \mu_2$ if and only if the $(1 - \alpha)$-level upper confidence bound for $\mu_1 - \mu_2$ is less than or equal to 0. In each case, illustrate the preceding relationship by obtaining the appropriate upper confidence bound and comparing the result to the conclusion of the hypothesis test in the specified exercise.
a. Exercise 10.81   **b.** Exercise 10.82

10.103 Right-Tailed Hypothesis Tests and CIs. If the assumptions for a nonpooled *t*-interval are satisfied, the formula for a $(1 - \alpha)$-level lower confidence bound for the difference, $\mu_1 - \mu_2$, between two population means is

$$(\bar{x}_1 - \bar{x}_2) - t_\alpha \cdot \sqrt{(s_1^2/n_1) + (s_2^2/n_2)}.$$

For a right-tailed hypothesis test at the significance level α, the null hypothesis H_0: $\mu_1 = \mu_2$ will be rejected in favor of the alternative hypothesis H_a: $\mu_1 > \mu_2$ if and only if the $(1 - \alpha)$-level lower confidence bound for $\mu_1 - \mu_2$ is greater than or equal to 0. In each case, illustrate the preceding relationship by obtaining the appropriate lower confidence bound and comparing the result to the conclusion of the hypothesis test in the specified exercise.
a. Exercise 10.80   **b.** Exercise 10.83

10.4   The Mann–Whitney Test*

To begin this section, we first formalize what it means for two or more distributions to have the *same shape*, as commonly used in nonparametric statistics.

> **DEFINITION 10.1**
>
> ### Distributions of the Same Shape
>
> We say that two or more distributions have the **same shape** if they are identical except possibly for the locations of their centers.[†]

For instance, two normal distributions with equal standard deviations have the same shape, regardless of their means. And, conversely, two normal distributions with different standard deviations have different shapes, regardless of their means. In short, *two normal distributions have the same shape if and only if they have equal standard deviations.*

[†]Observe that, in the context of our earlier discussions of distribution shape, Definition 10.1 means "same shape and spread."

Introducing the Mann–Whitney Test

Thus far, we have developed two procedures for performing a hypothesis test to compare the means of two populations: the pooled and nonpooled t-tests. Both tests require simple random samples, independent samples, and normal populations or large samples. The pooled t-test also requires equal population standard deviations.

As we have just seen, two normal distributions have the same shape if and only if they have equal standard deviations. Consequently, the pooled t-test applies when the two distributions (one for each population) of the variable under consideration are normal and have the same shape; the nonpooled t-test applies when the two distributions are normal, even if they don't have the same shape.

Another procedure for performing a hypothesis test based on independent simple random samples to compare the means of two populations is the **Mann–Whitney test.** This nonparametric test, introduced by Wilcoxon and further developed by Mann and Whitney, is also commonly referred to as the **Wilcoxon rank-sum test** or the **Mann–Whitney–Wilcoxon test.**

The Mann–Whitney test applies when the two distributions of the variable under consideration have the same shape, but it does not require that they be normal or have any other specific shape. See Fig. 10.9.

FIGURE 10.9

Appropriate procedure for comparing two population means based on independent simple random samples

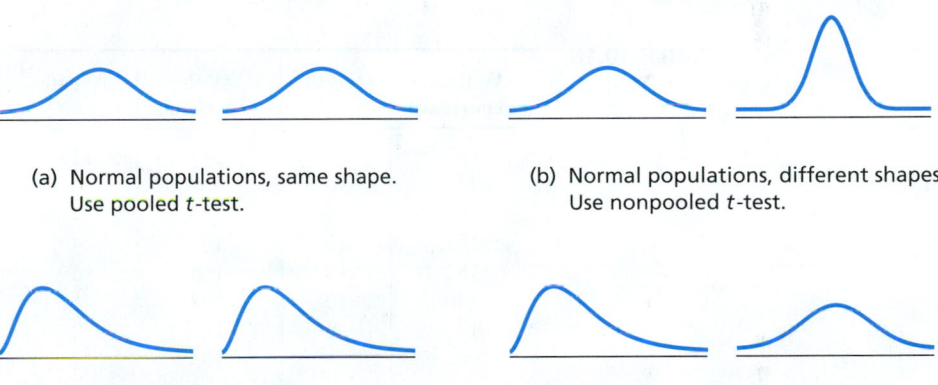

(a) Normal populations, same shape. Use pooled t-test.

(b) Normal populations, different shapes. Use nonpooled t-test.

(c) Nonnormal populations, same shape. Use Mann–Whitney test.

(d) Not both normal populations, different shapes. Use nonpooled t-test for large samples; otherwise, consult a statistician.

■■ ■ **EXAMPLE 10.9** **Introducing the Mann–Whitney Test**

TABLE 10.9

Times, in minutes, required to learn how to use the system

Without experience	With experience
139	142
118	109
164	130
151	107
182	155
140	88
134	95
	104

Computer-System Training A nationwide shipping firm purchased a new computer system to track its packages. Independent samples of employees with and without experience in this type of computer system were randomly selected and the times required to learn how to use the new system were measured. The times, in minutes, are given in Table 10.9.

At the 5% significance level, do the data provide sufficient evidence to conclude that the mean learning time for all employees without experience exceeds the mean learning time for all employees with experience?

Solution Let μ_1 and μ_2 denote the mean learning times for all employees without experience and with experience, respectively. Then the null and alternative hypotheses are, respectively,

H_0: $\mu_1 = \mu_2$ (mean time for inexperienced employees is not greater)

H_a: $\mu_1 > \mu_2$ (mean time for inexperienced employees is greater).

FIGURE 10.10

Back-to-back stem-and-leaf
diagram of the two learning-time
samples in Table 10.9

Without experience		With experience
	8	8
	9	5
	10	4 7 9
8	11	
	12	
9 4	13	0
0	14	2
1	15	5
4	16	
	17	
2	18	

To use the Mann–Whitney test, the learning-time distributions for employees without and with experience should have the same shape. If they do, then the distributions of the two samples in Table 10.9 should also have the same shape, roughly.

To check this condition, we constructed Fig. 10.10, a **back-to-back stem-and-leaf diagram** of the two samples in Table 10.9. In such a diagram, the leaves for the first sample are on the left, the stems are in the middle, and the leaves for the second sample are on the right. The stem-and-leaf diagrams in Fig. 10.10 have roughly the same shape and so do not reveal any obvious violations of the same-shape condition.[†]

To apply the Mann–Whitney test, we first rank all the data from both samples combined. (Referring to Fig. 10.10 is helpful in ranking the data.) The ranking, depicted in Table 10.10, shows, for instance, that the first employee without experience had the ninth-shortest learning time among all 15 employees in the two samples combined.

The idea behind the Mann–Whitney test is simple: If the sum of the ranks for the sample of employees without experience is too large, we conclude that the null hypothesis is false and, therefore, that the mean learning time for all employees without experience exceeds that for all employees with experience. From Table 10.10, the sum of the ranks for the sample of employees without experience, denoted M, is

$$9 + 6 + 14 + 12 + 15 + 10 + 8 = 74.$$

TABLE 10.10

Results of ranking the combined
data from Table 10.9

Without experience	Overall rank	With experience	Overall rank
139	9	142	11
118	6	109	5
164	14	130	7
151	12	107	4
182	15	155	13
140	10	88	1
134	8	95	2
		104	3

To decide whether $M = 74$ is large enough to reject the null hypothesis, we need to first discuss some preliminary material.

Using the Mann–Whitney Table[‡]

Table VI in Appendix A gives values of M_α for a Mann–Whitney test.[§] The size of the sample from Population 2 is given in the leftmost column of Table VI, the values of α in the next column, and the size of the sample from Population 1 along the top. As expected, the symbol M_α denotes the M-value with area (percentage, probability) α to its right.

We can express the critical value(s) for a Mann–Whitney test at the significance level α as follows:

- For a two-tailed test, the critical values are the M-values with area $\alpha/2$ to its left (or, equivalently, area $1 - \alpha/2$ to its right) and area $\alpha/2$ to its right, which are $M_{1-\alpha/2}$ and $M_{\alpha/2}$, respectively. See Fig. 10.11(a).

[†]For ease in explaining the Mann–Whitney test, we have chosen an example in which the sample sizes are very small. However, very small sample sizes make effectively checking the same-shape condition difficult, so proceed cautiously when dealing with very small samples.

[‡]We can use the Mann–Whitney table to estimate the P-value of a Mann–Whitney test. However, because doing so can be awkward or tedious, using statistical software is preferable. Thus, those concentrating on the P-value approach to hypothesis testing can skip to the subsection "Performing the Mann–Whitney Test" on page 474.

[§]Actually, the α-levels in Table VI are only approximate, but are used in practice.

- For a left-tailed test, the critical value is the M-value with area α to its left or, equivalently, area $1 - \alpha$ to its right, which is $M_{1-\alpha}$. See Fig. 10.11(b).
- For a right-tailed test, the critical value is the M-value with area α to its right, which is M_α. See Fig. 10.11(c).

FIGURE 10.11

Critical value(s) for a Mann–Whitney test at the significance level α if the test is (a) two tailed, (b) left tailed, or (c) right tailed

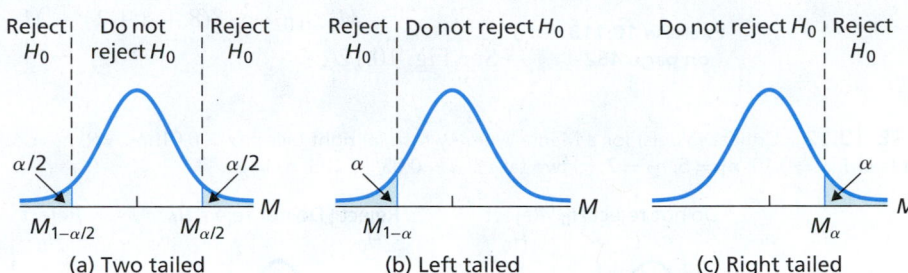

(a) Two tailed   (b) Left tailed   (c) Right tailed

Note the following:

- A critical value from Table VI is to be included as part of the rejection region.
- Although the variable M is discrete, we drew the "histograms" in Fig. 10.11 in the shape of a normal curve. This approach is acceptable because M is close to normally distributed except for very small sample sizes. We use this graphical convention throughout this section.

The distribution of the variable M is symmetric about $n_1(n_1 + n_2 + 1)/2$. This characteristic implies that the M-value with area A to its left (or, equivalently, area $1 - A$ to its right) equals $n_1(n_1 + n_2 + 1)$ minus the M-value with area A to its right. In symbols,

$$M_{1-A} = n_1(n_1 + n_2 + 1) - M_A. \tag{10.2}$$

Referring to Fig. 10.11, we see that by using Equation (10.2) and Table VI, we can determine the critical value for a left-tailed Mann–Whitney test and the critical values for a two-tailed Mann–Whitney test. The next example illustrates the use of Table VI to determine critical values for a Mann–Whitney test.

■■■ **EXAMPLE 10.10**   **Using the Mann–Whitney Table**

In each case, use Table VI to determine the critical value(s) for a Mann–Whitney test. Sketch graphs to illustrate your results.

a. $n_1 = 9, n_2 = 6$; significance level $= 0.01$; right tailed
b. $n_1 = 5, n_2 = 7$; significance level $= 0.10$; left tailed
c. $n_1 = 8, n_2 = 4$; significance level $= 0.05$; two tailed

Solution   In solving these problems, it helps to refer to Fig. 10.11.

a. The critical value for a right-tailed test at the 1% significance level is $M_{0.01}$. To find the critical value, we use Table VI. First we go down the leftmost column, labeled n_2, to "6." Then, going across the row for α labeled 0.01 to the column labeled "9," we reach 92, the required critical value. See Fig. 10.12(a) on the next page.

b. The critical value for a left-tailed test at the 10% significance level is $M_{1-0.10}$. To find the critical value, we use Table VI and Equation (10.2). First we go down the leftmost column, labeled n_2, to "7." Then, going across the row for α labeled 0.10 to the column labeled "5," we reach 41; thus $M_{0.10} = 41$. Now we apply Equation (10.2) and the result just obtained to get

$$M_{1-0.10} = 5(5 + 7 + 1) - M_{0.10} = 65 - 41 = 24,$$

which is the required critical value. See Fig. 10.12(b).

You try it!

Exercise 10.113 on page 482

c. The critical values for a two-tailed test at the 5% significance level are $M_{1-0.05/2}$ and $M_{0.05/2}$, that is, $M_{1-0.025}$ and $M_{0.025}$. First we use Table VI to find $M_{0.025}$. We go down the leftmost column, labeled n_2, to "4." Then, going across the row for α labeled 0.025 to the column labeled "8," we reach 64; thus $M_{0.025} = 64$. Now we apply Equation (10.2) and the result just obtained to get $M_{1-0.025}$:

$$M_{1-0.025} = 8(8 + 4 + 1) - M_{0.025} = 104 - 64 = 40.$$

See Fig. 10.12(c).

FIGURE 10.12 Critical value(s) for a Mann–Whitney test: (a) right tailed, $\alpha = 0.01$, $n_1 = 9$, $n_2 = 6$; (b) left tailed, $\alpha = 0.10$, $n_1 = 5$, $n_2 = 7$; (c) two tailed, $\alpha = 0.05$, $n_1 = 8$, $n_2 = 4$

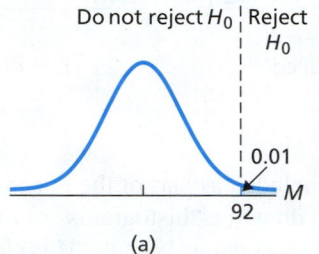
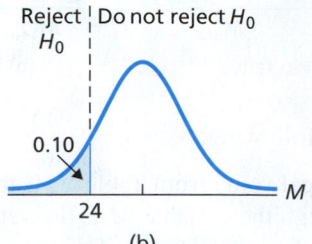
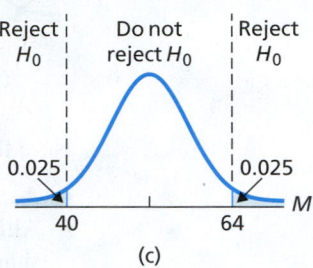

Performing the Mann–Whitney Test

Procedure 10.5 provides a step-by-step method for performing a Mann–Whitney test. Observe that we often use the phrase **same-shape populations** to indicate that the two distributions (one for each population) of the variable under consideration have the same shape.

> **Note:** When there are ties in the sample data, ranks are assigned in the same way as in the Wilcoxon signed-rank test. Namely, if two or more observations are tied, each is assigned the mean of the ranks they would have had if there had been no ties.

EXAMPLE 10.11   The Mann–Whitney Test

Computer-System Training Let's complete the hypothesis test of Example 10.9. Independent simple random samples of employees with and without computer-system experience were obtained. The employees selected were timed to see how long it would take them to learn how to use a certain computer system.

The times, in minutes, are given in Table 10.9 on page 471. At the 5% significance level, do the data provide sufficient evidence to conclude that the mean learning time for employees without experience exceeds that for employees with experience?

Solution We apply Procedure 10.5.

Step 1 State the null and alternative hypotheses.

Let μ_1 and μ_2 denote the mean learning times for all employees without and with experience, respectively. Then the null and alternative hypotheses are, respectively,

H_0: $\mu_1 = \mu_2$ (mean time for inexperienced employees is not greater)

H_a: $\mu_1 > \mu_2$ (mean time for inexperienced employees is greater).

Note that the hypothesis test is right tailed.

■■■■ **PROCEDURE 10.5** Mann–Whitney Test

Purpose To perform a hypothesis test to compare two population means, μ_1 and μ_2

Assumptions
1. Simple random samples
2. Independent samples
3. Same-shape populations

Step 1 The null hypothesis is H_0: $\mu_1 = \mu_2$, and the alternative hypothesis is

$$H_a\text{: } \mu_1 \neq \mu_2 \quad \text{or} \quad H_a\text{: } \mu_1 < \mu_2 \quad \text{or} \quad H_a\text{: } \mu_1 > \mu_2$$
$$\text{(Two tailed)} \qquad \text{(Left tailed)} \qquad \text{(Right tailed)}$$

Step 2 Decide on the significance level, α.

Step 3 Compute the value of the test statistic

$$M = \text{sum of the ranks for sample data from Population 1}$$

and denote that value M_0. To do so, construct a work table of the following form.

Sample from Population 1	Overall rank	Sample from Population 2	Overall rank
.	.	.	.
.	.	.	.
.	.	.	.

CRITICAL-VALUE APPROACH	OR	P-VALUE APPROACH

Step 4 The critical value(s) are

$$M_{1-\alpha/2} \text{ and } M_{\alpha/2} \quad \text{or} \quad M_{1-\alpha} \quad \text{or} \quad M_\alpha$$
$$\text{(Two tailed)} \qquad \text{(Left tailed)} \qquad \text{(Right tailed)}$$

Use Table VI to find the critical value(s). For a left-tailed or two-tailed test, you will also need the relation $M_{1-A} = n_1(n_1 + n_2 + 1) - M_A$.

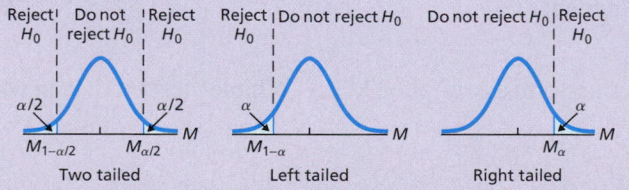

Step 5 If the value of the test statistic falls in the rejection region, reject H_0; otherwise, do not reject H_0.

Step 4 Obtain the *P*-value by using technology.

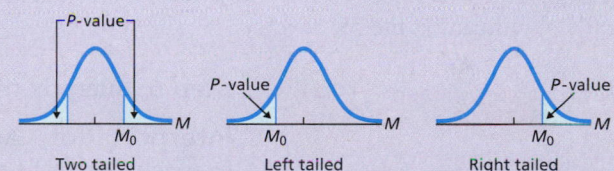

Step 5 If $P \leq \alpha$, reject H_0; otherwise, do not reject H_0.

Step 6 Interpret the results of the hypothesis test.

Step 2 Decide on the significance level, α.

We are to perform the test at the 5% significance level; so, $\alpha = 0.05$.

Step 3 Compute the value of the test statistic

$$M = \text{sum of the ranks for sample data from Population 1.}$$

From the second column of Table 10.10 on page 472, we see that

$$M = 9 + 6 + 14 + 12 + 15 + 10 + 8 = 74.$$

CRITICAL-VALUE APPROACH	OR	P-VALUE APPROACH

Step 4 The critical value for a right-tailed test is M_α. Use Table VI to find the critical value.

From Table 10.9 on page 471, we see that $n_1 = 7$ and $n_2 = 8$. The critical value for a right-tailed test at the 5% significance level is $M_{0.05}$. To find the critical value, we use Table VI. First we go down the leftmost column, labeled n_2, to "8." Then, going across the row for α labeled 0.05 to the column labeled "7," we reach 71, the required critical value. See Fig. 10.13A.

FIGURE 10.13A

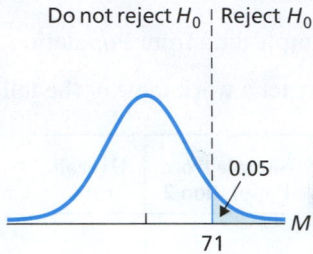

Do not reject H_0 | Reject H_0

0.05

71

M

Step 5 If the value of the test statistic falls in the rejection region, reject H_0; otherwise, do not reject H_0.

From Step 3, the value of the test statistic is $M = 74$. Figure 10.13A shows that this value falls in the rejection region. Thus we reject H_0. The test results are statistically significant at the 5% level.

Step 4 Obtain the *P*-value by using technology.

Using technology, we find that the *P*-value for the hypothesis test is $P = 0.02$, as shown in Fig. 10.13B.

FIGURE 10.13B

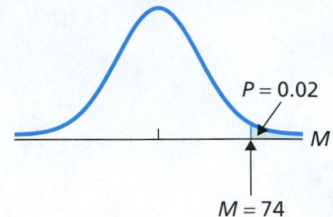

$P = 0.02$

M

$M = 74$

Step 5 If $P \le \alpha$, reject H_0; otherwise, do not reject H_0.

From Step 4, $P = 0.02$. Because the *P*-value is less than the specified significance level of 0.05, we reject H_0. The test results are statistically significant at the 5% level and (see Table 9.8 on page 386) provide strong evidence against the null hypothesis.

StatCrunch

Report 10.5

You try it!

Exercise 10.123 on page 482

Step 6 Interpret the results of the hypothesis test.

Interpretation At the 5% significance level, the data provide sufficient evidence to conclude that the mean learning time for employees without experience exceeds that for employees with experience.

In the next example, we perform a two-tailed Mann–Whitney test for data in which there are ties.

EXAMPLE 10.12   The Mann–Whitney Test

TABLE 10.11

Results of Elmendorf tear test on two different vinyl floor coverings (data in grams)

Brand A		Brand B	
2288	2384	2592	2384
2368	2304	2512	2432
2528	2240	2576	2112
2144	2208	2176	2288
2160	2112	2304	2752

Elmendorf Tear Strength Manufacturers use the Elmendorf tear test to evaluate material strength for various manufactured products. In the article "Using Repeatability and Reproducibility Studies to Evaluate a Destructive Test Method" (*Quality Engineering*, Vol. 10(2), pp. 283–290), A. Phillips et al. investigated that test.

In one aspect of the study, the researchers randomly and independently obtained the data shown in Table 10.11 on Elmendorf tear strength, in grams, for two different brands of vinyl floor covering. At the 5% significance level, do the data provide sufficient evidence to conclude that the mean tear strengths differ for the two vinyl floor coverings? Use the Mann–Whitney test.

Solution Graphical data analyses (not shown) suggest that we can reasonably assume that the distributions of tear strength for the two vinyl floor coverings have the same shape. Hence, we can apply Procedure 10.5 to carry out the required hypothesis test.

Step 1 State the null and alternative hypotheses.

Let μ_1 and μ_2 denote the mean tear strengths for Brand A and Brand B, respectively. Then the null and alternative hypotheses are, respectively,

$$H_0: \mu_1 = \mu_2 \text{ (mean tear strengths are equal)}$$
$$H_a: \mu_1 \neq \mu_2 \text{ (mean tear strengths are different).}$$

Note that the hypothesis test is two tailed.

Step 2 Decide on the significance level, α.

We are to perform the test at the 5% significance level, so $\alpha = 0.05$.

Step 3 Compute the value of the test statistic

$$M = \text{sum of the ranks for sample data from Population 1.}$$

We first construct the following work table. Note that, in several instances, there are ties in the data.

Brand A	Overall rank	Brand B	Overall rank
2288	8.5	2592	19
2384	13.5	2384	13.5
2368	12	2512	16
2304	10.5	2432	15
2528	17	2576	18
2240	7	2112	1.5
2144	3	2176	5
2208	6	2288	8.5
2160	4	2304	10.5
2112	1.5	2752	20

Referring now to the second column of the preceding table, we find that the value of the test statistic is

$$M = 8.5 + 13.5 + 12 + \cdots + 4 + 1.5 = 83.$$

CRITICAL-VALUE APPROACH	OR	P-VALUE APPROACH

Step 4 The critical values for a two-tailed test are $M_{1-\alpha/2}$ and $M_{\alpha/2}$. Use Table VI and the relation $M_{1-A} = n_1(n_1 + n_2 + 1) - M_A$ to find the critical values.

From Table 10.11, we see that $n_1 = 10$ and $n_2 = 10$. The critical values for a two-tailed test at the 5% significance level are $M_{1-0.05/2}$ and $M_{0.05/2}$, that is, $M_{1-0.025}$ and $M_{0.025}$. First we use Table VI to find $M_{0.025}$. We go down the leftmost column, labeled n_2, to "10." Then, going across the row for α labeled 0.025 to the column labeled "10," we reach 131; thus $M_{0.025} = 131$. Now we apply the aforementioned relation and the result just obtained to get $M_{1-0.025}$:

$$M_{1-0.025} = 10(10 + 10 + 1) - M_{0.025}$$
$$= 210 - 131 = 79.$$

See Fig. 10.14A.

Step 4 Obtain the P-value by using technology.

Using technology, we find that the P-value for the hypothesis test is $P = 0.104$, as shown in Fig. 10.14B.

FIGURE 10.14B

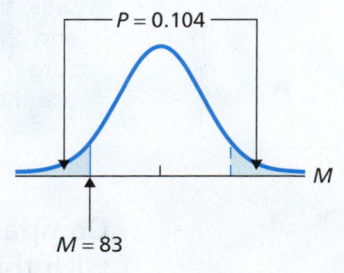

| CRITICAL-VALUE APPROACH | OR | *P*-VALUE APPROACH |

FIGURE 10.14A

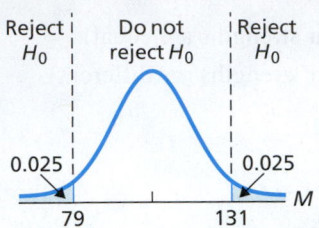

Step 5 **If the value of the test statistic falls in the rejection region, reject H_0; otherwise, do not reject H_0.**

The value of the test statistic is $M = 83$, as found in Step 3, which does not fall in the rejection region shown in Fig. 10.14A. Thus we do not reject H_0. The test results are not statistically significant at the 5% level.

Step 5 **If $P \leq \alpha$, reject H_0; otherwise, do not reject H_0.**

From Step 4, $P = 0.104$. Because the P-value exceeds the specified significance level of 0.05, we do not reject H_0. The test results are not statistically significant at the 5% level and (see Table 9.8 on page 386) provide at most weak evidence against the null hypothesis.

Step 6 **Interpret the results of the hypothesis test.**

Interpretation   At the 5% significance level, the data do not provide sufficient evidence to conclude that the mean tear strengths differ for the two brands of vinyl floor covering.

Comparing the Mann–Whitney Test and the Pooled *t*-Test

In Section 10.2, you learned how to perform a pooled *t*-test to compare two population means when the variable under consideration is normally distributed on each of the two populations and the population standard deviations are equal. Because two normal distributions with equal standard deviations have the same shape, you can also use the Mann–Whitney test to perform such a hypothesis test.

Under conditions of normality, the pooled *t*-test is more powerful than the Mann–Whitney test but, surprisingly, not much more powerful. However, if the two distributions of the variable under consideration have the same shape but are not normal, the Mann–Whitney test is usually more powerful than the pooled *t*-test, often considerably so.

KEY FACT 10.5

The Mann–Whitney Test Versus the Pooled *t*-Test

Suppose that the distributions of a variable of two populations have the same shape and that you want to compare, using independent simple random samples, the two population means. When deciding between the pooled *t*-test and the Mann–Whitney test, follow these guidelines: If you are reasonably sure that the two distributions are normal, use the pooled *t*-test; otherwise, use the Mann–Whitney test.

Comparing Two Population Medians with the Mann–Whitney Procedure

The Mann–Whitney test can be used to compare two population medians as well as two population means. To use Procedure 10.5 to compare two population medians, simply replace μ_1 with η_1 and μ_2 with η_2.

In some of the exercises at the end of this section, you will be asked to use the Mann–Whitney test to perform hypothesis tests for comparing two population medians.

Alternate Version of the Mann–Whitney Test

Assumption 3 of the Mann–Whitney test is that the two distributions (one for each population) of the variable under consideration have the same shape, which, for brevity, we refer to as "same-shape populations." This condition is often difficult to check and is frequently not satisfied.

An alternate version of the Mann–Whitney test is available that does not require the condition of same-shape populations. However, the null and alternative hypotheses for this different version are not stated in terms of means or medians but rather in terms of distributions. In other words, we can drop the "same-shape populations" condition if we change the hypotheses to those in terms of distributions instead of means or medians.

The null hypothesis for the alternate version of the Mann–Whitney test is that the distributions of the variable on the two populations are identical. The alternative hypothesis is that the distributions differ in some way:

- *Two tailed:* The distribution of the variable on the first population has either systematically smaller values or systematically larger values than that of the variable on the second population.
- *Left tailed:* The distribution of the variable on the first population has systematically smaller values than that of the variable on the second population.
- *Right tailed:* The distribution of the variable on the first population has systematically larger values than that of the variable on the second population.

Note that, if the "same-shape populations" condition does in fact hold, then the alternate version of the Mann–Whitney test is equivalent to the one in terms of means or medians (Procedure 10.5 on page 475).

THE TECHNOLOGY CENTER

Most statistical technologies have programs that automatically perform a Mann–Whitney test. In this subsection, we present output and step-by-step instructions for such programs.

As you will see, different programs may report slightly different P-values for a Mann–Whitney test. These differences are due to the fact that different programs may use different methods for obtaining or approximating such P-values.

Note to TI-83/84 Plus users: At the time of this writing, the TI-83/84 Plus does not have a built-in program for conducting a Mann–Whitney test. However, we have written a TI program called MANNWHIT for performing that test. It is located in the TI Programs section on the WeissStats site. Your instructor can show you how to download the program to your calculator. *Warning:* Any data that you may have previously stored in Lists 1–6 will be erased during program execution, so copy those data to other lists prior to program execution if you want to retain them.

EXAMPLE 10.13   Using Technology to Conduct a Mann–Whitney Test

Computer-System Training Table 10.9 on page 471 shows the times, in minutes, required to learn how to use a computer system for independent samples of employees without and with computer-system experience. Use Minitab, Excel, or the

TI-83/84 Plus to decide, at the 5% significance level, whether the data provide sufficient evidence to conclude that the mean learning time for employees without experience exceeds that for employees with experience.

Solution Let μ_1 and μ_2 denote the mean learning times for all employees without and with experience, respectively. We want to perform the hypothesis test

H_0: $\mu_1 = \mu_2$ (mean time for inexperienced employees is not greater)

H_a: $\mu_1 > \mu_2$ (mean time for inexperienced employees is greater)

at the 5% significance level. Note that the hypothesis test is right tailed.

We applied the Mann–Whitney test programs to the data, resulting in Output 10.3. Steps for generating that output are presented in Instructions 10.3.

Note to Minitab users: Like many statistical technologies, Minitab gives the results of the Mann–Whitney test in terms of medians, but those results can also be interpreted in terms of means. Observe also that Minitab uses the letter W (for Wilcoxon) instead of M to designate the test statistic (sum of the ranks for the sample data from Population 1).

Note to Excel users: As usual, we have presented only the essential portions of the actual output. We also point out that Excel uses a statistic, called the Mann−Whitney U statistic, instead of the statistic M (sum of the ranks for the sample data from Population 1). The two statistics are related by the equation $U = M - n_1(n_1 + 1)/2$.

OUTPUT 10.3 Mann–Whitney test on the learning-time data

MINITAB

Mann-Whitney Test and CI: WITHOUT, WITH

```
           N   Median
WITHOUT    7   140.00
WITH       8   108.00


Point estimate for η1 - η2 is 31.50
95.7 Percent CI for η1 - η2 is (3.99,56.00)
W = 74.0
Test of η1 = η2 vs η1 > η2 is significant at 0.0214
```

TI-83/84 PLUS

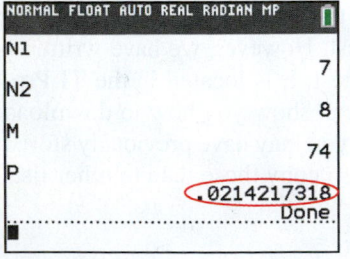

```
NORMAL FLOAT AUTO REAL RADIAN MP
N1
                              7
N2
                              8
M
                             74
P
              .0214217318
                          Done
■
```

EXCEL

Hypothesized difference (D): 0			
Significance level (%): 5			
p-value: Exact p-value			
Summary statistics:			
Variable	Observations	Mean	Std. deviation
WITHOUT	7	146.8571	21.0272
WITH	8	116.2500	23.5781
Mann-Whitney test / Upper-tailed test:			
U	46.0000		
Expected value	28.0000		
Variance (U)	74.6667		
p-value (one-tailed)	0.0200		
alpha	0.05		

As shown in Output 10.3, the *P*-value for the hypothesis test is 0.02. Because the *P*-value is less than the specified significance level of 0.05, we reject H_0. At the 5% significance level, the data provide sufficient evidence to conclude that the mean learning time for employees without experience exceeds that for employees with experience.

INSTRUCTIONS 10.3 Steps for generating Output 10.3

MINITAB

1 Store the data from Table 10.9 in columns named WITHOUT and WITH
2 Choose **Stat ➤ Nonparametrics ➤ Mann-Whitney...**
3 Press the F3 key to reset the dialog box
4 Specify WITHOUT in the **First Sample** text box
5 Specify WITH in the **Second Sample** text box
6 Click the arrow button at the right of the **Alternative** drop-down list box and select **greater than**
7 Click **OK**

EXCEL

1 Store the data from Table 10.9 in columns named WITHOUT and WITH
2 Choose **XLSTAT ➤ Nonparametric tests ➤ Comparison of two samples (Wilcoxon, Mann-Whitney, ...)**
3 Click the reset button in the lower left corner of the dialog box
4 Click in the **Sample 1** selection box and then select the column of the worksheet that contains the WITHOUT data
5 Click in the **Sample 2** selection box and then select the column of the worksheet that contains the WITH data

6 In the **Data format** list, select the **One column per sample** option button
7 Click the **Options** tab
8 Click the arrow button at the right of the **Alternative hypothesis** drop-down list box and select **Sample 1 – Sample 2 > D**
9 Type **5** in the **Significance level (%)** text box
10 Check the **Exact p-value** check box
11 Click **OK**
12 Click the **Continue** button in the **XLSTAT – Selections** dialog box

TI-83/84 PLUS

1 Store the data from Table 10.9 in lists named WOUT and WITH
2 Press **PRGM**
3 Arrow down to MANNWHIT and press **ENTER** twice
4 Press **2ND ➤ LIST**, arrow down to WOUT, and press **ENTER** twice
5 Press **2ND ➤ LIST**, arrow down to WITH, and press **ENTER** twice
6 Type **1** for **TYPE** and press **ENTER**

Exercises 10.4

Understanding the Concepts and Skills

10.104 State the conditions that are required for using the Mann–Whitney test.

10.105 Suppose that, for two populations, the distributions of the variable under consideration have the same shape. Further suppose that you want to perform a hypothesis test based on independent random samples to compare the two population means. If the distributions of the variable are normal, would you use the pooled t-test or the Mann–Whitney test? Explain your answer.

10.106 Suppose that, for two populations, the distributions of the variable under consideration have the same shape. Further suppose that you want to perform a hypothesis test based on independent random samples to compare the two population means. If the distributions of the variable are not normal, would you use the pooled t-test or the Mann–Whitney test? Explain your answer.

10.107 Suppose that you want to perform a hypothesis test based on independent random samples to compare the means of two populations. For each part, decide whether you would use the pooled t-test, the nonpooled t-test, the Mann–Whitney test, or none of these tests if preliminary data analyses of the samples suggest that the two distributions of the variable under consideration are
a. normal but do not have the same shape.
b. not normal but have the same shape.
c. not normal and do not have the same shape; both sample sizes are large.

10.108 Suppose that you want to perform a hypothesis test based on independent random samples to compare the means of two

populations. For each part, decide whether you would use the pooled t-test, the nonpooled t-test, the Mann–Whitney test, or none of these tests if preliminary data analyses of the samples suggest that the two distributions of the variable under consideration are
a. normal and have the same shape.
b. not normal and do not have the same shape; one of the sample sizes is large and the other is small.
c. different, one being normal and the other not; both sample sizes are large.

10.109 Suppose that you want to perform a hypothesis test based on independent random samples to compare the means of two populations. You know that the two distributions of the variable under consideration have the same shape and may be normal. You take the two samples and find that the data for one of the samples contain outliers. Which procedure would you use? Explain your answer.

10.110 Part of conducting a Mann–Whitney test involves ranking all the data from both samples combined. Explain how to deal with ties.

10.111 Why do two normal distributions that have equal standard deviations have the same shape?

10.112 Fill in the blank: The Mann–Whitney test can be used to compare two population means. That test can also be used to compare two population _____.

Exercises 10.113–10.116 pertain to critical values for a Mann–Whitney test. Use Table VI in Appendix A to determine the critical value(s) in each case. For a left-tailed or two-tailed test, you will also need the relation $M_{1-A} = n_1(n_1 + n_2 + 1) - M_A$.

10.113 $n_1 = 8, n_2 = 9$; Significance level $= 0.05$
a. Right-tailed test **b.** Left-tailed test
c. Two-tailed test

10.114 $n_1 = 8, n_2 = 9$; Significance level $= 0.01$
a. Right-tailed test **b.** Left-tailed test
c. Two-tailed test

10.115 $n_1 = 9, n_2 = 8$; Significance level $= 0.10$
a. Right-tailed test **b.** Left-tailed test
c. Two-tailed test

10.116 $n_1 = 9, n_2 = 8$; Significance level $= 0.05$
a. Right-tailed test **b.** Left-tailed test
c. Two-tailed test

In each of Exercises 10.117–10.122, the null hypothesis is $H_0: \mu_1 = \mu_2$ and the alternative hypothesis is as specified. We have provided data from independent simple random samples from the two populations under consideration. In each case, use the Mann–Whitney test to perform the required hypothesis test at the 10% significance level.

10.117 $H_a: \mu_1 > \mu_2$

Sample 1				Sample 2			
4	3	5	3	3	4	2	2	4

10.118 $H_a: \mu_1 < \mu_2$

Sample 1					Sample 2			
5	5	6	5	10	6	8	7	3

10.119 $H_a: \mu_1 \neq \mu_2$

Sample 1					Sample 2			
8	2	4	7	8	8	14	6	14	10

10.120 $H_a: \mu_1 > \mu_2$

Sample 1					Sample 2			
12	9	8	5	10	9	5	4	9	3

10.121 $H_a: \mu_1 < \mu_2$

Sample 1						Sample 2				
5	1	5	8	7	5	10	7	6	6	11

10.122 $H_a: \mu_1 \neq \mu_2$

Sample 1				Sample 2				
10	11	5	6	4	12	6	5	5	3

Applying the Concepts and Skills

In each of Exercises 10.123–10.128, use the Mann–Whitney test to perform the required hypothesis test.

10.123 Wing Stroke Frequency. T. Casey et al. investigated wing stroke frequencies among two species of Euglossine bees, Friese and

Cockerell, in the paper "Flight Energetics of Euglossine Bees in Relation to Morphology and Wing Stroke Frequency" (*Journal of Experimental Biology*, Vol. 116, Issue 1, pp. 271–289). Following are the wing stroke frequencies, in beats per second, for samples of each species.

Friese		Cockerell		
188	235	180	182	169
190	225	178	185	180

At the 5% significance level, do the data provide sufficient evidence to conclude that a difference exists in the mean wing stroke frequencies of the two species of Euglossine bees?

10.124 Mandate Perceptions. L. Grossback et al. examined mandate perceptions and their causes in the paper "Comparing Competing Theories on the Causes of Mandate Perceptions" (*American Journal of Political Science*, Vol. 49, Issue 2, pp. 406–419). Following are data on the percentage of members in each chamber of Congress who reacted to mandates in various years.

Senate					House			
21	38	40	39	27	30.3	41.1	15.6	10.1
27	17	25	25		23.9	15.2	11.7	

At the 10% significance level, do the data provide sufficient evidence to conclude that, on average, the percentage of senators who react to mandate perceptions each year exceeds that of representatives?

10.125 Math and Chemistry. A college chemistry instructor was concerned about the detrimental effects of poor mathematics background on her students. She randomly selected 15 students and divided them according to math background. Their semester averages were the following.

Fewer than 2 years of high school algebra		Two or more years of high school algebra		
58	61	84	92	75
81	64	67	83	81
74	43	65	52	74

At the 5% significance level, do the data provide sufficient evidence to conclude that, in this teacher's chemistry courses, students with fewer than 2 years of high school algebra have a lower mean semester average than those with 2 or more years?

10.126 Picoplankton in the Bay. Picoplankton are micron-sized, single-cell algae that are an integral component of aquatic ecosystems, both in estuarine and open ocean waters. In the paper "Spatial and Temporal Variability of Picocyanobacteria *Synechococcus* sp. in San Francisco Bay" (*Limnology and Oceanography*, Vol. 45(3), pp. 695–702), X. Ning et al. examined the spatial and temporal dynamics of picoplankton populations in the diverse estuarine environment of San Francisco Bay. Oceanographers classify the Bay into three spatial regions: North, Central, and South. Independent samples of picoplankton in the North and South Bays yielded the following data on concentration in units of 10^7 cells per liter.

North	16.2	11.2	24.8	36.4	15.0	23.6	12.1
South	9.8	18.7	26.0	7.4	15.0		

At the 5% significance level, do the data provide sufficient evidence to conclude that the mean concentrations of the picoplankton populations differ between the North and South Bays?

10.127 Weekly Earnings. The Bureau of Labor Statistics publishes data on weekly earnings of full-time wage and salary workers in *Employment and Earnings*. Independent random samples of male and female workers gave the data on weekly earnings, in dollars, found in the following table.

Men		Women	
924	575	2078	358
2621	415	2193	374
1888	405	594	1181
386	816	375	1445
		510	412

At the 5% significance level, do the data provide sufficient evidence to conclude that the median weekly earnings of male full-time wage and salary workers exceeds the median weekly earnings of female full-time wage and salary workers?

10.128 SAT Essay Scores. The SAT is a standardized test required for most college admissions in the United States. It is owned, published, and developed by the College Board and results are reported in *College-Bound Seniors Total Group Profile Report*. In 2006, an essay section with possible scores ranging from 2 to 12 was added to the test. The following essay scores are from samples of students who took their SAT during either their freshman or senior year.

Freshman					Senior				
8	9	8	6	6	8	6	7	9	9
4	9	6	8	5	10	8	6	5	8

At the 1% significance level, is there sufficient evidence to conclude that the median SAT essay score is lower for students who take the SAT during their freshman year than for those who take it during their senior year?

10.129 Doing Time. The Federal Bureau of Prisons publishes data in *Prison Statistics* on the times served by prisoners released from federal institutions for the first time. Independent random samples of released prisoners in the fraud and firearms offense categories yielded the following information on time served, in months.

Fraud		Firearms	
3.6	17.9	25.5	23.8
5.3	5.9	10.4	17.9
10.7	7.0	18.4	21.9
8.5	13.9	19.6	13.3
11.8	16.6	20.9	16.1

a. Do the data provide sufficient evidence to conclude that the mean time served for fraud is less than that for firearms offenses? Perform a Mann–Whitney test at a significance level of 0.05.
b. The hypothesis test in part (a) was done in Exercise 10.45 with the pooled *t*-test. The assumption there is that times served for both offense categories are normally distributed and have equal standard deviations. If that in fact is true, why can you use a Mann–Whitney test to compare the means? Is the pooled *t*-test or the Mann–Whitney test better in this case? Explain your answers.

10.130 Mutual Funds. A *mutual fund* is a professionally managed investment that can be sold to the general public. The following table lists the 3-month rates of return, in percent, for samples of healthcare-related and technology-related mutual funds, as reported by Morningstar, an independent investment research company.

Healthcare					Technology				
9.7	9.5	7.6	6.8	6.7	6.5	6.4	6.4	6.2	4.9
6.7	6.6	4.6	3.9	3.9	4.6	4.6	4.2	3.6	3.2
3.7	3.3	3.0	0.8	−9.9	2.9	2.8	2.7	2.5	

a. Use the technology of your choice to obtain normal probability plots and boxplots for the two samples.
b. Is it reasonable to use the pooled *t*-test to perform a hypothesis test here? Explain your answer.
c. Is it reasonable to use the Mann–Whitney test to perform a hypothesis test here? Explain your answer.

10.131 Weekly Earnings. Refer to Exercise 10.127.
a. Use the technology of your choice to obtain normal probability plots and boxplots for the two samples.
b. Is it reasonable to use the pooled *t*-test to perform the hypothesis test required in Exercise 10.127? Explain your answer.
c. Is it reasonable to use the Mann–Whitney test to perform the hypothesis test required in Exercise 10.127? Explain your answer.

Working with Large Data Sets

10.132 Gender and Direction. In the paper "The Relation of Sex and Sense of Direction to Spatial Orientation in an Unfamiliar Environment" (*Journal of Environmental Psychology*, Vol. 20, pp. 17–28), J. Sholl et al. published the results of examining the sense of direction of 30 male and 30 female students. After being taken to an unfamiliar wooded park, the students were given some spatial orientation tests, including pointing to south, which tested their absolute frame of reference. The students pointed by moving a pointer attached to a 360° protractor. The absolute pointing errors, in degrees, are provided on the WeissStats site.
a. Use the Mann–Whitney test to decide whether, on average, males have a better sense of direction and, in particular, a better frame of reference than females. Perform the test with $\alpha = 0.01$.
b. Obtain boxplots and normal probability plots for both samples.
c. In Exercise 10.46, you used the pooled *t*-test to conduct the hypothesis test. Based on your graphs in part (b), which test is more appropriate, the pooled *t*-test or the Mann–Whitney test? Explain your answer.

10.133 Formaldehyde Exposure. One use of the chemical formaldehyde is to preserve animal specimens. In the article "Exposure to Formaldehyde Among Animal Health Students" (*American Industrial Hygiene Association Journal*, Vol. 63, pp. 647–650), A. Dufresne et al. examined student exposure to formaldehyde. In the course of their lab work, 18 students at each of two animal health training centers were exposed to formaldehyde. Testing equipment recorded the total amount of formaldehyde, in milligrams per milliliter (mg/mL), to which each student was exposed. The results are presented on the WeissStats site.
a. Obtain boxplots and normal probability plots of the two samples.
b. Based on your results from part (a), given the choice between using a pooled *t*-test or a Mann–Whitney test, which would you choose? Explain your answer.

c. Use the test that you chose in part (b) to decide whether the data provide sufficient evidence to conclude that there is a difference in median formaldehyde exposure in the two labs. Perform the required hypothesis test at the 5% significance level.

10.134 Teacher Salaries. The National Education Association collects data on teacher salaries and publishes results in *Estimates of School Statistics Database*. Independent samples of 100 secondary school teachers and 125 elementary school teachers yielded the data, in thousands of dollars, on annual salaries as presented on the WeissStats site.

a. Obtain histograms, boxplots, and normal probability plots of the two samples.

b. Based on your results from part (a), given the choice between using a pooled *t*-test or a Mann–Whitney test, which would you choose? Explain your answer.

c. Use both the pooled *t*-test and the Mann–Whitney test to decide, at the 5% significance level, whether the data provide sufficient evidence to conclude that the average salary of secondary school teachers exceeds that of elementary school teachers. Compare the results of the two tests.

Extending the Concepts and Skills

Normal Approximation for *M*. The Mann–Whitney table, Table VI, stops at $n_1 = 10$ and $n_2 = 10$. For larger samples, a normal approximation can be used.

Normal Approximation for *M*

Suppose that the two distributions of the variable under consideration have the same shape. Then, for samples of sizes n_1 and n_2,

- $\mu_M = n_1(n_1 + n_2 + 1)/2$,
- $\sigma_M = \sqrt{n_1 n_2(n_1 + n_2 + 1)/12}$, and
- *M* is approximately normally distributed for $n_1 \geq 10$ and $n_2 \geq 10$.

Thus, for sample sizes of 10 or more, the standardized variable

$$z = \frac{M - n_1(n_1 + n_2 + 1)/2}{\sqrt{n_1 n_2(n_1 + n_2 + 1)/12}}$$

has approximately the standard normal distribution.

10.135 Large-Sample Mann–Whitney Test. Formulate a hypothesis-testing procedure for a Mann–Whitney test that uses the test statistic *z* given in the preceding box.

10.136 Doing Time. Refer to Exercise 10.129.

a. Use your procedure from Exercise 10.135 to perform the hypothesis test.

b. Compare your result in part (a) to the one you obtained in Exercise 10.129(a), where you didn't use the normal approximation.

10.137 The Distribution of *M*. In this exercise, you are to obtain the distribution of the variable *M* when the sample sizes are both 3. Doing so enables you to see how the Mann–Whitney table is constructed. All possible ranks for the data are displayed in the

following table; the letter *A* stands for a member from Population 1, and the letter *B* stands for a member from Population 2.

a. Complete the table. (*Hint:* There are 20 rows.)

b. If the null hypothesis, $H_0: \mu_1 = \mu_2$, is true, what percentages of samples will match any given row of the table? (*Hint:* The answer is the same for all rows.)

c. Use the answer from part (b) to obtain the distribution of *M* when $n_1 = 3$ and $n_2 = 3$.

d. Draw a relative-frequency histogram of the distribution obtained in part (c).

e. Use your histogram from part (d) to obtain the entry in Table VI for $n_1 = 3$, $n_2 = 3$, and $\alpha = 0.10$.

Rank						
1	2	3	4	5	6	*M*
A	A	A	B	B	B	6
A	A	B	A	B	B	7
A	A	B	B	A	B	8
.	.	.	.	.	.	.
.	.	.	.	.	.	.
.	.	.	.	.	.	.
B	B	A	B	A	A	14
B	B	B	A	A	A	15

10.138 Transformations. Often data do not satisfy the conditions for use of any of the standard hypothesis-testing procedures that we have discussed—the pooled *t*-test, nonpooled *t*-test, or Mann–Whitney test. However, by making a suitable *transformation,* you can often obtain data that do satisfy the assumptions of one or more of these standard tests.

In the paper "A Bayesian Analysis of a Multiplicative Treatment Effect in Weather Modification" (*Technometrics*, Vol. 17, pp. 161–166), J. Simpson et al. presented the results of a study on cloud seeding with silver nitrate. The rainfall amounts, in acre-feet, for unseeded and seeded clouds are provided on the WeissStats site. Suppose that you want to perform a hypothesis test to decide whether cloud seeding with silver nitrate increases rainfall.

a. Obtain boxplots and normal probability plots for both samples.

b. Is use of the pooled *t*-test appropriate? Why or why not?

c. Is use of the nonpooled *t*-test appropriate? Why or why not?

d. Is use of the Mann–Whitney test appropriate? Why or why not?

e. Now transform each sample by taking logarithms. That is, for each observation, *x*, obtain log *x*.

f. Obtain boxplots and normal probability plots for both transformed samples.

g. Is use of the pooled *t*-test on the transformed data appropriate? Why or why not?

h. Is use of the nonpooled *t*-test on the transformed data appropriate? Why or why not?

i. Is use of the Mann–Whitney test on the transformed data appropriate? Why or why not?

j. Which of the three procedures would you use to conduct the hypothesis test for the transformed data? Explain your answer.

k. Use the test you designated in part (j) to conduct the hypothesis test for the transformed data.

l. What conclusions can you draw?

10.5 Inferences for Two Population Means, Using Paired Samples

So far, we have compared the means of two populations by using independent samples. In this section and Section 10.6, we compare such means by using a *paired sample*. A paired sample may be appropriate when the members of the two populations have a natural pairing.

Each pair in a **paired sample** consists of a member of one population and that member's corresponding member in the other population. With a **simple random paired sample,** each possible paired sample is equally likely to be the one selected. Example 10.14 provides an unrealistically simple illustration of paired samples, but it will help you understand the concept.

EXAMPLE 10.14   Introducing Random Paired Samples

Husbands and Wives Let's consider two small populations, one consisting of five married women and the other of their five husbands, as shown in the following figure. The arrows in the figure indicate the married couples, which constitute the pairs for these two populations.

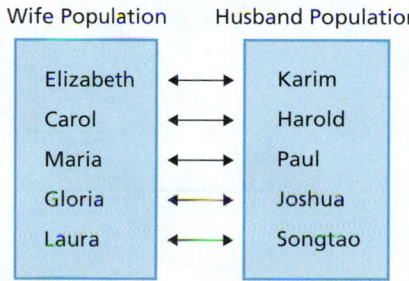

Suppose that we take a paired sample of size 3 (i.e., a sample of three pairs) from these two populations.

a. List the possible paired samples.
b. If a paired sample is selected at random (simple random paired sample), find the chance of obtaining any particular paired sample.

Solution We designated a wife–husband pair by using the first letter of each name. For example, (E, K) represents the couple Elizabeth and Karim.

a. There are 10 possible paired samples of size 3, as displayed in Table 10.12.
b. For a simple random paired sample of size 3, each of the 10 possible paired samples listed in Table 10.12 is equally likely to be the one selected. Therefore the chance of obtaining any particular paired sample of size 3 is $\frac{1}{10}$. ◼

TABLE 10.12
Possible paired samples of size 3 from the wife and husband populations

Paired sample
(E, K), (C, H), (M, P)
(E, K), (C, H), (G, J)
(E, K), (C, H), (L, S)
(E, K), (M, P), (G, J)
(E, K), (M, P), (L, S)
(E, K), (G, J), (L, S)
(C, H), (M, P), (G, J)
(C, H), (M, P), (L, S)
(C, H), (G, J), (L, S)
(M, P), (G, J), (L, S)

The previous example provides a concrete illustration of paired samples and emphasizes that, for simple random paired samples of any given size, each possible paired sample is equally likely to be the one selected. In practice, we neither obtain the number of possible paired samples nor explicitly compute the chance of selecting a particular paired sample. However, these concepts underlie the methods we do use.

Comparing Two Population Means, Using a Paired Sample

We are now ready to examine a process for comparing the means of two populations by using a paired sample.

■■■ **EXAMPLE 10.15** **Comparing Two Means, Using a Paired Sample**

Ages of Married People The U.S. Census Bureau publishes information on the ages of married people in *Current Population Reports*. Suppose that we want to decide whether, in the United States, the mean age of married men differs from the mean age of married women.

a. Formulate the problem statistically by posing it as a hypothesis test.
b. Explain the basic idea for carrying out the hypothesis test.
c. Suppose that 10 married heterosexual couples in the United States are selected at random and that the ages, in years, of the people chosen are as shown in the second and third columns of Table 10.13. Discuss the use of these data to make a decision concerning the hypothesis test.

TABLE 10.13

Ages, in years, of a random sample of 10 married couples

Couple	Husband	Wife	Difference, d
1	59	53	6
2	21	22	−1
3	33	36	−3
4	78	74	4
5	70	64	6
6	33	35	−2
7	68	67	1
8	32	28	4
9	54	41	13
10	52	44	8
			36

Solution

a. To formulate the problem statistically, we first note that we have one variable—namely, age—and two populations:

Population 1: All married men
Population 2: All married women.

Let μ_1 and μ_2 denote the means of the variable "age" for Population 1 and Population 2, respectively:

μ_1 = mean age of all married men
μ_2 = mean age of all married women.

We want to perform the hypothesis test

H_0: $\mu_1 = \mu_2$ (mean ages of married men and women are the same)
H_a: $\mu_1 \neq \mu_2$ (mean ages of married men and women differ).

b. Independent samples could be used to carry out the hypothesis test: Take independent simple random samples of, say, 10 married men and 10 married women and then apply a pooled or nonpooled t-test to the age data obtained. However, in this case, a paired sample is more appropriate. Here, a pair consists of a married (heterosexual) couple. The variable we analyze is the difference between the ages of the husband and wife in a couple. By using a paired sample, we can remove an extraneous source of variation: the variation in the ages among married couples. The sampling error thus made in estimating the difference between the population means will generally be smaller and, therefore, we are more likely to detect differences between the population means when such differences exist.

c. The last column of Table 10.13 contains the difference, d, between the ages of each of the 10 couples sampled. We refer to each difference as a **paired**

difference because it is the difference of a pair of observations. For example, in the first couple, the husband is 59 years old and the wife is 53 years old, giving a paired difference of 6 years, meaning that the husband is 6 years older than his wife.

If the null hypothesis of equal mean ages is true, the paired differences of the ages for the married couples sampled should average about 0; that is, the sample mean, $\bar{d}$, of the paired differences should be roughly 0. If $\bar{d}$ is too much different from 0, we would take this as evidence that the null hypothesis is false. From the last column of Table 10.13, we find that the sample mean of the paired differences is

$$\bar{d} = \frac{\Sigma d_i}{n} = \frac{36}{10} = 3.6.$$

The question now is, can this difference of 3.6 years be reasonably attributed to sampling error, or is the difference large enough to indicate that the two populations have different means? To answer that question, we need to know the distribution of the variable $\bar{d}$, which we discuss next.

The Paired t-Statistic

Suppose that x is a variable on each of two populations whose members can be paired. For each pair, we let d denote the difference between the values of the variable x on the members of the pair. We call d the **paired-difference variable.**

It can be shown that the mean of the paired differences equals the difference between the two population means. In symbols,

$$\mu_d = \mu_1 - \mu_2.$$

Furthermore, if d is normally distributed, we can apply this equation and our knowledge of the studentized version of a sample mean (Key Fact 8.5 on page 353) to obtain Key Fact 10.6.

KEY FACT 10.6

Distribution of the Paired t-Statistic

Suppose that x is a variable on each of two populations whose members can be paired. Further suppose that the paired-difference variable d is normally distributed. Then, for paired samples of size n, the variable

$$t = \frac{\bar{d} - (\mu_1 - \mu_2)}{s_d/\sqrt{n}}$$

has the t-distribution with df $= n - 1$.

Note: We use the phrase **normal differences** as an abbreviation of "the paired-difference variable is normally distributed."

Hypothesis Tests for the Means of Two Populations, Using a Paired Sample

We now present a hypothesis-testing procedure based on a paired sample for comparing the means of two populations when the paired-difference variable is normally distributed. In light of Key Fact 10.6, for a hypothesis test that has null hypothesis H_0: $\mu_1 = \mu_2$, we can use the variable

$$t = \frac{\bar{d}}{s_d/\sqrt{n}}$$

as the test statistic and obtain the critical value(s) or P-value from the t-table, Table IV.

We call this hypothesis-testing procedure the **paired *t*-test.** Note that the paired *t*-test is simply the one-mean *t*-test applied to the paired-difference variable with null hypothesis H_0: $\mu_d = 0$. Procedure 10.6 provides a step-by-step method for performing a paired *t*-test by using either the critical-value approach or the *P*-value approach.

■■■ PROCEDURE 10.6   Paired *t*-Test

Purpose   To perform a hypothesis test to compare two population means, μ_1 and μ_2

Assumptions
1.   Simple random paired sample
2.   Normal differences or large sample

Step 1   The null hypothesis is H_0: $\mu_1 = \mu_2$, and the alternative hypothesis is

$$H_a: \mu_1 \neq \mu_2 \quad\text{or}\quad H_a: \mu_1 < \mu_2 \quad\text{or}\quad H_a: \mu_1 > \mu_2$$
$$\text{(Two tailed)} \qquad\qquad \text{(Left tailed)} \qquad\qquad \text{(Right tailed)}$$

Step 2   Decide on the significance level, α.

Step 3   Compute the value of the test statistic

$$t = \frac{\bar{d}}{s_d/\sqrt{n}}$$

and denote that value t_0.

| **CRITICAL-VALUE APPROACH** | OR | **P-VALUE APPROACH** |

Step 4   The critical value(s) are

$$\pm t_{\alpha/2} \quad\text{or}\quad -t_{\alpha} \quad\text{or}\quad t_{\alpha}$$
$$\text{(Two tailed)} \qquad \text{(Left tailed)} \qquad \text{(Right tailed)}$$

with df $= n - 1$. Use Table IV to find the critical value(s).

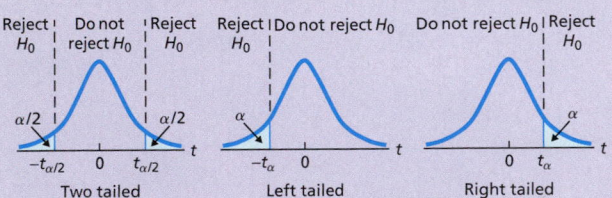

Step 5   If the value of the test statistic falls in the rejection region, reject H_0; otherwise, do not reject H_0.

Step 4   The *t*-statistic has df $= n - 1$. Use Table IV to estimate the *P*-value, or obtain it exactly by using technology.

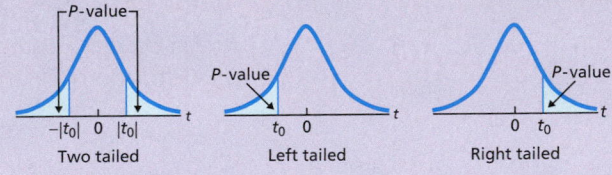

Step 5   If $P \leq \alpha$, reject H_0; otherwise, do not reject H_0.

Step 6   Interpret the results of the hypothesis test.

Note: The hypothesis test is exact for normal differences and is approximately correct for large samples and nonnormal differences.

Properties and guidelines for use of the paired *t*-test are the same as those given for the one-mean *z*-test in Key Fact 9.7 on page 389 when applied to paired differences. In particular, the paired *t*-test is robust to moderate violations of the normality assumption but, even for large samples, can sometimes be unduly affected by outliers because the sample mean and sample standard deviation are not resistant to outliers. Here are two other important points:

- Do not apply the paired t-test to independent samples, and, likewise, do not apply a pooled or nonpooled t-test to a paired sample.
- The normality assumption for a paired t-test refers to the distribution of the paired-difference variable, not to the two distributions of the variable under consideration.

■■■■ **EXAMPLE 10.16**   **The Paired t-Test**

Ages of Married People   We now return to the hypothesis test posed in Example 10.15. A random sample of 10 married couples gave the data on ages, in years, shown in the second and third columns of Table 10.13 on page 486. At the 5% significance level, do the data provide sufficient evidence to conclude that the mean age of married men differs from the mean age of married women?

Solution   First, we check the two conditions required for using the paired t-test, as listed in Procedure 10.6.

- Assumption 1 is satisfied because we have a simple random paired sample. Each pair consists of a married couple.
- Because the sample size, $n = 10$, is small, we need to examine issues of normality and outliers. (See the first bulleted item in Key Fact 9.7 on page 389.) To do so, we construct in Fig. 10.15 a normal probability plot for the sample of paired differences in the last column of Table 10.13. This plot reveals no outliers and is roughly linear. So we can consider Assumption 2 satisfied.

    From the preceding items, we see that the paired t-test can be used to conduct the required hypothesis test. We apply Procedure 10.6.

FIGURE 10.15

Normal probability plot of the paired differences in Table 10.13

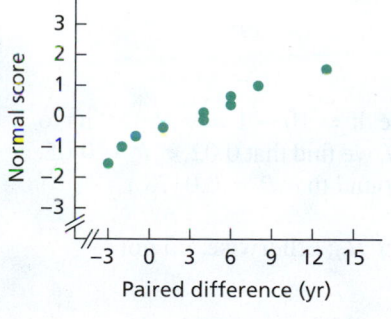

Paired difference (yr)

Step 1   State the null and alternative hypotheses.

Let μ_1 denote the mean age of all married men, and let μ_2 denote the mean age of all married women. Then the null and alternative hypotheses are, respectively,

$$H_0: \mu_1 = \mu_2 \text{ (mean ages are equal)}$$
$$H_a: \mu_1 \neq \mu_2 \text{ (mean ages differ)}.$$

Note that the hypothesis test is two tailed.

Step 2   Decide on the significance level, α.

We are to perform the test at the 5% significance level, so $\alpha = 0.05$.

Step 3   Compute the value of the test statistic

$$t = \frac{\bar{d}}{s_d / \sqrt{n}}.$$

The paired differences (d-values) of the sample pairs are shown in the last column of Table 10.13. We need to determine the sample mean and sample standard deviation of those paired differences. We do so in the usual manner:

$$\bar{d} = \frac{\Sigma d_i}{n} = \frac{36}{10} = 3.6,$$

and

$$s_d = \sqrt{\frac{\Sigma d_i^2 - (\Sigma d_i)^2/n}{n-1}} = \sqrt{\frac{352 - (36)^2/10}{10-1}} = 4.97.$$

Consequently, the value of the test statistic is

$$t = \frac{\overline{d}}{s_d/\sqrt{n}} = \frac{3.6}{4.97/\sqrt{10}} = 2.291.$$

| CRITICAL-VALUE APPROACH | OR | P-VALUE APPROACH |

CRITICAL-VALUE APPROACH

Step 4 The critical values for a two-tailed test are $\pm t_{\alpha/2}$ with df $= n - 1$. Use Table IV to find the critical values.

We have $n = 10$ and $\alpha = 0.05$. Table IV reveals that, for df $= 10 - 1 = 9$, $\pm t_{0.05/2} = \pm t_{0.025} = \pm 2.262$, as shown in Fig. 10.16A.

FIGURE 10.16A

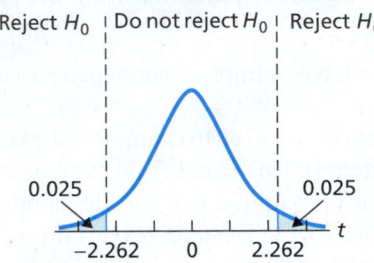

Reject H_0 | Do not reject H_0 | Reject H_0

0.025   0.025
            -2.262   0   2.262

Step 5 If the value of the test statistic falls in the rejection region, reject H_0; otherwise, do not reject H_0.

From Step 3, the value of the test statistic is $t = 2.291$, which falls in the rejection region depicted in Fig.10.16A. Thus we reject H_0. The test results are statistically significant at the 5% level.

P-VALUE APPROACH

Step 4 The t-statistic has df $= n - 1$. Use Table IV to estimate the P-value, or obtain it exactly by using technology.

From Step 3, the value of the test statistic is $t = 2.291$. The test is two tailed, so the P-value is the probability of observing a value of t of 2.291 or greater in magnitude if the null hypothesis is true. That probability equals the shaded area shown in Fig. 10.16B.

FIGURE 10.16B

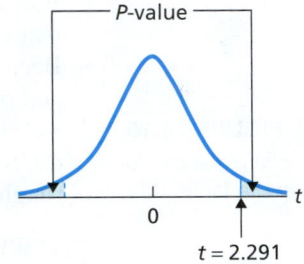

P-value

0
$t = 2.291$

Because $n = 10$, we have df $= 10 - 1 = 9$. Referring to Fig. 10.16B and Table IV, we find that $0.02 < P < 0.05$. (Using technology, we found that $P = 0.0478$.)

Step 5 If $P \leq \alpha$, reject H_0; otherwise, do not reject H_0.

From Step 4, $0.02 < P < 0.05$. Because the P-value is less than the specified significance level of 0.05, we reject H_0. The test results are statistically significant at the 5% level and (see Table 9.8 on page 386) provide strong evidence against the null hypothesis.

StatCrunch

Report 10.6

You try it!

Exercise 10.159 on page 496

Step 6 Interpret the results of the hypothesis test.

Interpretation At the 5% significance level, the data provide sufficient evidence to conclude that the mean age of married men differs from the mean age of married women.

Confidence Intervals for the Difference between the Means of Two Populations, Using a Paired Sample

We can also use Key Fact 10.6 on page 487 to derive a confidence-interval procedure for the difference between two population means. We call that confidence-interval procedure the **paired *t*-interval procedure**.

■ ■■ **PROCEDURE 10.7**   Paired *t*-Interval Procedure

Purpose To find a confidence interval for the difference between two population means, μ_1 and μ_2

Assumptions
1. Simple random paired sample
2. Normal differences or large sample

Step 1   For a confidence level of $1 - \alpha$, use Table IV to find $t_{\alpha/2}$ with df $= n - 1$.

Step 2   The endpoints of the confidence interval for $\mu_1 - \mu_2$ are

$$\bar{d} \pm t_{\alpha/2} \cdot \frac{s_d}{\sqrt{n}}.$$

Step 3   Interpret the confidence interval.

Note: The confidence interval is exact for normal differences and is approximately correct for large samples and nonnormal differences.

■ ■■ **EXAMPLE 10.17**   The Paired *t*-Interval Procedure

Ages of Married People Use the age data in the second and third columns of Table 10.13 on page 486 to obtain a 95% confidence interval for the difference, $\mu_1 - \mu_2$, between the mean ages of married men and married women.

Solution   We apply Procedure 10.7.

Step 1 For a confidence level of $1 - \alpha$, use Table IV to find $t_{\alpha/2}$ with df $= n - 1$.

For a 95% confidence interval, $\alpha = 0.05$. From Table IV, we determine that, for df $= n - 1 = 10 - 1 = 9$, we have $t_{\alpha/2} = t_{0.05/2} = t_{0.025} = 2.262$.

Step 2 The endpoints of the confidence interval for $\mu_1 - \mu_2$ are

$$\bar{d} \pm t_{\alpha/2} \cdot \frac{s_d}{\sqrt{n}}.$$

From Step 1, $t_{\alpha/2} = 2.262$ and $n = 10$. And, from Example 10.16, $\bar{d} = 3.6$ and $s_d = 4.97$. So, the endpoints of the confidence interval for $\mu_1 - \mu_2$ are

$$3.6 \pm 2.262 \cdot \frac{4.97}{\sqrt{10}},$$

or 0.04 to 7.16.

Step 3 Interpret the confidence interval.

Interpretation   We can be 95% confident that the difference between the mean ages of married men and married women is somewhere between 0.04 years and 7.16 years. In other words (see page 443), we can be 95% confident that the mean age of married men exceeds the mean age of married women by somewhere between 0.04 years and 7.16 years.

StatCrunch

Report 10.7

You try it!

Exercise 10.165 on page 496

Paired versus Independent Samples

Suppose that we want to perform a hypothesis test or obtain a confidence interval in order to compare two population means. As you know, two sampling designs that are candidates for the comparison are paired samples and independent samples.

When a paired sample is possible, it is often preferable to independent samples. Indeed, as we mentioned in the solution to part (b) of Example 10.15, by using a paired sample, we can frequently remove extraneous sources of variation. The sampling error thus made in estimating the difference between the population means will generally be smaller and, therefore, we are more likely to detect differences between the population means when such differences exist.

The paired-sample design is preferable when, for the variable under consideration, there is a positive correlation between the population pairs; that is, when population pairs tend to both have above-average values of the variable or below-average values of the variable. This situation is precisely the case for ages of husbands and wives (Example 10.15)—older husbands tend to be married to older wives.

THE TECHNOLOGY CENTER

Most statistical technologies have programs that automatically perform paired t-procedures. In this subsection, we present output and step-by-step instructions for such programs.

EXAMPLE 10.18 **Using Technology to Conduct Paired t-Procedures**

Ages of Married People The second and third columns of Table 10.13 on page 486 give the ages of 10 randomly selected married couples. Use Minitab, Excel, or the TI-83/84 Plus to perform the hypothesis test in Example 10.16 and obtain the confidence interval required in Example 10.17.

Solution Let μ_1 denote the mean age of all married men, and let μ_2 denote the mean age of all married women. The task in Example 10.16 is to perform the hypothesis test

$$H_0: \mu_1 = \mu_2 \text{ (mean ages are equal)}$$
$$H_a: \mu_1 \neq \mu_2 \text{ (mean ages differ)}$$

at the 5% significance level; the task in Example 10.17 is to obtain a 95% confidence interval for $\mu_1 - \mu_2$.

We applied the paired t-procedures programs to the data, resulting in Output 10.4. Steps for generating that output are presented in Instructions 10.4. *Note to Excel users:* For brevity, we have presented only the essential portions of the actual output.

As shown in Output 10.4, the P-value for the hypothesis test is about 0.048. Because the P-value is less than the specified significance level of 0.05, we reject H_0. Output 10.4 also shows that a 95% confidence interval for the difference between the means is from 0.04 to 7.16.

OUTPUT 10.4 Paired t-procedures on the age data

MINITAB

```
Paired T-Test and CI: HUSBAND, WIFE

Paired T for HUSBAND - WIFE

              N    Mean   StDev   SE Mean
HUSBAND      10   50.00   19.30    6.10
WIFE         10   46.40   17.47    5.52
Difference   10    3.60    4.97    1.57


95% CI for mean difference: (0.04, 7.16)
T-Test of mean difference = 0 (vs ≠ 0): T-Value = 2.29  P-Value = 0.048
```

OUTPUT 10.4 (cont.)   Paired *t*-procedures on the age data

EXCEL

Hypothesized difference (D): 0			
Significance level (%): 5			
Summary statistics:			
Variable	Observations	Mean	Std. deviation
HUSBAND	10	50.0000	19.2988
WIFE	10	46.4000	17.4687

t-test for two paired samples / Two-tailed test:

95% confidence interval on the difference between the means:
(0.0439, 7.1561)

Difference	3.6000
t (Observed value)	2.2901
|t| (Critical value)	2.2622
DF	9
p-value (Two-tailed)	0.0478
alpha	0.05

TI-83/84 PLUS

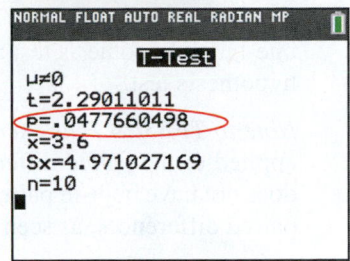

Using **T-Test**

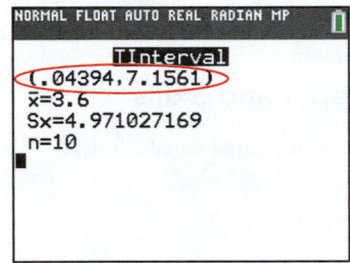

Using **TInterval**

INSTRUCTIONS 10.4   Steps for generating Output 10.4

MINITAB

1 Store the age data from the second and third columns of Table 10.13 in columns named HUSBAND and WIFE
2 Choose **Stat ➤ Basic Statistics ➤ Paired t...**
3 Press the F3 key to reset the dialog box
4 Click in the **Sample 1** text box and specify HUSBAND
5 Click in the **Sample 2** text box and specify WIFE
6 Click the **Options...** button
7 Click in the **Confidence level** text box and type 95
8 Click the arrow button at the right of the **Alternative hypothesis** drop-down list box and select **Difference ≠ hypothesized difference**
9 Click **OK** twice

EXCEL

1 Store the age data from the second and third columns of Table 10.13 in columns named HUSBAND and WIFE
2 Choose **XLSTAT ➤ Parametric tests ➤ Two-sample t-test and z-test**
3 Click the reset button in the lower left corner of the dialog box
4 Click in the **Sample 1** selection box and then select the column of the worksheet that contains the HUSBAND data
5 Click in the **Sample 2** selection box and then select the column of the worksheet that contains the WIFE data
6 In the **Data format** list, select the **Paired samples** option button
7 Click the **Options** tab
8 Click the arrow button at the right of the **Alternative hypothesis** drop-down list box and select **Mean 1 – Mean 2 ≠ D**
9 Type 5 in the **Significance level (%)** text box

10 Click **OK**
11 Click the **Continue** button in the **XLSTAT – Selections** dialog box

TI-83/84 PLUS

Store the age data from the second and third columns of Table 10.13 in lists named HUSB and WIFE.

FOR THE PAIRED DIFFERENCES:
1 Press **2nd ➤ LIST**, arrow down to HUSB, and press **ENTER**
2 Press −
3 Press **2nd ➤ LIST**, arrow down to WIFE, and press **ENTER**
4 Press **STO ▸**
5 Press **2nd ➤ A-LOCK**, type DIFF, and press **ENTER**

FOR THE HYPOTHESIS TEST:
1 Press **STAT**, arrow over to **TESTS**, and press **2**
2 Highlight **Data** and press **ENTER**
3 Press the down-arrow key, type 0 for μ_0, and press **ENTER**
4 Press **2nd ➤ LIST**, arrow down to DIFF, and press **ENTER** twice
5 Type 1 for **Freq** and then press **ENTER**
6 Highlight ≠ μ_0 and press **ENTER**
7 Arrow down to **Calculate** and press **ENTER**

FOR THE CI:
1 Press **STAT**, arrow over to **TESTS**, and press **8**
2 Highlight **Data** and press **ENTER**
3 Press the down-arrow key
4 Press **2nd ➤ LIST**, arrow down to DIFF, and press **ENTER** twice
5 Type 1 for **Freq** and then press **ENTER**
6 Type .95 for **C-Level** and press **ENTER** twice

Note to Minitab and Excel users: As we have noted, Minitab and Excel compute a two-sided confidence interval for a two-tailed test and a one-sided confidence interval for a one-tailed test. To perform a one-tailed hypothesis test and obtain a two-sided confidence interval, apply Minitab's or Excel's paired *t*-procedure twice: once for the one-tailed hypothesis test and once for the confidence interval specifying a two-tailed hypothesis test.

Note to TI-83/84 Plus users: The paired *t*-procedures are just one-mean *t*-procedures applied to the paired differences. Since, at the time of this writing, the TI-83/84 Plus does not have built-in paired *t*-procedures, we applied its one-mean *t*-procedures to the paired differences, as seen in Instructions 10.4.

Exercises 10.5

Understanding the Concepts and Skills

10.139 State one possible advantage of using paired samples instead of independent samples.

10.140 What constitutes each pair in a paired sample?

10.141 State the two conditions required for performing a paired *t*-procedure. How important are those conditions?

10.142 Provide an example (different from the ones considered in this section) of a procedure based on a paired sample being more appropriate than one based on independent samples.

In Exercises 10.143–10.148, hypothesis tests are proposed. For each hypothesis test,
a. *identify the variable.*
b. *identify the two populations.*
c. *identify the pairs.*
d. *identify the paired-difference variable.*
e. *determine the null and alternative hypotheses.*
f. *classify the hypothesis test as two tailed, left tailed, or right tailed.*

10.143 **TV Viewing.** The A. C. Nielsen Company collects data on the TV viewing habits of Americans and publishes the information in *Nielsen Report on Television*. Suppose that you want to use a paired sample to decide whether the mean viewing times of married men and married women differ.

10.144 **Self-Reported Weight.** The article "Accuracy of Self-Reported Height and Weight in a Community-Based Sample of Older African Americans and Whites" (*Journal of Gerontology Series A: Biological Sciences and Medical Sciences*, Vol. 65A, No. 10, pp. 1123–1129) by G. Fillenbaum et al. explores the relationship between measured and self-reported height and weight. The authors sampled African American and White women and men older than 70 years of age. A hypothesis test is to be performed to decide whether, on average, self-reported weight is less than measured weight for the aforementioned age group.

10.145 **Hypnosis and Pain.** In the paper "An Analysis of Factors That Contribute to the Efficacy of Hypnotic Analgesia" (*Journal of Abnormal Psychology*, Vol. 96, No. 1, pp. 46–51), D. Price and J. Barber examined the effects of hypnosis on pain. They measured response to pain using a visual analogue scale (VAS), in centimeters, where higher VAS indicates greater pain. VAS sensory ratings were made before and after hypnosis on each of 16 subjects. A hypothesis test is to be performed to decide whether, on average, hypnosis reduces pain.

10.146 **Sports Stadiums and Home Values.** In the paper "Housing Values Near New Sporting Stadiums" (*Land Economics*, Vol. 81, Issue 3, pp. 379–395), C. Tu examined the effects of construction of new sports stadiums on home values. Suppose that you want to use a paired sample to decide whether construction of new sports stadiums affects the mean price of neighboring homes.

10.147 **Breastmilk and Antioxidants.** There is convincing evidence that breastmilk containing antioxidants is important in the prevention of diseases in infants. Researchers A. Xavier et al. studied the effects of storing breastmilk on antioxidant levels in the article "Total Antioxidant Concentrations of Breastmilk—An Eye-Opener to the Negligent" (*Journal of Health, Population and Nutrition*, Vol. 29, No. 6, pp. 605–611). Samples of breastmilk were taken from women and divided into fresh samples that were immediately tested and the remaining samples that were stored in the refrigerator and tested after 48 hours. A hypothesis test is to be performed to decide whether, on average, stored breastmilk has a lower total antioxidant capacity.

10.148 **Fiber Density.** In the article "Comparison of Fiber Counting by TV Screen and Eyepieces of Phase Contrast Microscopy" (*American Industrial Hygiene Association Journal*, Vol. 63, pp. 756–761), I. Moa et al. reported on determining fiber density by two different methods. The fiber density of 10 samples with varying fiber density was obtained by using both an eyepiece method and a TV-screen method. A hypothesis test is to be performed to decide whether, on average, the eyepiece method gives a greater fiber density reading than the TV-screen method.

In Exercises 10.149–10.154, the null hypothesis is $H_0: \mu_1 = \mu_2$ and the alternative hypothesis is as specified. We have provided data from a simple random paired sample from the two populations under consideration. In each case, use the paired t-test to perform the required hypothesis test at the 10% significance level.

10.149 $H_a: \mu_1 \neq \mu_2$

Pair	Observation from	
	Population 1	**Population 2**
1	13	11
2	16	15
3	13	10
4	14	8
5	12	8
6	8	9
7	17	14

10.150 $H_a: \mu_1 < \mu_2$

	Observation from	
Pair	Population 1	Population 2
1	7	13
2	4	9
3	10	6
4	0	2
5	20	19
6	−1	5
7	12	10

10.151 $H_a: \mu_1 > \mu_2$

	Observation from	
Pair	Population 1	Population 2
1	7	3
2	4	5
3	9	8
4	7	2
5	19	16
6	12	12
7	13	18
8	5	11

10.152 $H_a: \mu_1 \neq \mu_2$

	Observation from	
Pair	Population 1	Population 2
1	10	12
2	8	7
3	13	11
4	13	16
5	17	15
6	12	9
7	12	12
8	11	7

10.153 $H_a: \mu_1 < \mu_2$

	Observation from	
Pair	Population 1	Population 2
1	15	18
2	22	25
3	15	17
4	27	24
5	24	30
6	23	23
7	8	10
8	20	27
9	2	3

10.154 $H_a: \mu_1 > \mu_2$

	Observation from	
Pair	Population 1	Population 2
1	40	32
2	30	29
3	34	36
4	22	18
5	35	31
6	26	26
7	26	25
8	27	25
9	11	15
10	35	31

Applying the Concepts and Skills

Preliminary data analyses indicate that use of a paired t-test is reasonable in Exercises 10.155–10.160. Perform each hypothesis test by using either the critical-value approach or the P-value approach.

10.155 **Zea Mays.** Charles Darwin, author of *Origin of Species*, investigated the effect of cross-fertilization on the heights of plants. In one study he planted 15 pairs of *Zea mays* plants. Each pair consisted of one cross-fertilized plant and one self-fertilized plant grown in the same pot. The following table gives the height differences, in eighths of an inch, for the 15 pairs. Each difference is obtained by subtracting the height of the self-fertilized plant from that of the cross-fertilized plant.

49	−67	8	16	6
23	28	41	14	29
56	24	75	60	−48

a. Identify the variable under consideration.
b. Identify the two populations.
c. Identify the paired-difference variable.
d. Are the numbers in the table paired differences? Why or why not?
e. At the 5% significance level, do the data provide sufficient evidence to conclude that the mean heights of cross-fertilized and self-fertilized *Zea mays* differ? (*Note:* $\bar{d} = 20.93$ and $s_d = 37.74$.)
f. Repeat part (e) at the 1% significance level.

10.156 **Sleep.** In 1908, W. S. Gosset published "The Probable Error of a Mean" (*Biometrika*, Vol. 6, pp. 1–25). In this pioneering paper, published under the pseudonym "Student," he introduced what later became known as Student's *t*-distribution. Gosset used the following data set, which gives the additional sleep in hours obtained by 10 patients who used laevohysocyamine hydrobromide.

1.9	0.8	1.1	0.1	−0.1
4.4	5.5	1.6	4.6	3.4

a. Identify the variable under consideration.
b. Identify the two populations.
c. Identify the paired-difference variable.
d. Are the numbers in the table paired differences? Why or why not?
e. At the 5% significance level, do the data provide sufficient evidence to conclude that laevohysocyamine hydrobromide is effective in increasing sleep? (*Note:* $\bar{d} = 2.33$ and $s_d = 2.002$.)
f. Repeat part (e) at the 1% significance level.

10.157 Anorexia Treatment. Anorexia nervosa is a serious eating disorder, particularly among young women. The following data provide the weights, in pounds, of 17 anorexic young women before and after receiving a family therapy treatment for anorexia nervosa. [SOURCE: D. Hand et al. (ed.) *A Handbook of Small Data Sets*, London: Chapman & Hall, 1994; raw data from B. Everitt (personal communication)]

Before	After	Before	After	Before	After
83.3	94.3	76.9	76.8	82.1	95.5
86.0	91.5	94.2	101.6	77.6	90.7
82.5	91.9	73.4	94.9	83.5	92.5
86.7	100.3	80.5	75.2	89.9	93.8
79.6	76.7	81.6	77.8	86.0	91.7
87.3	98.0	83.8	95.2		

Does family therapy appear to be effective in helping anorexic young women gain weight? Perform the appropriate hypothesis test at the 5% significance level.

10.158 Measuring Treadwear. R. Stichler et al. compared two methods of measuring treadwear in their paper "Measurement of Treadwear of Commercial Tires" (*Rubber Age*, Vol. 73:2). Eleven tires were each measured for treadwear by two methods, one based on weight and the other on groove wear. The data, in thousands of miles, are as follows.

Weight method	Groove method	Weight method	Groove method
30.5	28.7	24.5	16.1
30.9	25.9	20.9	19.9
31.9	23.3	18.9	15.2
30.4	23.1	13.7	11.5
27.3	23.7	11.4	11.2
20.4	20.9		

At the 5% significance level, do the data provide sufficient evidence to conclude that, on average, the two measurement methods give different results?

10.159 Glaucoma and Corneal Thickness. Glaucoma is a leading cause of blindness in the United States. N. Ehlers measured the corneal thickness of eight patients who had glaucoma in one eye but not in the other. The results of the study were published as the paper "On Corneal Thickness and Intraocular Pressure, II" (*Acta Opthalmologica*, Vol. 48, pp. 1107–1112). The data on corneal thickness, in microns, are shown in the following table.

Patient	Normal	Glaucoma
1	484	488
2	478	478
3	492	480
4	444	426
5	436	440
6	398	410
7	464	458
8	476	460

At the 10% significance level, do the data provide sufficient evidence to conclude that mean corneal thickness is greater in normal eyes than in eyes with glaucoma?

10.160 Cooling Down. Cooling down with a cold drink before exercise in the heat is believed to help an athlete perform. Researcher J. Dugas explored the difference between cooling down with an ice slurry (slushy) and with cold water in the article "Ice Slurry Ingestion Increases Running Time in the Heat" (*Clinical Journal of Sports Medicine*, Vol. 21, No. 6, pp. 541–542). Ten male participants drank a flavored ice slurry and ran on a treadmill in a controlled hot and humid environment. Days later, the same participants drank cold water and ran on a treadmill in the same hot and humid environment. The following table shows the times, in minutes, it took to fatigue on the treadmill for both the ice slurry and the cold water.

Subject	Cold Water	Ice Slurry
1	52	56
2	37	43
3	44	52
4	51	58
5	34	38
6	38	45
7	41	45
8	50	58
9	29	34
10	38	44

At the 1% significance level, do the data provide sufficient evidence to conclude that, on average, cold water is less effective than ice slurry for optimizing athletic performance in the heat? (*Note:* The mean and standard deviation of the paired differences are −5.9 minutes and 1.60 minutes, respectively.)

In Exercises 10.161–10.166, apply Procedure 10.7 on page 491 to obtain the required confidence interval. Interpret your result in each case.

10.161 Zea Mays. Refer to Exercise 10.155.
a. Determine a 95% confidence interval for the difference between the mean heights of cross-fertilized and self-fertilized *Zea mays*.
b. Repeat part (a) for a 99% confidence level.

10.162 Sleep. Refer to Exercise 10.156.
a. Determine a 90% confidence interval for the additional sleep that would be obtained, on average, by using laevohysocyamine hydrobromide.
b. Repeat part (a) for a 98% confidence level.

10.163 Anorexia Treatment. Refer to Exercise 10.157 and find a 90% confidence interval for the weight gain that would be obtained, on average, by using the family therapy treatment.

10.164 Measuring Treadwear. Refer to Exercise 10.158 and find a 95% confidence interval for the mean difference in measurement by the weight and groove methods.

10.165 Glaucoma and Corneal Thickness. Refer to Exercise 10.159 and obtain an 80% confidence interval for the difference between the mean corneal thickness of normal eyes and that of eyes with glaucoma.

10.166 Cooling Down. Refer to Exercise 10.160 and find a 98% confidence interval for the difference between the mean times to fatigue on a treadmill in a hot and humid environment after cooling down with cold water and after cooling down with an ice slurry.

In each of Exercises 10.167–10.169, use the technology of your choice to perform the required tasks.

10.167 Font Readability. In the online paper "A Comparison of Two Computer Fonts: Serif versus Ornate Sans Serif" (*Usability News*, Issue 5.3), researchers S. Morrison and J. Noyes studied whether the type of font used in a document affects reading speed or comprehension. The fonts used for the comparisons were the serif font Times New Roman (TNR) and a more ornate sans serif font called Gigi. The following table gives the times, in seconds, that it took each of the 25 participants to read paragraphs in the TNR and Gigi fonts.

TNR	Gigi	TNR	Gigi	TNR	Gigi
27.00	32.88	16.43	18.15	34.20	37.83
23.20	23.00	20.75	27.18	25.01	29.90
23.00	37.87	31.41	34.00	12.93	21.82
33.76	43.53	24.69	26.50	15.84	20.41
19.42	21.06	24.60	27.00	21.03	27.03
15.56	23.00	22.42	29.30	26.23	32.03
19.13	23.40	18.41	22.24	13.60	20.02
19.41	24.49	24.56	31.44		
22.28	22.76	16.75	17.95		

Suppose that you want to perform a hypothesis test to determine whether, on average, people read faster with the TNR font than with the Gigi font. Conduct preliminary graphical data analyses to decide whether applying the paired *t*-test is reasonable. Explain your decision.

10.168 Tobacco Mosaic Virus. To assess the effects of two different strains of the tobacco mosaic virus, W. Youden and H. Beale randomly selected eight tobacco leaves. Half of each leaf was subjected to one of the strains of tobacco mosaic virus and the other half to the other strain. The researchers then counted the number of local lesions apparent on each half of each leaf. The results of their study were published in the paper "A Statistical Study of the Local Lesion Method for Estimating Tobacco Mosaic Virus" (*Contributions to Boyce Thompson Institute*, Vol. 6, p. 437). Here are the data.

Leaf	1	2	3	4	5	6	7	8
Virus 1	31	20	18	17	9	8	10	7
Virus 2	18	17	14	11	10	7	5	6

Suppose that you want to perform a hypothesis test to determine whether a difference exists between the mean numbers of local lesions resulting from the two viral strains. Conduct preliminary graphical analyses to decide whether applying the paired *t*-test is reasonable. Explain your decision.

10.169 Antiviral Therapy. In the article "Improved Outcome for Children With Disseminated Adenoviral Infection Following Allogeneic Stem Cell Transplantation" (*British Journal of Haematology*, Vol. 130, Issue 4, p. 595), B. Kampmann et al. examined children who received stem cell transplants and subsequently became infected with a variety of ailments. A new antiviral therapy was administered to 11 patients. Their absolute lymphocyte counts (ABS lymphs) ($\times 10^9$/L) at onset and resolution were as shown in the following table.

Onset	Resolution	Onset	Resolution
0.08	0.59	0.31	0.38
0.02	0.37	0.23	0.39
0.03	0.07	0.09	0.02
0.64	0.81	0.10	0.38
0.03	0.76	0.04	0.60
0.15	0.44		

a. Obtain normal probability plots and boxplots of the onset data, the resolution data, and the paired differences of those data.
b. Based on your results from part (a), is applying a one-mean *t*-procedure to the onset data reasonable?
c. Based on your results from part (a), is applying a one-mean *t*-procedure to the resolution data reasonable?
d. Based on your results from part (a), is applying a paired *t*-procedure to the data reasonable?
e. What do your answers from parts (b)–(d) imply about the conditions for using a paired *t*-procedure?

Working with Large Data Sets

10.170 Faculty Salaries. The American Association of University Professors (AAUP) conducts salary studies of college professors and publishes its findings in *AAUP Annual Report on the Economic Status of the Profession*. In Example 10.3 on pages 449–451, we performed a hypothesis test based on independent samples to decide whether mean salaries differ for faculty in private and public institutions. Now you are to perform that same hypothesis test based on a paired sample. Pairs were formed by matching faculty in private and public institutions by rank and specialty. A random sample of 30 pairs yielded the data, in thousands of dollars, presented on the WeissStats site. Use the technology of your choice to do the following tasks.

a. Decide, at the 5% significance level, whether the data provide sufficient evidence to conclude that mean salaries differ for faculty in private institutions and public institutions. Use the paired *t*-test.
b. Compare your result in part (a) to the one obtained in Example 10.3.
c. Repeat both the pooled *t*-test of Example 10.3 and the paired *t*-test of part (a), using a 1% significance level, and compare your results.
d. Which test do you think is preferable here: the pooled *t*-test or the paired *t*-test? Explain your answer.
e. Find and interpret a 95% confidence interval for the difference between the mean salaries of faculty in private and public institutions. Use the paired *t*-interval procedure.
f. Compare your result in part (e) to the one obtained in Example 10.4 on page 452.
g. Obtain a normal probability plot and a boxplot of the paired differences.
h. Based on your graphs from part (g), do you think that applying paired *t*-procedures here is reasonable?

10.171 Marriage Ages. In the *Statistics Norway* on-line article "The Times They Are a Changing," J. Kristiansen discussed the changes in age at the time of marriage in Norway. The ages, in years, at the time of marriage for 75 Norwegian couples are presented on the WeissStats site. Use the technology of your choice to do the following.

a. Decide, at the 1% significance level, whether the data provide sufficient evidence to conclude that the mean age of Norwegian men at the time of marriage exceeds that of Norwegian women.

b. Find and interpret a 99% confidence interval for the difference between the mean ages at the time of marriage for Norwegian men and women.

c. Remove the two paired-difference (potential) outliers and repeat parts (a) and (b). Compare your results to those in parts (a) and (b).

10.172 Storm Hydrology and Clear Cutting. In the document "Peak Discharge from Unlogged and Logged Watersheds," J. Jones and G. Grant compiled (paired) data on peak discharge from storms in two watersheds, one unlogged and one logged (100% clear-cut). If there is an effect due to clear-cutting, one would expect that the runoff would be greater in the logged area than in the unlogged area. The runoffs, in cubic meters per second per square kilometer ($m^3/s/km^2$), are provided on the WeissStats site. Use the technology of your choice to do the following.

a. Formulate the null and alternative hypotheses to reflect the expectation expressed above.

b. Perform the required hypothesis test at the 1% significance level.

c. Obtain and interpret a 99% confidence interval for the difference between mean runoffs in the logged and unlogged watersheds.

d. Construct a histogram of the sample data to identify the approximate shape of the paired-difference variable.

e. Based on your result from part (d), do you think that applying the paired t-procedures in parts (b) and (c) is reasonable? Explain your answer.

Extending the Concepts and Skills

10.173 Explain exactly how a paired t-test can be formulated as a one-mean t-test. (*Hint:* Work solely with the paired-difference variable.)

10.174 A hypothesis test, based on a paired sample, is to be performed to compare the means of two populations. The sample of 15 paired differences contains an outlier but otherwise is approximately bell shaped. Assuming that removal of the outlier would not be legitimate, would use of the paired t-test or a nonparametric test be better? Explain your answer.

10.175 Gasoline Additive. This exercise shows what can happen when a hypothesis-testing procedure designed for use with independent samples is applied to perform a hypothesis test on a paired sample. The gas mileages, in miles per gallon (mpg), of 10 randomly selected cars, both with and without a new gasoline additive, are shown in the following table.

With additive	Without additive
25.7	24.9
20.0	18.8
28.4	27.7
13.7	13.0
18.8	17.8
12.5	11.3
28.4	27.8
8.1	8.2
23.1	23.1
10.4	9.9

a. Apply the paired t-test to decide, at the 5% significance level, whether the gasoline additive is effective in increasing gas mileage.

b. Apply the pooled t-test to the sample data to perform the hypothesis test.

c. Why is performing the hypothesis test the way you did in part (b) inappropriate?

d. Compare your result in parts (a) and (b).

10.176 Permutation Tests. With the advent of high-speed computing, new procedures have been developed that permit statistical inferences to be performed under less restrictive conditions than those of classical procedures. **Permutation tests** constitute one such collection of new procedures. To perform a permutation test to compare two population means using paired samples, we proceed as follows.

1. For each pair, switch or don't switch the two observations with probability 0.5. This procedure yields a new paired sample.
2. Compute the mean of the new paired differences.
3. Repeat steps 1 and 2 a large number (hundreds or thousands) of times.
4. The distribution of the resulting paired-difference means provides an estimate of the sampling distribution of the sample mean of paired differences when the null hypothesis of equal population means is true. This estimate is called a **permutation distribution.**
5. The (estimated) P-value of the hypothesis test equals the proportion of values of the permutation distribution that are as extreme as or more extreme than the observed mean of the paired differences.

Refer to Example 10.16 on pages 489–490. Use the technology of your choice to conduct a permutation test and compare your results with those found by using the paired t-test. Discuss any discrepancy that you encounter.

10.6 The Paired Wilcoxon Signed-Rank Test*

In Section 10.5, we discussed the paired t-procedures, which provide methods for comparing two population means using paired samples. An assumption for use of those procedures is that the paired-difference variable is (approximately) normally distributed or that the sample size is large. For a small or moderate sample size where the distribution of the paired-difference variable is far from normal, a paired t-procedure is inappropriate and a nonparametric procedure should be used instead.

For instance, if the distribution of the paired-difference variable is symmetric (but not necessarily normal), we can perform a hypothesis test to compare the means of the two populations by applying the Wilcoxon signed-rank test (Procedure 9.3 on page 412) to the sample of paired differences. In this context, the Wilcoxon signed-rank test is called the **paired Wilcoxon signed-rank test.**

Procedure 10.8 provides the steps for performing a paired Wilcoxon signed-rank test. Note that we use the phrase **symmetric differences** as shorthand for "the paired-difference variable has a symmetric distribution."

■■■■ PROCEDURE 10.8 Paired Wilcoxon Signed-Rank Test

Purpose To perform a hypothesis test to compare two population means, μ_1 and μ_2

Assumptions
1. Simple random paired sample
2. Symmetric differences

Step 1 The null hypothesis is H_0: $\mu_1 = \mu_2$, and the alternative hypothesis is

$$H_a: \mu_1 \neq \mu_2 \qquad \text{or} \qquad H_a: \mu_1 < \mu_2 \qquad \text{or} \qquad H_a: \mu_1 > \mu_2$$
$$\text{(Two tailed)} \qquad\qquad \text{(Left tailed)} \qquad\qquad \text{(Right tailed)}.$$

Step 2 Decide on the significance level, α.

Step 3 Compute the value of the test statistic

$$W = \text{sum of the positive ranks}$$

and denote that value W_0. To do so, first calculate the paired differences of the sample pairs, next discard all paired differences that equal 0 and reduce the sample size accordingly, and then construct a work table of the following form.

Paired difference d	$|d|$	Rank of $|d|$	Signed rank R
.	.	.	.
.	.	.	.
.	.	.	.

CRITICAL-VALUE APPROACH	OR	P-VALUE APPROACH

Step 4 The critical value(s) are

$$W_{1-\alpha/2} \text{ and } W_{\alpha/2} \qquad \text{or} \qquad W_{1-\alpha} \qquad \text{or} \qquad W_{\alpha}$$
$$\text{(Two tailed)} \qquad\qquad \text{(Left tailed)} \qquad\qquad \text{(Right tailed)}.$$

Use Table V to find the critical value(s). For a left-tailed or two-tailed test, you will also need the relation $W_{1-A} = n(n+1)/2 - W_A$.

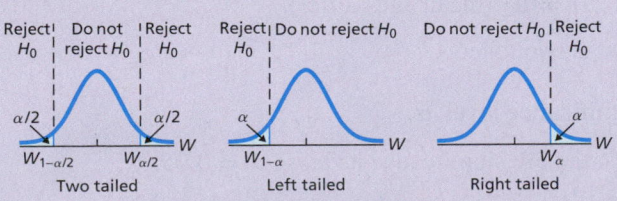

Step 5 If the value of the test statistic falls in the rejection region, reject H_0; otherwise, do not reject H_0.

Step 4 Obtain the *P*-value by using technology.

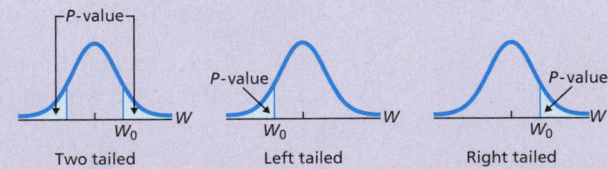

Step 5 If $P \leq \alpha$, reject H_0; otherwise, do not reject H_0.

Step 6 Interpret the results of the hypothesis test.

In Example 10.16 on pages 489–490, we used a paired *t*-test to decide whether a difference exists in the mean ages of married men and married women. Now we do so by using the paired Wilcoxon signed-rank test.

■■■■ EXAMPLE 10.19 The Paired Wilcoxon Signed-Rank Test

Ages of Married People The U.S. Census Bureau publishes information on the ages of married people in *Current Population Reports*. A random sample of 10 married couples gave the data on ages, in years, shown in the second and third columns of Table 10.14. The fourth column shows the paired differences, obtained by subtracting the age of each wife from that of her husband.

TABLE 10.14

Ages, in years, of a random sample of 10 married couples

Couple	Husband	Wife	Difference, *d*
1	59	53	6
2	21	22	−1
3	33	36	−3
4	78	74	4
5	70	64	6
6	33	35	−2
7	68	67	1
8	32	28	4
9	54	41	13
10	52	44	8

At the 5% significance level, do the data provide sufficient evidence to conclude that the mean age of married men differs from the mean age of married women?

Solution First, we check the two conditions required for using the paired Wilcoxon signed-rank test, as listed in Procedure 10.8.

- Assumption 1 is satisfied because we have a simple random paired sample. Each pair consists of a married couple.
- Figure 10.17 shows a stem-and-leaf diagram for the sample of paired differences in the last column of Table 10.14. Because the diagram is roughly symmetric, we can consider Assumption 2 satisfied.

FIGURE 10.17

Stem-and-leaf diagram (using five lines per stem) of the paired differences in Table 10.14

From the preceding items, we see that the paired Wilcoxon signed-rank test can be used to conduct the required hypothesis test. We apply Procedure 10.8.

Step 1 State the null and alternative hypotheses.

Let μ_1 denote the mean age of all married men, and let μ_2 denote the mean age of all married women. Then the null and alternative hypotheses are, respectively,

$$H_0: \mu_1 = \mu_2 \text{ (mean ages are equal)}$$
$$H_a: \mu_1 \neq \mu_2 \text{ (mean ages differ)}.$$

Note that the hypothesis test is two tailed.

Step 2 Decide on the significance level, α.

We are to perform the test at the 5% significance level, so $\alpha = 0.05$.

Step 3 Compute the value of the test statistic

$$W = \text{sum of the positive ranks.}$$

The paired differences (*d*-values) are shown in the fourth column of Table 10.14. We note that none of the paired differences equal 0 and proceed to construct the following work table. Observe that, in several instances, ties occur among the absolute paired differences ($|d|$-values). To deal with such ties, we proceed in the usual manner. Specifically, if two or more absolute paired differences are tied, each is assigned the mean of the ranks they would have had if there had been no ties. For instance, the second and seventh paired differences (-1 and 1) both have the smallest absolute paired difference, each of which is assigned rank $(1 + 2)/2$ or 1.5, as shown in the third column of the table.

Paired difference d	$|d|$	Rank of $|d|$	Signed rank R
6	6	7.5	7.5
−1	1	1.5	−1.5
−3	3	4	−4
4	4	5.5	5.5
6	6	7.5	7.5
−2	2	3	−3
1	1	1.5	1.5
4	4	5.5	5.5
13	13	10	10
8	8	9	9

Referring to the last column of the preceding table, we find that the value of the test statistic is

$$W = 7.5 + 5.5 + 7.5 + 1.5 + 5.5 + 10 + 9 = 46.5.$$

CRITICAL-VALUE APPROACH	OR	P-VALUE APPROACH

Step 4 The critical values for a two-tailed test are $W_{1-\alpha/2}$ and $W_{\alpha/2}$. Use Table V and the relation $W_{1-A} = n(n + 1)/2 - W_A$ to find the critical values.

From Table 10.14, we see that $n = 10$. The critical values for a two-tailed test at the 5% significance level are $W_{1-0.05/2}$ and $W_{0.05/2}$, that is, $W_{1-0.025}$ and $W_{0.025}$. First we use Table V to find $W_{0.025}$. We go down the outside columns, labeled n, to "10." Then, going across that row to the column labeled $W_{0.025}$, we reach 47; thus $W_{0.025} = 47$. Now we apply the aforementioned relation and the result just obtained to get $W_{1-0.025}$:

$$W_{1-0.025} = 10(10 + 1)/2 - W_{0.025} = 55 - 47 = 8.$$

See Fig. 10.18A.

FIGURE 10.18A

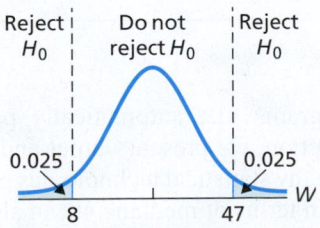

Step 5 If the value of the test statistic falls in the rejection region, reject H_0; otherwise, do not reject H_0.

The value of the test statistic is $W = 46.5$, as found in Step 3, which does not fall in the rejection region shown in Fig. 10.18A. Thus we do not reject H_0. The test results are not statistically significant at the 5% level.

Step 4 Obtain the P-value by using technology.

Using technology, we find that the P-value for the hypothesis test is $P = 0.059$, as shown in Fig. 10.18B.

FIGURE 10.18B

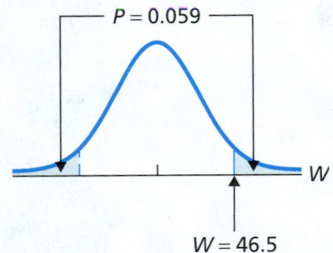

Step 5 If $P \le \alpha$, reject H_0; otherwise, do not reject H_0.

From Step 4, $P = 0.059$. Because the P-value exceeds the specified significance level of 0.05, we do not reject H_0. The test results are not statistically significant at the 5% level but (see Table 9.8 on page 386) the data do nonetheless provide moderate evidence against the null hypothesis.

StatCrunch

Report 10.8

Exercise 10.191
on page 506

Step 6 Interpret the results of the hypothesis test.

Interpretation At the 5% significance level, the data do not provide sufficient evidence to conclude that the mean age of married men differs from the mean age of married women.

It is interesting to note that, although we reject the null hypothesis of equal mean ages at the 5% significance level by using the paired t-test (Example 10.16), we do not reject it by using the paired Wilcoxon signed-rank test (Example 10.19). Nonetheless, the evidence against the null hypothesis is comparable with both tests: $P = 0.048$ and $P = 0.059$, respectively.

Comparing the Paired Wilcoxon Signed-Rank Test and the Paired t-Test

As we demonstrated in Section 10.5, a paired t-test can be used to conduct a hypothesis test to compare two population means when we have a paired sample and the paired-difference variable is normally distributed. Because normally distributed variables have symmetric distributions, we can also use the paired Wilcoxon signed-rank test to perform such a hypothesis test.

For a normally distributed paired-difference variable, the paired t-test is more powerful than the paired Wilcoxon signed-rank test because it is designed expressly for such paired-difference variables; surprisingly, though, the paired t-test is not much more powerful than the paired Wilcoxon signed-rank test. However, if the paired-difference variable has a symmetric distribution but is not normally distributed, the paired Wilcoxon signed-rank test is usually more powerful than the paired t-test and is often considerably more powerful.

KEY FACT 10.7

Paired Wilcoxon Signed-Rank Test Versus the Paired t-Test

Suppose that you want to perform a hypothesis test using a paired sample to compare the means of two populations. When deciding between the paired t-test and the paired Wilcoxon signed-rank test, follow these guidelines:

- If you are reasonably sure that the paired-difference variable is normally distributed, use the paired t-test.
- If you are not reasonably sure that the paired-difference variable is normally distributed but are reasonably sure that it has a symmetric distribution, use the paired Wilcoxon signed-rank test.

▉ ▉ ▉ THE TECHNOLOGY CENTER

Most statistical technologies have programs that automatically perform a paired Wilcoxon signed-rank test. In this subsection, we present output and step-by-step instructions for such programs. Although many statistical technologies present the output of the paired Wilcoxon signed-rank test in terms of medians, it can also be interpreted in terms of means.

As you will see, different programs may report slightly different P-values for a paired Wilcoxon signed-rank test. These differences are due to the fact that different programs may use different methods for obtaining or approximating such P-values.

Note to Minitab users: At the time of this writing, Minitab does not have a built-in program for a paired Wilcoxon signed-rank test. You can conduct such a test, however, by applying Minitab's (one-sample) Wilcoxon signed-rank test to the sample of paired differences, using the null hypothesis $H_0: \mu_d = 0$.

Note to TI-83/84 Plus users: At the time of this writing, the TI-83/84 Plus does not have a built-in program for conducting a paired Wilcoxon signed-rank test. However, we have written a TI program called PDWILCOX for performing that test. It is located in the TI Programs section on the WeissStats site. Your instructor can show you how to download the program to your calculator. *Warning:* Any data that you may have previously stored in Lists 1–6 will be erased during program execution, so copy those data to other lists prior to program execution if you want to retain them.

EXAMPLE 10.20   **Using Technology to Conduct Paired Wilcoxon Signed-Rank Test**

Ages of Married People   The second and third columns of Table 10.14 on page 500 give the ages of 10 randomly selected married couples. Use Minitab, Excel, or the TI-83/84 Plus to decide, at the 5% significance level, whether the data provide sufficient evidence to conclude that the mean age of married men differs from the mean age of married women.

Solution   Let μ_1 denote the mean age of all married men, and let μ_2 denote the mean age of all married women. We want to perform the hypothesis test

$$H_0: \ \mu_1 = \mu_2 \text{ (mean ages are equal)}$$
$$H_a: \ \mu_1 \neq \mu_2 \text{ (mean ages differ)},$$

at the 5% significance level.

We applied the Wilcoxon signed-rank programs to the data, resulting in Output 10.5. Steps for generating that output are presented in Instructions 10.5 on the following page. *Note to Excel users:* For brevity, we have presented only the essential portions of the actual output.

OUTPUT 10.5   Paired Wilcoxon signed-rank test on the age data

MINITAB

Wilcoxon Signed Rank Test: DIFFERENCE

Test of median = 0.000000 versus median ≠ 0.000000

	N	N for Test	Wilcoxon Statistic	P	Estimated Median
DIFFERENCE	10	10	46.5	0.059	3.500

EXCEL

Hypothesized difference (D): 0		
Significance level (%): 5		
p-value: Exact p-value		

Summary statistics:		

Variable	Observations	Mean	Std. deviation
HUSBAND	10	50.0000	19.2988
WIFE	10	46.4000	17.4687

Wilcoxon signed-rank test / Two-tailed test:	
V	46.5000
Expected value	27.5000
Variance (V)	95.8750
p-value (Two-tailed)	0.0588
alpha	0.05

TI-83/84 PLUS

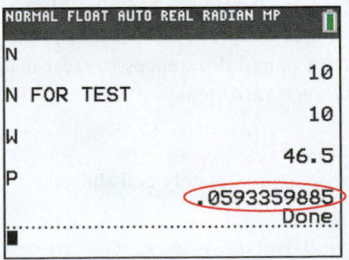

```
NORMAL FLOAT AUTO REAL RADIAN MP

N
              10
N FOR TEST
              10
W
            46.5
P
     .0593359885
              Done
■
```

As shown in Output 10.5, the *P*-value for the hypothesis test is 0.059. Because the *P*-value exceeds the specified significance level of 0.05, we do not reject H_0. At the 5% significance level, the data do not provide sufficient evidence to conclude that the mean age of married men differs from the mean age of married women.

INSTRUCTIONS 10.5 Steps for generating Output 10.5

MINITAB

1 Store the data from the second and third columns of Table 10.14 in columns named HUSBAND and WIFE
2 Choose **Calc ➤ Calculator. . .**
3 Type <u>DIFFERENCE</u> in the **Store result in variable** text box
4 Specify 'HUSBAND'–'WIFE' in the **Expression** text box and click **OK**
5 Choose **Stat ➤ Nonparametrics ➤ 1-Sample Wilcoxon...**
6 Press the F3 key to reset the dialog box
7 Specify DIFFERENCE in the **Variables** text box
8 Select the **Test median** option button
9 Click the arrow button at the right of the **Alternative** drop-down list box and select **not equal**
10 Click **OK**

EXCEL

1 Store the age data from the second and third columns of Table 10.14 in columns named HUSBAND and WIFE
2 Choose **XLSTAT ➤ Nonparametric tests ➤ Comparison of two samples (Wilcoxon, Mann-Whitney, . . .)**
3 Click the reset button in the lower left corner of the dialog box
4 Click in the **Sample 1** selection box and then select the column of the worksheet that contains the HUSBAND data

5 Click in the **Sample 2** selection box and then select the column of the worksheet that contains the WIFE data
6 Uncheck the **Sign test** check box
7 Click the **Options** tab
8 Click the arrow button at the right of the **Alternative hypothesis** drop-down list box and select **Sample 1 – Sample 2 ≠ D**
9 Type <u>5</u> in the **Significance level (%)** text box
10 Check the **Exact p-value** check box
11 Click **OK**
12 Click the **Continue** button in the **XLSTAT – Selections** dialog box

TI-83/84 PLUS

1 Store the data from Table 10.14 in lists named HUSB and WIFE
2 Press **PRGM**
3 Arrow down to PDWILCOX and press **ENTER** twice
4 Press **2ND ➤ LIST**, arrow down to HUSB, and press **ENTER** twice
5 Press **2ND ➤ LIST**, arrow down to WIFE, and press **ENTER** twice
6 Type <u>0</u> for **TYPE** and press **ENTER**

Exercises 10.6

Understanding the Concepts and Skills

10.177 Suppose that you want to perform a hypothesis test based on a simple random paired sample to compare the means of two populations and you know that the paired-difference variable has a symmetric distribution that is far from normal.
a. Is use of the paired t-test acceptable if the sample size is small or moderate? Why or why not?
b. Is use of the paired t-test acceptable if the sample size is large? Why or why not?
c. Is use of the paired Wilcoxon signed-rank test acceptable? Why or why not?
d. If both the paired t-test and the paired Wilcoxon signed-rank test are acceptable, which test is preferable? Explain your answer.

10.178 A hypothesis test based on a simple random paired sample is to be performed to compare the means of two populations. The sample of 15 paired differences contains an outlier but otherwise is approximately bell shaped. Assuming that removing the outlier is not legitimate, which test is better to use—the paired t-test or the paired Wilcoxon signed-rank test? Explain your answer.

10.179 Suppose that you want to perform a hypothesis test based on a simple random paired sample to compare the means of two populations. For each part, decide whether you would use the paired t-test,

the paired Wilcoxon signed-rank test, or neither of these tests. Preliminary data analyses of the sample of paired differences suggest that the distribution of the paired-difference variable is
a. approximately normal.
b. highly skewed; the sample size is 20.
c. symmetric bimodal.

10.180 Suppose that you want to perform a hypothesis test based on a simple random paired sample to compare the means of two populations. For each part, decide whether you would use the paired t-test, the paired Wilcoxon signed-rank test, or neither of these tests. Preliminary data analyses of the sample of paired differences suggest that the distribution of the paired-difference variable is
a. uniform.
b. not symmetric; the sample size is 132.
c. moderately skewed but otherwise approximately bell shaped.

*In Exercises **10.181–10.186**, the null hypothesis is H_0: $\mu_1 = \mu_2$ and the alternative hypothesis is as specified. We have provided data from a simple random paired sample from the two populations under consideration. In each case, use the paired Wilcoxon signed-rank test to perform the required hypothesis test at the 10% significance level. (Note: These problems were presented as Exercises 10.149–10.154 in Section 10.5, where they were to be solved by using the paired t-test.)*

10.181 $H_a: \mu_1 \neq \mu_2$

	Observation from	
Pair	Population 1	Population 2
1	13	11
2	16	15
3	13	10
4	14	8
5	12	8
6	8	9
7	17	14

10.182 $H_a: \mu_1 < \mu_2$

	Observation from	
Pair	Population 1	Population 2
1	7	13
2	4	9
3	10	6
4	0	2
5	20	19
6	−1	5
7	12	10

10.183 $H_a: \mu_1 > \mu_2$

	Observation from	
Pair	Population 1	Population 2
1	7	3
2	4	5
3	9	8
4	7	2
5	19	16
6	12	12
7	13	18
8	5	11

10.184 $H_a: \mu_1 \neq \mu_2$

	Observation from	
Pair	Population 1	Population 2
1	10	12
2	8	7
3	13	11
4	13	16
5	17	15
6	12	9
7	12	12
8	11	7

10.185 $H_a: \mu_1 < \mu_2$

	Observation from	
Pair	Population 1	Population 2
1	15	18
2	22	25
3	15	17
4	27	24
5	24	30
6	23	23
7	8	10
8	20	27
9	2	3

10.186 $H_a: \mu_1 > \mu_2$

	Observation from	
Pair	Population 1	Population 2
1	40	32
2	30	29
3	34	36
4	22	18
5	35	31
6	26	26
7	26	25
8	27	25
9	11	15
10	35	31

Applying the Concepts and Skills

*Exercises **10.187–10.192** repeat Exercises 10.155–10.160 of Section 10.5. There, you applied the paired t-test to solve each problem. Now solve each problem by applying the paired Wilcoxon signed-rank test.*

10.187 *Zea Mays.* Charles Darwin, author of *Origin of Species*, investigated the effect of cross-fertilization on the heights of plants. In one study he planted 15 pairs of *Zea mays* plants. Each pair consisted of one cross-fertilized plant and one self-fertilized plant grown in the same pot. The following table gives the height differences, in eighths of an inch, for the 15 pairs. Each difference is obtained by subtracting the height of the self-fertilized plant from that of the cross-fertilized plant.

49	−67	8	16	6
23	28	41	14	29
56	24	75	60	−48

a. At the 5% significance level, do the data provide sufficient evidence to conclude that the mean heights of cross-fertilized and self-fertilized *Zea mays* differ?
b. Repeat part (a) at the 1% significance level.

10.188 **Sleep.** In 1908, W. S. Gosset published "The Probable Error of a Mean" (*Biometrika*, Vol. 6, pp. 1–25). In this pioneering paper, published under the pseudonym "Student," he introduced what later became known as Student's t-distribution. Gosset used the following data set, which gives the additional sleep in hours obtained by 10 patients who used laevohysocyamine hydrobromide.

1.9	0.8	1.1	0.1	−0.1
4.4	5.5	1.6	4.6	3.4

a. At the 5% significance level, do the data provide sufficient evidence to conclude that laevohysocyamine hydrobromide is effective in increasing sleep?
b. Repeat part (a) at the 1% significance level.

10.189 Anorexia Treatment. Anorexia nervosa is a serious eating disorder, particularly among young women. The following data provide the weights, in pounds, of 17 anorexic young women before and after receiving a family therapy treatment for anorexia nervosa. [SOURCE: D. Hand et al., ed., *A Handbook of Small Data Sets*, London: Chapman & Hall, 1994; raw data from B. Everitt (personal communication)]

Before	After	Before	After	Before	After
83.3	94.3	76.9	76.8	82.1	95.5
86.0	91.5	94.2	101.6	77.6	90.7
82.5	91.9	73.4	94.9	83.5	92.5
86.7	100.3	80.5	75.2	89.9	93.8
79.6	76.7	81.6	77.8	86.0	91.7
87.3	98.0	83.8	95.2		

Does family therapy appear to be effective in helping anorexic young women gain weight? Perform the appropriate hypothesis test at the 5% significance level.

10.190 Measuring Treadwear. R. Stichler et al. compared two methods of measuring treadwear in their paper "Measurement of Treadwear of Commercial Tires" (*Rubber Age*, 73:2). Eleven tires were each measured for treadwear by two methods, one based on weight and the other on groove wear. The following are the data, in thousands of miles.

Weight method	Groove method	Weight method	Groove method
30.5	28.7	24.5	16.1
30.9	25.9	20.9	19.9
31.9	23.3	18.9	15.2
30.4	23.1	13.7	11.5
27.3	23.7	11.4	11.2
20.4	20.9		

At the 5% significance level, do the data provide sufficient evidence to conclude that, on average, the two measurement methods give different results?

10.191 Glaucoma and Corneal Thickness. Glaucoma is a leading cause of blindness in the United States. N. Ehlers measured the corneal thickness of eight patients who had glaucoma in one eye but not in the other. The results of the study were published in the paper "On Corneal Thickness and Intraocular Pressure, II" (*Acta Opthalmologica*, Vol. 48, pp. 1107–1112). The following are the data on corneal thickness, in microns.

Patient	Normal	Glaucoma
1	484	488
2	478	478
3	492	480
4	444	426
5	436	440
6	398	410
7	464	458
8	476	460

At the 10% significance level, do the data provide sufficient evidence to conclude that mean corneal thickness is greater in normal eyes than in eyes with glaucoma?

10.192 Cooling Down. Cooling down with a cold drink before exercise in the heat is believed to help an athlete perform. Researcher J. Dugas explored the difference between cooling down with an ice slurry (slushy) and with cold water in the article "Ice Slurry Ingestion Increases Running Time in the Heat" (*Clinical Journal of Sports Medicine*, Vol. 21, No. 6, pp. 541–542). Ten male participants drank a flavored ice slurry and ran on a treadmill in a controlled hot and humid environment. Days later, the same participants drank cold water and ran on a treadmill in the same hot and humid environment. The following table shows the times, in minutes, it took to fatigue on the treadmill for both the ice slurry and the cold water.

Subject	Cold Water	Ice Slurry
1	52	56
2	37	43
3	44	52
4	51	58
5	34	38
6	38	45
7	41	45
8	50	58
9	29	34
10	38	44

At the 1% significance level, do the data provide sufficient evidence to conclude that, on average, cold water is less effective than ice slurry for optimizing athletic performance in the heat?

In each of Exercises 10.193–10.195, use the technology of your choice to perform the required tasks.

10.193 Font Readability. In the online paper "A Comparison of Two Computer Fonts: Serif versus Ornate Sans Serif" (*Usability News*, Issue 5.3), researchers S. Morrison and J. Noyes studied whether the type of font used in a document affects reading speed or comprehension. The fonts used for the comparisons were the serif font Times New Roman (TNR) and a more ornate sans serif font called Gigi. There were 10 substitution words used for testing the comprehensibility of the two fonts. The substitution words were inappropriate to the context of the passage and varied grammatically from the original words in the paragraphs. The following table gives the number of inappropriate words out of the 10 that were identified in the TNR and Gigi fonts by each of the 25 participants.

TNR	Gigi	TNR	Gigi	TNR	Gigi
8	9	8	8	9	6
10	7	10	9	8	5
5	4	8	7	9	8
10	8	8	6	8	5
8	8	10	9	9	6
10	8	10	6	9	4
9	7	9	8	10	8
10	10	9	3		
9	2	7	6		

Suppose that you want to perform a hypothesis test to determine whether, on average, people have better comprehension with the TNR font than with the Gigi font. Conduct preliminary graphical data analyses to decide whether applying the paired Wilcoxon signed-rank test is reasonable. Explain your decision.

10.194 Tobacco Mosaic Virus. To assess the effects of two different strains of the tobacco mosaic virus, W. Youden and H. Beale randomly selected eight tobacco leaves. Half of each leaf was subjected to one of the strains of tobacco mosaic virus and the other half to the other strain. The researchers then counted the number of local lesions apparent on each half of each leaf. The results of their study were published in the paper "A Statistical Study of the Local Lesion Method for Estimating Tobacco Mosaic Virus" (*Contributions to Boyce Thompson Institute*, Vol. 6, p. 437). Here are the data.

Leaf	1	2	3	4	5	6	7	8
Virus 1	31	20	18	17	9	8	10	7
Virus 2	18	17	14	11	10	7	5	6

Suppose that you want to perform a hypothesis test to determine whether a difference exists between the mean numbers of local lesions resulting from the two viral strains. Conduct preliminary graphical analyses to decide whether applying the paired Wilcoxon signed-rank test is reasonable. Explain your decision.

10.195 Consonantal Inventory Size. In the article "Intervocalic Consonants in the Speech of Typically Developing Children: Emergence and Early Use" (*Clinical Linguistics and Phonetics*, Vol. 16, Issue 3, pp. 155–168), C. Stoel-Gammon examined the development of intervocalic consonants (consonants appearing between two vowels) by children during the first years of life. The following data provide word-initial and word-final consonantal inventory sizes for nine children at age 21 months.

Child	1	2	3	4	5	6	7	8	9
Initial	16	14	13	12	12	11	8	7	6
Final	4	10	0	7	7	6	3	4	6

Suppose that you want to use these data to perform a hypothesis test to determine whether mean word-initial consonantal inventory size is greater than mean word-final consonantal inventory size. Conduct preliminary graphical data analyses to decide whether it is reasonable to apply the

a. paired *t*-test.
b. paired Wilcoxon signed-rank test.

Working with Large Data Sets

10.196 Faculty Salaries. The American Association of University Professors (AAUP) conducts salary studies of college professors and publishes its findings in *AAUP Annual Report on the Economic Status of the Profession*. Pairs were formed by matching faculty in private and public institutions by rank and specialty. A random sample of 30 pairs yielded the data, in thousands of dollars, presented on the WeissStats site. Use the technology of your choice to do the following.

a. Apply the paired Wilcoxon signed-rank test to decide, at the 5% significance level, whether the data provide sufficient evidence to conclude that mean salaries differ for faculty in private and public institutions.
b. Compare your result in part (a) to the one obtained in Exercise 10.170 on page 497, where the paired *t*-test was used.
c. Which test do you think is preferable: the paired *t*-test or the paired Wilcoxon signed-rank test? Explain your answer.

10.197 Marriage Ages. In the *Statistics Norway* on-line article "The Times They Are a Changing," J. Kristiansen discussed the changes in age at the time of marriage in Norway. The ages, in years, at the time of marriage for 75 Norwegian couples is presented on the WeissStats site. Use the technology of your choice to do the following.

a. Apply the paired Wilcoxon signed-rank test to decide, at the 1% significance level, whether the data provide sufficient evidence to conclude that the mean age of Norwegian men at the time of marriage exceeds that of Norwegian women.
b. Compare your result in part (a) to the one obtained in Exercise 10.171 on pages 497–498, where the paired *t*-test was used.
c. Which test do you think is preferable: the paired *t*-test or the paired Wilcoxon signed-rank test? Explain your answer.

10.198 Storm Hydrology and Clear Cutting. In the document "Peak Discharge from Unlogged and Logged Watersheds," J. Jones and G. Grant compiled (paired) data on peak discharge from storms in two watersheds, one unlogged and one logged (100% clear-cut). If there is an effect due to clear-cutting, one would expect that the runoff would be greater in the logged area than in the unlogged area. The runoffs, in cubic meters per second per square kilometer ($m^3/s/km^2$), are provided on the WeissStats site. Use the technology of your choice to do the following.

a. Formulate the null and alternative hypotheses to reflect the expectation expressed above.
b. Apply the paired Wilcoxon signed-rank test to perform the required hypothesis test at the 1% significance level.
c. Compare your result in part (b) to the one obtained in Exercise 10.172 on page 498, where the paired *t*-test was used.
d. Which test do you think is preferable: the paired *t*-test or the paired Wilcoxon signed-rank test? Explain your answer.

Extending the Concepts and Skills

10.199 Explain why the paired Wilcoxon signed-rank test is simply a Wilcoxon signed-rank test on the sample of paired differences with null hypothesis H_0: $\mu_d = 0$.

Paired Sign Test. Recall that the paired Wilcoxon signed-rank test, which can be used to perform a hypothesis test to compare two population medians, requires that the paired-difference variable, d, has a symmetric distribution. If that is not the case, the **paired sign test** can be used instead. Technically, like the paired Wilcoxon signed-rank test, use of the paired sign test requires that the paired-difference variable has a continuous distribution. In practice, however, that restriction is usually ignored.

The null hypothesis for a paired sign test is H_0: $\eta_d = 0$, that is, the median of the population of paired differences is 0. If the null hypothesis is true, the probability is 0.5 that an observed paired difference exceeds 0. Therefore, in a simple random sample of size n, the number of paired differences, s, that exceed 0 has a binomial distribution with parameters n and 0.5.

To perform a paired sign test, first assign a "+" sign to each paired difference that exceeds 0 and then obtain the number of "+" signs, which we denote s_0. The P-value for the hypothesis test can be found by applying Exercise 9.71 on page 387 and obtaining the required binomial probability.

10.200 Assuming that the null hypothesis H_0: $\eta_d = 0$ is true, answer the following questions.
a. Why is the probability that an observed paired difference exceeds 0 equal to 0.5?
b. In a simple random sample of size n, why does the number of paired differences that exceed 0 have a binomial distribution with parameters n and 0.5?

10.201 The paired sign test can be used whether or not the paired-difference variable has a symmetric distribution.

a. If the distribution is in fact symmetric, the paired Wilcoxon signed-rank test is preferable. Why do you think that is so?
b. What advantage does the paired sign test have over the paired Wilcoxon signed-rank test?

10.202 Explain how to proceed with a paired sign test if one or more of the paired differences equals 0.

In Exercises 10.203–10.208, do the following.
a. *Apply the paired sign test to the specified exercise.*
b. *Compare your result in part (a) to that obtained by using the paired Wilcoxon signed-rank test earlier in this exercise section. Pay particular attention to the P-values.*
Note: If the paired-difference variable has a symmetric distribution, then $\eta_d = \mu_1 - \mu_2$.

10.203 *Zea Mays.* Exercise 10.187.

10.204 **Sleep.** Exercise 10.188.

10.205 **Anorexia Treatment.** Exercise 10.189.

10.206 **Measuring Treadwear.** Exercise 10.190.

10.207 **Glaucoma and Corneal Thickness.** Exercise 10.191.

10.208 **Cooling Down.** Exercise 10.192.

CHAPTER IN REVIEW

You Should Be Able to

1. use and understand the formulas in this chapter.

2. perform inferences based on independent simple random samples to compare the means of two populations when the population standard deviations are unknown but are assumed to be equal.

3. perform inferences based on independent simple random samples to compare the means of two populations when the population standard deviations are unknown but are not assumed to be equal.

*4. perform a hypothesis test based on independent simple random samples to compare the means of two populations when

the distributions of the variable under consideration have the same shape.

5. perform inferences based on a simple random paired sample to compare the means of two populations.

*6. perform a hypothesis test based on a simple random paired sample to compare the means of two populations when the paired-difference variable has a symmetric distribution.

**7. decide which procedure should be used to perform an inference to compare the means of two populations.

Key Terms

back-to-back stem-and-leaf
    diagram,* 472
independent samples, 439
independent simple random
    samples, 439
M_α,* 472
Mann–Whitney test,* 475
Mann–Whitney–Wilcoxon test,* 471
nonpooled t-interval procedure, 462
nonpooled t-test, 459

normal differences, 487
paired difference, 486
paired-difference variable, 487
paired sample, 485
paired t-interval procedure, 491
paired t-test, 488
paired Wilcoxon signed-rank test,* 499
pool, 447
pooled sample standard
    deviation (s_p), 447

pooled t-interval procedure, 452
pooled t-test, 448
same-shape distributions,* 470
same-shape populations,* 474
sampling distribution of the difference
    between two sample means, 444
simple random paired sample, 485
symmetric differences,* 499
Wilcoxon rank-sum test,* 471

REVIEW PROBLEMS

Understanding the Concepts and Skills

1. Discuss the basic strategy for comparing the means of two populations based on independent simple random samples.

2. Discuss the basic strategy for comparing the means of two populations based on a simple random paired sample.

3. Regarding the pooled and nonpooled t-procedures,
a. what is the difference in assumptions between the two procedures?
b. how important is the assumption of independent simple random samples for these procedures?
c. how important is the normality assumption for these procedures?

4. Suppose that the variable under consideration is normally distributed on each of the two populations and that you are going to use independent simple random samples to compare the population means. Fill in the blank and explain your answer: Unless you are quite sure that the _____ are equal, the nonpooled t-procedures should be used instead of the pooled t-procedures.

*5. Suppose that independent simple random samples are taken from two populations to compare their means. Further suppose that the two distributions of the variable under consideration have the same shape.
a. Would the nonpooled t-test ever be the procedure of choice in these circumstances? Explain your answer.
b. Under what conditions would the pooled t-test be preferable to the Mann–Whitney test? Explain your answer.

6. Explain one possible advantage of using a paired sample instead of independent samples.

*7. Suppose that a simple random paired sample is taken from two populations to compare their means. Further suppose that the distribution of the paired-difference variable has a symmetric distribution. Under what conditions would the paired t-test be preferable to the paired Wilcoxon signed-rank test? Explain your answer.

Applying the Concepts and Skills

8. Grip and Leg Strength. In the paper, "Sex Differences in Static Strength and Fatigability in Three Different Muscle Groups" (*Research Quarterly for Exercise and Sport*, Vol. 61(3), pp. 238–242), J. Misner et al. published results of a study on grip and leg strength of males and females. The following data, in newtons, is based on their measurements of right-leg strength.

Male			Female		
2632	1796	2256	1344	1351	1369
2235	2298	1917	2479	1573	1665
1105	1926	2644	1791	1866	1544
1569	3129	2167	2359	1694	2799
1977			1868	2098	

Preliminary data analyses indicate that you can reasonably presume leg strength is normally distributed for both males and females and that the standard deviations of leg strength are approximately equal. At the 5% significance level, do the data provide sufficient evidence to conclude that mean right-leg strength of males exceeds that of females? (*Note:* $\bar{x}_1 = 2127$, $s_1 = 513$, $\bar{x}_2 = 1843$, and $s_2 = 446$.)

9. Grip and Leg Strength. Refer to Problem 8. Determine a 90% confidence interval for the difference between the mean right-leg strengths of males and females. Interpret your result.

10. Cottonmouth Litter Size. In the article "The Eastern Cottonmouth (*Agkistrodon piscivorus*) at the Northern Edge of Its Range" (*Journal of Herpetology*, Vol. 29, No. 3, pp. 391–398), C. Blem and L. Blem examined the reproductive characteristics of the eastern cottonmouth. The data in the following table, based on the results of the researchers' study, give the number of young per litter for 24 female cottonmouths in Florida and 44 female cottonmouths in Virginia.

Florida			Virginia					
8	6	7	5	12	7	7	6	8
7	4	3	12	9	7	4	9	6
1	7	5	12	7	5	6	10	3
6	6	5	10	8	8	12	5	6
6	8	5	10	11	3	8	4	5
5	7	4	7	6	11	7	6	8
6	6	5	8	14	8	7	11	7
5	5	4	5	4				

Preliminary data analyses indicate that you can reasonably presume that litter sizes of cottonmouths in both states are approximately normally distributed. At the 1% significance level, do the data provide sufficient evidence to conclude that, on average, the number of young per litter of cottonmouths in Florida is less than that in Virginia? Do not assume that the population standard deviations are equal. (*Note:* $\bar{x}_1 = 5.46$, $s_1 = 1.59$, $\bar{x}_2 = 7.59$, and $s_2 = 2.68$.)

11. Cottonmouth Litter Size. Refer to Problem 10. Find a 98% confidence interval for the difference between the mean litter sizes of cottonmouths in Florida and Virginia. Interpret your result.

*12. Home Prices. The National Association of Realtors publishes information on the cost of existing single-family homes in *Median Sales Price of Existing Single-Family Homes for Metropolitan Areas*. Independent random samples of 10 homes each in Atlantic City and Las Vegas yielded the following data on home prices in thousands of dollars.

Atlantic City		Las Vegas	
207.2	166.0	177.65	182.75
186.2	229.6	165.95	162.15
596.3	223.4	418.15	125.85
265.2	370.7	121.25	488.95
209.3	275.1	300.65	129.75

At the 5% significance level, can you conclude that the median costs for existing single-family homes differ in Atlantic City and Las Vegas? (*Note:* Preliminary data analyses suggest that you can reasonably presume that the cost distributions for the two cities have roughly the same shape but that those distributions are right skewed.)

13. Ecosystem Response. In the on-line paper "Changes in Lake Ice: Ecosystem Response to Global Change" (*Teaching Issues and Experiments in Ecology*, tiee.ecoed.net, Vol. 3), R. Bohanan et al. questioned whether there is evidence for global warming in long-term data on changes in dates of ice cover in three Wisconsin Lakes. The following table gives data, for a sample of eight years, on the number of days that ice stayed on two lakes in Madison, Wisconsin—Lake Mendota and Lake Monona.

Year	Mendota	Monona
1	119	107
2	115	108
3	53	52
4	108	108
5	74	85
6	47	47
7	102	96
8	87	91

a. Obtain a normal probability plot and boxplot of the paired differences.

b. Based on your results from part (a), is performing a paired t-test on the data reasonable? Explain your answer.

c. At the 10% significance level, do the data provide sufficient evidence to conclude that a difference exists in the mean length of time that ice stays on these two lakes?

14. Ecosystem Response. Refer to Problem 13, and find a 90% confidence interval for the difference in the mean lengths of time that ice stays on the two lakes. Interpret your result.

*15. Antidepressants and Lipid Profile. In the online article "Comparing the Effects of Fluoxetine and Imipramine on Total Cholesterol, Triglyceride, and Weight in Patients with Major Depression" (*Journal of Pharmaceutical Sciences*, Vol. 21, No. 4), E. Ananloo et al. study the effect of antidepressants on lipid profiles. Nineteen females who suffered from depression were given the drug fluoxetine and their total cholesterol level, in milligrams per deciliter (mg/dL), was measured before and after a 4-week testing period. The findings are summarized in the table below.

Patient	Before	After	Patient	Before	After
1	212	194	11	137	129
2	142	159	12	212	212
3	200	202	13	195	164
4	180	160	14	175	149
5	115	105	15	176	180
6	184	134	16	225	214
7	211	219	17	233	189
8	202	179	18	185	231
9	165	137	19	228	187
10	173	175			

At the 5% significance level, is there sufficient evidence to conclude that, for females who suffer from depression, the mean total cholesterol level is lower after a 4-week treatment of the antidepressant fluoxetine? Apply the paired Wilcoxon signed-rank test.

Working with Large Data Sets

16. Drink and Be Merry? In the paper, "Drink and Be Merry? Gender, Life Satisfaction, and Alcohol Consumption Among College Students" (*Psychology of Addictive Behaviors*, Vol. 19, Issue 2, pp. 184–191), J. Murphy et al. examined the impact of alcohol use and alcohol-related problems on several domains of life satisfaction (LS) in a sample of 353 college students. All LS items were rated on a 7-point Likert scale that ranged from 1 (strongly disagree)

to 7 (strongly agree). On the WeissStats site you will find data for dating satisfaction, based on the results of the study. Use the technology of your choice to do the following.

a. Obtain normal probability plots, boxplots, and the standard deviations for the two samples.

b. Based on your results from part (a), which are preferable here, pooled or nonpooled t-procedures? Explain your reasoning.

c. At the 5% significance level, do the data provide sufficient evidence to conclude that a difference exists in mean dating satisfaction of male and female college students?

d. Determine a 95% confidence interval for the difference between mean dating satisfaction of male and female college students.

e. Are your procedures in parts (c) and (d) justified? Explain your answer.

* 17. Drink and Be Merry? Refer to Problem 16. Use the technology of your choice to do the following.

a. Obtain histograms of the two samples.

b. Based on your histograms in part (a), do you think that conducting a Mann–Whitney test is reasonable here?

c. Apply the Mann–Whitney test to decide, at the 5% significance level, whether a difference exists in mean dating satisfaction for male and female college students.

d. Compare your result from part (c) to that of Problem 16(c).

18. Insulin and BMD. I. Ertuğrul et al. conducted a study to determine the association between insulin growth factor 1 (IGF-1) and bone mineral density (BMD) in men over 65 years of age. The researchers published their results in the paper "Relationship Between Insulin-Like Growth Factor 1 and Bone Mineral Density in Men Aged over 65 Years" (*Medical Principles and Practice*, Vol. 12, pp. 231–236). Forty-one men over 65 years old were enrolled in the study, as was a control group consisting of 20 younger men, ages 19–62 years. On the WeissStats site, we provide data on IGF-1 levels (in ng/mL), based on the results of the study. Use the technology of your choice to do the following.

a. Obtain normal probability plots, boxplots, and the standard deviations for the two samples.

b. Based on your results from part (a), which are preferable here, pooled or nonpooled t-procedures? Explain your reasoning.

c. At the 1% significance level, do the data provide sufficient evidence to conclude that, on average, men over 65 have a lower IGF-1 level than younger men?

d. Find and interpret a 99% confidence interval for the difference between the mean IGF-1 levels of men over 65 and younger men.

e. Are your procedures in parts (c) and (d) justified? Explain your answer.

19. Weekly Earnings. The Bureau of Labor Statistics publishes data on weekly earnings of full-time wage and salary workers in *Employment and Earnings*. Male and female workers were paired according to occupation and experience. Their weekly earnings, in dollars, are provided on the WeissStats site. Use the technology of your choice to do the following.

a. Apply the paired t-test to decide, at the 5% significance level, whether the data provide sufficient evidence to conclude that, on average, the weekly earnings of male full-time wage and salary workers exceed those of women.

b. Find and interpret a 90% confidence interval for the difference between the mean weekly earnings of male and female full-time wage and salary workers. Use the paired t-interval procedure.

c. Obtain a normal probability plot, boxplot, and stem-and-leaf diagram of the paired differences.

d. Based on your results in part (c), are your procedures in parts (a) and (b) justified? Explain your answer.

* **20. Weekly Earnings.** Refer to Problem 19. Use the technology of your choice to do the following.

a. Apply the paired Wilcoxon signed-rank test to decide, at the 5% significance level, whether the data provide sufficient evi-

dence to conclude that, on average, the weekly earnings of male full-time wage and salary workers exceed those of women.

b. Compare your result in part (a) to that in Problem 19(a).

c. Obtain a histogram, boxplot, and stem-and-leaf diagram of the paired differences.

d. Based on your results in part (c), is your procedure in part (a) justified? Explain your answer.

FOCUSING ON DATA ANALYSIS

UWEC UNDERGRADUATES

Recall from Chapter 1 (see page 34) that the Focus database and Focus sample contain information on the undergraduate students at the University of Wisconsin - Eau Claire (UWEC). Now would be a good time for you to review the discussion about these data sets.

Open the Focus sample worksheet (FocusSample) in the technology of your choice and then do the following.

a. Obtain normal probability plots, boxplots, and the sample standard deviations of the ACT composite scores for the sampled males and the sampled females.

b. At the 5% significance level, do the data provide sufficient evidence to conclude that mean ACT composite scores differ for male and female UWEC undergraduates? Justify the use of the procedure you chose to carry out the hypothesis test.

c. Determine and interpret a 95% confidence interval for the difference between the mean ACT composite scores of male and female UWEC undergraduates.

d. Repeat parts (a)–(c) for cumulative GPA.

e. Obtain a normal probability plot, boxplot, and histogram of the paired differences of the ACT English scores and ACT math scores for the sampled UWEC undergraduates.

f. At the 5% significance level, do the data provide sufficient evidence to conclude that, for UWEC undergraduates, the mean ACT English score is less than the mean ACT math score? Justify the use of the procedure you chose to carry out the hypothesis test.

g. Repeat part (f) at the 10% significance level.

h. Find and interpret a 90% confidence interval for the difference between the mean ACT English score and the mean ACT math score of UWEC undergraduates.

i. Repeat part (h), using an 80% confidence level.

CASE STUDY DISCUSSION

DEXAMETHASONE THERAPY AND IQ

On page 439, we presented cognition data obtained by researchers studying the effects of early postnatal dexamethasone therapy for the prevention or treatment of chronic lung disease of prematurity. The study included 146 school-age children of which 72 had been in a dexamethasone group and 74 had been in a control group. As infants, all of the children in both groups had severe respiratory distress syndrome (RDS) due to insufficiently developed lungs.

To measure cognition, the researchers employed the Wechsler Intelligence Scale for Children, Third Edition (WISC-III). Use the data in the table to perform the following statistical inferences for each of the three WISC-III main scales and each of

the four WISC-III subscales. So, you will need do parts (a) and (b) seven times.

a. At the 5% significance level, do the data provide sufficient evidence to conclude that a difference exists between mean scores of school-age children who as (premature) infants receive dexamethasone therapy and those who do not?

b. Determine and interpret a 95% confidence interval for the difference between mean scores of school-age children who as (premature) infants do not receive dexamethasone therapy and those who do.

c. Discuss your results in detail.

BIOGRAPHY

GERTRUDE COX: SPREADING THE GOSPEL ACCORDING TO ST. GERTRUDE

Gertrude Mary Cox was born on January 13, 1900, in Dayton, Iowa, the daughter of John and Emmaline Cox. She graduated from Perry High School, Perry, Iowa, in 1918. Between 1918 and 1925, Cox prepared to become a deaconess in the Methodist

Episcopal Church. However, in 1925, she decided to continue her education at Iowa State College in Ames, where she studied mathematics and statistics.

In 1929 and 1931, Cox received a B.S. and an M.S., respectively. Her work was directed by George W. Snedecor, and her degree was the first master's degree in statistics given by the Department of Mathematics at Iowa State.

From 1931 to 1933, Cox studied psychological statistics at the University of California at Berkeley. Snedecor meanwhile had established a new Statistical Laboratory at Iowa State, and in 1933 he asked her to be his assistant. This position launched her internationally influential career in statistics. Cox worked in the lab until she became assistant professor at Iowa State in 1939.

In 1940, the committee in charge of filling a newly created position as head of the department of experimental statistics at North Carolina State College in Raleigh asked Snedecor for recommendations; he first named several male statisticians, then wrote, "... but if you would consider a woman for this position I would recommend Gertrude Cox of my staff." They did consider a woman, and Cox accepted their offer.

In 1945, Cox organized and became director of the Institute of Statistics, which combined the teaching of statistics at the University of North Carolina and North Carolina State. Work conferences that Cox organized established the Institute as an international center for statistics. She also developed statistical programs at institutions throughout the South, referred to as "spreading the gospel according to St. Gertrude."

Cox's area of expertise was experimental design. She, with W. G. Cochran, wrote *Experimental Designs* (1950), recognized as the classic textbook on design and analysis of replicated experiments.

From 1960 to 1964, Cox was director of the Statistics Section of the Research Triangle Institute in Durham, North Carolina. She then retired, working only as a consultant. She died on October 17, 1978, in Durham.

Inferences for Population Standard Deviations*

CHAPTER OBJECTIVES

So far, our study of inferential statistics has focused on inferences for population means. Now we will focus on inferences for population standard deviations (or variances).

For example, in Chapter 9, we discussed the problem of deciding whether the mean net weight of bags of pretzels being packaged by a machine equals the advertised weight of 454 grams. This decision involves a hypothesis test for a population mean.

We should also be concerned with the variation in weights from bag to bag. If the variation is too large, many bags will contain either considerably more or considerably less than they should. To investigate the variation, we can perform a hypothesis test or construct a confidence interval for the standard deviation of the weights. These are inferences for one population standard deviation.

In addition, we might want to compare two different machines for packaging the pretzels to see whether one provides a smaller variation in weights than the other. We could do so by using inferences for two population standard deviations.

In this chapter, we discuss inferences for one population standard deviation and for two population standard deviations.

CASE STUDY

Speaker Woofer Driver Manufacturing

Speaker driver manufacturing is an important industry in many countries. In Taiwan, for example, more than 100 companies or factories produce and supply parts and driver units for speakers.

An essential component in driver units is the rubber edge, which affects aspects of sound quality such as musical image and clarity. And an important characteristic of the rubber edge is weight. Generally, each process for manufacturing rubber edges calls for a *production weight specification* that consists of a lower specification limit (LSL), a target weight (*T*), and an upper specification limit (USL).

The (population) mean and standard deviation of the weights of the rubber edges actually produced are called, respectively, the process mean (μ) and process standard deviation (σ). An on-target process ($\mu = T$) is called *super* if (USL − LSL)/(6σ) > 2 or, equivalently, if $\sigma <$ (USL − LSL)/12.

In the paper "Multiprocess Performance Analysis: A Case Study" (*Quality Engineering*, Vol. 10, No. 1, pp. 1–8), W. Pearn and K. Chen investigated a rubber-edge manufacturing process at Bopro, a company located in Taipei, Taiwan. The following table, adapted from the researchers' paper, provides weight data for the process.

In this chapter, you will study inferences for population standard deviations. In the Case Study Discussion at the end of the chapter, you will use these weight data to determine process capability.

17.59	17.63	17.68	17.57	17.70	17.77	17.54	17.65	17.49	17.60
17.61	17.72	17.68	17.69	17.57	17.66	17.55	17.80	17.67	17.53
17.71	17.54	17.68	17.75	17.46	17.82	17.62	17.53	17.47	17.50
17.74	17.65	17.68	17.68	17.71	17.64	17.65	17.62	17.56	17.60
17.51	17.70	17.47	17.57	17.55	17.63	17.44	17.60	17.63	17.59
17.69	17.53	17.59	17.57	17.49	17.52	17.71	17.56	17.49	17.58

11.1 Inferences for One Population Standard Deviation*

Recall that standard deviation is a measure of the variation (or spread) of a data set. Also recall that, for a variable x, the standard deviation of all possible observations for the entire population is called the *population standard deviation* or *standard deviation of the variable x*. It is denoted σ_x or, when no confusion will arise, simply σ.

Suppose that we want to obtain information about a population standard deviation. If the population is small, we can often determine σ exactly by first taking a census and then computing σ from the population data. However, if the population is large, which is usually the case, a census is generally not feasible, and we must use inferential methods to obtain the required information about σ.

In this section, we describe how to perform hypothesis tests and construct confidence intervals for the standard deviation of a normally distributed variable. Such inferences are based on a distribution called the *chi-square distribution*. Chi (pronounced "kī") is a Greek letter whose lowercase form is χ.

FIGURE 11.1

χ^2-curves for df = 5, 10, and 19

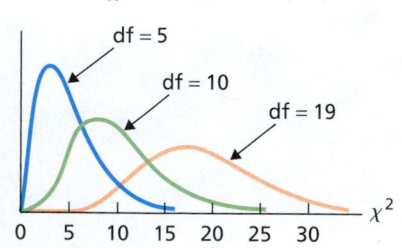

The Chi-Square Distribution

A variable has a **chi-square distribution** if its distribution has the shape of a special type of right-skewed curve, called a **chi-square (χ^2) curve.** Actually, there are infinitely many chi-square distributions, and we identify the chi-square distribution (and χ^2-curve) in question by its number of degrees of freedom, just as we did for t-distributions. Figure 11.1 shows three χ^2-curves and illustrates some basic properties of χ^2-curves.

KEY FACT 11.1

Basic Properties of χ^2-Curves

Property 1: The total area under a χ^2-curve equals 1.

Property 2: A χ^2-curve starts at 0 on the horizontal axis and extends indefinitely to the right, approaching, but never touching, the horizontal axis as it does so.

Property 3: A χ^2-curve is right skewed.

Property 4: As the number of degrees of freedom becomes larger, χ^2-curves look increasingly like normal curves.

Percentages (and probabilities) for a variable having a chi-square distribution are equal to areas under its associated χ^2-curve. To perform a hypothesis test or construct a confidence interval for a population standard deviation, we need to know how to find the χ^2-value that corresponds to a specified area under a χ^2-curve. Table VII in Appendix A provides χ^2-values corresponding to several areas for various degrees of freedom.

The χ^2-table (Table VII) is similar to the t-table (Table IV). The two outside columns of Table VII, labeled df, display the number of degrees of freedom. As expected, the symbol χ_α^2 denotes the χ^2-value having area α to its right under a χ^2-curve. Thus the column headed $\chi_{0.995}^2$, for example, contains χ^2-values having area 0.995 to their right.

■■■ **EXAMPLE 11.1** Finding the χ^2-Value Having a Specified Area to Its Right

For a χ^2-curve with 12 degrees of freedom, find $\chi_{0.025}^2$; that is, find the χ^2-value having area 0.025 to its right, as shown in Fig. 11.2(a).

FIGURE 11.2
Finding the χ^2-value having area 0.025 to its right

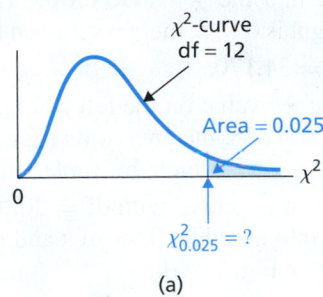
(a)

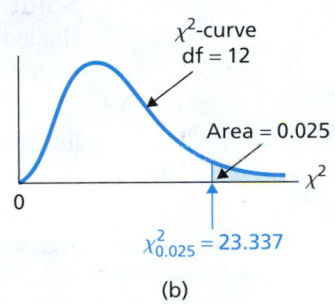
(b)

You try it!

Exercise 11.5 on page 524

Solution To find this χ^2-value, we use Table VII. The number of degrees of freedom is 12, so we first go down the outside columns, labeled df, to "12." Then, going across that row to the column labeled $\chi_{0.025}^2$, we reach 23.337. This number is the χ^2-value having area 0.025 to its right, as shown in Fig. 11.2(b). In other words, for a χ^2-value with df = 12, $\chi_{0.025}^2 = 23.337$. ■

■■■ **EXAMPLE 11.2** Finding the χ^2-Value Having a Specified Area to Its Left

Determine the χ^2-value having area 0.05 to its left for a χ^2-curve with df = 7, as depicted in Fig. 11.3(a).

FIGURE 11.3
Finding the χ^2-value having area 0.05 to its left

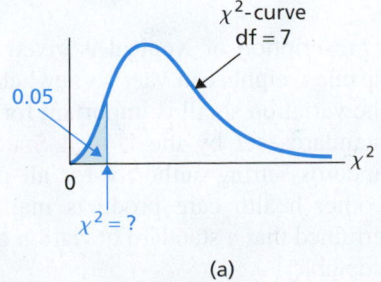
(a)

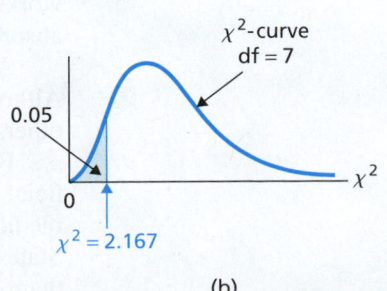
(b)

You try it!

Exercise 11.9 on page 524

Solution Because the total area under a χ^2-curve equals 1 (Property 1 of Key Fact 11.1), the unshaded area in Fig. 11.3(a) must equal $1 - 0.05 = 0.95$. Thus the required χ^2-value is $\chi_{0.95}^2$. From Table VII with df = 7, $\chi_{0.95}^2 = 2.167$. So, for a χ^2-curve with df = 7, the χ^2-value having area 0.05 to its left is 2.167, as shown in Fig. 11.3(b). ■

■■□■ **EXAMPLE 11.3** **Finding the χ^2-Values for a Specified Area**

For a χ^2-curve with df = 20, determine the two χ^2-values that divide the area under the curve into a middle 0.95 area and two outside 0.025 areas, as shown in Fig. 11.4(a).

FIGURE 11.4

Finding the two χ^2-values that divide the area under the curve into a middle 0.95 area and two outside 0.025 areas

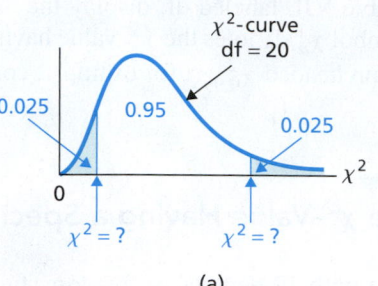

(a)

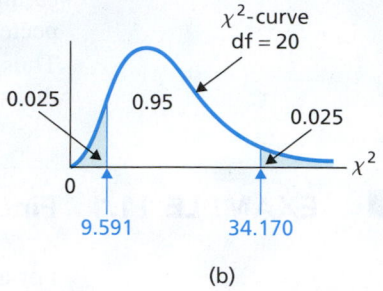

(b)

You try it!

Exercise 11.11 on page 524

Solution First, we find the χ^2-value on the right in Fig. 11.4(a). Because the shaded area on the right is 0.025, the χ^2-value on the right is $\chi^2_{0.025}$. From Table VII with df = 20, $\chi^2_{0.025} = 34.170$.

Next, we find the χ^2-value on the left in Fig. 11.4(a). Because the area to the left of that χ^2-value is 0.025, the area to its right is $1 - 0.025 = 0.975$. Hence the χ^2-value on the left is $\chi^2_{0.975}$, which, by Table VII, equals 9.591 for df = 20.

Consequently, for a χ^2-curve with df = 20, the two χ^2-values that divide the area under the curve into a middle 0.95 area and two outside 0.025 areas are 9.591 and 34.170, as shown in Fig. 11.4(b). ■

The Logic Behind Hypothesis Tests for One Population Standard Deviation

We illustrate the logic behind hypothesis tests for one population standard deviation in the next example.

■■□■ **EXAMPLE 11.4** **Hypothesis Tests for a Population Standard Deviation**

Xenical Capsules Xenical is used to treat obesity in people with risk factors such as diabetes, high blood pressure, and high cholesterol or triglycerides. Xenical works in the intestines, where it blocks some of the fat a person eats from being absorbed.

A standard prescription of Xenical is given in 120-milligram (mg) capsules. Although the capsule weights can vary somewhat from 120 mg and also from each other, keeping the variation small is important for various medical reasons.

Based on standards set by the United States Pharmacopeia (USP)—an official public standards-setting authority for all prescription and over-the-counter medicines and other health care products manufactured or sold in the United States—we determined that a standard deviation of Xenical capsule weights of less than 2 mg is acceptable.[†]

a. Formulate statistically the problem of deciding whether the standard deviation of Xenical capsule weights is less than 2.0 mg.

[†]See Exercise 11.40 for an explanation of how that information could be obtained.

b. Explain the basic idea for carrying out the hypothesis test.
c. In the paper "HPLC Analysis of Orlistat and Its Application to Drug Quality Control Studies" (*Chemical & Pharmaceutical Bulletin*, Vol. 55, No. 2, pp. 251–254), E. Souri et al. studied various properties of Xenical. A sample of 10 Xenical capsules had the weights shown in Table 11.1. Discuss the use of these data to make a decision concerning the hypothesis test.

TABLE 11.1

Weights (mg) of 10 Xenical capsules

120.94	118.58	119.41	120.23
121.13	118.22	119.71	121.09
120.56	119.11		

Solution

a. We want to perform the hypothesis test

$$H_0: \sigma = 2.0 \text{ mg (too much weight variation)}$$

$$H_a: \sigma < 2.0 \text{ mg (not too much weight variation).}$$

If the null hypothesis can be rejected, we can be confident that the variation in capsule weights is acceptable.[†]

b. Roughly speaking, the hypothesis test can be carried out in the following manner:

1. Take a random sample of Xenical capsules.
2. Find the standard deviation, s, of the weights of the capsules sampled.
3. If s is "too much smaller" than 2.0 mg, reject the null hypothesis in favor of the alternative hypothesis; otherwise, do not reject the null hypothesis.

c. The sample standard deviation of the capsule weights in Table 11.1 is

$$s = \sqrt{\frac{\Sigma x_i^2 - (\Sigma x_i)^2/n}{n-1}} = \sqrt{\frac{143765.3242 - (1198.98)^2/10}{9}} = 1.055 \text{ mg.}$$

Is this value of s "too much smaller" than 2.0 mg, suggesting that the null hypothesis be rejected? Or can the difference between $s = 1.055$ mg and the null hypothesis value of $\sigma = 2.0$ mg be attributed to sampling error? To answer these questions, we need to know the distribution of the variable s, that is, the distribution of all possible sample standard deviations that could be obtained by sampling 10 Xenical capsules. We examine that distribution and then return to complete the hypothesis test. ■

Sampling Distribution of the Sample Standard Deviation

Recall that to perform a hypothesis test with null hypothesis $H_0: \mu = \mu_0$ for the mean, μ, of a normally distributed variable, we do not use the variable $\bar{x}$ as the test statistic; rather, we use the variable

$$t = \frac{\bar{x} - \mu_0}{s/\sqrt{n}}.$$

Similarly, when performing a hypothesis test with null hypothesis $H_0: \sigma = \sigma_0$ for the standard deviation, σ, of a normally distributed variable, we do not use the variable s as the test statistic; rather, we use a modified version of that variable:

$$\chi^2 = \frac{n-1}{\sigma_0^2}s^2.$$

This variable has a chi-square distribution.

[†]Another approach would be to let the null hypothesis be $H_0: \sigma = 2.0$ mg (not too much weight variation) and the alternative hypothesis be $H_a: \sigma > 2.0$ mg (too much weight variation). Then rejection of the null hypothesis would indicate that the variation in capsule weights is unacceptable.

KEY FACT 11.2

The Sampling Distribution of the Sample Standard Deviation[†]

Suppose that a variable of a population is normally distributed with standard deviation σ. Then, for samples of size n, the variable

$$\chi^2 = \frac{n-1}{\sigma^2}s^2$$

has the chi-square distribution with $n-1$ degrees of freedom.

APPLET

Applet 11.1

■■ ■ **EXAMPLE 11.5**

The Sampling Distribution of the Sample Standard Deviation

Xenical Capsules In Example 11.4, suppose that the capsule weights are normally distributed with mean 120 mg and standard deviation 2.0 mg. Then, according to Key Fact 11.2, for samples of size 10, the variable

$$\chi^2 = \frac{n-1}{\sigma^2}s^2 = \frac{10-1}{(2.0)^2}s^2 = 2.25\,s^2$$

has a chi-square distribution with 9 degrees of freedom. Use simulation to make that fact plausible.

OUTPUT 11.1

Histogram of χ^2 for 1000 samples of 10 capsule weights with superimposed χ^2-curve

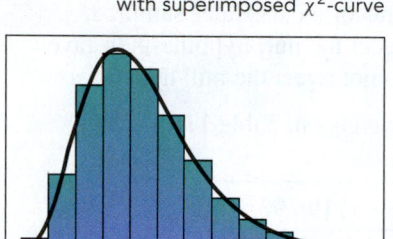

0 2 4 6 8 10 12 14 16 18 20 22 24
CHISQ

Solution We first simulated 1000 samples of 10 capsule weights each, that is, 1000 samples of 10 observations each of a normally distributed variable with mean 120 and standard deviation 2.0. Then, for each of those 1000 samples, we determined the sample standard deviation, s, and obtained the value of the variable χ^2 displayed above. Output 11.1 shows a histogram of those 1000 values of χ^2, which is shaped like the superimposed χ^2-curve with df = 9. ■

Hypothesis Tests for a Population Standard Deviation

In light of Key Fact 11.2, for a hypothesis test with null hypothesis H_0: $\sigma = \sigma_0$, we can use the variable

$$\chi^2 = \frac{n-1}{\sigma_0^2}s^2$$

as the test statistic and obtain the critical value(s) from the χ^2-table, Table VII. We call this hypothesis-testing procedure the **one-standard-deviation χ^2-test.**[‡]

Procedure 11.1 gives a step-by-step method for performing a one-standard-deviation χ^2-test by using either the critical-value approach or the P-value approach. For the P-value approach, we could use Table VII to estimate the P-value, but to do so is awkward and tedious; thus, we recommend using statistical software.

Unlike the z-tests and t-tests for one and two population means, the one-standard-deviation χ^2-test is not robust to moderate violations of the normality assumption. In fact, it is so nonrobust that many statisticians advise against its use unless there is considerable evidence that the variable under consideration is normally distributed or very nearly so.

Consequently, before applying Procedure 11.1, construct a normal probability plot. If the plot creates any doubt about the normality of the variable under consideration, do not use Procedure 11.1.

[†] Strictly speaking, the sampling distribution presented here is not the sampling distribution of the sample standard deviation but is the sampling distribution of a multiple of the sample variance.

[‡] The one-standard-deviation χ^2-test is also known as the **χ^2-test for one population standard deviation.** This test is often formulated in terms of variance instead of standard deviation.

▪▪▪ PROCEDURE 11.1   One-Standard-Deviation χ^2-Test

Purpose   To perform a hypothesis test for a population standard deviation, σ

Assumptions
1.   Simple random sample
2.   Normal population

Step 1   The null hypothesis is H_0: $\sigma = \sigma_0$, and the alternative hypothesis is

$$H_a: \sigma \neq \sigma_0 \quad \text{or} \quad H_a: \sigma < \sigma_0 \quad \text{or} \quad H_a: \sigma > \sigma_0$$
$$\text{(Two tailed)} \qquad \text{(Left tailed)} \qquad \text{(Right tailed)}.$$

Step 2   Decide on the significance level, α.

Step 3   Compute the value of the test statistic

$$\chi^2 = \frac{n-1}{\sigma_0^2}s^2$$

and denote that value χ_0^2.

CRITICAL-VALUE APPROACH	OR	P-VALUE APPROACH

Step 4   The critical value(s) are

$$\chi_{1-\alpha/2}^2 \text{ and } \chi_{\alpha/2}^2 \quad \text{or} \quad \chi_{1-\alpha}^2 \quad \text{or} \quad \chi_\alpha^2$$
$$\text{(Two tailed)} \qquad \text{(Left tailed)} \qquad \text{(Right tailed)}$$

with df $= n - 1$. Use Table VII to find the critical value(s).

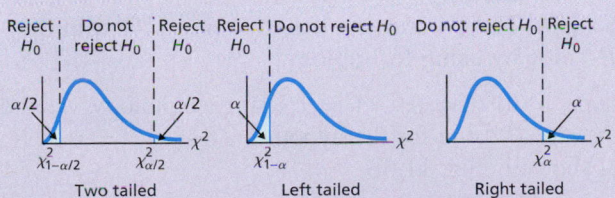

Step 5   If the value of the test statistic falls in the rejection region, reject H_0; otherwise, do not reject H_0.

Step 4   The χ^2-statistic has df $= n - 1$. Obtain the P-value by using technology.

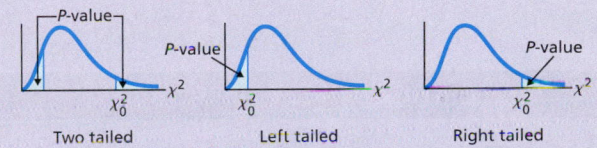

Step 5   If $P \leq \alpha$, reject H_0; otherwise, do not reject H_0.

Step 6   Interpret the results of the hypothesis test.

We note that nonparametric procedures, which do not require normality, have been developed to perform inferences for a population standard deviation. If you have doubts about the normality of the variable under consideration, you can often use one of those procedures to perform a hypothesis test or find a confidence interval for a population standard deviation.

▪▪▪ EXAMPLE 11.6   The One-Standard-Deviation χ^2-Test

Xenical Capsules   We can now complete the hypothesis test proposed in Example 11.4. A sample of 10 Xenical capsules have the weights, in milligrams (mg), shown in Table 11.2 on the next page. At the 5% significance level, do the data provide sufficient evidence to conclude that the standard deviation of the weights of all Xenical capsules is less than 2.0 mg?

TABLE 11.2

Weights (mg) of 10 Xenical capsules

120.94	118.58	119.41	120.23
121.13	118.22	119.71	121.09
120.56	119.11		

Solution To begin, we construct Fig. 11.5, which is a normal probability plot for the data in Table 11.2. Because the plot is reasonably linear, we can use Procedure 11.1 to perform the required hypothesis test.[†]

Step 1 **State the null and alternative hypotheses.**

Let σ denote the population standard deviation of Xenical capsule weights. Then the null and alternative hypotheses are, respectively,

$$H_0: \ \sigma = 2.0 \text{ mg (too much weight variation)}$$
$$H_a: \ \sigma < 2.0 \text{ mg (not too much weight variation)}.$$

FIGURE 11.5

Normal probability plot for the weights in Table 11.2

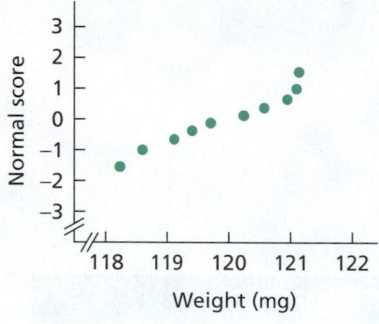

Step 2 **Decide on the significance level, α.**

The test is to be performed at the 5% level of significance, or $\alpha = 0.05$.

Step 3 **Compute the value of the test statistic**

$$\chi^2 = \frac{n-1}{\sigma_0^2} s^2.$$

First, we find the sample variance, s^2. From Table 11.2,

$$s^2 = \frac{\Sigma x_i^2 - (\Sigma x_i)^2/n}{n-1} = \frac{143765.3242 - (1198.98)^2/10}{9} = 1.113.$$

Because $n = 10$ and $\sigma_0 = 2.0$, the value of the test statistic is

$$\chi^2 = \frac{n-1}{\sigma_0^2} s^2 = \frac{10-1}{(2.0)^2} \cdot 1.113 = 2.504.$$

CRITICAL-VALUE APPROACH	OR	P-VALUE APPROACH

Step 4 The critical value for a left-tailed test is $\chi^2_{1-\alpha}$ with df $= n - 1$. Use Table VII to find the critical value.

We have $\alpha = 0.05$. Also, $n = 10$, so df $= 10 - 1 = 9$. In Table VII, we determine that the critical value is $\chi^2_{1-\alpha} = \chi^2_{1-0.05} = \chi^2_{0.95} = 3.325$, as shown in Fig. 11.6A.

FIGURE 11.6A

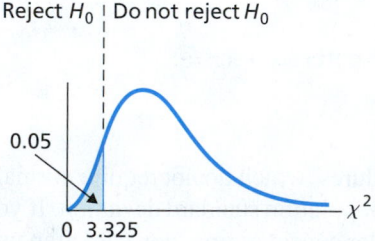

Step 5 **If the value of the test statistic falls in the rejection region, reject H_0; otherwise, do not reject H_0.**

From Step 3, the value of the test statistic is $\chi^2 = 2.504$, which falls in the rejection region shown in Fig. 11.6A. Thus we reject H_0. The test results are statistically significant at the 5% level.

Step 4 The χ^2-statistic has df $= n - 1$. Obtain the P-value by using technology.

For $n = 10$, df $= 10 - 1 = 9$. Using technology, we find that the P-value for the hypothesis test is $P = 0.0193$, as shown in Fig. 11.6B.

FIGURE 11.6B

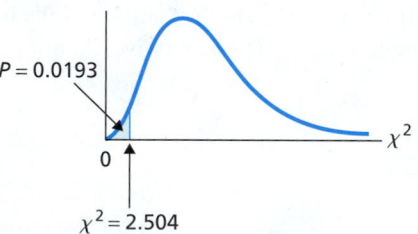

Step 5 **If $P \leq \alpha$, reject H_0; otherwise, do not reject H_0.**

From Step 4, $P = 0.0193$. Because the P-value is less than the specified significance level of 0.05, we reject H_0. The test results are statistically significant at the 5% level and (see Table 9.8 on page 386) provide strong evidence against the null hypothesis.

[†]Some statisticians might regard the plot sufficiently nonlinear to require the use of a nonparametric method instead of Procedure 11.1.

StatCrunch

Report 11.1

Exercise 11.21
on page 524

Step 6 Interpret the results of the hypothesis test.

Interpretation At the 5% significance level, the data provide sufficient evidence to conclude that the standard deviation of the weights of all Xenical capsules is less than 2.0 mg. Evidently, the variation in capsule weights is acceptable according to United States Pharmacopeia standards.

Confidence Intervals for a Population Standard Deviation

Using Key Fact 11.2 on page 518, we can also obtain a confidence-interval procedure for a population standard deviation. We call this procedure the **one-standard-deviation χ^2-interval procedure** and present it as Procedure 11.2.[†] Like the one-standard-deviation χ^2-test, this procedure is not at all robust to violations of the normality assumption.

■■■ PROCEDURE 11.2 One-Standard-Deviation χ^2-Interval Procedure

Purpose To find a confidence interval for a population standard deviation, σ

Assumptions
1. Simple random sample
2. Normal population

Step 1 For a confidence level of $1 - \alpha$, use Table VII to find $\chi^2_{1-\alpha/2}$ and $\chi^2_{\alpha/2}$ with df $= n - 1$.

Step 2 The confidence interval for σ is from

$$\sqrt{\frac{n-1}{\chi^2_{\alpha/2}}} \cdot s \quad \text{to} \quad \sqrt{\frac{n-1}{\chi^2_{1-\alpha/2}}} \cdot s,$$

where $\chi^2_{1-\alpha/2}$ and $\chi^2_{\alpha/2}$ are found in Step 1, n is the sample size, and s is computed from the sample data obtained.

Step 3 Interpret the confidence interval.

■■■ EXAMPLE 11.7 The One-Standard-Deviation χ^2-Interval Procedure

Xenical Capsules Use the sample data in Table 11.2 to determine a 90% confidence interval for the standard deviation, σ, of the weights of all Xenical capsules.

Solution We apply Procedure 11.2.

Step 1 For a confidence level of $1 - \alpha$, use Table VII to find $\chi^2_{1-\alpha/2}$ and $\chi^2_{\alpha/2}$ with df $= n - 1$.

For a 90% confidence interval, the confidence level is $0.90 = 1 - 0.10$, and so $\alpha = 0.10$. Also, for $n = 10$, df $= 9$. In Table VII, we find that

$$\chi^2_{1-\alpha/2} = \chi^2_{1-0.10/2} = \chi^2_{0.95} = 3.325$$

and

$$\chi^2_{\alpha/2} = \chi^2_{0.10/2} = \chi^2_{0.05} = 16.919.$$

[†] The one-standard-deviation χ^2-interval procedure is also known as the **χ^2-interval procedure for one population standard deviation.** This confidence-interval procedure is often formulated in terms of variance instead of standard deviation.

Step 2 The confidence interval for σ is from

$$\sqrt{\frac{n-1}{\chi^2_{\alpha/2}}} \cdot s \quad \text{to} \quad \sqrt{\frac{n-1}{\chi^2_{1-\alpha/2}}} \cdot s.$$

We have $n = 10$, and from Step 1, $\chi^2_{1-\alpha/2} = 3.325$ and $\chi^2_{\alpha/2} = 16.919$. Also, we found in Example 11.4 that $s = 1.055$ mg. So, a 90% confidence interval for σ is from

$$\sqrt{\frac{10-1}{16.919}} \cdot 1.055 \quad \text{to} \quad \sqrt{\frac{10-1}{3.325}} \cdot 1.055,$$

or 0.77 to 1.74.

Step 3 Interpret the confidence interval.

Interpretation We can be 90% confident that the standard deviation of the weights of all Xenical capsules is somewhere between 0.77 mg and 1.74 mg. ■

StatCrunch

Report 11.2

You try it!

Exercise 11.27
on page 525

THE TECHNOLOGY CENTER

Some statistical technologies have programs that automatically perform one-standard-deviation χ^2-procedures, but others do not. In this subsection, we present output and step-by-step instructions for such programs.

Note to Excel users: At the time of this writing, Excel does not have a built-in program for one-standard-deviation χ^2-procedures.

Note to TI-83/84 Plus users:

- At the time of this writing, the TI-83/84 Plus does not have a built-in program for one-standard-deviation χ^2-procedures. However, TI programs called STDEVHT and STDEVINT, supplied in the TI Programs section on the WeissStats site, allow you to perform those procedures. Your instructor can show you how to download the programs to your calculator.
- *Warning:* Any data that you may have previously stored in List 1 will be erased during program execution, so copy those data to another list prior to program execution if you want to retain them.

EXAMPLE 11.8 **Using Technology to Conduct One-Standard-Deviation χ^2-Procedures**

Xenical Capsules Table 11.2 on page 520 gives the weights, in milligrams, of a sample of 10 Xenical capsules. Use Minitab or the TI-83/84 Plus to perform the hypothesis test in Example 11.6 and obtain the confidence interval in Example 11.7.

Solution Let σ denote the population standard deviation of weights of all Xenical capsules. The task in Example 11.6 is to perform the hypothesis test

$$H_0: \quad \sigma = 2.0 \text{ mg (too much weight variation)}$$
$$H_a: \quad \sigma < 2.0 \text{ mg (not too much weight variation)}$$

at the 5% significance level; the task in Example 11.7 is to find a 90% confidence interval for σ. We applied the appropriate Minitab and TI-83/84 Plus programs to the data, resulting in Output 11.2. Steps for generating that output are presented in Instructions 11.1. *Note to Minitab users:* For brevity, we have presented only the essential portions of the actual output.

OUTPUT 11.2   One-standard-deviation χ^2-test and interval on the weight data

<table>
<tr><td>

MINITAB

```
[FOR THE HYPOTHESIS TEST]
```

Test and CI for One Variance: WEIGHT

```
Null hypothesis        σ = 2
Alternative hypothesis σ < 2

The chi-square method is only for the normal distribution.
The Bonett method is for any continuous distribution.


Statistics

Variable   N  StDev  Variance
WEIGHT    10  1.06    1.11

Tests

                      Test
Variable  Method   Statistic  DF  P-Value
WEIGHT    Chi-Square   2.51    9   0.019
          Bonett        -      -   0.013
```

```
[FOR THE CONFIDENCE INTERVAL]
```

Test and CI for One Variance: WEIGHT

```
90% Confidence Intervals

                  CI for        CI for
Variable  Method   StDev        Variance
WEIGHT  Chi-Square (0.77, 1.74) (0.59, 3.01)
        Bonett     (0.81, 1.64) (0.66, 2.70)
```

</td><td>

TI-83/84 PLUS

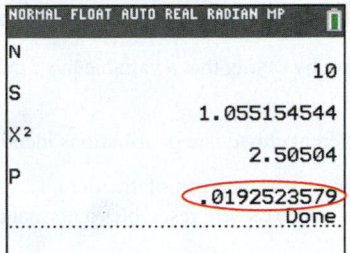

Using the **STDEVHT** program

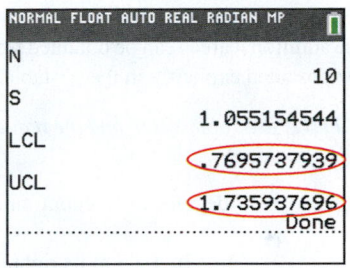

Using the **STDEVINT** program

</td></tr>
</table>

As shown in Output 11.2, the *P*-value for the hypothesis test is 0.019. Because the *P*-value is less than the specified significance level of 0.05, we reject H_0. Output 11.2 also shows that a 90% confidence interval for σ is from 0.77 mg to 1.74 mg.

INSTRUCTIONS 11.1   Steps for generating Output 11.2

<table>
<tr><td>

MINITAB

Store the data from Table 11.2 in a column named WEIGHT.

FOR THE HYPOTHESIS TEST:
1. Choose **Stat ➤ Basic Statistic ➤ 1 Variance...**
2. Press the F3 key to reset the dialog box
3. Click in the text box directly below the **One or more samples, each in a column** drop-down list box and specify WEIGHT
4. Check the **Perform hypothesis test** check box
5. Type 2 in the **Value** text box
6. Click the **Options...** button
7. Click the arrow button at the right of the **Alternative hypothesis** drop-down list box and select **Standard deviation < hypothesized standard deviation**
8. Click **OK** twice

FOR THE CI:
1. Repeat steps 1–3 from the hypothesis-test instructions
2. Click the **Options...** button
3. Click in the **Confidence level** text box and type 90
4. Click **OK** twice

</td><td>

TI-83/84 PLUS

Store the data from Table 11.2 in a list named WT.

FOR THE HYPOTHESIS TEST:
1. Press **PRGM**
2. Arrow down to STDEVHT and press **ENTER** twice
3. Type 1 for **TYPE** and press **ENTER**
4. Press **2ND ➤ LIST**, arrow down to WT, and press **ENTER** twice
5. Type 2 for **SIGMA0** and press **ENTER**
6. Type –1 for **TYPE** and press **ENTER**

FOR THE CI:
1. Press **PRGM**
2. Arrow down to STDEVINT and press **ENTER** twice
3. Type 1 for **TYPE** and press **ENTER**
4. Press **2ND ➤ LIST**, arrow down to WT, and press **ENTER** twice
5. Type .90 for **C-LEVEL** and press **ENTER**

</td></tr>
</table>

Exercises 11.1

Understanding the Concepts and Skills

11.1 What is meant by saying that a variable has a chi-square distribution?

11.2 How are different chi-square distributions identified?

11.3 Two χ^2-curves have degrees of freedom 12 and 20, respectively. Which curve more closely resembles a normal curve? Explain your answer.

11.4 The t-table has entries for areas of 0.10, 0.05, 0.025, 0.01, and 0.005. In contrast, the χ^2-table has entries for those areas and for 0.995, 0.99, 0.975, 0.95, and 0.90. Explain why the t-values corresponding to these additional areas can be obtained from the existing t-table but must be provided explicitly in the χ^2-table.

In Exercises 11.5–11.12, use Table VII to find the required χ^2-values. Illustrate your work graphically.

11.5 For a χ^2-curve with 19 degrees of freedom, find the χ^2-value that has area
a. 0.025 to its right.   **b.** 0.95 to its right.

11.6 For a χ^2-curve with 22 degrees of freedom, find the χ^2-value that has area
a. 0.01 to its right.   **b.** 0.995 to its right.

11.7 For a χ^2-curve with df = 10, determine
a. $\chi^2_{0.05}$.   **b.** $\chi^2_{0.975}$.

11.8 For a χ^2-curve with df = 4, determine
a. $\chi^2_{0.005}$.   **b.** $\chi^2_{0.99}$.

11.9 Consider a χ^2-curve with df = 8. Obtain the χ^2-value that has area
a. 0.01 to its left.   **b.** 0.95 to its left.

11.10 Consider a χ^2-curve with df = 16. Obtain the χ^2-value that has area
a. 0.025 to its left.   **b.** 0.975 to its left.

11.11 Determine the two χ^2-values that divide the area under the curve into a middle 0.95 area and two outside 0.025 areas for a χ^2-curve with
a. df = 5.   **b.** df = 26.

11.12 Determine the two χ^2-values that divide the area under the curve into a middle 0.90 area and two outside 0.05 areas for a χ^2-curve with
a. df = 11.   **b.** df = 28.

11.13 When you use chi-square procedures to make inferences about a population standard deviation, why should the variable under consideration be normally distributed or nearly so?

11.14 Give two situations in which making an inference about a population standard deviation would be important.

In each of Exercises 11.15–11.20, we have provided a sample standard deviation and sample size. In each case, use the one-standard-deviation χ^2-test and the one-standard-deviation χ^2-interval procedure to conduct the required hypothesis test and obtain the specified confidence interval.

11.15 $s = 3$ and $n = 10$
a. H_0: $\sigma = 4$, H_a: $\sigma < 4$, $\alpha = 0.05$
b. 90% confidence interval

11.16 $s = 2$ and $n = 10$
a. H_0: $\sigma = 4$, H_a: $\sigma < 4$, $\alpha = 0.05$
b. 90% confidence interval

11.17 $s = 7$ and $n = 26$
a. H_0: $\sigma = 5$, H_a: $\sigma > 5$, $\alpha = 0.01$
b. 98% confidence interval

11.18 $s = 6$ and $n = 26$
a. H_0: $\sigma = 5$, H_a: $\sigma > 5$, $\alpha = 0.01$
b. 98% confidence interval

11.19 $s = 5$ and $n = 20$
a. H_0: $\sigma = 6$, H_a: $\sigma \neq 6$, $\alpha = 0.05$
b. 95% confidence interval

11.20 $s = 8$ and $n = 20$
a. H_0: $\sigma = 6$, H_a: $\sigma \neq 6$, $\alpha = 0.05$
b. 95% confidence interval

Applying the Concepts and Skills

Preliminary data analyses and other information suggest that you can reasonably assume that the variables under consideration in Exercises 11.21–11.26 are normally distributed. In each case, use either the critical-value approach or the P-value approach to perform the required hypothesis test.

11.21 Grey-Seal Nursing. The average lactation (nursing) period of all earless seals is 23 days. Grey seals are one of several types of earless seals. The length of time that a female grey seal nurses her pup is studied by S. Twiss et al. in the article "Variation in Female Grey Seal (*Halichoerus grypus*) Reproductive Performance Correlates to Proactive-Reactive Behavioural Types" (*PLOS ONE* 7(11): e49598. doi:10.1371/journal.pone.0049598). A sample of 14 female grey seals had the following lactation periods, in days.

20.2	20.9	20.6	23.6	19.6	15.9	19.8
15.4	21.4	19.5	17.4	21.9	22.3	16.4

In Exercise 9.84, you were asked to use these data to decide whether the mean lactation period of grey seals differs from 23 days. There, you were to assume that the population standard deviation of lactation periods for grey seals is 3.0 days. At the 10% significance level, do the data provide evidence against that assumption? (*Note: s = 2.501.*)

11.22 EPA Gas Mileage Estimates. Gas mileage estimates for cars and light-duty trucks are determined and published by the U.S. Environmental Protection Agency (EPA). According to the EPA, "...the mileages obtained by most drivers will be within plus or minus 15 percent of the [EPA] estimates...." The mileage estimate given for one model is 23 mpg on the highway. If the EPA claim is true, the standard deviation of mileages should be about $0.15 \cdot 23/3 = 1.15$ mpg. A random sample of 12 cars of this model yields the following highway mileages.

24.1	23.3	22.5	23.2
22.3	21.1	21.4	23.4
23.5	22.8	24.5	24.3

a. At the 5% significance level, do the data suggest that the standard deviation of highway mileages for all cars of this model is different from 1.15 mpg? (*Note: s = 1.071.*)

b. Why it is useful to know the standard deviation of the gas mileages as well as the mean gas mileage?

11.23 Process Capability. R. Morris and E. Watson studied various aspects of process capability in the paper "Determining Process Capability in a Chemical Batch Process" (*Quality Engineering*, Vol. 10(2), pp. 389–396). In one part of the study, the researchers compared the variability in product (as measured by standard deviation) of a particular piece of equipment to a known analytic capability to decide whether product consistency could be improved. The following data were obtained for 10 batches of product.

30.1	30.7	30.2	29.3	31.0
29.6	30.4	31.2	28.8	29.8

At the 1% significance level, do the data provide sufficient evidence to conclude that the product variability for this piece of equipment exceeds the analytic capability of 0.27? (*Note: s = 0.756.*)

11.24 Premade Pizza. Homestyle Pizza of Camp Verde, Arizona, provides baking instructions for its premade pizzas. According to the instructions, the average baking time is 12 to 18 minutes. If the times are normally distributed, the standard deviation of the times should be approximately 1 minute. A random sample of 15 pizzas yielded the following baking times to the nearest tenth of a minute.

15.4	15.1	14.0	15.8	16.0
13.7	15.6	11.6	14.8	12.8
17.6	15.1	16.4	13.1	15.3

At the 1% significance level, do the data provide sufficient evidence to conclude that the standard deviation of baking times exceeds 1 minute? (*Note:* The sample standard deviation of the 15 baking times is 1.54 minutes.)

11.25 Dispensing Coffee. A coffee machine is supposed to dispense 6 fluid ounces (fl oz) of coffee into a paper cup. In reality, the amounts dispensed vary from cup to cup. However, if the machine is working properly, most of the cups will contain within 10% of the advertised 6 fl oz. In other words, the standard deviation of the amounts dispensed should be less than 0.2 fl oz. A random sample of 15 cups provided the following data, in fluid ounces.

5.90	5.82	6.20	6.09	5.93
6.18	5.99	5.79	6.28	6.16
6.00	5.85	6.13	6.09	6.18

a. At the 5% significance level, do the data provide sufficient evidence to conclude that the standard deviation of the amounts being dispensed is less than 0.2 fl oz? (*Note: s = 0.154.*)

b. Why is it important that the standard deviation of the amounts of coffee being dispensed not be too large?

11.26 Counting Production. In Issue 10 of *STATS* from Iowa State University, data were published from an experiment that examined the effects of machine adjustment on bolt production. An electronic counter records the number of bolts passing it on a conveyer belt and stops the run when the count reaches a preset number. The following data give the times, in seconds, needed to count 20 bolts for eight different runs.

10.78	9.39	9.84	13.94
12.33	7.32	7.91	15.58

Do the data provide sufficient evidence to conclude that the standard deviation in the time needed to count 20 bolts exceeds 2 seconds? Use $\alpha = 0.05$. (*Note:* The sample standard deviation of the eight times is 2.8875 seconds.)

In Exercises 11.27–11.32, use Procedure 11.2 on page 521 to obtain the required confidence interval.

11.27 Grey-Seal Nursing. Refer to Exercise 11.21 and find a 90% confidence interval for the standard deviation of lactation periods of grey seals.

11.28 EPA Gas Mileage Estimates. Refer to Exercise 11.22 and find a 95% confidence interval for the standard deviation of highway gas mileages for all cars of the model in question.

11.29 Process Capability. Refer to Exercise 11.23 and determine a 98% confidence interval for the product variability of the piece of equipment under consideration.

11.30 Premade Pizza. Refer to Exercise 11.24 and determine a 98% confidence interval for the standard deviation of baking times.

11.31 Dispensing Coffee. Refer to Exercise 11.25 and obtain a 90% confidence interval for the standard deviation of the amounts of coffee being dispensed.

11.32 Counting Production. Refer to Exercise 11.26 and obtain a 90% confidence interval for the standard deviation of the times needed to count 20 bolts.

In each of Exercises 11.33–11.36, use the technology of your choice to decide whether applying one-standard-deviation χ^2-procedures appears reasonable. Explain your answers.

11.33 Military Assistance Loans. The annual update of *U.S. Overseas Loans and Grants*, informally known as the "Greenbook," contains data on U.S. government monetary economic and military assistance loans. The following table shows military assistance loans, in thousands of dollars, to a sample of 10 countries, as reported by the U.S. Agency for International Development.

102	280	33	1643	177
69	180	89	205	695

11.34 Positively Selected Genes. R. Nielsen et al. compared 13,731 annotated genes from humans with their chimpanzee orthologs to identify genes that show evidence of positive selection. The researchers published their findings in "A Scan for Positively Selected Genes in the Genomes of Humans and Chimpanzees" (*PLOS Biology*, Vol. 3, Issue 6, pp. 976–985). A simple random sample of 14 tissue types yielded the following number of genes.

66	47	43	101	201	83	93
82	120	64	244	51	70	14

11.35 Total Solar Irradiance. The amount of solar energy transmitted through each square meter of space prior to entry in the earth's atmosphere, as measured by satellites, is called the total solar irradiance (TSI). A random sample of daily TSI measurements, in watts per square meter, from the World Radiation Center in Switzerland is as follows.

1365.9587	1365.9485	1365.9308	1365.8971
1365.8925	1365.8803	1365.8829	1365.8680
1365.8817	1365.8631	1365.8836	1365.8846
1365.9098	1365.9680	1365.9091	1365.9294

11.36 Plesiadapis cookei. One extinct relative of primates that lived in North America about 60 million years ago is called *Plesiadapis cookei*. Dental characteristics of *P. cookei* were compared to those of other primate species in the article "Evidence of Dietary Differentiation among Late Paleocene–Early Eocene Plesiadapids" (*American Journal of Physical Anthropology*, Vol. 142, No. 2, pp. 194–210) by D. Boyer et al. The following table gives the dentary depth, in millimeters, for a sample of molars from the skulls of 18 *P. cookei* specimens.

18.12	19.48	19.36	15.94	15.83	19.70
15.76	17.00	13.96	16.55	15.70	17.83
13.25	16.12	18.13	14.02	14.04	16.20

Working with Large Data Sets

11.37 Body Temperature. A study by researchers at the University of Maryland addressed the question of whether the mean body temperature of humans is 98.6°F. The results of the study by P. Mackowiak et al. appeared in the article "A Critical Appraisal of 98.6°F, the Upper Limit of the Normal Body Temperature, and Other Legacies of Carl Reinhold August Wunderlich" (*Journal of the American Medical Association*, Vol. 268, pp. 1578–1580). Among other data, the researchers obtained the body temperatures of 93 healthy humans, as provided on the WeissStats site. Use the technology of your choice to do the following.

a. Obtain a normal probability plot, boxplot, histogram, and stem-and-leaf diagram of the data.

b. Based on your results from part (a), can you reasonably apply one-standard-deviation χ^2-procedures to the data? Explain your reasoning.

c. In Exercise 9.91, you were asked to use these data to decide whether mean body temperature of healthy humans differs from 98.6°F. There, you were to assume that the population standard deviation of body temperatures for healthy humans is 0.63°F. At the 5% significance level, do the data provide evidence against that assumption?

d. Find and interpret a 95% confidence interval for the population standard deviation of body temperatures for healthy humans.

11.38 Dexamethasone and IQ. In the paper "Outcomes at School Age After Postnatal Dexamethasone Therapy for Lung Disease of Prematurity" (*New England Journal of Medicine*, Vol. 350, No. 13, pp. 1304–1313), T. Yeh et al. studied the outcomes at school age in children who had participated in a double-blind, placebo-controlled trial of early postnatal dexamethasone therapy for the prevention of chronic lung disease of prematurity. All of the infants in the study had had severe respiratory distress syndrome requiring mechanical ventilation shortly after birth. On the WeissStats site, we provide the school-age IQs of the 74 children in the control group, based on the study results. Use the technology of your choice to do the following.

a. Obtain a normal probability plot, boxplot, histogram, and stem-and-leaf diagram of the data.

b. Based on your results from part (a), can you reasonably apply one-standard-deviation χ^2-procedures to the data? Explain your reasoning.

c. Overall, IQs of school-age children have a standard deviation of 16. At the 1% significance level, do the data provide sufficient evidence to conclude that IQs of school-age children in similar postnatal circumstances as those in the control group of this study have a smaller standard deviation than that of school-age children in general?

d. Find and interpret a 99% confidence interval for the standard deviation of IQs of all school-age children in similar postnatal circumstances as those in the control group of this study.

11.39 Forearm Length. In 1903, K. Pearson and A. Lee published the paper "On the Laws of Inheritance in Man. I. Inheritance of Physical Characters" (*Biometrika*, Vol. 2, pp. 357–462). The article examined and presented data on forearm length, in inches, for a sample of 140 men, which we provide on the WeissStats site. Use the technology of your choice to do the following.

a. Obtain a normal probability plot, boxplot, and histogram of the data.

b. Based on your results from part (a), can you reasonably apply one-standard-deviation χ^2-procedures to the data? Explain your reasoning.

c. If you answered "yes" to part (b), determine and interpret a 95% confidence interval for the standard deviation of men's forearm length.

Extending the Concepts and Skills

11.40 Xenical Capsules. In Example 11.4 on page 516, we stated that, based on standards set by the United States Pharmacopeia (USP), a standard deviation of Xenical capsule weights of less than 2 mg is acceptable. We now ask you to obtain that result. In doing so, we presume that weights of Xenical capsules are normally distributed with a mean of 120 mg.

a. According to USP, the requirements for weight variation of capsules are met if each of the individual weights is within the limits of 90% and 110% of the mean weight. Find the lower and upper weight limits in order for USP requirements to be met.

b. Using statistical software, find the percentage of all possible observations of a normally distributed variable that lie within six standard deviations to either side of the mean.

c. Show that, if $\sigma < 2$, then fewer than two of every billion Xenical capsules will have weights that violate USP requirements. (*Hint:* First determine the value of σ for which six standard deviations to either side of the mean give the lower and upper weight limits for USP requirements to be met.)

d. Explain why a standard deviation of Xenical capsule weights of less than 2 mg is reasonably acceptable with respect to USP requirements.

11.41 Intelligence Quotients. Measured on the Stanford Revision of the Binet–Simon Intelligence Scale, intelligence quotients (IQs) are known to be normally distributed with a mean of 100 and a standard deviation of 16. Use the technology of your choice to do the following.

a. Simulate 1000 samples of four IQs each.

b. Determine the sample standard deviation of each of the 1000 samples.

c. Obtain the following quantity for each of the 1000 samples:

$$\frac{n-1}{\sigma^2}s^2 = \frac{4-1}{16^2}s^2.$$

d. Obtain a histogram of the 1000 values found in part (c).

e. Theoretically, what is the distribution of the variable in part (c)?

f. Compare your answers from parts (d) and (e).

11.2 Inferences for Two Population Standard Deviations, Using Independent Samples*

In Section 11.1, we discussed hypothesis tests and confidence intervals for one population standard deviation. We now introduce hypothesis tests and confidence intervals for two population standard deviations. More precisely, we examine inferences to compare the standard deviations of one variable of two different populations. Such inferences are based on a distribution called the *F-distribution*, named in honor of Sir Ronald Fisher.

The *F*-Distribution

A variable is said to have an **F-distribution** if its distribution has the shape of a special type of right-skewed curve, called an **F-curve.** Actually, there are infinitely many *F*-distributions, and we identify the *F*-distribution (and *F*-curve) in question by its number of degrees of freedom, just as we did for *t*-distributions and chi-square distributions.

An *F*-distribution, however, has two numbers of degrees of freedom instead of one. Figure 11.7 depicts two different *F*-curves; one has df $= (10, 2)$, and the other has df $= (9, 50)$.

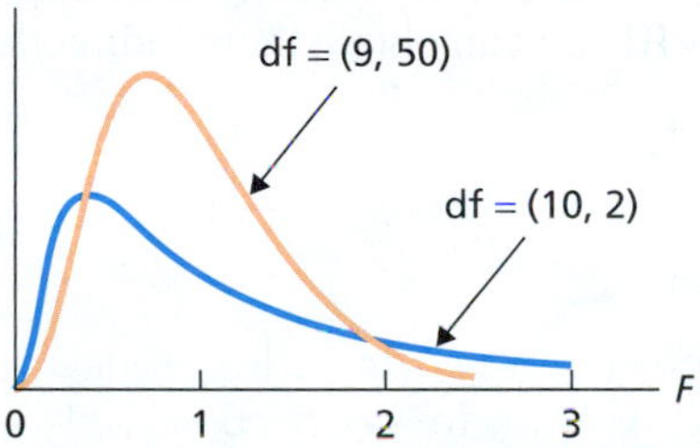

The first number of degrees of freedom for an *F*-curve is called the **degrees of freedom for the numerator** and the second the **degrees of freedom for the denominator.** (This terminology will become clear shortly.) Thus, for the *F*-curve in Fig. 11.7 with df $= (10, 2)$, we have

$$\mathrm{df} \;=\; (10, 2).$$

Degrees of freedom for the numerator   Degrees of freedom for the denominator

FIGURE 11.7

Two different *F*-curves

df = (9, 50)

df = (10, 2)

KEY FACT 11.3

Basic Properties of *F*-Curves

Property 1: The total area under an *F*-curve equals 1.

Property 2: An *F*-curve starts at 0 on the horizontal axis and extends indefinitely to the right, approaching, but never touching, the horizontal axis as it does so.

Property 3: An *F*-curve is right skewed.

Percentages (and probabilities) for a variable having an *F*-distribution equal areas under its associated *F*-curve. To perform a hypothesis test or construct a confidence interval for comparing two population standard deviations, we need to know how to find the *F*-value that corresponds to a specified area under an *F*-curve. The symbol F_α denotes the *F*-value having area α to its right.

Table VIII in Appendix A gives *F*-values with areas 0.005, 0.01, 0.025, 0.05, and 0.10 to their right for various degrees of freedom. The degrees of freedom for the denominator (dfd) are displayed in the outside columns of the table; the values of α are in the next columns; and the degrees of freedom for the numerator (dfn) are along the top.

EXAMPLE 11.9   Finding the *F*-Value Having a Specified Area to Its Right

For an *F*-curve with df $= (4, 12)$, find $F_{0.05}$; that is, find the *F*-value having area 0.05 to its right, as shown in Fig. 11.8(a).

FIGURE 11.8
Finding the *F*-value having
area 0.05 to its right

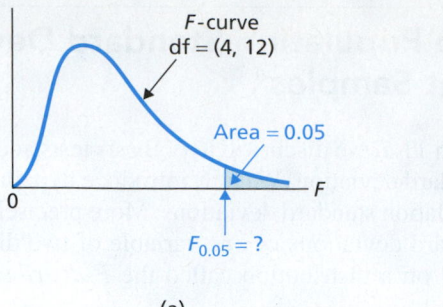
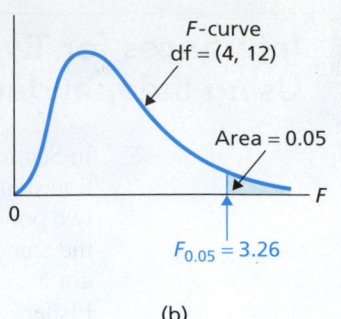

(a)   (b)

Solution To obtain the *F*-value, we use Table VIII. In this case, $\alpha = 0.05$, the degrees of freedom for the numerator is 4, and the degrees of freedom for the denominator is 12.

> *You try it!*
>
> Exercise 11.47
> on page 538

We first go down the dfd column to "12." Next, we concentrate on the row for α labeled 0.05. Then, going across that row to the column labeled "4," we reach 3.26. This number is the *F*-value having area 0.05 to its right, as shown in Fig. 11.8(b). In other words, for an *F*-curve with df = (4, 12), $F_{0.05} = 3.26$.

In many statistical analyses that involve the *F*-distribution, we also need to determine *F*-values having areas 0.005, 0.01, 0.025, 0.05, and 0.10 to their left. Although such *F*-values aren't available directly from Table VIII, we can obtain them indirectly from the table by using Key Fact 11.4.

KEY FACT 11.4

Reciprocal Property of *F*-Curves

For an *F*-curve with df = (ν_1, ν_2), the *F*-value having area α to its left equals the reciprocal of the *F*-value having area α to its right for an *F*-curve with df = (ν_2, ν_1).

EXAMPLE 11.10 **Finding the *F*-Value Having a Specified Area to Its Left**

For an *F*-curve with df = (60, 8), find the *F*-value having area 0.05 to its left.

Solution We apply Key Fact 11.4. Accordingly, the required *F*-value is the reciprocal of the *F*-value having area 0.05 to its right for an *F*-curve with df = (8, 60). From Table VIII, this latter *F*-value equals 2.10. Consequently, the required *F*-value is $\frac{1}{2.10}$, or 0.48, as shown in Fig. 11.9.

FIGURE 11.9
Finding the *F*-value having
area 0.05 to its left

> *You try it!*
>
> Exercise 11.51
> on page 538

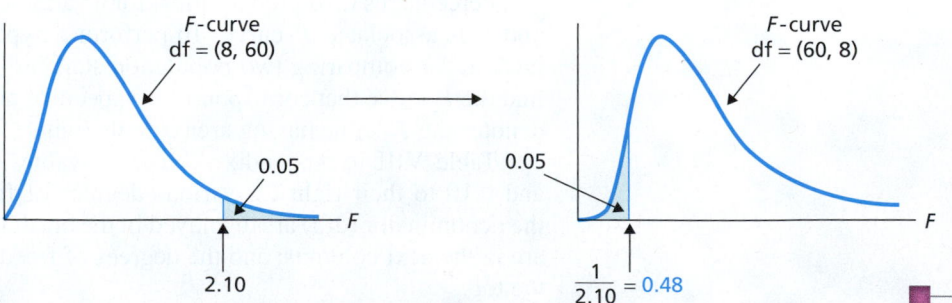

EXAMPLE 11.11 **Finding the *F*-Values for a Specified Area**

For an *F*-curve with df = (9, 8), determine the two *F*-values that divide the area under the curve into a middle 0.95 area and two outside 0.025 areas, as shown in Fig. 11.10(a).

FIGURE 11.10

Finding the two F-values that divide the area under the curve into a middle 0.95 area and two outside 0.025 areas

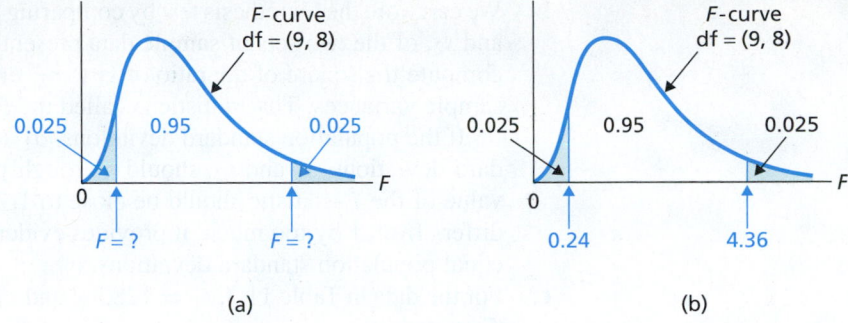

(a)   (b)

Solution First, we find the F-value on the right in Fig. 11.10(a). Because the shaded area on the right is 0.025, the F-value on the right is $F_{0.025}$. From Table VIII with df $= (9, 8)$, $F_{0.025} = 4.36$.

Next, we find the F-value on the left in Fig. 11.10(a). By Key Fact 11.4, that F-value is the reciprocal of the F-value having area 0.025 to its right for an F-curve with df $= (8, 9)$. From Table VIII, we find that this latter F-value equals 4.10. Thus the F-value on the left in Fig. 11.10(a) is $\frac{1}{4.10}$, or 0.24.

Consequently, for an F-curve with df $= (9, 8)$, the two F-values that divide the area under the curve into a middle 0.95 area and two outside 0.025 areas are 0.24 and 4.36, as shown in Fig. 11.10(b).

You try it!

Exercise 11.53 on page 538

The Logic Behind Hypothesis Tests for Comparing Two Population Standard Deviations

We illustrate the logic behind hypothesis tests for comparing two population standard deviations in the next example.

■■■ EXAMPLE 11.12   Hypothesis Tests for Two Population Standard Deviations

Elmendorf Tear Strength Variation within a method used for testing a product is an essential factor in deciding whether the method should be employed. Indeed, when the variation of such a test is high, ascertaining the true quality of a product is difficult.

Manufacturers use the Elmendorf tear test to evaluate material strength for various manufactured products. In the article "Using Repeatability and Reproducibility Studies to Evaluate a Destructive Test Method" (*Quality Engineering*, Vol. 10(2), pp. 283–290), A. Phillips et al. investigated the variation of that test. In one aspect of the study, the researchers randomly and independently obtained the data shown in Table 11.3 on Elmendorf tear strength, in grams, of two different vinyl floor coverings.

Suppose that we want to decide whether the standard deviations of tear strength differ between the two vinyl floor coverings.

a. Formulate the problem statistically by posing it as a hypothesis test.
b. Explain the basic idea for carrying out the hypothesis test.
c. Discuss the use of the data in Table 11.3 to make a decision concerning the hypothesis test.

TABLE 11.3

Results of Elmendorf tear test on two different vinyl floor coverings (data in grams)

Brand A		Brand B	
2288	2384	2592	2384
2368	2304	2512	2432
2528	2240	2576	2112
2144	2208	2176	2288
2160	2112	2304	2752

Solution

a. We want to perform the hypothesis test

$$H_0: \sigma_1 = \sigma_2 \text{ (standard deviations of tear strength are the same)}$$
$$H_a: \sigma_1 \neq \sigma_2 \text{ (standard deviations of tear strength are different)},$$

where σ_1 and σ_2 denote the population standard deviations of tear strength for Brand A and Brand B, respectively.

b. We carry out the hypothesis test by comparing the sample standard deviations, s_1 and s_2, of the two sets of sample data presented in Table 11.3. Specifically, we compute the square of the ratio of s_1 to s_2, or, equivalently, the quotient of the sample variances. That statistic is called the **F-statistic.**

If the population standard deviations, σ_1 and σ_2, are equal, the sample standard deviations, s_1 and s_2, should be roughly the same, which means that the value of the F-statistic should be close to 1. When the value of the F-statistic differs from 1 by too much, it provides evidence against the null hypothesis of equal population standard deviations.

c. For the data in Table 11.3, $s_1 = 128.3$ g and $s_2 = 199.7$ g. Thus the value of the F-statistic is

$$F = \frac{s_1^2}{s_2^2} = \frac{128.3^2}{199.7^2} = 0.413.$$

Does this value of F differ from 1 by enough to conclude that the null hypothesis of equal population standard deviations is false? To answer that question, we need to know the distribution of the F-statistic. We discuss that distribution and then return to complete the hypothesis test.

The Distribution of the *F*-Statistic

To perform hypothesis tests and obtain confidence intervals for two population standard deviations, we need Key Fact 11.5.

KEY FACT 11.5

> ### Distribution of the *F*-Statistic for Comparing Two Population Standard Deviations
>
> Suppose that the variable under consideration is normally distributed on each of two populations. Then, for independent samples of sizes n_1 and n_2 from the two populations, the variable
>
> $$F = \frac{s_1^2/\sigma_1^2}{s_2^2/\sigma_2^2}$$
>
> has the *F*-distribution with df = $(n_1 - 1, n_2 - 1)$.

■■■ **EXAMPLE 11.13**   **The Distribution of the *F*-Statistic**

Elmendorf Tear Strength In Example 11.12, suppose that the Elmendorf tear strengths for Brands A and B vinyl floor coverings are normally distributed with means 2275 g and 2405 g, respectively, and equal standard deviations 168 g. Then, according to Key Fact 11.5, for independent random samples, each of size 10, from Brands A and B, the variable $F = s_1^2/s_2^2$ has the F-distribution with df = $(9, 9)$. Use simulation to make that fact plausible.

OUTPUT 11.3

Histogram of F for 1000 independent samples with superimposed F-curve

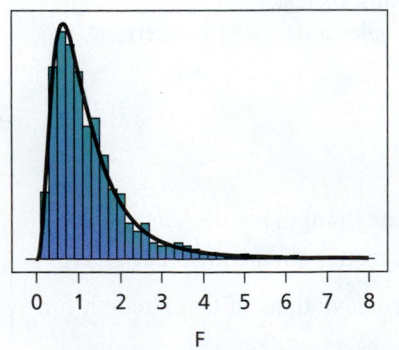

Solution   We first simulated 1000 samples of 10 tear strengths each for Brand A vinyl floor covering, that is, 1000 samples of 10 observations each of a normally distributed variable with mean 2275 and standard deviation 168. Next we simulated 1000 samples of 10 tear strengths each for Brand B vinyl floor covering, that is, 1000 samples of 10 observations each of a normally distributed variable with mean 2405 and standard deviation 168. Then, for each of the 1000 pairs of samples from the two brands, we determined the sample standard deviations, s_1 and s_2, and obtained the value of the variable $F = s_1^2/s_2^2$.

Output 11.3 shows a histogram of those 1000 values of F, which is shaped like the superimposed F-curve with df = $(9, 9)$.

Hypothesis Tests for Two Population Standard Deviations

In light of Key Fact 11.5, for a hypothesis test with null hypothesis H_0: $\sigma_1 = \sigma_2$ (population standard deviations are equal), we can use the variable

$$F = \frac{s_1^2}{s_2^2}$$

as the test statistic and obtain the critical value(s) from the F-table, Table VIII. We call this hypothesis-testing procedure the **two-standard-deviations F-test.**[†]

Procedure 11.3 gives a step-by-step method for performing a two-standard-deviations F-test by using either the critical-value approach or the P-value approach.

PROCEDURE 11.3 Two-Standard-Deviations F-Test

Purpose To perform a hypothesis test to compare two population standard deviations, σ_1 and σ_2

Assumptions
1. Simple random samples
2. Independent samples
3. Normal populations

Step 1 The null hypothesis is H_0: $\sigma_1 = \sigma_2$, and the alternative hypothesis is

$$H_a: \sigma_1 \neq \sigma_2 \quad \text{or} \quad H_a: \sigma_1 < \sigma_2 \quad \text{or} \quad H_a: \sigma_1 > \sigma_2$$
$$\text{(Two tailed)} \qquad \text{(Left tailed)} \qquad \text{(Right tailed)}.$$

Step 2 Decide on the significance level, α.

Step 3 Compute the value of the test statistic

$$F = \frac{s_1^2}{s_2^2}$$

and denote that value F_0.

CRITICAL-VALUE APPROACH	OR	P-VALUE APPROACH

CRITICAL-VALUE APPROACH

Step 4 The critical value(s) are

$$F_{1-\alpha/2} \text{ and } F_{\alpha/2} \quad \text{or} \quad F_{1-\alpha} \quad \text{or} \quad F_{\alpha}$$
$$\text{(Two tailed)} \qquad \text{(Left tailed)} \qquad \text{(Right tailed)}$$

with df $= (n_1 - 1, n_2 - 1)$. Use Table VIII to find the critical value(s).

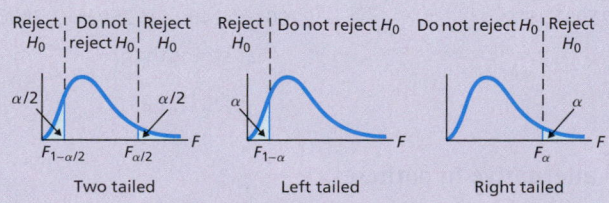

Step 5 If the value of the test statistic falls in the rejection region, reject H_0; otherwise, do not reject H_0.

P-VALUE APPROACH

Step 4 The F-statistic has df $= (n_1 - 1, n_2 - 1)$. Obtain the P-value by using technology.

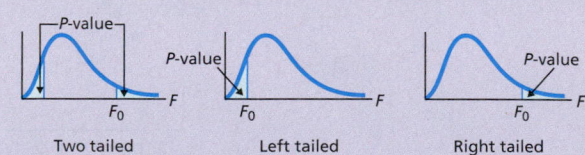

Step 5 If $P \leq \alpha$, reject H_0; otherwise, do not reject H_0.

Step 6 Interpret the results of the hypothesis test.

[†]The two-standard-deviations F-test is also known as the **F-test for two population standard deviations** and the **two-sample F-test.** This test is often formulated in terms of variances instead of standard deviations.

For the P-value approach, we could use Table VIII to estimate the P-value, but to do so is awkward and tedious; thus, we recommend using statistical software.

Unlike the z-tests and t-tests for one and two population means, the two-standard-deviations F-test is not robust to moderate violations of the normality assumption. In fact, it is so nonrobust that many statisticians advise against its use unless there is considerable evidence that the variable under consideration is normally distributed, or very nearly so, on each population.

Consequently, before applying Procedure 11.3, construct a normal probability plot of each sample. If either plot creates any doubt about the normality of the variable under consideration, do not use Procedure 11.3.

We note that nonparametric procedures, which do not require normality, have been developed to perform inferences for comparing two population standard deviations. If you have doubts about the normality of the variable on the two populations under consideration, you can often use one of those procedures to perform a hypothesis test or find a confidence interval for two population standard deviations.

■■■■ **EXAMPLE 11.14**

The Two-Standard-Deviations *F*-Test

TABLE 11.4

Results of Elmendorf tear test on two different vinyl floor coverings (data in grams)

Brand A		Brand B	
2288	2384	2592	2384
2368	2304	2512	2432
2528	2240	2576	2112
2144	2208	2176	2288
2160	2112	2304	2752

Elmendorf Tear Strength We can now complete the hypothesis test proposed in Example 11.12. Independent random samples of two vinyl floor coverings yield the data on Elmendorf tear strength repeated here in Table 11.4. At the 5% significance level, do the data provide sufficient evidence to conclude that the population standard deviations of tear strength differ for the two vinyl floor coverings?

Solution To begin, we construct normal probability plots for the two samples in Table 11.4, shown in Fig. 11.11. The plots suggest that we can reasonably presume that tear strength is normally distributed for each brand of vinyl flooring. Hence we can use Procedure 11.3 to perform the required hypothesis test.

FIGURE 11.11 Normal probability plots of the sample data for (a) Brand A and (b) Brand B

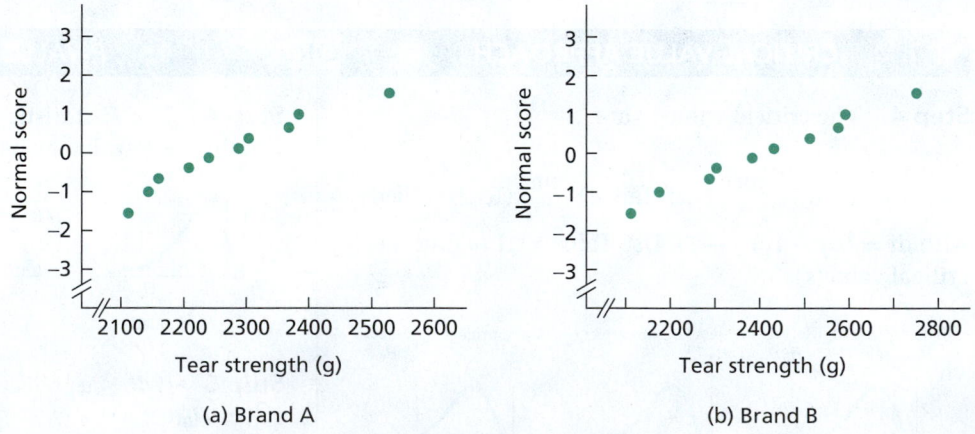

(a) Brand A

(b) Brand B

Step 1 State the null and alternative hypotheses.

Let σ_1 and σ_2 denote the population standard deviations of tear strength for Brand A and Brand B, respectively. Then the null and alternative hypotheses are, respectively,

$$H_0\text{: } \sigma_1 = \sigma_2 \text{ (standard deviations of tear strength are the same)}$$
$$H_a\text{: } \sigma_1 \neq \sigma_2 \text{ (standard deviations of tear strength are different).}$$

Note that the hypothesis test is two tailed.

Step 2 Decide on the significance level, α.

The test is to be performed at the 5% level of significance, or $\alpha = 0.05$.

Step 3 Compute the value of the test statistic $F = s_1^2/s_2^2$.

We computed the value of the test statistic at the end of Example 11.12, where we found that $F = 0.413$.

CRITICAL-VALUE APPROACH	OR	P-VALUE APPROACH

Step 4 The critical values for a two-tailed test are $F_{1-\alpha/2}$ and $F_{\alpha/2}$ with df $= (n_1 - 1, n_2 - 1)$. Use Table VIII to find the critical values.

We have $\alpha = 0.05$. Also, $n_1 = 10$ and $n_2 = 10$, so df $= (9, 9)$. Therefore the critical values are $F_{1-\alpha/2} = F_{1-0.05/2} = F_{0.975}$ and $F_{\alpha/2} = F_{0.05/2} = F_{0.025}$. From Table VIII, $F_{0.025} = 4.03$. To obtain $F_{0.975}$, we first note that it is the F-value having area 0.025 to its left. Applying the reciprocal property of F-curves (see page 528), we conclude that $F_{0.975}$ equals the reciprocal of the F-value having area 0.025 to its right for an F-curve with df $= (9, 9)$. (We switched the degrees of freedom, but because they are the same, the difference isn't apparent.) Thus $F_{0.975} = \frac{1}{4.03} = 0.25$. Figure 11.12A summarizes our results.

FIGURE 11.12A

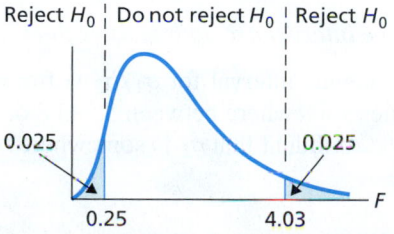

Step 5 If the value of the test statistic falls in the rejection region, reject H_0; otherwise, do not reject H_0.

From Step 3, the value of the test statistic is $F = 0.413$. This value does not fall in the rejection region shown in Fig. 11.12A, so we do not reject H_0. The test results are not statistically significant at the 5% level.

Step 4 The F-statistic has df $= (n_1 - 1, n_2 - 1)$. Obtain the P-value by using technology.

We have $n_1 = 10$ and $n_2 = 10$, so df $= (9, 9)$. Using technology, we find that the P-value for the hypothesis test is $P = 0.204$, as depicted in Fig. 11.12B.

FIGURE 11.12B

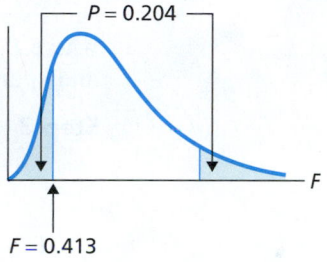

Step 5 If $P \leq \alpha$, reject H_0; otherwise, do not reject H_0.

From Step 4, $P = 0.204$. Because the P-value exceeds the specified significance level of 0.05, we do not reject H_0. The test results are not statistically significant at the 5% level and (see Table 9.8 on page 386) provide at most weak evidence against the null hypothesis.

Step 6 Interpret the results of the hypothesis test.

Interpretation At the 5% significance level, the data do not provide sufficient evidence to conclude that the population standard deviations of tear strength differ for the two vinyl floor coverings.

StatCrunch
Report 11.3

You try it!

Exercise 11.63 on page 538

Confidence Intervals for Two Population Standard Deviations

Using Key Fact 11.5 on page 530, we can also obtain a confidence-interval procedure, Procedure 11.4, for the ratio of two population standard deviations. We call it the **two-standard-deviations F-interval procedure.**[†] Like the two-standard-deviations F-test, this procedure is not at all robust to violations of the normality assumption.

[†]The two-standard-deviations F-interval procedure is also known as the **F-interval procedure for two population standard deviations** and the **two-sample F-interval procedure.** This confidence-interval procedure is often formulated in terms of variances instead of standard deviations.

▪▪▪▪ **PROCEDURE 11.4** **Two-Standard-Deviations *F*-Interval Procedure**

Purpose To find a confidence interval for the ratio of two population standard deviations, σ_1 and σ_2

Assumptions
1. Simple random samples
2. Independent samples
3. Normal populations

Step 1 For a confidence level of $1 - \alpha$, use Table VIII to find $F_{1-\alpha/2}$ and $F_{\alpha/2}$ with df $= (n_1 - 1, n_2 - 1)$.

Step 2 The confidence interval for σ_1/σ_2 is from

$$\frac{1}{\sqrt{F_{\alpha/2}}} \cdot \frac{s_1}{s_2} \quad \text{to} \quad \frac{1}{\sqrt{F_{1-\alpha/2}}} \cdot \frac{s_1}{s_2},$$

where $F_{1-\alpha/2}$ and $F_{\alpha/2}$ are found in Step 1, n_1 and n_2 are the sample sizes, and s_1 and s_2 are computed from the sample data obtained.

Step 3 Interpret the confidence interval.

To interpret confidence intervals for the ratio, σ_1/σ_2, of two population standard deviations, considering three cases is helpful.

Case 1: The endpoints of the confidence interval are both greater than 1.

To illustrate, suppose that a 95% confidence interval for σ_1/σ_2 is from 5 to 8. Then we can be 95% confident that σ_1/σ_2 lies somewhere between 5 and 8 or, equivalently, $5\sigma_2 < \sigma_1 < 8\sigma_2$. Thus, we can be 95% confident that σ_1 is somewhere between 5 and 8 times greater than σ_2.

Case 2: The endpoints of the confidence interval are both less than 1.

To illustrate, suppose that a 95% confidence interval for σ_1/σ_2 is from 0.5 to 0.8. Then we can be 95% confident that σ_1/σ_2 lies somewhere between 0.5 and 0.8 or, equivalently, $0.5\sigma_2 < \sigma_1 < 0.8\sigma_2$. Thus, noting that $1/0.5 = 2$ and $1/0.8 = 1.25$, we can be 95% confident that σ_1 is somewhere between 1.25 and 2 times less than σ_2.

Case 3: One endpoint of the confidence interval is less than 1 and the other is greater than 1.

To illustrate, suppose that a 95% confidence interval for σ_1/σ_2 is from 0.5 to 8. Then we can be 95% confident that σ_1/σ_2 lies somewhere between 0.5 and 8 or, equivalently, $0.5\sigma_2 < \sigma_1 < 8\sigma_2$. Thus, we can be 95% confident that σ_1 is somewhere between 2 times less than and 8 times greater than σ_2.

▪▪▪ **EXAMPLE 11.15** **The Two-Standard-Deviations *F*-Interval Procedure**

Elmendorf Tear Strength Use the sample data in Table 11.4 on page 532 to determine a 95% confidence interval for the ratio, σ_1/σ_2, of the standard deviations of tear strength for Brand A and Brand B vinyl floor coverings.

Solution As found in Example 11.14, we can reasonably presume that tear strengths are normally distributed for both Brand A and Brand B vinyl floor coverings. Consequently, we can apply Procedure 11.4 to obtain the required confidence interval.

Step 1 For a confidence level of $1 - \alpha$, use Table VIII to find $F_{1-\alpha/2}$ and $F_{\alpha/2}$ with df $= (n_1 - 1, n_2 - 1)$.

We want to obtain a 95% confidence interval; consequently, $\alpha = 0.05$. Hence we need to find $F_{0.975}$ and $F_{0.025}$ for df $= (n_1 - 1, n_2 - 1) = (9, 9)$. We did so earlier (Example 11.14, Step 4 of the critical-value approach), where we determined that $F_{0.975} = 0.25$ and $F_{0.025} = 4.03$.

Step 2 The confidence interval for σ_1/σ_2 is from

$$\frac{1}{\sqrt{F_{\alpha/2}}} \cdot \frac{s_1}{s_2} \quad \text{to} \quad \frac{1}{\sqrt{F_{1-\alpha/2}}} \cdot \frac{s_1}{s_2}.$$

For the data in Table 11.4, $s_1 = 128.3$ g and $s_2 = 199.7$ g. From Step 1, we know that $F_{0.975} = 0.25$ and $F_{0.025} = 4.03$. Consequently, the required 95% confidence interval is from

$$\frac{1}{\sqrt{4.03}} \cdot \frac{128.3}{199.7} \quad \text{to} \quad \frac{1}{\sqrt{0.25}} \cdot \frac{128.3}{199.7},$$

or 0.32 to 1.28.

Step 3 Interpret the confidence interval.

Interpretation We can be 95% confident that the ratio of the standard deviations of tear strength for Brand A and Brand B vinyl floor coverings is somewhere between 0.32 and 1.28 (i.e., $0.32\sigma_2 < \sigma_1 < 1.28\sigma_2$). In other words, we can be 95% confident that the standard deviation of tear strength for Brand A is somewhere between 3.125 times less than and 1.28 times greater than that of Brand B.

StatCrunch

Report 11.4

Exercise 11.69
on page 540

THE TECHNOLOGY CENTER

Most statistical technologies have programs that automatically perform two-standard-deviations F-procedures. In this subsection, we present output and step-by-step instructions for such programs.

Note to TI-83/84 Plus users:

- At the time of this writing, the TI-83/84 Plus does not have a built-in program for a two-standard-deviations F-interval procedure. However, a TI program, FINT, for that procedure is supplied in the TI Programs section on the WeissStats site. Your instructor can show you how to download the program to your calculator.

- *Warning:* Any data that you may have previously stored in Lists 1 and 2 will be erased during program execution, so copy those data to other lists prior to program execution if you want to retain them.

EXAMPLE 11.16 Using Technology to Conduct Two-Standard-Deviations *F*-Procedures

Elmendorf Tear Strength Table 11.4 on page 532 gives the Elmendorf tear strengths, in grams, for independent random samples of Brand A and Brand B vinyl floor coverings. Use Minitab, Excel, or the TI-83/84 Plus to perform the hypothesis test in Example 11.14 and obtain the confidence interval in Example 11.15.

Solution Let σ_1 and σ_2 denote the population standard deviations of tear strength for Brands A and B, respectively. The task in Example 11.14 is to perform the hypothesis test

$$H_0:\ \sigma_1 = \sigma_2 \ \text{(standard deviations of tear strength are the same)}$$
$$H_a:\ \sigma_1 \neq \sigma_2 \ \text{(standard deviations of tear strength are different)}$$

at the 5% significance level; the task in Example 11.15 is to find a 95% confidence interval for σ_1/σ_2.

We applied the two-standard-deviations F-procedures programs to the data, resulting in Output 11.4. Steps for generating that output are presented in Instructions 11.2. *Note to Excel users:* For brevity, we have presented only the essential portions of the actual output.

As shown in Output 11.4, the P-value for the hypothesis test is 0.204. Because the P-value exceeds the specified significance level of 0.05, we do not reject H_0. Output 11.4 also shows that a 95% confidence interval for σ_1/σ_2 is from 0.32 to 1.29.

Note to Excel users: The confidence interval provided in the Excel output is for the ratio of the two population variances, not the ratio of the two population standard deviations. To get the confidence interval for the latter ratio, simply take the square root of each of the endpoints of the confidence interval for the former ratio. In this case, we take the square root of 0.1026 and 1.6628, which gives 0.32 and 1.29, respectively. ∎

OUTPUT 11.4   Two-standard-deviations F-test and interval on the tear-strength data

MINITAB

Test and CI for Two Variances: BRAND A, BRAND B

```
Method

Null hypothesis          σ(BRAND A) / σ(BRAND B) = 1
Alternative hypothesis   σ(BRAND A) / σ(BRAND B) ≠ 1
Significance level       α = 0.05

F method was used. This method is accurate for normal data only.


Statistics

                                       95% CI for
Variable   N    StDev    Variance        StDevs
BRAND A   10   128.322   16466.489   ( 88.264, 234.266)
BRAND B   10   199.669   39867.733   (137.339, 364.518)

Ratio of standard deviations = 0.643
Ratio of variances = 0.413


95% Confidence Intervals

                            CI for
           CI for StDev    Variance
Method        Ratio         Ratio
F         (0.320, 1.290)  (0.103, 1.663)


Tests

                     Test
Method  DF1  DF2  Statistic  P-Value
F        9    9     0.41      0.204
```

OUTPUT 11.4 (cont.)   Two-standard-deviations *F*-test and interval on the tear-strength data

EXCEL

Summary statistics:			
Variable	Observations	Mean	Std. deviation
BRAND A	10	2273.6000	128.3218
BRAND B	10	2412.8000	199.6691

Fisher's F-test / Two-tailed test:

95% confidence interval on the ratio of variances:

(0.1026, 1.6628)

Ratio	0.4130
F (Observed value)	0.4130
F (Critical value)	4.0260
DF1	9
DF2	9
p-value (Two-tailed)	0.2039
alpha	0.05

TI-83/84 PLUS

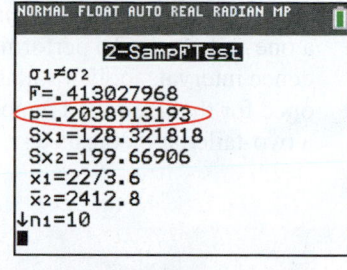

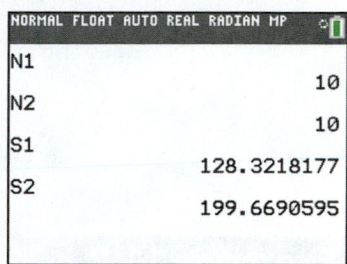

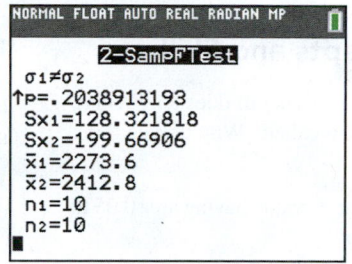

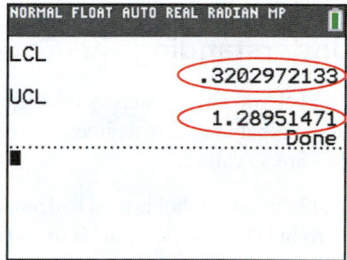

Using **2-SampFTest**   Using the **FINT** program

INSTRUCTIONS 11.2   Steps for generating Output 11.4

MINITAB

1. Store the tear-strength data from Table 11.4 in columns named BRAND A and BRAND B
2. Choose **Stat ➤ Basic Statistics ➤ 2 Variances...**
3. Press the F3 key to reset the dialog box
4. Select **Each sample is in its own column** from the drop-down list box
5. Click in the **Sample 1** text box and specify 'BRAND A'
6. Click in the **Sample 2** text box and specify 'BRAND B'
7. Click the **Options...** button
8. Click in the **Confidence level** text box and type 95
9. Click the arrow button at the right of the **Alternative hypothesis** drop-down list box and select **Ratio ≠ hypothesized ratio**
10. Check the **Use test and confidence intervals based on normal distribution** check box
11. Click **OK**
12. Click the **Graphs...** button and uncheck the **Summary plot** check box
13. Click **OK** twice

EXCEL

1. Store the tear-strength data from Table 11.4 in columns named BRAND A and BRAND B
2. Choose **XLSTAT ➤ Parametric tests ➤ Two-sample comparison of variances**
3. Click the reset button in the lower left corner of the dialog box
4. Click in the **Sample 1** selection box and then select the column of the worksheet that contains the BRAND A data
5. Click in the **Sample 2** selection box and then select the column of the worksheet that contains the BRAND B data
6. Click the **Options** tab

7. Click the arrow button at the right of the **Alternative hypothesis** drop-down list box and select **Variance 1 / Variance 2 ≠ R**
8. Type 5 in the **Significance level (%)** text box
9. Click **OK**
10. Click the **Continue** button in the **XLSTAT – Selections** dialog box

TI-83/84 PLUS

Store the tear-strength data from Table 11.4 in lists named BRNDA and BRNDB

FOR THE HYPOTHESIS TEST:
1. Press **STAT**, arrow over to **TESTS**, arrow down to **2-SampFTest**, and press **ENTER**
2. Highlight **Data** and press **ENTER**
3. Press the down-arrow key
4. Press **2nd ➤ LIST**, arrow down to BRNDA, and press **ENTER** twice
5. Press **2nd ➤ LIST**, arrow down to BRNDB, and press **ENTER** twice
6. Type 1 for **Freq1**, press **ENTER**, type 1 for **Freq2**, and press **ENTER**
7. Highlight ≠ σ2 and press **ENTER**
8. Arrow down to **Calculate** and press **ENTER**

FOR THE CI:
1. Press **PRGM**
2. Arrow down to FINT and press **ENTER** twice
3. Type 1 for **TYPE** and press **ENTER**
4. Press **2ND ➤ LIST**, arrow down to BRNDA, and press **ENTER** twice
5. Press **2ND ➤ LIST**, arrow down to BRNDB, and press **ENTER** twice
6. Type .95 for **C-LEVEL** and press **ENTER**
7. After the screen changes, press **ENTER** when ready

Note to Minitab and Excel users: Although Minitab and Excel simultaneously perform a hypothesis test and obtain a confidence interval, the type of confidence interval found depends on the type of hypothesis test. Specifically, Minitab and Excel compute a two-sided confidence interval for a two-tailed test and a one-sided confidence interval for a one-tailed test. To perform a one-tailed hypothesis test and obtain a two-sided confidence interval, apply Minitab's or Excel's two-standard-deviations F-procedure twice: once for the one-tailed hypothesis test and once for the confidence interval specifying a two-tailed hypothesis test.

Exercises 11.2

Understanding the Concepts and Skills

11.42 How many numbers of degrees of freedom does an F-curve have? What are those degrees of freedom called? Why do you think they are so called?

11.43 What symbol is used to denote the F-value having area 0.05 to its right? 0.025 to its right? α to its right?

11.44 Using the F_α-notation, identify the F-value having area 0.975 to its left.

11.45 An F-curve has df $= (12, 7)$. What is the number of degrees of freedom for the
a. numerator? **b.** denominator?

11.46 An F-curve has df $= (8, 19)$. What is the number of degrees of freedom for the
a. denominator? **b.** numerator?

In Exercises 11.47–11.54, use Table VIII and, if necessary, the reciprocal property of F-curves to find the required F-values. Illustrate your work graphically.

11.47 An F-curve has df $= (24, 30)$. In each case, find the F-value that has the specified area to its right.
a. 0.05 **b.** 0.01 **c.** 0.025

11.48 An F-curve has df $= (12, 5)$. In each case, find the F-value that has the specified area to its right.
a. 0.01 **b.** 0.05 **c.** 0.005

11.49 For an F-curve with df $= (20, 21)$, find
a. $F_{0.01}$. **b.** $F_{0.05}$. **c.** $F_{0.10}$.

11.50 For an F-curve with df $= (6, 10)$, find
a. $F_{0.05}$. **b.** $F_{0.01}$. **c.** $F_{0.025}$.

11.51 Consider an F-curve with df $= (6, 8)$. Determine the F-value that has area
a. 0.01 to its left. **b.** 0.95 to its left.

11.52 Consider an F-curve with df $= (15, 5)$. Determine the F-value that has area
a. 0.025 to its left. **b.** 0.975 to its left.

11.53 Determine the two F-values that divide the area under the curve into a middle 0.95 area and two outside 0.025 areas for an F-curve with
a. df $= (7, 4)$. **b.** df $= (12, 20)$.

11.54 Determine the two F-values that divide the area under the curve into a middle 0.90 area and two outside 0.05 areas for an F-curve with
a. df $= (10, 8)$. **b.** df $= (12, 12)$.

11.55 In using F-procedures to make inferences for two population standard deviations, why should the distributions (one for each population) of the variable under consideration be normally distributed or nearly so?

11.56 Give two situations in which comparing two population standard deviations would be important.

In each of Exercises 11.57–11.62, we have provided the sample standard deviations and sample sizes for independent simple random samples from two populations. In each case, use the two-standard-deviations F-test and the two-standard-deviations F-interval procedure to conduct the required hypothesis test and obtain the specified confidence interval.

11.57 $s_1 = 19.4$, $n_1 = 31$, $s_2 = 10.5$, $n_2 = 16$
a. right-tailed test, $\alpha = 0.01$ **b.** 98% confidence interval

11.58 $s_1 = 12.04$, $n_1 = 25$, $s_2 = 11.25$, $n_2 = 21$
a. right-tailed test, $\alpha = 0.10$ **b.** 80% confidence interval

11.59 $s_1 = 28.82$, $n_1 = 8$, $s_2 = 38.97$, $n_2 = 13$
a. left-tailed test, $\alpha = 0.10$ **b.** 80% confidence interval

11.60 $s_1 = 38.2$, $n_1 = 6$, $s_2 = 84.7$, $n_2 = 16$
a. left-tailed test, $\alpha = 0.05$ **b.** 90% confidence interval

11.61 $s_1 = 14.5$, $n_1 = 11$, $s_2 = 30.4$, $n_2 = 9$
a. two-tailed test, $\alpha = 0.05$ **b.** 95% confidence interval

11.62 $s_1 = 74.8$, $n_1 = 7$, $s_2 = 30.4$, $n_2 = 9$
a. two-tailed test, $\alpha = 0.01$ **b.** 99% confidence interval

Applying the Concepts and Skills

Preliminary data analyses and other information indicate that you can reasonably assume that, in Exercises 11.63–11.68, the variable under consideration is normally distributed on both populations. For each exercise, use either the critical-value approach or the P-value approach to perform the required hypothesis test.

11.63 Algebra Exam Scores. One year at Arizona State University, the algebra course director decided to experiment with a new teaching method that might reduce variability in final-exam scores by eliminating lower scores. The director randomly divided the algebra students who were registered for class at 9:40 A.M. into two groups. One of the groups, called the control group, was taught the usual algebra course; the other group, called the experimental group, was taught by the new teaching method. Both classes covered the same material, took the same unit quizzes, and took the same final exam at the same time. The final-exam scores (out of 40 possible) for the two groups are shown in the following table.

Control						Experimental			
36	35	35	33	32	32	36	35	35	31
31	29	29	28	28	28	30	29	27	27
27	27	27	26	26	25	26	23	21	21
24	24	24	23	20	20	35	32	28	28
19	19	18	18	18	17	25	23	21	19
17	16	15	15	15	15				
14	11	10	9	4					

Do the data provide sufficient evidence to conclude that there is less variation among final-exam scores when the new teaching method is used rather than the usual one? Perform an F-test at the 5% significance level. (*Note:* $s_1 = 7.813$ and $s_2 = 5.286$.)

11.64 Pulmonary Hypertension. In the paper "Persistent Pulmonary Hypertension of the Neonate and Asymmetric Growth Restriction" (*Obstetrics & Gynecology*, Vol. 91, No. 3, pp. 336–341), M. Williams et al. reported on a study of characteristics of neonates. Infants treated for pulmonary hypertension, called the PH group, were compared with those not so treated, called the control group. One of the characteristics measured was head circumference. The following data, in centimeters (cm), are based on the results obtained by the researchers.

PH		Control				
33.9	35.1	35.2	35.6	36.7	35.1	36.0
33.4	34.5	33.4	31.3	33.5	35.8	36.3
37.9	31.3	34.3	33.1	32.4	35.1	33.6
32.5	32.9	31.8	34.1	35.2	34.8	34.5
36.3	34.2	31.6	31.9	31.9	32.8	34.0

Do the data provide sufficient evidence to conclude that variation in head circumference differs among neonates treated for pulmonary hypertension and those not so treated? Perform an F-test at the 5% significance level. (*Note:* $s_1 = 1.907$ and $s_2 = 1.594$.)

11.65 Chronic Hemodialysis and Anxiety. Patients who undergo chronic hemodialysis often experience severe anxiety. Videotapes of progressive relaxation exercises were shown to one group of patients and neutral videotapes to another group. Then both groups took the State-Trait Anxiety Inventory, a psychiatric questionnaire used to measure anxiety, where higher scores correspond to higher anxiety. In the paper "The Effectiveness of Progressive Relaxation in Chronic Hemodialysis Patients" (*Journal of Chronic Diseases*, 35(10)), R. Alarcon et al. presented the results of the study. The following data are based on those results.

Relaxation tapes				Neutral tapes			
30	41	28	14	36	44	47	45
40	36	38	24	50	54	54	45
61	36	24	45	50	46	28	35
38	43	32	28	42	35	32	43
37	34	20	23	41	33	35	36
34	47	25	31	32	17	45	
39	14	43	40	24	46		
29	21	40					

Do the data provide sufficient evidence to conclude that variation in anxiety-test scores differs between patients who are shown videotapes of progressive relaxation exercises and those who are shown neutral

videotapes? Perform an F-test at the 10% significance level. (*Note:* $s_1 = 10.154$ and $s_2 = 9.197$.)

11.66 Stuttering. Researchers S. Reilly et al. studied the indicators of stuttering onset and the potential effects that stuttering may have on a child in the article "Natural History of Stuttering to 4 Years of Age: A Prospective Community-Based Study" (*Pediatrics*, Vol. 132, Issue 3, pp. 460–467). The following table provides the scores on the Communication and Symbolic Behavior Scale (CSBS), an assessment tool used to identify communication disorders, for random samples of stuttering and nonstuttering children.

Stutterers					Nonstutterers			
120	109	77	119	72	101	93	95	116
84	122	108	127		126	111	99	92
97	107	108	105		81	123	107	75

At the 10% significance level, do the data provide sufficient evidence to conclude that, in this age group, there is more variation in CSBS scores for stutterers than for nonstutterers? (*Note:* $s_1 = 17.326$ and $s_2 = 15.745$.)

11.67 A Better Golf Tee? An independent golf equipment testing facility compared the difference in the performance of golf balls hit off a regular 2-3/4" wooden tee to those hit off a 3" Stinger Competition golf tee. A Callaway Great Big Bertha driver with 10 degrees of loft was used for the test, and a robot swung the club head at approximately 95 miles per hour. Data on ball velocity (in miles per hour) with each type of tee are as follows.

Stinger	Regular
$\bar{x}_1 = 128.83$	$\bar{x}_2 = 127.01$
$s_1 = 0.410$	$s_2 = 0.894$
$n_1 = 30$	$n_2 = 30$

At the 1% significance level, do the data provide sufficient evidence to conclude that the standard deviation of velocity is less with the Stinger tee than with the regular tee? (*Note:* For df = (29, 29), $F_{0.01} = 2.42$.)

11.68 Nitrogen and Seagrass. The seagrass *Thalassia testudinum* is an integral part of the Texas coastal ecosystem. Essential to the growth of *T. testudinum* is ammonium. Researchers K. Lee and K. Dunton of the Marine Science Institute of the University of Texas at Austin noticed that the seagrass beds in Corpus Christi Bay (CCB) were taller and thicker than those in Lower Laguna Madre (LLM). They compared the sediment ammonium concentrations in the two locations and published their findings in *Marine Ecology Progress Series* (Vol. 196, pp. 39–48). The summary statistics on sediment ammonium concentrations, in micromoles, obtained by the researchers are shown in the following table.

LLM	CCB
$\bar{x}_1 = 24.3$	$\bar{x}_2 = 115.1$
$s_1 = 10.5$	$s_2 = 79.4$
$n_1 = 19$	$n_2 = 51$

At the 1% significance level, do the data provide sufficient evidence to conclude that the standard deviation of sediment ammonium concentrations is less in LLM seagrass beds than in CCB seagrass beds? (*Note:* For df = (50, 18), $F_{0.01} = 2.78$.)

In each of Exercises 11.69–11.74, use Procedure 11.4 on page 534 to obtain the required confidence interval.

11.69 Algebra Exam Scores. Refer to Exercise 11.63, and find a 90% confidence interval for the ratio of the population standard deviations of final-exam scores for students taught by the conventional method and for students taught by the new method. (*Note:* For df = (19, 40), $F_{0.05} = 1.85$.)

11.70 Pulmonary Hypertension. Refer to Exercise 11.64, and find a 95% confidence interval for the ratio of the population standard deviations of head circumferences for neonates treated for pulmonary hypertension and those not so treated.

11.71 Chronic Hemodialysis and Anxiety. Refer to Exercise 11.65, and determine a 90% confidence interval for the ratio of the population standard deviations of scores for patients who are shown videotapes of progressive relaxation exercises and those who are shown neutral videotapes.

11.72 Stuttering. Refer to Exercise 11.66 and find an 80% confidence interval for the ratio of the population standard deviations of CSBS scores for stutterers and nonstutterers in the age group under consideration. (*Note:* For df = (11, 12), $F_{0.10} = 2.17$.)

11.73 A Better Golf Tee? Refer to Exercise 11.67, and obtain a 98% confidence interval for the ratio of the population standard deviations of ball velocity for the Stinger tee and the regular tee.

11.74 Nitrogen and Seagrass. Refer to Exercise 11.68, and obtain a 98% confidence interval for the ratio of the population standard deviations of sediment ammonium concentrations for LLM seagrass beds and CCB seagrass beds. (*Note:* For df = (18, 50), $F_{0.01} = 2.32$.)

Working with Large Data Sets

11.75 The Etruscans. Anthropologists are still trying to unravel the mystery of the origins of the Etruscan empire, a highly advanced Italic civilization formed around the eighth century B.C. in central Italy. Were they native to the Italian peninsula or, as many aspects of their civilization suggest, did they migrate from the East by land or sea? The maximum head breadth, in millimeters, of 70 modern Italian male skulls and that of 84 preserved Etruscan male skulls were analyzed to help researchers decide whether the Etruscans were native to Italy. The resulting data can be found on the WeissStats site. [SOURCE: N. Barnicot and D. Brothwell, "The Evaluation of Metrical Data in the Comparison of Ancient and Modern Bones." In *Medical Biology and Etruscan Origins*, G. E. W. Wolstenholme and C. M. O'Connor, eds., Little, Brown & Co., 1959] Use the technology of your choice to solve parts (a)–(c).
a. Perform a two-standard-deviations *F*-test at the 5% significance level to decide whether the data provide sufficient evidence to conclude that variations in skull measurements differ between the two populations.
b. Use the two-standard-deviations *F*-interval procedure to determine a 95% confidence interval for the ratio of the standard deviations of skull measurements of the two populations.
c. Obtain a normal probability plot for each sample.
d. In light of your plots in part (c), does conducting the inferences you did in parts (a) and (b) seem reasonable? Explain your answer.

11.76 Dexamethasone Therapy and IQ. In the paper "Outcomes at School Age after Postnatal Dexamethasone Therapy for Lung Disease of Prematurity" (*New England Journal of Medicine*, Vol. 350, No. 13, pp. 1304–1313), T. Yeh et al. studied the outcomes at school age in children who had participated in a double-blind, placebo-controlled trial of early postnatal dexamethasone therapy for the prevention of chronic lung disease of prematurity. All of the infants in the study had had severe respiratory distress syndrome requiring mechanical ventilation shortly after birth. On the WeissStats site, we provide the school-age IQs of the 72 children in the dexamethasone group and the 74 children in the control group, based on the study results. Use the technology of your choice to solve parts (a)–(c).
a. Do the data provide sufficient evidence to conclude that the early postnatal dexamethasone therapy increases the variation in IQ? Perform a two-standard-deviations *F*-test at the 10% significance level.
b. Use the two-standard-deviations *F*-interval procedure to find an 80% confidence interval for the ratio of the population standard deviations of IQs with and without early postnatal dexamethasone therapy.
c. Obtain a normal probability plot for each sample.
d. In light of your plots in part (c), does conducting the inferences you did in parts (a) and (b) seem reasonable? Explain your answer.

11.77 RBC Transfusions. In the article "Reduction in Red Blood Cells Transfusions Among Preterm Infants: Results of a Randomized Trial With an In-Line Blood Gas and Chemistry Monitor" (*Pediatrics*, Vol. 115, Issue 5, pp. 1299–1306), J. Widness et al. examined extremely premature infants who develop anemia caused by intensive laboratory blood testing and multiple red blood cell (RBC) transfusions. The goal of the study was to reduce the number of RBC transfusions. Two groups were studied, a control group and a monitor group (which used the in-line blood gas and chemistry monitor). Data on hemoglobin level, in grams per liter (g/L), based on the results of the study, are provided on the WeissStats site. Use the technology of your choice to solve parts (a)–(c).
a. Do the data provide sufficient evidence to conclude that the variation in hemoglobin level is less without the inline blood gas and chemistry monitor? Perform a two-standard-deviations *F*-test at the 5% significance level.
b. Use the two-standard-deviations *F*-interval procedure to determine a 90% confidence interval for the ratio of the population standard deviations of hemoglobin levels with and without the inline blood gas and chemistry monitor.
c. Obtain a normal probability plot for each sample.
d. In light of your plots in part (c), does conducting the inferences you did in parts (a) and (b) seem reasonable? Explain your answer.

Extending the Concepts and Skills

11.78 Simulation. Use the technology of your choice to conduct the simulation discussed in Example 11.13 on page 530.

11.79 Elmendorf Tear Strength. Refer to Example 11.14 on pages 532–533. Use Table VIII to show that the *P*-value for the hypothesis test exceeds 0.20.

CHAPTER IN REVIEW

You Should Be Able to

1. use and understand the formulas in this chapter.

2. state the basic properties of χ^2-curves.

3. use the chi-square table, Table VII.

4. perform a hypothesis test for a population standard deviation when the variable under consideration is normally distributed.

5. obtain a confidence interval for a population standard deviation when the variable under consideration is normally distributed.

6. state the basic properties of F-curves.

7. apply the reciprocal property of F-curves.

8. use the F-table, Table VIII.

9. perform a hypothesis test to compare two population standard deviations when the variable under consideration is normally distributed on both populations.

10. find a confidence interval for the ratio of two population standard deviations when the variable under consideration is normally distributed on both populations.

Key Terms

χ^2_α, *515*
chi-square (χ^2) curve, *514*
chi-square distribution, *514*
degrees of freedom for the
   denominator, *527*

degrees of freedom for the
   numerator, *527*
F_α, *527*
F-curve, *527*
F-distribution, *527*
F-statistic, *530*

one-standard-deviation χ^2-interval
   procedure, *521*
one-standard-deviation χ^2-test, *519*
two-standard-deviations F-interval
   procedure, *534*
two-standard-deviations F-test, *531*

REVIEW PROBLEMS

Understanding the Concepts and Skills

1. What distribution is used in this chapter to make inferences for one population standard deviation?

2. Fill in the blanks.
a. A χ^2-curve is _____ skewed.
b. A χ^2-curve looks increasingly like a _____ curve as the number of degrees of freedom becomes larger.

3. When you use the one-standard-deviation χ^2-test or χ^2-interval procedure, what assumption must be met by the variable under consideration? How important is that assumption?

In each of Problems 4–8, use Table VII in Appendix A to find the required chi-square value(s) for a χ^2-curve with 17 degrees of freedom.

4. $\chi^2_{0.99}$

5. $\chi^2_{0.01}$

6. The χ^2-value with area 0.05 to its right

7. The χ^2-value with area 0.05 to its left

8. The two χ^2-values that divide the area under the curve into a middle 0.95 area and two outside 0.025 areas

9. What distribution is used in this chapter to make inferences for two population standard deviations?

10. Fill in the blanks:
a. An F-curve is _____ skewed.
b. For an F-curve with df $= (14, 5)$, the F-value having area 0.05 to its left equals the _____ of the F-value having area 0.05 to its right for an F-curve with df $= (\_\_\_\_, \_\_\_\_)$.
c. The observed value of a variable having an F-distribution must be greater than or equal to _____.

11. When you use the two-standard-deviations F-test, what assumption must be met by the variable under consideration? How important is that assumption?

In each of Problems 12–16, use Table VIII in Appendix A to find the required F-value(s) for an F-curve with df $= (4, 8)$.

12. $F_{0.01}$

13. $F_{0.99}$

14. The F-value with area 0.05 to its right

15. The F-value with area 0.05 to its left

16. The two F-values that divide the area under the curve into a middle 0.95 area and two outside 0.025 areas

Applying the Concepts and Skills

17. Intelligence Quotients. IQs measured on the Stanford Revision of the Binet–Simon Intelligence Scale are supposed to have a standard deviation of 16 points. Twenty-five randomly selected people were given the IQ test; here are the data that were obtained.

91	96	106	116	97
102	96	124	115	121
95	111	105	101	86
88	129	112	82	98
104	118	127	66	102

Preliminary data analyses and other information indicate the reasonableness of presuming that IQs measured on the Stanford Revision of the Binet–Simon Intelligence Scale are normally distributed.

a. Do the data provide sufficient evidence to conclude that IQs measured on this scale have a standard deviation different from 16 points? Perform the required hypothesis test at the 10% significance level. (*Note:* $s = 15.006$.)

b. How crucial is the normality assumption for the hypothesis test you performed in part (a)? Explain your answer.

18. Bouncing Blocks. Researcher P. Paronuzzi modeled the effects of a rock wall collapsing on a slope in the article "Rockfall-Induced Block Propagation on a Soil Slope, Northern Italy" (*Environmental Geology*, Vol. 58, No. 7, pp. 1451–1466). One of the variables measured was the distance that a rock block rebounded after bouncing down the slope. The following table lists the rebound distances, in meters, for a sample of 16 simulations of a rock bouncing down a soil slope with a 20° angle.

12.65	9.08	5.25	2.12	4.52	2.75	6.10	11.16
7.87	2.94	6.96	4.71	7.46	6.51	6.51	2.31

Find a 99% confidence interval for the standard deviation of the rebound distances of the bouncing rocks. (*Note:* $s = 3.054$ m)

19. Skinfold Thickness. A study entitled "Body Composition of Elite Class Distance Runners" was conducted by M. Pollock et al. to decide whether elite distance runners are thinner than other people. Their results were published in *The Marathon: Physiological, Medical, Epidemiological, and Psychological Studies*, P. Milvey (ed.), New York: New York Academy of Sciences, 1977, p. 366. The researchers measured the skinfold thickness, an indirect indicator of body fat, of runners and nonrunners in the same age group. The data, in millimeters (mm), shown in the following table are based on the skinfold thickness measurements on the thighs of the people sampled.

Runners			Others			
7.3	6.7	8.7	24.0	19.9	7.5	18.4
3.0	5.1	8.8	28.0	29.4	20.3	19.0
7.8	3.8	6.2	9.3	18.1	22.8	24.2
5.4	6.4	6.3	9.6	19.4	16.3	16.3
3.7	7.5	4.6	12.4	5.2	12.2	15.6

a. For an F-test to compare the standard deviations of skinfold thickness of runners and others, identify the appropriate F-distribution.

b. At the 1% significance level, do the data provide sufficient evidence to conclude that runners have less variability in skinfold thickness than others? (*Note:* $s_1 = 1.798$ and $s_2 = 6.606$. For df = (19, 14), $F_{0.01} = 3.53$.)

c. What assumption about skinfold thickness did you make in carrying out the hypothesis test in part (b)? How would you check that assumption?

d. In addition to the assumption on skinfold thickness discussed in part (c), what other assumptions are required for performing the two-standard-deviations F-test?

20. Laughter. The effect of laughter was studied in the article "Energy Expenditure of Genuine Laughter" (*International Journal of Obesity*, Vol. 31, pp. 131–137) by M. Buchowski et al. Subjects were split into groups of two males, one male and one female, or two females. The subjects viewed film clips intended to evoke laughter. The statistics in the first row of the following table provide the standard deviations of heart-rate increase, in beats per minute, for the two same-gender groups.

Male-male	Female-female
$s_1 = 4.76$	$s_2 = 3.72$
$n_1 = 8$	$n_2 = 20$

Find a 90% confidence interval for the ratio of the population standard deviations of heart-rate increase when laughing for male-male and female-female dyads. (*Note:* For df = (19, 7), $F_{0.05} = 3.46$.)

Working with Large Data Sets

21. Body Mass Index. Body mass index (BMI) is a measure of body fat based on height and weight. According to the document *Dietary Guidelines for Americans* published by the U.S. Department of Agriculture and the U.S. Department of Health and Human Services, for adults, a BMI of greater than 25 indicates an above healthy weight (i.e., overweight or obese). The BMIs of 75 randomly selected U.S. adults provided the data on the WeissStats site. Use the technology of your choice to do the following.

a. Obtain a normal probability plot, a boxplot, and a histogram of the data.

b. Based on your graphs from part (a), is it reasonable to apply one-standard-deviation χ^2-procedures to the data? Explain your answer.

c. In Problem 40 of Chapter 9, we applied the one-mean z-test to the data, assuming a standard deviation of 5.0 for the BMIs of all U.S. adults. At the 5% significance level, do the data provide evidence against that assumption?

22. Body Mass Index. Refer to Problem 21, and find a 95% confidence interval for the standard deviation of BMIs for all U.S. adults.

23. Gender and Direction. In the paper "The Relation of Sex and Sense of Direction to Spatial Orientation in an Unfamiliar Environment" (*Journal of Environmental Psychology*, Vol. 20, pp. 17–28), J. Sholl et al. published the results of examining the sense of direction of 30 male and 30 female students. After being taken to an unfamiliar wooded park, the students were given a number of spatial orientation tests, including pointing to south, which tested their absolute frame of reference. To point south, the students moved a pointer attached to a 360° protractor. The absolute pointing errors, in degrees, for students who rated themselves with a good sense of direction (GSOD) and those who rated themselves with a poor sense of direction (PSOD) are provided on the WeissStats site. Can you reasonably apply the two-standard-deviations F-test to compare the variation in pointing errors between people who rate themselves with a good sense of direction and those who rate themselves with a poor sense of direction? Explain your answer.

24. Microwave Popcorn. Two brands of microwave popcorn, which we will call Brand A and Brand B, were compared for consistency in popping time. The popping times, in seconds, for 30 bags of each brand are provided on the WeissStats site. Use the technology of your choice to do the following.

a. Obtain normal probability plots and boxplots, and histograms for the two data sets.

b. Based on your graphs from part (a), do you think it reasonable to perform a two-standard-deviations F-test on the data? Explain your answer.

c. At the 5% significance level, do the data provide sufficient evidence to conclude that Brand B has a more consistent popping time than Brand A?

d. Find a 90% confidence interval for the ratio of the standard deviations of popping times for Brand A and Brand B.

FOCUSING ON DATA ANALYSIS

UWEC UNDERGRADUATES

Recall from Chapter 1 (see page 34) that the Focus database and Focus sample contain information on the undergraduate students at the University of Wisconsin - Eau Claire (UWEC). Now would be a good time for you to review the discussion about these data sets.

Open the Focus sample worksheet (FocusSample) in the technology of your choice and then do the following.

a. At the 5% significance level, do the data provide sufficient evidence to conclude that the standard deviation of ACT composite scores of all UWEC undergraduates differs from 3 points?

b. Determine and interpret a 95% confidence interval for the standard deviation of ACT composite scores of all UWEC undergraduates.

c. Obtain a normal probability plot and a boxplot of the ACT composite scores of the sampled UWEC undergraduates.

d. Based on your results from part (c), do you think that performing the inferences in parts (a) and (b) is reasonable? Explain your answer.

e. At the 5% significance level, do the data provide sufficient evidence to conclude that the standard deviations of ACT English scores and ACT math scores differ for UWEC undergraduates?

f. Determine and interpret a 95% confidence interval for the ratio of the standard deviation of ACT English scores to the standard deviation of ACT math scores for UWEC undergraduates.

g. Obtain normal probability plots and boxplots of the ACT English scores and the ACT math scores of the sampled UWEC undergraduates.

h. Based on your results from part (g), do you think that performing the inference in parts (e) and (f) is reasonable? Explain your answer.

CASE STUDY DISCUSSION

SPEAKER WOOFER DRIVER MANUFACTURING

At the beginning of this chapter, we discussed rubber-edge manufacturing for speaker woofer drivers and a criterion for classifying process capability.

Recall that each process for manufacturing rubber edges requires a production weight specification that consists of a lower specification limit (LSL), a target weight (T), and an upper specification limit (USL). The actual mean and standard deviation of the weights of the rubber edges being produced are called the process mean (μ) and process standard deviation (σ). A process that is on target ($\mu = T$) is called *super* if $\sigma < (\text{USL} - \text{LSL})/12$.

The table on page 514 provides data on rubber-edge weight for a sample of 60 observations. Use those data and the procedures discussed in this chapter to solve the following problems:

a. Find a 99% confidence interval for the process standard deviation.

b. The process under consideration is known to be on target, and its production weight specification is LSL = 16.72, $T = 17.60$, and USL = 18.48. Do the data provide sufficient evidence to conclude that the process is super? Perform the required hypothesis test at the 1% significance level.

c. Obtain a normal probability plot of the data.

d. Based on your plot in part (c), was conducting the inferences that you did in parts (a) and (b) reasonable? Explain your answer.

BIOGRAPHY

W. EDWARDS DEMING: TRANSFORMING INDUSTRY WITH SQC

William Edwards Deming was born on October 14, 1900, in Sioux City, Iowa. Shortly after his birth, his father secured homestead land and moved the family first to Cody, Wyoming, and then to Powell, Wyoming.

Deming obtained a B.S. in physics at the University of Wyoming in 1921, a master's degree in physics and mathematics at the University of Colorado in 1924, and a doctorate in mathematical physics at Yale University in 1928.

While working for various federal agencies during the next decade, Deming became an expert on sampling and quality control. In 1939, he accepted the position of head mathematician and advisor in sampling at the U.S. Census Bureau. Deming began the use of sampling at the Census Bureau and, expanding the work of Walter A. Shewhart (later known as the father of statistical quality control, or SQC), also applied statistical methods of quality control to provide reliability and quality to the nonmanufacturing environment.

In 1946, Deming left the Census Bureau, joined the Graduate School of Business Administration at New York University, and offered his services to the private sector as a consultant in statistical studies. It was in this last-named capacity that Deming transformed industry in Japan. Deming began his long association with Japanese businesses in 1947 when the U.S. War Department engaged him to instruct Japanese industrialists in statistical quality control methods. The reputation of Japan's goods changed from definitely shoddy to amazingly excellent over the next two decades as the businessmen of Japan implemented Deming's teachings.

More than 30 years passed before Deming's methods gained widespread recognition by the business community in the United States. Finally, in 1980, as the result of the NBC white paper *If Japan Can, Why Can't We?*, in which Deming's role was publicized, executives of major corporations (among them, Ford Motor Company) contracted with Deming to improve the quality of U.S. goods.

Deming maintained an intense work schedule throughout his 80s, giving 4-day managerial seminars, teaching classes at NYU, sponsoring clinics for statisticians, and consulting with businesses internationally. His last book, *The New Economics*, was published in 1993. Dr. Deming died at his home in Washington, D.C., on December 20, 1993.

Inferences for Population Proportions

CHAPTER OBJECTIVES

In Chapters 8–10, we discussed methods for finding confidence intervals and performing hypothesis tests for one or two population means. Now we describe how to conduct those inferences for one or two population proportions.

A *population proportion* is the proportion (percentage) of a population that has a specified attribute. For example, if the population under consideration consists of all Americans and the specified attribute is "retired," the population proportion is the proportion of all Americans who are retired.

In Section 12.1, we begin by introducing notation and terminology needed to perform proportion inferences; then we discuss confidence intervals for one population proportion. Next, in Section 12.2, we examine a method for conducting a hypothesis test for one population proportion.

In Section 12.3, we investigate how to perform a hypothesis test to compare two population proportions and how to construct a confidence interval for the difference between two population proportions.

CASE STUDY

Arrested Youths

In a *New York Times* article titled "Many in U.S. Are Arrested by Age 23, Study Finds," E. Goode noted that, by age 23, almost a third of Americans have been arrested (excluding arrests for minor traffic violations). The article reports on a recent study by R. Brame et al. published as "Cumulative Prevalence of Arrest from Ages 8 to 23 in a National Sample" (*Pediatrics*, Vol. 129, No. 1, pp. 21–27). That study analyzed self-reported, arrest-history data from the *National Longitudinal Survey of Youth*.

The results of the study suggest a substantial increase in the prevalence of youth (children and young adults) arrests since the last national study, which occurred in the 1960s. Furthermore, according to S. Bushway, a criminologist at the University of Albany and a co-author of the aforementioned paper, "This estimate provides a real sense that the proportion of people who have criminal history records is sizable and perhaps much larger than most people would expect."

Professor Brame, lead author of the paper and a criminologist at the University of North Carolina at Charlotte, told the reporter that he hoped the research would be useful to physicians in helping their young patients recover from being arrested. More generally, the researchers concluded that "At a minimum, being arrested for criminal activity signifies increased risk of unhealthy

lifestyle, violence involvement, and violent victimization. Incorporating this insight into regular clinical assessment could yield significant benefits for patients and the larger community."

After studying the inferential methods discussed in this chapter, you will be asked to conduct statistical analyses on arrests of American youths based on data presented in the researchers' paper.

12.1 Confidence Intervals for One Population Proportion

Statisticians often need to determine the proportion (percentage) of a population that has a specified attribute. Some examples are

- the percentage of U.S. adults who have health insurance
- the percentage of cars in the United States that are imports
- the percentage of U.S. adults who favor stricter clean air health standards
- the percentage of Canadian women in the labor force.

In the first case, the population consists of all U.S. adults and the specified attribute is "has health insurance." For the second case, the population consists of all cars in the United States and the specified attribute is "is an import." The population in the third case is all U.S. adults and the specified attribute is "favors stricter clean air health standards." In the fourth case, the population consists of all Canadian women and the specified attribute is "is in the labor force."

We know that it is often impractical or impossible to take a census of a large population. In practice, therefore, we use data from a sample to make inferences about the population proportion. We introduce proportion notation and terminology in the next example.

EXAMPLE 12.1   Proportion Notation and Terminology

Playing Hooky From Work Many employers are concerned about the problem of employees who call in sick when they are not ill. The Hilton Hotels Corporation commissioned a survey to investigate this issue. One question asked the respondents whether they call in sick at least once a year when they simply need time to relax. For brevity, we use the phrase *play hooky* to refer to that practice.

The survey polled 1010 randomly selected U.S. employees. The proportion of the 1010 employees sampled who play hooky was used to estimate the proportion of all U.S. employees who play hooky. Discuss the statistical notation and terminology used in this and similar studies on proportions.

Solution   We use p to denote the proportion of all U.S. employees who play hooky; it represents the **population proportion** and is the parameter whose value is to be estimated. The proportion of the 1010 U.S. employees sampled who play hooky is designated $\hat{p}$ (read "p hat") and represents a **sample proportion;** it is the statistic used to estimate the unknown population proportion, p.

Although unknown, the population proportion, p, is a fixed number. In contrast, the sample proportion, $\hat{p}$, is a variable; its value varies from sample to sample. For instance, if 202 of the 1010 employees sampled play hooky, then

$$\hat{p} = \frac{202}{1010} = 0.2,$$

that is, 20.0% of the employees sampled play hooky. If 184 of the 1010 employees sampled play hooky, however, then

$$\hat{p} = \frac{184}{1010} = 0.182,$$

You try it!

Exercise 12.9(a)–(b)
on page 553

that is, 18.2% of the employees sampled play hooky.

These two calculations also reveal how to compute a sample proportion: Divide the number of employees sampled who play hooky, denoted x, by the total number of employees sampled, n. In symbols, $\hat{p} = x/n$. We generalize these new concepts below.

DEFINITION 12.1

Population Proportion and Sample Proportion

Consider a population in which each member either has or does not have a specified attribute. Then we use the following notation and terminology.

Population proportion, p: The proportion (percentage) of the entire population that has the specified attribute.

Sample proportion, $\hat{p}$: The proportion (percentage) of a sample from the population that has the specified attribute.

FORMULA 12.1

? What Does It Mean?

A sample proportion is obtained by dividing the number of members sampled that have the specified attribute by the total number of members sampled.

Sample Proportion

A sample proportion, $\hat{p}$, is computed by using the formula

$$\hat{p} = \frac{x}{n},$$

where x denotes the number of members in the sample that have the specified attribute and, as usual, n denotes the sample size.

Note: For convenience, we sometimes refer to x (the number of members in the sample that have the specified attribute) as the **number of successes** and to $n - x$ (the number of members in the sample that do not have the specified attribute) as the **number of failures.** In this context, the words *success* and *failure* may not have their ordinary meanings.

TABLE 12.1

Correspondence between notations for means and proportions

	Parameter	Statistic
Means	μ	$\bar{x}$
Proportions	p	$\hat{p}$

Table 12.1 shows the correspondence between the notation for means and the notation for proportions. Recall that a sample mean, $\bar{x}$, can be used to make inferences about a population mean, μ. Similarly, a sample proportion, $\hat{p}$, can be used to make inferences about a population proportion, p.

The Sampling Distribution of the Sample Proportion

To make inferences about a population mean, μ, we must know the sampling distribution of the sample mean, that is, the distribution of the variable $\bar{x}$. The same is true for proportions: To make inferences about a population proportion, p, we need to know the **sampling distribution of the sample proportion,** that is, the distribution of the variable $\hat{p}$.

Because a proportion can always be regarded as a mean, we can use our knowledge of the sampling distribution of the sample mean to derive the sampling distribution of the sample proportion. In practice, the sample size usually is large, so we concentrate on that case.

KEY FACT 12.1

What Does It Mean?

If n is large, the possible sample proportions for samples of size n have approximately a normal distribution with mean p and standard deviation $\sqrt{p(1-p)/n}$.

The Sampling Distribution of the Sample Proportion

For samples of size n,

- the mean of $\hat{p}$ equals the population proportion: $\mu_{\hat{p}} = p$ (i.e., the sample proportion is an unbiased estimator of the population proportion);
- the standard deviation of $\hat{p}$ equals the square root of the product of the population proportion and one minus the population proportion divided by the sample size: $\sigma_{\hat{p}} = \sqrt{p(1-p)/n}$; and
- $\hat{p}$ is approximately normally distributed for large n.

APPLET

Applet 12.1

The accuracy of the normal approximation depends on n and p. If p is close to 0.5, the approximation is quite accurate, even for moderate n. The farther p is from 0.5, the larger n must be for the approximation to be accurate. As a rule of thumb, we use the normal approximation when np and $n(1 - p)$ *are both 5 or greater*.[†] In this chapter, when we say that n is large, we mean that np and $n(1 - p)$ are both 5 or greater.

■■ ■ EXAMPLE 12.2

The Sampling Distribution of the Sample Proportion

OUTPUT 12.1

Histogram of $\hat{p}$ for 2000 samples of size 1010 with superimposed normal curve

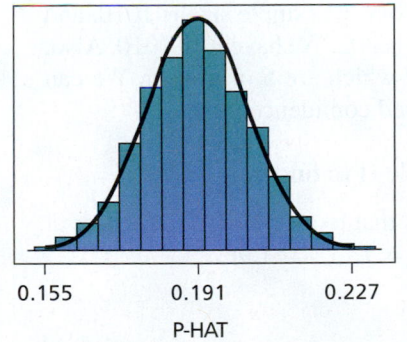

0.155   0.191   0.227

P-HAT

Playing Hooky From Work In Example 12.1, suppose that 19.1% of all U.S. employees play hooky, that is, that the population proportion is $p = 0.191$. Then, according to Key Fact 12.1, for samples of size 1010, the variable $\hat{p}$ is approximately normally distributed with mean $\mu_{\hat{p}} = p = 0.191$ and standard deviation

$$\sigma_{\hat{p}} = \sqrt{\frac{p(1-p)}{n}} = \sqrt{\frac{0.191(1-0.191)}{1010}} = 0.012.$$

Use simulation to make that fact plausible.

Solution We first simulated 2000 samples of 1010 U.S. employees each. Then, for each of those 2000 samples, we found the sample proportion, $\hat{p}$, of those who play hooky. Output 12.1 shows a histogram of those 2000 values of $\hat{p}$, which is shaped like the superimposed normal curve with parameters 0.191 and 0.012. ■

Large-Sample Confidence Intervals for a Population Proportion

Procedure 12.1 gives a step-by-step method for finding a confidence interval for a population proportion. We call this method the **one-proportion z-interval procedure**.[‡] It is based on Key Fact 12.1 and is derived in a way similar to the one-mean z-interval procedure (Procedure 8.1 on page 339).

[†]Another commonly used rule of thumb is that np and $n(1 - p)$ are both 10 or greater; still another is that $np(1 - p)$ is 25 or greater. However, our rule of thumb, which is less conservative than either of those two, is consistent with the conditions required for performing a chi-square goodness-of-fit test (discussed in Chapter 13).

[‡]The one-proportion z-interval procedure is also known as the **one-sample z-interval procedure for a population proportion** and the **one-variable proportion interval procedure**.

■■■ PROCEDURE 12.1 One-Proportion z-Interval Procedure

Purpose To find a confidence interval for a population proportion, p

Assumptions
1. Simple random sample
2. The number of successes, x, and the number of failures, $n - x$, are both 5 or greater.

Step 1 For a confidence level of $1 - \alpha$, use Table II to find $z_{\alpha/2}$.

Step 2 The confidence interval for p is from

$$\hat{p} - z_{\alpha/2} \cdot \sqrt{\hat{p}(1 - \hat{p})/n} \quad \text{to} \quad \hat{p} + z_{\alpha/2} \cdot \sqrt{\hat{p}(1 - \hat{p})/n},$$

where $z_{\alpha/2}$ is found in Step 1, n is the sample size, and $\hat{p} = x/n$ is the sample proportion.

Step 3 Interpret the confidence interval.

APPLET

Applet 12.2

Note: As stated in Assumption 2 of Procedure 12.1, a condition for using that procedure is that "the number of successes, x, and the number of failures, $n - x$, are both 5 or greater." We can restate this condition as "$n\hat{p}$ and $n(1 - \hat{p})$ are both 5 or greater," which, for an unknown p, corresponds to the rule of thumb for using the normal approximation given after Key Fact 12.1.

■■■ EXAMPLE 12.3 The One-Proportion z-Interval Procedure

Playing Hooky From Work A poll was taken of 1010 U.S. employees. The employees sampled were asked whether they "play hooky," that is, call in sick at least once a year when they simply need time to relax; 202 responded "yes." Use these data to find a 95% confidence interval for the proportion, p, of all U.S. employees who play hooky.

Solution The attribute in question is "plays hooky," the sample size is 1010, and the number of employees sampled who play hooky is 202. We have $n = 1010$. Also, $x = 202$ and $n - x = 1010 - 202 = 808$, both of which are 5 or greater. We can therefore apply Procedure 12.1 to obtain the required confidence interval.

Step 1 For a confidence level of $1 - \alpha$, use Table II to find $z_{\alpha/2}$.

We want a 95% confidence interval, which means that $\alpha = 0.05$. In Table II or at the bottom of Table IV, we find that $z_{\alpha/2} = z_{0.05/2} = z_{0.025} = 1.96$.

Step 2 The confidence interval for p is from

$$\hat{p} - z_{\alpha/2} \cdot \sqrt{\hat{p}(1 - \hat{p})/n} \quad \text{to} \quad \hat{p} + z_{\alpha/2} \cdot \sqrt{\hat{p}(1 - \hat{p})/n}.$$

We have $n = 1010$ and, from Step 1, $z_{\alpha/2} = 1.96$. Also, because 202 of the 1010 employees sampled play hooky, $\hat{p} = x/n = 202/1010 = 0.2$. Consequently, a 95% confidence interval for p is from

$$0.2 - 1.96 \cdot \sqrt{(0.2)(1 - 0.2)/1010} \quad \text{to} \quad 0.2 + 1.96 \cdot \sqrt{(0.2)(1 - 0.2)/1010},$$

or

$$0.2 - 0.025 \quad \text{to} \quad 0.2 + 0.025,$$

or 0.175 to 0.225.

StatCrunch

Report 12.1

You try it!

Exercise 12.51 on page 554

Step 3 Interpret the confidence interval.

Interpretation We can be 95% confident that the percentage of all U.S. employees who play hooky is somewhere between 17.5% and 22.5%.

Margin of Error

In Sections 8.2 and 8.3, we discussed the *margin of error* in estimating an unknown population mean by a sample mean. In general, the **margin of error** of a point estimate indicates the accuracy with which it estimates the value of the unknown parameter in question.

As we noted previously, most confidence intervals that we encounter in our study of statistics have endpoints of the form

point estimate ± margin of error.

For the one-proportion z-interval procedure (Procedure 12.1), the point estimate is the sample proportion, $\hat{p}$. Referring now to Step 2 of Procedure 12.1, we see that the margin of error for a one-proportion z-interval is $z_{\alpha/2} \cdot \sqrt{\hat{p}(1-\hat{p})/n}$, which we denote by the letter E. Formula 12.2 summarizes our discussion.

FORMULA 12.2

What Does It Mean?

The margin of error for the estimate of a population proportion indicates the accuracy with which a sample proportion estimates the unknown population proportion at the specified confidence level.

Margin of Error for the Estimate of p

The **margin of error** for the estimate of p is

$$E = z_{\alpha/2} \cdot \sqrt{\hat{p}(1-\hat{p})/n}.$$

In Example 12.3, the margin of error is

$$E = z_{\alpha/2} \cdot \sqrt{\hat{p}(1-\hat{p})/n} = 1.96 \cdot \sqrt{(0.2)(1-0.2)/1010} = 0.025,$$

which can also be obtained by taking one-half the length of the confidence interval: $(0.225 - 0.175)/2 = 0.025$. Therefore we can be 95% confident that the error in estimating the proportion, p, of all U.S. employees who play hooky by the proportion, 0.2, of those in the sample who play hooky is at most 0.025, that is, plus or minus 2.5 percentage points.

On the one hand, given a confidence interval, we can find the margin of error by taking half the length of the confidence interval. On the other hand, given the sample proportion and the margin of error, we can determine the confidence interval—its endpoints are $\hat{p} \pm E$.

Most newspaper and magazine polls provide the sample proportion and the margin of error associated with a 95% confidence interval. For example, a survey of U.S. women conducted by Gallup for the CNBC cable network stated, "36% of those polled believe their gender will hurt them; the margin of error for the poll is plus or minus 4 percentage points."

Translated into our terminology, $\hat{p} = 0.36$ and $E = 0.04$. Thus the confidence interval has endpoints $\hat{p} \pm E = 0.36 \pm 0.04$, or 0.32 to 0.40. As a result, we can be 95% confident that the percentage of all U.S. women who believe that their gender will hurt them is somewhere between 32% and 40%.

Determining the Required Sample Size

If the margin of error and confidence level are specified in advance, then we must determine the sample size required to meet those specifications. Solving for n in the formula for the margin of error, we get

$$n = \hat{p}(1-\hat{p})\left(\frac{z_{\alpha/2}}{E}\right)^2. \tag{12.1}$$

This formula cannot be used to obtain the required sample size because the sample proportion, $\hat{p}$, is not known prior to sampling.

There are two ways around this problem. To begin, we examine the graph of $\hat{p}(1-\hat{p})$ versus $\hat{p}$ shown in Fig. 12.1. The graph reveals that the largest $\hat{p}(1-\hat{p})$ can be is 0.25, which occurs when $\hat{p} = 0.5$. The farther $\hat{p}$ is from 0.5, the smaller will be the value of $\hat{p}(1-\hat{p})$.

Because the largest possible value of $\hat{p}(1-\hat{p})$ is 0.25, the most conservative approach for determining sample size is to use that value in Equation (12.1). The sample

FIGURE 12.1

Graph of $\hat{p}(1-\hat{p})$ versus $\hat{p}$

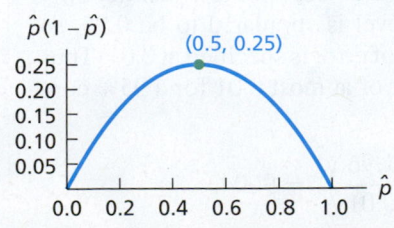

size obtained then will generally be larger than necessary and the margin of error less than required. Nonetheless, this approach guarantees that the specifications will at least be met.

However, because sampling tends to be time consuming and expensive, we usually do not want to take a larger sample than necessary. If we can make an educated guess for the observed value of $\hat{p}$—say, from a previous study or theoretical considerations—we can use that guess to obtain a more realistic sample size.

In this same vein, if we have in mind a likely range for the observed value of $\hat{p}$, then, in light of Fig. 12.1, we should take as our educated guess for $\hat{p}$ the value in the range closest to 0.5. In either case, we should be aware that, if the observed value of $\hat{p}$ is closer to 0.5 than is our educated guess, the margin of error will be larger than desired.

FORMULA 12.3

Sample Size for Estimating p

- A $(1 - \alpha)$-level confidence interval for a population proportion that has a margin of error of at most E can be obtained by choosing

$$n = 0.25 \left(\frac{z_{\alpha/2}}{E} \right)^2$$

rounded up to the nearest whole number.
- If you can make an educated guess, $\hat{p}_g$ (g for guess), for the observed value of $\hat{p}$, then you should instead choose

$$n = \hat{p}_g (1 - \hat{p}_g) \left(\frac{z_{\alpha/2}}{E} \right)^2$$

rounded up to the nearest whole number.
- If you have in mind a likely range for the observed value of $\hat{p}$, then you should apply the preceding formula with your educated guess for the observed value of $\hat{p}$ being the value in the range closest to 0.5.

EXAMPLE 12.4   Sample Size for Estimating p

Playing Hooky From Work   Consider again the problem of estimating the proportion of all U.S. employees who play hooky.

a.   Obtain a sample size that will ensure a margin of error of at most 0.01 for a 95% confidence interval.
b.   Find a 95% confidence interval for p if, for a sample of the size determined in part (a), the proportion of those who play hooky is 0.194.
c.   Determine the margin of error for the estimate in part (b), and compare it to the margin of error specified in part (a).
d.   Repeat parts (a)–(c) if the proportion of those sampled who play hooky can reasonably be presumed to be between 0.1 and 0.3.
e.   Compare the results obtained in parts (a)–(c) with those obtained in part (d).

Solution

a.   We apply the first equation in Formula 12.3. To do so, we must identify $z_{\alpha/2}$ and the margin of error, E. The confidence level is stipulated to be 0.95, so $z_{\alpha/2} = z_{0.05/2} = z_{0.025} = 1.96$, and the margin of error is specified at 0.01. Thus a sample size that will ensure a margin of error of at most 0.01 for a 95% confidence interval is

$$n = 0.25 \left(\frac{z_{\alpha/2}}{E} \right)^2 = 0.25 \left(\frac{1.96}{0.01} \right)^2 = 9604.$$

Interpretation If we take a sample of 9604 U.S. employees, the margin of error for our estimate of the proportion of all U.S. employees who play hooky will be 0.01 or less—that is, plus or minus at most 1 percentage point.

b. We find, by applying Procedure 12.1 (page 548) with $\alpha = 0.05$, $n = 9604$, and $\hat{p} = 0.194$, that a 95% confidence interval for p has endpoints

$$0.194 \pm 1.96 \cdot \sqrt{(0.194)(1 - 0.194)/9604},$$

or 0.194 ± 0.008, or 0.186 to 0.202.

Interpretation Based on a sample of 9604 U.S. employees, we can be 95% confident that the percentage of all U.S. employees who play hooky is somewhere between 18.6% and 20.2%.

c. The margin of error for the estimate in part (b) is 0.008. Not surprisingly, this is less than the margin of error of 0.01 specified in part (a).

d. If we can reasonably presume that the proportion of those sampled who play hooky will be between 0.1 and 0.3, we use the second equation in Formula 12.3, with $\hat{p}_g = 0.3$ (the value in the range closest to 0.5), to determine the sample size:

$$n = \hat{p}_g(1 - \hat{p}_g)\left(\frac{z_{\alpha/2}}{E}\right)^2 = (0.3)(1 - 0.3)\left(\frac{1.96}{0.01}\right)^2 = 8068 \text{ (rounded up)}.$$

Applying Procedure 12.1 with $\alpha = 0.05$, $n = 8068$, and $\hat{p} = 0.194$, we find that a 95% confidence interval for p has endpoints

$$0.194 \pm 1.96 \cdot \sqrt{(0.194)(1 - 0.194)/8068},$$

or 0.194 ± 0.009, or 0.185 to 0.203.

Interpretation Based on a sample of 8068 U.S. employees, we can be 95% confident that the percentage of all U.S. employees who play hooky is somewhere between 18.5% and 20.3%. The margin of error for the estimate is 0.009.

e. By using the educated guess for $\hat{p}$ in part (d), we reduced the required sample size by more than 1500 (from 9604 to 8068). Moreover, only 0.1% (0.001) of accuracy was lost—the margin of error rose from 0.008 to 0.009. The risk of using the guess 0.3 for $\hat{p}$ is that, if the observed value of $\hat{p}$ had turned out to be larger than 0.3 (but smaller than 0.7), the achieved margin of error would have exceeded the specified 0.01.

You try it!

Exercise 12.59 on page 555

The One-Proportion Plus-Four z-Interval Procedure

The confidence interval for a population proportion presented in Procedure 12.1 on page 548 does not always provide reasonably good accuracy, even for relatively large samples. As a consequence, more accurate methods have been developed. One such method is called the **one-proportion plus-four z-interval procedure.**[†] We will examine the one-proportion plus-four z-interval procedure in the exercises.

■■■■ THE TECHNOLOGY CENTER

Most statistical technologies have programs that automatically perform the one-proportion z-interval procedure. In this subsection, we present output and step-by-step instructions for such programs.

[†] See "Approximate Is Better than 'Exact' for Interval Estimation of Binomial Proportions" (*The American Statistician*, Vol. 52, No. 2, pp. 119–126) by A. Agresti and B. Coull, and "Simple and Effective Confidence Intervals for Proportions and Differences of Proportions Result from Adding Two Successes and Two Failures" (*The American Statistician*, Vol. 54, No. 4, pp. 280–288) by A. Agresti and B. Caffo.

EXAMPLE 12.5 Using Technology to Obtain a One-Proportion *z*-Interval

Playing Hooky From Work Of 1010 randomly selected U.S. employees asked whether they play hooky from work, 202 said they do. Use Minitab, Excel, or the TI-83/84 Plus to find a 95% confidence interval for the proportion, *p*, of all U.S. employees who play hooky.

Solution We applied the one-proportion *z*-interval programs to the data, resulting in Output 12.2. Steps for generating that output are presented in Instructions 12.1. *Note to Excel users:* For brevity, we have presented only the essential portions of the actual output.

OUTPUT 12.2 One-proportion *z*-interval on the data on playing hooky from work

MINITAB

Test and CI for One Proportion

```
Sample    X    N    Sample p        95% CI
1        202  1010  0.200000   (0.175331, 0.224669)

Using the normal approximation.
```

EXCEL

Frequency: 202			
Sample size: 1010			
95% confidence interval on the proportion (Wald):			
(0.1753, 0.2247)			

TI-83/84 PLUS

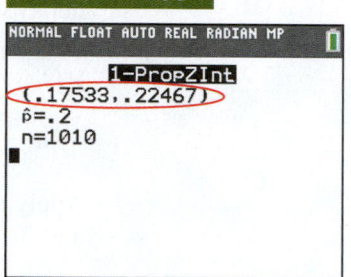

```
NORMAL FLOAT AUTO REAL RADIAN MP
            1-PropZInt
    (.17533,.22467)
p̂=.2
n=1010
```

As shown in Output 12.2, the required 95% confidence interval is from 0.175 to 0.225. We can be 95% confident that the percentage of all U.S. employees who play hooky is somewhere between 17.5% and 22.5%.

INSTRUCTIONS 12.1 Steps for generating Output 12.2

MINITAB

1 Choose **Stat ➤ Basic Statistics ➤ 1 Proportion...**
2 Press the F3 key to reset the dialog box
3 Select **Summarized data** from the drop-down list box
4 Click in the **Number of events** text box and type 202
5 Click in the **Number of trials** text box and type 1010
6 Click the **Options...** button
7 Click in the **Confidence level** text box and type 95
8 Select **Normal approximation** from the **Method** drop-down list box
9 Click **OK** twice

EXCEL

1 Choose **XLSTAT ➤ Parametric tests ➤ Tests for one proportion**

2 Click the reset button in the lower left corner of the dialog box
3 Type 202 in the **Frequency** text box
4 Type 1010 in the **Sample size** text box
5 Click the **Options** tab
6 Type 5 in the **Significance level (%)** text box
7 Click **OK**

TI-83/84 PLUS

1 Press **STAT**, arrow over to **TESTS**, and press **ALPHA ➤ A**
2 Type 202 for **x** and press **ENTER**
3 Type 1010 for **n** and press **ENTER**
4 Type .95 for **C-Level** and press **ENTER** twice

Exercises 12.1

Understanding the Concepts and Skills

12.1 In a newspaper or magazine of your choice, find a statistical study that contains an estimated population proportion.

12.2 Why is statistical inference generally used to obtain information about a population proportion?

12.3 Is a population proportion a parameter or a statistic? What about a sample proportion? Explain your answers.

12.4 Regarding a population proportion:
a. What is it?
b. What symbol is used for it?

12.5 Regarding a sample proportion:
a. What is it?
b. What symbol is used for it?

12.6 Regarding the phrase "number of successes":
a. For what is it an abbreviation?
b. What symbol is used for it?

12.7 For what is the phrase "number of failures" an abbreviation?

12.8 Explain the relationships among the sample proportion, the number of successes in the sample, and the sample size.

12.9 This exercise involves the use of an unrealistically small population to provide a concrete illustration for the exact distribution of a sample proportion. A population consists of three men and two women. The first names of the men are Jose, Pete, and Carlo; the first names of the women are Gail and Frances. Suppose that the specified attribute is "female."
a. Determine the population proportion, p.
b. The first column of the following table provides the possible samples of size 2, where each person is represented by the first letter of his or her first name; the second column gives the number of successes—the number of females obtained—for each sample; and the third column shows the sample proportion. Complete the table.

Sample	Number of females x	Sample proportion $\hat{p}$
J, G	1	0.5
J, P	0	0.0
J, C	0	0.0
J, F	1	0.5
G, P		
G, C		
G, F		
P, C		
P, F		
C, F		

c. Construct a dotplot for the sampling distribution of the proportion for samples of size 2. Mark the position of the population proportion on the dotplot.
d. Use the third column of the table to obtain the mean of the variable $\hat{p}$.
e. Compare your answers from parts (a) and (d). Why are they the same?

12.10 Repeat parts (b)–(e) of Exercise 12.9 for samples of size 1.

12.11 Repeat parts (b)–(e) of Exercise 12.9 for samples of size 3. (There are 10 possible samples.)

12.12 Repeat parts (b)–(e) of Exercise 12.9 for samples of size 4. (There are five possible samples.)

12.13 Repeat parts (b)–(e) of Exercise 12.9 for samples of size 5.

12.14 Prerequisite to this exercise are Exercises 12.9–12.13. What do your graphs in parts (c) of those exercises illustrate about the impact of increasing sample size on sampling error? Explain your answer.

12.15 NBA Draft Picks. From Wikipedia's on-line document "List of First Overall NBA Draft Picks," we found that, since 1947, 10.4% of the number-one draft picks in the National Basketball Association have been other than U.S. nationals.
a. Identify the population.
b. Identify the specified attribute.
c. Is the proportion 0.104 (10.4%) a population proportion or a sample proportion? Explain your answer.

12.16 Staying Single. From the U.S. Census Bureau document *America's Families and Living Arrangements* and an article in *Time* magazine, we found that, in 1963, 83.0% of American women between the ages of 25 and 54 were married, compared to 64.6% in 2010.
a. For 2010, identify the population.
b. For 2010, identify the specified attribute.
c. Under what circumstances is the proportion 0.646 a population proportion?
d. Under what circumstances is the proportion 0.646 a sample proportion?

12.17 Random Drug Testing. A *Harris Poll* asked Americans whether states should be allowed to conduct random drug tests on elected officials. Of 21,355 respondents, 79% said "yes."
a. Determine the margin of error for a 99% confidence interval.
b. Without doing any calculations, indicate whether the margin of error is larger or smaller for a 90% confidence interval. Explain your answer.

12.18 Genetic Binge Eating. According to an article in *Science News*, binge eating has been associated with a mutation of the gene for a brain protein called melanocortin 4 receptor (MC4R). In one study, F. Horber of the Hirslanden Clinic in Zurich and his colleagues genetically analyzed the blood of 469 obese people and found that 24 carried a mutated MC4R gene. Suppose that you want to estimate the proportion of all obese people who carry a mutated MC4R gene.
a. Determine the margin of error for a 90% confidence interval.
b. Without doing any calculations, indicate whether the margin of error is larger or smaller for a 95% confidence interval. Explain your answer.

In each of Exercises 12.19–12.24, we have given a likely range for the observed value of a sample proportion $\hat{p}$.
a. Based on the given range, identify the educated guess that should be used for the observed value of $\hat{p}$ to calculate the required sample size for a prescribed confidence level and margin of error.
b. Identify the observed values of the sample proportion that will yield a larger margin of error than the one specified if the educated guess is used for the sample-size computation.

12.19 0.2 to 0.4   **12.20** 0.4 to 0.7

12.21 0.2 or less

12.22 0.7 or greater

12.23 0.4 or greater

12.24 0.7 or less

In each of Exercises 12.25–12.30, we have given the number of successes and the sample size for a simple random sample from a population. In each case, do the following tasks.

a. *Determine the sample proportion.*

b. *Decide whether using the one-proportion z-interval procedure is appropriate.*

c. *If appropriate, use the one-proportion z-interval procedure to find the confidence interval at the specified confidence level.*

d. *If appropriate, find the margin of error for the estimate of p and express the confidence interval in terms of the sample proportion and the margin of error.*

12.25 $x = 8, n = 40$, 95% level

12.26 $x = 10, n = 40$, 90% level

12.27 $x = 35, n = 50$, 99% level

12.28 $x = 40, n = 50$, 95% level

12.29 $x = 16, n = 20$, 90% level

12.30 $x = 3, n = 100$, 99% level

In each of Exercises 12.31–12.36, we have specified a margin of error and a confidence level. For each exercise, obtain a sample size that will ensure a margin of error of at most the one specified.

12.31 margin of error = 0.01; confidence level = 95%

12.32 margin of error = 0.02; confidence level = 95%

12.33 margin of error = 0.02; confidence level = 90%

12.34 margin of error = 0.01; confidence level = 90%

12.35 margin of error = 0.03; confidence level = 99%

12.36 margin of error = 0.04; confidence level = 99%

In Exercises 12.37–12.42, we have specified the margin of errors and confidence levels from Exercises 12.31–12.36, respectively. Additionally, we have, in each case, provided an educated guess for the observed value of the sample proportion. For each exercise,

a. *obtain a sample size that will ensure a margin of error of at most the one specified (provided of course that the observed value of the sample proportion is further from 0.5 than the educated guess).*

b. *compare your answer to the corresponding one from Exercises 12.31–12.36 and explain the reason for the difference, if any.*

12.37 margin of error = 0.01; confidence level = 95%; educated guess = 0.3

12.38 margin of error = 0.02; confidence level = 95%; educated guess = 0.6

12.39 margin of error = 0.02; confidence level = 90%; educated guess = 0.1

12.40 margin of error = 0.01; confidence level = 90%; educated guess = 0.9

12.41 margin of error = 0.03; confidence level = 99%; educated guess = 0.5

12.42 margin of error = 0.04; confidence level = 99%; educated guess = 0.5

In each of Exercises 12.43–12.48, we have specified a margin of error, a confidence level, and a likely range for the observed value of the sample proportion. For each exercise, obtain a sample size that will ensure a margin of error of at most the one specified (provided of course that the observed value of the sample proportion is further from 0.5 than the educated guess).

12.43 margin of error = 0.01; confidence level = 95%; likely range = 0.2 to 0.4

12.44 margin of error = 0.02; confidence level = 95%; likely range = 0.4 to 0.7

12.45 margin of error = 0.02; confidence level = 90%; likely range = 0.2 or less

12.46 margin of error = 0.01; confidence level = 90%; likely range = 0.7 or greater

12.47 margin of error = 0.03; confidence level = 99%; likely range = 0.4 or greater

12.48 margin of error = 0.04; confidence level = 99%; likely range = 0.7 or less

Applying the Concepts and Skills

In Exercises 12.49–12.54, use Procedure 12.1 on page 548 to find the required confidence interval. Be sure to check the conditions for using that procedure.

12.49 **Online Clothing Shopping.** In a HuffPost Style and YouGov poll, 1000 U.S. adults were asked about their online vs. in-store clothes shopping. One finding was that 32% of respondents never clothes-shop online. Find and interpret a 95% confidence interval for the proportion of all U.S. adults who never clothes-shop online.

12.50 **Life Support.** In 2005, the Terri Schiavo case focused national attention on the issue of withdrawal of life support from terminally ill patients or those in a vegetative state. A *Harris Poll* of 1010 U.S. adults was conducted by telephone on April 5–10, 2005. Of those surveyed, 140 had experienced the death of at least one family member or close friend within the last 10 years who died after the removal of life support. Find and interpret a 90% confidence interval for the proportion of all U.S. adults who had experienced the death of at least one family member or close friend within the last 10 years after life support had been withdrawn.

12.51 **Asthmatics and Sulfites.** In the article "Explaining an Unusual Allergy," appearing on the Everyday Health Network, Dr. A. Feldweg explained that allergy to sulfites is usually seen in patients with asthma. The typical reaction is a sudden increase in asthma symptoms after eating a food containing sulfites. Studies are performed to estimate the percentage of the nation's 10 million asthmatics who are allergic to sulfites. In one survey, 38 of 500 randomly selected U.S. asthmatics were found to be allergic to sulfites. Find and interpret a 95% confidence interval for the proportion, p, of all U.S. asthmatics who are allergic to sulfites.

12.52 **Drinking Habits.** In a nationwide survey, conducted by Pulse Opinion Research, LLC for Rasmussen Reports, 1000 American adults were asked, among other things, whether they drink alcoholic beverages at least once a week; 38% said "yes." Determine and interpret a 95% confidence interval for the proportion, p, of all American adults who drink alcoholic beverages at least once a week.

12.53 **Factory Farming Funk.** During one year, the U.S. Environmental Protection Agency reported that concentrated animal feeding operations (CAFOs) dump 2 trillion pounds of waste into the environment annually, contaminating the ground water in 17 states and polluting more than 35,000 miles of our nation's rivers. In a survey of 1000 registered voters by Snell, Perry and Associates, 80% favored the creation of standards to limit such pollution and, in general, viewed CAFOs unfavorably. Find and interpret a 99% confidence interval for the percentage of all registered voters who favor the creation of standards on CAFO pollution and, in general, view CAFOs unfavorably.

12.54 The Nipah Virus. During one year, Malaysia was the site of an encephalitis outbreak caused by the Nipah virus, a paramyxovirus that appears to spread from pigs to workers on pig farms. As reported by K. Goh et al. in the paper "Clinical Features of Nipah Virus Encephalitis among Pig Farmers in Malaysia" (*New England Journal of Medicine*, Vol. 342, No. 17, pp. 1229–1235), neurologists from the University of Malaysia found that, among 94 patients infected with the Nipah virus, 30 died from encephalitis. Find and interpret a 90% confidence interval for the percentage of Malaysians infected with the Nipah virus who would die from encephalitis.

12.55 Literate Adults. Suppose that you have been hired to estimate the percentage of adults in your state who are literate. You take a random sample of 100 adults and find that 96 are literate. You then obtain a 95% confidence interval of

$$0.96 \pm 1.96 \cdot \sqrt{(0.96)(0.04)/100},$$

or 0.922 to 0.998. From it you conclude that you can be 95% confident that the percentage of all adults in your state who are literate is somewhere between 92.2% and 99.8%. Is anything wrong with this reasoning?

12.56 IMR in Singapore. The infant mortality rate (IMR) is the number of infant deaths per 1000 live births. Suppose that you have been commissioned to estimate the IMR in Singapore. From a random sample of 1109 live births in Singapore, you find that 0.361% of them resulted in infant deaths. You next find a 90% confidence interval:

$$0.00361 \pm 1.645 \cdot \sqrt{(0.00361)(0.99639)/1109},$$

or 0.000647 to 0.00657. You then conclude, "I can be 90% confident that the IMR in Singapore is somewhere between 0.647 and 6.57." How did you do?

12.57 Bank Breakup. In a nationwide survey, conducted by Pulse Opinion Research, LLC for Rasmussen Reports, a sample of American adults were asked whether they favor a plan to break up the 12 megabanks, which currently control about 69% of the banking industry; 50% of those sampled responded in the affirmative. According to the report, "the margin of sampling error is +/−3 percentage points with a 95% level of confidence." Find and interpret a 95% confidence interval for the percentage of all American adults who favor a plan to break up the 12 megabanks.

12.58 Online Tax Returns. According to the Internal Revenue Service, among people entitled to tax refunds, those who file online receive their refunds twice as fast as paper filers. A study conducted by International Communications Research (ICR) of Media, Pennsylvania, found that 57% of those polled said that they are not worried about the privacy of their financial information when filing their tax returns online. The survey had a margin of error of plus or minus 3 percentage points (for a 0.95 confidence level). Use this information to determine a 95% confidence interval for the percentage of all people who are not worried about the privacy of their financial information when filing their tax returns online.

12.59 Asthmatics and Sulfites. Refer to Exercise 12.51.
a. Determine the margin of error for the estimate of p.
b. Obtain a sample size that will ensure a margin of error of at most 0.01 for a 95% confidence interval without making a guess for the observed value of $\hat{p}$.
c. Find a 95% confidence interval for p if, for a sample of the size determined in part (b), the proportion of asthmatics sampled who are allergic to sulfites is 0.071.
d. Determine the margin of error for the estimate in part (c) and compare it to the margin of error specified in part (b).
e. Repeat parts (b)–(d) if you can reasonably presume that the proportion of asthmatics sampled who are allergic to sulfites will be at most 0.10.
f. Compare the results you obtained in parts (b)–(d) with those obtained in part (e).

12.60 Drinking Habits. Refer to Exercise 12.52.
a. Find the margin of error for the estimate of p.
b. Obtain a sample size that will ensure a margin of error of at most 0.02 for a 95% confidence interval without making a guess for the observed value of $\hat{p}$.
c. Find a 95% confidence interval for p if, for a sample of the size determined in part (b), 36.0% of those sampled drink alcoholic beverages at least once a week.
d. Determine the margin of error for the estimate in part (c) and compare it to the margin of error specified in part (b).
e. Repeat parts (b)–(d) if you can reasonably presume that the percentage of adults sampled who drink alcoholic beverages at least once a week will be at most 40%.
f. Compare the results you obtained in parts (b)–(d) with those obtained in part (e).

12.61 Factory Farming Funk. Refer to Exercise 12.53.
a. Find the margin of error for the estimate of the percentage.
b. Obtain a sample size that will ensure a margin of error of at most 1.5 percentage points for a 99% confidence interval without making a guess for the observed value of $\hat{p}$.
c. Find a 99% confidence interval for p if, for a sample of the size determined in part (b), 82.2% of the registered voters sampled favor the creation of standards on CAFO pollution and, in general, view CAFOs unfavorably.
d. Determine the margin of error for the estimate in part (c) and compare it to the margin of error specified in part (b).
e. Repeat parts (b)–(d) if you can reasonably presume that the percentage of registered voters sampled who favor the creation of standards on CAFO pollution and, in general, view CAFOs unfavorably will be between 75% and 85%.
f. Compare the results you obtained in parts (b)–(d) with those obtained in part (e).

12.62 The Nipah Virus. Refer to Exercise 12.54.
a. Find the margin of error for the estimate of the percentage.
b. Obtain a sample size that will ensure a margin of error of at most 5 percentage points for a 90% confidence interval without making a guess for the observed value of $\hat{p}$.
c. Find a 90% confidence interval for p if, for a sample of the size determined in part (b), 28.8% of the sampled Malaysians infected with the Nipah virus die from encephalitis.
d. Find the margin of error for the estimate in part (c) and compare it to the margin of error specified in part (b).
e. Repeat parts (b)–(d) if you can reasonably presume that the percentage of sampled Malaysians infected with the Nipah virus who would die from encephalitis would be between 25% and 40%.

f. Compare the results you obtained in parts (b)–(d) with those obtained in part (e).

12.63 Product Response Rate. A company manufactures goods that are sold exclusively by mail order. The director of market research needed to test market a new product. She planned to send brochures to a random sample of households and use the proportion of orders obtained as an estimate of the true proportion, known as the *product response rate*. The results of the market research were to be utilized as a primary source for advance production planning, so the director wanted the figures she presented to be as accurate as possible. Specifically, she wanted to be 95% confident that the estimate of the product response rate would be accurate to within 1%.

a. Without making any assumptions, determine the sample size required.

b. Historically, product response rates for products sold by this company have ranged from 0.5% to 4.9%. If the director had been willing to assume that the sample product response rate for this product would also fall in that range, find the required sample size.

c. Compare the results from parts (a) and (b).

d. Discuss the possible consequences if the assumption made in part (b) turns out to be incorrect.

12.64 Indicted Governor. On Thursday, June 13, 1996, then-Arizona Governor Fife Symington was indicted on 23 counts of fraud and extortion. Just hours after the federal prosecutors announced the indictment, several polls were conducted of Arizonans asking whether they thought Symington should resign. A poll conducted by Research Resources, Inc., that appeared in the *Phoenix Gazette*, revealed that 58% of Arizonans felt that Symington should resign; it had a margin of error of plus or minus 4.9 percentage points. Another poll, conducted by Phoenix-based Behavior Research Center and appearing in the *Tempe Daily News*, reported that 54% of Arizonans felt that Symington should resign; it had a margin of error of plus or minus 4.4 percentage points. Can the conclusions of both polls be correct? Explain your answer.

12.65 President's Job Rating. A poll conducted by Gallup in December 2013 asked a sample of American adults whether they approved of the way President Obama was doing his job; 42% said yes, with a margin of error of plus or minus 3 percentage points. During that same time period, Quinnipiac University asked the same question of a sample of American adults; 38% said yes, with a margin of error of plus or minus 2 percentage points. Can the conclusions of both polls be correct? Explain your answer.

Extending the Concepts and Skills

12.66 What important theorem in statistics implies that, for a large sample size, the possible sample proportions of that size have approximately a normal distribution?

12.67 In discussing the sample size required for obtaining a confidence interval with a prescribed confidence level and margin of error, we made the following statement: "If we have in mind a likely range for the observed value of $\hat{p}$, then, in light of Fig. 12.1, we should take as our educated guess for $\hat{p}$ the value in the range closest to 0.5." Explain why.

12.68 In discussing the sample size required for obtaining a confidence interval with a prescribed confidence level and margin of error, we made the following statement: "...we should be aware that, if the observed value of $\hat{p}$ is closer to 0.5 than is our educated guess, the margin of error will be larger than desired." Explain why.

One-Proportion Plus-Four z-Interval Procedure. To obtain a plus-four z-interval for a population proportion, we first add two successes and two failures to our data (hence, the term "plus four") and then apply Procedure 12.1 on page 548 to the new data. In other words, in place of $\hat{p}$ (which is x/n), we use $\tilde{p} = (x + 2)/(n + 4)$. Consequently, for a confidence level of $1 - \alpha$, the endpoints of the plus-four z-interval are

$$\tilde{p} \pm z_{\alpha/2} \cdot \sqrt{\tilde{p}(1 - \tilde{p})/(n + 4)}.$$

As a rule of thumb, the one-proportion plus-four z-interval procedure should be used only with confidence levels of 90% or greater and sample sizes of 10 or more.

In each of Exercises 12.69–12.74, we have given the number of successes and the sample size for a simple random sample from a population. In each case,

a. use the one-proportion plus-four z-interval procedure to find the required confidence interval.

b. compare your result with the corresponding confidence interval found in Exercises 12.25–12.30, if finding such a confidence interval was appropriate.

12.69 $x = 8, n = 40$, 95% level

12.70 $x = 10, n = 40$, 90% level

12.71 $x = 35, n = 50$, 99% level

12.72 $x = 40, n = 50$, 95% level

12.73 $x = 16, n = 20$, 90% level

12.74 $x = 3, n = 100$, 99% level

In each of Exercises 12.75–12.78, use the one-proportion plus-four z-interval procedure to find the required confidence interval. Interpret your results.

12.75 Working with Millions. A poll by Gallup asked, "If you won 10 million dollars in the lottery, would you continue to work or stop working?" Of the 1039 American adults surveyed, 707 said that they would continue working. Obtain a 95% confidence interval for the proportion of all American adults who would continue working if they won 10 million dollars in the lottery.

12.76 Social Networking. A Pew Internet & American Life Project examined Internet social networking. Among a sample of 929 online adults 18–29 years old, 836 said they use social networking sites. Determine a 95% confidence interval for the percentage of all online adults 18–29 years old who use social networking sites.

12.77 Breast-Feeding. In a *New York Times* article "More Mothers Breast-Feed, in First Months at Least," G. Harris reported that 77% of new mothers breast-feed their infants at least briefly, the highest rate seen in the United States in more than a decade. His report was based on data for 434 infants from the *National Health and Nutrition Examination Survey*, which involved in-person interviews and physical examinations. Find a 90% confidence interval for the percentage of all new mothers who breast-feed their infants at least briefly.

12.78 Offshore Drilling. In the February 2013 article "Offshore Drilling Support High as Deepwater Horizon Oil Spill Trial Opens," E. Swanson reported on a HuffPost and YouGov poll that asked Americans what they think about increased offshore drilling for oil and natural gas. Of the 1000 U.S. adults surveyed, 280 said that they were opposed. Find a 99% confidence interval for the proportion of all U.S. adults who, at the time, opposed increased offshore drilling for oil and natural gas.

12.2   Hypothesis Tests for One Population Proportion

In Section 12.1, we showed how to obtain confidence intervals for a population proportion. Now we show how to perform hypothesis tests for a population proportion. This procedure is actually a special case of the one-mean z-test.

From Key Fact 12.1 on page 547, we deduce that, for large n, the standardized version of $\hat{p}$,

$$z = \frac{\hat{p} - p}{\sqrt{p(1 - p)/n}},$$

has approximately the standard normal distribution. Consequently, to perform a large-sample hypothesis test with null hypothesis H_0: $p = p_0$, we can use the variable

$$z = \frac{\hat{p} - p_0}{\sqrt{p_0(1 - p_0)/n}}$$

as the test statistic and obtain the critical value(s) or P-value from the standard normal table, Table II.

APPLET

Applet 12.3

We call this hypothesis-testing procedure the **one-proportion z-test.**[†] Procedure 12.2 provides a step-by-step method for performing a one-proportion z-test by using either the critical-value approach or the P-value approach.

■■ ■ **PROCEDURE 12.2**   One-Proportion z-Test

Purpose To perform a hypothesis test for a population proportion, p

Assumptions
1. Simple random sample
2. Both np_0 and $n(1 - p_0)$ are 5 or greater

Step 1   The null hypothesis is H_0: $p = p_0$, and the alternative hypothesis is

H_a: $p \neq p_0$	or	H_a: $p < p_0$	or	H_a: $p > p_0$
(Two tailed)		(Left tailed)		(Right tailed)

Step 2   Decide on the significance level, α.

Step 3   Compute the value of the test statistic

$$z = \frac{\hat{p} - p_0}{\sqrt{p_0(1 - p_0)/n}}$$

and denote that value z_0.

CRITICAL-VALUE APPROACH	**OR**	***P*-VALUE APPROACH**

Step 4   The critical value(s) are

$\pm z_{\alpha/2}$	or	$-z_\alpha$	or	z_α
(Two tailed)		(Left tailed)		(Right tailed)

Use Table II to find the critical value(s).

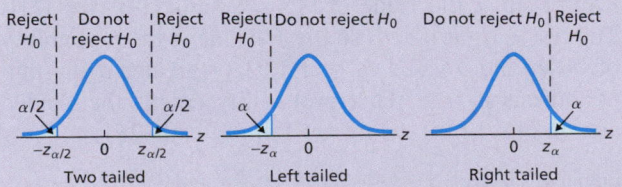

Step 5   If the value of the test statistic falls in the rejection region, reject H_0; otherwise, do not reject H_0.

Step 4   Use Table II to obtain the P-value.

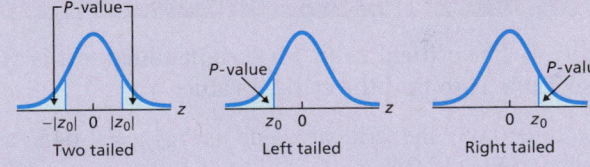

Step 5   If $P \leq \alpha$, reject H_0; otherwise, do not reject H_0.

Step 6   Interpret the results of the hypothesis test.

[†]The one-proportion z-test is also known as the **one-sample z-test for a population proportion** and the **one-variable proportion test.**

■■■   **EXAMPLE 12.6**   **The One-Proportion z-Test**

Economic Powerhouse In early February 2013, Gallup conducted a national poll of 1015 U.S. adults that asked which country is the leading economic power in the world. Of those sampled, 538 said China. At the 5% significance level, do the data provide sufficient evidence to conclude that, at the time, a majority (more than 50%) of U.S. adults considered China the leading economic power in the world?

Solution Because $n = 1015$ and $p_0 = 0.50$ (50%), we have

$$np_0 = 1015 \cdot 0.50 = 507.5 \quad \text{and} \quad n(1 - p_0) = 1015 \cdot (1 - 0.50) = 507.5.$$

As both np_0 and $n(1 - p_0)$ are 5 or greater, we can apply Procedure 12.2.

Step 1 **State the null and alternative hypotheses.**

Let p denote the proportion of all U.S. adults who considered China the leading economic power in the world. Then the null and alternative hypotheses are, respectively,

$$H_0: \ p = 0.50 \text{ (it is not true that a majority considered China top)}$$
$$H_a: \ p > 0.50 \text{ (a majority considered China top)}.$$

Note that the hypothesis test is right tailed.

Step 2 **Decide on the significance level, α.**

We are to perform the hypothesis test at the 5% significance level; so, $\alpha = 0.05$.

Step 3 **Compute the value of the test statistic**

$$z = \frac{\hat{p} - p_0}{\sqrt{p_0(1 - p_0)/n}}.$$

We have $n = 1015$ and $p_0 = 0.50$. The number of U.S. adults surveyed who considered China the leading economic power in the world is 538. Therefore the proportion of those surveyed who considered China the leading economic power in the world is $\hat{p} = x/n = 538/1015 = 0.530$ (53.0%). So, the value of the test statistic is

$$z = \frac{0.530 - 0.50}{\sqrt{(0.50)(1 - 0.50)/1015}} = 1.91.$$

CRITICAL-VALUE APPROACH	OR	P-VALUE APPROACH

CRITICAL-VALUE APPROACH

Step 4 The critical value for a right-tailed test is z_α. Use Table II to find the critical value.

For $\alpha = 0.05$, the critical value is $z_{0.05} = 1.645$, as shown in Fig. 12.2A.

FIGURE 12.2A

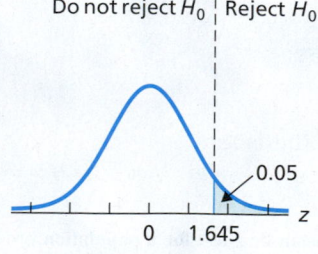

P-VALUE APPROACH

Step 4 Use Table II to obtain the P-value.

From Step 3, the value of the test statistic is $z = 1.91$. The test is right tailed, so the P-value is the probability of observing a value of z of 1.91 or greater if the null hypothesis is true. That probability equals the shaded area in Fig. 12.2B, which by Table II is 0.0281.

FIGURE 12.2B

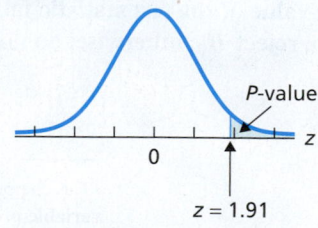

CRITICAL-VALUE APPROACH	OR	P-VALUE APPROACH

Step 5 If the value of the test statistic falls in the rejection region, reject H_0; otherwise, do not reject H_0.

From Step 3, the value of the test statistic is $z = 1.91$, which, as Fig. 12.2A shows, falls in the rejection region. Thus we reject H_0. The test results are statistically significant at the 5% level.

Step 5 If $P \leq \alpha$, reject H_0; otherwise, do not reject H_0.

From Step 4, $P = 0.0281$. Because the P-value is less than the specified significance level of 0.05, we reject H_0. The test results are statistically significant at the 5% level and (see Table 9.8 on page 386) the data provide strong evidence against the null hypothesis.

 StatCrunch
Report 12.2

 **You try it!**

Exercise 12.85 on page 561

Step 6 Interpret the results of the hypothesis test.

Interpretation At the 5% significance level, the data provide sufficient evidence to conclude that a majority of U.S. adults considered China the leading economic power in the world.

 # THE TECHNOLOGY CENTER

Most statistical technologies have programs that automatically perform the one-proportion z-test. In this subsection, we present output and step-by-step instructions for such programs.

EXAMPLE 12.7   Using Technology to Conduct a One-Proportion z-Test

Economic Powerhouse Of 1015 U.S. adults who were asked which country is the leading economic power in the world, 538 said China. Use Minitab, Excel, or the TI-83/84 Plus to decide, at the 5% significance level, whether the data provide sufficient evidence to conclude that a majority of U.S. adults considered China the leading economic power in the world.

Solution Let p denote the proportion of all U.S. adults who considered China the leading economic power in the world. The task is to perform the hypothesis test

H_0: $p = 0.50$ (it is not true that a majority considered China top)
H_a: $p > 0.50$ (a majority considered China top)

at the 5% significance level. Note that the hypothesis test is right tailed.

We applied the one-proportion z-test programs to the data, resulting in Output 12.3. Steps for generating that output are presented in Instructions 12.2 on the next page. *Note to Excel users:* For brevity, we have presented only the essential portions of the actual output.

OUTPUT 12.3   One-proportion z-test on the data on the leading economic power in the world

MINITAB

Test and CI for One Proportion

Test of p = 0.5 vs p > 0.5

Sample	X	N	Sample p	95% Lower Bound	Z-Value	P-Value
1	538	1015	0.530049	0.504281	1.91	0.028

Using the normal approximation.

OUTPUT 12.3 (cont.)
One-proportion z-test on the
data on the leading economic
power in the world

EXCEL		
Frequency: 538		
Sample size: 1015		
Test proportion: 0.5		
Hypothesized difference (D): 0		
Variance (confidence interval): Sample		
Significance level (%): 5		
z-test for one proportion / Upper-tailed test:		
Difference	0.0300	
z (Observed value)	1.9147	
z (Critical value)	1.6449	
p-value (one-tailed)	0.0278	
alpha	0.05	

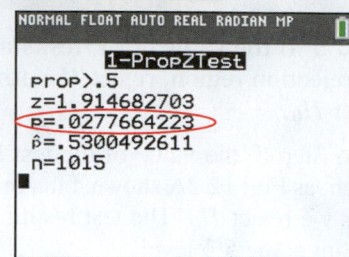

```
NORMAL FLOAT AUTO REAL RADIAN MP

         1-PropZTest
prop>.5
z=1.914682703
p=.0277664223
p̂=.5300492611
n=1015
■
```

As shown in Output 12.3, the *P*-value for the hypothesis test is 0.028. Because the *P*-value is less than the specified significance level of 0.05, we reject H_0. At the 5% significance level, the data provide sufficient evidence to conclude that a majority of U.S. adults considered China the leading economic power in the world.

INSTRUCTIONS 12.2   Steps for generating Output 12.3

MINITAB

1. Choose **Stat ➤ Basic Statistics ➤ 1 Proportion...**
2. Press the F3 key to reset the dialog box
3. Select **Summarized data** from the drop-down list box
4. Click in the **Number of events** text box and type 538
5. Click in the **Number of trials** text box and type 1015
6. Check the **Perform hypothesis test** check box
7. Click in the **Hypothesized proportion** text box and type 0.50
8. Click the **Options...** button
9. Click the arrow button at the right of the **Alternative hypothesis** drop-down list box and select **Proportion > hypothesized proportion**
10. Select **Normal approximation** from the **Method** drop-down list box
11. Click **OK** twice

EXCEL

1. Choose **XLSTAT ➤ Parametric tests ➤ Tests for one proportion**

2. Click the reset button in the lower left corner of the dialog box
3. Type 538 in the **Frequency** text box
4. Type 1015 in the **Sample size** text box
5. Type 0.5 in the **Test proportion** text box
6. Click the **Options** tab
7. Click the arrow button at the right of the **Alternative hypothesis** drop-down list box and select **Proportion – Test proportion > D**
8. Type 5 in the **Significance level (%)** text box
9. Click **OK**

TI-83/84 PLUS

1. Press **STAT**, arrow over to **TESTS**, and press **5**
2. Type 0.50 for **p₀** and press **ENTER**
3. Type 538 for **x** and press **ENTER**
4. Type 1015 for **n** and press **ENTER**
5. Highlight **> p₀** and press **ENTER**
6. Arrow down to **Calculate**, and press **ENTER**

Exercises 12.2

Understanding the Concepts and Skills

In each of Exercises 12.79–12.84, we have given the number of successes and the sample size for a simple random sample from a population. In each case, do the following.

a. *Determine the sample proportion.*
b. *Decide whether using the one-proportion z-test is appropriate.*
c. *If appropriate, use the one-proportion z-test to perform the specified hypothesis test.*

12.79 $x = 8, n = 40, H_0: p = 0.3, H_a: p < 0.3, \alpha = 0.10$

12.80 $x = 10, n = 40, H_0: p = 0.3, H_a: p < 0.3, \alpha = 0.05$

12.81 $x = 35, n = 50, H_0: p = 0.6, H_a: p > 0.6, \alpha = 0.05$

12.82 $x = 40, n = 50, H_0: p = 0.6, H_a: p > 0.6, \alpha = 0.01$

12.83 $x = 16, n = 20, H_0: p = 0.7, H_a: p \neq 0.7, \alpha = 0.05$

12.84 $x = 3, n = 100, H_0: p = 0.04, H_a: p \neq 0.04, \alpha = 0.10$

Applying the Concepts and Skills

In Exercises 12.85–12.94, use Procedure 12.2 on page 557 to perform an appropriate hypothesis test. Be sure to check the conditions for using that procedure.

12.85 Government Surveillance. Gallup conducted a national survey of 1008 American adults, asking "As you may know, as part of its efforts to investigate terrorism, a federal government agency obtained records from larger U.S. telephone and Internet companies in order to compile telephone call logs and Internet communications. Based on what you have heard or read about the program, would you say that you approve or disapprove of this government program?" Of those surveyed, 534 said they disapprove.
a. Determine and interpret the sample proportion.
b. At the 5% significance level, do the data provide sufficient evidence to conclude that a majority (more than 50%) of American adults disapprove of this government surveillance program?

12.86 Christmas Presents. The *Arizona Republic* conducted a telephone poll of 758 Arizona adults who celebrate Christmas. The question asked was, "In your family, do you open presents on Christmas Eve or Christmas Day?" Of those surveyed, 394 said they wait until Christmas Day.
a. Determine and interpret the sample proportion.
b. At the 5% significance level, do the data provide sufficient evidence to conclude that a majority (more than 50%) of Arizona families who celebrate Christmas wait until Christmas Day to open their presents?

12.87 Third Party. In 2010, shortly before that year's midterm elections, when Americans were dissatisfied with government and the Tea Party movement was emerging as a political force, 58% of American adults thought that a third major party was needed. In a recent Gallup poll, 60% of 1028 Americans felt the same way. At the 10% significance level, do the data provide sufficient evidence to conclude that the percentage of all American adults who now think that a third major party is needed has changed from that in 2010?

12.88 Families in Poverty. In 2012, 15.9% of all U.S. families had incomes below the poverty level, as reported by the U.S. Census Bureau in *American Community Survey*. During that same year, of 400 randomly selected Wyoming families, 50 had incomes below the poverty level. At the 5% significance level, do the data provide sufficient evidence to conclude that, in 2012, the percentage of families with incomes below the poverty level was lower among those living in Wyoming than among all U.S. families?

12.89 Labor Union Support. Labor Day was created by the U.S. labor movement over 100 years ago. It was subsequently adopted by most states as an official holiday. In a poll by Gallup, 1003 randomly selected adults were asked whether they approve of labor unions; 65% said yes.
a. In 1936, about 72% of Americans approved of labor unions. At the 5% significance level, do the data provide sufficient evidence to conclude that the percentage of Americans who approve of labor unions now has decreased since 1936?

b. In 1963, roughly 67% of Americans approved of labor unions. At the 5% significance level, do the data provide sufficient evidence to conclude that the percentage of Americans who approve of labor unions now has decreased since 1963?

12.90 An Edge in Roulette? Of the 38 numbers on an American roulette wheel, 18 are red, 18 are black, and 2 are green. If the wheel is balanced, the probability of the ball landing on red is $\frac{18}{38} = 0.474$. A gambler has been studying a roulette wheel. If the wheel is out of balance, he can improve his odds of winning. The gambler observes 200 spins of the wheel and finds that the ball lands on red 93 times. At the 10% significance level, do the data provide sufficient evidence to conclude that the ball is not landing on red the correct percentage of the time for a balanced wheel?

12.91 Economic Stimulus. In a national poll, 1053 U.S. adults were asked, "As you may know, Congress is considering a new economic stimulus package of at least 800 billion dollars. Do you favor or oppose Congress passing this legislation?" Of those sampled, 548 favored passage.
a. At the 5% significance level, do the data provide sufficient evidence to conclude that a majority (more than 50%) of U.S. adults favored passage?
b. The headline on the website featuring the survey read, "In U.S., Slim Majority Supports Economic Stimulus Plan." In view of your result from part (a), discuss why the headline might be misleading.
c. How could the headline be made more precise?

12.92 Delayed Perinatal Stroke. In the article "Prothrombotic Factors in Children With Stroke or Porencephaly" (*Pediatrics Journal*, Vol. 116, Issue 2, pp. 447–453), J. Lynch et al. compared differences and similarities in children with arterial ischemic stroke and porencephaly. Three classification categories were used: perinatal stroke, delayed perinatal stroke, and childhood stroke. Of 59 children, 25 were diagnosed with delayed perinatal stroke. At the 5% significance level, do the data provide sufficient evidence to conclude that delayed perinatal stroke does not comprise one-third of the cases among the three categories?

12.93 Washing Up. A Harris Interactive survey found that 92.0% of 1001 American adults said they always wash up after using the bathroom.
a. At the 5% significance level, do the data provide sufficient evidence to conclude that more than 9 of 10 Americans always wash up after using the bathroom?
b. Repeat part (a), using a 1% level of significance.

12.94 Drowning Deaths. In the article "Drowning Deaths of Zero to Five Year Old Children in Victorian Dams, 1989–2001" (*Australian Journal of Rural Health*, Vol. 13, Issue 5, pp. 300–308), L. Bugeja and R. Franklin examined drowning deaths of young children in Victorian dams to identify common contributing factors and develop strategies for future prevention. Of 11 young children who drowned in Victorian dams located on farms, 5 were girls. At the 5% significance level, do the data provide sufficient evidence to conclude that, of all young children drowning in Victorian dams located on farms, less than half are girls?

12.3 Inferences for Two Population Proportions

In Sections 12.1 and 12.2, you studied inferences for one population proportion. Now we examine inferences for comparing two population proportions. In this case, we have two populations and one specified attribute; the problem is to compare the proportion of one population that has the specified attribute to the proportion of the other population that has the specified attribute. We begin by discussing hypothesis testing.

■■■ **EXAMPLE 12.8**   Hypothesis Tests for Two Population Proportions

Eating Out Vegetarian Zogby International surveyed 1181 U.S. adults to gauge the demand for vegetarian meals in restaurants. The study, commissioned by the Vegetarian Resource Group and published in the *Vegetarian Journal*, polled independent random samples of 747 men and 434 women. Of those sampled, 276 men and 195 women said that they sometimes order a dish without meat, fish, or fowl when they eat out.

Suppose we want to use the data to decide whether, in the United States, the percentage of men who sometimes order a dish without meat, fish, or fowl is smaller than the percentage of women who sometimes order a dish without meat, fish, or fowl.

a. Formulate the problem statistically by posing it as a hypothesis test.
b. Explain the basic idea for carrying out the hypothesis test.
c. Discuss the use of the data to make a decision concerning the hypothesis test.

Solution

a. The specified attribute is "sometimes orders a dish without meat, fish, or fowl," which we abbreviate throughout this section as "sometimes orders veg." The two populations are

$$\text{Population 1: All U.S. men}$$
$$\text{Population 2: All U.S. women.}$$

Let p_1 and p_2 denote the population proportions for the two populations:

$$p_1 = \text{proportion of all U.S. men who sometimes order veg}$$
$$p_2 = \text{proportion of all U.S. women who sometimes order veg.}$$

We want to perform the hypothesis test

$$H_0: p_1 = p_2 \text{ (percentage for men is not less than that for women)}$$
$$H_a: p_1 < p_2 \text{ (percentage for men is less than that for women).}$$

b. Roughly speaking, we can carry out the hypothesis test as follows:

1. Compute the proportion of the men sampled who sometimes order veg, $\hat{p}_1$, and compute the proportion of the women sampled who sometimes order veg, $\hat{p}_2$.
2. If $\hat{p}_1$ is too much smaller than $\hat{p}_2$, reject H_0; otherwise, do not reject H_0.

c. To use the data to make a decision concerning the hypothesis test, we apply the two steps just listed. For the first step, we note that 276 of the 747 men sampled sometimes order veg and 195 of the 434 women sampled sometimes order veg, $x_1 = 276$, $n_1 = 747$, $x_2 = 195$, and $n_2 = 434$. Hence,

$$\hat{p}_1 = \frac{x_1}{n_1} = \frac{276}{747} = 0.369 \ (36.9\%)$$

and

$$\hat{p}_2 = \frac{x_2}{n_2} = \frac{195}{434} = 0.449 \ (44.9\%).$$

For the second step, we must decide whether the sample proportion $\hat{p}_1 = 0.369$ is less than the sample proportion $\hat{p}_2 = 0.449$ by a sufficient amount to warrant rejecting the null hypothesis in favor of the alternative hypothesis. To make that decision, we need to know the distribution of the difference between two sample proportions.

The Sampling Distribution of the Difference Between Two Sample Proportions for Large and Independent Samples

Let's begin by summarizing the required notation in Table 12.2.

TABLE 12.2

Notation for parameters and statistics when two population proportions are being considered

	Population 1	Population 2
Population proportion	p_1	p_2
Sample size	n_1	n_2
Number of successes	x_1	x_2
Sample proportion	$\hat{p}_1$	$\hat{p}_2$

You try it!

Exercise 12.97 on page 570

Recall that the *number of successes* refers to the number of members sampled that have the specified attribute. Consequently, we compute the sample proportions by using the formulas

$$\hat{p}_1 = \frac{x_1}{n_1} \quad \text{and} \quad \hat{p}_2 = \frac{x_2}{n_2}.$$

Armed with the notation in Table 12.2, we now describe the **sampling distribution of the difference between two sample proportions**.

KEY FACT 12.2

? What Does It Mean?

For large independent samples, the possible differences between two sample proportions have approximately a normal distribution with mean $p_1 - p_2$ and standard deviation $\sqrt{p_1(1-p_1)/n_1 + p_2(1-p_2)/n_2}$.

The Sampling Distribution of the Difference Between Two Sample Proportions for Independent Samples

For independent samples of sizes n_1 and n_2 from the two populations,

- $\mu_{\hat{p}_1 - \hat{p}_2} = p_1 - p_2$ (i.e., the difference between sample proportions is an unbiased estimator of the difference between population proportions),
- $\sigma_{\hat{p}_1 - \hat{p}_2} = \sqrt{p_1(1-p_1)/n_1 + p_2(1-p_2)/n_2}$, and
- $\hat{p}_1 - \hat{p}_2$ is approximately normally distributed for large n_1 and n_2.

Large-Sample Hypothesis Tests for Two Population Proportions, Using Independent Samples

Now we can develop a hypothesis-testing procedure for comparing two population proportions. Our immediate goal is to identify a variable that we can use as the test statistic. From Key Fact 12.2, we know that, for large, independent samples, the standardized variable

$$z = \frac{(\hat{p}_1 - \hat{p}_2) - (p_1 - p_2)}{\sqrt{p_1(1-p_1)/n_1 + p_2(1-p_2)/n_2}} \tag{12.2}$$

has approximately the standard normal distribution.

The null hypothesis for a hypothesis test to compare two population proportions is

$$H_0: p_1 = p_2 \text{ (population proportions are equal)}.$$

If the null hypothesis is true, then $p_1 - p_2 = 0$, and, consequently, the variable in Equation (12.2) becomes

$$z = \frac{\hat{p}_1 - \hat{p}_2}{\sqrt{p(1-p)/n_1 + p(1-p)/n_2}}, \tag{12.3}$$

where p denotes the common value of p_1 and p_2. Factoring $p(1-p)$ out of the denominator of Equation (12.3) yields the variable

$$z = \frac{\hat{p}_1 - \hat{p}_2}{\sqrt{p(1-p)}\sqrt{(1/n_1) + (1/n_2)}}. \tag{12.4}$$

However, because p is unknown, we cannot use this variable as the test statistic.

Consequently, we must estimate p by using sample information. The best estimate of p is obtained by pooling the data to get the proportion of successes in both samples combined; that is, we estimate p by

$$\hat{p}_p = \frac{x_1 + x_2}{n_1 + n_2}.$$

We call $\hat{p}_p$ the **pooled sample proportion.**

Replacing p in Equation (12.4) with its estimate $\hat{p}_p$ yields the variable

$$\frac{\hat{p}_1 - \hat{p}_2}{\sqrt{\hat{p}_p(1 - \hat{p}_p)}\sqrt{(1/n_1) + (1/n_2)}},$$

which can be used as the test statistic and, like the variable in Equation (12.4), has approximately the standard normal distribution for large samples if the null hypothesis is true. Hence we have Procedure 12.3, the **two-proportions z-test.**

▪▪▪▪ PROCEDURE 12.3 Two-Proportions z-Test

Purpose To perform a hypothesis test to compare two population proportions, p_1 and p_2

Assumptions
1. Simple random samples
2. Independent samples
3. $x_1, n_1 - x_1, x_2$, and $n_2 - x_2$ are all 5 or greater

Step 1 The null hypothesis is $H_0: p_1 = p_2$, and the alternative hypothesis is

$$H_a: p_1 \neq p_2 \quad \text{or} \quad H_a: p_1 < p_2 \quad \text{or} \quad H_a: p_1 > p_2$$
$$\text{(Two tailed)} \qquad \text{(Left tailed)} \qquad \text{(Right tailed)}.$$

Step 2 Decide on the significance level, α.

Step 3 Compute the value of the test statistic

$$z = \frac{\hat{p}_1 - \hat{p}_2}{\sqrt{\hat{p}_p(1 - \hat{p}_p)}\sqrt{(1/n_1) + (1/n_2)}},$$

where $\hat{p}_p = (x_1 + x_2)/(n_1 + n_2)$. Denote the value of the test statistic z_0.

CRITICAL-VALUE APPROACH	OR	P-VALUE APPROACH

CRITICAL-VALUE APPROACH

Step 4 The critical value(s) are

$$\pm z_{\alpha/2} \quad \text{or} \quad -z_\alpha \quad \text{or} \quad z_\alpha$$
$$\text{(Two tailed)} \qquad \text{(Left tailed)} \qquad \text{(Right tailed)}.$$

Use Table II to find the critical value(s).

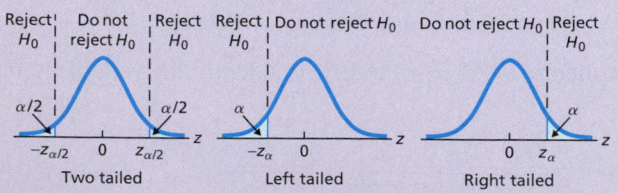

Step 5 If the value of the test statistic falls in the rejection region, reject H_0; otherwise, do not reject H_0.

P-VALUE APPROACH

Step 4 Use Table II to obtain the P-value.

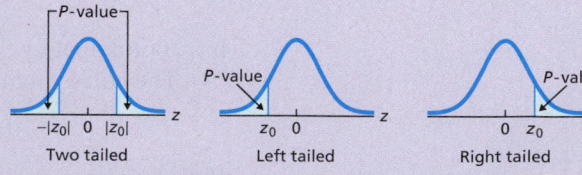

Step 5 If $P \leq \alpha$, reject H_0; otherwise, do not reject H_0.

Step 6 Interpret the results of the hypothesis test.

Note the following:

- The two-proportions z-test is also known as the **two-sample z-test for two population proportions** and the **two-variable proportions test**.
- Procedure 12.3 and its confidence-interval counterpart (Procedure 12.4 on page 566) also apply to designed experiments with two treatments.

EXAMPLE 12.9 The Two-Proportions z-Test

Eating Out Vegetarian Let's solve the problem posed in Example 12.8: Do the data from the Zogby International poll provide sufficient evidence to conclude that the percentage of U.S. men who sometimes order veg is smaller than the percentage of U.S. women who sometimes order veg? Use a 5% level of significance.

Solution We apply Procedure 12.3, noting first that the assumptions for its use are satisfied.

Step 1 State the null and alternative hypotheses.

Let p_1 and p_2 denote the proportions of all U.S. men and all U.S. women who sometimes order veg, respectively. The null and alternative hypotheses are

$$H_0: p_1 = p_2 \text{ (percentage for men is not less than that for women)}$$
$$H_a: p_1 < p_2 \text{ (percentage for men is less than that for women)}.$$

Note that the hypothesis test is left tailed.

Step 2 Decide on the significance level, α.

The test is to be performed at the 5% significance level, or $\alpha = 0.05$.

Step 3 Compute the value of the test statistic

$$z = \frac{\hat{p}_1 - \hat{p}_2}{\sqrt{\hat{p}_p(1 - \hat{p}_p)}\sqrt{(1/n_1) + (1/n_2)}},$$

where $\hat{p}_p = (x_1 + x_2)/(n_1 + n_2)$.

We first obtain $\hat{p}_1$, $\hat{p}_2$, and $\hat{p}_p$. Because 276 of the 747 men sampled and 195 of the 434 women sampled sometimes order veg, $x_1 = 276$, $n_1 = 747$, $x_2 = 195$, and $n_2 = 434$. Therefore,

$$\hat{p}_1 = \frac{x_1}{n_1} = \frac{276}{747} = 0.369, \qquad \hat{p}_2 = \frac{x_2}{n_2} = \frac{195}{434} = 0.449,$$

and

$$\hat{p}_p = \frac{x_1 + x_2}{n_1 + n_2} = \frac{276 + 195}{747 + 434} = \frac{471}{1181} = 0.399.$$

Consequently, the value of the test statistic is

$$z = \frac{\hat{p}_1 - \hat{p}_2}{\sqrt{\hat{p}_p(1 - \hat{p}_p)}\sqrt{(1/n_1) + (1/n_2)}}$$

$$= \frac{0.369 - 0.449}{\sqrt{(0.399)(1 - 0.399)}\sqrt{(1/747) + (1/434)}} = -2.71.$$

CRITICAL-VALUE APPROACH	OR	P-VALUE APPROACH

Step 4 The critical value for a left-tailed test is $-z_\alpha$. Use Table II to find the critical value.

For $\alpha = 0.05$, we find from Table II that the critical value is $-z_{0.05} = -1.645$, as shown in Fig. 12.3A.

FIGURE 12.3A

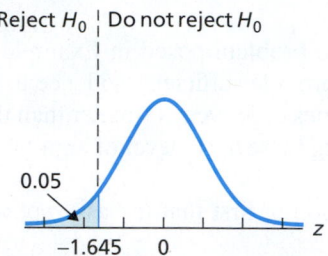

Step 5 If the value of the test statistic falls in the rejection region, reject H_0; otherwise, do not reject H_0.

From Step 3, the value of the test statistic is $z = -2.71$, which, as Fig. 12.3A shows, falls in the rejection region. Thus we reject H_0. The test results are statistically significant at the 5% level.

Step 4 Use Table II to obtain the P-value.

From Step 3, the value of the test statistic is $z = -2.71$. The test is left tailed, so the P-value is the probability of observing a value of z of -2.71 or less if the null hypothesis is true. That probability equals the shaded area in Fig. 12.3B, which, by Table II, is 0.0034.

FIGURE 12.3B

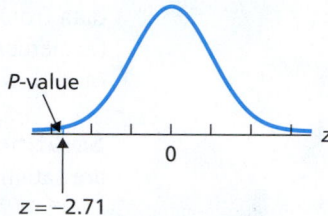

Step 5 If $P \le \alpha$, reject H_0; otherwise, do not reject H_0.

From Step 4, $P = 0.0034$. Because the P-value is less than the specified significance level of 0.05, we reject H_0. The test results are statistically significant at the 5% level and (see Table 9.8 on page 386) provide very strong evidence against the null hypothesis.

Report 12.3

Exercise 12.107
on page 570

Step 6 Interpret the results of the hypothesis test.

Interpretation   At the 5% significance level, the data provide sufficient evidence to conclude that, in the United States, the percentage of men who sometimes order veg is smaller than the percentage of women who sometimes order veg.

Large-Sample Confidence Intervals for the Difference Between Two Population Proportions

We can also use Key Fact 12.2 on page 563 to derive a confidence-interval procedure for the difference between two population proportions, called the **two-proportions** **z-interval procedure.**

■■■■ PROCEDURE 12.4   Two-Proportions z-Interval Procedure

Purpose   To find a confidence interval for the difference between two population proportions, p_1 and p_2

Assumptions
1.   Simple random samples
2.   Independent samples
3. $x_1, n_1 - x_1, x_2,$ and $n_2 - x_2$ are all 5 or greater

Step 1   For a confidence level of $1 - \alpha$, use Table II to find $z_{\alpha/2}$.

Step 2   The endpoints of the confidence interval for $p_1 - p_2$ are

$$(\hat{p}_1 - \hat{p}_2) \pm z_{\alpha/2} \cdot \sqrt{\hat{p}_1(1 - \hat{p}_1)/n_1 + \hat{p}_2(1 - \hat{p}_2)/n_2}.$$

Step 3   Interpret the confidence interval.

Note the following:

- The two-proportions z-interval procedure is also known as the **two-sample z-interval procedure for two population proportions** and the **two-variable proportions interval procedure.**
- Guidelines for interpreting confidence intervals for the difference, $p_1 - p_2$, between two population proportions are similar to those for interpreting confidence intervals for the difference, $\mu_1 - \mu_2$, between two population means, as described on page 443.

■■■■ **EXAMPLE 12.10** **The Two-Proportions z-Interval Procedure**

Eating Out Vegetarian Refer to Example 12.9, and find a 90% confidence interval for the difference, $p_1 - p_2$, between the proportions of U.S. men and U.S. women who sometimes order veg.

Solution We apply Procedure 12.4, noting first that the conditions for its use are met.

Step 1 **For a confidence level of $1 - \alpha$, use Table II to find $z_{\alpha/2}$.**

For a 90% confidence interval, we have $\alpha = 0.10$. From Table II, we determine that $z_{\alpha/2} = z_{0.10/2} = z_{0.05} = 1.645$.

Step 2 **The endpoints of the confidence interval for $p_1 - p_2$ are**

$$(\hat{p}_1 - \hat{p}_2) \pm z_{\alpha/2} \cdot \sqrt{\hat{p}_1(1 - \hat{p}_1)/n_1 + \hat{p}_2(1 - \hat{p}_2)/n_2}.$$

From Step 1, $z_{\alpha/2} = 1.645$. As we found in Example 12.9, $\hat{p}_1 = 0.369$, $n_1 = 747$, $\hat{p}_2 = 0.449$, and $n_2 = 434$. Therefore the endpoints of the 90% confidence interval for $p_1 - p_2$ are

$$(0.369 - 0.449) \pm 1.645 \cdot \sqrt{(0.369)(1 - 0.369)/747 + (0.449)(1 - 0.449)/434},$$

or -0.080 ± 0.049, or -0.129 to -0.031.

Step 3 **Interpret the confidence interval.**

Interpretation We can be 90% confident that, in the United States, the difference between the proportions of men and women who sometimes order veg is somewhere between -0.129 and -0.031. In other words, we can be 90% confident that the percentage of U.S. men who sometimes order veg is less than the percentage of U.S. women who sometimes order veg by somewhere between 3.1 and 12.9 percentage points.

StatCrunch

You try it!

Report 12.4

Exercise 12.113 on page 571

The Two-Proportions Plus-Four z-Interval Procedure

The confidence interval for the difference between two population proportions presented in Procedure 12.4 does not always provide reasonably good accuracy, even for relatively large samples. As a consequence, more accurate methods have been developed. One such method is called the **two-proportions plus-four z-interval procedure.**[†] We will examine the two-proportions plus-four z-interval procedure in the exercises.

[†]See "Simple and Effective Confidence Intervals for Proportions and Differences of Proportions Result from Adding Two Successes and Two Failures" (*The American Statistician*, Vol. 54, No. 4, pp. 280–288) by A. Agresti and B. Caffo.

THE TECHNOLOGY CENTER

Most statistical technologies have programs that automatically perform two-proportions z-procedures. In this subsection, we present output and step-by-step instructions for such programs.

EXAMPLE 12.11 ## Using Technology to Conduct Two-Proportions z-Procedures

Eating Out Vegetarian Independent random samples of 747 U.S. men and 434 U.S. women were taken. Of those sampled, 276 men and 195 women said that they sometimes order veg. Use Minitab, Excel, or the TI-83/84 Plus to perform the hypothesis test in Example 12.9 and obtain the confidence interval in Example 12.10.

Solution Let p_1 and p_2 denote the proportions of all U.S. men and all U.S. women who sometimes order veg, respectively. The task in Example 12.9 is to perform the hypothesis test

H_0: $p_1 = p_2$ (percentage for men is not less than that for women)

H_a: $p_1 < p_2$ (percentage for men is less than that for women)

at the 5% significance level; the task in Example 12.10 is to obtain a 90% confidence interval for $p_1 - p_2$.

We applied the two-proportions z-procedures programs to the data, resulting in Output 12.4. Steps for generating that output are presented in Instructions 12.3. *Note to Excel users:* For brevity, we have presented only the essential portions of the actual output.

As shown in Output 12.4, the P-value for the hypothesis test is 0.003. Because the P-value is less than the specified significance level of 0.05, we reject H_0. Output 12.4 also shows that a 90% confidence interval for the difference between the population proportions is from -0.129 to -0.031.

OUTPUT 12.4   Two-proportions z-test and z-interval on the ordering-vegetarian data

> **MINITAB**

```
[FOR THE HYPOTHESIS TEST]
Test and CI for Two Proportions

Sample    X    N   Sample p
1        276  747  0.369478
2        195  434  0.449309


Difference = p (1) - p (2)
Estimate for difference:  -0.0798308
95% upper bound for difference:  -0.0309817
Test for difference = 0 (vs < 0):  Z = -2.70  P-Value = 0.003

[FOR THE CONFIDENCE INTERVAL]
Test and CI for Two Proportions

Sample    X    N   Sample p
1        276  747  0.369478
2        195  434  0.449309


Difference = p (1) - p (2)
Estimate for difference:  -0.0798308
90% CI for difference: (-0.128680, -0.0309817)
Test for difference = 0 (vs ≠ 0):  Z = -2.69  P-Value = 0.007
```

OUTPUT 12.4 (cont.)
Two-proportions z-test and z-interval on the ordering-vegetarian data

EXCEL		
[FOR THE HYPOTHESIS TEST]		
Frequency 1: 276		
Sample size 1: 747		
Frequency 2: 195		
Sample size 2: 434		
Hypothesized difference (D): 0		
Variance: pq(1/n1+1/n2)		
Significance level (%): 5		
z-test for two proportions / Lower-tailed test:		
Difference	-0.0798	
z (Observed value)	-2.7012	
z (Critical value)	-1.6449	
p-value (one-tailed)	0.0035	
alpha	0.05	
[FOR THE CONFIDENCE INTERVAL]		
Variance: p1q1/n1+p2q2/n2		
Significance level (%): 10		
z-test for two proportions / Two-tailed test:		
90% confidence interval on the difference between the proportions:		
(-0.1287 , -0.0310)		

TI-83/84 PLUS
NORMAL FLOAT AUTO REAL RADIAN MP
2-PropZTest
P1<P2
z=-2.701227783
p=.0034542498
p̂1=.3694779116
p̂2=.4493087558
p̂=.3988145639
n1=747
n2=434

Using **2-PropZTest**

NORMAL FLOAT AUTO REAL RADIAN MP
2-PropZInt
(-.1287,-.031)
p̂1=.3694779116
p̂2=.4493087558
n1=747
n2=434

Using **2-PropZInt**

INSTRUCTIONS 12.3   Steps for generating Output 12.4

MINITAB

FOR THE HYPOTHESIS TEST:
1. Choose **Stat ➤ Basic Statistics ➤ 2 Proportions…**
2. Press the F3 key to reset the dialog box
3. Select **Summarized data** from the drop-down list box
4. Click in the **Number of events** text box for **Sample 1** and type 276
5. Click in the **Number of trials** text box for **Sample 1** and type 747
6. Click in the **Number of events** text box for **Sample 2** and type 195
7. Click in the **Number of trials** text box for **Sample 2** and type 434
8. Click the **Options…** button
9. Click the arrow button at the right of the **Alternative hypothesis** drop-down list box and select **Difference < hypothesized difference**
10. Click the arrow button at the right of the **Test method** drop-down list box and select **Use the pooled estimate of the proportion**
11. Click **OK** twice

FOR THE CI:
1. Repeat steps 1–8 from the hypothesis-test instructions
2. Click in the **Confidence level** text box and type 90
3. Click **OK** twice

EXCEL

FOR THE HYPOTHESIS TEST:
1. Choose **XLSTAT ➤ Parametric tests ➤ Tests for two proportions**
2. Click the reset button in the lower left corner of the dialog box

3. Type 276 in the **Frequency 1** text box
4. Type 747 in the **Sample size 1** text box
5. Type 195 in the **Frequency 2** text box
6. Type 434 in the **Sample size 2** text box
7. Click the **Options** tab
8. Click the arrow button at the right of the **Alternative hypothesis** drop-down list box and select **Proportion 1 – Proportion 2 < D**
9. Type 5 in the **Significance level (%)** text box
10. Select the second option button in the **Variance** list
11. Click **OK**

FOR THE CI:
1. Repeat steps 1–7 from the hypothesis-test instructions
2. Type 10 in the **Significance level (%)** text box
3. Click **OK**

TI-83/84 PLUS

FOR THE HYPOTHESIS TEST:
1. Press **STAT**, arrow over to **TESTS**, and press **6**
2. Type 276 for **x1** and press **ENTER**
3. Type 747 for **n1** and press **ENTER**
4. Type 195 for **x2** and press **ENTER**
5. Type 434 for **n2** and press **ENTER**
6. Highlight **<p2** and press **ENTER**
7. Arrow down to **Calculate** and press **ENTER**

FOR THE CI:
1. Press **STAT**, arrow over to **TESTS**, and press **ALPHA ➤ B**
2. Type 276 for **x1** and press **ENTER**
3. Type 747 for **n1** and press **ENTER**
4. Type 195 for **x2** and press **ENTER**
5. Type 434 for **n2** and press **ENTER**
6. Type .90 for **C-Level** and press **ENTER** twice

Note to Minitab and Excel users: Although Minitab and Excel simultaneously perform a hypothesis test and obtain a confidence interval, the type of confidence interval found depends on the type of hypothesis test. Specifically, Minitab and Excel compute a two-sided confidence interval for a two-tailed test and a one-sided confidence interval for a one-tailed test. To perform a one-tailed hypothesis test and obtain a two-sided confidence interval, apply Minitab's or Excel's two-proportions z-procedure twice: once for the one-tailed hypothesis test and once for the confidence interval specifying a two-tailed hypothesis test. For Excel, be sure to select the second option button in the **Variance** list for a hypothesis test and the first option button (the default) in the **Variance** list for a confidence interval.

Exercises 12.3

Understanding the Concepts and Skills

12.95 Explain the basic idea for performing a hypothesis test, based on independent samples, to compare two population proportions.

12.96 Kids Attending Church. In an *ABC Global Kids Study*, conducted by Roper Starch Worldwide, Inc., estimates were made in various countries of the percentage of children who attend church at least once a week. Two of the countries in the survey were the United States and Germany. Considering these two countries only,
a. identify the specified attribute.
b. identify the two populations.
c. What are the two population proportions under consideration?

12.97 Sunscreen Use. Industry Research polled teenagers on sunscreen use. The survey revealed that 46% of teenage girls and 30% of teenage boys regularly use sunscreen before going out in the sun.
a. Identify the specified attribute.
b. Identify the two populations.
c. Are the proportions 0.46 (46%) and 0.30 (30%) sample proportions or population proportions? Explain your answer.

12.98 Consider a hypothesis test for two population proportions with the null hypothesis $H_0: p_1 = p_2$. What parameter is being estimated by the
a. sample proportion $\hat{p}_1$?
b. sample proportion $\hat{p}_2$?
c. pooled sample proportion $\hat{p}_p$?

12.99 Of the quantities p_1, p_2, x_1, x_2, $\hat{p}_1$, $\hat{p}_2$, and $\hat{p}_p$,
a. which represent parameters and which represent statistics?
b. which are fixed numbers and which are variables?

In each of Exercises 12.100–12.105, we have provided the numbers of successes and the sample sizes for independent simple random samples from two populations. In each case, do the following tasks.
a. Determine the sample proportions.
b. Decide whether using the two-proportions z-procedures is appropriate. If so, also do parts (c) and (d).
c. Use the two-proportions z-test to conduct the required hypothesis test.
d. Use the two-proportions z-interval procedure to find the specified confidence interval.

12.100 $x_1 = 10, n_1 = 20, x_2 = 18, n_2 = 30$;
left-tailed test, $\alpha = 0.10$; 80% confidence interval

12.101 $x_1 = 18, n_1 = 40, x_2 = 30, n_2 = 40$;
left-tailed test, $\alpha = 0.10$; 80% confidence interval

12.102 $x_1 = 14, n_1 = 20, x_2 = 8, n_2 = 20$;
right-tailed test, $\alpha = 0.05$; 90% confidence interval

12.103 $x_1 = 15, n_1 = 20, x_2 = 18, n_2 = 30$;
right-tailed test, $\alpha = 0.05$; 90% confidence interval

12.104 $x_1 = 18, n_1 = 30, x_2 = 10, n_2 = 20$;
two-tailed test, $\alpha = 0.05$; 95% confidence interval

12.105 $x_1 = 30, n_1 = 80, x_2 = 15, n_2 = 20$;
two-tailed test, $\alpha = 0.05$; 95% confidence interval

Applying the Concepts and Skills

In Exercises 12.106–12.111, use either the critical-value approach or the P-value approach to perform the required hypothesis test.

12.106 Vasectomies and Prostate Cancer. In the United States, approximately 450,000 vasectomies are performed each year. In this surgical procedure for contraception, the tube carrying sperm from the testicles is cut and tied. Several studies have been conducted to analyze the relationship between vasectomies and prostate cancer. The results of one such study by E. Giovannucci et al. appeared in the paper "A Retrospective Cohort Study of Vasectomy and Prostate Cancer in U.S. Men" (*Journal of the American Medical Association*, Vol. 269(7), pp. 878–882). Of 21,300 men who had not had a vasectomy, 69 were found to have prostate cancer; of 22,000 men who had had a vasectomy, 113 were found to have prostate cancer.
a. At the 1% significance level, do the data provide sufficient evidence to conclude that men who have had a vasectomy are at greater risk of having prostate cancer? Consider men who had had a vasectomy Population 2.
b. Is this study a designed experiment or an observational study? Explain your answer.
c. In view of your answers to parts (a) and (b), could you reasonably conclude that having a vasectomy causes an increased risk of prostate cancer? Explain your answer.

12.107 Folic Acid and Birth Defects. For several years, evidence had been mounting that folic acid reduces major birth defects. A. Czeizel and I. Dudas of the National Institute of Hygiene in Budapest directed a study that provided the strongest evidence to date. Their results were published in the paper "Prevention of the First Occurrence of Neural-Tube Defects by Periconceptional Vitamin Supplementation" (*New England Journal of Medicine*, Vol. 327(26), p. 1832). For the study, the doctors enrolled women prior to conception and divided them randomly into two groups. One group, consisting of 2701 women, took daily multivitamins containing 0.8 mg of folic acid; the other group, consisting of 2052 women, received only trace elements. Major birth defects occurred in 35 cases when the women took folic acid and in 47 cases when the women did not.
a. At the 1% significance level, do the data provide sufficient evidence to conclude that women who take folic acid are at lesser risk of having children with major birth defects?

b. Is this study a designed experiment or an observational study? Explain your answer.

c. In view of your answers to parts (a) and (b), could you reasonably conclude that taking folic acid causes a reduction in major birth defects? Explain your answer.

12.108 Racial Crossover. In the paper "The Racial Crossover in Comorbidity, Disability, and Mortality" (*Demography*, Vol. 37(3), pp. 267–283), N. Johnson investigated the health of independent random samples of white and African-American elderly (aged 70 years or older). Of the 4989 white elderly surveyed, 529 had at least one stroke, whereas 103 of the 906 African-American elderly surveyed reported at least one stroke. At the 5% significance level, do the data suggest that there is a difference in stroke incidence between white and African-American elderly?

12.109 Buckling Up. The National Highway Traffic Safety Administration collects data on seat-belt use and publishes results in the document *Occupant Restraint Use*. Of 1000 drivers 16–24 years old, 79% said that they buckle up, whereas 924 of 1100 drivers 25–69 years old said that they did. At the 1% significance level, do the data provide sufficient evidence to conclude that there is a difference in seat-belt use between the two age groups?

12.110 Ballistic Fingerprinting. Guns make unique markings on bullets they fire and their shell casings. These markings are called *ballistic fingerprints*. An *ABCNEWS Poll* examined the opinions of Americans on the enactment of a law "…that would require every gun sold in the United States to be test-fired first, so law enforcement would have its fingerprint in case it were ever used in a crime." The following problem is based on the results of that poll. Independent simple random samples were taken of 537 women and 495 men. When asked whether they support a ballistic fingerprinting law, 446 of the women and 307 of the men said "yes." At the 1% significance level, do the data provide sufficient evidence to conclude that women tend to favor ballistic fingerprinting more than men?

12.111 Body Mass Index. Body mass index (BMI) is a measure of body fat based on height and weight. According to the document *Dietary Guidelines for Americans*, published by the U.S. Department of Agriculture and the U.S. Department of Health and Human Services, for adults, a BMI of greater than 25 indicates an above healthy weight (i.e., overweight or obese). Of 750 randomly selected adults whose highest degree is a bachelor's, 386 have an above healthy weight; and of 500 randomly selected adults with a graduate degree, 237 have an above healthy weight.

a. What assumptions are required for using the two-proportions *z*-test here?

b. Apply the two-proportions *z*-test to determine, at the 5% significance level, whether the percentage of adults who have an above healthy weight is greater for those whose highest degree is a bachelor's than for those with a graduate degree.

In Exercises 12.112–12.117, apply Procedure 12.4 on page 566 to find the required confidence interval.

12.112 Vasectomies and Prostate Cancer. Refer to Exercise 12.106 and determine and interpret a 98% confidence interval for the difference between the prostate cancer rates of men who have had a vasectomy and those who have not.

12.113 Folic Acid and Birth Defects. Refer to Exercise 12.107 and determine and interpret a 98% confidence interval for the difference between the rates of major birth defects for babies born to women who have taken folic acid and those born to women who have not.

12.114 Racial Crossover. Refer to Exercise 12.108 and find and interpret a 95% confidence interval for the difference between the stroke incidences of white and African-American elderly.

12.115 Buckling Up. Refer to Exercise 12.109 and find and interpret a 99% confidence interval for the difference between the proportions of seat-belt users for drivers in the age groups 16–24 years and 25–69 years.

12.116 Ballistic Fingerprinting. Refer to Exercise 12.110 and find and interpret a 98% confidence interval for the difference between the percentages of women and men who favor ballistic fingerprinting.

12.117 Body Mass Index. Refer to Exercise 12.111 and find and interpret a 90% confidence interval for the difference between the percentages of adults in the two degree categories who have an above healthy weight.

12.118 Hormone Therapy and Dementia. An issue of *Science News* (Vol. 163, No. 22, pp. 341–342) reported that the Women's Health Initiative cast doubts on the benefit of hormone-replacement therapy. Researchers randomly divided 4532 healthy women over the age of 65 years into two groups. One group, consisting of 2229 women, received hormone-replacement therapy; the other group, consisting of 2303 women, received placebo. Over 5 years, 40 of the women receiving the hormone-replacement therapy were diagnosed with dementia, compared with 21 of those getting placebo.

a. At the 5% significance level, do the data provide sufficient evidence to conclude that healthy women over 65 years old who take hormone-replacement therapy are at greater risk for dementia than those who do not?

b. Determine and interpret a 90% confidence interval for the difference in dementia risk rates for healthy women over 65 years old who take hormone-replacement therapy and those who do not.

12.119 Women in the Labor Force. The Organization for Economic Cooperation and Development (OECD) summarizes data on labor-force participation rates in *OECD in Figures*. Independent simple random samples were taken of 300 U.S. women and 250 Canadian women. Of the U.S. women, 215 were found to be in the labor force; of the Canadian women, 186 were found to be in the labor force.

a. At the 5% significance level, do the data suggest that there is a difference between the labor-force participation rates of U.S. and Canadian women?

b. Find and interpret a 95% confidence interval for the difference between the labor-force participation rates of U.S. and Canadian women.

12.120 Neutropenia. Neutropenia is an abnormally low number of neutrophils (a type of white blood cell) in the blood. Chemotherapy often reduces the number of neutrophils to a level that makes patients susceptible to fever and infections. G. Bucaneve et al. published a study of such cancer patients in the paper "Levofloxacin to Prevent Bacterial Infection in Patients With Cancer and Neutropenia" (*New England Journal of Medicine*, Vol. 353, No. 10, pp. 977–987). For the study, 375 patients were randomly assigned to receive a daily dose of levofloxacin, and 363 were given placebo. In the group receiving levofloxacin, fever was present in 243 patients for the duration of neutropenia, whereas fever was experienced by 308 patients in the placebo group.

a. At the 1% significance level, do the data provide sufficient evidence to conclude that levofloxacin is effective in reducing the occurrence of fever in such patients?

b. Find a 98% confidence level for the difference in the proportions of such cancer patients who would experience fever for the duration of neutropenia.

12.121 Is College Worth It? In the *New York Times* article "College Graduates Fare Well in Jobs Market, Even through Recession," C. Rampell noted that college graduates have suffered through the recession and lackluster recovery with remarkable resilience. Of a random sample of 1020 college graduates, 35 were unemployed; and of a random sample of 1008 high-school graduates (no college), 69 were unemployed.

a. At the 1% significance level, do the data provide sufficient evidence to conclude that college graduates have a lower unemployment rate than high-school graduates?

b. Find and interpret a 98% confidence interval for the difference in unemployment rates of college and high-school graduates.

Extending the Concepts and Skills

Two-Proportions Plus-Four z-Interval Procedure. To obtain a plus-four z-interval for the difference between two population proportions, we first add one success and one failure to each of our two samples of data (hence, the term "plus four") and then apply Procedure 12.4 on page 566 to the new data. In other words, in place of $\hat{p}_1$ (which is x_1/n_1), we use $\tilde{p}_1 = (x_1 + 1)/(n_1 + 2)$, and in place of $\hat{p}_2$ (which is x_2/n_2), we use $\tilde{p}_2 = (x_2 + 1)/(n_2 + 2)$. Thus, for a confidence level of $1 - \alpha$, the plus-four z-interval for $p_1 - p_2$ has endpoints

$$(\tilde{p}_1 - \tilde{p}_2) \pm z_{\alpha/2} \cdot \sqrt{\frac{\tilde{p}_1(1 - \tilde{p}_1)}{n_1 + 2} + \frac{\tilde{p}_2(1 - \tilde{p}_2)}{n_2 + 2}}.$$

As a rule of thumb, the two-proportions plus-four z-interval procedure can be used when both sample sizes are 5 or more.

In each of Exercises 12.122–12.127, we have given the numbers of successes and the sample sizes for simple random samples for independent random samples from two populations. In each case,

a. *use the two-proportions plus-four z-interval procedure to find the required confidence interval for the difference between the two population proportions.*

b. *compare your result with the corresponding confidence interval found in parts (d) of Exercises 12.100–12.105, if finding such a confidence interval was appropriate.*

12.122 $x_1 = 10, n_1 = 20, x_2 = 18, n_2 = 30$; 80% confidence interval

12.123 $x_1 = 18, n_1 = 40, x_2 = 30, n_2 = 40$; 80% confidence interval

12.124 $x_1 = 14, n_1 = 20, x_2 = 8, n_2 = 20$; 90% confidence interval

12.125 $x_1 = 15, n_1 = 20, x_2 = 18, n_2 = 30$; 90% confidence interval

12.126 $x_1 = 18, n_1 = 30, x_2 = 10, n_2 = 20$; 95% confidence interval

12.127 $x_1 = 30, n_1 = 80, x_2 = 15, n_2 = 20$; 95% confidence interval

In each of Exercises 12.128–12.131, use the two-proportions plus-four z-interval procedure to find the required confidence interval. Interpret your results.

12.128 Confidence in Banks. Since 1973, Gallup has asked Americans how much confidence they have in a variety of U.S. institutions. One question asked of those polled is whether they have a great deal of confidence in banks. In 2007, of a random sample of 1008 adult Americans, 413 said yes; and, in 2013, of a random sample of 1529 adult Americans, 398 said yes. For the two years, find and interpret a 95% confidence interval for the difference between the percentages of adult Americans who had a great deal of confidence in banks.

12.129 Unemployment Rates. The Organization for Economic Cooperation and Development (OECD) conducts studies on unemployment rates by country and publishes its findings in the document *Main Economic Indicators.* Independent random samples of 100 and 75 people in the civilian labor forces of Finland and Denmark, respectively, revealed 7 and 3 unemployed, respectively. Find a 95% confidence interval for the difference between the unemployment rates in Finland and Denmark.

12.130 Federal Gas Tax. The Quinnipiac University Poll conducts nationwide surveys as a public service and for research. In one poll, participants were asked whether they thought eliminating the federal gas tax for the summer months is a good idea. The following problems are based on the results of that poll.

a. Of 611 Republicans, 275 thought it a good idea, and, of 872 Democrats, 366 thought it a good idea. Obtain a 90% confidence interval for the difference between the proportions of Republicans and Democrats who think that eliminating the federal gas tax for the summer months is a good idea.

b. Of 907 women, 417 thought it a good idea, and, of 838 men, 310 thought it a good idea. Obtain a 90% confidence interval for the difference between the percentages of women and men who think that eliminating the federal gas tax for the summer months is a good idea.

12.131 Blockers and Cancer. A *Wall Street Journal* article, titled "Hypertension Drug Linked to Cancer," reported on a study of several types of high-blood-pressure drugs and links to cancer. For one type, called calcium-channel blockers, 27 of 202 elderly patients taking the drug developed cancer. For another type, called beta-blockers, 28 of 424 other elderly patients developed cancer. Find a 90% confidence interval for the difference between the cancer rates of elderly people taking calcium-channel blockers and those taking beta-blockers. *Note:* The results of this study were challenged and questioned by several sources that claimed, for example, that the study was flawed and that several other studies have suggested that calcium-channel blockers are safe.

Margin of Error and Sample Size for $p_1 - p_2$. Exercises 12.132–12.137 examine the margin of error and sample size for estimating the difference between two population proportions.

12.132 Obtain a formula for the margin of error, E, in estimating the difference between two population proportions by referring to Step 2 of Procedure 12.4 on page 566.

12.133 Use your result from Exercise 12.132 to show that a $(1 - \alpha)$-level confidence interval for the difference between two population proportions that has a margin of error of at most E can be obtained by choosing

$$n_1 = n_2 = 0.5 \left(\frac{z_{\alpha/2}}{E}\right)^2$$

rounded up to the nearest whole number.

12.134 Suppose that you can make reasonably good educated guesses, $\hat{p}_{1g}$ and $\hat{p}_{2g}$, for the observed values of $\hat{p}_1$ and $\hat{p}_2$.

a. Use your result from Exercise 12.132 to show that a $(1 - \alpha)$-level confidence interval for the difference between two population proportions that has an approximate margin of error of E can be obtained by choosing

$$n_1 = n_2 = \left(\hat{p}_{1g}(1 - \hat{p}_{1g}) + \hat{p}_{2g}(1 - \hat{p}_{2g})\right) \left(\frac{z_{\alpha/2}}{E}\right)^2$$

rounded up to the nearest whole number. *Note:* If you know likely ranges instead of exact educated guesses for the observed values of the two sample proportions, use the values in the ranges closest to 0.5 as the educated guesses.

b. Explain why the formula in part (a) yields smaller (or at worst the same) sample sizes than the formula in Exercise 12.133.

c. When reasonably good educated guesses for the observed values of $\hat{p}_1$ and $\hat{p}_2$ can be made, explain why choosing the sample sizes by using the formula in part (a) is preferable to choosing them by using the formula in Exercise 12.133.

12.135 Eating Out Vegetarian. Refer to the study on ordering vegetarian considered in Examples 12.8–12.10.

a. Obtain the margin of error for the estimate of the difference between the proportions of men and women who sometimes order veg by taking half the length of the confidence interval found in Example 12.10 on page 567. Interpret your answer in words.

b. Obtain the margin of error for the estimate of the difference between the proportions of men and women who sometimes order veg by applying Exercise 12.132.

12.136 Eating Out Vegetarian. Refer to the study on ordering vegetarian considered in Examples 12.8–12.10.

a. Without making a guess for the observed values of the sample proportions, find the common sample size that will ensure a margin of error of at most 0.01 for a 90% confidence interval. *Hint:* Use Exercise 12.133.

b. Find a 90% confidence interval for $p_1 - p_2$ if, for samples of the size determined in part (a), 38.3% of the men and 43.7% of the women sometimes order veg.

c. Determine the margin of error for the estimate in part (b), and compare it to the required margin of error specified in part (a).

12.137 Eating Out Vegetarian. Repeat Exercise 12.136 by applying the formula in Exercise 12.134(a) if you can reasonably presume that at most 41% of the men sampled and at most 49% of the women sampled will be people who sometimes order veg. Compare the results obtained in Exercise 12.136 to those obtained here.

CHAPTER IN REVIEW

You Should Be Able to

1. use and understand the formulas in this chapter.

2. determine a large-sample confidence interval for a population proportion.

3. compute the margin of error for the estimate of a population proportion.

4. understand the relationship between the sample size, confidence level, and margin of error for a confidence interval for a population proportion.

5. determine the sample size required for a specified confidence level and margin of error for the estimate of a population proportion.

6. perform a large-sample hypothesis test for a population proportion.

7. perform large-sample inferences (hypothesis tests and confidence intervals) to compare two population proportions.

8. understand the relationship between the sample sizes, confidence level, and margin of error for a confidence interval for the difference between two population proportions.

9. determine the sample sizes required for a specified confidence level and margin of error for the estimate of the difference between two population proportions.

Key Terms

margin of error, *549*
number of failures, *546*
number of successes, *546*
one-proportion plus-four z-interval procedure, *551*
one-proportion z-interval procedure, *548*

one-proportion z-test, *557*
pooled sample proportion ($\hat{p}_p$), *564*
population proportion (p), *546*
sample proportion ($\hat{p}$), *546*
sampling distribution of the difference between two sample proportions, *563*

sampling distribution of the sample proportion, *547*
two-proportions plus-four z-interval procedure, *567*
two-proportions z-interval procedure, *566*
two-proportions z-test, *564*

REVIEW PROBLEMS

Understanding the Concepts and Skills

1. Medical Marijuana? An international poll of physicians was conducted on the *New England Journal of Medicine* website asking "Do you believe that the overall medicinal benefits of marijuana outweigh the risks and potential harms?" Identify the
a. specified attribute.   **b.** population.

c. population proportion.

d. Of the 1446 physicians who responded, 76% said yes. Is the proportion 0.76 (76%) a sample proportion or a population proportion? Explain your answer.

2. Why is a sample proportion generally used to estimate a population proportion instead of obtaining the population proportion directly?

3. Explain what each phrase means in the context of inferences for a population proportion.

a. Number of successes

b. Number of failures

4. Fill in the blanks.

a. The mean of all possible sample proportions is equal to the _____.

b. For large samples, the possible sample proportions have approximately a _____ distribution.

c. A rule of thumb for using a normal distribution to approximate the distribution of all possible sample proportions is that both _____ and _____ are _____ or greater.

5. What does the margin of error for the estimate of a population proportion tell you?

6. Holiday Blues. A poll was conducted by Opinion Research Corporation to estimate the proportions of men and women who get the "holiday blues." Identify the

a. specified attribute. **b.** two populations.

c. two population proportions.

d. two sample proportions.

e. According to the poll, 34% of men and 44% of women get the "holiday blues." Are the proportions 0.34 and 0.44 sample proportions or population proportions? Explain your answer.

7. Suppose that you are using independent samples to compare two population proportions. Fill in the blanks.

a. The mean of all possible differences between the two sample proportions equals the _____.

b. For large samples, the possible differences between the two sample proportions have approximately a _____ distribution.

Applying the Concepts and Skills

8. Smallpox Vaccine. *ABCNEWS.com* published the results of a poll that asked U.S. adults whether they would get a smallpox shot if it were available. Sampling, data collection, and tabulation were done by TNS Intersearch of Horsham, Pennsylvania. When the risk of the vaccine was described in detail, 4 in 10 of those surveyed said they would take the smallpox shot. According to the article, "the results have a three-point margin of error" (for a 0.95 confidence level). Use the information provided to obtain a 95% confidence interval for the percentage of all U.S. adults who would take a smallpox shot, knowing the risk of the vaccine.

9. Getting a Job. The National Association of Colleges and Employers sponsors the *Graduating Student and Alumni Survey*. Part of the survey gauges student optimism in landing a job after graduation. According to one year's survey results, published in *American Demographics*, among the 1218 respondents, 733 said that they expected difficulty finding a job. *Note:* In this problem and the next, round your proportion answers to four decimal places.

a. Use these data to find and interpret a 95% confidence interval for the proportion of students who expect difficulty finding a job.

b. Find the margin of error for the estimate of p.

c. Express the confidence interval in the form point estimate ± margin of error.

10. Getting a Job. Refer to Problem 9.

a. Determine a sample size that will ensure a margin of error of at most 0.02 for a 95% confidence interval without making a guess for the observed value of $\hat{p}$.

b. Find a 95% confidence interval for p if, for a sample of the size determined in part (a), 58.7% of those surveyed say that they expect difficulty finding a job.

c. Determine the margin of error for the estimate in part (b), and compare it to the required margin of error specified in part (a).

d. Repeat parts (a)–(c) if you can reasonably presume that the percentage of those surveyed who say that they expect difficulty finding a job will be at least 56%.

e. Compare the results obtained in parts (a)–(c) with those obtained in part (d).

11. Justice in the Courts? In an issue of *Parade Magazine*, the editors reported on a national survey on law and order. One question asked of the 2512 U.S. adults who took part was whether they believed that juries "almost always" convict the guilty and free the innocent. Only 578 said that they did. At the 5% significance level, do the data provide sufficient evidence to conclude that less than one in four Americans believe that juries "almost always" convict the guilty and free the innocent?

12. Height and Breast Cancer. In the article "Height and Weight at Various Ages and Risk of Breast Cancer" (*Annals of Epidemiology*, Vol. 2, pp. 597–609), L. Brinton and C. Swanson discussed the relationship between height and breast cancer. The study, sponsored by the National Cancer Institute, took 5 years and involved more than 1500 women with breast cancer and 2000 women without breast cancer; it revealed a trend between height and breast cancer: "...taller women have a 50 to 80 percent greater risk of getting breast cancer than women who are closer to 5 feet tall." Christine Swanson, a nutritionist who was involved with the study, added, "...height may be associated with the culprit, ...but no one really knows" the exact relationship between height and the risk of breast cancer.

a. Classify this study as either an observational study or a designed experiment. Explain your answer.

b. Interpret the statement made by Christine Swanson in light of your answer to part (a).

13. Smartphone Ownership. The Pew Internet & American Life Project conducted a survey of smartphone ownership. One aspect of the study involved the gender of smartphone owners. Of 1029 sampled men, 607 owned a smartphone; and of 1223 sampled women, 648 owned a smartphone. At the 1% significance level, do the data provide sufficient evidence to conclude that a difference exists in the percentages of smartphone owners between men and women?

14. Smartphone Ownership. Refer to Problem 13.

a. Find a 99% confidence interval for the difference, $p_1 - p_2$, between the proportions of men and women smartphone owners.

b. Interpret your answer in part (a) in terms of the difference in percentages of men and women smartphone owners.

15. Bulletproof Vests. In the *New York Times* article "A Common Police Vest Fails the Bulletproof Test," E. Lichtblau reported on a U.S. Department of Justice study of 103 bulletproof vests containing a fiber known as Zylon. In ballistics tests, only 4 of these vests produced acceptable safety outcomes (and resulted in immediate changes in federal safety guidelines). Find a 95% confidence interval for the proportion of all such vests that would produce acceptable safety outcomes by using the

a. one-proportion z-interval procedure.

b. one-proportion plus-four z-interval procedure. (See page 556 for the details of this procedure.)

c. Explain the large discrepancy between the two methods.

d. Which confidence interval would you use? Explain your answer.

FOCUSING ON DATA ANALYSIS

UWEC UNDERGRADUATES

Recall from Chapter 1 (see page 34) that the Focus database and Focus sample contain information on the undergraduate students at the University of Wisconsin - Eau Claire (UWEC). Now would be a good time for you to review the discussion about these data sets.

Open the Focus sample worksheet (FocusSample) in the technology of your choice and then do the following.

a. At the 5% significance level, do the data provide sufficient evidence to conclude that more than half of UWEC undergraduates are females?

b. Repeat part (a) using a 1% significance level.

c. Determine and interpret a 95% confidence interval for the percentage of UWEC undergraduates who are females.

d. At the 5% significance level, do the data provide sufficient evidence to conclude that a difference exists in the percentages of females among resident and nonresident UWEC undergraduates?

e. Repeat part (d) using a 10% significance level.

f. Determine and interpret a 95% confidence interval for the difference between the percentages of females among resident and nonresident UWEC undergraduates.

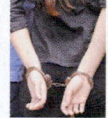

CASE STUDY DISCUSSION

ARRESTED YOUTHS

As you learned on page 544, recent research by R. Brame et al. indicates that almost a third of Americans have been arrested by age 23, and this figure excludes arrests for minor traffic violations. Now that you have studied inferences for proportions, we ask you to conduct statistical analyses based on the results obtained by the researchers. A random sample of 6157 Americans revealed that 1858 had been arrested by the time they were 23 years old.

a. Obtain and interpret a point estimate for the percentage of Americans who have been arrested by age 23.

b. Assuming a simple random sample, determine and interpret a 95% confidence interval for the percentage of Americans who have been arrested by age 23.

c. At the 1% significance level, do the data provide sufficient evidence to conclude that the percentage of Americans who have been arrested by age 23 is less than a third?

d. At the 1% significance level, do the data provide sufficient evidence to conclude that the (current) percentage of Americans who have been arrested by age 23 exceeds the 1965 percentage of 22%?

BIOGRAPHY

ABRAHAM DE MOIVRE: PAVING THE WAY FOR PROPORTION INFERENCES

Abraham de Moivre was born in Vitry-le-Francois, France, on May 26, 1667, the son of a country surgeon. He was educated in the Catholic school in his village and at the Protestant Academy at Sedan. In 1684, he went to Paris to study under Jacques Ozanam.

In late 1685, de Moivre, a French Huguenot (Protestant), was imprisoned in Paris because of his religion. (In October, 1685, Louis XIV revoked an edict that had allowed Protestantism in addition to the Catholicism favored by the French Court.) The duration of his incarceration is unclear, but de Moivre was probably jailed 1 to 3 years. In any case, upon his release he fled to London, where he began tutoring students in mathematics.

In London, de Moivre mastered Sir Isaac Newton's *Principia* and became a close friend of Newton's and of Edmond Halley's, an English astronomer (in whose honor, incidentally, Halley's Comet is named). In Newton's later years, he would refuse to take new students, saying, "Go to Mr. de Moivre; he knows these things better than I do."

De Moivre's contributions to probability theory, mathematics, and statistics range from the definition of statistical independence to analytical trigonometric formulas to his major discovery:

the normal approximation to the binomial distribution—of monumental importance in its own right, precursor to the central limit theorem, and fundamental to proportion inferences. The definition of statistical independence appeared in *The Doctrine of Chances,* published in 1718 and dedicated to Newton; the normal approximation to the binomial distribution was contained in a Latin pamphlet published in 1733. Many of his other papers were published in *Philosophical Transactions of the Royal Society.*

De Moivre also did research on the analysis of mortality statistics and the theory of annuities. In 1725, the first edition of his *Annuities on Lives*, in which he derived annuity formulas and addressed other annuity problems, was published.

De Moivre was elected to the Royal Society in 1697, to the Berlin Academy of Sciences in 1735, and to the Paris Academy in 1754. Despite his obvious talents as a mathematician and his many champions, he was never able to obtain a position in any of England's universities. Instead, he had to rely on his meager earnings as a tutor in mathematics and a consultant on gambling and insurance, supplemented by the sales of his books. De Moivre died in London on November 27, 1754.

Chi-Square Procedures

CHAPTER OBJECTIVES

The statistical-inference techniques presented so far have dealt exclusively with hypothesis tests and confidence intervals for population parameters, such as population means and population proportions. In this chapter, we consider three widely used inferential procedures that are not concerned with population parameters. These three procedures are often called **chi-square procedures** because they rely on a distribution called the *chi-square distribution*, which we discuss in Section 13.1.

In Section 13.2, we present the chi-square goodness-of-fit test, a hypothesis test that can be used to make inferences about the distribution of a variable. For instance, we could apply that test to a sample of university students to decide whether the political preference distribution of all university students differs from that of the population as a whole.

In Section 13.3, as a preliminary to the study of our second chi-square procedure, we discuss contingency tables and related topics. Next, in Section 13.4, we present the chi-square independence test, a hypothesis test used to decide whether an association exists between two variables of a population. For instance, we could apply that test to a sample of U.S. adults to decide whether an association exists between annual income and educational level for all U.S. adults.

Then, in Section 13.5, we examine the chi-square homogeneity test, a hypothesis test used to decide whether a difference exists among the distributions of a variable of two or more populations. For instance, we could apply that test to decide whether race distributions differ in the four U.S. regions.

CASE STUDY

Eye and Hair Color

Statistically speaking, does eye color depend on hair color? In other words, is there an association between those two characteristics?

We would think so, but how do we establish our conjecture?

In the article "Graphical Display of Two-Way Contingency Tables" (*The American Statistician*, Vol. 28, No. 1, pp. 9–12), R. Snee presented sample data on hair color and eye color among 592 people. The data, which are provided on the WeissStats site, were collected as part of a class project by students in an elementary statistics course taught by Snee at the University of Delaware.

From the (raw) data on the WeissStats site, we constructed

the following two-way table, which gives a frequency distribution for the cross-classified data. For instance, the table shows that 16 of the 592 people sampled have blonde hair and green eyes.

We can use the frequencies in this table to perform a hypothesis test to decide whether an association exists between eye color and hair color. After studying the inferential methods discussed in this chapter, you will be asked to do just that.

| | Hair color | | | | |
	Black	Blonde	Brown	Red	Total
Blue	20	94	84	17	215
Brown	68	7	119	26	220
Green	5	16	29	14	64
Hazel	15	10	54	14	93
Total	108	127	286	71	592

(Eye color labels the rows.)

13.1 The Chi-Square Distribution

FIGURE 13.1

χ^2-curves for df = 5, 10, and 19

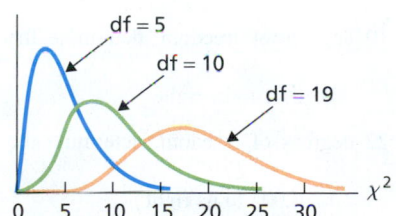

The statistical-inference procedures discussed in this chapter rely on a distribution called the *chi-square distribution*. Chi (pronounced "kī") is a Greek letter whose lowercase form is χ.

A variable has a **chi-square distribution** if its distribution has the shape of a special type of right-skewed curve, called a **chi-square (χ^2) curve**. Actually, there are infinitely many chi-square distributions, and we identify the chi-square distribution (and χ^2-curve) in question by its number of degrees of freedom, just as we did for *t*-distributions. Figure 13.1 shows three χ^2-curves and illustrates some basic properties of χ^2-curves.

KEY FACT 13.1

Basic Properties of χ^2-Curves

Property 1: The total area under a χ^2-curve equals 1.

Property 2: A χ^2-curve starts at 0 on the horizontal axis and extends indefinitely to the right, approaching, but never touching, the horizontal axis.

Property 3: A χ^2-curve is right skewed.

Property 4: As the number of degrees of freedom becomes larger, χ^2-curves look increasingly like normal curves.

Using the χ^2-Table

Percentages (and probabilities) for a variable that has a chi-square distribution are equal to areas under its associated χ^2-curve. To perform a chi-square test, we need to know how to find the χ^2-value that has a specified area to its right. Table VII in Appendix A provides χ^2-values that correspond to several areas.

The χ^2-table (Table VII) is similar to the *t*-table (Table IV). The two outside columns of Table VII, labeled df, display the number of degrees of freedom. As expected, the symbol χ^2_α denotes the χ^2-value that has area α to its right under a χ^2-curve. Thus the column headed $\chi^2_{0.05}$, for example, contains χ^2-values that have area 0.05 to their right.

■■■ **EXAMPLE 13.1** Finding the χ^2-Value Having a Specified Area to Its Right

For a χ^2-curve with 12 degrees of freedom, find $\chi^2_{0.025}$; that is, find the χ^2-value that has area 0.025 to its right, as shown in Fig. 13.2(a).

FIGURE 13.2
Finding the χ^2-value that has area 0.025 to its right

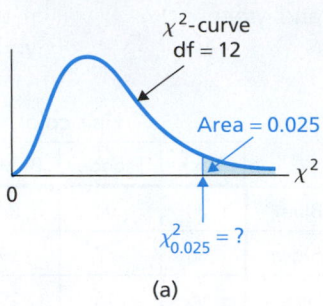

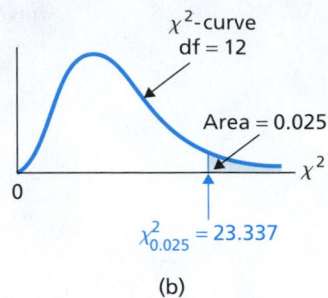

(a)

(b)

You try it!

Exercise 13.5 on page 578

Solution To find this χ^2-value, we use Table VII. The number of degrees of freedom is 12, so we first go down the outside columns, labeled df, to "12." Then, going across that row to the column labeled $\chi^2_{0.025}$, we reach 23.337. This number is the χ^2-value having area 0.025 to its right, as shown in Fig. 13.2(b). In other words, for a χ^2-curve with df = 12, $\chi^2_{0.025} = 23.337$. ■

Exercises 13.1

Understanding the Concepts and Skills

13.1 What is meant by saying that a variable has a chi-square distribution?

13.2 How do you identify different chi-square distributions?

13.3 Consider two χ^2-curves with degrees of freedom 12 and 20, respectively. Which one more closely resembles a normal curve? Explain your answer.

13.4 The t-table has entries for areas of 0.10, 0.05, 0.025, 0.01, and 0.005. In contrast, the χ^2-table has entries for those areas and for 0.995, 0.99, 0.975, 0.95, and 0.90. Explain why the t-values corresponding to these additional areas can be obtained from the existing t-table but must be provided explicitly in the χ^2-table.

In Exercises 13.5–13.8, use Table VII to find the required χ^2-values. Illustrate your work graphically.

13.5 For a χ^2-curve with 19 degrees of freedom, determine the χ^2-value that has area
a. 0.025 to its right.   **b.** 0.01 to its right.

13.6 For a χ^2-curve with 22 degrees of freedom, determine the χ^2-value that has area
a. 0.01 to its right.   **b.** 0.005 to its right.

13.7 For a χ^2-curve with df = 10, determine
a. $\chi^2_{0.05}$.   **b.** $\chi^2_{0.025}$.

13.8 For a χ^2-curve with df = 4, determine
a. $\chi^2_{0.005}$.   **b.** $\chi^2_{0.01}$.

13.2 Chi-Square Goodness-of-Fit Test

Our first chi-square procedure is called the **chi-square goodness-of-fit test.** We can use this procedure to perform a hypothesis test about the distribution of a qualitative (categorical) variable or a discrete quantitative variable that has only finitely many possible values. We introduce and explain the reasoning behind the chi-square goodness-of-fit test next.

■■■ **EXAMPLE 13.2** Introduces the Chi-Square Goodness-of-Fit Test

Violent Crimes The Federal Bureau of Investigation (FBI) compiles data on crimes and crime rates and publishes the information in *Crime in the United States*. A violent crime is classified by the FBI as murder, forcible rape, robbery, or aggravated assault. Table 13.1 gives a relative-frequency distribution for (reported) violent crimes in 2010. For instance, in 2010, 29.5% of violent crimes were robberies.

TABLE 13.1

Distribution of violent crimes
in the United States, 2010

Type of violent crime	Relative frequency
Murder	0.012
Forcible rape	0.068
Robbery	0.295
Agg. assault	0.625
	1.000

TABLE 13.2

Sample results for 500 randomly
selected violent-crime
reports from last year

Type of violent crime	Frequency
Murder	3
Forcible rape	36
Robbery	170
Agg. assault	291
	500

TABLE 13.3

Expected frequencies if last year's
violent-crime distribution is the same
as the 2010 distribution

Type of violent crime	Expected frequency
Murder	6.0
Forcible rape	34.0
Robbery	147.5
Agg. assault	312.5

A simple random sample of 500 violent-crime reports from last year yielded the frequency distribution shown in Table 13.2. Suppose that we want to use the data in Tables 13.1 and 13.2 to decide whether last year's distribution of violent crimes is changed from the 2010 distribution.

a. Formulate the problem statistically by posing it as a hypothesis test.
b. Explain the basic idea for carrying out the hypothesis test.
c. Discuss the details for making a decision concerning the hypothesis test.

Solution

a. The population is last year's (reported) violent crimes. The variable is "type of violent crime," and its possible values are murder, forcible rape, robbery, and aggravated assault. We want to perform the following hypothesis test.

H_0: Last year's violent-crime distribution is the same as the 2010 distribution.

H_a: Last year's violent-crime distribution is different from the 2010 distribution.

b. The idea behind the chi-square goodness-of-fit test is to compare the **observed frequencies** in the second column of Table 13.2 to the frequencies that would be expected—the **expected frequencies**—if last year's violent-crime distribution is the same as the 2010 distribution. If the observed and expected frequencies match fairly well (i.e., each observed frequency is roughly equal to its corresponding expected frequency), we do not reject the null hypothesis; otherwise, we reject the null hypothesis.

c. To formulate a precise procedure for carrying out the hypothesis test, we need to answer two questions:

1. What frequencies should we expect from a random sample of 500 violent-crime reports from last year if last year's violent-crime distribution is the same as the 2010 distribution?
2. How do we decide whether the observed and expected frequencies match fairly well?

The first question is easy to answer, which we illustrate with robberies. If last year's violent-crime distribution is the same as the 2010 distribution, then, according to Table 13.1, 29.5% of last year's violent crimes would have been robberies. Therefore, in a random sample of 500 violent-crime reports from last year, we would expect about 29.5% of the 500 to be robberies. In other words, we would expect the number of robberies to be $500 \cdot 0.295$, or 147.5.

In general, we compute each expected frequency, denoted E, by using the formula

$$E = np,$$

where n is the sample size and p is the appropriate relative frequency from the second column of Table 13.1. Using this formula, we calculated the expected frequencies for all four types of violent crime. The results are displayed in the second column of Table 13.3.

The second column of Table 13.3 answers the first question. It gives the frequencies that we would expect if last year's violent-crime distribution is the same as the 2010 distribution.

The second question—whether the observed and expected frequencies match fairly well—is harder to answer. We need to calculate a number that measures the goodness of fit.

In Table 13.4 (next page), the second column repeats the observed frequencies from the second column of Table 13.2. The third column of Table 13.4 repeats the expected frequencies from the second column of Table 13.3.

TABLE 13.4

Calculating the goodness of fit

Type of violent crime x	Observed frequency O	Expected frequency E	Difference $O - E$	Square of difference $(O - E)^2$	Chi-square subtotal $(O - E)^2/E$
Murder	3	6.0	−3.0	9.00	1.500
Forcible rape	36	34.0	2.0	4.00	0.118
Robbery	170	147.5	22.5	506.25	3.432
Agg. assault	291	312.5	−21.5	462.25	1.479
	500	500.0	0		6.529

To measure the goodness of fit of the observed and expected frequencies, we look at the differences, $O - E$, shown in the fourth column of Table 13.4. Summing these differences to obtain a measure of goodness of fit isn't very useful because the sum is 0. Instead, we square each difference (shown in the fifth column) and then divide by the corresponding expected frequency. Doing so gives the values $(O - E)^2/E$, called **chi-square subtotals,** shown in the sixth column. The sum of the chi-square subtotals,

$$\Sigma(O - E)^2/E = 6.529,$$

is the statistic used to measure the goodness of fit of the observed and expected frequencies.[†]

If the null hypothesis is true, the observed and expected frequencies should be roughly equal, resulting in a small value of the test statistic, $\Sigma(O - E)^2/E$. So, large values of $\Sigma(O - E)^2/E$ provide evidence against the null hypothesis.

As we have seen, $\Sigma(O - E)^2/E = 6.529$. Can this value be reasonably attributed to sampling error, or is it large enough to suggest that the null hypothesis is false? To answer this question, we need to know the distribution of the test statistic $\Sigma(O - E)^2/E$. ■

First we present the formula for expected frequencies in a chi-square goodness-of-fit test, as discussed in the preceding example, and then we provide the distribution of the test statistic for a chi-square goodness-of-fit test.

FORMULA 13.1

? **What Does It Mean?**

To obtain an expected frequency, multiply the sample size by the null-hypothesis relative frequency.

Expected Frequencies for a Goodness-of-Fit Test

In a chi-square goodness-of-fit test, the expected frequency for each possible value of the variable is found by using the formula

$$E = np,$$

where n is the sample size and p is the relative frequency (or probability) given for the value in the null hypothesis.

KEY FACT 13.2

? **What Does It Mean?**

To obtain a chi-square subtotal, square the difference between an observed and expected frequency and divide the result by the expected frequency. Adding the chi-square subtotals gives the χ^2-statistic, which has approximately a chi-square distribution.

Distribution of the χ^2-Statistic for a Goodness-of-Fit Test

For a chi-square goodness-of-fit test, the test statistic

$$\chi^2 = \Sigma(O - E)^2/E$$

has approximately a chi-square distribution if the null hypothesis is true. The number of degrees of freedom is 1 less than the number of possible values for the variable under consideration.

[†]Using subscripts alone or both subscripts and indices, we would write $\Sigma(O - E)^2/E$ as

$$\Sigma(O_i - E_i)^2/E_i \quad \text{or} \quad \sum_{i=1}^{c}(O_i - E_i)^2/E_i,$$

where c denotes the number of possible values for the variable, in this case, four ($c = 4$).

Procedure for the Chi-Square Goodness-of-Fit Test

In light of Key Fact 13.2, we now present, in Procedure 13.1, a step-by-step method for conducting a chi-square goodness-of-fit test. Because the null hypothesis is rejected only when the test statistic is too large, a chi-square goodness-of-fit test is always right tailed.

■■■ PROCEDURE 13.1   **Chi-Square Goodness-of-Fit Test**

Purpose To perform a hypothesis test for the distribution of a variable

Assumptions

1. All expected frequencies are 1 or greater
2. At most 20% of the expected frequencies are less than 5
3. Simple random sample

Step 1   The null and alternative hypotheses are, respectively,

   H_0: The variable has the specified distribution

   H_a: The variable does not have the specified distribution.

Step 2   Decide on the significance level, α.

Step 3   Compute the value of the test statistic

$$\chi^2 = \Sigma(O - E)^2/E,$$

where O and E represent observed and expected frequencies, respectively. Denote the value of the test statistic χ_0^2.

CRITICAL-VALUE APPROACH	OR	P-VALUE APPROACH

Step 4   The critical value is χ_α^2 with df $= c - 1$, where c is the number of possible values for the variable. Use Table VII to find the critical value.

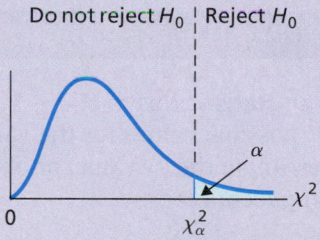

Step 5   If the value of the test statistic falls in the rejection region, reject H_0; otherwise, do not reject H_0.

Step 4   The χ^2-statistic has df $= c - 1$, where c is the number of possible values for the variable. Use Table VII to estimate the P-value, or obtain it exactly by using technology.

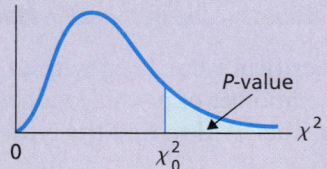

Step 5   If $P \leq \alpha$, reject H_0; otherwise, do not reject H_0.

Step 6   Interpret the results of the hypothesis test.

Note: Regarding Assumptions 1 and 2, in many texts the rule given is that all expected frequencies be 5 or greater. However, research by the noted statistician W. G. Cochran shows that the "rule of 5" is too restrictive. See, for instance, W. G. Cochran, "Some Methods for Strengthening the Common χ^2 Tests" (*Biometrics*, Vol. 10, No. 4, pp. 417–451).

■■■ EXAMPLE 13.3   **The Chi-Square Goodness-of-Fit Test**

Violent Crimes We can now complete the hypothesis test introduced in Example 13.2. Table 13.5 repeats the relative-frequency distribution for violent crimes in the United States in 2010.

TABLE 13.5

Distribution of violent crimes in the United States, 2010

Type of violent crime	Relative frequency
Murder	0.012
Forcible rape	0.068
Robbery	0.295
Agg. assault	0.625

A random sample of 500 violent-crime reports from last year yielded the frequency distribution shown in Table 13.6. At the 5% significance level, do the data provide sufficient evidence to conclude that last year's violent-crime distribution is different from the 2010 distribution?

Solution We displayed the expected frequencies in Table 13.3 on page 579. From the second column of that table, we see that the expected-frequency conditions, Assumptions 1 and 2 of Procedure 13.1, are satisfied because all of the expected frequencies exceed 5. Hence, we can apply Procedure 13.1 to perform the required hypothesis test.

Step 1 State the null and alternative hypotheses.

The null and alternative hypotheses are, respectively,

H_0: Last year's violent-crime distribution is the same as the 2010 distribution.

H_a: Last year's violent-crime distribution is different from the 2010 distribution.

TABLE 13.6

Sample results for 500 randomly selected violent-crime reports from last year

Type of violent crime	Observed frequency
Murder	3
Forcible rape	36
Robbery	170
Agg. assault	291

Step 2 Decide on the significance level, α.

We are to perform the test at the 5% significance level, so $\alpha = 0.05$.

Step 3 Compute the value of the test statistic

$$\chi^2 = \Sigma(O - E)^2/E,$$

where O and E represent observed and expected frequencies, respectively.

We already calculated the value of the test statistic in Table 13.4 on page 580:

$$\chi^2 = \Sigma(O - E)^2/E = 6.529,$$

to three decimal places.

CRITICAL-VALUE APPROACH	OR	P-VALUE APPROACH

Step 4 The critical value is χ_α^2 with df $= c - 1$, where c is the number of possible values for the variable. Use Table VII to find the critical value.

From Step 2, $\alpha = 0.05$. The variable is "type of violent crime." There are four types of violent crime, so $c = 4$. In Table VII, we find that, for df $= c - 1 = 4 - 1 = 3$, $\chi_{0.05}^2 = 7.815$, as shown in Fig. 13.3A.

FIGURE 13.3A

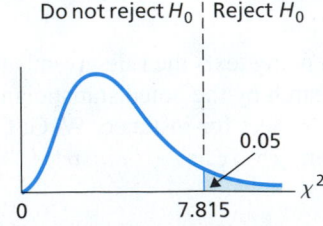

Do not reject H_0 | Reject H_0

0.05

0   7.815 χ^2

Step 4 The χ^2-statistic has df $= c - 1$, where c is the number of possible values for the variable. Use Table VII to estimate the P-value, or obtain it exactly by using technology.

From Step 3, the value of the test statistic is $\chi^2 = 6.529$. The test is right tailed, so the P-value is the probability of observing a value of χ^2 of 6.529 or greater if the null hypothesis is true. That probability equals the shaded area in Fig. 13.3B.

FIGURE 13.3B

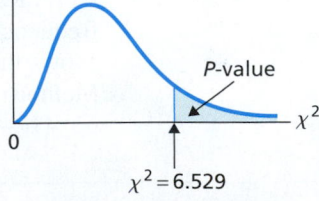

P-value

0          $\chi^2 = 6.529$ χ^2

The variable is "type of violent crime." Because there are four types of violent crime, $c = 4$. Referring to Fig. 13.3B and Table VII with df $= c - 1 = 4 - 1 = 3$, we find that $0.05 < P < 0.10$. (Using technology, we obtain $P = 0.089$.)

| CRITICAL-VALUE APPROACH | OR | P-VALUE APPROACH |

Step 5 If the value of the test statistic falls in the rejection region, reject H_0; otherwise, do not reject H_0.

From Step 3, the value of the test statistic is $\chi^2 = 6.529$. Because it does not fall in the rejection region, as shown in Fig. 13.3A, we do not reject H_0. The test results are not statistically significant at the 5% level.

Step 5 If $P \leq \alpha$, reject H_0; otherwise, do not reject H_0.

From Step 4, $0.05 < P < 0.10$. Because the P-value exceeds the specified significance level of 0.05, we do not reject H_0. The test results are not statistically significant at the 5% level, but (see Table 9.8 on page 386) the data do provide moderate evidence against the null hypothesis.

StatCrunch

Report 13.1

You try it!

Exercise 13.25
on page 586

Step 6 Interpret the results of the hypothesis test.

Interpretation At the 5% significance level, the data do not provide sufficient evidence to conclude that last year's violent-crime distribution differs from the 2010 distribution.

THE TECHNOLOGY CENTER

Most statistical technologies have programs that automatically perform a chi-square goodness-of-fit test, but others do not. In this subsection, we present output and step-by-step instructions for such programs.

Note to TI-83/84 Plus users:

- The TI-83 Plus does not have a built-in program for performing a chi-square goodness-of-fit test. And, although the TI-84 Plus does have such a built-in program, χ^2GOF-Test, that program applies to other types of chi-square goodness-of-fit tests in addition to the one discussed in this section. As a consequence, χ^2GOF-Test requires the user to enter items that would be unnecessary to enter for the specific test discussed here.
- In view of the above facts, we have written a TI program called CHIGFT specifically for the chi-square goodness-of-fit test discussed in this section. You can find that program in the TI Programs section on the WeissStats site and your instructor can show you how to download the program to your calculator.
- *Warning:* Any data that you may have previously stored in Lists 1–3 will be erased during program execution, so copy those data to other lists prior to program execution if you want to retain them.

EXAMPLE 13.4   Using Technology to Perform a Goodness-of-Fit Test

Violent Crimes Table 13.5 shows a relative-frequency distribution for violent crimes in the United States in 2010, and Table 13.6 gives a frequency distribution for a random sample of 500 violent-crime reports from last year. Use Minitab, Excel, or the TI-83/84 Plus to decide, at the 5% significance level, whether the data provide sufficient evidence to conclude that last year's violent-crime distribution is different from the 2010 distribution.

Solution We want to perform the hypothesis test

H_0: Last year's violent-crime distribution is the same as the 2010 distribution

H_a: Last year's violent-crime distribution is different from the 2010 distribution

at the 5% significance level.

We applied the chi-square goodness-of-fit programs to the data, resulting in Output 13.1. Steps for generating that output are presented in Instructions 13.1. *Note to Excel users:* For brevity, we have presented only the essential portions of the actual output.

OUTPUT 13.1
Goodness-of-fit test on the violent-crime data

MINITAB

Chi-Square Goodness-of-Fit Test for Observed Counts in Variable: O

Using category names in CRIME

Category	Observed	Test Proportion	Expected	Contribution to Chi-Sq
Murder	3	0.012	6.0	1.50000
Forcible rape	36	0.068	34.0	0.11765
Robbery	170	0.295	147.5	3.43220
Agg. assault	291	0.625	312.5	1.47920

N	DF	Chi-Sq	P-Value
500	3	6.52905	0.089

EXCEL

Significance level (%): 5	
Chi-square test:	
Chi-square (Observed value)	6.5291
Chi-square (Critical value)	7.8147
DF	3
p-value	0.0885
alpha	0.05

TI-83/84 PLUS

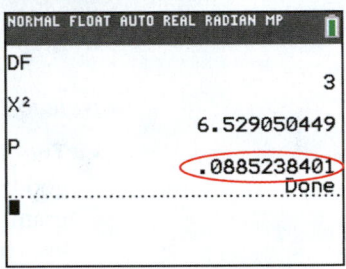

Using the **CHIGFT** program

As shown in Output 13.1, the *P*-value for the hypothesis test is 0.089. Because the *P*-value exceeds the specified significance level of 0.05, we do not reject H_0. At the 5% significance level, the data do not provide sufficient evidence to conclude that last year's violent-crime distribution differs from the 2010 distribution.

INSTRUCTIONS 13.1   Steps for generating Output 13.1

MINITAB

1 Store the violent-crime types, year-2010 relative frequencies, and observed frequencies in columns named CRIME, P, and O, respectively
2 Choose **Stat ➤ Tables ➤ Chi-Square Goodness-of-Fit Test (One Variable)...**
3 Press the F3 key to reset the dialog box
4 Specify O in the **Observed counts** text box
5 Specify CRIME in the **Category names (optional)** text box
6 Select the **Specific proportions** option button from the **Test** list
7 Specify P in the **Specific proportions** text box
8 Click the **Graphs...** button and uncheck all the check boxes
9 Click **OK** twice

EXCEL

1 Store the violent-crime types, year-2010 relative frequencies, and observed frequencies in columns named CRIME, P, and O, respectively
2 Choose **XLSTAT ➤ Parametric tests ➤ Multinomial goodness of fit test**
3 Click the reset button in the lower left corner of the dialog box
4 Click in the **Frequencies** selection box and then select the column of the worksheet that contains the O data
5 Click in the **Expected proportions** selection box and then select the column of the worksheet that contains the P data
6 Type **5** in the **Significance level (%)** text box
7 Click **OK**
8 Click the **Continue** button in the **XLSTAT – Selections** dialog box

(continued)

Exercises 13.2

Understanding the Concepts and Skills

13.9 Why is the phrase "goodness of fit" used to describe the type of hypothesis test considered in this section?

13.10 Are the observed frequencies variables? What about the expected frequencies? Explain your answers.

In each of Exercises 13.11–13.16, we have given the relative frequencies for the null hypothesis of a chi-square goodness-of-fit test and the sample size. In each case, decide whether Assumptions 1 and 2 for using that test are satisfied.

13.11 Sample size: $n = 100$.
Relative frequencies: 0.65, 0.30, 0.05.

13.12 Sample size: $n = 50$.
Relative frequencies: 0.65, 0.30, 0.05.

13.13 Sample size: $n = 50$.
Relative frequencies: 0.20, 0.20, 0.25, 0.30, 0.05.

13.14 Sample size: $n = 50$.
Relative frequencies: 0.22, 0.21, 0.25, 0.30, 0.02.

13.15 Sample size: $n = 50$.
Relative frequencies: 0.22, 0.22, 0.25, 0.30, 0.01.

13.16 Sample size: $n = 100$.
Relative frequencies: 0.44, 0.25, 0.30, 0.01.

13.17 Primary Heating Fuel. According to *Current Housing Reports*, published by the U.S. Census Bureau, the primary heating fuel for all occupied housing units is distributed as follows.

Primary heating fuel	Percentage
Utility gas	50.8
Fuel oil, kerosene	7.9
Electricity	33.9
Bottled, tank, or LPG	5.2
Wood and other fuel	1.9
None	0.3

Suppose that you want to determine whether the distribution of primary heating fuel for occupied housing units built after 2010 differs from that of all occupied housing units. To decide, you take a random sample of housing units built after 2010 and obtain a frequency distribution of their primary heating fuel.

a. Identify the population and variable under consideration here.

b. For each of the following sample sizes, determine whether conducting a chi-square goodness-of-fit test is appropriate and explain your answers: 200; 300; 400.

c. Strictly speaking, what is the smallest sample size for which conducting a chi-square goodness-of-fit test is appropriate?

In each of Exercises 13.18–13.23, we have provided a distribution and the observed frequencies of the values of a variable from a simple random sample of a population. In each case, use the chi-square goodness-of-fit test to decide, at the specified significance level, whether the distribution of the variable differs from the given distribution.

13.18 Distribution: 0.2, 0.4, 0.3, 0.1
Observed frequencies: 39, 78, 64, 19
Significance level = 0.05

13.19 Distribution: 0.2, 0.4, 0.3, 0.1
Observed frequencies: 85, 215, 130, 70
Significance level = 0.05

13.20 Distribution: 0.2, 0.1, 0.1, 0.3, 0.3
Observed frequencies: 29, 13, 5, 25, 28
Significance level = 0.10

13.21 Distribution: 0.2, 0.1, 0.1, 0.3, 0.3
Observed frequencies: 9, 7, 1, 12, 21
Significance level = 0.10

13.22 Distribution: 0.5, 0.3, 0.2
Observed frequencies: 45, 39, 16
Significance level = 0.01

13.23 Distribution: 0.5, 0.3, 0.2
Observed frequencies: 147, 115, 88
Significance level = 0.01

Applying the Concepts and Skills

In each of Exercises 13.24–13.33, apply the chi-square goodness-of-fit test, using either the critical-value approach or the P-value approach, to perform the required hypothesis test.

13.24 Population by Region. According to the U.S. Census Bureau publication *Demographic Profiles*, a relative-frequency distribution of the U.S. resident population by region in 2000 was as follows.

Region	Northeast	Midwest	South	West
Rel. freq.	0.190	0.229	0.356	0.225

A simple random sample of this year's U.S. residents gave the following frequency distribution.

Region	Northeast	Midwest	South	West
Frequency	259	306	568	367

a. Identify the population and variable under consideration here.

b. At the 5% significance level, do the data provide sufficient evidence to conclude that this year's resident population distribution by region has changed from the 2000 distribution?

13.25 Freshmen Politics. The Higher Education Research Institute of the University of California, Los Angeles, publishes information on characteristics of incoming college freshmen in *The American Freshman*. In 2000, 27.7% of incoming freshmen characterized their political views as liberal, 51.9% as moderate, and 20.4% as conservative. For this year, a random sample of 500 incoming college freshmen yielded the following frequency distribution for political views.

Political view	Frequency
Liberal	147
Moderate	237
Conservative	116

a. Identify the population and variable under consideration here.
b. At the 5% significance level, do the data provide sufficient evidence to conclude that this year's distribution of political views for incoming college freshmen has changed from the 2000 distribution?
c. Repeat part (b), using a significance level of 10%.

13.26 Road Rage. The report *Controlling Road Rage: A Literature Review and Pilot Study* was prepared for the AAA Foundation for Traffic Safety by D. Rathbone and J. Huckabee. The authors discussed the results of a literature review and pilot study on how to prevent aggressive driving and road rage. *Road rage* is defined as "...an incident in which an angry or impatient motorist or passenger intentionally injures or kills another motorist, passenger, or pedestrian, or attempts or threatens to injure or kill another motorist, passenger, or pedestrian." One aspect of the study was to investigate road rage as a function of the day of the week. The following table provides a frequency distribution for the days on which 69 road-rage incidents occurred.

Day	Frequency
Sunday	5
Monday	5
Tuesday	11
Wednesday	12
Thursday	11
Friday	18
Saturday	7

At the 5% significance level, do the data provide sufficient evidence to conclude that road-rage incidents are more likely to occur on some days than on others?

13.27 M&M Colors. Observing that the proportion of blue M&Ms in his bowl of candy appeared to be less than that of the other colors, R. Fricker, Jr., decided to compare the color distribution in randomly chosen bags of M&Ms to the theoretical distribution reported by M&M/MARS consumer affairs. Fricker published his findings in "The Mysterious Case of the Blue M&Ms" (*Chance*, Vol. 9(4), pp. 19–22). The following table gives the theoretical distribution.

Color	Percentage
Brown	30
Yellow	20
Red	20
Orange	10
Green	10
Blue	10

For his study, Fricker bought three bags of M&Ms from local stores and counted the number of each color. The average number of each color in the three bags was distributed as follows.

Color	Frequency
Brown	152
Yellow	114
Red	106
Orange	51
Green	43
Blue	43

Do the data provide sufficient evidence to conclude that the color distribution of M&Ms differs from that reported by M&M/MARS consumer affairs? Use $\alpha = 0.05$.

13.28 An Edge in Roulette? An American roulette wheel contains 18 red numbers, 18 black numbers, and 2 green numbers. The following table shows the frequency with which the ball landed on each color in 200 trials.

Number	Red	Black	Green
Frequency	88	102	10

At the 5% significance level, do the data suggest that the wheel is out of balance?

13.29 Loaded Die? A gambler thinks a die may be loaded, that is, that the six numbers are not equally likely. To test his suspicion, he rolled the die 150 times and obtained the data shown in the following table.

Number	1	2	3	4	5	6
Frequency	23	26	23	21	31	26

Do the data provide sufficient evidence to conclude that the die is loaded? Perform the hypothesis test at the 0.05 level of significance.

13.30 Bottled Water. A project exploring the bottled-water phenomenon and preference of water types was conducted by researchers M. Lunsford and A. Fink in the article "Water Taste Test Data" (*Journal of Statistics Education*, Vol. 18, No. 1). One hundred nine subjects participated in double-blind taste tests of three different bottled water brands (Fiji, Aquafina, and Sam's Choice) and tap water. Twelve people preferred the tap water, 27 Aquafina, 44 Fiji, and 26 Sam's Choice. At the 5% significance level, do the data provide sufficient evidence to conclude that the four different water types are not equally likely in preference?

13.31 World Series. The World Series in baseball is won by the first team to win four games (ignoring the 1903 and 1919–1921 World Series, when it was a best of nine). Thus it takes at least four games and no more than seven games to establish a winner. If two teams are evenly matched, the probabilities of the series lasting 4, 5, 6, or 7 games are as given in the second column of the following table. From the document *World Series History* on the Baseball Almanac website, as of November 2013, the actual numbers of times that the series lasted 4, 5, 6, or 7 games are as shown in the third column of the table.

Games	Probability	Actual
4	0.1250	21
5	0.2500	24
6	0.3125	24
7	0.3125	36

a. At the 5% significance level, do the data provide sufficient evidence to conclude that World Series teams are not evenly matched?

b. Repeat part (a) at the 10% significance level.

c. Discuss the appropriateness of using the chi-square goodness-of-fit test here.

13.32 Variegated Plants. *Arabidopsis* is a genus of flowering plant related to cabbage. A variegated mutant of the *Arabidopsis* has yellow streaks or marks. E. Miura et al. studied the origin of this variegated mutant in the article "The Balance between Protein Synthesis and Degradation in Chloroplasts Determines Leaf Variegation in *Arabidopsis yellow variegated* Mutants" (*The Plant Cell*, Vol. 19, No. 4, pp. 1313–1328). In a second-generation cross of variegated plants, 216 were variegated and 84 were normal. Genetics predicts that 75% of crossed variegated plants would be variegated and 25% would be normal. At the 10% significance level, do the data provide sufficient evidence to conclude that the second generation of crossed variegated plants does not follow the genetic predictions?

13.33 Girls and Boys. One probability model for child gender is that a boy or a girl is equally likely to be born. If that model is correct, then, for a two-child family, the probabilities are 0.25, 0.50, and 0.25 of two girls, one girl and one boy, and two boys, respectively. W. Stansfield and M. Carlton examined data collected in the *National Health Interview Study* on two-child families in the article "The Most Widely Published Gender Problem in Human Genetics" (*Human Biology*, Vol. 81, No. 1, pp. 3–11). Of 42,888 families with exactly two children, 9,523 had two girls, 22,031 had one girl and one boy, and 11,334 had two boys.

a. At the 1% significance level, do the data provide sufficient evidence to conclude that the distribution of genders in two-children families differs from the distribution predicted by the model described?

b. In view of your result from part (a), what conclusion can you draw?

Extending the Concepts and Skills

13.34 Table 13.4 on page 580 showed the calculated sums of the observed frequencies, the expected frequencies, and their differences. Strictly speaking, those sums are not needed. However, they serve as a check for computational errors.

a. In general, what common value should the sum of the observed frequencies and the sum of the expected frequencies equal? Explain your answer.

b. Fill in the blank. The sum of the differences between each observed and expected frequency should equal _____.

c. Suppose that you are conducting a chi-square goodness-of-fit test. If the sum of the expected frequencies does not equal the sample size, what do you conclude?

d. Suppose that you are conducting a chi-square goodness-of-fit test. If the sum of the expected frequencies equals the sample size, can you conclude that you made no error in calculating the expected frequencies? Explain your answer.

13.35 The chi-square goodness-of-fit test provides a method for performing a hypothesis test about the distribution of a variable that has c possible values. If the number of possible values is 2, that is, $c = 2$, the chi-square goodness-of-fit test is equivalent to a procedure that you studied earlier.

a. Which procedure is that? Explain your answer.

b. Suppose that you want to perform a hypothesis test to decide whether the proportion of a population that has a specified attribute is different from p_0. Discuss the method for performing such a test if you use (1) the one-proportion z-test (page 557) or (2) the chi-square goodness-of-fit test.

13.3   Contingency Tables; Association

Before we present our next chi-square procedure, we need to discuss two prerequisite concepts: *contingency tables* and *association*.

Contingency Tables

In Section 2.2, you learned how to group data from one variable into a frequency distribution. Data from one variable of a population are called **univariate data.**

Now, we show how to simultaneously group data from two variables into a frequency distribution. Data from two variables of a population are called **bivariate data,** and a frequency distribution for bivariate data is called a **contingency table** or **two-way table,** also known as a **cross-tabulation table** or **cross tabs.**

■■■   **EXAMPLE 13.5   Introducing Contingency Tables**

Political Party and Class Level In Example 2.5 on page 42, we considered univariate data on political party affiliation for the students in Professor Weiss's introductory statistics course. Now, we consider bivariate data on political party affiliation and class level for those same students, as shown in Table 13.7 (next page). Group these bivariate data into a contingency table.

TABLE 13.7

Political party affiliation and class level for students in introductory statistics

Student	Political party	Class level	Student	Political party	Class level
1	Democratic	Freshman	21	Democratic	Junior
2	Other	Junior	22	Democratic	Senior
3	Democratic	Senior	23	Republican	Freshman
4	Other	Sophomore	24	Democratic	Sophomore
5	Democratic	Sophomore	25	Democratic	Senior
6	Republican	Sophomore	26	Republican	Sophomore
7	Republican	Junior	27	Republican	Junior
8	Other	Freshman	28	Other	Junior
9	Other	Sophomore	29	Other	Junior
10	Republican	Sophomore	30	Democratic	Sophomore
11	Republican	Sophomore	31	Republican	Sophomore
12	Republican	Junior	32	Democratic	Junior
13	Republican	Sophomore	33	Republican	Junior
14	Democratic	Junior	34	Other	Senior
15	Republican	Sophomore	35	Other	Sophomore
16	Republican	Senior	36	Republican	Freshman
17	Democratic	Sophomore	37	Republican	Freshman
18	Democratic	Junior	38	Republican	Freshman
19	Other	Senior	39	Democratic	Junior
20	Republican	Sophomore	40	Republican	Senior

Solution   A contingency table must accommodate each possible pair of values for the two variables. The contingency table for these two variables has the form shown in Table 13.8. The small boxes inside the rectangle formed by the heavy lines are called **cells,** which hold the frequencies.

TABLE 13.8

Preliminary contingency table for political party affiliation and class level

Party		Freshman	Sophomore	Junior	Senior	Total
	Democratic	I	IIII	IHI	III	
	Republican	IIII	IHI III	IIII	II	
	Other	I	III	III	II	
	Total					

To complete the contingency table, we first go through the data in Table 13.7 and place a tally mark in the appropriate cell of Table 13.8 for each student. For instance, the first student is both a Democrat and a freshman, so this calls for a tally mark in the upper left cell of Table 13.8. The results of the tallying procedure are shown in Table 13.8. Replacing the tallies in Table 13.8 by the frequencies (counts of the tallies), we obtain the required contingency table, as shown in Table 13.9.

TABLE 13.9

Contingency table for political party affiliation and class level

Party		Freshman	Sophomore	Junior	Senior	Total
	Democratic	1	4	5	3	13
	Republican	4	8	4	2	18
	Other	1	3	3	2	9
	Total	6	15	12	7	40

The upper left cell of Table 13.9 shows that one student in the course is both a Democrat and a freshman. The cell diagonally below and to the right of that cell shows that eight students in the course are both Republicans and sophomores.

You try it!

Exercise 13.43(a)
on page 593

According to the first row total, 13 ($1 + 4 + 5 + 3$) of the students are Democrats. Similarly, the third column total shows that 12 of the students are juniors. The lower right corner gives the total number of students in the course, 40. You can find that total by summing the row totals, the column totals, or the frequencies in the 12 cells.

Grouping bivariate data into a contingency table by hand, as we did in Example 13.5, is a useful teaching tool. In practice, however, computers are almost always used to accomplish such tasks.

Association between Variables

Next, we need to discuss the concept of *association* between two variables. We do so for variables that are either categorical or quantitative with only finitely many possible values. Roughly speaking, two variables of a population are associated if knowing the value of one of the variables imparts information about the value of the other variable.

■■ ■ **EXAMPLE 13.6**   Introduces Association between Variables

Political Party and Class Level In Example 13.5, we presented data on political party affiliation and class level for the students in Professor Weiss's introductory statistics course. Consider those students a population of interest.

a. Find the distribution of political party affiliation within each class level.
b. Use the result of part (a) to decide whether the variables "political party affiliation" and "class level" are associated.
c. What would it mean if the variables "political party affiliation" and "class level" were not associated?
d. Explain how a *segmented bar graph* represents whether the variables "political party affiliation" and "class level" are associated.
e. Discuss another method for deciding whether the variables "political party affiliation" and "class level" are associated.

Solution

a. To obtain the distribution of political party affiliation within each class level, divide each entry in a column of the contingency table in Table 13.9 by its column total. Table 13.10 shows the results.

TABLE 13.10

Conditional distributions of political party affiliation by class level

		Class level				
		Freshman	Sophomore	Junior	Senior	**Total**
Party	Democratic	0.167	0.267	0.417	0.429	0.325
	Republican	0.667	0.533	0.333	0.286	0.450
	Other	0.167	0.200	0.250	0.286	0.225
	Total	1.000	1.000	1.000	1.000	1.000

The first column of Table 13.10 gives the distribution of political party affiliation for freshmen: 16.7% are Democrats, 66.7% are Republicans, and 16.7% are Other. This distribution is called the **conditional distribution** of the variable "political party affiliation" corresponding to the value "freshman" of the variable "class level"; or, more simply, the conditional distribution of political party affiliation for freshmen.

Similarly, the second, third, and fourth columns give the conditional distributions of political party affiliation for sophomores, juniors, and seniors, respectively. The "Total" column provides the (unconditional) distribution of political party affiliation for the entire population, which, in this context, is called the **marginal distribution** of the variable "political party affiliation." This distribution is the same as the one we found in Example 2.6 (Table 2.3 on page 44).

b. Table 13.10 reveals that the variables "political party affiliation" and "class level" are associated because knowing the value of the variable "class level" imparts information about the value of the variable "political party affiliation." For instance, as shown in Table 13.10, if we do not know the class level of a student in the course, there is a 32.5% chance that the student is a Democrat. If we know that the student is a junior, however, there is a 41.7% chance that the student is a Democrat.

c. If the variables "political party affiliation" and "class level" were not associated, the four conditional distributions of political party affiliation would be the same as each other and as the marginal distribution of political party affiliation; in other words, all five columns of Table 13.10 would be identical.

d. A **segmented bar graph** lets us visualize the concept of association. The first four bars of the segmented bar graph in Fig. 13.4 show the conditional distributions of political party affiliation for freshmen, sophomores, juniors, and seniors, respectively, and the fifth bar gives the marginal distribution of political party affiliation. This segmented bar graph is derived from Table 13.10.

FIGURE 13.4

Segmented bar graph for the conditional distributions and marginal distribution of political party affiliation

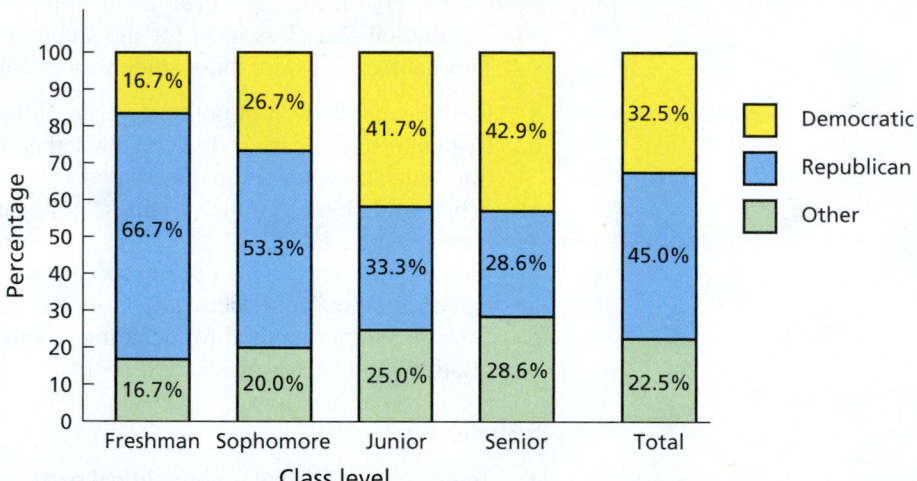

If political party affiliation and class level were not associated, the four bars displaying the conditional distributions of political party affiliation would be the same as each other and as the bar displaying the marginal distribution of political party affiliation; in other words, all five bars in Fig. 13.4 would be identical. That political party affiliation and class level are in fact associated is illustrated by the nonidentical bars.

e. Alternatively, we could decide whether the two variables are associated by obtaining the conditional distribution of class level within each political party affiliation. The conclusion regarding association (or nonassociation) will be the same, regardless of which variable's conditional distributions we examined.

StatCrunch

Report 13.3

You try it!

Exercise 13.43(b)–(d) on page 593

DEFINITION 13.1

? **What Does It Mean?**

Roughly speaking, two variables of a population are associated if knowing the value of one variable imparts information about the value of the other variable.

Association between Variables

We say that two variables of a population are **associated** (or that an **association** exists between the two variables) if the conditional distributions of one variable given the other are not identical.

Note: Two associated variables are also called **statistically dependent variables.** Similarly, two nonassociated variables are often called **statistically independent variables.**

In the preceding example, we illustrated how to determine whether two variables of a population are associated by simply comparing conditional distributions of one variable given the other—if those distributions are identical, the variables are not associated; otherwise, they are associated. This comparison method works only with population data, that is, when we have bivariate data for the entire population.

If we have bivariate data for only a sample of the population, then we must apply inferential methods to decide whether the two variables are associated. One such inferential method is discussed in the next section.

 ## THE TECHNOLOGY CENTER

Some statistical technologies have programs that automatically group bivariate data into a contingency table and also obtain conditional and marginal distributions. In this subsection, we present output and step-by-step instructions for such programs. (*Note to TI-83/84 Plus users:* At the time of this writing, the TI-83/84 Plus does not have a built-in program for conducting these analyses.)

EXAMPLE 13.7   Using Technology to Group Bivariate Data

Political Party and Class Level Table 13.7 on page 588 gives the political party affiliations and class levels for the students in Professor Weiss's introductory statistics course. Use Minitab or Excel to group these data into a contingency table.

Solution We applied the bivariate grouping programs to the data, resulting in Output 13.2. Steps for generating that output are presented in Instructions 13.2.

OUTPUT 13.2   Contingency table for political-party and class-level data

MINITAB

```
Tabulated Statistics: PARTY, CLASS

Rows: PARTY   Columns: CLASS

            Freshman   Junior   Senior   Sophomore   All

Democratic       1        5        3          4       13
Other            1        3        2          3        9
Republican       4        4        2          8       18
All              6       12        7         15       40

Cell Contents:        Count
```

EXCEL

Results for the variables PARTY and CLASS:

Observed frequencies (PARTY / CLASS):

	Freshman	Junior	Senior	Sophomore	Total
Democratic	1	5	3	4	13
Other	1	3	2	3	9
Republican	4	4	2	8	18
Total	6	12	7	15	40

INSTRUCTIONS 13.2   Steps for generating Output 13.2

MINITAB

1 Store the political-party and class-level data from Table 13.7 in columns named PARTY and CLASS, respectively
2 Choose **Stat ➤ Tables ➤ Cross Tabulation and Chi-Square...**
3 Press the F3 key to reset the dialog box
4 Specify PARTY in the **Rows** text box
5 Specify CLASS in the **Columns** text box
6 Click **OK**

EXCEL

1 Store the political-party and class-level data from Table 13.7 in columns named PARTY and CLASS, respectively
2 Choose **XLSTAT ➤ Correlation/Association tests ➤ Tests on contingency tables (Chi-square...)**
3 Click the reset button in the lower left corner of the dialog box
4 In the **Data format** list, select the **Qualitative variables** option button
5 Click in the **Row variable(s)** selection box and then select the column of the worksheet that contains the PARTY data
6 Click in the **Column variable(s)** selection box and then select the column of the worksheet that contains the CLASS data
7 Click the **Outputs** tab, uncheck all check boxes, and then check the **Observed frequencies** check box
8 Click **OK**
9 Click the **Continue** button in the **XLSTAT – Selections** dialog box

Compare Output 13.2 to Table 13.9 on page 588. (*Note to Minitab users:* By using **Column ➤ Value Order...**, available from the **Editor** menu when in the Worksheet window, you can order the rows and columns of the Minitab output to match that in Table 13.9.)

EXAMPLE 13.8 **Using Technology to Get Conditional and Marginal Distributions**

Political Party and Class Level Table 13.7 on page 588 gives the political party affiliations and class levels for the students in Professor Weiss's introductory statistics course. Use Minitab or Excel to determine the conditional distribution of party within each class level and the marginal distribution of party.

Solution We applied the appropriate programs to the data, resulting in Output 13.3. Steps for generating that output are presented in Instructions 13.3.

OUTPUT 13.3 Conditional distribution of party within each class level and marginal distribution of party

MINITAB

```
Tabulated Statistics: PARTY, CLASS

Rows: PARTY    Columns: CLASS

            Freshman   Junior   Senior   Sophomore    All

Democratic     16.67    41.67    42.86       26.67   32.50
Other          16.67    25.00    28.57       20.00   22.50
Republican     66.67    33.33    28.57       53.33   45.00
All           100.00   100.00   100.00      100.00  100.00

Cell Contents:      % of Column
```

EXCEL

Results for the variables PARTY and CLASS:

Proportions / Column (PARTY / CLASS):

	Freshman	Junior	Senior	Sophomore	Total
Democratic	0.1667	0.4167	0.4286	0.2667	0.3250
Other	0.1667	0.2500	0.2857	0.2000	0.2250
Republican	0.6667	0.3333	0.2857	0.5333	0.4500
Total	1	1	1	1	1

Compare Output 13.3 to Table 13.10 on page 589. (*Note to Minitab users:* By using **Column ➤ Value Order...**, available from the **Editor** menu when in the Worksheet window, you can order the rows and columns of the Minitab output to match that in Table 13.10.)

INSTRUCTIONS 13.3 Steps for generating Output 13.3

MINITAB

1. Store the political-party and class-level data from Table 13.7 in columns named PARTY and CLASS, respectively
2. Choose **Stat ➤ Tables ➤ Cross Tabulation and Chi-Square...**
3. Press the F3 key to reset the dialog box
4. Specify PARTY in the **Rows** text box
5. Specify CLASS in the **Columns** text box
6. In the **Display** list, check only the **Column percents** check box
7. Click **OK**

EXCEL

1. Store the political-party and class-level data from Table 13.7 in columns named PARTY and CLASS, respectively
2. Choose **XLSTAT ➤ Correlation/Association tests ➤ Tests on contingency tables (Chi-square...)**
3. Click the reset button in the lower left corner of the dialog box
4. In the **Data format** list, select the **Qualitative variables** option button
5. Click in the **Row variable(s)** selection box and then select the column of the worksheet that contains the PARTY data
6. Click in the **Column variable(s)** selection box and then select the column of the worksheet that contains the CLASS data
7. Click the **Outputs** tab, uncheck all check boxes, and then check the **Proportions / Column** check box
8. Click **OK**
9. Click the **Continue** button in the **XLSTAT – Selections** dialog box

Exercises 13.3

Understanding the Concepts and Skills

13.36 Identify the type of table that is used to group bivariate data.

13.37 What are the small boxes inside the heavy lines of a contingency table called?

13.38 Suppose that bivariate data are to be grouped into a contingency table. Determine the number of cells that the contingency table will have if the numbers of possible values for the two variables are
a. two and three.
b. four and three.
c. m and n.

13.39 Identify three ways in which the total number of observations of bivariate data can be obtained from the frequencies in a contingency table.

13.40 Presidential Election. According to *Dave Leip's Atlas of U.S. Presidential Elections*, in the 2012 presidential election, 51.01% of those voting voted for the Democratic candidate (Barack H. Obama), whereas 57.50% of those voting who lived in Illinois did so. For that presidential election, does an association exist between the variables "party of presidential candidate voted for" and "state of residence" for those who voted? Explain your answer.

13.41 Physician Specialty. According to the document *Physician Specialty Data Book*, published by the Association of American Medical Colleges, in 2010, 12.9% of active male physicians specialized in internal medicine and 15.3% of active female physicians specialized in internal medicine. Does an association exist between the variables "gender" and "specialty" for active physicians in 2010? Explain your answer.

Applying the Concepts and Skills

Table 13.11 provides data on gender, class level, and college for the students in one section of the course *Introduction to Computer Science* during one semester at Arizona State University. In the table, we use the abbreviations BUS for Business, ENG for Engineering and Applied Sciences, and LIB for Liberal Arts and Sciences.

TABLE 13.11

Gender, class level, and college for students in Introduction to Computer Science

Gender	Class	College	Gender	Class	College
M	Junior	ENG	F	Soph	BUS
M	Soph	ENG	F	Junior	ENG
F	Senior	BUS	M	Junior	LIB
F	Junior	BUS	F	Junior	BUS
M	Junior	ENG	M	Soph	BUS
F	Junior	LIB	M	Junior	BUS
M	Senior	LIB	M	Soph	ENG
M	Soph	ENG	M	Junior	ENG
M	Junior	ENG	M	Junior	ENG
M	Soph	ENG	M	Soph	LIB
F	Soph	BUS	F	Senior	ENG
F	Junior	BUS	F	Senior	BUS
M	Junior	ENG			

In Exercises 13.42–13.44, use the data in Table 13.11.

13.42 Gender and Class Level. Refer to Table 13.11. Consider the variables "gender" and "class level."
a. Group the bivariate data for these two variables into a contingency table.
b. Determine the conditional distribution of gender within each class level and the marginal distribution of gender.
c. Determine the conditional distribution of class level within each gender and the marginal distribution of class level.
d. Does an association exist between the variables "gender" and "class level" for this population? Explain your answer.

13.43 Gender and College. Refer to Table 13.11. Consider the variables "gender" and "college."
a. Group the bivariate data for these two variables into a contingency table.
b. Determine the conditional distribution of gender within each college and the marginal distribution of gender.
c. Determine the conditional distribution of college within each gender and the marginal distribution of college.
d. Does an association exist between the variables "gender" and "college" for this population? Explain your answer.

13.44 Class Level and College. Refer to Table 13.11. Consider the variables "class level" and "college."
a. Group the bivariate data for these two variables into a contingency table.
b. Determine the conditional distribution of class level within each college and the marginal distribution of class level.
c. Determine the conditional distribution of college within each class level and the marginal distribution of college.
d. Does an association exist between the variables "class level" and "college" for this population? Explain your answer.

Table 13.12 provides hypothetical data on political party affiliation and class level for the students in a night-school course.

TABLE 13.12

Political party affiliation and class level for the students in a night-school course (hypothetical data)

Party	Class	Party	Class	Party	Class
Rep	Jun	Rep	Soph	Rep	Jun
Dem	Soph	Other	Jun	Rep	Soph
Dem	Jun	Dem	Soph	Rep	Soph
Other	Jun	Rep	Soph	Rep	Fresh
Dem	Jun	Dem	Sen	Rep	Soph
Dem	Fresh	Rep	Jun	Rep	Jun
Dem	Soph	Dem	Jun	Rep	Sen
Dem	Sen	Dem	Jun	Rep	Jun
Other	Sen	Rep	Sen	Dem	Soph
Dem	Fresh	Rep	Fresh	Rep	Jun
Rep	Jun	Rep	Jun	Other	Jun
Rep	Jun	Dem	Jun	Dem	Jun
Dem	Sen	Rep	Sen	Other	Soph
Rep	Jun	Rep	Sen	Rep	Sen
Dem	Sen	Rep	Sen	Other	Soph
Rep	Jun	Dem	Soph	Rep	Soph
Rep	Soph	Other	Fresh	Other	Soph
Rep	Fresh	Rep	Soph	Other	Sen
Rep	Jun	Other	Jun	Rep	Soph
Dem	Soph	Dem	Jun	Dem	Jun

In Exercises 13.45 and 13.46, use the data in Table 13.12.

13.45 Party and Class. Refer to Table 13.12.

a. Group the bivariate data for the two variables into a contingency table.

b. Determine the conditional distribution of political party affiliation within each class level.

c. Are the variables "political party affiliation" and "class level" for this population of night-school students associated? Explain your answer.

d. Without doing any further calculation, determine the marginal distribution of political party affiliation.

e. Without doing further calculation, respond true or false to the following statement and explain your answer: "The conditional distributions of class level within political party affiliations are identical to each other and to the marginal distribution of class level."

13.46 Party and Class. Refer to Table 13.12.

a. If you have not done Exercise 13.45, group the bivariate data for the two variables into a contingency table.

b. Determine the conditional distribution of class level within each political party affiliation.

c. Are the variables "political party affiliation" and "class level" for this population of night-school students associated? Explain your answer.

d. Without doing any further calculation, determine the marginal distribution of class level.

e. Without doing further calculation, respond true or false to the following statement and explain your answer: "The conditional distributions of political party affiliation within class levels are identical to each other and to the marginal distribution of political party affiliation."

13.47 AIDS Cases. According to the Centers for Disease Control and Prevention publication *HIV Surveillance Report*, the number of AIDS cases in the United States in 2011, by region and race, is as shown in the following contingency table.

		Race			
		White	Black	Other	Total
Region	Northeast	1,100	2,493		5,117
	Midwest	1,137		504	
	South	2,761	7,848		
	West	1,700	764	1,766	4,230
	Total			6,052	25,435

a. How many cells does this contingency table have?

b. Fill in the missing entries.

c. What was the total number of AIDS cases in the United States in 2011?

d. How many AIDS cases were whites?

e. How many AIDS cases were Southerners?

f. How many AIDS cases were black Westerners?

13.48 School Enrollment. The *Current Population Survey* from the U.S. Census Bureau provides data on school enrollment. The following contingency table shows school enrollment below postsecondary, in thousands, by level and gender.

		Gender		
		Male	Female	Total
Level	Nursery	2,331	2,377	4,708
	Kindergarten		1,997	
	Elementary	16,504	15,734	
	High school	8,393		16,446
	Total	29,363		

a. How many cells does this contingency table have?

b. Fill in the missing entries.

c. What is the total enrollment?

d. How many students are female?

e. How many students are in high school?

f. How many students are male kindergarteners?

13.49 Education of Prisoners. In the article "Education and Correctional Populations" (*Bureau of Justice Statistics Special Report*, NCJ 195670), C. Harlow examined the educational attainment of prisoners by type of prison facility. The following contingency table was adapted from Table 1 of the article. Frequencies are in thousands, rounded to the nearest hundred.

		Prison facility			
		State	Federal	Local	Total
Educational attainment	8th grade or less	149.9	10.6	66.0	226.5
	Some high school	269.1	12.9	168.2	450.2
	GED	300.8	20.1	71.0	391.9
	High school diploma	216.4	24.0	130.4	370.8
	Postsecondary	95.0	14.0	51.9	160.9
	College grad or more	25.3	7.2	16.1	48.6
	Total	1056.5	88.8	503.6	1648.9

How many prisoners

a. are in state facilities?

b. have at least a college education?

c. are in federal facilities and have at most an 8th-grade education?

d. are in federal facilities or have at most an 8th-grade education?

e. in local facilities have a postsecondary educational attainment?

f. who have a postsecondary educational attainment are in local facilities?

g. are not in federal facilities?

13.50 U.S. Hospitals. The American Hospital Association publishes information about U.S. hospitals and nursing homes in *Hospital Statistics*. The following contingency table provides a cross-classification of U.S. hospitals and nursing homes by type of facility and number of beds.

Number of beds

		24 or fewer	25–74	75 or more	Total
Facility	General	260	1586	3557	5403
	Psychiatric	24	242	471	737
	Chronic	1	3	22	26
	Tuberculosis	0	2	2	4
	Other	25	177	208	410
	Total	310	2010	4260	6580

Tenure

		Owner	Renter	Total
Foundation	Full basement	22,987	3,301	26,288
	Partial basement	7,516	990	8,506
	Crawl space	14,289	3,477	17,766
	Concrete slab	21,013	5,651	26,664
	Other	949	334	1,283
	Total	66,754	13,753	80,507

In the following questions, the term *hospital* refers to either a hospital or nursing home.

a. How many hospitals have at least 75 beds?
b. How many hospitals are psychiatric facilities?
c. How many hospitals are psychiatric facilities with at least 75 beds?
d. How many hospitals either are psychiatric facilities or have at least 75 beds?
e. How many general facilities have between 25 and 74 beds?
f. How many hospitals with between 25 and 74 beds are chronic facilities?
g. How many hospitals have more than 24 beds?

13.51 Farms. The U.S. Department of Agriculture publishes information about U.S. farms in *Census of Agriculture*. A joint frequency distribution for number of farms, by acreage and tenure of operator, is provided in the following contingency table. Frequencies are in thousands.

Tenure of operator

		Full owner	Part owner	Tenant	Total
Acreage	Under 50		70	44	
	50–179	492	130	38	660
	180–499	198			368
	500–999	51	84	14	149
	1000 & over	41	114	17	172
	Total	1521	541		

a. Fill in the six missing entries.
b. How many cells does this contingency table have?
c. How many farms have under 50 acres?
d. How many farms are tenant operated?
e. How many farms are operated by part owners and have between 500 acres and 999 acres, inclusive?
f. How many farms are not full-owner operated?
g. How many tenant-operated farms have 180 acres or more?

13.52 Housing Foundations. The U.S. Census Bureau publishes information about housing units in *American Housing Survey for the United States*. The following table cross-classifies single-unit occupied housing units by foundation type and tenure of occupier. The frequencies are in thousands.

For single-unit occupied housing units:
a. How many have crawl spaces?
b. How many are owner occupied?
c. How many are rented and have full basements?
d. How many have a basement?
e. How many are either owner occupied or have a concrete slab?

13.53 AIDS Cases. Refer to Exercise 13.47. For AIDS cases in the United States in 2011, solve the following problems:
a. Find and interpret the conditional distributions of region by race.
b. Find and interpret the marginal distribution of region.
c. Are the variables "region" and "race" associated? Explain your answer.
d. What percentage of AIDS cases were in the South?
e. What percentage of AIDS cases among whites were in the South?
f. Without doing further calculations, respond true or false to the following statement and explain your answer: "The conditional distributions of race by region are not identical."
g. Find and interpret the marginal distribution of race and the conditional distributions of race by region.

13.54 School Enrollment. Refer to Exercise 13.48. For students below postsecondary, solve the following problems.
a. Determine the conditional distribution of level for each gender.
b. Determine the marginal distribution of level.
c. Are the variables "gender" and "level" associated? Explain your answer.
d. Find the percentage of students in nursery school.
e. Find the percentage of females in nursery school.
f. Without doing any further calculations, respond true or false to the following statement and explain your answer: "The conditional distributions of gender within levels are not identical."
g. Determine and interpret the marginal distribution of gender and the conditional distributions of gender within levels.

13.55 Education of Prisoners. Refer to Exercise 13.49.
a. Find the conditional distribution of educational attainment within each type of prison facility.
b. Does an association exist between educational attainment and type of prison facility for prisoners? Explain your answer.
c. Determine the marginal distribution of educational attainment for prisoners.
d. Construct a segmented bar graph for the conditional distributions of educational attainment and marginal distribution of educational attainment that you obtained in parts (a) and (c), respectively. Interpret the graph in light of your answer to part (b).
e. Without doing any further calculations, respond true or false to the following statement and explain your answer: "The conditional

distributions of facility type within educational attainment categories are identical."

f. Determine the marginal distribution of facility type and the conditional distributions of facility type within educational attainment categories.

g. Find the percentage of prisoners who are in federal facilities.

h. Find the percentage of prisoners with at most an 8th-grade education who are in federal facilities.

i. Find the percentage of prisoners in federal facilities who have at most an 8th-grade education.

13.56 U.S. Hospitals. Refer to Exercise 13.50.

a. Determine the conditional distribution of number of beds within each facility type.

b. Does an association exist between facility type and number of beds for U.S. hospitals? Explain your answer.

c. Determine the marginal distribution of number of beds for U.S. hospitals.

d. Construct a segmented bar graph for the conditional distributions and marginal distribution of number of beds. Interpret the graph in light of your answer to part (b).

e. Without doing any further calculations, respond true or false to the following statement and explain your answer: "The conditional distributions of facility type within number-of-beds categories are identical."

f. Determine the marginal distribution of facility type and the conditional distributions of facility type within number-of-beds categories.

g. What percentage of hospitals are general facilities?

h. What percentage of hospitals that have at least 75 beds are general facilities?

i. What percentage of general facilities have at least 75 beds?

Working with Large Data Sets

In each of Exercises 13.57–13.59, use the technology of your choice to solve the specified problems.

13.57 Governors. The National Governors Association publishes information on U.S. governors in *Governors' Political Affiliations & Terms of Office*. Based on that document, we obtained the data on region of residence and political party given on the WeissStats site.

a. Group the bivariate data for these two variables into a contingency table.

b. Determine the conditional distribution of region within each party and the marginal distribution of region.

c. Determine the conditional distribution of party within each region and the marginal distribution of party.

d. Are the variables "region" and "party" for U.S. governors associated? Explain your answer.

13.58 Motorcycle Accidents. The *Scottish Executive, Analytical Services Division Transport Statistics*, compiles information on motorcycle accidents in Scotland. During one year, data on the number of motorcycle accidents, by day of the week and type of road (built-up or non–built-up), are as presented on the WeissStats site.

a. Group the bivariate data for these two variables into a contingency table.

b. Determine the conditional distribution of day of the week within each type-of-road category and the marginal distribution of day of the week.

c. Determine the conditional distribution of type of road within each day of the week and the marginal distribution of type of road.

d. Does an association exist between the variables "day of the week" and "type of road" for these motorcycle accidents? Explain your answer.

13.59 Senators. The U.S. Congress, Joint Committee on Printing, provides information on the composition of Congress in *Congressional Directory*. On the WeissStats site, we present data on party and class for the senators in the 113th Congress.

a. Group the bivariate data for these two variables into a contingency table.

b. Determine the conditional distribution of party within each class and the marginal distribution of party.

c. Determine the conditional distribution of class within each party and the marginal distribution of class.

d. Are the variables "party" and "class" for U.S. senators in the 113th Congress associated? Explain your answer.

Extending the Concepts and Skills

13.60 In this exercise, you are to consider two variables, x and y, defined on a hypothetical population. Following are the conditional distributions of the variable y corresponding to each value of the variable x.

			x	
		A	B	C	Total
y	0	0.316	0.316	0.316	
	1	0.422	0.422	0.422	
	2	0.211	0.211	0.211	
	3	0.047	0.047	0.047	
	4	0.004	0.004	0.004	
	Total	1.000	1.000	1.000	

a. Are the variables x and y associated? Explain your answer.

b. Determine the marginal distribution of y.

c. Can you determine the marginal distribution of x? Explain your answer.

13.61 Age and Gender. The U.S. Census Bureau publishes census data on the resident population of the United States in *Current Population Survey*. According to that document, 7.3% of male residents are in the age group 20–24 years.

a. If no association exists between age group and gender, what percentage of the resident population would be in the age group 20–24 years? Explain your answer.

b. If no association exists between age group and gender, what percentage of female residents would be in the age group 20–24 years? Explain your answer.

c. There are about 158 million female residents of the United States. If no association exists between age group and gender, how many female residents would there be in the age group 20–24 years?

d. In fact, there are some 10.9 million female residents in the age group 20–24 years. Given this number and your answer to part (c), what do you conclude?

13.4 Chi-Square Independence Test

In Section 13.3, you learned how to determine whether an association exists between two variables of a population if you have the bivariate data for the entire population. However, because, in most cases, data for an entire population are not available, you must usually apply inferential methods to decide whether an association exists between two variables.

One of the most commonly used procedures for making such decisions is the **chi-square independence test.** In the next example, we introduce and explain the reasoning behind the chi-square independence test.

EXAMPLE 13.9 Introducing the Chi-Square Independence Test

Marital Status and Drinking A national survey was conducted to obtain information on the alcohol consumption patterns of U.S. adults by marital status. A random sample of 1772 residents 18 years old and older yielded the data displayed in Table 13.13.[†]

TABLE 13.13

Contingency table of marital status and alcohol consumption for 1772 randomly selected U.S. adults

	Drinks per month			
Marital status	Abstain	1–60	Over 60	Total
Single	67	213	74	354
Married	411	633	129	1173
Widowed	85	51	7	143
Divorced	27	60	15	102
Total	590	957	225	1772

Suppose we want to use the data in Table 13.13 to decide whether marital status and alcohol consumption are associated.

a. Formulate the problem statistically by posing it as a hypothesis test.
b. Explain the basic idea for carrying out the hypothesis test.
c. Develop a formula for computing the expected frequencies.
d. Construct a table that provides both the observed frequencies in Table 13.13 and the expected frequencies.
e. Discuss the details for making a decision concerning the hypothesis test.

Solution

a. For a chi-square independence test, the null hypothesis is that the two variables are not associated; the alternative hypothesis is that the two variables are associated. Thus, we want to perform the following hypothesis test.

H_0: Marital status and alcohol consumption are not associated.

H_a: Marital status and alcohol consumption are associated.

b. The idea behind the chi-square independence test is to compare the observed frequencies in Table 13.13 with the frequencies we would expect if the null hypothesis of nonassociation is true. The test statistic for making the comparison has the same form as the one used for the goodness-of-fit test: $\chi^2 = \Sigma(O - E)^2/E$, where O represents observed frequency and E represents expected frequency.

[†]Adapted from research by W. Clark and L. Midanik. In: National Institute on Alcohol Abuse and Alcoholism, *Alcohol Consumption and Related Problems: Alcohol and Health Monograph 1* (DHHS Pub. No. (ADM) 82–1190).

c. To develop a formula for computing the expected frequencies, consider, for instance, the cell of Table 13.13 corresponding to "Married *and* Abstain," the cell in the second row and first column. We note that the population proportion of all adults who abstain can be estimated by the sample proportion of the 1772 adults sampled who abstain, that is, by

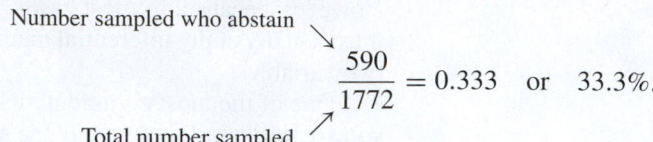

$$\underset{\text{Total number sampled}}{\overset{\text{Number sampled who abstain}}{\frac{590}{1772}}} = 0.333 \quad \text{or} \quad 33.3\%.$$

If no association exists between marital status and alcohol consumption (i.e., if H_0 is true), then the proportion of married adults who abstain is the same as the proportion of all adults who abstain. Therefore, of the 1173 married adults sampled, we would expect about

$$\frac{590}{1772} \cdot 1173 = 390.6$$

to abstain from alcohol.

Let's rewrite the left side of this expected-frequency computation in a slightly different way. By using algebra and referring to Table 13.13, we obtain

$$\text{Expected frequency} = \frac{590}{1772} \cdot 1173$$

$$= \frac{1173 \cdot 590}{1772}$$

$$= \frac{(\text{Row total}) \cdot (\text{Column total})}{\text{Sample size}}.$$

If we let R denote "Row total" and C denote "Column total," we can write this equation as

$$E = \frac{R \cdot C}{n}, \tag{13.1}$$

where, as usual, E denotes expected frequency and n denotes sample size.

d. Using Equation (13.1), we can calculate the expected frequencies for all the cells in Table 13.13. For the cell in the upper right corner of the table, we get

$$E = \frac{R \cdot C}{n} = \frac{354 \cdot 225}{1772} = 44.9.$$

In Table 13.14, we have modified Table 13.13 by including each expected frequency beneath the corresponding observed frequency. Table 13.14 shows,

TABLE 13.14

Observed and expected frequencies for marital status and alcohol consumption (expected frequencies printed below observed frequencies)

Marital status		Drinks per month			
		Abstain	1–60	Over 60	**Total**
	Single	67 117.9	213 191.2	74 44.9	354
	Married	411 390.6	633 633.5	129 148.9	1173
	Widowed	85 47.6	51 77.2	7 18.2	143
	Divorced	27 34.0	60 55.1	15 13.0	102
	Total	590	957	225	1772

for instance, that of the adults sampled, 74 were observed to be single and consume more than 60 drinks per month, whereas if marital status and alcohol consumption are not associated, the expected frequency is 44.9.

e. If the null hypothesis of nonassociation is true, the observed and expected frequencies should be approximately equal, which would result in a relatively small value of the test statistic, $\chi^2 = \Sigma(O - E)^2/E$. Consequently, if χ^2 is too large, we reject the null hypothesis and conclude that an association exists between marital status and alcohol consumption. From Table 13.14, we find that

$$\chi^2 = \Sigma(O - E)^2/E$$

$$= (67 - 117.9)^2/117.9 + (213 - 191.2)^2/191.2 + (74 - 44.9)^2/44.9$$

$$+ (411 - 390.6)^2/390.6 + (633 - 633.5)^2/633.5 + (129 - 148.9)^2/148.9$$

$$+ (85 - 47.6)^2/47.6 + (51 - 77.2)^2/77.2 + (7 - 18.2)^2/18.2$$

$$+ (27 - 34.0)^2/34.0 + (60 - 55.1)^2/55.1 + (15 - 13.0)^2/13.0$$

$$= 94.4.$$

Can this value be reasonably attributed to sampling error, or is it large enough to indicate that marital status and alcohol consumption are associated? Before we can answer that question, we must know the distribution of the χ^2-statistic.

First we present the formula for expected frequencies in a chi-square independence test, as discussed in the preceding example.

FORMULA 13.2

What Does It Mean?

To obtain an expected frequency, multiply the row total by the column total and divide by the sample size.

Expected Frequencies for an Independence Test

In a chi-square independence test, the expected frequency for each cell is found by using the formula

$$E = \frac{R \cdot C}{n},$$

where R is the row total, C is the column total, and n is the sample size.

Now we provide the distribution of the test statistic for a chi-square independence test.

KEY FACT 13.3

What Does It Mean?

To obtain a chi-square subtotal, square the difference between an observed and expected frequency and divide the result by the expected frequency. Adding the chi-square subtotals gives the χ^2-statistic, which has approximately a chi-square distribution.

Distribution of the χ^2-Statistic for a Chi-Square Independence Test

For a chi-square independence test, the test statistic

$$\chi^2 = \Sigma(O - E)^2/E$$

has approximately a chi-square distribution if the null hypothesis of non-association is true. The number of degrees of freedom is $(r - 1)(c - 1)$, where r and c are the number of possible values for the two variables under consideration.

Procedure for the Chi-Square Independence Test

In light of Key Fact 13.3, we present, in Procedure 13.2 (next page), a step-by-step method for conducting a chi-square independence test by using either the critical-value approach or the P-value approach. Because the null hypothesis is rejected only when the test statistic is too large, a chi-square independence test is always right tailed.

▪▪▪ PROCEDURE 13.2   Chi-Square Independence Test

Purpose To perform a hypothesis test to decide whether two variables are associated

Assumptions
1.   All expected frequencies are 1 or greater
2.   At most 20% of the expected frequencies are less than 5
3.   Simple random sample

Step 1   The null and alternative hypotheses are, respectively,

H_0: The two variables are not associated
H_a: The two variables are associated.

Step 2   Decide on the significance level, α.

Step 3   Compute the value of the test statistic

$$\chi^2 = \Sigma(O - E)^2/E,$$

where O and E represent observed and expected frequencies, respectively. Denote the value of the test statistic χ_0^2.

CRITICAL-VALUE APPROACH	OR	P-VALUE APPROACH

Step 4   The critical value is χ_α^2 with df $= (r-1)\times$ $(c-1)$, where r and c are the number of possible values for the two variables. Use Table VII to find the critical value.

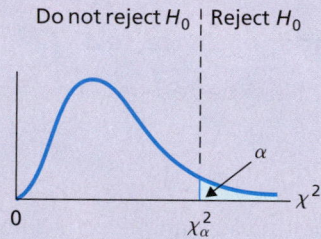

Step 5   If the value of the test statistic falls in the rejection region, reject H_0; otherwise, do not reject H_0.

Step 4   The χ^2-statistic has df $= (r-1)(c-1)$, where r and c are the number of possible values for the two variables. Use Table VII to estimate the P-value, or obtain it exactly by using technology.

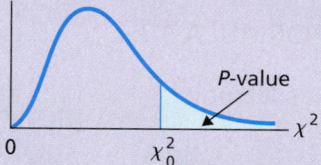

Step 5   If $P \le \alpha$, reject H_0; otherwise, do not reject H_0.

Step 6   Interpret the results of the hypothesis test.

Note: Regarding Assumptions 1 and 2, in many texts the rule given is that all expected frequencies be 5 or greater. However, research by the noted statistician W. G. Cochran shows that the "rule of 5" is too restrictive. See, for instance, W. G. Cochran, "Some Methods for Strengthening the Common χ^2 Tests" (*Biometrics*, Vol. 10, No. 4, pp. 417–451).

▪▪▪ EXAMPLE 13.10   The Chi-Square Independence Test

Marital Status and Drinking A random sample of 1772 U.S. adults yielded the data on marital status and alcohol consumption displayed in Table 13.13 on page 597. At the 5% significance level, do the data provide sufficient evidence to conclude that an association exists between marital status and alcohol consumption?

Solution   We calculated the expected frequencies earlier and displayed them in Table 13.14 (page 598) below the observed frequencies. From Table 13.14, we see that the expected-frequency conditions, Assumptions 1 and 2 of Procedure 13.2, are satisfied because all of the expected frequencies exceed 5. Consequently, we can apply Procedure 13.2 to perform the required hypothesis test.

Step 1 **State the null and alternative hypotheses.**

The null and alternative hypotheses are, respectively,

H_0: Marital status and alcohol consumption are not associated

H_a: Marital status and alcohol consumption are associated.

Step 2 **Decide on the significance level, α.**

The test is to be performed at the 5% significance level, so $\alpha = 0.05$.

Step 3 **Compute the value of the test statistic**

$$\chi^2 = \Sigma(O - E)^2/E,$$

where O and E represent observed and expected frequencies, respectively.

We did this earlier (see page 599) and found that $\chi^2 = 94.4$.

| CRITICAL-VALUE APPROACH | OR | P-VALUE APPROACH |

Step 4 The critical value is χ_α^2 with df $= (r - 1)\times (c - 1)$, where r and c are the number of possible values for the two variables. Use Table VII to find the critical value.

The number of marital status categories is four, and the number of drinks-per-month categories is three. Hence $r = 4, c = 3$, and

$$df = (r - 1)(c - 1) = 3 \cdot 2 = 6.$$

For $\alpha = 0.05$, Table VII reveals that the critical value is $\chi_{0.05}^2 = 12.592$, as shown in Fig. 13.5A.

FIGURE 13.5A

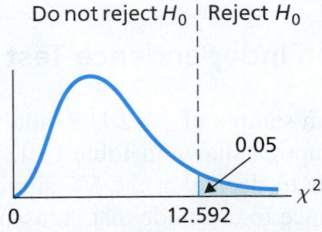

Step 5 If the value of the test statistic falls in the rejection region, reject H_0; otherwise, do not reject H_0.

From Step 3, we see that the value of the test statistic is $\chi^2 = 94.4$, which falls in the rejection region, as shown in Fig. 13.5A. Thus we reject H_0. The test results are statistically significant at the 5% level.

Step 4 The χ^2-statistic has df $= (r - 1)(c - 1)$, where r and c are the number of possible values for the two variables. Use Table VII to estimate the P-value, or obtain it exactly by using technology.

From Step 3, we see that the value of the test statistic is $\chi^2 = 94.4$. Because the test is right tailed, the P-value is the probability of observing a value of χ^2 of 94.4 or greater if the null hypothesis is true. That probability equals the shaded area shown in Fig. 13.5B.

FIGURE 13.5B

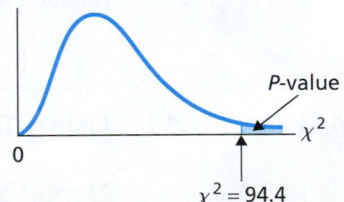

The number of marital status categories is four, and the number of drinks-per-month categories is three. Hence $r = 4, c = 3$, and

$$df = (r - 1)(c - 1) = 3 \cdot 2 = 6.$$

From Fig. 13.5B and Table VII with df $= 6$, we find that $P < 0.005$. (Using technology, we determined that $P = 0.000$ to three decimal places.)

Step 5 If $P \leq \alpha$, reject H_0; otherwise, do not reject H_0.

From Step 4, $P < 0.005$. Because the P-value is less than the specified significance level of 0.05, we reject H_0. The test results are statistically significant at the 5% level and (see Table 9.8 on page 386) provide very strong evidence against the null hypothesis.

StatCrunch
Report 13.4

Exercise 13.79
on page 604

Step 6 Interpret the results of the hypothesis test.

Interpretation At the 5% significance level, the data provide sufficient evidence to conclude that marital status and alcohol consumption are associated.

Concerning the Assumptions

In Procedure 13.2, we made two assumptions about expected frequencies:

1. All expected frequencies are 1 or greater.
2. At most 20% of the expected frequencies are less than 5.

What can we do if one or both of these assumptions are violated? Three approaches are possible. We can combine rows or columns to increase the expected frequencies in those cells in which they are too small; we can eliminate certain rows or columns in which the small expected frequencies occur; or we can increase the sample size.

What Does It Mean?

Association does not imply causation!

Association and Causation

Two variables may be associated without being causally related. In Example 13.10, we concluded that the variables marital status and alcohol consumption are associated. This result means that knowing the marital status of a person imparts information about the alcohol consumption of that person, and vice versa. It does not necessarily mean, however, for instance, that being single *causes* a person to drink more.

Although we must keep in mind that association does not imply causation, we must also note that, if two variables are not associated, there is no point in looking for a causal relationship. In other words, association is a necessary but not sufficient condition for causation.

THE TECHNOLOGY CENTER

Most statistical technologies have programs that automatically perform a chi-square independence test. In this subsection, we present output and step-by-step instructions for such programs.

EXAMPLE 13.11   Using Technology to Perform an Independence Test

Marital Status and Drinking A random sample of 1772 U.S. adults yielded the data on marital status and alcohol consumption shown in Table 13.13 on page 597. Use Minitab, Excel, or the TI-83/84 Plus to decide, at the 5% significance level, whether the data provide sufficient evidence to conclude that an association exists between marital status and alcohol consumption.

Solution We want to perform the hypothesis test

H_0: Marital status and alcohol consumption are not associated

H_a: Marital status and alcohol consumption are associated

at the 5% significance level.

We applied the chi-square independence test programs to the data, resulting in Output 13.4. Steps for generating that output are presented in Instructions 13.4. *Note to Excel users:* For brevity, we have presented only the essential portions of the actual output.

As shown in Output 13.4, the *P*-value for the hypothesis test is 0.000 to three decimal places. Because the *P*-value is less than the specified significance level of 0.05, we reject H_0. At the 5% significance level, the data provide sufficient evidence to conclude that marital status and alcohol consumption are associated.

OUTPUT 13.4
Chi-square independence test on the data on marital status and alcohol consumption

MINITAB

Chi-Square Test for Association: MARITAL STATUS, DRINKS PER MONTH

Rows: MARITAL STATUS   Columns: DRINKS PER MONTH

	Abstain	1-60	Over 60	All
Single	67	213	74	354
	117.87	191.18	44.95	
Married	411	633	129	1173
	390.56	633.50	148.94	
Widowed	85	51	7	143
	47.61	77.23	18.16	
Divorced	27	60	15	102
	33.96	55.09	12.95	
All	590	957	225	1772

Cell Contents: Count
                    Expected count

Pearson Chi-Square = 94.269, DF = 6, P-Value = 0.000
Likelihood Ratio Chi-Square = 93.096, DF = 6, P-Value = 0.000

EXCEL

Test of independence between the rows and the columns (Chi-square):	
Chi-square (Observed value)	94.2688
Chi-square (Critical value)	12.5916
DF	6
p-value	< 0.0001
alpha	0.05

TI-83/84 PLUS

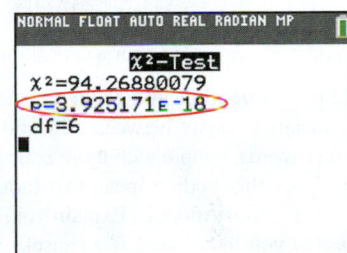

NORMAL FLOAT AUTO REAL RADIAN MP

χ^2-Test
χ^2=94.26880079
p=3.925171E-18
df=6

INSTRUCTIONS 13.4   Steps for generating Output 13.4

MINITAB

1. Store the marital-status categories from Table 13.13 in a column named MARITAL STATUS
2. Store the cell data from Table 13.13 in columns named Abstain, 1–60, and Over 60
3. Choose **Stat ➤ Tables ➤ Chi-Square Test for Association...**
4. Press the F3 key to reset the dialog box
5. Select **Summarized data in a two-way table** from the drop-down list box
6. Specify Abstain, '1–60', and 'Over 60' in the **Columns containing the table** text box
7. Specify 'MARITAL STATUS' in the **Rows** text box
8. Type DRINKS PER MONTH in the **Columns** text box
9. Click **OK**

EXCEL

1. Store the marital-status categories from Table 13.13 in a column named MARITAL STATUS
2. Directly to the right of the column entered at step 1, store the cell data from Table 13.13 in columns named Abstain, 1–60, and Over 60
3. Choose **XLSTAT ➤ Correlation/Association tests ➤ Tests on contingency tables (Chi-square...)**

4. Click the reset button in the lower left corner of the dialog box
5. Click in the **Contingency table** selection box and then select the columns of the worksheet that contain the data stored in steps 1 and 2
6. Click the **Options** tab
7. Check the **Chi-square test** check box
8. Type 5 in the **Significance level (%)** text box
9. Click the **Outputs** tab and then uncheck all check boxes
10. Click **OK**
11. Click the **Continue** button in the **XLSTAT – Selections** dialog box

TI-83/84 PLUS

1. Press **2nd ➤ MATRIX**, arrow over to **EDIT**, and press **1**
2. Type 4 and press **ENTER**
3. Type 3 and press **ENTER**
4. Enter the cell data from Table 13.13, pressing **ENTER** after each entry
5. Press **STAT**, arrow over to **TESTS**, and press **ALPHA ➤ C**
6. Press **2nd ➤ MATRIX**, press **1**, and press **ENTER**
7. Press **2nd ➤ MATRIX**, press **2**, and press **ENTER**
8. Arrow down to **Calculate**, and press **ENTER**

Note to Excel users: You can obtain results in addition to those shown in Output 13.4 by checking boxes in the **Outputs** tab as desired. For instance, to get tables of the observed and expected frequencies, check the **Observed frequencies** and **Theoretical frequencies** check boxes, respectively.

Exercises 13.4

Understanding the Concepts and Skills

13.62 To decide whether two variables of a population are associated, we usually need to resort to inferential methods such as the chi-square independence test. Why?

13.63 Step 1 of Procedure 13.2 gives generic statements for the null and alternative hypotheses of a chi-square independence test. Use the terms *statistically dependent* and *statistically independent,* introduced on page 590, to restate those hypotheses.

13.64 In Example 13.9, we made the following statement: If no association exists between marital status and alcohol consumption, the proportion of married adults who abstain is the same as the proportion of all adults who abstain. Explain why that statement is true.

13.65 A chi-square independence test is to be conducted to decide whether an association exists between two variables of a population. One variable has six possible values, and the other variable has four. What is the degrees of freedom for the χ^2-statistic?

13.66 We stated earlier that, if two variables are not associated, there is no point in looking for a causal relationship. Why is that so?

13.67 **Education and Salary.** Studies have shown that a positive association exists between educational level and annual salary; in other words, people with more education tend to make more money.
a. Does this finding mean that more education *causes* a person to make more money? Explain your answer.
b. Do you think there is a causal relationship between educational level and annual salary? Explain your answer.

13.68 Identify three techniques that can be tried as a remedy when one or more of the expected-frequency assumptions for a chi-square independence test are violated.

In each of Exercises 13.69–13.74, we have given the number of possible values for two variables of a population. For each exercise, determine the maximum number of expected frequencies that can be less than 5 in order that Assumption 2 of Procedure 13.2 on page 600 be satisfied. Note: The number of cells for a contingency table with m rows and n columns is m · n.

13.69 four and five

13.70 five and three

13.71 two and three

13.72 four and three

13.73 two and two

13.74 six and seven

In each of Exercises 13.75–13.78, we have presented a contingency table that gives a cross-classification of a random sample of values for two variables, x and y, of a population. For each exercise, perform the following tasks.
a. *Find the expected frequencies. Note: You will first need to compute the row totals, column totals, and grand total.*
b. *Determine the value of the chi-square statistic.*
c. *Decide at the 5% significance level whether the data provide sufficient evidence to conclude that the two variables are associated.*

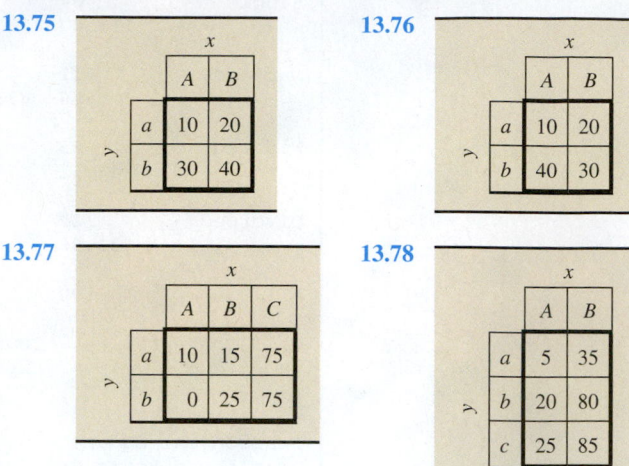

13.75

	x	
	A	B
y a	10	20
b	30	40

13.76

	x	
	A	B
y a	10	20
b	40	30

13.77

	x		
	A	B	C
y a	10	15	75
b	0	25	75

13.78

	x	
	A	B
y a	5	35
b	20	80
c	25	85

Applying the Concepts and Skills

In Exercises 13.79–13.86, use either the critical-value approach or the P-value approach to perform a chi-square independence test, provided the conditions for using the test are met.

13.79 **Siskel and Ebert.** In the classic TV show *Sneak Previews*, originally hosted by the late Gene Siskel and Roger Ebert, the two Chicago movie critics reviewed the week's new movie releases and then rated them thumbs up (positive), mixed, or thumbs down (negative). These two critics often saw the merits of a movie differently. In general, however, were the ratings given by Siskel and Ebert associated? The answer to this question was the focus of the paper "Evaluating Agreement and Disagreement Among Movie Reviewers" by A. Agresti and L. Winner that appeared in *Chance* (Vol. 10(2), pp. 10–14). The following contingency table summarizes the ratings by Siskel and Ebert for 160 movies.

		Ebert's rating			
		Thumbs down	Mixed	Thumbs up	Total
Siskel's rating	Thumbs down	24	8	13	45
	Mixed	8	13	11	32
	Thumbs up	10	9	64	83
	Total	42	30	88	160

At the 1% significance level, do the data provide sufficient evidence to conclude that an association exists between the ratings of Siskel and Ebert?

13.80 **Diabetes in Native Americans.** Preventable chronic diseases are increasing rapidly in Native American populations, particularly diabetes. F. Gilliland et al. examined the diabetes issue in the paper

"Preventative Health Care among Rural American Indians in New Mexico" (*Preventative Medicine*, Vol. 28, pp. 194–202). Following is a contingency table showing cross-classification of educational attainment and diabetic state for a sample of 1273 Native Americans (HS is high school).

	Diabetic state		
	Diabetes	No diabetes	**Total**
Less than HS	33	218	251
HS grad	25	389	414
Some college	20	393	413
College grad	17	178	195
Total	95	1178	1273

(Left label: **Education**)

At the 1% significance level, do the data provide sufficient evidence to conclude that an association exists between educational level and diabetic state for Native Americans?

13.81 Learning at Home. M. Stuart et al. studied various aspects of grade-school children and their mothers and reported their findings in the article "Learning to Read at Home and at School" (*British Journal of Educational Psychology*, 68(1), pp. 3–14). The researchers gave a questionnaire to parents of 66 children in kindergarten through second grade. Two social-class groups, middle and working, were identified based on the mother's occupation.

a. One of the questions dealt with the children's knowledge of nursery rhymes. The following data were obtained.

	Nursery-rhyme knowledge		
	A few	Some	Lots
Middle	4	13	15
Working	5	11	18

(Left label: **Social class**)

Are Assumptions 1 and 2 satisfied for a chi-square independence test? If so, conduct the test at the 1% significance level and interpret your results.

b. Another question dealt with whether the parents played "I Spy" games with their children. The following data were obtained.

	Frequency of games		
	Never	Sometimes	Often
Middle	2	8	22
Working	11	10	13

(Left label: **Social class**)

Are Assumptions 1 and 2 satisfied for a chi-square independence test? If so, conduct the test at the 1% significance level and interpret your results.

13.82 Deceptive News. In the article "When News Reporters Deceive: The Production of Stereotypes" (*Journalism & Mass Communication Quarterly*, Vol. 84, No. 2, pp. 281–298), researchers D. La-

sorsa and J. Dai investigated the relationship between authenticity and tone for news stories. Tone was measured by using a method of coding sentences and was categorized as positive, neutral, or negative. A sample of stories yielded the following data.

	Authenticity		
	Authentic	Deceptive	**Total**
Negative	59	111	170
Neutral	49	61	110
Positive	20	11	31
Total	128	183	311

(Left label: **Tone**)

At the 5% significance level, do the data provide sufficient evidence to conclude that an association exists between authenticity and tone?

13.83 Fear of Gangs. In the article "Growing Pains and Fear of Gangs" (*Applied Psychology in Criminal Justice*, Vol. 5, No. 2, pp. 139–164), B. Brown and W. Benedict examined the relationship between worry about a gang attack and actually being a victim of a gang attack. Interviews of a sample of high school students yielded the following contingency table.

	Worry		
	Yes	Not	**Total**
Yes	18	21	39
No	22	152	174
Total	40	173	213

(Left label: **Victim**)

At the 1% significance level, do the data provide sufficient evidence to conclude that an association exists between worry about a gang attack and actually being a victim of a gang attack?

13.84 HPV Vaccine. In the article "Correlates for Completion of 3-dose Regimen of HPV Vaccine in Female Members of a Managed Care Organization" (*Mayo Clinical Proceedings*, Vol. 84, pp. 864–870), C. Chao et al. examined factors that may influence whether young female patients complete a three-injection sequence of the Gardasil quadrivalent human papillomavirus vaccine (HPV4). HPV is a virus that has been linked to the development of cervical cancer. The following contingency table summarizes the data obtained for completion of treatment versus practice type.

	Completion		
	Yes	No	**Total**
Pediatric	162	353	515
Family	106	259	365
OB/GYN	201	332	533
Total	469	944	1413

(Left label: **Practice**)

a. At the 5% significance level, do the data provide sufficient evidence to conclude that an association exists between completion of treatment and practice type?
b. Repeat part (a) at the 1% significance level.

13.85 Religion. A worldwide poll on religion was conducted by WIN-Gallup International and published as the document *Global Index of Religiosity and Atheism*. One question involved religious belief and educational attainment. The following data is based on the answers to that question.

Religiosity	Basic (Education)	Secondary	Advanced	Total
Religious	77	149	78	304
Not religious	23	56	36	115
Atheist	8	24	29	61
Don't know	6	15	8	29
Total	114	244	151	509

a. At the 5% significance level, do the data provide sufficient evidence to conclude that an association exists between religiosity and education?
b. Repeat part (a) at the 1% significance level.

13.86 BMD and Depression. In the paper "Depression and Bone Mineral Density: Is There a Relationship in Elderly Asian Men?" (*Osteoporosis International*, Vol. 16, pp. 610–615), S. Wong et al. published results of their study on bone mineral density (BMD) and depression for 1999 Hong Kong men aged 65 to 92 years. Here are the cross-classified data.

BMD	Depressed (Depression)	Not depressed	Total
Osteoporitic	3	35	38
Low BMD	69	533	602
Normal	97	1262	1359
Total	169	1830	1999

At the 1% significance level, do the data provide sufficient evidence to conclude that BMD and depression are statistically dependent for elderly Asian men?

Job Satisfaction. A *CNN/USA TODAY* poll conducted by Gallup asked a sample of employed Americans the following question: "Which do you enjoy more, the hours when you are on your job, or the hours when you are not on your job?" The responses to this question were cross-tabulated against several characteristics, among which were gender, age, type of community, educational attainment, income, and type of employer. The data are provided on the WeissStats site. In each of Exercises 13.87–13.92, use the technology of your choice to decide, at the 5% significance level, whether an association exists between the specified pair of variables.

13.87 gender and response (to the question)

13.88 age and response

13.89 type of community and response

13.90 educational attainment and response

13.91 income and response

13.92 type of employer and response

13.5 Chi-Square Homogeneity Test

The purpose of a **chi-square homogeneity test** is to compare the distributions of a variable of two or more populations. As a special case, it can be used to decide whether a difference exists among two or more population proportions.

For a chi-square homogeneity test, the null hypothesis is that the distributions of the variable are the same for all the populations, and the alternative hypothesis is that the distributions of the variable are not all the same (i.e., the distributions differ for at least two of the populations).

When the populations under consideration have the same distribution for a variable, they are said to be **homogeneous** with respect to the variable; otherwise, they are said to be **nonhomogeneous** with respect to the variable. Using this terminology, we can state the null and alternative hypotheses for a chi-square homogeneity test simply as follows:

H_0: The populations are homogeneous with respect to the variable
H_a: The populations are nonhomogeneous with respect to the variable.

The assumptions for use of the chi-square homogeneity test are simple random samples, independent samples, and the same two expected-frequency assumptions required for performing a chi-square independence test.

Although the context of and assumptions for the chi-square homogeneity test differ from those of the chi-square independence test, the steps for carrying out the two tests are the same. In particular, the test statistics for the two tests are identical.

As with a chi-square independence test, the observed frequencies for a chi-square homogeneity test are arranged in a contingency table. Moreover, the expected frequencies are computed in the same way.

FORMULA 13.3

What Does It Mean?

To obtain an expected frequency, multiply the row total by the column total and divide by the sample size

Expected Frequencies for a Homogeneity Test

In a chi-square homogeneity test, the expected frequency for each cell is found by using the formula

$$E = \frac{R \cdot C}{n},$$

where R is the row total, C is the column total, and n is the sample size.

The distribution of the test statistic for a chi-square homogeneity test is presented in Key Fact 13.4.

KEY FACT 13.4

What Does It Mean?

To obtain a chi-square subtotal, square the difference between an observed and expected frequency and divide the result by the expected frequency. Adding the chi-square subtotals gives the χ^2-statistic, which has approximately a chi-square distribution.

Distribution of the χ^2-Statistic for a Chi-Square Homogeneity Test

For a chi-square homogeneity test, the test statistic

$$\chi^2 = \Sigma(O - E)^2 / E$$

has approximately a chi-square distribution if the null hypothesis of homogeneity is true. The number of degrees of freedom is $(r - 1)(c - 1)$, where r is the number of populations and c is the number of possible values for the variable under consideration.

Procedure for the Chi-Square Homogeneity Test

In light of Key Fact 13.4, we present, in Procedure 13.3 (next page), a step-by-step method for conducting a chi-square homogeneity test by using either the critical-value approach or the P-value approach. Because the null hypothesis is rejected only when the test statistic is too large, a chi-square homogeneity test is always right tailed.

◼◼◼ **EXAMPLE 13.12**   ## The Chi-Square Homogeneity Test

Region and Educational Attainment The U.S. Census Bureau compiles data on the resident population by region and educational attainment. Results are published in *Current Population Survey*. Independent simple random samples of (adult) residents in the four U.S. regions gave the following data on educational attainment (HS is high school; Assoc's is Associate's). At the 5% significance level, do the data provide sufficient evidence to conclude that a difference exists in educational-attainment distributions among residents of the four U.S. regions?

TABLE 13.15

Sample data for educational attainment in the four U.S. regions

		Educational attainment						
		Not HS grad	HS grad	Some college	Assoc's degree	Bachelor's degree	Advanced degree	**Total**
Region	Northeast	7	13	7	4	10	6	47
	Midwest	5	18	13	6	9	4	55
	South	11	30	14	7	19	10	91
	West	8	16	13	2	10	8	57
	Total	31	77	47	19	48	28	250

Solution We first calculate the expected frequencies by using Formula 13.3. Doing so, we obtain Table 13.16 shown at the bottom of the next page, which displays the expected frequencies below the observed frequencies from Table 13.15.

■ ■ ■ ■ **PROCEDURE 13.3** Chi-Square Homogeneity Test

Purpose To perform a hypothesis test to compare the distributions of a variable of two or more populations

Assumptions

1. All expected frequencies are 1 or greater
2. At most 20% of the expected frequencies are less than 5
3. Simple random samples
4. Independent samples

Step 1 The null and alternative hypotheses are, respectively,

H_0: The populations are homogeneous with respect to the variable

H_a: The populations are nonhomogeneous with respect to the variable.

Step 2 Decide on the significance level, α.

Step 3 Compute the value of the test statistic

$$\chi^2 = \Sigma(O - E)^2/E,$$

where O and E represent observed and expected frequencies, respectively. Denote the value of the test statistic χ_0^2.

CRITICAL-VALUE APPROACH	OR	P-VALUE APPROACH

Step 4 The critical value is χ_α^2 with df $= (r - 1) \times (c - 1)$, where r is the number of populations and c is the number of possible values for the variable. Use Table VII to find the critical value.

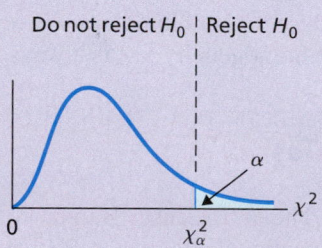

Do not reject H_0 | Reject H_0

Step 5 If the value of the test statistic falls in the rejection region, reject H_0; otherwise, do not reject H_0.

Step 4 The χ^2-statistic has df $= (r - 1)(c - 1)$, where r is the number of populations and c is the number of possible values for the variable. Use Table VII to estimate the P-value, or obtain it exactly by using technology.

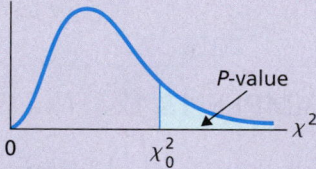

P-value

Step 5 If $P \le \alpha$, reject H_0; otherwise, do not reject H_0.

Step 6 Interpret the results of the hypothesis test.

TABLE 13.16

Observed and expected frequencies for the data in Table 13.15

		Educational attainment						
		Not HS grad	HS grad	Some college	Assoc's degree	Bachelor's degree	Advanced degree	Total
Region	Northeast	7 5.8	13 14.5	7 8.8	4 3.6	10 9.0	6 5.3	47
	Midwest	5 6.8	18 16.9	13 10.3	6 4.2	9 10.6	4 6.2	55
	South	11 11.3	30 28.0	14 17.1	7 6.9	19 17.5	10 10.2	91
	West	8 7.1	16 17.6	13 10.7	2 4.3	10 10.9	8 6.4	57
	Total	31	77	47	19	48	28	250

We see from Table 13.16 that all of the expected frequencies are 1 or greater; hence, Assumption 1 of Procedure 13.3 is satisfied. We also see from Table 13.16 that three of the expected frequencies are less than 5. Noting that there are 24 cells, we conclude that 3/24, or 12.5%, of the expected frequencies are less than 5; hence, Assumption 2 of Procedure 13.3 is satisfied. Consequently, we can apply Procedure 13.3 to perform the required hypothesis test.

Step 1 State the null and alternative hypotheses.

The null and alternative hypotheses are, respectively,

H_0: The residents of the four U.S. regions are homogeneous with respect to educational attainment

H_a: The residents of the four U.S. regions are nonhomogeneous with respect to educational attainment.

Step 2 Decide on the significance level, α.

We are to perform the test at the 5% significance level, so $\alpha = 0.05$.

Step 3 Compute the value of the test statistic

$$\chi^2 = \Sigma(O - E)^2/E,$$

where O and E represent observed and expected frequencies, respectively.

The observed and expected frequencies are displayed in Table 13.16. Using them, we compute the value of the test statistic:

$$\chi^2 = (7 - 5.8)^2/5.8 + (13 - 14.5)^2/14.5 + \cdots + (8 - 6.4)^2/6.4 = 7.386.$$

CRITICAL-VALUE APPROACH	OR	P-VALUE APPROACH

Step 4 The critical value is χ^2_α with df $= (r - 1) \times (c - 1)$, where r is the number of populations and c is the number of possible values for the variable. Use Table VII to find the critical value.

The populations are the residents of the four U.S. regions; hence, $r = 4$. The variable has six possible values, namely, the six educational-attainment categories; hence, $c = 6$. Consequently, we have

$$df = (r - 1)(c - 1) = 3 \cdot 5 = 15.$$

For $\alpha = 0.05$, Table VII reveals that the critical value is $\chi^2_{0.05} = 24.996$, as shown in Fig. 13.6A.

FIGURE 13.6A

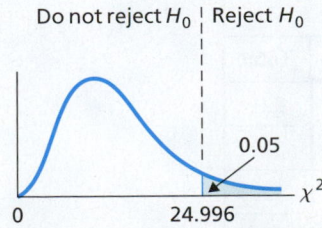

Do not reject H_0 | Reject H_0

0.05

0   24.996 χ^2

Step 4 The χ^2-statistic has df $= (r - 1)(c - 1)$, where r is the number of populations and c is the number of possible values for the variable. Use Table VII to estimate the P-value, or obtain it exactly by using technology.

From Step 3, we see that the value of the test statistic is $\chi^2 = 7.386$. Because the test is right tailed, the P-value is the probability of observing a value of χ^2 of 7.386 or greater if the null hypothesis is true. That probability equals the shaded area shown in Fig. 13.6B.

FIGURE 13.6B

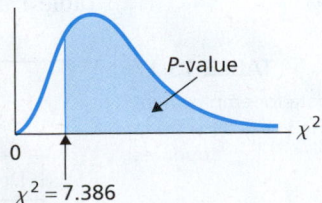

P-value

0

$\chi^2 = 7.386$

The populations are the residents of the four U.S. regions; hence, $r = 4$. The variable has six possible values, namely, the six educational-attainment categories; hence, $c = 6$. Consequently, we have

$$df = (r - 1)(c - 1) = 3 \cdot 5 = 15$$

From Fig. 13.6B and Table VII with df $= 15$, we find that $0.90 < P < 0.95$. (Using technology, $P = 0.946$.)

CRITICAL-VALUE APPROACH	OR	P-VALUE APPROACH

Step 5 If the value of the test statistic falls in the rejection region, reject H_0; otherwise, do not reject H_0.

The value of the test statistic is $\chi^2 = 7.386$, as found in Step 3, which does not fall in the rejection region shown in Fig. 13.6A. Thus we do not reject H_0. The test results are not statistically significant at the 5% level.

Step 5 If $P \leq \alpha$, reject H_0; otherwise, do not reject H_0.

From Step 4, $0.90 < P < 0.95$. Because the P-value exceeds the specified significance level of 0.05, we do not reject H_0. The test results are not statistically significant at the 5% level and (see Table 9.8 on page 386) provide virtually no evidence against the null hypothesis.

StatCrunch

Report 13.5

You try it!

Exercise 13.103 on page 612

Step 6 Interpret the results of the hypothesis test.

Interpretation At the 5% significance level, the data do not provide sufficient evidence to conclude that a difference exists in educational-attainment distributions among residents of the four U.S. regions.

Comparing Several Population Proportions

As we mentioned, a special use of the chi-square homogeneity test is for comparing several population proportions. Recall that a population proportion is the proportion of an entire population that has a specified attribute.

In these circumstances, the variable has two possible values, namely, "the specified attribute" and "not the specified attribute." Furthermore, the distribution of such a variable is completely determined by the proportion of the population that has the specified attribute, that is, by the population proportion, p. (Why is that so?)

Consequently, populations are homogeneous with respect to such a variable if and only if the population proportions are equal. Hence, in this case, we can state the respective null and alternative hypotheses for a chi-square homogeneity test as follows:

H_0: $p_1 = p_2 = \cdots = p_r$ (population proportions are all equal)

H_a: Not all the population proportions are equal.

In other words, if a variable has only two possible values, then the chi-square homogeneity test provides a procedure for comparing several population proportions.

EXAMPLE 13.13   The Chi-Square Homogeneity Test

Scandinavian Unemployment Rates The Organization for Economic Cooperation and Development compiles information on unemployment rates of selected countries and publishes its findings in *Main Economic Indicators*. Independent simple random samples from the civilian labor forces of the five Scandinavian countries—Denmark, Norway, Sweden, Finland, and Iceland—yielded the data in Table 13.17 on employment status.

TABLE 13.17

Sample data for employment status in the five Scandinavian countries

		Status		
		Unemployed	Employed	Total
Country	Denmark	12	309	321
	Norway	7	265	272
	Sweden	32	498	530
	Finland	21	286	307
	Iceland	1	69	70
	Total	73	1427	1500

Do the data provide sufficient evidence to conclude that a difference exists in the unemployment rates of the five Scandinavian countries?

Solution Let p_1, p_2, p_3, p_4, and p_5 denote the population proportions of the unemployed people in the civilian labor forces of Denmark, Norway, Sweden, Finland, and Iceland, respectively. We want to perform the following hypothesis test.

H_0: $p_1 = p_2 = p_3 = p_4 = p_5$ (unemployment rates are equal)

H_a: Not all the unemployment rates are equal.

Proceeding in the usual manner, we first computed the expected frequencies by using Formula 13.3 on page 607. We found that all of the expected frequencies are 1 or greater; hence, Assumption 1 of Procedure 13.3 is satisfied. We also found that one of the expected frequencies is less than 5. Noting that there are 10 cells, we conclude that 1/10, or 10%, of the expected frequencies are less than 5; hence, Assumption 2 of Procedure 13.3 is satisfied. Consequently, we can apply Procedure 13.3 to perform the required hypothesis test.

We have df $= (5 - 1)(2 - 1) = 4$ and, using the formula $\chi^2 = \Sigma(O - E)^2/E$, we find that the value of the test statistic is $\chi^2 = 9.912$.

Critical-value approach: From Table VII, the critical value for a test at the 5% significance level is 9.488. Because the value of the test statistic exceeds the critical value, we reject H_0.

P-value approach: From Table VII, we find that $0.025 < P < 0.05$. (Using technology, we get $P = 0.042$.) Because the P-value is less than the specified significance level of 0.05, we reject H_0. Furthermore, Table 9.8 on page 386 shows that the data provide strong evidence against the null hypothesis.

You try it!

Exercise 13.105 on page 613

Interpretation At the 5% significance level, the data provide sufficient evidence to conclude that a difference exists in the unemployment rates of the five Scandinavian countries. ■

The Chi-Square Homogeneity Test and the Two-Proportions z-Test

When $r = 2$ (i.e., there are two populations under consideration), the respective null and alternative hypotheses of a chi-square homogeneity test for comparing population proportions can be reexpressed as follows:

H_0: $p_1 = p_2$

H_a: $p_1 \neq p_2$.

However, these are the null and alternative hypotheses of a two-tailed test for comparing two population proportions. As you know, we can also use the two-proportions z-test (Procedure 12.3 on page 564) to conduct such a hypothesis test. The question now is whether these two tests yield the same results. In fact, they always do; that is, the chi-square homogeneity test for comparing two population proportions and the two-tailed two-proportions z-test are equivalent.[†]

THE TECHNOLOGY CENTER

As we have seen, although the chi-square homogeneity test and the chi-square independence test are used for quite different purposes, the procedures for carrying them out are essentially identical. Hence, to use technology to perform a chi-square homogeneity test, we can apply the same method used for a chi-square independence test, as described in The Technology Center on pages 602–604.

[†]See, for instance, the paper "Equivalence of Different Statistical Tests for Common Problems" (*The AMATYC Review*, Vol. 4, No. 2, pp. 5–13) by M. Hassett and N. Weiss.

Exercises 13.5

Understanding the Concepts and Skills

13.93 Although the chi-square homogeneity test and the chi-square independence test use essentially the same basic procedure, the context of the two tests are quite different. Explain the difference.

13.94 For what purpose is a chi-square homogeneity test used?

13.95 Consider a variable of several populations. Define the terms *homogeneous* and *nonhomogeneous* in this context.

13.96 State the null and alternative hypotheses for a chi-square homogeneity test
a. without using the terms homogeneous and nonhomogeneous.
b. using the terms homogeneous and nonhomogeneous.

13.97 Fill in the blank: If a variable has only two possible values, the chi-square homogeneity test provides a procedure for comparing several population _____.

13.98 If a variable of two populations has only two possible values, the chi-square homogeneity test is equivalent to a two-tailed test that we discussed in an earlier chapter. What test is that?

13.99 A chi-square homogeneity test is to be conducted to decide whether a difference exists among the distributions of a variable of six populations. The variable has five possible values. What are the degrees of freedom for the χ^2-statistic?

13.100 A chi-square homogeneity test is to be conducted to decide whether four populations are nonhomogeneous with respect to a variable that has eight possible values. What are the degrees of freedom for the χ^2-statistic?

Applying the Concepts and Skills

In Exercises 13.101–13.106, use either the critical-value approach or the P-value approach to perform a chi-square homogeneity test, provided the conditions for using the test are met.

13.101 Self-Concept and Sightedness. Self-concept can be defined as the general view of oneself in terms of personal value and capabilities. A study of whether visual impairment affects self-concept was reported in the article "An Exploration into Self Concept: A Comparative Analysis between the Adolescents Who Are Sighted and Blind in India" (*British Journal of Visual Impairment*, Vol. 30, No. 1, pp. 31–41) by S. Halder and P. Datta. Independent random samples of sighted and blind Indian adolescents gave the following data on self-concept.

	Self-concept			
Sightedness		High	Moderate	Low	Total
	Sighted	13	73	14	100
	Blind	3	40	17	60
	Total	16	113	31	160

a. At the 5% significance level, do the data provide sufficient evidence to conclude that a difference exists in self-concept distributions between sighted and blind Indian adolescents?
b. Repeat part (a) at the 1% significance level.

13.102 Slot Machines. In the article "The Influence of Theme as Slot Machine Attribute on Casino Gamers Decision-Making" (*American Journal of Applied Sciences*, Vol. 10, No. 7, pp. 734–739), E. Wannenburg et al. explore the effects of theme on slot-machine gamers. Independent random samples of male and female slot gamers were asked whether the theme of a slot machine plays a role in their slot-machine game selection. The responses are summarized in the following table.

		Gender		
Response		Male	Female	**Total**
	Disagree	57	54	111
	Neutral	33	37	70
	Agree	181	322	503
	Total	271	413	684

At the 1% significance level, do the data provide sufficient evidence to conclude that there is a difference between male and female slot gamers with regard to the way they perceive theme as a slot machine attribute?

13.103 Region and Race. The U.S. Census Bureau compiles data on the U.S. population by region and race and publishes its findings in *Current Population Reports*. Independent simple random samples of residents in the four U.S. regions gave the following data on race.

		Race			
		White	Black	Other	**Total**
Region	Northeast	93	14	6	113
	Midwest	118	14	4	136
	South	167	42	7	216
	West	113	7	15	135
	Total	491	77	32	600

At the 1% significance level, do the data provide sufficient evidence to conclude that a difference exists in race distributions among the four U.S. regions?

13.104 State of the Union. The Quinnipiac University Poll conducts nationwide surveys as a public service and for research. This problem is based on the results of one such poll. Independent simple random samples of 300 residents each in red (predominantly Republican), blue (predominantly Democratic), and purple (mixed) states were asked how satisfied they were with the way things are going today. The following table summarizes the responses.

	State classification			
Satisfaction level		Red	Blue	Purple	**Total**
	Very satisfied	9	3	9	21
	Somewhat satisfied	48	48	36	132
	Somewhat dissatisfied	114	114	108	336
	Very dissatisfied	129	135	147	411
	Total	300	300	300	900

At the 10% significance level, do the data provide sufficient evidence to conclude that the satisfaction-level distributions differ among residents of red, blue, and purple states?

13.105 Gun Control. The Quinnipiac University Poll conducts nationwide surveys as a public service and for research. This problem is based on the results of one such poll that asked independent random samples of American adults in urban, suburban, and rural regions, "Do you support or oppose requiring background checks for all gun buyers?" Here are the results.

	Region			
Response		Urban	Suburban	Rural	**Total**
	Support	335	348	318	1001
	Oppose	35	23	50	108
	Total	370	371	368	1109

At the 1% significance level, do the data provide sufficient evidence to conclude that a difference exists in the proportions of supporters among the three regions?

13.106 Scoliosis. Scoliosis is a condition involving curvature of the spine. In a study by A. Nachemson and L. Peterson, reported in the *Journal of Bone and Joint Surgery* (Vol. 77, Issue 6, pp. 815–822), 286 girls aged 10 to 15 years were followed to determine the effects of observation only (129 patients), an underarm plastic brace (111 patients), and nighttime surface electrical stimulation (46 patients). A treatment was deemed to have failed if the curvature of the spine increased by 6° on two successive examinations. The following table summarizes the results obtained by the researchers.

	Result		
Treatment		Not failure	Failure	**Total**
	Brace	94	17	111
	Stimulation	24	22	46
	Observation	71	58	129
	Total	189	97	286

At the 5% significance level, do the data provide sufficient evidence to conclude that a difference in failure rate exists among the three types of treatments?

*In each of Exercises **13.107** and **13.108**,*
a. *use the two-proportions z-test (Procedure 12.3 on page 564) to perform the required hypothesis test.*
b. *use the chi-square homogeneity test to perform the required hypothesis test.*
c. *compare your results in parts (a) and (b).*
d. *explain what principle is being illustrated.*

13.107 Vegetarians. A poll conducted by Gallup asked American adults about vegetarianism. This problem is based on that poll. Of independent random samples of 500 men and 512 women, 20 of the men and 36 of the women said they were vegetarians. At the 5% significance level do the data provide sufficient evidence to conclude that a difference exists in the proportions of male and female vegetarians?

13.108 Fatty Acids and Allergies. P. Noakes et al. researched the effects of fatty acids found in oily fish on lowering the risk of allergic disease in the article "Increased Intake of Oily Fish in Pregnancy: Effects on Neonatal Immune Responses and on Clinical Outcomes in Infants at 6 Mo." (*American Journal of Clinical Nutrition*, Vol. 95, No. 2, pp. 395–404). Pregnant women were randomly assigned to continue their habitual diet (control group), which was low in oily fish, or to consume two portions of salmon per week (treatment group). Their infants were clinically evaluated at 6 months of age and the frequency of many different symptoms was recorded. Of the 37 infants in the control group, 12 had symptoms of dry skin; and of the 45 infants in the experimental group, 14 had symptoms of dry skin. At the 5% significance level, do the data provide sufficient evidence to conclude that a difference exists in the proportions of infants who have symptoms of dry skin at 6 months between those whose mothers continue their habitual diet and those whose mothers consume two portions of salmon per week?

CHAPTER IN REVIEW

You Should Be Able to

1. use and understand the formulas in this chapter.

2. identify the basic properties of χ^2-curves.

3. use the chi-square table, Table VII.

4. explain the reasoning behind the chi-square goodness-of-fit test.

5. perform a chi-square goodness-of-fit test.

6. group bivariate data into a contingency table.

7. find and graph marginal and conditional distributions.

8. decide whether an association exists between two variables of a population, given bivariate data for the entire population.

9. explain the reasoning behind the chi-square independence test.

10. perform a chi-square independence test to decide whether an association exists between two variables of a population, given bivariate data for a sample of the population.

11. perform a chi-square homogeneity test to compare the distributions of a variable of two or more populations.

Key Terms

associated variables, *590*
association, *590*
bivariate data, *587*
cells, *588*
χ^2_α, *577*
chi-square (χ^2) curve, *577*
chi-square distribution, *577*
chi-square goodness-of-fit test, *581*
chi-square homogeneity test, *608*

chi-square independence test, *600*
chi-square procedures, *576*
chi-square subtotals, *580*
conditional distribution, *589*
contingency table, *587*
cross tabs, *587*
cross-tabulation table, *587*
expected frequencies, *579*
homogeneous, *606*

marginal distribution, *590*
nonhomogeneous, *606*
observed frequencies, *579*
segmented bar graph, *590*
statistically dependent variables, *590*
statistically independent variables, *590*
two-way table, *587*
univariate data, *587*

Summary of Chi-Square Procedures

The following table compares the three chi-square procedures discussed in this chapter by describing the situations in which they are used.

Note: All three chi-square procedures can also be used with discrete quantitative variables that have only a finite number of possible values.

Type	Description	Procedure to use
Goodness-of-fit test	Perform a hypothesis test for the distribution of one qualitative variable on one population	13.1 (page 581)
Independence test	Perform a hypothesis test to decide whether two qualitative variables on one population are associated	13.2 (page 600)
Homogeneity test	Perform a hypothesis test to compare the distributions of one qualitative variable on two or more populations	13.3 (page 608)

REVIEW PROBLEMS

Understanding the Concepts and Skills

1. How do you distinguish among the infinitely many different chi-square distributions and their corresponding χ^2-curves?

2. Regarding a χ^2-curve:
a. At what point on the horizontal axis does the curve begin?
b. Classify its shape as symmetric, left skewed, or right skewed.
c. As the number of degrees of freedom increases, a χ^2-curve begins to look like another type of curve. What type of curve is that?

3. Recall that the number of degrees of freedom for the *t*-distribution used in a one-mean *t*-test depends on the sample size. Is that true for the chi-square distribution used in a chi-square
a. goodness-of-fit test?
b. independence test?
c. homogeneity test?
Explain your answers.

4. Explain why a chi-square goodness-of-fit test, a chi-square independence test, or a chi-square homogeneity test is always right tailed.

5. If the observed and expected frequencies for a chi-square goodness-of-fit test, a chi-square independence test, or a chi-square homogeneity test matched perfectly, what would be the value of the test statistic?

6. Regarding the expected-frequency assumptions for a chi-square goodness-of-fit test, a chi-square independence test, or a chi-square homogeneity test,
a. state them.
b. how important are they?

7. **Ancestry and Region.** The U.S. Census Bureau collects information on the U.S. population by ancestry and region of residence and publishes the results in *American Community Survey*. According to that document, 18% of the population resides in the Northeast.
a. If ancestry and region of residence are not associated, what percentage of Americans of Irish ancestry would reside in the Northeast?
b. There are roughly 37 million Americans of Irish ancestry. If ancestry and region of residence are not associated, how many Americans of Irish ancestry would reside in the Northeast?
c. There are, in fact, 9.25 million Americans of Irish ancestry who reside in the Northeast. Given this information and your answer to part (b), what can you conclude?

8. Suppose that you have bivariate data for an entire population.
a. How would you decide whether an association exists between the two variables under consideration?
b. Assuming that you make no calculation mistakes, could your conclusion be in error? Explain your answer.

9. Suppose that you have bivariate data for a sample of a population.
   a. How would you decide whether an association exists between the two variables under consideration?
   b. Assuming that you make no calculation mistakes, could your conclusion be in error? Explain your answer.

10. Consider a χ^2-curve with 17 degrees of freedom. Use Table VII to determine
   a. $\chi^2_{0.10}$.
   b. $\chi^2_{0.01}$.
   c. the χ^2-value that has area 0.05 to its right.

Applying the Concepts and Skills

11. **Educational Attainment.** The U.S. Census Bureau compiles census data on educational attainment of Americans. From the document *Current Population Survey*, we obtained the 2010 distribution of educational attainment for U.S. adults 25 years old and older. Here is that distribution.

Highest level	Percentage
Not HS graduate	12.9
HS graduate	31.2
Some college	16.8
Associate's degree	9.1
Bachelor's degree	19.5
Advanced degree	10.5

A random sample of 500 U.S. adults (25 years old and older) taken this year gave the following frequency distribution.

Highest level	Frequency
Not HS graduate	45
HS graduate	163
Some college	76
Associate's degree	53
Bachelor's degree	102
Advanced degree	61

Decide, at the 5% significance level, whether this year's distribution of educational attainment differs from the 2010 distribution.

12. **Presidents.** From the *Information Please Almanac*, we compiled the following table on U.S. region of birth and political party of the first 44 U.S. presidents. The table uses these abbreviations: F = Federalist, DR = Democratic-Republican, D = Democratic, W = Whig, R = Republican, U = Union; NE = Northeast, MW = Midwest, SO = South, WE = West.

Region	Party	Region	Party	Region	Party
SO	F	SO	R	MW	R
NE	F	SO	U	NE	D
SO	DR	MW	R	MW	D
SO	DR	MW	R	SO	R
SO	DR	MW	R	NE	D
NE	DR	NE	R	SO	D
SO	D	NE	D	WE	R
NE	D	MW	R	MW	R
SO	W	NE	D	SO	D
SO	W	MW	R	MW	R
SO	D	NE	R	NE	R
SO	W	MW	R	SO	D
NE	W	SO	D	NE	R
NE	D	MW	R	WE	D
NE	D	NE	R		

a. What is the population under consideration?
b. What are the two variables under consideration?
c. Group the bivariate data for the variables "birth region" and "party" into a contingency table.

13. **Presidents.** Refer to Problem 12.
a. Find the conditional distributions of party by birth region and the marginal distribution of party.
b. Does an association exist between the variables "birth region" and "party" for the U.S. presidents? Explain your answer.
c. What percentage of presidents are Republicans?
d. If no association existed between birth region and party, what percentage of presidents born in the South would be Republicans?
e. In reality, what percentage of presidents born in the South are Republicans?

14. **Presidents.** Refer to Problem 12.
a. Find the conditional distributions of birth region by party and the marginal distribution of birth region.
b. Does an association exist between the variables "birth region" and "party" for the U.S. presidents? Explain your answer.
c. What percentage of presidents were born in the South?
d. If no association existed between birth region and party, what percentage of Republican presidents would have been born in the South?
e. In reality, what percentage of Republican presidents were born in the South?

15. **Hospitals.** From data in *Hospital Statistics*, published by the American Hospital Association, we obtained the following contingency table for U.S. hospitals and nursing homes by type of facility and type of control. We used the abbreviations Gov for Government, Prop for Proprietary, and NP for nonprofit.

		Control			
		Gov	Prop	NP	Total
Facility	General	1697	660	3046	5403
	Psychiatric	266	358	113	737
	Chronic	21	1	4	26
	Tuberculosis	3	0	1	4
	Other	59	148	203	410
	Total	2046	1167	3367	6580

In the following questions, the term *hospital* refers to either a hospital or nursing home:
a. How many hospitals are government controlled?
b. How many hospitals are psychiatric facilities?
c. How many hospitals are government controlled psychiatric facilities?
d. How many general facilities are nonprofit?
e. How many hospitals are not under proprietary control?
f. How many hospitals are either general facilities or under proprietary control?

16. **Hospitals.** Refer to Problem 15.
a. Obtain the conditional distribution of control type within each facility type.
b. Does an association exist between facility type and control type? Explain your answer.
c. Find the marginal distribution of control type.

d. Construct a segmented bar graph for the conditional distributions and marginal distribution of control type. Interpret the graph in light of your answer to part (b).

17. Hospitals. Refer to Problems 15 and 16.
a. In view of your answer to Problem 16(b), without doing any further calculations, respond true or false to the following statement and explain your answer: "The conditional distributions of facility type within control types are identical."
b. Obtain the conditional distributions of facility type within control types.
c. Find the marginal distribution of facility type.

18. Hospitals. Refer to Problems 15–17.
a. What percentage of hospitals are under proprietary control?
b. What percentage of psychiatric hospitals are under proprietary control?
c. What percentage of hospitals are psychiatric hospitals?
d. What percentage of hospitals under proprietary control are psychiatric hospitals?

19. Internet Use. In the article "Happier and Less Isolated: Internet Use in Old Age" (*Journal of Poverty & Social Justice*, Vol. 21, Issue 1, pp. 33–45), researcher O. Lelkes explores the impact of Internet use. The following problem is based on the article. A random sample of adults yielded the following data on age and Internet usage.

Usage	Age: 18–24	25–64	65+	Total
Never	6	38	31	75
Sometimes	14	31	4	49
Every day	50	50	5	105
Total	70	119	40	229

At the 1% significance level, do the data provide sufficient evidence to conclude that an association exists between age and Internet usage?

20. Income and Residence. The U.S. Census Bureau compiles information on money income of people by type of residence and publishes its finding in *Current Population Reports*. Independent simple random samples of people residing inside principal cities (IPC), outside principal cities but within metropolitan areas (OPC), and outside metropolitan areas (OMA), gave the following data on income level.

Income	Residence: IPC	OPC	OMA	Total
Under $15,000	75	106	46	227
$15,000–$34,999	106	161	61	328
$35,000–$74,999	98	183	52	333
$75,000 and over	48	102	14	164
Total	327	552	173	1052

a. Identify the populations under consideration.
b. Identify the variable under consideration.
c. At the 5% significance level, do the data provide sufficient evidence to conclude that people residing in the three types of residence are nonhomogeneous with respect to income level?

21. Economy in Recession? The Quinnipiac University Poll conducts nationwide surveys as a public service and for research. This problem is based on the results of one such poll. Independent simple random samples of registered Democrats, Republicans, and Independents were asked, "Do you think the United States economy is in a recession now?" Of the 628 Democrats sampled, 220 responded "yes," as did 349 of the 471 Republicans sampled and 342 of the 646 Independents sampled. At the 1% significance level, do the data provide sufficient evidence to conclude that a difference exists in the percentages of registered Democrats, Republicans, and Independents who thought the U.S. economy was in a recession at the time?

Working with Large Data Sets

22. Yakashba Estates. The document *Arizona Residential Property Valuation System*, published by the Arizona Department of Revenue, describes how county assessors use computerized systems to value single-family residential properties for property tax purposes. On the WeissStats site are data on lot size (in acres) and house size (in square feet) for homes in the Yakashba Estates, a private community in Prescott, AZ. We used the following codings for lot size and home size.

Lot size		House size	
Size (acres)	*Coding*	*Size (sq. ft.)*	*Coding*
Under 2.25	L1	Under 3000	H1
2.25–2.49	L2	3000–3999	H2
2.50–2.74	L3	4000 & over	H3
2.75 & over	L4		

Use the technology of your choice to do the following tasks for the coded variables.
a. Group the bivariate data for the variables "lot size" and "house size" into a contingency table.
b. Find the conditional distributions of lot size by house size and the marginal distribution of lot size.
c. Find the conditional distributions of house size by lot size and the marginal distribution of house size.
d. Does an association exist between the variables "lot size" and "house size" for homes in the Yakashba Estates? Explain your answer.

23. Withholding Treatment. Several years ago, a poll by Gallup asked 1528 adults the following question: "The New Jersey Supreme Court recently ruled that all life-sustaining medical treatment may be withheld or withdrawn from terminally ill patients, provided that is what the patients want or would want if they were able to express their wishes. Would you like to see such a ruling in the state in which you live, or not?" The data on the WeissStats site give the responses by opinion and educational level. Use the technology of your choice to decide, at the 1% significance level, whether the data provide sufficient evidence to conclude that opinion on this issue and educational level are associated.

FOCUSING ON DATA ANALYSIS

UWEC UNDERGRADUATES

Recall from Chapter 1 (see page 34) that the Focus database and Focus sample contain information on the undergraduate students at the University of Wisconsin - Eau Claire (UWEC). Now would be a good time for you to review the discussion about these data sets.

Open the Focus sample worksheet (FocusSample) in the technology of your choice. In each part, apply the chi-square independence test to decide, at the 5% significance level, whether the data provide sufficient evidence to conclude that an association exists between the indicated variables for the population of all UWEC undergraduates. Be sure to check whether the assumptions for performing each test are satisfied. Interpret your results.

a. sex and classification
b. sex and residency
c. sex and college
d. classification and residency
e. classification and college
f. college and residency

CASE STUDY DISCUSSION

EYE AND HAIR COLOR

At the beginning of this chapter, we presented a cross-classification of data on eye color and hair color collected as part of a class project by students in an elementary statistics course at the University of Delaware.

a. Explain what it would mean for an association to exist between eye color and hair color.
b. Do you think that an association exists between eye color and hair color? Explain your answer.

c. Use the data provided in the contingency table to decide, at the 5% significance level, whether an association exists between eye color and hair color.
d. The raw data on eye color and hair color are provided on the WeissStats site. Use the technology of your choice to group the bivariate data into a contingency table. Compare your results with the table presented on page 577.

BIOGRAPHY

KARL PEARSON: THE FOUNDING DEVELOPER OF CHI-SQUARE TESTS

Karl Pearson was born on March 27, 1857, in London, the second son of William Pearson, a prominent lawyer, and his wife, Fanny Smith. Karl Pearson's early education took place at home. At the age of 9, he was sent to University College School in London, where he remained for the next 7 years. Because of ill health, Pearson was then privately tutored for a year. He received a scholarship at King's College, Cambridge, in 1875. There he earned a B.A. (with honors) in mathematics in 1879 and an M.A. in law in 1882. He then studied physics and metaphysics in Heidelberg, Germany.

In addition to his expertise in mathematics, law, physics, and metaphysics, Pearson was competent in literature and knowledgeable about German history, folklore, and philosophy. He was also considered somewhat of a political radical because of his interest in the ideas of Karl Marx and the rights of women.

In 1884, Pearson was appointed Goldsmid Professor of Applied Mathematics and Mechanics at University College; from 1891–1894, he was also a lecturer in geometry at Gresham College, London. In 1911, he gave up the Goldsmid chair to become the first Galton Professor of Eugenics at University College. Pearson was elected to the Royal Society—a prestigious association of scientists—in 1896 and was awarded the society's Darwin Medal in 1898.

Pearson really began his pioneering work in statistics in 1893, mainly through an association with Walter Weldon (a zoology professor at University College), Francis Edgeworth (a professor of logic at University College), and Sir Francis Galton (see the Chapter 15 Biography). An analysis of published data on roulette wheels at Monte Carlo led to Pearson's discovery of the chi-square goodness-of-fit test. He also coined the term *standard deviations,* introduced his amazingly diverse skew curves, and developed the most widely used measure of correlation, the correlation coefficient.

Pearson, Weldon, and Galton cofounded the statistical journal *Biometrika,* of which Pearson was editor from 1901 to 1936 and a major contributor. Pearson retired from University College in 1933. He died in London on April 27, 1936.

CHAPTER

14

Descriptive Methods in Regression and Correlation

CHAPTER OUTLINE

CHAPTER OBJECTIVES

We often want to know whether two or more variables are related and, if they are, how they are related. In Sections 13.3 and 13.4, we examined relationships between two qualitative (categorical) variables. In this chapter, we discuss relationships between two quantitative variables.

Linear regression and *correlation* are two commonly used methods for examining the relationship between quantitative variables and for making predictions. We discuss descriptive methods in linear regression and correlation in this chapter and consider inferential methods in Chapter 15.

To prepare for our discussion of linear regression, we review linear equations with one independent variable in Section 14.1. In Section 14.2, we explain how to determine the *regression equation,* the equation of the line that best fits a set of data points.

In Section 14.3, we examine the coefficient of determination, a descriptive measure of the utility of the regression equation for making predictions. In Section 14.4, we discuss the linear correlation coefficient, which provides a descriptive measure of the strength of the linear relationship between two quantitative variables.

CASE STUDY

Healthcare: Spending and Outcomes

It goes without saying that healthcare is an essential element of any society. The Organisation for Economic Co-operation and Development (OECD) publishes annual data on healthcare, among other important topics. OECD's origins date back to 1960, but its roots are directly post World War II. Currently, there are 34 member countries, but OECD also works closely with emerging giants like China, India, and Brazil and developing economies in Africa, Asia, Latin America, and the Caribbean.

In many ways, the United States is a "healthcare outlier." For instance, in 2011, the per capita health expenditure in the United States was $8508, over two and two-thirds times that of the average of the other 33 member countries. In addition, as a percentage of gross domestic product (GDP), total healthcare expenditures in the United States were 17.7%, almost twice that of the average of the others. Moreover, the OECD reported that the United

618

States ranks relatively poorly among those countries on measures such as life expectancy and infant mortality.

The following table provides selected OECD healthcare data for the 34 member countries for the year 2011 or, in a few cases, the nearest year to that. Note that

% GDP is an abbreviation for "healthcare expenditures as a percentage of gross domestic product," LE for "life expectancy" (at birth), and IMR for "infant mortality rate" (number of deaths of babies under 1 year of age per 1000 live births).

Country	% GDP	LE	IMR	Country	% GDP	LE	IMR
Australia	8.9	82.0	3.8	Japan	9.6	82.7	2.3
Austria	10.8	81.1	3.6	Korea	7.4	81.1	3.0
Belgium	10.5	80.5	3.3	Luxembourg	8.2	81.1	4.3
Canada	11.2	81.0	4.9	Mexico	6.2	74.2	13.6
Chile	7.5	78.3	7.4	Netherlands	11.9	81.3	3.6
Czech Republic	7.5	78.0	2.7	New Zealand	10.3	81.2	5.5
Denmark	11.1	79.9	3.6	Norway	9.3	81.4	2.4
Estonia	5.9	76.3	2.5	Poland	6.9	76.9	4.7
Finland	9.0	80.6	2.4	Portugal	10.2	80.8	3.1
France	11.6	82.2	3.5	Slovak Republic	7.9	76.1	4.9
Germany	11.3	80.8	3.6	Slovenia	8.9	80.1	2.9
Greece	9.1	80.8	3.4	Spain	9.3	82.4	3.2
Hungary	7.9	75.0	4.9	Sweden	9.5	81.9	2.1
Iceland	9.0	82.4	0.9	Switzerland	11.0	82.8	3.8
Ireland	8.9	80.6	3.5	Turkey	6.1	74.6	7.7
Israel	7.7	81.8	3.5	United Kingdom	9.4	81.1	4.3
Italy	9.2	82.7	3.4	United States	17.7	78.7	6.1

After studying the techniques in this chapter, you will be able to conduct statistical analyses on the data to see for yourself some of the relationships between healthcare spending and its outcomes.

Linear Equations with One Independent Variable

To understand linear regression, let's first review linear equations with one independent variable. The general form of a **linear equation** with one independent variable can be written as

$$y = b_0 + b_1 x,$$

where b_0 and b_1 are constants (fixed numbers), x is the independent variable, and y is the dependent variable.[†]

The graph of a linear equation with one independent variable is a **straight line,** or simply a **line;** furthermore, any nonvertical line can be represented by such an equation. Examples of linear equations with one independent variable are $y = 4 + 0.2x$, $y = -1.5 - 2x$, and $y = -3.4 + 1.8x$. The graphs of these three linear equations are shown in Fig. 14.1 on the following page.

Linear equations with one independent variable occur frequently in applications of mathematics to many different fields, including the management, life, and social sciences, as well as the physical and mathematical sciences.

[†]You may be familiar with the form $y = mx + b$ instead of the form $y = b_0 + b_1 x$. Statisticians prefer the latter form because it allows a smoother transition to multiple regression, in which there is more than one independent variable. Material on multiple regression is provided in the modules *Multiple Regression Analysis* and *Model Building in Regression*, available in the Regression-ANOVA Modules section of the WeissStats site.

FIGURE 14.1

Graphs of three linear equations

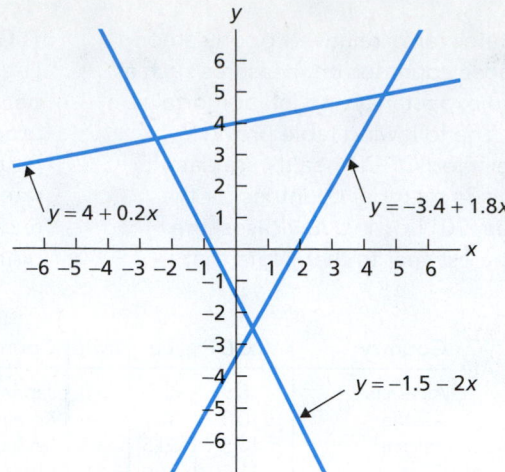

$y = 4 + 0.2x$

$y = -3.4 + 1.8x$

$y = -1.5 - 2x$

■■■ **EXAMPLE 14.1**   **Linear Equations**

TABLE 14.1

Times and costs for five word-processing jobs

Time (hr) x	Cost ($) y
5.0	125
7.5	175
15.0	325
20.0	425
22.5	475

FIGURE 14.2

Graph of $y = 25 + 20x$, obtained from the points displayed in Table 14.1

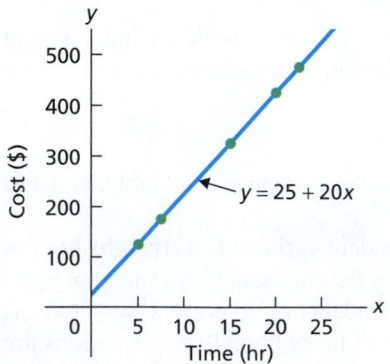

$y = 25 + 20x$

You try it!

Exercise 14.23 on page 623

Word-Processing Costs CJ² Business Services offers its clients word processing at a rate of $20 per hour plus a $25 disk charge. The total cost to a customer depends, of course, on the number of hours needed to complete the job.

a. Find the equation that expresses the total cost in terms of the number of hours needed to complete the job.

b. Determine b_0 and b_1.

c. Construct a table that gives the costs for jobs that take the following numbers of hours: 5, 7.5, 15, 20, and 22.5.

d. Draw the graph of the equation found in part (a) by plotting the points from the table constructed in part (c).

e. Visually estimate the cost of a 10-hour job from the graph constructed in part (d). Then calculate that cost exactly by using the equation obtained in part (a).

Solution

a. Because the rate for word processing is $20 per hour, a job that takes x hours will cost $20x$ plus the $25 disk charge. Hence the total cost, y, of a job that takes x hours is $y = 25 + 20x$. This equation gives us the exact cost for a job if we know the number of hours required.

b. The equation $y = 25 + 20x$ is linear; here $b_0 = 25$ and $b_1 = 20$.

c. A job that takes 5 hours will cost $y = 25 + 20 \cdot 5 = \$125$; a job that takes 7.5 hours will cost $y = 25 + 20 \cdot 7.5 = \$175$. Proceeding similarly, we get the costs for the other three numbers of hours. See Table 14.1.

d. As we noted, the graph of a linear equation, such as $y = 25 + 20x$, is a line. To obtain the graph of $y = 25 + 20x$, we first plot the points displayed in Table 14.1 and then connect them with a line, as shown in Fig. 14.2.

e. Figure 14.2 shows that a 10-hour job will cost somewhere between $200 and $300. Using the equation from part (a), we find that the exact cost is $y = 25 + 20 \cdot 10 = \$225$. ■

Intercept and Slope

For a linear equation $y = b_0 + b_1 x$, the number b_0 is the y-value of the point of intersection of the line and the y-axis. The number b_1 measures the steepness of the line; more precisely, b_1 indicates how much the y-value changes when the x-value increases by 1 unit. Figure 14.3 illustrates these relationships.

FIGURE 14.3

Graph of $y = b_0 + b_1 x$

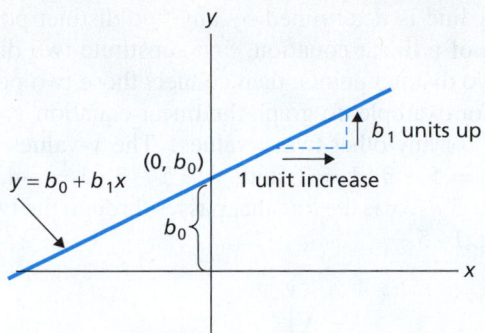

The numbers b_0 and b_1 have special names that reflect these geometric inter-pretations.

DEFINITION 14.1

What Does It Mean?

The y-intercept of a line is where it intersects the y-axis. The slope of a line measures its steepness.

y-Intercept and Slope

For a linear equation $y = b_0 + b_1 x$, the number b_0 is called the **y-intercept** and the number b_1 is called the **slope.**

In the next example, we apply the concepts of y-intercept and slope to the illustration of word-processing costs.

■■■   **EXAMPLE 14.2** y-Intercept and Slope

Word-Processing Costs In Example 14.1, we found the linear equation that ex-presses the total cost, y, of a word-processing job in terms of the number of hours, x, required to complete the job. The equation is $y = 25 + 20x$.

a. Determine the y-intercept and slope of that linear equation.
b. Interpret the y-intercept and slope in terms of the graph of the equation.
c. Interpret the y-intercept and slope in terms of word-processing costs.

Solution

a. The y-intercept for the equation is $b_0 = 25$, and the slope is $b_1 = 20$.
b. The y-intercept $b_0 = 25$ is the y-value where the line intersects the y-axis, as shown in Fig. 14.4. The slope $b_1 = 20$ indicates that the y-value increases by 20 units for every increase in x of 1 unit.

FIGURE 14.4

Graph of $y = 25 + 20x$

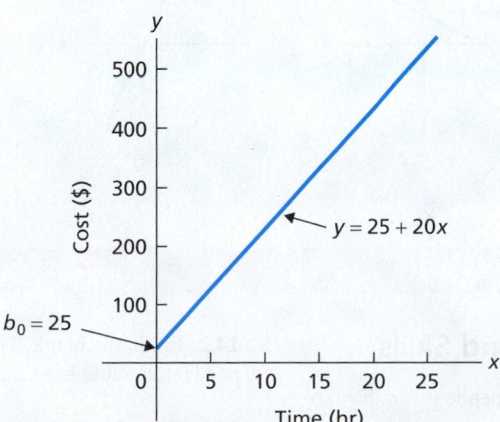

You try it!

Exercise 14.27
on page 623

c. The y-intercept $b_0 = 25$ represents the total cost of a job that takes 0 hours. In other words, the y-intercept of \$25 is a fixed cost that is charged no matter how long the job takes. The slope $b_1 = 20$ represents the cost per hour of \$20; it is the amount that the total cost goes up for every additional hour the job takes. **■**

A line is determined by any two distinct points that lie on it. Thus, to draw the graph of a linear equation, first substitute two different x-values into the equation to get two distinct points; then connect those two points with a line.

For example, to graph the linear equation $y = 5 - 3x$, we can use the x-values 1 and 3 (or any other two x-values). The y-values corresponding to those two x-values are $y = 5 - 3 \cdot 1 = 2$ and $y = 5 - 3 \cdot 3 = -4$, respectively. Therefore the graph of $y = 5 - 3x$ is the line that passes through the two points $(1, 2)$ and $(3, -4)$, as shown in Fig. 14.5.

FIGURE 14.5

Graph of $y = 5 - 3x$

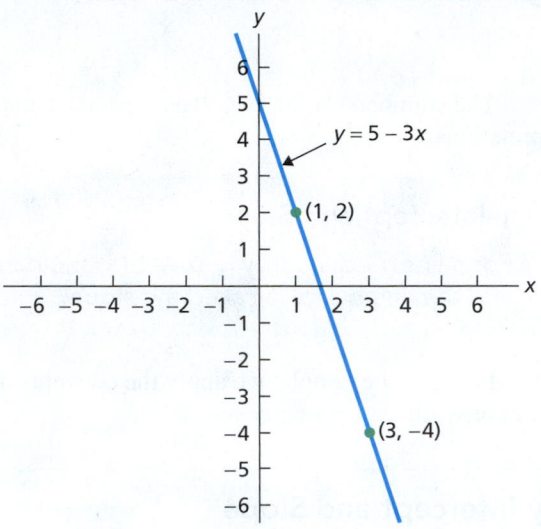

Note that the line in Fig. 14.5 slopes downward—the y-values decrease as x increases—because the slope of the line is negative: $b_1 = -3 < 0$. Now look at the line in Fig. 14.4 on page 621, the graph of the linear equation $y = 25 + 20x$. That line slopes upward—the y-values increase as x increases—because the slope of the line is positive: $b_1 = 20 > 0$.

KEY FACT 14.1

Graphical Interpretation of Slope

The graph of the linear equation $y = b_0 + b_1 x$ slopes upward if $b_1 > 0$, slopes downward if $b_1 < 0$, and is horizontal if $b_1 = 0$, as shown in Fig. 14.6.

FIGURE 14.6

Graphical interpretation of slope

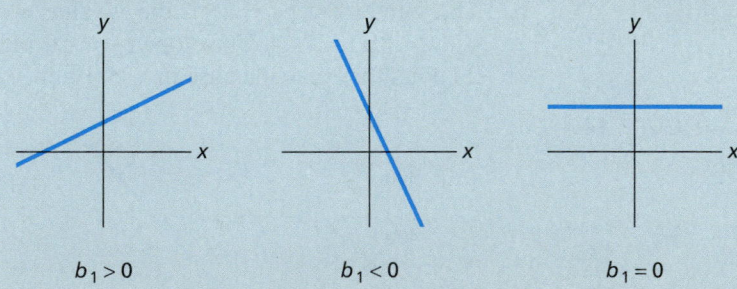

Exercises 14.1

Understanding the Concepts and Skills

14.1 Regarding linear equations with one independent variable, answer the following questions:
a. What is the general form of such an equation?
b. In your expression in part (a), which letters represent constants and which represent variables?
c. In your expression in part (a), which letter represents the independent variable and which represents the dependent variable?

14.2 Fill in the blank. The graph of a linear equation with one independent variable is a _____.

14.3 Consider the linear equation $y = b_0 + b_1 x$.
a. Identify and give the geometric interpretation of b_0.
b. Identify and give the geometric interpretation of b_1.

14.4 Answer true or false to each statement, and explain your answers.

a. The graph of a linear equation slopes upward unless the slope is 0.
b. The value of the y-intercept has no effect on the direction that the graph of a linear equation slopes.

In Exercises 14.5–14.14, we give linear equations. For each equation,
a. find the y-intercept and slope.
b. determine whether the line slopes upward, slopes downward, or is horizontal, without graphing the equation.
c. use two points to graph the equation.

14.5 $y = 3 + 4x$ **14.6** $y = -1 + 2x$

14.7 $y = 6 - 7x$ **14.8** $y = -8 - 4x$

14.9 $y = 0.5x - 2$ **14.10** $y = -0.75x - 5$

14.11 $y = 2$ **14.12** $y = -3x$

14.13 $y = 1.5x$ **14.14** $y = -3$

In Exercises 14.15–14.22, we identify the y-intercepts and slopes, respectively, of lines. For each line,
a. determine whether it slopes upward, slopes downward, or is horizontal, without graphing the equation.
b. find its equation.
c. use two points to graph the equation.

14.15 5 and 2   **14.16** -3 and 4

14.17 -2 and -3 **14.18** 0.4 and 1

14.19 0 and -0.5 **14.20** -1.5 and 0

14.21 3 and 0   **14.22** 0 and 3

Applying the Concepts and Skills

14.23 Rental-Car Costs. During one month, the Avis Rent-A-Car rate for renting a Buick LeSabre in Mobile, Alabama, was $68.22 per day plus 25¢ per mile. For a 1-day rental, let x denote the number of miles driven and let y denote the total cost, in dollars.
a. Find the equation that expresses y in terms of x.
b. Determine b_0 and b_1.
c. Construct a table similar to Table 14.1 on page 620 for the x-values 50, 100, and 250 miles.
d. Draw the graph of the equation that you determined in part (a) by plotting the points from part (c) and connecting them with a line.
e. Apply the graph from part (d) to estimate visually the cost of driving the car 150 miles. Then calculate that cost exactly by using the equation from part (a).

14.24 Air-Conditioning Repairs. Richard's Heating and Cooling in Prescott, Arizona, charges $55 per hour plus a $30 service charge. Let x denote the number of hours required for a job, and let y denote the total cost to the customer.
a. Find the equation that expresses y in terms of x.
b. Determine b_0 and b_1.
c. Construct a table similar to Table 14.1 on page 620 for the x-values 0.5, 1, and 2.25 hours.
d. Draw the graph of the equation that you determined in part (a) by plotting the points from part (c) and connecting them with a line.
e. Apply the graph from part (d) to estimate visually the cost of a job that takes 1.75 hours. Then calculate that cost exactly by using the equation from part (a).

14.25 Measuring Temperature. The two most commonly used scales for measuring temperature are the Fahrenheit and Celsius scales. If you let y denote Fahrenheit temperature and x denote

Celsius temperature, you can express the relationship between those two scales with the linear equation $y = 32 + 1.8x$.
a. Determine b_0 and b_1.
b. Find the Fahrenheit temperatures corresponding to the Celsius temperatures $-40°$, $0°$, $20°$, and $100°$.
c. Graph the linear equation $y = 32 + 1.8x$, using the four points found in part (b).
d. Apply the graph obtained in part (c) to estimate visually the Fahrenheit temperature corresponding to a Celsius temperature of $28°$. Then calculate that temperature exactly by using the linear equation $y = 32 + 1.8x$.

14.26 A Law of Physics. A ball is thrown straight up in the air with an initial velocity of 64 feet per second (ft/sec). According to the laws of physics, if you let y denote the velocity of the ball after x seconds, $y = 64 - 32x$.
a. Determine b_0 and b_1 for this linear equation.
b. Determine the velocity of the ball after 1, 2, 3, and 4 sec.
c. Graph the linear equation $y = 64 - 32x$, using the four points obtained in part (b).
d. Use the graph from part (c) to estimate visually the velocity of the ball after 1.5 sec. Then calculate that velocity exactly by using the linear equation $y = 64 - 32x$.

In each of Exercises 14.27–14.30,
a. find the y-intercept and slope of the specified linear equation.
b. explain what the y-intercept and slope represent in terms of the graph of the equation.
c. explain what the y-intercept and slope represent in terms relating to the application.

14.27 Rental-Car Costs. Refer to Exercise 14.23. The linear equation is $y = 68.22 + 0.25x$.

14.28 Air-Conditioning Repairs. Refer to Exercise 14.24. The linear equation is $y = 30 + 55x$.

14.29 Measuring Temperature. Refer to Exercise 14.25. The linear equation is $y = 32 + 1.8x$.

14.30 A Law of Physics. Refer to Exercise 14.26. The linear equation is $y = 64 - 32x$.

Extending the Concepts and Skills

14.31 Hooke's Law. According to *Hooke's law* for springs, developed by Robert Hooke (1635–1703), the force exerted by a spring that has been compressed to a length x is given by the formula $F = -k(x - x_0)$, where x_0 is the natural length of the spring and k is a constant, called the *spring constant*. A certain spring exerts a force of 32 lb when compressed to a length of 2 ft and a force of 16 lb when compressed to a length of 3 ft. For this spring, find the following.
a. The linear equation that relates the force exerted to the length compressed
b. The spring constant
c. The natural length of the spring

14.32 Road Grade. The *grade* of a road is defined as the distance it rises (or falls) to the distance it runs horizontally, usually expressed as a percentage. Consider a road with positive grade, g. Suppose that you begin driving on that road at an altitude a_0.
a. Find the linear equation that expresses the altitude, a, when you have driven a distance, d, along the road. (*Hint:* Draw a graph and apply the Pythagorean Theorem.)

b. Identify and interpret the y-intercept and slope of the linear equation in part (a).

c. Apply your results in parts (a) and (b) to a road with a 5% grade and an initial altitude of 1 mile. Express your answer for the slope to four decimal places.

d. For the road in part (c), what altitude will you reach after driving 10 miles along the road?

e. For the road in part (c), how far along the road must you drive to reach an altitude of 3 miles?

14.33 Vertical Lines. In this section, we stated that any nonvertical line can be described by an equation of the form $y = b_0 + b_1 x$.

a. Explain in detail why a vertical line can't be expressed in this form.

b. What is the form of the equation of a vertical line?

c. Does a vertical line have a slope? Explain your answer.

14.2   The Regression Equation

In Examples 14.1 and 14.2, we discussed the linear equation $y = 25 + 20x$, which expresses the total cost, y, of a word-processing job in terms of the time in hours, x, required to complete it. Given the amount of time required, x, we can use the equation to determine the *exact* cost of the job, y.

Real-life applications are seldom as simple as the word-processing example, in which one variable (cost) can be predicted exactly in terms of another variable (time required). Rather, we must often rely on rough predictions.

For instance, we cannot predict the exact asking price, y, of a particular make and model of car just by knowing its age, x. Indeed, even for a fixed age, say, 3 years old, price varies from car to car. We must be content with making a rough prediction for the price of a 3-year-old car of the particular make and model or with an estimate of the mean price of all such 3-year-old cars.

As you know, when analyzing data, it is essential to first construct a graph of the data. In this chapter and the next, a particularly useful graphic is a **scatterplot** (or **scatter diagram**).

DEFINITION 14.2

Scatterplot

A **scatterplot** is a graph of data from two quantitative variables of a population. In a scatterplot, we use a horizontal axis for the observations of one variable and a vertical axis for the observations of the other variable. Each pair of observations is then plotted as a point.

Note: Data from two quantitative variables of a population are called **bivariate quantitative data.**

■■■ **EXAMPLE 14.3**   Scatterplot

Age and Price of Orions Table 14.2 displays data on age and price for a sample of cars of a particular make and model. We refer to the car as the Orion, but the data, obtained from the *Asian Import* edition of the *Auto Trader* magazine, are for a real car. Ages are in years; prices are in hundreds of dollars, rounded to the nearest hundred dollars. Construct a scatterplot for these age and price data.

StatCrunch

Report 14.1

Solution Here the population under consideration consists of all Orions and the two quantitative variables are age and price. For the scatterplot, we will use a horizontal axis for ages and a vertical axis for prices. Each pair of observations (in this case, each age-price observation) is plotted as a point.

For instance, the second car in Table 14.2 is 4 years old and has a price of 103 ($10,300). We plot this age–price observation as the point (4, 103), shown in magenta in Fig. 14.7. Continuing in this way for the other 10 cars, we get the complete scatterplot of the age and price data, as depicted in Fig. 4.7. ■

As we see from Fig. 14.7, although the age–price data points do not fall exactly on a line, they appear to cluster about a line. We want to fit a line to the data points and use that line to predict the price of an Orion based on its age.

TABLE 14.2 Age and price data for a sample of 11 Orions

Car	Age (yr) x	Price ($100) y
1	5	85
2	4	103
3	6	70
4	5	82
5	5	89
6	5	98
7	6	66
8	6	95
9	2	169
10	7	70
11	7	48

FIGURE 14.7 Scatterplot for the age and price data of Orions from Table 14.2

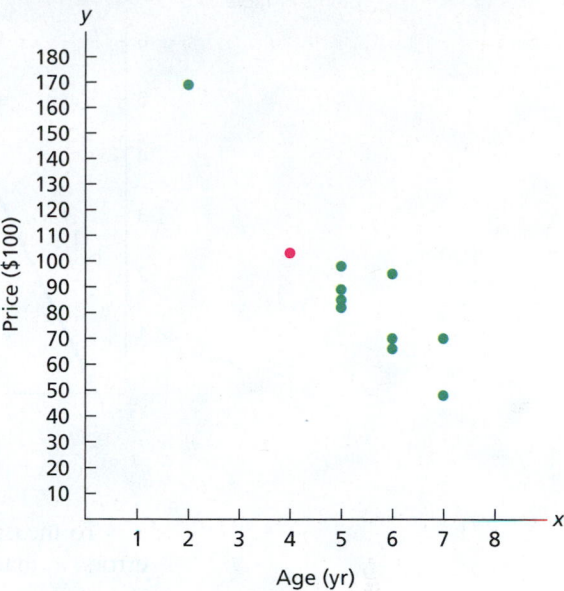

Because we could draw many different lines through the cluster of data points, we need a method to choose the "best" line. The method, called the *least-squares criterion,* is based on an analysis of the errors made in using a line to fit the data points. To introduce the least-squares criterion, we use a very simple data set in Example 14.4. We return to the Orion data soon.

EXAMPLE 14.4 Introducing the Least-Squares Criterion

TABLE 14.3
Three data points

x	y
1	3
2	1
3	5

Consider the problem of fitting a line to the three data points in Table 14.3, whose scatterplot is shown in Fig. 14.8. Many (in fact, infinitely many) lines can "fit" those three data points. Two of the many possibilities, which we call Line A ($y = -1 + 2x$) and Line B ($y = 1 + x$), are shown in Figs. 14.9(a) and 14.9(b) on the following page. Determine which of those two lines fits the data better.

Solution To avoid confusion, we use $\hat{y}$ to denote the x-value predicted by a line for a value of x. For instance, the y-value predicted by Line A for $x = 1$ is

$$\hat{y} = -1 + 2 \cdot 1 = 1,$$

and the y-value predicted by Line B for $x = 1$ is

$$\hat{y} = 1 + 1 = 2.$$

FIGURE 14.8
Scatterplot for the data points in Table 14.3

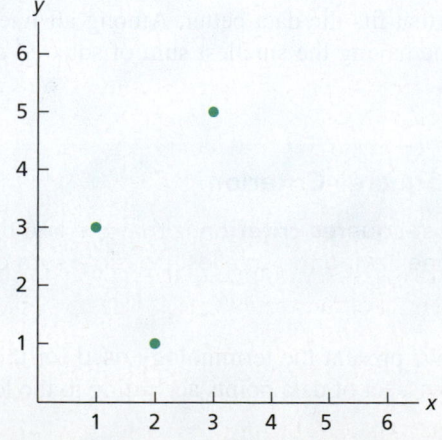

FIGURE 14.9
Two possible lines to fit the data
points in Table 14.3

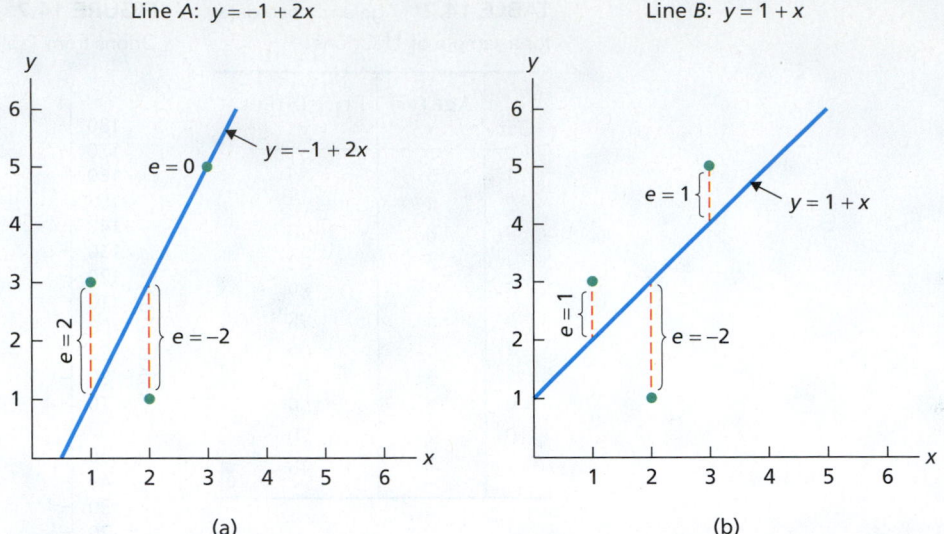

To measure quantitatively how well a line fits the data, we first consider the errors, e, made in using the line to predict the y-values of the data points. For instance, as we have just demonstrated, Line B predicts a y-value of $\hat{y} = 2$ when $x = 1$. The actual y-value for $x = 1$ is $y = 3$ (see Table 14.3). So, the error made in using Line B to predict the y-value of the data point $(1, 3)$ is

$$e = y - \hat{y} = 3 - 2 = 1,$$

as seen in Fig. 14.9(b).

In general, an **error**, e, is the signed vertical distance from the line to a data point. The fourth column of Table 14.4(a) shows the errors made by Line A for all three data points; the fourth column of Table 14.4(b) shows the same for Line B.

TABLE 14.4
Determining how well the data
points in Table 14.3 are fit
by (a) Line A and (b) Line B

Line A: $y = -1 + 2x$

x	y	$\hat{y}$	e	e^2
1	3	1	2	4
2	1	3	-2	4
3	5	5	0	0
				8

(a)

Line B: $y = 1 + x$

x	y	$\hat{y}$	e	e^2
1	3	2	1	1
2	1	3	-2	4
3	5	4	1	1
				6

(b)

You try it!

Exercise 14.43
on page 635

To decide which line, Line A or Line B, fits the data better, we first compute the sum of the squared errors, Σe_i^2, in the final columns of Tables 14.4(a) and 14.4(b). The line having the smaller sum of squared errors, in this case Line B, is the one that fits the data better. Among all lines, the **least-squares criterion** is that the line having the smallest sum of squared errors is the one that fits the data best.

KEY FACT 14.2

Least-Squares Criterion

The **least-squares criterion** is that the line that best fits a set of data points is the one having the smallest possible sum of squared errors.

Next we present the terminology used for the line (and corresponding equation) that best fits a set of data points according to the least-squares criterion.

DEFINITION 14.3

Regression Line and Regression Equation

Regression line: The line that best fits a set of data points according to the least-squares criterion.

Regression equation: The equation of the regression line.

APPLET

Applet 14.1

Although the least-squares criterion states the property that the regression line for a set of data points must satisfy, it does not tell us how to find that line. This task is accomplished by Formula 14.1. In preparation, we introduce some notation that will be used throughout our study of regression and correlation.

DEFINITION 14.4

Notation Used in Regression and Correlation

For a set of n data points, the defining and computing formulas for S_{xx}, S_{xy}, and S_{yy} are as follows.

Quantity	Defining formula	Computing formula
S_{xx}	$\Sigma(x_i - \bar{x})^2$	$\Sigma x_i^2 - (\Sigma x_i)^2/n$
S_{xy}	$\Sigma(x_i - \bar{x})(y_i - \bar{y})$	$\Sigma x_i y_i - (\Sigma x_i)(\Sigma y_i)/n$
S_{yy}	$\Sigma(y_i - \bar{y})^2$	$\Sigma y_i^2 - (\Sigma y_i)^2/n$

FORMULA 14.1

Regression Equation

The regression equation for a set of n data points is $\hat{y} = b_0 + b_1 x$, where

$$b_1 = \frac{S_{xy}}{S_{xx}} \quad \text{and} \quad b_0 = \bar{y} - b_1\bar{x} = \frac{1}{n}(\Sigma y_i - b_1 \Sigma x_i).$$

These two equations give the slope and y-intercept of the regression line, respectively.

Note: Although we have not used S_{yy} in Formula 14.1, we will use it later in this chapter when we study correlation.

In the next example, we will find the regression equation for the three-point data set considered earlier in this section. We'll use the defining formulas so that you can see how they work.

EXAMPLE 14.5   **The Regression Equation**

Consider again the three-point data set shown in Table 14.3, which we repeat in the first two columns of Table 14.5.

a. Determine the regression equation for the data. Use the defining formulas in Definition 14.4 for obtaining S_{xx} and S_{xy}.

b. Graph the regression equation and the data points.

TABLE 14.5

Table for finding S_{xx} and S_{yy} for the data in Table 14.3 by using the defining formulas

x	y	$x - \bar{x}$	$y - \bar{y}$	$(x - \bar{x})^2$	$(x - \bar{x})(y - \bar{y})$
1	3	-1	0	1	0
2	1	0	-2	0	0
3	5	1	2	1	2
				2	2

Solution We first note that $\bar{x} = 2$ and $\bar{y} = 3$. Now we construct Table 14.5, which provides the required quantities for applying the defining formulas in Definition 14.4 to get S_{xx} and S_{xy}.

a. Referring now to Formula 14.1 and the defining formulas in Definition 14.4, we get, in view of the last row of Table 14.5, that

$$b_1 = \frac{S_{xy}}{S_{xx}} = \frac{\Sigma(x_i - \bar{x})(y_i - \bar{y})}{\Sigma(x_i - \bar{x})^2} = \frac{2}{2} = 1$$

and

$$b_0 = \bar{y} - b_1\bar{x} = 3 - 1 \cdot 2 = 1.$$

Hence, the regression equation is

$$\hat{y} = b_0 + b_1 x = 1 + 1 \cdot x,$$

that is, $\hat{y} = 1 + x$.

b. A graph of the regression equation and the data points is shown in Fig. 14.10.

FIGURE 14.10

Regression line and data points

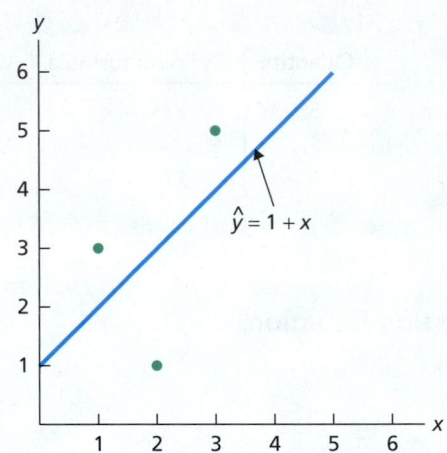

You try it!

Exercise 14.55 on page 636

Note that the regression line is just Line B of Example 14.4 (see pages 625–626). In that example, we showed that Line B has a smaller sum of squared errors than Line A and, hence, is the one that fits the data better. Because we have now seen that Line B is actually the regression line for the data, we can in fact say that it fits the data better than any other straight line. ◼

In practice, a regression equation is almost always found by using statistical software. When a regression equation is obtained by hand, the computing formulas for S_{xx} and S_{xy} are usually applied to avoid rounding error. We now return to the Orion data to illustrate this latter technique.

◼◼◼ **EXAMPLE 14.6** **The Regression Equation**

Age and Price of Orions In the first two columns of Table 14.6, we repeat our data on age and price for a sample of 11 Orions.

a. Determine the regression equation for the data. Use the computing formulas in Definition 14.4 (page 627) for obtaining S_{xx} and S_{xy}.
b. Graph the regression equation and the data points.
c. Describe the apparent relationship between age and price of Orions.
d. Interpret the slope of the regression line in terms of prices for Orions.
e. Use the regression equation to predict the price of a 3-year-old Orion and a 4-year-old Orion.

Solution

a. We first construct Table 14.6, which provides the required quantities for applying the computing formulas in Definition 14.4 to get S_{xx} and S_{xy}. Now, in

TABLE 14.6

Table for computing the regression equation for the Orion data

Age (yr)	Price ($100)		
x	y	xy	x^2
5	85	425	25
4	103	412	16
6	70	420	36
5	82	410	25
5	89	445	25
5	98	490	25
6	66	396	36
6	95	570	36
2	169	338	4
7	70	490	49
7	48	336	49
58	975	4732	326

view of the first equation in Formula 14.1 (page 627), the computing formulas in Definition 14.4, and the last row of Table 14.6, we see that the slope of the regression line is

$$b_1 = \frac{S_{xy}}{S_{xx}} = \frac{\Sigma x_i y_i - (\Sigma x_i)(\Sigma y_i)/n}{\Sigma x_i^2 - (\Sigma x_i)^2/n} = \frac{4732 - (58)(975)/11}{326 - (58)^2/11} = -20.26.$$

Now applying the second equation in Formula 14.1, we find that the y-intercept is

$$b_0 = \frac{1}{n}(\Sigma y_i - b_1 \Sigma x_i) = \frac{1}{11}[975 - (-20.26) \cdot 58] = 195.47.$$

So the regression equation is $\hat{y} = 195.47 - 20.26x$.

Note: The usual warnings about rounding apply. When computing the slope, b_1, of the regression line, do not round until the computation is finished. When computing the y-intercept, b_0, do not use the rounded value of b_1; instead, keep full calculator accuracy.

b. To graph the regression equation, we need to substitute two different x-values in the regression equation to obtain two distinct points. Let's use the x-values 2 and 8. The corresponding y-values are

$$\hat{y} = 195.47 - 20.26 \cdot 2 = 154.95 \quad \text{and} \quad \hat{y} = 195.47 - 20.26 \cdot 8 = 33.39.$$

Therefore, the regression line goes through the two points $(2, 154.95)$ and $(8, 33.39)$. In Fig. 14.11, we plotted these two points with open dots. Drawing a line through the two open dots yields the regression line, the graph of the regression equation. Figure 14.11 also shows the data points from the first two columns of Table 14.6.

FIGURE 14.11

Regression line and data points for Orion data

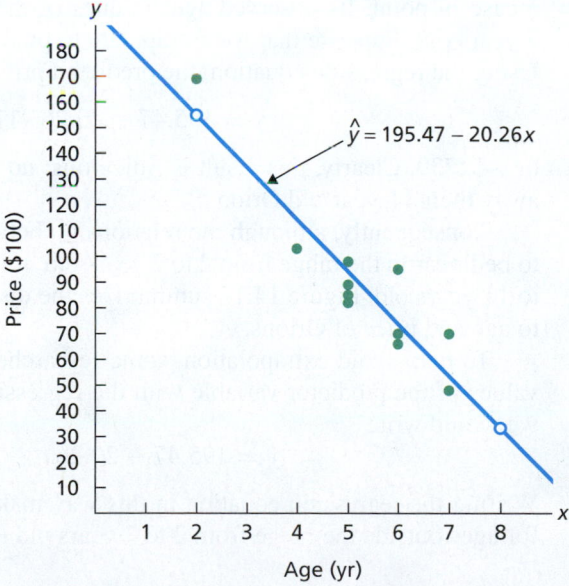

c. Because the slope of the regression line is negative, price tends to decrease as age increases, which is no particular surprise.

d. Because x represents age in years and y represents price in hundreds of dollars, the slope of -20.26 indicates that Orions depreciate an estimated $2026 per year, at least in the 2- to 7-year-old range.

e. For a 3-year-old Orion, $x = 3$, and the regression equation yields the predicted price of

$$\hat{y} = 195.47 - 20.26 \cdot 3 = 134.69.$$

Similarly, the predicted price for a 4-year-old Orion is

$$\hat{y} = 195.47 - 20.26 \cdot 4 = 114.43.$$

StatCrunch

Report 14.2

You try it!

Exercise 14.59
on page 636

Interpretation The estimated price of a 3-year-old Orion is $13,469, and the estimated price of a 4-year-old Orion is $11,443.

We discuss questions concerning the accuracy and reliability of such predictions later in this chapter and also in Chapter 15.

Predictor Variable and Response Variable

For a linear equation $y = b_0 + b_1x$, y is the dependent variable and x is the independent variable. However, in the context of regression analysis, we usually call y the **response variable** and x the **predictor variable** or **explanatory variable** (because it is used to predict or explain the values of the response variable). For the Orion example, then, age is the predictor variable and price is the response variable.

DEFINITION 14.5

> ### Response Variable and Predictor Variable
>
> **Response variable:** The variable to be measured or observed.
>
> **Predictor variable:** A variable used to predict or explain the values of the response variable.

Extrapolation

Suppose that a scatterplot indicates a linear relationship between two variables. Then, within the range of the observed values of the predictor variable, we can reasonably use the regression equation to make predictions for the response variable. However, to do so outside that range, which is called **extrapolation,** may not be reasonable because the linear relationship between the predictor and response variables may not hold there.

Grossly incorrect predictions can result from extrapolation. The Orion example is a case in point. Its observed ages (values of the predictor variable) range from 2 to 7 years old. Suppose that we extrapolate to predict the price of an 11-year-old Orion. Using the regression equation, the predicted price is

$$\hat{y} = 195.47 - 20.26 \cdot 11 = -27.39,$$

or −$2739. Clearly, this result is ridiculous: no one is going to pay us $2739 to take away their 11-year-old Orion.

Consequently, although the relationship between age and price of Orions appears to be linear in the range from 2 to 7 years old, it is definitely not so in the range from 2 to 11 years old. Figure 14.12 summarizes the discussion on extrapolation as it applies to age and price of Orions.

To help avoid extrapolation, some researchers include the range of the observed values of the predictor variable with the regression equation. For the Orion example, we would write

$$\hat{y} = 195.47 - 20.26x, \qquad 2 \le x \le 7.$$

Writing the regression equation in this way makes clear that using it to predict price for ages outside the range from 2 to 7 years old is extrapolation.

Outliers and Influential Observations

Recall that an outlier is an observation that lies outside the overall pattern of the data. In the context of regression, an **outlier** is a data point that lies far from the regression line, relative to the other data points. Figure 14.11 on page 629 shows that the Orion data have no outliers.

An outlier can sometimes have a significant effect on a regression analysis. Thus, as usual, we need to identify outliers and remove them from the analysis when appropriate—for example, if we find that an outlier is a measurement or recording error.

APPLET

Applet 14.2

We must also watch for *influential observations*. In regression analysis, an **influential observation** is a data point whose removal causes the regression equation (and

FIGURE 14.12

Extrapolation in the Orion example

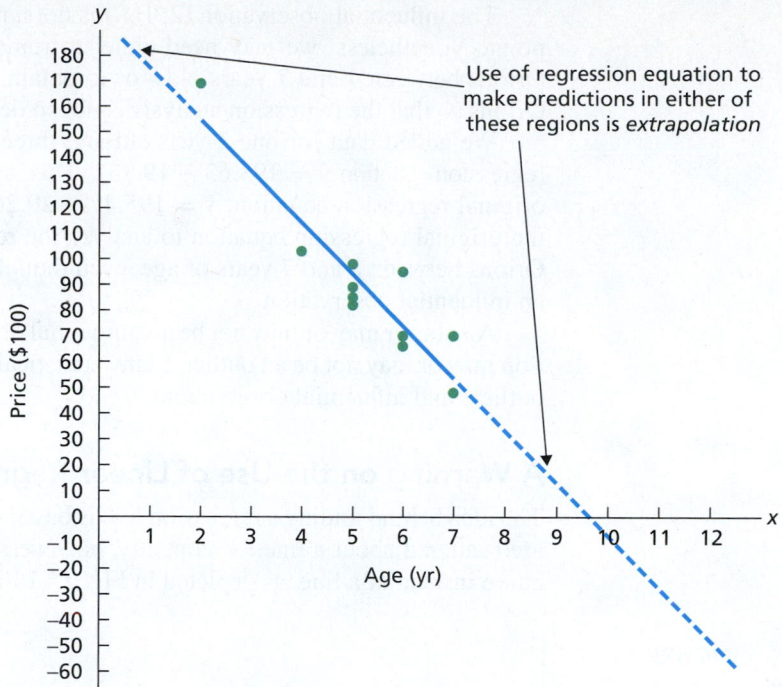

line) to change considerably. A data point separated in the *x*-direction from the other data points is often an influential observation because the regression line is "pulled" toward such a data point without counteraction by other data points.

If an influential observation is due to a measurement or recording error, or if for some other reason it clearly does not belong in the data set, it can be removed without further consideration. However, if no explanation for the influential observation is apparent, the decision whether to retain it is often difficult and calls for a judgment by the researcher.

For the Orion data, Fig. 14.11 on page 629 (or Table 14.6 on page 629) shows that the data point (2, 169) might be an influential observation because the age of 2 years appears separated from the other observed ages. Removing that data point and recalculating the regression equation yields $\hat{y} = 160.33 - 14.24x$. Figure 14.13 reveals that this equation differs markedly from the regression equation based on the full data set. The data point (2, 169) is indeed an influential observation.

FIGURE 14.13

Regression lines with and without the influential observation removed

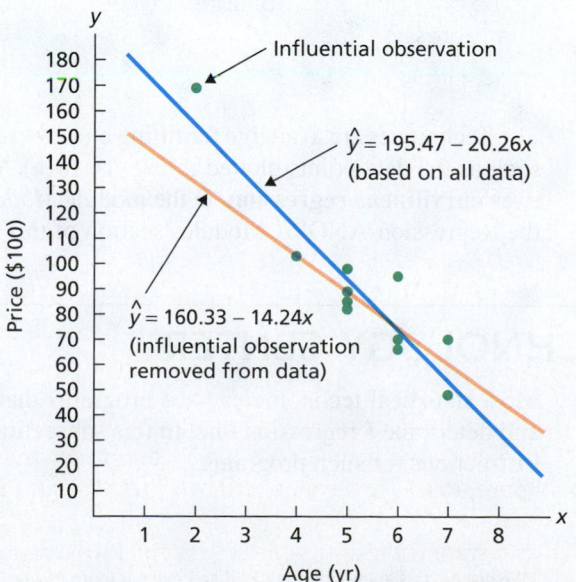

The influential observation (2, 169) is not a recording error; it is a legitimate data point. Nonetheless, we may need either to remove it—thus limiting the analysis to Orions between 4 and 7 years old—or to obtain additional data on 2- and 3-year-old Orions so that the regression analysis is not so dependent on one data point.

We added data for one 2-year-old and three 3-year-old Orions and obtained the regression equation $\hat{y} = 193.63 - 19.93x$. This regression equation differs little from our original regression equation, $\hat{y} = 195.47 - 20.26x$. Therefore we could justify using the original regression equation to analyze the relationship between age and price of Orions between 2 and 7 years of age, even though the corresponding data set contains an influential observation.

An outlier may or may not be an influential observation, and an influential observation may or may not be an outlier. Many statistical software packages identify potential outliers and influential observations.

A Warning on the Use of Linear Regression

The idea behind finding a regression line is based on the assumption that the data points are scattered about a line.[†] Frequently, however, the data points are scattered about a curve instead of a line, as depicted in Fig. 14.14(a).

FIGURE 14.14

(a) Data points scattered about a curve; (b) inappropriate line fit to the data points

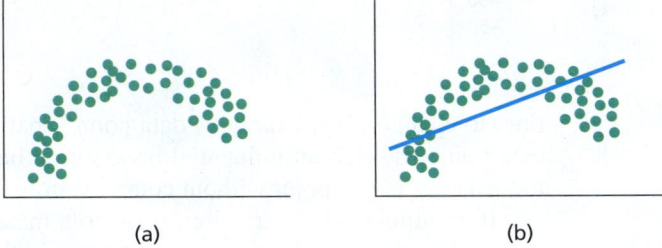

(a)   (b)

One can still compute the values of b_0 and b_1 to obtain a regression line for these data points. The result, however, will yield an inappropriate fit by a line, as shown in Fig. 14.14(b), when in fact a curve should be used. For instance, the regression line suggests that y-values in Fig. 14.14(a) will keep increasing when they have actually begun to decrease.

KEY FACT 14.3

> ### Criterion for Finding a Regression Line
>
> Before finding a regression line for a set of data points, draw a scatterplot. If the data points do not appear to be scattered about a line, do not determine a regression line.

Techniques are available for fitting curves to data points that show a curved pattern, such as the data points plotted in Fig. 14.14(a). We discuss those techniques, referred to as **curvilinear regression,** in the module *Model Building in Regression*, available in the Regression-ANOVA Modules section of the WeissStats site.

 # THE TECHNOLOGY CENTER

Most statistical technologies have programs that automatically generate a scatterplot and determine a regression line. In this subsection, we present output and step-by-step instructions for such programs.

[†]We discuss this assumption in detail and make it more precise in Section 15.1.

EXAMPLE 14.7   Using Technology to Obtain a Scatterplot

Age and Price of Orions Use Minitab, Excel, or the TI-83/84 Plus to obtain a scatterplot for the age and price data in Table 14.2 on page 625.

Solution We applied the scatterplot programs to the data, resulting in Output 14.1. Steps for generating that output are presented in Instructions 14.1.

OUTPUT 14.1   Scatterplots for the age and price data of 11 Orions

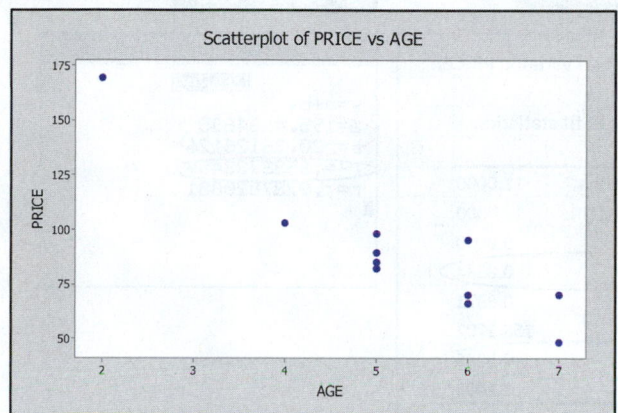

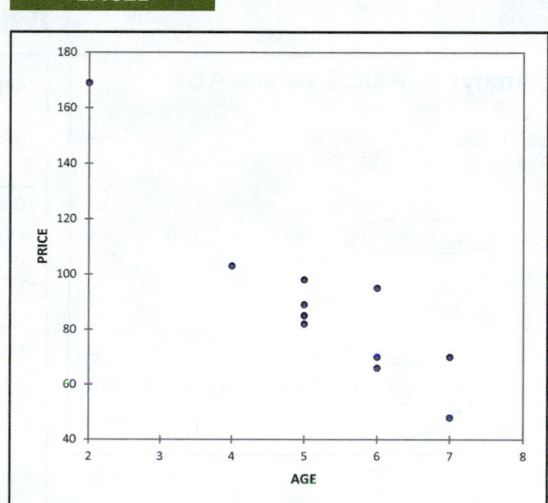

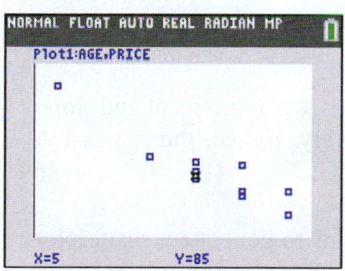

As shown in Output 14.1, the data points are scattered about a line. So, we can reasonably find a regression line for these data.

INSTRUCTIONS 14.1   Steps for generating Output 14.1

MINITAB

1 Store the age and price data from Table 14.2 in columns named AGE and PRICE, respectively
2 Choose **Graph ➤ Scatterplot...**
3 Click **OK**
4 Press the F3 key to reset the dialog box
5 Specify PRICE in the **Y variables** text box
6 Specify AGE in the **X variables** text box
7 Click **OK**

EXCEL

1 Store the age and price data from Table 14.2 in columns named AGE and PRICE, respectively
2 Choose **XLSTAT ➤ Visualizing data ➤ Scatter plots**
3 Click the reset button in the lower left corner of the dialog box
4 Click in the **X** selection box and then select the column of the worksheet that contains the AGE data

5 Click in the **Y** selection box and then select the column of the worksheet that contains the PRICE data
6 Click **OK**
7 Click the **Continue** button in the **XLSTAT – Selections** dialog box

TI-83/84 PLUS

1 Store the age and price data from Table 14.2 in lists named AGE and PRICE, respectively
2 Ensure that all stat plots and all **Y =** functions are off
3 Press **2nd ➤ STAT PLOT** and then press **ENTER** twice
4 Arrow to the first graph icon and press **ENTER**
5 Press the down-arrow key
6 Press **2nd ➤ LIST**, arrow down to AGE, and press **ENTER** twice
7 Press **2nd ➤ LIST**, arrow down to PRICE, and press **ENTER** twice
8 Press **ZOOM**, then **9**, and then **TRACE**

EXAMPLE 14.8 Using Technology to Obtain a Regression Line

Age and Price of Orions Use Minitab, Excel, or the TI-83/84 Plus to determine the regression equation for the age and price data in Table 14.2 on page 625.

Solution We applied the regression programs to the data, resulting in Output 14.2. Steps for generating that output are presented in Instructions 14.2. *Note to Excel users:* For brevity, we have presented only the essential portions of the actual output.

OUTPUT 14.2 Regression analysis on the age and price data of 11 Orions

MINITAB

Regression Analysis: PRICE versus AGE

The regression equation is
PRICE = 195.5 - 20.26 AGE

S = 12.5766   R-Sq = 85.3%   R-Sq(adj) = 83.7%

Analysis of Variance

Source	DF	SS	MS	F	P
Regression	1	8285.01	8285.01	52.38	0.000
Error	9	1423.53	158.17		
Total	10	9708.55			

EXCEL

Regression of variable PRICE:

Goodness of fit statistics:

Observations	11.0000
Sum of weights	11.0000
DF	9.0000
R^2	0.8534
Adjusted R^2	0.8371
MSE	158.1702
RMSE	12.5766
DW	1.2488

Equation of the model:

PRICE = 195.46847-20.26126*AGE

TI-83/84 PLUS

```
NORMAL FLOAT AUTO REAL RADIAN MP
            LinReg
   y=a+bx
   a=195.4684685
   b=-20.26126126
   r²=.8533733464
   r=-.9237820881
```

As shown in Output 14.2 (see the items circled in red), the y-intercept and slope of the regression line are 195.5 and -20.26, respectively. Hence, the regression equation is $\hat{y} = 195.5 - 20.26x$.

We can also use Minitab, Excel, or the TI-83/84 Plus to generate a scatterplot of the age and price data with a superimposed regression line, similar to the graph in Fig. 14.11 on page 629. Note that Minitab's **Fitted Line Plot** automatically provides a scatterplot of the data with a superimposed regression line. Users of Excel or the TI-83/84 Plus can consult their instructors.

INSTRUCTIONS 14.2 Steps for generating Output 14.2

MINITAB

1. Store the age and price data from Table 14.2 in columns named AGE and PRICE, respectively
2. Choose **Stat ➤ Regression ➤ Fitted Line Plot...**
3. Press the F3 key to reset the dialog box
4. Specify PRICE in the **Response (Y)** text box
5. Specify AGE in the **Predictor (X)** text box
6. Click **OK**
7. Choose **Window ➤ Session**

EXCEL

1. Store the age and price data from Table 14.2 in columns named AGE and PRICE, respectively
2. Choose **XLSTAT ➤ Modeling data ➤ Linear regression**

3. Click the reset button in the lower left corner of the dialog box
4. In the **Y / Dependent variables** list, click in the **Quantitative** selection box and then select the column of the worksheet that contains the PRICE data
5. In the **X / Explanatory variables** list, click in the **Quantitative** selection box and then select the column of the worksheet that contains the AGE data
6. Click the **Outputs** tab and then uncheck all the check boxes therein
7. Click the **Charts** tab and then uncheck the **Regression charts** check box
8. Click **OK**
9. Click the **Continue** button in the **XLSTAT – Selections** dialog box

(continued)

Exercises 14.2

Understanding the Concepts and Skills

14.34 Regarding a scatterplot,
a. identify one of its uses.
b. what property should it have to obtain a regression line for the data?

14.35 Regarding the criterion used to decide on the line that best fits a set of data points,
a. what is that criterion called?
b. specifically, what is the criterion?

14.36 Regarding the line that best fits a set of data points,
a. what is that line called?
b. what is the equation of that line called?

14.37 Regarding the two variables under consideration in a regression analysis,
a. what is the dependent variable called?
b. what is the independent variable called?

14.38 Using the regression equation to make predictions for values of the predictor variable outside the range of the observed values of the predictor variable is called _____.

14.39 Fill in the blanks.
a. In the context of regression, an _____ is a data point that lies far from the regression line, relative to the other data points.
b. In regression analysis, an _____ is a data point whose removal causes the regression equation to change considerably.

14.40 For which of the following sets of data points can you reasonably determine a regression line? Explain your answer.

14.41 For which of the following sets of data points can you reasonably determine a regression line? Explain your answer.

In each of Exercises 14.42–14.45, we have presented two linear equations and a set of data points. For each exercise,
a. *plot the data points and the first linear equation on one graph and the data points and the second linear equation on another.*
b. *construct tables for x, y, ŷ, e, and e^2 like Table 14.4 (page 626).*
c. *determine which line fits the set of data points better, according to the least-squares criterion.*

14.42 Line A: $y = 9 - 2x$, Line B: $y = 6 - x$

Data points:

x	1	3	4	4
y	8	0	3	1

14.43 Line A: $y = -1 + 3x$, Line B: $y = 1 + 2x$

Data points:

x	1	2	3
y	4	3	8

14.44 Line A: $y = 1.5 + 0.5x$, Line B: $y = 1.125 + 0.375x$

Data points:

x	1	1	5	5
y	1	3	2	4

14.45 Line A: $y = 3 - 0.6x$, Line B: $y = 4 - x$

Data points:

x	0	2	2	5	6
y	4	2	0	-2	1

14.46 For a data set consisting of two data points:
a. Identify the regression line.
b. What is the sum of squared errors for the regression line? Explain your answer.

14.47 Refer to Exercise 14.46. For each of the following sets of data points, determine the regression equation both without and with the use of Formula 14.1 on page 627.
a.

x	2	4
y	1	3

b.

x	1	5
y	3	-3

In each of Exercises 14.48–14.57,
a. *find the regression equation for the data points. Use the defining formulas in Definition 14.4 to obtain S_{xx} and S_{xy}.*
b. *graph the regression equation and the data points.*

14.48

x	2	4	3
y	3	5	7

14.49

x	3	1	2
y	-4	0	-5

14.50

x	0	4	3	1	2
y	1	9	8	4	3

14.51

x	3	4	1	2
y	4	5	0	-1

14.52

x	1	3	4	4
y	13	-1	3	5

14.53

x	2	2	3	4	4
y	3	4	0	2	1

14.54 The data points in Exercise 14.42

14.55 The data points in Exercise 14.43

14.56 The data points in Exercise 14.44

14.57 The data points in Exercise 14.45

Applying the Concepts and Skills

In each of Exercises 14.58–14.63,
a. *find the regression equation for the data points.*
b. *graph the regression equation and the data points.*
c. *describe the apparent relationship between the two variables under consideration.*
d. *interpret the slope of the regression line.*
e. *identify the predictor and response variables.*
f. *identify outliers and potential influential observations.*
g. *predict the values of the response variable for the specified values of the predictor variable, and interpret your results.*

14.58 Tax Efficiency. *Tax efficiency* is a measure, ranging from 0 to 100, of how much tax due to capital gains stock or mutual funds investors pay on their investments each year; the higher the tax efficiency, the lower is the tax. In the article "At the Mercy of the Manager" (*Financial Planning*, Vol. 30(5), pp. 54–56), C. Israelsen examined the relationship between investments in mutual fund portfolios and their associated tax efficiencies. The following table shows percentage of investments in energy securities (x) and tax efficiency (y) for 10 mutual fund portfolios. For part (g), predict the tax efficiency of a mutual fund portfolio with 5.0% of its investments in energy securities and one with 7.4% of its investments in energy securities.

x	3.1	3.2	3.7	4.3	4.0	5.5	6.7	7.4	7.4	10.6
y	98.1	94.7	92.0	89.8	87.5	85.0	82.0	77.8	72.1	53.5

14.59 Corvette Prices. The *Kelley Blue Book* provides information on wholesale and retail prices of cars. Following are age and price data for 10 randomly selected Corvettes between 1 and 6 years old.

Here, x denotes age, in years, and y denotes price, in hundreds of dollars. For part (g), predict the prices of a 2-year-old Corvette and a 3-year-old Corvette.

x	6	6	6	2	2	5	4	5	1	4
y	290	280	295	425	384	315	355	328	425	325

14.60 Custom Homes. Hanna Properties specializes in custom-home resales in the Equestrian Estates, an exclusive subdivision in Phoenix, Arizona. A random sample of nine custom homes currently listed for sale provided the following information on size and price. Here, x denotes size, in hundreds of square feet, rounded to the nearest hundred, and y denotes price, in thousands of dollars, rounded to the nearest thousand. For part (g), predict the price of a 2600-sq. ft. home in the Equestrian Estates.

x	26	27	33	29	29	34	30	40	22
y	540	555	575	577	606	661	738	804	496

14.61 Plant Emissions. Plants emit gases that trigger the ripening of fruit, attract pollinators, and cue other physiological responses. N. Agelopolous et al. examined factors that affect the emission of volatile compounds by the potato plant *Solanum tuberosum* and published their findings in the paper "Factors Affecting Volatile Emissions of Intact Potato Plants, *Solanum tuberosum*: Variability of Quantities and Stability of Ratios" (*Journal of Chemical Ecology*, Vol. 26, No. 2, pp. 497–511). The volatile compounds analyzed were hydrocarbons used by other plants and animals. Following are data on plant weight (x), in grams, and quantity of volatile compounds emitted (y), in hundreds of nanograms, for 11 potato plants. For part (g), predict the quantity of volatile compounds emitted by a potato plant that weighs 75 grams.

x	57	85	57	65	52	67	62	80	77	53	68
y	8.0	22.0	10.5	22.5	12.0	11.5	7.5	13.0	16.5	21.0	12.0

14.62 Crown-Rump Length. In the article "The Human Vomeronasal Organ. Part II: Prenatal Development" (*Journal of Anatomy*, Vol. 197, Issue 3, pp. 421–436), T. Smith and K. Bhatnagar examined the controversial issue of the human vomeronasal organ, regarding its structure, function, and identity. The following table shows the age of fetuses (x), in weeks, and length of crown-rump (y), in millimeters. For part (g), predict the crown-rump length of a 19-week-old fetus.

x	10	10	13	13	18	19	19	23	25	28
y	66	66	108	106	161	166	177	228	235	280

14.63 Study Time and Score. An instructor at Arizona State University asked a random sample of eight students to record their study times in a beginning calculus course. She then made a table for total hours studied (x) over 2 weeks and test score (y) at the end of the 2 weeks. Here are the results. For part (g), predict the score of a student who studies for 15 hours.

x	10	15	12	20	8	16	14	22
y	92	81	84	74	85	80	84	80

14.64 Tax Efficiency. In Exercise 14.58, you determined a regression equation that relates the variables percentage of investments in energy securities and tax efficiency for mutual fund portfolios.

a. Should that regression equation be used to predict the tax efficiency of a mutual fund portfolio with 6.4% of its investments in energy securities? with 15% of its investments in energy securities? Explain your answers.

b. For which percentages of investments in energy securities is use of the regression equation to predict tax efficiency reasonable?

14.65 Corvette Prices. In Exercise 14.59, you determined a regression equation that can be used to predict the price of a Corvette, given its age.

a. Should that regression equation be used to predict the price of a 4-year-old Corvette? a 10-year-old Corvette? Explain your answers.

b. For which ages is use of the regression equation to predict price reasonable?

14.66 Anscombe's Quartet. In the article "Graphs in Statistical Analysis" (*American Statistician*, Vol. 27, Issue 1, pp 17–21), F. Anscombe presented four sets of data points with almost identical basic statistical properties (means, standard deviations, regression lines, etc.) but quite different scatterplots. We have provided Anscombe's four sets of data points on the WeissStats site. Use the technology of your choice to solve the following problems.

a. Obtain the mean and standard deviation of each set of x-values. Compare your results.

b. Obtain the mean and standard deviation of each set of y-values. Compare your results.

c. Find the regression equation for each set of data points. Compare your results.

d. Draw a scatterplot with superimposed regression line for each set of data points.

e. Discuss your results in part (d) with respect to the importance of plotting data before analyzing it and to the effect of outliers.

14.67 Study Time and Score. The negative relation between study time and test score found in Exercise 14.63 has been discovered by many investigators. Provide a possible explanation for it.

14.68 Age and Price of Orions. In Table 14.2, we provided data on age and price for a sample of 11 Orions between 2 and 7 years old. On the WeissStats site, we have given the ages and prices for a sample of 31 Orions between 1 and 11 years old.

a. Obtain a scatterplot for the data.

b. Is it reasonable to find a regression line for the data? Explain your answer.

14.69 Wasp Mating Systems. In the paper "Mating System and Sex Allocation in the Gregarious Parasitoid *Cotesia glomerata*" (*Animal Behaviour*, Vol. 66, pp. 259–264), H. Gu and S. Dorn reported on various aspects of the mating system and sex allocation strategy of the wasp *C. glomerata*. One part of the study involved the investigation of the percentage of male wasps dispersing before mating in relation to the brood sex ratio (proportion of males). The data obtained by the researchers are on the WeissStats site.

a. Obtain a scatterplot for the data.

b. Is it reasonable to find a regression line for the data? Explain your answer.

Working with Large Data Sets

In Exercises 14.70–14.80, use the technology of your choice to do the following tasks.

a. Obtain a scatterplot for the data.

b. Decide whether finding a regression line for the data is reasonable. If so, then also do parts (c)–(f).

c. Determine and interpret the regression equation for the data.

d. Identify potential outliers and influential observations.

e. In case a potential outlier is present, remove it and discuss the effect.

f. In case a potential influential observation is present, remove it and discuss the effect.

14.70 Birdies and Score. How important are birdies (a score of one under par on a given golf hole) in determining the final total score of a woman golfer? From the *U.S. Women's Open* website, we obtained data on number of birdies during a tournament and final score for 63 women golfers. The data are presented on the WeissStats site.

14.71 U.S. Presidents. The *Information Please Almanac* provides data on the ages at inauguration and of death for the presidents of the United States. We give those data on the WeissStats site for those presidents who are not still living at the time of this writing.

14.72 Movie Grosses. Box Office Mojo collects and posts data on movie grosses. For a random sample of 50 movies, we obtained both the domestic (U.S.) and overseas grosses, in millions of dollars. The data are presented on the WeissStats site.

14.73 Acreage and Value. The document *Arizona Residential Property Valuation System*, published by the Arizona Department of Revenue, describes how county assessors use computerized systems to value single-family residential properties for property tax purposes. On the WeissStats site are data on lot size (in acres) and assessed value (in thousands of dollars) for a sample of homes in a particular area.

14.74 Home Size and Value. On the WeissStats site are data on home size (in square feet) and assessed value (in thousands of dollars) for the same homes as in Exercise 14.73.

14.75 High and Low Temperature. The National Oceanic and Atmospheric Administration publishes temperature information of cities around the world in *Climates of the World*. A random sample of 50 cities gave the data on average high and low temperatures in January shown on the WeissStats site.

14.76 PCBs and Pelicans. Polychlorinated biphenyls (PCBs), industrial pollutants, are known to be carcinogens and a great danger to natural ecosystems. As a result of several studies, PCB production was banned in the United States in 1979 and by the Stockholm Convention on Persistent Organic Pollutants in 2001. One study, published in 1972 by R. Risebrough, is titled "Effects of Environmental Pollutants Upon Animals Other Than Man" (*Proceedings of the 6th Berkeley Symposium on Mathematics and Statistics, VI*, University of California Press, pp. 443–463). In that study, 60 Anacapa pelican eggs were collected and measured for their shell thickness, in millimeters (mm), and concentration of PCBs, in parts per million (ppm). The data are on the WeissStats site.

14.77 More Money, More Beer? Does a higher state per capita income equate to a higher per capita beer consumption? From the document *Survey of Current Business*, published by the U.S. Bureau of Economic Analysis, and from the *Brewer's Almanac*, published by the Beer Institute, we obtained data on personal income per capita, in thousands of dollars, and per capita beer consumption, in gallons, for the 50 states and Washington, D.C. Those data are provided on the WeissStats site.

14.78 Gas Guzzlers. The magazine *Consumer Reports* publishes information on automobile gas mileage and variables that affect gas mileage. In one issue, data on gas mileage (in miles per gallon) and engine displacement (in liters) were published for 121 vehicles. Those data are available on the WeissStats site.

14.79 Top Wealth Managers. An issue of *BARRON'S* presented information on top wealth managers in the United States, based on individual clients with accounts of $1 million or more. Data were given for various variables, two of which were number of private client managers and private client assets. Those data are provided on the WeissStats site, where private client assets are in billions of dollars.

14.80 Shortleaf Pines. The ability to estimate the volume of a tree based on a simple measurement, such as the tree's diameter, is important to the lumber industry, ecologists, and conservationists. Data on volume, in cubic feet, and diameter at breast height, in inches, for 70 shortleaf pines were reported in C. Bruce and F. X. Schumacher's *Forest Mensuration* (New York: McGraw-Hill, 1935) and analyzed by A. C. Akinson in the article "Transforming Both Sides of a Tree" (*The American Statistician*, Vol. 48, pp. 307–312). The data are presented on the WeissStats site.

Extending the Concepts and Skills

14.81 Sample Covariance. For a set of n data points, the **sample covariance**, s_{xy}, is given by

$$s_{xy} = \underbrace{\frac{\Sigma(x_i - \bar{x})(y_i - \bar{y})}{n-1}}_{\text{Defining formula}} = \underbrace{\frac{\Sigma x_i y_i - (\Sigma x_i)(\Sigma y_i)/n}{n-1}}_{\text{Computing formula}}. \quad (14.1)$$

The sample covariance can be used as an alternative method for finding the slope and y-intercept of a regression line. The formulas are

$$b_1 = s_{xy}/s_x^2 \quad \text{and} \quad b_0 = \bar{y} - b_1\bar{x}, \quad (14.2)$$

where s_x denotes the sample standard deviation of the x-values.

a. Use Equation (14.1) to determine the sample covariance of the data points in Exercise 14.45.

b. Use Equation (14.2) and your answer from part (a) to find the regression equation. Compare your result to that found in Exercise 14.57.

14.82 Time Series. A collection of observations of a variable y taken at regular intervals over time is called a **time series.** Economic data and electrical signals are examples of time series. We can think of a time series as providing data points (x_i, y_i), where x_i is the ith observation time and y_i is the observed value of y at time x_i. If a time series exhibits a linear trend, we can find that trend by determining the regression equation for the data points. We can then use the regression equation for forecasting purposes.

As an illustration, consider the data on the WeissStats site that shows the U.S. population, in millions of persons, for the years 1990–2013, as provided by the U.S. Census Bureau.

a. Use the technology of your choice to obtain a scatterplot of the data.

b. Use the technology of your choice to find the regression equation.

c. Use your result from part (b) to forecast the U.S. population for the years 2014 and 2015.

14.3 The Coefficient of Determination

In Example 14.6, we determined the regression equation, $\hat{y} = 195.47 - 20.26x$, for data on age and price of a sample of 11 Orions, where x represents age, in years, and $\hat{y}$ represents predicted price, in hundreds of dollars. We also applied the regression equation to predict the price of a 4-year-old Orion:

$$\hat{y} = 195.47 - 20.26 \cdot 4 = 114.43,$$

or $11,443. But how valuable are such predictions? Is the regression equation useful for predicting price, or could we do just as well by ignoring age?

In general, several methods exist for evaluating the utility of a regression equation for making predictions. One method is to determine the percentage of variation in the observed values of the response variable that is explained by the regression (or predictor variable), as discussed below. To find this percentage, we need to define two measures of variation: (1) the total variation in the observed values of the response variable and (2) the amount of variation in the observed values of the response variable that is explained by the regression.

Sums of Squares and Coefficient of Determination

To measure the total variation in the observed values of the response variable, we use the sum of squared deviations of the observed values of the response variable from the mean of those values. This measure of variation is called the **total sum of squares, *SST*.** Thus, $SST = \Sigma(y_i - \bar{y})^2$. If we divide SST by $n - 1$, we get the sample variance of the observed values of the response variable. So, SST really is a measure of total variation.

To measure the amount of variation in the observed values of the response variable that is explained by the regression, we first look at a particular observed value of the response variable, say, corresponding to the data point (x_i, y_i), as shown in Fig. 14.15.

The total variation in the observed values of the response variable is based on the deviation of each observed value from the mean value, $y_i - \bar{y}$. As shown in Fig. 14.15, each such deviation can be decomposed into two parts: the deviation explained by the

FIGURE 14.15 Decomposing the deviation of an observed y-value from the mean into the deviations explained and not explained by the regression

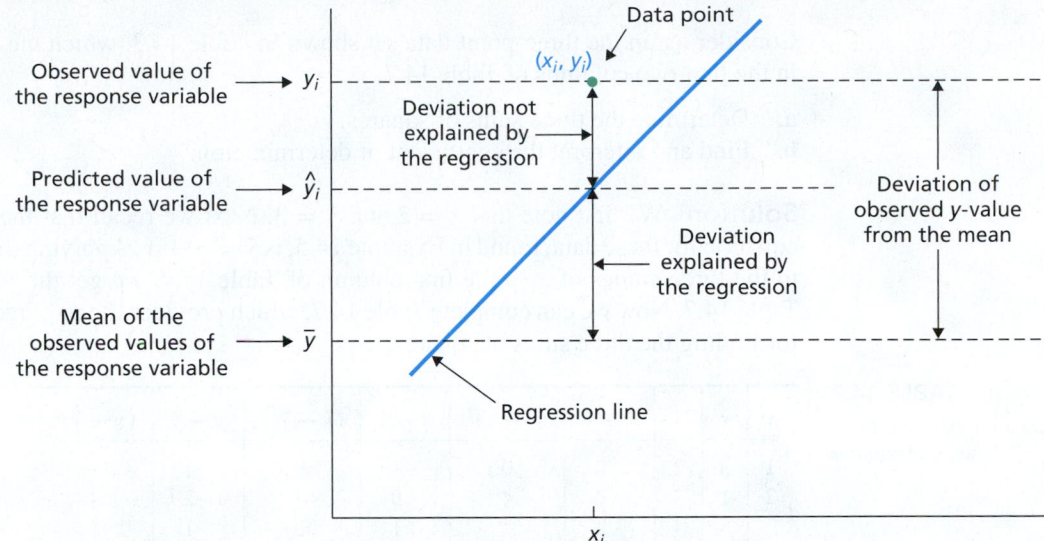

regression line, $\hat{y}_i - \bar{y}$, and the remaining unexplained deviation, $y_i - \hat{y}_i$. Hence the amount of variation (squared deviation) in the observed values of the response variable that is explained by the regression is $\Sigma(\hat{y}_i - \bar{y})^2$. This measure of variation is called the **regression sum of squares, SSR.** Thus, $SSR = \Sigma(\hat{y}_i - \bar{y})^2$.

Using the total sum of squares and the regression sum of squares, we can determine the percentage of variation in the observed values of the response variable that is explained by the regression, namely, SSR/SST. This quantity is called the **coefficient of determination** and is denoted r^2. Thus, $r^2 = SSR/SST$.

Before applying the coefficient of determination, let's consider the remaining deviation portrayed in Fig. 14.15: the deviation not explained by the regression, $y_i - \hat{y}_i$. The amount of variation (squared deviation) in the observed values of the response variable that is not explained by the regression is $\Sigma(y_i - \hat{y}_i)^2$. This measure of variation is called the **error sum of squares, SSE.** Thus, $SSE = \Sigma(y_i - \hat{y}_i)^2$.

DEFINITION 14.6

Sums of Squares in Regression

Total sum of squares, SST: The total variation in the observed values of the response variable: $SST = \Sigma(y_i - \bar{y})^2$.

Regression sum of squares, SSR: The variation in the observed values of the response variable explained by the regression: $SSR = \Sigma(\hat{y}_i - \bar{y})^2$.

Error sum of squares, SSE: The variation in the observed values of the response variable not explained by the regression: $SSE = \Sigma(y_i - \hat{y}_i)^2$.

DEFINITION 14.7

? **What Does It Mean?**

The coefficient of determination is a descriptive measure of the utility of the regression equation for making predictions.

Coefficient of Determination

The **coefficient of determination, r^2,** is the proportion of variation in the observed values of the response variable explained by the regression. Thus,

$$r^2 = \frac{SSR}{SST}.$$

The coefficient of determination, r^2, always lies between 0 and 1. A value of r^2 near 0 suggests that the regression equation is not very useful for making predictions, whereas a value of r^2 near 1 suggests that the regression equation is quite useful for making predictions.

■■■ **EXAMPLE 14.9** **Sums of Squares and Coefficient of Determination**

Consider again the three-point data set shown in Table 14.3, which we repeat here in the first two columns of Table 14.7.

a. Determine the three sums of squares.
b. Find and interpret the coefficient of determination.

Solution We first note that $\bar{x} = 2$ and $\bar{y} = 3$. Next we recall that the regression equation for these data, found in Example 14.5, is $\hat{y} = 1 + x$. Applying that equation to the three values of x in the first column of Table 14.7, we get the $\hat{y}$-column of Table 14.7. Now we can complete Table 14.7, which provides the required quantities for finding the three sums of squares.

TABLE 14.7

Table for finding the three sums of squares

x	y	$\hat{y}$	$y - \bar{y}$	$(y - \bar{y})^2$	$\hat{y} - \bar{y}$	$(\hat{y} - \bar{y})^2$	$y - \hat{y}$	$(y - \hat{y})^2$
1	3	2	0	0	-1	1	1	1
2	1	3	-2	4	0	0	-2	4
3	5	4	2	4	1	1	1	1
				8		2		6

a. Referring to Definition 14.6 and the last row of Table 14.7, we get that

$$SST = \Sigma(y - \bar{y})^2 = 8,$$
$$SSR = \Sigma(\hat{y} - \bar{y})^2 = 2,$$
$$SSE = \Sigma(y - \hat{y})^2 = 6.$$

For the three-point data set shown in the first two columns of Table 14.7, the total sum of squares is 8, the regression sum of squares is 2, and the error sum of squares is 6.

b. Referring to Definition 14.7 and the results of part (a), we see that

$$r^2 = \frac{SSR}{SST} = \frac{2}{8} = 0.25.$$

You try it!

Exercise 14.95 on page 643

For the three-point data set shown in the first two columns of Table 14.7, the coefficient of determination is 0.25. Thus, 25% of the variation in the observed y-values is explained by the regression (i.e., by the linear relationship between the x-values and the y-values). ■

The Regression Identity

For the three-point data set shown in the first two columns of Table 14.7, we found that $SST = 8$, $SSR = 2$, and $SSE = 6$. Because $8 = 2 + 6$, we see that $SST = SSR + SSE$. This equation is always true and is called the **regression identity.**

KEY FACT 14.4

? What Does It Mean?

The total variation in the observed values of the response variable can be partitioned into two components, one representing the variation explained by the regression and the other representing the variation not explained by the regression.

Regression Identity

The total sum of squares equals the regression sum of squares plus the error sum of squares: $SST = SSR + SSE$.

Because of the regression identity, we can also express the coefficient of determination in terms of the total sum of squares and the error sum of squares:

$$r^2 = \frac{SSR}{SST} = \frac{SST - SSE}{SST} = 1 - \frac{SSE}{SST}.$$

This formula shows that, when expressed as a percentage, we can also interpret the coefficient of determination as the percentage reduction obtained in the total squared error by using the regression equation instead of the mean, $\bar{y}$, to predict the observed values of the response variable. See Exercise 14.117 (page 644).

Computing Formulas for the Sums of Squares

Calculating the three sums of squares—*SST*, *SSR*, and *SSE*—with the defining formulas is usually time consuming and can lead to significant roundoff error unless full accuracy is retained. For those reasons, we generally use computing formulas or statistical software to find the sums of squares.

To obtain the computing formulas for the sums of squares, we first note that they can be expressed as

$$SST = S_{yy}, \qquad SSR = \frac{S_{xy}^2}{S_{xx}}, \qquad \text{and} \qquad SSE = S_{yy} - \frac{S_{xy}^2}{S_{xx}},$$

where S_{xx}, S_{xy}, and S_{yy} are given in Definition 14.4 on page 627. Referring again to that definition, we get Formula 14.2.

FORMULA 14.2

Computing Formulas for the Sums of Squares

The computing formulas for the three sums of squares are

$$SST = \Sigma y_i^2 - (\Sigma y_i)^2/n, \quad SSR = \frac{[\Sigma x_i y_i - (\Sigma x_i)(\Sigma y_i)/n]^2}{\Sigma x_i^2 - (\Sigma x_i)^2/n},$$

and $SSE = SST - SSR$.

■■■ **EXAMPLE 14.10**   **The Coefficient of Determination**

Age and Price of Orions The scatterplot and regression line for the age and price data of 11 Orions are repeated in Fig. 14.16.

FIGURE 14.16

Scatterplot and regression line for Orion data

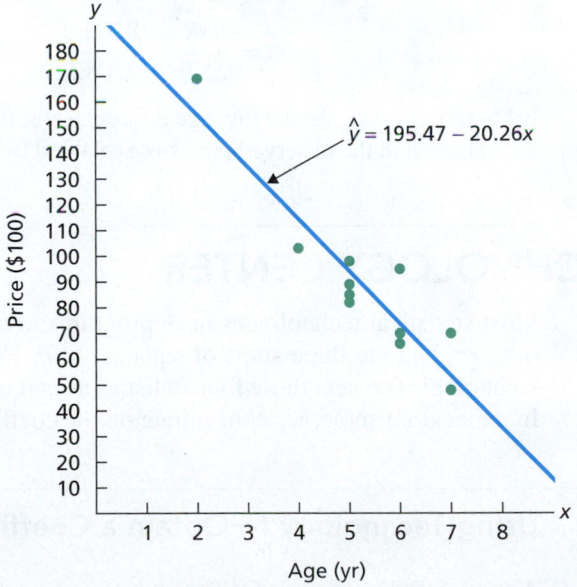

$\hat{y} = 195.47 - 20.26x$

The scatterplot reveals that the prices of the 11 Orions vary widely, ranging from a low of 48 ($4800) to a high of 169 ($16,900). But Fig. 14.16 also shows that much of the price variation is "explained" by the regression (or age); that is, the regression line, with age as the predictor variable, predicts a sizeable portion of the type of variation found in the prices. Make this qualitative statement precise by finding and interpreting the coefficient of determination for the Orion data.

Solution To determine the coefficient of determination, we first need to find the total sum of squares (*SST*) and regression sum of squares (*SSR*). We do so by applying the computing formulas in Formula 14.2. And to apply the computing formulas, we need a table of values for x (age), y (price), xy, x^2, y^2, and their sums, as shown in Table 14.8.

TABLE 14.8

Table for finding *SST* and *SSR* for the Orion data by using the computing formulas

Age (yr) x	Price ($100) y	xy	x^2	y^2
5	85	425	25	7,225
4	103	412	16	10,609
6	70	420	36	4,900
5	82	410	25	6,724
5	89	445	25	7,921
5	98	490	25	9,604
6	66	396	36	4,356
6	95	570	36	9,025
2	169	338	4	28,561
7	70	490	49	4,900
7	48	336	49	2,304
58	975	4732	326	96,129

Using the last row of Table 14.8 and Formula 14.2, we can now find *SST* and *SSR*. The total sum of squares is

$$SST = \Sigma y_i^2 - (\Sigma y_i)^2/n = 96{,}129 - (975)^2/11 = 9708.5,$$

and the regression sum of squares is

$$SSR = \frac{[\Sigma x_i y_i - (\Sigma x_i)(\Sigma y_i)/n]^2}{\Sigma x_i^2 - (\Sigma x_i)^2/n} = \frac{[4732 - (58)(975)/11]^2}{326 - (58)^2/11} = 8285.0.$$

From *SST* and *SSR*, we compute the coefficient of determination, the percentage of variation in the observed prices explained by the regression (i.e., by the linear relationship between age and price for the sampled Orions):

$$r^2 = \frac{SSR}{SST} = \frac{8285.0}{9708.5} = 0.853 \quad (85.3\%).$$

StatCrunch

Report 14.3

You try it!

Exercise 14.99
on page 644

Interpretation Evidently, age is quite useful for predicting price because 85.3% of the variation in the observed prices is explained by the regression of price on age. ◾

THE TECHNOLOGY CENTER

Most statistical technologies have programs to compute the coefficient of determination, r^2, and the three sums of squares, *SST*, *SSR*, and *SSE*. In fact, many statistical technologies present those four statistics as part of the output for a regression equation. In the next example, we concentrate on the coefficient of determination.

EXAMPLE 14.11 **Using Technology to Obtain a Coefficient of Determination**

Age and Price of Orions The age and price data for a sample of 11 Orions are given in Table 14.2 on page 625. Use Minitab, Excel, or the TI-83/84 Plus to obtain the coefficient of determination, r^2, for those data.

Solution In Section 14.2, we used the three statistical technologies to find the regression equation for the age and price data. The results, displayed in Output 14.2 on page 634, also give the coefficient of determination. See the items circled in blue. Thus, to three decimal places, $r^2 = 0.853$. ◾

We can also use Minitab, Excel, or the TI-83/84 Plus to get the three sums of squares: regression sum of squares (*SSR*), error sum of squares (*SSE*), and total sum of squares (*SST*). To do so, proceed as follows.

- *Minitab:* The three sums of squares can be found in Minitab's standard regression output, as discussed in Example 14.8 on page 634. See the SS column of the Analysis of Variance table at the bottom of the output.
- *Excel:* You can get the three sums of squares by checking the **Analysis of variance** check box in step 6 of Instructions 14.2 on page 634. In the resulting output, the three sums of squares will appear in the Sum of squares column of the Analysis of variance table. Note that Excel uses the term "Model" instead of "Regression" and the term "Corrected Total" instead of "Total."
- *TI-83/84 Plus:* You can get the three sums of squares by running the TI program REGSS, which can be found in the TI Programs section on the WeissStats site.

Exercises 14.3

Understanding the Concepts and Skills

14.83 In this section, we introduced a descriptive measure of the utility of the regression equation for making predictions.
a. Identify the term and symbol for that descriptive measure.
b. Provide an interpretation.

In each of Exercises 14.84–14.86, fill in the blanks.

14.84 A measure of total variation in the observed values of the response variable is the _____. The mathematical abbreviation for it is _____.

14.85 A measure of the amount of variation in the observed values of the response variable explained by the regression is the _____. The mathematical abbreviation for it is _____.

14.86 A measure of the amount of variation in the observed values of the response variable not explained by the regression is the _____. The mathematical abbreviation for it is _____.

14.87 For a regression analysis, $SST = 8291.0$ and $SSR = 7626.6$.
a. Obtain and interpret the coefficient of determination.
b. Determine SSE.

In Exercises 14.88–14.97, we repeat the data and provide the regression equations for Exercises 14.48–14.57. In each exercise,
a. compute the three sums of squares, SST, SSR, and SSE, using the defining formulas (Definition 14.6 on page 639).
b. verify the regression identity, $SST = SSR + SSE$.
c. compute the coefficient of determination.
d. determine the percentage of variation in the observed values of the response variable that is explained by the regression.
e. state how useful the regression equation appears to be for making predictions.

14.88

x	2	4	3
y	3	5	7

$\hat{y} = 2 + x$

14.89

x	3	1	2
y	−4	0	−5

$\hat{y} = 1 - 2x$

14.90

x	0	4	3	1	2
y	1	9	8	4	3

$\hat{y} = 1 + 2x$

14.91

x	3	4	1	2
y	4	5	0	−1

$\hat{y} = -3 + 2x$

14.92

x	1	3	4	4
y	13	−1	3	5

$\hat{y} = 14 - 3x$

14.93

x	2	2	3	4	4
y	3	4	0	2	1

$\hat{y} = 5 - x$

14.94

x	1	3	4	4
y	8	0	3	1

$\hat{y} = 9 - 2x$

14.95

x	1	2	3
y	4	3	8

$\hat{y} = 1 + 2x$

14.96

x	1	1	5	5
y	1	3	2	4

$\hat{y} = 1.75 + 0.25x$

14.97

x	0	2	2	5	6
y	4	2	0	−2	1

$\hat{y} = 2.875 - 0.625x$

Applying the Concepts and Skills

For Exercises 14.98–14.103,
a. compute SST, SSR, and SSE, using Formula 14.2 on page 641.
b. compute the coefficient of determination, r^2.
c. determine the percentage of variation in the observed values of the response variable explained by the regression, and interpret your answer.
d. state how useful the regression equation appears to be for making predictions.

14.98 Tax Efficiency. Following are the data on percentage of investments in energy securities and tax efficiency from Exercise 14.58.

x	3.1	3.2	3.7	4.3	4.0	5.5	6.7	7.4	7.4	10.6
y	98.1	94.7	92.0	89.8	87.5	85.0	82.0	77.8	72.1	53.5

14.99 Corvette Prices. Following are the age and price data for Corvettes from Exercise 14.59:

x	6	6	6	2	2	5	4	5	1	4
y	290	280	295	425	384	315	355	328	425	325

14.100 Custom Homes. Following are the size and price data for custom homes from Exercise 14.60.

x	26	27	33	29	29	34	30	40	22
y	540	555	575	577	606	661	738	804	496

14.101 Plant Emissions. Following are the data on plant weight and quantity of volatile emissions from Exercise 14.61.

x	57	85	57	65	52	67	62	80	77	53	68
y	8.0	22.0	10.5	22.5	12.0	11.5	7.5	13.0	16.5	21.0	12.0

14.102 Crown-Rump Length. Following are the data on age and crown-rump length for fetuses from Exercise 14.62.

x	10	10	13	13	18	19	19	23	25	28
y	66	66	108	106	161	166	177	228	235	280

14.103 Study Time and Score. Following are the data on study time and score for calculus students from Exercise 14.63.

x	10	15	12	20	8	16	14	22
y	92	81	84	74	85	80	84	80

Working with Large Data Sets

In Exercises 14.104–14.115, use the technology of your choice to perform the following tasks.
a. Decide whether finding a regression line for the data is reasonable. If so, then also do parts (b)–(d).
b. Obtain the coefficient of determination.
c. Determine the percentage of variation in the observed values of the response variable explained by the regression, and interpret your answer.
d. State how useful the regression equation appears to be for making predictions.

14.104 Birdies and Score. The data from Exercise 14.70 for number of birdies during a tournament and final score for 63 women golfers are on the WeissStats site.

14.105 U.S. Presidents. The data from Exercise 14.71 for the ages at inauguration and of death for the presidents of the United States are on the WeissStats site.

14.106 Movie Grosses. The data from Exercise 14.72 on domestic and overseas grosses for a random sample of 50 movies are on the WeissStats site.

14.107 Acreage and Value. The data from Exercise 14.73 for lot size (in acres) and assessed value (in thousands of dollars) for a sample of homes in a particular area are on the WeissStats site.

14.108 Home Size and Value. The data from Exercise 14.74 for home size (in square feet) and assessed value (in thousands of dollars) for the same homes as in Exercise 14.107 are on the WeissStats site.

14.109 High and Low Temperature. The data from Exercise 14.75 for average high and low temperatures in January for a random sample of 50 cities are on the WeissStats site.

14.110 PCBs and Pelicans. The data for shell thickness and concentration of PCBs for 60 Anacapa pelican eggs from Exercise 14.76 are on the WeissStats site.

14.111 More Money, More Beer? The data for per capita income and per capita beer consumption for the 50 states and Washington, D.C., from Exercise 14.77 are on the WeissStats site.

14.112 Gas Guzzlers. The data for gas mileage and engine displacement for 121 vehicles from Exercise 14.78 are provided on the WeissStats site.

14.113 Shortleaf Pines. The data from Exercise 14.80 for volume, in cubic feet, and diameter at breast height, in inches, for 70 shortleaf pines are on the WeissStats site.

14.114 Body Fat. In the paper "Total Body Composition by Dual-Photon (^{153}Gd) Absorptiometry" (*American Journal of Clinical Nutrition*, Vol. 40, pp. 834–839), R. Mazess et al. studied methods for quantifying body composition. Eighteen randomly selected adults were measured for percentage of body fat, using dual-photon absorptiometry. Each adult's age and percentage of body fat are shown on the WeissStats site.

14.115 Estriol Level and Birth Weight. J. Greene and J. Touchstone conducted a study on the relationship between the estriol levels of pregnant women and the birth weights of their children. Their findings, "Urinary Tract Estriol: An Index of Placental Function," were published in the *American Journal of Obstetrics and Gynecology* (Vol. 85(1), pp. 1–9). The data from the study are provided on the WeissStats site, where estriol levels are in mg/24 hr and birth weights are in hectograms.

Extending the Concepts and Skills

14.116 What can you say about *SSE*, *SSR*, and the utility of the regression equation for making predictions if
a. $r^2 = 1$?   b. $r^2 = 0$?

14.117 As we noted, because of the regression identity, we can express the coefficient of determination in terms of the total sum of squares and the error sum of squares as $r^2 = 1 - SSE/SST$.
a. Explain why this formula shows that the coefficient of determination can also be interpreted as the percentage reduction obtained in the total squared error by using the regression equation instead of the mean, $\bar{y}$, to predict the observed values of the response variable.
b. Refer to Exercise 14.99. What percentage reduction is obtained in the total squared error by using the regression equation instead of the mean of the observed prices to predict the observed prices?

14.4 Linear Correlation

We often hear statements pertaining to the correlation or lack of correlation between two variables: "There is a positive correlation between advertising expenditures and sales" or "IQ and alcohol consumption are uncorrelated." In this section, we explain the meaning of such statements.

Several statistics can be used to measure the correlation between two quantitative variables. The statistic most commonly used is the **linear correlation coefficient, r,** which is also called the **Pearson product moment correlation coefficient** in honor of its developer, Karl Pearson.

DEFINITION 14.8

? What Does It Mean?

The linear correlation coefficient is a descriptive measure of the strength and direction of the linear (straight-line) relationship between two variables.

Linear Correlation Coefficient

For a set of n data points, the **linear correlation coefficient, r,** is defined by

$$r = \frac{\frac{1}{n-1}\Sigma(x_i - \bar{x})(y_i - \bar{y})}{s_x s_y},$$

where s_x and s_y denote the sample standard deviations of the x-values and y-values, respectively.

Note: Because of the division by the sample standard deviations, s_x and s_y, in the defining formula for r, we can show that r *is independent of the choice of units and always lies between* -1 *and* 1.

■■■ **EXAMPLE 14.12**   **The Linear Correlation Coefficient**

Consider again the three-point data set shown in Table 14.3, which we repeat here in the first two columns of Table 14.9. Find the linear correlation coefficient of these data.

Solution   We first note that $\bar{x} = 2$ and $\bar{y} = 3$. Now we construct Table 14.9, which provides the required quantities for determining the linear correlation coefficient, r.

TABLE 14.9

Table for finding the linear correlation coefficient

x	y	$x - \bar{x}$	$y - \bar{y}$	$(x - \bar{x})(y - \bar{y})$	$(x - \bar{x})^2$	$(y - \bar{y})^2$
1	3	-1	0	0	1	0
2	1	0	-2	0	0	4
3	5	1	2	2	1	4
				2	2	8

From the defining formula for the sample standard deviation and the last two entries in the last row of Table 14.9, we get that

$$s_x = \sqrt{\frac{\Sigma(x_i - \bar{x})^2}{n - 1}} = \sqrt{\frac{2}{3 - 1}} = \sqrt{1} = 1$$

and

$$s_y = \sqrt{\frac{\Sigma(y_i - \bar{y})^2}{n - 1}} = \sqrt{\frac{8}{3 - 1}} = \sqrt{4} = 2.$$

Also, from the last entry in the fifth column of Table 14.9, we get that

$$\frac{1}{n - 1}\Sigma(x_i - \bar{x})(y_i - \bar{y}) = \frac{1}{3 - 1} \cdot 2 = 1.$$

Exercise 14.141
on page 651

Referring now to Definition 14.8, we find that

$$r = \frac{\frac{1}{n-1}\Sigma(x_i - \bar{x})(y_i - \bar{y})}{s_x s_y} = \frac{1}{1 \cdot 2} = 0.5.$$

For the three-point data set shown in the first two columns of Table 14.9, the linear correlation coefficient is 0.5.

Understanding the Linear Correlation Coefficient

As we have already noted, the linear correlation coefficient, r, is independent of the choice of units and always lies between -1 and 1. We now discuss some other important properties of the linear correlation coefficient, r. Keep in mind that r measures the strength of the *linear* relationship between two variables and that the following properties of r are meaningful only when the data points are scattered about a line.

- *r reflects the slope of the scatterplot.* The linear correlation coefficient is positive when the scatterplot shows a positive slope and is negative when the scatterplot shows a negative slope. To demonstrate why this property is true, we refer to Definition 14.8 and to Fig. 14.17, where we have drawn a coordinate system with a second set of axes centered at point $(\bar{x}, \bar{y})$.

  If the scatterplot shows a positive slope, the data points, on average, will lie either in Region I or Region III. For such a data point, the deviations from the means, $x_i - \bar{x}$ and $y_i - \bar{y}$, will either both be positive or both be negative. This condition implies that, on average, the product $(x_i - \bar{x})(y_i - \bar{y})$ will be positive and consequently that the correlation coefficient will be positive.

  If the scatterplot shows a negative slope, the data points, on average, will lie either in Region II or Region IV. For such a data point, one of the deviations from the mean will be positive and the other negative. This condition implies that, on average, the product $(x_i - \bar{x})(y_i - \bar{y})$ will be negative and consequently that the correlation coefficient will be negative.

- *The magnitude of r indicates the strength of the linear relationship.* A value of r close to -1 or to 1 indicates a strong linear relationship between the variables and that the variable x is a good linear predictor of the variable y (i.e., the regression equation is extremely useful for making predictions). A value of r near 0 indicates at most a weak linear relationship between the variables and that the variable x is a poor linear predictor of the variable y (i.e., the regression equation is either useless or not very useful for making predictions).

- *The sign of r suggests the type of linear relationship.* A positive value of r suggests that the variables are **positively linearly correlated,** meaning that y tends to increase linearly as x increases, with the tendency being greater the closer that r is to 1. A negative value of r suggests that the variables are **negatively linearly correlated,** meaning that y tends to decrease linearly as x increases, with the tendency being greater the closer that r is to -1.

- *The sign of r and the sign of the slope of the regression line are identical.* If r is positive, so is the slope of the regression line (i.e., the regression line slopes upward); if r is negative, so is the slope of the regression line (i.e., the regression line slopes downward).

  To graphically portray the meaning of the linear correlation coefficient, we present various degrees of linear correlation in Fig. 14.18.

  If r is close to ± 1, the data points are clustered closely about the regression line, as shown in Fig. 14.18(b) and (e). If r is farther from ± 1, the data points are more widely scattered about the regression line, as shown in Fig. 14.18(c) and (f). If r is near 0, the data points are essentially scattered about a horizontal line, as shown in Fig. 14.18(g), indicating at most a weak linear relationship between the variables.

FIGURE 14.17

Coordinate system with a second set of axes centered at $(\bar{x}, \bar{y})$

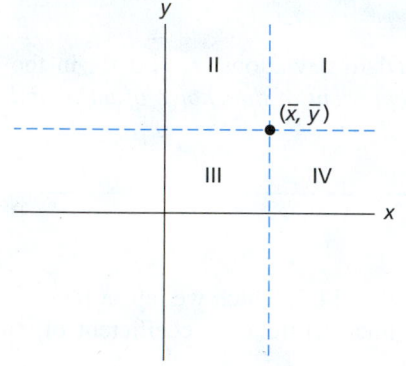

APPLET

Applet 14.3

Computing Formula for a Linear Correlation Coefficient

Calculating a linear correlation coefficient, r, by using the defining formula (Definition 14.8 on page 645) is usually time consuming and can lead to significant roundoff

FIGURE 14.18
Various degrees of linear correlation

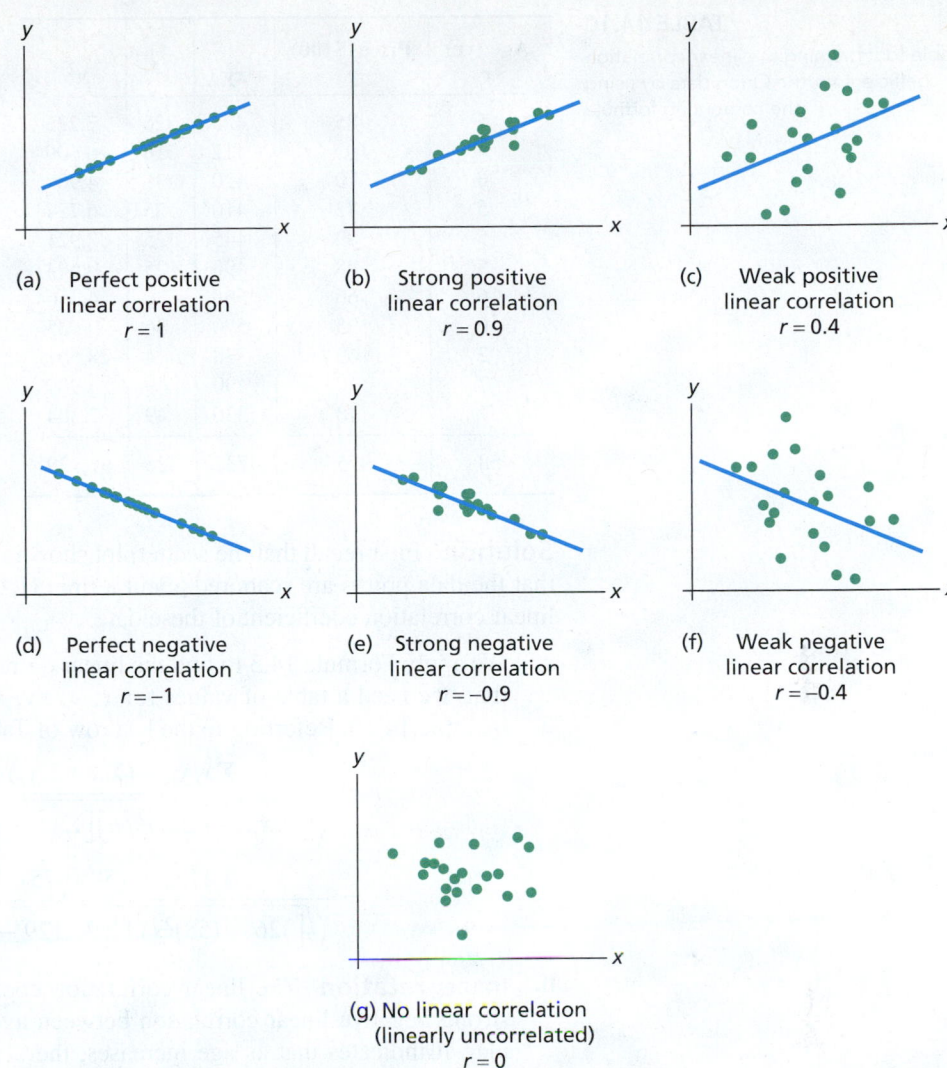

(a) Perfect positive
     linear correlation
     $r = 1$

(b) Strong positive
     linear correlation
     $r = 0.9$

(c) Weak positive
     linear correlation
     $r = 0.4$

(d) Perfect negative
     linear correlation
     $r = -1$

(e) Strong negative
     linear correlation
     $r = -0.9$

(f) Weak negative
     linear correlation
     $r = -0.4$

(g) No linear correlation
     (linearly uncorrelated)
     $r = 0$

error unless full accuracy is retained. For those reasons, we generally use a computing formula or statistical software to find r.

Using algebra, we can show that the linear correlation coefficient can be expressed as $r = S_{xy}/\sqrt{S_{xx}S_{yy}}$, where S_{xx}, S_{xy}, and S_{yy} are given in Definition 14.4 on page 627. Referring again to that definition, we get Formula 14.3.

FORMULA 14.3

Computing Formula for a Linear Correlation Coefficient

The computing formula for a linear correlation coefficient is

$$r = \frac{\sum x_i y_i - (\sum x_i)(\sum y_i)/n}{\sqrt{[\sum x_i^2 - (\sum x_i)^2/n][\sum y_i^2 - (\sum y_i)^2/n]}}.$$

EXAMPLE 14.13   The Linear Correlation Coefficient

Age and Price of Orions The age and price data for a sample of 11 Orions are repeated in the first two columns of Table 14.10 on the next page.

a. Compute the linear correlation coefficient, r, of the data.
b. Interpret the value of r obtained in part (a) in terms of the linear relationship between the variables age and price of Orions.
c. Discuss the graphical implications of the value of r.

TABLE 14.10

Table for obtaining the linear correlation coefficient for the Orion data by using the computing formula

Age (yr) x	Price ($100) y	xy	x^2	y^2
5	85	425	25	7,225
4	103	412	16	10,609
6	70	420	36	4,900
5	82	410	25	6,724
5	89	445	25	7,921
5	98	490	25	9,604
6	66	396	36	4,356
6	95	570	36	9,025
2	169	338	4	28,561
7	70	490	49	4,900
7	48	336	49	2,304
58	975	4732	326	96,129

Solution   First recall that the scatterplot shown in Fig. 14.7 on page 625 indicates that the data points are scattered about a line. Hence it is meaningful to obtain the linear correlation coefficient of these data.

a.   We apply Formula 14.3 to find the linear correlation coefficient. To accomplish that, we need a table of values for x, y, xy, x^2, y^2, and their sums, as shown in Table 14.10. Referring to the last row of Table 14.10, we get

$$r = \frac{\Sigma x_i y_i - (\Sigma x_i)(\Sigma y_i)/n}{\sqrt{[\Sigma x_i^2 - (\Sigma x_i)^2/n][\Sigma y_i^2 - (\Sigma y_i)^2/n]}}$$

$$= \frac{4732 - (58)(975)/11}{\sqrt{[326 - (58)^2/11][96,129 - (975)^2/11]}} = -0.924.$$

b.   **Interpretation**   The linear correlation coefficient, $r = -0.924$, suggests a strong negative linear correlation between age and price of Orions. In particular, it indicates that as age increases, there is a strong tendency for price to decrease, which is not surprising. It also implies that the regression equation, $\hat{y} = 195.47 - 20.26x$, is extremely useful for making predictions.

StatCrunch

You try it!

Report 14.4

Exercise 14.145 on page 652

c.   Because the correlation coefficient, $r = -0.924$, is quite close to -1, the data points should be clustered closely about the regression line. Figure 14.16 on page 641 shows that to be the case.   ■

Relationship between the Correlation Coefficient and the Coefficient of Determination

In Section 14.3, we discussed the coefficient of determination, r^2, a descriptive measure of the utility of the regression equation for making predictions. In this section, we introduced the linear correlation coefficient, r, as a descriptive measure of the strength of the linear relationship between two variables.

We expect the strength of the linear relationship also to indicate the usefulness of the regression equation for making predictions. In other words, there should be a relationship between the linear correlation coefficient and the coefficient of determination—and there is. The relationship is precisely the one suggested by the notation used.

KEY FACT 14.5

Relationship between the Correlation Coefficient and the Coefficient of Determination

The coefficient of determination equals the square of the linear correlation coefficient.

In Example 14.13, we found that the linear correlation coefficient for the data on age and price of a sample of 11 Orions is $r = -0.924$. From this result and Key Fact 14.5, we can easily obtain the coefficient of determination: $r^2 = (-0.924)^2 = 0.854$. As expected, this value is the same (except for roundoff error) as the value we found for r^2 on page 642 by using the defining formula $r^2 = SSR/SST$. In general, we can find the coefficient of determination either by using the defining formula or by first finding the linear correlation coefficient and then squaring the result.

Likewise, we can find the linear correlation coefficient, r, either by using Definition 14.8 (or Formula 14.3) or from the coefficient of determination, r^2, provided we also know the direction of the regression line. Specifically, the square root of r^2 gives the magnitude of r; the sign of r is the same as that of the slope of the regression line.

Warnings on the Use of the Linear Correlation Coefficient

Because the linear correlation coefficient describes the strength of the *linear* relationship between two variables, it should be used as a descriptive measure only when a scatterplot indicates that the data points are scattered about a line.

For instance, in general, we cannot say that a value of r near 0 implies that there is no relationship between the two variables under consideration, nor can we say that a value of r near ± 1 implies that a linear relationship exists between the two variables. Such statements are meaningful only when a scatterplot indicates that the data points are scattered about a line. See Exercises 14.151–14.153 for more on these issues.

When using the linear correlation coefficient, you must also watch for outliers and influential observations. Such data points can sometimes unduly affect r because sample means and sample standard deviations are not resistant to outliers and other extreme values.

Correlation and Causation

Two variables may have a high correlation without being causally related. For example, Table 14.11 displays data on total pari-mutuel turnover (money wagered) at U.S. racetracks and college enrollment for five randomly selected years. [SOURCE: *National Association of State Racing Commissioners* and *National Center for Education Statistics*]

TABLE 14.11

Pari-mutuel turnover and college enrollment for five randomly selected years

Pari-mutuel turnover ($ millions) x	College enrollment (thousands) y
5,977	8,581
7,862	11,185
10,029	11,260
11,677	12,372
11,888	12,426

The linear correlation coefficient of the data points in Table 14.11 is $r = 0.931$, suggesting a strong positive linear correlation between pari-mutuel wagering and college enrollment. But this result doesn't mean that a causal relationship exists between the two variables, such as that when people go to racetracks they are somehow inspired to go to college. On the contrary, we can only infer that the two variables have a strong tendency to increase (or decrease) simultaneously and that total pari-mutuel turnover is a good predictor of college enrollment.

Two variables may be strongly correlated because they are both associated with other variables, called **lurking variables,** that cause changes in the two variables under consideration. For example, a study showed that teachers' salaries and the dollar amount of liquor sales are positively linearly correlated. A possible explanation for this curious fact might be that both variables are tied to other variables, such as the rate of inflation, that pull them along together.

? What Does It Mean?

Correlation does not imply causation!

THE TECHNOLOGY CENTER

Most statistical technologies have programs that automatically determine a linear correlation coefficient. In this subsection, we present output and step-by-step instructions for such programs.

EXAMPLE 14.14 Using Technology to Find a Linear Correlation Coefficient

Age and Price of Orions Use Minitab, Excel, or the TI-83/84 Plus to determine the linear correlation coefficient of the age and price data in Table 14.2 on page 625.

Solution We applied the linear correlation coefficient programs to the data, resulting in Output 14.3. Steps for generating that output are presented in Instructions 14.3.

OUTPUT 14.3 Linear correlation coefficient for the age and price data of 11 Orions

MINITAB

Correlation: AGE, PRICE

Pearson correlation of AGE and PRICE = -0.924

EXCEL

Correlation matrix (Pearson):

Variables	AGE	PRICE
AGE	1	-0.9238
PRICE	-0.9238	1

TI-83/84 PLUS

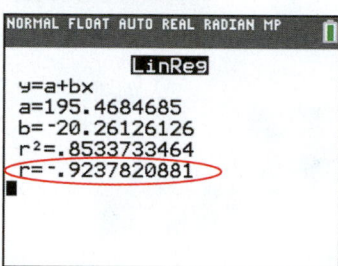

As shown in Output 14.3, the linear correlation coefficient for the age and price data is -0.924.

INSTRUCTIONS 14.3 Steps for generating Output 14.3

MINITAB

1. Store the age and price data from Table 14.2 in columns named AGE and PRICE, respectively
2. Choose **Stat ➤ Basic Statistics ➤ Correlation...**
3. Press the F3 key to reset the dialog box
4. Specify AGE and PRICE in the **Variables** text box
5. Uncheck the **Display p-values** check box
6. Click **OK**

EXCEL

1. Store the age and price data from Table 14.2 in columns named AGE and PRICE, respectively
2. Choose **XLSTAT ➤ Correlation/Association tests ➤ Correlation tests**
3. Click the reset button in the lower left corner of the dialog box
4. Click in the **Observations/variables table** selection box and then select the columns of the worksheet that contain the AGE and PRICE data
5. Click the **Outputs** tab and then uncheck all the check boxes therein
6. Click **OK**

7. Click the **Continue** button in the **XLSTAT – Selections** dialog box

TI-83/84 PLUS

1. Store the age and price data from Table 14.2 in lists named AGE and PRICE, respectively
2. Press **2nd ➤ CATALOG** and then press **D**
3. Arrow down to **DiagnosticOn** and press **ENTER** twice
4. Press **STAT**, arrow over to **CALC**, and press **8**

FOR THE TI-84 PLUS C
5. Press **2nd ➤ LIST**, arrow down to AGE, and press **ENTER** twice
6. Press **2nd ➤ LIST**, arrow down to PRICE, and press **ENTER** twice
7. Press **CLEAR**, arrow down to **Calculate**, and press **ENTER**

FOR THE TI-83/84 PLUS
5. Press **2nd ➤ LIST**, arrow down to AGE, and press **ENTER**
6. Press **, ➤ 2nd ➤ LIST**, arrow down to PRICE, and press **ENTER** twice

Exercises 14.4

Understanding the Concepts and Skills

14.118 What is one purpose of the linear correlation coefficient?

14.119 The linear correlation coefficient is also known by another name. What is it?

In Exercises 14.120–14.127, fill in the blanks.

14.120 The symbol that is used for the linear correlation coefficient is _____.

14.121 A value of r close to ± 1 indicates that there is a _____ linear relationship between the variables.

14.122 A value of r close to _____ indicates that there is either no linear relationship between the variables or a weak one.

14.123 A value of r close to _____ indicates that the regression equation is extremely useful for making predictions.

14.124 A value of r close to 0 indicates that the regression equation is either useless or _____ for making predictions.

14.125 If y tends to increase linearly as x increases, the variables are _____ linearly correlated.

14.126 If y tends to decrease linearly as x increases, the variables are _____ linearly correlated.

14.127 If there is no linear relationship between x and y, the variables are linearly _____.

In each of Exercises 14.128–14.130, determine whether r is positive, negative, or zero.

14.128   **14.129**   **14.130**

14.131 Answer true or false to the following statement and provide a reason for your answer: If there is a very strong positive correlation between two variables, a causal relationship exists between the two variables.

14.132 The linear correlation coefficient of a set of data points is 0.846.
a. Is the slope of the regression line positive or negative? Explain your answer.
b. Determine the coefficient of determination.

14.133 The coefficient of determination of a set of data points is 0.709 and the slope of the regression line is -3.58. Determine the linear correlation coefficient of the data.

In Exercises 14.134–14.143, we repeat data from exercises in Section 14.2. For each exercise, determine the linear correlation coefficient by using
a. *Definition 14.8 on page 645.*
b. *Formula 14.3 on page 647.*
Compare your answers in parts (a) and (b).

14.134

x	2	4	3
y	3	5	7

14.135

x	3	1	2
y	-4	0	-5

14.136

x	0	4	3	1	2
y	1	9	8	4	3

14.137

x	3	4	1	2
y	4	5	0	-1

14.138

x	1	3	4	4
y	13	-1	3	5

14.139

x	2	2	3	4	4
y	3	4	0	2	1

14.140

x	1	3	4	4
y	8	0	3	1

14.141

x	1	2	3
y	4	3	8

14.142

x	1	1	5	5
y	1	3	2	4

14.143

x	0	2	2	5	6
y	4	2	0	-2	1

Applying the Concepts and Skills

In Exercises 14.144–14.149, we repeat data from exercises in Section 14.2. For each exercise here,
a. *obtain the linear correlation coefficient.*
b. *interpret the value of r in terms of the linear relationship between the two variables in question.*
c. *discuss the graphical interpretation of the value of r and verify that it is consistent with the graph you obtained in the corresponding exercise in Section 14.2.*
d. *square r and compare the result with the value of the coefficient of determination you obtained in the corresponding exercise in Section 14.3.*

14.144 **Tax Efficiency.** Following are the data on percentage of investments in energy securities and tax efficiency from Exercises 14.58 and 14.98.

x	3.1	3.2	3.7	4.3	4.0	5.5	6.7	7.4	7.4	10.6
y	98.1	94.7	92.0	89.8	87.5	85.0	82.0	77.8	72.1	53.5

14.145 Corvette Prices. Following are the age and price data for Corvettes from Exercises 14.59 and 14.99.

x	6	6	6	2	2	5	4	5	1	4
y	290	280	295	425	384	315	355	328	425	325

14.146 Custom Homes. Following are the size and price data for custom homes from Exercises 14.60 and 14.100.

x	26	27	33	29	29	34	30	40	22
y	540	555	575	577	606	661	738	804	496

14.147 Plant Emissions. Following are the data on plant weight and quantity of volatile emissions from Exercises 14.61 and 14.101.

x	57	85	57	65	52	67	62	80	77	53	68
y	8.0	22.0	10.5	22.5	12.0	11.5	7.5	13.0	16.5	21.0	12.0

14.148 Crown-Rump Length. Following are the data on age and crown-rump length for fetuses from Exercises 14.62 and 14.102.

x	10	10	13	13	18	19	19	23	25	28
y	66	66	108	106	161	166	177	228	235	280

14.149 Study Time and Score. Following are the data on study time and score for calculus students from Exercises 14.63 and 14.103.

x	10	15	12	20	8	16	14	22
y	92	81	84	74	85	80	84	80

14.150 Height and Score. A random sample of 10 students was taken from an introductory statistics class. The following data were obtained, where x denotes height, in inches, and y denotes score on the final exam.

x	71	68	71	65	66	68	68	64	62	65
y	87	96	66	71	71	55	83	67	86	60

a. What sort of value of r would you expect to find for these data? Explain your answer.
b. Compute r.

14.151 Consider the following set of data points.

x	−3	−2	−1	0	1	2	3
y	9	4	1	0	1	4	9

a. Compute the linear correlation coefficient, r.
b. Can you conclude from your answer in part (a) that the variables x and y are unrelated? Explain your answer.
c. Draw a scatterplot for the data.
d. Is use of the linear correlation coefficient as a descriptive measure for the data appropriate? Explain your answer.
e. Show that the data are related by the quadratic equation $y = x^2$. Graph that equation and the data points.

In each of Exercises 14.152 and 14.153, perform the following tasks.
a. Compute the linear correlation coefficient, r.
b. Can you conclude from your answer in part (a) that the variables x and y are linearly related? Explain your answer.
c. Draw a scatterplot for the data.
d. Is use of the linear correlation coefficient as a descriptive measure for the data appropriate? Explain your answer.
e. Show that the data are related by the equation provided.
f. Graph the equation and the data points.

14.152

x	−5	−3	−1	1	3	5
y	−32	−60	−80	−92	−96	−92

Equation: $y = x^2 - 6x - 87$

14.153

x	0	1	2	3	4	5	6
y	0	1	4	9	16	25	36

Equation: $y = x^2$

Working with Large Data Sets

In Exercises 14.154–14.166, use the technology of your choice to
a. decide whether use of the linear correlation coefficient as a descriptive measure for the data is appropriate. If so, then also do parts (b) and (c).
b. obtain the linear correlation coefficient.
c. interpret the value of r in terms of the linear relationship between the two variables in question.

14.154 Birdies and Score. The data from Exercise 14.70 for number of birdies during a tournament and final score for 63 women golfers are on the WeissStats site.

14.155 U.S. Presidents. The data from Exercise 14.71 for the ages at inauguration and of death for the presidents of the United States are on the WeissStats site.

14.156 Movie Grosses. The data from Exercise 14.72 on domestic and overseas grosses for a random sample of 50 movies are on the WeissStats site.

14.157 Acreage and Value. The data from Exercise 14.73 for lot size (in acres) and assessed value (in thousands of dollars) for a sample of homes in a particular area are on the WeissStats site.

14.158 Home Size and Value. The data from Exercise 14.74 for home size (in square feet) and assessed value (in thousands of dollars) for the same homes as in Exercise 14.157 are on the WeissStats site.

14.159 High and Low Temperature. The data from Exercise 14.75 for average high and low temperatures in January for a random sample of 50 cities are on the WeissStats site.

14.160 PCBs and Pelicans. The data on shell thickness and concentration of PCBs for 60 Anacapa pelican eggs from Exercise 14.76 are on the WeissStats site.

14.161 More Money, More Beer? The data for per capita income and per capita beer consumption for the 50 states and Washington, D.C., from Exercise 14.77 are on the WeissStats site.

14.162 Gas Guzzlers. The data for gas mileage and engine displacement for 121 vehicles from Exercise 14.78 are provided on the WeissStats site.

14.163 Shortleaf Pines. The data from Exercise 14.80 for volume, in cubic feet, and diameter at breast height, in inches, for 70 shortleaf pines are on the WeissStats site.

14.164 Body Fat. The data from Exercise 14.114 for age and percentage of body fat for 18 randomly selected adults are on the WeissStats site.

14.165 Estriol Level and Birth Weight. The data for estriol levels of pregnant women and birth weights of their children from Exercise 14.115 are on the WeissStats site.

14.166 Fiber Density. In the article "Comparison of Fiber Counting by TV Screen and Eyepieces of Phase Contrast Microscopy" (*American Industrial Hygiene Association Journal*, Vol. 63, pp. 756–761), I. Moa et al. reported on determining fiber density by two different methods. Twenty samples of varying fiber density were each counted by 10 viewers by means of an eyepiece method and a television-screen method to determine the relationship between the counts done by each method. The results, in fibers per square millimeter, are presented on the WeissStats site.

Extending the Concepts and Skills

14.167 The coefficient of determination of a set of data points is 0.716.
a. Can you determine the linear correlation coefficient? If yes, obtain it. If no, why not?
b. Can you determine whether the slope of the regression line is positive or negative? Why or why not?
c. If we tell you that the slope of the regression line is negative, can you determine the linear correlation coefficient? If yes, obtain it. If no, why not?

d. If we tell you that the slope of the regression line is positive, can you determine the linear correlation coefficient? If yes, obtain it. If no, why not?

14.168 Country Music Blues. A Knight-Ridder News Service article in an issue of the *Wichita Eagle* discussed a study on the relationship between country music and suicide. The results of the study, coauthored by S. Stack and J. Gundlach, appeared as the paper "The Effect of Country Music on Suicide" (*Social Forces*, Vol. 71, Issue 1, pp. 211–218). According to the article, "...analysis of 49 metropolitan areas shows that the greater the airtime devoted to country music, the greater the white suicide rate." (Suicide rates in the black population were found to be uncorrelated with the amount of country music airtime.)
a. Use the terminology introduced in this section to describe the statement quoted above.
b. One of the conclusions stated in the journal article was that country music "nurtures a suicidal mood" by dwelling on marital status and alienation from work. Is this conclusion warranted solely on the basis of the positive correlation found between airtime devoted to country music and white suicide rate? Explain your answer.

Rank Correlation. The **rank correlation coefficient,** r_s, is a nonparametric alternative to the linear correlation coefficient. It was developed by Charles Spearman (1863–1945) and therefore is also known as the **Spearman rank correlation coefficient.** To determine the rank correlation coefficient, we first rank the x-values among themselves and the y-values among themselves, and then we compute the linear correlation coefficient of the rank pairs. An advantage of the rank correlation coefficient over the linear correlation coefficient is that the former can be used to describe the strength of a positive or negative nonlinear (as well as linear) relationship between two variables. Ties are handled as usual.

In each of Exercises 14.169 and 14.170,
a. construct a scatterplot for the data.
b. decide whether using the rank correlation coefficient is reasonable.
c. decide whether using the linear correlation coefficient is reasonable.
d. find and interpret the rank correlation coefficient.

14.169 Exercise 14.149

14.170 Exercise 14.163 (Use technology.)

CHAPTER IN REVIEW

You Should Be Able to

1. use and understand the formulas in this chapter.

2. define and apply the concepts related to linear equations with one independent variable.

3. explain the least-squares criterion.

4. obtain and graph the regression equation for a set of data points, interpret the slope of the regression line, and use the regression equation to make predictions.

5. define and use the terminology *predictor variable* and *response variable*.

6. understand the concept of extrapolation.

7. identify outliers and influential observations.

8. know when obtaining a regression line for a set of data points is appropriate.

9. calculate and interpret the three sums of squares, *SST, SSE,* and *SSR*, and the coefficient of determination, r^2.

10. find and interpret the linear correlation coefficient, *r*.

11. identify the relationship between the linear correlation coefficient and the coefficient of determination.

Key Terms

bivariate quantitative data, *624*
coefficient of determination (r^2), *639*
curvilinear regression, *632*
error, *626*
error sum of squares (*SSE*), *639*
explanatory variable, *630*
extrapolation, *630*
influential observation, *630*
least-squares criterion, *626*
line, *619*
linear correlation coefficient (r), *645*

linear equation, *619*
lurking variables, *649*
negatively linearly correlated
   variables, *646*
outlier, *630*
Pearson product moment correlation
   coefficient, *645*
positively linearly correlated
   variables, *646*
predictor variable, *630*
regression equation, *627*

regression identity, *640*
regression line, *627*
regression sum of squares
   (*SSR*), *639*
response variable, *630*
scatter diagram, *624*
scatterplot, *624*
slope, *621*
straight line, *619*
total sum of squares (*SST*), *639*
y-intercept, *621*

REVIEW PROBLEMS

Understanding the Concepts and Skills

1. For a linear equation $y = b_0 + b_1 x$, identify the
a. independent variable.   **b.** dependent variable.
c. slope.   **d.** y-intercept.

2. Consider the linear equation $y = 4 - 3x$.
a. At what y-value does its graph intersect the y-axis?
b. At what x-value does its graph intersect the y-axis?
c. What is its slope?
d. By how much does the y-value on the line change when the x-value increases by 1 unit?
e. By how much does the y-value on the line change when the x-value decreases by 2 units?

In Problems 3–5, answer true or false to each statement. Explain your answers.

3. The y-intercept of a line has no effect on the steepness of the line.

4. A horizontal line has no slope.

5. If a line has a positive slope, y-values on the line decrease as the x-values decrease.

6. What kind of plot is useful for deciding whether finding a regression line for a set of data points is reasonable?

7. Identify one use of a regression equation.

8. Regarding the variables in a regression analysis,
a. what is the independent variable called?
b. what is the dependent variable called?

In each of Problems 9–11, fill in the blank.

9. Based on the least-squares criterion, the line that best fits a set of data points is the one with the _____ possible sum of squared errors.

10. The line that best fits a set of data points according to the least-squares criterion is called the _____ line.

11. Using a regression equation to make predictions for values of the predictor variable outside the range of the observed values of the predictor variable is called _____.

12. In the context of regression analysis, what is an
a. outlier?   **b.** influential observation?

13. Identify a use of the coefficient of determination as a descriptive measure.

14. For each of the sums of squares in regression, state its name and what it measures.
a. *SST*   **b.** *SSR*   **c.** *SSE*

In each of Problems 15–18, fill in the blank.

15. One use of the linear correlation coefficient is as a descriptive measure of the strength of the _____ relationship between two variables.

16. A positive linear relationship between two variables means that one variable tends to increase linearly as the other _____.

17. A value of r close to -1 suggests a strong _____ linear relationship between the variables.

18. A value of r close to _____ suggests at most a weak linear relationship between the variables.

19. Answer true or false to the following statement, and explain your answer: A strong correlation between two variables doesn't necessarily mean that they're causally related.

Applying the Concepts and Skills

20. Equipment Depreciation. A small company has purchased a computer system for $7200 and plans to depreciate the value of the equipment by $1200 per year for 6 years. Let x denote the age of the equipment, in years, and y denote the value of the equipment, in hundreds of dollars.
a. Find the equation that expresses y in terms of x.
b. Find the y-intercept, b_0, and slope, b_1, of the linear equation in part (a).
c. Without graphing the equation in part (a), decide whether the line slopes upward, slopes downward, or is horizontal.
d. Find the value of the computer equipment after 2 years; after 5 years.
e. Obtain the graph of the equation in part (a) by plotting the points from part (d) and connecting them with a line.
f. Use the graph from part (e) to visually estimate the value of the equipment after 4 years. Then calculate that value exactly, using the equation from part (a).

21. Graduation Rates. Graduation rate—the percentage of entering freshmen attending full time and graduating within 5 years— and what influences it is a concern in U.S. colleges and universities.

U.S. News and World Report's "College Guide" provides data on graduation rates for colleges and universities as a function of the percentage of freshmen in the top 10% of their high school class, total spending per student, and student-to-faculty ratio. A random sample of 10 universities gave the following data on student-to-faculty ratio (S/F ratio) and graduation rate (Grad rate).

S/F ratio x	Grad rate y	S/F ratio x	Grad rate y
16	45	17	46
20	55	17	50
17	70	17	66
19	50	10	26
22	47	18	60

a. Draw a scatterplot of the data.
b. Is finding a regression line for the data reasonable? Explain your answer.
c. Determine the regression equation for the data, and draw its graph on the scatterplot you drew in part (a).
d. Describe the apparent relationship between student-to-faculty ratio and graduation rate.
e. What does the slope of the regression line represent in terms of student-to-faculty ratio and graduation rate?
f. Use the regression equation to predict the graduation rate of a university having a student-to-faculty ratio of 17.
g. Identify outliers and potential influential observations.

22. Graduation Rates. Refer to Problem 21.
a. Determine *SST*, *SSR*, and *SSE* by using the computing formulas.
b. Obtain the coefficient of determination.
c. Obtain the percentage of the total variation in the observed graduation rates that is explained by student-to-faculty ratio (i.e., by the regression line).
d. State how useful the regression equation appears to be for making predictions.

23. Graduation Rates. Refer to Problem 21.
a. Compute the linear correlation coefficient, *r*.
b. Interpret your answer from part (a) in terms of the linear relationship between student-to-faculty ratio and graduation rate.
c. Discuss the graphical implications of the value of the linear correlation coefficient, *r*.
d. Use your answer from part (a) to obtain the coefficient of determination.

Working with Large Data Sets

24. Exotic Plants. In the article "Effects of Human Population, Area, and Time on Non-native Plant and Fish Diversity in the United States" (*Biological Conservation*, Vol. 100, No. 2, pp. 243–252), M. McKinney investigated the relationship of various factors on the number of exotic plants in each state. On the WeissStats site, you will find the data on population (in millions), area (in thousands of square miles), and number of exotic plants for each state. Use the technology of your choice to determine the linear correlation coefficient between each of the following:
a. population and area
b. population and number of exotic plants
c. area and number of exotic plants
d. Interpret and explain the results you got in parts (a)–(c).

In Problems 25–27, use the technology of your choice to do the following tasks.
a. *Construct and interpret a scatterplot for the data.*
b. *Decide whether finding a regression line for the data is reasonable. If so, then also do parts (c)–(f).*
c. *Determine and interpret the regression equation.*
d. *Make the indicated predictions.*
e. *Compute and interpret the correlation coefficient.*
f. *Identify potential outliers and influential observations.*

25. IMR and Life Expectancy. From the *International Data Base*, published by the U.S. Census Bureau, we obtained data on infant mortality rate (IMR) and life expectancy (LE), in years, for a sample of 60 countries. The data are presented on the WeissStats site. For part (d), predict the life expectancy of a country with an IMR of 30.

26. High Temperature and Precipitation. The National Oceanic and Atmospheric Administration publishes temperature and precipitation information for cities around the world in *Climates of the World*. Data on average high temperature (in degrees Fahrenheit) in July and average precipitation (in inches) in July for 48 cities are on the WeissStats site. For part (d), predict the average July precipitation of a city with an average July temperature of 83°F.

27. U.S. Open. From the website Golf.com, part of Sports Illustrated Sites, we obtained the scores for the first and second rounds of the 2013 U.S. Open golf tournament. You will find those scores on the WeissStats site. For part (d), predict the second-round score of a golfer who got a 72 on the first round.

FOCUSING ON DATA ANALYSIS

UWEC UNDERGRADUATES

Recall from Chapter 1 (see page 34) that the Focus database and Focus sample contain information on the undergraduate students at the University of Wisconsin - Eau Claire (UWEC). Now would be a good time for you to review the discussion about these data sets.

Open the Focus sample worksheet (FocusSample) in the technology of your choice and do the following.

a. Find the linear correlation coefficient between cumulative GPA and high school percentile for the 200 UWEC undergraduate students in the Focus sample.
b. Repeat part (a) for cumulative GPA and each of ACT English score, ACT math score, and ACT composite score.

c. Among the variables high school percentile, ACT English score, ACT math score, and ACT composite score, identify the one that appears to be the best predictor of cumulative GPA. Explain your reasoning.

Now perform a regression analysis on cumulative GPA, using the predictor variable identified in part (c), as follows.

d. Obtain and interpret a scatterplot.
e. Find and interpret the regression equation.
f. Find and interpret the coefficient of determination.
g. Determine and interpret the three sums of squares *SSR*, *SSE*, and *SST*.

CASE STUDY DISCUSSION

HEALTHCARE: SPENDING AND OUTCOMES

At the beginning of this chapter, we presented selected OECD healthcare data for the 34 member countries for the year 2011 or, in a few cases, the nearest year to that. Recall that % GDP is an abbreviation for "healthcare expenditures as a percentage of gross domestic product," LE for "life expectancy" (at birth), and IMR for "infant mortality rate" (number of deaths of babies under 1 year of age per 1000 live births).

Now that you have studied regression and correlation, you can analyze the relationship between % GDP and LE and that between % GDP and IMR. We recommend that you use statistical software or a graphing calculator to solve the following problems, but they can also be done by hand.

a. Considering % GDP as the predictor variable and LE as the response variable, construct a scatterplot of the data.

b. Does the data point for the United States appear to be an outlier? Explain your answer and discuss it in the context of % GDP and LE.

c. Find the regression equation for the data, both with and without the data point for the United States. Plot the regression lines and the data points.

d. Is the data point for the United States an influential observation? Explain your answer and discuss it in the context of % GDP and LE.

e. Obtain the coefficient of determination, both with and without the data point for the United States. Interpret your results.

f. Determine the correlation coefficient, both with and without the data point for the United States. Interpret your results.

g. Repeat parts (a)–(f) with % GDP as the predictor variable and IMR as the response variable.

BIOGRAPHY

ADRIEN LEGENDRE: INTRODUCING THE METHOD OF LEAST SQUARES

Adrien-Marie Legendre was born in Paris, France, on September 18, 1752, the son of a moderately wealthy family. He studied at the Collège Mazarin and received degrees in mathematics and physics in 1770 at the age of 18.

Although Legendre's financial assets were sufficient to allow him to devote himself to research, he took a position teaching mathematics at the École Militaire in Paris from 1775 to 1780. In March 1783, he was elected to the Academie des Sciences in Paris, and, in 1787, he was assigned to a project undertaken jointly by the observatories at Paris and at Greenwich, England. At that time, he became a fellow of the Royal Society.

As a result of the French Revolution, which began in 1789, Legendre lost his "small fortune" and was forced to find work. He held various positions during the early 1790s, including commissioner of astronomical operations for the Academie des Sciences, Professor of Pure Mathematics at the Institut de Marat, and Head of the National Executive Commission of Public Instruction. During this same period, Legendre wrote a geometry book

that became the major text used in elementary geometry courses for nearly a century.

Legendre's major contribution to statistics was the publication, in 1805, of the first statement and the first application of the most widely used, nontrivial technique of statistics: the method of least squares. In his book, *The History of Statistics: The Measurement of Uncertainty Before 1900* (Cambridge, MA: Belknap Press of Harvard University Press, 1986), Stephen M. Stigler wrote "[Legendre's] presentation ... must be counted as one of the clearest and most elegant introductions of a new statistical method in the history of statistics."

Because Gauss also claimed the method of least squares, there was strife between the two men. Although evidence shows that Gauss was not successful in any communication of the method prior to 1805, his development of the method was crucial to its usefulness.

In 1813, Legendre was appointed Chief of the Bureau des Longitudes. He remained in that position until his death, following a long illness, in Paris on January 10, 1833.

Inferential Methods in Regression and Correlation

CHAPTER OBJECTIVES

In Chapter 14, you studied descriptive methods in regression and correlation. You discovered how to determine the regression equation for a set of data points and how to use that equation to make predictions. You also learned how to compute and interpret the coefficient of determination and the linear correlation coefficient for a set of data points.

In this chapter, you will study inferential methods in regression and correlation. In Section 15.1, we examine the conditions required for performing such inferences and the methods for checking whether those conditions are satisfied. In presenting the first inferential method, in Section 15.2, we show how to decide whether a regression equation is useful for making predictions.

In Section 15.3, we investigate two additional inferential methods: one for estimating the mean of the response variable corresponding to a particular value of the predictor variable, the other for predicting the value of the response variable for a particular value of the predictor variable. We also discuss, in Section 15.4, the use of the linear correlation coefficient of a set of data points to decide whether the two variables under consideration are linearly correlated and, if so, the nature of the linear correlation.

We also present, in Section 15.5 (on the WeissStats site), an inferential procedure for testing whether a variable is normally distributed.

CASE STUDY

Shoe Size and Height

Most of us have heard that tall people generally have larger feet than do short people. Is that really true, and, if so, what is the precise relationship between height and foot length?

To examine the relationship, Professor D. Young obtained data on shoe size and height for a sample of students at Arizona State University. We have displayed the results obtained by Professor Young in the following table, where height is measured in inches.

Shoe size	Height	Gender	Shoe size	Height	Gender
6.5	66.0	F	13.0	77	M
9.0	68.0	F	11.5	72	M
8.5	64.5	F	8.5	59	F
8.5	65.0	F	5.0	62	F
10.5	70.0	M	10.0	72	M
7.0	64.0	F	6.5	66	F
9.5	70.0	F	7.5	64	F
9.0	71.0	F	8.5	67	M
13.0	72.0	M	10.5	73	M
7.5	64.0	F	8.5	69	F
10.5	74.5	M	10.5	72	M
8.5	67.0	F	11.0	70	M
12.0	71.0	M	9.0	69	M
10.5	71.0	M	13.0	70	M

At the end of this chapter, you will be asked to use these data to make regression and correlation inferences for shoe size and height. In particular, you will perform a correlation analysis to decide whether shoe size (foot length) and height are positively correlated.

15.1 The Regression Model; Analysis of Residuals

Before we can perform statistical inferences in regression and correlation, we must know whether the variables under consideration satisfy certain conditions. In this section, we discuss those conditions and examine methods for deciding whether they hold.

The Regression Model

TABLE 15.1

Age and price data for a sample of 11 Orions

Age (yr) x	Price ($100) y
5	85
4	103
6	70
5	82
5	89
5	98
6	66
6	95
2	169
7	70
7	48

Let's return to the Orion illustration used throughout Chapter 14. In Table 15.1, we reproduce the data on age and price for a sample of 11 Orions.

With age as the predictor variable and price as the response variable, the regression equation for these data is $\hat{y} = 195.47 - 20.26x$, as we found in Chapter 14 on page 629. Recall that the regression equation can be used to predict the price of an Orion from its age. However, we cannot expect such predictions to be completely accurate because prices vary even for Orions of the same age.

For instance, the sample data in Table 15.1 include four 5-year-old Orions. Their prices are $8500, $8200, $8900, and $9800. We expect this variation in price for 5-year-old Orions because such cars generally have different mileages, interior conditions, paint quality, and so forth.

We use the population of all 5-year-old Orions to introduce some important regression terminology. The distribution of their prices is called the **conditional distribution** of the response variable "price" corresponding to the value 5 of the predictor variable "age." Likewise, their mean price is called the **conditional mean** of the response variable "price" corresponding to the value 5 of the predictor variable "age." Similar terminology applies to the standard deviation and other parameters.

Of course, there is a population of Orions for each age. The distribution, mean, and standard deviation of prices for that population are called the *conditional distribution, conditional mean,* and *conditional standard deviation,* respectively, of the response variable "price" corresponding to the value of the predictor variable "age."

The terminology of conditional distributions, means, and standard deviations is used in general for any predictor variable and response variable. In other words, we have the following definitions.

DEFINITION 15.1   **Conditional Distribution, Mean, and Standard Deviation**

Suppose that x and y are predictor and response variables, respectively, on a population. Let x_p denote a particular value of the predictor variable and consider the subpopulation consisting of all members of the population whose value of the predictor variable is x_p.

Conditional distribution of the response variable corresponding to x_p: The distribution of all possible values of the response variable on the aforementioned subpopulation.

Conditional mean of the response variable corresponding to x_p: The mean of all possible values of the response variable on the aforementioned subpopulation.

Conditional standard deviation of the response variable corresponding to x_p: The standard deviation of all possible values of the response variable on the aforementioned subpopulation.

Using the terminology presented in Definition 15.1, we can now state the conditions required for applying inferential methods in regression analysis.

KEY FACT 15.1

? What Does It Mean?

Assumptions 1–3 require that there are constants β_0, β_1, and σ so that, for each value x of the predictor variable, the conditional distribution of the response variable, y, is a normal distribution with mean $\beta_0 + \beta_1 x$ and standard deviation σ. These assumptions are often referred to as the **regression model**.

Assumptions (Conditions) for Regression Inferences

1. **Population regression line:** There are constants β_0 and β_1 such that, for each value x of the predictor variable, the conditional mean of the response variable is $\beta_0 + \beta_1 x$.

2. **Equal standard deviations:** The conditional standard deviations of the response variable are the same for all values of the predictor variable. We denote this common standard deviation σ.[†]

3. **Normal populations:** For each value of the predictor variable, the conditional distribution of the response variable is a normal distribution.

4. **Independent observations:** The observations of the response variable are independent of one another.

Note: We refer to the line $y = \beta_0 + \beta_1 x$—on which the conditional means of the response variable lie—as the **population regression line** and to its equation as the **population regression equation**. Observe that β_0 is the y-intercept of the population regression line and β_1 is its slope.

The inferential procedures in regression are robust to moderate violations of Assumptions 1–3 for regression inferences. In other words, the inferential procedures work reasonably well provided the variables under consideration don't violate any of those assumptions too badly.

■■■ **EXAMPLE 15.1**   **Assumptions for Regression Inferences**

Age and Price of Orions For Orions, with age as the predictor variable and price as the response variable, what would it mean for the regression-inference Assumptions 1–3 to be satisfied? Display those assumptions graphically.

[†]The condition of equal standard deviations is called **homoscedasticity.** When that condition fails, we have what is called **heteroscedasticity.**

Solution Satisfying regression-inference Assumptions 1–3 requires that there are constants β_0, β_1, and σ so that for each age, x, the prices of all Orions of that age are normally distributed with mean $\beta_0 + \beta_1 x$ and standard deviation σ. Thus the prices of all 2-year-old Orions must be normally distributed with mean $\beta_0 + \beta_1 \cdot 2$ and standard deviation σ, the prices of all 3-year-old Orions must be normally distributed with mean $\beta_0 + \beta_1 \cdot 3$ and standard deviation σ, and so on.

To display the assumptions for regression inferences graphically, let's first consider Assumption 1. This assumption requires that for each age, the mean price of all Orions of that age lies on the line $y = \beta_0 + \beta_1 x$, as shown in Fig. 15.1.

FIGURE 15.1

Population regression line

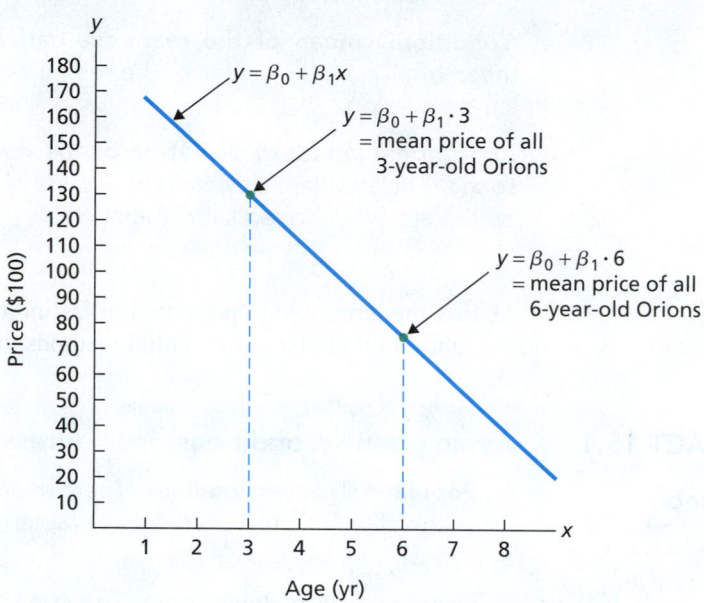

Assumptions 2 and 3 require that the price distributions for the various ages of Orions are all normally distributed with the same standard deviation, σ. Figure 15.2 illustrates those two assumptions for the price distributions of 2-, 5-, and 7-year-old Orions. The shapes of the three normal curves in Fig. 15.2 are identical because normal distributions that have the same standard deviation have the same shape.

FIGURE 15.2

Price distributions for 2-, 5-, and 7-year-old Orions under Assumptions 2 and 3 (The means shown for the three normal distributions reflect Assumption 1)

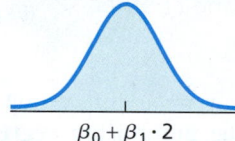 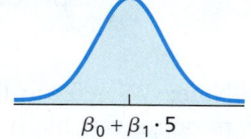 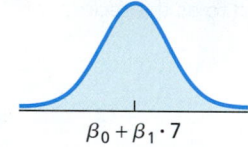

Prices of 2-year-old Orions   Prices of 5-year-old Orions   Prices of 7-year-old Orions

You try it!

Exercise 15.23 on page 668

Assumptions 1–3 for regression inferences, as they pertain to the variables age and price of Orions, can be portrayed graphically by combining Figs. 15.1 and 15.2 into a three-dimensional graph, as shown in Fig. 15.3. Whether those assumptions actually hold remains to be seen.

Estimating the Regression Parameters

Suppose that we are considering two variables, x and y, for which the assumptions for regression inferences are met. Then there are constants β_0, β_1, and σ so that, for each value x of the predictor variable, the conditional distribution of the response variable is a normal distribution with mean $\beta_0 + \beta_1 x$ and standard deviation σ.

Because the parameters β_0, β_1, and σ are usually unknown, we must estimate them from sample data. We use the y-intercept and slope of a sample regression line as point estimates of the y-intercept and slope, respectively, of the population regression line; that is, we use b_0 (the y-intercept of a sample regression line) to estimate β_0 (the y-intercept of the population regression line) and we use b_1 (the slope of a sample

FIGURE 15.3   Graphical portrayal of Assumptions 1–3 for regression inferences pertaining to age and price of Orions

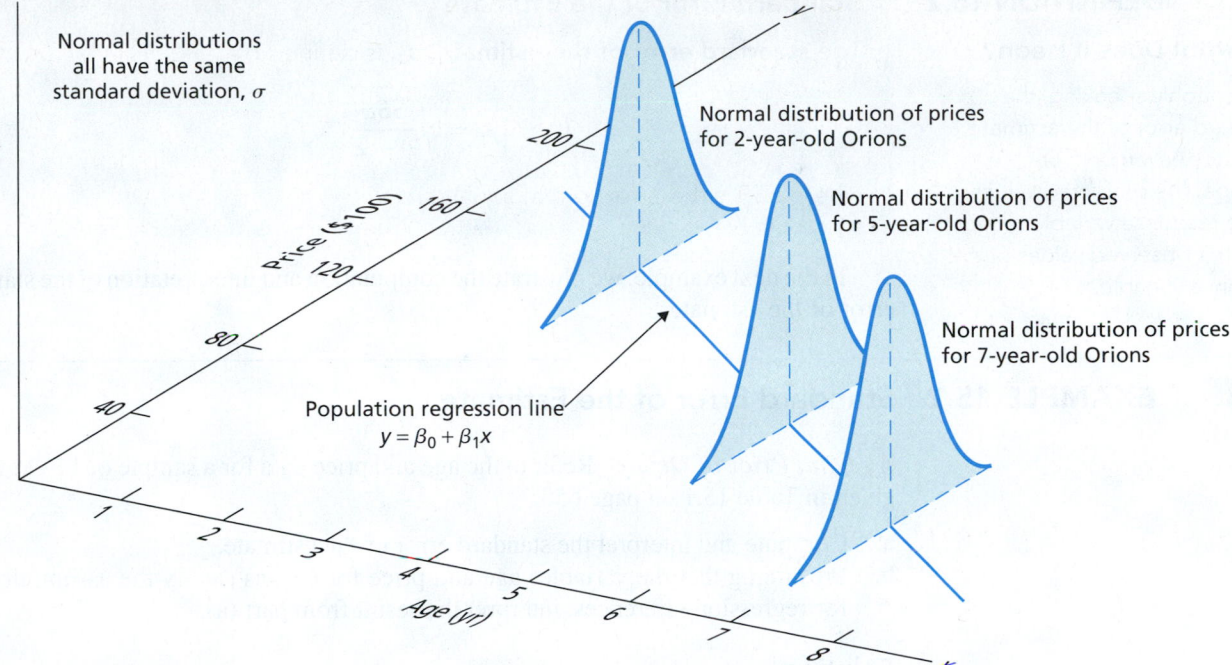

regression line) to estimate β_1 (the slope of the population regression line). We note that b_0 is an unbiased estimator of β_0 and that b_1 is an unbiased estimator of β_1.

Equivalently, we use a sample regression line to estimate the unknown population regression line. Of course, a sample regression line ordinarily will not be the same as the population regression line, just as a sample mean generally will not equal the population mean. In Fig. 15.4, we illustrate this situation for the Orion example. Although the population regression line is unknown, we have drawn it to illustrate the difference between the population regression line and a sample regression line.

FIGURE 15.4

Population regression line and sample regression line for age and price of Orions

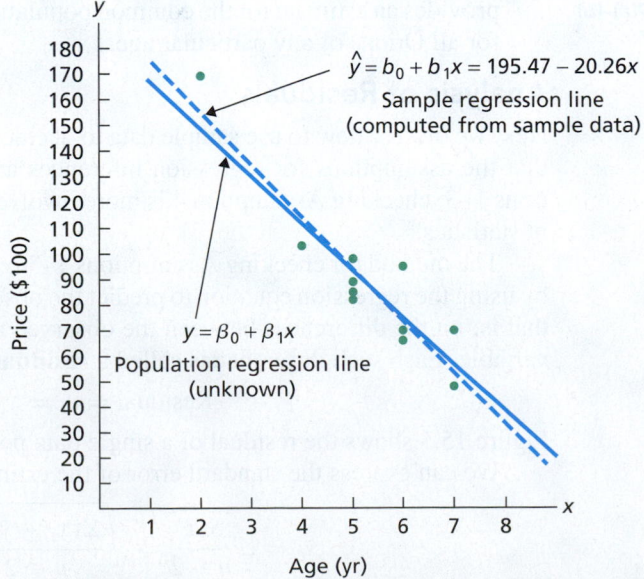

In Fig. 15.4, the sample regression line (the dashed line) is the best approximation that can be made to the population regression line (the solid line) by using the sample data in Table 15.1 on page 658. A different sample of Orions would almost certainly yield a different sample regression line.

The statistic used to obtain a point estimate for the common conditional standard deviation σ is called the **standard error of the estimate.**

DEFINITION 15.2

What Does It Mean?

Roughly speaking, the standard error of the estimate indicates how much, on average, the predicted values of the response variable differ from the observed values of the response variable.

Standard Error of the Estimate

The **standard error of the estimate, s_e,** is defined by

$$s_e = \sqrt{\frac{SSE}{n-2}},$$

where SSE is the error sum of squares.

In the next example, we illustrate the computation and interpretation of the standard error of the estimate.

■■■   **EXAMPLE 15.2**   **Standard Error of the Estimate**

Age and Price of Orions Refer to the age and price data for a sample of 11 Orions given in Table 15.1 on page 658.

a.   Compute and interpret the standard error of the estimate.
b.   Presuming that the variables age and price for Orions satisfy the assumptions for regression inferences, interpret the result from part (a).

Solution

a.   On page 642, we found that $SST = 9708.5$ and $SSR = 8285.0$. So, by the regression identity, $SSE = 9708.5 - 8285.0 = 1423.5$. Thus,

$$s_e = \sqrt{\frac{SSE}{n-2}} = \sqrt{\frac{1423.5}{11-2}} = 12.58.$$

Interpretation Roughly speaking, the predicted price of an Orion in the sample differs, on average, from the observed price by $1258.

StatCrunch

Report 15.1

Exercise 15.29(a)–(b) on page 669

b.   Presuming that the variables age and price for Orions satisfy the assumptions for regression inferences, the standard error of the estimate, $s_e = 12.58$, or $1258, provides an estimate for the common population standard deviation, σ, of prices for all Orions of any particular age.

Analysis of Residuals

Next we discuss how to use sample data to decide whether we can reasonably presume that the assumptions for regression inferences are met. We concentrate on Assumptions 1–3; checking Assumption 4 is more involved and is best left for a second course in statistics.

The method for checking Assumptions 1–3 relies on an analysis of the errors made by using the regression equation to predict the observed values of the response variable, that is, on the differences between the observed and predicted values of the response variable. Each such difference is called a **residual**, generically denoted **e**. Thus,

$$\text{Residual} = e_i = y_i - \hat{y}_i.$$

Figure 15.5 shows the residual of a single data point.

We can express the standard error of the estimate in terms of the residuals:

$$s_e = \sqrt{\frac{SSE}{n-2}} = \sqrt{\frac{\Sigma(y_i - \hat{y}_i)^2}{n-2}} = \sqrt{\frac{\Sigma e_i^2}{n-2}}. \tag{15.1}$$

We can show that the sum of the residuals is always 0, which, in turn, implies that $\bar{e} = 0$. Consequently, the standard error of the estimate is essentially the same as the standard deviation of the residuals.[†] Thus the standard error of the estimate is sometimes called the **residual standard deviation**.

[†]The exact standard deviation of the residuals is obtained by dividing by $n - 1$ instead of $n - 2$.

FIGURE 15.5

Residual of a data point

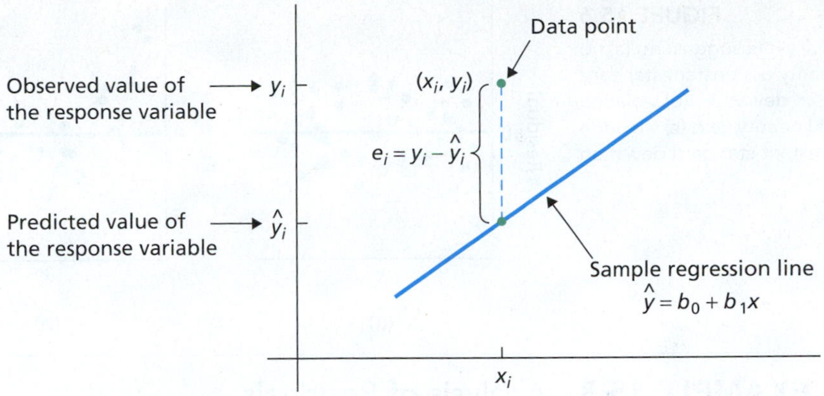

We can analyze the residuals to decide whether Assumptions 1–3 for regression inferences are met because those assumptions can be translated into conditions on the residuals. To show how, let's consider a sample of data points obtained from two variables that satisfy the assumptions for regression inferences.

In light of Assumption 1, the data points should be scattered about the (sample) regression line, which means that the residuals should be scattered about the x-axis. In light of Assumption 2, the variation of the observed values of the response variable should remain approximately constant from one value of the predictor variable to the next, which means the residuals should fall roughly in a horizontal band. In light of Assumption 3, for each value of the predictor variable, the distribution of the corresponding observed values of the response variable should be approximately bell shaped, which implies that the horizontal band should be centered and symmetric about the x-axis.

Furthermore, considering all four regression assumptions simultaneously, we can regard the residuals as independent observations of a variable having a normal distribution with mean 0 and standard deviation σ. Thus a normal probability plot of the residuals should be roughly linear.

KEY FACT 15.2

Residual Analysis for the Regression Model

If the assumptions for regression inferences are met, the following two conditions should hold:

- A plot of the residuals against the observed values of the predictor variable should fall roughly in a horizontal band centered and symmetric about the x-axis.
- A normal probability plot of the residuals should be roughly linear.

Failure of either of these two conditions casts doubt on the validity of one or more of the assumptions for regression inferences for the variables under consideration.

A plot of the residuals against the observed values of the predictor variable, which for brevity we call a **residual plot,** provides approximately the same information as does a scatterplot of the data points. However, a residual plot makes spotting patterns such as curvature and nonconstant standard deviation easier.

To illustrate the use of residual plots for regression diagnostics, let's consider the three plots in Fig. 15.6 on the next page.

- Fig. 15.6(a): In this plot, the residuals are scattered about the x-axis (residuals $= 0$) and fall roughly in a horizontal band, so Assumptions 1 and 2 appear to be met.
- Fig. 15.6(b): This plot suggests that the relation between the variables is curved, indicating that Assumption 1 may be violated.
- Fig. 15.6(c): This plot suggests that the conditional standard deviations increase as x increases, indicating that Assumption 2 may be violated.

FIGURE 15.6

Residual plots suggesting (a) no violation of linearity or constant standard deviation, (b) violation of linearity, and (c) violation of constant standard deviation

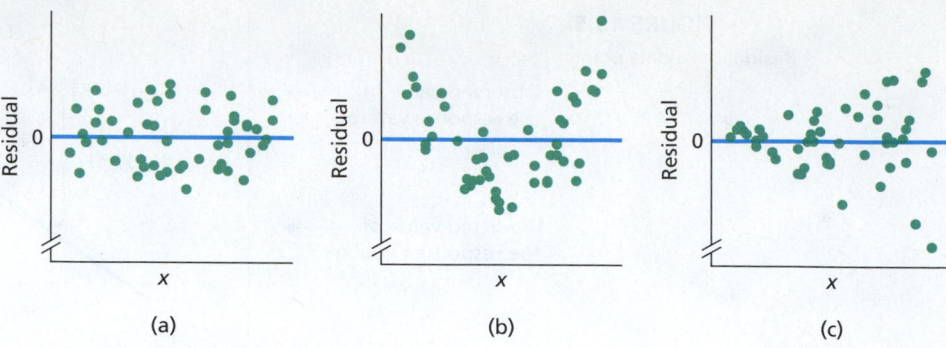

(a)   (b)   (c)

EXAMPLE 15.3   Analysis of Residuals

Age and Price of Orions   The age and price data for a sample of 11 Orions are repeated in the first two columns of Table 15.2. Perform a residual analysis to decide whether we can reasonably consider the assumptions for regression inferences met by the variables age and price of Orions.

Solution   We must first determine the residuals. Each residual is the difference of the observed price (y) and the predicted price ($\hat{y}$). We find each predicted price by substituting each age in the first column of Table 15.2 into the regression equation, $\hat{y} = 195.47 - 20.26x$. The results are shown in the third column of Table 15.2.

TABLE 15.2

Table for obtaining the residuals for the Orion data

Age (yr)	Price ($100)	Predicted price	Residual
x	y	$\hat{y}$	e
5	85	94.17	−9.17
4	103	114.43	−11.43
6	70	73.91	−3.91
5	82	94.17	−12.17
5	89	94.17	−5.17
5	98	94.17	3.83
6	66	73.91	−7.91
6	95	73.91	21.09
2	169	154.95	14.05
7	70	53.65	16.35
7	48	53.65	−5.65

Now we obtain the residuals by subtracting the predicted prices in the third column of Table 15.2 from the observed prices in the second column of Table 15.2. The results are shown in the fourth column of Table 15.2.

We can now perform the required residual analysis by applying the criteria presented in Key Fact 15.2 on page 663. Figure 15.7(a) shows a plot of the residuals against age, and Fig. 15.7(b) shows a normal probability plot for the residuals.

FIGURE 15.7

(a) Residual plot (b) normal probability plot for residuals

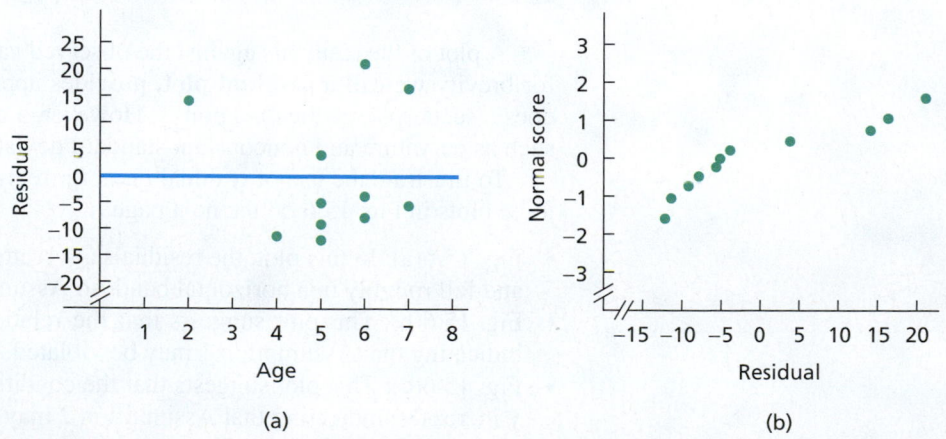

(a)   (b)

StatCrunch

Report 15.2

Exercise 15.29(c)–(d)
on page 669

Taking into account the small sample size, we can say that the residuals fall roughly in a horizontal band that is centered and symmetric about the *x*-axis. We can also say that the normal probability plot for the residuals is (very) roughly linear, although the departure from linearity is sufficient for some concern.[†]

Interpretation There are no obvious violations of the assumptions for regression inferences for the variables age and price of 2- to 7-year-old Orions.

THE TECHNOLOGY CENTER

Most statistical technologies provide the standard error of the estimate as part of their regression analysis output. For instance, consider the Minitab and Excel regression analysis in Output 14.2 on page 634 for the age and price data of 11 Orions. The items circled in green give the standard error of the estimate, so $s_e = 12.58$. As you can see, instead of the notation s_e, Minitab uses S and Excel uses RMSE (root mean square error).

Although the TI-83/84 Plus does not provide the standard error of the estimate in the output of its basic regression procedure, **LinReg(a+bx)**, it can easily be obtained after applying that procedure. Specifically, the TI-83/84 Plus automatically stores the residuals in a list named RESID. In view of the last expression in Equation (15.1) on page 662, we can then determine the standard error of the estimate by using the calculator.

We can also use statistical technology to obtain a residual plot and a normal probability plot of the residuals. The next example illustrates how this is done.

EXAMPLE 15.4 Using Technology to Obtain Plots of Residuals

Age and Price of Orions Use Minitab, Excel, or the TI-83/84 Plus to obtain a residual plot and a normal probability plot of the residuals for the age and price data of Orions given in Table 15.1 on page 658.

Solution We applied the plots-of-residuals programs to the data, resulting in Output 15.1. Steps for generating that output are presented in Instructions 15.1 on the following page.

OUTPUT 15.1 Residual plots and normal probability plots of the residuals for the age and price data of 11 Orions

MINITAB

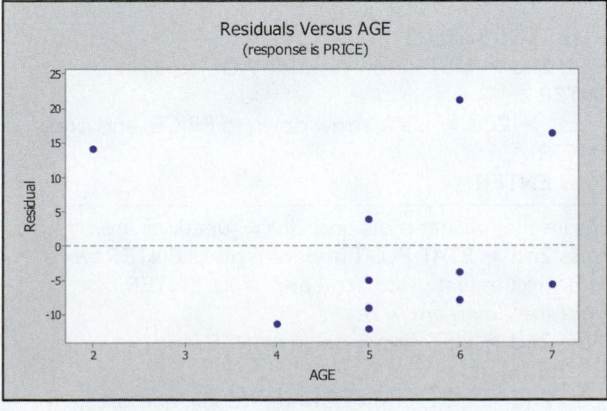

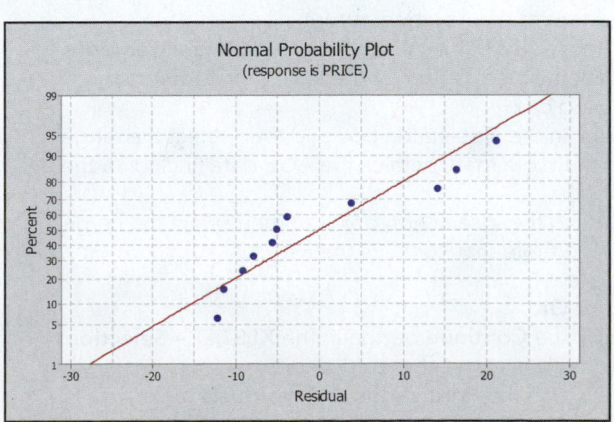

[†]Recall, though, that the inferential procedures in regression analysis are robust to moderate violations of Assumptions 1–3 for regression inferences.

OUTPUT 15.1 (cont.) Residual plots and normal probability plots of the residuals for the age and price data of 11 Orions

EXCEL

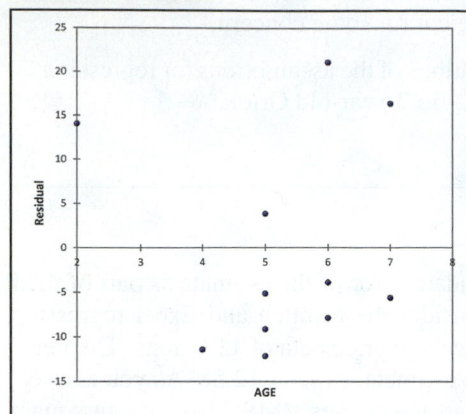

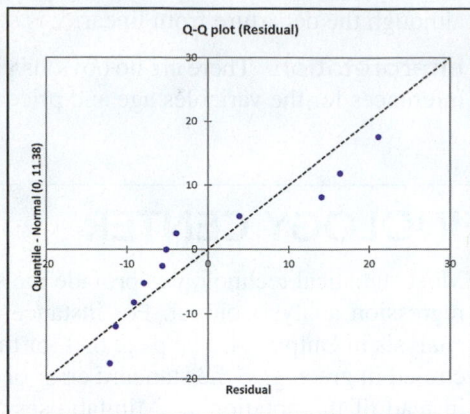

TI-83/84 PLUS

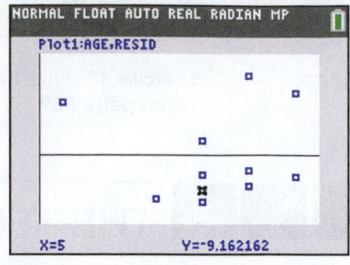

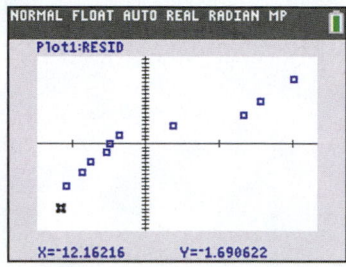

INSTRUCTIONS 15.1 Steps for generating Output 15.1

MINITAB

1 Store the age and price data from Table 15.1 in columns named AGE and PRICE, respectively
2 Choose **Stat ➤ Regression ➤ Fitted Line Plot…**
3 Press the F3 key to reset the dialog box
4 Specify PRICE in the **Response (Y)** text box
5 Specify AGE in the **Predictor (X)** text box
6 Click the **Graphs…** button
7 In the **Individual plots** list, check the **Normal plot of residuals** check box
8 Click in the **Residuals versus the variables** text box and specify AGE
9 Click **OK** twice

EXCEL

1 Store the age and price data from Table 15.1 in columns named AGE and PRICE, respectively
2 Repeat steps 2–9 of Instructions 14.2 on page 634 except that, in step 6, check (only) the **Predictions and residuals** and **X** check boxes
3 Go to the **Predictions and residuals** table at the bottom of the resulting output
4 Choose **XLSTAT ➤ Visualizing data ➤ Scatter plots**
5 Click the reset button in the lower left corner of the dialog box
6 Click in the **X** selection box and then select the range of the table that contains the AGE data (including the label)
7 Click in the **Y** selection box and then select the range of the table that contains the Residual data (including the label)
8 Click **OK**
9 Click the **Continue** button in the **XLSTAT – Selections** dialog box to get the residual plot
10 Return to the **Predictions and residuals** table
11 Choose **XLSTAT ➤ Visualizing data ➤ Univariate plots**
12 Click the reset button in the lower left corner of the dialog box

13 Select the range of the table that contains the Residual data (including the label)
14 Click the **Options** tab and uncheck the **Descriptive statistics** check box
15 Click the **Charts (1)** tab, uncheck the **Box plots** check box, and check the **Normal Q-Q plots** check box
16 Click **OK**
17 Click the **Continue** button in the **XLSTAT – Selections** dailog box to get the normal probability plot of the residuals

TI-83/84 PLUS

1 Store the age and price data from Table 15.1 in lists named AGE and PRICE, respectively
2 Press **STAT**, arrow over to **CALC**, and press **8**

FOR THE TI-84 PLUS *C*
3 Press **2nd ➤ LIST**, arrow down to AGE, and press **ENTER** twice
4 Press **2nd ➤ LIST**, arrow down to PRICE, and press **ENTER** twice
5 Press **CLEAR**, arrow down to **Calculate**, and press **ENTER**

FOR THE TI-83/84 PLUS
3 Press **2nd ➤ LIST**, arrow down to AGE, and press **ENTER**
4 Press **, ➤ 2nd ➤ LIST**, arrow down to PRICE, and press **ENTER**
5 Press **ENTER**

6 Ensure that all stat plots and all **Y=** functions are off
7 Press **2nd ➤ STAT PLOT** and then press **ENTER** twice
8 Arrow to the first graph icon and press **ENTER**
9 Press the down-arrow key
10 Press **2nd ➤ LIST**, arrow down to AGE, and press **ENTER** twice
11 Press **2nd ➤ LIST**, arrow down to RESID, and press **ENTER** twice

(continued)

Exercises 15.1

Understanding the Concepts and Skills

15.1 Suppose that x and y are predictor and response variables, respectively, of a population. Consider the population that consists of all members of the original population that have a specified value of the predictor variable. The distribution, mean, and standard deviation of the response variable for this population are called the _____, _____, and _____, respectively, corresponding to the specified value of the predictor variable.

15.2 State the four conditions required for making regression inferences.

In Exercises 15.3–15.6, assume that the variables under considera-tion satisfy the assumptions for regression inferences.

15.3 Fill in the blanks.
a. The line $y = \beta_0 + \beta_1 x$ is called the _____.
b. The common conditional standard deviation of the response variable is denoted _____.
c. For $x = 6$, the conditional distribution of the response variable is a _____ distribution having mean _____ and standard deviation _____.

15.4 What statistic is used to estimate
a. the y-intercept of the population regression line?
b. the slope of the population regression line?
c. the common conditional standard deviation, σ, of the response variable?

15.5 Based on a sample of data points, what is the best estimate of the population regression line?

15.6 Regarding the standard error of the estimate,
a. give two interpretations of it.
b. identify another name used for it, and explain the rationale for that name.
c. which one of the three sums of squares figures in its computation?

15.7 The difference between an observed value and a predicted value of the response variable is called a _____.

15.8 Identify two graphs used in a residual analysis to check the Assumptions 1–3 for regression inferences, and explain the reasoning behind their use.

15.9 Which graph used in a residual analysis provides roughly the same information as a scatterplot? What advantages does it have over a scatterplot?

15.10 Figure 15.8 shows three residual plots and a normal probabil-ity plot of residuals. For each part, decide whether the graph suggests violation of one or more of the assumptions for regression inferences. Explain your answers.

15.11 Figure 15.9 on the next page shows three residual plots and a normal probability plot of residuals. For each part, decide whether the graph suggests violation of one or more of the assumptions for regression inferences. Explain your answers.

FIGURE 15.8
Plots for Exercise 15.10

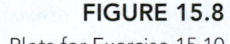
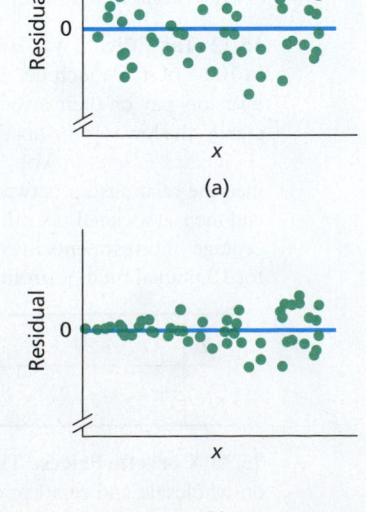

(a)

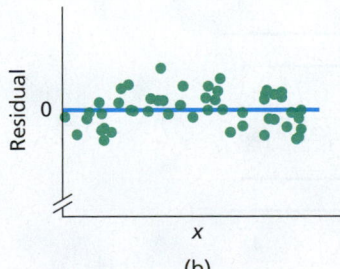

(b)

(c)

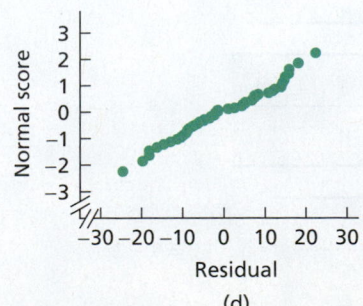

(d)

FIGURE 15.9
Plots for Exercise 15.11

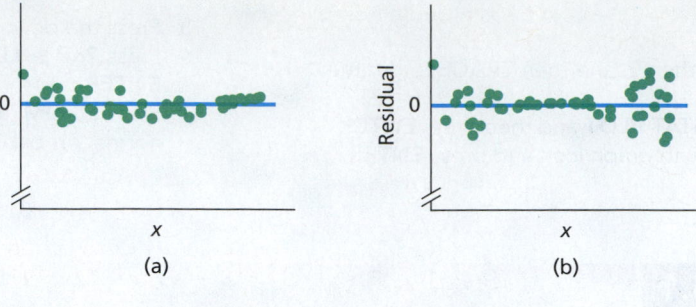

(a)   (b)

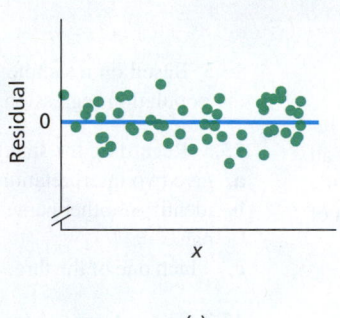   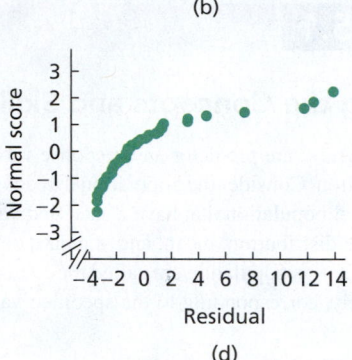

(c)   (d)

In Exercises 15.12–15.21, we repeat the data and provide the sample regression equations for Exercises 14.48–14.57.
a. Determine the standard error of the estimate.
b. Construct a residual plot.
c. Construct a normal probability plot of the residuals.

15.12

x	2	4	3
y	3	5	7

$\hat{y} = 2 + x$

15.13

x	3	1	2
y	−4	0	−5

$\hat{y} = 1 - 2x$

15.14

x	0	4	3	1	2
y	1	9	8	4	3

$\hat{y} = 1 + 2x$

15.15

x	3	4	1	2
y	4	5	0	−1

$\hat{y} = -3 + 2x$

15.16

x	1	3	4	4
y	13	−1	3	5

$\hat{y} = 14 - 3x$

15.17

x	2	2	3	4	4
y	3	4	0	2	1

$\hat{y} = 5 - x$

15.18

x	1	3	4	4
y	8	0	3	1

$\hat{y} = 9 - 2x$

15.19

x	1	2	3
y	4	3	8

$\hat{y} = 1 + 2x$

15.20

x	1	1	5	5
y	1	3	2	4

$\hat{y} = 1.75 + 0.25x$

15.21

x	0	2	2	5	6
y	4	2	0	−2	1

$\hat{y} = 2.875 - 0.625x$

Applying the Concepts and Skills

In Exercises 15.22–15.27, we repeat the information from Exercises 14.58–14.63. For each exercise here, discuss what satisfying Assumptions 1–3 for regression inferences by the variables under consideration would mean.

15.22 Tax Efficiency. *Tax efficiency* is a measure—ranging from 0 to 100—of how much tax due to capital gains stock or mutual funds investors pay on their investments each year; the higher the tax efficiency, the lower is the tax. The paper "At the Mercy of the Manager" (*Financial Planning*, Vol. 30(5), pp. 54–56) by C. Israelsen examined the relationship between investments in mutual fund portfolios and their associated tax efficiencies. The following table shows percentage of investments in energy securities (x) and tax efficiency (y) for 10 mutual fund portfolios.

x	3.1	3.2	3.7	4.3	4.0	5.5	6.7	7.4	7.4	10.6
y	98.1	94.7	92.0	89.8	87.5	85.0	82.0	77.8	72.1	53.5

15.23 Corvette Prices. The *Kelley Blue Book* provides information on wholesale and retail prices of cars. Following are age and price data for 10 randomly selected Corvettes between 1 and 6 years old.

Here, x denotes age, in years, and y denotes price, in hundreds of dollars.

x	6	6	6	2	2	5	4	5	1	4
y	290	280	295	425	384	315	355	328	425	325

15.24 Custom Homes. Hanna Properties specializes in custom-home resales in the Equestrian Estates, an exclusive subdivision in Phoenix, Arizona. A random sample of nine custom homes currently listed for sale provided the following information on size and price. Here, x denotes size, in hundreds of square feet, rounded to the nearest hundred, and y denotes price, in thousands of dollars, rounded to the nearest thousand.

x	26	27	33	29	29	34	30	40	22
y	540	555	575	577	606	661	738	804	496

15.25 Plant Emissions. Plants emit gases that trigger the ripening of fruit, attract pollinators, and cue other physiological responses. N. Agelopolous et al. examined factors that affect the emission of volatile compounds by the potato plant *Solanum tuberosum* and published their findings in the paper "Factors Affecting Volatile Emissions of Intact Potato Plants, *Solanum tuberosum*: Variability of Quantities and Stability of Ratios" (*Journal of Chemical Ecology*, Vol. 26(2), pp. 497–511). The volatile compounds analyzed were hydrocarbons used by other plants and animals. Following are data on plant weight (x), in grams, and quantity of volatile compounds emitted (y), in hundreds of nanograms, for 11 potato plants.

x	57	85	57	65	52	67	62	80	77	53	68
y	8.0	22.0	10.5	22.5	12.0	11.5	7.5	13.0	16.5	21.0	12.0

15.26 Crown-Rump Length. In the article "The Human Vomeronasal Organ. Part II: Prenatal Development" (*Journal of Anatomy*, Vol. 197, Issue 3, pp. 421–436), T. Smith and K. Bhatnagar examined the controversial issue of the human vomeronasal organ, regarding its structure, function, and identity. The following table shows the age of fetuses (x), in weeks, and length of crown-rump (y), in millimeters.

x	10	10	13	13	18	19	19	23	25	28
y	66	66	108	106	161	166	177	228	235	280

15.27 Study Time and Score. An instructor at Arizona State University asked a random sample of eight students to record their study times in a beginning calculus course. She then made a table for total hours studied (x) over 2 weeks and test score (y) at the end of the 2 weeks. Here are the results.

x	10	15	12	20	8	16	14	22
y	92	81	84	74	85	80	84	80

In Exercises 15.28–15.33,
a. compute the standard error of the estimate and interpret your answer.
b. interpret your result from part (a) if the assumptions for regression inferences hold.

c. obtain a residual plot and a normal probability plot of the residuals.
d. decide whether you can reasonably consider Assumptions 1–3 for regression inferences to be met by the variables under consideration. (The answer here is subjective, especially in view of the extremely small sample sizes.)

15.28 Tax Efficiency. Use the data on percentage of investments in energy securities and tax efficiency from Exercise 15.22.

15.29 Corvette Prices. Use the age and price data for Corvettes from Exercise 15.23.

15.30 Custom Homes. Use the size and price data for custom homes from Exercise 15.24.

15.31 Plant Emissions. Use the data on plant weight and quantity of volatile emissions from Exercise 15.25.

15.32 Crown-Rump Length. Use the data on age of fetuses and length of crown-rump from Exercise 15.26.

15.33 Study Time and Score. Use the data on total hours studied over 2 weeks and test score at the end of the 2 weeks from Exercise 15.27.

Working with Large Data Sets

In Exercises 15.34–15.43, use the technology of your choice to
a. obtain and interpret the standard error of the estimate.
b. obtain a residual plot and a normal probability plot of the residuals.
c. decide whether you can reasonably consider Assumptions 1–3 for regression inferences met by the two variables under consideration.

15.34 Birdies and Score. How important are birdies (a score of one under par on a given hole) in determining the final total score of a woman golfer? From the *U.S. Women's Open* website, we obtained data on number of birdies during a tournament and final score for 63 women golfers. The data are presented on the WeissStats site.

15.35 U.S. Presidents. The *Information Please Almanac* provides data on the ages at inauguration and of death for the presidents of the United States. We give those data on the WeissStats site for those presidents who are not still living at the time of this writing.

15.36 Movie Grosses. Box Office Mojo collects and posts data on movie grosses. For a random sample of 50 movies, we obtained both the domestic (U.S.) and overseas grosses, in millions of dollars. The data are presented on the WeissStats site.

15.37 Acreage and Value. The document *Arizona Residential Property Valuation System*, published by the Arizona Department of Revenue, describes how county assessors use computerized systems to value single-family residential properties for property tax purposes. On the WeissStats site are data on lot size (in acres) and assessed value (in thousands of dollars) for a sample of homes in a particular area.

15.38 Home Size and Value. On the WeissStats site are data on home size (in square feet) and assessed value (in thousands of dollars) for the same homes as in Exercise 15.37.

15.39 High and Low Temperature. The National Oceanic and Atmospheric Administration publishes temperature information of cities around the world in *Climates of the World*. A random sample of 50 cities gave the data on average high and low temperatures in January shown on the WeissStats site.

15.40 PCBs and Pelicans. Polychlorinated biphenyls (PCBs), industrial pollutants, are known to be carcinogens and a great danger to natural ecosystems. As a result of several studies, PCB production was banned in the United States in 1979 and by the Stockholm Convention on Persistent Organic Pollutants in 2001. One study, published in 1972 by R. Risebrough, is titled "Effects of Environmental Pollutants Upon Animals Other Than Man" (*Proceedings of the 6th Berkeley Symposium on Mathematics and Statistics, VI*, University of California Press, pp. 443–463). In that study, 60 Anacapa pelican eggs were collected and measured for their shell thickness, in millimeters (mm), and concentration of PCBs, in parts per million (ppm). The data are on the WeissStats site.

15.41 Gas Guzzlers. The magazine *Consumer Reports* publishes information on automobile gas mileage and variables that affect gas mileage. In one issue, data on gas mileage (in mpg) and engine displacement (in liters, L) were published for 121 vehicles. Those data are stored on the WeissStats site.

15.42 Estriol Level and Birth Weight. J. Greene and J. Touchstone conducted a study on the relationship between the estriol levels of pregnant women and the birth weights of their children. Their findings, "Urinary Tract Estriol: An Index of Placental Function," were published in the *American Journal of Obstetrics and Gynecology* (Vol. 85(1), pp. 1–9). The data points are provided on the WeissStats site, where estriol levels are in mg/24 hr and birth weights are in hectograms (hg).

15.43 Shortleaf Pines. The ability to estimate the volume of a tree based on a simple measurement, such as the diameter of the tree, is important to the lumber industry, ecologists, and conservationists. Data on volume, in cubic feet, and diameter at breast height, in inches, for 70 shortleaf pines was reported in C. Bruce and F. X. Schumacher's *Forest Mensuration* (New York: McGraw-Hill, 1935) and analyzed by A. C. Akinson in the article "Transforming Both Sides of a Tree" (*The American Statistician*, Vol. 48, pp. 307–312). The data are provided on the WeissStats site.

15.2 Inferences for the Slope of the Population Regression Line

In this section and the next, we examine several inferential procedures used in regression analysis. Strictly speaking, these inferential techniques require that the assumptions given in Key Fact 15.1 on page 659 be satisfied. However, as we noted earlier, these techniques are robust to moderate violations of those assumptions.

The first inferential methods we present concern the slope, β_1, of the population regression line. To begin, we consider hypothesis testing.

Hypothesis Tests for the Slope of the Population Regression Line

TABLE 15.3

Age and price data for a sample of 11 Orions

Age (yr) x	Price ($100) y
5	85
4	103
6	70
5	82
5	89
5	98
6	66
6	95
2	169
7	70
7	48

Suppose that the variables x and y satisfy the assumptions for regression inferences. Then, for each value x of the predictor variable, the conditional distribution of the response variable is a normal distribution with mean $\beta_0 + \beta_1 x$ and standard deviation σ.

Of particular interest is whether the slope, β_1, of the population regression line equals 0. If $\beta_1 = 0$, then, for each value x of the predictor variable, the conditional distribution of the response variable is a normal distribution having mean $\beta_0 \ (= \beta_0 + 0 \cdot x)$ and standard deviation σ. Because x does not appear in either of those two parameters, it is useless as a predictor of y.[†]

Hence, we can decide whether x is useful as a (linear) predictor of y—that is, whether the regression equation has utility—by performing the hypothesis test

H_0: $\beta_1 = 0$ (x is not useful for predicting y)

H_a: $\beta_1 \neq 0$ (x is useful for predicting y).

We base hypothesis tests for β_1 (the slope of the population regression line) on the statistic b_1 (the slope of a sample regression line). To explain how this method works, let's return to the Orion illustration. The data on age and price for a sample of 11 Orions are repeated in Table 15.3.

[†]Although x alone may not be useful for predicting y, it may be useful in conjunction with another variable or variables. Thus, in this section, when we say that x is not useful for predicting y, we really mean that the regression equation with x as the only predictor variable is not useful for predicting y. Conversely, although x alone may be useful for predicting y, it may not be useful in conjunction with another variable or variables. Thus, in this section, when we say that x is useful for predicting y, we really mean that the regression equation with x as the only predictor variable is useful for predicting y.

With age as the predictor variable and price as the response variable, the regression equation for these data is $\hat{y} = 195.47 - 20.26x$, as we found in Chapter 14. In particular, the slope, b_1, of the sample regression line is -20.26.

We now consider all possible samples of 11 Orions whose ages are the same as those given in the first column of Table 15.3. For such samples, the slope, b_1, of the sample regression line varies from one sample to another and is therefore a variable. Its distribution is called the **sampling distribution of the slope of the regression line.** From the assumptions for regression inferences, we can show that this distribution is a normal distribution whose mean is the slope, β_1, of the population regression line. More generally, we have Key Fact 15.3.

KEY FACT 15.3

? **What Does It Mean?**

For fixed values of the predictor variable, the slopes of all possible sample regression lines have a normal distribution with mean β_1 and standard deviation $\sigma/\sqrt{S_{xx}}$.

The Sampling Distribution of the Slope of the Regression Line

Suppose that the variables x and y satisfy the four assumptions for regression inferences. Then, for samples of size n, each with the same values $x_1, x_2, \ldots, x_n$ for the predictor variable, the following properties hold for the slope, b_1, of the sample regression line:

- The mean of b_1 equals the slope of the population regression line; that is, we have $\mu_{b_1} = \beta_1$ (i.e., the slope of the sample regression line is an unbiased estimator of the slope of the population regression line).
- The standard deviation of b_1 is $\sigma_{b_1} = \sigma/\sqrt{S_{xx}}$.
- The variable b_1 is normally distributed.

As a consequence of Key Fact 15.3, the standardized variable

$$z = \frac{b_1 - \beta_1}{\sigma/\sqrt{S_{xx}}}$$

has the standard normal distribution. But this variable cannot be used as a basis for the required test statistic because the common conditional standard deviation, σ, is unknown. We therefore replace σ with its sample estimate s_e, the standard error of the estimate. As you might suspect, the resulting variable has a t-distribution.

KEY FACT 15.4

t-Distribution for Inferences for β_1

Suppose that the variables x and y satisfy the four assumptions for regression inferences. Then, for samples of size n, each with the same values $x_1, x_2, \ldots, x_n$ for the predictor variable, the variable

$$t = \frac{b_1 - \beta_1}{s_e/\sqrt{S_{xx}}}$$

has the t-distribution with df $= n - 2$.

In light of Key Fact 15.4, for a hypothesis test with the null hypothesis $H_0\colon \beta_1 = 0$, we can use the variable

$$t = \frac{b_1}{s_e/\sqrt{S_{xx}}}$$

as the test statistic and obtain the critical values or P-value from the t-table, Table IV in Appendix A. We call this hypothesis-testing procedure the **regression t-test.** Procedure 15.1 provides a step-by-step method for performing a regression t-test by using either the critical-value approach or the P-value approach. *Note:* By "the four assumptions for regression inferences," we mean the four conditions stated in Key Fact 15.1 on page 659.

███ ■ **PROCEDURE 15.1** Regression *t*-Test

Purpose To perform a hypothesis test to decide whether a predictor variable is useful for making predictions

Assumptions The four assumptions for regression inferences

Step 1 The null and alternative hypotheses are, respectively,

$H_0: \beta_1 = 0$ (predictor variable is not useful for making predictions)
$H_a: \beta_1 \neq 0$ (predictor variable is useful for making predictions).

Step 2 Decide on the significance level, α.

Step 3 Compute the value of the test statistic

$$t = \frac{b_1}{s_e/\sqrt{S_{xx}}}$$

and denote that value t_0.

CRITICAL-VALUE APPROACH	OR	P-VALUE APPROACH

Step 4 The critical values are $\pm t_{\alpha/2}$ with df $= n - 2$. Use Table IV to find the critical values.

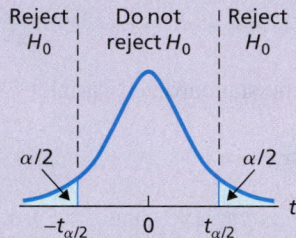

Step 5 If the value of the test statistic falls in the rejection region, reject H_0; otherwise, do not reject H_0.

Step 4 The *t*-statistic has df $= n - 2$. Use Table IV to estimate the *P*-value, or obtain it exactly by using technology.

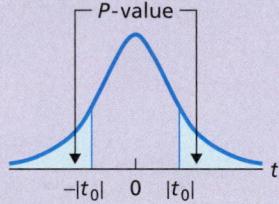

Step 5 If $P \leq \alpha$, reject H_0; otherwise, do not reject H_0.

Step 6 Interpret the results of the hypothesis test.

███ ■ **EXAMPLE 15.5** The Regression *t*-Test

Age and Price of Orions The data on age and price for a sample of 11 Orions are displayed in Table 15.3 on page 670. At the 5% significance level, do the data provide sufficient evidence to conclude that age is useful as a (linear) predictor of price for Orions?

Solution As we discovered in Example 15.3, we can reasonably consider the assumptions for regression inferences to be satisfied by the variables age and price for Orions, at least for Orions between 2 and 7 years old. So we apply Procedure 15.1 to carry out the required hypothesis test.

Step 1 State the null and alternative hypotheses.

Let β_1 denote the slope of the population regression line that relates price to age for Orions. Then the null and alternative hypotheses are, respectively,

$H_0: \beta_1 = 0$ (age is not useful for predicting price)
$H_a: \beta_1 \neq 0$ (age is useful for predicting price).

Step 2 Decide on the significance level, α.

We are to perform the hypothesis test at the 5% significance level, or $\alpha = 0.05$.

Step 3 Compute the value of the test statistic

$$t = \frac{b_1}{s_e/\sqrt{S_{xx}}}.$$

In Example 14.6 on page 629, we found that $b_1 = -20.26$, $\Sigma x_i^2 = 326$, and $\Sigma x_i = 58$. Also, in Example 15.2 on page 662, we determined that $s_e = 12.58$. Therefore, because $n = 11$, the value of the test statistic is

$$t = \frac{b_1}{s_e/\sqrt{S_{xx}}} = \frac{b_1}{s_e/\sqrt{\Sigma x_i^2 - (\Sigma x_i)^2/n}}$$

$$= \frac{-20.26}{12.58/\sqrt{326 - (58)^2/11}} = -7.235.$$

CRITICAL-VALUE APPROACH	OR	P-VALUE APPROACH

Step 4 The critical values are $\pm t_{\alpha/2}$ with df $= n - 2$. Use Table IV to find the critical values.

From Step 2, $\alpha = 0.05$. For $n = 11$, df $= n - 2 = 11 - 2 = 9$. Using Table IV, we find that the critical values are $\pm t_{\alpha/2} = \pm t_{0.025} = \pm 2.262$, as depicted in Fig. 15.10A.

FIGURE 15.10A

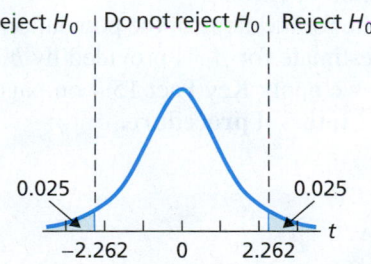

Reject H_0 | Do not reject H_0 | Reject H_0

0.025   0.025

-2.262 0   2.262

Step 5 If the value of the test statistic falls in the rejection region, reject H_0; otherwise, do not reject H_0.

The value of the test statistic, found in Step 3, is $t = -7.235$. Because this value falls in the rejection region, we reject H_0. The test results are statistically significant at the 5% level.

Step 4 The t-statistic has df $= n - 2$. Use Table IV to estimate the P-value or obtain it exactly by using technology.

The value of the test statistic, found in Step 3, is $t = -7.235$. Because the test is two tailed, the P-value is the probability of observing a value of t of 7.235 or greater in magnitude if the null hypothesis is true. That probability equals the shaded area shown in Fig. 15.10B.

FIGURE 15.10B

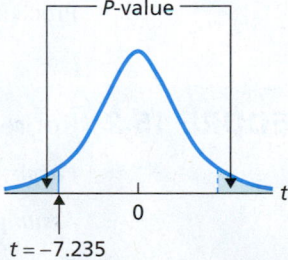

P-value

0

$t = -7.235$

For $n = 11$, df $= 11 - 2 = 9$. Referring to Fig. 15.10B and to Table IV with df $= 9$, we find that $P < 0.01$. (Using technology, we obtain $P = 0.0000488$.)

Step 5 If $P \le \alpha$, reject H_0; otherwise, do not reject H_0.

From Step 4, $P < 0.01$. Because the P-value is less than the specified significance level of 0.05, we reject H_0. The test results are statistically significant at the 5% level and (see Table 9.8 on page 386) provide very strong evidence against the null hypothesis.

StatCrunch

Report 15.3

You try it!

Exercise 15.59
on page 678

Step 6 Interpret the results of the hypothesis test.

Interpretation At the 5% significance level, the data provide sufficient evidence to conclude that the slope of the population regression line is not 0 and hence that age is useful as a (linear) predictor of price for Orions.

Other Procedures for Testing Utility of the Regression

We use Procedure 15.1 on page 672, which is based on the statistic b_1, to perform a hypothesis test to decide whether the slope of the population regression line is not 0 or, equivalently, whether the regression equation is useful for making predictions.

In Section 14.3, we introduced the coefficient of determination, r^2, as a descriptive measure of the utility of the regression equation for making predictions. We should therefore also be able to use the statistic r^2 as a basis for performing a hypothesis test to decide whether the regression equation is useful for making predictions—and indeed we can. However, we do not cover the hypothesis test based on r^2 because it is equivalent to the hypothesis test based on b_1.

We can also use the linear correlation coefficient, r, introduced in Section 14.4, as a basis for performing a hypothesis test to decide whether the regression equation is useful for making predictions. That test too is equivalent to the hypothesis test based on b_1, but, because it has other uses, we discuss it in Section 15.4.

Confidence Intervals for the Slope of the Population Regression Line

Recall that the slope of a line represents the change in the dependent variable, y, resulting from an increase in the independent variable, x, by 1 unit. Also recall that the population regression line, whose slope is β_1, gives the conditional means of the response variable. Therefore β_1 represents the change in the conditional mean of the response variable for each increase in the value of the predictor variable by 1 unit.

For instance, consider the variables age and price of Orions. In this case, β_1 is the amount that the mean price decreases for every increase in age by 1 year. In other words, β_1 is the mean yearly depreciation of Orions.

Consequently, obtaining an estimate for the slope of the population regression line is worthwhile. We know that a point estimate for β_1 is provided by b_1. To determine a confidence-interval estimate for β_1, we apply Key Fact 15.4 on page 671 to obtain Procedure 15.2, called the **regression t-interval procedure.**

■■■■ PROCEDURE 15.2   Regression t-Interval Procedure

Purpose To find a confidence interval for the slope, β_1, of the population regression line

Assumptions The four assumptions for regression inferences

Step 1 For a confidence level of $1 - \alpha$, use Table IV to find $t_{\alpha/2}$ with df $= n - 2$.

Step 2 The endpoints of the confidence interval for β_1 are

$$b_1 \pm t_{\alpha/2} \cdot \frac{s_e}{\sqrt{S_{xx}}}.$$

Step 3 Interpret the confidence interval.

■■■ EXAMPLE 15.6   The Regression t-Interval Procedure

Age and Price of Orions Use the data in Table 15.3 on page 670 to determine a 95% confidence interval for the slope of the population regression line that relates price to age for Orions.

Solution We apply Procedure 15.2.

Step 1 For a confidence level of $1 - \alpha$, use Table IV to find $t_{\alpha/2}$ with df $= n - 2$.

For a 95% confidence interval, $\alpha = 0.05$. Because $n = 11$, df $= 11 - 2 = 9$. From Table IV, $t_{\alpha/2} = t_{0.05/2} = t_{0.025} = 2.262$.

Step 2 The endpoints of the confidence interval for β_1 are

$$b_1 \pm t_{\alpha/2} \cdot \frac{s_e}{\sqrt{S_{xx}}}.$$

From Example 14.6, $b_1 = -20.26$, $\Sigma x_i^2 = 326$, and $\Sigma x_i = 58$. Also, from Example 15.2, $s_e = 12.58$. Hence the endpoints of the confidence interval for β_1 are

$$-20.26 \pm 2.262 \cdot \frac{12.58}{\sqrt{326 - (58)^2/11}},$$

or -20.26 ± 6.33, or -26.59 to -13.93.

Step 3 Interpret the confidence interval.

StatCrunch

Report 15.4

Exercise 15.65
on page 678

Interpretation We can be 95% confident that the slope of the population regression line is somewhere between -26.59 and -13.93. In other words, we can be 95% confident that the yearly decrease in mean price for Orions is somewhere between $1393 and $2659.

THE TECHNOLOGY CENTER

Most statistical technologies can be used to perform regression t-procedures. In this subsection, we present output and step-by-step instructions for such programs.

Note to TI-83 Plus users: Although the TI-84 Plus has a built-in regression t-interval procedure, the TI-83 Plus does not. Thus, we have supplied a TI program called REGTINT that can be used with the TI-83 Plus to perform the regression t-interval procedure. You can find that program in the TI Programs section on the WeissStats site. Your instructor can show you how to download the program to your calculator. *Warning:* Any data that you may have previously stored in Lists 1 and 2 will be erased during program execution, so copy those data to other lists prior to program execution if you want to retain them.

EXAMPLE 15.7   Using Technology to Conduct Regression t-Procedures

Age and Price of Orions Table 15.3 on page 670 gives the age and price data for a sample of 11 Orions. Use Minitab, Excel, or the TI-83/84 Plus to perform the hypothesis test in Example 15.5 and obtain the confidence interval required in Example 15.6.

Solution Let β_1 denote the slope of the population regression line that relates price to age for Orions. The task in Example 15.5 is to perform the hypothesis test

$$H_0: \beta_1 = 0 \text{ (age is not useful for predicting price)}$$

$$H_a: \beta_1 \neq 0 \text{ (age is useful for predicting price)}$$

at the 5% significance level; the task in Example 15.6 is to obtain a 95% confidence interval for β_1.

We applied the regression t-procedures to the data, resulting in Output 15.2. Steps for generating that output are presented in Instructions 15.2. *Note to Excel users:* For brevity, we have presented only the essential portions of the actual output.

OUTPUT 15.2   Regression *t*-procedures for the age and price data of 11 Orions

MINITAB

Regression Analysis: PRICE versus AGE

Coefficients

Term	Coef	SE Coef	95% CI	T-Value	P-Value	VIF
Constant	195.5	15.2	(161.0, 229.9)	12.83	0.000	
AGE	-20.26	2.80	(-26.59, -13.93)	-7.24	0.000	1.00

TI-83/84 PLUS

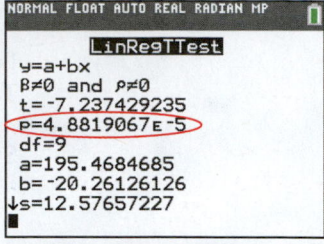

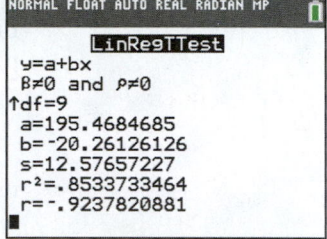

Using **LinRegTTest**

EXCEL

Regression of variable PRICE:							
Model parameters:							
Source	Value	Standard error	t	Pr > |t|	Lower bound (95%)	Upper bound (95%)	
Intercept	195.4685	15.2403	12.8257	< 0.0001	160.9924	229.9445	
AGE	-20.2613	2.7995	-7.2374	< 0.0001	-26.5942	-13.9283	

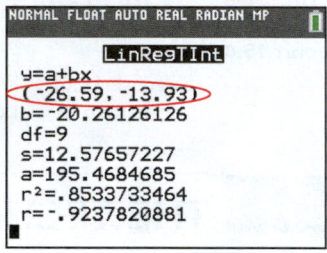

Using **LinRegTInt**

As shown in Output 15.2, $P = 0.000$ (to three decimal places). Because the P-value is less than the specified significance level of 0.05, we reject H_0. At the 5% significance level, the data provide sufficient evidence to conclude that the slope of the population regression line is not 0 and hence that age is useful as a (linear) predictor of price for Orions.

Output 15.2 also shows that a 95% confidence interval for β_1 is from -26.59 to -13.93. We can be 95% confident that the slope of the population regression line is somewhere between -26.59 and -13.93. In other words, we can be 95% confident that the yearly decrease in mean price for Orions is somewhere between \$1393 and \$2659.

INSTRUCTIONS 15.2   Steps for generating Output 15.2

MINITAB

1 Store the age and price data from Table 15.3 in columns named AGE and PRICE, respectively
2 Choose **Stat ➤ Regression ➤ Regression ➤ Fit Regression Model...**
3 Press the F3 key to reset the dialog box
4 Specify PRICE in the **Responses** text box
5 Specify AGE in the **Continuous predictors** text box
6 Click the **Options...** button
7 Type 95 in the **Confidence level for all intervals** text box
8 Click **OK**
9 Click the **Results...** button
10 Click the arrow button at the right of the **Display of results** drop-down list box and select **Expanded tables**
11 Uncheck all check boxes except for **Coefficients**
12 Click **OK** twice

EXCEL

1 Store the age and price data from Table 15.3 in columns named AGE and PRICE, respectively
2 Choose **XLSTAT ➤ Modeling data ➤ Linear regression**
3 Click the reset button in the lower left corner of the dialog box
4 In the **Y / Dependent variables** list, click in the **Quantitative** selection box and then select the column of the worksheet that contains the PRICE data
5 In the **X / Explanatory variables** list, click in the **Quantitative** selection box and then select the column of the worksheet that contains the AGE data
6 Click the **Options** tab and then type 95 in the **Confidence interval (%)** text box
7 Click the **Outputs** tab and then uncheck all the check boxes therein

(continued)

EXCEL

8 Click the **Charts** tab and then uncheck the **Regression charts** check box
9 Click **OK**
10 Click the **Continue** button in the **XLSTAT – Selections** dialog box

TI-83/84 PLUS

Store the age and price data from Table 15.3 in lists named AGE and PRICE, respectively

FOR THE HYPOTHESIS TEST:
1 Press **STAT**, arrow over to **TESTS**, arrow down to **LinRegTTest**, and press **ENTER**
2 Press **2nd ➤ LIST**, arrow down to AGE, and press **ENTER** twice

3 Press **2nd ➤ LIST**, arrow down to PRICE, and press **ENTER** twice
4 Type 1 for **Freq** and press **ENTER**
5 Highlight ≠0 and press **ENTER**
6 Arrow down to **Calculate** and press **ENTER**

FOR THE CI:
1 Press **STAT**, arrow over to **TESTS**, arrow down to **LinRegTInt**, and press **ENTER**
2 Press **2nd ➤ LIST**, arrow down to AGE, and press **ENTER** twice
3 Press **2nd ➤ LIST**, arrow down to PRICE, and press **ENTER** twice
4 Type 1 for **Freq** and press **ENTER**
5 Type .95 for **C-Level** and press **ENTER**
6 Arrow down to **Calculate** and press **ENTER**

Exercises 15.2

Understanding the Concepts and Skills

15.44 Explain why the predictor variable is useless as a predictor of the response variable if the slope of the population regression line is 0.

15.45 For two variables satisfying Assumptions 1–3 for regression inferences, the population regression equation is $y = 20 - 3.5x$. For samples of size 10 and given values of the predictor variable, the distribution of slopes of all possible sample regression lines is a _____ distribution with mean _____.

15.46 Consider the standardized variable

$$z = \frac{b_1 - \beta_1}{\sigma/\sqrt{S_{xx}}}.$$

a. Identify its distribution.
b. Why can't it be used as the test statistic for a hypothesis test concerning β_1?
c. What statistic is used? What is the distribution of that statistic?

15.47 In this section, we used the statistic b_1 as a basis for conducting a hypothesis test to decide whether a regression equation is useful for prediction. Identify two other statistics that can be used as a basis for such a test.

In Exercises 15.48–15.57, we repeat the information from Exercises 15.12–15.21.
a. Decide, at the 10% significance level, whether the data provide sufficient evidence to conclude that x is useful for predicting y.
b. Find a 90% confidence interval for the slope of the population regression line.

15.48

x	2	4	3
y	3	5	7

$\hat{y} = 2 + x$

15.49

x	3	1	2
y	−4	0	−5

$\hat{y} = 1 - 2x$

15.50

x	0	4	3	1	2
y	1	9	8	4	3

$\hat{y} = 1 + 2x$

15.51

x	3	4	1	2
y	4	5	0	−1

$\hat{y} = -3 + 2x$

15.52

x	1	3	4	4
y	13	−1	3	5

$\hat{y} = 14 - 3x$

15.53

x	2	2	3	4	4
y	3	4	0	2	1

$\hat{y} = 5 - x$

15.54

x	1	3	4	4
y	8	0	3	1

$\hat{y} = 9 - 2x$

15.55

x	1	2	3
y	4	3	8

$\hat{y} = 1 + 2x$

15.56

x	1	1	5	5
y	1	3	2	4

$\hat{y} = 1.75 + 0.25x$

15.57

x	0	2	2	5	6
y	4	2	0	−2	1

$\hat{y} = 2.875 - 0.625x$

Applying the Concepts and Skills

In Exercises 15.58–15.63, we repeat the information from Exercises 15.22–15.27. Presuming that the assumptions for regression inferences are met, decide at the specified significance level whether the data provide sufficient evidence to conclude that the predictor variable is useful for predicting the response variable.

15.58 Tax Efficiency. Following are the data on percentage of investments in energy securities and tax efficiency from Exercise 15.22. Use $\alpha = 0.05$.

x	3.1	3.2	3.7	4.3	4.0	5.5	6.7	7.4	7.4	10.6
y	98.1	94.7	92.0	89.8	87.5	85.0	82.0	77.8	72.1	53.5

15.59 Corvette Prices. Following are the age and price data for Corvettes from Exercise 15.23. Use $\alpha = 0.10$.

x	6	6	6	2	2	5	4	5	1	4
y	290	280	295	425	384	315	355	328	425	325

15.60 Custom Homes. Following are the size and price data for custom homes from Exercise 15.24. Use $\alpha = 0.01$.

x	26	27	33	29	29	34	30	40	22
y	540	555	575	577	606	661	738	804	496

15.61 Plant Emissions. Following are the data on plant weight and quantity of volatile emissions from Exercise 15.25. Use $\alpha = 0.05$.

x	57	85	57	65	52	67	62	80	77	53	68
y	8.0	22.0	10.5	22.5	12.0	11.5	7.5	13.0	16.5	21.0	12.0

15.62 Crown-Rump Length. Following are the data on age of fetuses and length of crown-rump from Exercise 15.26. Use $\alpha = 0.10$.

x	10	10	13	13	18	19	19	23	25	28
y	66	66	108	106	161	166	177	228	235	280

15.63 Study Time and Score. Following are the data on total hours studied over 2 weeks and test score at the end of the 2 weeks from Exercise 15.27. Use $\alpha = 0.01$.

x	10	15	12	20	8	16	14	22
y	92	81	84	74	85	80	84	80

In each of Exercises 15.64–15.69, apply Procedure 15.2 on page 674 to find and interpret a confidence interval, at the specified confidence level, for the slope of the population regression line that relates the response variable to the predictor variable.

15.64 Tax Efficiency. Refer to Exercise 15.58; 95%.

15.65 Corvette Prices. Refer to Exercise 15.59; 90%.

15.66 Custom Homes. Refer to Exercise 15.60; 99%.

15.67 Plant Emissions. Refer to Exercise 15.61; 95%.

15.68 Crown-Rump Length. Refer to Exercise 15.62; 90%.

15.69 Study Time and Score. Refer to Exercise 15.63; 99%.

Working with Large Data Sets

In Exercises 15.70–15.80, use the technology of your choice to do the following tasks.
a. Decide whether you can reasonably apply the regression t-test. If so, then also do part (b).
b. Decide, at the 5% significance level, whether the data provide sufficient evidence to conclude that the predictor variable is useful for predicting the response variable.

15.70 Birdies and Score. The data from Exercise 15.34 for number of birdies during a tournament and final score for 63 women golfers are on the WeissStats site.

15.71 U.S. Presidents. The data from Exercise 15.35 for the ages at inauguration and of death for the presidents of the United States are on the WeissStats site.

15.72 Movie Grosses. The data from Exercise 15.36 on domestic and overseas grosses for a random sample of 50 movies are on the WeissStats site.

15.73 Acreage and Value. The data from Exercise 15.37 for lot size (in acres) and assessed value (in thousands of dollars) for a sample of homes in a particular area are on the WeissStats site.

15.74 Home Size and Value. The data from Exercise 15.38 for home size (in square feet) and assessed value (in thousands of dollars) for the same homes as in Exercise 15.73 are on the WeissStats site.

15.75 High and Low Temperature. The data from Exercise 15.39 for average high and low temperatures in January for a random sample of 50 cities are on the WeissStats site.

15.76 PCBs and Pelicans. Use the data points given on the WeissStats site for shell thickness and concentration of PCBs for 60 Anacapa pelican eggs referred to in Exercise 15.40.

15.77 Gas Guzzlers. Use the data on the WeissStats site for gas mileage and engine displacement for 121 vehicles referred to in Exercise 15.41.

15.78 Estriol Level and Birth Weight. Use the data on the WeissStats site for estriol levels of pregnant women and birth weights of their children referred to in Exercise 15.42.

15.79 Shortleaf Pines. The data from Exercise 15.43 for volume, in cubic feet, and diameter at breast height, in inches, for 70 shortleaf pines are on the WeissStats site.

15.80 Body Fat. In the paper "Total Body Composition by Dual-Photon (^{153}Gd) Absorptiometry" (*American Journal of Clinical Nutrition*, Vol. 40, pp. 834–839), R. Mazess et al. studied methods for quantifying body composition. Eighteen randomly selected adults were measured for percentage of body fat, using dual-photon absorptiometry. Each adult's age and percentage of body fat are shown on the WeissStats site.

15.3 Estimation and Prediction

In this section, we examine how a sample regression equation can be used to make two important inferences:

- Estimate the conditional mean of the response variable corresponding to a particular value of the predictor variable.
- Predict the value of the response variable for a particular value of the predictor variable.

We again use the Orion data to illustrate the pertinent ideas. In doing so, we presume that the assumptions for regression inferences (Key Fact 15.1 on page 659) are satisfied by the variables age and price for Orions. Example 15.3 on pages 664–665 shows that to presume so is not unreasonable.

EXAMPLE 15.8   Estimating Conditional Means in Regression

TABLE 15.4

Age and price data for a sample of 11 Orions

Age (yr) x	Price ($100) y
5	85
4	103
6	70
5	82
5	89
5	98
6	66
6	95
2	169
7	70
7	48

 StatCrunch

Report 15.5

 You try it!

Exercise 15.93(a) on page 686

Age and Price of Orions   Use the data on age and price for a sample of 11 Orions, repeated in Table 15.4, to estimate the mean price of all 3-year-old Orions.

Solution   By Assumption 1 for regression inferences, the population regression line gives the mean prices for the various ages of Orions. In particular, the mean price of all 3-year-old Orions is $\beta_0 + \beta_1 \cdot 3$. Because β_0 and β_1 are unknown, we estimate the mean price of all 3-year-old Orions ($\beta_0 + \beta_1 \cdot 3$) by the corresponding value on the sample regression line, namely, $b_0 + b_1 \cdot 3$.

Recalling that the sample regression equation for the age and price data in Table 15.4 is $\hat{y} = 195.47 - 20.26x$, we estimate that the mean price of all 3-year-old Orions is

$$\hat{y} = 195.47 - 20.26 \cdot 3 = 134.69,$$

or $13,469. Note that the estimate for the mean price of all 3-year-old Orions is the same as the predicted price for a 3-year-old Orion. Both are obtained by substituting $x = 3$ into the sample regression equation.

Confidence Intervals for Conditional Means in Regression

The estimate of $13,469 for the mean price of all 3-year-old Orions found in the previous example is a point estimate. Providing a confidence-interval estimate for the mean price of all 3-year-old Orions would be more informative.

To that end, consider all possible samples of 11 Orions whose ages are the same as those given in the first column of Table 15.4. For such samples, the predicted price of a 3-year-old Orion varies from one sample to another and is therefore a variable. Using the assumptions for regression inferences, we can show that its distribution is a normal distribution whose mean equals the mean price of all 3-year-old Orions. More generally, we have Key Fact 15.5.

KEY FACT 15.5

Distribution of the Predicted Value of a Response Variable

Suppose that the variables x and y satisfy the four assumptions for regression inferences. Let x_p denote a particular value of the predictor variable, and let $\hat{y}_p$ be the corresponding value predicted for the response variable by the sample regression equation; that is, $\hat{y}_p = b_0 + b_1 x_p$. Then, for samples of size n, each with the same values $x_1, x_2, \ldots, x_n$ for the predictor variable, the following properties hold for $\hat{y}_p$.

- The mean of $\hat{y}_p$ equals the conditional mean of the response variable corresponding to the value x_p of the predictor variable: $\mu_{\hat{y}_p} = \beta_0 + \beta_1 x_p$.
- The standard deviation of $\hat{y}_p$ is

$$\sigma_{\hat{y}_p} = \sigma \sqrt{\frac{1}{n} + \frac{(x_p - \bar{x})^2}{S_{xx}}}.$$

- The variable $\hat{y}_p$ is normally distributed.

In particular, for fixed values of the predictor variable, the possible predicted values of the response variable corresponding to x_p have a normal distribution with mean $\beta_0 + \beta_1 x_p$.

In light of Key Fact 15.5, if we standardize the variable $\hat{y}_p$, the resulting variable has the standard normal distribution. However, because the standardized variable contains the unknown parameter σ, it cannot be used as a basis for a confidence-interval formula. Therefore we replace σ by its estimate s_e, the standard error of the estimate. The resulting variable has a t-distribution.

KEY FACT 15.6 | **t-Distribution for Confidence Intervals for Conditional Means in Regression**

Suppose that the variables x and y satisfy the four assumptions for regression inferences. Then, for samples of size n, each with the same values $x_1, x_2, \ldots, x_n$ for the predictor variable, the variable

$$t = \frac{\hat{y}_p - (\beta_0 + \beta_1 x_p)}{s_e \sqrt{\dfrac{1}{n} + \dfrac{(x_p - \bar{x})^2}{S_{xx}}}}$$

has the t-distribution with df $= n - 2$.

Recalling that $\beta_0 + \beta_1 x_p$ is the conditional mean of the response variable corresponding to the value x_p of the predictor variable, we can apply Key Fact 15.6 to derive a confidence-interval procedure for means in regression. We call that procedure the **conditional mean t-interval procedure.**

PROCEDURE 15.3 | **Conditional Mean t-Interval Procedure**

Purpose To find a confidence interval for the conditional mean of the response variable corresponding to a particular value of the predictor variable, x_p

Assumptions The four assumptions for regression inferences

Step 1 For a confidence level of $1 - \alpha$, use Table IV to find $t_{\alpha/2}$ with df $= n - 2$.

Step 2 Compute the point estimate, $\hat{y}_p = b_0 + b_1 x_p$.

Step 3 The endpoints of the confidence interval for the conditional mean of the response variable are

$$\hat{y}_p \pm t_{\alpha/2} \cdot s_e \sqrt{\frac{1}{n} + \frac{(x_p - \bar{x})^2}{S_{xx}}}.$$

Step 4 Interpret the confidence interval.

EXAMPLE 15.9 | **The Conditional Mean t-Interval Procedure**

Age and Price of Orions Use the sample data in Table 15.4 on page 679 to obtain a 95% confidence interval for the mean price of all 3-year-old Orions.

Solution We apply Procedure 15.3.

Step 1 For a confidence level of $1 - \alpha$, use Table IV to find $t_{\alpha/2}$ with df $= n - 2$.

We want a 95% confidence interval, or $\alpha = 0.05$. Because $n = 11$, we have df $= 9$. From Table IV, $t_{\alpha/2} = t_{0.05/2} = t_{0.025} = 2.262$.

Step 2 Compute the point estimate, $\hat{y}_p = b_0 + b_1 x_p$.

Here, $x_p = 3$ (3-year-old Orions). From Example 15.8, the point estimate for the mean price of all 3-year-old Orions is

$$\hat{y}_p = 195.47 - 20.26 \cdot 3 = 134.69.$$

Step 3 The endpoints of the confidence interval for the conditional mean of the response variable are

$$\hat{y}_p \pm t_{\alpha/2} \cdot s_e \sqrt{\frac{1}{n} + \frac{(x_p - \bar{x})^2}{S_{xx}}}.$$

In Example 14.6, we found that $\Sigma x_i = 58$ and $\Sigma x_i^2 = 326$; in Example 15.2, we determined that $s_e = 12.58$. Also, from Step 1, $t_{\alpha/2} = 2.262$ and, from Step 2, $\hat{y}_p = 134.69$. Consequently, the endpoints of the confidence interval for the conditional mean are

$$134.69 \pm 2.262 \cdot 12.58 \sqrt{\frac{1}{11} + \frac{(3 - 58/11)^2}{326 - (58)^2/11}},$$

or 134.69 ± 16.76, or 117.93 to 151.45.

StatCrunch

Report 15.6

Exercise 15.93(b)
on page 686

Step 4 Interpret the confidence interval.

Interpretation We can be 95% confident that the mean price of all 3-year-old Orions is somewhere between $11,793 and $15,145. ∎

Prediction Intervals

A primary use of a sample regression equation is to make predictions. As we have seen, for the Orion data in Table 15.4 on page 679, the sample regression equation is $\hat{y} = 195.47 - 20.26x$. Substituting $x = 3$ into that equation, we get the predicted price for a 3-year-old Orion of 134.69, or $13,469. Because the prices of such cars vary, finding a **prediction interval** for the price of a 3-year-old Orion makes more sense than giving a single predicted value.[†]

To that end, we first recall that, from the assumptions for regression inferences, the price of a 3-year-old Orion has a normal distribution with mean $\beta_0 + \beta_1 \cdot 3$ and standard deviation σ. Because β_0 and β_1 are unknown, we estimate the mean price by its point estimate $b_0 + b_1 \cdot 3$, which is also the predicted price of a 3-year-old Orion.

Thus, to find a prediction interval, we need the distribution of the difference between the price of a 3-year-old Orion and the predicted price of a 3-year-old Orion. Using the assumptions for regression inferences, we can show that this distribution is normal. More generally, we have Key Fact 15.7.

KEY FACT 15.7

Distribution of the Difference between the Observed and Predicted Values of the Response Variable

Suppose that the variables x and y satisfy the four assumptions for regression inferences. Let x_p denote a particular value of the predictor variable, and let $\hat{y}_p$ be the corresponding value predicted for the response variable by the sample regression equation. Furthermore, let y_p be an independently observed value of the response variable corresponding to the value x_p of the predictor variable. Then, for samples of size n, each with the same values $x_1, x_2, \ldots, x_n$ for the predictor variable, the following properties hold for $y_p - \hat{y}_p$, the difference between the observed and predicted values.

- The mean of $y_p - \hat{y}_p$ equals zero: $\mu_{y_p - \hat{y}_p} = 0$.
- The standard deviation of $y_p - \hat{y}_p$ is

$$\sigma_{y_p - \hat{y}_p} = \sigma \sqrt{1 + \frac{1}{n} + \frac{(x_p - \bar{x})^2}{S_{xx}}}.$$

- The variable $y_p - \hat{y}_p$ is normally distributed.

In particular, for fixed values of the predictor variable, the possible differences between the observed and predicted values of the response variable corresponding to x_p have a normal distribution with a mean of 0.

[†]Prediction intervals are similar to confidence intervals. The term *confidence* is usually reserved for interval estimates of parameters, such as the mean price of all 3-year-old Orions. The term *prediction* is used for interval estimates of variables, such as the price of a 3-year-old Orion.

In light of Key Fact 15.7, if we standardize the variable $y_p - \hat{y}_p$, the resulting variable has the standard normal distribution. However, because the standardized variable contains the unknown parameter σ, it cannot be used as a basis for a prediction-interval formula. So we replace σ by its estimate s_e, the standard error of the estimate. The resulting variable has a t-distribution.

KEY FACT 15.8

t-Distribution for Prediction Intervals in Regression

Suppose that the variables x and y satisfy the four assumptions for regression inferences. Then, for samples of size n, each with the same values $x_1, x_2, \ldots, x_n$ for the predictor variable, the variable

$$t = \frac{y_p - \hat{y}_p}{s_e\sqrt{1 + \dfrac{1}{n} + \dfrac{(x_p - \bar{x})^2}{S_{xx}}}}$$

has the t-distribution with df $= n - 2$.

Using Key Fact 15.8, we can derive a prediction-interval procedure, called the **predicted value t-interval procedure.**

■■■■ **PROCEDURE 15.4**   **Predicted Value *t*-Interval Procedure**

Purpose To find a prediction interval for the value of the response variable corresponding to a particular value of the predictor variable, x_p

Assumptions The four assumptions for regression inferences

Step 1   For a prediction level of $1 - \alpha$, use Table IV to find $t_{\alpha/2}$ with df $= n - 2$.

Step 2   Compute the predicted value, $\hat{y}_p = b_0 + b_1 x_p$.

Step 3   The endpoints of the prediction interval for the value of the response variable are

$$\hat{y}_p \pm t_{\alpha/2} \cdot s_e \sqrt{1 + \frac{1}{n} + \frac{(x_p - \bar{x})^2}{S_{xx}}}.$$

Step 4   Interpret the prediction interval.

■■■ **EXAMPLE 15.10**   **The Predicted Value *t*-Interval Procedure**

Age and Price of Orions Using the sample data in Table 15.4 on page 679, find a 95% prediction interval for the price of a 3-year-old Orion.

Solution   We apply Procedure 15.4.

Step 1 For a prediction level of $1 - \alpha$, use Table IV to find $t_{\alpha/2}$ with df $= n - 2$.

We want a 95% prediction interval, or $\alpha = 0.05$. Also, because $n = 11$, we have df $= 9$. From Table IV, $t_{\alpha/2} = t_{0.05/2} = t_{0.025} = 2.262$.

Step 2 Compute the predicted value, $\hat{y}_p = b_0 + b_1 x_p$.

As previously shown, the sample regression equation for the data in Table 15.4 is $\hat{y} = 195.47 - 20.26x$. Therefore, the predicted price for a 3-year-old Orion is

$$\hat{y}_p = 195.47 - 20.26 \cdot 3 = 134.69.$$

Step 3 The endpoints of the prediction interval for the value of the response variable are

$$\hat{y}_p \pm t_{\alpha/2} \cdot s_e \sqrt{1 + \frac{1}{n} + \frac{(x_p - \bar{x})^2}{S_{xx}}}.$$

From Example 14.6, $\Sigma x_i = 58$ and $\Sigma x_i^2 = 326$; from Example 15.2, we know that $s_e = 12.58$. Also, $n = 11$, $t_{\alpha/2} = 2.262$, $x_p = 3$, and $\hat{y}_p = 134.69$. Consequently, the endpoints of the prediction interval are

$$134.69 \pm 2.262 \cdot 12.58 \sqrt{1 + \frac{1}{11} + \frac{(3 - 58/11)^2}{326 - (58)^2/11}},$$

or 134.69 ± 33.02, or 101.67 to 167.71.

Step 4 Interpret the prediction interval.

Interpretation We can be 95% certain that the price of a 3-year-old Orion will be somewhere between $10,167 and $16,771.

We just demonstrated that a 95% prediction interval for the observed price of a 3-year-old Orion is from $10,167 to $16,771. In Example 15.9, we found that a 95% confidence interval for the mean price of all 3-year-old Orions is from $11,793 to $15,145. We show both intervals in Fig. 15.11.

StatCrunch
Report 15.7

You try it!

Exercise 15.93(d) on page 686

FIGURE 15.11
Prediction and confidence intervals for 3-year-old Orions

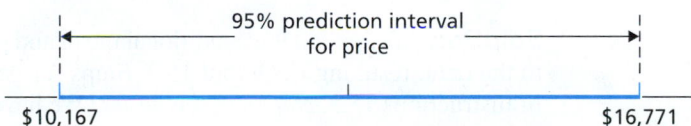

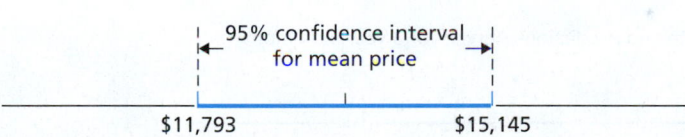

What Does It Mean?

More error is involved in predicting the price of a single 3-year-old Orion than in estimating the mean price of all 3-year-old Orions.

Note that the prediction interval is wider than the confidence interval, a result to be expected, for the following reason: The error in the estimate of the mean price of all 3-year-old Orions is due only to the fact that the population regression line is being estimated by a sample regression line, whereas the error in the prediction of the price of one particular 3-year-old Orion is due to the error in estimating the mean price plus the variation in prices of 3-year-old Orions.

Multiple Regression

In Chapter 14 and in this chapter, we examined descriptive and inferential methods for **simple linear regression,** where one predictor variable is used to predict a response variable by using a straight-line fit. However, we often want to use more than one predictor variable in a regression analysis—so-called **multiple regression analysis**—or use a model other than a straight-line fit.

For instance, we have been using the variable "age" as a single predictor for the price of an Orion. Using, in addition, the variable "mileage" (i.e., number of miles driven) might improve our predictions. In other words, it might be preferable to use both age and mileage to predict the price of an Orion. This is an example of a multiple regression analysis with two predictor variables.

We cover multiple regression and model building in the optional modules *Multiple Regression Analysis* (Module A) and *Model Building in Regression* (Module B). These two modules are located in the Regression-ANOVA Modules section on the WeissStats site.

THE TECHNOLOGY CENTER

Most statistical technologies have programs that automatically perform conditional mean and predicted value t-interval procedures. In this subsection, we present output and step-by-step instructions for such programs.

Note to TI-83/84 Plus users: At the time of this writing, the TI-83/84 Plus does not have a built-in program for performing conditional mean and predicted value t-interval procedures. However, we have written a TI program called REGEST for performing those procedures. It is located in the TI Programs section on the WeissStats site. Your instructor can show you how to download the program to your calculator. *Warning:* Any data that you may have previously stored in Lists 1 and 2 will be erased during program execution, so copy those data to other lists prior to program execution if you want to retain them.

EXAMPLE 15.11 Using Technology to Obtain Conditional Mean and Predicted Value *t*-Intervals

Age and Price of Orions Table 15.4 on page 679 gives the age and price data for a sample of 11 Orions. Use Minitab, Excel, or the TI-83/84 Plus to determine a 95% confidence interval for the mean price of all 3-year-old Orions and a 95% prediction interval for the price of a 3-year-old Orion.

Solution We applied the conditional mean and predicted value t-interval programs to the data, resulting in Output 15.3. Steps for generating that output are presented in Instructions 15.3. *Note to Excel users:* We have presented only the portion of the regression output essential to the confidence and prediction intervals.

OUTPUT 15.3 Confidence and prediction intervals for 3-year-old Orions

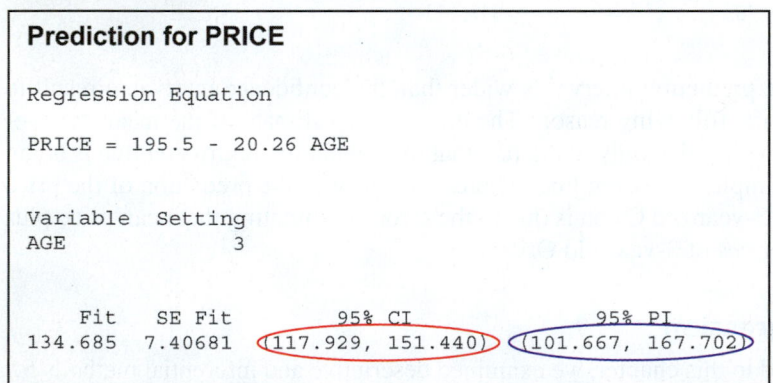

MINITAB

Prediction for PRICE

Regression Equation

PRICE = 195.5 - 20.26 AGE

Variable	Setting
AGE	3

Fit	SE Fit	95% CI	95% PI
134.685	7.40681	(117.929, 151.440)	(101.667, 167.702)

EXCEL

Regression of variable PRICE:

Predictions for the new observations:

Observation	AGE	Pred(PRICE)	Std. dev. on pred. (Mean)	Lower bound 95% (Mean)	Upper bound 95% (Mean)
PredObs1	3.0000	134.6847	7.4068	117.9293	151.4401

Std. dev. on pred. (Observation)	Lower bound 95% (Observation)	Upper bound 95% (Observation)
14.5956	101.6672	167.7022

OUTPUT 15.3 (cont.)   Confidence and prediction intervals for 3-year-old Orions

TI-83/84 PLUS

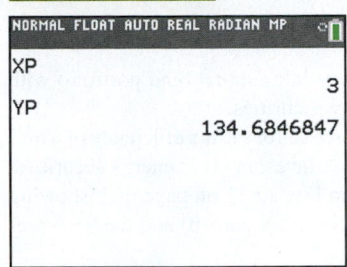

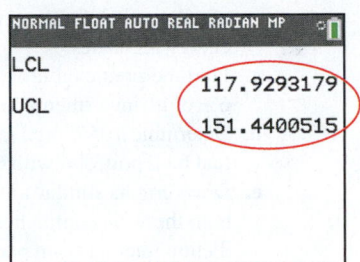

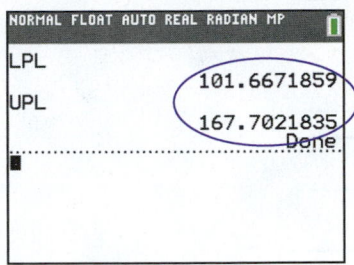

In Output 15.3, the items that are circled in red and blue give the required 95% confidence and prediction intervals, respectively, in hundreds of dollars.

INSTRUCTIONS 15.3   Steps for generating Output 15.3

MINITAB

1 Store the age and price data from Table 15.4 in columns named AGE and PRICE, respectively
2 Choose **Stat ➤ Regression ➤ Regression ➤ Fit Regression Model...**
3 Press the F3 key to reset the dialog box
4 Specify PRICE in the **Responses** text box
5 Specify AGE in the **Continuous predictors** text box
6 Click the **Results...** button
7 Uncheck all check boxes
8 Click **OK** twice
9 Choose **Stat ➤ Regression ➤ Regression ➤ Predict...**
10 Press the F3 key to reset the dialog box
11 Click in the first text box in the **AGE** list and type 3
12 Click the **Options...** button
13 Type 95 in the **Confidence level** text box
14 Click **OK** twice

EXCEL

1 Store the age and price data from Table 15.4 in columns named AGE and PRICE, respectively
2 Type 3 in any cell in another column
3 Choose **XLSTAT ➤ Modeling data ➤ Linear regression**
4 Click the reset button in the lower left corner of the dialog box
5 In the **Y / Dependent variables** list, click in the **Quantitative** selection box and then select the column of the worksheet that contains the PRICE data

6 In the **X / Explanatory variables** list, click in the **Quantitative** selection box and then select the column of the worksheet that contains the AGE data
7 Click the **Options** tab and then type 95 in the **Confidence interval (%)** text box
8 Click the **Prediction** tab, check the **Prediction** check box, and then select the cell in the worksheet that contains the 3 from step 2
9 Click the **Outputs** tab and then check only the **Predictions and residuals** and **X** check boxes
10 Click the **Charts** tab and then uncheck the **Regression charts** check box
11 Click **OK**
12 Click the **Continue** button in the **XLSTAT – Selections** dialog box

TI-83/84 PLUS

1 Store the age and price data from Table 15.4 in lists named AGE and PRICE, respectively
2 Press **PRGM**, arrow down to REGEST, and press **ENTER** twice
3 Press **2nd ➤ LIST**, arrow down to AGE, and press **ENTER** twice
4 Press **2nd ➤ LIST**, arrow down to PRICE, and press **ENTER** twice
5 Type .95 for **LEVEL** and then press **ENTER**
6 Type 3 for **XP** and then press **ENTER**
7 After the screen changes, press **ENTER** when ready
8 Again, after the screen changes, press **ENTER** when ready

Exercises 15.3

Understanding the Concepts and Skills

15.81 Without doing any calculations, fill in the blank. Based on the sample data in Table 15.4, the predicted price for a 4-year-old Orion is $11,443. A point estimate for the mean price of all 4-year-old Orions, based on the same sample data, is _____.

In Exercises 15.82–15.91, we repeat the data from Exercises 15.12–15.21 and specify a value of the predictor variable.

a. Determine a point estimate for the conditional mean of the response variable corresponding to the specified value of the predictor variable.

b. Find a 95% confidence interval for the conditional mean of the response variable corresponding to the specified value of the predictor variable.

c. Determine the predicted value of the response variable corresponding to the specified value of the predictor variable.

d. *Find a 95% prediction interval for the value of the response variable corresponding to the specified value of the predictor variable.*

15.82

x	2	4	3
y	3	5	7

$x = 3$

15.83

x	3	1	2
y	−4	0	−5

$x = 2$

15.84

x	0	4	3	1	2
y	1	9	8	4	3

$x = 1$

15.85

x	3	4	1	2
y	4	5	0	−1

$x = 4$

15.86

x	1	3	4	4
y	13	−1	3	5

$x = 2$

15.87

x	2	2	3	4	4
y	3	4	0	2	1

$x = 3$

15.88

x	1	3	4	4
y	8	0	3	1

$x = 3$

15.89

x	1	2	3
y	4	3	8

$x = 2$

15.90

x	1	1	5	5
y	1	3	2	4

$x = 2$

15.91

x	0	2	2	5	6
y	4	2	0	−2	1

$x = 3$

Applying the Concepts and Skills

In Exercises 15.92–15.97, presume that the assumptions for regression inferences are met.

15.92 **Tax Efficiency.** Following are the data on percentage of investments in energy securities and tax efficiency from Exercise 15.22.

x	3.1	3.2	3.7	4.3	4.0	5.5	6.7	7.4	7.4	10.6
y	98.1	94.7	92.0	89.8	87.5	85.0	82.0	77.8	72.1	53.5

a. Obtain a point estimate for the mean tax efficiency of all mutual fund portfolios with 6% of their investments in energy securities.
b. Determine a 95% confidence interval for the mean tax efficiency of all mutual fund portfolios with 6% of their investments in energy securities.
c. Find the predicted tax efficiency of a mutual fund portfolio with 6% of its investments in energy securities.
d. Determine a 95% prediction interval for the tax efficiency of a mutual fund portfolio with 6% of its investments in energy securities.
e. Draw graphs similar to those in Fig. 15.11 on page 683, showing both the 95% confidence interval from part (b) and the 95% prediction interval from part (d).
f. Why is the prediction interval wider than the confidence interval?

15.93 **Corvette Prices.** Following are the age and price data for Corvettes from Exercise 15.23.

x	6	6	6	2	2	5	4	5	1	4
y	290	280	295	425	384	315	355	328	425	325

a. Obtain a point estimate for the mean price of all 4-year-old Corvettes.
b. Determine a 90% confidence interval for the mean price of all 4-year-old Corvettes.
c. Find the predicted price of a 4-year-old Corvette.
d. Determine a 90% prediction interval for the price of a 4-year-old Corvette.
e. Draw graphs similar to those in Fig. 15.11 on page 683, showing both the 90% confidence interval from part (b) and the 90% prediction interval from part (d).
f. Why is the prediction interval wider than the confidence interval?

15.94 **Custom Homes.** Following are the size and price data for custom homes from Exercise 15.24.

x	26	27	33	29	29	34	30	40	22
y	540	555	575	577	606	661	738	804	496

a. Determine a point estimate for the mean price of all 2800-sq. ft. Equestrian Estate homes.
b. Find a 99% confidence interval for the mean price of all 2800-sq. ft. Equestrian Estate homes.
c. Find the predicted price of a 2800-sq. ft. Equestrian Estate home.
d. Determine a 99% prediction interval for the price of a 2800-sq. ft. Equestrian Estate home.

15.95 **Plant Emissions.** Following are the data on plant weight and quantity of volatile emissions from Exercise 15.25.

x	57	85	57	65	52	67	62	80	77	53	68
y	8.0	22.0	10.5	22.5	12.0	11.5	7.5	13.0	16.5	21.0	12.0

a. Obtain a point estimate for the mean quantity of volatile emissions of all (*Solanum tuberosum*) plants that weigh 60 g.
b. Find a 95% confidence interval for the mean quantity of volatile emissions of all plants that weigh 60 g.
c. Find the predicted quantity of volatile emissions for a plant that weighs 60 g.
d. Determine a 95% prediction interval for the quantity of volatile emissions for a plant that weighs 60 g.

15.96 Crown-Rump Length. Following are the data on age of fetuses and length of crown-rump from Exercise 15.26.

x	10	10	13	13	18	19	19	23	25	28
y	66	66	108	106	161	166	177	228	235	280

a. Determine a point estimate for the mean crown-rump length of all 19-week-old fetuses.
b. Find a 90% confidence interval for the mean crown-rump length of all 19-week-old fetuses.
c. Find the predicted crown-rump length of a 19-week-old fetus.
d. Determine a 90% prediction interval for the crown-rump length of a 19-week-old fetus.

15.97 Study Time and Score. Following are the data on total hours studied over 2 weeks and test score at the end of the 2 weeks from Exercise 15.27.

x	10	15	12	20	8	16	14	22
y	92	81	84	74	85	80	84	80

a. Determine a point estimate for the mean test score of all beginning calculus students who study for 15 hours.
b. Find a 99% confidence interval for the mean test score of all beginning calculus students who study for 15 hours.
c. Find the predicted test score of a beginning calculus student who studies for 15 hours.
d. Determine a 99% prediction interval for the test score of a beginning calculus student who studies for 15 hours.

Working with Large Data Sets

In Exercises 15.98–15.108, use the technology of your choice to do the following tasks.
a. Decide whether you can reasonably apply the conditional mean and predicted value t-interval procedures to the data. If so, then also do parts (b)–(f).
b. Determine and interpret a point estimate for the conditional mean of the response variable corresponding to the specified value of the predictor variable.
c. Find and interpret a 95% confidence interval for the conditional mean of the response variable corresponding to the specified value of the predictor variable.
d. Determine and interpret the predicted value of the response variable corresponding to the specified value of the predictor variable.
e. Find and interpret a 95% prediction interval for the value of the response variable corresponding to the specified value of the predictor variable.
f. Compare and discuss the differences between the confidence interval that you obtained in part (c) and the prediction interval that you obtained in part (e).

15.98 Birdies and Score. The data from Exercise 15.34 for number of birdies during a tournament and final score of 63 women golfers are on the WeissStats site. Specified value of the predictor variable: 12 birdies.

15.99 U.S. Presidents. The data from Exercise 15.35 for the ages at inauguration and of death of the presidents of the United States are on the WeissStats site. Specified value of the predictor variable: 53 years.

15.100 Movie Grosses. The data from Exercise 15.36 on domestic and overseas grosses for a random sample of 50 movies are on the WeissStats site. Specified value of the predictor variable: $100 million.

15.101 Acreage and Value. The data from Exercise 15.37 for lot size (in acres) and assessed value (in thousands of dollars) of a sample of homes in a particular area are on the WeissStats site. Specified value of the predictor variable: 2.5 acres.

15.102 Home Size and Value. The data from Exercise 15.38 for home size (in square feet) and assessed value (in thousands of dollars) for the same homes as in Exercise 15.101 are on the WeissStats site. Specified value of the predictor variable: 3000 sq. ft.

15.103 High and Low Temperature. The data from Exercise 15.39 for average high and low temperatures in January of a random sample of 50 cities are on the WeissStats site. Specified value of the predictor variable: 55°F.

15.104 PCBs and Pelicans. The data from Exercise 15.40 for shell thickness and concentration of PCBs of 60 Anacapa pelican eggs are on the WeissStats site. Specified value of the predictor variable: 220 ppm.

15.105 Gas Guzzlers. The data from Exercise 15.41 for gas mileage and engine displacement of 121 vehicles are on the WeissStats site. Specified value of the predictor variable: 3.0 L.

15.106 Estriol Level and Birth Weight. The data from Exercise 15.42 for estriol levels of pregnant women and birth weights of their children are on the WeissStats site. Specified value of the predictor variable: 18 mg/24 hr.

15.107 Shortleaf Pines. The data from Exercise 15.43 for volume, in cubic feet, and diameter at breast height, in inches, of 70 shortleaf pines are on the WeissStats site. Specified value of the predictor variable: 11 inches.

15.108 Body Fat. The data from Exercise 15.80 for age and body fat of 18 randomly selected adults are on the WeissStats site. Specified value of the predictor variable: 30 years.

Extending the Concepts and Skills

Margin of Error in Regression. In Exercises 15.109 and 15.110, you will examine the magnitude of the margin of error of confidence intervals and prediction intervals in regression as a function of how far the specified value of the predictor variable is from the mean of the observed values of the predictor variable.

15.109 Age and Price of Orions. Refer to the data on age and price of a sample of 11 Orions given in Table 15.4 on page 679.
a. For each age between 2 and 7 years, obtain a 95% confidence interval for the mean price of all Orions of that age. Plot the confidence intervals against age and discuss your results.
b. Determine the margin of error for each confidence interval that you obtained in part (a). Plot the margins of error against age and discuss your results.
c. Repeat parts (a) and (b) for prediction intervals.

15.110 Refer to the confidence interval and prediction interval formulas in Procedures 15.3 and 15.4, respectively.
a. Explain why, for a fixed confidence level, the margin of error for the estimate of the conditional mean of the response variable increases as the value of the predictor variable moves farther from the mean of the observed values of the predictor variable.
b. Explain why, for a fixed prediction level, the margin of error for the estimate of the predicted value of the response variable increases as the value of the predictor variable moves farther from the mean of the observed values of the predictor variable.

15.4 Inferences in Correlation

Frequently, we want to decide whether two variables are linearly correlated, that is, whether there is a linear relationship between the two variables. In the context of regression, we can make that decision by performing a hypothesis test for the slope of the population regression line, as discussed in Section 15.2.

Alternatively, we can perform a hypothesis test for the **population linear correlation coefficient, ρ** (rho). This parameter measures the linear correlation of all possible pairs of observations of two variables in the same way that a sample linear correlation coefficient, r, measures the linear correlation of a sample of pairs. Thus, ρ actually describes the strength of the linear relationship between two variables; r is only an estimate of ρ obtained from sample data.

The population linear correlation coefficient of two variables x and y always lies between -1 and 1. Values of ρ near -1 or 1 indicate a strong linear relationship between the variables, whereas values of ρ near 0 indicate a weak linear relationship between the variables. Note the following:

- If $\rho = 0$, the variables are **linearly uncorrelated,** meaning that there is no linear relationship between the variables.
- If $\rho > 0$, the variables are **positively linearly correlated,** meaning that y tends to increase linearly as x increases (and vice versa), with the tendency being greater the closer ρ is to 1.
- If $\rho < 0$, the variables are **negatively linearly correlated,** meaning that y tends to decrease linearly as x increases (and vice versa), with the tendency being greater the closer ρ is to -1.
- If $\rho \neq 0$, the variables are **linearly correlated.** Linearly correlated variables are either positively linearly correlated or negatively linearly correlated.

As we mentioned, a sample linear correlation coefficient, r, is an estimate of the population linear correlation coefficient, ρ. Consequently, we can use r as a basis for performing a hypothesis test for ρ. To do so, we require the following fact.

KEY FACT 15.9

t-Distribution for a Correlation Test

Suppose that the variables x and y satisfy the four assumptions for regression inferences and that $\rho = 0$. Then, for samples of size n, the variable

$$t = \frac{r}{\sqrt{\dfrac{1 - r^2}{n - 2}}}$$

has the t-distribution with df $= n - 2$.

In light of Key Fact 15.9, for a hypothesis test with the null hypothesis H_0: $\rho = 0$, we can use the variable

$$t = \frac{r}{\sqrt{\dfrac{1 - r^2}{n - 2}}}$$

as the test statistic and obtain the critical values or P-value from the t-table, Table IV. We call this hypothesis-testing procedure the **correlation t-test.** Procedure 15.5 provides a step-by-step method for performing a correlation t-test by using either the critical-value approach or the P-value approach.

■■■ **PROCEDURE 15.5**   Correlation *t*-Test

Purpose To perform a hypothesis test for a population linear correlation coefficient, ρ

Assumptions The four assumptions for regression inferences

Step 1   The null hypothesis is H_0: $\rho = 0$, and the alternative hypothesis is

$$H_a: \rho \neq 0 \qquad H_a: \rho < 0 \qquad H_a: \rho > 0$$
$$\text{(Two tailed)} \quad \text{or} \quad \text{(Left tailed)} \quad \text{or} \quad \text{(Right tailed)}.$$

Step 2   Decide on the significance level, α.

Step 3   Compute the value of the test statistic

$$t = \frac{r}{\sqrt{\dfrac{1 - r^2}{n - 2}}}$$

and denote that value t_0.

CRITICAL-VALUE APPROACH	OR	P-VALUE APPROACH

Step 4   The critical value(s) are

$$\pm t_{\alpha/2} \qquad -t_\alpha \qquad t_\alpha$$
$$\text{(Two tailed)} \quad \text{or} \quad \text{(Left tailed)} \quad \text{or} \quad \text{(Right tailed)}$$

with df $= n - 2$. Use Table IV to find the critical value(s).

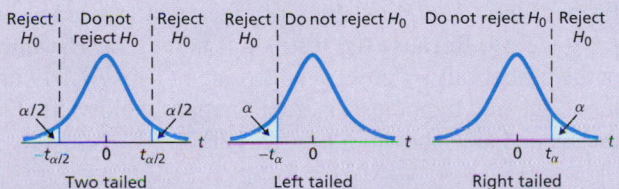

Step 5   If the value of the test statistic falls in the rejection region, reject H_0; otherwise, do not reject H_0.

Step 4   The *t*-statistic has df $= n - 2$. Use Table IV to estimate the *P*-value, or obtain it exactly by using technology.

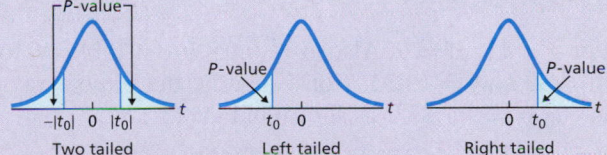

Step 5   If $P \leq \alpha$, reject H_0; otherwise, do not reject H_0.

Step 6   Interpret the results of the hypothesis test.

■■■ **EXAMPLE 15.12**   The Correlation *t*-Test

TABLE 15.5

Age and price data for a sample of 11 Orions

Age (yr) x	Price ($100) y
5	85
4	103
6	70
5	82
5	89
5	98
6	66
6	95
2	169
7	70
7	48

Age and Price of Orions The data on age and price for a sample of 11 Orions are repeated in Table 15.5. At the 5% significance level, do the data provide sufficient evidence to conclude that age and price of Orions are negatively linearly correlated?

Solution As we learned in Example 15.3 on pages 664–665, considering that the assumptions for regression inferences are met by the variables age and price for Orions is not unreasonable, at least for Orions between 2 and 7 years old. Consequently, we apply Procedure 15.5 to carry out the required hypothesis test.

Step 1   **State the null and alternative hypotheses.**

Let ρ denote the population linear correlation coefficient for the variables age and price of Orions. Then the null and alternative hypotheses are, respectively,

H_0: $\rho = 0$ (age and price are linearly uncorrelated)
H_a: $\rho < 0$ (age and price are negatively linearly correlated).

Note that the hypothesis test is left tailed.

Step 2 Decide on the significance level, α.

We are to use $\alpha = 0.05$.

Step 3 Compute the value of the test statistic

$$t = \frac{r}{\sqrt{\dfrac{1 - r^2}{n - 2}}}.$$

In Example 14.13 on page 648, we found that $r = -0.924$, so the value of the test statistic is

$$t = \frac{-0.924}{\sqrt{\dfrac{1 - (-0.924)^2}{11 - 2}}} = -7.249.$$

CRITICAL-VALUE APPROACH	OR	P-VALUE APPROACH

Step 4 The critical value for a left-tailed test is $-t_\alpha$ with df $= n - 2$. Use Table IV to find the critical value.

For $n = 11$, df $= 9$. Also, $\alpha = 0.05$. From Table IV, for df $= 9$, $t_{0.05} = 1.833$. Consequently, the critical value is $-t_{0.05} = -1.833$, as shown in Fig. 15.12A.

FIGURE 15.12A

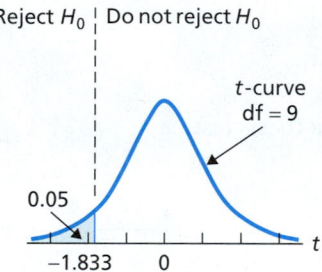

Step 5 If the value of the test statistic falls in the rejection region, reject H_0; otherwise, do not reject H_0.

The value of the test statistic, found in Step 3, is $t = -7.249$. Figure 15.12A shows that this value falls in the rejection region, so we reject H_0. The test results are statistically significant at the 5% level.

Step 4 The t-statistic has df $= n - 2$. Use Table IV to estimate the P-value or obtain it exactly by using technology.

From Step 3, the value of the test statistic is $t = -7.249$. Because the test is left tailed, the P-value is the probability of observing a value of t of -7.249 or less if the null hypothesis is true. That probability equals the shaded area shown in Fig. 15.12B.

FIGURE 15.12B

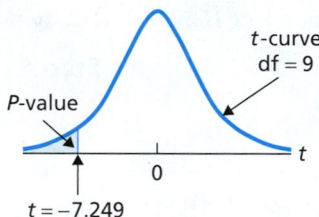

For $n = 11$, df $= 9$. Referring now to Fig. 15.12B and Table IV, we find that $P < 0.005$. (Using technology, we obtain $P = 0.0000244$.)

Step 5 If $P \leq \alpha$, reject H_0; otherwise, do not reject H_0.

From Step 4, $P < 0.005$. Because the P-value is less than the specified significance level of 0.05, we reject H_0. The test results are statistically significant at the 5% level and (see Table 9.8 on page 386) provide very strong evidence against the null hypothesis.

StatCrunch

Report 15.8

Exercise 15.129 on page 693

You try it!

Step 6 Interpret the results of the hypothesis test.

Interpretation At the 5% significance level, the data provide sufficient evidence to conclude that age and price of Orions are negatively linearly correlated. Prices for 2- to 7-year-old Orions tend to decrease linearly with increasing age. ■

THE TECHNOLOGY CENTER

Most statistical technologies have programs that automatically perform a correlation *t*-test. In this subsection, we present output and step-by-step instructions for such programs.

Note to Minitab and Excel users: At the time of this writing, Minitab and Excel do only a two-tailed correlation *t*-test. However, we can get a one-tailed *P*-value from the provided two-tailed *P*-value by using the result of Exercise 9.71 on page 387. This result implies, for instance, that if the sign of the sample linear correlation coefficient is in the same direction as the alternative hypothesis, then the one-tailed *P*-value equals one-half of the two-tailed *P*-value.

EXAMPLE 15.13 Using Technology to Conduct a Correlation *t*-Test

Age and Price of Orions Table 15.5 on page 689 gives the age and price data for a sample of 11 Orions. Use Minitab, Excel, or the TI-83/84 Plus to decide, at the 5% significance level, whether the data provide sufficient evidence to conclude that age and price of Orions are negatively linearly correlated.

Solution Let ρ denote the population linear correlation coefficient for the variables age and price of Orions. We want to perform the hypothesis test

$$H_0: \rho = 0 \text{ (age and price are linearly uncorrelated)}$$
$$H_a: \rho < 0 \text{ (age and price are negatively linearly correlated)}$$

at the 5% significance level. Note that the hypothesis test is left tailed.

We applied the correlation *t*-test programs to the data, resulting in Output 15.4. Steps for generating that output are presented in Instructions 15.4 (next page).

OUTPUT 15.4 Correlation *t*-test on the Orion data

MINITAB

Correlation: AGE, PRICE

Pearson correlation of AGE and PRICE = -0.924
P-Value = 0.000

EXCEL

Correlation matrix (Pearson):		
Variables	AGE	PRICE
AGE	1	-0.9238
PRICE	-0.9238	1
p-values:		
Variables	AGE	PRICE
AGE	0	<0.0001
PRICE	<0.0001	0

TI-83/84 PLUS

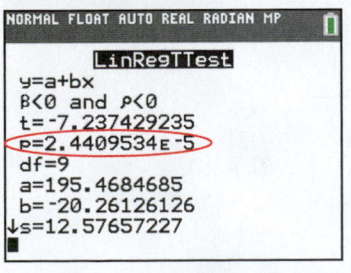

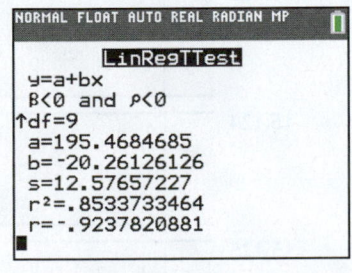

As shown in Output 15.4, the *P*-value is less than the specified significance level of 0.05, so we reject H_0. At the 5% significance level, the data provide sufficient evidence to conclude that age and price of Orions are negatively linearly correlated.

INSTRUCTIONS 15.4 Steps for generating Output 15.4

MINITAB

1 Store the age and price data from Table 15.5 in columns named AGE and PRICE, respectively
2 Choose **Stat ➤ Basic Statistics ➤ Correlation...**
3 Press the F3 key to reset the dialog box
4 Specify AGE and PRICE in the **Variables** text box
5 Click **OK**

EXCEL

1 Store the age and price data from Table 15.5 in columns named AGE and PRICE, respectively
2 Choose **XLSTAT ➤ Correlation/Association tests ➤ Correlation tests**
3 Click the reset button in the lower left corner of the dialog box
4 Click in the **Observations/variables table** selection box and then select the columns of the worksheet that contain the AGE and PRICE data
5 Type 5 in the **Significance level (%)** text box

6 Click the **Outputs** tab and then check only the **Correlations** and **p-values** check boxes
7 Click **OK**
8 Click the **Continue** button in the **XLSTAT – Selections** dialog box

TI-83/84 PLUS

1 Store the age and price data from Table 15.5 in lists named AGE and PRICE, respectively
2 Press **STAT**, arrow over to **TESTS**, arrow down to **LinRegTTest**, and press **ENTER**
3 Press **2nd ➤ LIST**, arrow down to AGE, and press **ENTER** twice
4 Press **2nd ➤ LIST**, arrow down to PRICE, and press **ENTER** twice
5 Type 1 for **Freq** and press **ENTER**
6 Highlight <0 and press **ENTER**
7 Arrow down to **Calculate** and press **ENTER**

Exercises 15.4

Understanding the Concepts and Skills

15.111 Identify the statistic used to estimate the population linear correlation coefficient.

15.112 Suppose that, for a sample of pairs of observations from two variables, the linear correlation coefficient, r, is positive. Does this result necessarily imply that the variables are positively linearly correlated? Explain.

15.113 Suppose that, for a sample of pairs of observations from two variables, the linear correlation coefficient, r, is negative. Does this result necessarily imply that the variables are negatively linearly correlated? Explain.

15.114 Is ρ a parameter or a statistic? What about r? Explain your answers.

In each of Exercises 15.115–15.117, fill in the blank.

15.115 If $\rho = 0$, then the two variables under consideration are linearly _____.

15.116 If two variables are positively linearly correlated, one of the variables tends to increase as the other _____.

15.117 If two variables are _____ linearly correlated, one of the variables tends to decrease as the other increases.

In Exercises 15.118–15.127, we repeat the data from Exercises 15.12–15.21 and specify an alternative hypothesis for a correlation t-test. For each exercise, decide, at the 10% significance level, whether the data provide sufficient evidence to reject the null hypothesis in favor of the alternative hypothesis.

15.118

x	2	4	3
y	3	5	7

$H_a: \rho > 0$

15.119

x	3	1	2
y	−4	0	−5

$H_a: \rho < 0$

15.120

x	0	4	3	1	2
y	1	9	8	4	3

$H_a: \rho \neq 0$

15.121

x	3	4	1	2
y	4	5	0	−1

$H_a: \rho > 0$

15.122

x	1	3	4	4
y	13	−1	3	5

$H_a: \rho \neq 0$

15.123

x	2	2	3	4	4
y	3	4	0	2	1

$H_a: \rho \neq 0$

15.124

x	1	3	4	4
y	8	0	3	1

$H_a: \rho < 0$

15.125

x	1	2	3
y	4	3	8

$H_a: \rho > 0$

15.126

x	1	1	5	5
y	1	3	2	4

$H_a: \rho < 0$

15.127

x	0	2	2	5	6
y	4	2	0	−2	1

$H_a: \rho \neq 0$

Applying the Concepts and Skills

In Exercises 15.128–15.133, we repeat the information from Exercises 15.22–15.27. Presuming that the assumptions for regression inferences are met, perform the required correlation t-tests, using either the critical-value approach or the P-value approach.

15.128 Tax Efficiency. Following are the data on percentage of investments in energy securities and tax efficiency from Exercise 15.22.

x	3.1	3.2	3.7	4.3	4.0	5.5	6.7	7.4	7.4	10.6
y	98.1	94.7	92.0	89.8	87.5	85.0	82.0	77.8	72.1	53.5

At the 2.5% significance level, do the data provide sufficient evidence to conclude that percentage of investments in energy securities and tax efficiency are negatively linearly correlated for mutual fund portfolios?

15.129 Corvette Prices. Following are the age and price data for Corvettes from Exercise 15.23.

x	6	6	6	2	2	5	4	5	1	4
y	290	280	295	425	384	315	355	328	425	325

At the 5% level of significance, do the data provide sufficient evidence to conclude that age and price of Corvettes are negatively linearly correlated?

15.130 Custom Homes. Following are the size and price data for custom homes from Exercise 15.24.

x	26	27	33	29	29	34	30	40	22
y	540	555	575	577	606	661	738	804	496

At the 0.5% significance level, do the data provide sufficient evidence to conclude that, for custom homes in the Equestrian Estates, size and price are positively linearly correlated?

15.131 Plant Emissions. Following are the data on plant weight and quantity of volatile emissions from Exercise 15.25.

x	57	85	57	65	52	67	62	80	77	53	68
y	8.0	22.0	10.5	22.5	12.0	11.5	7.5	13.0	16.5	21.0	12.0

Do the data suggest that, for the potato plant *Solanum tuberosum*, weight and quantity of volatile emissions are linearly correlated? Use $\alpha = 0.05$.

15.132 Crown-Rump Length. Following are the data on age of fetuses and length of crown-rump from Exercise 15.26.

x	10	10	13	13	18	19	19	23	25	28
y	66	66	108	106	161	166	177	228	235	280

At the 10% significance level, do the data provide sufficient evidence to conclude that age and crown-rump length are linearly correlated?

15.133 Study Time and Score. Following are the data on total hours studied over 2 weeks and test score at the end of the 2 weeks from Exercise 15.27.

x	10	15	12	20	8	16	14	22
y	92	81	84	74	85	80	84	80

a. At the 1% significance level, do the data provide sufficient evidence to conclude that a negative linear correlation exists between study time and test score for beginning calculus students?
b. Repeat part (a) using a 5% significance level.

Working with Large Data Sets

In each of Exercises 15.134–15.144, use the technology of your choice to decide whether you can reasonably apply the correlation t-test. If so, perform and interpret the required correlation t-test(s) at the 5% significance level.

15.134 Birdies and Score. The data from Exercise 15.34 for number of birdies during a tournament and final score of 63 women golfers are on the WeissStats site. Do the data provide sufficient evidence to conclude that, for women golfers, number of birdies and score are negatively linearly correlated?

15.135 U.S. Presidents. The data from Exercise 15.35 for the ages at inauguration and of death of the presidents of the United States are on the WeissStats site. Do the data provide sufficient evidence to conclude that, for U.S. presidents, age at inauguration and age at death are positively linearly correlated?

15.136 Movie Grosses. The data from Exercise 15.36 on domestic and overseas grosses for a random sample of 50 movies are on the WeissStats site. Do the data provide sufficient evidence to conclude that domestic and overseas grosses are positively linearly correlated?

15.137 Acreage and Value. The data from Exercise 15.37 for lot size (in acres) and assessed value (in thousands of dollars) of a sample of homes in a particular area are on the WeissStats site. Do the data provide sufficient evidence to conclude that, for homes in this particular area, lot size and assessed value are positively linearly correlated?

15.138 Home Size and Value. The data from Exercise 15.38 for home size (in square feet) and assessed value (in thousands of dollars) for the same homes as in Exercise 15.137 are on the WeissStats site. Do the data provide sufficient evidence to conclude that, for homes in this particular area, home size and assessed value are positively linearly correlated?

15.139 High and Low Temperature. The data from Exercise 15.39 for average high and low temperatures in January of a random sample of 50 cities are on the WeissStats site. Do the data provide sufficient evidence to conclude that, for cities, average high and low temperatures in January are linearly correlated?

15.140 PCBs and Pelicans. The data from Exercise 15.40 for shell thickness and concentration of PCBs of 60 Anacapa pelican eggs are on the WeissStats site. Do the data provide sufficient evidence to conclude that concentration of PCBs and shell thickness are linearly correlated for Anacapa pelican eggs?

15.141 Gas Guzzlers. The data from Exercise 15.41 for gas mileage and engine displacement of 121 vehicles are on the WeissStats site.

Do the data provide sufficient evidence to conclude that engine displacement and gas mileage are negatively linearly correlated?

15.142 Estriol Level and Birth Weight. The data from Exercise 15.42 for estriol levels of pregnant women and birth weights of their children are on the WeissStats site. Do the data provide sufficient evidence to conclude that estriol level and birth weight are positively linearly correlated?

15.143 Shortleaf Pines. The data from Exercise 15.43 for volume, in cubic feet, and diameter at breast height, in inches, of 70 shortleaf pines are on the WeissStats site. Do the data provide sufficient evidence to conclude that diameter at breast height and volume are positively linearly correlated for shortleaf pines?

15.144 Body Fat. The data from Exercise 15.80 for age and body fat of 18 randomly selected adults are on the WeissStats site.
a. Do the data provide sufficient evidence to conclude that, for adults, age and percentage of body fat are positively linearly correlated?
b. Remove the potential outlier and repeat part (a).
c. Compare your results with and without the removal of the potential outlier and state your conclusions.

CHAPTER IN REVIEW

You Should Be Able to

1. use and understand the formulas in this chapter.

2. state the assumptions for regression inferences.

3. understand the difference between the population regression line and a sample regression line.

4. estimate the regression parameters β_0, β_1, and σ.

5. determine the standard error of the estimate.

6. perform a residual analysis to check the assumptions for regression inferences.

7. perform a hypothesis test to decide whether the slope, β_1, of the population regression line is not 0 and hence whether x is useful for predicting y.

8. obtain a confidence interval for β_1.

9. determine a point estimate and a confidence interval for the conditional mean of the response variable corresponding to a particular value of the predictor variable.

10. determine a predicted value and a prediction interval for the response variable corresponding to a particular value of the predictor variable.

11. understand the difference between the population correlation coefficient and a sample correlation coefficient.

12. perform a hypothesis test for a population linear correlation coefficient.

Key Terms

conditional distribution, *659*
conditional mean, *659*
conditional mean *t*-interval procedure, *680*
conditional standard deviation, *659*
correlation *t*-test, *689*
linearly correlated variables, *688*
linearly uncorrelated variables, *688*
multiple regression analysis, *683*

negatively linearly correlated variables, *688*
population linear correlation coefficient (ρ), *688*
population regression equation, *659*
population regression line, *659*
positively linearly correlated variables, *688*
predicted value *t*-interval procedure, *682*
prediction interval, *681*

regression model, *659*
regression *t*-interval procedure, *674*
regression *t*-test, *672*
residual (*e*), *662*
residual plot, *663*
residual standard deviation, *662*
sampling distribution of the slope of the regression line, *671*
simple linear regression, *683*
standard error of the estimate (s_e), *662*

REVIEW PROBLEMS

Understanding the Concepts and Skills

1. Suppose that x and y are two variables of a population with x a predictor variable and y a response variable.
a. The distribution of all possible values of the response variable y corresponding to a particular value of the predictor variable x is called a _____ distribution of the response variable.
b. State the four assumptions for regression inferences.

2. Suppose that x and y are two variables of a population and that the assumptions for regression inferences are met with x as the predictor variable and y as the response variable.

a. What statistic is used to estimate the slope of the population regression line?
b. What statistic is used to estimate the y-intercept of the population regression line?
c. What statistic is used to estimate the common conditional standard deviation of the response variable corresponding to fixed values of the predictor variable?

3. What two plots did we use in this chapter to decide whether we can reasonably presume that the assumptions for regression inferences are met by two variables of a population? What properties should those plots have?

4. Regarding analysis of residuals, decide in each case which assumption for regression inferences may be violated.
  a. A residual plot—that is, a plot of the residuals against the observed values of the predictor variable—shows curvature.
  b. A residual plot becomes wider with increasing values of the predictor variable.
  c. A normal probability plot of the residuals shows extreme curvature.
  d. A normal probability plot of the residuals shows outliers but is otherwise roughly linear.

5. Suppose that you perform a hypothesis test for the slope of the population regression line with the null hypothesis $H_0: \beta_1 = 0$ and the alternative hypothesis $H_a: \beta_1 \neq 0$. If you reject the null hypothesis, what can you say about the utility of the regression equation for making predictions?

6. Identify three statistics that can be used as a basis for testing the utility of a regression.

7. For a particular value of a predictor variable, is there a difference between the predicted value of the response variable and the point estimate for the conditional mean of the response variable? Explain your answer.

8. Generally speaking, what is the difference between a confidence interval and a prediction interval?

9. Fill in the blank: $\bar{x}$ is to μ as r is to _____.

10. Identify the relationship between two variables and the terminology used to describe that relationship if
  a. $\rho > 0$.   **b.** $\rho = 0$.   **c.** $\rho < 0$.

Applying the Concepts and Skills

11. Graduation Rates. Graduation rate—the percentage of entering freshmen attending full time and graduating within 5 years—and what influences it is a concern in U.S. colleges and universities. *U.S. News and World Report*'s "College Guide" provides data on graduation rates for colleges and universities as a function of the percentage of freshmen in the top 10% of their high school class, total spending per student, and student-to-faculty ratio. A random sample of 10 universities gave the following data on student-to-faculty ratio (S/F ratio) and graduation rate (Grad rate).

S/F ratio x	Grad rate y	S/F ratio x	Grad rate y
16	45	17	46
20	55	17	50
17	70	17	66
19	50	10	26
22	47	18	60

Discuss what satisfying the assumptions for regression inferences would mean with student-to-faculty ratio as the predictor variable and graduation rate as the response variable.

12. Graduation Rates. Refer to Problem 11.
  a. Determine the regression equation for the data.
  b. Compute and interpret the standard error of the estimate.
  c. Presuming that the assumptions for regression inferences are met, interpret your answer to part (b).

13. Graduation Rates. Refer to Problems 11 and 12. Perform a residual analysis to decide whether considering the assumptions for regression inferences to be met by the variables student-to-faculty ratio and graduation rate is reasonable.

For Problems 14–17, presume that the variables student-to-faculty ratio and graduation rate satisfy the assumptions for regression inferences.

14. Graduation Rates. Refer to Problems 11 and 12.
  a. At the 5% significance level, do the data provide sufficient evidence to conclude that student-to-faculty ratio is useful as a predictor of graduation rate?
  b. Determine a 95% confidence interval for the slope, β_1, of the population regression line that relates graduation rate to student-to-faculty ratio. Interpret your answer.

15. Graduation Rates. Refer to Problems 11 and 12.
  a. Find a point estimate for the mean graduation rate of all universities that have a student-to-faculty ratio of 17.
  b. Determine a 95% confidence interval for the mean graduation rate of all universities that have a student-to-faculty ratio of 17.

16. Graduation Rates. Refer to Problems 11 and 12.
  a. Find the predicted graduation rate for a university that has a student-to-faculty ratio of 17.
  b. Find a 95% prediction interval for the graduation rate of a university that has a student-to-faculty ratio of 17.
  c. Explain why the prediction interval in part (b) is wider than the confidence interval in Problem 15(b).

17. Graduation Rates. Refer to Problem 11. At the 2.5% significance level, do the data provide sufficient evidence to conclude that the variables student-to-faculty ratio and graduation rate are positively linearly correlated?

Working with Large Data Sets

In Problems 18–20, use the technology of your choice to
  a. determine the sample regression equation.
  b. find and interpret the standard error of the estimate.
  c. decide, at the 5% significance level, whether the data provide sufficient evidence to conclude that the predictor variable is useful for predicting the response variable.
  d. determine and interpret a point estimate for the conditional mean of the response variable corresponding to the specified value of the predictor variable.
  e. find and interpret a 95% confidence interval for the conditional mean of the response variable corresponding to the specified value of the predictor variable.
  f. determine and interpret the predicted value of the response variable corresponding to the specified value of the predictor variable.
  g. find and interpret a 95% prediction interval for the value of the response variable corresponding to the specified value of the predictor variable.
  h. compare and discuss the differences between the confidence interval that you obtained in part (e) and the prediction interval that you obtained in part (g).
  i. perform and interpret the required correlation t-test at the 5% significance level.
  j. perform a residual analysis to decide whether making the preceding inferences is reasonable. Explain your answer.

18. IMR and Life Expectancy. From the *International Data Base*, published by the U.S. Census Bureau, we obtained data on infant mortality rate (IMR) and life expectancy (LE), in years, for a sample of 60 countries. The data are presented on the WeissStats site.

- For the estimations and predictions, use an IMR of 30.
- For the correlation test, decide whether IMR and life expectancy are negatively linearly correlated.

19. High Temperature and Precipitation. The National Oceanic and Atmospheric Administration publishes temperature and precipitation information for cities around the world in *Climates of the World*. Data on average high temperature (in degrees Fahrenheit) in July and average precipitation (in inches) in July for 48 cities are on the WeissStats site.

- For the estimations and predictions, use an average July temperature of 83°F.
- For the correlation test, decide whether average high temperature in July and average precipitation in July are linearly correlated.

20. U.S. Open. From the website Golf.com, part of Sports Illustrated Sites, we obtained the scores for the first and second rounds of the 2013 U.S. Open golf tournament. You will find those scores on the WeissStats site.

- For the estimations and predictions, use a first-round score of 72.
- For the correlation test, decide whether first-round and second-round scores are positively linearly correlated.

FOCUSING ON DATA ANALYSIS

UWEC UNDERGRADUATES

Recall from Chapter 1 (see page 34) that the Focus database and Focus sample contain information on the undergraduate students at the University of Wisconsin - Eau Claire (UWEC). Now would be a good time for you to review the discussion about these data sets.

Open the Focus sample worksheet (FocusSample) in the technology of your choice and do the following.

a. Perform a residual analysis to decide whether considering the assumptions for regression inferences met by the variables high school percentile and cumulative GPA appears reasonable.
b. With high school percentile as the predictor variable and cumulative GPA as the response variable, determine and interpret the standard error of the estimate.

c. At the 5% significance level, do the data provide sufficient evidence to conclude that high school percentile is useful for predicting cumulative GPA of UWEC undergraduates?
d. Determine a point estimate for the mean cumulative GPA of all UWEC undergraduates who had high school percentiles of 74.
e. Find a 95% confidence interval for the mean cumulative GPA of all UWEC undergraduates who had high school percentiles of 74.
f. Determine the predicted cumulative GPA of a UWEC undergraduate who had a high school percentile of 74.
g. Find a 95% prediction interval for the cumulative GPA of a UWEC undergraduate who had a high school percentile of 74.
h. At the 5% significance level, do the data provide sufficient evidence to conclude that high school percentile and cumulative GPA are positively linearly correlated?

CASE STUDY DISCUSSION

SHOE SIZE AND HEIGHT

At the beginning of this chapter, we presented data on shoe size and height for a sample of students at Arizona State University. Using the regression and correlation techniques that you learned in Chapter 14 and this chapter, solve the following problems. *Note:* We recommend that you use statistical software or a graphing calculator to solve these problems, but they can also be done by hand.

a. Separate the data in the table on page 658 into two tables, one for males and the other for females. Parts (b)–(j) are for the male data.
b. Determine the sample regression equation with shoe size as the predictor variable for height.
c. Perform a residual analysis to decide whether considering Assumptions 1–3 for regression inferences to be satisfied by the variables shoe size and height appears reasonable.
d. Find and interpret the standard error of the estimate.
e. Determine the *P*-value for a test of whether shoe size is useful for predicting height. Then refer to Table 9.8 on page 386 to assess the evidence in favor of utility.

f. Find a point estimate for the mean height of all males who wear a size $10\frac{1}{2}$ shoe.
g. Obtain a 95% confidence interval for the mean height of all males who wear a size $10\frac{1}{2}$ shoe. Interpret your answer.
h. Determine the predicted height of a male who wears a size $10\frac{1}{2}$ shoe.
i. Find a 95% prediction interval for the height of a male who wears a size $10\frac{1}{2}$ shoe. Interpret your answer.
j. At the 5% significance level, do the data provide sufficient evidence to conclude that shoe size and height are positively linearly correlated?
k. Repeat parts (b)–(j) for the unabridged data on shoe size and height for females. Do the estimation and prediction problems for a size 8 shoe.
l. Repeat part (k) for the data on shoe size and height for females with the outlier removed. Compare your results with those obtained in part (k).

BIOGRAPHY

SIR FRANCIS GALTON: DISCOVERER OF REGRESSION AND CORRELATION

Francis Galton was born on February 16, 1822, into a wealthy Quaker family of bankers and gunsmiths on his father's side and as a cousin of Charles Darwin on his mother's side. Although his IQ was estimated to be about 200, his formal education was unfinished.

He began training in medicine in Birmingham and London but quit when, in his words, "A passion for travel seized me as if I had been a migratory bird." After a tour through Germany and southeastern Europe, he went to Trinity College in Cambridge to study mathematics. He left Cambridge in his third year, broken from overwork. He recovered quickly and resumed his medical studies in London. However, his father died before he had finished medical school and left to him, at 22, "a sufficient fortune to make me independent of the medical profession."

Galton held no professional or academic positions; nearly all his experiments were conducted at his home or performed by friends. He was curious about almost everything, and carried out research in fields that included meteorology, biology, psychology, statistics, and genetics.

The origination of the concepts of regression and correlation, developed by Galton as tools for measuring the influence of heredity, are summed up in his work *Natural Inheritance.* He discovered regression during experiments with sweet-pea seeds to determine the law of inheritance of size. He made his other great discovery, correlation, while applying his techniques to the problem of measuring the degree of association between the sizes of two different body organs of an individual.

In his later years, Galton was associated with Karl Pearson, who became his champion and an extender of his ideas. Pearson was the first holder of the chair of eugenics at University College in London, which Galton had endowed in his will. Galton was knighted in 1909. He died in Haslemere, Surrey, England, in 1911.

CHAPTER 16

Analysis of Variance (ANOVA)

CHAPTER OBJECTIVES

In Chapter 10, you studied inferential methods for comparing the means of two populations. Now you will study **analysis of variance,** or **ANOVA,** which provides methods for comparing the means of more than two populations. For instance, you could use ANOVA to compare the mean energy consumption by households among the four U.S. regions. Just as there are several different procedures for comparing two population means, there are several different ANOVA procedures.

In Section 16.1, to prepare for the study of ANOVA, we consider the *F*-distribution. Next, in Section 16.2, we introduce one-way analysis of variance and examine the logic behind it. Then we discuss the one-way ANOVA procedure itself in Section 16.3.

If you conduct a one-way ANOVA and decide that the population means are not all equal, you may then want to know which means are different, which mean is largest, and, in general, the relation among all the means. *Multiple comparison* methods, which we discuss in Section 16.4, are used to tackle these types of questions.

In Section 16.5, we investigate the Kruskal–Wallis test. This hypothesis-testing procedure is a generalization of the Mann–Whitney test to more than two populations and provides a nonparametric alternative to one-way ANOVA.

CASE STUDY

Self-Perception and Physical Activity

In the paper "Aspects of Self Differ among Physically Active and Inactive Youths" (*International Journal of Public Health*, Vol. 56, Issue 3, pp. 311–318), Z. Veselska et al.

explored connections between aspects of self-perception and physical activity among adolescents. The study involved a sample of adolescents from the Slovak and Czech Republics with ages ranging roughly between 12 and 16.

The youths completed the Self-competence/Self-liking Scale, the Rosenberg's Self-esteem Scale, and the Self-efficacy Scale and answered a question regarding their physical activity. The respondents were divided into three physical-activity categories: no physical activity (none), infrequent physical activity (infrequent), and everyday physical activity (everyday).

The results were used by the researchers to examine the connections between self-perception and frequency of physical activity both for boys and girls combined and for boys and girls separately. In this case study, we will restrict our analyses to the data on boys and girls separately, whose summary statistics are presented in the following table.

Aspect of self	Boys None (n = 34) $\bar{x}$	s	Infrequent (n = 139) $\bar{x}$	s	Everyday (n = 61) $\bar{x}$	s	Girls None (n = 53) $\bar{x}$	s	Infrequent (n = 156) $\bar{x}$	s	Everyday (n = 37) $\bar{x}$	s
Self-esteem												
Positive	14.3	2.76	15.0	2.12	15.1	2.70	12.7	2.88	14.3	2.28	14.6	2.19
Negative	12.4	2.23	12.2	2.59	11.8	2.47	14.1	2.69	12.4	2.72	12.7	2.68
Self-liking	32.1	6.44	33.8	5.18	35.6	6.97	28.6	7.93	32.6	6.56	33.6	6.93
Self-competence	33.4	6.42	35.4	5.64	36.0	6.70	30.9	6.86	34.2	5.87	35.5	5.61
Self-efficacy												
General	56.4	8.50	57.7	7.86	59.1	10.15	51.9	7.89	58.5	9.47	59.8	7.83
Social	18.7	3.23	19.8	3.19	20.7	3.73	19.1	4.49	20.7	3.74	21.4	3.34

After studying the inferential methods in this chapter, you will be able to conduct statistical analyses on these data to compare aspects of self-perception among youths in the three physical-activity categories.

16.1   The *F*-Distribution

Analysis-of-variance procedures rely on a distribution called the *F-distribution,* named in honor of Sir Ronald Fisher. See the biography at the end of this chapter for more information about Fisher.

A variable is said to have an **F-distribution** if its distribution has the shape of a special type of right-skewed curve, called an **F-curve.** There are infinitely many *F*-distributions, and we identify an *F*-distribution (and *F*-curve) by its number of degrees of freedom, just as we did for *t*-distributions and chi-square distributions.

An *F*-distribution, however, has two numbers of degrees of freedom instead of one. Figure 16.1 depicts two different *F*-curves; one has df = (10, 2), and the other has df = (9, 50).

The first number of degrees of freedom for an *F*-curve is called the **degrees of freedom for the numerator,** and the second is called the **degrees of freedom for the denominator.** (The reason for this terminology will become clear in Section 16.3.) Thus, for the *F*-curve in Fig. 16.1 with df = (10, 2), we have

$$\text{df} = (10, 2)$$

Degrees of freedom for the numerator ↗   ↖ Degrees of freedom for the denominator

FIGURE 16.1

Two different *F*-curves

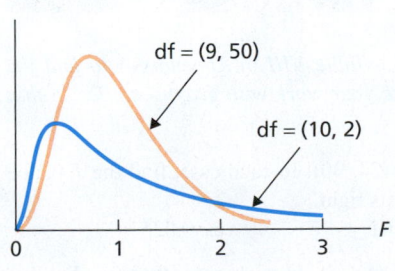

KEY FACT 16.1

Basic Properties of *F*-Curves

Property 1: The total area under an *F*-curve equals 1.

Property 2: An *F*-curve starts at 0 on the horizontal axis and extends indefinitely to the right, approaching, but never touching, the horizontal axis as it does so.

Property 3: An *F*-curve is right skewed.

Using the F-Table

Percentages (and probabilities) for a variable having an F-distribution are equal to areas under its associated F-curve. To perform an ANOVA test, we need to know how to find the F-value having a specified area to its right. The symbol F_α denotes the F-value having area α to its right.

Table VIII in Appendix A provides F-values corresponding to several areas for various degrees of freedom. The degrees of freedom for the denominator (dfd) are displayed in the outside columns of the table, the values of α in the next columns, and the degrees of freedom for the numerator (dfn) along the top.

■■ ■■ **EXAMPLE 16.1** **Finding the F-Value Having a Specified Area to Its Right**

For an F-curve with df $= (4, 12)$, find $F_{0.05}$; that is, find the F-value having area 0.05 to its right, as shown in Fig. 16.2(a).

FIGURE 16.2

Finding the F-value having area 0.05 to its right

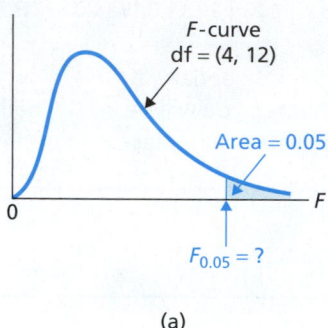

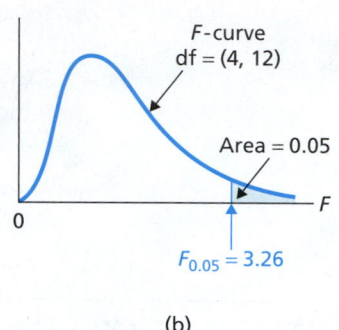

(a) (b)

Solution We use Table VIII to find the F-value. In this case, $\alpha = 0.05$, the degrees of freedom for the numerator is 4, and the degrees of freedom for the denominator is 12.

You try it!

Exercise 16.7 on page 700

We first go down the dfd column to "12." Next, we concentrate on the row for α labeled 0.05. Then, going across that row to the column labeled "4," we reach 3.26. This number is the F-value having area 0.05 to its right, as shown in Fig. 16.2(b). In other words, for an F-curve with df $= (4, 12)$, $F_{0.05} = 3.26$. ■

Exercises 16.1

Understanding the Concepts and Skills

16.1 How do we identify an F-distribution and its corresponding F-curve?

16.2 How many degrees of freedom does an F-curve have? What are those degrees of freedom called?

16.3 What symbol is used to denote the F-value having area 0.05 to its right? 0.025 to its right? α to its right?

16.4 Using the F_α-notation, identify the F-value having area 0.975 to its left.

16.5 An F-curve has df $= (12, 7)$. What is the number of degrees of freedom for the
a. numerator? **b.** denominator?

16.6 An F-curve has df $= (8, 19)$. What is the number of degrees of freedom for the
a. denominator? **b.** numerator?

In Exercises 16.7–16.10, use Table VIII in Appendix A to find the required F-values. Illustrate your work with graphs similar to that shown in Fig. 16.2.

16.7 An F-curve has df $= (24, 30)$. In each case, find the F-value having the specified area to its right.
a. 0.05 **b.** 0.01 **c.** 0.025

16.8 An F-curve has df $= (12, 5)$. In each case, find the F-value having the specified area to its right.
a. 0.01 **b.** 0.05 **c.** 0.005

16.9 For an F-curve with df $= (20, 21)$, find
a. $F_{0.01}$. **b.** $F_{0.05}$. **c.** $F_{0.10}$.

16.10 For an F-curve with df $= (6, 10)$, find
a. $F_{0.05}$. **b.** $F_{0.01}$. **c.** $F_{0.025}$.

16.2   One-Way ANOVA: The Logic

In Chapter 10, you learned how to compare two population means, that is, the means of a single variable for two different populations. You studied various methods for making such comparisons, one being the pooled *t*-procedure.

Analysis of variance (ANOVA) provides methods for comparing several population means, that is, the means of a single variable for several populations. In this section and Section 16.3, we present the simplest kind of ANOVA, **one-way analysis of variance.** This type of ANOVA is called *one-way* analysis of variance because it compares the means of a variable for populations that result from a classification by *one* other variable, called the **factor.** The possible values of the factor are referred to as the **levels** of the factor.

For example, suppose that you want to compare the mean energy consumption by households among the four regions of the United States. The variable under consideration is "energy consumption," and there are four populations: households in the Northeast, Midwest, South, and West. The four populations result from classifying households in the United States by the factor "region," whose levels are Northeast, Midwest, South, and West.

One-way analysis of variance is the generalization to more than two populations of the pooled *t*-procedure (i.e., both procedures give the same results when applied to two populations). As in the pooled *t*-procedure, we make the following assumptions.

KEY FACT 16.2

Assumptions (Conditions) for One-Way ANOVA

1. **Simple random samples:** The samples taken from the populations under consideration are simple random samples.
2. **Independent samples:** The samples taken from the populations under consideration are independent of one another.
3. **Normal populations:** For each population, the variable under consideration is normally distributed.
4. **Equal standard deviations:** The standard deviations of the variable under consideration are the same for all the populations.

Regarding Assumptions 1 and 2, we note that one-way ANOVA can also be used as a method for comparing several means with a designed experiment. In addition, like the pooled *t*-procedure, one-way ANOVA is robust to moderate violations of Assumption 3 (normal populations) and is also robust to moderate violations of Assumption 4 (equal standard deviations) provided the sample sizes are roughly equal.

How can the conditions of normal populations and equal standard deviations be checked? Normal probability plots of the sample data are effective in detecting gross violations of normality. Checking equal population standard deviations, however, can be difficult, especially when the sample sizes are small; as a rule of thumb, you can consider that condition met if *the ratio of the largest to the smallest sample standard deviation is less than 2.* We call that rule of thumb the **rule of 2.**

Another way to assess the normality and equal-standard-deviations assumptions is to perform a **residual analysis.** In ANOVA, the **residual** of an observation is the difference between the observation and the mean of the sample containing it. If the normality and equal-standard-deviations assumptions are met, a normal probability plot of (all) the residuals should be roughly linear. Moreover, a plot of the residuals against the sample means should fall roughly in a horizontal band centered and symmetric about the horizontal axis.

The Logic Behind One-Way ANOVA

The reason for the word *variance* in *analysis of variance* is that the procedure for comparing the means analyzes the variation in the sample data. To examine how this procedure works, let's suppose that independent random samples are taken from two populations—say, Populations 1 and 2—with means μ_1 and μ_2. Further, let's suppose that the means of the two samples are $\bar{x}_1 = 20$ and $\bar{x}_2 = 25$. Can we reasonably conclude from these statistics that $\mu_1 \neq \mu_2$, that is, that the population means are different? To answer this question, we must consider the variation within the samples.

Suppose, for instance, that the sample data are as displayed in Table 16.1 and depicted in Fig. 16.3.

TABLE 16.1

Sample data from Populations 1 and 2

Sample from Population 1	21	37	11	20	8	23
Sample from Population 2	24	31	29	40	9	17

FIGURE 16.3

Dotplots for sample data in Table 16.1

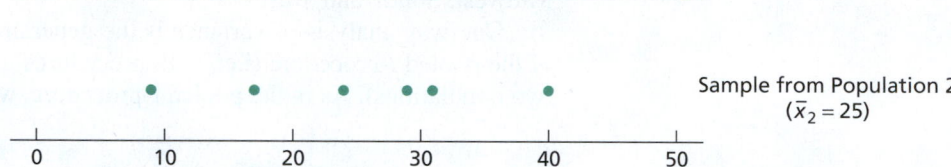

For these two samples, $\bar{x}_1 = 20$ and $\bar{x}_2 = 25$. But here we cannot infer that $\mu_1 \neq \mu_2$ because it is not clear whether the difference between the sample means is due to a difference between the population means or to the variation within the populations.

However, suppose that the sample data are as displayed in Table 16.2 and depicted in Fig. 16.4.

> **What Does It Mean?**
>
> Intuitively speaking, because the variation between the sample means is not large relative to the variation within the samples, we cannot conclude that $\mu_1 \neq \mu_2$.

TABLE 16.2

Sample data from Populations 1 and 2

Sample from Population 1	21	21	20	18	20	20
Sample from Population 2	25	28	25	24	24	24

FIGURE 16.4

Dotplots for sample data in Table 16.2

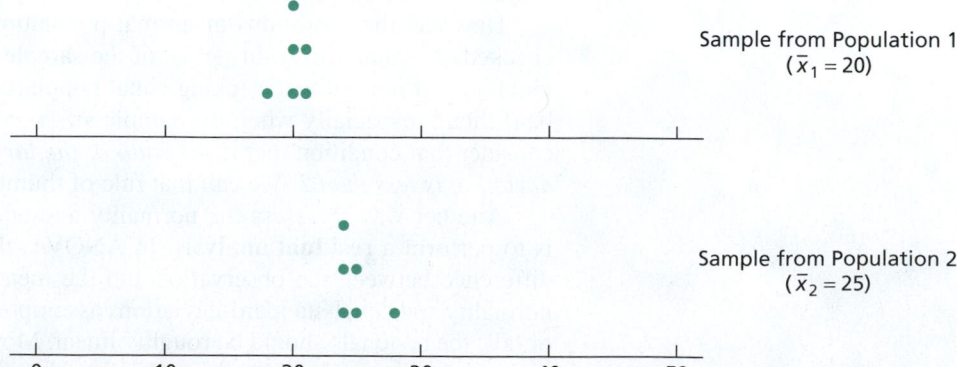

Again, for these two samples, $\bar{x}_1 = 20$ and $\bar{x}_2 = 25$. But this time, we *can* infer that $\mu_1 \neq \mu_2$ because it seems clear that the difference between the sample means is due to a difference between the population means, not to the variation within the populations.

> **What Does It Mean?**
>
> Intuitively speaking, because the variation between the sample means is large relative to the variation within the samples, we can conclude that $\mu_1 \neq \mu_2$.

The preceding two illustrations reveal the basic idea for performing a one-way analysis of variance to compare the means of several populations:

1. Take independent simple random samples from the populations.
2. Compute the sample means.
3. If the variation among the sample means is large relative to the variation within the samples, conclude that the means of the populations are not all equal.

To make this process precise, we need quantitative measures of the variation among the sample means and the variation within the samples. We also need an objective method for deciding whether the variation among the sample means is large relative to the variation within the samples.

Mean Squares and *F*-Statistic in One-Way ANOVA

As before, when dealing with several populations, we use subscripts on parameters and statistics. Thus, for Population j, we use μ_j, $\bar{x}_j$, s_j, and n_j to denote the population mean, sample mean, sample standard deviation, and sample size, respectively.

We first consider the measure of variation among the sample means. In hypothesis tests for two population means, we measure the variation between the two sample means by calculating their difference, $\bar{x}_1 - \bar{x}_2$. When more than two populations are involved, we cannot measure the variation among the sample means simply by taking a difference. However, we can measure that variation by computing the standard deviation or variance of the sample means or by computing any descriptive statistic that measures variation.

In one-way ANOVA, we measure the variation among the sample means by a weighted average of their squared deviations about the mean, $\bar{x}$, of all the sample data. That measure of variation is called the **treatment mean square, *MSTR*,** and is defined as

$$MSTR = \frac{SSTR}{k - 1}, \tag{16.1}$$

where k denotes the number of populations being sampled and

$$SSTR = n_1(\bar{x}_1 - \bar{x})^2 + n_2(\bar{x}_2 - \bar{x})^2 + \cdots + n_k(\bar{x}_k - \bar{x})^2. \tag{16.2}$$

The quantity **SSTR** is called the **treatment sum of squares.**

We note that *MSTR* is similar to the sample variance of the sample means. In fact, if all the sample sizes are identical, then *MSTR* equals that common sample size times the sample variance of the sample means.

Next we consider the measure of variation within the samples. This measure is the pooled estimate of the common population variance, σ^2. It is called the **error mean square, *MSE*,** and is defined as

$$MSE = \frac{SSE}{n - k}, \tag{16.3}$$

where n denotes the total number of observations and

$$SSE = (n_1 - 1)s_1^2 + (n_2 - 1)s_2^2 + \cdots + (n_k - 1)s_k^2. \tag{16.4}$$

The quantity **SSE** is called the **error sum of squares.**[†‡]

Finally, we consider how to compare the variation among the sample means, *MSTR*, to the variation within the samples, *MSE*. To do so, we use the statistic $F = MSTR/MSE$, which we refer to as the **F-statistic.** Large values of F indicate that the variation among the sample means is large relative to the variation within the samples and hence that the null hypothesis of equal population means should be rejected.

What Does It Mean?

MSTR measures the variation among the sample means.

What Does It Mean?

MSE measures the variation within the samples.

What Does It Mean?

The *F*-statistic compares the variation among the sample means to the variation within the samples.

[†]The terms **treatment** and **error** arose from the fact that many ANOVA techniques were first developed to analyze agricultural experiments. In any case, the treatments refer to the different populations, and the errors pertain to the variation within the populations.

[‡]For two populations (i.e., $k = 2$), *MSE* is the pooled variance, s_p^2, defined in Section 10.2 on page 447.

> **DEFINITION 16.1**
>
> ### Mean Squares and *F*-Statistic in One-Way ANOVA
>
> **Treatment mean square, *MSTR*:** The variation among the sample means: $MSTR = SSTR/(k-1)$, where $SSTR$ is the treatment sum of squares and k is the number of populations under consideration.
>
> **Error mean square, *MSE*:** The variation within the samples: $MSE = SSE/(n-k)$, where SSE is the error sum of squares and n is the total number of observations.
>
> **F-statistic, *F*:** The ratio of the variation among the sample means to the variation within the samples: $F = MSTR/MSE$.

■■ ■■ **EXAMPLE 16.2** **Introducing One-Way ANOVA**

Energy Consumption The Energy Information Administration gathers data on residential energy consumption and expenditures and publishes its findings in *Residential Energy Consumption Survey*. Suppose that we want to decide whether a difference exists in mean annual energy consumption by households among the four U.S. regions.

Let μ_1, μ_2, μ_3, and μ_4 denote last year's mean energy consumptions by households in the Northeast, Midwest, South, and West, respectively. Then the hypotheses to be tested are

H_0: $\mu_1 = \mu_2 = \mu_3 = \mu_4$ (mean energy consumptions are all equal)

H_a: Not all the means are equal (i.e., at least two of the means differ).

The basic strategy for carrying out this hypothesis test follows the three steps mentioned on page 703 and is illustrated in Fig. 16.5.

Step 1. Independently and randomly take samples of households in the four U.S. regions.

Step 2. Compute last year's mean energy consumptions, $\bar{x}_1$, $\bar{x}_2$, $\bar{x}_3$, and $\bar{x}_4$, of the four samples.

Step 3. Reject the null hypothesis if the variation among the sample means is large relative to the variation within the samples; otherwise, do not reject the null hypothesis.

FIGURE 16.5 Process for comparing four population means

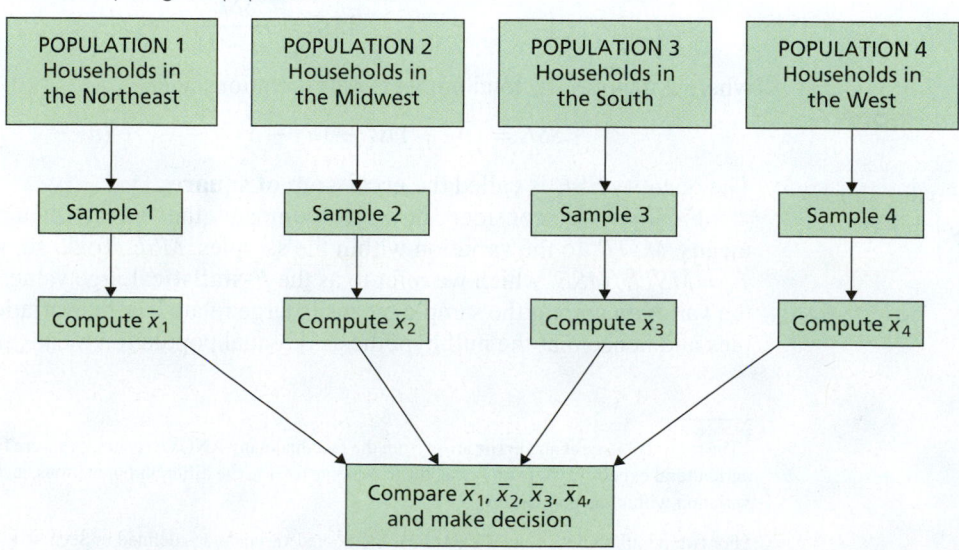

In Steps 1 and 2, we obtain the sample data and compute the sample means. Suppose that the results of those steps are as shown in Table 16.3, where the data are displayed to the nearest 10 million BTU.

TABLE 16.3

Samples and their means of last year's energy consumptions for households in the four U.S. regions

Northeast	Midwest	South	West
13	15	5	8
8	10	11	10
11	16	9	6
12	11	5	5
11	13		7
	10		
11.0	12.5	7.5	7.2	← *Means*

In Step 3, we compare the variation among the four sample means (see bottom of Table 16.3) to the variation within the samples. To accomplish that, we need to compute the treatment mean square (*MSTR*), the error mean square (*MSE*), and the *F*-statistic.

We'll require, among other things, the number of populations under consideration (k), the total number of observations (n), and the overall mean ($\bar{x}$). Referring to Table 16.3, we see that

$$k = 4,$$

$$n = 5 + 6 + 4 + 5 = 20,$$

and

$$\bar{x} = \frac{\Sigma x_i}{n} = \frac{13 + 8 + 11 + \cdots + 5 + 7}{20} = \frac{196}{20} = 9.8.$$

To compute *MSTR*, we first need to find *SSTR*. In view of Equation (16.2) on page 703, we construct the following table by referring to Table 16.3.

TABLE 16.4

Table for computing *SSTR* for the energy-consumption data

Region	Size n_j	Mean $\bar{x}_j$	$\bar{x}_j - \bar{x}$	$(\bar{x}_j - \bar{x})^2$	$n_j(\bar{x}_j - \bar{x})^2$
Northeast	5	11.0	1.2	1.44	7.20
Midwest	6	12.5	2.7	7.29	43.74
South	4	7.5	−2.3	5.29	21.16
West	5	7.2	−2.6	6.76	33.80
					105.90

From the final column of Table 16.4, we see that

$$SSTR = n_1(\bar{x}_1 - \bar{x})^2 + n_2(\bar{x}_2 - \bar{x})^2 + n_3(\bar{x}_3 - \bar{x})^2 + n_4(\bar{x}_4 - \bar{x})^2 = 105.9.$$

Therefore, from Definition 16.1 or Equation (16.1) on page 703,

$$MSTR = \frac{SSTR}{k-1} = \frac{105.9}{4-1} = 35.3.$$

To compute *MSE*, we first need to find *SSE*. Referring to Table 16.3, we computed the sample variance of each sample:

$$s_1^2 = 3.5, \quad s_2^2 = 6.7, \quad s_3^2 = 9.0, \quad \text{and} \quad s_4^2 = 3.7.$$

In view of Equation (16.4) on page 703, we construct the following table.

TABLE 16.5

Table for computing *SSE* for the energy-consumption data

Region	Size n_j	Variance s_j^2	$n_j - 1$	$(n_j - 1)s_j^2$
Northeast	5	3.5	4	14.0
Midwest	6	6.7	5	33.5
South	4	9.0	3	27.0
West	5	3.7	4	14.8
				89.3

From the final column of Table 16.5, we see that

$$SSE = (n_1 - 1)s_1^2 + (n_2 - 1)s_2^2 + (n_3 - 1)s_3^2 + (n_4 - 1)s_4^2 = 89.3.$$

Therefore, from Definition 16.1 or Equation (16.3) on page 703,

$$MSE = \frac{SSE}{n - k} = \frac{89.3}{20 - 4} = 5.581.$$

Finally, we determine F. As $MSTR = 35.3$ and $MSE = 5.581$, the value of the F-statistic is

$$F = \frac{MSTR}{MSE} = \frac{35.3}{5.581} = 6.32.$$

Is this value of F large enough to conclude that the null hypothesis of equal population means is false? To answer that question, we need to know the distribution of the F-statistic, which we discuss in Section 16.3.

You try it!

Exercise 16.25 on page 707

Exercises 16.2

Understanding the Concepts and Skills

16.11 One-way ANOVA is a procedure for comparing the means of several populations. It is the generalization of what procedure for comparing the means of two populations?

16.12 If we define $s = \sqrt{MSE}$, of which parameter is s an estimate?

16.13 Explain the reason for the word *variance* in the phrase *analysis of variance*.

16.14 For a one-way ANOVA test, suppose that, in reality, the null hypothesis is false. Does that mean that no two of the populations have the same mean? If not, what does it mean?

Regarding one-way ANOVA, fill in the blanks in each of Exercises 16.15–16.17.

16.15 A measure of variation among the sample means is called the _____. The mathematical abbreviation for it is _____.

16.16 A measure of variation within the samples is called the _____. The mathematical abbreviation for it is _____.

16.17 To compare the variation among the sample means to the variation within the samples, we use the ratio of *MSTR* to _____. This ratio is called the _____.

16.18 Explain the logic behind one-way ANOVA.

16.19 What does the term *one-way* signify in the phrase *one-way ANOVA*?

16.20 Figure 16.6 shows side-by-side boxplots of independent samples from three normally distributed populations having equal standard deviations. Based on these boxplots, would you be inclined to reject the null hypothesis of equal population means? Why?

FIGURE 16.6

Side-by-side boxplots for Exercise 16.20

16.21 Figure 16.7 shows side-by-side boxplots of independent samples from three normally distributed populations having equal standard deviations. Based on these boxplots, would you be inclined to reject the null hypothesis of equal population means? Why?

16.22 Discuss two methods for checking the assumptions of normal populations and equal standard deviations for a one-way ANOVA.

16.23 In one-way ANOVA, what is the residual of an observation?

FIGURE 16.7

Side-by-side boxplots for Exercise 16.21

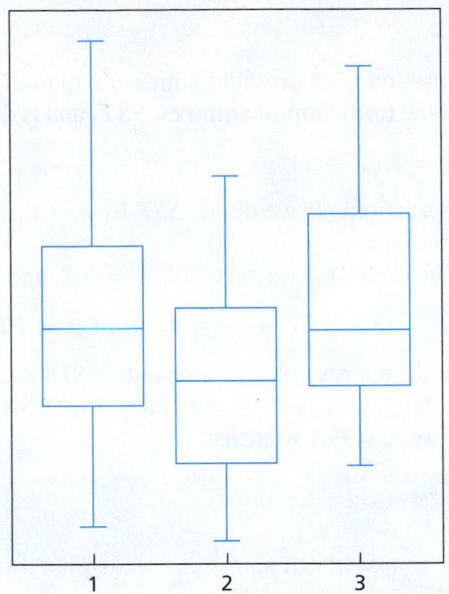

In Exercises 16.24–16.29, we have provided data from independent simple random samples from several populations. In each case, determine the following items.
a. SSTR   **b.** MSTR   **c.** SSE   **d.** MSE   **e.** F

16.24

Sample 1	Sample 2	Sample 3
1	10	4
9	4	16
	8	10
	6	
	2	

16.25

Sample 1	Sample 2	Sample 3
8	2	4
4	1	3
6	3	6
		3

16.26

Sample 1	Sample 2	Sample 3	Sample 4
6	9	4	8
3	5	4	4
3	7	2	6
	8	2	
	6	3	

16.27

Sample 1	Sample 2	Sample 3	Sample 4	Sample 5
7	5	6	3	7
4	9	7	7	9
5	4	5	7	11
4		4	4	
		8	4	

16.28

Sample 1	Sample 2	Sample 3	Sample 4	Sample 5
4	8	9	4	3
2	5	6	0	6
3	5	9	2	9

16.29

Sample 1	Sample 2	Sample 3	Sample 4
11	9	16	5
6	2	10	1
7	4	10	3

Extending the Concepts and Skills

16.30 Show that, for two populations, $MSE = s_p^2$, where s_p^2 is the pooled variance defined in Section 10.2 on page 447. Conclude that $\sqrt{MSE}$ is the pooled sample standard deviation, s_p.

16.31 Suppose that the variable under consideration is normally distributed on each of two populations and that the population standard deviations are equal. Further suppose that you want to perform a hypothesis test to decide whether the populations have different means, that is, whether $\mu_1 \neq \mu_2$. If independent simple random samples are used, identify two hypothesis-testing procedures that you can use to carry out the hypothesis test.

16.3 One-Way ANOVA: The Procedure

In this section, we present a step-by-step procedure for performing a one-way ANOVA to compare the means of several populations. To begin, we need to identify the distribution of the variable $F = MSTR/MSE$, introduced in Section 16.2.

KEY FACT 16.3

Distribution of the F-Statistic for One-Way ANOVA

Suppose that the variable under consideration is normally distributed on each of k populations and that the population standard deviations are equal. Then, for independent samples from the k populations, the variable

$$F = \frac{MSTR}{MSE}$$

has the F-distribution with df $= (k - 1, n - k)$ if the null hypothesis of equal population means is true. Here, n denotes the total number of observations.

Although we have now covered all the elements required to formulate a procedure for performing a one-way ANOVA, we still need to consider two additional concepts.

One-Way ANOVA Identity

First, we define another sum of squares—one that provides a measure of total variation among all the sample data. It is called the **total sum of squares, *SST*,** and is defined by

$$SST = \Sigma(x_i - \bar{x})^2,$$

where the sum extends over all n observations. If we divide *SST* by $n - 1$, we get the sample variance of all the observations.

For the energy consumption data in Table 16.3 on page 705, $\bar{x} = 9.8$, and therefore

$$SST = \Sigma(x_i - \bar{x})^2 = (13 - 9.8)^2 + (8 - 9.8)^2 + \cdots + (7 - 9.8)^2 = 195.2.$$

In Section 16.2, we found that, for the energy consumption data, $SSTR = 105.9$ and $SSE = 89.3$. Because $195.2 = 105.9 + 89.3$, we have $SST = SSTR + SSE$. This equation is always true and is called the **one-way ANOVA identity.**

KEY FACT 16.4

One-Way ANOVA Identity

The total sum of squares equals the treatment sum of squares plus the error sum of squares: $SST = SSTR + SSE$.

Note: The one-way ANOVA identity shows that the total variation among all the observations can be partitioned into two components. The partitioning of the total variation among all the observations into two or more components is fundamental not only in one-way ANOVA but also in all types of ANOVA.

We provide a graphical representation of the one-way ANOVA identity in Fig. 16.8.

FIGURE 16.8

Partitioning of the total sum of squares into the treatment sum of squares and the error sum of squares

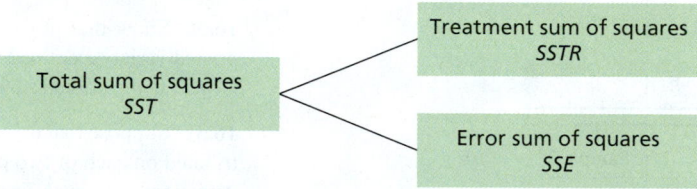

One-Way ANOVA Tables

To organize and summarize the quantities required for performing a one-way analysis of variance, we use a **one-way ANOVA table.** The general format of such a table is as shown in Table 16.6.

TABLE 16.6

ANOVA table format for a one-way analysis of variance

Source	df	SS	MS = SS/df	F-statistic
Treatment	$k - 1$	$SSTR$	$MSTR = \dfrac{SSTR}{k-1}$	$F = \dfrac{MSTR}{MSE}$
Error	$n - k$	SSE	$MSE = \dfrac{SSE}{n-k}$	
Total	$n - 1$	SST		

For the energy consumption data in Table 16.3, we have already computed all quantities appearing in the one-way ANOVA table. See Table 16.7.

TABLE 16.7

One-way ANOVA table for the energy consumption data

Source	df	SS	MS = SS/df	F-statistic
Treatment	3	105.9	35.3	6.32
Error	16	89.3	5.581	
Total	19	195.2		

Performing a One-Way ANOVA

To perform a one-way ANOVA, we need to determine the three sums of squares, *SST*, *SSTR*, and *SSE*. We can do so by using the defining formulas introduced earlier. Generally, however, when calculating by hand from the raw data, computing formulas are more accurate and easier to use. Both sets of formulas are presented next.

FORMULA 16.1

Sums of Squares in One-Way ANOVA

For a one-way ANOVA of *k* population means, the defining and computing formulas for the three sums of squares are as follows.

Sum of squares	Defining formula	Computing formula
Total, *SST*	$\Sigma(x_i - \bar{x})^2$	$\Sigma x_i^2 - (\Sigma x_i)^2/n$
Treatment, *SSTR*	$\Sigma n_j(\bar{x}_j - \bar{x})^2$	$\Sigma(T_j^2/n_j) - (\Sigma x_i)^2/n$
Error, *SSE*	$\Sigma(n_j - 1)s_j^2$	$SST - SSTR$

In this table, we used the notation

$$n = \text{total number of observations}$$
$$\bar{x} = \text{mean of all } n \text{ observations;}$$

and, for $j = 1, 2, \ldots, k,$

$$n_j = \text{size of sample from Population } j$$
$$\bar{x}_j = \text{mean of sample from Population } j$$
$$s_j^2 = \text{variance of sample from Population } j$$
$$T_j = \text{sum of sample data from Population } j.$$

Note that a summation involving a subscript *i* is over all *n* observations; one involving a subscript *j* is over the *k* populations.

Keep the following facts in mind when you use Formula 16.1.

- Only two of the three sums of squares need ever be calculated; the remaining one can always be found by using the one-way ANOVA identity.
- When using the computing formulas, the most efficient formula for calculating the sum of all *n* observations is $\Sigma x_i = \Sigma T_j$.

Procedure 16.1 on the following page gives a step-by-step method for conducting a **one-way ANOVA test** by using either the critical-value approach or the *P*-value approach. Because the null hypothesis is rejected only when the test statistic, *F*, is too large, a one-way ANOVA test is always right tailed.

■■■ **EXAMPLE 16.3**   **The One-Way ANOVA Test**

Energy Consumption Recall that independent simple random samples of households in the four U.S. regions yielded the data on last year's energy consumptions shown in Table 16.8. At the 5% significance level, do the data provide sufficient

TABLE 16.8

Last year's energy consumptions for samples of households in the four U.S. regions

Northeast	Midwest	South	West
13	15	5	8
8	10	11	10
11	16	9	6
12	11	5	5
11	13		7
	10		

■■■ **PROCEDURE 16.1** **One-Way ANOVA Test**

Purpose To perform a hypothesis test to compare k population means, $\mu_1, \mu_2, \ldots, \mu_k$

Assumptions
1. Simple random samples
2. Independent samples
3. Normal populations
4. Equal population standard deviations

Step 1 The null and alternative hypotheses are, respectively,

$$H_0: \mu_1 = \mu_2 = \cdots = \mu_k$$
$$H_a: \text{Not all the means are equal.}$$

Step 2 Decide on the significance level, α.

Step 3 Compute the value of the test statistic

$$F = \frac{MSTR}{MSE}$$

and denote that value F_0. To do so, construct a one-way ANOVA table:

Source	df	SS	MS = SS/df	F-statistic
Treatment	$k-1$	SSTR	$MSTR = \dfrac{SSTR}{k-1}$	$F = \dfrac{MSTR}{MSE}$
Error	$n-k$	SSE	$MSE = \dfrac{SSE}{n-k}$	
Total	$n-1$	SST		

CRITICAL-VALUE APPROACH	OR	P-VALUE APPROACH

Step 4 The critical value is F_α with $df = (k-1, n-k)$. Use Table VIII to find the critical value.

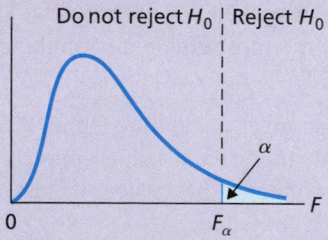

Step 5 If the value of the test statistic falls in the rejection region, reject H_0; otherwise, do not reject H_0.

Step 4 The F-statistic has $df = (k-1, n-k)$. Use Table VIII to estimate the P-value or obtain it exactly by using technology.

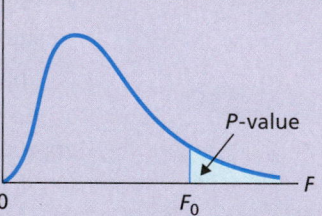

Step 5 If $P \le \alpha$, reject H_0; otherwise, do not reject H_0.

Step 6 Interpret the results of the hypothesis test.

evidence to conclude that a difference exists in last year's mean energy consumption by households among the four U.S. regions?

Solution First, we check the four conditions required for performing a one-way ANOVA test, as listed in Procedure 16.1.

• The samples are given as simple random samples and, therefore, Assumption 1 is satisfied.

- The samples are given as independent samples and, therefore, Assumption 2 is satisfied.
- Normal probability plots of the four samples, presented in Fig. 16.9, show no outliers and are roughly linear, indicating no gross violations of the normality assumption; thus we can consider Assumption 3 satisfied.
- The sample standard deviations of the four samples are 1.87, 2.59, 3.00, and 1.92, respectively. The ratio of the largest to the smallest standard deviation is $3.00/1.87 = 1.60$, which is less than 2. Thus, by the rule of 2, we can consider Assumption 4 satisfied.

FIGURE 16.9

Normal probability plots of the energy-consumption data: (a) Northeast, (b) Midwest, (c) South, (d) West

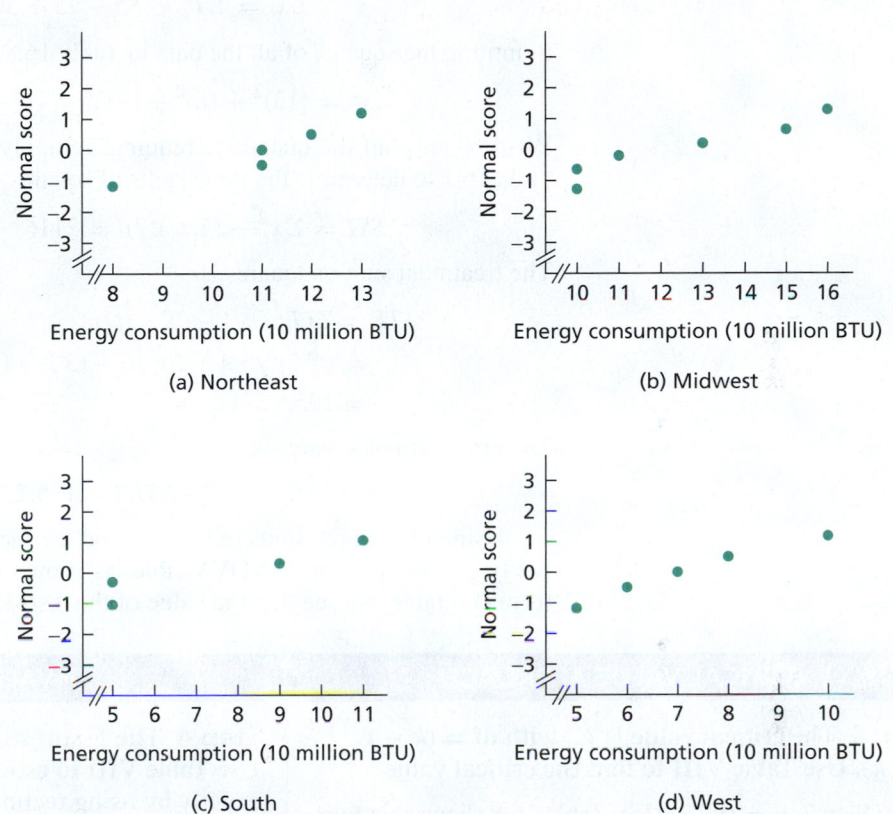

As it is reasonable to presume that the four assumptions for performing a one-way ANOVA test are satisfied, we now apply Procedure 16.1 to carry out the required hypothesis test.

Step 1 State the null and alternative hypotheses.

Let μ_1, μ_2, μ_3, and μ_4 denote last year's mean energy consumptions for households in the Northeast, Midwest, South, and West, respectively. Then the null and alternative hypotheses are, respectively,

H_0: $\mu_1 = \mu_2 = \mu_3 = \mu_4$ (mean energy consumptions are equal)

H_a: Not all the means are equal.

Step 2 Decide on the significance level, α.

We are to perform the test at the 5% significance level; so, $\alpha = 0.05$.

Step 3 Compute the value of the test statistic

$$F = \frac{MSTR}{MSE}.$$

To begin, we need to determine the three sums of squares: SST, SSTR, and SSE. Although we obtained these sums earlier by using the defining formulas, we find them again to illustrate use of the computing formulas. Referring to Table 16.8 and the notation in Formula 16.1 (both on page 709), we find that

$$k = 4$$

$$n_1 = 5 \quad n_2 = 6 \quad n_3 = 4 \quad n_4 = 5$$

$$T_1 = 55 \quad T_2 = 75 \quad T_3 = 30 \quad T_4 = 36$$

and

$$n = \Sigma n_j = 5 + 6 + 4 + 5 = 20$$

$$\Sigma x_i = \Sigma T_j = 55 + 75 + 30 + 36 = 196.$$

Summing the squares of all the data in Table 16.8, we get

$$\Sigma x_i^2 = (13)^2 + (8)^2 + (11)^2 + \cdots + (5)^2 + (7)^2 = 2116.$$

We now have all the quantities required to apply the computing formulas in Formula 16.1 to determine the three sums of squares. The total sum of squares is

$$SST = \Sigma x_i^2 - (\Sigma x_i)^2/n = 2116 - (196)^2/20 = 195.2.$$

The treatment sum of squares is

$$\begin{aligned} SSTR &= \Sigma(T_j^2/n_j) - (\Sigma x_i)^2/n \\ &= \big((55)^2/5 + (75)^2/6 + (30)^2/4 + (36)^2/5\big) - (196)^2/20 \\ &= 105.9. \end{aligned}$$

The error sum of squares is

$$SSE = SST - SSTR = 195.2 - 105.9 = 89.3.$$

Using these three sums of squares and the fact that $k = 4$ and $n = 20$, we now easily get the one-way ANOVA table, as shown in Table 16.7 on page 708. And, from that table, we see that the value of the test statistic is $F = 6.32$.

CRITICAL-VALUE APPROACH	OR	P-VALUE APPROACH

Step 4 The critical value is F_α with df = $(k - 1, n - k)$. Use Table VIII to find the critical value.

From Step 2, $\alpha = 0.05$. Also, Table 16.8 shows that four populations are under consideration, or $k = 4$, and that the number of observations total 20, or $n = 20$. Hence, df = $(k - 1, n - k) = (4 - 1, 20 - 4) = (3, 16)$. From Table VIII, the critical value is $F_{0.05} = 3.24$, as shown in Fig. 16.10A.

FIGURE 16.10A

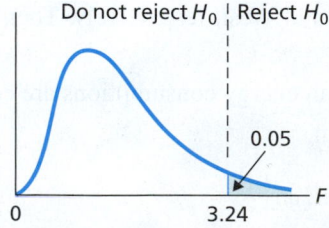

Step 4 The F-statistic has df = $(k - 1, n - k)$. Use Table VIII to estimate the P-value or obtain it exactly by using technology.

From Step 3, the value of the test statistic is $F = 6.32$. Because the test is right tailed, the P-value is the probability of observing a value of F of 6.32 or greater if the null hypothesis is true. That probability equals the shaded area in Fig. 16.10B.

FIGURE 16.10B

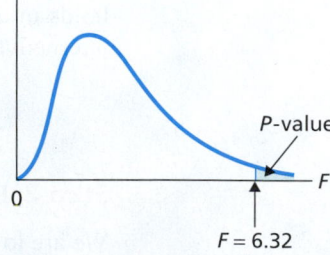

From Table 16.8, four populations are under consideration, or $k = 4$, and the number of observations total 20, or $n = 20$. Thus, we have df = $(k - 1, n - k) = (4 - 1, 20 - 4) = (3, 16)$. Referring to Fig. 16.10B and to Table VIII with df = $(3, 16)$, we find $P < 0.005$. (Using technology, we get $P = 0.00495$.)

CRITICAL-VALUE APPROACH	OR	*P*-VALUE APPROACH

Step 5 If the value of the test statistic falls in the rejection region, reject H_0; otherwise, do not reject H_0.

From Step 3, the value of the test statistic is $F = 6.32$, which, as Fig. 16.10A shows, falls in the rejection region. Thus we reject H_0. The test results are statistically significant at the 5% level.

Step 5 If $P \leq \alpha$, reject H_0; otherwise, do not reject H_0.

From Step 4, $P < 0.005$. Because the P-value is less than the specified significance level of 0.05, we reject H_0. The test results are statistically significant at the 5% level and (see Table 9.8 on page 386) provide very strong evidence against the null hypothesis.

Step 6 Interpret the results of the hypothesis test.

Interpretation At the 5% significance level, the data provide sufficient evidence to conclude that a difference exists in last year's mean energy consumption by households among the four U.S. regions. Evidently, at least two of the regions have different mean energy consumptions.

StatCrunch

Report 16.1   Exercise 16.49 on page 716

Using Summary Statistics in One-Way ANOVA

Journal articles and other sources frequently only provide summary statistics of data. To perform a one-way ANOVA with summary statistics, we need the sample sizes, sample means, and sample standard deviations.

We can determine the mean of all the observations from the individual sample means by using the formula

$$\bar{x} = \frac{n_1 \bar{x}_1 + n_2 \bar{x}_2 + \cdots + n_k \bar{x}_k}{n_1 + n_2 + \cdots + n_k}.$$

Note that, if all the sample sizes are equal, then the mean of all the observations is just the mean of the sample means.

Using the summary statistics and the preceding formula, we can apply the defining formulas for *SSTR* and *SSE* in Formula 16.1 on page 709 and the one-way ANOVA identity to obtain the three sums of squares and, subsequently, the value of the F-statistic. Exercises 16.60–16.63 provide practice for performing a one-way ANOVA given only the required summary statistics.

Other Types of ANOVA

We can consider one-way ANOVA to be a method for comparing the means of populations classified according to one factor. Put another way, it is a method for analyzing the effect of one factor on the mean of the variable under consideration, called the **response variable.**

For instance, in Example 16.3, we compared last year's mean energy consumption by households among the four U.S. regions (Northeast, Midwest, South, and West). Here, the factor is "region," and the response variable is "energy consumption." One-way ANOVA permits us to analyze the effect of region on mean energy consumption.

Other ANOVA procedures provide methods for comparing the means of populations classified according to two or more factors. Put another way, these are methods for simultaneously analyzing the effect of two or more factors on the mean of a response variable.

For example, suppose that you want to consider the effect of "region" and "home type" (the two factors) on energy consumption (the response variable). Two-way ANOVA permits you to determine simultaneously whether region affects mean energy consumption, whether home type affects mean energy consumption, and whether region and home type interact in their effect on mean energy consumption (e.g., whether the effect of home type on mean energy consumption depends on region).

Two-way ANOVA and other ANOVA procedures, such as randomized block ANOVA, are treated in detail in the module *Design of Experiments and Analysis of Variance* in the Regression-ANOVA Modules section on the WeissStats site.

THE TECHNOLOGY CENTER

Most statistical technologies have programs that automatically perform a one-way analysis of variance. In this subsection, we present output and step-by-step instructions for such programs.

EXAMPLE 16.4   Using Technology to Conduct a One-Way ANOVA Test

Energy Consumption Table 16.8 on page 709 shows last year's energy consumptions for independent random samples of households in the four U.S. regions. Use Minitab, Excel, or the TI-83/84 Plus to decide, at the 5% significance level, whether the data provide sufficient evidence to conclude that a difference exists in last year's mean energy consumption by households among the four U.S. regions.

Solution Let μ_1, μ_2, μ_3, and μ_4 denote last year's mean energy consumptions for households in the Northeast, Midwest, South, and West, respectively. We want to perform the hypothesis test

$$H_0: \mu_1 = \mu_2 = \mu_3 = \mu_4 \text{ (mean energy consumptions are equal)}$$

$$H_a: \text{Not all the means are equal}$$

at the 5% significance level.

We applied the one-way ANOVA programs to the data, resulting in Output 16.1. Steps for generating that output are presented in Instructions 16.1 on the next page. *Note to Excel users:* For brevity, we have presented only the essential portions of the actual output.

OUTPUT 16.1   One-way ANOVA test on the energy consumption data

MINITAB

One-way ANOVA: ENERGY versus REGION

Analysis of Variance

Source	DF	Adj SS	Adj MS	F-Value	P-Value
REGION	3	105.90	35.300	6.32	0.005
Error	16	89.30	5.581		
Total	19	195.20			

EXCEL

Analysis of variance:

Source	DF	Sum of squares	Mean squares	F	Pr > F
Model	3	105.9000	35.3000	6.3247	0.0049
Error	16	89.3000	5.5813		
Corrected Total	19	195.2000			

TI-83/84 PLUS

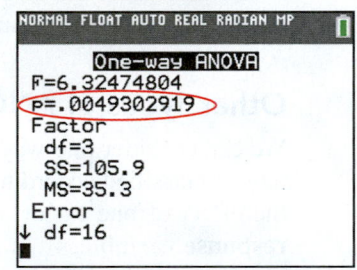

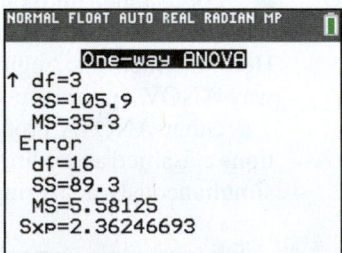

As shown in Output 16.1, the *P*-value for the hypothesis test is about 0.005. Because the *P*-value is less than the specified significance level of 0.05, we reject H_0. At the 5% significance level, the data provide sufficient evidence to conclude that last year's mean energy consumptions for households in the four U.S. regions are not all the same.

INSTRUCTIONS 16.1 Steps for generating Output 16.1

MINITAB

1. Store all 20 energy consumptions from Table 16.8 in a column named ENERGY
2. Store the regions corresponding to the energy consumptions in a column named REGION
3. Choose **Stat ➤ ANOVA ➤ One-Way...**
4. Press the F3 key to reset the dialog box
5. Specify ENERGY in the **Response** text box
6. Specify REGION in the **Factor** text box
7. Click the **Graphs...** button, uncheck the **Interval plot** check box, and click **OK**
8. Click the **Results...** button, uncheck all but the **Analysis of variance** check box, and click **OK** twice

EXCEL

1. Store all 20 energy consumptions from Table 16.8 in a column named ENERGY
2. Store the regions corresponding to the energy consumptions in a column named REGION
3. Choose **XLSTAT ➤ Modeling data ➤ ANOVA**
4. Click the reset button in the lower left corner of the dialog box
5. In the **Y / Dependent variables** list, click in the **Quantitative** selection box and then select the column of the worksheet that contains the ENERGY data

6. In the **X / Explanatory variables** list, click in the **Qualitative** selection box and then select the column of the worksheet that contains the REGION data
7. Click the **Outputs** tab and then uncheck all the check boxes except for the **Analysis of variance** check box
8. Click the **Charts** tab and then uncheck the **Regression charts** and **Means charts** check boxes
9. Click **OK**
10. Click the **Continue** button in the **XLSTAT – Selections** dialog box

TI-83/84 PLUS

1. Store the four samples from Table 16.8 in lists named NE, MW, SO, and WE
2. Press **STAT**, arrow over to **TESTS**, arrow down to **ANOVA(**, and press **ENTER**
3. Press **2nd ➤ LIST**, arrow down to NE and press **ENTER**
4. Press **, ➤ 2nd ➤ LIST**, arrow down to MW, and press **ENTER**
5. Press **, ➤ 2nd ➤ LIST**, arrow down to SO, and press **ENTER**
6. Press **, ➤ 2nd ➤ LIST**, arrow down to WE, and press **ENTER**
7. Press **)** and then **ENTER**

Exercises 16.3

Understanding the Concepts and Skills

16.32 Suppose that a one-way ANOVA is being performed to compare the means of three populations and that the sample sizes are 10, 12, and 15. Determine the degrees of freedom for the F-statistic.

16.33 We stated earlier that a one-way ANOVA test is always right tailed because the null hypothesis is rejected only when the test statistic, F, is too large. Why is the null hypothesis rejected only when F is too large?

16.34 Following are the notations for the three sums of squares. State the name of each sum of squares and the source of variation each sum of squares represents.
a. *SSE* b. *SSTR* c. *SST*

16.35 State the one-way ANOVA identity, and interpret its meaning with regard to partitioning the total variation in the data.

16.36 True or false: If you know any two of the three sums of squares, *SST*, *SSTR*, and *SSE*, you can determine the remaining one. Explain your answer.

16.37 In each part, specify what type of analysis you might use.
a. To study the effect of one factor on the mean of a response variable
b. To study the effect of two factors on the mean of a response variable

In Exercises 16.38–16.41, fill in the missing entries in the partially completed one-way ANOVA tables.

16.38

Source	df	SS	MS = SS/df	F-statistic
Treatment	2		21.652	
Error		84.400		
Total	14			

16.39

Source	df	SS	MS = SS/df	F-statistic
Treatment		2.124	0.708	0.75
Error	20			
Total				

16.40

Source	df	SS	MS = SS/df	F-statistic
Treatment	4			
Error	20		6.76	
Total		173.04		

16.41

Source	df	SS	MS = SS/df	F-statistic
Treatment			1.4	
Error	12		0.9	
Total	14			

In Exercises 16.42–16.47, we provide data from independent simple random samples from several populations. In each case,

a. *compute SST, SSTR, and SSE by using the computing formulas given in Formula 16.1 on page 709.*
b. *compare your results in part (a) for SSTR and SSE with those you obtained in Exercises 16.24–16.29, where you employed the defining formulas.*
c. *construct a one-way ANOVA table.*
d. *decide, at the 5% significance level, whether the data provide sufficient evidence to conclude that the means of the populations from which the samples were drawn are not all the same.*

16.42

Sample 1	Sample 2	Sample 3
1	10	4
9	4	16
	8	10
	6	
	2	

16.43

Sample 1	Sample 2	Sample 3
8	2	4
4	1	3
6	3	6
		3

16.44

Sample 1	Sample 2	Sample 3	Sample 4
6	9	4	8
3	5	4	4
3	7	2	6
	8	2	
	6	3	

16.45

Sample 1	Sample 2	Sample 3	Sample 4	Sample 5
7	5	6	3	7
4	9	7	7	9
5	4	5	7	11
4		4	4	
		8	4	

16.46

Sample 1	Sample 2	Sample 3	Sample 4	Sample 5
4	8	9	4	3
2	5	6	0	6
3	5	9	2	9

16.47

Sample 1	Sample 2	Sample 3	Sample 4
11	9	16	5
6	2	10	1
7	4	10	3

Applying the Concepts and Skills

In Exercises 16.48–16.53, apply Procedure 16.1 on page 710 to perform a one-way ANOVA test by using either the critical-value approach or the P-value approach.

16.48 Movie Guide. Movie fans use the annual *Leonard Maltin Movie Guide* for facts, cast members, and reviews of more than 21,000 films. The movies are rated from 4 stars (4*), indicating a very good movie, to 1 star (1*), which Leonard Maltin refers to as a BOMB. The following table gives the running times, in minutes, of a random sample of films listed in one year's guide.

1* or 1.5*	2* or 2.5*	3* or 3.5*	4*
75	97	101	101
95	70	89	135
84	105	97	93
86	119	103	117
58	87	86	126
85	95	100	119

At the 1% significance level, do the data provide sufficient evidence to conclude that a difference exists in mean running times among films in the four rating groups? (*Note:* $T_1 = 483$, $T_2 = 573$, $T_3 = 576$, $T_4 = 691$, and $\Sigma x_i^2 = 232{,}117$.)

16.49 Copepod Cuisine. Copepods are tiny crustaceans that are an essential link in the estuarine food web. Marine scientists G. Weiss et al. at the Chesapeake Biological Laboratory in Maryland designed an experiment to determine whether dietary lipid (fat) content is important in the population growth of a Chesapeake Bay copepod. Their findings were published as the paper "Development and Lipid Composition of the Harpacticoid Copepod Nitocra Spinipes Reared on Different Diets" (*Marine Ecology Progress Series*, Vol. 132, pp. 57–61). Independent random samples of copepods were placed in containers containing lipid-rich diatoms, bacteria, or leafy macroalgae. There were 12 containers total with four replicates per diet. Five gravid (egg-bearing) females were placed in each container. After 14 days, the number of copepods in each container were as follows.

Diatoms	Bacteria	Macroalgae
426	303	277
467	301	324
438	293	302
497	328	272

At the 5% significance level, do the data provide sufficient evidence to conclude that a difference exists in mean number of copepods among the three different diets? (*Note:* $T_1 = 1828$, $T_2 = 1225$, $T_3 = 1175$, and $\Sigma x_i^2 = 1{,}561{,}154$.)

16.50 In Section 16.2, we considered two hypothetical examples to explain the logic behind one-way ANOVA. Now, you are to further examine those examples.
a. Refer to Table 16.1 on page 702. Perform a one-way ANOVA on the data and compare your conclusion to that stated in the corresponding "What Does it Mean?" box. Use $\alpha = 0.05$.
b. Repeat part (a) for the data in Table 16.2 on page 702.

16.51 Staph Infections. In the article "Using EDE, ANOVA and Regression to Optimize Some Microbiology Data" (*Journal of Statistics Education*, Vol. 12, No. 2, online), N. Binnie analyzed bacteria-culture data collected by G. Cooper at the Auckland University of Technology. Five strains of cultured *Staphylococcus aureus*—bacteria that cause staph infections—were observed for 24 hours at 27°C. The following table reports bacteria counts, in millions, for different cases from each of the five strains.

Strain A	Strain B	Strain C	Strain D	Strain E
9	3	10	14	33
27	32	47	18	43
22	37	50	17	28
30	45	52	29	59
16	12	26	20	31

At the 5% significance level, do the data provide sufficient evidence to conclude that a difference exists in mean bacteria counts among the five strains of *Staphylococcus aureus*? (*Note:* $T_1 = 104$, $T_2 = 129$, $T_3 = 185$, $T_4 = 98$, $T_5 = 194$, and $\Sigma x_i^2 = 25{,}424$.)

16.52 Permeation Sampling. Permeation sampling is a method of sampling air in buildings for pollutants. It can be used over a long period of time and is not affected by humidity, air currents, or temperature. In the paper "Calibration of Permeation Passive Samplers With Silicone Membranes Based on Physicochemical Properties of the Analytes" (*Analytical Chemistry*, Vol. 75, No. 13, pp. 3182–3192), B. Zabiegata et al. obtained calibration constants experimentally for samples of compounds in each of four compound groups. The following data summarize their results.

Esters	Alcohols	Aliphatic hydrocarbons	Aromatic hydrocarbons
0.185	0.185	0.230	0.166
0.155	0.160	0.184	0.144
0.131	0.142	0.160	0.117
0.103	0.122	0.132	0.072
0.064	0.117	0.100	
	0.115	0.064	
	0.110		
	0.095		
	0.085		
	0.075		

At the 5% significance level, do the data provide sufficient evidence to conclude that a difference exists in mean calibration constant among the four compound groups? (*Note:* $T_1 = 0.638$, $T_2 = 1.206$, $T_3 = 0.870$, $T_4 = 0.499$, and $\Sigma x_i^2 = 0.456919$.)

16.53 Laptop Battery Life. *Consumer Reports* publishes reviews and comparisons of products based on results from its laboratory. Data from their website gave the following table for battery lives, in hours, for samples of laptops made by four different computer companies. The four brands are Apple, Dell, Samsung, and Toshiba, but we have not used the names and have permuted the order.

Brand A	Brand B	Brand C	Brand D
7.25	12.00	11.50	8.00
9.50	7.25	9.50	7.50
6.25	11.75	10.00	5.50
8.00	15.00	5.00	4.75
8.25	8.50	9.50	6.25
10.25	8.75	5.25	6.00
6.25	10.00	9.75	4.75
5.75	10.25	8.25	4.50
		5.25	
		11.50	

At the 5% significance level, do the data provide sufficient evidence to conclude that a difference exists in mean battery life among the four

brands? (*Note:* $T_1 = 61.50$, $T_2 = 83.50$, $T_3 = 85.50$, $T_4 = 47.25$, and $\Sigma x_i^2 = 2483.8125$.)

In Exercises 16.54–16.59, use the technology of your choice to
a. *conduct a one-way ANOVA test on the data.*
b. *interpret your results from part (a).*
c. *decide whether presuming that the assumptions of normal populations and equal population standard deviations are met is reasonable.*

16.54 Empty Stomachs. In the publication "How Often Do Fishes 'Run on Empty'?" (*Ecology*, Vol. 83, No 8, pp. 2145–2151), D. Arrington et al. examined almost 37,000 fish of 254 species from the waters of Africa, South and Central America, and North America to determine the percentage of fish with empty stomachs. The fish were classified as piscivores (fish-eating), invertivores (invertibrate-eating), omnivores (anything-eating) and algivores/detritivores (eating algae and other organic matter). For those fish in African waters, the data on the WeissStats site give the proportions of each species of fish with empty stomachs. At the 1% significance level, do the data provide sufficient evidence to conclude that a difference exists in the mean percentages of fish with empty stomachs among the four different types of feeders?

16.55 Monthly Rents. The U.S. Census Bureau collects data on monthly rents of newly completed apartments and publishes the results in *Current Housing Reports*. Independent random samples of newly completed apartments in the four U.S. regions yielded the data on monthly rents, in dollars, given on the WeissStats site. At the 5% significance level, do the data provide sufficient evidence to conclude that a difference exists in mean monthly rents among newly completed apartments in the four U.S. regions?

16.56 Ground Water. The U.S. Geological Survey, in cooperation with the Florida Department of Environmental Protection, investigated the effects of waste disposal practices on ground water quality at five poultry farms in north-central Florida. At one site, they drilled four monitoring wells, numbered 1, 2, 3, and 4. Over a period of 9 months, water samples were collected from the last three wells and analyzed for a variety of chemicals, including potassium, chlorides, nitrates, and phosphorus. The concentrations, in milligrams per liter, are provided on the WeissStats site. For each of the four chemicals, decide whether the data provide sufficient evidence to conclude that a difference exists in mean concentration among the three wells. Use $\alpha = 0.01$. [SOURCE: USGS Water Resources Investigations Report 95-4064, *Effects of Waste-Disposal Practices on Ground-Water Quality at Five Poultry (Broiler) Farms in North-Central Florida*, H. Hatzell, U.S. Geological Survey]

16.57 Rock Sparrows. Rock Sparrows breeding in northern Italy are the subject of a long-term ecology and conservation study due to their wide variety of breeding patterns. Both males and females have a yellow patch on their breasts that is thought to play a significant role in their sexual behavior. A. Pilastro et al. conducted an experiment in which they increased or reduced the size of a female's breast patch by dying feathers at the edge of a patch and then observed several characteristics of the behavior of the male. Their results were published in the paper "Male Rock Sparrows Adjust Their Breeding Strategy According to Female Ornamentation: Parental or Mating Investment?" (*Animal Behaviour*, Vol. 66, Issue 2, pp. 265–271). Eight mating pairs were observed in each of three groups: a reduced-patch-size group, a control group, and an enlarged-patch-size group. The data on the WeissStats site, based on the results reported by the researchers, give the number of minutes per hour that males sang in the vicinity of the nest after the patch size manipulation was done on the females.

At the 1% significance level, do the data provide sufficient evidence to conclude that a difference exists in the mean singing rates among male Rock Sparrows exposed to the three types of breast treatments?

16.58 Artificial Teeth: Wear. In a study by J. Zeng et al., three materials for making artificial teeth—Endura, Duradent, and Duracross—were tested for wear. Their results were published as the paper "In Vitro Wear Resistance of Three Types of Composite Resin Denture Teeth" (*Journal of Prosthetic Dentistry*, Vol. 94, Issue 5, pp. 453–457). Using a machine that simulated grinding by two right first molars at 60 strokes per minute for a total of 50,000 strokes, the researchers measured the volume of material worn away, in cubic millimeters. Six pairs of teeth were tested for each material. The data on the WeissStats site are based on the results obtained by the researchers. At the 5% significance level, do the data provide sufficient evidence to conclude that there is a difference in mean wear among the three materials?

16.59 Artificial Teeth: Hardness. In a study by J. Zeng et al., three materials for making artificial teeth—Endura, Duradent, and Duracross—were tested for hardness. Their results were published as the paper "In Vitro Wear Resistance of Three Types of Composite Resin Denture Teeth" (*Journal of Prosthetic Dentistry*, Vol. 94, Issue 5, pp. 453–457). The Vickers microhardness (VHN) of the occlusal surfaces was measured with a load of 50 grams and a loading time of 30 seconds. Six pairs of teeth were tested for each material. The data on the WeissStats site are based on the results obtained by the researchers. At the 5% significance level, do the data provide sufficient evidence to conclude that there is a difference in mean hardness among the three materials?

In Exercises 16.60–16.63, refer to the discussion of using summary statistics in one-way ANOVA on page 713. Note: We have provided values of F_α not given in Table VIII.

16.60 Breast Milk and IQ. Considerable controversy exists over whether long-term neurodevelopment is affected by nutritional factors in early life. A. Lucas and R. Morley summarized their findings on that question for preterm babies in the publication "Breast Milk and Subsequent Intelligence Quotient in Children Born Preterm" (*The Lancet*, Vol. 339, Issue 8788, pp. 261–264). The researchers analyzed IQ data on children at age $7\frac{1}{2}$–8 years. The mothers of the children in the study had chosen whether to provide their infants with breast milk within 72 hours of delivery. The researchers used the following designations. Group I: mothers declined to provide breast milk; Group IIa: mothers had chosen but were unable to provide breast milk; and Group IIb: mothers had chosen and were able to provide breast milk. Here are the summary statistics on IQ.

Group	n_j	$\bar{x}_j$	s_j
I	90	92.8	15.2
IIa	17	94.8	19.0
IIb	193	103.7	15.3

At the 1% significance level, do the data provide sufficient evidence to conclude that a difference exists in mean IQ at age $7\frac{1}{2}$–8 years for preterm children among the three groups? *Note:* For the degrees of freedom in this exercise:

α	0.10	0.05	0.025	0.01	0.005
F_α	2.32	3.03	3.74	4.68	5.39

16.61 Denosumab and Osteoporosis. A clinical study was conducted to see whether an antibody called denosumab is effective in treatment of osteoporosis of postmenopausal women, as reported in the article "Denosumab in Postmenopausal Women with Low Bone Mineral Density (*New England Journal of Medicine*, Vol. 354, No. 8, pp. 821–831) by M. McClung et al. Postmenopausal women with osteoporosis were randomly assigned into groups that received either a placebo, or a six-month regimen of Denosumab at doses of 14 mg, 60 mg, 100 mg, or 210 mg. The following table provides summary statistics for the body-mass indexes (BMI) of the women in each treatment group.

Treatment	n_j	$\bar{x}_j$	s_j
Placebo	46	25.9	4.3
14 mg	54	25.8	5.3
60 mg	47	27.5	5.8
100 mg	42	26.0	4.6
210 mg	47	25.9	4.3

At the 10% significance level, do the data provide sufficient evidence to conclude that a difference exists in mean BMI for women in the five different treatment groups? *Note:* For the degrees of freedom in this exercise:

α	0.10	0.05	0.025	0.01	0.005
F_α	1.97	2.41	2.84	3.40	3.82

16.62 Minke Whales. Entanglement of marine mammals in fishing gear is a global issue and a significant threat to minke whales, in particular. In the article "Fishing Gears Involved in Entanglements of Minke Whales (*Balaenoptera acutorostrata*) in the East Sea of Korea" (*Marine Mammal Science*, Vol. 26, No. 2, pp. 282–295), K. Song et al. studied the body lengths of minke whales entangled near Korea. The following table provides summary statistics for the body lengths of minke whales entangled at different ocean depths. Both variables, body length and ocean depth, are in meters.

Depth	n_j	$\bar{x}_j$	s_j
0–49	39	4.49	0.80
50–99	20	5.06	0.79
100–149	28	5.75	1.20
150–199	14	5.99	0.97

At the 1% significance level, do the data provide sufficient evidence to conclude that a difference exists in mean body length among minke whales entangled at the four different depths? *Note:* For the degrees of freedom in this exercise:

α	0.10	0.05	0.025	0.01	0.005
F_α	2.14	2.70	3.25	3.99	4.55

16.63 Starting Salaries. The National Association of Colleges and Employers (NACE) conducts surveys on salary offers to college graduates by field and degree. Results are published in *Salary Survey*. The following table provides summary statistics for starting salaries, in thousands of dollars, to samples of bachelor's-degree graduates in six fields.

Field	n_j	$\bar{x}_j$	s_j
Business	46	55.1	5.6
Communications	11	44.6	4.7
Computer Science	30	59.1	4.0
Education	11	40.6	5.0
Engineering	44	62.6	5.7
Math & Sciences	18	43.0	4.8

At the 1% significance level, do the data provide sufficient evidence to conclude that a difference exists in mean starting salaries among bachelor's-degree candidates in the six fields? *Note:* For the degrees of freedom in this exercise:

α	0.10	0.05	0.025	0.01	0.005
F_α	1.89	2.27	2.65	3.14	3.50

Working with Large Data Sets

In Exercises 16.64–16.72, use the technology of your choice to do the following tasks.
a. *Obtain individual normal probability plots and the standard deviations of the samples.*
b. *Perform a residual analysis.*
c. *Use your results from parts (a) and (b) to decide whether conducting a one-way ANOVA test on the data is reasonable. If so, also do parts (d) and (e).*
d. *Use a one-way ANOVA test to decide, at the 5% significance level, whether the data provide sufficient evidence to conclude that a difference exists among the means of the populations from which the samples were taken.*
e. *Interpret your results from part (d).*

16.64 Daily TV Viewing Time. Nielsen Media Research collects information on daily TV viewing time, in hours, and publishes its findings in *Time Spent Viewing*. The WeissStats site provides data on daily viewing times of independent simple random samples of men, women, teens, and children.

16.65 Fish of Lake Laengelmaevesi. An article by J. Puranen of the Department of Statistics, University of Helsinki, discussed a classic study on several variables of seven different species of fish caught in Lake Laengelmaevesi, Finland. On the WeissStats site, we present the data on weight (in grams) and length (in centimeters) from the nose to the beginning of the tail for four of the seven species. Perform the required parts for both the weight and length data.

16.66 Popular Diets. In the article "Comparison of the Atkins, Ornish, Weight Watchers, and Zone Diets for Weight Loss and Heart Disease Risk Reduction" (*Journal of the American Medical Association*, Vol. 293, No. 1, pp. 43–53), M. Dansinger et al. conducted a randomized trial to assess the effectiveness of four popular diets for weight loss. Overweight adults with average body mass index of 35 and ages 22–72 years participated in the randomized trial for 1 year. The weight losses, in kilograms, based on the results of the experiment are given on the WeissStats site. Negative losses are gains. WW = Weight Watchers.

16.67 Cuckoo Care. Many species of cuckoos are brood parasites. The females lay their eggs in the nests of smaller bird species that then raise the young cuckoos at the expense of their own young. The question might be asked, "Do the cuckoos lay the same size eggs regardless of the size of the bird whose nest they use?" Data on

the lengths, in millimeters, of cuckoo eggs found in the nests of six bird species—Meadow Pipit, Tree Pipit, Hedge Sparrow, Robin, Pied Wagtail, and Wren—are provided on the WeissStats site. These data were collected by the late O. Latter in 1902 and used by L. Tippett in his text *The Methods of Statistics* (New York: Wiley, 1952, p. 176).

16.68 Doing Time. The Federal Bureau of Prisons publishes data in *Statistical Report* on the times served by prisoners released from federal institutions for the first time. Independent simple random samples of released prisoners for five different offense categories yielded the data on time served, in months, shown on the WeissStats site.

16.69 Book Prices. The R. R. Bowker Company collects data on book prices and publishes its findings in *The Bowker Annual Library and Book Trade Almanac*. Independent simple random samples of hardcover books in law, science, medicine, and technology gave the data, in dollars, on the WeissStats site.

16.70 Magazine Ads. Advertising researchers F. Shuptrine and D. McVicker wanted to determine whether there were significant differences in the readability of magazine advertisements. Thirty magazines were classified based on their educational level—high, mid, or low—and then three magazines were randomly selected from each level. From each magazine, six advertisements were randomly chosen and examined for readability. In this particular case, readability was characterized by the numbers of words, sentences, and words of three syllables or more in each ad. The researchers published their findings in the paper "Readability Levels of Magazine Ads" (*Journal of Advertising Research*, Vol. 21, No. 5, pp. 45–51). The number of words of three syllables or more in each ad are provided on the WeissStats site.

16.71 Sickle Cell Disease. A study by E. Anionwu et al., published as the paper "Sickle Cell Disease in a British Urban Community" (*British Medical Journal*, Vol. 282, pp. 283–286), measured the steady-state hemoglobin levels of patients with three different types of sickle cell disease: HB SS, HB ST, and HB SC. The data are presented on the WeissStats site.

16.72 Prolonging Life. Vitamin C (ascorbate) boosts the human immune system and is effective in preventing a variety of illnesses. In a study by E. Cameron and L. Pauling, published as the paper "Supplemental Ascorbate in the Supportive Treatment of Cancer: Reevaluation of Prolongation of Survival Times in Terminal Human Cancer" (*Proceedings of the National Academy of Science USA*, Vol. 75, No. 9, pp. 4538–4542), patients in advanced stages of cancer were given a vitamin C supplement. Patients were grouped according to the organ affected by cancer: stomach, bronchus, colon, ovary, or breast. The study yielded the survival times, in days, given on the WeissStats site.

Extending the Concepts and Skills

16.73 On page 713, we discussed how to use summary statistics (sample sizes, sample means, and sample standard deviations) to conduct a one-way ANOVA.
a. Verify the formula presented there for obtaining the mean of all the observations, namely,

$$\bar{x} = \frac{n_1\bar{x}_1 + n_2\bar{x}_2 + \cdots + n_k\bar{x}_k}{n_1 + n_2 + \cdots + n_k}.$$

b. Show that, if all the sample sizes are equal, then the mean of all the observations is just the mean of the sample means.
c. Explain in detail how to obtain the value of the F-statistic from the summary statistics.

Confidence Intervals in One-Way ANOVA. Assume that the conditions for one-way ANOVA are satisfied, and let $s = \sqrt{MSE}$. Then we have the following confidence-interval formulas.

- A $(1 - \alpha)$-level confidence interval for any particular population mean, say, μ_i, has endpoints

$$\bar{x}_i \pm t_{\alpha/2} \cdot \frac{s}{\sqrt{n_i}}.$$

- A $(1 - \alpha)$-level confidence interval for the difference between any two particular population means, say, μ_i and μ_j, has endpoints

$$(\bar{x}_i - \bar{x}_j) \pm t_{\alpha/2} \cdot s\sqrt{(1/n_i) + (1/n_j)}.$$

In both formulas, $df = n - k$, where, as usual, k denotes the number of populations and n denotes the total number of observations. Apply these formulas in Exercise 16.74.

16.74 Monthly Rents. Refer to Exercise 16.55.
a. Find and interpret a 95% confidence interval for the mean monthly rent of newly completed apartments in the Midwest.
b. Find and interpret a 95% confidence interval for the difference between the mean monthly rents of newly completed apartments in the Northeast and South.
c. What assumptions are you making in solving parts (a) and (b)?

16.75 Monthly Rents. Refer to Exercise 16.74. Suppose that you have obtained a 95% confidence interval for each of the two differences, $\mu_1 - \mu_2$ and $\mu_1 - \mu_3$. Can you be 95% confident of both results simultaneously, that is, that both differences are contained in their corresponding confidence intervals? Explain your answer.

16.4 Multiple Comparisons*[†]

Suppose that you perform a one-way ANOVA and reject the null hypothesis. Then you can conclude that the means of the populations under consideration are not all the same. Once you make that decision, you may also want to know which means are different, which mean is largest, or, more generally, the relation among all the means. Methods for dealing with these problems are called **multiple comparisons.**

In this book, we discuss the **Tukey multiple-comparison method.** Other commonly used multiple-comparison methods are the Bonferroni method, the Fisher method, and the Scheffé method.

One approach for implementing multiple comparisons is to determine confidence intervals for the differences between all possible pairs of population means. Two means are declared different if the confidence interval for their difference does not contain 0. [Recall from Chapter 10 (see page 453) that if a confidence interval for the difference between two population means does not contain 0, then we can reject the null hypothesis that the two means are equal in favor of the alternative hypothesis that they are different.]

? What Does It Mean?

It is at the family confidence level that we can be confident in the truth of our conclusions when comparing all the population means simultaneously.

In multiple comparisons, we must distinguish between the *individual confidence level* and the *family confidence level*. The **individual confidence level** is the confidence we have that any particular confidence interval contains the difference between the corresponding population means. The **family confidence level** is the confidence we have that all the confidence intervals contain the differences between the corresponding population means.

The Studentized Range Distribution

The Tukey multiple-comparison method is based on the **studentized range distribution,** also known as the **q-distribution.** A variable has a q-distribution if its distribution has the shape of a special type of right-skewed curve, called a **q-curve.** There are infinitely many q-distributions (and q-curves); a particular one is identified by two parameters, which we denote κ (kappa) and ν (nu).

Percentages and probabilities for a variable having a q-distribution equal areas under its associated q-curve. To perform a Tukey multiple comparison, we need to know how to find the q-value having a specified area to its right. We use the symbol q_α to denote the q-value having area α to its right. Values of $q_{0.01}$ and $q_{0.05}$ are presented in Tables IX and X, respectively, in Appendix A.

[†]If you plan to study the module *Design of Experiments and Analysis of Variance* (Module C) in the Regression-ANOVA Modules section on the WeissStats site, you should cover this section.

■■■ **EXAMPLE 16.5** **Finding the *q*-Value Having a Specified Area to Its Right**

For the *q*-curve with parameters $\kappa = 4$ and $\nu = 16$, find $q_{0.05}$; that is, find the *q*-value having area 0.05 to its right, as shown in Fig. 16.11(a).

FIGURE 16.11

Finding the *q*-value having area 0.05 to its right

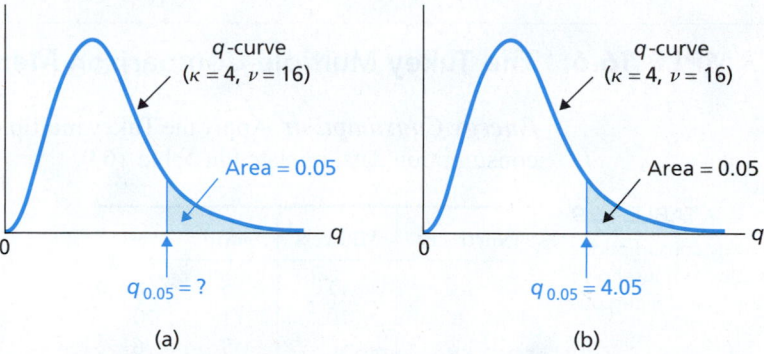

(a)   (b)

Exercise 16.83 on page 725

Solution To obtain the *q*-value in question, we use Table X with $\kappa = 4$ and $\nu = 16$. We first go down the outside columns to the row labeled "16." Then, going across that row to the column labeled "4," we reach 4.05. This number is the *q*-value having area 0.05 to its right, as shown in Fig. 16.11(b). In other words, for a *q*-curve with parameters $\kappa = 4$ and $\nu = 16$, $q_{0.05} = 4.05$. ■

The Tukey Multiple-Comparison Method

The formulas used in the Tukey multiple-comparison method for obtaining confidence intervals for the differences between means are similar to the pooled *t*-interval formula (Procedure 10.2 on page 452). The essential difference is that, in the Tukey multiple-comparison method, we consult a *q*-table instead of a *t*-table.

Procedure 16.2 provides a step-by-step method for performing a Tukey multiple comparison. Note that the assumptions for its use are the same as those for a one-way ANOVA test.

■■■ **PROCEDURE 16.2** **Tukey Multiple-Comparison Method**

Purpose To determine the relationship among k population means $\mu_1, \mu_2, \ldots, \mu_k$

Assumptions
1. Simple random samples
2. Independent samples
3. Normal populations
4. Equal population standard deviations

Step 1 Decide on the family confidence level, $1 - \alpha$.

Step 2 Find q_α for the *q*-curve with parameters $\kappa = k$ and $\nu = n - k$, where n is the total number of observations.

Step 3 Obtain the endpoints of the confidence interval for $\mu_i - \mu_j$:

$$(\bar{x}_i - \bar{x}_j) \pm \frac{q_\alpha}{\sqrt{2}} \cdot s\sqrt{(1/n_i) + (1/n_j)},$$

where $s = \sqrt{MSE}$. Do so for all possible pairs of means with $i < j$.

Step 4 Declare two population means different if the confidence interval for their difference does not contain 0; otherwise, do not declare the two population means different.

Step 5 Summarize the results in Step 4 by ranking the sample means from smallest to largest and by connecting with lines those whose population means were not declared different.

Step 6 Interpret the results of the multiple comparison.

Performing a Tukey Multiple Comparison

In Example 16.3 (pages 709–713), we used a one-way ANOVA test to conclude, at the 5% significance level, that at least two of the four U.S. regions have different mean household energy consumptions. The Tukey multiple-comparison method allows us to elaborate on this conclusion.

■■■■   **EXAMPLE 16.6**   The Tukey Multiple-Comparison Method

Energy Consumption Apply the Tukey multiple-comparison method to the energy consumption data, repeated in Table 16.9.

TABLE 16.9

Last year's energy consumptions for samples of households in the four U.S. regions

Northeast	Midwest	South	West
13	15	5	8
8	10	11	10
11	16	9	6
12	11	5	5
11	13		7
	10		

Solution In the solution to Example 16.3 (beginning on page 710), we showed that it is reasonable to presume that the four assumptions for performing a one-way ANOVA test are satisfied. Because the assumptions for conducting a Tukey multiple comparison are identical to those for performing a one-way ANOVA, we see that it is reasonable to apply Procedure 16.2.

Step 1 Decide on the family confidence level, $1 - \alpha$.

As we have done previously in this application, we use $\alpha = 0.05$, so the family confidence level is 0.95 (95%).

Step 2 Find q_α for the q-curve with parameters $\kappa = k$ and $v = n - k$, where n is the total number of observations.

From Table 16.9, $\kappa = k = 4$ and $v = n - k = 20 - 4 = 16$. In Table X, we find that $q_\alpha = q_{0.05} = 4.05$.

Step 3 Obtain the endpoints of the confidence interval for $\mu_i - \mu_j$:

$$(\bar{x}_i - \bar{x}_j) \pm \frac{q_\alpha}{\sqrt{2}} \cdot s\sqrt{(1/n_i) + (1/n_j)},$$

where $s = \sqrt{MSE}$. Do so for all possible pairs of means with $i < j$.

Table 16.10 gives the means and sizes for the sample data in Table 16.9.

TABLE 16.10

Sample means and sample sizes for the energy consumption data

Region	Northeast	Midwest	South	West
j	1	2	3	4
$\bar{x}_j$	11.0	12.5	7.5	7.2
n_j	5	6	4	5

From Step 2, $q_\alpha = 4.05$. Also, on page 706, we found that $MSE = 5.581$ for the energy consumption data. Now, we are ready to obtain the required confidence intervals. The endpoints of the confidence interval for $\mu_1 - \mu_2$ are

$$(11.0 - 12.5) \pm \frac{4.05}{\sqrt{2}} \cdot \sqrt{5.581}\sqrt{(1/5) + (1/6)},$$

or -5.60 to 2.60.

Likewise, the endpoints of the confidence interval for $\mu_1 - \mu_3$ are

$$(11.0 - 7.5) \pm \frac{4.05}{\sqrt{2}} \cdot \sqrt{5.581}\sqrt{(1/5) + (1/4)},$$

or -1.04 to 8.04.

In a similar way, we find the remaining confidence intervals. All six are displayed in Table 16.11.

TABLE 16.11

Simultaneous 95% confidence intervals for the differences between the energy consumption means. The number used to represent a region is shown parenthetically following the region.

	Northeast (1)	Midwest (2)	South (3)
Midwest (2)	(−5.60, 2.60)		
South (3)	(−1.04, 8.04)	(0.63, 9.37)	
West (4)	(−0.48, 8.08)	(1.20, 9.40)	(−4.24, 4.84)

Each entry in Table 16.11 is the confidence interval for the difference between the mean labeled by the column and the mean labeled by the row. For instance, the entry in the column labeled "Midwest (2)" and the row labeled "West (4)" is $(1.20, 9.40)$. So the confidence interval for the difference, $\mu_2 - \mu_4$, between last year's mean energy consumptions for households in the Midwest and West is from 1.20 to 9.40.

Step 4 Declare two population means different if the confidence interval for their difference does not contain 0; otherwise, do not declare the two population means different.

Referring to Table 16.11, we see that we can declare the means μ_2 and μ_3 different and the means μ_2 and μ_4 different; all other pairs of means are not declared different.

Step 5 Summarize the results in Step 4 by ranking the sample means from smallest to largest and by connecting with lines those whose population means were not declared different.

In light of Table 16.10, Step 4, and the numbering used to represent the U.S. regions (shown parenthetically), we obtain the following diagram.

West (4)	South (3)	Northeast (1)	Midwest (2)
7.2	7.5	11.0	12.5

Step 6 Interpret the results of the multiple comparison.

StatCrunch
Report 16.2

Exercise 16.95 on page 726

Interpretation Referring to the diagram in Step 5, we conclude that last year's mean energy consumption in the Midwest exceeds that in the West and South and that no other means can be declared different. All of this can be said with 95% confidence, the family confidence level. ◼

THE TECHNOLOGY CENTER

Some statistical technologies have programs that automatically perform a Tukey multiple comparison. In this subsection, we present output and step-by-step instructions for such programs. (*Note to TI-83/84 Plus users:* At the time of this writing, the TI-83/84 Plus does not have a built-in program for conducting a Tukey multiple comparison. However, a TI program, TUKEY, to help with the analysis is located in the TI Programs section on the WeissStats site.)

EXAMPLE 16.7   Using Technology to Conduct a Tukey Multiple Comparison

Energy Consumption Table 16.9 on page 722 shows last year's energy consumptions for independent simple random samples of households in the four U.S. regions. Apply Minitab or Excel to conduct a Tukey multiple comparison, using a 95% family confidence level.

Solution We applied the Tukey multiple-comparison programs to the data. Output 16.2 shows only the portion of the output essential to the Tukey multiple comparison. Steps for generating that output are presented in Instructions 16.2 on this and the next page.

OUTPUT 16.2   Tukey multiple comparison on the energy-consumption data

MINITAB

One-way ANOVA: ENERGY versus REGION

Tukey Pairwise Comparisons

Grouping Information Using the Tukey Method and 95% Confidence

REGION	N	Mean	Grouping	
Midwest	6	12.50	A	
Northeast	5	11.000	A	B
South	4	7.50		B
West	5	7.200		B

Means that do not share a letter are significantly different.

EXCEL

REGION / Tukey (HSD) / Analysis of the differences between the categories with a confidence interval of 95%:

Category	LS means	Groups	
Midwest	12.5000	A	
Northeast	11.0000	A	B
South	7.5000		B
West	7.2000		B

In Output 16.2, means that do not share a common letter are declared different. Thus, we see that last year's mean energy consumption in the Midwest exceeds that in the West and South and that no other means can be declared different. All of this can be said with 95% confidence, the family confidence level.

INSTRUCTIONS 16.2   Steps for generating Output 16.2

MINITAB

1 Store all 20 energy consumptions from Table 16.9 in a column named ENERGY
2 Store the regions corresponding to the energy consumptions in a column named REGION
3 Choose **Stat ➤ ANOVA ➤ One-Way...**
4 Press the F3 key to reset the dialog box
5 Specify ENERGY in the **Response** text box
6 Specify REGION in the **Factor** text box
7 Click the **Comparisons...** button

8 Type 5 (100 minus the family confidence level expressed as a percentage) in the **Error rate for comparisons** text box
9 In the **Comparison procedures assuming equal variances** check-box list, check the **Tukey** check box
10 In the **Results** list, uncheck the **Interval plot for differences of means** check box
11 Click **OK**
12 Click the **Graphs...** button, uncheck the **Interval plot** check box, and click **OK**
13 Click the **Results...** button, uncheck all the check boxes, and click **OK** twice

(continued)

EXCEL

1. Store all 20 energy consumptions from Table 16.9 in a column named ENERGY
2. Store the regions corresponding to the energy consumptions in a column named REGION
3. Choose **XLSTAT ➤ Modeling data ➤ ANOVA**
4. Click the reset button in the lower left corner of the dialog box
5. In the **Y / Dependent variables** list, click in the **Quantitative** selection box and then select the column of the worksheet that contains the ENERGY data
6. In the **X / Explanatory variables** list, click in the **Qualitative** selection box and then select the

column of the worksheet that contains the REGION data
7. Click the **Options** tab and then type 95 in the **Confidence interval (%)** text box
8. Click the **Outputs** tab and uncheck all check boxes
9. Click the **Multiple comparisons** subtab
10. Check the **Pairwise comparisons** check box
11. In the **Pairwise comparisons** list box, ensure that only **Tukey (HSD)** is checked
12. Click the **Charts** tab and then uncheck the **Regression charts** and **Means charts** check boxes
13. Click **OK**
14. Click the **Continue** button in the **XLSTAT – Selections** dialog box

Exercises 16.4

Understanding the Concepts and Skills

16.76 What is the purpose of doing a multiple comparison?

16.77 Fill in the blank: If a confidence interval for the difference between two population means does not contain _____, we can reject the null hypothesis that the two means are equal in favor of the alternative hypothesis that the two means are different; and vice versa.

16.78 Explain the difference between the family confidence level and the individual confidence level.

16.79 Regarding family and individual confidence levels, answer the following questions and explain your answers.
a. Which is smaller for multiple comparisons involving three or more means, the family confidence level or the individual confidence level?
b. For multiple comparisons involving two means, what is the relationship between the family confidence level and the individual confidence level?

16.80 What is the name of the distribution on which the Tukey multiple-comparison method is based? What is its abbreviation?

16.81 The parameter ν for the q-curve in a Tukey multiple comparison equals one of the degrees of freedom for the F-curve in a one-way ANOVA. Which one?

16.82 Explain the essential difference between obtaining a confidence interval by using the pooled t-interval procedure and obtaining a confidence interval by using the Tukey multiple-comparison procedure.

16.83 Determine the following for a q-curve with parameters $\kappa = 6$ and $\nu = 13$.
a. The q-value having area 0.05 to its right   b. $q_{0.01}$

16.84 Determine the following for a q-curve with parameters $\kappa = 8$ and $\nu = 20$.
a. The q-value having area 0.01 to its right   b. $q_{0.05}$

16.85 Find the following for a q-curve with parameters $\kappa = 9$ and $\nu = 30$.
a. The q-value having area 0.01 to its right   b. $q_{0.05}$

16.86 Find the following for a q-curve with parameters $\kappa = 4$ and $\nu = 11$.
a. The q-value having area 0.05 to its right   b. $q_{0.01}$

16.87 Suppose that you conduct a one-way ANOVA test and find that the test is not statistically significant at the 5% level. If you subsequently perform a Tukey multiple comparison at a family confidence level of 0.95, what will be the results? Explain your answer.

In Exercises 16.88–16.93, we repeat the data from Exercises 16.42–16.47 of Section 16.3 for independent simple random samples from several populations. In each case, conduct a Tukey multiple comparison at the 95% family confidence level. Interpret your results.

16.88

Sample 1	Sample 2	Sample 3
1	10	4
9	4	16
	8	10
	6	
	2	

16.89

Sample 1	Sample 2	Sample 3
8	2	4
4	1	3
6	3	6
		3

16.90

Sample 1	Sample 2	Sample 3	Sample 4
6	9	4	8
3	5	4	4
3	7	2	6
	8	2	
	6	3	

16.91

Sample 1	Sample 2	Sample 3	Sample 4	Sample 5
7	5	6	3	7
4	9	7	7	9
5	4	5	7	11
4		4	4	
		8	4	

16.92

Sample 1	Sample 2	Sample 3	Sample 4	Sample 5
4	8	9	4	3
2	5	6	0	6
3	5	9	2	9

16.93

Sample 1	Sample 2	Sample 3	Sample 4
11	9	16	5
6	2	10	1
7	4	10	3

Applying the Concepts and Skills

In Exercises 16.94–16.99, use Procedure 16.2 on page 721 to perform a Tukey multiple comparison at the specified family confidence level.

16.94   Movie Guide. Following are the data from Exercise 16.48 on running times, in minutes, for random samples of films in four rating groups. Use a family confidence level of 0.99.

1* or 1.5*	2* or 2.5*	3* or 3.5*	4*
75	97	101	101
95	70	89	135
84	105	97	93
86	119	103	117
58	87	86	126
85	95	100	119

16.95   Copepod Cuisine. Following are the data on the number of copepods in each of 12 containers after 14 days for three different diets from Exercise 16.49. Use a family confidence level of 0.95.

Diatoms	Bacteria	Macroalgae
426	303	277
467	301	324
438	293	302
497	328	272

16.96 From Exercise 16.50: In Section 16.2, we considered two hypothetical examples to explain the logic for one-way ANOVA.
a. Refer to Table 16.1 on page 702 (95% family confidence level).
b. Refer to Table 16.2 on page 702 (95% family confidence level).

16.97   Staph Infections. Following are the data from Exercise 16.51 on bacteria counts, in millions, for different cases from each of five strains of cultured *Staphylococcus aureus*.

Strain A	Strain B	Strain C	Strain D	Strain E
9	3	10	14	33
27	32	47	18	43
22	37	50	17	28
30	45	52	29	59
16	12	26	20	31

a. Use a 95% family confidence level.
b. Without doing any further work or referring to Exercise 16.51, decide at the 5% significance level whether the data provide sufficient evidence to conclude that a difference exists in mean bacteria counts among the five strains of *Staphylococcus aureus*. Explain your reasoning.

16.98   Permeation Sampling. Following are the data from Exercise 16.52 on experimentally obtained calibration constants for samples of compounds in each of four compound groups.

Esters	Alcohols	Aliphatic hydrocarbons	Aromatic hydrocarbons
0.185	0.185	0.230	0.166
0.155	0.160	0.184	0.144
0.131	0.142	0.160	0.117
0.103	0.122	0.132	0.072
0.064	0.117	0.100	
	0.115	0.064	
	0.110		
	0.095		
	0.085		
	0.075		

a. Use a 95% family confidence level.
b. Without doing any further work or referring to Exercise 16.52, decide at the 5% significance level whether the data provide sufficient evidence to conclude that a difference exists in mean calibration constant among the four compound groups. Explain your reasoning.

16.99   Laptop Battery Life. In the following table are the data from Exercise 16.53 on battery lives, in hours, for samples of laptops made by four different computer companies. The four brands are Apple, Dell, Samsung, and Toshiba, but we have not used the names and have permuted the order. Use a 95% family confidence level.

Brand A	Brand B	Brand C	Brand D
7.25	12.00	11.50	8.00
9.50	7.25	9.50	7.50
6.25	11.75	10.00	5.50
8.00	15.00	5.00	4.75
8.25	8.50	9.50	6.25
10.25	8.75	5.25	6.00
6.25	10.00	9.75	4.75
5.75	10.25	8.25	4.50
		5.25	
		11.50	

In Exercises 16.100–16.105, use the technology of your choice to perform and interpret a Tukey multiple comparison at the specified family confidence level. All data sets are on the WeissStats site.

16.100   Empty Stomachs. The data from Exercise 16.54 on the proportions of fish with empty stomachs among four species in African waters. Use a 99% family confidence level.

16.101   Monthly Rents. The data from Exercise 16.55 on monthly rents, in dollars, for independent random samples of newly completed apartments in the four U.S. regions. Use a 95% family confidence level.

16.102   Ground Water. The data from Exercise 16.56 on the concentrations, in milligrams per liter, of each of four chemicals among three different wells. Use a 99% family confidence level.

16.103   Rock Sparrows. The data from Exercise 16.57 on the number of minutes per hour that male Rock Sparrows sang in the vicinity of the nests after patch-size manipulations were done on three different groups of females. Use a 99% family confidence level.

16.104   Artificial Teeth: Wear. The data from Exercise 16.58 on the volume of material worn away, in cubic millimeters, among three different materials for making artificial teeth. Use a 95% family confidence level.

16.105 Artificial Teeth: Hardness. The data from Exercise 16.59 on the Vickers microhardness (VHN) of the occlusal surfaces among three different materials for making artificial teeth. Use a 95% family confidence level.

In Exercises 16.106–16.109, use Procedure 16.2 on page 721 to perform a Tukey multiple comparison at the specified family confidence level. Note: We have provided values of q_α not given in Table IX or X.

16.106 Breast Milk and IQ. Following are summary statistics from Exercise 16.60 on IQ for samples of children at age 7½–8 years who were born preterm. The researchers used the following designations. Group I: mothers declined to provide breast milk; Group IIa: mothers had chosen but were unable to provide breast milk; and Group IIb: mothers had chosen and were able to provide breast milk. Use a family confidence level of 0.99. Here $q_\alpha = 4.15$.

Group	n_j	$\bar{x}_j$	s_j
I	90	92.8	15.2
IIa	17	94.8	19.0
IIb	193	103.7	15.3

16.107 Denosumab and Osteoporosis. In the following table are summary statistics from Exercise 16.61 on body-mass indexes (BMI) of the women in five denosumab treatment groups. Use a family confidence level of 0.90. Here $q_\alpha = 3.50$.

Treatment	n_j	$\bar{x}_j$	s_j
Placebo	46	25.9	4.3
14 mg	54	25.8	5.3
60 mg	47	27.5	5.8
100 mg	42	26.0	4.6
210 mg	47	25.9	4.3

16.108 Minke Whales. In the following table are summary statistics from Exercise 16.62 on body lengths, in meters, of minke whales entangled at four different ocean depths, in meters. Use a 99% family confidence level. Here $q_\alpha = 4.52$.

Depth	n_j	$\bar{x}_j$	s_j
0–49	39	4.49	0.80
50–99	20	5.06	0.79
100–149	28	5.75	1.20
150–199	14	5.99	0.97

16.109 Starting Salaries. Following are summary statistics from Exercise 16.63 on starting salaries, in thousands of dollars, for samples of bachelor's-degree graduates in six fields. Use a family confidence level of 0.99. Here $q_\alpha = 4.85$.

Field	n_j	$\bar{x}_j$	s_j
Business	46	55.1	5.6
Communications	11	44.6	4.7
Computer Science	30	59.1	4.0
Education	11	40.6	5.0
Engineering	44	62.6	5.7
Math & Sciences	18	43.0	4.8

Working with Large Data Sets

In Exercises 16.110–16.118, we repeat information from Exercises 16.64–16.72, where you were asked to decide whether conducting a one-way ANOVA test on the data is reasonable. For those exercises where it is, use the technology of your choice to perform and interpret a Tukey multiple comparison at the 95% family confidence level. All data sets are on the WeissStats site.

16.110 Daily TV Viewing Time. The data from Exercise 16.64 on the daily TV viewing times, in hours, of independent simple random samples of men, women, teens, and children.

16.111 Fish of Lake Laengelmaevesi. The data from Exercise 16.65 on weight (in grams) and length (in centimeters) from the nose to the beginning of the tail for four species of fish caught in Lake Laengelmaevesi, Finland. Consider both the weight and length data for possible analysis.

16.112 Popular Diets. The data from Exercise 16.66 on weight losses, in kilograms, over a 1-year period of four popular diets. Recall that negative losses are gains and that WW = Weight Watchers.

16.113 Cuckoo Care. The data from Exercise 16.67 on the lengths, in millimeters, of cuckoo eggs found in the nests of six bird species.

16.114 Doing Time. The data from Exercise 16.68 on times served, in months, of independent simple random samples of released prisoners among five different offense categories.

16.115 Book Prices. The data from Exercise 16.69 on book prices, in dollars, for independent random samples of hardcover books in law, science, medicine, and technology.

16.116 Magazine Ads. The data from Exercise 16.70 on the number of words of three syllables or more in advertisements from magazines of three different educational levels.

16.117 Sickle Cell Disease. The data from Exercise 16.71 on the steady-state hemoglobin levels of patients with three different types of sickle cell disease.

16.118 Prolonging Life. The data from Exercise 16.72 on the survival times, in days, among samples of patients in advanced stages of cancer, grouped by the affected organ, who were given a vitamin C supplement.

Extending the Concepts and Skills

16.119 Explain why the family confidence level, not the individual confidence level, is the appropriate level for comparing all population means simultaneously.

16.120 In Step 3 of Procedure 16.2, we obtain confidence intervals only when $i < j$. Explain how to determine the remaining confidence intervals from those obtained.

16.121 Energy Consumption. Apply Table 16.11 on page 723 and your answer from Exercise 16.120 to determine the remaining six confidence intervals for the differences between the energy consumption means.

16.5 The Kruskal–Wallis Test*

In this section, we examine the **Kruskal–Wallis test,** a nonparametric alternative to the one-way ANOVA procedure discussed in Section 16.3. The Kruskal–Wallis test applies when the distributions (one for each population) of the variable under consideration have the same shape in the sense of Definition 10.1 on page 470; it does not require that the distributions be normal or have any other specific shape.

Like the Mann–Whitney test, the Kruskal–Wallis test is based on ranks. When ties occur, ranks are assigned in the same way as in the Mann–Whitney test: *If two or more observations are tied, each is assigned the mean of the ranks they would have had if there were no ties.*

■■■ **EXAMPLE 16.8** Introducing the Kruskal–Wallis Test

Vehicle Miles The Federal Highway Administration conducts annual surveys on motor vehicle travel by type of vehicle and publishes its findings in *Highway Statistics*. Independent simple random samples of cars, buses, and trucks yielded the data on number of thousands of miles driven last year shown in Table 16.12.

Suppose that we want to use the sample data in Table 16.12 to decide whether a difference exists in last year's mean number of miles driven among cars, buses, and trucks.

a. Formulate the problem statistically by posing it as a hypothesis test.
b. Is it appropriate to apply the one-way ANOVA test here? What about the Kruskal–Wallis test?
c. Explain the basic idea for carrying out a Kruskal–Wallis test.
d. Discuss the use of the sample data in Table 16.12 to make a decision concerning the hypothesis test.

TABLE 16.12

Number of miles driven (1000s) last year for independent samples of cars, buses, and trucks

Cars	Buses	Trucks
19.9	1.8	24.6
15.3	7.2	37.0
2.2	7.2	21.2
6.8	6.5	23.6
34.2	13.3	23.0
8.3	25.4	15.3
12.0		57.1
7.0		14.5
9.5		26.0
1.1		

Solution

a. Let μ_1, μ_2, and μ_3 denote last year's mean number of miles driven for cars, buses, and trucks, respectively. Then the null and alternative hypotheses are, respectively,

$$H_0: \mu_1 = \mu_2 = \mu_3 \text{ (mean miles driven are equal)}$$

$$H_a: \text{Not all the means are equal.}$$

b. We constructed stem-and-leaf diagrams of the three samples, as shown in Fig. 16.12. These diagrams suggest that the distributions of miles driven have roughly the same shape for cars, buses, and trucks but that those distributions are far from normal. Thus, although the one-way ANOVA test of Section 16.3 is probably inappropriate, the Kruskal–Wallis procedure appears suitable.[†]

c. To apply the Kruskal–Wallis test, we first rank the data from all three samples combined, as shown in Table 16.13.

The idea behind the Kruskal–Wallis test is simple: If the null hypothesis of equal population means is true, the means of the ranks for the three samples should be roughly equal. Put another way, if the variation among the mean ranks for the three samples is too large, we have evidence against the null hypothesis.

To measure the variation among the mean ranks, we use the treatment sum of squares, *SSTR,* computed for the ranks. To decide whether that quantity is too large, we compare it to the variance of all the ranks, which can be expressed as $SST/(n-1)$, where *SST* is the total sum of squares for the ranks and n is the

FIGURE 16.12

Stem-and-leaf diagrams of the three samples in Table 16.12

```
0 | 1 2        0 | 1          1 | 4
0 | 6 7 8 9    0 | 6 7 7      1 | 5
1 | 2          1 | 3          2 | 1 3 3 4
1 | 5 9        1 |            2 | 6
2 |            2 |            3 |
2 |            2 | 5          3 | 7
3 | 4          4 |
3 |            4 |
                              5 |
                              5 | 7
```

(a) Cars   (b) Buses   (c) Trucks

[†]To explain the Kruskal–Wallis test, we have chosen an example with very small sample sizes. However, because having very small sample sizes makes effectively checking the same-shape condition difficult, proceed cautiously when dealing with them.

TABLE 16.13

Results of ranking the combined data from Table 16.12

Cars	Rank	Buses	Rank	Trucks	Rank
19.9	16	1.8	2	24.6	20
15.3	14.5	7.2	7.5	37.0	24
2.2	3	7.2	7.5	21.2	17
6.8	5	6.5	4	23.6	19
34.2	23	13.3	12	23.0	18
8.3	9	25.4	21	15.3	14.5
12.0	11			57.1	25
7.0	6			14.5	13
9.5	10			26.0	22
1.1	1				
	9.850|	9.000|	19.167	← *Mean ranks*

What Does It Mean?

The K-statistic is the ratio of the variation among the mean ranks to the variation of all the ranks.

total number of observations.[†] More precisely, the test statistic for a Kruskal–Wallis test, denoted K, is

$$K = \frac{SSTR}{SST/(n-1)}. \qquad (16.5)$$

Large values of K indicate that the variation among the mean ranks is large (relative to the variance of all the ranks) and hence that the null hypothesis of equal population means should be rejected.

d. For the ranks in Table 16.13, we find that $SSTR = 537.475$, $SST = 1299$, and $n = 25$. Thus the value of the test statistic is

$$K = \frac{SSTR}{SST/(n-1)} = \frac{537.475}{1299/24} = 9.930.$$

Is this value of K large enough to conclude that the null hypothesis of equal population means is false? To answer this question, we need to know the distribution of the variable K.

KEY FACT 16.5

Distribution of the K-Statistic for a Kruskal–Wallis Test

Suppose that the k distributions (one for each population) of the variable under consideration have the same shape. Then, for independent samples from the k populations, the variable

$$K = \frac{SSTR}{SST/(n-1)}$$

has approximately a chi-square distribution with $df = k - 1$ if the null hypothesis of equal population means is true. Here, n denotes the total number of observations.

Note the following:

- A rule of thumb for using the chi-square distribution as an approximation to the true distribution of K is that all sample sizes should be 5 or greater. Although we adopt that rule of thumb, some statisticians consider it too restrictive. Instead, they regard the chi-square approximation to be adequate unless $k = 3$ and none of the sample sizes exceed 5.

[†]Recall from Sections 16.2 and 16.3 that the treatment sum of squares, *SSTR*, is a measure of variation among means and that the total sum of squares, *SST*, is a measure of variation among all the data. The defining and computing formulas for *SSTR* and *SST* are given in Formula 16.1 on page 709. For the Kruskal–Wallis test, we apply those formulas to the ranks of the sample data, not to the sample data themselves.

• Notations other than K, most often H, are used for the Kruskal–Wallis test statistic. We have chosen the letter K instead of H to avoid confusion with the H used for "hypothesis."

Computing Formula for K

> ### What Does It Mean?
>
> This is the computing formula for K, used for hand calculations.

Usually, an easier way to compute the test statistic K by hand from the raw data is to apply the computing formula

$$K = \frac{12}{n(n+1)} \sum_{j=1}^{k} \frac{R_j^2}{n_j} - 3(n+1), \qquad (16.6)$$

where R_1 denotes the sum of the ranks for the sample data from Population 1, R_2 denotes the sum of the ranks for the sample data from Population 2, and so on.

Strictly speaking, the computing formula for K is equivalent to the defining formula for K only if no ties occur. In practice, however, the computing formula provides a sufficiently accurate approximation unless the number of ties is relatively large.

Performing the Kruskal–Wallis Test

Procedure 16.3 provides a step-by-step method for conducting a Kruskal–Wallis test by using either the critical-value approach or the P-value approach. Because the null hypothesis is rejected only when the test statistic, K, is too large, a Kruskal–Wallis test is always right tailed.

Although the Kruskal–Wallis test can be used to compare several population medians as well as several population means, we state Procedure 16.3 in terms of population means. To apply the procedure for population medians, simply replace μ_1 by η_1, μ_2 by η_2, and so on.

Regarding the third and fourth assumptions of Procedure 16.3, note the following:

• **Assumption 3:** For brevity, we use the phrase "same-shape populations" to indicate that the k distributions of the variable under consideration have the same shape.
• **Assumption 4:** This assumption is necessary only when we are using the chi-square distribution as an approximation to the distribution of K. Tables of critical values for K are available in cases where Assumption 4 fails.

■■■   **EXAMPLE 16.9**   The Kruskal–Wallis Test

Vehicle Miles We now complete the hypothesis test introduced in Example 16.8. Independent simple random samples of cars, buses, and trucks gave the data on number of thousands of miles driven last year shown in Table 16.14. At the 5% significance level, do the data provide sufficient evidence to conclude that a difference exists in last year's mean number of miles driven among cars, buses, and trucks?

TABLE 16.14

Number of miles driven (1000s) last year for independent samples of cars, buses, and trucks

Cars	Buses	Trucks
19.9	1.8	24.6
15.3	7.2	37.0
2.2	7.2	21.2
6.8	6.5	23.6
34.2	13.3	23.0
8.3	25.4	15.3
12.0		57.1
7.0		14.5
9.5		26.0
1.1		

Solution We apply Procedure 16.3.

Step 1   State the null and alternative hypotheses.

Let μ_1, μ_2, and μ_3 denote last year's mean number of miles driven for cars, buses, and trucks, respectively. Then the null and alternative hypotheses are, respectively,

$$H_0: \mu_1 = \mu_2 = \mu_3 \text{ (mean miles driven are equal)}$$
$$H_a: \text{Not all the means are equal.}$$

Step 2   Decide on the significance level, α.

We are to perform the test at the 5% significance level; so, $\alpha = 0.05$.

■■■■ **PROCEDURE 16.3** **Kruskal–Wallis Test**

Purpose To perform a hypothesis test to compare k population means, $\mu_1, \mu_2, \ldots, \mu_k$

Assumptions

1. Simple random samples
2. Independent samples
3. Same-shape populations
4. All sample sizes are 5 or greater

Step 1   The null and alternative hypotheses are, respectively,

$$H_0: \mu_1 = \mu_2 = \cdots = \mu_k$$
$$H_a: \text{Not all the means are equal.}$$

Step 2   Decide on the significance level, α.

Step 3   Compute the value of the test statistic

$$K = \frac{12}{n(n+1)} \sum_{j=1}^{k} \frac{R_j^2}{n_j} - 3(n+1)$$

and denote that value K_0. Here, n is the total number of observations and $R_1, R_2, \ldots, R_k$ denote the sums of the ranks for the sample data from Populations $1, 2, \ldots, k$, respectively. To obtain K, first construct a work table to rank the data from all the samples combined.

CRITICAL-VALUE APPROACH	OR	P-VALUE APPROACH

Step 4   The critical value is χ_α^2 with df $= k - 1$. Use Table VII to find the critical value.

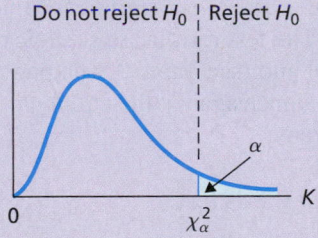

Step 5   If the value of the test statistic falls in the rejection region, reject H_0; otherwise, do not reject H_0.

Step 4   The K-statistic has df $= k - 1$. Use Table VII to estimate the P-value or obtain it exactly by using technology.

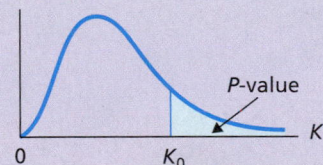

Step 5   If $P \le \alpha$, reject H_0; otherwise, do not reject H_0.

Step 6   Interpret the results of the hypothesis test.

Step 3   Compute the value of the test statistic

$$K = \frac{12}{n(n+1)} \sum_{j=1}^{k} \frac{R_j^2}{n_j} - 3(n+1).$$

We have $n = 10 + 6 + 9 = 25$. Summing the second, fourth, and sixth columns of the work table in Table 16.13 on page 729 yields $R_1 = 98.5$, $R_2 = 54.0$, and $R_3 = 172.5$. Thus the value of the test statistic is

$$K = \frac{12}{25(25+1)} \left(\frac{98.5^2}{10} + \frac{54.0^2}{6} + \frac{172.5^2}{9} \right) - 3(25+1) = 9.923.$$

CRITICAL-VALUE APPROACH	OR	P-VALUE APPROACH

Step 4 The critical value is χ_α^2 with df $= k - 1$. Use Table VII to find the critical value.

We have $k = 3$—the three types of vehicles—hence df $= 3 - 1 = 2$. From Table VII, the critical value is $\chi_{0.05}^2 = 5.991$, as shown in Fig. 16.13A.

FIGURE 16.13A

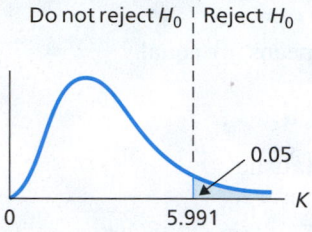

Step 5 If the value of the test statistic falls in the rejection region, reject H_0; otherwise, do not reject H_0.

From Step 3, we see that the value of the test statistic is $K = 9.923$. Figure 16.13A shows that this value falls in the rejection region. Thus we reject H_0. The test results are statistically significant at the 5% level.

Step 4 The K-statistic has df $= k - 1$. Use Table VII to estimate the P-value or obtain it exactly by using technology.

From Step 3, we see that the value of the test statistic is $K = 9.923$. Because the test is right tailed, the P-value is the probability of observing a value of K of 9.923 or greater if the null hypothesis is true. That probability equals the shaded area shown in Fig. 16.13B.

FIGURE 16.13B

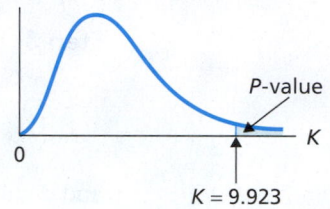

We have $k = 3$—the three types of vehicles—hence df $= 3 - 1 = 2$. Referring to Fig. 16.13B and Table VII, we find that $0.005 < P < 0.01$. (Using technology, we obtain $P = 0.007$.)

Step 5 If $P \leq \alpha$, reject H_0; otherwise, do not reject H_0.

From Step 4, $0.005 < P < 0.01$. Because the P-value is less than the specified significance level of 0.05, we reject H_0. The test results are statistically significant at the 5% level and (see Table 9.8 on page 386) provide very strong evidence against the null hypothesis of equal population means.

StatCrunch

Report 16.3

You try it!

Exercise 16.149 on page 736

Step 6 Interpret the results of the hypothesis test.

Interpretation At the 5% significance level, the data provide sufficient evidence to conclude that a difference exists in last year's mean number of miles driven among cars, buses, and trucks.

Comparison of the Kruskal–Wallis Test and the One-Way ANOVA Test

In Section 16.3, you learned how to perform a one-way ANOVA test to compare k population means when the variable under consideration is normally distributed on each of the k populations and the population standard deviations are equal. Because normal distributions with equal standard deviations have the same shape, you can also use the Kruskal–Wallis test to perform such a hypothesis test.

Under conditions of normality, the one-way ANOVA test is more powerful (but not much more powerful) than the Kruskal–Wallis test. However, if the distributions of the variable under consideration have the same shape but are not normal, the Kruskal–Wallis test is usually more powerful than the one-way ANOVA test, often considerably so.

KEY FACT 16.6

> ### The Kruskal–Wallis Test Versus the One-Way ANOVA Test
>
> Suppose that the distributions of a variable of several populations have the same shape and that you want to compare the population means, using independent simple random samples. When deciding between the one-way ANOVA test and the Kruskal–Wallis test, follow these guidelines: If you are reasonably sure that the distributions are normal, use the one-way ANOVA test; otherwise, use the Kruskal–Wallis test.

Alternate Version of the Kruskal-Wallis Test

Assumption 3 of the Kruskal–Wallis test is that the distributions (one for each population) of the variable under consideration have the same shape, which, for brevity, we refer to as "same-shape populations." This condition is often difficult to check and is frequently not satisfied.

An alternate version of the Kruskal–Wallis test is available that does not require the condition of same-shape populations. However, the null and alternative hypotheses for this different version are not stated in terms of means or medians but rather in terms of distributions. In other words, we can drop the "same-shape populations" condition if we change the hypotheses to those in terms of distributions instead of means or medians.

The null hypothesis for the alternate version of the Kruskal–Wallis test is that the distribution of the variable is the same on all the populations. The alternative hypothesis is that the distribution of the variable has systematically larger values in some populations than in others.

Note that, if the "same-shape populations" condition does in fact hold, then the alternate version of the Kruskal–Wallis test is equivalent to the one in terms of means or medians (Procedure 16.3 on page 731).

THE TECHNOLOGY CENTER

Most statistical technologies have programs that automatically perform a Kruskal–Wallis test. In this subsection, we present output and step-by-step instructions for such programs.

Note to TI-83/84 Plus users: At the time of this writing, the TI-83/84 Plus does not have a built-in program for conducting a Kruskal–Wallis test. However, we have written a TI program called KWTEST for performing that test. It is located in the TI Programs section on the WeissStats site. Your instructor can show you how to download the program to your calculator. *Warning:* Any data that you may have previously stored in Lists 1–5 will be erased during program execution, so copy those data to other lists prior to program execution if you want to retain them.

EXAMPLE 16.10   Using Technology to Perform a Kruskal–Wallis Test

Vehicle Miles Table 16.14 on page 730 shows last year's number of thousands of miles driven for independent simple random samples of cars, buses, and trucks. Use Minitab, Excel, or the TI-83/84 Plus to decide, at the 5% significance level, whether the data provide sufficient evidence to conclude that a difference exists in last year's mean number of miles driven among cars, buses, and trucks.

Solution Let μ_1, μ_2, and μ_3 denote last year's mean number of miles driven for cars, buses, and trucks, respectively. The task is to use the Kruskal–Wallis procedure to perform the hypothesis test

$$H_0: \mu_1 = \mu_2 = \mu_3 \text{ (mean miles driven are equal)}$$

H_a: Not all the means are equal

at the 5% significance level.

We applied the Kruskal–Wallis test programs to the data, resulting in Output 16.3. Steps for generating that output are presented in Instructions 16.3. *Note to Excel users:* For brevity, we have presented only the essential portion of the output.

OUTPUT 16.3 Kruskal–Wallis test on the miles-driven data

MINITAB

Kruskal-Wallis Test: MILES versus VEHICLE

```
Kruskal-Wallis Test on MILES

VEHICLE    N   Median   Ave Rank      Z
Buses      6    7.200        9.0  -1.53
Cars      10    8.900        9.8  -1.75
Trucks     9   23.600       19.2   3.14
Overall   25                13.0

H = 9.92  DF = 2  P = 0.007
H = 9.93  DF = 2  P = 0.007  (adjusted for ties)
```

TI-83/84 PLUS

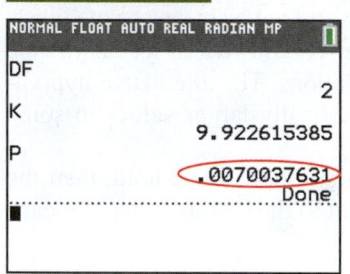

EXCEL

Significance level (%): 5			
p-value: Asymptotic p-value			
Summary statistics:			
Variable	Observations	Mean	Std. deviation
MILES | Cars	10	11.6300	9.7274
MILES | Buses	6	10.2333	8.2812
MILES | Trucks	9	26.9222	13.0722
Kruskal-Wallis test (MILES):			
K (Observed value)	9.9303		
K (Critical value)	5.9915		
DF	2		
p-value (Two-tailed)	0.0070		
alpha	0.05		

As shown in Output 16.3, the P-value for the hypothesis test is 0.007. Because the P-value is less than the specified significance level of 0.05, we reject H_0. At the 5% significance level, the data provide sufficient evidence to conclude that a difference exists in last year's mean number of miles driven among cars, buses, and trucks.

INSTRUCTIONS 16.3 Steps for generating Output 16.3

MINITAB

1. Store all 25 mileages from Table 16.14 in a column named MILES
2. Store the corresponding vehicle types in a column named VEHICLE
3. Choose **Stat ➤ Nonparametrics ➤ Kruskal-Wallis...**
4. Press the F3 key to reset the dialog box
5. Specify MILES in the **Response** text box
6. Specify VEHICLE in the **Factor** text box
7. Click **OK**

EXCEL

1. Store all 25 mileages from Table 16.14 in a column named MILES
2. Store the corresponding vehicle types in a column named VEHICLE
3. Choose **XLSTAT ➤ Nonparametric tests ➤ Comparison of k samples (Kruskal-Wallis, Friedman, . . .)**
4. Click the reset button in the lower left corner of the dialog box
5. In the **Data format** list, select the **One column per variable** option button

6. Click in the **Data** selection box and then select the column of the worksheet that contains the MILES data
7. Click in the **Sample identifiers** selection box and then select the column of the worksheet that contains the VEHICLE data
8. Click the **Options** tab and then type 5 in the **Significance level (%)** text box
9. Click **OK**
10. Click the **Continue** button in the **XLSTAT – Selections** dialog box

TI-83/84 PLUS

1. Store all 25 mileages from Table 16.14 in a list named MILES
2. Store the corresponding vehicle types in a list named VEHIC, using the coding 1 for Cars, 2 for Buses, and 3 for Trucks
3. Press **PRGM**
4. Arrow down to KWTEST and press **ENTER** twice
5. Press **2ND ➤ LIST**, arrow down to MILES, and press **ENTER** twice
6. Press **2ND ➤ LIST**, arrow down to VEHIC, and press **ENTER** twice

Exercises 16.5

Understanding the Concepts and Skills

16.122 Of what test is the Kruskal–Wallis test a nonparametric version?

16.123 State the conditions required for performing a Kruskal–Wallis test.

16.124 In the Kruskal–Wallis test, how should you deal with tied ranks?

16.125 Fill in the blank: If the null hypothesis of equal population means is true, the sample mean ranks should be roughly _____.

Regarding a Kruskal–Wallis test, fill in the blanks in each of Exercises 16.126–16.128.

16.126 The measure of variation among the sample mean ranks is _____ computed for the _____.

16.127 The measure of total variation of all the ranks is the _____ of all the ranks, which can be expressed as _____ $/(n - 1)$, where _____ is the total sum of squares computed for the ranks.

16.128 To decide whether the variation among sample mean ranks is large enough to warrant rejection of the null hypothesis of equal population means, we use the ratio of _____ to _____. This ratio is called the _____ -statistic.

16.129 For a Kruskal–Wallis test to compare five population means, what is the approximate distribution of K?

In each of Exercises 16.130–16.133, suppose that you want to perform a hypothesis test to compare several population means, using independent samples. In each case, decide whether you would use the one-way ANOVA test, the Kruskal–Wallis test, or neither of these tests.

16.130 Preliminary data analyses of the samples suggest that the distributions of the variable are not normal but have the same shape.

16.131 Preliminary data analyses of the samples suggest that the distributions of the variable are normal and have the same shape.

16.132 Preliminary data analyses of the samples suggest that the distributions of the variable are not normal and have quite different shapes.

16.133 Preliminary data analyses of the samples suggest that the distributions of the variable are normal but have quite different shapes.

In each of Exercises 16.134–16.137, we provide independent simple random samples from several populations where, in each case, all the observations are distinct values. For each exercise, perform the following tasks.
a. Compute the Kruskal–Wallis test statistic, K, by using the defining formula, given in Key Fact 16.5 on page 729. Use the defining formulas when obtaining SSTR and SST for the ranks.
b. Compute the Kruskal–Wallis test statistic, K, by using the computing formula, given in Equation (16.6) on page 730.
c. Compare your results in parts (a) and (b).
d. Explain the fact that parts (a)–(c) are illustrating.

16.134

Sample 1	Sample 2	Sample 3
51	9	1
3	73	19
	33	

16.135

Sample 1	Sample 2	Sample 3
31	49	97
7	17	71
	1	

16.136

Sample 1	Sample 2	Sample 3	Sample 4
10	6	12	13
8	9	11	15
	1		14
			4

16.137

Sample 1	Sample 2	Sample 3	Sample 4
14	2	11	9
4	16	10	12
13			8
			7

In Exercises 16.138–16.143, we repeat the data from Exercises 16.42–16.47 of Section 16.3 for independent simple random samples from several populations. In each case, determine the
a. total number of observations, n.
b. number of populations, k.
c. rank sums, $R_1, R_2, \ldots, R_k$.

16.138

Sample 1	Sample 2	Sample 3
1	10	4
9	4	16
	8	10
	6	
	2	

16.139

Sample 1	Sample 2	Sample 3
8	2	4
4	1	3
6	3	6
		3

16.140

Sample 1	Sample 2	Sample 3	Sample 4
6	9	4	8
3	5	4	4
3	7	2	6
	8	2	
	6	3	

16.141

Sample 1	Sample 2	Sample 3	Sample 4	Sample 5
7	5	6	3	7
4	9	7	7	9
5	4	5	7	11
4		4	4	
		8	4	

16.142

Sample 1	Sample 2	Sample 3	Sample 4	Sample 5
4	8	9	4	3
2	5	6	0	6
3	5	9	2	9

16.143

Sample 1	Sample 2	Sample 3	Sample 4
11	9	16	5
6	2	10	1
7	4	10	3

Applying the Concepts and Skills

*In Exercises **16.144–16.149**, perform a Kruskal–Wallis test by using either the critical-value approach or the P-value approach.*

16.144 **Entertainment Expenditures.** The Bureau of Labor Statistics conducts surveys on consumer expenditures for various types of entertainment and publishes its findings in *Consumer Expenditure Survey*. Independent random samples yielded the following data, in dollars, on last year's expenditures for three entertainment categories.

Fees and admissions	TV, radio, and sound equipment	Other equipment and services
303	230	130
242	1878	381
152	526	1423
625	130	161
241	600	205
1333	130	1154
739	692	1759
430		232
		1368

At the 5% significance level, do the data provide sufficient evidence to conclude that a difference exists in last year's mean expenditures among the three entertainment categories?

16.145 **TV-Commercial Placement.** In the article "Violence and Sex Impair Memory for Television Ads" (*Journal of Applied Psychology*, Vol. 87, No. 3, pp. 557–564), B. Bushman and A. Bonacci studied whether the type of television program in which commercials are placed affects a viewer's ability to remember the product. Participants watched a violent, sexually explicit, or neutral TV program that contained nine advertisements. The next day, participants were asked to recall the advertised brands. The number of brands the participants were able to recall are presented in the following table.

Violent	Sexually explicit	Neutral
0	0	4
0	2	1
2	2	2
1	2	3
1	0	2
0	1	7

At the 5% significance level, do the data provide sufficient evidence to conclude that a difference exists in the median number of advertised brands a viewer is able to recall among the three TV program types?

16.146 **Ages of Car Buyers.** Information on characteristics of new-car buyers appears in *Buyers of New Cars*, a publication of Newsweek, Inc. Independent random samples of new-car buyers

yielded the data on age of purchaser, in years, by origin of car purchased, shown in the following table. Do the data provide sufficient evidence to conclude that a difference exists in the median ages of buyers of new domestic, Asian, and European cars? Use $\alpha = 0.05$.

Domestic	Asian	European
41	78	72
42	42	42
51	51	58
47	45	39
33	21	67
83	24	39
35	21	45
69	39	27
50	45	33
60	30	55

16.147 **Home Size.** The U.S. Census Bureau publishes information on the sizes of housing units in *Current Housing Reports*. Independent random samples of single-family detached homes (including mobile homes) in the four U.S. regions yielded the following data on square footage.

Northeast	Midwest	South	West
3182	2115	1591	1345
2130	2413	1354	694
1781	1639	722	2789
2989	1691	2135	1649
1581	1655	1982	2203
2149	1605	1639	2068
2286	3361	642	1565
1293	2058	1513	1655

At the 10% significance level, do the data provide sufficient evidence to conclude that a difference exists in median square footage of single-family detached homes among the four U.S. regions?

16.148 **Free Lunch.** In the publication "What Makes a High School Great?" (*Newsweek*, May 8, 2006, pp. 50–60), B. Kantrowitz and P. Wingert looked for America's best high schools. One relevant variable is the percentage of the student body that is eligible for free and reduced lunches, an indicator of socioeconomic status. A percentage of 40% or more generally indicates a high concentration of children in poverty. The following table provides the percentages for independent simple random samples of high schools from the four U.S. regions.

Northeast	South	Midwest	West
15.3	41.8	12.4	26.0
3.3	18.4	7.2	45.6
13.2	11.0	1.0	3.3
10.0	18.0	2.0	45.0
5.9	30.7	58.0	10.0
16.0	50.0	2.0	26.8
1.0	7.4	2.8	10.0
40.5	8.0	13.0	11.0
	6.0		

At the 5% significance level, do the data provide sufficient evidence to conclude that a difference exists in mean percent eligibility for free and reduced lunches among the four regions of the United States?

16.149 **Spider Mites.** The reproduction capability of spider mites infected with a bacteria called *Cardinium* was studied in the article "*Cardinium* Symbionts Cause Cytoplasmic Incompatibility in Spider

Mites" (*Heredity*, Vol. 98, No. 1, pp. 13–20) by T. Gotoh et al. Infected mites were crossed with mites treated with antibiotics. The following table shows the number of eggs per female for each combination tested. Here we use "A" for antibiotic-treated, "I" for infected, "M" for male, and "F" for female.

IF & IM	IF & AM	AF & IM	AF & AM
18	26	11	19
19	21	18	22
22	19	22	16
19	19	17	20
22	20	13	18
18	17	19	19
25	17	5	17
	22		14

At the 10% significance level, do the data provide sufficient evidence to conclude that a difference exists in mean number of eggs among the four combinations?

16.150 Movie Guide. Following are the data from Exercise 16.48 on running times, in minutes, for random samples of films in four rating groups.

1* or 1.5*	2* or 2.5*	3* or 3.5*	4*
75	97	101	101
95	70	89	135
84	105	97	93
86	119	103	117
58	87	86	126
85	95	100	119

a. Use the Kruskal–Wallis test to decide, at the 1% significance level, whether the data provide sufficient evidence to conclude that a difference exists in mean running times among films in the four rating groups.
b. The hypothesis test in part (a) was done in Exercise 16.48 by using the one-way ANOVA test. The assumption there is that running times in the four rating groups are normally distributed and have equal standard deviations. Presuming that to be true, why is performing a Kruskal–Wallis test to compare the means permissible? In this case, is use of the one-way ANOVA test or the Kruskal–Wallis test better? Explain your answers.

16.151 Staph Infections. Following are the data from Exercise 16.51 on bacteria counts, in millions, for different cases from each of five strains of cultured *Staphylococcus aureus*.

Strain A	Strain B	Strain C	Strain D	Strain E
9	3	10	14	33
27	32	47	18	43
22	37	50	17	28
30	45	52	29	59
16	12	26	20	31

a. Use the Kruskal–Wallis test to decide, at the 5% significance level, whether the data provide sufficient evidence to conclude that a difference exists in mean bacteria counts among the five strains of *Staphylococcus aureus*.
b. The hypothesis test in part (a) was done in Exercise 16.51 by using the one-way ANOVA test. The assumption there is that bacteria counts in the five strains are normally distributed and have equal standard deviations. Presuming that to be true, why is performing a Kruskal–Wallis test to compare the means permissible? In this

case, is use of the one-way ANOVA test or the Kruskal–Wallis test better? Explain your answers.

In Exercises 16.152–16.157, use the technology of your choice to
a. *conduct a Kruskal–Wallis test on the data at the specified significance level.*
b. *interpret your results from part (a).*
Note: All data sets are on the WeissStats site.

16.152 Empty Stomachs. The data from Exercise 16.54 on the proportions of fish with empty stomachs among four species in African waters. Use a 1% significance level.

16.153 Monthly Rents. The data from Exercise 16.55 on monthly rents, in dollars, for independent random samples of newly completed apartments in the four U.S. regions. Use a 5% significance level.

16.154 Ground Water. The data from Exercise 16.56 on the concentrations, in milligrams per liter, of each of four chemicals among three different wells. Use a 1% significance level.

16.155 Rock Sparrows. The data from Exercise 16.57 on the number of minutes per hour that male Rock Sparrows sang in the vicinity of the nests after patch-size manipulations were done on three different groups of females. Use a 1% significance level.

16.156 Artificial Teeth: Wear. The data from Exercise 16.58 on the volume of material worn away, in cubic millimeters, among three different materials for making artificial teeth. Use a 5% significance level.

16.157 Artificial Teeth: Hardness. The data from Exercise 16.59 on the Vickers microhardness (VHN) of the occlusal surfaces among three different materials for making artificial teeth. Use a 5% significance level.

Working with Large Data Sets

In Exercises 16.158–16.166, use the technology of your choice to do the following tasks.
a. *Decide whether conducting a Kruskal–Wallis test on the data is reasonable. If so, also do parts (b)–(d).*
b. *Use a Kruskal–Wallis test to decide, at the 5% significance level, whether the data provide sufficient evidence to conclude that a difference exists among the means of the populations from which the samples were taken.*
c. *Interpret your results from part (b).*
d. *If a one-way ANOVA test was performed on the data in Section 16.3, compare your results there to those obtained here.*
Note: All data sets are on the WeissStats site.

16.158 Daily TV Viewing Time. The data from Exercise 16.64 on the daily TV viewing times, in hours, of independent simple random samples of men, women, teens, and children.

16.159 Fish of Lake Laengelmaevesi. The data from Exercise 16.65 on weight (in grams) and length (in centimeters) from the nose to the beginning of the tail for four species of fish caught in Lake Laengelmaevesi, Finland. Consider both the weight and length data for possible analysis.

16.160 Popular Diets. The data from Exercise 16.66 on weight losses, in kilograms, over a 1-year period of four popular diets. Recall that negative losses are gains and that WW = Weight Watchers.

16.161 Cuckoo Care. The data from Exercise 16.67 on the lengths, in millimeters, of cuckoo eggs found in the nests of six bird species.

16.162 Doing Time. The data from Exercise 16.68 on times served, in months, of independent simple random samples of released prisoners among five different offense categories.

16.163 Book Prices. The data from Exercise 16.69 on book prices, in dollars, for independent random samples of hardcover books in law, science, medicine, and technology.

16.164 Magazine Ads. The data from Exercise 16.70 on the number of words of three syllables or more in advertisements from magazines of three different educational levels.

16.165 Sickle Cell Disease. The data from Exercise 16.71 on the steady-state hemoglobin levels of patients with three different types of sickle cell disease.

16.166 Prolonging Life. The data from Exercise 16.72 on the survival times, in days, among samples of patients in advanced stages of

cancer, grouped by the affected organ, who were given a vitamin C supplement.

Extending the Concepts and Skills

16.167 Vehicle Miles. In this section, we illustrated the Kruskal–Wallis test with data on miles driven by samples of cars, buses, and trucks. The value of the test statistic K was computed on page 729 to be 9.930, whereas on page 731, we found its value to be 9.923.
a. Explain the discrepancy between the two values of K.
b. Does the difference in the values affect our conclusion in the hypothesis test we conducted? Explain your answer.
c. Does the difference in the values affect our estimate of the P-value of the hypothesis test? Explain your answer.

CHAPTER IN REVIEW

You Should Be Able to

1. use and understand the formulas in this chapter.
2. use the F-table, Table VIII in Appendix A.
3. explain the essential ideas behind a one-way ANOVA.
4. state and check the assumptions required for a one-way ANOVA.
5. obtain the sums of squares for a one-way ANOVA by using the defining formulas.
6. obtain the sums of squares for a one-way ANOVA by using the computing formulas.
7. compute the mean squares and the F-statistic for a one-way ANOVA.
8. construct a one-way ANOVA table.
9. perform a one-way ANOVA test.
*10. use the q-tables, Tables IX and X in Appendix A.
*11. perform a multiple comparison by using the Tukey method.
*12. perform a Kruskal–Wallis test.

Key Terms

analysis of variance (ANOVA), *698*
degrees of freedom for the denominator, *699*
degrees of freedom for the numerator, *699*
error, *703*
error mean square (MSE), *704*
error sum of squares (SSE), *703*
F_α, *700*
F-curve, *699*
F-distribution, *699*
F-statistic, *704*

factor, *701*
family confidence level,* *720*
individual confidence level,* *720*
Kruskal–Wallis test,* *731*
levels, *701*
multiple comparisons,* *720*
one-way analysis of variance, *701*
one-way ANOVA identity, *708*
one-way ANOVA table, *708*
one-way ANOVA test, *710*
q_α,* *720*
q-curve,* *720*

q-distribution,* *720*
residual, *701*
residual analysis, *701*
response variable, *713*
rule of 2, *701*
studentized range distribution,* *720*
total sum of squares (SST), *708*
treatment, *703*
treatment mean square (MSTR), *704*
treatment sum of squares (SSTR), *703*
Tukey multiple-comparison method,* *721*

REVIEW PROBLEMS

Understanding the Concepts and Skills

1. For what is one-way ANOVA used?

2. State the four assumptions for one-way ANOVA, and explain how those assumptions can be checked.

3. On what distribution does one-way ANOVA rely?

4. Suppose that you want to compare the means of three populations by using one-way ANOVA. If the sample sizes are 5, 6, and 6, determine the degrees of freedom for the appropriate F-curve.

5. In one-way ANOVA, identify a statistic that measures
a. the variation among the sample means.
b. the variation within the samples.

6. In one-way ANOVA,
a. list and interpret the three sums of squares.
b. state the one-way ANOVA identity and interpret its meaning with regard to partitioning the total variation among all the data.

7. For a one-way ANOVA,
a. identify one purpose of one-way ANOVA tables.
b. construct a generic one-way ANOVA table.

*8. Explain in detail the purpose of conducting a multiple comparison.

*9. Explain the difference between the individual confidence level and the family confidence level. Which confidence level is appropriate for multiple comparisons? Explain your answer.

*10. On what distribution does the Tukey multiple-comparison procedure rely?

*11. Consider a Tukey multiple comparison of four population means with a family confidence level of 0.95. Is the individual confidence level smaller or larger than 0.95? Explain your answer.

*12. Suppose that you want to compare the means of three populations by using the Tukey multiple-comparison procedure. If the sample sizes are 5, 6, and 6, determine the parameters for the appropriate q-curve.

*13. Identify a nonparametric alternative to the one-way ANOVA procedure.

*14. Identify the distribution used as an approximation to the true distribution of the K-statistic for a Kruskal–Wallis test.

*15. Explain the logic of a Kruskal–Wallis test.

*16. Suppose that you want to compare the means of several populations, using independent samples. If given the choice between using the one-way ANOVA test and the Kruskal–Wallis test, which would you choose if outliers occur in the sample data? Explain your answer.

In Problems 17–21, consider an F-curve with df = (2, 14).

17. Identify the degrees of freedom for the numerator.

18. Identify the degrees of freedom for the denominator.

19. Determine $F_{0.05}$.

20. Find the F-value with area 0.01 to its right.

21. Find the F-value with area 0.05 to its right.

*22. Consider a q-curve with parameters 3 and 14.
a. Determine $q_{0.05}$.
b. Find the q-value with area 0.01 to its right.

23. Consider the following hypothetical samples.

A	B	C
1	0	3
3	6	12
5	2	6
	5	3
	2	

a. Obtain the sample mean and sample variance of each of the three samples.
b. Obtain SST, SSTR, and SSE by using the defining formulas and verify that the one-way ANOVA identity holds.
c. Obtain SST, SSTR, and SSE by using the computing formulas.
d. Construct the one-way ANOVA table.

Applying the Concepts and Skills

24. **Losses to Robbery.** The Federal Bureau of Investigation conducts surveys to obtain information on the value of losses from various types of robberies. Results of the surveys are published in *Population-at-Risk Rates and Selected Crime Indicators*. Independent simple random samples of reports for three types of robberies—highway, gas station, and convenience store—gave the following data, in dollars, on value of losses.

Highway	Gas station	Convenience store
952	1298	844
996	1195	921
839	1174	880
1088	1113	706
1024	953	602
	1280	614

a. What does *MSTR* measure?
b. What does *MSE* measure?
c. Suppose that you want to perform a one-way ANOVA to compare the mean losses among the three types of robberies. What conditions are necessary? How crucial are those conditions?

25. **Losses to Robbery.** Refer to Problem 24.
a. Obtain individual normal probability plots and the standard deviations of the samples.
b. Perform a residual analysis.
c. Decide whether presuming that the assumptions of normal populations and equal standard deviations are met is reasonable.

26. **Losses to Robbery.** Refer to Problem 24. At the 5% significance level, do the data provide sufficient evidence to conclude that a difference in mean losses exists among the three types of robberies? Use one-way ANOVA to perform the required hypothesis test. (*Note:* $T_1 = 4899$, $T_2 = 7013$, $T_3 = 4567$, and $\Sigma x^2 = 16,683,857$.)

*27. **Losses to Robbery.** Refer to Problem 24.
a. Apply the Tukey multiple-comparison method to the data. Use a family confidence level of 0.95.
b. Interpret your results from part (a).

*28. **Losses to Robbery.** Refer to Problem 24.
a. At the 5% significance level, do the data provide sufficient evidence to conclude that a difference in mean losses exists among the three types of robberies? Use the Kruskal–Wallis procedure to perform the required hypothesis test.
b. The hypothesis test in part (a) was done in Problem 26 by using the one-way ANOVA procedure. The assumptions in that exercise are that, for the three types of robberies, losses are normally distributed and have equal standard deviations. Presuming that to be true, why is performing a Kruskal–Wallis test to compare the means permissible? In this case, is use of the one-way ANOVA test or the Kruskal–Wallis test better? Explain your answers.
c. Compare your hypothesis-testing results with the Kruskal–Wallis test to those of the one-way ANOVA test.

29. **Foot-pressure Angle.** Genu valgum, commonly known as "knee-knock," is a condition in which the knees angle in and touch one another when standing. Genu varum, commonly known as "bow-legged," is a condition in which the knees angle out and the legs bow when standing. In the article "Frontal Plane Knee Angle Affects Dynamic Postural Control Strategy during Unilateral Stance" (*Medicine and Science in Sports & Exercise*, Vol. 34, No. 7, pp. 1150–1157), J. Nyland et al. studied patients with and without these conditions. One aspect of the study was to see whether patients with genu valgum or genu varum had a different angle of foot pressure when standing. The following table provides summary statistics for the angle, in degrees, of anterior-posterior center of foot pressure for patients that have genu valgum, genu varum, or neither condition.

Condition	n_j	$\bar{x}_j$	s_j
Genu varum	11	60.6	8
Genu valgum	16	60.7	7
Neither	29	54.2	6

At the 1% significance level, do the data provide sufficient evidence to conclude that a difference exists in the mean angle of anterior-posterior center of foot pressure among people in the three condition groups? *Note:* For the degrees of freedom in this exercise:

α	0.10	0.05	0.025	0.01	0.005
F_α	2.41	3.17	3.96	5.03	5.87

Use the technology of your choice to solve Problems 30 and 31.

30. Smoking and Cotinine. Smoking during pregnancy is hazardous to both the mother and baby. Passive smoking, or inhalation of second-hand smoke, is also a concern. In the article "Detection of Cotinine in Neonate Meconium as a Marker for Nicotine Exposure in Utero" (*Eastern Mediterranean Health Journal*, Vol. 10, No. 1/2, pp. 96–105), N. Sherif et al. studied whether active-smoking or passive-smoking mothers were passing along harmful chemicals to their babies. The level of cotinine in a newborn's first meconium, which is the infant's stool, is reported on the WeissStats site for independent samples of active-, passive-, and non-smoking mothers. Cotinine levels are measured in nanograms per milliliter (ng/mL). Decide whether presuming that the assumptions of normal populations and equal population standard deviations are met is reasonable.

***31. Smoking and Cotinine.** Refer to Problem 30. Decide whether conducting a Kruskal–Wallis test on the data is reasonable.

Working with Large Data Sets

In Problems 32–34, use the technology of your choice to do the following tasks.
a. Obtain individual normal probability plots and the standard deviations of the samples.
b. Perform a residual analysis.
c. Use your results from parts (a) and (b) to decide whether conducting a one-way ANOVA test on the data is reasonable. If so, also do parts (d)–(f).
d. Use a one-way ANOVA test to decide, at the 5% significance level, whether the data provide sufficient evidence to conclude that a difference exists among the means of the populations from which the samples were taken.
e. Interpret your results from part (d).

*f. If the result of the one-way ANOVA test is statistically significant, perform and interpret a Tukey multiple comparison.

32. Weight Loss and BMI. In the paper "Voluntary Weight Reduction in Older Men Increases Hip Bone Loss: The Osteoporotic Fractures in Men Study" (*Journal of Clinical Endocrinology & Metabolism*, Vol. 90, Issue 4, pp. 1998–2004), K. Ensrud et al. reported on the effect of voluntary weight reduction on hip bone loss in older men. In the study, 1342 older men participated in two physical examinations an average of 1.8 years apart. After the second exam, they were categorized into three groups according to their change in weight between exams: weight loss of more than 5%, weight gain of more than 5%, and stable weight (between 5% loss and 5% gain). For purposes of the hip bone density study, other characteristics were compared, one such being body mass index (BMI). On the WeissStats site, we provide the BMI data for the three groups, based on the results obtained by the researchers.

33. Weight Loss and Leg Power. Another characteristic compared in the hip bone density study discussed in Problem 32 was Maximum Nottingham leg power, in watts. On the WeissStats site, we provide the leg-power data for the three groups, based on the results obtained by the researchers.

34. Income by Age. The U.S. Census Bureau collects information on incomes of employed persons and publishes the results in *Historical Income Tables*. Independent simple random samples of 100 employed persons in each of four age groups gave the data on annual income, in thousands of dollars, presented on the WeissStats site.

In Problems 35–37, refer to the specified problem and use the technology of your choice to do the following tasks.
a. Decide whether conducting a Kruskal–Wallis test on the data is reasonable. If so, also do parts (b)–(d).
b. Use a Kruskal–Wallis test to decide, at the 5% significance level, whether the data provide sufficient evidence to conclude that a difference exists among the means of the populations from which the samples were taken.
c. Interpret your results from part (b).
d. If a one-way ANOVA test was performed on the data, compare the results of that test with those of the Kruskal–Wallis test, paying particular attention to the P-values.
Note: All data sets are on the WeissStats site.

***35. Weight Loss and BMI.** The data from Problem 32 on the BMIs of three weight-loss groups of older men.

***36. Weight Loss and Leg Power.** The data from Problem 33 on the Maximum Nottingham leg power of three weight-loss groups of older men.

***37. Income by Age.** The data from Problem 34 on the annual incomes, in thousands of dollars, for independent simple random samples of 100 employed persons in each of four age groups.

FOCUSING ON DATA ANALYSIS

UWEC UNDERGRADUATES

Recall from Chapter 1 (see page 34) that the Focus database and Focus sample contain information on the undergraduate students at the University of Wisconsin - Eau Claire (UWEC). Now would be a good time for you to review the discussion about these data sets.

Open the Focus sample worksheet (FocusSample) in the technology of your choice and do the following.

a. At the 10% significance level, do the data provide sufficient evidence to conclude that a difference exists among mean cumulative GPA for freshmen, sophomores, juniors, and seniors at UWEC? Use the one-way ANOVA procedure.

b. Obtain individual normal probability plots and the sample standard deviations of the GPAs of the sampled students in each class level. Based on your results, decide whether conducting a one-way ANOVA test on the data is reasonable.

c. Perform a residual analysis of the GPAs by class level. Based on your results, decide whether conducting a one-way ANOVA test on the data is reasonable.

*d. Conduct and interpret a Tukey multiple comparison corresponding to the ANOVA test in part (a). Use a 90% family confidence level.

*e. Repeat part (a), using the Kruskal–Wallis test. Compare your results with those of the one-way ANOVA test.

f. Repeat parts (a)–(c) for mean cumulative GPA by college.

CASE STUDY DISCUSSION

SELF-PERCEPTION AND PHYSICAL ACTIVITY

As you learned at the beginning of this chapter, Z. Veselska et al. explored connections between aspects of self-perception and physical activity among adolescents. The study involved a sample of adolescents from the Slovak and Czech Republics with ages ranging roughly between 12 and 16.

The youths sampled completed several self-perception questionnaires and also answered a question about their physical activity. The respondents were divided into three physical-activity categories: no physical activity (none), infrequent physical activity (infrequent), and everyday physical activity (everyday). Summary statistics for the researchers' findings for boys and girls separately are presented in the table on page 699.

First consider self-liking.

a. Assuming that self-liking scores for boys are approximately normally distributed in each physical-activity category, can we reasonably presume that the conditions for performing a one-way ANOVA are met? (*Hint:* Rule of 2)

b. Perform a one-way ANOVA to decide, at the 5% significance level, whether the data provide sufficient evidence to conclude that a difference exists in mean self-liking scores among adolescent boys in the three physical-activity categories. Interpret your result.

*c. If appropriate, conduct a Tukey multiple comparison of the three mean self-liking scores and interpret your results. Use a family confidence level of 0.95.

d. Repeat parts (a)–(c) for girls.

Next consider self-competence.

e. Assuming that self-competence scores for boys are approximately normally distributed in each physical-activity category, can we reasonably presume that the conditions for performing a one-way ANOVA are met?

f. Perform a one-way ANOVA to decide, at the 5% significance level, whether the data provide sufficient evidence to conclude that a difference exists in mean self-competence scores among adolescent boys in the three physical-activity categories. Interpret your result.

*g. If appropriate, conduct a Tukey multiple comparison of the three mean self-competence scores and interpret your results. Use a family confidence level of 0.95.

h. Repeat parts (e)–(g) for girls.

Now select any one of the remaining aspects-of-self categories and repeat parts (a)–(d).

BIOGRAPHY

SIR RONALD FISHER: MR. ANOVA

Ronald Fisher was born on February 17, 1890, in London, England, a surviving twin in a family of eight children; his father was a prominent auctioneer. Fisher graduated from Cambridge in 1912 with degrees in mathematics and physics.

From 1912 to 1919, Fisher worked at an investment house, did farm chores in Canada, and taught high school. In 1919, he took a position as a statistician at Rothamsted Experimental Station in Harpenden, West Hertford, England. His charge was to sort and reassess a 66-year accumulation of data on manurial field trials and weather records.

Fisher's work at Rothamsted during the next 15 years earned him the reputation as the leading statistician of his day and as a top-ranking geneticist. It was there, in 1925, that he published *Statistics for Research Workers,* a book that remained in print for 50 years. Fisher made important contributions to analysis of

variance (ANOVA), exact tests of significance for small samples, and maximum-likelihood solutions. He developed experimental designs to address issues in biological research, such as small samples, variable materials, and fluctuating environments.

Fisher has been described as "slight, bearded, eloquent, reactionary, and quirkish; genial to his disciples and hostile to his dissenters." He was also a prolific writer—over a span of 50 years, he wrote an average of one paper every 2 months!

In 1933, Fisher became Galton Professor of Eugenics at University College in London and, in 1943, Balfour Professor of Genetics at Cambridge. In 1952, he was knighted. Fisher "retired" in 1959, moved to Australia, and spent the last 3 years of his life working at the Division of Mathematical Statistics of the Commonwealth Scientific and Industrial Research Organization. He died in 1962 in Adelaide, Australia.

APPENDIX A

Statistical Tables

TABLE OUTLINE

TABLE I
Random numbers

Line number	Column number									
	00–09		10–19		20–29		30–39		40–49	
00	15544	80712	97742	21500	97081	42451	50623	56071	28882	28739
01	01011	21285	04729	39986	73150	31548	30168	76189	56996	19210
02	47435	53308	40718	29050	74858	64517	93573	51058	68501	42723
03	91312	75137	86274	59834	69844	19853	06917	17413	44474	86530
04	12775	08768	80791	16298	22934	09630	98862	39746	64623	32768
05	31466	43761	94872	92230	52367	13205	38634	55882	77518	36252
06	09300	43847	40881	51243	97810	18903	53914	31688	06220	40422
07	73582	13810	57784	72454	68997	72229	30340	08844	53924	89630
08	11092	81392	58189	22697	41063	09451	09789	00637	06450	85990
09	93322	98567	00116	35605	66790	52965	62877	21740	56476	49296
10	80134	12484	67089	08674	70753	90959	45842	59844	45214	36505
11	97888	31797	95037	84400	76041	96668	75920	68482	56855	97417
12	92612	27082	59459	69380	98654	20407	88151	56263	27126	63797
13	72744	45586	43279	44218	83638	05422	00995	70217	78925	39097
14	96256	70653	45285	26293	78305	80252	03625	40159	68760	84716
15	07851	47452	66742	83331	54701	06573	98169	37499	67756	68301
16	25594	41552	96475	56151	02089	33748	65289	89956	89559	33687
17	65358	15155	59374	80940	03411	94656	69440	47156	77115	99463
18	09402	31008	53424	21928	02198	61201	02457	87214	59750	51330
19	97424	90765	01634	37328	41243	33564	17884	94747	93650	77668

TABLE II

Areas under the
standard normal curve

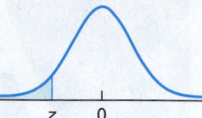

0.09	0.08	0.07	0.06	0.05	0.04	0.03	0.02	0.01	0.00	z
									0.0000†	−3.9
0.0001	0.0001	0.0001	0.0001	0.0001	0.0001	0.0001	0.0001	0.0001	0.0001	−3.8
0.0001	0.0001	0.0001	0.0001	0.0001	0.0001	0.0001	0.0001	0.0001	0.0001	−3.7
0.0001	0.0001	0.0001	0.0001	0.0001	0.0001	0.0001	0.0001	0.0002	0.0002	−3.6
0.0002	0.0002	0.0002	0.0002	0.0002	0.0002	0.0002	0.0002	0.0002	0.0002	−3.5
0.0002	0.0003	0.0003	0.0003	0.0003	0.0003	0.0003	0.0003	0.0003	0.0003	−3.4
0.0003	0.0004	0.0004	0.0004	0.0004	0.0004	0.0004	0.0005	0.0005	0.0005	−3.3
0.0005	0.0005	0.0005	0.0006	0.0006	0.0006	0.0006	0.0006	0.0007	0.0007	−3.2
0.0007	0.0007	0.0008	0.0008	0.0008	0.0008	0.0009	0.0009	0.0009	0.0010	−3.1
0.0010	0.0010	0.0011	0.0011	0.0011	0.0012	0.0012	0.0013	0.0013	0.0013	−3.0
0.0014	0.0014	0.0015	0.0015	0.0016	0.0016	0.0017	0.0018	0.0018	0.0019	−2.9
0.0019	0.0020	0.0021	0.0021	0.0022	0.0023	0.0023	0.0024	0.0025	0.0026	−2.8
0.0026	0.0027	0.0028	0.0029	0.0030	0.0031	0.0032	0.0033	0.0034	0.0035	−2.7
0.0036	0.0037	0.0038	0.0039	0.0040	0.0041	0.0043	0.0044	0.0045	0.0047	−2.6
0.0048	0.0049	0.0051	0.0052	0.0054	0.0055	0.0057	0.0059	0.0060	0.0062	−2.5
0.0064	0.0066	0.0068	0.0069	0.0071	0.0073	0.0075	0.0078	0.0080	0.0082	−2.4
0.0084	0.0087	0.0089	0.0091	0.0094	0.0096	0.0099	0.0102	0.0104	0.0107	−2.3
0.0110	0.0113	0.0116	0.0119	0.0122	0.0125	0.0129	0.0132	0.0136	0.0139	−2.2
0.0143	0.0146	0.0150	0.0154	0.0158	0.0162	0.0166	0.0170	0.0174	0.0179	−2.1
0.0183	0.0188	0.0192	0.0197	0.0202	0.0207	0.0212	0.0217	0.0222	0.0228	−2.0
0.0233	0.0239	0.0244	0.0250	0.0256	0.0262	0.0268	0.0274	0.0281	0.0287	−1.9
0.0294	0.0301	0.0307	0.0314	0.0322	0.0329	0.0336	0.0344	0.0351	0.0359	−1.8
0.0367	0.0375	0.0384	0.0392	0.0401	0.0409	0.0418	0.0427	0.0436	0.0446	−1.7
0.0455	0.0465	0.0475	0.0485	0.0495	0.0505	0.0516	0.0526	0.0537	0.0548	−1.6
0.0559	0.0571	0.0582	0.0594	0.0606	0.0618	0.0630	0.0643	0.0655	0.0668	−1.5
0.0681	0.0694	0.0708	0.0721	0.0735	0.0749	0.0764	0.0778	0.0793	0.0808	−1.4
0.0823	0.0838	0.0853	0.0869	0.0885	0.0901	0.0918	0.0934	0.0951	0.0968	−1.3
0.0985	0.1003	0.1020	0.1038	0.1056	0.1075	0.1093	0.1112	0.1131	0.1151	−1.2
0.1170	0.1190	0.1210	0.1230	0.1251	0.1271	0.1292	0.1314	0.1335	0.1357	−1.1
0.1379	0.1401	0.1423	0.1446	0.1469	0.1492	0.1515	0.1539	0.1562	0.1587	−1.0
0.1611	0.1635	0.1660	0.1685	0.1711	0.1736	0.1762	0.1788	0.1814	0.1841	−0.9
0.1867	0.1894	0.1922	0.1949	0.1977	0.2005	0.2033	0.2061	0.2090	0.2119	−0.8
0.2148	0.2177	0.2206	0.2236	0.2266	0.2296	0.2327	0.2358	0.2389	0.2420	−0.7
0.2451	0.2483	0.2514	0.2546	0.2578	0.2611	0.2643	0.2676	0.2709	0.2743	−0.6
0.2776	0.2810	0.2843	0.2877	0.2912	0.2946	0.2981	0.3015	0.3050	0.3085	−0.5
0.3121	0.3156	0.3192	0.3228	0.3264	0.3300	0.3336	0.3372	0.3409	0.3446	−0.4
0.3483	0.3520	0.3557	0.3594	0.3632	0.3669	0.3707	0.3745	0.3783	0.3821	−0.3
0.3859	0.3897	0.3936	0.3974	0.4013	0.4052	0.4090	0.4129	0.4168	0.4207	−0.2
0.4247	0.4286	0.4325	0.4364	0.4404	0.4443	0.4483	0.4522	0.4562	0.4602	−0.1
0.4641	0.4681	0.4721	0.4761	0.4801	0.4840	0.4880	0.4920	0.4960	0.5000	−0.0

Second decimal place in z

† For $z \leq -3.90$, the areas are 0.0000 to four decimal places.

TABLE II (cont.)
Areas under the
standard normal curve

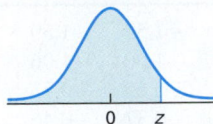

z	0.00	0.01	0.02	0.03	0.04	0.05	0.06	0.07	0.08	0.09
				Second decimal place in z						
0.0	0.5000	0.5040	0.5080	0.5120	0.5160	0.5199	0.5239	0.5279	0.5319	0.5359
0.1	0.5398	0.5438	0.5478	0.5517	0.5557	0.5596	0.5636	0.5675	0.5714	0.5753
0.2	0.5793	0.5832	0.5871	0.5910	0.5948	0.5987	0.6026	0.6064	0.6103	0.6141
0.3	0.6179	0.6217	0.6255	0.6293	0.6331	0.6368	0.6406	0.6443	0.6480	0.6517
0.4	0.6554	0.6591	0.6628	0.6664	0.6700	0.6736	0.6772	0.6808	0.6844	0.6879
0.5	0.6915	0.6950	0.6985	0.7019	0.7054	0.7088	0.7123	0.7157	0.7190	0.7224
0.6	0.7257	0.7291	0.7324	0.7357	0.7389	0.7422	0.7454	0.7486	0.7517	0.7549
0.7	0.7580	0.7611	0.7642	0.7673	0.7704	0.7734	0.7764	0.7794	0.7823	0.7852
0.8	0.7881	0.7910	0.7939	0.7967	0.7995	0.8023	0.8051	0.8078	0.8106	0.8133
0.9	0.8159	0.8186	0.8212	0.8238	0.8264	0.8289	0.8315	0.8340	0.8365	0.8389
1.0	0.8413	0.8438	0.8461	0.8485	0.8508	0.8531	0.8554	0.8577	0.8599	0.8621
1.1	0.8643	0.8665	0.8686	0.8708	0.8729	0.8749	0.8770	0.8790	0.8810	0.8830
1.2	0.8849	0.8869	0.8888	0.8907	0.8925	0.8944	0.8962	0.8980	0.8997	0.9015
1.3	0.9032	0.9049	0.9066	0.9082	0.9099	0.9115	0.9131	0.9147	0.9162	0.9177
1.4	0.9192	0.9207	0.9222	0.9236	0.9251	0.9265	0.9279	0.9292	0.9306	0.9319
1.5	0.9332	0.9345	0.9357	0.9370	0.9382	0.9394	0.9406	0.9418	0.9429	0.9441
1.6	0.9452	0.9463	0.9474	0.9484	0.9495	0.9505	0.9515	0.9525	0.9535	0.9545
1.7	0.9554	0.9564	0.9573	0.9582	0.9591	0.9599	0.9608	0.9616	0.9625	0.9633
1.8	0.9641	0.9649	0.9656	0.9664	0.9671	0.9678	0.9686	0.9693	0.9699	0.9706
1.9	0.9713	0.9719	0.9726	0.9732	0.9738	0.9744	0.9750	0.9756	0.9761	0.9767
2.0	0.9772	0.9778	0.9783	0.9788	0.9793	0.9798	0.9803	0.9808	0.9812	0.9817
2.1	0.9821	0.9826	0.9830	0.9834	0.9838	0.9842	0.9846	0.9850	0.9854	0.9857
2.2	0.9861	0.9864	0.9868	0.9871	0.9875	0.9878	0.9881	0.9884	0.9887	0.9890
2.3	0.9893	0.9896	0.9898	0.9901	0.9904	0.9906	0.9909	0.9911	0.9913	0.9916
2.4	0.9918	0.9920	0.9922	0.9925	0.9927	0.9929	0.9931	0.9932	0.9934	0.9936
2.5	0.9938	0.9940	0.9941	0.9943	0.9945	0.9946	0.9948	0.9949	0.9951	0.9952
2.6	0.9953	0.9955	0.9956	0.9957	0.9959	0.9960	0.9961	0.9962	0.9963	0.9964
2.7	0.9965	0.9966	0.9967	0.9968	0.9969	0.9970	0.9971	0.9972	0.9973	0.9974
2.8	0.9974	0.9975	0.9976	0.9977	0.9977	0.9978	0.9979	0.9979	0.9980	0.9981
2.9	0.9981	0.9982	0.9982	0.9983	0.9984	0.9984	0.9985	0.9985	0.9986	0.9986
3.0	0.9987	0.9987	0.9987	0.9988	0.9988	0.9989	0.9989	0.9989	0.9990	0.9990
3.1	0.9990	0.9991	0.9991	0.9991	0.9992	0.9992	0.9992	0.9992	0.9993	0.9993
3.2	0.9993	0.9993	0.9994	0.9994	0.9994	0.9994	0.9994	0.9995	0.9995	0.9995
3.3	0.9995	0.9995	0.9995	0.9996	0.9996	0.9996	0.9996	0.9996	0.9996	0.9997
3.4	0.9997	0.9997	0.9997	0.9997	0.9997	0.9997	0.9997	0.9997	0.9997	0.9998
3.5	0.9998	0.9998	0.9998	0.9998	0.9998	0.9998	0.9998	0.9998	0.9998	0.9998
3.6	0.9998	0.9998	0.9999	0.9999	0.9999	0.9999	0.9999	0.9999	0.9999	0.9999
3.7	0.9999	0.9999	0.9999	0.9999	0.9999	0.9999	0.9999	0.9999	0.9999	0.9999
3.8	0.9999	0.9999	0.9999	0.9999	0.9999	0.9999	0.9999	0.9999	0.9999	0.9999
3.9	1.0000[†]									

[†] For $z \geq 3.90$, the areas are 1.0000 to four decimal places.

TABLE III
Normal scores

Ordered position	n								
	5	6	7	8	9	10	11	12	13
1	−1.18	−1.28	−1.36	−1.43	−1.50	−1.55	−1.59	−1.64	−1.68
2	−0.50	−0.64	−0.76	−0.85	−0.93	−1.00	−1.06	−1.11	−1.16
3	0.00	−0.20	−0.35	−0.47	−0.57	−0.65	−0.73	−0.79	−0.85
4	0.50	0.20	0.00	−0.15	−0.27	−0.37	−0.46	−0.53	−0.60
5	1.18	0.64	0.35	0.15	0.00	−0.12	−0.22	−0.31	−0.39
6		1.28	0.76	0.47	0.27	0.12	0.00	−0.10	−0.19
7			1.36	0.85	0.57	0.37	0.22	0.10	0.00
8				1.43	0.93	0.65	0.46	0.31	0.19
9					1.50	1.00	0.73	0.53	0.39
10						1.55	1.06	0.79	0.60
11							1.59	1.11	0.85
12								1.64	1.16
13									1.68

TABLE III   (cont.)
Normal scores

Ordered position	n								
	14	15	16	17	18	19	20	21	22
1	−1.71	−1.74	−1.77	−1.80	−1.82	−1.85	−1.87	−1.89	−1.91
2	−1.20	−1.24	−1.28	−1.32	−1.35	−1.38	−1.40	−1.43	−1.45
3	−0.90	−0.94	−0.99	−1.03	−1.06	−1.10	−1.13	−1.16	−1.18
4	−0.66	−0.71	−0.76	−0.80	−0.84	−0.88	−0.92	−0.95	−0.98
5	−0.45	−0.51	−0.57	−0.62	−0.66	−0.70	−0.74	−0.78	−0.81
6	−0.27	−0.33	−0.39	−0.45	−0.50	−0.54	−0.59	−0.63	−0.66
7	−0.09	−0.16	−0.23	−0.29	−0.35	−0.40	−0.45	−0.49	−0.53
8	0.09	0.00	−0.08	−0.15	−0.21	−0.26	−0.31	−0.36	−0.40
9	0.27	0.16	0.08	0.00	−0.07	−0.13	−0.19	−0.24	−0.28
10	0.45	0.33	0.23	0.15	0.07	0.00	−0.06	−0.12	−0.17
11	0.66	0.51	0.39	0.29	0.21	0.13	0.06	0.00	−0.06
12	0.90	0.71	0.57	0.45	0.35	0.26	0.19	0.12	0.06
13	1.20	0.94	0.76	0.62	0.50	0.40	0.31	0.24	0.17
14	1.71	1.24	0.99	0.80	0.66	0.54	0.45	0.36	0.28
15		1.74	1.28	1.03	0.84	0.70	0.59	0.49	0.40
16			1.77	1.32	1.06	0.88	0.74	0.63	0.53
17				1.80	1.35	1.10	0.92	0.78	0.66
18					1.82	1.38	1.13	0.95	0.81
19						1.85	1.40	1.16	0.98
20							1.87	1.43	1.18
21								1.89	1.45
22									1.91

TABLE III (cont.)
Normal scores

Ordered position	n							
	23	24	25	26	27	28	29	30
1	−1.93	−1.95	−1.97	−1.98	−2.00	−2.01	−2.03	−2.04
2	−1.48	−1.50	−1.52	−1.54	−1.56	−1.58	−1.59	−1.61
3	−1.21	−1.24	−1.26	−1.28	−1.30	−1.32	−1.34	−1.36
4	−1.01	−1.04	−1.06	−1.09	−1.11	−1.13	−1.15	−1.17
5	−0.84	−0.87	−0.90	−0.93	−0.95	−0.98	−1.00	−1.02
6	−0.70	−0.73	−0.76	−0.79	−0.82	−0.84	−0.87	−0.89
7	−0.57	−0.60	−0.63	−0.66	−0.69	−0.72	−0.75	−0.77
8	−0.44	−0.48	−0.52	−0.55	−0.58	−0.61	−0.64	−0.67
9	−0.33	−0.37	−0.41	−0.44	−0.48	−0.51	−0.54	−0.57
10	−0.22	−0.26	−0.30	−0.34	−0.38	−0.41	−0.44	−0.47
11	−0.11	−0.15	−0.20	−0.24	−0.28	−0.31	−0.35	−0.38
12	0.00	−0.05	−0.10	−0.14	−0.18	−0.22	−0.26	−0.29
13	0.11	0.05	0.00	−0.05	−0.09	−0.13	−0.17	−0.21
14	0.22	0.15	0.10	0.05	0.00	−0.04	−0.09	−0.12
15	0.33	0.26	0.20	0.14	0.09	0.04	0.00	−0.04
16	0.44	0.37	0.30	0.24	0.18	0.13	0.09	0.04
17	0.57	0.48	0.41	0.34	0.28	0.22	0.17	0.12
18	0.70	0.60	0.52	0.44	0.38	0.31	0.26	0.21
19	0.84	0.73	0.63	0.55	0.48	0.41	0.35	0.29
20	1.01	0.87	0.76	0.66	0.58	0.51	0.44	0.38
21	1.21	1.04	0.90	0.79	0.69	0.61	0.54	0.47
22	1.48	1.24	1.06	0.93	0.82	0.72	0.64	0.57
23	1.93	1.50	1.26	1.09	0.95	0.84	0.75	0.67
24		1.95	1.52	1.28	1.11	0.98	0.87	0.77
25			1.97	1.54	1.30	1.13	1.00	0.89
26				1.98	1.56	1.32	1.15	1.02
27					2.00	1.58	1.34	1.17
28						2.01	1.59	1.36
29							2.03	1.61
30								2.04

TABLE IV
Values of t_α

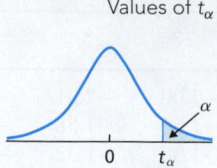

df	$t_{0.10}$	$t_{0.05}$	$t_{0.025}$	$t_{0.01}$	$t_{0.005}$	df
1	3.078	6.314	12.706	31.821	63.657	1
2	1.886	2.920	4.303	6.965	9.925	2
3	1.638	2.353	3.182	4.541	5.841	3
4	1.533	2.132	2.776	3.747	4.604	4
5	1.476	2.015	2.571	3.365	4.032	5
6	1.440	1.943	2.447	3.143	3.707	6
7	1.415	1.895	2.365	2.998	3.499	7
8	1.397	1.860	2.306	2.896	3.355	8
9	1.383	1.833	2.262	2.821	3.250	9
10	1.372	1.812	2.228	2.764	3.169	10
11	1.363	1.796	2.201	2.718	3.106	11
12	1.356	1.782	2.179	2.681	3.055	12
13	1.350	1.771	2.160	2.650	3.012	13
14	1.345	1.761	2.145	2.624	2.977	14
15	1.341	1.753	2.131	2.602	2.947	15
16	1.337	1.746	2.120	2.583	2.921	16
17	1.333	1.740	2.110	2.567	2.898	17
18	1.330	1.734	2.101	2.552	2.878	18
19	1.328	1.729	2.093	2.539	2.861	19
20	1.325	1.725	2.086	2.528	2.845	20
21	1.323	1.721	2.080	2.518	2.831	21
22	1.321	1.717	2.074	2.508	2.819	22
23	1.319	1.714	2.069	2.500	2.807	23
24	1.318	1.711	2.064	2.492	2.797	24
25	1.316	1.708	2.060	2.485	2.787	25
26	1.315	1.706	2.056	2.479	2.779	26
27	1.314	1.703	2.052	2.473	2.771	27
28	1.313	1.701	2.048	2.467	2.763	28
29	1.311	1.699	2.045	2.462	2.756	29
30	1.310	1.697	2.042	2.457	2.750	30
31	1.309	1.696	2.040	2.453	2.744	31
32	1.309	1.694	2.037	2.449	2.738	32
33	1.308	1.692	2.035	2.445	2.733	33
34	1.307	1.691	2.032	2.441	2.728	34
35	1.306	1.690	2.030	2.438	2.724	35
36	1.306	1.688	2.028	2.434	2.719	36
37	1.305	1.687	2.026	2.431	2.715	37
38	1.304	1.686	2.024	2.429	2.712	38
39	1.304	1.685	2.023	2.426	2.708	39
40	1.303	1.684	2.021	2.423	2.704	40
41	1.303	1.683	2.020	2.421	2.701	41
42	1.302	1.682	2.018	2.418	2.698	42
43	1.302	1.681	2.017	2.416	2.695	43
44	1.301	1.680	2.015	2.414	2.692	44
45	1.301	1.679	2.014	2.412	2.690	45
46	1.300	1.679	2.013	2.410	2.687	46
47	1.300	1.678	2.012	2.408	2.685	47
48	1.299	1.677	2.011	2.407	2.682	48
49	1.299	1.677	2.010	2.405	2.680	49

TABLE IV (cont.)
Values of t_α

df	$t_{0.10}$	$t_{0.05}$	$t_{0.025}$	$t_{0.01}$	$t_{0.005}$	df
50	1.299	1.676	2.009	2.403	2.678	50
51	1.298	1.675	2.008	2.402	2.676	51
52	1.298	1.675	2.007	2.400	2.674	52
53	1.298	1.674	2.006	2.399	2.672	53
54	1.297	1.674	2.005	2.397	2.670	54
55	1.297	1.673	2.004	2.396	2.668	55
56	1.297	1.673	2.003	2.395	2.667	56
57	1.297	1.672	2.002	2.394	2.665	57
58	1.296	1.672	2.002	2.392	2.663	58
59	1.296	1.671	2.001	2.391	2.662	59
60	1.296	1.671	2.000	2.390	2.660	60
61	1.296	1.670	2.000	2.389	2.659	61
62	1.295	1.670	1.999	2.388	2.657	62
63	1.295	1.669	1.998	2.387	2.656	63
64	1.295	1.669	1.998	2.386	2.655	64
65	1.295	1.669	1.997	2.385	2.654	65
66	1.295	1.668	1.997	2.384	2.652	66
67	1.294	1.668	1.996	2.383	2.651	67
68	1.294	1.668	1.995	2.382	2.650	68
69	1.294	1.667	1.995	2.382	2.649	69
70	1.294	1.667	1.994	2.381	2.648	70
71	1.294	1.667	1.994	2.380	2.647	71
72	1.293	1.666	1.993	2.379	2.646	72
73	1.293	1.666	1.993	2.379	2.645	73
74	1.293	1.666	1.993	2.378	2.644	74
75	1.293	1.665	1.992	2.377	2.643	75
80	1.292	1.664	1.990	2.374	2.639	80
85	1.292	1.663	1.988	2.371	2.635	85
90	1.291	1.662	1.987	2.368	2.632	90
95	1.291	1.661	1.985	2.366	2.629	95
100	1.290	1.660	1.984	2.364	2.626	100
200	1.286	1.653	1.972	2.345	2.601	200
300	1.284	1.650	1.968	2.339	2.592	300
400	1.284	1.649	1.966	2.336	2.588	400
500	1.283	1.648	1.965	2.334	2.586	500
600	1.283	1.647	1.964	2.333	2.584	600
700	1.283	1.647	1.963	2.332	2.583	700
800	1.283	1.647	1.963	2.331	2.582	800
900	1.282	1.647	1.963	2.330	2.581	900
1000	1.282	1.646	1.962	2.330	2.581	1000
2000	1.282	1.646	1.961	2.328	2.578	2000

1.282	1.645	1.960	2.326	2.576
$z_{0.10}$	$z_{0.05}$	$z_{0.025}$	$z_{0.01}$	$z_{0.005}$

TABLE V
Values of W_α

n	$W_{0.10}$	$W_{0.05}$	$W_{0.025}$	$W_{0.01}$	$W_{0.005}$	n
7	22	24	26	28	—	7
8	28	30	32	34	36	8
9	34	37	39	42	43	9
10	41	44	47	50	52	10
11	48	52	55	59	61	11
12	56	61	64	68	71	12
13	65	70	74	78	81	13
14	74	79	84	89	92	14
15	83	90	95	100	104	15
16	94	100	106	112	117	16
17	104	112	118	125	130	17
18	116	124	131	138	143	18
19	128	136	144	152	158	19
20	140	150	158	167	173	20

TABLE VI
Values of M_α

n_2	α	n_1							
		3	4	5	6	7	8	9	10
	0.10	14	20	27	36	45	55	66	78
	0.05	15	21	29	37	46	57	68	80
3	0.025	—	22	30	38	48	58	70	82
	0.01	—	—	—	39	49	59	71	83
	0.005	—	—	—	—	—	60	72	85
	0.10	16	23	31	40	49	60	72	85
	0.05	17	24	32	41	51	62	74	87
4	0.025	18	25	33	43	53	64	76	89
	0.01	—	26	35	44	54	65	78	91
	0.005	—	—	—	45	55	66	79	93
	0.10	18	26	34	44	54	65	78	91
	0.05	20	27	36	46	56	68	80	94
5	0.025	21	28	37	47	58	70	83	96
	0.01	—	30	39	49	60	72	85	99
	0.005	—	—	40	50	61	73	86	101
	0.10	21	29	38	48	59	71	84	98
	0.05	22	30	40	50	61	73	87	101
6	0.025	23	32	41	52	63	76	89	103
	0.01	24	33	43	54	65	78	92	106
	0.005	—	34	44	55	67	80	94	108
	0.10	23	31	41	52	63	76	89	104
	0.05	24	33	43	54	66	79	93	107
7	0.025	26	35	45	56	68	81	95	110
	0.01	27	36	47	58	71	84	98	114
	0.005	—	37	48	60	72	86	101	116
	0.10	25	34	44	56	68	81	95	110
	0.05	27	36	47	58	71	84	99	114
8	0.025	28	38	49	61	73	87	102	117
	0.01	29	39	51	63	76	90	105	121
	0.005	30	40	52	65	78	92	108	124
	0.10	27	37	48	60	72	86	101	116
	0.05	29	39	50	63	76	90	105	121
9	0.025	31	41	53	65	78	93	108	124
	0.01	32	43	55	68	81	96	112	129
	0.005	33	44	56	70	84	99	114	131
	0.10	29	40	51	64	77	91	106	123
	0.05	31	42	54	67	80	95	111	127
10	0.025	33	44	56	69	83	98	114	131
	0.01	34	46	59	72	87	102	119	136
	0.005	36	48	61	74	89	105	121	139

TABLE VII
Values of χ_α^2

df	$\chi_{0.995}^2$	$\chi_{0.99}^2$	$\chi_{0.975}^2$	$\chi_{0.95}^2$	$\chi_{0.90}^2$
1	0.000	0.000	0.001	0.004	0.016
2	0.010	0.020	0.051	0.103	0.211
3	0.072	0.115	0.216	0.352	0.584
4	0.207	0.297	0.484	0.711	1.064
5	0.412	0.554	0.831	1.145	1.610
6	0.676	0.872	1.237	1.635	2.204
7	0.989	1.239	1.690	2.167	2.833
8	1.344	1.646	2.180	2.733	3.490
9	1.735	2.088	2.700	3.325	4.168
10	2.156	2.558	3.247	3.940	4.865
11	2.603	3.053	3.816	4.575	5.578
12	3.074	3.571	4.404	5.226	6.304
13	3.565	4.107	5.009	5.892	7.042
14	4.075	4.660	5.629	6.571	7.790
15	4.601	5.229	6.262	7.261	8.547
16	5.142	5.812	6.908	7.962	9.312
17	5.697	6.408	7.564	8.672	10.085
18	6.265	7.015	8.231	9.390	10.865
19	6.844	7.633	8.907	10.117	11.651
20	7.434	8.260	9.591	10.851	12.443
21	8.034	8.897	10.283	11.591	13.240
22	8.643	9.542	10.982	12.338	14.041
23	9.260	10.196	11.689	13.091	14.848
24	9.886	10.856	12.401	13.848	15.659
25	10.520	11.524	13.120	14.611	16.473
26	11.160	12.198	13.844	15.379	17.292
27	11.808	12.879	14.573	16.151	18.114
28	12.461	13.565	15.308	16.928	18.939
29	13.121	14.256	16.047	17.708	19.768
30	13.787	14.953	16.791	18.493	20.599
40	20.707	22.164	24.433	26.509	29.051
50	27.991	29.707	32.357	34.764	37.689
60	35.534	37.485	40.482	43.188	46.459
70	43.275	45.442	48.758	51.739	55.329
80	51.172	53.540	57.153	60.391	64.278
90	59.196	61.754	65.647	69.126	73.291
100	67.328	70.065	74.222	77.930	82.358

TABLE VII (cont.)
Values of χ_α^2

$\chi_{0.10}^2$	$\chi_{0.05}^2$	$\chi_{0.025}^2$	$\chi_{0.01}^2$	$\chi_{0.005}^2$	df
2.706	3.841	5.024	6.635	7.879	1
4.605	5.991	7.378	9.210	10.597	2
6.251	7.815	9.348	11.345	12.838	3
7.779	9.488	11.143	13.277	14.860	4
9.236	11.070	12.833	15.086	16.750	5
10.645	12.592	14.449	16.812	18.548	6
12.017	14.067	16.013	18.475	20.278	7
13.362	15.507	17.535	20.090	21.955	8
14.684	16.919	19.023	21.666	23.589	9
15.987	18.307	20.483	23.209	25.188	10
17.275	19.675	21.920	24.725	26.757	11
18.549	21.026	23.337	26.217	28.300	12
19.812	22.362	24.736	27.688	29.819	13
21.064	23.685	26.119	29.141	31.319	14
22.307	24.996	27.488	30.578	32.801	15
23.542	26.296	28.845	32.000	34.267	16
24.769	27.587	30.191	33.409	35.718	17
25.989	28.869	31.526	34.805	37.156	18
27.204	30.143	32.852	36.191	38.582	19
28.412	31.410	34.170	37.566	39.997	20
29.615	32.671	35.479	38.932	41.401	21
30.813	33.924	36.781	40.290	42.796	22
32.007	35.172	38.076	41.638	44.181	23
33.196	36.415	39.364	42.980	45.559	24
34.382	37.653	40.647	44.314	46.928	25
35.563	38.885	41.923	45.642	48.290	26
36.741	40.113	43.195	46.963	49.645	27
37.916	41.337	44.461	48.278	50.994	28
39.087	42.557	45.722	49.588	52.336	29
40.256	43.773	46.979	50.892	53.672	30
51.805	55.759	59.342	63.691	66.767	40
63.167	67.505	71.420	76.154	79.490	50
74.397	79.082	83.298	88.381	91.955	60
85.527	90.531	95.023	100.424	104.213	70
96.578	101.879	106.628	112.328	116.320	80
107.565	113.145	118.135	124.115	128.296	90
118.499	124.343	129.563	135.811	140.177	100

TABLE VIII
Values of F_α

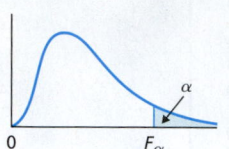

		dfn								
dfd	α	1	2	3	4	5	6	7	8	9
	0.10	39.86	49.50	53.59	55.83	57.24	58.20	58.91	59.44	59.86
	0.05	161.45	199.50	215.71	224.58	230.16	233.99	236.77	238.88	240.54
1	0.025	647.79	799.50	864.16	899.58	921.85	937.11	948.22	956.66	963.28
	0.01	4052.2	4999.5	5403.4	5624.6	5763.6	5859.0	5928.4	5981.1	6022.5
	0.005	16211	20000	21615	22500	23056	23437	23715	23925	24091
	0.10	8.53	9.00	9.16	9.24	9.29	9.33	9.35	9.37	9.38
	0.05	18.51	19.00	19.16	19.25	19.30	19.33	19.35	19.37	19.38
2	0.025	38.51	39.00	39.17	39.25	39.30	39.33	39.36	39.37	39.39
	0.01	98.50	99.00	99.17	99.25	99.30	99.33	99.36	99.37	99.39
	0.005	198.50	199.00	199.17	199.25	199.30	199.33	199.36	199.37	199.39
	0.10	5.54	5.46	5.39	5.34	5.31	5.28	5.27	5.25	5.24
	0.05	10.13	9.55	9.28	9.12	9.01	8.94	8.89	8.85	8.81
3	0.025	17.44	16.04	15.44	15.10	14.88	14.73	14.62	14.54	14.47
	0.01	34.12	30.82	29.46	28.71	28.24	27.91	27.67	27.49	27.35
	0.005	55.55	49.80	47.47	46.19	45.39	44.84	44.43	44.13	43.88
	0.10	4.54	4.32	4.19	4.11	4.05	4.01	3.98	3.95	3.94
	0.05	7.71	6.94	6.59	6.39	6.26	6.16	6.09	6.04	6.00
4	0.025	12.22	10.65	9.98	9.60	9.36	9.20	9.07	8.98	8.90
	0.01	21.20	18.00	16.69	15.98	15.52	15.21	14.98	14.80	14.66
	0.005	31.33	26.28	24.26	23.15	22.46	21.97	21.62	21.35	21.14
	0.10	4.06	3.78	3.62	3.52	3.45	3.40	3.37	3.34	3.32
	0.05	6.61	5.79	5.41	5.19	5.05	4.95	4.88	4.82	4.77
5	0.025	10.01	8.43	7.76	7.39	7.15	6.98	6.85	6.76	6.68
	0.01	16.26	13.27	12.06	11.39	10.97	10.67	10.46	10.29	10.16
	0.005	22.78	18.31	16.53	15.56	14.94	14.51	14.20	13.96	13.77
	0.10	3.78	3.46	3.29	3.18	3.11	3.05	3.01	2.98	2.96
	0.05	5.99	5.14	4.76	4.53	4.39	4.28	4.21	4.15	4.10
6	0.025	8.81	7.26	6.60	6.23	5.99	5.82	5.70	5.60	5.52
	0.01	13.75	10.92	9.78	9.15	8.75	8.47	8.26	8.10	7.98
	0.005	18.63	14.54	12.92	12.03	11.46	11.07	10.79	10.57	10.39
	0.10	3.59	3.26	3.07	2.96	2.88	2.83	2.78	2.75	2.72
	0.05	5.59	4.74	4.35	4.12	3.97	3.87	3.79	3.73	3.68
7	0.025	8.07	6.54	5.89	5.52	5.29	5.12	4.99	4.90	4.82
	0.01	12.25	9.55	8.45	7.85	7.46	7.19	6.99	6.84	6.72
	0.005	16.24	12.40	10.88	10.05	9.52	9.16	8.89	8.68	8.51
	0.10	3.46	3.11	2.92	2.81	2.73	2.67	2.62	2.59	2.56
	0.05	5.32	4.46	4.07	3.84	3.69	3.58	3.50	3.44	3.39
8	0.025	7.57	6.06	5.42	5.05	4.82	4.65	4.53	4.43	4.36
	0.01	11.26	8.65	7.59	7.01	6.63	6.37	6.18	6.03	5.91
	0.005	14.69	11.04	9.60	8.81	8.30	7.95	7.69	7.50	7.34

TABLE VIII (cont.)
Values of F_α

10	12	15	20	24	30	40	60	120	α	dfd
60.19	60.71	61.22	61.74	62.00	62.26	62.53	62.79	63.06	0.10	
241.88	243.91	245.95	248.01	249.05	250.10	251.14	252.20	253.25	0.05	
968.63	976.71	984.87	993.10	997.25	1001.41	1005.60	1009.80	1014.02	0.025	1
6055.8	6106.3	6157.3	6208.7	6234.6	6260.6	6286.7	631.9	6339.4	0.01	
24224	24426	24630	24836	24940	25044	25148	25253	25359	0.005	
9.39	9.41	9.42	9.44	9.45	9.46	9.47	9.47	9.48	0.10	
19.40	19.41	19.43	19.45	19.45	19.46	19.47	19.48	19.49	0.05	
39.40	39.41	39.43	39.45	39.46	39.46	39.47	39.48	39.49	0.025	2
99.40	99.42	99.43	99.45	99.46	99.47	99.47	99.48	99.49	0.01	
199.40	199.42	199.43	199.45	199.46	199.47	199.47	199.48	199.49	0.005	
5.23	5.22	5.20	5.18	5.18	5.17	5.16	5.15	5.14	0.10	
8.79	8.74	8.70	8.66	8.64	8.62	8.59	8.57	8.55	0.05	
14.42	14.34	14.25	14.17	14.12	14.08	14.04	13.99	13.95	0.025	3
27.23	27.05	26.87	26.69	26.60	26.50	26.41	26.32	26.22	0.01	
43.69	43.39	43.08	42.78	42.62	42.47	42.31	42.15	41.99	0.005	
3.92	3.90	3.87	3.84	3.83	3.82	3.80	3.79	3.78	0.10	
5.96	5.91	5.86	5.80	5.77	5.75	5.72	5.69	5.66	0.05	
8.84	8.75	8.66	8.56	8.51	8.46	8.41	8.36	8.31	0.025	4
14.55	14.37	14.20	14.02	13.93	13.84	13.75	13.65	13.56	0.01	
20.97	20.70	20.44	20.17	20.03	19.89	19.75	19.61	19.47	0.005	
3.30	3.27	3.24	3.21	3.19	3.17	3.16	3.14	3.12	0.10	
4.74	4.68	4.62	4.56	4.53	4.50	4.46	4.43	4.40	0.05	
6.62	6.52	6.43	6.33	6.28	6.23	6.18	6.12	6.07	0.025	5
10.05	9.89	9.72	9.55	9.47	9.38	9.29	9.20	9.11	0.01	
13.62	13.38	13.15	12.90	12.78	12.66	12.53	12.40	12.27	0.005	
2.94	2.90	2.87	2.84	2.82	2.80	2.78	2.76	2.74	0.10	
4.06	4.00	3.94	3.87	3.84	3.81	3.77	3.74	3.70	0.05	
5.46	5.37	5.27	5.17	5.12	5.07	5.01	4.96	4.90	0.025	6
7.87	7.72	7.56	7.40	7.31	7.23	7.14	7.06	6.97	0.01	
10.25	10.03	9.81	9.59	9.47	9.36	9.24	9.12	9.00	0.005	
2.70	2.67	2.63	2.59	2.58	2.56	2.54	2.51	2.49	0.10	
3.64	3.57	3.51	3.44	3.41	3.38	3.34	3.30	3.27	0.05	
4.76	4.67	4.57	4.47	4.41	4.36	4.31	4.25	4.20	0.025	7
6.62	6.47	6.31	6.16	6.07	5.99	5.91	5.82	5.74	0.01	
8.38	8.18	7.97	7.75	7.64	7.53	7.42	7.31	7.19	0.005	
2.54	2.50	2.46	2.42	2.40	2.38	2.36	2.34	2.32	0.10	
3.35	3.28	3.22	3.15	3.12	3.08	3.04	3.01	2.97	0.05	
4.30	4.20	4.10	4.00	3.95	3.89	3.84	3.78	3.73	0.025	8
5.81	5.67	5.52	5.36	5.28	5.20	5.12	5.03	4.95	0.01	
7.21	7.01	6.81	6.61	6.50	6.40	6.29	6.18	6.06	0.005	

TABLE VIII   (cont.)
Values of F_α

dfd	α	1	2	3	4	5	6	7	8	9
						dfn				
	0.10	3.36	3.01	2.81	2.69	2.61	2.55	2.51	2.47	2.44
	0.05	5.12	4.26	3.86	3.63	3.48	3.37	3.29	3.23	3.18
9	0.025	7.21	5.71	5.08	4.72	4.48	4.32	4.20	4.10	4.03
	0.01	10.56	8.02	6.99	6.42	6.06	5.80	5.61	5.47	5.35
	0.005	13.61	10.11	8.72	7.96	7.47	7.13	6.88	6.69	6.54
	0.10	3.29	2.92	2.73	2.61	2.52	2.46	2.41	2.38	2.35
	0.05	4.96	4.10	3.71	3.48	3.33	3.22	3.14	3.07	3.02
10	0.025	6.94	5.46	4.83	4.47	4.24	4.07	3.95	3.85	3.78
	0.01	10.04	7.56	6.55	5.99	5.64	5.39	5.20	5.06	4.94
	0.005	12.83	9.43	8.08	7.34	6.87	6.54	6.30	6.12	5.97
	0.10	3.23	2.86	2.66	2.54	2.45	2.39	2.34	2.30	2.27
	0.05	4.84	3.98	3.59	3.36	3.20	3.09	3.01	2.95	2.90
11	0.025	6.72	5.26	4.63	4.28	4.04	3.88	3.76	3.66	3.59
	0.01	9.65	7.21	6.22	5.67	5.32	5.07	4.89	4.74	4.63
	0.005	12.23	8.91	7.60	6.88	6.42	6.10	5.86	5.68	5.54
	0.10	3.18	2.81	2.61	2.48	2.39	2.33	2.28	2.24	2.21
	0.05	4.75	3.89	3.49	3.26	3.11	3.00	2.91	2.85	2.80
12	0.025	6.55	5.10	4.47	4.12	3.89	3.73	3.61	3.51	3.44
	0.01	9.33	6.93	5.95	5.41	5.06	4.82	4.64	4.50	4.39
	0.005	11.75	8.51	7.23	6.52	6.07	5.76	5.52	5.35	5.20
	0.10	3.14	2.76	2.56	2.43	2.35	2.28	2.23	2.20	2.16
	0.05	4.67	3.81	3.41	3.18	3.03	2.92	2.83	2.77	2.71
13	0.025	6.41	4.97	4.35	4.00	3.77	3.60	3.48	3.39	3.31
	0.01	9.07	6.70	5.74	5.21	4.86	4.62	4.44	4.30	4.19
	0.005	11.37	8.19	6.93	6.23	5.79	5.48	5.25	5.08	4.94
	0.10	3.10	2.73	2.52	2.39	2.31	2.24	2.19	2.15	2.12
	0.05	4.60	3.74	3.34	3.11	2.96	2.85	2.76	2.70	2.65
14	0.025	6.30	4.86	4.24	3.89	3.66	3.50	3.38	3.29	3.21
	0.01	8.86	6.51	5.56	5.04	4.69	4.46	4.28	4.14	4.03
	0.005	11.06	7.92	6.68	6.00	5.56	5.26	5.03	4.86	4.72
	0.10	3.07	2.70	2.49	2.36	2.27	2.21	2.16	2.12	2.09
	0.05	4.54	3.68	3.29	3.06	2.90	2.79	2.71	2.64	2.59
15	0.025	6.20	4.77	4.15	3.80	3.58	3.41	3.29	3.20	3.12
	0.01	8.68	6.36	5.42	4.89	4.56	4.32	4.14	4.00	3.89
	0.005	10.80	7.70	6.48	5.80	5.37	5.07	4.85	4.67	4.54
	0.10	3.05	2.67	2.46	2.33	2.24	2.18	2.13	2.09	2.06
	0.05	4.49	3.63	3.24	3.01	2.85	2.74	2.66	2.59	2.54
16	0.025	6.12	4.69	4.08	3.73	3.50	3.34	3.22	3.12	3.05
	0.01	8.53	6.23	5.29	4.77	4.44	4.20	4.03	3.89	3.78
	0.005	10.58	7.51	6.30	5.64	5.21	4.91	4.69	4.52	4.38

TABLE VIII (cont.)
Values of F_α

			dfn								
10	12	15	20	24	30	40	60	120	α	dfd
2.42	2.38	2.34	2.30	2.28	2.25	2.23	2.21	2.18	0.10	
3.14	3.07	3.01	2.94	2.90	2.86	2.83	2.79	2.75	0.05	
3.96	3.87	3.77	3.67	3.61	3.56	3.51	3.45	3.39	0.025	9
5.26	5.11	4.96	4.81	4.73	4.65	4.57	4.48	4.40	0.01	
6.42	6.23	6.03	5.83	5.73	5.62	5.52	5.41	5.30	0.005	
2.32	2.28	2.24	2.20	2.18	2.16	2.13	2.11	2.08	0.10	
2.98	2.91	2.85	2.77	2.74	2.70	2.66	2.62	2.58	0.05	
3.72	3.62	3.52	3.42	3.37	3.31	3.26	3.20	3.14	0.025	10
4.85	4.71	4.56	4.41	4.33	4.25	4.17	4.08	4.00	0.01	
5.85	5.66	5.47	5.27	5.17	5.07	4.97	4.86	4.75	0.005	
2.25	2.21	2.17	2.12	2.10	2.08	2.05	2.03	2.00	0.10	
2.85	2.79	2.72	2.65	2.61	2.57	2.53	2.49	2.45	0.05	
3.53	3.43	3.33	3.23	3.17	3.12	3.06	3.00	2.94	0.025	11
4.54	4.40	4.25	4.10	4.02	3.94	3.86	3.78	3.69	0.01	
5.42	5.24	5.05	4.86	4.76	4.65	4.55	4.45	4.34	0.005	
2.19	2.15	2.10	2.06	2.04	2.01	1.99	1.96	1.93	0.10	
2.75	2.69	2.62	2.54	2.51	2.47	2.43	2.38	2.34	0.05	
3.37	3.28	3.18	3.07	3.02	2.96	2.91	2.85	2.79	0.025	12
4.30	4.16	4.01	3.86	3.78	3.70	3.62	3.54	3.45	0.01	
5.09	4.91	4.72	4.53	4.43	4.33	4.23	4.12	4.01	0.005	
2.14	2.10	2.05	2.01	1.98	1.96	1.93	1.90	1.88	0.10	
2.67	2.60	2.53	2.46	2.42	2.38	2.34	2.30	2.25	0.05	
3.25	3.15	3.05	2.95	2.89	2.84	2.78	2.72	2.66	0.025	13
4.10	3.96	3.82	3.66	3.59	3.51	3.43	3.34	3.25	0.01	
4.82	4.64	4.46	4.27	4.17	4.07	3.97	3.87	3.76	0.005	
2.10	2.05	2.01	1.96	1.94	1.91	1.89	1.86	1.83	0.10	
2.60	2.53	2.46	2.39	2.35	2.31	2.27	2.22	2.18	0.05	
3.15	3.05	2.95	2.84	2.79	2.73	2.67	2.61	2.55	0.025	14
3.94	3.80	3.66	3.51	3.43	3.35	3.27	3.18	3.09	0.01	
4.60	4.43	4.25	4.06	3.96	3.86	3.76	3.66	3.55	0.005	
2.06	2.02	1.97	1.92	1.90	1.87	1.85	1.82	1.79	0.10	
2.54	2.48	2.40	2.33	2.29	2.25	2.20	2.16	2.11	0.05	
3.06	2.96	2.86	2.76	2.70	2.64	2.59	2.52	2.46	0.025	15
3.80	3.67	3.52	3.37	3.29	3.21	3.13	3.05	2.96	0.01	
4.42	4.25	4.07	3.88	3.79	3.69	3.58	3.48	3.37	0.005	
2.03	1.99	1.94	1.89	1.87	1.84	1.81	1.78	1.75	0.10	
2.49	2.42	2.35	2.28	2.24	2.19	2.15	2.11	2.06	0.05	
2.99	2.89	2.79	2.68	2.63	2.57	2.51	2.45	2.38	0.025	16
3.69	3.55	3.41	3.26	3.18	3.10	3.02	2.93	2.84	0.01	
4.27	4.10	3.92	3.73	3.64	3.54	3.44	3.33	3.22	0.005	

TABLE VIII (cont.)
Values of F_α

dfd	α	dfn								
		1	*2*	*3*	*4*	*5*	*6*	*7*	*8*	*9*
	0.10	3.03	2.64	2.44	2.31	2.22	2.15	2.10	2.06	2.03
	0.05	4.45	3.59	3.20	2.96	2.81	2.70	2.61	2.55	2.49
17	*0.025*	6.04	4.62	4.01	3.66	3.44	3.28	3.16	3.06	2.98
	0.01	8.40	6.11	5.18	4.67	4.34	4.10	3.93	3.79	3.68
	0.005	10.38	7.35	6.16	5.50	5.07	4.78	4.56	4.39	4.25
	0.10	3.01	2.62	2.42	2.29	2.20	2.13	2.08	2.04	2.00
	0.05	4.41	3.55	3.16	2.93	2.77	2.66	2.58	2.51	2.46
18	*0.025*	5.98	4.56	3.95	3.61	3.38	3.22	3.10	3.01	2.93
	0.01	8.29	6.01	5.09	4.58	4.25	4.01	3.84	3.71	3.60
	0.005	10.22	7.21	6.03	5.37	4.96	4.66	4.44	4.28	4.14
	0.10	2.99	2.61	2.40	2.27	2.18	2.11	2.06	2.02	1.98
	0.05	4.38	3.52	3.13	2.90	2.74	2.63	2.54	2.48	2.42
19	*0.025*	5.92	4.51	3.90	3.56	3.33	3.17	3.05	2.96	2.88
	0.01	8.18	5.93	5.01	4.50	4.17	3.94	3.77	3.63	3.52
	0.005	10.07	7.09	5.92	5.27	4.85	4.56	4.34	4.18	4.04
	0.10	2.97	2.59	2.38	2.25	2.16	2.09	2.04	2.00	1.96
	0.05	4.35	3.49	3.10	2.87	2.71	2.60	2.51	2.45	2.39
20	*0.025*	5.87	4.46	3.86	3.51	3.29	3.13	3.01	2.91	2.84
	0.01	8.10	5.85	4.94	4.43	4.10	3.87	3.70	3.56	3.46
	0.005	9.94	6.99	5.82	5.17	4.76	4.47	4.26	4.09	3.96
	0.10	2.96	2.57	2.36	2.23	2.14	2.08	2.02	1.98	1.95
	0.05	4.32	3.47	3.07	2.84	2.68	2.57	2.49	2.42	2.37
21	*0.025*	5.83	4.42	3.82	3.48	3.25	3.09	2.97	2.87	2.80
	0.01	8.02	5.78	4.87	4.37	4.04	3.81	3.64	3.51	3.40
	0.005	9.83	6.89	5.73	5.09	4.68	4.39	4.18	4.01	3.88
	0.10	2.95	2.56	2.35	2.22	2.13	2.06	2.01	1.97	1.93
	0.05	4.30	3.44	3.05	2.82	2.66	2.55	2.46	2.40	2.34
22	*0.025*	5.79	4.38	3.78	3.44	3.22	3.05	2.93	2.84	2.76
	0.01	7.95	5.72	4.82	4.31	3.99	3.76	3.59	3.45	3.35
	0.005	9.73	6.81	5.65	5.02	4.61	4.32	4.11	3.94	3.81
	0.10	2.94	2.55	2.34	2.21	2.11	2.05	1.99	1.95	1.92
	0.05	4.28	3.42	3.03	2.80	2.64	2.53	2.44	2.37	2.32
23	*0.025*	5.75	4.35	3.75	3.41	3.18	3.02	2.90	2.81	2.73
	0.01	7.88	5.66	4.76	4.26	3.94	3.71	3.54	3.41	3.30
	0.005	9.63	6.73	5.58	4.95	4.54	4.26	4.05	3.88	3.75
	0.10	2.93	2.54	2.33	2.19	2.10	2.04	1.98	1.94	1.91
	0.05	4.26	3.40	3.01	2.78	2.62	2.51	2.42	2.36	2.30
24	*0.025*	5.72	4.32	3.72	3.38	3.15	2.99	2.87	2.78	2.70
	0.01	7.82	5.61	4.72	4.22	3.90	3.67	3.50	3.36	3.26
	0.005	9.55	6.66	5.52	4.89	4.49	4.20	3.99	3.83	3.69

TABLE VIII (cont.)
Values of F_α

10	12	15	20	24	30	40	60	120	α	dfd
2.00	1.96	1.91	1.86	1.84	1.81	1.78	1.75	1.72	0.10	
2.45	2.38	2.31	2.23	2.19	2.15	2.10	2.06	2.01	0.05	
2.92	2.82	2.72	2.62	2.56	2.50	2.44	2.38	2.32	0.025	17
3.59	3.46	3.31	3.16	3.08	3.00	2.92	2.83	2.75	0.01	
4.14	3.97	3.79	3.61	3.51	3.41	3.31	3.21	3.10	0.005	
1.98	1.93	1.89	1.84	1.81	1.78	1.75	1.72	1.69	0.10	
2.41	2.34	2.27	2.19	2.15	2.11	2.06	2.02	1.97	0.05	
2.87	2.77	2.67	2.56	2.50	2.44	2.38	2.32	2.26	0.025	18
3.51	3.37	3.23	3.08	3.00	2.92	2.84	2.75	2.66	0.01	
4.03	3.86	3.68	3.50	3.40	3.30	3.20	3.10	2.99	0.005	
1.96	1.91	1.86	1.81	1.79	1.76	1.73	1.70	1.67	0.10	
2.38	2.31	2.23	2.16	2.11	2.07	2.03	1.98	1.93	0.05	
2.82	2.72	2.62	2.51	2.45	2.39	2.33	2.27	2.20	0.025	19
3.43	3.30	3.15	3.00	2.92	2.84	2.76	2.67	2.58	0.01	
3.93	3.76	3.59	3.40	3.31	3.21	3.11	3.00	2.89	0.005	
1.94	1.89	1.84	1.79	1.77	1.74	1.71	1.68	1.64	0.10	
2.35	2.28	2.20	2.12	2.08	2.04	1.99	1.95	1.90	0.05	
2.77	2.68	2.57	2.46	2.41	2.35	2.29	2.22	2.16	0.025	20
3.37	3.23	3.09	2.94	2.86	2.78	2.69	2.61	2.52	0.01	
3.85	3.68	3.50	3.32	3.22	3.12	3.02	2.92	2.81	0.005	
1.92	1.87	1.83	1.78	1.75	1.72	1.69	1.66	1.62	0.10	
2.32	2.25	2.18	2.10	2.05	2.01	1.96	1.92	1.87	0.05	
2.73	2.64	2.53	2.42	2.37	2.31	2.25	2.18	2.11	0.025	21
3.31	3.17	3.03	2.88	2.80	2.72	2.64	2.55	2.46	0.01	
3.77	3.60	3.43	3.24	3.15	3.05	2.95	2.84	2.73	0.005	
1.90	1.86	1.81	1.76	1.73	1.70	1.67	1.64	1.60	0.10	
2.30	2.23	2.15	2.07	2.03	1.98	1.94	1.89	1.84	0.05	
2.70	2.60	2.50	2.39	2.33	2.27	2.21	2.14	2.08	0.025	22
3.26	3.12	2.98	2.83	2.75	2.67	2.58	2.50	2.40	0.01	
3.70	3.54	3.36	3.18	3.08	2.98	2.88	2.77	2.66	0.005	
1.89	1.84	1.80	1.74	1.72	1.69	1.66	1.62	1.59	0.10	
2.27	2.20	2.13	2.05	2.01	1.96	1.91	1.86	1.81	0.05	
2.67	2.57	2.47	2.36	2.30	2.24	2.18	2.11	2.04	0.025	23
3.21	3.07	2.93	2.78	2.70	2.62	2.54	2.45	2.35	0.01	
3.64	3.47	3.30	3.12	3.02	2.92	2.82	2.71	2.60	0.005	
1.88	1.83	1.78	1.73	1.70	1.67	1.64	1.61	1.57	0.10	
2.25	2.18	2.11	2.03	1.98	1.94	1.89	1.84	1.79	0.05	
2.64	2.54	2.44	2.33	2.27	2.21	2.15	2.08	2.01	0.025	24
3.17	3.03	2.89	2.74	2.66	2.58	2.49	2.40	2.31	0.01	
3.59	3.42	3.25	3.06	2.97	2.87	2.77	2.66	2.55	0.005	

TABLE VIII (cont.)

Values of F_α

dfd	α					dfn				
		1	*2*	*3*	*4*	*5*	*6*	*7*	*8*	*9*
	0.10	2.92	2.53	2.32	2.18	2.09	2.02	1.97	1.93	1.89
	0.05	4.24	3.39	2.99	2.76	2.60	2.49	2.40	2.34	2.28
25	*0.025*	5.69	4.29	3.69	3.35	3.13	2.97	2.85	2.75	2.68
	0.01	7.77	5.57	4.68	4.18	3.85	3.63	3.46	3.32	3.22
	0.005	9.48	6.60	5.46	4.84	4.43	4.15	3.94	3.78	3.64
	0.10	2.91	2.52	2.31	2.17	2.08	2.01	1.96	1.92	1.88
	0.05	4.23	3.37	2.98	2.74	2.59	2.47	2.39	2.32	2.27
26	*0.025*	5.66	4.27	3.67	3.33	3.10	2.94	2.82	2.73	2.65
	0.01	7.72	5.53	4.64	4.14	3.82	3.59	3.42	3.29	3.18
	0.005	9.41	6.54	5.41	4.79	4.38	4.10	3.89	3.73	3.60
	0.10	2.90	2.51	2.30	2.17	2.07	2.00	1.95	1.91	1.87
	0.05	4.21	3.35	2.96	2.73	2.57	2.46	2.37	2.31	2.25
27	*0.025*	5.63	4.24	3.65	3.31	3.08	2.92	2.80	2.71	2.63
	0.01	7.68	5.49	4.60	4.11	3.78	3.56	3.39	3.26	3.15
	0.005	9.34	6.49	5.36	4.74	4.34	4.06	3.85	3.69	3.56
	0.10	2.89	2.50	2.29	2.16	2.06	2.00	1.94	1.90	1.87
	0.05	4.20	3.34	2.95	2.71	2.56	2.45	2.36	2.29	2.24
28	*0.025*	5.61	4.22	3.63	3.29	3.06	2.90	2.78	2.69	2.61
	0.01	7.64	5.45	4.57	4.07	3.75	3.53	3.36	3.23	3.12
	0.005	9.28	6.44	5.32	4.70	4.30	4.02	3.81	3.65	3.52
	0.10	2.89	2.50	2.28	2.15	2.06	1.99	1.93	1.89	1.86
	0.05	4.18	3.33	2.93	2.70	2.55	2.43	2.35	2.28	2.22
29	*0.025*	5.59	4.20	3.61	3.27	3.04	2.88	2.76	2.67	2.59
	0.01	7.60	5.42	4.54	4.04	3.73	3.50	3.33	3.20	3.09
	0.005	9.23	6.40	5.28	4.66	4.26	3.98	3.77	3.61	3.48
	0.10	2.88	2.49	2.28	2.14	2.05	1.98	1.93	1.88	1.85
	0.05	4.17	3.32	2.92	2.69	2.53	2.42	2.33	2.27	2.21
30	*0.025*	5.57	4.18	3.59	3.25	3.03	2.87	2.75	2.65	2.57
	0.01	7.56	5.39	4.51	4.02	3.70	3.47	3.30	3.17	3.07
	0.005	9.18	6.35	5.24	4.62	4.23	3.95	3.74	3.58	3.45
	0.10	2.79	2.39	2.18	2.04	1.95	1.87	1.82	1.77	1.74
	0.05	4.00	3.15	2.76	2.53	2.37	2.25	2.17	2.10	2.04
60	*0.025*	5.29	3.93	3.34	3.01	2.79	2.63	2.51	2.41	2.33
	0.01	7.08	4.98	4.13	3.65	3.34	3.12	2.95	2.82	2.72
	0.005	8.49	5.79	4.73	4.14	3.76	3.49	3.29	3.13	3.01
	0.10	2.75	2.35	2.13	1.99	1.90	1.82	1.77	1.72	1.68
	0.05	3.92	3.07	2.68	2.45	2.29	2.18	2.09	2.02	1.96
120	*0.025*	5.15	3.80	3.23	2.89	2.67	2.52	2.39	2.30	2.22
	0.01	6.85	4.79	3.95	3.48	3.17	2.96	2.79	2.66	2.56
	0.005	8.18	5.54	4.50	3.92	3.55	3.28	3.09	2.93	2.81

TABLE VIII (cont.)
Values of F_α

				dfn						
10	12	15	20	24	30	40	60	120	α	dfd
1.87	1.82	1.77	1.72	1.69	1.66	1.63	1.59	1.56	0.10	
2.24	2.16	2.09	2.01	1.96	1.92	1.87	1.82	1.77	0.05	
2.61	2.51	2.41	2.30	2.24	2.18	2.12	2.05	1.98	0.025	25
3.13	2.99	2.85	2.70	2.62	2.54	2.45	2.36	2.27	0.01	
3.54	3.37	3.20	3.01	2.92	2.82	2.72	2.61	2.50	0.005	
1.86	1.81	1.76	1.71	1.68	1.65	1.61	1.58	1.54	0.10	
2.22	2.15	2.07	1.99	1.95	1.90	1.85	1.80	1.75	0.05	
2.59	2.49	2.39	2.28	2.22	2.16	2.09	2.03	1.95	0.025	26
3.09	2.96	2.81	2.66	2.58	2.50	2.42	2.33	2.23	0.01	
3.49	3.33	3.15	2.97	2.87	2.77	2.67	2.56	2.45	0.005	
1.85	1.80	1.75	1.70	1.67	1.64	1.60	1.57	1.53	0.10	
2.20	2.13	2.06	1.97	1.93	1.88	1.84	1.79	1.73	0.05	
2.57	2.47	2.36	2.25	2.19	2.13	2.07	2.00	1.93	0.025	27
3.06	2.93	2.78	2.63	2.55	2.47	2.38	2.29	2.20	0.01	
3.45	3.28	3.11	2.93	2.83	2.73	2.63	2.52	2.41	0.005	
1.84	1.79	1.74	1.69	1.66	1.63	1.59	1.56	1.52	0.10	
2.19	2.12	2.04	1.96	1.91	1.87	1.82	1.77	1.71	0.05	
2.55	2.45	2.34	2.23	2.17	2.11	2.05	1.98	1.91	0.025	28
3.03	2.90	2.75	2.60	2.52	2.44	2.35	2.26	2.17	0.01	
3.41	3.25	3.07	2.89	2.79	2.69	2.59	2.48	2.37	0.005	
1.83	1.78	1.73	1.68	1.65	1.62	1.58	1.55	1.51	0.10	
2.18	2.10	2.03	1.94	1.90	1.85	1.81	1.75	1.70	0.05	
2.53	2.43	2.32	2.21	2.15	2.09	2.03	1.96	1.89	0.025	29
3.00	2.87	2.73	2.57	2.49	2.41	2.33	2.23	2.14	0.01	
3.38	3.21	3.04	2.86	2.76	2.66	2.56	2.45	2.33	0.005	
1.82	1.77	1.72	1.67	1.64	1.61	1.57	1.54	1.50	0.10	
2.16	2.09	2.01	1.93	1.89	1.84	1.79	1.74	1.68	0.05	
2.51	2.41	2.31	2.20	2.14	2.07	2.01	1.94	1.87	0.025	30
2.98	2.84	2.70	2.55	2.47	2.39	2.30	2.21	2.11	0.01	
3.34	3.18	3.01	2.82	2.73	2.63	2.52	2.42	2.30	0.005	
1.71	1.66	1.60	1.54	1.51	1.48	1.44	1.40	1.35	0.10	
1.99	1.92	1.84	1.75	1.70	1.65	1.59	1.53	1.47	0.05	
2.27	2.17	2.06	1.94	1.88	1.82	1.74	1.67	1.58	0.025	60
2.63	2.50	2.35	2.20	2.12	2.03	1.94	1.84	1.73	0.01	
2.90	2.74	2.57	2.39	2.29	2.19	2.08	1.96	1.83	0.005	
1.65	1.60	1.55	1.48	1.45	1.41	1.37	1.32	1.26	0.10	
1.91	1.83	1.75	1.66	1.61	1.55	1.50	1.43	1.35	0.05	
2.16	2.05	1.94	1.82	1.76	1.69	1.61	1.53	1.43	0.025	120
2.47	2.34	2.19	2.03	1.95	1.86	1.76	1.66	1.53	0.01	
2.71	2.54	2.37	2.19	2.09	1.98	1.87	1.75	1.61	0.005	

TABLE IX
Values of $q_{0.01}$

v	κ									v
	2	3	4	5	6	7	8	9	10	
1	90.0	135	164	186	202	216	227	237	246	1
2	14.0	19.0	22.3	24.7	26.6	28.2	29.5	30.7	31.7	2
3	8.26	10.6	12.2	13.3	14.2	15.0	15.6	16.2	16.7	3
4	6.51	8.12	9.17	9.96	10.6	11.1	11.5	11.9	12.3	4
5	5.70	6.97	7.80	8.42	8.91	9.32	9.67	9.97	10.2	5
6	5.24	6.33	7.03	7.56	7.97	8.32	8.61	8.87	9.10	6
7	4.95	5.92	6.54	7.01	7.37	7.68	7.94	8.17	8.37	7
8	4.74	5.63	6.20	6.63	6.96	7.24	7.47	7.68	7.87	8
9	4.60	5.43	5.96	6.35	6.66	6.91	7.13	7.32	7.49	9
10	4.48	5.27	5.77	6.14	6.43	6.67	6.87	7.05	7.21	10
11	4.39	5.14	5.62	5.97	6.25	6.48	6.67	6.84	6.99	11
12	4.32	5.04	5.50	5.84	6.10	6.32	6.51	6.67	6.81	12
13	4.26	4.96	5.40	5.73	5.98	6.19	6.37	6.53	6.67	13
14	4.21	4.89	5.32	5.63	5.88	6.08	6.26	6.41	6.54	14
15	4.17	4.83	5.25	5.56	5.80	5.99	6.16	6.31	6.44	15
16	4.13	4.78	5.19	5.49	5.72	5.92	6.08	6.22	6.35	16
17	4.10	4.74	5.14	5.43	5.66	5.85	6.01	6.15	6.27	17
18	4.07	4.70	5.09	5.38	5.60	5.79	5.94	6.08	6.20	18
19	4.05	4.67	5.05	5.33	5.55	5.73	5.89	6.02	6.14	19
20	4.02	4.64	5.02	5.29	5.51	5.69	5.84	5.97	6.09	20
24	3.96	4.54	4.91	5.17	5.37	5.54	5.69	5.81	5.92	24
30	3.89	4.45	4.80	5.05	5.24	5.40	5.54	5.65	5.76	30
40	3.82	4.37	4.70	4.93	5.11	5.27	5.39	5.50	5.60	40
60	3.76	4.28	4.60	4.82	4.99	5.13	5.25	5.36	5.45	60
120	3.70	4.20	4.50	4.71	4.87	5.01	5.12	5.21	5.30	120
∞	3.64	4.12	4.40	4.60	4.76	4.88	4.99	5.08	5.16	∞

TABLE X
Values of $q_{0.05}$

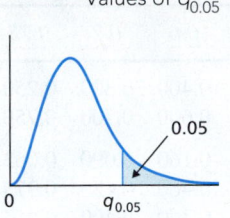

					κ						
ν	2	3	4	5	6	7	8	9	10	ν
1	18.0	27.0	32.8	37.1	40.4	43.1	45.4	47.4	49.1	1
2	6.08	8.33	9.80	10.9	11.7	12.4	13.0	13.5	14.0	2
3	4.50	5.91	6.82	7.50	8.04	8.48	8.85	9.18	9.46	3
4	3.93	5.04	5.76	6.29	6.71	7.05	7.35	7.60	7.83	4
5	3.64	4.60	5.22	5.67	6.03	6.33	6.58	6.80	6.99	5
6	3.46	4.34	4.90	5.30	5.63	5.90	6.12	6.32	6.49	6
7	3.34	4.16	4.68	5.06	5.36	5.61	5.82	6.00	6.16	7
8	3.26	4.04	4.53	4.89	5.17	5.40	5.60	5.77	5.92	8
9	3.20	3.95	4.41	4.76	5.02	5.24	5.43	5.59	5.74	9
10	3.15	3.88	4.33	4.65	4.91	5.12	5.30	5.46	5.60	10
11	3.11	3.82	4.26	4.57	4.82	5.03	5.20	5.35	5.49	11
12	3.08	3.77	4.20	4.51	4.75	4.95	5.12	5.27	5.39	12
13	3.06	3.73	4.15	4.45	4.69	4.88	5.05	5.19	5.32	13
14	3.03	3.70	4.11	4.41	4.64	4.83	4.99	5.13	5.25	14
15	3.01	3.67	4.08	4.37	4.59	4.78	4.94	5.08	5.20	15
16	3.00	3.65	4.05	4.33	4.56	4.74	4.90	5.03	5.15	16
17	2.98	3.63	4.02	4.30	4.52	4.70	4.86	4.99	5.11	17
18	2.97	3.61	4.00	4.28	4.49	4.67	4.82	4.96	5.07	18
19	2.96	3.59	3.98	4.25	4.47	4.65	4.79	4.92	5.04	19
20	2.95	3.58	3.96	4.23	4.45	4.62	4.77	4.90	5.01	20
24	2.92	3.53	3.90	4.17	4.37	4.54	4.68	4.81	4.92	24
30	2.89	3.49	3.85	4.10	4.30	4.46	4.60	4.72	4.82	30
40	2.86	3.44	3.79	4.04	4.23	4.39	4.52	4.63	4.73	40
60	2.83	3.40	3.74	3.98	4.16	4.31	4.44	4.55	4.65	60
120	2.80	3.36	3.68	3.92	4.10	4.24	4.36	4.47	4.56	120
∞	2.77	3.31	3.63	3.86	4.03	4.17	4.29	4.39	4.47	∞

TABLE XI

Binomial probabilities:

$$\binom{n}{x} p^x (1-p)^{n-x}$$

							p						
n	x	0.1	0.2	0.25	0.3	0.4	0.5	0.6	0.7	0.75	0.8	0.9
1	0	0.900	0.800	0.750	0.700	0.600	0.500	0.400	0.300	0.250	0.200	0.100
	1	0.100	0.200	0.250	0.300	0.400	0.500	0.600	0.700	0.750	0.800	0.900
2	0	0.810	0.640	0.563	0.490	0.360	0.250	0.160	0.090	0.063	0.040	0.010
	1	0.180	0.320	0.375	0.420	0.480	0.500	0.480	0.420	0.375	0.320	0.180
	2	0.010	0.040	0.063	0.090	0.160	0.250	0.360	0.490	0.563	0.640	0.810
3	0	0.729	0.512	0.422	0.343	0.216	0.125	0.064	0.027	0.016	0.008	0.001
	1	0.243	0.384	0.422	0.441	0.432	0.375	0.288	0.189	0.141	0.096	0.027
	2	0.027	0.096	0.141	0.189	0.288	0.375	0.432	0.441	0.422	0.384	0.243
	3	0.001	0.008	0.016	0.027	0.064	0.125	0.216	0.343	0.422	0.512	0.729
4	0	0.656	0.410	0.316	0.240	0.130	0.063	0.026	0.008	0.004	0.002	0.000
	1	0.292	0.410	0.422	0.412	0.346	0.250	0.154	0.076	0.047	0.026	0.004
	2	0.049	0.154	0.211	0.265	0.346	0.375	0.346	0.265	0.211	0.154	0.049
	3	0.004	0.026	0.047	0.076	0.154	0.250	0.346	0.412	0.422	0.410	0.292
	4	0.000	0.002	0.004	0.008	0.026	0.063	0.130	0.240	0.316	0.410	0.656
5	0	0.590	0.328	0.237	0.168	0.078	0.031	0.010	0.002	0.001	0.000	0.000
	1	0.328	0.410	0.396	0.360	0.259	0.156	0.077	0.028	0.015	0.006	0.000
	2	0.073	0.205	0.264	0.309	0.346	0.312	0.230	0.132	0.088	0.051	0.008
	3	0.008	0.051	0.088	0.132	0.230	0.312	0.346	0.309	0.264	0.205	0.073
	4	0.000	0.006	0.015	0.028	0.077	0.156	0.259	0.360	0.396	0.410	0.328
	5	0.000	0.000	0.001	0.002	0.010	0.031	0.078	0.168	0.237	0.328	0.590
6	0	0.531	0.262	0.178	0.118	0.047	0.016	0.004	0.001	0.000	0.000	0.000
	1	0.354	0.393	0.356	0.303	0.187	0.094	0.037	0.010	0.004	0.002	0.000
	2	0.098	0.246	0.297	0.324	0.311	0.234	0.138	0.060	0.033	0.015	0.001
	3	0.015	0.082	0.132	0.185	0.276	0.313	0.276	0.185	0.132	0.082	0.015
	4	0.001	0.015	0.033	0.060	0.138	0.234	0.311	0.324	0.297	0.246	0.098
	5	0.000	0.002	0.004	0.010	0.037	0.094	0.187	0.303	0.356	0.393	0.354
	6	0.000	0.000	0.000	0.001	0.004	0.016	0.047	0.118	0.178	0.262	0.531
7	0	0.478	0.210	0.133	0.082	0.028	0.008	0.002	0.000	0.000	0.000	0.000
	1	0.372	0.367	0.311	0.247	0.131	0.055	0.017	0.004	0.001	0.000	0.000
	2	0.124	0.275	0.311	0.318	0.261	0.164	0.077	0.025	0.012	0.004	0.000
	3	0.023	0.115	0.173	0.227	0.290	0.273	0.194	0.097	0.058	0.029	0.003
	4	0.003	0.029	0.058	0.097	0.194	0.273	0.290	0.227	0.173	0.115	0.023
	5	0.000	0.004	0.012	0.025	0.077	0.164	0.261	0.318	0.311	0.275	0.124
	6	0.000	0.000	0.001	0.004	0.017	0.055	0.131	0.247	0.311	0.367	0.372
	7	0.000	0.000	0.000	0.000	0.002	0.008	0.028	0.082	0.133	0.210	0.478

Answers to Selected Exercises

NOTE:

- This appendix contains answers to most of the odd-numbered Understanding the Concepts and Skills and Applying the Concepts and Skills section exercises and to most of the Understanding the Concepts and Skills and Applying the Concepts and Skills review problems.

- Most of the numerical answers presented here were obtained by using a computer. If you solve a problem by hand and do some intermediate rounding or use provided summary statistics, your answer may differ slightly from the one given in this appendix.

- The *Student's Solutions Manual* contains detailed, worked-out solutions to the odd-numbered section exercises (Understanding the Concepts and Skills, Applying the Concepts and Skills, Working with Large Data Sets, and Extending the Concepts and Skills) and all review problems.

Chapter 1

Exercises 1.1

1.1 See Definition 1.2 on page 3.

1.3 Descriptive statistics includes the construction of graphs, charts, and tables and the calculation of various descriptive measures such as averages, measures of variation, and percentiles.

1.5
a. In an observational study, researchers simply observe characteristics and take measurements, as in a sample survey.
b. In a designed experiment, researchers impose treatments and controls and then observe characteristics and take measurements.

1.7 Inferential   **1.9** Descriptive   **1.11** Descriptive

1.13 a. Inferential
b. The sample consists of those U.S. adults who were interviewed; the population consists of all U.S. adults.

1.15 a. Descriptive   **b.** Inferential   **1.17** Designed experiment

1.19 Observational study   **1.21** Designed experiment

Exercises 1.2

1.27 Conducting a census may be time consuming, costly, impractical, or even impossible.

1.29 Because the sample will be used to draw conclusions about the entire population.

1.31
a. In probability sampling, a random device—such as tossing a coin, consulting a table of random numbers, or employing a random number generator—is used to decide which members of the population will constitute the sample instead of leaving such decisions to human judgment.
b. No. Because probability sampling uses a random device, it is possible to obtain a nonrepresentative sample.
c. Probability sampling helps eliminate unintentional selection bias and permits the researcher to control the chance of obtaining a nonrepresentative sample. Also, use of probability sampling guarantees that the techniques of inferential statistics can be applied.

1.33 Simple random sampling   **1.35** SRS

1.37 a.

1, 2	1, 3	1, 4	2, 3	2, 4	3, 4

b. $\frac{1}{6}$ **c.** 1, 4

1.39 a. 43, 45, 1, 42, 37, 47   **b.** Answers will vary.

1.41 Dentists form a high-income group whose incomes are not representative of the incomes of Seattle residents in general.

1.43 a.

G, L, S	G, L, A	G, I, T	G, S, A	G, S, T
G, A, T	L, S, A	L, S, T	L, A, T	S, A, T

b. $\frac{1}{10}, \frac{1}{10}, \frac{1}{10}$

1.45 a.

E, M, P, L	E, M, P, A	E, M, P, B	E, M, L, A	E, M, L, B
E, M, A, B	E, P, L, A	E, P, L, B	E, P, A, B	E, L, A, B
M, P, L, A	M, P, L, B	M, P, A, B	M, L, A, B	P, L, A, B

b. Write the initials of the six artists on separate pieces of paper, place the six slips of paper in a box, and then, while blindfolded, pick four of the slips of paper.

c. $\frac{1}{15}, \frac{1}{15}$

1.47 a.

F, T	F, G	F, H	F, L	F, B	F, A	T, G
T, H	T, L	T, B	T, A	G, H	G, L	G, B
G, A	H, L	H, B	H, A	L, B	L, A	B, A

b. $\frac{1}{21}, \frac{1}{21}$

1.49 a.

452	16	343	242	428	378	163	182	293	422

b. Answers will vary.

1.51 a. Answers will vary.   **b.** Answers will vary.

Exercises 1.3

1.55 cyclical pattern   **1.57** homogeneous

1.59 a. Answers will vary.   **b.** 10, 84, 158, 232, 306

1.61 a. Answers will vary.   **b.** 1–10, 21–30

1.63 a. 6, 4, 8, 2   **b.** Answers will vary.   **1.65** Stratified sampling

1.67 Systematic random sampling   **1.69** Cluster sampling

1.71 a. Answers will vary.   **b.** Systematic random sampling
c. Answers will vary.

1.73
a. Number the suites from 1 to 48, use a table of random numbers to randomly
select 3 of the 48 suites, and take as the sample the 24 dormitory residents
living in the 3 suites obtained.
b. Probably not, because friends often have similar opinions.
c. Proportional allocation dictates that the number of freshmen, sophomores,
juniors, and seniors selected be 8, 7, 6, and 3, respectively. Thus a stratified
sample of 24 dormitory residents can be obtained as follows: Number the
freshman dormitory residents from 1 through 128 and use a table of random
numbers to randomly select 8 of the 128 freshman dormitory residents;
number the sophomore dormitory residents from 1 through 112 and use a
table of random numbers to randomly select 7 of the 112 sophomore
dormitory residents; and so on.

1.75 a. Answers will vary.
b. The representatives numbered 12, 41, 70, 99, 128, 157, 186, 215, 244, 273,
302, 331, 360, 389, and 418

Exercises 1.4

1.81 a. The individuals or items on which the experiment is performed
b. Subject

1.83 See Definition 1.6 on page 26.   **1.85** 4

1.87 a. In the table, TT = Treatment.

		B			
		b_1	b_2	b_3	b_4
A	a_1	TT 1	TT 2	TT 3	TT 4
	a_2	TT 5	TT 6	TT 7	TT 8
	a_3	TT 9	TT 10	TT 11	TT 12

b. 12   **c.** Yes, there are $3 \times 4 = 12$ treatments.

1.89 $m \times n$

1.91 a. Three   **b.** The pharmacologic therapy alone group
c. Two. Pharmacologic therapy with a pacemaker; pharmacologic therapy
with a pacemaker–defibrillator combination
d. 304 in the pharmacologic therapy alone group; 608 each in the
pharmacologic therapy with a pacemaker group and the pharmacologic
therapy with a pacemaker–defibrillator combination group

1.93 a. The drivers in the study   **b.** Detection distance, in feet
c. Sign size and sign material
d. Sign size has three levels: small, medium, and large. Sign material has three
levels: 1, 2, and 3.
e. The nine different possible combinations of the three sign sizes and the
three sign materials

1.95 a. The female lions
b. Whether or not (yes or no) the female lions approached a male dummy
c. Mane length and mane color
d. Mane length has two levels: long and short. Mane color has two levels:
blonde and dark.
e. The four different possible combinations of the two mane lengths and the
two mane colors

1.97 a. The children in the study   **b.** IQ score at school age
c. Dexamethasone usage
d. Dexamethasone usage has two levels: not used and used.
e. Same as the two levels of dexamethasone usage

1.99 a. Randomized block design
b. The two genders (female and male preschoolers)

Review Problems for Chapter 1

1. Answers will vary.

2. It is almost always necessary to invoke techniques of descriptive
statistics to organize and summarize the information obtained from a
sample before carrying out an inferential analysis.

3. **a.** In an observational study, researchers simply observe characteristics
and take measurements, as in a sample survey. In a designed
experiment, researchers impose treatments and controls and then
observe characteristics and take measurements.
b. Observational studies can reveal only association, whereas designed
experiments can help establish causation.

4. A literature search

5. **a.** A representative sample is a sample that reflects as closely as possible
the relevant characteristics of the population under consideration.
b. In probability sampling, a random device, such as tossing a coin or
consulting a table of random numbers, is used to decide which
members of the population will constitute the sample instead of
leaving such decisions to human judgement.
c. Simple random sampling is a sampling procedure for which each
possible sample of a given size is equally likely to be the one obtained
from the population.

6. Only (b)

7. See Section 1.3 and, in particular, Procedure 1.1 on page 17,
Procedure 1.2 on page 18, and Procedure 1.3 on page 21.

8. See Key Fact 1.1 on page 26   9. Descriptive

10. **a.** Descriptive   **b.** Inferential   **11.** Inferential

12. **a.** Descriptive   **b.** Observational   **13.** Observational

14. Designed experiment

15. No, because parents of students at Yale tend to have higher incomes than
parents of college students in general.

16. **a.**

H, Z, C	H, Z, A	H, Z, J	H, C, A	H, C, J
H, A, J	Z, C, A	Z, C, J	Z, A, J	C, A, J

   b. $\frac{1}{10}, \frac{1}{10}, \frac{1}{10}$ **c.** Answers will vary.   **d.** Answers will vary.

17. **a.** Number the athletes from 1 to 100, use Table I to obtain 15 different
numbers between 1 and 100, and take as the sample the 15 athletes
who are numbered with the numbers obtained.
b. 82, 8, 16, 1, 47, 94, 97, 74, 52, 76, 98, 3, 89, 41, 63
c. Answers will vary.

18. The statement is a disclaimer as to the validity of the survey. Because the
results reflect only responses of Internet users, they cannot be regarded as
representative of the public in general. Moreover, because the sample
was not chosen at random from Internet users, but rather was obtained
only from volunteers, the results cannot even be considered
representative of Internet users.

19. Answers will vary.

20. The study is observational and, hence, can reveal only association. A
designed experiment and possibly other information are needed to try to
establish causation.

21. **a.** Answers will vary.
b. Yes, unless for some reason there is a cyclical pattern in the listing of
the athletes.

22. Proportional allocation dictates that the sample sizes for the strata be
9, 22, 19, 11, 9, and 10, respectively. So, to obtain the required stratified
sample, select a simple random sample of 9 water samples from the first
stratum, a simple random sample of 22 water samples from the second
stratum, and so forth. Then combine all six samples.

23. **a.** Designed experiment
b. The treatment group consists of the 158 patients who took AVONEX.
The control group consists of the 143 patients who were given
placebo. The treatments are AVONEX and placebo.

24. **a.** The tomato plants in the study (Some might say the plots of land are
the experimental units.)
b. Yield of tomato plants   **c.** Tomato variety and planting density
d. Tomato variety has four levels: Harvester, Pusa Early Dwarf, Ife
No. 1, and Ibadan Local. Planting density has four levels: 10,000,
20,000, 30,000, and 40,000 plants/ha.

e. The 16 different possible combinations of the four tomato varieties and the four planting densities

25. a. The children on the panel
 b. Whether the bottle is opened or not (yes or no)
 c. Design type **d.** The three design types **e.** The three design types

26. Completely randomized design

27. a. Completely randomized design
 b. Randomized block design; the six different car models
 c. The randomized block design in part (b)

Chapter 2

Exercises 2.1

2.1 Answers will vary. **2.3** See Definition 2.2 on page 38.

2.5 Qualitative variable

2.7 a. Quantitative, continuous; time at which an earthquake occurred
b. Quantitative, continuous; magnitude of an earthquake on the Richter scale
c. Quantitative, continuous; depth, in kilometers, at which an earthquake occurred
d. Quantitative, discrete; number of stations that reported the activity on an earthquake.
e. Qualitative; region where an earthquake occurred

2.9
a. Quantitative, discrete; rank of a deceased celebrity by 2012 earnings
b. Qualitative; name of a deceased celebrity
c. Quantitative, discrete or continuous; 2012 earnings, in millions of dollars, of a deceased celebrity

2.11 a. Qualitative; product type
b. Quantitative, discrete; units shipped, in millions
c. Quantitative, discrete or continuous; retail value of shipments, in millions of dollars

2.13 Quantitative, qualitative, qualitative, and quantitative

2.15 Quantitative, qualitative, qualitative, and quantitative

Exercises 2.2

2.17 A frequency distribution of qualitative data is a listing of the distinct values and their frequencies. A frequency distribution is useful for organizing qualitative data so that the data are more compact and easier to understand.

2.19 a. True **b.** False
c. Relative frequencies always lie between 0 and 1 and hence provide a standard for comparison.

2.21 a.

Category	Frequency
A	3
B	1
C	1

b.

Category	Relative frequency
A	0.6
B	0.2
C	0.2

c. Data

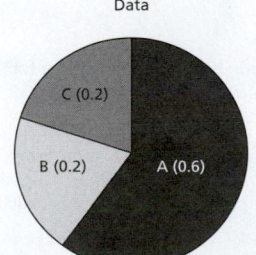

d. Data

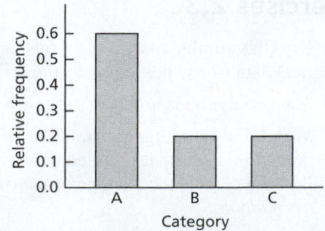

2.23 a.

Category	Frequency
A	4
B	3
C	1
D	2

b.

Category	Relative frequency
A	0.4
B	0.3
C	0.1
D	0.2

c. Data

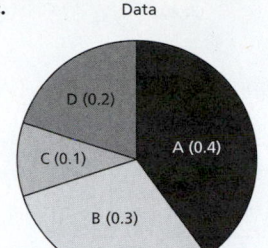

d. Data

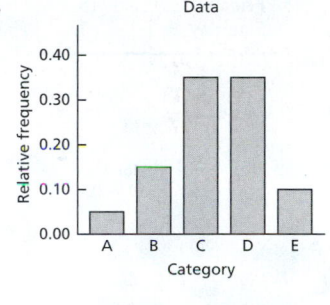

2.25 a.

Category	Frequency
A	1
B	3
C	7
D	7
E	2

b.

Category	Relative frequency
A	0.05
B	0.15
C	0.35
D	0.35
E	0.10

c. Data

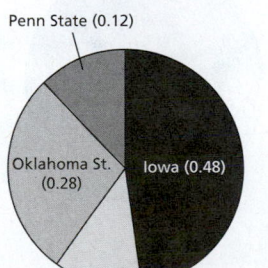

d. Data

2.27 a.

Champion	Freq.
Iowa	12
Minnesota	3
Oklahoma St.	7
Penn State	3

b.

Champion	Rel. freq.
Iowa	0.48
Minnesota	0.12
Oklahoma St.	0.28
Penn State	0.12

c. NCAA Wrestling Champs

d. NCAA Wrestling Champs

2.29 a.

Class	Frequency
Freshman	6
Sophomore	15
Junior	12
Senior	7

b.

Class	Relative frequency
Freshman	0.150
Sophomore	0.375
Junior	0.300
Senior	0.175

c.

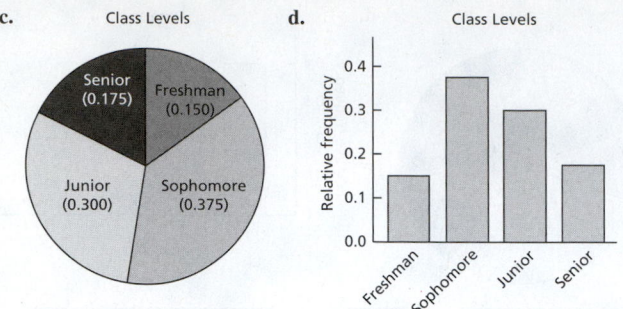

Class Levels

d.

Class Levels

2.31 a.

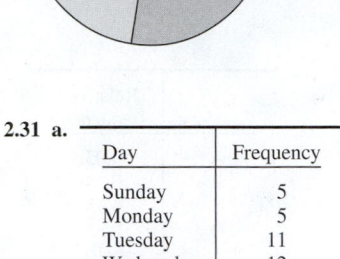

Day	Frequency
Sunday	5
Monday	5
Tuesday	11
Wednesday	12
Thursday	11
Friday	18
Saturday	7

b.

Day	Relative frequency
Sunday	0.072
Monday	0.072
Tuesday	0.159
Wednesday	0.174
Thursday	0.159
Friday	0.261
Saturday	0.101

c.

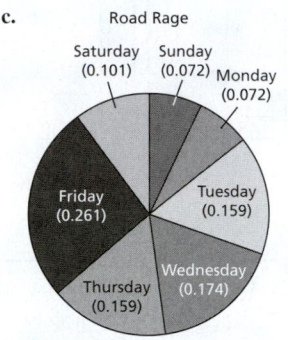

Road Rage

d.

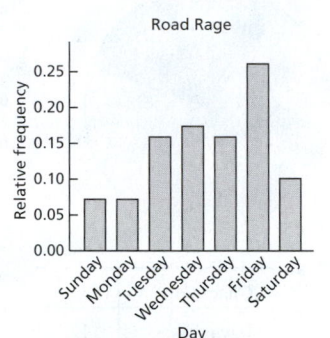

Road Rage

2.33 a.

Color	Rel. freq.
Brown	0.299
Yellow	0.224
Red	0.208
Orange	0.100
Green	0.084
Blue	0.084

b.

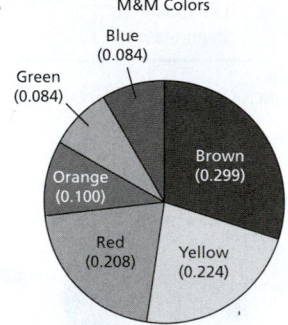

M&M Colors

c.

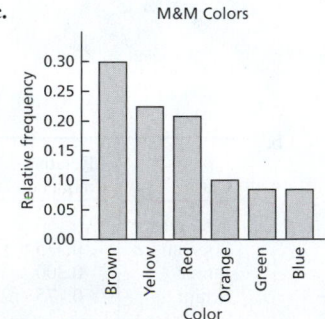

M&M Colors

2.35 a.

Rank	Rel. freq.
Professor	0.236
Associate professor	0.207
Assistant professor	0.430
Instructor	0.104
Other	0.024

b.

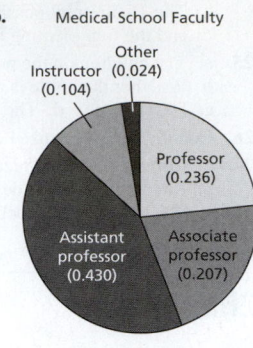

Medical School Faculty

c.

Medical School Faculty

2.37 a.

Number	Relative frequency
Red	0.44
Black	0.51
Green	0.05

b.

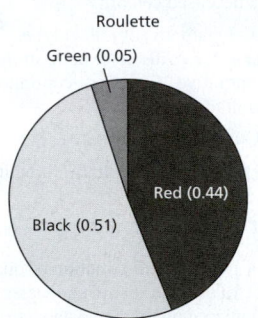

Roulette

c.

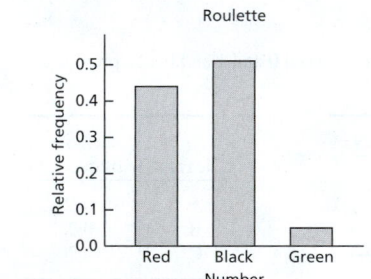

Roulette

Exercises 2.3

2.43 No. Class limits, marks, cutpoints, and midpoints make sense only for numerical data (for which doing arithmetic is meaningful).

2.45 The two methods are limit grouping and cutpoint grouping.

2.47 With limit grouping, the "middle" of a class is the average of the two class limits of the class; it is called the *class mark*. With cutpoint grouping, the "middle" of a class is the average of the two cutpoints of the class; it is called the *class midpoint*.

2.49 Answers will vary.   **2.51** Answers will vary.

2.53 Reconstruct the stem-and-leaf diagram, using more lines per stem.

2.55 Limit grouping.   **2.57** Single-value grouping.

2.59 Cutpoint grouping.

2.61 a.

Class	Frequency
1	2
2	2
3	5
4	1

b.

Class	Relative frequency
1	0.2
2	0.2
3	0.5
4	0.1

c.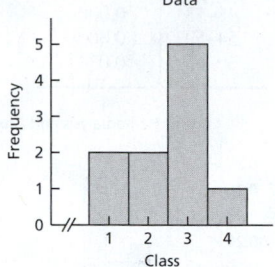

d.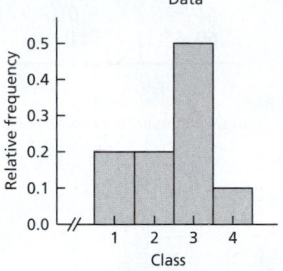

2.63 a.

Class	Frequency
0	1
1	1
2	4
3	8
4	6

b.

Class	Relative frequency
0	0.05
1	0.05
2	0.20
3	0.40
4	0.30

c.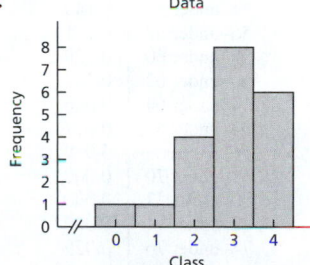

d.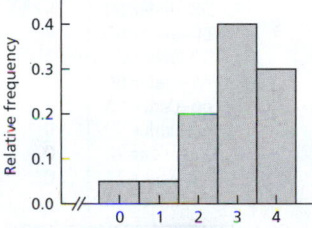

2.65 a.

Class	Frequency
0–4	5
5–9	5
10–14	1
15–19	4
20–24	2
25–29	3

b.

Class	Relative frequency
0–4	0.25
5–9	0.25
10–14	0.05
15–19	0.20
20–24	0.10
25–29	0.15

c.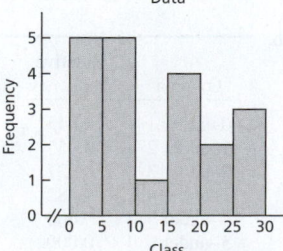

d.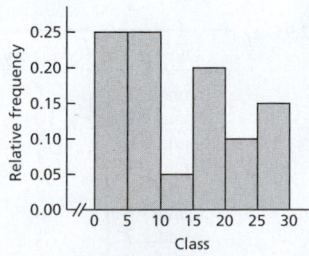

2.67 a.

Class	Frequency
50–59	6
60–69	3
70–79	6
80–89	7
90–99	3

b.

Class	Relative frequency
50–59	0.24
60–69	0.12
70–79	0.24
80–89	0.28
90–99	0.12

c.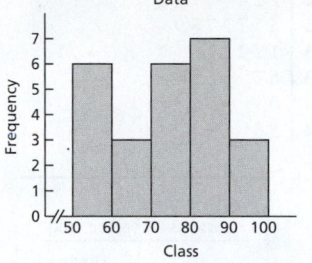

d.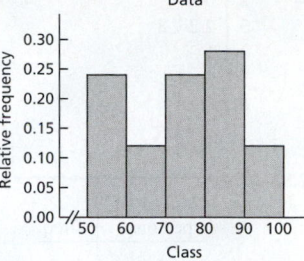

2.69 a.

Class	Frequency
40–under 46	3
46–under 52	6
52–under 58	10
58–under 64	0
64–under 70	1

b.

Class	Relative frequency
40–under 46	0.15
46–under 52	0.30
52–under 58	0.50
58–under 64	0.00
64–under 70	0.05

c.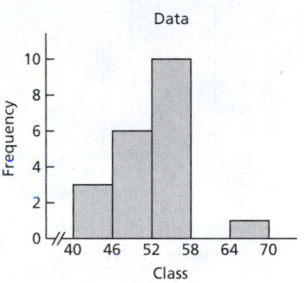

d.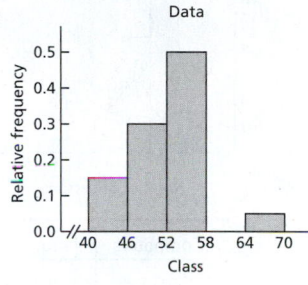

2.71 a.

Class	Freq.
25–under 28	1
28–under 31	2
31–under 34	5
34–under 37	7
37–under 40	5
40–under 43	4
43–under 46	1

b.

Class	Rel. freq.
25–under 28	0.04
28–under 31	0.08
31–under 34	0.20
34–under 37	0.28
37–under 40	0.20
40–under 43	0.16
43–under 46	0.04

c.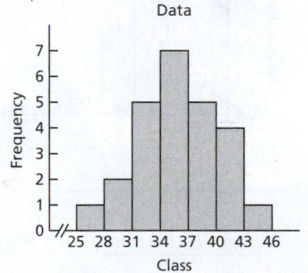

d.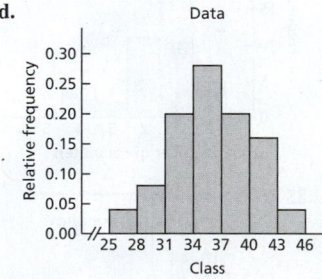

2.73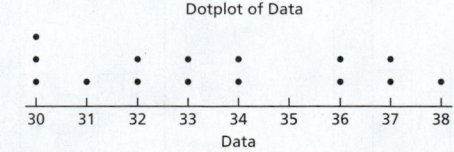

2.75

2.77

2	0 1
3	2 2 7 8
4	1 3
5	5
6	2

2.79

2	2 2 4
2	5 5 7 7 7 8 9
3	1 2 2 3 4
3	6 7
4	0
4	5 6

2.81 a.

Number of persons	Freq.
1	7
2	13
3	9
4	5
5	4
6	1
7	1

b.

Number of persons	Rel. freq.
1	0.175
2	0.325
3	0.225
4	0.125
5	0.100
6	0.025
7	0.025

c.

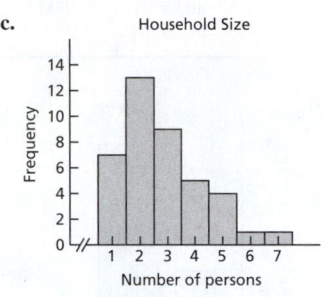

d.

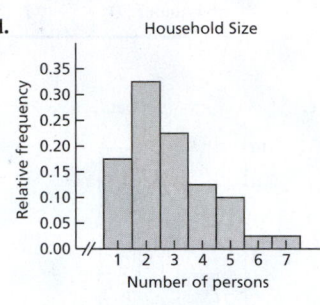

2.83 a.

Computers	Freq.
0	5
1	22
2	11
3	2
4	4
5	1

b.

Computers	Rel. freq.
0	0.111
1	0.489
2	0.244
3	0.044
4	0.089
5	0.022

c.

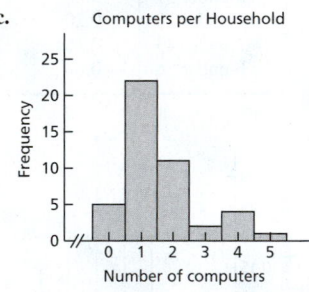

d.

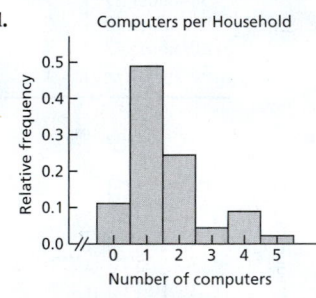

2.85 a.

Age	Frequency
40–44	4
45–49	3
50–54	4
55–59	8
60–64	2

b.

Age	Relative frequency
40–44	0.190
45–49	0.143
50–54	0.190
55–59	0.381
60–64	0.095

c.

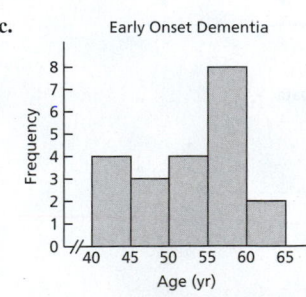

d.

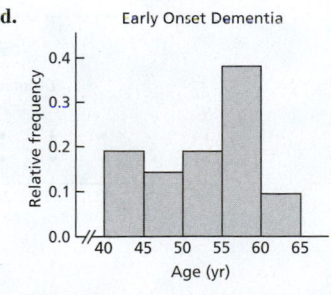

2.87 a.

Anxiety	Frequency
12–17	2
18–23	3
24–29	6
30–35	5
36–41	10
42–47	4
48–53	0
54–59	0
60–65	1

b.

Anxiety	Relative frequency
12–17	0.065
18–23	0.097
24–29	0.194
30–35	0.161
36–41	0.323
42–47	0.129
48–53	0.000
54–59	0.000
60–65	0.032

c.

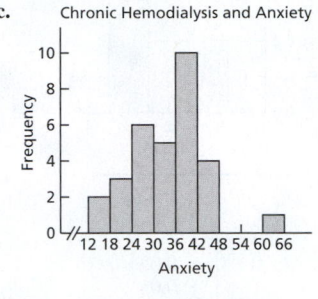

d.

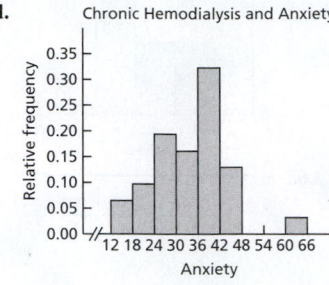

2.89 a.

Speed	Freq.
52–under 54	2
54–under 56	5
56–under 58	6
58–under 60	8
60–under 62	7
62–under 64	3
64–under 66	2
66–under 68	1
68–under 70	0
70–under 72	0
72–under 74	0
74–under 76	1

b.

Speed	Rel. freq.
52–under 54	0.057
54–under 56	0.143
56–under 58	0.171
58–under 60	0.229
60–under 62	0.200
62–under 64	0.086
64–under 66	0.057
66–under 68	0.029
68–under 70	0.000
70–under 72	0.000
72–under 74	0.000
74–under 76	0.029

c.

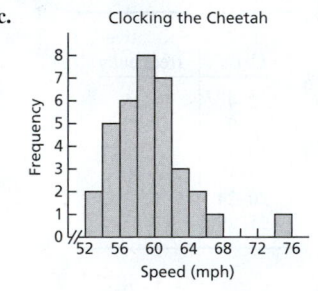

d.

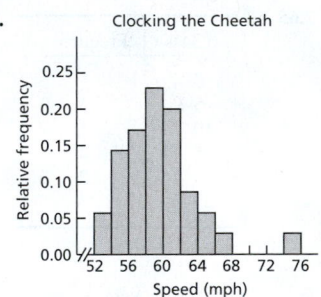

2.91 a.

Oxygen	Frequency
0–under 1	1
1–under 2	10
2–under 3	5
3–under 4	4
4–under 5	0
5–under 6	0
6–under 7	1
7–under 8	1

b.

Oxygen	Relative frequency
0–under 1	0.045
1–under 2	0.455
2–under 3	0.227
3–under 4	0.182
4–under 5	0.000
5–under 6	0.000
6–under 7	0.045
7–under 8	0.045

c.

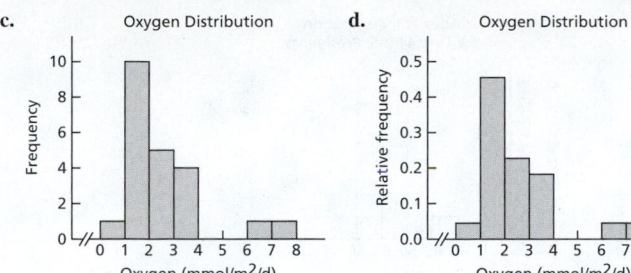

d.

2.101 a.

```
6 | 8
7 | 1 1 1 1 1 1 1 1
7 | 2 2 2 2 2 2 2 2 2 2 3 3 3 3 3
7 | 4 4 4 4 5 5 5 5 5
7 | 6 6 6 6 6 7 7 7 7 7 7
7 | 8
8 | 0
```

b. Using one or two lines per stem would give an insufficient number of stems (i.e., lines).

2.103 a. 20%   **b.** 25%   **c.** 7   **2.105 a.** 1   **b.** 3   **c.** 8

Exercises 2.4

2.123 *Sample data* are the values of a variable for a sample of the population.

2.125 A *sample distribution* is the distribution of sample data.

2.127 *Distribution of a variable* is another name for population distribution.

2.129 Roughly a bell shape (See Key Fact 2.1 on page 80.)

2.131 Answers will vary.   **2.133 a.** Unimodal   **b.** Symmetric

2.135 a. Unimodal   **b.** Not symmetric   **c.** Left skewed

2.137 a. Unimodal   **b.** Not symmetric   **c.** Left skewed

2.139 a. Multimodal   **b.** Not symmetric

2.141 Right skewed (Symmetric is also an acceptable answer.)

2.143 Left skewed   **2.145** Symmetric   **2.147** Left skewed

2.149 Right skewed

2.151 a. Year 1: unimodal. Year 2: unimodal
b. Year 1: not symmetric. Year 2: not symmetric
c. Year 1: right skewed. Year 2: right skewed
d. Although both distributions are right skewed, their centers are different and there is much more variation (spread) in Year 1 than in Year 2.

Exercises 2.5

2.163 a. Part of the vertical axis of the graph has been cut off, or truncated.
b. It may allow relevant information to be conveyed more easily.
c. Start the axis at 0 and put slashes in the axis to indicate that part of the axis is missing.

2.165
c. They give the misleading impression that the district average is much greater relative to the national average than it actually is.

2.167 a. It is a truncated graph.
b.

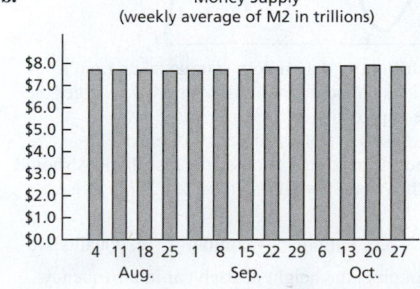

c.

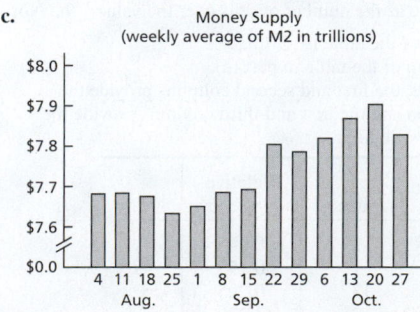

2.93

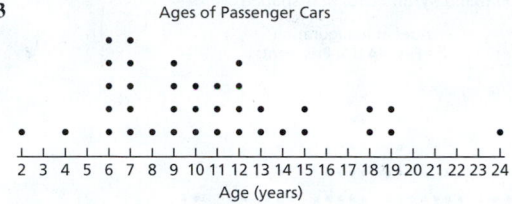

2.95 a.

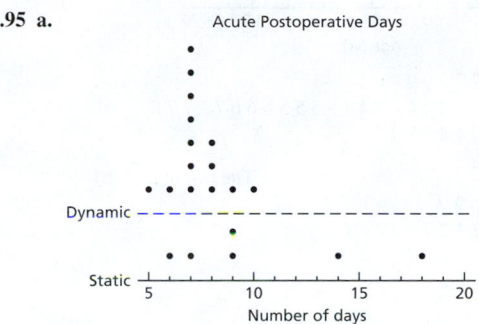

b. For these data, the number of acute postoperative days is, on average, less with the dynamic system than with the static system. Also, more variation exists in the number of acute postoperative days with the static system than with the dynamic system.

2.97

```
1 | 2 3 8
2 | 1 6 7 8 8 9 9
3 | 3 4 4 5 9
4 | 0 4
```

2.99 a.

```
0 | 2 2 3 4 7 9 9
1 | 1 1 1 4 5 5 6 6 6 8 9
2 | 0 2 3 4 7 9
3 | 0 0 4 5 5 5
4 | 1 9
5 | 5
6 | 9
7 | 9
8 |
9 | 3
```

b.

```
0 | 2 2 3 4
0 | 7 9 9
1 | 1 1 1 4
1 | 5 5 6 6 6 8 9
2 | 0 2 3 4
2 | 7 9
3 | 0 0 4
3 | 5 5 5
4 | 1
4 | 9
5 |
5 | 5
6 |
6 | 9
7 |
7 | 9
8 |
8 |
9 | 3
```

c. The stem-and-leaf diagram in part (a) (one line per stem) is more useful; the one in part (b) (two lines per stem) has an unnecessarily large number of stems (i.e., lines).

2.169 b. That happiness has dropped by roughly 25%.
c. About 4.5%　**d.** Because it is a truncated graph
e. Start the vertical scale of the graph at 0 instead of 4.5, or use some method (such as slashes) to warn the reader that the vertical scale has been modified.

Review Problems for Chapter 2

1. **a.** A characteristic that varies from one person or thing to another
   b. Quantitative variables and qualitative (or categorical) variables
   c. Discrete variables and continuous variables

2. **a.** Values of a variable　**b.** By the type of variable

3. **a.** A listing of the distinct values and their frequencies
   b. A listing of the distinct values and their relative frequencies

4. We construct frequency or relative-frequency distributions of quantitative data by treating the classes of the quantitative data as the distinct values of qualitative data.

5. Pie charts and bar charts

6. To avoid confusing bar graphs with histograms

7. Answers will vary.

8. When grouping discrete data in which there are only a small number of distinct observations

9. **a.** 11.5　**b.** 15 and 20　**c.** The fourth class

10. **a.** 6 and 10　**b.** 13　**c.** 16 and 20　**d.** The fifth class

11. **a.** 10　**b.** 20　**c.** 25 and 35　**d.** The third class

12. **a.** 6 and 14　**b.** 18　**c.** 22 and 30　**d.** The third class

13. **a.** The bar for a class extends horizontally from the lower limit of the class to the lower limit of the next higher class.
   b. The bar for a class extends horizontally from the lower cutpoint of the class to the lower cutpoint of the next higher class.
   c. The bar for a class is centered horizontally over the mark of the class.
   d. The bar for a class is centered horizontally over the midpoint of the class.

14.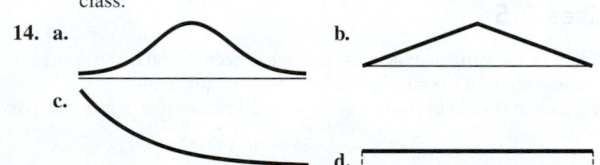

15. Answers will vary, but here is one possibility:

16. **a.** Left skewed. The distribution of a random sample taken from a population approximates the population distribution. The larger the sample, the better the approximation tends to be.
   b. No. Sample distributions vary from sample to sample.
   c. Yes. Left skewed. The shapes of the two sample distributions should be similar to that of the population distribution and hence to each other.

17. **a.** Discrete quantitative　**b.** Continuous quantitative　**c.** Qualitative

18. **a.** With single-value grouping, the height of each bar in a frequency histogram is the same as the number of dots over the value.　**b.** No.

19. **a.** See the first column of the table in part (c).
   b. See the fourth column of the table in part (c).
   c. In the following table, the first and second columns provide the frequency distribution and the first and third columns provide the relative-frequency distribution.

Age at inauguration	Frequency	Relative frequency	Mark
40–44	2	0.045	42
45–49	7	0.159	47
50–54	13	0.295	52
55–59	12	0.273	57
60–64	7	0.159	62
65–69	3	0.068	67

d.

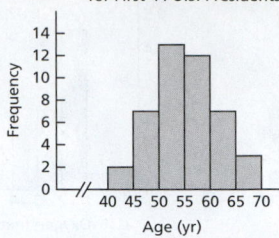

Ages at Inauguration for First 44 U.S. Presidents

e. Unimodal and symmetric; bell shaped

20.

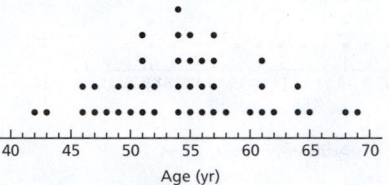

Ages at Inauguration for First 44 U.S. Presidents

21. **a.**

```
4 | 2 3 6 6 7 7 8 9 9
5 | 0 0 1 1 1 1 2 2 4 4 4 4 4 5 5 5 5 5 6 6 6 7 7 7 7 8
6 | 0 1 1 1 2 4 4 5 8 9
```

b.

```
4 | 2 3
4 | 6 6 7 7 8 9 9
5 | 0 0 1 1 1 1 2 2 4 4 4 4 4
5 | 5 5 5 5 6 6 6 7 7 7 7 8
6 | 0 1 1 1 2 4 4
6 | 5 8 9
```

c. The one in part (b)

22. **a.**

```
3 | 4 7 7 8
4 | 0 4 6 7
5 | 4 4 6 6 7 7 7 8
6 | 0
```

b.

```
3 | 4
3 | 7 7 8
4 | 0 4
4 | 6 7
5 | 4 4
5 | 6 6 7 7 7 8
6 | 0
```

c. The stem-and-leaf diagram in part (a) is only moderately useful because there are so few stems. The one in part (b) is a better stem-and-leaf diagram for these data.

23. **a.**

Number busy	Frequency	Relative frequency
0	1	0.04
1	2	0.08
2	2	0.08
3	4	0.16
4	5	0.20
5	7	0.28
6	4	0.16

b.

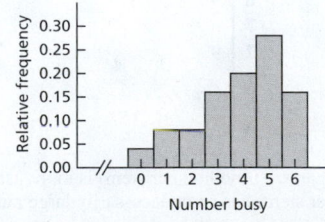

Busy Tellers

c. Unimodal　**d.** Left skewed

e.

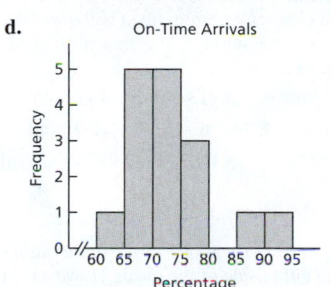

Busy Bank Tellers

f. They have identical shapes.

24. a. See the first column of the table in part (c).
b. See the fourth column of the table in part (c).
c. In the following table, the first and second columns provide the frequency distribution and the first and third columns provide the relative-frequency distribution.

Percentage on time	Frequency	Relative frequency	Midpoint
60–under 65	1	0.063	62.5
65–under 70	5	0.313	67.5
70–under 75	5	0.313	72.5
75–under 80	3	0.188	77.5
80–under 85	0	0.000	82.5
85–under 90	1	0.063	87.5
90–under 95	1	0.063	92.5

d.

On-Time Arrivals

e.
```
6 | 2
6 | 6 6 9
7 | 0 0 1 1 3 3 4
7 | 6 7 8
8 |
8 | 8
9 | 3
```

f.
```
6 | 1
6 | 5 6 9 9 9
7 | 0 1 2 3 3
7 | 6 7 7
8 |
8 | 7
9 | 3
```
g. The one in part (f)

25. a.

Oldest Players

b. Bimodal (or multimodal) and roughly symmetric

26. a.

Evidence	Relative frequency
Firearm	0.084
Magazine	0.047
Live cartridge	0.794
Spent cartridge casing	0.075

b.

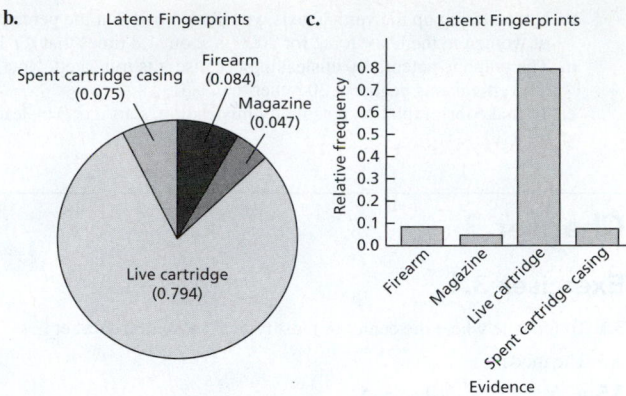

Latent Fingerprints

c. Latent Fingerprints

27. a. The population consists of the states in the United States; the variable under consideration is division.
b. In the following table, the first and second columns provide the frequency distribution and the first and third columns provide the relative-frequency distribution.

Division	Frequency	Relative frequency
East North Central	5	0.10
East South Central	4	0.08
Middle Atlantic	3	0.06
Mountain	8	0.16
New England	6	0.12
Pacific	5	0.10
South Atlantic	8	0.16
West North Central	7	0.14
West South Central	4	0.08

c.

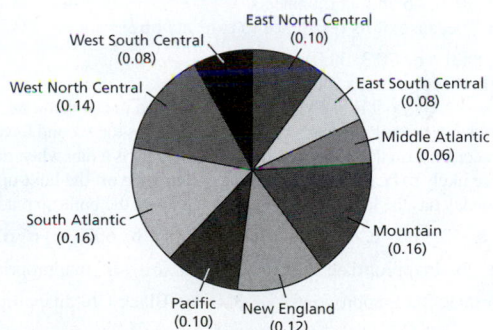

U. S. Divisions

d.

U.S. Divisions

28. a. To warn the reader that part of it has been removed
b. To enable the reader to see differences among the amounts of CO_2 that can be kept in different geological spaces without causing misinterpretation

29. b. Having followed the directions in part (a), you might conclude that the percentage of women in the labor force for 2000 is about 3.5 times that for 1960.

c. Not covering up the vertical axis, you would find that the percentage of women in the labor force for 2000 is about 1.8 times that for 1960.
d. The graph is potentially misleading because it is truncated. Note that the vertical axis begins at 30 rather than at 0.
e. To make the graph less potentially misleading, start it at 0 instead of at 30.

Chapter 3

Exercises 3.1

3.1 To indicate where the center or most typical value of a data set lies

3.3 The mode

3.5 a. Mean $= 5$; median $= 5$.
b. Mean $= 15$; median $= 5$. The median is a better measure of center because it is not influenced by the one unusually large value, 99.
c. Resistance

3.7 Median. Unlike the mean, the median is not affected strongly by the relatively few homes that have extremely large or small floor spaces.

3.9 a. 3 **b.** 4 **c.** no mode   **3.11 a.** 2.75 **b.** 3 **c.** 4

3.13 a. 5 **b.** 4 **c.** no mode

3.15
a. Σ is an abbreviation for the phrase "the sum of" and indicates that you should sum the values that follow it.
b. n represents the sample size (number of observations).
c. $\bar{x}$ denotes a sample mean.

3.17 a. 5 **b.** 27 **c.** 5.4

3.19 a. 7.3 days **b.** 6.0 days **c.** 5, 6, 11 days

3.21 a. 78.4 tornadoes **b.** 77.0 tornadoes **c.** no mode

3.23 a. \$34.71 billion **b.** \$29.45 billion **c.** \$31.0 billion

3.25 a. 14.6 songs **b.** 14.0 songs **c.** 14 songs

3.27 a. 292.8, 83.0, 46 cremation burials
b. The median, because of its resistance to extreme observations

3.29 a. 88.4, 88.0 **b.** 70.3, 59.0
c. Friday for built-up; Sunday for non–built-up
d. Friday is a work day, so it is likely that people involved in accidents are commuters using the built-up roads; note that Friday has the second lowest number of accidents on the non--built-up roads. Sunday is a day when riders may be more likely to be cruising around the countryside off the built-up roads; note that Sunday has the lowest number of accidents on the built-up roads.

3.31 a. 10 **b.** 23.3 hr **c.** 2.33 hr   **3.33 a.** 9 **b.** 607 yr **c.** 67.4 yr

3.35 a. Iowa **b.** Inappropriate   **3.37 a.** Harvard **b.** Inappropriate

3.39 a. Moderate **b.** Inappropriate   **3.41 a.** Black **b.** Inappropriate

Exercises 3.2

3.57 To indicate the amount of variation in a data set   **3.59** The mean

3.61 a. 2.7 **b.** 31.6 **c.** Resistance

3.63 a. 45 years **b.** 19.9 years **c.** 19.9 years   **3.65 a.** 5 **b.** 2.6

3.67 a. 3 **b.** 1.5   **3.69 a.** 8 **b.** 3.4

3.71 range $= 6$ days; $s = 2.6$ days

3.73 range $= 202$ tornadoes; $s = 53.9$ tornadoes

3.75 range $= \$41$ billion; $s = \$13.00$ billion

3.77 range $= 24$ songs; $s = 5.5$ songs

3.79 a. 586.3 cremation burials **b.** No, because of its lack of resistance

3.81 a. Non–built-up
b. Built-up: range $= 34$ accidents; $s = 12.8$ accidents
Non–built-up: range $= 49$ accidents; $s = 19.8$ accidents

Exercises 3.3

3.97 Because, in general, it provides better estimates

3.99
a. At least 93.75% of the observations in any data set lie within four standard deviations to either side of the mean.

b. At least 84% of the observations in any data set lie within 2.5 standard deviations to either side of the mean.

3.101 At least 96% of the observations in any data set lie within five standard deviations to either side of the mean.

3.103 a.

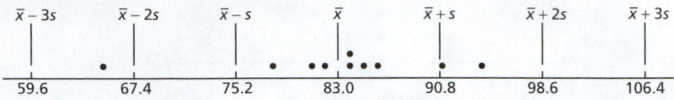

$\bar{x}-3s$	$\bar{x}-2s$	$\bar{x}-s$	$\bar{x}$	$\bar{x}+s$	$\bar{x}+2s$	$\bar{x}+3s$
59.6	67.4	75.2	83.0	90.8	98.6	106.4

b. Chebyshev's rule with $k = 2$ says that at least 75% of the observations lie within two standard deviations to either side of the mean. However, as we see from the graph in part (a), in this case, nine of the ten, or 90% of the observations actually lie within two standard deviations to either side of the mean.
c. Chebyshev's rule with $k = 3$ says that at least 89% of the observations lie within three standard deviations to either side of the mean. However, as we see from the graph in part (a), in this case, all (100%) of the observations actually lie within three standard deviations to either side of the mean.

3.105 a. Appropriate **b.** Not appropriate **c.** Not appropriate

3.107
a. Chebyshev's rule says that, for any data set, at least 75% of the observations lie within two standard deviations to either side of the mean; the empirical rule says that, for a data set with roughly a bell-shaped distribution, approximately 95% of the observations lie within two standard deviations to either side of the mean.
b. Chebyshev's rule says that, for any data set, at least 89% of the observations lie within three standard deviations to either side of the mean; the empirical rule says that, for a data set with roughly a bell-shaped distribution, approximately 99.7% of the observations lie within three standard deviations to either side of the mean.

3.109 a. 10; 40 **b.** 75   **3.111** 96%   **3.113** 60   **3.115** 45

3.117 45   **3.119** 134   **3.121 a.** 20; 30 **b.** 15; 35 **c.** 10; 40

3.123 95%   **3.125** 76   **3.127** 34   **3.129** 249 (249.25 rounded)

3.131 102

3.133
a. Chebyshev's rule with $k = 2$ says that at least 75% of the observations lie within two standard deviations to either side of the mean. However, 31 of the 33, or 93.9% of the observations actually lie within two standard deviations to either side of the mean.
b. Chebyshev's rule with $k = 3$ says that at least 89% of the observations lie within three standard deviations to either side of the mean. However, all (100%) of the observations actually lie within three standard deviations to either side of the mean.
c. Chebyshev's rule gives only a minimum for the percentage of observations that lie within a specified number of standard deviations to either side of the mean; the actual percentage will usually be higher.

3.135 a.

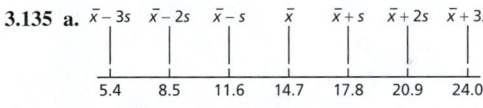

$\bar{x}-3s$	$\bar{x}-2s$	$\bar{x}-s$	$\bar{x}$	$\bar{x}+s$	$\bar{x}+2s$	$\bar{x}+3s$
5.4	8.5	11.6	14.7	17.8	20.9	24.0

b. At least 34 of the 45 adult females in the sample have daily iron intakes between 8.5 mg and 20.9 mg.
c. At least 41 of the 45 adult females in the sample have daily iron intakes between 5.4 mg and 24.0 mg.

3.137 a.

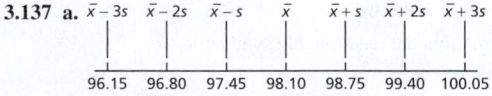

$\bar{x}-3s$	$\bar{x}-2s$	$\bar{x}-s$	$\bar{x}$	$\bar{x}+s$	$\bar{x}+2s$	$\bar{x}+3s$
96.15	96.80	97.45	98.10	98.75	99.40	100.05

b. At least 70 of the 93 healthy humans in the sample had body temperatures between 96.80°F and 99.40°F.
c. At least 83 of the 93 healthy humans in the sample had body temperatures between 96.15°F and 100.05°F.

3.139 a. Yes, the distribution of the data is close to bell shaped.
b. 68%, 95%, and 99.7% **c.** 73.3%, 95%, and 100%
d. Comparing the percentages obtained in parts (b) and (c), we see that, in this case, the empirical rule provides good estimates of the actual percentages. In view of the answer to part (a), this result is not surprising.

3.141 a.

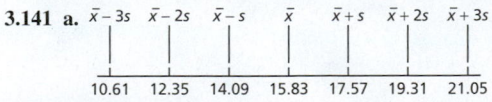

$\bar{x}-3s$	$\bar{x}-2s$	$\bar{x}-s$	$\bar{x}$	$\bar{x}+s$	$\bar{x}+2s$	$\bar{x}+3s$
10.61	12.35	14.09	15.83	17.57	19.31	21.05

b. Approximately 55 (55.08 rounded) of the 81 fifth-grade classes sampled have student-to-faculty ratios between 14.09 and 17.57.

c. Approximately 77 (76.95 rounded) of the 81 fifth-grade classes sampled have student-to-faculty ratios between 12.35 and 19.31.

d. Approximately 81 (80.757 rounded) of the 81 fifth-grade classes sampled have student-to-faculty ratios between 10.61 and 21.05.

3.143 a.

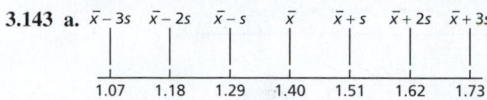

$\bar{x}-3s$ $\bar{x}-2s$ $\bar{x}-s$ $\bar{x}$ $\bar{x}+s$ $\bar{x}+2s$ $\bar{x}+3s$
1.07 1.18 1.29 1.40 1.51 1.62 1.73

b. Approximately 153 of the 225 Swedish men sampled have brain weights between 1.29 kg and 1.51 kg.

c. Approximately 214 (213.75 rounded) of the 225 Swedish men sampled have brain weights between 1.18 kg and 1.62 kg.

d. Approximately 224 (224.325 rounded) of the 225 Swedish men sampled have brain weights between 1.07 kg and 1.73 kg.

Exercises 3.4

3.149 The median and interquartile range are resistant measures, whereas the mean and standard deviation are not.

3.151 No. It may, for example, be an indication of skewness.

3.153 a. A measure of variation
b. Roughly, the range of the middle 50% of the observations

3.155 When both the minimum and maximum observations lie within the lower and upper limits

3.157 Q_1, Q_3

3.159 a. $Q_1 = 1.5, Q_2 = 2.5, Q_3 = 3.5$ **b.** 2 **c.** 1, 1.5, 2.5, 3.5, 4

3.161 a. $Q_1 = 2, Q_2 = 3, Q_3 = 4$ **b.** 2 **c.** 1, 2, 3, 4, 5

3.163 a. $Q_1 = 2, Q_2 = 3.5, Q_3 = 5$ **b.** 3 **c.** 1, 2, 3.5, 5, 6

3.165 a. $Q_1 = 2.5, Q_2 = 4, Q_3 = 5.5$ **b.** 3 **c.** 1, 2.5, 4, 5.5, 7

Note: If you use technology to obtain your results for Exercises 3.167–3.177, they may differ from those presented here because different technologies often use different rules for computing quartiles.

3.167 Units are in games.
a. $Q_1 = 73.5, Q_2 = 79, Q_3 = 80$ **b.** 6.5 **c.** 45, 73.5, 79, 80, 82
d. 45 and 48 **e.**

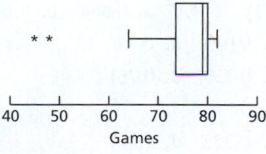

40 50 60 70 80 90
Games

3.169 Units are in days.
a. $Q_1 = 4, Q_2 = 7, Q_3 = 12$ **b.** 8 **c.** 1, 4, 7, 12, 55 **d.** 55
e.

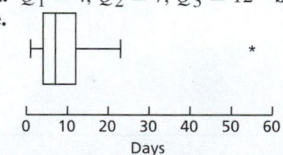

0 10 20 30 40 50 60
Days

3.171 Units are in kilograms per hectare per year.
a. $Q_1 = 88, Q_2 = 131.5, Q_3 = 154$ **b.** 66
c. 57, 88, 131.5, 154, 175 **d.** No potential outliers
e.

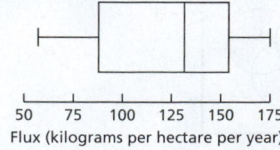

50 75 100 125 150 175
Flux (kilograms per hectare per year)

3.173 Units are in thousands of dollars.
a. $Q_1 = 660, Q_2 = 1800, Q_3 = 4749.5$ **b.** 4089.5
c. 21, 660, 1800, 4749.5, 17,341 **d.** 11,189 and 17,341

e.

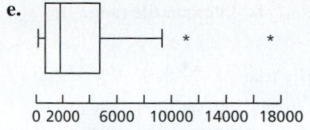

0 2000 6000 10000 14000 18000
Capital spending ($1000s)

3.175 Units are in centimeters (cm).
a. $Q_1 = 26.70, Q_2 = 27.12, Q_3 = 28.84$ **b.** 2.14
c. 26.09, 26.70, 27.12, 28.84, 29.40 **d.** No potential outliers
e.

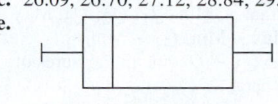

26.0 26.5 27.0 27.5 28.0 28.5 29.0 29.5
Diameter (cm)

3.177 a. $Q_1 = 8$ cigs/day, $Q_2 = 9$ cigs/day, $Q_3 = 10$ cigs/day
b. The quartiles for this data set are not particularly useful because of its small range and the relatively large number of identical values. Note, for instance, that Q_3 and Max are equal.

3.179 The weight losses for the two groups are, on average, roughly the same. However, there is less variation in the weight losses of Group 1 than of Group 2.

3.181 On average, the hemoglobin levels for HB SC and HB ST are roughly the same, and both exceed that for HB SS. Also, the variation in hemoglobin levels appears to be greatest for HB ST and least for HB SC.

Exercises 3.5

3.193 To describe the entire population

3.195 a. 0, 1
b. the number of standard deviations that the observation is from the mean, that is, how far the observation is from the mean in units of standard deviation
c. above (greater than); below (less than)

3.197 Parameter. A descriptive measure of a population is a parameter.

3.199 a. $\mu = 75.0$ in. **b.** $\sigma = 5.6$ in. **3.201 a.** 3 **b.** 2.2

3.203 a. 2.75 **b.** 1.3 **3.205 a.** 5 **b.** 3.0

3.207 a. The variable is age and the population consists of all U.S. residents.
b. Median = 37.0 yr. Statistic. $M = 37.0$ yr. **c.** Parameter. $\eta = 37.2$ yr.

3.209 a. $\mu = 97.0$ mph **b.** $\sigma = 13.1$ mph **c.** $\eta = 95.0$ mph
d. Modes = 80, 90, and 115 mph **e.** IQR = 25.0 mph

3.211 a. 465.4 cases; 265.8 cases
b. The standard deviation is smaller for Orlando because there is less variation in the numbers of cases for Orlando.
c. 67.9 cases; 167.1 cases **d.** Yes.

3.213 a. $z = (x - 32.9)/17.9$ **b.** 0; 1 **c.** 2.70; −0.68
d. The time served of 81.3 months is 2.70 standard deviations above the mean time served of 32.9 months; the time served of 20.8 months is 0.68 standard deviations below the mean time served of 32.9 months.

3.215 a. $z = (x - 6.71)/0.67$
b. −2.25; 2.07. The thumb length of 5.2 mm is 2.25 standard deviations below the mean thumb length of 6.71 mm; the thumb length of 8.1 mm is 2.07 standard deviations above the mean thumb length of 6.71 mm.

3.217 a. −3.13
b. Yes. Assuming the advertised claim is correct, the three-standard-deviations rule implies that your car's mileage is lower than most other cars of that model.

Review Problems for Chapter 3

1. **a.** Numbers that are used to describe data sets are called descriptive measures.
   b. Descriptive measures that indicate where the center or most typical value of a data set lies are called measures of center.
   c. Descriptive measures that indicate the amount of variation, or spread, in a data set are called measures of variation.

2. Mean and median. The median is a resistant measure, whereas the mean is not. The mean takes into account the actual numerical value of all observations, whereas the median does not.

3. The mode   **4. a.** Standard deviation **b.** Interquartile range

5. a. $\bar{x}$ **b.** s **c.** μ **d.** σ

6. a. Not necessarily true **b.** Necessarily true   **7.** three

8. a. At least 97.2% of the observations in any data set lie within six standard deviations to either side of the mean.
b. At least 55.6% of the observations in any data set lie within 1.5 standard deviations to either side of the mean.

9. 66 (65.25 rounded up)   **10.** 68%   **11.** 144 (144.4 rounded)

12. a. Minimum, quartiles, and maximum; that is, Min, Q_1, Q_2, Q_3, Max
b. Q_2 can be used to describe center. Max − Min, Q_1 − Min, Max − Q_3, Q_2 − Q_1, Q_3 − Q_2, and Q_3 − Q_1 are all measures of variation for different portions of the data.
c. Boxplot

13. a. An outlier is an observation that falls well outside the overall pattern of the data.
b. First, determine the lower and upper limits—the numbers 1.5 IQRs below the first quartile and 1.5 IQRs above the third quartile, respectively. Observations that lie outside the lower and upper limits—either below the lower limit or above the upper limit— are potential outliers.

14. a. Subtract from x its mean and then divide by its standard deviation.
b. The z-score of an observation gives the number of standard deviations that the observation is from the mean, that is, how far the observation is from the mean in units of standard deviation.
c. The observation is 2.9 standard deviations above the mean. It is larger than most of the other observations.

15. a. 2.35 drinks; 2.0 drinks; 1, 2 drinks **b.** Answers will vary.

16. The median, because it is resistant to outliers and other extreme values.

17. The mode; neither the mean nor the median can be used as a measure of center for qualitative data.

18. 30.53 mm; 32.50 mm; 33 mm

19. a. $\bar{x} = 45.7$ kg **b.** Range $= 17$ kg **c.** $s = 5.0$ kg

20. a. $\bar{x}-3s$ $\bar{x}-2s$ $\bar{x}-s$ $\bar{x}$ $\bar{x}+s$ $\bar{x}+2s$ $\bar{x}+3s$

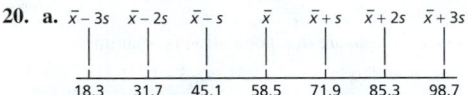

18.3 31.7 45.1 58.5 71.9 85.3 98.7

b. 18.3 yr, 98.7 yr

21. a. 7662 (7661.25 rounded up) **b.** 9.8; 41.6

22. a. 47.5; 79.1 **b.** 99.7 **c.** 170

23. a. $Q_1 = 48.0$ yr, $Q_2 = 59.5$ yr, $Q_3 = 68.5$ yr
b. 20.5 yr; roughly speaking, the middle 50% of the ages has a range of 20.5 yr.
c. 31, 48.0, 59.5, 68.5, 79 yr
d. Lower limit: 17.25 yr. Upper limit: 99.25 yr.
e. No potential outliers **f.**

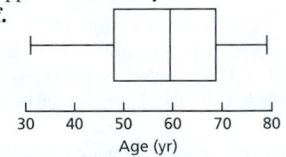

30 40 50 60 70 80
Age (yr)

24. Units are in millimoles per square meter per day.
a. 0.7, 1.50, 1.95, 3.30, 7.6 **b.** 6.7 and 7.6
c.

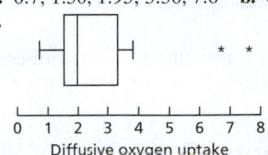

0 1 2 3 4 5 6 7 8
Diffusive oxygen uptake

25. On average, the driving distances with the Stinger tee are greater than with the regular tee. Moreover, there is somewhat more variation in driving distances with the Stinger tee than with the regular tee.

26. a. 54.9 minutes **b.** 16.1 minutes

27. a. 20.34 thousand students **b.** 6.41 thousand students
c. $z = (x - 20.34)/6.41$ **d.** 0; 1

e.

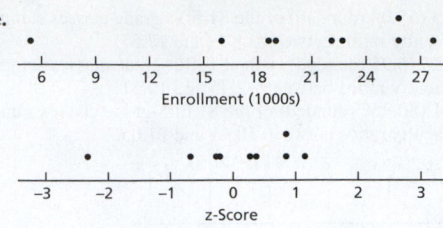

6 9 12 15 18 21 24 27
Enrollment (1000s)

−3 −2 −1 0 1 2 3
z-Score

f. 1.15; −0.27. The enrollment at Los Angeles is 1.15 standard deviations above the UC campuses' mean enrollment of 20.34 thousand students; the enrollment at Riverside is 0.27 standard deviations below that mean.

28. a. A sample mean **b.** $\bar{x}$ **c.** A statistic

Chapter 4

Exercises 4.1

4.1 An experiment is an action whose outcome cannot be predicted with certainty. An event is some specified result that may or may not occur when an experiment is performed.

4.3 There is no difference.   **4.5** 0.25

4.7 The probability of an event is the proportion of times it occurs in a large number of repetitions of the experiment.

4.9
a. In a large number of Texas hold'em hands, a person will be dealt a pocket pair about 5.9% of the time.
b. If the experiment of tossing a balanced dime three times is repeated many times, the percentage of experiments in which all three tosses come up heads will be approximately 12.5%.

4.11 The number in part (b). Because the probability of an event must always be between 0 and 1, inclusive.

4.13 a.

G, L, S, A	G, L, S, T	G, L, A, T	G, S, A, T	L, S, A, T

b. 0.2 **c.** 0.6 **d.** 0.8

4.15 a. 1/4 **b.** 7/12 **c.** 2/3   **4.17 a.** 0.644 **b.** 0.102 **c.** 0.356

4.19 a. 0.178 **b.** 0.720 **c.** 0.015 **d.** 0 **e.** 1

4.21 a. 0.189 **b.** 0.176 **c.** 0.239 **d.** 0.761

4.23 a. 0.131 **b.** 0.389 **c.** 0.868

4.25 a. 0.139 **b.** 0.500 **c.** 0.222 **d.** 0.111

4.27 a. The event in part (e) is certain; the event in part (d) is impossible.
b. The certain event has probability 1; the impossible event has probability 0.

4.29 Answers will vary.   **4.31** 333.5

Exercises 4.2

4.41 Venn diagrams

Note: In the each of the answers to Exercises 4.43–4.45, the shaded region represents the required event.

4.43 a.   **b.**

4.45 a.   **b.**

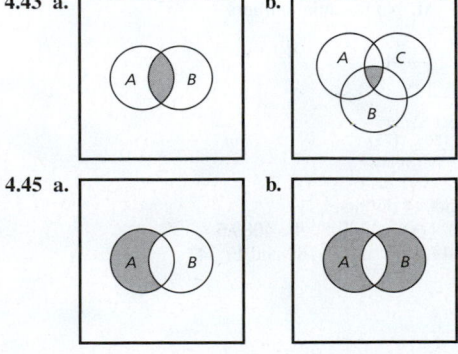

4.47 a. 6, 7, 8, 9, 10   **b.** 1, 2, 3   **c.** 2, 3, 4, 5

4.49 Three events are mutually exclusive if no two of them have outcomes in common.

4.51 False. Events A, B, and C may or may not be mutually exclusive.

4.53 $A = $ [dice: 4, 5, 6] $B = $ [dice: 4, 5, 6]

$C = $ [dice: 1, 2] $D = $ [dice: 1]

4.55 A: JM, WM, JS, WS, JH, WH, JW, WJ; B: HM, HS, HJ, HW;
C: MW, SW, HW, JW; D: MS, SM, HM, MH, SH, HS

4.57 a. (not A) = [dice: 1, 3, 5]

The event that the die comes up odd

b. (A & B) = [dice: 4, 6]

The event that the die comes up 4 or 6

c. (B or C) = [dice: 1, 2, 4, 5, 6]

The event that the die does not come up 3

4.59
a. (not A) = MS, SM, HM, MH, SH, HS, MJ, SJ, HJ, MW, SW, HW; the event that a female is appointed chairperson.
b. (B & D) = HM, HS; the event that Holly is appointed chairperson and either Maria or Susan is appointed secretary.
c. (B or C) = HM, HS, HJ, HW, MW, SW, JW; the event that either Holly is appointed chairperson or Will is appointed secretary (or both).

4.61
a. (not C) is the event that the state has a diabetes prevalence percentage of less than 6% or at least 13%; two states satisfy that property.
b. (A & B) is the event that the state has a diabetes prevalence percentage of at least 8%, but less than 7%, which is impossible; no states satisfy that property.
c. (C or D) is the event that the state has a diabetes prevalence percentage of less than 13%; 49 states satisfy that property.
d. (C & B) is the event that the state has a diabetes prevalence percentage of at least 6% but less than 7%; five states satisfy that property.

4.63
a. (not A) is the event that the World Series was decided in five or more games; 84 World Series satisfy that property.
b. (A & B) is the event that the World Series was decided in four games; 21 World Series satisfy that property.
c. (A or C) is the event that the World Series was decided in either four or seven games; 57 World Series satisfy that property.
d. (A & C) is the event that the World Series was decided in both four and seven games, which is impossible; no World Series satisfy that property.

4.65
a. (not A) is the event the unit has at least five rooms; 95,344 thousand units have that property.
b. (A & B) is the event the unit has two, three, or four rooms; 36,473 thousand units have that property.
c. (C or D) is the event the unit has at least five rooms; 95,344 thousand units have that property. (*Note:* From part (a), (not A) = (C or D).)

4.67 a. No   **b.** Yes   **c.** No   **d.** Yes, events B, C, and D. No.

4.69 A and C; A and D; C and D; A, C, and D

4.71
a. (1,0), (1,1), (2,0), (2,1), (2,2), (3,0), (3,1), (3,2), (3,3), (4,0), (4,1), (4,2), (4,3), (4,4), (5,0), (5,1), (5,2), (5,3), (5,4), (5,5), (6,0), (6,1), (6,2), (6,3), (6,4), (6,5), (6,6)
b. (1,0), (2,0), (2,2), (3,0), (3,2), (4,0), (4,2), (4,4), (5,0), (5,2), (5,4), (6,0), (6,2), (6,4), (6,6)

4.73
a. Suppose that A occurs. Because A and (not B) are mutually exclusive, it follows that (not B) does not occur, that is, that B occurs.
b. Suppose that A occurs. Because B occurs whenever A occurs, it follows that B occurs and, hence, that (not B) does not occur. Consequently, A and (not B) are mutually exclusive.

Exercises 4.3

4.77 0.5; $P(R) = 0.5$ **4.79** 0.65   **4.81** 0.65   **4.83** 0.45

4.85 a. No, because $P(A \& B) \neq 0$.   **b.** 4/15

4.87 a. 0.78   **b.** $S = (A \text{ or } B \text{ or } C)$ **c.** 0.11, 0.30, 0.37   **d.** 0.78

4.89 a. 0.267   **b.** 0.169   **c.** 0.088   **4.91 a.** 0.411   **b.** 0.829   **c.** 0.731

4.93 a. 0.89   **b.** 0.78   **4.95 a.** 0.969   **b.** 0.969

4.97 a. 0.167, 0.056, 0.028, 0.056, 0.028, 0.139, 0.167   **b.** 0.223   **c.** 0.112
d. 0.278   **e.** 0.278

4.99 93.1%

Exercises 4.4

4.103 Summing the row totals, summing the column totals, or summing the frequencies in the cells

4.105 a. univariate   **b.** bivariate

4.107
a. From left to right and top to bottom, the missing entries are 7, 8, 15, 11, and 14.
b. 0.44, 0.60, and 0.32
c.

	C_1	C_2	$P(R_i)$
R_1	0.12	0.28	0.40
R_2	0.32	0.28	0.60
$P(C_j)$	0.44	0.56	1.00

4.109
a. From left to right and top to bottom, the missing entries are 6, 22, 10, 13, 17, and 50.
b. 0.34, 0.56, and 0.10
c.

	C_1	C_2	C_3	$P(R_i)$
R_1	0.24	0.08	0.12	0.44
R_2	0.10	0.20	0.26	0.56
$P(C_j)$	0.34	0.28	0.38	1.00

4.111 a. 12   **b.** 62   **c.** 18   **d.** 41   **e.** 11

4.113
a. From left to right and top to bottom, the missing entries are 1,482, 99, 67, 17,470, and 50,109.
b. 50,109   **c.** 124,976   **d.** 306,052   **e.** 18,319   **f.** 306,901

4.115 a. 32   **b.** 23   **c.** 14
d. D_1 is the event that one of these teachers selected at random has only a bachelor's degree; (D_2 & F_2) is the event that one of these teachers selected at random has a master's degree but didn't offer field trips.
e. 0.549; 0.098

4.117
a. The player has between 6 and 10 years of experience; the player weighs between 200 and 300 lb; the player weighs less than 200 lb and has between 1 and 5 years of experience.
b. 0.194; 0.661; 0.081
c.

		Years of experience				
		Rookie Y_1	1–5 Y_2	6–10 Y_3	10+ Y_4	$P(W_i)$
Weight (lb)	Under 200 W_1	0.048	0.081	0.000	0.000	0.129
	200–300 W_2	0.177	0.339	0.113	0.032	0.661
	Over 300 W_3	0.065	0.065	0.081	0.000	0.210
	$P(Y_j)$	0.290	0.484	0.194	0.032	1.000

4.119 a. (i) M_2; (ii) G_3; (iii) (M_1 & G_1)   **b.** 0.050; 0.005; 0.364

c.

	Pay grade			
Marital status	Enlisted G_1	Officer G_2	Warrant G_3	**Total**
Single w/o children M_1	36.4	4.3	0.0	40.7
Single with children M_2	4.6	0.5	0.0	5.0
Joint service marriage M_3	4.5	0.8	0.0	5.4
Civilian marriage M_4	38.5	9.9	0.4	48.8
Total	84.0	15.4	0.5	100.0

Exercises 4.5

4.125 The conditional probability of tossing a head on the second toss, given that a head occurred on the first toss, equals the unconditional probability of tossing a head on the second toss.

4.127 0.4   **4.129** 21/50 or 0.42   **4.131** 5/8; 4/7   **4.133** 0.333; 0.167

4.135 a. 0.077   **b.** 0.333   **c.** 0.077   **d.** 0   **e.** 0.231   **f.** 1   **g.** 0.231
h. 0.167

4.137 a. 0.178   **b.** 0.179   **c.** 0.277
d. 17.8% of U.S. housing units have exactly four rooms; of those U.S. housing units with at least two rooms, 17.9% have exactly four rooms; of those U.S. housing units with at least two rooms, 27.7% have at most four rooms.

4.139 a. 0.290   **b.** 0.129   **c.** 0.375   **d.** 0.167
e. 29.0% of the players are rookies; 12.9% of the players weigh under 200 lb; 37.5% of the players who weigh under 200 lb are rookies; 16.7% of the rookies weigh under 200 lb.

4.141 a. 0.375   **b.** 0.106   **c.** 0.283   **d.** 0.283

4.143 a. 0.564   **b.** 0.391   **c.** 0.274   **d.** 0.701   **e.** 0.486
f. 56.4% of U.S. adults own a smartphone; 39.1% of U.S. adults are college grads; 27.4% of U.S. adults own a smartphone and are college grads; of U.S. adults who are college grads, 70.1% own a smartphone; of U.S. adults who own a smartphone, 48.6% are college grads.

4.145 a. 0.441   **b.** 0.686   **c.** 0.022   **d.** 0.133   **e.** 0.574

4.147 31.4%   **4.149** 63.8%   **4.151 a.** 0.5   **b.** 0.333

Exercises 4.6

4.157 0.24   **4.159** 1/5 or 0.2   **4.161** 0.42

4.163 5/14 or 0.357   **4.165** not independent   **4.167** independent

4.169 not possible to tell   **4.171** not possible to tell

4.173 independent   **4.175** not independent   **4.177** 0.12

4.179 0.229; 22.9% of U.S. adults are women who suffer from holiday depression.

4.181 a. 0.167   **b.** 0.4   **c.** 0.067   **d.** 0.067   **e.** 0.2

4.183 a. 0.054   **b.** 0.135   **d.** 0.115

4.185 a. 0.408   **b.** 0.370   **c.** No   **d.** No

4.187 a. 0.471; 0.187; 0.097
b. Not independent because $0.097 \neq 0.471 \cdot 0.187$

4.189 a. 0.5, 0.5, 0.375   **b.** 0.5   **c.** Yes.   **d.** 0.25   **e.** No.

4.191 a. 0.006   **b.** 0.005

4.193 a. 0.928   **b.** 0.072
c. There was a 7.2% chance that at least one "criticality 1" item would fail; in the long run, at least one "criticality 1" item will fail in 7.2 out of every 100 such missions.

4.195 a. 0.0239   **b.** 0.0222

4.197 a. 4.9%
b. Whether a nut is defective is independent of whether a bolt is defective

4.199 13/25 (a 52% chance)

Exercises 4.7

4.207 At least one of the four events must occur when the experiment is performed.

4.209 No. No.   **4.211 a.** 0.74   **b.** 0.432   **4.213 a.** 0.54   **b.** 0.370

4.215 a. $P(R_3)$ **b.** $P(S \mid R_3)$ **c.** $P(R_3 \mid S)$

4.217 a. 43.1%   **b.** 33%   **c.** 39.8%

4.219 a. 0.060   **b.** 0.112   **c.** 0.263

4.221 a. 52.1%   **b.** 57.9%   **c.** 31.7%   **4.223 a.** 34.0%   **b.** 35.3%

Exercises 4.8

4.229 Counting rules are techniques for determining the number of ways something can happen without directly listing all the possibilities. They are important because most often the number of possibilities is so large that a direct listing is impractical.

4.231
a. A permutation of r objects from a collection of m objects is any ordered arrangement of r of the m objects.
b. A combination of r objects from a collection of m objects is any unordered arrangement of r of the m objects.
c. Order matters in permutations but not in combinations.

4.233 40   **4.235** 1; 120; 720

4.237 a. 24   **b.** 32,760   **c.** 30   **d.** 1   **e.** 40,320

4.239 a. 4   **b.** 1365   **c.** 15   **d.** 1   **e.** 1   **4.241** $m!$

4.243 12,103,014   **4.245 b.** 15   **c.** 15

4.247 a. 80,000   **b.** 9000   **c.** 720,000,000

4.249 a. 900   **b.** 9,000,000   **c.** 8,100,000,000   **4.251** 657,720

4.253 60   **4.255** 5040   **4.257** $_{18}C_3 = 816$ **4.259** 45

4.261 a. 2,598,960   **b.** 24   **c.** 3744   **d.** 0.00144   **4.263** 0.278

4.265 a. 0.243   **b.** 0.972   **c.** 0.271

4.267 a. 0.00000000386   **b.** 0.00000135   **c.** 0.00000141   **d.** 0.0680

Review Problems for Chapter 4

1. It enables you to evaluate and control the likelihood that a statistical inference is correct. More generally, probability theory provides the mathematical basis for inferential statistics.

2. **a.** The experiment has a finite number of possible outcomes, all equally likely.
   b. The probability of an event equals the ratio of the number of ways that the event can occur to the total number of possible outcomes.

3. It is the proportion of times the event occurs in a large number of repetitions of the experiment.

4. (b) and (c), because the probability of an event must always be between 0 and 1, inclusive.

5. Venn diagrams

6. Two or more events are said to be mutually exclusive if at most one of them can occur when the experiment is performed, that is, if no two of them have outcomes in common.

7. **a.** $P(E)$ **b.** $P(E) = 0.436$ **8. a.** False   **b.** True

9. It is sometimes easier to compute the probability that an event does not occur than the probability that it does occur.

10. **a.** Univariate   **b.** Bivariate   **c.** Contingency table, or two-way table

11. Marginal   **12. a.** $P(B \mid A)$ **b.** A

13. Directly or using the conditional probability rule

14. The joint probability equals the product of the marginal probabilities.

15. Exhaustive   **16.** See Key Fact 4.2 on page 209.   **17.** 0.9

18. 0.6   **19.** 0.7   **20.** 0.167; 0.5   **21. a.** No.   **b.** Yes.

22. 0.28   **23.** 0.18   **24. a.** 0.606   **b.** 0.66   **25.** 72

26. 1; 1; 6; 24   **27.** 120

28. a.

abc	abd	acd	bcd
acb	adb	adc	bdc
bac	bad	cad	cbd
bca	bda	cda	cdb
cba	dab	dac	dbc
cab	dba	dca	dcb

b. $\{a, b, c\}, \{a, b, d\}, \{a, c, d\}, \{b, c, d\}$ **c.** 24; 4 **d.** 24; 4

29. a. 0.43 **b.** 0.57 **c.** 0.1

30. a. 0.184 **b.** 0.397 **c.** 0.184, 0.160, 0.131, 0.102, 0.078, 0.217, 0.128

31. a. (not J) is the event that the return shows an AGI of at least $100K. There are 18,227 thousand such returns.
b. (H & I) is the event that the return shows an AGI of between $20K and $50K. There are 44,251 thousand such returns.
c. (H or K) is the event that the return shows an AGI of at least $20K. There are 93,404 thousand such returns.
d. (H & K) is the event that the return shows an AGI of between $50K and $100K. There are 30,926 thousand such returns.

32. a. Not mutually exclusive **b.** Mutually exclusive
c. Mutually exclusive **d.** Not mutually exclusive

33. a. 0.528, 0.655, 0.872, 0.345
b. $H = (C$ or D or E or $F)$
$I = (A$ or B or C or D or $E)$
$J = (A$ or B or C or D or E or $F)$
$K = (F$ or $G)$
c. 0.528, 0.655, 0.872, 0.345

34. a. 0.128, 0.311, 0.656, 0.217 **b.** 0.872 **c.** 0.656
d. They are the same.

35. a. 6 **b.** 15,980 thousand **c.** 66,947 thousand **d.** 6059 thousand

36. a. L_3 is the event that the student selected is in college; T_1 is the event that the student selected attends a public school; (T_1 & L_3) is the event that the student selected attends a public college.
b. 0.281; 0.856; 0.203. 28.1% of students attend college, 85.6% attend public schools; 20.3% attend public colleges.
c.

Level	Type		$P(L_i)$
	Public T_1	Private T_2	
Elementary L_1	0.463	0.052	0.515
High school L_2	0.190	0.014	0.204
College L_3	0.203	0.077	0.281
$P(T_j)$	0.856	0.144	1.000

d. 0.934 **e.** 0.934 **f.** They are the same.

37. a. 0.238; 23.8% of students attending public schools are in college.
b. 0.237 **c.** Discrepancy is due to roundoff error.

38. a. 0.144, 0.069
b. No, because $P(T_2 \mid L_2) \neq P(T_2)$; 6.9% of high school students attend private schools, whereas 14.4% of all students attend private schools.
c. No, because both events can occur if the student selected is any one of the 1103 thousand students who attend a private high school.

39. a. 0.023 **b.** 0.309 **d.** 0.451 **40. a.** 47.4% **b.** 20.4% **c.** 32.3%

41. a. 0.686 **b.** 0.068 **c.** 0.271 **42. a.** 0.79 **b.** 0.554 **c.** 0.178

43. a. 66 **b.** 1320 **c.** 28; 336

44. a. 635,013,559,600 **b.** 0.213 **c.** 0.00045 **d.** 0.032 **e.** 0.013

45. 4,426,165,368

46. a. All households that own a DVD player also own a TV. **b.** 84.0%
c. The percentage of non-TV households that own a DVD player

Chapter 5

Exercises 5.1

5.1 a. probability **b.** probability

5.3 $\{X = 3\}$ is the event that the student has three siblings; $P(X = 3)$ is the probability of the event that the student has three siblings.

5.5 The probability distribution of the random variable

5.7 a.

x	1	2	3
$P(X = x)$	0.3	0.4	0.3

b. $\{X = 2\}; \{X \leq 2\}; \{X > 2\}$ **c.** 0.4; 0.7; 0.3
d.

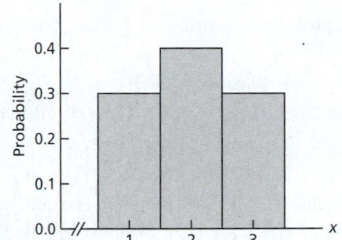

5.9 a.

y	0	1	4	6
$P(Y = y)$	0.36	0.28	0.16	0.20

b. $\{Y = 3\}; \{Y < 3\}; \{Y \geq 3\}$ **c.** 0.00; 0.64; 0.36

5.11 a. 2, 4, 5, 6, 7, 8 **b.** $\{X = 7\}$
c. 0.022. 2.2% of the shuttle missions between April 12, 1981 and July 8, 2011 had a crew size of 4.
d.

x	2	4	5	6	7	8
$P(X = x)$	0.030	0.022	0.267	0.207	0.467	0.007

e.

5.13 a. $\{Y \geq 1\}$ **b.** $\{Y = 3\}$ **c.** $\{2 \leq Y \leq 4\}$ **d.** 0.815 **e.** 0.093
f. 0.408

5.15 a. 2, 3, 4, 5, 6, 7, 8, 9, 10, 11, 12 **b.** $\{Y = 7\}$ **c.** $\frac{1}{6}$
d.

y	2	3	4	5	6	7	8	9	10	11	12
$P(Y = y)$	$\frac{1}{36}$	$\frac{1}{18}$	$\frac{1}{12}$	$\frac{1}{9}$	$\frac{5}{36}$	$\frac{1}{6}$	$\frac{5}{36}$	$\frac{1}{9}$	$\frac{1}{12}$	$\frac{1}{18}$	$\frac{1}{36}$

f. $\frac{2}{9}$ **g.** $\frac{1}{9}$

5.17 a.

s	0	5	10
$P(S = s)$	0.651	0.262	0.087

b. 0.262; 0.349; 0.913; 0.087; 1; 0

5.19 a. Because only a sample of litter sizes were observed. **b.** 0.758

Exercises 5.2

5.25 The mean of a variable of a finite population (population mean)

5.27 a. 2 **b.** 0.8 **5.29 a.** 2.12 **b.** 2.4

5.31 a. 6.05 crew members **b.** 1.2 crew members

5.33 a. 1.937 hurricanes **b.** 1.7 hurricanes

5.35 a. 7 **b.** 2.4 **5.37 a.** 2.18 points **b.** 3.2 points

5.39 b. −0.052 **c.** 5.2¢ **d.** $5.20, $52 **5.41 a.** $760 **b.** $810

5.43 a. $\mu_W = 0.25, \sigma_W = 0.536$ **b.** 0.25 **c.** 62.5

Exercises 5.3

5.49 A trial

5.51 In n Bernoulli trials, the number of outcomes that contain exactly x successes equals the binomial coefficient $\binom{n}{x}$.

5.53 The binomial distribution is the probability distribution for the number of successes in a sequence of Bernoulli trials.

5.55 Answers will vary. **5.57** 6; 5040; 40,320; 362,880

5.59 a. 10 **b.** 35 **c.** 120 **d.** 792 **5.61 a.** 4 **b.** 15 **c.** 56 **d.** 84

5.63 a. $p = 0.5$ **b.** $p < 0.5$

5.65
a. Each trial consists of observing whether a child with pinworm is cured by treatment with pyrantel pamoate and has two possible outcomes: cured or not cured. The trials are independent. The success probability is 0.9; that is, $p = 0.9$.

b.

Outcome	Probability
sss	$(0.9)(0.9)(0.9) = 0.729$
ssf	$(0.9)(0.9)(0.1) = 0.081$
sfs	$(0.9)(0.1)(0.9) = 0.081$
sff	$(0.9)(0.1)(0.1) = 0.009$
fss	$(0.1)(0.9)(0.9) = 0.081$
fsf	$(0.1)(0.9)(0.1) = 0.009$
ffs	$(0.1)(0.1)(0.9) = 0.009$
fff	$(0.1)(0.1)(0.1) = 0.001$

d. ssf, sfs, fss
e. 0.081. Because each probability is obtained by multiplying two success probabilities of 0.9 and one failure probability of 0.1.
f. 0.243 **g.**

x	0	1	2	3
$P(X = x)$	0.001	0.027	0.243	0.729

5.67 a. 0.265 **b.** 0.265 **5.69 a.** 0.234 **b.** 0.234

5.71 a. 0.396 **b.** 0.396

5.73 The appropriate binomial probability formula is

$$P(X = x) = \binom{3}{x}(0.9)^x(0.1)^{3-x}.$$

Applying this formula for $x = 0, 1, 2$, and 3, gives the same result as in part (g) of Exercise 5.65.

5.75 0.246

5.77 a. 0.161 **b.** 0.332 **c.** 0.468 **d.** 0.821
e.

x	0	1	2	3	4	5
$P(X = x)$	0.004	0.040	0.161	0.328	0.332	0.135

f. Left skewed **g.** $P(X = x)$

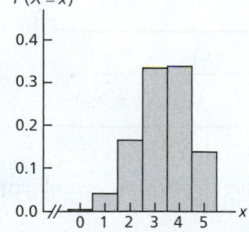

h. $\mu = 3.35$ times; $\sigma = 1.1$ times. **i.** $\mu = 3.35$ times; $\sigma = 1.1$ times.
j. On average, the favorite will finish in the money 3.35 times for every 5 races.

5.79 a. 0.279; 0.685; 0.594 **b.** 0.720
c. 3.2 traffic fatalities; on average, 3.2 of every 8 traffic fatalities involve an intoxicated or alcohol-impaired driver or nonoccupant.
d. 1.4 traffic fatalities

5.81 a. 0.118; 0.946; 0.172 **b.** 0.979; 0.121 **c.** 0.054
d.

x	$P(X = x)$	x	$P(X = x)$
0	0.021	5	0.118
1	0.100	6	0.042
2	0.216	7	0.010
3	0.272	8	0.001
4	0.219	9	0.000

e. Because the sampling is done without replacement from a finite population. Hypergeometric distribution.

5.83 a.

x	$P(X = x)$	x	$P(X = x)$
0	0.517	4	0.002
1	0.358	5	0.000
2	0.106	6	0.000
3	0.017	7	0.000

b. 0.63; on average, we would expect about 0.63 of seven youths in grades 3–8 to be PVGU.
c. Yes, because if the percentage of PVGUs in grades 3–8 today is the same as in 2011, there is only a 1.9% chance that three or more of the seven youths selected would be PVGU.
d. Probably not, because if the percentage of PVGUs in grades 3–8 today is the same as in 2011, there is a 12.5% chance that two or more of the seven youths selected would be PVGU.

Exercises 5.4

5.91 (1) To model the frequency with which a specified event occurs during a particular period of time; (2) to approximate binomial probabilities

5.93 They are the same. **5.95** np

5.97 a. 0.175; 0.040; 0.875 **b.** 5; 2.2

5.99 a. 0.157; 0.401; 0.848 **b.** 4.7; 2.2

5.101 a. 0.195 **b.** 0.102 **c.** 0.704
d.

Particles y	Probability $P(Y = y)$	Particles y	Probability $P(Y = y)$
0	0.021	7	0.054
1	0.081	8	0.026
2	0.156	9	0.011
3	0.201	10	0.004
4	0.195	11	0.002
5	0.151	12	0.000
6	0.097		

f. 3.87 particles

5.103 a. 0.497 **b.** 0.966 **c.** 0.498
d. 0.7 wars; on average, 0.7 wars begin during a calendar year. **e.** 0.8 wars

5.105 a. 1.2 cherries
b.

Cherries	Rel. freq.
0	0.314
1	0.343
2	0.229
3	0.057
4	0.057

c.

Cherries	Probability
0	0.301
1	0.361
2	0.217
3	0.087
4	0.026

5.107 0.758 **5.109 a.** 6.667 **b.** 0.352; 0.923

5.111 a. 0.526 **b.** 69,077,553

Review Problems for Chapter 5

1. **a.** random variable   **b.** can be listed

2. The possible values and corresponding probabilities of the discrete random variable

3. Probability histogram   **4.** 1

5. **a.** $P(X = 2) = 0.386$ **b.** 38.6%   **c.** 19.3; 193   **6.** 3.6

7. X, because it has a smaller standard deviation, therefore less variation.

8. 1, 6, 24, 5040   **9. a.** 56   **b.** 56   **c.** 1   **d.** 45   **e.** 91,390   **f.** 1

10. Each trial has the same two possible outcomes; the trials are independent; the probability of a success remains the same from trial to trial.

11. The binomial distribution is the probability distribution for the number of successes in a finite sequence of Bernoulli trials.

12. 120

13. **a.** $p = 0.493$ **b.**

Outcome	Probability
sss	$(0.493)(0.493)(0.493) = 0.120$
ssf	$(0.493)(0.493)(0.507) = 0.123$
sfs	$(0.493)(0.507)(0.493) = 0.123$
sff	$(0.493)(0.507)(0.507) = 0.127$
fss	$(0.507)(0.493)(0.493) = 0.123$
fsf	$(0.507)(0.493)(0.507) = 0.127$
ffs	$(0.507)(0.507)(0.493) = 0.127$
fff	$(0.507)(0.507)(0.507) = 0.130$

   d. *ssf, sfs, fss*
   e. 0.123. Each probability is obtained by multiplying two success probabilities of 0.493 and one failure probability of 0.507.
   f. 0.369   **g.**

y	0	1	2	3
$P(Y = y)$	0.130	0.381	0.369	0.120

   h. Binomial with parameters $n = 3$ and $p = 0.493$

14. **a.** $p > 0.5$ **b.** $p = 0.5$

15. Substitute the binomial (or Poisson) probability formula into the formulas for the mean and standard deviation of a discrete random variable and then simplify mathematically.

16. **a.** Binomial distribution   **b.** Hypergeometric distribution
   c. When the sample size does not exceed 5% of the population size because, under this condition, there is little difference between sampling with and without replacement.

17. **a.** 1, 2, 3, 4   **b.** $\{X = 3\}$
   c. 0.292; 29.2% of undergraduates at ASU are juniors.
   d.

x	1	2	3	4
$P(X = x)$	0.163	0.188	0.292	0.357

   e. $P(X=x)$

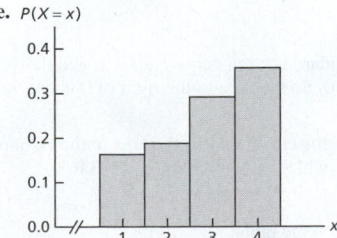

18. **a.** $\{Y = 4\}$ **b.** $\{Y \geq 4\}$ **c.** $\{2 \leq Y \leq 4\}$ **d.** $\{Y \geq 1\}$ **e.** 0.174
   f. 0.322   **g.** 0.646   **h.** 0.948

19. **a.** 2.817 lines   **b.** 2.817 lines   **c.** 1.5 lines

20.

y	0	1	2	3
$P(Y = y)$	0.130	0.380	0.370	0.120

The discrepancy between the probability distribution found here and the one found in Problem 13(g) is due to rounding error in the latter problem.

21. **a.** 0.087   **b.** 0.604   **c.** 12.75; 1.4

22. **a.** 0.346; 0.476; 0.872
   b.

x	0	1	2	3	4
$P(X = x)$	0.026	0.154	0.346	0.346	0.130

   c. Left skewed, because $p > 0.5$

23. **a.** $P(X=x)$

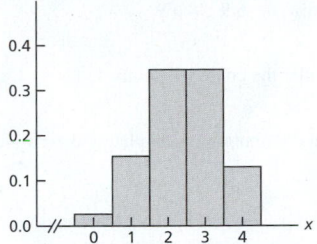

   b. The probability distribution is only approximately correct because the sampling is without replacement; a hypergeometric distribution.
   c. Yes, because the sample size certainly does not exceed 5% of the population size.

24. **a.** 0.266   **b.** 0.099   **c.** 0.826

25. **a.**

x	$P(X = x)$	x	$P(X = x)$
0	0.174	5	0.024
1	0.304	6	0.007
2	0.266	7	0.002
3	0.155	8	0.000
4	0.068		

   b. $P(X=x)$

   c. Right skewed. Yes, all Poisson distributions are right skewed.
   d. $\mu = 1.75$ calls; on average, there are 1.75 calls per minute to a wrong number.
   e. $\sigma = 1.3$ calls

26. **a.** 3   **b.** 0.616   **c.** 0.950

27. **a.** $n = 100$ and $p = 0.015$ **b.** $\lambda = 1.5$
   c. & d.

x	Binomial probability	Poisson approximation
0	0.2206	0.2231
1	0.3360	0.3347
2	0.2532	0.2510
3	0.1260	0.1255
4	0.0465	0.0471
5	0.0136	0.0141
6	0.0033	0.0035
7	0.0007	0.0008
8	0.0001	0.0001
9	0.0000	0.0000

28. **a.** 0.1260, 0.1255   **b.** 0.4393, 0.4377   **c.** 0.9358, 0.9343
   d. 0.1902, 0.1911

Chapter 6

Exercises 6.1

6.1 A density curve of a variable is a smooth curve with which we can identify the shape of the distribution of the variable.

6.3 They are equal (at least approximately) when the area is expressed as a percentage.

6.5 4   **6.7 a.** 32.4%   **b.** 67.6%   **6.9** 58.6%

6.11 a. 0.284   **b.** 0.716

6.13 No, because the total area under the curve is 0.9, not 1.

6.15 Roughly bell shaped

6.17 They are the same. A normal distribution is completely determined by the mean and standard deviation.

6.19
a. True. They have the same spread because their standard deviations are equal.
b. False. A normal distribution is centered at its mean, which is different for these two distributions.

6.21 True. The spread of a normal distribution is completely determined by its standard deviation.

6.23 a.

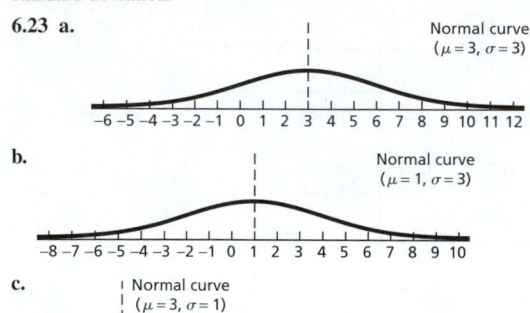

b.

c.

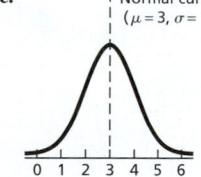

6.25 They are equal. They are approximately equal.

6.27 62.27%

6.29 a.

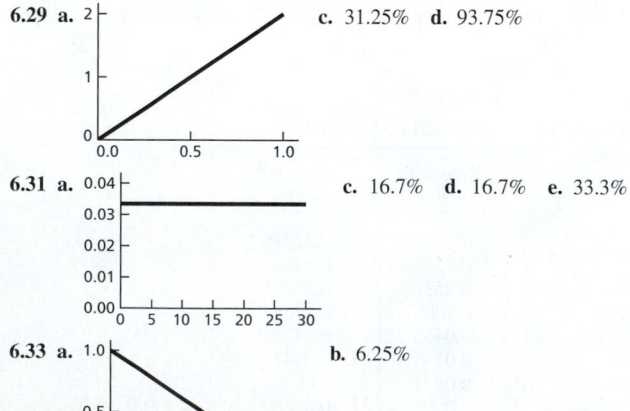

c. 31.25%   **d.** 93.75%

6.31 a.

c. 16.7%   **d.** 16.7%   **e.** 33.3%

6.33 a. 1.0

b. 6.25%

6.35 a. 55.70%
b. 0.5570; this is only an estimate because the distribution of heights is only approximately normally distributed.

6.37 a.

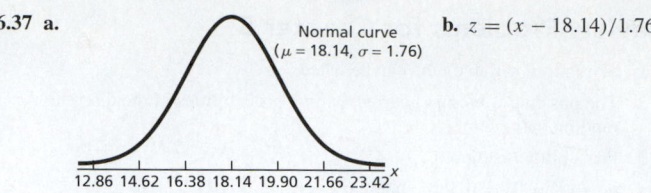

Normal curve $(\mu = 18.14, \sigma = 1.76)$

b. $z = (x - 18.14)/1.76$

c. Standard normal distribution   **d.** $-1.22; -0.65$ **e.** right; 0.49

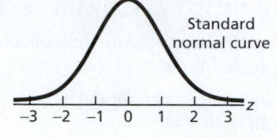

Standard normal curve

6.39 a.

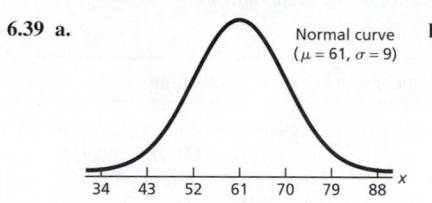

Normal curve $(\mu = 61, \sigma = 9)$

b. $z = (x - 61)/9$

c. Standard normal distribution; see the graph in the answer to Exercise 6.37(c).
d. $-1.22; 1$ **e.** left; 1.56

6.41 a.

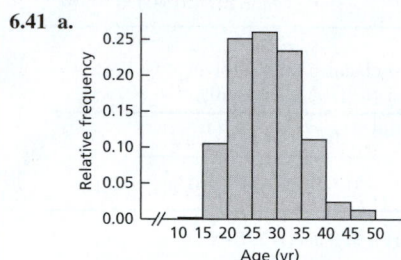

b. Yes, because the age distribution is shaped roughly like a normal curve.

6.43 a.

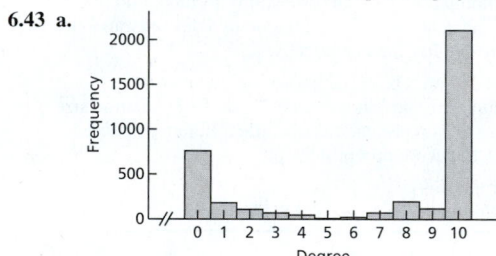

b. No, because the degree-of-cloudiness distribution has a shape far different from that of a normal curve.

Exercises 6.2

6.51 The total area under the standard normal curve equals 1, and the standard normal curve is symmetric about 0. So the area to the right of 0 is one-half of 1, or 0.5.

6.53 0.3336. The total area under the curve is 1, so the area to the right of 0.43 equals 1 minus the area to its left, which is $1 - 0.6664 = 0.3336$.

6.55 99.74%

6.57 a. Read the area directly from the table.
b. Subtract the table area from 1.
c. Subtract the smaller table area from the larger.

6.59 a. 0.9875   **b.** 0.0594   **c.** 0.5   **d.** 0.0000 (to four decimal places)

6.61 a. 0.8577   **b.** 0.2743   **c.** 0.5   **d.** 0.0000 (to four decimal places)

6.63 a. 0.9105   **b.** 0.0440   **c.** 0.2121   **d.** 0.1357

6.65 a. 0.0645   **b.** 0.7975

6.67 a. 0.7994   **b.** 0.8990   **c.** 0.0500   **d.** 0.0198

6.69 a. 0.6826 **b.** 0.9544

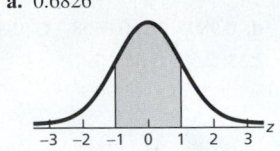

c. 0.9974

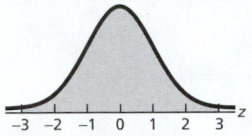

6.71 −1.96 **6.73** 0.67 **6.75** −1.645 **6.77** 0.44
6.79 a. 1.88 **b.** 2.575 **6.81** ±1.645
6.83 The four missing entries are 1.645, 1.96, 2.33, and 2.575.

Exercises 6.3

6.87 The z-scores corresponding to the x-values that lie two standard deviations below and above the mean are −2 and 2, respectively.

Note: In the remainder of this chapter, your answers may vary from those given here depending on whether you use Table II or technology.

6.89 a. 68.53% **b.** 69.15% **c.** 15.87%
6.91 a. 6.69% **b.** 50% **c.** 99.38%
6.93 a. 4.66, 6, 7.34 **b.** 8.08 **c.** 5.22 **d.** 2.08, 9.92
6.95 a. 7.99, 10, 12.01 **b.** 11.56 **c.** 11.17 **d.** 2.28, 17.73
6.97 a. 14.66% **b.** 31.21% **c.** 16.96 mm, 18.14 mm, 19.32 mm
d. 21.04 mm; 95% of adult male *G. mollicoma* have carapace lengths less than 21.04 mm and 5% have carapace lengths greater than 21.04 mm.
6.99 a. 73.01% **b.** 94.06%
c. 58.8 minutes; 40% of finishers in the New York City 10-km run have times less than 58.8 minutes and 60% have times greater than 58.8 minutes.
d. 68.6 minutes; 80% of finishers in the New York City 10-km run have times less than 68.6 minutes and 20% have times greater than 68.6 minutes.
6.101 a. 96.34% **b.** 21.19% **6.103 a.** 68% **b.** 95% **c.** 99.7%
6.105 a. 1.29 kg; 1.51 kg **b.** 1.18 kg; 1.62 kg **c.** 1.07 kg; 1.73 kg
d. See the graphs shown in Fig. A.1.
6.107 a. (i) 11.70% (ii) 12.23% **b.** (i) 39.83% (ii) 39.14%
6.109 a. 100% **b.** 61.41% **c.** No.

Exercises 6.4

6.117 Decisions about whether a variable is normally distributed often are important in subsequent analyses—from percentage or percentile calculations to statistical inferences.

6.119 In a normal probability plot, outliers lie outside the overall pattern formed by the other points in the plot.

6.121 The variable under consideration is approximately normally distributed.

6.123 The variable under consideration is not approximately normally distributed.

6.125 The variable under consideration is not approximately normally distributed.

6.127 a.

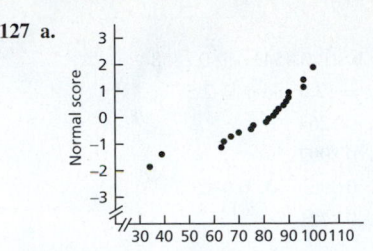

b. 34 and 39 are outliers.

c. Final-exam scores in this introductory statistics class do not appear to be normally distributed.

6.129 a. **b.** No outliers

c. It appears plausible that finishing times for the winners of 1-mile thoroughbred horse races are (approximately) normally distributed.

6.131 a. **b.** No outliers

c. It appears plausible that the average times spent per user per month from January to June of the year in question are (approximately) normally distributed.

6.133 a. **b.** 6.7 and 7.6 are outliers.

c. Diffusive oxygen uptakes in surface sediments from central Sagami Bay do not appear to be normally distributed.

Exercises 6.5

6.141 It is not practical to use the binomial probability formula when the number of trials is very large.
6.143 between 7.5 and 8.5 **6.145** between −0.5 and 6.5
6.147 between −0.5 and 7.5 **6.149** between 7.5 and 10.5
6.151 between 6.5 and 9.5 **6.153** between 6.5 and 10.5
6.155 between 7.5 and 9.5 **6.157** between 10.5 and $n + 0.5$

FIGURE A.1 Graphs for Exercise 6.105(d)

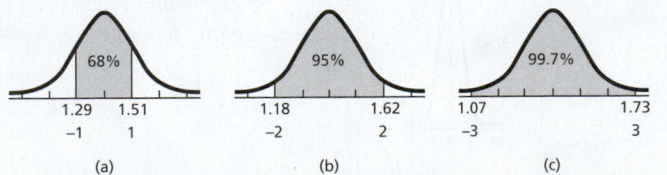

6.159 between 9.5 and $n + 0.5$

6.161 a. (i) 0.4512 (ii) 0.8907 **b.** (i) 0.4544 (ii) 0.8858

6.163 The one with parameters $\mu = 12.5$ and $\sigma = 2.5$

6.165 a. 0.1492 **b.** 0.3135 **c.** 0.3264

6.167 a. 0.0824 **b.** 0.4458 **c.** 0.8997

6.169 a. 0.1288 **b.** 0.2191 **c.** 0.9956 **d.** 0.0363

6.171 a. 0.0558 **b.** 0.4721 **c.** 0.4721

Review Problems for Chapter 6

1. A density curve of a variable is a smooth curve with which we can identify the shape of the distribution of the variable. For a variable with a density curve, the percentage of all possible observations of the variable that lie within any specified range equals (at least approximately) the corresponding area under the density curve, expressed as a percentage.

2. 25; 50 **3.** 36.4%; 63.6% **4.** 27.2%

5. a. 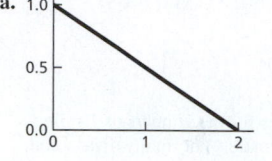 **c.** 31.25% **d.** 6.25%

6. It appears again and again in both theory and practice.

7. a. A variable is said to be normally distributed if its distribution has the shape of a normal curve.
b. If a variable of a population is normally distributed and is the only variable under consideration, common practice is to say that the population is a normally distributed population.
c. The parameters for a normal curve are the corresponding mean and standard deviation of the variable.

8. a. False
b. True. A normal distribution is completely determined by its mean and standard deviation.

9. They are the same when areas are expressed as percentages.

10. Standard normal distribution **11 a.** True **b.** True

12. a. The second curve **b.** The first and second curves
c. The first and third curves **d.** The third curve **e.** The fourth curve

13. Key Fact 6.4 on page 269, which states that the standardized version of a normally distributed variable has the standard normal distribution

14. a. Read the area directly from the table.
b. Subtract the table area from 1.
c. Subtract the smaller table area from the larger.

15. a. Locate the table entry closest to the specified area and read the corresponding z-score.
b. Locate the table entry closest to 1 minus the specified area and read the corresponding z-score.

16. The z-score having area α to its right under the standard normal curve

17. See Key Fact 6.6 on page 282.

18. The observations expected for a sample of the same size from a variable that has the standard normal distribution

19. linear (i.e., a straight line)

20. a.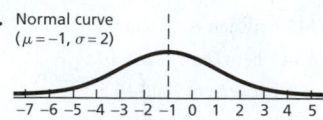
b. 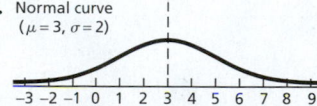 **c.**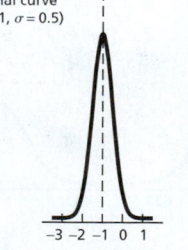

21. a. 0.1469 **b.** 0.1469 **c.** 0.7062

22. a. 0.0013 **b.** 0.2709 **c.** 0.1305 **d.** 0.9803 **e.** 0.0668 **f.** 0.8426

23. a. −0.52 **b.** 1.28 **c.** 1.96; 1.645; 2.33; 2.575 **d.** ±2.575

24. a. 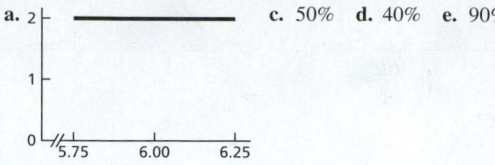 **c.** 50% **d.** 40% **e.** 90%

25. a. 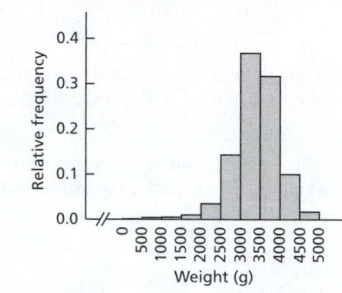 **b.** $z = (x - 18.8)/1.1$
c. Standard normal distribution **d.** 0.8115 **e.** left; −2.55

26. a.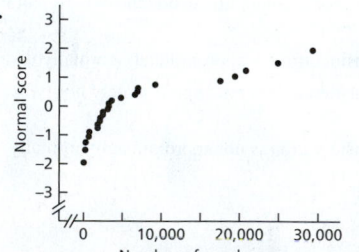
b. No, because the histogram is left skewed.

27. a. 59.87% **b.** 73.33% **c.** 2.28%

28. a. 144.1 points, 150 points, 155.9 points; 25% of the scores are less than 144.1 points, 25% are between 144.1 points and 150 points, 25% are between 150 points and 155.9 points, and 25% exceed 155.9 points.
b. 170.4 points; 99% of the scores are less than 170.4 points and 1% are greater than 170.4 points.

29. a. 141.25 points; 158.75 points **b.** 132.50 points; 167.50 points
c. 123.75 points; 176.25 points

30. a. **b.** No outliers

c. It appears plausible that lengths of adult male Northwest Atlantic cod in Nova Scotia, Canada, are (approximately) normally distributed.

31. a. **b.** No outliers

c. The numbers of employees of publicly traded mortgage industry companies do not appear to be normally distributed.

32. a. 0.0076 **b.** 0.9505 **c.** 0.9988

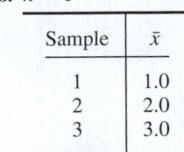

Chapter 7

Exercises 7.1

7.1 Generally, sampling is less costly and can be done more quickly than a census.

7.3 a. 2
b. $n = 1$

Sample	$\bar{x}$
1	1.0
2	2.0
3	3.0

$n = 2$

Sample	$\bar{x}$
1, 2	1.5
1, 3	2.0
2, 3	2.5

$n = 3$

Sample	$\bar{x}$
1, 2, 3	2.0

For the dotplots, see part (c).
c.

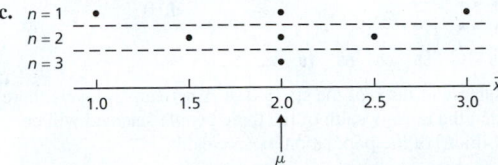

d. 1/3; 1/3; 1   **e.** 1/3; 1; 1

7.5 a. 2.5
b. $n = 1$

Sample	$\bar{x}$
1	1.0
2	2.0
3	3.0
4	4.0

$n = 2$

Sample	$\bar{x}$
1, 2	1.5
1, 3	2.0
1, 4	2.5
2, 3	2.5
2, 4	3.0
3, 4	3.5

$n = 3$

Sample	$\bar{x}$
1, 2, 3	2.0
1, 2, 4	2.3
1, 3, 4	2.7
2, 3, 4	3.0

$n = 4$

Sample	$\bar{x}$
1, 2, 3, 4	2.5

For the dotplots, see part (c).
c.

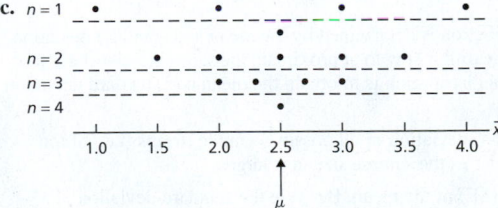

d. 0; 1/3; 0; 1   **e.** 1/2; 2/3; 1; 1

7.7 a. 3
b. $n = 1$

Sample	$\bar{x}$
1	1.0
2	2.0
3	3.0
4	4.0
5	5.0

$n = 2$

Sample	$\bar{x}$
1, 2	1.5
1, 3	2.0
1, 4	2.5
1, 5	3.0
2, 3	2.5
2, 4	3.0
2, 5	3.5
3, 4	3.5
3, 5	4.0
4, 5	4.5

$n = 3$

Sample	$\bar{x}$
1, 2, 3	2.0
1, 2, 4	2.3
1, 2, 5	2.7
1, 3, 4	2.7
1, 3, 5	3.0
1, 4, 5	3.3
2, 3, 4	3.0
2, 3, 5	3.3
2, 4, 5	3.7
3, 4, 5	4.0

$n = 4$

Sample	$\bar{x}$
1, 2, 3, 4	2.50
1, 2, 3, 5	2.75
1, 2, 4, 5	3.00
1, 3, 4, 5	3.25
2, 3, 4, 5	3.50

$n = 5$

Sample	$\bar{x}$
1, 2, 3, 4, 5	3

For the dotplots, see part (c).
c.

d. 1/5; 1/5; 1/5; 1/5; 1   **e.** 1/5; 3/5; 3/5; 1; 1

7.9 a. 3.5
b. $n = 1$

Sample	$\bar{x}$
1	1.0
2	2.0
3	3.0
4	4.0
5	5.0
6	6.0

$n = 2$

Sample	$\bar{x}$
1, 2	1.5
1, 3	2.0
1, 4	2.5
1, 5	3.0
1, 6	3.5
2, 3	2.5
2, 4	3.0
2, 5	3.5
2, 6	4.0
3, 4	3.5
3, 5	4.0
3, 6	4.5
4, 5	4.5
4, 6	5.0
5, 6	5.5

$n = 3$

Sample	$\bar{x}$
1, 2, 3	2.0
1, 2, 4	2.3
1, 2, 5	2.7
1, 2, 6	3.0
1, 3, 4	2.7
1, 3, 5	3.0
1, 3, 6	3.3
1, 4, 5	3.3
1, 4, 6	3.7
1, 5, 6	4.0
2, 3, 4	3.0
2, 3, 5	3.3
2, 3, 6	3.7
2, 4, 5	3.7
2, 4, 6	4.0
2, 5, 6	4.3
3, 4, 5	4.0
3, 4, 6	4.3
3, 5, 6	4.7
4, 5, 6	5.0

$n = 4$

Sample	$\bar{x}$
1, 2, 3, 4	2.50
1, 2, 3, 5	2.75
1, 2, 3, 6	3.00
1, 2, 4, 5	3.00
1, 2, 4, 6	3.25
1, 2, 5, 6	3.50
1, 3, 4, 5	3.25
1, 3, 4, 6	3.50
1, 3, 5, 6	3.75
1, 4, 5, 6	4.00
2, 3, 4, 5	3.50
2, 3, 4, 6	3.75
2, 3, 5, 6	4.00
2, 4, 5, 6	4.25
3, 4, 5, 6	4.50

$n = 5$

Sample	$\bar{x}$
1, 2, 3, 4, 5	3.0
1, 2, 3, 4, 6	3.2
1, 2, 3, 5, 6	3.4
1, 2, 4, 5, 6	3.6
1, 3, 4, 5, 6	3.8
2, 3, 4, 5, 6	4.0

$n = 6$

Sample	$\bar{x}$
1, 2, 3, 4, 5, 6	3.5

For the dotplots, see part (c).

c. $n = 1$

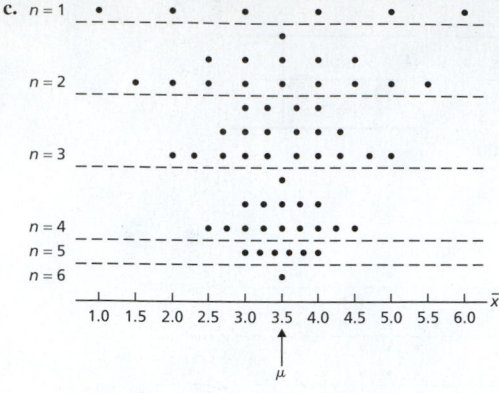

d. 0; 1/5; 0; 1/5; 0; 1   **e.** 1/3; 7/15; 3/5; 11/15; 1; 1

7.11 a. $\mu = 78.6$ inches

b.

Sample	Heights	$\bar{x}$	Sample	Heights	$\bar{x}$
B, W	83, 76	79.5	W, C	76, 74	75.0
B, J	83, 80	81.5	W, H	76, 80	78.0
B, C	83, 74	78.5	J, C	80, 74	77.0
B, H	83, 80	81.5	J, H	80, 80	80.0
W, J	76, 80	78.0	C, H	74, 80	77.0

c.

d. 0

e. 0.4. If a random sample of two players is taken, there is a 40% chance that the mean height of the two players selected will be within 1 inch of the population mean height.

7.13 b.

Sample	Heights	$\bar{x}$	Sample	Heights	$\bar{x}$
B, W, J	83, 76, 80	79.7	B, C, H	83, 74, 80	79.0
B, W, C	83, 76, 74	77.7	W, J, C	76, 80, 74	76.7
B, W, H	83, 76, 80	79.7	W, J, H	76, 80, 80	78.7
B, J, C	83, 80, 74	79.0	W, C, H	76, 74, 80	76.7
B, J, H	83, 80, 80	81.0	J, C, H	80, 74, 80	78.0

c.

d. 0

e. 0.5. If a random sample of three players is taken, there is a 50% chance that the mean height of the three players selected will be within 1 inch of the population mean height.

7.15 b.

Sample	Heights	$\bar{x}$
B, W, J, C, H	83, 76, 80, 74, 80	78.6

c.

d. 1

e. 1. If a random sample of five players is taken, there is a 100% chance that the mean height of the five players selected will be within 1 inch of the population mean height.

7.17 a. $\mu = \$46.5$ billion

b.

Sample	Wealth	$\bar{x}$	Sample	Wealth	$\bar{x}$
G, B	72, 59	65.5	B, W	59, 35	47.0
G, E	72, 41	56.5	E, C	41, 36	38.5
G, C	72, 36	54.0	E, D	41, 36	38.5
G, D	72, 36	54.0	E, W	41, 35	38.0
G, W	72, 35	53.5	C, D	36, 36	36.0
B, E	59, 41	50.0	C, W	36, 35	35.5
B, C	59, 36	47.5	D, W	36, 35	35.5
B, D	59, 36	47.5			

c.

d. 0

e. 0.2. If a random sample of two of the six richest Americans is taken, there is a 20% chance that the mean wealth of the two people selected will be within 3 (i.e., $3 billion) of the population mean wealth.

7.19 b.

Sample	Wealth	$\bar{x}$	Sample	Wealth	$\bar{x}$
G, B, E	72, 59, 41	57.3	B, E, C	59, 41, 36	45.3
G, B, C	72, 59, 36	55.7	B, E, D	59, 41, 36	45.3
G, B, D	72, 59, 36	55.7	B, E, W	59, 41, 35	45.0
G, B, W	72, 59, 35	55.3	B, C, D	59, 36, 36	43.7
G, E, C	72, 41, 36	49.7	B, C, W	59, 36, 35	43.3
G, E, D	72, 41, 36	49.7	B, D, W	59, 36, 35	43.3
G, E, W	72, 41, 35	49.3	E, C, D	41, 36, 36	37.7
G, C, D	72, 36, 36	48.0	E, C, W	41, 36, 35	37.3
G, C, W	72, 36, 35	47.7	E, D, W	41, 36, 35	37.3
G, D, W	72, 36, 35	47.7	C, D, W	36, 36, 35	35.7

c.

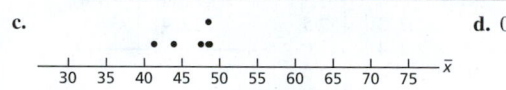

d. 0

e. 0.4. If a random sample of three of the six richest Americans is taken, there is a 40% chance that the mean wealth of the three people selected will be within 3 (i.e., $3 billion) of the population mean wealth.

7.21 b.

Sample	Wealth	$\bar{x}$
G, B, E, C, D	72, 59, 41, 36, 36	48.8
G, B, E, C, W	72, 59, 41, 36, 35	48.6
G, B, E, D, W	72, 59, 41, 36, 35	48.6
G, B, C, D, W	72, 59, 36, 36, 35	47.6
G, E, C, D, W	72, 41, 36, 36, 35	44.0
B, E, C, D, W	59, 41, 36, 36, 35	41.4

c.

d. 0

e. 0.833. If a random sample of five of the six richest Americans is taken, there is an 83.3% chance that the mean wealth of the five people selected will be within 3 (i.e., $3 billion) of the population mean wealth.

7.23 Sampling error tends to be smaller for large samples than for small samples.

Exercises 7.2

7.27 A normal distribution is determined by the mean and standard deviation. Hence a first step in learning how to approximate the sampling distribution of the mean by a normal distribution is to obtain the mean and standard deviation of the variable $\bar{x}$.

7.29 Yes. The standard deviation of all possible sample means (i.e., of the variable $\bar{x}$) gets smaller as the sample size gets larger.

7.31 Standard error (SE) of the mean. Because the standard deviation of $\bar{x}$ determines the amount of sampling error to be expected when a population mean is estimated by a sample mean.

7.33
a. Applying Definition 3.11 on page 140 and the answers to Exercise 7.3(b), we find that, for each sample size, $\mu_{\bar{x}} = 2$.
b. Applying Formula 7.1 on page 314 and the answer to Exercise 7.3(a), we find that, for each sample size, $\mu_{\bar{x}} = \mu = 2$.

7.35
a. Applying Definition 3.11 on page 140 and the answers to Exercise 7.5(b), we find that, for each sample size, $\mu_{\bar{x}} = 2.5$.
b. Applying Formula 7.1 on page 314 and the answer to Exercise 7.5(a), we find that, for each sample size, $\mu_{\bar{x}} = \mu = 2.5$.

7.37
a. Applying Definition 3.11 on page 140 and the answers to Exercise 7.7(b), we find that, for each sample size, $\mu_{\bar{x}} = 3$.
b. Applying Formula 7.1 on page 314 and the answer to Exercise 7.7(a), we find that, for each sample size, $\mu_{\bar{x}} = \mu = 3$.

7.39
a. Applying Definition 3.11 on page 140 and the answers to Exercise 7.9(b), we find that, for each sample size, $\mu_{\bar{x}} = 3.5$.
b. Applying Formula 7.1 on page 314 and the answer to Exercise 7.9(a), we find that, for each sample size, $\mu_{\bar{x}} = \mu = 3.5$.

7.41 a. $\mu = 78.6$ inches   **b.** $\mu_{\bar{x}} = 78.6$ inches   **c.** $\mu_{\bar{x}} = \mu = 78.6$ inches

7.43 b. $\mu_{\bar{x}} = 78.6$ inches   **c.** $\mu_{\bar{x}} = \mu = 78.6$ inches

7.45 b. $\mu_{\bar{x}} = 78.6$ inches   **c.** $\mu_{\bar{x}} = \mu = 78.6$ inches

7.47
a. The population consists of all babies. The variable is birth weight.
b. 3369 g; 41.1 g   **c.** 3369 g; 29.1 g

7.49
a. $\mu_{\bar{x}} = \$65{,}100$, $\sigma_{\bar{x}} = \$1018.2$. For samples of 50 new mobile homes, the mean and standard deviation of all possible sample mean prices are $65,100 and $1018.2, respectively.
b. $\mu_{\bar{x}} = \$65{,}100$, $\sigma_{\bar{x}} = \$720.0$. For samples of 100 new mobile homes, the mean and standard deviation of all possible sample mean prices are $65,100 and $720.0, respectively.

7.51 a. 437 days   **b.** ±598.5 days

Exercises 7.3

7.61
a. Approximately normally distributed with a mean of 100 and a standard deviation of 4
b. None
c. No. Because the distribution of the variable under consideration is not specified, a sample size of at least 30 is needed to apply Key Fact 7.4.

7.63 a. Normal with mean μ and standard deviation $\sigma/\sqrt{n}$
b. No. Because the variable under consideration is normally distributed.
c. μ and $\sigma/\sqrt{n}$
d. Essentially, no. For any variable, the mean of $\bar{x}$ equals the population mean, and the standard deviation of $\bar{x}$ equals (at least approximately) the population standard deviation divided by the square root of the sample size.

7.65
a. All four graphs are centered at the same place because $\mu_{\bar{x}} = \mu$ and normal distributions are centered at their means.
b. Because $\sigma_{\bar{x}} = \sigma/\sqrt{n}$, $\sigma_{\bar{x}}$ decreases as n increases. This fact results in a diminishing of the spread because the spread of a distribution is determined by its standard deviation. As a consequence, the larger the sample size, the greater is the likelihood for small sampling error.
c. If the variable under consideration is normally distributed, so is the sampling distribution of the mean, regardless of sample size.
d. The central limit theorem indicates that, if the sample size is relatively large, the sampling distribution of the mean is approximately a normal distribution, regardless of the distribution of the variable under consideration.

7.67
a. A normal distribution with a mean of 1.40 and a standard deviation of 0.064. Thus, for samples of three Swedish men, the possible sample mean brain weights have a normal distribution with a mean of 1.40 kg and a standard deviation of 0.064 kg.
b. A normal distribution with a mean of 1.40 and a standard deviation of 0.032. Thus, for samples of 12 Swedish men, the possible sample mean brain weights have a normal distribution with a mean of 1.40 kg and a standard deviation of 0.032 kg.
c.

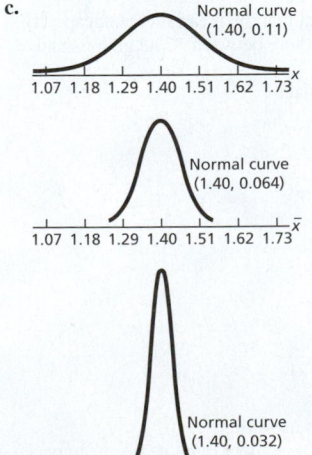

d. 88.12%. Chances are 88.12% that the sampling error made in estimating the mean brain weight of all Swedish men by that of a sample of three Swedish men will be at most 0.1 kg.
e. 99.82%. Chances are 99.82% that the sampling error made in estimating the mean brain weight of all Swedish men by that of a sample of 12 Swedish men will be at most 0.1 kg.

7.69
a. Approximately a normal distribution with a mean of 55.4 thousand and a standard deviation of 1.15 thousand. Thus, for samples of 64 classroom teachers in the public school system, the possible sample mean annual salaries are approximately normally distributed with a mean of $55.4 thousand and a standard deviation of $1.15 thousand.
b. Approximately a normal distribution with a mean of 55.4 thousand and a standard deviation of 0.575 thousand. Thus, for samples of 256 classroom teachers in the public school system, the possible sample mean annual salaries are approximately normally distributed with a mean of $55.4 thousand and a standard deviation of $0.575 thousand.
c. No. Because, in each case, the sample size exceeds 30.
d. 0.6156   **e.** 0.9182

7.71 Let μ denote the mean length of hospital stay on the intervention ward.
a. Approximately a normal distribution with mean μ and standard deviation 0.93 days.
b. No, because the sample size is well in excess of 30.
c. 0.9684

7.73 0.9476. There is about a 94.8% chance that the sampling error made in estimating the mean tariff rate of all railroad shipments of ethanol by that of a sample of 500 such shipments will be at most $100.

7.75 11.70%. Here we assume that the calcium intakes of adults with incomes below the poverty level are (approximately) normally distributed.

7.77 0.0012. Here we assume that the post-work heart rate for casting workers is (approximately) normally distributed.

Review Problems for Chapter 7

1. Sampling error is the error resulting from using a sample to estimate a population characteristic.

2. The distribution of a statistic (i.e., of all possible observations of the statistic for samples of a given size) is called the sampling distribution of the statistic.

3. Sampling distribution of the sample mean; distribution of the variable $\bar{x}$

4. The possible sample means cluster closer around the population mean as the sample size increases. Thus, the larger the sample size, the smaller the sampling error tends to be in estimating a population mean, μ, by a sample mean, $\bar{x}$.

5. **a.** $\mu = \$18$ thousand
   b. The completed table is as follows.

Sample	Salaries	$\bar{x}$
A, B, C, D	8, 12, 16, 20	14
A, B, C, E	8, 12, 16, 24	15
A, B, C, F	8, 12, 16, 28	16
A, B, D, E	8, 12, 20, 24	16
A, B, D, F	8, 12, 20, 28	17
A, B, E, F	8, 12, 24, 28	18
A, C, D, E	8, 16, 20, 24	17
A, C, D, F	8, 16, 20, 28	18
A, C, E, F	8, 16, 24, 28	19
A, D, E, F	8, 20, 24, 28	20
B, C, D, E	12, 16, 20, 24	18
B, C, D, F	12, 16, 20, 28	19
B, C, E, F	12, 16, 24, 28	20
B, D, E, F	12, 20, 24, 28	21
C, D, E, F	16, 20, 24, 28	22

c.

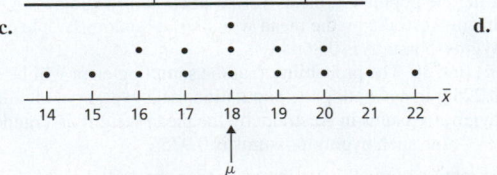

d. $\dfrac{7}{15}$

6. a. $18 thousand. For samples of four officers from the six, the mean of all possible sample mean monthly salaries equals $18 thousand.
   b. Yes. Because $\mu_{\bar{x}} = \mu$ and, from Problem 5(a), $\mu = \$18$ thousand.

7. a. For a normally distributed variable, the sampling distribution of the sample mean is a normal distribution, regardless of the sample size. Also, we know that $\mu_{\bar{x}} = \mu$. Consequently, because the normal curve for a normally distributed variable is centered at the mean, all three curves are centered at the same place.
   b. Curve B. Because $\sigma_{\bar{x}} = \sigma/\sqrt{n}$, the larger the sample size, the smaller is the value of $\sigma_{\bar{x}}$ and hence the smaller is the spread of the normal curve for $\bar{x}$. Thus, Curve B, which has the smaller spread, corresponds to the larger sample size.
   c. Because $\sigma_{\bar{x}} = \sigma/\sqrt{n}$ and the spread of a normal curve is determined by the standard deviation, different sample sizes result in normal curves with different spreads.
   d. Curve B. The smaller the value of $\sigma_{\bar{x}}$, the smaller the sampling error tends to be.
   e. Because the variable under consideration is normally distributed and, hence, so is the sampling distribution of the sample mean, regardless of sample size.

8. a. The error resulting from using the mean income tax, $\bar{x}$ of the 308,946 tax returns selected as an estimate of the mean income tax, μ, of all 2010 tax returns.
   b. $88
   c. No, not necessarily. However, increasing the sample size from 308,946 to 400,000 would increase the likelihood for smaller sampling error.
   d. Increase the sample size.

9. a. The population consists of all new cars sold in the United States during the year in question. The variable is the amount spent on a new car.
   b. $30,803; $1442.5   **c.** $30,803; $1020.0
   d. Smaller, because $\sigma_{\bar{x}} = \sigma/\sqrt{n}$ and hence $\sigma_{\bar{x}}$ decreases with increasing sample size.

10. a. False   **b.** Not possible to tell   **c.** True

11. a. False   **b.** True   **c.** True

12. a. See the first graph that follows.
   b. Normal distribution with a mean of 8.5 g and a standard deviation of 0.15 g, as shown in the second graph that follows.
   c. Normal distribution with a mean of 8.5 g and a standard deviation of 0.10 g, as shown in the third graph that follows.

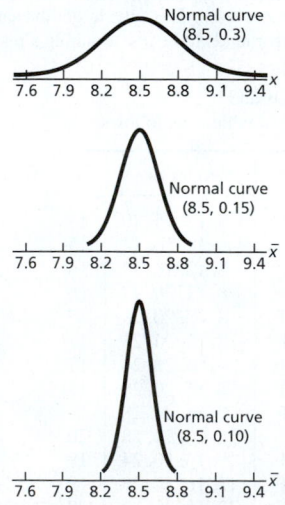

13. a. 86.64%   **b.** 0.8664
   c. The probability that the sampling error will be at most 0.225 g in estimating the population mean weight of all adult male pygmy-possums in Australia by the mean weight of a random sample of four such pygmy-possums is 0.8664.
   d. 97.56%. 0.9756. The probability that the sampling error will be at most 0.225 g in estimating the population mean weight of all adult male pygmy-possums in Australia by the mean weight of a random sample of nine such pygmy-possums is 0.9756.

14. a. Approximately normally distributed with mean 4.60 and standard deviation 0.021.

   b. Approximately normally distributed with mean 4.60 and standard deviation 0.015.
   c. No, because, in each case, the sample size exceeds 30.

15. a. 0.6212
   b. No. Because the sample size is large and therefore $\bar{x}$ is approximately normally distributed, regardless of the distribution of life insurance amounts. Yes.
   c. 0.9946

16. a. No. If the manufacturer's claim is correct, the probability that the paint life for a randomly selected house painted with this paint will be 4.5 years or less is 0.1587; that is, such an event would occur roughly 16% of the time.
   b. Yes. If the manufacturer's claim is correct, the probability that the mean paint life for 10 randomly selected houses painted with this paint will be 4.5 years or less is 0.0008; that is, such an event would occur less than 0.1% of the time.
   c. No. If the manufacturer's claim is correct, the probability that the mean paint life for 10 randomly selected houses painted with this paint will be 4.9 years or less is 0.2643; that is, such an event would occur roughly 26% of the time.

17. a. 5.82%
   b. No, because the distribution of the degree of cloudiness is far from normally distributed.

Chapter 8

Exercises 8.1

8.1 Point estimate

8.3 The margin of error indicates the accuracy of our estimate for the value of the unknown parameter.

8.5 950   **8.7** 23   **8.9** 9.7

8.11 a. 19 to 21
   b. 1. We can be 95% confident that the population mean, μ, is within 1 of the sample mean, 20.
   c. 20 ± 1

8.13 a. 28.4 to 31.6   **b.** 1.6   **c.** 30 ± 1.6

8.15 a. 47.5 to 52.5   **b.** 2.5   **c.** 50 ± 2.5

8.17 a. $26,326.9
   b. No. It is unlikely that a sample mean, $\bar{x}$, will exactly equal the population mean, μ; some sampling error is to be anticipated.

8.19 a. $22,704.5 to $29,949.3
   b. We can be 95% confident that the mean cost, μ, of all recent U.S. weddings is somewhere between $22,704.5 and $29,949.3.
   c. It may or may not, but we can be 95% confident that it does.

8.21
   a. 19.00 gallons. Based on the sample data, the mean fuel tank capacity of all 2003 automobile models is estimated to be 19.00 gallons.
   b. 17.82 to 20.18. We can be 95% confident that the mean fuel tank capacity of all 2003 automobile models is somewhere between 17.82 gallons and 20.18 gallons.
   c. Obtain a normal probability plot of the data.
   d. No. Because the sample size is large.

8.23 a.

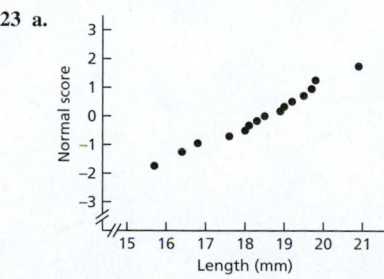

   b. Yes, the plot is roughly linear and shows no outliers.
   c. 17.52 mm to 19.34 mm. We can be 95% confident that the mean carapace length of all adult male Brazilian giant tawny red tarantulas is somewhere between 17.52 mm and 19.34 mm.
   d. Yes. No.

Exercises 8.2

8.27 a. Confidence level $= 0.90$; $\alpha = 0.10$
b. Confidence level $= 0.99$; $\alpha = 0.01$

8.29
a. Saying that the CI is exact means that the true confidence level is equal to $1 - \alpha$.
b. Saying that the CI is approximately correct means that the true confidence level is only approximately equal to $1 - \alpha$.

8.31 The variable under consideration is normally distributed on the population of interest.

8.33 A statistical procedure is said to be *robust* if it is insensitive to departures from the assumptions on which it is based.

Note: In obtaining the answers to Exercises 8.35–8.39, we referred to Key Fact 8.1 on page 340.

8.35 Reasonable   **8.37** Reasonable   **8.39** Reasonable

8.41 a 95% confidence level

8.43 The margin of error equals the standard error of the mean multiplied by $z_{\alpha/2}$.

8.45 Decreases the margin of error and, hence, increases the accuracy of estimating a population mean by a sample mean

8.47 Increases the margin of error and, hence, decreases the accuracy of estimating a population mean by a sample mean

8.49 a. 6.8   **b.** 49.4 to 56.2   **8.51 a.** 10   **b.** 50 to 70

8.53 True   **8.55** False   **8.57** False   **8.59** True

8.61
a. The sample size (number of observations) cannot be fractional; it must be a whole number.
b. The number resulting from Formula 8.2 is the smallest value that will provide the required margin of error. If that value were rounded down, the sample size thus obtained would be insufficient to ensure the required margin of error.

8.63 a. 19.0 to 21.0   **b.** 1.0   **c.** 1.0

8.65 a. 28.7 to 31.3   **b.** 1.3   **c.** 1.3

8.67 a. 46.8 to 53.2   **b.** 3.2   **c.** 3.2

8.69 $5.389 million to $7.274 million. We can be 95% confident that the mean amount of all venture-capital investments in the fiber optics business sector is somewhere between $5.389 million and $7.274 million.

8.71 0.251 ppm to 0.801 ppm. We can be 99% confident that the mean cadmium level of all *Boletus pinicola* mushrooms is somewhere between 0.251 ppm and 0.801 ppm.

8.73 18.8 to 48.0 months. We can be 95% confident that the mean duration of imprisonment, μ, of all East German political prisoners with chronic PTSD is somewhere between 18.8 and 48.0 months.

8.75 a. $5.093 million to $7.570 million
b. It is longer because the confidence level is greater.
c.

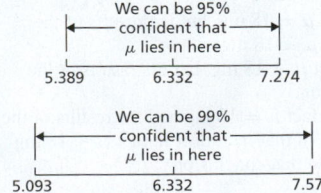

d. The 95% CI is a more accurate estimate of μ because it is narrower than the 99% CI.

8.77 a. 96.99 days to 108.45 days   **b.** 97.92 days to 107.52 days
c.

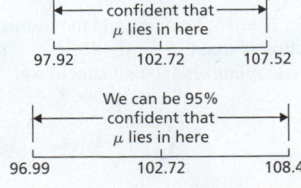

d. The 90% CI is a more accurate estimate of μ because it is narrower than the 95% CI.

8.79 a. 5.73 days   **b.** 4.80 days
c. The margin of error for the 90% CI is smaller than that for the 95% CI.
d. For a fixed sample size, decreasing the confidence level decreases the margin of error.

8.81 a. 93.98 days to 116.88 days
b.

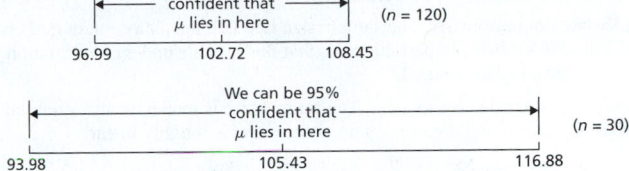

c. The margins of error for the two 95% CIs are 11.45 days and 5.73 days for $n = 30$ and $n = 120$, respectively.
d. For a fixed confidence level, increasing the sample size decreases the margin of error.

8.83
a. 60.93 to 65.63. We can be 95% confident that the mean price, μ, of all new mobile homes is somewhere between $60,930 and $65,630.
b. The confidence interval found in Example 8.2(c) is from 60.88 to 65.68 (i.e., $60,880 to $65,680). The discrepancy in the two confidence intervals is due to the difference of using 2 (from Property 2 of the empirical rule) and 1.96 (from $z_{0.025}$).

8.85 a. 33.1 cm to 35.3 cm   **b.** 1.1 cm
c. We can be 90% confident that the error made in estimating μ by $\bar{x}$ is at most 1.1 cm.
d. 68

8.87 a. $0.94 million   **b.** $0.9424 million

8.89 a. 14.6 months
b. We can be 95% confident that the error made in estimating μ by $\bar{x}$ is at most 14.6 months.
c. 82 prisoners   **d.** 24.3 months to 48.1 months

8.91 Referring to the second bulleted item in Key Fact 8.1 on page 340, we see that applying the z-interval procedure is not reasonable because the data contains an outlier, namely, 360 lumens.

8.93 a. 276.8 months to 303.1 months   **c.** 272.0 months to 299.0 months
d. Although removal of the outlier (374) does not appreciably affect the confidence interval, using the z-interval procedure here is not advisable because the sample size is moderate and the data contain an outlier.

8.95 0.79 year

8.97 a. 172
b. Because σ is unknown; because the sample size for the prior study is 36, which is at least 30

Exercises 8.3

8.109 a. $z = 1$ **b.** $t = 1.333$

8.111 a. The standard normal distribution   **b.** t-distribution with df $= 11$

8.113 The variation in the possible values of the standardized version is due solely to the variation of sample means, whereas that of the studentized version is due to the variation of both sample means and sample standard deviations.

8.115 a. 1.440   **b.** 2.447   **c.** 3.143

8.117 a. 1.323   **b.** 2.518   **c.** -2.080 **d.** ± 1.721

8.119 Yes. Because the sample size exceeds 30 and there are no outliers.

8.121 $E = t_{\alpha/2} \cdot s/\sqrt{n}$ **8.123 a.** 19.0 to 21.0   **b.** 1.0   **c.** 1.0

8.125 a. 28.6 to 31.4   **b.** 1.4   **c.** 1.4

8.127 a. 46.3 to 53.7   **b.** 3.7   **c.** 3.7

8.129 24.9 minutes to 31.1 minutes. We can be 90% confident that the mean commute time of all commuters in Washington, D.C., is somewhere between 24.9 minutes and 31.1 minutes.

8.131
a. 0.90 hr to 3.76 hr. We can be 95% confident that the additional sleep that would be obtained on average for all people using laevohysocyamine hydrobromide is somewhere between 0.90 hr and 3.76 hr.

b. It appears so because, based on the confidence interval, we can be 95% confident that the mean additional sleep is somewhere between 0.90 hr and 3.76 hr and that, in particular, the mean is positive.

8.133

a. 1225.0 chips per bag to 1298.2 chips per bag. We can be 95% confident that the mean number of chips per bag for all 18-ounce bags of Chips Ahoy! Cookies is somewhere between 1225.0 and 1298.2.

b. Yes, because, from part (a), we can be confident that the average bag contains at least 1225 chocolate chips.

8.135 No, not reasonable. The sample size is small, the data contain outliers, and a normal probability plot indicates that the variable under consideration is far from normally distributed.

8.137 Yes, it appears reasonable. The sample size is moderate and a normal probability plot of the data shows no outliers and is roughly linear.

Review Problems for Chapter 8

1. A point estimate of a parameter is the value of a statistic that is used to estimate the parameter; it consists of a single number, or point. A confidence-interval estimate of a parameter consists of an interval of numbers obtained from a point estimate of the parameter and a percentage that specifies how confident we are that the parameter lies in the interval.

2. False. The mean of the population may or may not lie somewhere between 33.8 and 39.0, but we can be 95% confident that it does.

3. No. See the guidelines in Key Fact 8.1 on page 340.

4. Roughly 950 intervals would actually contain μ.

5. Look at graphical displays of the data to ascertain whether the conditions required for using the procedure appear to be satisfied.

6. **a.** The accuracy of the estimate would decrease because the CI would be wider for a sample of size 50.
   b. The accuracy of the estimate would increase because the CI would be narrower for a 90% confidence level.

7. **a.** Because the length of a CI is twice the margin of error, the length of the CI is 21.4.
   b. 64.5 to 85.9 **c.** 75.2 ± 10.7

8. **a.** 6.58 **b.** The sample mean, $\bar{x}$

9. **a.** $z = -0.77$ **b.** $t = -0.605$

10. **a.** Standard normal distribution
    b. t-distribution with 14 degrees of freedom

11. From Property 4 of Key Fact 8.6 (page 354), as the number of degrees of freedom becomes larger, t-curves look increasingly like the standard normal curve. So the curve that is closer to the standard normal curve has the larger degrees of freedom.

12. t-interval procedure   13. z-interval procedure

14. z-interval procedure   15. Neither procedure

16. z-interval procedure   17. Neither procedure

18. **a.** 2.101 **b.** 1.734 **c.** -1.330 **d.** ± 2.878

19. 54.3 yr to 62.8 yr

20. Part (c) provides the correct interpretation of the statement in quotes.

21. **a.** 8.09 yr to 10.21 yr. We can be 90% confident that the mean sentence length of all federally sentenced adult male prisoners is somewhere between 8.09 yr and 10.21 yr.
    b. The data contain no outliers.

22. **a.** 1.06 yr
    b. We can be 90% confident that the mean sentence length of all federally sentenced adult male prisoners is within 1.06 yr of the sample mean sentence length of 9.15 yr.
    c. 3203 **d.** 9.60 yr to 10.60 yr

23. **a.** 81.69 mm Hg to 90.30 mm Hg. We can be 95% confident that the mean arterial blood pressure of all children of diabetic mothers is somewhere between 81.69 mm Hg and 90.30 mm Hg.
    c. Yes, the sample size is moderate, none of the graphs show any outliers, and the normal probability plot is linear.

24. **a.** $1880.1 to $2049.4. We can be 90% confident that the mean price of all one-half-carat diamonds is somewhere between $1880.1 and $2049.4.

c. This one is a tough call, but using the t-interval procedure is probably reasonable. The sample size is moderate and, although the boxplot shows a potential outlier (1442), the other three plots suggest that the potential outlier may, in fact, not be an outlier. Furthermore, the normal probability plot is roughly linear.

25. No, not reasonable. The sample size is small, but the data indicate that the population under consideration is far from normally distributed.

Chapter 9

Exercises 9.1

9.1 A hypothesis is a statement that something is true.

9.3 The decision criterion provides an objective method for deciding whether the null hypothesis should be rejected in favor of the alternative hypothesis.

9.5 a. $H_a: \mu \neq \mu_0$; two tailed **b.** $H_a: \mu < \mu_0$; left tailed
c. $H_a: \mu > \mu_0$; right tailed

9.7

a. No. A Type I error occurs when a true null hypothesis is rejected, which is impossible if the null hypothesis is in fact false.

b. Yes. If the (false) null hypothesis is not rejected, a Type II error will be made.

9.9 True. Because the significance level, α, is the probability of making a Type I error, it is unlikely that a true null hypothesis will be rejected if the hypothesis test is conducted at a small significance level.

9.11 The two types of incorrect decisions are a Type I error (rejection of a true null hypothesis) and a Type II error (nonrejection of a false null hypothesis). The probabilities of these two errors are denoted α and β, respectively.

9.13 It should be small (close to 0).

9.15 Let μ denote the mean cadmium level in *Boletus pinicola* mushrooms.
a. $H_0: \mu = 0.5$ ppm **b.** $H_a: \mu > 0.5$ ppm **c.** Right-tailed test

9.17 Let μ denote the mean iron intake (per day) of all adult females under the age of 51.
a. $H_0: \mu = 18$ mg **b.** $H_a: \mu < 18$ mg **c.** Left-tailed test

9.19 Let μ denote the mean length of imprisonment for motor-vehicle-theft offenders in Sydney, Australia.
a. $H_0: \mu = 16.7$ months **b.** $H_a: \mu \neq 16.7$ months **c.** Two-tailed test

9.21 Let μ denote the mean body temperature of all healthy humans.
a. $H_0: \mu = 98.6°F$ **b.** $H_a: \mu \neq 98.6°F$ **c.** Two-tailed test

9.23

a. A Type I error would occur if in fact $\mu = 0.5$ ppm, but the results of the sampling lead to the conclusion that $\mu > 0.5$ ppm.

b. A Type II error would occur if in fact $\mu > 0.5$ ppm, but the results of the sampling fail to lead to that conclusion.

c. A correct decision would occur if in fact $\mu = 0.5$ ppm and the results of the sampling do not lead to the rejection of that fact; or if in fact $\mu > 0.5$ ppm and the results of the sampling lead to that conclusion.

d. Correct decision **e.** Type II error

9.25

a. A Type I error would occur if in fact $\mu = 18$ mg, but the results of the sampling lead to the conclusion that $\mu < 18$ mg.

b. A Type II error would occur if in fact $\mu < 18$ mg, but the results of the sampling fail to lead to that conclusion.

c. A correct decision would occur if in fact $\mu = 18$ mg and the results of the sampling do not lead to the rejection of that fact; or if in fact $\mu < 18$ mg and the results of the sampling lead to that conclusion.

d. Type I error **e.** Correct decision

9.27

a. A Type I error would occur if in fact $\mu = 16.7$ months, but the results of the sampling lead to the conclusion that $\mu \neq 16.7$ months.

b. A Type II error would occur if in fact $\mu \neq 16.7$ months, but the results of the sampling fail to lead to that conclusion.

c. A correct decision would occur if in fact $\mu = 16.7$ months and the results of the sampling do not lead to the rejection of that fact; or if in fact $\mu \neq 16.7$ months and the results of the sampling lead to that conclusion.

d. Correct decision **e.** Type II error

9.29
a. A Type I error would occur if in fact $\mu = 98.6°F$, but the results of the sampling lead to the conclusion that $\mu \neq 98.6°F$.
b. A Type II error would occur if in fact $\mu \neq 98.6°F$, but the results of the sampling fail to lead to that conclusion.
c. A correct decision would occur if in fact $\mu = 98.6°F$ and the results of the sampling do not lead to the rejection of that fact; or if in fact $\mu \neq 98.6°F$ and the results of the sampling lead to that conclusion.
d. Type I error   **e.** Correct decision

Exercises 9.2

9.31 A statistic calculated from the data that is used as a basis for deciding whether the null hypothesis should be rejected

9.33 The set of values for the test statistic that leads to nonrejection of the null hypothesis

9.35 a. $z \geq 1.645$ **b.** $z < 1.645$ **c.** $z = 1.645$ **d.** $\alpha = 0.05$
e.   **f.** Right-tailed test

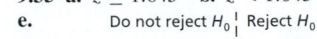
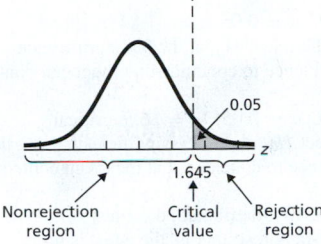

9.37 a. $z \leq -2.33$ **b.** $z > -2.33$ **c.** $z = -2.33$ **d.** $\alpha = 0.01$
e.   **f.** Left-tailed test

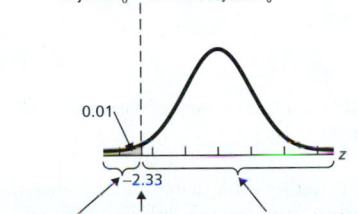

9.39 a. $z \leq -1.645$ or $z \geq 1.645$ **b.** $-1.645 < z < 1.645$
c. $z = \pm 1.645$ **d.** $\alpha = 0.10$
e.   **f.** Two-tailed test

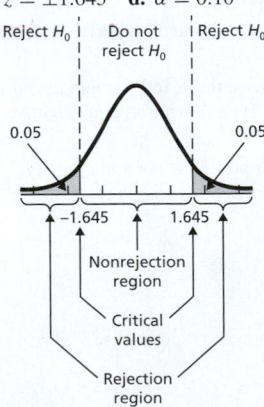

9.41 Critical values: $\pm z_{0.05} = \pm 1.645$

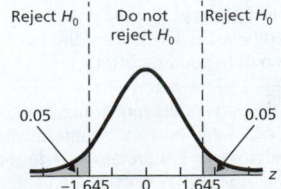

9.43 Critical value: $-z_{0.01} = -2.33$

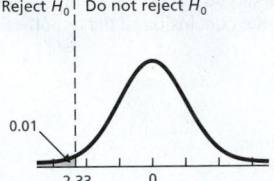

9.45 Critical value: $z_{0.01} = 2.33$

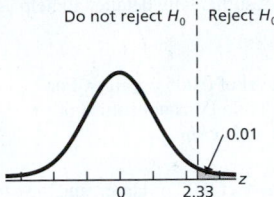

Exercises 9.3

9.47 (1) It allows you to assess significance at any desired level.
(2) It permits you to evaluate the strength of the evidence against the null hypothesis.

9.51 a. Do not reject the null hypothesis.   **b.** Reject the null hypothesis.
c. Reject the null hypothesis.

9.53 A P-value of 0.02 provides stronger evidence against the null hypothesis because it reflects an observed value of the test statistic that is more inconsistent with the null hypothesis.

9.55 Moderate   **9.57** Strong   **9.59** Weak or none

9.61 Very strong   **9.63 a.** 0.0212; reject H_0 **b.** 0.6217; do not reject H_0

9.65 a. 0.2296; do not reject H_0 **b.** 0.8770; do not reject H_0

9.67 a. 0.0970; do not reject H_0 **b.** 0.6030; do not reject H_0

Exercises 9.4

9.73 Inappropriate   **9.75** Appropriate

> *Note:* Throughout this answer section, we provide both the critical values and P-values for hypothesis-test exercises and problems. If you are concentrating on the critical-value approach, you can ignore the P-value information. Likewise, if you are concentrating on the P-value approach, you can ignore the critical-value information.

9.77 $z = -2.83$; critical value $= -1.645$; $P = 0.002$; reject H_0

9.79 $z = 1.94$; critical value $= 1.645$; $P = 0.026$; reject H_0

9.81 $z = 1.22$; critical values $= \pm 1.96$; $P = 0.221$; do not reject H_0

9.83 H_0: $\mu = 0.5$ ppm, H_a: $\mu > 0.5$ ppm; $\alpha = 0.05$; $z = 0.24$; critical value $= 1.645$; $P = 0.404$; do not reject H_0; at the 5% significance level, the data do not provide sufficient evidence to conclude that the mean cadmium level in *Boletus pinicola* mushrooms is greater than the government's recommended limit of 0.5 ppm.

9.85 H_0: $\mu = 18$ mg, H_a: $\mu < 18$ mg; $\alpha = 0.01$; $z = -5.30$; critical value $= -2.33$; $P = 0.000$; reject H_0; at the 1% significance level, the data provide sufficient evidence to conclude that adult females under the age of 51 years are, on average, getting less than the RDA of 18 mg of iron.

9.87 H_0: $\mu = 16.7$ months, H_a: $\mu \neq 16.7$ months; $\alpha = 0.05$; $z = 1.83$; critical values $= \pm 1.96$; $P = 0.067$; do not reject H_0; at the 5% significance level, the data do not provide sufficient evidence to conclude that the mean length of imprisonment for motor-vehicle-theft offenders in Sydney differs from the national mean in Australia.

9.89
a. At the 5% significance level, the data do not provide sufficient evidence to conclude that, on average, the net percentage gain for jobs exceeds 0.2.
c. Removing the potential outlier (-1.1), we conclude, at the 5% significance level, that, on average, the net percentage gain for jobs exceeds 0.2.

d. The sample size is moderate, there is a potential outlier in the data, and the variable under consideration appears to be left skewed. Furthermore, removal of the potential outlier affects the conclusion of the hypothesis test. Using the z-test here is not advisable.

Exercises 9.5

9.99 a. No **b.** A nonparametric procedure

9.101 a. $0.01 < P < 0.025$
b. We can reject H_0 at any significance level of 0.025 or larger, and we cannot reject H_0 at any significance level of 0.01 or smaller. For significance levels between 0.01 and 0.025, Table IV is not sufficiently detailed to help us to decide whether to reject H_0.

9.103 a. $P < 0.005$
b. We can reject H_0 at any significance level of 0.005 or larger. For significance levels smaller than 0.005, Table IV is not sufficiently detailed to help us to decide whether to reject H_0.

9.105 a. $0.01 < P < 0.02$
b. We can reject H_0 at any significance level of 0.02 or larger, and we cannot reject H_0 at any significance level of 0.01 or smaller. For significance levels between 0.01 and 0.02, Table IV is not sufficiently detailed to help us to decide whether to reject H_0.

> *Note to users of P-values:* Throughout this answer section, we provide, for hypothesis-test exercises and problems, both estimated P-values (using Appendix A tables) and exact P-values (using technology). The exact P-values are shown parenthetically and are usually given to three decimal places.

9.107 $t = -2.83$; critical value $= -1.696$; $P < 0.005$ ($P = 0.004$); reject H_0

9.109 $t = 1.94$; critical value $= 1.761$; $0.025 < P < 0.05$ ($P = 0.037$); reject H_0

9.111 $t = 1.22$; critical values $= \pm 2.069$; $P > 0.20$ ($P = 0.233$); do not reject H_0

9.113 $H_0: \mu = 4.55$ hr, $H_a: \mu \neq 4.55$ hr; $\alpha = 0.10$; $t = 0.41$; critical values $= \pm 1.729$; $P > 0.20$ ($P = 0.687$); do not reject H_0; at the 10% significance level, the data do not provide sufficient evidence to conclude that the amount of television watched per day last year by the average person differed from that in 2005.

9.115 $H_0: \mu = 45°$, $H_a: \mu > 45°$; $\alpha = 0.05$; $t = 1.784$; critical value $= 1.729$; $0.025 < P < 0.05$ ($P = 0.045$); reject H_0; at the 5% significance level, the data provide sufficient evidence to conclude that, on average, the angle between the body and head of an alligator during a death roll is greater than 45°.

9.117 $H_0: \mu = 0.9$, $H_a: \mu < 0.9$; $\alpha = 0.01$; $t = -23.703$; critical value $= -2.347$; $P < 0.005$ ($P = 0.000$); reject H_0; at the 1% significance level, the data provide sufficient evidence to conclude that, on average, women with peripheral arterial disease have an unhealthy ABI.

9.119 Yes, it appears reasonable. The sample size is moderate, and a normal probability plot shows no outliers and is (very) roughly linear.

9.121 No, not reasonable. The sample size is only moderate, and it appears that the variable under consideration is highly right skewed and hence far from normally distributed.

Exercises 9.6

9.129 Technically, inferential methods that are not concerned with parameters are called nonparametric methods. Current statistical practice is to refer to most methods that can be applied without assuming normality as nonparametric methods.

9.131 Because the D-value for such an observation equals 0, a sign cannot be attached to the rank of $|D|$.

9.133 t-test   **9.135** Wilcoxon signed-rank test   **9.137** Neither

9.139 a. 30 **b.** 6 **c.** 4, 32   **9.141 a.** 128 **b.** 62 **c.** 54, 136

9.143 $W = 29$; critical value $= 28$; $P = 0.071$; reject H_0

9.145 $W = 4.5$; critical values $= 4$ and 24; $P = 0.128$; do not reject H_0

9.147 $W = 20$; critical value $= 11$; $P = 0.406$; do not reject H_0

9.149 $H_0: \mu = 124.9$ days, $H_a: \mu < 124.9$ days; $\alpha = 0.05$; $W = 3$; critical value $= 6$; $P = 0.021$; reject H_0; at the 5% significance level, the data provide

sufficient evidence to conclude that the average number of ice days is less now than in the late 1800s.

9.151 $H_0: \eta = 37.2$ yr, $H_a: \eta > 37.2$ yr; $\alpha = 0.01$; $W = 35$; critical value $= 50$; $P = 0.238$; do not reject H_0; at the 1% significance level, the data do not provide sufficient evidence to conclude that the median age of today's U.S. residents has increased from the 2010 median age of 37.2 yr.

9.153 $H_0: \mu = \$18{,}000$, $H_a: \mu < \$18{,}000$; $\alpha = 0.01$; $W = 0$; critical value $= 5$; $P = 0.003$; reject H_0; at the 1% significance level, the data provide sufficient evidence to conclude that the mean purchase price from private parties for 2-year-old Ford Mustang coupes is less than the fair purchase price from dealers.

9.155
a. $H_0: \mu = 45°$, $H_a: \mu > 45°$; $\alpha = 0.05$; $W = 150$; critical value $= 150$; $P = 0.048$; reject H_0; at the 5% significance level, the data provide sufficient evidence to conclude that, on average, the angle between the body and head of an alligator during a death roll is greater than 45°.
b. Because a normal distribution is symmetric

9.157
a. $H_0: \mu = 310$ mL, $H_a: \mu < 310$ mL; $\alpha = 0.05$; $t = -1.845$; critical value $= -1.753$; $0.025 < P < 0.05$; reject H_0; at the 5% significance level, the data provide sufficient evidence to conclude that the mean content is less than advertised.
b. $H_0: \mu = 310$ mL, $H_a: \mu < 310$ mL; $\alpha = 0.05$; $W = 36.5$; critical value $= 36$; $P = 0.054$; do not reject H_0; at the 5% significance level, the data do not provide sufficient evidence to conclude that the mean content is less than advertised.
c. Assuming that the contents are normally distributed, the t-test is more powerful than the Wilcoxon signed-rank test; that is, the t-test is more likely to detect a false null hypothesis.

9.159 Not reasonable. It appears that HIC levels for small SUVs are not symmetric, but rather right skewed.

Exercises 9.7

9.179 Because partial information, obtained from a sample, is used to draw conclusions about the entire population

9.181
a. The probability of making a Type I error (rejecting a true null hypothesis), also known as the significance level of the hypothesis test
b. The probability of making a Type II error (not rejecting a false null hypothesis)
c. The power of the hypothesis test (the probability of not making a Type II error or, equivalently, of rejecting a false null hypothesis)

9.183 The power curve provides a visual display of the overall effectiveness of the hypothesis test.

9.185 Decreasing the significance level of a hypothesis test without changing the sample size increases the probability of a Type II error or, equivalently, decreases the power.

9.187 Obtaining Type II error probabilities or powers is computationally intensive. Moreover, determining those quantities by hand can result in substantial roundoff error.

9.189 a.

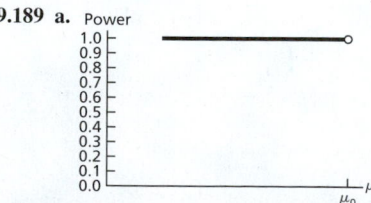

b. The ideal power curve for a left-tailed test portrays that the power (i.e., the probability of rejecting a false null hypothesis) is 1 whenever the true value of μ is less than μ_0 (i.e., whenever the null hypothesis is false).

> *Note:* The answers obtained to many of the parts in the remaining exercises of Section 9.7 may vary depending on when and how much intermediate rounding is done. We used statistical software to get the answers to most parts of each of these exercises.

9.191 a. 0.05 **b.** 0.8803 **c.** Power

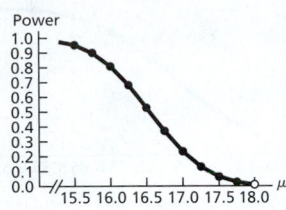

9.193 a. 0.01 **b.** 0.0478 **c.** Power

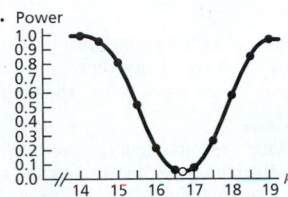

9.195 a. 0.05 **b.** 0.0055 **c.** Power

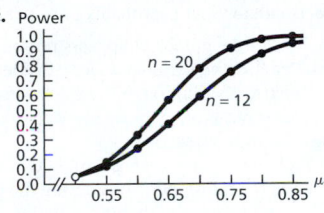

9.197 a. 0.05 **b.** 0.8509 **c.** Power

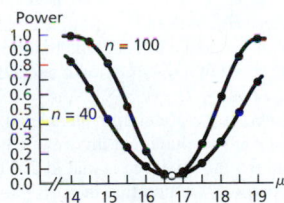

For a fixed significance level, increasing the sample size increases the power.

9.199 a. 0.05 **b.** 0.1878 **c.** Power

For a fixed significance level, decreasing the sample size decreases the power.

Review Problems for Chapter 9

1. **a.** The null hypothesis is a hypothesis to be tested.
   b. The alternative hypothesis is a hypothesis to be considered as an alternate to the null hypothesis.
   c. The test statistic is the statistic used as a basis for deciding whether the null hypothesis should be rejected.
   d. The significance level of a hypothesis test is the probability of making a Type I error, that is, of rejecting a true null hypothesis.

2. **a.** The weight of a package of Tide is a variable. A particular package may weigh slightly more or less than the marked weight. The mean weight of all packages produced on any specified day (the population mean weight for that day) exceeds the marked weight.
   b. The null hypothesis would be that the population mean weight for a specified day equals the marked weight; the alternative hypothesis would be that the population mean weight for the specified day exceeds the marked weight.
   c. The null hypothesis would be that the population mean weight for a specified day equals the marked weight of 76 oz; the alternative hypothesis would be that the population mean weight for the specified day exceeds the marked weight of 76 oz. In statistical terminology, the hypothesis test would be $H_0: \mu = 76$ oz and $H_a: \mu > 76$ oz, where μ is the mean weight of all packages produced on the specified day.

3. **a.** Obtain the data from a random sample of the population or from a designed experiment. If the data are consistent with the null hypothesis, do not reject the null hypothesis; if the data are inconsistent with the null hypothesis, reject the null hypothesis and conclude that the alternative hypothesis is true.

b. We establish a precise criterion for deciding whether to reject the null hypothesis prior to obtaining the data.

4. Two-tailed test, $H_a: \mu \neq \mu_0$. Used when the primary concern is deciding whether a population mean, μ, is different from a specified value μ_0.
   Left-tailed test, $H_a: \mu < \mu_0$. Used when the primary concern is deciding whether a population mean, μ, is less than a specified value μ_0.
   Right-tailed test, $H_a: \mu > \mu_0$. Used when the primary concern is deciding whether a population mean, μ, is greater than a specified value μ_0.

5. **a.** A Type I error is the incorrect decision of rejecting a true null hypothesis. A Type II error is the incorrect decision of not rejecting a false null hypothesis.
   b. α and β, respectively **c.** A Type I error **d.** A Type II error

6. It increases.

7. **a.** The rejection region is the set of values for the test statistic that leads to rejection of the null hypothesis.
   b. The nonrejection region is the set of values for the test statistic that leads to nonrejection of the null hypothesis.
   c. The critical values are the values of the test statistic that separate the rejection and nonrejection regions.

8. True

9. It must be chosen so that, if the null hypothesis is true, the probability equals 0.05 that the test statistic will fall in the rejection region, in this case, to the left of the critical value.

10. **a.** 2.33 **b.** −2.33 **c.** −2.575 and 2.575

11. **a.** $z \geq 1.28$ **b.** $z < 1.28$ **c.** $z = 1.28$ **d.** $\alpha = 0.10$
    e.  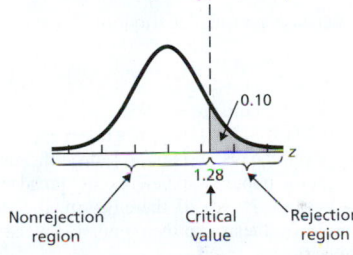 **f.** Right tailed

12. See Table 9.5 on page 380.

13. The *P*-value of a hypothesis test is the probability of getting sample data at least as inconsistent with the null hypothesis (and supportive of the alternative hypothesis) as the sample data actually obtained.

14. True

15. If the *P*-value is less than or equal to the specified significance level, reject the null hypothesis; otherwise, do not reject the null hypothesis. In other words, if $P \leq \alpha$, reject H_0; otherwise, do not reject H_0.

16. Because it is the smallest significance level for which the observed sample data result in rejection of the null hypothesis.

17. To determine the *P*-value of a hypothesis test, we assume that the null hypothesis is true and compute the probability of observing a value of the test statistic as extreme as or more extreme than that observed. By *extreme* we mean "far from what we would expect to observe if the null hypothesis is true."

18. **a.** 0.1056; do not reject H_0 **b.** 0.0091; reject H_0
    c. 0.0672; do not reject H_0

19. See Table 9.7 on page 386. **20.** Moderate

21. **a.** The true significance level equals α.
    b. The true significance level only approximately equals α.

22. The results of a hypothesis test are statistically significant if the null hypothesis is rejected at the specified significance level. Statistical significance means that the data provide sufficient evidence to conclude that the truth is different from the stated null hypothesis. It does not necessarily mean that the difference is important in any practical sense.

23. **a.** Assumptions: simple random sample; normal population or large sample; σ unknown. Test statistic: $t = (\bar{x} - \mu_0)/(s/\sqrt{n})$.
    b. Assumptions: simple random sample; normal population or large sample; σ known. Test statistic: $z = (\bar{x} - \mu_0)/(\sigma/\sqrt{n})$.
    c. Assumptions: simple random sample; symmetric population. Test statistic: $W =$ sum of the positive ranks.

25. a. The probability of rejecting a false null hypothesis **b.** It increases.

26. Let μ denote last year's mean cheese consumption by Americans.
a. H_0: $\mu = 33$ lb **b.** H_a: $\mu > 33$ lb **c.** Right tailed

27. a. A Type I error would occur if in fact $\mu = 33$ lb, but the results of the sampling lead to the conclusion that $\mu > 33$ lb.
b. A Type II error would occur if in fact $\mu > 33$ lb, but the results of the sampling fail to lead to that conclusion.
c. A correct decision would occur if in fact $\mu = 33$ lb and the results of the sampling do not lead to the rejection of that fact; or if in fact $\mu > 33$ lb and the results of the sampling lead to that conclusion.
d. Type I error **e.** Correct decision

28. a. H_0: $\mu = 33$ lb, H_a: $\mu > 33$ lb; $\alpha = 0.10$; $z = 1.54$; critical value $= 1.28$; $P = 0.061$; reject H_0; at the 10% significance level, the data provide sufficient evidence to conclude that last year's mean cheese consumption for all Americans has increased over the 2010 mean of 33 lb.
b. A Type I error because, given that the null hypothesis was rejected, the only error that could be made is the error of rejecting a true null hypothesis.

29. H_0: $\mu = \$468$, H_a: $\mu < \$468$; $\alpha = 0.05$; $t = -0.519$; critical value $= -1.796$; $P > 0.10$ ($P = 0.307$); do not reject H_0; at the 5% significance level, the data do not provide sufficient evidence to conclude that last year's mean value lost to purse snatching has decreased from the 2012 mean of $468.

30. a. H_0: $\mu = \$468$, H_a: $\mu < \$468$; $\alpha = 0.05$; $W = 35$; critical value $= 17$; $P = 0.392$; do not reject H_0; at the 5% significance level, the data do not provide sufficient evidence to conclude that last year's mean value lost to purse snatching has decreased from the 2012 mean of $468.
b. It is symmetric. **c.** Because a normal distribution is symmetric

31. t-test

32. a. 0 points
b. H_0: $\mu = 0$ points, H_a: $\mu \neq 0$ points; $\alpha = 0.05$; $t = -0.843$; critical values $= \pm 1.96$; $P > 0.20$ ($P = 0.400$); do not reject H_0.
c. At the 5% significance level, the data do not provide sufficient evidence to conclude that the population mean point-spread error differs from 0. In fact, because $P > 0.20$, there is virtually no evidence against the null hypothesis that the population mean point-spread error equals 0.

33. *Note:* The answers obtained to many of the parts of this problem may vary depending on when and how much intermediate rounding is done. We used statistical software to get the answers to most parts of this problem.
a. 0.10
b. Approximately normal with a mean of 36.5 and a standard deviation of $6.9/\sqrt{35} \approx 1.17$
c. 0.0428
d. Approximately normal with the specified mean and a standard deviation of $6.9/\sqrt{35} \approx 1.17$. The Type II error probabilities, β, are shown in the table in part (e).
e.

μ	β	Power	μ	β	Power
33.5	0.8031	0.1969	35.5	0.1944	0.8056
34.0	0.6643	0.3357	36.0	0.0984	0.9016
34.5	0.4982	0.5018	36.5	0.0428	0.9572
35.0	0.3324	0.6676	37.0	0.0159	0.9841

f. Power

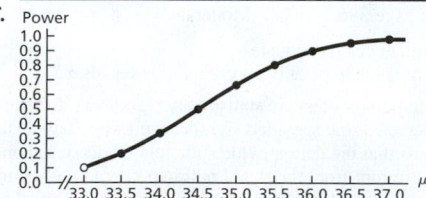

g. Approximately normal with a mean of 36.5 and a standard deviation of $6.9/\sqrt{60} \approx 0.89$
h. 0.0041
i. Approximately normal with the specified mean and a standard deviation of $6.9/\sqrt{60} \approx 0.89$. The Type II error probabilities, β, are shown in the table in part (j).

j.

μ	β	Power	μ	β	Power
33.5	0.7643	0.2357	35.5	0.0636	0.9364
34.0	0.5631	0.4369	36.0	0.0185	0.9815
34.5	0.3437	0.6563	36.5	0.0041	0.9959
35.0	0.1676	0.8324	37.0	0.0007	0.9993

k. Power

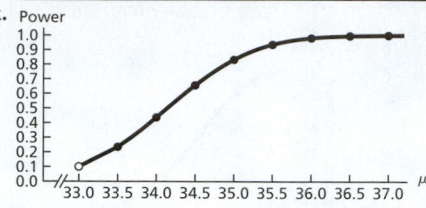

l. For a fixed significance level, increasing the sample size increases the power.

34. It is probably okay to use the z-test because the sample size is large and σ is known. However, it does appear from the normal probability plot that there may be outliers, so one should proceed cautiously in using the z-test.

35. It appears that the variable under consideration is far from being normally distributed and, in fact, has a left-skewed distribution. However, the sample size is large and the plots reveal no outliers. Keeping in mind that σ is unknown, it is probably reasonable to use the t-test.

36. a. In view of the graphs, it appears reasonable to assume that, in Problem 34, the variable under consideration has (approximately) a symmetric distribution, but not so in Problem 35. Consequently, it would be reasonable to use the Wilcoxon signed-rank test in the first case, but not the second.
b. In Problem 34, it is a tough call between the Wilcoxon signed-rank test and the z-test, but, considering the possible outliers, the Wilcoxon signed-rank test is probably the better one to use.

37. a. H_0: $\mu = \$239$, H_a: $\mu > \$239$; $\alpha = 0.05$; $t = 1.608$; $P = 0.069$; do not reject H_0; at the 5% significance level, the data do not provide sufficient evidence to conclude that this year's average cost for a private room in a nursing home exceeds that in 2011.
b. H_0: $\mu = \$239$, H_a: $\mu > \$239$; $\alpha = 0.05$; $W = 53.5$; $P = 0.038$; reject H_0; at the 5% significance level, the data provide sufficient evidence to conclude that this year's average cost for a private room in a nursing home exceeds that in 2011.
c. From part (c), we find that the variable under consideration appears to be symmetric, but that the observation $140 is an outlier. This explains the discrepancy between the results of the two tests. In view of the small sample size, the Wilcoxon signed-rank test is preferable to the t-test.

Chapter 10

Exercises 10.1

10.1 Answers will vary.

10.3 a. μ_1, σ_1, μ_2, and σ_2 are parameters; $\bar{x}_1, s_1, \bar{x}_2$, and s_2 are statistics.
b. μ_1, σ_1, μ_2, and σ_2 are fixed numbers; $\bar{x}_1, s_1, \bar{x}_2$, and s_2 are variables.

10.5 a. H_0: $\mu_1 = \mu_2$; H_a: $\mu_1 \neq \mu_2$ **b.** Two tailed

10.7 a. H_0: $\mu_1 = \mu_2$; H_a: $\mu_1 > \mu_2$ **b.** Right tailed

10.9 a. H_0: $\mu_1 = \mu_2$; H_a: $\mu_1 < \mu_2$ **b.** Left tailed

10.11 So that you can determine whether the observed difference between the two sample means can be reasonably attributed to sampling error or whether that difference suggests that the null hypothesis of equal population means should be rejected in favor of the alternative hypothesis

10.13 We can be 95% confident that $\mu_1 - \mu_2$ lies somewhere between 15 and 20. Equivalently, we can be 95% confident that μ_1 is somewhere between 15 and 20 greater than μ_2.

10.15 We can be 90% confident that $\mu_1 - \mu_2$ lies somewhere between -10 and -5. Equivalently, we can be 90% confident that μ_1 is somewhere between 5 and 10 less than μ_2.

10.17 We can be 99% confident that $\mu_1 - \mu_2$ lies somewhere between -20 and 15. Equivalently, we can be 99% confident that μ_1 is somewhere between 20 less than and 15 more than μ_2.

10.19 a. 0 and 5 **b.** No. **c.** No.

10.21 a. 0 and 5 **b.** Yes. **c.** 95.44%

10.23 Let μ_1 and μ_2 denote the mean salaries of faculty in private and public institutions, respectively. The null and alternative hypotheses are $H_0: \mu_1 = \mu_2$ and $H_a: \mu_1 > \mu_2$, respectively.

10.25 a. Systolic blood pressure
b. ODM adolescents and ONM adolescents
c. Let μ_1 and μ_2 denote the mean systolic blood pressures of ODM adolescents and ONM adolescents, respectively. The null and alternative hypotheses are $H_0: \mu_1 = \mu_2$ and $H_a: \mu_1 > \mu_2$, respectively.
d. Right tailed

10.27 a. Last year's vehicle miles of travel (VMT)
b. Households in the Midwest and households in the South
c. Let μ_1 and μ_2 denote last year's mean VMT for households in the Midwest and South, respectively. The null and alternative hypotheses are $H_0: \mu_1 = \mu_2$ and $H_a: \mu_1 \neq \mu_2$, respectively.
d. Two tailed

10.29 a. Operative time
b. Dynamic-system operations and static-system operations
c. Let μ_1 and μ_2 denote the mean operative times with the dynamic and static systems, respectively. The null and alternative hypotheses are $H_0: \mu_1 = \mu_2$ and $H_a: \mu_1 < \mu_2$, respectively.
d. Left tailed

Exercises 10.2

10.33
a. Simple random samples, independent samples, normal populations or large samples, and equal population standard deviations
b. Simple random samples and independent samples are essential assumptions. Moderate violations of the normality assumption are permissible even for small or moderate size samples. Moderate violations of the equal-standard-deviations requirement are not serious provided the two sample sizes are roughly equal.

Note: From the instructions for Exercises 10.35–10.38, the only assumption for pooled t-procedures we need to address is that of equal population standard deviations.

10.35 No, not reasonable, because the sample standard deviations suggest that the two population standard deviations differ and the sample sizes are not roughly equal.

10.37 Yes, because the sample standard deviations are close to being equal, suggesting that assuming the population standard deviations are equal is reasonable.

10.39
a. $t = -2.487$; critical values $= \pm 2.048$; $0.01 < P < 0.02$ $(P = 0.019)$; reject H_0
b. -3.65 to -0.35

10.41
a. $t = 1.057$; critical value $= 1.714$; $P > 0.10$ $(P = 0.151)$; do not reject H_0
b. -1.24 to 5.24

10.43
a. $t = -2.631$; critical value $= -1.692$; $0.005 < P < 0.01$ $(P = 0.006)$; reject H_0
b. -6.57 to -1.43

10.45 $H_0: \mu_1 = \mu_2$, $H_a: \mu_1 < \mu_2$; $\alpha = 0.05$; $t = -4.058$; critical value $= -1.734$; $P < 0.005$ $(P = 0.000)$; reject H_0; at the 5% significance level, the data provide sufficient evidence to conclude that the mean time served for fraud is less than that for firearms offenses.

10.47 $H_0: \mu_1 = \mu_2$, $H_a: \mu_1 > \mu_2$; $\alpha = 0.05$; $t = 0.520$; critical value $= 1.711$; $P > 0.10$ $(P = 0.304)$; do not reject H_0; at the 5% significance level, the data do not provide sufficient evidence to conclude that drinking fortified orange juice reduces PTH level more than drinking unfortified orange juice.

10.49 $H_0: \mu_1 = \mu_2$, $H_a: \mu_1 \neq \mu_2$; $\alpha = 0.01$; $t = 8.834$; critical values $= \pm 2.601$; $P < 0.01$ $(P = 0.000)$; reject H_0; at the 1% significance level, the data provide sufficient evidence to conclude that a difference exists in mean spleen lengths of male and female Nigerians.

10.51 -12.36 to -4.96 months. We can be 90% confident that the difference between the mean times served by prisoners in the fraud and firearms offense categories is somewhere between -12.36 months and -4.96 months. In other words, we can be 90% confident that mean time served by prisoners in the fraud offense category is somewhere between 4.96 months and 12.36 months less than that served by prisoners in the firearms offense category.

10.53 -16.92 pg/mL to 31.72 pg/mL. We can be 90% confident that the difference between the mean reductions in PTH levels for fortified and unfortified orange juice is somewhere between -16.92 pg/mL and 31.72 pg/mL. In other words, we can be 90% confident that the mean reduction in PTH level for fortified orange juice is somewhere between 16.92 pg/mL less than and 31.72 pg/mL more than that for unfortified orange juice.

10.55 0.71 cm to 1.29 cm. We can be 99% confident that the difference between mean spleen lengths of Nigerian males and females is somewhere between 0.71 cm and 1.29 cm. In other words, we can be 99% confident that the mean spleen length of Nigerian males is somewhere between 0.71 cm and 1.29 cm greater than that of Nigerian females.

Exercises 10.3

10.67 Pooled t-test   **10.69** Nonpooled t-test

10.71 a. Pooled t-test **b.** Nonpooled t-test **c.** Neither **d.** Neither

Note: Answers for exercises that require nonpooled t-procedures may vary depending on whether you use statistical software. Furthermore, discrepancies may occur among results provided by statistical technologies because some round the number of degrees of freedom and others do not.

10.73
a. $t = -1.438$; critical values $= \pm 2.101$; $0.10 < P < 0.20$ $(P = 0.167)$; do not reject H_0
b. -4.92 to 0.92

10.75
a. $t = 1.107$; critical value $= 1.717$; $P > 0.10$ $(P = 0.140)$; do not reject H_0
b. -1.10 to 5.10

10.77
a. $t = -2.782$; critical value $= -1.711$; $0.005 < P < 0.01$ $(P = 0.0051)$; reject H_0
b. -6.46 to -1.54

10.79 $H_0: \mu_1 = \mu_2$, $H_a: \mu_1 \neq \mu_2$; $\alpha = 0.10$; $t = 1.791$; critical values $= \pm 1.677$; $0.05 < P < 0.10$ $(P = 0.080)$; reject H_0; at the 10% significance level, the data provide sufficient evidence to conclude that a difference exists in the mean age at arrest of East German prisoners with chronic PTSD and remitted PTSD.

10.81 $H_0: \mu_1 = \mu_2$, $H_a: \mu_1 < \mu_2$; $\alpha = 0.05$; $t = -1.651$; critical value $= -2.015$; $0.05 < P < 0.10$ $(P = 0.080)$; do not reject H_0; at the 5% significance level, the data do not provide sufficient evidence to conclude that the mean number of acute postoperative days in the hospital is smaller with the dynamic system than with the static system.

10.83 $H_0: \mu_1 = \mu_2$, $H_a: \mu_1 > \mu_2$; $\alpha = 0.01$; $t = 3.863$; critical value $= 2.552$; $P < 0.005$ $(P = 0.001)$; reject H_0; at the 1% significance level, the data provide sufficient evidence to conclude that dopamine activity is higher, on average, in psychotic patients.

10.85 0.2 yr to 7.2 yr. We can be 90% confident that the difference between the mean ages at arrest of East German prisoners with chronic PTSD and remitted PTSD is somewhere between 0.2 yr and 7.2 yr. In other words, we can be 90% confident that the mean age at arrest of East German prisoners with chronic PTSD is somewhere between 0.2 yr and 7.2 yr greater than that of those with remitted PTSD.

10.87 -6.97 days to 0.69 days. We can be 90% confident that the difference between the mean number of acute postoperative days in the hospital with the dynamic and static systems is somewhere between -6.97 days and 0.69 days. In other words, we can be 90% confident that the mean number of acute postoperative days in the hospital with the dynamic system is somewhere between 6.97 days less than and 0.69 days more than that with the static system.

10.89 0.00266 to 0.01301 nmol/mL-hr/mg. We can be 98% confident that the difference between the mean dopamine activities of psychotic and nonpsychotic patients is somewhere between 0.00266 nmol/mL-hr/mg and 0.01301 nmol/mL-hr/mg. In other words, we can be 98% confident that the mean dopamine activities of psychotic patients exceeds that of nonpsychotic patients by somewhere between 0.00266 nmol/mL-hr/mg and 0.01301 nmol/mL-hr/mg.

10.91

a. Nonpooled t-procedures because the sample standard deviations indicate that the population standard deviations are far from equal and the sample sizes are quite different.

b. No, because a normal probability plot for the males' data is far from linear and indicates the presence of outliers.

10.93

a. Nonpooled t-procedures. The sample standard deviations of the two samples are 4.68% and 1.50%, thus suggesting that the population standard deviations are not equal.

b. No, because the healthcare data contain an outlier.

10.95

a. $H_0: \mu_1 = \mu_2$, $H_a: \mu_1 < \mu_2$; $\alpha = 0.05$; $t = -2.02$; critical value $= -1.734$; $0.025 < P < 0.05$ ($P = 0.029$); reject H_0; at the 5% significance level, the data provide sufficient evidence to conclude that the mean operative time is less with the dynamic system than with the static system.

b. The null hypothesis is rejected using both the nonpooled t-test and the pooled t-test. However, the nonpooled t-test provides very strong evidence against the null hypothesis, whereas the pooled t-test provides (only) strong evidence against the null hypothesis.

d. The nonpooled t-test, because the sample standard deviations suggest that the population standard deviations are not equal.

Exercises 10.4

10.105 Pooled t-test

10.107 a. Nonpooled t-test **b.** Mann–Whitney test **c.** Nonpooled t-test

10.109 Mann–Whitney test

10.111 Because the shape of a normal distribution is determined by its standard deviation

10.113 a. 90 **b.** 54 **c.** 51, 93   **10.115 a.** 95 **b.** 67 **c.** 63, 99

10.117 $M = 24$; critical value $= 26$; $P = 0.196$ (0.186 adjusted for ties); do not reject H_0

10.119 $M = 19$; critical values $= 19$ and 36; $P = 0.095$ (0.090 adjusted for ties); reject H_0

10.121 $M = 26.5$; critical value $= 28$; $P = 0.050$ (0.048 adjusted for ties); reject H_0

10.123 $H_0: \mu_1 = \mu_2$, $H_a: \mu_1 \neq \mu_2$; $\alpha = 0.05$; $M = 34$; critical values $= 12$ and 32; $P = 0.014$; reject H_0; at the 5% significance level, the data provide sufficient evidence to conclude that a difference exists in the mean wing stroke frequencies of the two species of Euglossine bees.

10.125 $H_0: \mu_1 = \mu_2$, $H_a: \mu_1 < \mu_2$; $\alpha = 0.05$; $M = 33$; critical value $= 33$; $P = 0.044$; reject H_0; at the 5% significance level, the data provide sufficient evidence to conclude that, in this teacher's chemistry courses, students with fewer than 2 years of high school algebra have a lower mean semester average than those with two or more years.

10.127 $H_0: \eta_1 = \eta_2$, $H_a: \eta_1 > \eta_2$; $\alpha = 0.05$; $M = 81$; critical value $= 95$; $P = 0.345$; do not reject H_0; at the 5% significance level, the data do not provide sufficient evidence to conclude that the median weekly earnings of male full-time wage and salary workers exceeds the median weekly earnings of female full-time wage and salary workers.

10.129

a. $H_0: \mu_1 = \mu_2$, $H_a: \mu_1 < \mu_2$; $\alpha = 0.05$; $M = 65.5$; critical value $= 83$; $P = 0.002$; reject H_0; at the 5% significance level, the data provide sufficient evidence to conclude that the mean time served for fraud is less than that for firearms offenses.

b. Because two normal distributions with equal standard deviations have the same shape. The pooled t-test is better because, in the normal case, it is more powerful than the Mann–Whitney test.

10.131

b. No, because the sample sizes are small and the normality assumption appears to be violated.

c. Yes, because presuming that the two distributions of the variable under consideration have the same shape appears reasonable.

Exercises 10.5

10.139 By using a paired sample, extraneous sources of variation can be removed. The sampling error thus made in estimating the difference between the population means will generally be smaller. As a result, detecting differences between the population means is more likely when such differences exist.

10.141 Simple random paired sample, and normal differences or large sample. The simple-random-paired-sample assumption is essential. Moderate violations of the normal-differences assumption are permissible even for small or moderate size samples.

10.143 a. TV viewing time **b.** Married men and married women

c. Married couples

d. The difference between the TV viewing times of a married couple

e. Let μ_1 and μ_2 denote the mean TV viewing times of married men and married women, respectively. The null and alternative hypotheses are $H_0: \mu_1 = \mu_2$ and $H_a: \mu_1 \neq \mu_2$, respectively.

f. Two tailed

10.145 a. Response to pain (VAS sensory rating)

b. All people when not under hypnosis and all people when under hypnosis

c. Each pair consists of a person when not under hypnosis and the same person when under hypnosis.

d. The difference between the response to pain (VAS sensory rating) of a person when not under hypnosis and the same person when under hypnosis

e. Let μ_1 and μ_2 denote the mean response to pain of all people when not under hypnosis and all people when under hypnosis, respectively. The null and alternative hypotheses are $H_0: \mu_1 = \mu_2$ and $H_a: \mu_1 > \mu_2$, respectively.

f. Right tailed

10.147 a. Antioxidant capacity

b. All stored breastmilk and all fresh breastmilk

c. Each pair consists of stored breastmilk and fresh breastmilk from the same woman.

d. The difference between the antioxidant capacity of stored breastmilk and fresh breastmilk from a woman

e. Let μ_1 and μ_2 denote the mean antioxidant capacities of stored breastmilk and fresh breastmilk, respectively. The null and alternative hypotheses are $H_0: \mu_1 = \mu_2$ and $H_a: \mu_1 < \mu_2$, respectively.

f. Left tailed

10.149 $t = 3.057$; critical values $= \pm 1.943$; $0.02 < P < 0.05$ ($P = 0.022$); reject H_0

10.151 $t = 0.088$; critical value $= 1.415$; $P > 0.10$ ($P = 0.466$); do not reject H_0

10.153 $t = -2.333$; critical value $= -1.397$; $0.01 < P < 0.025$ ($P = 0.024$); reject H_0

10.155 a. Height (of *Zea mays*)

b. Cross-fertilized *Zea mays* and self-fertilized *Zea mays*

c. The difference between the heights of a cross-fertilized *Zea mays* and a self-fertilized *Zea mays* grown in the same pot

d. Yes. Because each number is the difference between the heights of a cross-fertilized *Zea mays* and a self-fertilized *Zea mays* grown in the same pot

e. $H_0: \mu_1 = \mu_2$, $H_a: \mu_1 \neq \mu_2$; $\alpha = 0.05$; $t = 2.148$; critical values $= \pm 2.145$; $0.02 < P < 0.05$ ($P = 0.0497$); reject H_0; at the 5% significance level, the data provide sufficient evidence to conclude that the mean heights of cross-fertilized and self-fertilized *Zea mays* differ.

f. $H_0: \mu_1 = \mu_2$, $H_a: \mu_1 \neq \mu_2$; $\alpha = 0.01$; $t = 2.148$; critical values $= \pm 2.977$; $0.02 < P < 0.05$ ($P = 0.0497$); do not reject H_0; at the 1% significance level, the data do not provide sufficient evidence to conclude that the mean heights of cross-fertilized and self-fertilized *Zea mays* differ.

10.157 $H_0: \mu_1 = \mu_2$, $H_a: \mu_1 < \mu_2$; $\alpha = 0.05$; $t = -4.185$; critical value $= -1.746$; $P < 0.005$ ($P = 0.000$); reject H_0; at the 5% significance level, the data provide sufficient evidence to conclude that family therapy is effective in helping anorexic young women gain weight.

10.159 $H_0: \mu_1 = \mu_2$, $H_a: \mu_1 > \mu_2$; $\alpha = 0.10$; $t = 1.053$; critical value $= 1.415$; $P > 0.10$ ($P = 0.164$); do not reject H_0; at the 10% significance level, the data do not provide sufficient evidence to conclude that mean corneal thickness is greater in normal eyes than in eyes with glaucoma.

10.161

a. 0.03 to 41.84 eighths of an inch. We can be 95% confident that the difference between the mean heights of cross-fertilized and self-fertilized *Zea mays* is somewhere between 0.03 and 41.84 eighths of an inch. In other words, we can be 95% confident that the mean height of cross-fertilized *Zea mays* exceeds that of self-fertilized *Zea mays* by somewhere between 0.03 eighth of an inch and 41.84 eighths of an inch.

b. −8.08 to 49.94 eighths of an inch. We can be 99% confident that the difference between the mean heights of cross-fertilized and self-fertilized *Zea mays* is somewhere between −8.08 and 49.94 eighths of an inch.

In other words, we can be 99% confident that the mean height of cross-fertilized *Zea mays* is somewhere between 8.08 eighths of an inch less than and 49.94 eighths of an inch more than that of self-fertilized *Zea mays*.

10.163 -10.30 lb to -4.23 lb. We can be 90% confident that the weight gain that would be obtained, on average, by using the family therapy treatment is somewhere between 4.23 lb and 10.30 lb.

10.165 -1.4 microns to 9.4 microns. We can be 80% confident that the difference between the mean corneal thickness of normal eyes and that of eyes with glaucoma is somewhere between -1.4 microns and 9.4 microns. In other words, we can be 80% confident that the mean corneal thickness of normal eyes is somewhere between 1.4 microns less than and 9.4 microns more than that of eyes with glaucoma.

10.167 Graphical data analyses suggest that the paired difference of the third pair of observations is an outlier, albeit a mild one. Thus, although a tough call, we would probably not recommend use of the paired *t*-test.

10.169
b. The normal probability plot of the onset data indicates extreme deviation from normality. Therefore, in view of the small sample size, applying a one-mean *t*-procedure is not reasonable.
c. The normal probability plot of the resolution data is only roughly linear, and the boxplot suggests a potential outlier. Therefore, in view of the small sample size, applying a one-mean *t*-procedure is probably not reasonable.
d. Neither the normal probability plot nor the boxplot of the paired differences suggests the presence of outliers, and, furthermore, the normal probability plot of the paired differences is quite linear. Therefore, applying a paired *t*-procedure is reasonable.
e. Whether applying a paired *t*-procedure is reasonable depends on the properties of the paired-difference variable and not on those of the individual variables that constitute the paired-difference variable.

Exercises 10.6

10.177
a. No. Because the paired-difference variable is far from normally distributed and the sample size is not large
b. Yes. Because the sample size is large
c. Yes. Because both assumptions required for a paired Wilcoxon signed-rank test are satisfied
d. The paired Wilcoxon signed-rank test because it is usually more powerful than the paired *t*-test when the paired-difference variable is not normally distributed.

10.179 a. Paired *t*-test **b.** Neither of the two tests
c. Paired Wilcoxon signed-rank test

10.181 $W = 26.5$; critical values $= 4$ and 24; $P = 0.043$; reject H_0

10.183 $W = 14$; critical value $= 22$; $P = 0.534$; do not reject H_0

10.185 $W = 5$; critical value $= 8$; $P = 0.040$; reject H_0

10.187
a. $H_0: \mu_1 = \mu_2$, $H_a: \mu_1 \neq \mu_2$; $\alpha = 0.05$; $W = 96$; critical values $= 25$ and 95; $P = 0.044$; reject H_0; at the 5% significance level, the data provide sufficient evidence to conclude that the mean heights of cross-fertilized and self-fertilized *Zea mays* differ.
b. $H_0: \mu_1 = \mu_2$, $H_a: \mu_1 \neq \mu_2$; $\alpha = 0.01$; $W = 96$; critical values $= 16$ and 104; $P = 0.044$; do not reject H_0; at the 1% significance level, the data do not provide sufficient evidence to conclude that the mean heights of cross-fertilized and self-fertilized *Zea mays* differ.

10.189 $H_0: \mu_1 = \mu_2$, $H_a: \mu_1 < \mu_2$; $\alpha = 0.05$; $W = 11$; critical value $= 41$; $P = 0.001$; reject H_0; at the 5% significance level, the data provide sufficient evidence to conclude that family therapy is effective in helping anorexic young women gain weight.

10.191 $H_0: \mu_1 = \mu_2$, $H_a: \mu_1 > \mu_2$; $\alpha = 0.10$; $W = 20.5$; critical value $= 22$; $P = 0.155$; do not reject H_0; at the 10% significance level, the data do not provide sufficient evidence to conclude that mean corneal thickness is greater in normal eyes than in eyes with glaucoma.

10.193 Although a tough call, applying the paired Wilcoxon signed-rank test is probably reasonable.

10.195
a. A normal probability plot suggests that the paired-difference variable is not normally distributed. Therefore, in view of the small sample size, applying the paired *t*-test is not reasonable.
b. Graphical analyses suggest that the paired-difference variable is roughly symmetric. So, using the paired Wilcoxon signed-rank test is reasonable.

Review Problems for Chapter 10

1. Independently and randomly take samples from the two populations; compute the two sample means; compare the two sample means; and make the decision.

2. Randomly take a paired sample from the two populations; calculate the paired differences of the sample pairs; compute the mean of the sample of paired differences; compare that sample mean to 0; and make the decision.

3. **a.** The pooled *t*-procedures require equal population standard deviations, whereas the nonpooled *t*-procedures do not.
   b. It is essential that the assumption of independence be satisfied.
   c. For very small sample sizes, the normality assumption is essential for both *t*-procedures. However, for larger samples, the normality assumption is less important.

4. population standard deviations

5. **a.** No. If the two distributions are normal and have the same shape, they have equal population standard deviations; in this case, the pooled *t*-test is preferred. If the two distributions are nonnormal but have the same shape, the Mann–Whitney test is preferred.
   b. The two distributions are normal. In this case, the pooled *t*-test is more powerful than the Mann–Whitney test.

6. By using a paired sample, extraneous sources of variation can be removed. As a consequence, the sampling error made in estimating the difference between the population means will generally be smaller. This fact, in turn, makes it more likely that differences between the population means will be detected when such differences exist.

7. If the paired-difference variable is normally distributed, it would be preferable to use the paired *t*-test because, in that case, it is more powerful than the paired Wilcoxon signed-rank test.

8. $H_0: \mu_1 = \mu_2$, $H_a: \mu_1 > \mu_2$; $\alpha = 0.05$; $t = 1.538$; critical value $= 1.708$; $0.05 < P < 0.10$ ($P = 0.068$); do not reject H_0; at the 5% significance level, the data do not provide sufficient evidence to conclude that the mean right-leg strength of males exceeds that of females.

9. -31.3 to 599.3 newtons (N). We can be 90% confident that the difference between the mean right-leg strengths of males and females is somewhere between -31.3 N and 599.3 N. In other words, we can be 90% confident that the mean right-leg strength of males is somewhere between 31.3 N less than and 599.3 N more than that of females.

10. $H_0: \mu_1 = \mu_2$, $H_a: \mu_1 < \mu_2$; $\alpha = 0.01$; $t = -4.118$; critical value $= -2.385$; $P < 0.005$ ($P = 0.000$); reject H_0; at the 1% significance level, the data provide sufficient evidence to conclude that, on average, the number of young per litter of cottonmouths in Florida is less than that in Virginia.

11. -3.4 to -0.9 young per litter. We can be 98% confident that the difference between the mean litter sizes of cottonmouths in Florida and Virginia is somewhere between -3.4 and -0.9. With 98% confidence, we can say that, on average, cottonmouths in Virginia have somewhere between 0.9 and 3.4 more young per litter than those in Florida.

12. $H_a: \eta_1 = \eta_2$, $H_a: \eta_1 \neq \eta_2$; $\alpha = 0.05$; $M = 127$; critical values $= 79$ and 131; $P = 0.104$; do not reject H_0; at the 5% significance level, the data do not provide sufficient evidence to conclude that the median costs for existing single-family homes differ in Atlantic City and Las Vegas.

13. **b.** Yes, the normal probability plot is quite linear, and neither that plot nor the boxplot reveals any outliers.
   c. $H_0: \mu_1 = \mu_2$, $H_a: \mu_1 \neq \mu_2$; $\alpha = 0.10$; $t = 0.55$; critical values $= \pm 1.895$; $P > 0.20$ ($P = 0.600$); do not reject H_0; at the 10% significance level, the data do not provide sufficient evidence to conclude that a difference exists in the mean length of time that ice stays on these two lakes.

14. -3.4 to 6.1 days. We can be 90% confident that the difference in the mean lengths of time that ice stays on the two lakes is somewhere between -3.4 and 6.1 days. In other words, we can be 90% confident that the mean length of time that ice stays on Lake Mendota is somewhere between 3.4 days less than and 6.1 days more than that on Lake Monona.

15. $H_0: \mu_1 = \mu_2$, $H_a: \mu_1 > \mu_2$; $\alpha = 0.05$; $W = 135.5$; critical value $= 124$; $P = 0.016$; reject H_0; at the 5% significance level, the data provide sufficient evidence to conclude that, for females who suffer from depression, the mean total cholesterol level is lower after a 4-week treatment of the antidepressant fluoxetine.

Chapter 11

Exercises 11.1

11.1 A variable is said to have a chi-square distribution if its distribution has the shape of a special type of right-skewed curve, called a chi-square curve.

11.3 The χ^2-curve with 20 degrees of freedom more closely resembles a normal curve. As the number of degrees of freedom becomes larger, χ^2-curves look increasingly like normal curves.

11.5 a. 32.852   **b.** 10.117   **11.7 a.** 18.307   **b.** 3.247

11.9 a. 1.646   **b.** 15.507   **11.11 a.** 0.831, 12.833   **b.** 13.844, 41.923

11.13 Because the procedures are based on the assumption that the variable under consideration is normally distributed and are nonrobust to violations of that assumption

11.15 a. $\chi^2 = 5.062$; critical value = 3.325; $P = 0.171$; do not reject H_0
b. 2.19 to 4.94

11.17 a. $\chi^2 = 49$; critical value = 44.314; $P = 0.003$; reject H_0
b. 5.26 to 10.31

11.19
a. $\chi^2 = 13.194$; critical values = 8.907 and 32.852; $P = 0.343$; do not reject H_0
b. 3.80 to 7.30

11.21 H_0: $\sigma = 3.0$ days, H_a: $\sigma \neq 3.0$ days; $\alpha = 0.10$; $\chi^2 = 9.035$; critical values = 5.892 and 22.362; $P = 0.459$; do not reject H_0; at the 10% significance level, the data do not provide sufficient evidence to conclude that the population standard deviation of lactation periods for grey seals differs from 3.0 days.

11.23 H_0: $\sigma = 0.27$, H_a: $\sigma > 0.27$; $\alpha = 0.01$; $\chi^2 = 70.631$; critical value = 21.666; $P = 0.000$; reject H_0; at the 1% significance level, the data provide sufficient evidence to conclude that the product variability for this piece of equipment exceeds the analytical capability of 0.27.

11.25
a. H_0: $\sigma = 0.2$ fl oz, H_a: $\sigma < 0.2$ fl oz; $\alpha = 0.05$; $\chi^2 = 8.317$; critical value = 6.571; $P = 0.128$; do not reject H_0; at the 5% significance level, the data do not provide sufficient evidence to conclude that the standard deviation of the amounts being dispensed is less than 0.2 fl oz.
b. To ensure that the amount of coffee being dispensed does not vary too much from cup to cup (nor from the advertised weight of 6 fl oz)

11.27 1.91 to 3.71 days. We can be 90% confident that the standard deviation of lactation periods of grey seals is somewhere between 1.91 and 3.71 days.

11.29 0.49 to 1.57. We can be 98% confident that the product variability for this piece of equipment is somewhere between 0.49 and 1.57.

11.31 0.119 to 0.225 fl oz. We can be 90% confident that the standard deviation of the amounts of coffee being dispensed is somewhere between 0.119 and 0.225 fl oz.

11.33 A normal probability plot of the data suggests that the variable under consideration is far from normally distributed. So, using one-standard-deviation χ^2-procedures is not reasonable.

11.35 This one is a tough call. However, as a normal probability plot of the data is only roughly linear, using one-standard-deviation χ^2-procedures here is probably not reasonable.

Exercises 11.2

11.43 $F_{0.05}$; $F_{0.025}$; F_α **11.45 a.** 12   **b.** 7

11.47 a. 1.89   **b.** 2.47   **c.** 2.14   **11.49 a.** 2.88   **b.** 2.10   **c.** 1.78

11.51 a. 0.12   **b.** 3.58   **11.53 a.** 0.18, 9.07   **b.** 0.33, 2.68

11.55 Because the procedures are based in part on the assumption that the variable under consideration is normally distributed on each population and are nonrobust to violations of that assumption

11.57 a. $F = 3.41$; critical value = 3.21; $P = 0.007$; reject H_0
b. 1.03 to 3.04

11.59 a. $F = 0.55$; critical value = 0.37; $P = 0.216$; do not reject H_0
b. 0.49 to 1.21

11.61 a. $F = 0.23$; critical values = 0.26 and 4.30; $P = 0.032$; reject H_0
b. 0.23 to 0.94

11.63 H_0: $\sigma_1 = \sigma_2$, H_a: $\sigma_1 > \sigma_2$; $F = 2.19$; $\alpha = 0.05$; critical value = 2.03; $P = 0.035$; reject H_0; at the 5% significance level, the data provide sufficient evidence to conclude that there is less variation among final-exam scores using the new teaching method.

11.65 H_0: $\sigma_1 = \sigma_2$, H_a: $\sigma_1 \neq \sigma_2$; $F = 1.22$; $\alpha = 0.10$; critical values = 0.53 and 1.94; $P = 0.624$; do not reject H_0; at the 10% significance level, the data do not provide sufficient evidence to conclude that the variation in anxiety-test scores differs between patients seeing videotapes showing progressive relaxation exercises and those seeing neutral videotapes.

11.67 H_0: $\sigma_1 = \sigma_2$, H_a: $\sigma_1 < \sigma_2$; $F = 0.21$; $\alpha = 0.01$; critical value = 0.41; $P = 0.000$; reject H_0; at the 1% significance level, the data provide sufficient evidence to conclude that the standard deviation of velocity is less with the Stinger tee than with the regular tee.

11.69 1.04 to 2.01. We can be 90% confident that the ratio of the population standard deviations of final-exam scores for students taught by the conventional method and those taught by the new method is somewhere between 1.04 and 2.01 (i.e., $1.04\sigma_2 < \sigma_1 < 2.01\sigma_2$). In other words, we can be 90% confident that the standard deviation of final-exam scores for students taught by the conventional method is somewhere between 1.04 and 2.01 times greater than that for those taught by the new method.

11.71 0.79 to 1.52. We can be 90% confident that the ratio of the population standard deviations of scores for patients who are shown videotapes of progressive relaxation exercises and those who are shown neutral videotapes is somewhere between 0.79 and 1.52 (i.e., $0.79\sigma_2 < \sigma_1 < 1.52\sigma_2$). In other words, we can be 90% confident that the standard deviation of scores for patients who are shown videotapes of progressive relaxation exercises is somewhere between 1.27 times less than and 1.52 times greater than that for those who are shown neutral videotapes.

11.73 0.295 to 0.714. We can be 98% confident that the ratio of the population standard deviations of ball velocity for the Stinger tee and the regular tee is somewhere between 0.295 and 0.714 (i.e., $0.295\sigma_2 < \sigma_1 < 0.714\sigma_2$). In other words, we can be 98% confident that the standard deviation of ball velocity for the Stinger tee is somewhere between 1.40 and 3.39 times less than that for the regular tee.

Review Problems for Chapter 11

1. Chi-square distribution   **2. a.** Right   **b.** Normal

3. The variable under consideration must be normally distributed or nearly so. It is very important because the procedures are nonrobust to violations of that assumption.

4. 6.408   **5.** 33.409   **6.** 27.587   **7.** 8.672

8. 7.564, 30.191   **9.** The F-distribution

10. a. Right   **b.** Reciprocal; 5, 14   **c.** 0

11. The distributions (one for each population) of the variable under consideration must be normally distributed or nearly so. It is very important because the procedure is nonrobust to violations of that assumption.

12. 7.01   **13.** 0.07   **14.** 3.84   **15.** 0.17   **16.** 0.11, 5.05

17. a. H_0: $\sigma = 16$ points, H_a: $\sigma \neq 16$ points; $\alpha = 0.10$; $\chi^2 = 21.110$; critical values = 13.848 and 36.415; $P = 0.736$; do not reject H_0; at the 10% significance level, the data do not provide sufficient evidence to conclude that IQs measured on this scale have a standard deviation different from 16 points.
b. It is essential because the one-standard-deviation χ^2-test is nonrobust to violations of that assumption.

18. 2.07 to 5.51 m. We can be 99% confident that the standard deviation of the rebound distances of the bouncing rocks is somewhere between 2.07 m and 5.51 m.

19. a. F-distribution with df = (14, 19)
b. H_0: $\sigma_1 = \sigma_2$, H_a: $\sigma_1 < \sigma_2$; $\alpha = 0.01$; $F = 0.07$; critical value = 0.28; $P = 0.000$; reject H_0; at the 1% significance level, the data provide sufficient evidence to conclude that runners have less variability in skinfold thickness than others.
c. Skinfold thickness is normally distributed for runners and for others. Construct normal probability plots of the two samples.
d. The samples from the two populations must be independent and simple random samples.

20. 0.802 to 2.378. We can be 90% confident that the ratio of the population standard deviations of heart-rate increase when laughing for male-male and female-female dyads is somewhere between 0.802 and 2.378 (i.e., $0.802\sigma_2 < \sigma_1 < 2.378\sigma_2$). In other words, we can be 90% confident that the population standard deviation of heart-rate increase when laughing for male-male dyads is somewhere between 1.247 times less than and 2.378 times greater than that for female-female dyads.

Chapter 12

Exercises 12.1

12.1 Answers will vary.

12.3 A population proportion is a parameter because it is a descriptive measure for a population. A sample proportion is a statistic because it is a descriptive measure for a sample.

12.5
a. The proportion (percentage) of a sample from the population that has the specified attribute
b. $\hat{p}$

12.7 The number of members in the sample that do not have the specified attribute

12.9 a. $p = 0.4$
b.

Sample	No. of females x	Sample proportion $\hat{p}$
J, G	1	0.5
J, P	0	0.0
J, C	0	0.0
J, F	1	0.5
G, P	1	0.5
G, C	1	0.5
G, F	2	1.0
P, C	0	0.0
P, F	1	0.5
C, F	1	0.5

c.

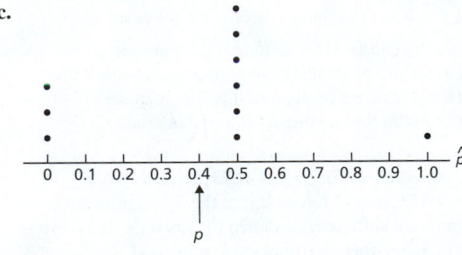

d. 0.4
e. They are the same because the mean of the variable $\hat{p}$ equals the population proportion; in symbols, $\mu_{\hat{p}} = p$.

12.11 b.

Sample	No. of females x	Sample proportion $\hat{p}$
J, P, C	0	0.00
J, P, G	1	0.33
J, P, F	1	0.33
J, C, G	1	0.33
J, C, F	1	0.33
J, G, F	2	0.67
P, C, G	1	0.33
P, C, F	1	0.33
P, G, F	2	0.67
C, G, F	2	0.67

c.

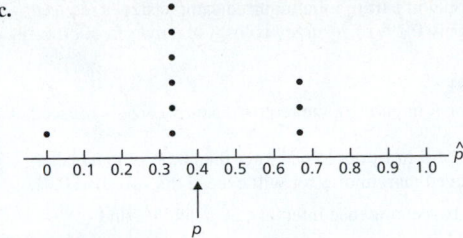

d. 0.4
e. They are the same because the mean of the variable $\hat{p}$ equals the population proportion; in symbols, $\mu_{\hat{p}} = p$.

12.13 b.

Sample	No. of females x	Sample proportion $\hat{p}$
J, P, C, G, F	2	0.4

c.

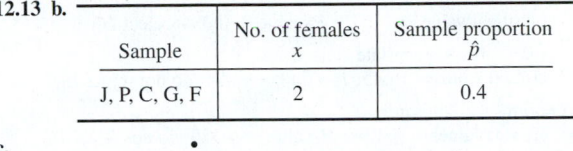

d. 0.4
e. They are the same because the mean of the variable $\hat{p}$ equals the population proportion; in symbols, $\mu_{\hat{p}} = p$.

12.15 a. The No. 1 draft picks in the NBA since 1947
b. Being other than a U.S. national
c. Population proportion. It is the proportion of the population of No. 1 draft picks in the NBA since 1947 who are other than U.S. nationals.

12.17 a. 0.00718 **b.** Smaller **12.19 a.** 0.4 **b.** $0.4 < \hat{p} < 0.6$

12.21 a. 0.2 **b.** $0.2 < \hat{p} < 0.8$ **12.23 a.** 0.5 **b.** None

12.25 a. $\hat{p} = 0.2$ **b.** Appropriate **c.** 0.076 to 0.324
d. 0.124; 0.2 ± 0.124

12.27 a. $\hat{p} = 0.7$ **b.** Appropriate **c.** 0.533 to 0.867
d. 0.167; 0.7 ± 0.167

12.29 a. $\hat{p} = 0.8$ **b.** Not appropriate

> *Note:* In each of Exercises 12.31–12.47, the answer you get can depend on how you obtain $z_{\alpha/2}$.

12.31 9604 **12.33** 1692 **12.35** 1844 **12.37** 8068 **12.39** 609

12.41 1844 **12.43** 9220 **12.45** 1083 **12.47** 1844

12.49 0.291 to 0.349. We can be 95% confident that the proportion of all U.S. adults who never clothes-shop online is somewhere between 0.291 and 0.349.

12.51 0.0528 to 0.0992. We can be 95% confident that the proportion of all U.S. asthmatics who are allergic to sulfites is somewhere between 0.0528 and 0.0992.

12.53 76.7% to 83.3%. We can be 99% confident that the percentage of all registered voters who favor the creation of standards on CAFO pollution and, in general, view CAFOs unfavorably is somewhere between 76.7% and 83.3%.

12.55 Yes. Procedure 12.1 was applied without checking one of the assumptions for its use; namely, that the number of successes, x, and the number of failures, $n - x$, are both 5 or greater. Because the number of failures here is only 4, Procedure 12.1 should not have been used.

12.57 47% to 53%. We can be 95% confident that the percentage of all American adults who favor a plan to break up the 12 megabanks is somewhere between 47% and 53%.

12.59 a. 0.0232 **b.** 9604 **c.** 0.0659 to 0.0761
d. 0.0051, which is less than 0.01
e. 3458; 0.0624 to 0.0796; 0.0086, which is less than 0.01
f. By using the guess for $\hat{p}$ in part (e), the required sample size is reduced by 6146. Moreover, only 0.35% of accuracy is lost—the margin of error rises from 0.0051 to 0.0086.

12.61 a. 3.3% (i.e., 3.3 percentage points) **b.** 7368 **c.** 0.811 to 0.833
d. 0.011, which is less than 0.015 (1.5%)
e. 5526; 0.809 to 0.835; 0.013, which is less than 0.015 (1.5%)

f. By using the guess for $\hat{p}$ in part (e), the required sample size is reduced by 1842. Moreover, only 0.2% of accuracy is lost—the margin of error rises from 0.011 to 0.013.

12.63 a. 9604   **b.** 1791

c. By using the guess for $\hat{p}$ in part (b), the required sample size is reduced by 7813.

d. If the observed value of $\hat{p}$ turns out to be larger than 0.049 (but smaller than 0.951), the achieved margin of error will exceed the specified 0.01.

12.65 Yes. Because the two confidence intervals, 39% to 45% and 36% to 40%, overlap

Exercises 12.2

12.79 a. $\hat{p} = 0.2$ **b.** Appropriate

c. $z = -1.38$; critical value $= -1.28$; P-value $= 0.084$; reject H_0

12.81 a. $\hat{p} = 0.7$ **b.** Appropriate

c. $z = 1.44$; critical value $= 1.645$; P-value $= 0.074$; do not reject H_0

12.83 a. $\hat{p} = 0.8$ **b.** Appropriate

c. $z = 0.98$; critical values $= \pm 1.96$; P-value $= 0.329$; do not reject H_0

12.85

a. 0.530. 53.0% of those sampled disapprove of this government surveillance program.

b. H_0: $p = 0.5$, H_a: $p > 0.5$; $\alpha = 0.05$; $z = 1.89$; critical value $= 1.645$; $P = 0.029$; reject H_0; at the 5% significance level, the data provide sufficient evidence to conclude that a majority of American adults disapprove of this government surveillance program.

12.87 H_0: $p = 0.58$, H_a: $p \neq 0.58$; $\alpha = 0.10$; $z = 1.31$; critical values $= \pm 1.645$; $P = 0.190$; do not reject H_0; at the 10% significance level, the data do not provide sufficient evidence to conclude that the percentage of all American adults who now think that a third major party is needed has changed from that in 2010.

12.89

a. H_0: $p = 0.72$, H_a: $p < 0.72$; $\alpha = 0.05$; $z = -4.93$; critical value $= -1.645$; $P = 0.000$; reject H_0; at the 5% significance level, the data provide sufficient evidence to conclude that the percentage of Americans who approve of labor unions now has decreased since 1936.

b. H_0: $p = 0.67$, H_a: $p < 0.67$; $\alpha = 0.05$; $z = -1.34$; critical value $= -1.645$; $P = 0.090$; do not reject H_0; at the 5% significance level, the data do not provide sufficient evidence to conclude that the percentage of Americans who approve of labor unions now has decreased since 1963.

12.91

a. H_0: $p = 0.5$, H_a: $p > 0.5$; $\alpha = 0.05$; $z = 1.33$; critical value $= 1.645$; $P = 0.093$; do not reject H_0; at the 5% significance level, the data do not provide sufficient evidence to conclude that a majority of U.S. adults favored passage.

b. From part (a), at the 5% significance level, we cannot conclude that "In U.S., Slim Majority Supports Economic Stimulus Plan."

c. It could state, for instance, that "Slim Majority (52.0%) of Respondents Supports Economic Stimulus Plan."

12.93

a. H_0: $p = 0.9$, H_a: $p > 0.9$; $\alpha = 0.05$; $z = 2.12$; critical value $= 1.645$; $P = 0.017$; reject H_0; at the 5% significance level, the data provide sufficient evidence to conclude that more than 9 of 10 Americans always wash up after using the bathroom.

b. H_0: $p = 0.9$, H_a: $p > 0.9$; $\alpha = 0.01$; $z = 2.12$; critical value $= 2.33$; $P = 0.017$; do not reject H_0; at the 1% significance level, the data do not provide sufficient evidence to conclude that more than 9 of 10 Americans always wash up after using the bathroom.

Exercises 12.3

12.95 For a two-tailed test, the basic strategy is as follows: (1) independently and randomly take samples from the two populations under consideration; (2) compute the sample proportions, $\hat{p}_1$ and $\hat{p}_2$; and (3) reject the null hypothesis if the sample proportions differ by too much—otherwise, do not reject the null hypothesis. The process is the same for a one-tailed test except that, for a left-tailed test, the null hypothesis is rejected only when $\hat{p}_1$ is too much smaller than $\hat{p}_2$, and, for a right-tailed test, the null hypothesis is rejected only when $\hat{p}_1$ is too much larger than $\hat{p}_2$.

12.97 a. Uses sunscreen before going out in the sun

b. Teenage girls and teenage boys

c. Sample proportions. Industry Research acquired those proportions by polling samples of the populations of all teenage girls and all teenage boys.

12.99 a. p_1 and p_2 are parameters, and the other quantities are statistics.

b. p_1 and p_2 are fixed numbers, and the other quantities are variables.

12.101 a. $\hat{p}_1 = 0.45$, $\hat{p}_2 = 0.75$ **b.** Appropriate

c. $z = -2.74$; critical value $= -1.28$; P-value $= 0.003$; reject H_0

d. -0.434 to -0.166

12.103 a. $\hat{p}_1 = 0.75$, $\hat{p}_2 = 0.60$ **b.** Appropriate

c. $z = 1.10$; critical value $= 1.645$; P-value $= 0.136$; do not reject H_0

d. -0.067 to 0.367

12.105 a. $\hat{p}_1 = 0.375$, $\hat{p}_2 = 0.750$ **b.** Appropriate

c. $z = -3.02$; critical values $= \pm 1.96$; P-value $= 0.003$; reject H_0

d. -0.592 to -0.158

12.107

a. H_0: $p_1 = p_2$, H_a: $p_1 < p_2$; $\alpha = 0.01$; $z = -2.61$; critical value $= -2.33$; $P = 0.005$; reject H_0; at the 1% significance level, the data provide sufficient evidence to conclude that women who take folic acid are at lesser risk of having children with major birth defects.

b. Designed experiment

c. Yes. Because for a designed experiment, it is reasonable to interpret statistical significance as a causal relationship.

12.109 H_0: $p_1 = p_2$, H_a: $p_1 \neq p_2$; $\alpha = 0.01$; $z = -2.95$; critical values $= \pm 2.576$; $P = 0.003$; reject H_0; at the 1% significance level, the data provide sufficient evidence to conclude that there is a difference in seat-belt use between drivers who are 16–24 years old and drivers who are 25–69 years old.

12.111

a. The samples must be independent simple random samples; of those sampled whose highest degree is a bachelor's, at least five must be overweight and at least five must not be overweight; and of those sampled with a graduate degree, at least five must be overweight and at least five must not be overweight.

b. H_0: $p_1 = p_2$, H_a: $p_1 > p_2$; $\alpha = 0.05$; $z = 1.41$; critical value $= 1.645$; $P = 0.079$; do not reject H_0; at the 5% significance level, the data do not provide sufficient evidence to conclude that the percentage who are overweight is greater for those whose highest degree is a bachelor's than for those with a graduate degree.

12.113 -0.0191 to -0.000746, or about -0.019 to -0.001. Roughly, we can be 98% confident that the rate of major birth defects for babies born to women who have taken folic acid is somewhere between 1 per 1000 and 19 per 1000 lower than for babies born to women who have not taken folic acid.

12.115 -0.0937 to -0.00628. We can be 99% confident that the proportion of seat-belt users in the age group 16–24 years is somewhere between 0.00628 and 0.0937 less than that for drivers in the age group 25–69 years.

12.117 -0.68% to 8.81%. We can be 90% confident that, among adults whose highest degree is a bachelor's, the percentage who have an above healthy weight is somewhere between 0.68 percentage points less than and 8.81 percentage points more than that among adults with a graduate degree.

12.119

a. H_0: $p_1 = p_2$, H_a: $p_1 \neq p_2$; $\alpha = 0.05$; $z = -0.72$; critical values $= \pm 1.96$; $P = 0.473$; do not reject H_0; at the 5% significance level, the data do not provide sufficient evidence to conclude that there is a difference between the labor-force participation rates of U.S. and Canadian women.

b. -0.102 to 0.047. We can be 95% confident that the labor-force participation rate of U.S. women is somewhere between 10.2 percentage points less than and 4.7 percentage points more than that of Canadian women.

12.121

a. H_0: $p_1 = p_2$, H_a: $p_1 < p_2$; $\alpha = 0.01$; $z = -3.48$; critical value $= -2.33$; $P = 0.000$; reject H_0; at the 1% significance level, the data provide sufficient evidence to conclude that college graduates have a lower unemployment rate than high-school graduates.

b. -0.057 to -0.011. We can be 98% confident that the unemployment rate for college graduates is somewhere between 1.1 and 5.7 percentage points lower than that of high-school graduates.

Review Problems for Chapter 12

1. a. Believing that the overall medicinal benefits of marijuana outweigh the risks and potential harms

b. All physicians

c. The proportion of all physicians that believe the overall medicinal benefits of marijuana outweigh the risks and potential harms

d. Sample proportion. It is the proportion of the responding physicians that believe the overall medicinal benefits of marijuana outweigh the risks and potential harms.

2. Generally, obtaining a sample proportion can be done more quickly and is less costly than obtaining the population proportion. Sampling is often the only practical way to proceed.

3. **a.** The number of members in the sample that have the specified attribute
   b. The number of members in the sample that do not have the specified attribute

4. **a.** Population proportion **b.** Normal **c.** np, $n(1-p)$, 5

5. The accuracy with which a sample proportion, $\hat{p}$, estimates the population proportion, p, at the specified confidence level

6. **a.** Getting the "holiday blues" **b.** All men, all women
   c. The proportion of all men who get the "holiday blues" and the proportion of all women who get the "holiday blues"
   d. The proportion of all sampled men who get the "holiday blues" and the proportion of all sampled women who get the "holiday blues"
   e. Sample proportions. The poll used samples of men and women to obtain the proportions.

7. **a.** Difference between the population proportions **b.** Normal

8. 37.0% to 40.0%

9. **a.** 0.5743 to 0.6293. We can be 95% confident that the proportion of students who expect difficulty finding a job is somewhere between 0.5743 and 0.6293.
   b. 0.0275 **c.** 0.6018 ± 0.0275

10. **a.** 2401 **b.** 0.5671 to 0.6065
    c. 0.0197, which is less than the required 0.02.
    d. 2367; 0.5670 to 0.6067; 0.0199 (or 0.0198)
    e. By using the guess of 0.56 for $\hat{p}$, the required sample size is reduced by 34 with virtually no sacrifice of accuracy.

11. H_0: $p = 0.25$, H_a: $p < 0.25$; $\alpha = 0.05$; $z = -2.30$; critical value $= -1.645$; $P = 0.011$; reject H_0; at the 5% significance level, the data provide sufficient evidence to conclude that less than one in four Americans believe that juries "almost always" convict the guilty and free the innocent.

12. **a.** Observational study
    b. Being observational, the study established only an association between height and breast cancer; no causal relationship can be inferred, although there may be one.

13. H_0: $p_1 = p_2$, H_a: $p_1 \neq p_2$; $\alpha = 0.01$; $z = 2.86$; critical values $= \pm 2.575$; $P = 0.004$; reject H_0; at the 1% significance level, the data provide sufficient evidence to conclude that a difference exists in the percentages of smartphone owners between men and women.

14. **a.** 0.006 to 0.114.
    b. We can be 99% confident that the difference in percentages of men and women smartphone owners is somewhere between 0.6 and 11.4 percentage points.

15. **a.** 0.00152 to 0.0761 **b.** 0.0125 to 0.0997
    c. The number of "successes" is less than 5, so the one-proportion z-interval procedure should not be used here and is unreliable. However, the requirements for use of the one-proportion plus-four z-interval procedure are met.
    d. The one-proportion plus-four z-interval procedure.

Chapter 13

Exercises 13.1

13.1 A variable is said to have a chi-square distribution if its distribution has the shape of a special type of right-skewed curve, called a chi-square curve.

13.3 The χ^2-curve with 20 degrees of freedom more closely resembles a normal curve. As the number of degrees of freedom becomes larger, χ^2-curves look increasingly like normal curves.

13.5 a. 32.852 **b.** 36.191   **13.7 a.** 18.307 **b.** 20.483

Exercises 13.2

13.9 Because the hypothesis test is carried out by determining how well the observed frequencies fit the expected frequencies

13.11 Both assumptions are satisfied.

13.13 Both assumptions are satisfied. Note that 20% of the expected frequencies are less than 5.

13.15 Assumption 2 is satisfied because only 20% of the expected frequencies are less than 5, but Assumption 1 fails because there is an expected frequency of 0.5 (which is less than 1).

13.17
a. The population consists of occupied housing units built after 2010; the variable is primary heating fuel.
b. In the following table, the first column gives the sample size, the second column shows the number of expected frequencies less than 1 and parenthetically whether Assumption 1 is satisfied, the third column shows the percentage of expected frequencies less than 5 and parenthetically whether Assumption 2 is satisfied, and the fourth column states whether both assumptions for a chi-square goodness-of-fit test are satisfied.

Sample size	Number less than 1	Percentage less than 5	Both satisfied?
200	1 (no)	33.3 (no)	no
300	1 (no)	16.7 (yes)	no
400	0 (yes)	16.7 (yes)	yes

c. 334

Note: In each of Exercises 13.19–13.23, the null hypothesis is that the variable has the distribution given in the problem statement and the alternative hypothesis is that the variable does not have that distribution.

13.19 $\chi^2 = 14.042$; critical value $= 7.815$; $P < 0.005$ ($P = 0.003$); reject H_0

13.21 $\chi^2 = 7.1$; critical value $= 7.779$; $P > 0.10$ ($P = 0.131$); do not reject H_0

13.23 $\chi^2 = 10.061$; critical value $= 9.210$; $0.005 < P < 0.01$ ($P = 0.007$); reject H_0

13.25
a. The population consists of all this year's incoming college freshmen in the United States; the variable is political view.
b. H_0: This year's distribution of political views for incoming college freshmen is the same as the 2000 distribution. H_a: This year's distribution of political views for incoming college freshmen has changed from the 2000 distribution. $\alpha = 0.05$; $\chi^2 = 4.394$; critical value $= 5.991$; $P > 0.10$ ($P = 0.111$); do not reject H_0; at the 5% significance level, the data do not provide sufficient evidence to conclude that this year's distribution of political views for incoming college freshmen has changed from the 2000 distribution.
c. Same null and alternative hypotheses as in part (b). $\alpha = 0.10$; $\chi^2 = 4.394$; critical value $= 4.605$; $P > 0.10$ ($P = 0.111$); do not reject H_0; at the 10% significance level, the data do not provide sufficient evidence to conclude that this year's distribution of political views for incoming college freshmen has changed from the 2000 distribution.

13.27 H_0: The color distribution of M&Ms is that reported by M&M/MARS consumer affairs. H_a: The color distribution of M&Ms differs from that reported by M&M/MARS consumer affairs. $\alpha = 0.05$; $\chi^2 = 4.091$; critical value $= 11.070$; $P > 0.10$ ($P = 0.536$); do not reject H_0; at the 5% significance level, the data do not provide sufficient evidence to conclude that the color distribution of M&Ms differs from that reported by M&M/MARS consumer affairs.

13.29 H_0: The die is not loaded. H_a: The die is loaded. $\alpha = 0.05$; $\chi^2 = 2.48$; critical value $= 11.070$; $P > 0.10$ ($P = 0.780$); do not reject H_0; at the 5% significance level, the data do not provide sufficient evidence to conclude that the die is loaded.

13.31
a. H_0: World Series teams are evenly matched. H_a: World Series teams are not evenly matched. $\alpha = 0.05$; $\chi^2 = 7.594$; critical value $= 7.815$; $0.05 < P < 0.10$ ($P = 0.055$); do not reject H_0; at the 5% significance level, the data do not provide sufficient evidence to conclude that World Series teams are not evenly matched.
b. Same null and alternative hypotheses as in part (a). $\alpha = 0.10$; $\chi^2 = 7.594$; critical value $= 6.251$; $0.05 < P < 0.10$ ($P = 0.055$); reject H_0; at the 10% significance level, the data provide sufficient evidence to conclude that World Series teams are not evenly matched.
c. The data are not from a simple random sample, so using the chi-square goodness-of-fit test here is inappropriate.

13.33

a. H_0: The distribution of genders in two-children families does not differ from the distribution predicted by the model described. H_a: The distribution of genders in two-children families differs from the distribution predicted by the model described. $\alpha = 0.01$; $\chi^2 = 185.080$; critical value $= 9.210$; $P < 0.005$ ($P = 0.000$); reject H_0; at the 1% significance level, the data provide sufficient evidence to conclude that the distribution of genders in two-children families differs from the distribution predicted by the model described.

b. Boys and girls are not equally likely to be born.

Exercises 13.3

13.37 Cells

13.39 Summing the row totals, summing the column totals, or summing the frequencies in the cells

13.41 Yes. If no association existed between "gender" and "specialty," the percentage of active male physicians who specialized in internal medicine would be identical to the percentage of active female physicians who specialized in internal medicine. As that is not the case, an association exists between the two variables.

13.43 a.

		Bus.	Engr.	Lib. Arts	Total
Gender	Male	2	10	3	15
	Female	7	2	1	10
	Total	9	12	4	25

College

b.

		Bus.	Engr.	Lib. Arts	Total
Gender	Male	0.222	0.833	0.750	0.600
	Female	0.778	0.167	0.250	0.400
	Total	1.000	1.000	1.000	1.000

College

c.

		Bus.	Engr.	Lib. Arts	Total
Gender	Male	0.133	0.667	0.200	1.000
	Female	0.700	0.200	0.100	1.000
	Total	0.360	0.480	0.160	1.000

College

d. Yes. The tables in parts (b) and (c) show that the conditional distributions of one variable given the other are not identical.

13.45

a.

		Fresh.	Soph.	Junior	Senior	Total
Party	Republican	3	9	12	6	30
	Democrat	2	6	8	4	20
	Other	1	3	4	2	10
	Total	6	18	24	12	60

Class

b.

		Fresh.	Soph.	Junior	Senior
Party	Republican	0.500	0.500	0.500	0.500
	Democrat	0.333	0.333	0.333	0.333
	Other	0.167	0.167	0.167	0.167
	Total	1.000	1.000	1.000	1.000

Class

c. No. The table in part (b) shows that the conditional distributions of political party affiliation within class levels are identical.

d. Republican 0.500, Democrat 0.333, Other 0.167, Total 1.000

e. True. From part (c), political party affiliation and class level are not associated. Therefore the conditional distributions of class level within political party affiliations are identical to each other and to the marginal distribution of class level.

13.47 a. 12

b. The missing entries, from left to right and top to bottom, are 1524, 1580, 3221, 2258, 12,867, 6698, and 12,685.

c. 25,435   **d.** 6698   **e.** 12,867   **f.** 764

13.49 a. 1056.5 thousand   **b.** 48.6 thousand   **c.** 10.6 thousand

d. 304.7 thousand   **e.** 51.9 thousand   **f.** 51.9 thousand

g. 1560.1 thousand

13.51

a. The missing entries, from left to right and top to bottom, are 739, 853, 143, 27, 140, and 2202.

b. 15   **c.** 853 thousand   **d.** 140 thousand   **e.** 84 thousand

f. 681 thousand   **g.** 58 thousand

13.53 a.

		White	Black	Other
Region	Northeast	0.164	0.197	0.252
	Midwest	0.170	0.125	0.083
	South	0.412	0.619	0.373
	West	0.254	0.060	0.292
	Total	1.000	1.000	1.000

Race

b. Northeast: 0.201; Midwest: 0.127; South: 0.506; West: 0.166

c. Yes. Because the conditional distributions of region within races are not identical.

d. 50.6%   **e.** 41.2%

f. True. Because by part (c), an association exists between the variables "region" and "race."

g.

		White	Black	Other	Total
Region	Northeast	0.215	0.487	0.298	1.000
	Midwest	0.353	0.491	0.156	1.000
	South	0.215	0.610	0.175	1.000
	West	0.402	0.181	0.417	1.000
	Total	0.263	0.499	0.238	1.000

Race

13.55 a.

		Prison facility		
		State	Federal	Local
Educational attainment	8th grade or less	0.142	0.119	0.131
	Some high school	0.255	0.145	0.334
	GED	0.285	0.226	0.141
	High school diploma	0.205	0.270	0.259
	Postsecondary	0.090	0.158	0.103
	College grad or more	0.024	0.081	0.032
	Total	1.000	1.000	1.000

b. Yes. Because the conditional distributions of educational attainment within type of prison facility categories are not identical.

c. 8th grade or less: 0.137; Some high school: 0.273; GED: 0.238; High school diploma: 0.225; Postsecondary: 0.098; College grad or more: 0.029

d.

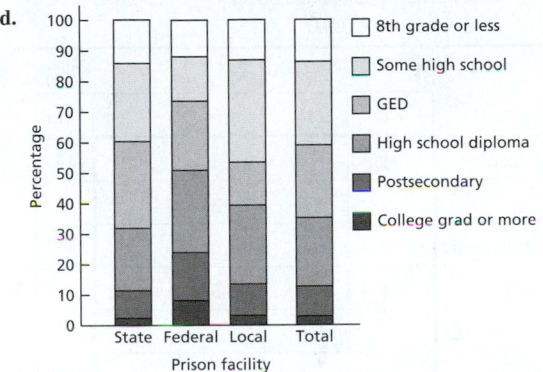

That the bars are not identical reflects the fact that there is an association between educational attainment and type of prison facility.

e. False. Because by part (b), there is an association between educational attainment and type of prison facility.

f.

		Prison facility			
		State	Federal	Local	**Total**
Educational attainment	8th grade or less	0.662	0.047	0.291	1.000
	Some high school	0.598	0.029	0.374	1.000
	GED	0.768	0.051	0.181	1.000
	High school diploma	0.584	0.065	0.352	1.000
	Postsecondary	0.590	0.087	0.323	1.000
	College grad or more	0.521	0.148	0.331	1.000
	Total	0.641	0.054	0.305	1.000

g. 5.4%   **h.** 4.7%   **i.** 11.9%

Exercises 13.4

13.63 H_0: The two variables under consideration are statistically independent. H_a: The two variables under consideration are statistically dependent.

13.65 15

13.67
a. No. Association in observational studies does not imply causation.
b. Answers will vary.

13.69 4   **13.71** 1   **13.73** 0

13.75 a.

	x	
	A	B
a	12	18
b	28	42

(y labels rows)

b. $\chi^2 = 0.794$
c. Critical value = 3.841; $P > 0.10$ ($P = 0.373$); do not reject H_0; at the 5% significance level, the data do not provide sufficient evidence to conclude that the two variables are associated.

13.77 a.

	x		
	A	B	C
a	5	20	75
b	5	20	75

(y labels rows)

b. $\chi^2 = 12.500$
c. Critical value = 5.991; $P < 0.005$ ($P = 0.002$); reject H_0; at the 5% significance level, the data provide sufficient evidence to conclude that the two variables are associated.

13.79 H_0: An association does not exist between the ratings of Siskel and Ebert. H_a: An association exists between the ratings of Siskel and Ebert. $\alpha = 0.01$; $\chi^2 = 45.357$; critical value = 13.277; $P < 0.005$ ($P = 0.000$); reject H_0; at the 1% significance level, the data provide sufficient evidence to conclude that an association exists between the ratings of Siskel and Ebert.

13.81
a. No. Assumption 2 fails because 33.3% of the expected frequencies are less than 5.
b. Yes. H_0: Social class and frequency of games are not associated. H_a: Social class and frequency of games are associated. $\alpha = 0.01$; $\chi^2 = 8.715$; critical value = 9.210, $0.01 < P < 0.025$ ($P = 0.013$); do not reject H_0; at the 1% significance level, the data do not provide sufficient evidence to conclude that an association exists between social class and frequency of games.

13.83 H_0: An association does not exist between worry about a gang attack and actually being a victim of a gang attack. H_a: An association exists between worry about a gang attack and actually being a victim of a gang attack. $\alpha = 0.01$; $\chi^2 = 23.455$; critical value = 6.635; $P < 0.005$ ($P = 0.000$); reject H_0; at the 1% significance level, the data provide sufficient evidence to conclude that an association exists between worry about a gang attack and actually being a victim of a gang attack.

13.85
a. H_0: An association does not exist between religiosity and education. H_a: An association exists between religiosity and education. $\alpha = 0.05$; $\chi^2 = 13.322$; critical value = 12.592; $0.025 < P < 0.05$ ($P = 0.038$); reject H_0; at the 5% significance level, the data provide sufficient evidence to conclude that an association exists between religiosity and education.
b. Same null and alternative hypotheses as in part (a). $\alpha = 0.01$; $\chi^2 = 13.322$; critical value = 16.812; $0.025 < P < 0.05$ ($P = 0.038$); do not reject H_0; at the 1% significance level, the data do not provide sufficient evidence to conclude that an association exists between religiosity and education.

13.87 H_0: An association does not exist between gender and response. H_a: An association exists between gender and response. $\alpha = 0.05$; $\chi^2 = 1.350$; $P = 0.509$; do not reject H_0; at the 5% significance level, the data do not provide sufficient evidence to conclude that an association exists between gender and response.

13.89 H_0: An association does not exist between type of community and response. H_a: An association exists between type of community and response. $\alpha = 0.05$; $\chi^2 = 6.952$; $P = 0.138$; do not reject H_0; at the 5% significance level, the data do not provide sufficient evidence to conclude that an association exists between type of community and response.

13.91 H_0: An association does not exist between income and response. H_a: An association exists between income and response. $\alpha = 0.05$; $\chi^2 = 16.634$; $P = 0.011$; reject H_0; at the 5% significance level, the data provide sufficient evidence to conclude that an association exists between income and response.

Exercises 13.5

13.93 The chi-square independence test is used to decide whether two variables of one one population are associated. The chi-square homogeneity test is used to decide whether the distributions of one variable of two or more populations differ.

13.95 When the populations under consideration have the same distribution for the variable, they are said to be *homogeneous* with respect to the variable; otherwise, they are said to be *nonhomogeneous* with respect to the variable.

13.97 proportions   **13.99** 20

13.101

a. H_0: A difference does not exist in self-concept distributions between sighted and blind Indian adolescents. H_a: A difference exists in self-concept distributions between sighted and blind Indian adolescents. $\alpha = 0.05$; $\chi^2 = 6.589$; critical value $= 5.991$; $0.025 < P < 0.05$ ($P = 0.037$); reject H_0; at the 5% significance level, the data provide sufficient evidence to conclude that a difference exists in self-concept distributions between sighted and blind Indian adolescents.

b. Same null and alternative hypotheses as in part (a). $\alpha = 0.01$; $\chi^2 = 6.589$; critical value $= 9.210$; $0.025 < P < 0.05$ ($P = 0.037$); do not reject H_0; at the 1% significance level, the data do not provide sufficient evidence to conclude that a difference exists in self-concept distributions between sighted and blind Indian adolescents.

13.103 H_0: No difference exists in race distributions among the four U.S. regions. H_a: A difference exists in race distributions among the four U.S. regions. $\alpha = 0.01$; $\chi^2 = 26.897$; critical value $= 16.812$; $P < 0.005$ ($P = 0.000$); reject H_0; at the 1% significance level, the data provide sufficient evidence to conclude that a difference exists in race distributions among the four U.S. regions.

13.105 H_0: $p_1 = p_2 = p_3$ (proportions of supporters are equal). H_a: Not all the proportions of supporters are equal. $\alpha = 0.01$; $\chi^2 = 11.520$; critical value $= 9.210$; $P < 0.005$ ($P = 0.003$); reject H_0; at the 1% significance level, the data provide sufficient evidence to conclude that a difference exists in the proportions of supporters among the three regions.

13.107

a. H_0: $p_1 = p_2$, H_a: $p_1 \neq p_2$; $\alpha = 0.05$; $z = -2.11$; critical values $= \pm 1.96$; $P = 0.035$; reject H_0; at the 5% significance level, the data provide sufficient evidence to conclude that a difference exists in the proportions of male and female vegetarians.

b. H_0: $p_1 = p_2$, H_a: $p_1 \neq p_2$; $\alpha = 0.05$; $\chi^2 = 4.446$; critical value $= 3.841$; $0.025 < P < 0.05$ ($P = 0.035$); reject H_0; at the 5% significance level, the data provide sufficient evidence to conclude that a difference exists in the proportions of male and female vegetarians.

c. The results are the same.

d. The chi-square homogeneity test for comparing two population proportions and the two-tailed two-proportions z-test are equivalent.

Review Problems for Chapter 13

1. By their degrees of freedom

2. **a.** 0 **b.** Right skewed **c.** Normal curve

3. **a.** No. The degrees of freedom for the chi-square goodness-of-fit test depends on the number of possible values for the variable under consideration, not on the sample size.
   b. No. The degrees of freedom for the chi-square independence test depends on the number of possible values for the two variables under consideration, not on the sample size.
   c. No. The degrees of freedom for the chi-square homogeneity test depends on the number of populations and the number of possible values for the variable under consideration, not on the sample size.

4. For all three tests, the null hypothesis is rejected only when the observed and expected frequencies match up poorly, which corresponds to large values of the chi-square test statistic. Thus all three tests are always right tailed.

5. 0

6. **a.** (1) All expected frequencies are 1 or greater. (2) At most 20% of the expected frequencies are less than 5.
   b. They are very important. If the assumptions are not met, the results could be invalid.

7. **a.** 18% **b.** Roughly 6.66 million
   c. Ancestry and region of residence are associated.

8. **a.** Obtain the conditional distribution of one of the variables for each possible value of the other variable. If all these conditional distributions are identical, no association exists between the two variables; otherwise, an association exists between the two variables.
   b. No. Because the data are for an entire population, no inference is being made from a sample to the population. The conclusion is a fact.

9. **a.** Perform a chi-square independence test.
   b. Yes. As in any inference, it is always possible that the conclusion is in error.

10. **a.** 24.769 **b.** 33.409 **c.** 27.587

11. H_0: This year's distribution of educational attainment is the same as the 2010 distribution. H_a: This year's distribution of educational attainment differs from the 2010 distribution. Assumptions 1 and 2 are satisfied because all expected frequencies are 5 or greater. $\alpha = 0.05$; $\chi^2 = 9.792$; critical value $= 11.070$; $0.05 < P < 0.10$ ($P = 0.081$); do not reject H_0; at the 5% significance level, the data do not provide sufficient evidence to conclude that this year's distribution of educational attainment differs from the 2010 distribution.

12. **a.** The first 44 presidents of the United States
    b. Region of birth and political party
    c.

	Region				
	NE	MW	SO	WE	Total
Federalist	1	0	1	0	2
DR	1	0	3	0	4
Democratic	7	1	6	1	15
Whig	1	0	3	0	4
Republican	5	10	2	1	18
Union	0	0	1	0	1
Total	15	11	16	2	44

(Party)

13. **a.**

	Region				
	NE	MW	SO	WE	Total
Federalist	0.067	0.000	0.063	0.000	0.046
DR	0.067	0.000	0.188	0.000	0.091
Democratic	0.467	0.091	0.375	0.500	0.341
Whig	0.067	0.000	0.188	0.000	0.091
Republican	0.333	0.909	0.125	0.500	0.409
Union	0.000	0.000	0.063	0.000	0.023
Total	1.000	1.000	1.000	1.000	1.000

(Party)

b. Yes. Because the conditional distributions of party within birth regions are not identical.
c. 40.9% **d.** 40.9% **e.** 12.5%

14. a.

		Region			
Party		NE	MW	SO	WE	Total
	Federalist	0.500	0.000	0.500	0.000	1.000
	DR	0.250	0.000	0.750	0.000	1.000
	Democratic	0.467	0.067	0.400	0.067	1.000
	Whig	0.250	0.000	0.750	0.000	1.000
	Republican	0.278	0.556	0.111	0.056	1.000
	Union	0.000	0.000	1.000	0.000	1.000
	Total	0.341	0.250	0.364	0.046	1.000

b. Yes. Because the conditional distributions of birth region within parties are not identical.
c. 36.4%   **d.** 36.4%   **e.** 11.1%

15. a. 2046   **b.** 737   **c.** 266   **d.** 3046   **e.** 5413   **f.** 5910

16. a.

	Control			
Facility	Gov	Prop	NP	Total
General	0.314	0.122	0.564	1.000
Psychiatric	0.361	0.486	0.153	1.000
Chronic	0.808	0.038	0.154	1.000
Tuberculosis	0.750	0.000	0.250	1.000
Other	0.144	0.361	0.495	1.000

b. Yes. Because the conditional distributions of control type within facility types are not identical.
c. Gov: 0.311; Prop: 0.177; NP: 0.512
d.

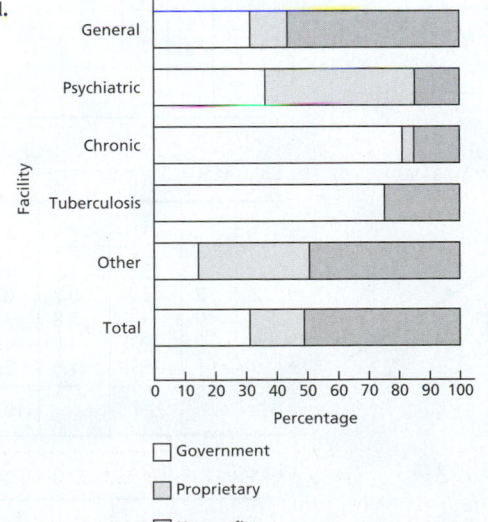

☐ Government
☐ Proprietary
■ Nonprofit

That the bars are not identical reflects the fact that an association exists between facility type and control type.

17. a. False. By Problem 16(b), an association exists between facility type and control type.

b.

	Control		
Facility	Gov	Prop	NP
General	0.829	0.566	0.905
Psychiatric	0.130	0.307	0.034
Chronic	0.010	0.001	0.001
Tuberculosis	0.001	0.000	0.000
Other	0.029	0.127	0.060
Total	1.000	1.000	1.000

c. General: 0.821; Psychiatric: 0.112; Chronic: 0.004; Tuberculosis: 0.001; Other: 0.062

18. a. 17.7%   **b.** 48.6%   **c.** 11.2%   **d.** 30.7%

19. H_0: An association does not exist between age and Internet usage. H_a: An association exists between age and Internet usage. $\alpha = 0.01$; $\chi^2 = 60.746$; critical value = 13.277; $P < 0.005$ ($P = 0.000$); reject H_0; at the 1% significance level, the data provide sufficient evidence to conclude that an association exists between age and Internet usage.

20. a. There are three populations here: People in the United States that reside inside principal cities, outside principal cities but within metropolitan areas, and outside metropolitan areas.
b. Income level
c. H_0: People residing in the three types of residence are homogeneous with respect to income level. H_a: People residing in the three types of residence are nonhomogeneous with respect to income level. $\alpha = 0.05$; $\chi^2 = 15.728$; critical value = 12.592; $0.01 < P < 0.025$ ($P = 0.015$); reject H_0; at the 5% significance level, the data provide sufficient evidence to conclude that people residing in the three types of residence are nonhomogeneous with respect to income level.

21. H_0: $p_1 = p_2 = p_3$ (percentages are equal). H_a: Not all the percentages are equal. $\alpha = 0.01$; $\chi^2 = 164.842$; critical value = 9.210; $P < 0.005$ ($P = 0.000$); reject H_0; at the 1% significance level, the data provide sufficient evidence to conclude that a difference exists in the percentages of registered Democrats, Republicans, and Independents who thought the U.S. economy was in a recession at the time.

Chapter 14

Exercises 14.1

14.1 a. $y = b_0 + b_1 x$
b. b_0 and b_1 represent constants; x and y represent variables.
c. x is the independent variable; y is the dependent variable.

14.3
a. The number b_0 is the y-intercept. It is the y-value of the point of intersection of the line and the y-axis.
b. The number b_1 is the slope. It measures the steepness of the line; more precisely, b_1 indicates how much the y-value changes (increases or decreases) when the x-value increases by 1 unit.

14.5 a. $b_0 = 3, b_1 = 4$ **b.** Slopes upward
c.

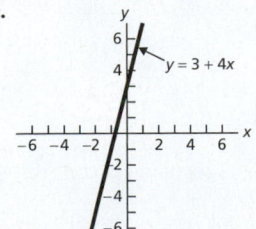

14.7 a. $b_0 = 6, b_1 = -7$ **b.** Slopes downward
c.

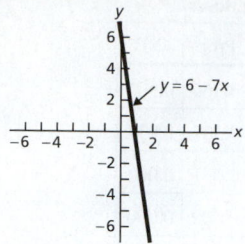
$y = 6 - 7x$

14.9 a. $b_0 = -2, b_1 = 0.5$ **b.** Slopes upward
14.11 a. $b_0 = 2, b_1 = 0$ **b.** Horizontal
14.13 a. $b_0 = 0, b_1 = 1.5$ **b.** Slopes upward
14.15 a. Slopes upward   **b.** $y = 5 + 2x$
c.

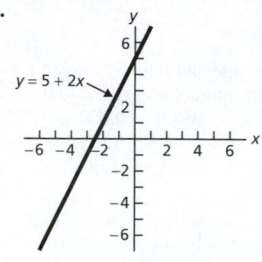
$y = 5 + 2x$

14.17 a. Slopes downward   **b.** $y = -2 - 3x$
c.

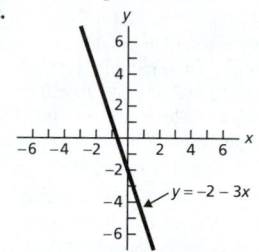
$y = -2 - 3x$

14.19 a. Slopes downward   **b.** $y = -0.5x$
14.21 a. Horizontal   **b.** $y = 3$
14.23 a. $y = 68.22 + 0.25x$ **b.** $b_0 = 68.22, b_1 = 0.25$
c.

x	50	100	250
y	80.72	93.22	130.72

d.

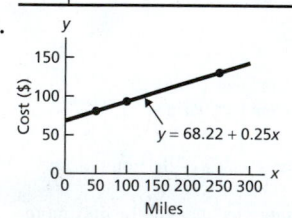
$y = 68.22 + 0.25x$

e. About $105; exact cost is $105.72
14.25 a. $b_0 = 32, b_1 = 1.8$ **b.** $-40, 32, 68, 212$
c.

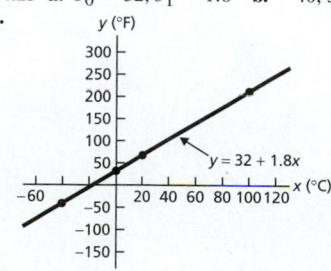
$y = 32 + 1.8x$

d. About $80°$ F; exact temperature is $82.4°$ F

14.27 a. $b_0 = 68.22, b_1 = 0.25$
b. The y-intercept $b_0 = 68.22$ gives the y-value at which the line $y = 68.22 + 0.25x$ intersects the y-axis. The slope $b_1 = 0.25$ indicates that the y-value increases by 0.25 unit for every increase in x of 1 unit.
c. The y-intercept $b_0 = 68.22$ is the cost (in dollars) for driving the car 0 miles. The slope $b_1 = 0.25$ represents the fact that the cost per mile is $0.25; it is the amount the total cost increases for each additional mile driven.

14.29 a. $b_0 = 32, b_1 = 1.8$
b. The y-intercept $b_0 = 32$ gives the y-value at which the line $y = 32 + 1.8x$ intersects the y-axis. The slope $b_1 = 1.8$ indicates that the y-value increases by 1.8 units for every increase in x of 1 unit.
c. The y-intercept $b_0 = 32$ is the Fahrenheit temperature corresponding to $0°$ C. The slope $b_1 = 1.8$ represents the fact that the Fahrenheit temperature increases by $1.8°$ for every increase of the Celsius temperature of $1°$.

Exercises 14.2

14.35 a. Least-squares criterion
b. The line that best fits a set of data points is the one having the smallest possible sum of squared errors.

14.37 a. Response variable   **b.** Predictor variable, or explanatory variable
14.39 a. Outlier   **b.** Influential observation
14.41 Only the second one
14.43 a. Line A: $y = -1 + 3x$ **b.** Line A: $y = -1 + 3x$

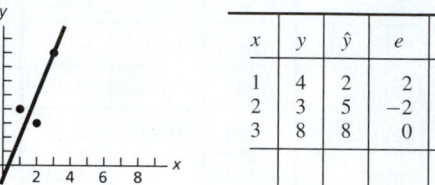

x	y	$\hat{y}$	e	e^2
1	4	2	2	4
2	3	5	-2	4
3	8	8	0	0
				8

Line B: $y = 1 + 2x$ Line B: $y = 1 + 2x$

x	y	$\hat{y}$	e	e^2
1	4	3	1	1
2	3	5	-2	4
3	8	7	1	1
				6

c. Line B

14.45 a. Line A: $y = 3 - 0.6x$ **b.** Line A: $y = 3 - 0.6x$

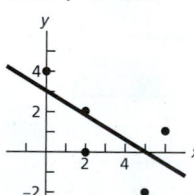

x	y	$\hat{y}$	e	e^2
0	4	3.0	1.0	1.00
2	2	1.8	0.2	0.04
2	0	1.8	-1.8	3.24
5	-2	0.0	-2.0	4.00
6	1	-0.6	1.6	2.56
				10.84

Line B: $y = 4 - x$ Line B: $y = 4 - x$

x	y	$\hat{y}$	e	e^2
0	4	4	0	0
2	2	2	0	0
2	0	2	-2	4
5	-2	-1	-1	1
6	1	-2	3	9
				14

c. Line A

14.47 a. $\hat{y} = -1 + x$ **b.** $\hat{y} = 4.5 - 1.5x$

14.49 a. $\hat{y} = 1 - 2x$
b.

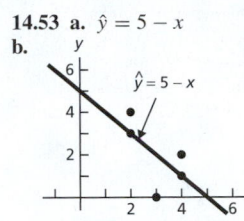

14.51 a. $\hat{y} = -3 + 2x$
b.

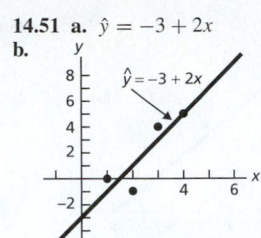

14.53 a. $\hat{y} = 5 - x$
b.

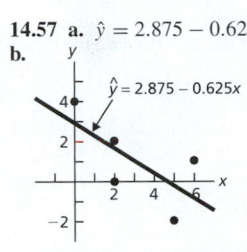

14.55 a. $\hat{y} = 1 + 2x$
b.

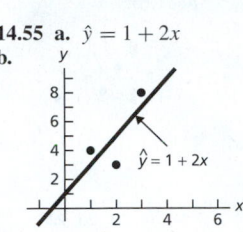

14.57 a. $\hat{y} = 2.875 - 0.625x$
b.

Note: Recall the second bulleted item on page A-23.

14.59 a. $\hat{y} = 456.6 - 27.9x$ **b.**

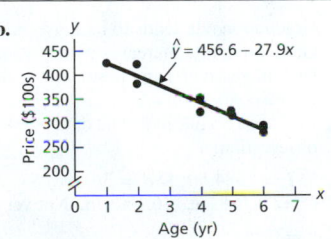

c. Price tends to decrease as age increases.
d. Corvettes depreciate an estimated \$2790 per year, at least in the 1- to 6-year-old range.
e. The predictor variable is age (in years); the response variable is price (in hundreds of dollars).
f. None **g.** \$40,080; \$37,289

14.61 a. $\hat{y} = 3.52 + 0.16x$ **b.**

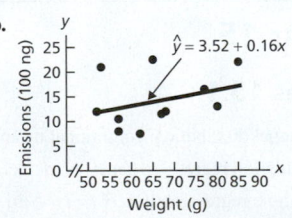

c. Quantity of volatile compounds emitted tends to increase as potato plant weight increases.
d. The quantity of volatile compounds emitted increases an estimated 16 nanograms for each increase in potato plant weight of 1 g.
e. The predictor variable is potato plant weight (in grams); the response variable is quantity of volatile compounds emitted (in hundreds of nanograms).
f. None **g.** 1574 nanograms

14.63 a. $\hat{y} = 94.9 - 0.8x$ **b.**

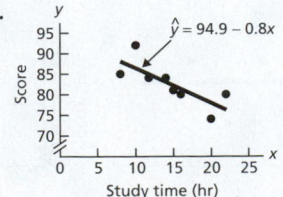

c. Test score in beginning calculus courses tends to decrease as study time increases.
d. Test score in beginning calculus courses decreases an estimated 0.8 point for each increase in study time of 1 hour.
e. The predictor variable is study time (in hours); the response variable is test score.
f. None **g.** 82.2 points

14.65
a. It is acceptable to use the regression equation to predict the price of a 4-year-old Corvette because that age lies within the range of ages in the sample data. It is not acceptable (and would be extrapolation) to use the regression equation to predict the price of a 10-year-old Corvette because that age lies outside the range of the ages in the sample data.
b. Ages between 1 and 6 years, inclusive

14.67 Answers will vary. One possible explanation is that students with an aptitude for calculus will not need to study as long to master the material.

14.69 a.

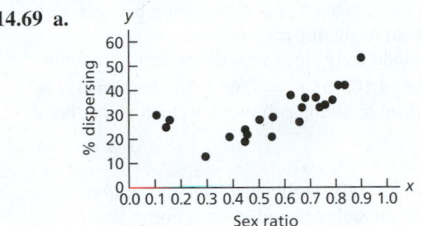

b. No, because the data points are scattered about a curve, not a line.

Exercises 14.3

14.83 a. The coefficient of determination, r^2
b. The proportion of variation in the observed values of the response variable explained by the regression

14.85 regression sum of squares; *SSR*

14.87
a. $r^2 = 0.920$; 92.0% of the variation in the observed values of the response variable is explained by the regression. The fact that r^2 is near 1 indicates that the regression equation is extremely useful for making predictions.
b. 664.4

14.89 a. $SST = 14$, $SSR = 8$, $SSE = 6$ **b.** $14 = 8 + 6$ **c.** $r^2 = 0.571$
d. 57.1% **e.** Moderately useful

14.91 a. $SST = 26$, $SSR = 20$, $SSE = 6$ **b.** $26 = 20 + 6$ **c.** $r^2 = 0.769$
d. 76.9% **e.** Useful

14.93 a. $SST = 10$, $SSR = 4$, $SSE = 6$ **b.** $10 = 4 + 6$ **c.** $r^2 = 0.4$
d. 40% **e.** Moderately useful

14.95 a. $SST = 14$, $SSR = 8$, $SSE = 6$ **b.** $14 = 8 + 6$ **c.** $r^2 = 0.571$
d. 57.1% **e.** Moderately useful

14.97 a. $SST = 20$, $SSR = 9.375$, $SSE = 10.625$ **b.** $20 = 9.375 + 10.625$
c. $r^2 = 0.469$ **d.** 46.9% **e.** Moderately useful

14.99 a. $SST = 25{,}681.6$, $SSR = 24{,}057.9$, $SSE = 1623.7$ **b.** 0.937
c. 93.7%; 93.7% of the variation in the price data is explained by age.
d. Extremely useful

14.101 a. $SST = 296.68$, $SSR = 32.52$, $SSE = 264.16$ **b.** 0.110
c. 11.0%; 11.0% of the variation in the quantity of volatile emissions is explained by potato plant weight.
d. Not very useful

14.103 a. $SST = 188.0$, $SSR = 112.9$, $SSE = 75.1$ **b.** 0.600
c. 60.0%; 60.0% of the variation in the score data is explained by study time.
d. Useful

Exercises 14.4

14.119 Pearson product moment correlation coefficient

14.121 strong **14.123** ± 1 **14.125** positively

14.127 uncorrelated **14.129** negative

14.131 False. Correlation does not imply causation.

14.133 $r = -0.842$ **14.135 a.** $r = -0.756$ **b.** $r = -0.756$

14.137 a. $r = 0.877$ **b.** $r = 0.877$

14.139 a. $r = -0.632$ **b.** $r = -0.632$

14.141 a. $r = 0.756$ **b.** $r = 0.756$

14.143 a. $r = -0.685$ **b.** $r = -0.685$

14.145 a. $r = -0.968$
b. Suggests an extremely strong negative linear relationship between age and price of Corvettes.
c. Data points are clustered closely about the regression line.
d. $r^2 = 0.937$. This value of r^2 is the same as the one obtained in Exercise 14.99(b).

14.147 a. $r = 0.331$
b. Suggests a weak positive linear relationship between potato plant weight and quantity of volatile emissions.
c. Data points are scattered widely about the regression line.
d. $r^2 = 0.110$. This value of r^2 is the same as the one obtained in Exercise 14.101(b).

14.149 a. $r = -0.775$
b. Suggests a moderately strong negative linear relationship between study time and score for students in beginning calculus courses.
c. Data points are clustered moderately closely about the regression line.
d. $r^2 = 0.601$. From Exercise 14.103(b), $r^2 = 0.600$. The discrepancy is due to the error resulting from rounding r to three decimal places before squaring.

14.151 a. $r = 0$
b. No. Only that there is no *linear* relationship between the variables
d. No. Because the data points are not scattered about a line
e. For each data point (x, y), the relation $y = x^2$ holds.

14.153 a. $r = 0.961$
b. No. A linear correlation coefficient close to ± 1 does not necessarily imply a linear relationship between two variables (unless the data points are scattered about a line, in which case, indications are that there is a strong linear relationship between the variables).
d. No, because the data points are not scattered about a line.
e. For each data point (x, y), the relation $y = x^2$ holds.

Review Problems for Chapter 14

1. a. x **b.** y **c.** b_1 **d.** b_0

2. a. $y = 4$ **b.** $x = 0$ **c.** -3 **d.** -3 units **e.** 6 units

3. True. The y-intercept indicates only where the line crosses the y-axis; that is, it is the y-value when $x = 0$.

4. False. Its slope is 0.

5. True. The statement is equivalent to saying: If a line has a positive slope, then y-values on the line increase as the x-values increase.

6. Scatterplot

7. Within the range of the observed values of the predictor variable, we can use the regression equation to make predictions for the response variable.

8. a. Predictor variable, or explanatory variable **b.** Response variable

9. smallest **10.** regression **11.** extrapolation

12. a. An outlier is a data point that lies far from the regression line, relative to the other data points.
b. An influential observation is a data point whose removal causes the regression equation (and regression line) to change considerably.

13. It is a descriptive measure of the utility of the regression equation for making predictions.

14. a. SST is the total sum of squares. It measures the variation in the observed values of the response variable.
b. SSR is the regression sum of squares. It measures the variation in the observed values of the response variable explained by the regression.
c. SSE is the error sum of squares. It measures the variation in the observed values of the response variable not explained by the regression.

15. linear **16.** increases **17.** negative **18.** 0 **19.** True

20. a. $y = 72 - 12x$ **b.** $b_0 = 72, b_1 = -12$
c. The line slopes downward because $b_1 < 0$. **d.** $4800; $1200

e.
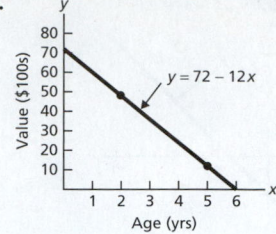

f. About $2500; exact value is $2400.

21. a.
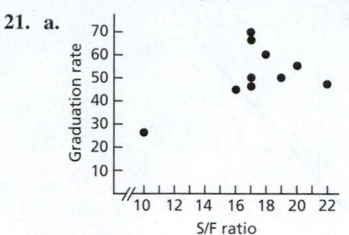

b. It is reasonable to find a regression line for the data because the data points appear to be scattered about a line.
c. $\hat{y} = 16.4 + 2.03x$

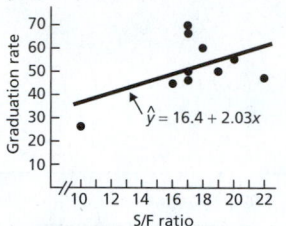

d. Graduation rate tends to increase as student-to-faculty ratio increases.
e. Graduation rate increases by an estimated 2.03 percentage points for each increase of 1 in the student-to-faculty ratio.
f. 50.9%
g. There are no outliers. The data point (10, 26) is a potential influential observation.

22. a. $SST = 1384.50$; $SSR = 361.66$; $SSE = 1022.84$
b. $r^2 = 0.261$ **c.** 26.1% **d.** Not very useful

23. a. $r = 0.511$
b. Suggests a moderately weak positive linear relationship between student-to-faculty ratio and graduation rate.
c. Data points are rather widely scattered about the regression line.
d. $r^2 = (0.511)^2 = 0.261$

Chapter 15

Exercises 15.1

15.1 Conditional distribution, conditional mean, conditional standard deviation

15.3 a. Population regression line **b.** σ **c.** Normal; $\beta_0 + 6\beta_1$; σ

15.5 The sample regression line, $\hat{y} = b_0 + b_1 x$ **15.7** Residual

15.9 A residual plot, that is, a plot of the residuals against the values of the predictor variable. A residual plot makes it easier to spot patterns such as curvature and nonconstant standard deviation than does a scatterplot.

15.11 Part (a) is a tough call, but the assumption of linearity (Assumption 1) may be violated, as may be the assumption of equal standard deviations (Assumption 2). In part (b), it appears that the assumption of equal standard deviations (Assumption 2) is violated. In part (d), it appears that the normality assumption (Assumption 3) is violated. Part (c) shows no violations.

15.13 a. 2.45
b.   **c.**
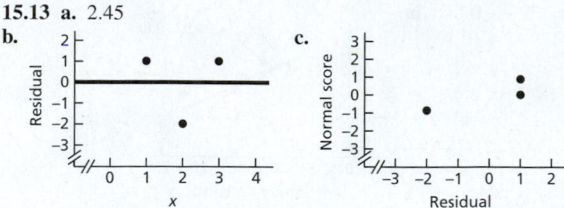

15.15 a. 1.73

b. **c.**

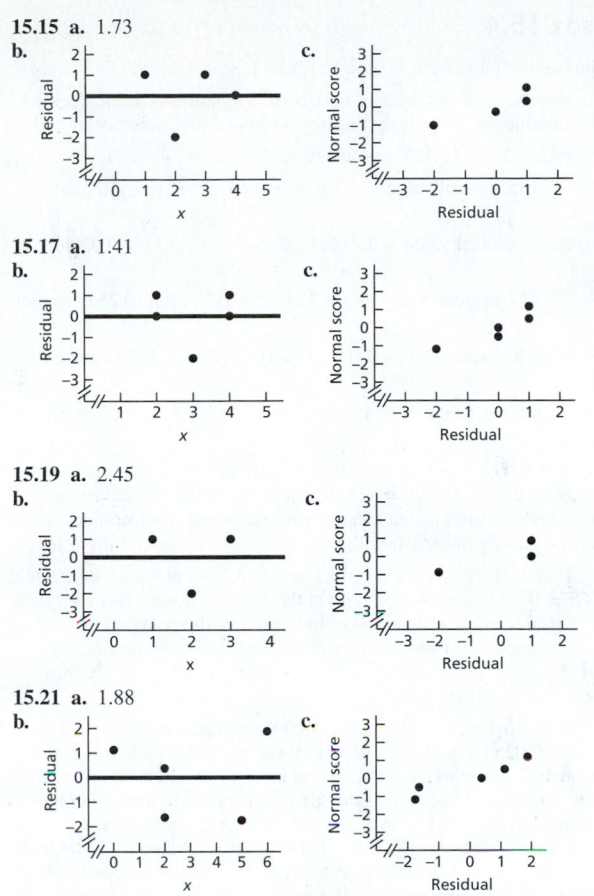

15.17 a. 1.41

b. **c.**

15.19 a. 2.45

b. **c.**

15.21 a. 1.88

b. **c.**

15.23 There are constants, β_0, β_1, and σ, such that, for each age, x, the prices of all Corvettes of that age are normally distributed with mean $\beta_0 + \beta_1 x$ and standard deviation σ.

15.25 There are constants, β_0, β_1, and σ, such that, for each weight, x, the quantities of volatile compounds emitted by all potato plants of that weight are normally distributed with mean $\beta_0 + \beta_1 x$ and standard deviation σ.

15.27 There are constants, β_0, β_1, and σ, such that, for each total number of hours studied, x, the test scores of all students in beginning calculus courses who study that number of hours are normally distributed with mean $\beta_0 + \beta_1 x$ and standard deviation σ.

15.29

a. $s_e = 14.25$; very roughly speaking, on average, the predicted price of a Corvette in the sample differs from the observed price by about \$1425.

b. Presuming that, for Corvettes, the variables age (x) and price (y) satisfy the assumptions for regression inferences, the standard error of the estimate, $s_e = 14.25$, provides an estimate for the common population standard deviation, σ, of prices (in hundreds of dollars) for all Corvettes of any particular age.

c. See Fig. A.4.   **d.** It appears reasonable.

FIGURE A.4

(a) Residual plot and (b) normal probability plot of residuals for Exercise 15.29(c)

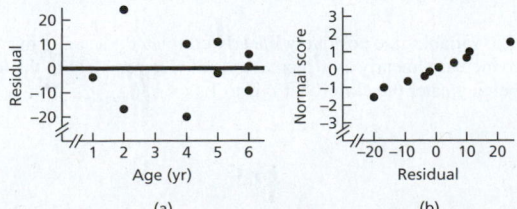

(a) (b)

15.31

a. $s_e = 5.42$; very roughly speaking, on average, the predicted quantity of volatile compounds emitted by a potato plant in the sample differs from the observed quantity by about 542 nanograms.

b. Presuming that, for potato plants, the variables weight (x) and quantity of volatile compounds emitted (y) satisfy the assumptions for regression

inferences, the standard error of the estimate, $s_e = 5.42$, provides an estimate for the common population standard deviation, σ, of quantities of volatile compounds emitted (in hundreds of nanograms) for all potato plants of any particular weight.

c. See Fig. A.6.

FIGURE A.6

(a) Residual plot and (b) normal probability plot of residuals for Exercise 15.31(c)

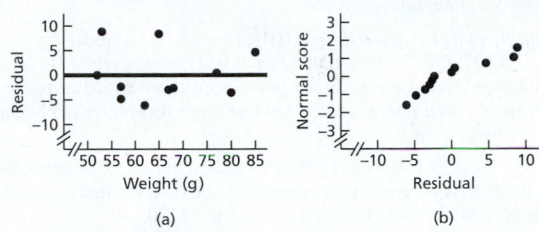

(a) (b)

d. Although Fig. A.6(b) shows some curvature, it is probably not sufficiently curved to call into question the validity of the normality assumption (Assumption 3).

15.33

a. $s_e = 3.54$; very roughly speaking, on average, the predicted test score of a student in the sample differs from the observed score by about 3.54 points.

b. Presuming that, for students in beginning calculus courses, the variables study time (x) and test score (y) satisfy the assumptions for regression inferences, the standard error of the estimate, $s_e = 3.54$, provides an estimate for the common population standard deviation, σ, of test scores for all students who study for any particular amount of time.

c. See Fig. A.8.   **d.** It appears reasonable.

FIGURE A.8

(a) Residual plot and (b) normal probability plot of residuals for Exercise 15.33(c)

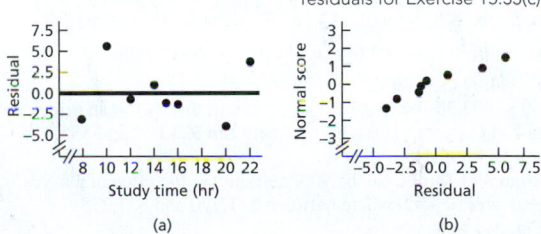

(a) (b)

Exercises 15.2

15.45 normal, -3.5

15.47 r^2, r

Note: In each of Exercises 15.49–15.57, the null hypothesis is that x is not useful for predicting y and the alternative hypothesis is that x is useful for predicting y.

15.49

a. $t = -1.155$; critical values $= \pm 6.314$; $P > 0.20$ ($P = 0.454$); do not reject H_0

b. -12.94 to 8.94

15.51

a. $t = 2.582$; critical values $= \pm 2.920$; $0.10 < P < 0.20$ ($P = 0.123$); do not reject H_0

b. -0.26 to 4.26

15.53

a. $t = -1.414$; critical values $= \pm 2.353$; $P > 0.20$ ($P = 0.252$); do not reject H_0

b. -2.66 to 0.66

15.55

a. $t = 1.155$; critical values $= \pm 6.314$; $P > 0.20$ ($P = 0.454$); do not reject H_0

b. -8.94 to 12.94

15.57

a. $t = -1.627$; critical values $= \pm 2.353$; $P > 0.20$ ($P = 0.202$); do not reject H_0

b. -1.53 to 0.28

15.59 $H_0: \beta_1 = 0$, $H_a: \beta_1 \neq 0$; $\alpha = 0.10$; $t = -10.887$; critical values $= \pm 1.860$; $P < 0.01$ ($P = 0.000$); reject H_0; at the 10% significance level, the data provide sufficient evidence to conclude that age is useful as a predictor of price for Corvettes.

15.61 $H_0: \beta_1 = 0$, $H_a: \beta_1 \neq 0$; $\alpha = 0.05$; $t = 1.053$; critical values $= \pm 2.262$; $P > 0.20$ ($P = 0.320$); do not reject H_0; at the 5% significance level, the data do not provide sufficient evidence to conclude that weight is useful as a predictor of quantity of volatile emissions for the potato plant *Solanum tuberosum*.

15.63 $H_0: \beta_1 = 0$, $H_a: \beta_1 \neq 0$; $\alpha = 0.01$; $t = -3.003$; critical values $= \pm 3.707$; $0.02 < P < 0.05$ ($P = 0.024$); do not reject H_0; at the 1% significance level, the data do not provide sufficient evidence to conclude that study time is useful as a predictor of test score for students in beginning calculus courses.

15.65 -32.7 to -23.1. We can be 90% confident that, for Corvettes, the decrease in mean price per 1-year increase in age (i.e., the mean annual depreciation) is somewhere between \$2310 and \$3270.

15.67 -0.19 to 0.51. We can be 95% confident that, for the potato plant *Solanum tuberosum*, the change in the mean quantity of volatile emissions per 1 g increase of weight is somewhere between -19 ng and 51 ng.

15.69 -1.89 to 0.20. We can be 99% confident that, for students in beginning calculus courses, the change in mean test score per increase of 1 hour studied is somewhere between -1.89 points and 0.20 points.

Exercises 15.3

15.81 \$11,443. A point estimate for the mean price is the same as the predicted price.

15.83 **a.** -3 **b.** -20.97 to 14.97 **c.** -3 **d.** -38.94 to 32.94

15.85 **a.** 5 **b.** -1.24 to 11.24 **c.** 5 **d.** -4.72 to 14.72

15.87 **a.** 2 **b.** -0.01 to 4.01 **c.** 2 **d.** -2.93 to 6.93

15.89 **a.** 5 **b.** -12.97 to 22.97 **c.** 5 **d.** -30.94 to 40.94

15.91 **a.** 1 **b.** -1.68 to 3.68 **c.** 1 **d.** -5.56 to 7.56

15.93 **a.** 344.99 (\$34,499)
b. 336.60 to 353.38. We can be 90% confident that the mean price of all 4-year-old Corvettes is somewhere between \$33,660 and \$35,338.
c. 344.99 (\$34,499)
d. 317.20 to 372.78. We can be 90% certain that the price of a 4-year-old Corvette will be somewhere between \$31,720 and \$37,278.
e. See Fig. A.10.
f. The error in the estimate of the mean price of all 4-year-old Corvettes is due only to the fact that the population regression line is being estimated by a sample regression line. In contrast, the error in the prediction of the price of a 4-year-old Corvette is due to the error in estimating the mean price plus the variation in prices of 4-year-old Corvettes.

15.95 **a.** 13.29 (1329 ng)
b. 9.09 to 17.50. We can be 95% confident that the mean quantity of volatile emissions of all plants that weigh 60 g is somewhere between 909 ng and 1750 ng.
c. 13.29 (1329 ng)
d. 0.34 to 26.25. We can be 95% certain that the quantity of volatile emissions of a plant that weighs 60 g will be somewhere between 34 ng and 2625 ng.

15.97 **a.** 82.2 points
b. 77.5 to 86.8. We can be 99% confident that the mean test score of all beginning calculus students who study for 15 hours is somewhere between 77.5 points and 86.8 points.
c. 82.2 points
d. 68.3 to 96.1. We can be 99% certain that the test score of a beginning calculus student who studies for 15 hours will be somewhere between 68.3 points and 96.1 points.

Exercises 15.4

15.111 The (sample) linear correlation coefficient, r

15.113 No. However, we can use r as a basis for a hypothesis test to decide whether the variables are negatively linearly correlated (i.e., whether $\rho < 0$).

15.115 uncorrelated　　**15.117** negatively

15.119 $t = -1.155$; critical value $= -3.078$; $P > 0.10$ ($P = 0.227$); do not reject H_0

15.121 $t = 2.582$; critical value $= 1.886$; $0.05 < P < 0.10$ ($P = 0.061$); reject H_0

15.123 $t = -1.414$; critical values $= \pm 2.353$; $P > 0.20$ ($P = 0.252$); do not reject H_0

15.125 $t = 1.155$; critical value $= 3.078$; $P > 0.10$ ($P = 0.227$); do not reject H_0

15.127 $t = -1.627$; critical values $= \pm 2.353$; $P > 0.20$ ($P = 0.202$); do not reject H_0

15.129 $H_0: \rho = 0$, $H_a: \rho < 0$; $\alpha = 0.05$; $t = -10.887$; critical value $= -1.860$; $P < 0.005$ ($P = 0.000$); reject H_0; at the 5% significance level, the data provide sufficient evidence to conclude that, for Corvettes, age and price are negatively linearly correlated.

15.131 $H_0: \rho = 0$, $H_a: \rho \neq 0$; $\alpha = 0.05$; $t = 1.053$; critical values $= \pm 2.262$; $P > 0.20$ ($P = 0.320$); do not reject H_0; at the 5% significance level, the data do not provide sufficient evidence to conclude that, for the potato plant *Solanum tuberosum*, weight and quantity of volatile emissions are linearly correlated.

15.133
a. $H_0: \rho = 0$, $H_a: \rho < 0$; $\alpha = 0.01$; $t = -3.003$; critical value $= -3.143$; $0.01 < P < 0.025$ ($P = 0.012$); do not reject H_0; at the 1% significance level, the data do not provide sufficient evidence to conclude that a negative linear correlation exists between study time and test score for beginning calculus students.
b. $H_0: \rho = 0$, $H_a: \rho < 0$; $\alpha = 0.05$; $t = -3.003$; critical value $= -1.943$; $0.01 < P < 0.025$ ($P = 0.012$); reject H_0; at the 5% significance level, the data provide sufficient evidence to conclude that a negative linear correlation exists between study time and test score for beginning calculus students.

Review Problems for Chapter 15

1. **a.** conditional **b.** See Key Fact 15.1 on page 659.

2. **a.** b_1 **b.** b_0 **c.** s_e

3. A residual plot (i.e., a plot of the residuals against the observed values of the predictor variable) and a normal probability plot of the residuals. A plot of the residuals against the observed values of the predictor variable should fall roughly in a horizontal band centered and symmetric about the x-axis. A normal probability plot of the residuals should be roughly linear.

4. **a.** Assumption 1 **b.** Assumption 2
   c. Assumption 3 **d.** Assumption 3

5. The regression equation is useful for making predictions.

6. b_1, r, r^2

7. No. Both equal the number obtained by substituting the specified value of the predictor variable into the sample regression equation.

8. The term *confidence* is usually reserved for interval estimates of parameters, whereas the term *prediction* is used for interval estimates of variables.

9. ρ

10. **a.** The variables are positively linearly correlated, meaning that y tends to increase linearly as x increases (and vice versa), with the tendency being greater the closer that ρ is to 1.

FIGURE A.10 90% confidence and prediction intervals for Exercise 15.93(e)

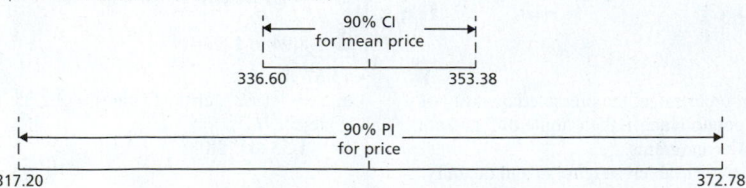

b. The variables are linearly uncorrelated, meaning that there is no linear relationship between the variables.

c. The variables are negatively linearly correlated, meaning that y tends to decrease linearly as x increases (and vice versa), with the tendency being greater the closer that ρ is to -1.

11. There are constants, β_0, β_1, and σ, such that, for each student-to-faculty ratio, x, the graduation rates for all universities with that student-to-faculty ratio are normally distributed with mean $\beta_0 + \beta_1 x$ and standard deviation σ.

12. a. $\hat{y} = 16.4 + 2.03x$

b. $s_e = 11.31\%$; very roughly speaking, on average, the predicted graduation rate for a university in the sample differs from the observed graduation rate by about 11.31 percentage points.

c. Presuming that, for universities, the variables student-to-faculty ratio (x) and graduation rate (y) satisfy the assumptions for regression inferences, the standard error of the estimate, $s_e = 11.31\%$, provides an estimate for the common population standard deviation, σ, of graduation rates for all universities with any particular student-to-faculty ratio.

13.

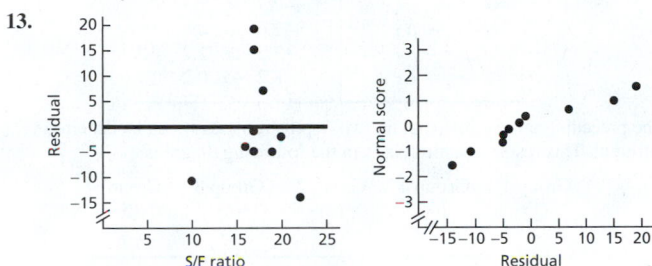

It appears reasonable.

14. a. H_0: $\beta_1 = 0$, H_a: $\beta_1 \neq 0$; $\alpha = 0.05$; $t = 1.682$; critical values $= \pm 2.306$; $0.10 < P < 0.20$ ($P = 0.131$); do not reject H_0; at the 5% significance level, the data do not provide sufficient evidence to conclude that, for universities, student-to-faculty ratio is useful as a predictor of graduation rate.

b. -0.75% to 4.80%. We can be 95% confident that, for universities, the change in mean graduation rate per increase by 1 in the student-to-faculty ratio is somewhere between -0.75 and 4.80 percentage points.

15. a. 50.9%

b. 42.6% to 59.2%. We can be 95% confident that the mean graduation rate of all universities that have a student-to-faculty ratio of 17 is somewhere between 42.6% and 59.2%.

16. a. 50.9%

b. 23.5% to 78.3%. We can be 95% certain that the observed graduation rate of a university that has a student-to-faculty ratio of 17 will be somewhere between 23.5% and 78.3%.

c. The error in the estimate of the mean graduation rate of all universities that have a student-to-faculty ratio of 17 is due only to the fact that the population regression line is being estimated by a sample regression line, whereas the error in the prediction of the observed graduation rate of a university that has a student-to-faculty ratio of 17 is due to the error in estimating the mean graduation rate plus the variation in graduation rates of universities that have a student-to-faculty ratio of 17.

17. H_0: $\rho = 0$, H_a: $\rho > 0$; $\alpha = 0.025$; $t = 1.682$; critical value $= 2.306$; $0.05 < P < 0.10$ ($P = 0.066$); do not reject H_0; at the 2.5% significance level, the data do not provide sufficient evidence to conclude that, for universities, the variables student-to-faculty ratio and graduation rate are positively linearly correlated.

Chapter 16

Exercises 16.1

16.1 By stating its two numbers of degrees of freedom

16.3 $F_{0.05}$, $F_{0.025}$, F_α **16.5 a.** 12 **b.** 7

16.7 a. 1.89 **b.** 2.47 **c.** 2.14   **16.9 a.** 2.88 **b.** 2.10 **c.** 1.78

Exercises 16.2

16.11 The pooled t-procedure of Section 10.2

16.13 The procedure for comparing the means analyzes the variation in the sample data.

16.15 treatment mean square; $MSTR$ **16.17** MSE; F-statistic

16.19 It signifies that the ANOVA compares the means of a variable for populations that result from a classification by *one* other variable (called the *factor*).

16.21 No. Because the variation among the sample means is not large relative to the variation within the samples

16.23 The difference between the observation and the mean of the sample containing it

16.25 a. 24 **b.** 12 **c.** 16 **d.** 2.29 **e.** 5.25

16.27 a. 36 **b.** 9 **c.** 52 **d.** 3.47 **e.** 2.60

16.29 a. 138 **b.** 46 **c.** 72 **d.** 9 **e.** 5.11

Exercises 16.3

16.33 A small value of F results when $SSTR$ is small relative to SSE, that is, when the variation among sample means is small relative to the variation within samples. This result describes what is expected when the null hypothesis is true; thus it doesn't constitute evidence against the null hypothesis. Only when the variation among sample means is large relative to the variation within samples (i.e., only when F is large), is there evidence that the null hypothesis is false.

16.35 $SST = SSTR + SSE$. The total variation among all the sample data can be partitioned into a component representing variation among the sample means and a component representing variation within the samples.

16.37 a. One-way ANOVA   **b.** Two-way ANOVA

16.39 The missing entries are as follows: In the first row, it is 3; in the second row, they are 18.880 and 0.944; and in the third row, they are 23 and 21.004.

16.41 The missing entries are as follows: In the first row, they are 2, 2.8, and 1.56; in the second row, it is 10.8.

16.43 a. 40, 24, 16   **b.** They are the same.

c.

Source	df	SS	MS	F
Treatment	2	24	12	5.25
Error	7	16	2.29	
Total	9	40		

d. H_0: $\mu_1 = \mu_2 = \mu_3$, H_a: Not all the means are equal. $\alpha = 0.05$; $F = 5.25$; critical value $= 4.74$; $0.025 < P < 0.05$ ($P = 0.040$); reject H_0.

16.45 a. 88, 36, 52   **b.** They are the same.

c.

Source	df	SS	MS	F
Treatment	4	36	9	2.60
Error	15	52	3.47	
Total	19	88		

d. H_0: $\mu_1 = \mu_2 = \mu_3 = \mu_4 = \mu_5$, H_a: Not all the means are equal. $\alpha = 0.05$; $F = 2.60$; critical value $= 3.06$; $0.05 < P < 0.10$ ($P = 0.079$); do not reject H_0.

16.47 a. 210, 138, 72   **b.** They are the same.

c.

Source	df	SS	MS	F
Treatment	3	138	46	5.11
Error	8	72	9	
Total	11	210		

d. H_0: $\mu_1 = \mu_2 = \mu_3 = \mu_4$, H_a: Not all the means are equal. $\alpha = 0.05$; $F = 5.11$; critical value $= 4.07$; $0.025 < P < 0.05$ ($P = 0.029$); reject H_0.

16.49 H_0: $\mu_1 = \mu_2 = \mu_3$, H_a: Not all the means are equal. $\alpha = 0.05$; $F = 54.58$; critical value $= 4.26$; $P < 0.005$ ($P = 0.000$); reject H_0; at the 5% significance level, the data provide sufficient evidence to conclude that a difference exists in mean number of copepods among the three different diets.

16.51 H_0: $\mu_1 = \mu_2 = \mu_3 = \mu_4 = \mu_5$, H_a: Not all the means are equal. $\alpha = 0.05$; $F = 2.23$; critical value = 2.87; $P > 0.10$ ($P = 0.103$); do not reject H_0; at the 5% significance level, the data do not provide sufficient evidence to conclude that a difference exists in mean bacteria counts among the five strains of *Staphylococcus aureus*.

16.53 H_0: $\mu_1 = \mu_2 = \mu_3 = \mu_4$, H_a: Not all the means are equal. $\alpha = 0.05$; $F = 6.60$; critical value = 2.92; $P < 0.005$ ($P = 0.001$); reject H_0; at the 5% significance level, the data provide sufficient evidence to conclude that a difference exists in mean battery life among the four brands.

16.55
a. H_0: $\mu_1 = \mu_2 = \mu_3 = \mu_4$, H_a: Not all the means are equal. $\alpha = 0.05$; $F = 7.54$; $P = 0.002$; reject H_0.
b. At the 5% significance level, the data provide sufficient evidence to conclude that a difference exists in mean monthly rents among newly completed apartments in the four U.S. regions.
c. It appears reasonable to presume that the assumptions of normal populations and equal population standard deviations are both met.

16.57
a. H_0: $\mu_1 = \mu_2 = \mu_3$, H_a: Not all the means are equal. $\alpha = 0.01$; $F = 6.09$; $P = 0.008$; reject H_0.
b. At the 1% significance level, the data provide sufficient evidence to conclude that a difference exists in the mean singing rates among male Rock Sparrows exposed to the three types of breast treatments.
c. It appears reasonable to presume that the assumption of normal populations is met, but the assumption of equal population standard deviations appears to be violated.

16.59
a. H_0: $\mu_1 = \mu_2 = \mu_3$, H_a: Not all the means are equal. $\alpha = 0.05$; $F = 114.71$; $P = 0.000$; reject H_0.
b. At the 5% significance level, the data provide sufficient evidence to conclude that there is a difference in mean hardness among the three materials.
c. The assumptions of normal populations and equal population standard deviations both appear to be violated.

16.61 H_0: $\mu_1 = \mu_2 = \mu_3 = \mu_4 = \mu_5$, H_a: Not all the means are equal. $\alpha = 0.10$; $F = 1.01$; critical value = 1.97; $P > 0.10$ ($P = 0.401$); do not reject H_0; at the 10% significance level, the data do not provide sufficient evidence to conclude that a difference exists in mean BMI for women in the five different treatment groups.

16.63 H_0: $\mu_1 = \mu_2 = \mu_3 = \mu_4 = \mu_5 = \mu_6$, H_a: Not all the means are equal. $\alpha = 0.01$; $F = 67.82$; critical value = 3.14; $P < 0.005$ ($P = 0.000$); reject H_0; at the 1% significance level, the data provide sufficient evidence to conclude that a difference exists in mean starting salaries among bachelor's-degree candidates in the six fields.

Exercises 16.4

16.77 0

16.79
a. The family confidence level. Because the family confidence level is the confidence we have that all the confidence intervals contain the differences between the corresponding population means, whereas the individual confidence level is the confidence we have that any particular confidence interval contains the difference between the corresponding population means.
b. They are identical.

16.81 Degrees of freedom for the denominator

16.83 a. 4.69 **b.** 5.98   **16.85 a.** 5.65 **b.** 4.72

16.87 No two population means will be declared different.

16.89 Family confidence level = 0.95; $q_{0.05} = 4.16$; simultaneous 95% confidence intervals are as follows.

Means difference	Confidence interval
$\mu_1 - \mu_2$	0.4 to 7.6
$\mu_1 - \mu_3$	−1.4 to 5.4
$\mu_2 - \mu_3$	−5.4 to 1.4

The preceding table shows that only μ_1 and μ_2 can be declared different. This result is summarized in the following diagram.

Group 2 (2)	Group 3 (3)	Group 1 (1)
2	4	6

Interpreting this diagram, we conclude with 95% confidence that the mean of Population 1 exceeds the mean of Population 2; no other population means can be declared different.

16.91 Family confidence level = 0.95; $q_{0.05} = 4.37$; simultaneous 95% confidence intervals are as follows.

Means difference	Confidence interval
$\mu_1 - \mu_2$	−5.4 to 3.4
$\mu_1 - \mu_3$	−4.9 to 2.9
$\mu_1 - \mu_4$	−3.9 to 3.9
$\mu_1 - \mu_5$	−8.4 to 0.4
$\mu_2 - \mu_3$	−4.2 to 4.2
$\mu_2 - \mu_4$	−3.2 to 5.2
$\mu_2 - \mu_5$	−7.7 to 1.7
$\mu_3 - \mu_4$	−2.6 to 4.6
$\mu_3 - \mu_5$	−7.2 to 1.2
$\mu_4 - \mu_5$	−8.2 to 0.2

The preceding table shows that no two population means can be declared different. This result is summarized in the following diagram.

Group 1 (1)	Group 4 (4)	Group 2 (2)	Group 3 (3)	Group 5 (5)
5	5	6	6	9

16.93 Family confidence level = 0.95; $q_{0.05} = 4.53$; simultaneous 95% confidence intervals are as follows.

Means difference	Confidence interval
$\mu_1 - \mu_2$	−4.8 to 10.8
$\mu_1 - \mu_3$	−11.8 to 3.8
$\mu_1 - \mu_4$	−2.8 to 12.8
$\mu_2 - \mu_3$	−14.8 to 0.8
$\mu_2 - \mu_4$	−5.8 to 9.8
$\mu_3 - \mu_4$	1.2 to 16.8

The preceding table shows that only μ_3 and μ_4 can be declared different. This result is summarized in the following diagram.

Group 4 (4)	Group 2 (2)	Group 1 (1)	Group 3 (3)
3	5	8	12

Interpreting this diagram, we conclude with 95% confidence that the mean of Population 3 exceeds the mean of Population 4; no other population means can be declared different.

16.95 Family confidence level = 0.95; $q_{0.05} = 3.95$; simultaneous 95% confidence intervals are as follows.

Means difference	Confidence interval
$\mu_1 - \mu_2$	102.2 to 199.3
$\mu_1 - \mu_3$	114.7 to 211.8
$\mu_2 - \mu_3$	−36.1 to 61.1

The preceding table shows that the following pairs of means can be declared different: μ_1 and μ_2, μ_1 and μ_3. This result is summarized in the following diagram.

Macroalgae (3)	Bacteria (2)	Diatoms (1)
293.75	306.25	457.00

Interpreting this diagram, we conclude with 95% confidence that the mean number of copepods is greater with the diatoms diet than with the other two diets; no other means can be declared different.

16.97
a. Family confidence level = 0.95; $q_{0.05} = 4.23$; simultaneous 95% confidence intervals are as follows.

Means difference	Confidence interval
$\mu_1 - \mu_2$	−30.5 to 20.5
$\mu_1 - \mu_3$	−41.7 to 9.3
$\mu_1 - \mu_4$	−24.3 to 26.7
$\mu_1 - \mu_5$	−43.5 to 7.5
$\mu_2 - \mu_3$	−36.7 to 14.3
$\mu_2 - \mu_4$	−19.3 to 31.7
$\mu_2 - \mu_5$	−38.5 to 12.5
$\mu_3 - \mu_4$	−8.1 to 42.9
$\mu_3 - \mu_5$	−27.3 to 23.7
$\mu_4 - \mu_5$	−44.7 to 6.3

The preceding table shows that no two population means can be declared different. This result is summarized in the following diagram.

Strain D	Strain A	Strain B	Strain C	Strain E
(4)	(1)	(2)	(3)	(5)
19.6	20.8	25.8	37.0	38.8

Interpreting this diagram, we conclude with 95% confidence that no two mean bacteria counts can be declared different.

b. Because the Tukey multiple comparison, performed using a 95% family confidence level, does not detect a difference between any two means, we deduce that, at the 5% significance level, the data do not provide sufficient evidence to conclude that a difference exists in mean bacteria counts among the five strains of *Staphylococcus aureus*.

16.99 Family confidence level = 0.95; $q_{0.05} = 3.85$; simultaneous 95% confidence intervals are as follows.

Means difference	Confidence interval
$\mu_1 - \mu_2$	−5.58 to 0.08
$\mu_1 - \mu_3$	−3.54 to 1.82
$\mu_1 - \mu_4$	−1.05 to 4.61
$\mu_2 - \mu_3$	−0.79 to 4.57
$\mu_2 - \mu_4$	1.70 to 7.36
$\mu_3 - \mu_4$	−0.04 to 5.33

The preceding table shows that only μ_2 and μ_4 can be declared different. This result is summarized in the following diagram.

Brand D	Brand A	Brand C	Brand B
(4)	(1)	(3)	(2)
5.91	7.69	8.55	10.44

Interpreting this diagram, we conclude with 95% confidence that the mean battery life for Brand B is greater than that for Brand D; no other mean battery lives can be declared different.

16.101 With 95% confidence, we conclude that the mean monthly rents for newly completed apartments in the Northeast and West exceed that for those in the Midwest; no other mean monthly rents can be declared different.

16.103 With 99% confidence, we conclude that the mean singing rate of male Rock Sparrows exposed to the enlarged breast treatment exceeds that of those exposed to the reduced breast treatment; no other mean singing rates can be declared different.

16.105 With 95% confidence, we conclude that the mean hardness of Duradent is less than that of Endura, which is less than that of Duracross.

16.107 Family confidence level = 0.90; $q_{0.10} = 3.50$; simultaneous 90% confidence intervals are as follows.

Means difference	Confidence interval
$\mu_1 - \mu_2$	−2.34 to 2.54
$\mu_1 - \mu_3$	−4.12 to 0.92
$\mu_1 - \mu_4$	−2.70 to 2.50
$\mu_1 - \mu_5$	−2.52 to 2.52
$\mu_2 - \mu_3$	−4.13 to 0.73
$\mu_2 - \mu_4$	−2.70 to 2.30
$\mu_2 - \mu_5$	−2.53 to 2.33
$\mu_3 - \mu_4$	−1.08 to 4.08
$\mu_3 - \mu_5$	−0.91 to 4.11
$\mu_4 - \mu_5$	−2.48 to 2.68

The preceding table shows that no two population means can be declared different. This result is summarized in the following diagram.

14 mg	Placebo	210 mg	100 mg	60 mg
(2)	(1)	(5)	(4)	(3)
25.8	25.9	25.9	26.0	27.5

Interpreting this diagram, we conclude with 90% confidence that no two mean body-mass indexes can be declared different.

16.109 Family confidence level = 0.99; $q_{0.01} = 4.85$; simultaneous 99% confidence intervals are as follows.

Means difference	Confidence interval
$\mu_1 - \mu_2$	4.54 to 16.46
$\mu_1 - \mu_3$	−8.17 to 0.17
$\mu_1 - \mu_4$	8.54 to 20.46
$\mu_1 - \mu_5$	−11.25 to −3.75
$\mu_1 - \mu_6$	7.16 to 17.04
$\mu_2 - \mu_3$	−20.76 to −8.24
$\mu_2 - \mu_4$	−3.58 to 11.58
$\mu_2 - \mu_5$	−23.99 to −12.01
$\mu_2 - \mu_6$	−5.20 to 8.40
$\mu_3 - \mu_4$	12.24 to 24.76
$\mu_3 - \mu_5$	−7.71 to 0.71
$\mu_3 - \mu_6$	10.80 to 21.40
$\mu_4 - \mu_5$	−27.99 to −16.01
$\mu_4 - \mu_6$	−9.20 to 4.40
$\mu_5 - \mu_6$	14.63 to 24.57

The preceding table shows that the following pairs of population means can be declared different: μ_1 and μ_2, μ_1 and μ_4, μ_1 and μ_5, μ_1 and μ_6, μ_2 and μ_3, μ_2 and μ_5, μ_3 and μ_4, μ_3 and μ_6, μ_4 and μ_5, μ_5 and μ_6. This result is summarized in the following diagram.

Ed	Math/Sci	Comm	Bus	CS	Engr
(4)	(6)	(2)	(1)	(3)	(5)
40.6	43.0	44.6	55.1	59.1	62.6

Interpreting this diagram, we conclude with 99% confidence that, among bachelor's-degree graduates in the six fields, the mean starting salary of education, math & sciences, and communications majors is less than that of business, computer science, and engineering majors; that of business majors is less than that of engineering majors; no other mean starting salaries can be declared different.

Exercises 16.5

16.123 Simple random samples, independent samples, same-shape populations, and all sample sizes are 5 or greater

16.125 equal   **16.127** variance; *SST*; *SST*

16.129 Chi-square with df = 4   **16.131** One-way ANOVA test

16.133 Neither of the tests

16.135 a. 3.75   **b.** 3.75   **c.** They are the same.
d. The computing formula for K is equivalent to the defining formula for K if no ties occur among the ranks.

16.137 a. 0.682   **b.** 0.682   **c.** They are the same.
d. The computing formula for K is equivalent to the defining formula for K if no ties occur among the ranks.

16.139 a. 10   **b.** 3   **c.** 25, 7, 23

16.141 a. 20   **b.** 5   **c.** 32, 32, 55.5, 38, 52.5

16.143 a. 12   **b.** 4   **c.** 24, 14, 31, 9

16.145 H_0: $\eta_1 = \eta_2 = \eta_3$, H_a: Not all the medians are equal. $\alpha = 0.05$; $K = 6.77$; critical value = 5.991; $0.025 < P < 0.05$ ($P = 0.034$); reject H_0; at the 5% significance level, the data provide sufficient evidence to conclude that a difference exists in the median number of advertised brands a viewer is able to recall among the three TV program types.

16.147 H_0: $\eta_1 = \eta_2 = \eta_3 = \eta_4$, H_a: Not all the medians are equal. $\alpha = 0.10$; $K = 6.37$; critical value = 7.815; $0.05 < P < 0.10$ ($P = 0.095$); reject H_0; at the 10% significance level, the data provide sufficient evidence to conclude that a difference exists in median square footage of single-family detached homes among the four U.S. regions.

16.149 H_0: $\mu_1 = \mu_2 = \mu_3 = \mu_4$, H_a: Not all the means are equal. $\alpha = 0.10$; $K = 6.02$; critical value $= 6.251$; $P > 0.10$ ($P = 0.111$); do not reject H_0; at the 10% significance level, the data do not provide sufficient evidence to conclude that a difference exists in mean number of eggs among the four combinations.

16.151

a. H_0: $\mu_1 = \mu_2 = \mu_3 = \mu_4 = \mu_5$, H_a: Not all the means are equal. $\alpha = 0.05$; $K = 7.19$; critical value $= 9.488$; $P > 0.10$ ($P = 0.126$); do not reject H_0; at the 5% significance level, the data do not provide sufficient evidence to conclude that a difference exists in mean bacteria counts among the five strains of *Staphylococcus aureus*.

b. Because normal distributions with equal standard deviations have the same shape. It is better to use the one-way ANOVA test because, when the assumptions for that test are met, it is more powerful than the Kruskal–Wallis test.

Note: In each of the answers to Exercises 16.153–16.157, we have provided the *P*-value obtained from the chi-square approximation to the distribution of the test statistic, *K*. For some of these exercises, using a *P*-value obtained from the exact distribution of *K* would be preferable. If your statistical technology has an option for the latter, use that option instead.

16.153

a. H_0: $\mu_1 = \mu_2 = \mu_3 = \mu_4$, H_a: Not all the means are equal. $\alpha = 0.05$; $K = 12.23$; $P = 0.007$; reject H_0.

b. At the 5% significance level, the data provide sufficient evidence to conclude that a difference exists in mean monthly rents among newly completed apartments in the four U.S. regions.

16.155

a. H_0: $\mu_1 = \mu_2 = \mu_3$, H_a: Not all the means are equal. $\alpha = 0.01$; $K = 9.86$; $P = 0.007$; reject H_0.

b. At the 1% significance level, the data provide sufficient evidence to conclude that a difference exists in the mean singing rates among male Rock Sparrows exposed to the three types of breast treatments.

16.157

a. H_0: $\mu_1 = \mu_2 = \mu_3$, H_a: Not all the means are equal. $\alpha = 0.05$; $K = 15.20$; $P = 0.0005$; reject H_0.

b. At the 5% significance level, the data provide sufficient evidence to conclude that there is a difference in mean hardness among the three materials.

Review Problems for Chapter 16

1. To compare the means of a variable for populations that result from a classification by one other variable (called the *factor*)

2. *Simple random samples:* Check by carefully studying the way the sampling was done. *Independent samples:* Check by carefully studying the way the sampling was done. *Normal populations:* Check by constructing normal probability plots. *Equal standard deviations:* As a rule of thumb, this assumption is considered to be satisfied if the ratio of the largest sample standard deviation to the smallest sample standard deviation is less than 2.

   Also, the normality and equal-standard-deviations assumptions can be assessed by performing a residual analysis.

3. The *F*-distribution   4. df $= (2, 14)$

5. **a.** *MSTR* (or *SSTR*)   **b.** *MSE* (or *SSE*)

6. **a.** The total sum of squares, *SST*, represents the total variation among all the sample data; the treatment sum of squares, *SSTR*, represents the variation among the sample means; and the error sum of squares, *SSE*, represents the variation within the samples.

   b. $SST = SSTR + SSE$; the one-way ANOVA identity shows that the total variation among all the sample data can be partitioned into a component representing variation among the sample means and a component representing variation within the samples.

7. **a.** For organizing and summarizing the quantities required for performing a one-way analysis of variance

   b.

Source	df	SS	MS = SS/df	F
Treatment	$k - 1$	SSTR	$MSTR = \dfrac{SSTR}{k-1}$	$F = \dfrac{MSTR}{MSE}$
Error	$n - k$	SSE	$MSE = \dfrac{SSE}{n-k}$	
Total	$n - 1$	SST		

8. Suppose that, in a one-way ANOVA, the null hypothesis of equal population means is rejected. The purpose of a multiple comparison is to then decide which means are different, which mean is largest, or, more generally, the relation among all the means.

9. The individual confidence level is the confidence we have that any particular confidence interval contains the difference between the corresponding population means; the family confidence level is the confidence we have that all the confidence intervals contain the differences between the corresponding population means. It is at the family confidence level that we can be confident in the truth of our conclusions when comparing all the population means simultaneously; thus the family confidence level is the appropriate one for multiple comparisons.

10. Studentized range distribution (or *q*-distribution)

11. Larger. We can be more confident about the truth of one of several statements than about the truth of all statements simultaneously.

12. $\kappa = 3$, $\nu = 14$ 13. Kruskal–Wallis test

14. Chi-square distribution with df $= k - 1$, where *k* is the number of populations under consideration

15. The Kruskal–Wallis test is based on ranks. If the null hypothesis of equal population means is true, the means of the ranks for the samples should be roughly equal. Put another way, an unduly large variation among the mean ranks provides evidence against the null hypothesis.

16. The Kruskal–Wallis test because, unlike the one-way ANOVA test, it is resistant to outliers and other extreme values.

17. 2   **18.** 14   **19.** 3.74   **20.** 6.51   **21.** 3.74

22. **a.** 3.70   **b.** 4.89

23. **a.** The sample means are 3, 3, and 6, respectively; the sample variances are 4, 6, and 18, respectively.

   b. $SST = 110$, $SSTR = 24$, $SSE = 86$; $110 = 24 + 86$

   c. $SST = 110$, $SSTR = 24$, $SSE = 86$

   d.

Source	df	SS	MS = SS/df	F
Treatment	2	24	12	1.26
Error	9	86	9.556	
Total	11	110		

24. **a.** The variation among the sample means

   b. The variation within the samples

   c. Simple random samples, independent samples, normal populations, and equal (population) standard deviations. One-way ANOVA is robust to moderate violations of the normality assumption. It is also reasonably robust to moderate violations of the equal-standard-deviations assumption if the sample sizes are roughly equal.

25. **a.** $s_1 = \$92.9$, $s_2 = \$126.1$, $s_3 = \$139.0$. Figure A.11 shows individual normal probability plots of the three samples.

   b. See Fig. A.12.

   c. Referring to the results of either part (a) or part (b), we conclude that presuming that the assumptions of normal populations and equal standard deviations are met is reasonable.

FIGURE A.11   Normal probability plots for Problem 25(a)

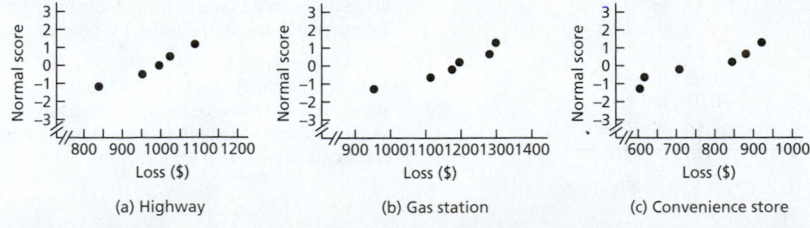

(a) Highway   (b) Gas station   (c) Convenience store

FIGURE A.12   (a) Residual plot and (b) normal probability plot of the residuals for Problem 25(b)

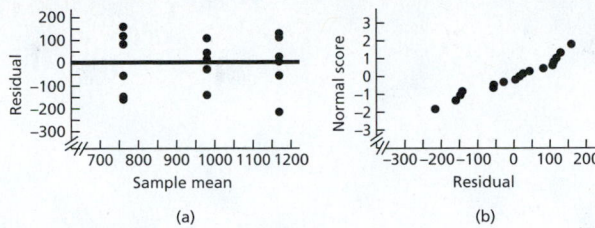

(a)   (b)

26. $H_0: \mu_1 = \mu_2 = \mu_3$, H_a: Not all the means are equal. $\alpha = 0.05$;
$F = 16.60$; critical value $= 3.74$; $P < 0.005$ ($P = 0.000$); reject H_0;
at the 5% significance level, the data provide sufficient evidence to conclude
that a difference in mean losses exists among the three types of robberies.

27. **a.** Family confidence level $= 0.95$; $q_{0.05} = 3.70$; simultaneous
95% confidence intervals are as follows.

Means difference	Confidence interval
$\mu_1 - \mu_2$	-383.3 to 5.2
$\mu_1 - \mu_3$	24.4 to 412.9
$\mu_2 - \mu_3$	222.4 to 592.9

b. The table in part (a) shows that the following pairs of means can be
declared different: μ_1 and μ_3, μ_2 and μ_3. This result is summarized
in the following diagram.

Convenience store (3)	Highway (1)	Gas station (2)
761.2	979.8	1168.8

Interpreting the diagram, we conclude with 95% confidence that the
mean loss due to convenience-store robberies is less than that of both
highway robberies and gas-station robberies; the mean losses due to
highway robberies and gas-station robberies cannot be declared different.

28. **a.** $H_0: \mu_1 = \mu_2 = \mu_3$, H_a: Not all the means are equal. $\alpha = 0.05$;
$K = 11.76$; critical value $= 5.991$; $P < 0.005$ ($P = 0.003$);
reject H_0; at the 5% significance level, the data provide sufficient
evidence to conclude that a difference in mean losses exists among
the three types of robberies.

b. Because normal distributions with equal standard deviations have the
same shape. It is better to use the one-way ANOVA test because,
when the assumptions for that test are met, it is more powerful than
the Kruskal–Wallis test.

c. Both tests reject the null hypothesis of equal population means.

29. $H_0: \mu_1 = \mu_2 = \mu_3$, H_a: Not all the means are equal. $\alpha = 0.01$;
$F = 6.49$; critical value $= 5.03$; $P < 0.005$ ($P = 0.003$); reject H_0;
at the 1% significance level, the data provide sufficient evidence to
conclude that a difference exists in the mean angle of anterior-posterior
center of foot pressure among people in the three condition groups.

30. Normal probability plots suggest that it is reasonable to presume that the
distributions of cotinine for the three populations under consideration are
all normal. However, the standard deviations of the three samples
are 121.2 ng/mL, 19.2 ng/mL, and 61.8 ng/mL. As the rule of 2 fails, it
appears that the assumption of equal population standard is violated.

31. Graphical analyses (e.g., dotplots, stem-and-leaf diagrams) suggest that
the distributions of cotinine for the three populations under consideration
do not have the same shape. Therefore, performing a Kruskal–Wallis test
on the data is not reasonable.

Index

Photo Credits

About the Author

p. v, Carol Weiss

Chapter 1

p. 1, United Archives GmbH/Alamy
p. 2, TSN/ZUMAPRESS/Newscom
p. 3, Bettmann/Corbis
p. 5, Sports Illustrated/Getty Images
p. 34 (top), United Archives GmbH/Alamy
p. 34 (bottom), Library of Congress

Chapter 2

p. 36, Jason DeCrow/AP Images
p. 91, Jason DeCrow/AP Images
p. 92, Library of Congress

Chapter 3

p. 93, Gamma/Gamma-Rapho/Getty Images
p. 154 (top), Gamma/Gamma-Rapho/Getty Images
p. 154 (bottom), Pearson Education

Chapter 4

p. 156, Krastiu Vasilev/Dreamstime
p. 222 (top), Krastiu Vasilev/Dreamstime
p. 222 (bottom), Pearson Education

Chapter 5

p. 223, Ed Reinke/AP Images
p. 261 (top), Ed Reinke/AP Images
p. 261 (bottom), Pearson Education

Chapter 6

p. 262, Holmes Garden Photos/Alamy
p. 306 (top), Holmes Garden Photos/Alamy
p. 306 (bottom), Library of Congress Prints and Photographs Division

Chapter 7

p. 307, S_oleg/ Shutterstock
p. 329 (top), S_oleg/Shutterstock
p. 329 (bottom), Pearson Education

Chapter 8

p. 331, Folio/Alamy
p. 366 (top), Folio/Alamy
p. 366 (bottom), Pearson Education

Chapter 9

p. 367, Erlucho/iStockphoto
p. 437 (top), Erlucho/iStockphoto
p. 437 (bottom), Pearson Education

Chapter 10

p. 438, Steve Lovegrove/Fotolia
p. 511 (top), Steve Lovegrove/Fotolia
p. 511 (bottom), North Carolina State University Archives

Chapter 11

p. 513, Manchan/Getty Images
p. 543 (top), Manchan/Getty Images
p. 543 (bottom), Pearson Education

Chapter 12

p. 544, Ivanff/Fotolia
p. 575 (top), Ivanff/Fotolia
p. 575 (bottom), Pearson Education

Chapter 13

p. 576, Shutterstock
p. 617 (top), Shutterstock
p. 617 (bottom), SPL/Photo Researchers Inc.

Chapter 14

p. 618, Wavebreakmedia/Shutterstock
p. 656 (top), Wavebreakmedia/Shutterstock
p. 656 (bottom), Pearson Education

Chapter 15

p. 657, Warren Goldswain/Fotolia
p. 696, Warren Goldswain/Fotolia
p. 697, Pearson Education

Chapter 16

p. 698, Gareth Boden/Pearson Education
p. 741 (top), Gareth Boden/Pearson Education
p. 741 (bottom), Pearson Education.

Indexes for
Case Studies & Biographical Sketches

TABLE IV

Values of t_α

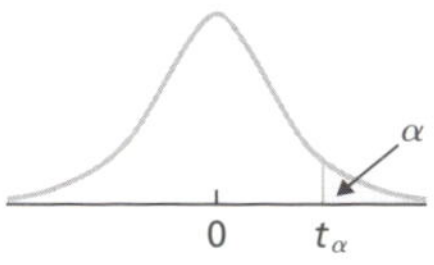

NOTE: *See the version of Table IV in Appendix A for additional values of t_α.*

df	$t_{0.10}$	$t_{0.05}$	$t_{0.025}$	$t_{0.01}$	$t_{0.005}$	df
1	3.078	6.314	12.706	31.821	63.657	1
2	1.886	2.920	4.303	6.965	9.925	2
3	1.638	2.353	3.182	4.541	5.841	3
4	1.533	2.132	2.776	3.747	4.604	4
5	1.476	2.015	2.571	3.365	4.032	5
6	1.440	1.943	2.447	3.143	3.707	6
7	1.415	1.895	2.365	2.998	3.499	7
8	1.397	1.860	2.306	2.896	3.355	8
9	1.383	1.833	2.262	2.821	3.250	9
10	1.372	1.812	2.228	2.764	3.169	10
11	1.363	1.796	2.201	2.718	3.106	11
12	1.356	1.782	2.179	2.681	3.055	12
13	1.350	1.771	2.160	2.650	3.012	13
14	1.345	1.761	2.145	2.624	2.977	14
15	1.341	1.753	2.131	2.602	2.947	15
16	1.337	1.746	2.120	2.583	2.921	16
17	1.333	1.740	2.110	2.567	2.898	17
18	1.330	1.734	2.101	2.552	2.878	18
19	1.328	1.729	2.093	2.539	2.861	19
20	1.325	1.725	2.086	2.528	2.845	20
21	1.323	1.721	2.080	2.518	2.831	21
22	1.321	1.717	2.074	2.508	2.819	22
23	1.319	1.714	2.069	2.500	2.807	23
24	1.318	1.711	2.064	2.492	2.797	24
25	1.316	1.708	2.060	2.485	2.787	25
26	1.315	1.706	2.056	2.479	2.779	26
27	1.314	1.703	2.052	2.473	2.771	27
28	1.313	1.701	2.048	2.467	2.763	28
29	1.311	1.699	2.045	2.462	2.756	29
30	1.310	1.697	2.042	2.457	2.750	30
35	1.306	1.690	2.030	2.438	2.724	35
40	1.303	1.684	2.021	2.423	2.704	40
50	1.299	1.676	2.009	2.403	2.678	50
60	1.296	1.671	2.000	2.390	2.660	60
70	1.294	1.667	1.994	2.381	2.648	70
80	1.292	1.664	1.990	2.374	2.639	80
90	1.291	1.662	1.987	2.369	2.632	90
100	1.290	1.660	1.984	2.364	2.626	100
1000	1.282	1.646	1.962	2.330	2.581	1000
2000	1.282	1.646	1.961	2.328	2.578	2000

1.282	1.645	1.960	2.326	2.576
$z_{0.10}$	$z_{0.05}$	$z_{0.025}$	$z_{0.01}$	$z_{0.005}$

TABLE II
Areas under the
standard normal curve

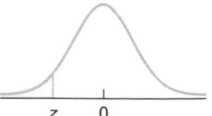

					Second decimal place in z					
0.09	0.08	0.07	0.06	0.05	0.04	0.03	0.02	0.01	0.00	z
									0.0000†	−3.9
0.0001	0.0001	0.0001	0.0001	0.0001	0.0001	0.0001	0.0001	0.0001	0.0001	−3.8
0.0001	0.0001	0.0001	0.0001	0.0001	0.0001	0.0001	0.0001	0.0001	0.0001	−3.7
0.0001	0.0001	0.0001	0.0001	0.0001	0.0001	0.0001	0.0001	0.0002	0.0002	−3.6
0.0002	0.0002	0.0002	0.0002	0.0002	0.0002	0.0002	0.0002	0.0002	0.0002	−3.5
0.0002	0.0003	0.0003	0.0003	0.0003	0.0003	0.0003	0.0003	0.0003	0.0003	−3.4
0.0003	0.0004	0.0004	0.0004	0.0004	0.0004	0.0004	0.0005	0.0005	0.0005	−3.3
0.0005	0.0005	0.0005	0.0006	0.0006	0.0006	0.0006	0.0006	0.0007	0.0007	−3.2
0.0007	0.0007	0.0008	0.0008	0.0008	0.0008	0.0009	0.0009	0.0009	0.0010	−3.1
0.0010	0.0010	0.0011	0.0011	0.0011	0.0012	0.0012	0.0013	0.0013	0.0013	−3.0
0.0014	0.0014	0.0015	0.0015	0.0016	0.0016	0.0017	0.0018	0.0018	0.0019	−2.9
0.0019	0.0020	0.0021	0.0021	0.0022	0.0023	0.0023	0.0024	0.0025	0.0026	−2.8
0.0026	0.0027	0.0028	0.0029	0.0030	0.0031	0.0032	0.0033	0.0034	0.0035	−2.7
0.0036	0.0037	0.0038	0.0039	0.0040	0.0041	0.0043	0.0044	0.0045	0.0047	−2.6
0.0048	0.0049	0.0051	0.0052	0.0054	0.0055	0.0057	0.0059	0.0060	0.0062	−2.5
0.0064	0.0066	0.0068	0.0069	0.0071	0.0073	0.0075	0.0078	0.0080	0.0082	−2.4
0.0084	0.0087	0.0089	0.0091	0.0094	0.0096	0.0099	0.0102	0.0104	0.0107	−2.3
0.0110	0.0113	0.0116	0.0119	0.0122	0.0125	0.0129	0.0132	0.0136	0.0139	−2.2
0.0143	0.0146	0.0150	0.0154	0.0158	0.0162	0.0166	0.0170	0.0174	0.0179	−2.1
0.0183	0.0188	0.0192	0.0197	0.0202	0.0207	0.0212	0.0217	0.0222	0.0228	−2.0
0.0233	0.0239	0.0244	0.0250	0.0256	0.0262	0.0268	0.0274	0.0281	0.0287	−1.9
0.0294	0.0301	0.0307	0.0314	0.0322	0.0329	0.0336	0.0344	0.0351	0.0359	−1.8
0.0367	0.0375	0.0384	0.0392	0.0401	0.0409	0.0418	0.0427	0.0436	0.0446	−1.7
0.0455	0.0465	0.0475	0.0485	0.0495	0.0505	0.0516	0.0526	0.0537	0.0548	−1.6
0.0559	0.0571	0.0582	0.0594	0.0606	0.0618	0.0630	0.0643	0.0655	0.0668	−1.5
0.0681	0.0694	0.0708	0.0721	0.0735	0.0749	0.0764	0.0778	0.0793	0.0808	−1.4
0.0823	0.0838	0.0853	0.0869	0.0885	0.0901	0.0918	0.0934	0.0951	0.0968	−1.3
0.0985	0.1003	0.1020	0.1038	0.1056	0.1075	0.1093	0.1112	0.1131	0.1151	−1.2
0.1170	0.1190	0.1210	0.1230	0.1251	0.1271	0.1292	0.1314	0.1335	0.1357	−1.1
0.1379	0.1401	0.1423	0.1446	0.1469	0.1492	0.1515	0.1539	0.1562	0.1587	−1.0
0.1611	0.1635	0.1660	0.1685	0.1711	0.1736	0.1762	0.1788	0.1814	0.1841	−0.9
0.1867	0.1894	0.1922	0.1949	0.1977	0.2005	0.2033	0.2061	0.2090	0.2119	−0.8
0.2148	0.2177	0.2206	0.2236	0.2266	0.2296	0.2327	0.2358	0.2389	0.2420	−0.7
0.2451	0.2483	0.2514	0.2546	0.2578	0.2611	0.2643	0.2676	0.2709	0.2743	−0.6
0.2776	0.2810	0.2843	0.2877	0.2912	0.2946	0.2981	0.3015	0.3050	0.3085	−0.5
0.3121	0.3156	0.3192	0.3228	0.3264	0.3300	0.3336	0.3372	0.3409	0.3446	−0.4
0.3483	0.3520	0.3557	0.3594	0.3632	0.3669	0.3707	0.3745	0.3783	0.3821	−0.3
0.3859	0.3897	0.3936	0.3974	0.4013	0.4052	0.4090	0.4129	0.4168	0.4207	−0.2
0.4247	0.4286	0.4325	0.4364	0.4404	0.4443	0.4483	0.4522	0.4562	0.4602	−0.1
0.4641	0.4681	0.4721	0.4761	0.4801	0.4840	0.4880	0.4920	0.4960	0.5000	−0.0

†For $z \leq -3.90$, the areas are 0.0000 to four decimal places.